DYNAMIC ASPECTS OF CELL SURFACE ORGANIZATION

contain authoritative and topical reviews by investigators who have contributed to progress in their respective research fields. While individual reviews will provide comprehensive coverage of specialized topics, all of the reviews published within each volume will be related to an overall common theme. This format represents a departure from that adopted by most of the existing series of "review" publications which usually provide heterogeneous collections of reviews on unrelated topics. While this latter format is considerably more convenient from an editorial standpoint, we feel that publication together of a number of related reviews will better serve the stated aims of this series – to bridge the information and specialization "gap" among investigators in related areas. Each volume will therefore present a fairly complete and critical survey of the more important and recent advances in well defined topics in biology and medicine. The level will be advanced, directed primarily to the needs of the research worker and graduate students.

Editorial policy will be to impose as few restrictions as possible on contributors. This is appropriate since the volumes published in this series will represent collections of review articles and will not be definitive monographs dealing with all aspects of the selected subject. Contributors will be encouraged, however, to provide comprehensive, critical reviews that attempt to integrate the available data into a broad conceptual framework. Emphasis will also be given to identification of major problems demanding further study and the possible avenues by which these might be investigated. Scope will also be offered for the presentation of new and challenging ideas and hypotheses for which complete evidence is still lacking.

The first four volumes of this series will be published within one year, after which volumes will appear at approximately one year intervals.

George Poste
Garth L. Nicolson
Editors

Contents of previous and forthcoming volumes

Volume 1
THE CELL SURFACE IN ANIMAL EMBRYOGENESIS AND DEVELOPMENT
Fertilization – R.B.L. Gwatkin (Rahway)
Cytokinesis in Animal Cells: New Answers to Old Questions – J.M. Arnold (Honolulu)
The Implanting Mouse Blastocyst – M.I. Sherman and L.R. Wudl (Nutley)
Cell Surface Antigens in Mammalian Development – M. Edidin (Baltimore)
The Transport of Molecules across Placental Membranes – R.K. Miller (Rochester), T.R. Koszalka and R.L. Brent (Philadelphia)
Mechanisms of Metazoan Cell Movements – J.P. Trinkaus (New Haven)
Inductive Tissue Interactions – L. Saxén, M. Karkinen-Jääskelainen, E. Lehtonen, S. Nordling and J. Wartiovaara (Helsinki)
Cell Coupling and Cell Communication during Embryogenesis – J.D. Sheridan (Minneapolis)
Transduction of Positional Information during Development – D. McMahon and C. West (Pasadena)
Cell Interactions in Vertebrate Limb Development – D.A. Ede (Glasgow)
Heart Development: Interactions Involved in Cardiac Morphogenesis – F.J. Manasek (Chicago)
Development and Differentiation of Lymphocytes – I. Goldschneider and R.W. Barton (Farmington)
In Vitro Analysis of Surface Specificity in Embryonic Cells – D. Maslow (Buffalo)

Volume 2
VIRUS INFECTION AND THE CELL SURFACE
Host and Tissue Specificities in Virus Infections of Animals – H. Smith (Birmingham)

Structure and Assembly of Viral Envelopes – R. Rott and H.-D. Klenk (Giessen)

Virus–Erythrocyte Membrane Interactions – T. Bächi (Zürich) and J.E. Deas and C. Howe (New Orleans)

Cell Fusion by Sendai Virus – Y. Hosaka and K. Shimizu (Sendai City)

The Interaction of Viruses with Model Membranes – J.M. Tiffany (Oxford)

Virus–Host Cell Interactions in "Slow" Virus Diseases of the Nervous System – L.P. Weiner and R.T. Johnson (Baltimore)

Surface Antigens of Virus-Infected Cells – W.H. Burns (Parkville) and A.C. Allison (Harrow)

T Cell Recognition of Virus-Infected Cells – R.V. Blanden, T.E. Pang and M.B.C. Dunlop (Canberra)

Host Antigens in Enveloped RNA Viruses – J. Lindenmann (Zürich)

Volume 4
MEMBRANE ASSEMBLY AND TURNOVER
The Golgi Apparatus and Membrane Biogenesis – J. Morré (West Lafayette, Ind.)

Biosynthesis of Plasma Membrane Proteins – G.M.W. Cook (Cambridge)

Turnover of Proteins of the Eukaryotic Cell Surface – J. Tweto and D. Doyle (Buffalo)

Origin and Fate of the Membranes of Secretion Granules and Synaptic Vesicles – E. Holtzman, S. Schacher, J. Evans and S. Teichberg (New York)

Cell Surface Changes in Phagocytosis – P. Elsbach (New York)

Plasma Membrane Assembly as Related to Cell Division – J. Bluemink and S.W. de Laat (Utrecht)

Biosynthesis and Assembly of Bacterial Cell Walls – J.M. Ghuysen (Liège)

The Bacterial Cell Surface in Growth and Division – L. Daneo-Moore and G.D. Shockman (Philadelphia)

The Synthesis and Assembly of Plant Cell Walls: Possible Control Mechanisms – D.H. Northcote (Cambridge)

Envelopes of Lipid Containing Viruses as Models for Membrane Assembly – L. Kääriäinen and O. Renkonen (Helsinki)

In Vitro and In Vivo Assembly of Bacteriophage PM2: A Model for Protein–Lipid Interactions – R.M. Franklin (Basel)

Local Differentiations of the Cell Surface of Ciliates: Their Determination, Effects and Genetics – T.M. Sonneborn (Bloomington, Ind.)

Volume 5
MEMBRANE FUSION
Calcium-Induced Structural Reorganization and Fusion of Natural and Model Membranes – G. Poste and D. Papahadjopoulos (Buffalo)

Membrane Interactions in Endocytosis – P. Edelson and Z. Cohn (New York)

Mechanisms of Mediator Release from Inflammatory Cells – P.M. Henson, D. Morrison and M. Ginsberg (La Jolla)

Ultrastructural Aspects of Exocytosis – L. Orci and A. Perrelet (Geneva)

Cytoplasmic Membranes and the Secretory Process – J. Meldolesi (Milan)

Fusion of Gametes: Vertebrates – J.M. Bedford and G.W. Cooper (New York)

Fusion of Gametes: Invertebrates – D. Epel (San Diego) and V. Vacquier (Davis)

Cell Surface Components and Fusion of Muscle Cells – R. Bischoff (St. Louis)

Macrophage Fusion In Vivo and In Vitro – J.M. Papadimitriou (Perth)

Cell Fusion and Somatic Incompatibility in Fungi – M. Carlile (London) and G. Gooday (Aberdeen)

Fusion of Plant Protoplasts – J.B. Power, K. Evans and E.C. Cocking (Nottingham)

Virus-Induced Cell Fusion – G. Poste (Buffalo) and C.A. Pasternak (London)

Chemically Induced Cell Fusion – J. Lucy (London)

Membranes of Ciliates: Morphology, Biochemistry and Fusion Behavior – R.D. Allen (Honolulu)

Problems in the Physical Interpretation of Membrane Interaction and Fusion – D. Gingell and L. Ginsberg (London)

Preface

There is now general acceptance of the view that the plasma membranes of most, if not all, cells are dynamic assemblies of molecules that are able to undergo rapid and reversible structural rearrangements in response to both intra- and extracellular stimuli. This concept is, however, of relatively recent origin. Until the mid-nineteen sixties, experimental observations on the morphology, chemical composition and function of biological membranes were interpreted against a conceptual background in which membrane components were thought to exist largely in rather rigid, ordered and static structural arrangements. In the late nineteen sixties and early seventies information began to accumulate from studies in a variety of disciplines which indicated that certain membrane components were free to move within the membrane and thus undergo topographic rearrangement. These findings were difficult to reconcile with the existing static models for membrane structure and prompted the formulation of new generalized concepts for membrane organization such as the "liquid crystalline" and "fluid mosaic" models which, for the first time, appeared to fit the growing number of observations on the dynamic properties of biological membranes. From these initial experimental observations less than a decade ago, information on membrane dynamics has grown at a remarkable pace so that today the literature on the subject is of voluminous proportion. Although many fundamental questions remain to be answered, and speculation currently surrounds interpretation of several aspects of the subject, sufficient information is already available to support a functional relationship between the topography and dynamics of plasma membrane macromolecules and the control of cell surface properties.

In its most simple form, the plasma membrane can be thought of as a two-dimensional solution of a mosaic of lipids and proteins. The lipids, arranged predominantly as a bilayer, exist in a "fluid" state while the proteins (and glycoproteins) are either inserted to varying depths into the lipid bilayer (integral proteins) or bound loosely to the surfaces of the bilayer (peripheral proteins). Both proteins and lipids can be organized asymmetrically within this type of structure, thus allowing specific classes of membrane components to be localized exclusively or predominantly at the inner or outer membrane surface. Another important feature is that the fluid state of the membrane lipids permits lateral diffusion of macromolecules within the plane of the membrane, thereby creating opportunities for rapid and reversible changes in membrane topography. The distinctive lateral mobilities of different membrane components, ranging from extremely rapid to relatively immobile, permits cells to maintain certain surface molecules in relatively ordered topographic arrays or "patterns", while allowing others to diffuse randomly and avoid ordered topographic restraint. The fluid nature of cell membranes also dictates that environmental changes in temperature, ionic strength, pressure, binding of material to membrane "receptors" and many other stimuli can induce drastic changes in the physical state of the membrane, leading to phase changes and separations whereby molecules are excluded away from or sequestered into specific membrane regions or domains. Finally, trans-membrane interaction of components within the plasma membrane with membrane-associated structures in the cytoplasm offers a potential mechanism for the transmission of "information" across the membrane in either direction via various combinations of cooperative, allosteric and transductive coupling mechanisms. These and other possible regulatory devices may well allow different cell types to react individually to the same stimuli, thereby creating opportunities for a large repertoire of cell-specific responses and membrane specialization based on a relatively restricted and conservative framework of membrane structure.

Membrane dynamics embraces a broad range of physical, chemical and biological processes. Comprehensive discussion of this subject demands consideration of the basic physico-chemical properties of the different molecules and macromolecules found in membranes, analysis of their functional interrelationships within membranes, characterization of the diverse factors that influence membrane organization in both physiological and pathological states and, finally, identification of relationships between the control of plasma membrane organization and changes in cell surface properties and cell behavior. Insight into these various aspects of membrane dynamics has come from work in many disciplines, from detailed observations at the molecular level as well as phenomenological descriptions of the behavior of intact cells, and from the application of numerous techniques, some old but improved, and some entirely new.

This volume, the third in the *Cell Surface Reviews* series, contains fourteen chapters which reflect the dramatic growth of information of the dynamic nature of membrane organization. No attempt has been made to cover all aspects of this vast subject. However, the reviews in this volume offer a broad perspective of current concepts of plasma membrane organization and the range of experimental strategies and techniques used to investigate this important structure. Discussion of the importance of the biosynthesis, assembly and turnover of plasma membrane components in cell surface dynamics has been largely excluded from this volume since the next volume in this series will be devoted entirely to this subject. Together, volumes 3 and 4 of *Cell Surface Reviews* provide an extensive and up-to-date discussion of what we feel to be the more important aspects of cell surface dynamics.

We thank the contributors for their authoritative and comprehensive chapters. We are also grateful to Judy Kaiser, Adele Brodginski, Shirley Guagliardi and Molly Terhaar for their assistance in preparing the edited manuscripts.

George Poste
Buffalo, New York
August, 1976

Garth L. Nicolson
Irvine, California
August, 1976

List of contributors

The numbers in parentheses indicate the page on which the authors' contributions begin.

Jan ANDERSSON (601), Department of Immunology, Biomedical Center, University of Uppsala, S-751 23 Uppsala, Sweden.

Robert W. BALDWIN (423), Cancer Research Campaign Laboratories, University of Nottingham, Nottingham NG7 2RD, England.

Matko CIKES (473), Swiss Institute for Experimental Cancer Research, Ludwig Institute for Cancer Research, Unit of Human Cancer Immunology, 1011 Lausanne, Switzerland.

David R. CRITCHLEY (307), Department of Biochemistry, University of Leicester, Leicester LE1 7RH, England.

Stefanello DE PETRIS (643), Basel Institute for Immunology, CH-4058 Basel, Switzerland.

Sten FRIBERG Jr. (473), Karolinska Hospital, Radiumhemmet, 104 01 Stockholm 60, Sweden.

Carl G. GAHMBERG (371), Department of Serology and Bacteriology, University of Helsinki, Haartmaninkatu 3, Helsinki 29, Finland.

Alan F. HORWITZ (295), Departments of Biochemistry and Biophysics and Human Genetics, University of Pennsylvania Medical School, Philadelphia, Pennsylvania 19174, U.S.A.

Robert HYMAN (513), Department of Cancer Biology, The Salk Institute for Biological Studies, P.O. Box 1809, San Diego, California 92112, U.S.A.

Tae H. JI (1), Division of Biochemistry, University of Wyoming, Laramie, Wyoming 82071, U.S.A.

J-C. KADER (127), Université de Paris VI, Uer 59-Physiologie Cellulaire, 12, Rue Cuvier, 75005 Paris, France.

Harold K. KIMELBERG (205), Division of Neurosurgery and Department of Biochemistry, Albany Medical College of Union University, Albany, New York 12208, U.S.A.

N. Scott McNUTT (75), Electron Microscopy Laboratory, Department of Pathology, Veterans Administration Hospital and University of California School of Medicine, San Francisco, California 94121, U.S.A.

Fritz MELCHERS (601), Basel Institute for Immunology, CH-4058 Basel, Switzerland.

Garth L. NICOLSON (1), Department of Developmental and Cell Biology, University of California, Irvine, California 92717, U.S.A.

George POSTE (1), Department of Experimental Pathology, Roswell Park Memorial Institute, Buffalo, New York 14263, U.S.A.

Michael R. PRICE (423), Cancer Research Campaign Laboratories, University of Nottingham, Nottingham NG7 2RD, England.

George F. SCHREINER (619), Department of Pathology, Harvard Medical School, Boston, Massachusetts 02115, U.S.A.

Karl-Gösta SUNDQVIST (551), Department of Immunology, Karolinska Institutet and Department of Immunology, The National Bacteriological Laboratory, S-105 21, Stockholm, Sweden.

Emil R. UNANUE (619), Department of Pathology, Harvard Medical School, Boston, Massachusetts 02115, U.S.A.

Michael G. VICKER (307), Department of Biochemistry, University of Leicester, Leicester LE1 7RH, England.

Contents

General preface v
Contents of previous and forthcoming volumes vii
Preface xi
List of contributors xv

1 The dynamics of cell membrane organization, by G.L. Nicolson, G. Poste and T.H. Ji

1. Introduction 1
   1.1. Some basic principles 1
   1.2. Some basic methodology 4
      1.2.1. Physical techniques 4
      1.2.2. Biological techniques 10
2. Dynamics of cell membrane components 14
   2.1. Lipid motion 14
      2.1.1. Lipid viscosity 15
      2.1.2. Lipid lateral motion 17
      2.1.3. Lipid perpendicular motion 19
      2.1.4. Lipid phase separation 21
   2.2. Protein and glycoprotein motion 26
      2.2.1. Lateral mobility of proteins and glycoproteins 26
      2.2.2. Ligand-induced redistribution of cell surface components 28
      2.2.3. Protein and glycoprotein turnover 31
3. Mechanisms of receptor control 34
   3.1. Planar (cis) control 34
      3.1.1. Planar or lateral associations 34
      3.1.2. Domain formation 37
   3.2. Trans-membrane control 38
      3.2.1. Peripheral membrane components 38
      3.2.2. Membrane-associated (cytoskeletal) components 40
4. Trans-membrane architecture 48

2 Freeze-fracture techniques and applications to the structural analysis of the mammalian plasma membrane, by N.S. McNutt

1. Introduction 75
2. Freeze-fracture technique 76
3. Interpretation of freeze-fracture images 80
4. Freeze-fracture images of plasma membranes 84
   4.1. The erythrocyte 84
   4.2. Other cells 95
5. Intercellular junctions 100
   5.1. Symmetrical junctions 101
      5.1.1. Adherens junctions 101
      5.1.2. Occludens junctions 103
      5.1.3. Gap junctions 107
         5.1.3.1. Small subunit gap junction 107
         5.1.3.2. Large subunit gap junctions 110
         5.1.3.3. Gap junction formation and degradation . . . 112
   5.2. Asymmetrical junctions 117
6. Conclusions 117

3 Exchange of phospholipids between membranes, by J-C. Kader

1. Introduction 127
2. The origin of the hypothesis for exchange of phospholipids within the cell . . 129
   2.1. In vivo experiments 129
   2.2. In vitro experiments 131
3. Exchange of phospholipids between membranes 132
   3.1. Identification of phospholipid exchange processes 132
   3.2. Distribution of phospholipid exchange processes 134
   3.3. Properties of phospholipid exchange between membranes . . 135
      3.3.1. Phospholipid exchange is time- and temperature-dependent . . 135
      3.3.2. Phospholipid exchange involves intact phospholipid molecules . . 135
      3.3.3. Phospholipid exchange is bidirectional 138
      3.3.4. Phospholipid exchange is energy-independent and does not require biologically active membranes 138
      3.3.5. Phospholipid exchange involves a major proportion of the phospholipid pool . . 138
      3.3.6. Phospholipid exchange does not involve all the phospholipid classes . . 139
      3.3.7. Phospholipid exchange is stimulated by the 105 000 g supernatant . . 139
4. Phospholipid exchange involving lipoproteins 140
   4.1. Lipoprotein–lipoprotein exchange 141
   4.2. Lipoproteins and cellular fractions 141
   4.3. Chylomicrons and serum lipoproteins 141
   4.4. Erythrocytes and plasma lipoproteins 142
   4.5. Tissues and the external medium 143
5. Involvement of proteins in the exchange of phospholipids between membranes . . 144
   5.1. Role of the pH 5.1 supernatant 144
      5.1.1. Generalization of the pH 5.1 supernatant stimulated exchange . . 144
      5.1.2. Which phospholipids are preferentially transferred? . . . 145
      5.1.3. Evidence for the protein nature of the active fraction . . 147
   5.2. Discovery of phospholipid exchange proteins (PLEP) 148
6. Phospholipid exchange in artificial (model) membranes 149
   6.1. Phospholipid exchange between liposomes and cell fractions . 149
   6.2. Exchange of phospholipids between liposomes 153

7. Purification and properties of phospholipid exchange proteins		154
7.1. Purification		155
7.1.1. Phosphatidylcholine exchange protein from beef liver (PC-PLEP)		155
7.1.2. PLEP from beef heart		157
7.1.3. PLEP from rat intestine		158
7.1.4. PLEP from beef brain (PI-PLEP)		158
7.1.5. PLEP from plant cytosols		159
7.2. Common properties of phospholipid exchange proteins		160
8. Do phospholipid exchange proteins act as phospholipid carriers?		163
8.1. Studies with phospholipid monolayers		163
8.2. Binding experiments		165
8.3. Net transfer		166
9. Nature of the interactions between phospholipid exchange proteins and bound phospholipids		168
9.1. Hydrophobic interactions		168
9.2. Electrostatic interactions		171
10. Control of phospholipid exchange by membrane surface charge and the mode of action of phospholipid exchange proteins		175
10.1. Liposome–liposome interactions		175
10.2. Liposome–mitochondria and liposome–microsome interactions		178
10.3. Mode of action of PLEP		181
11. Membrane asymmetry: implications for phospholipid exchange		182
11.1. Natural membranes		182
11.2. Model membranes		184
12. The physiological significance of phospholipid exchange		187
13. Unanswered questions		191

4 The influence of membrane fluidity on the activity of membrane-bound enzymes, by H.K. Kimelberg

1. Introduction		205
2. Lipid fluidity in model and biological membranes		209
2.1. Structure and formation of liposomes		209
2.2. Lipid fluidity and phase transitions in liposomes		210
2.3. Effects of cholesterol, divalent cations and pH on lipid fluidity		218
2.4. Effects of proteins on lipid fluidity		221
2.5. Lipid fluidity in biological membranes		224
3. Lipid fluidity and membrane enzymes		228
3.1. Lipid dependence of membrane enzymes		228
3.2. Membrane fluidity and Arrhenius plots of enzyme activity		231
3.3. Alterations in membrane lipid composition and enzyme activity		236
3.3.1. In *Escherichia coli*		236
3.3.2. *Mycoplasma* and *Acholeplasma*		238
3.3.3. Yeast		239
3.3.4. Effects of temperature acclimation in plants and poikilotherms		239
3.3.5. Diet-induced alterations		240
3.4. Lipid phase separation and enzyme activity		240
3.5. Correlation of Arrhenius plots of enzyme activity and lipid fluidity in natural membranes		244
3.5.1. Mitochondria		244
3.5.2. Sarcoplasmic reticulum and microsomal enzymes		248
3.5.3. Plasma membrane enzymes		249

3.6. Enzyme activity and lipid fluidity in reconstituted systems	256
3.7. Membrane microenvironments and lateral translational mobility	266
4. A regulatory role for membrane lipid fluidity?	269
4.1. Alterations in normal states	269
4.2. Alterations in disease	273
5. Concluding remarks	278

5 Manipulation of the lipid composition of cultured animal cells, by A.F. Horwitz

1. Introduction	295
2. Lipid metabolism	295
3. Lipid alterations	296
4. Some observations and generalizations	299
5. Conclusion	302

6 Glycolipids as membrane receptors important in growth regulation and cell–cell interactions, by D.R. Critchley and M.G. Vicker

1. Introduction	307
2. Glycolipid structure, biosynthesis and nomenclature	308
3. Glycolipids in normal and transformed cells	308
3.1. Enzymatic basis for change in glycolipid pattern on transformation	313
3.2. Subcellular distribution of glycolipids	316
3.3. Correlation between glycolipid changes and transformation	317
3.3.1. Generality of the effect of transformation on glycolipid pattern	317
3.3.2. Studies with viral mutants temperature sensitive for transformation	318
3.3.3. Studies with revertant cell lines	320
3.4. Molecular basis of the viral effect on glycolipid synthesis	321
3.5. Effects of exogenous glycolipids on cell growth	321
4. Variations in glycolipid pattern in normal cells	322
4.1. Growth dependent variation	322
4.1.1. Cell density-dependent glycolipids	322
4.1.2. Kinetics of the change in glycolipid pattern as related to cell density	324
4.2. Glycolipid metabolism as a function of cell cycle	326
4.3. Possible mechanisms to explain density-dependent glycolipids	327
4.3.1. Cell cycle effects	327
4.3.2. Synthesis of density-dependent glycolipids by transglycosylation	328
4.3.3. Enzyme induction	330
5. Exposure of glycolipids at the cell surface	330
5.1. In cultured cells	331
5.2. Possible mechanisms to explain alterations in exposure of glycolipids	334
6. Correlation between altered glycolipid pattern and malignancy	335
7. Glycolipids as cell surface receptors	338
7.1. Studies with cholera toxin	338
7.1.1. Binding specificity	338
7.1.2. Cholera toxin: interaction with adenyl cyclase in intact cells	341
7.1.3. Cholera toxin: interaction with adenyl cyclase in isolated membranes	343
7.2. Protein–glycolipid interactions in other systems	345

7.3.	Glycolipids as receptor for interferon	347
7.4.	Possible role of glycolipids as receptors for molecules involved in growth regulation	348
7.5.	Glycolipids as cell surface receptors involved in intercellular recognition	351

7 Cell surface proteins: changes during cell growth and malignant transformation, by C.G. Gahmberg

1.	Introduction	371
2.	The erythrocyte membrane as a model for the structure of mammalian cell surface membranes	372
	2.1. Lipids	372
	2.2. Proteins	373
	2.3. Carbohydrates	377
3.	Methods for studying cell surface proteins	378
	3.1. Isolation of plasma membranes	378
	3.2. Isolation of membrane proteins	380
	3.3. Polyacrylamide gel electrophoresis	382
	3.4. Specific labeling of the outer surface of the plasma membrane	384
	3.4.1. Labeling of cell surface carbohydrates	385
	3.4.1.1. The galactose oxidase method for labeling galactose and N-acetyl galactosamine	385
	3.4.1.2. The periodate-[^{3}H]NaBH$_4$ method for labeling sialic acids	386
	3.4.1.3. The cytidine monophosphate (CMP)-[^{14}C]sialic acid method for labeling cell surface glycoproteins	386
	3.4.2. Labeling of the polypeptide portion of cell surface glycoproteins	387
	3.4.2.1. The lactoperoxidase method for labeling cell surface proteins	387
	3.4.2.2. The pyridoxal phosphate-[^{3}H]NaBH$_4$ method for labeling cell surface proteins	387
	3.4.2.3. The [^{35}S]formyl methionyl sulfone methyl phosphate method for labeling cell surface proteins	388
	3.4.2.4. The [^{3}H]/[^{14}C]isothionyl acetimidate method for labeling cell surface proteins	388
	3.4.2.5. Photochemical labeling of cell surface proteins	388
	3.4.2.6. Labeling of cell surface proteins using transglutaminase	388
	3.5. Chemical analysis of membranes	389
	3.5.1. Proteins	389
	3.5.2. Carbohydrates	389
	3.6. Fluorescence microscopy and electron microscopy	390
4.	Studies of surface glycoproteins of normal and transformed cells	391
	4.1. External labeling of normal and malignant cells	391
	4.1.1. Fibronectin	391
	4.1.2. Other cell surface proteins	398
	4.2. Cell surface proteins and regulation of cell growth	399
	4.3. Cell surface antigens of normal and transformed cells	401
	4.4. Interaction of normal and transformed cells with lectins: insights into plasma membrane glycoprotein organization	402
	4.5. Glycopeptides and oligosaccharides from the surface of normal and transformed cells	404
	4.6. Transport of glucose and glucose analogs in normal and malignant cells	406
	4.7. Cell surface recognition and cellular adhesion	407

8 Shedding of tumor cell surface antigens, by M.R. Price and R.W. Baldwin

1. Introduction 423
2. Nature of tumor antigens and molecular expression at the cell surface 424
3. Tumor antigen shedding in vitro 432
   3.1. Background considerations 432
   3.2. Detection of antigens shed in vitro 435
   3.3. Induction of release of cell surface antigens by immune mechanisms 438
4. Detection of antigens shed in vivo 442
   4.1. Identification of tumor-associated antigens in body fluids 442
   4.2. Blocking factors 447
   4.3. Inhibitory factors 451
5. Immunobiological effects of circulating antigens 454
   5.1. Correlation of serum factors with tumor growth 454
   5.2. Immune responses to circulating tumor antigens in the tumor host 457
6. Conclusion 460

9 Expression of cell surface antigens on cultured tumor cells, by M. Cikes and S. Friberg Jr.

1. Introduction 473
2. Cell surface antigens in non-synchronized cultures 473
   2.1. Variations in cell surface antigen expression during a single growth cycle . . 473
   2.2. Cell cycle kinetics during a single growth cycle 477
3. Cell surface antigens in synchronized cultures 480
   3.1. Variations in cell surface antigen expression during the cell cycle 480
   3.2. Possible mechanisms of cell cycle-dependent antigen expression 484
4. Macromolecular synthesis and the expression of cell surface antigens 493
5. Culture- and transformation-induced alterations of cell surface antigen expression . 496
   5.1. Alterations of cell surface antigen expression in explanted cells 496
   5.2. Reversion of culture-induced antigenic changes by retransplantation of cultured cells into syngeneic hosts 498
   5.3. Alterations of cell surface antigen expression in transformed cells 499
6. Conclusions 500

10 Somatic genetic analysis of the surface antigens of murine lymphoid tumors, by R. Hyman

1. Rationale for the somatic analysis of surface antigens 513
2. Characteristics of murine lymphoid cell surface antigens which have been used in somatic genetic studies 517
   2.1. H-2 517
   2.2. TL 519
   2.3. Thy-1 522
3. Surface antigen variation in tumor cell populations 524
   3.1. Selection of tumor cell surface antigen variants 524
      3.1.1. Selection of H-2 antigen variants in vivo 524
      3.1.2. Selection of H-2 and HL-A variants in vitro 525
      3.1.3. Selection of variants for Thy-1 and TL from homozygous tumor cells . 526
   3.2. Are surface antigen variants of tumor cells mutants? 527
4. Genetic behavior of antigen loss variants 530

4.1. Antigen loss variants derived from heterozygous tumors	530
4.2. Antigen loss variants derived from homozygous tumors	532
4.3. Dominant "suppression" of antigen expression	536
5. Future prospects	538
5.1. Analysis of differentiation alloantigens	538
5.2. Tumor-associated antigens	538

11 Dynamics of antibody binding and complement interactions at the cell surface, by K.G. Sundqvist

1. Introduction	551
2. Cell surface molecules as antibody-binding structures	552
2.1. Accessibility	552
2.2. Valence	554
2.3. Number of antigenic sites per cell and antigenic density	554
2.4. Distribution	555
3. The precipitin reaction at the cell surface	556
3.1. General considerations of antibody structure and its implications for lattice formation	556
3.1.1. Multivalent interaction and affinity	557
3.1.2. Span and flexibility	558
3.2. Lattice formation by a single layer of antibody	558
3.2.1. Model systems	558
3.2.2. Experimental systems	560
3.3. Lattice formation by a double-layer of antibody	562
3.3.1. Model systems	562
3.3.2. Experimental systems	563
4. Prozone effects	565
4.1. Observations of prozone effect	565
4.2. Studies of the mechanism	565
5. Behaviour and fate of antibody following binding to the cell surface	571
5.1. Dissociation of antibodies from antigenic determinants	572
5.2. Release of antigen-antibody complexes from the cell surface	572
5.2.1. Demonstration of release and its differentiation from dissociation	572
5.2.2. Metabolic dependence of antibody release	573
5.2.3. Kinetics of release of antibodies from cells	573
5.3. Redistribution of plasma membrane components within the plane of the membrane	574
5.3.1. Characteristics of ligand-induced redistribution of surface antigens	575
5.3.2. Capping is influenced by the kind of antigen and the cell type	576
5.4. Endocytosis of ligand-receptor complexes	576
5.5. Mechanisms for the control of distribution and mobility of cell surface components and bound ligands	577
5.6. Factors determining the behaviour and fate of membrane-bound antibodies	579
5.6.1. The ligand	579
5.6.2. The cell	581
6. The interaction of complement with antibodies at the cell surface	582
6.1. Factors influencing the susceptibility of target cells to lysis by antibodies and complement	583
6.2. Cell cycle-dependent variation in cellular susceptibility to lysis by antibodies and complement	584
6.3. Influence of the mobility of membrane components on cellular susceptibility to lysis by antibodies and complement	585
6.4. Complement modulates the binding of antibodies and immune complexes to cells	589

12 Mitogen stimulation of B lymphocytes. A mitogen receptor complex which influences reactions leading to proliferation and differentiation, by J. Andersson and F. Melchers

1. Introduction	601
2. Lymphocyte heterogeneity	603
3. The small B lymphocyte	603
4. The mitogens	604
5. The induction of B lymphocytes by mitogens	606
6. Proliferation and maturation of B lymphocytes	608
7. The role of surface membrane-bound Ig in the induction of B cells to growth and differentiation	609
8. Mitogen-receptors on B cells	612

13 Structure and function of surface immunoglobulin of lymphocytes, by E.R. Unanue and G.F. Schreiner

1. Introduction	619
2. Detection of surface Ig	620
3. Class and antigen specificity of Ig	622
3.1. Class	622
3.2. Antibody specificity	623
3.3. Allelic exclusion	624
4. Topography and redistribution	626
5. Synthesis and dynamics of surface Ig	632
6. Summary	635

14 Distribution and mobility of plasma membrane components on lymphocytes, by S. de Petris

1. Introduction	643
2. General characteristics of redistribution of lymphocyte surface components	644
2.1. Lymphocyte surface components	644
2.2. Experimental methods	646
2.3. Early distribution studies of lymphocyte surface components	647
2.4. Basic redistribution phenomena	647
3. Normal distribution of membrane components	649
4. Metabolically independent redistribution: patching	653
5. Metabolically dependent redistribution: capping	659
5.1. Dependence on cell metabolism	660
5.2. Dependence on crosslinking	664
5.3. Kinetics of capping	667
5.4. Movement of surface molecules during capping	669
5.4.1. The polar movement	669
5.4.2. Independence of membrane components	670
5.4.3. Mechanism of segregation of membrane proteins	672
5.5. Cell movement, cell morphology and capping	675
5.5.1. Stimulation of motility and changes of cell shape	675
5.5.2. Relationship between cell movement and capping	677
5.5.3. Capping on cells bound to a solid substrate	680
5.6. Role of intracellular structures in capping	683

	5.6.1. Structural cellular components (microfilaments, microtubules)	683
	5.6.2. Evidence for a role of microfilaments	689
	5.6.3. Evidence for a role of microtubules	692
	5.6.4. Mechanisms of membrane–cytoplasm interaction	697
	5.7. Biological significance of capping	702
6.	Fate of labelled material	704
	6.1. Pinocytosis	704
	6.2. Shedding	709
	6.3. Antigenic modulation	710
	6.4. Resynthesis of surface components (recovery from modulation)	712
7.	Concluding remarks	713

Subject Index 729

The dynamics of cell membrane organization

Garth L. NICOLSON, George POSTE and Tae H. JI

1. Introduction

The exact molecular structure and organization of cell membranes has yet to be elucidated. We are, however, much closer than ever to determining at least the gross construction of certain cell membranes (reviews, Singer, 1974a; Steck, 1974; Nicolson, 1976a). In this article we will emphasize a number of general points concerning the dynamics of cell surface components which will serve as an introduction and background to the topics discussed in greater detail in the other chapters of this volume. No attempt will be made to provide a comprehensive survey of this enormous subject and, wherever possible, we have cited recent review articles which can be consulted for further information. Similarly, we will not dwell extensively on the various proposals concerning the precise structural organization of membranes. This is not because we feel that this information is any less important in achieving detailed insight into the structural and functional organization of the cell surface, but simply because there are many excellent reviews on membrane models to which the reader can refer (Table I). This chapter therefore represents a brief overview of the factors that are presently considered of importance in regulating plasma membrane organization and the dynamic properties of this important structure.

1.1. Some basic principles

Within the past few years a general consensus has been reached that the structure of different cellular membranes conforms to a number of basic principles (reviews, Korn, 1969a,b; Goldup et al., 1970; Singer, 1971, 1974a; Singer and Nicolson, 1972; Wallach, 1972; Bretscher, 1973; Overath and Träuble, 1973; Fox, 1975a). The first of these is that the major membrane lipids, such as the phospholipids, are arranged in a planar bilayer configuration which is predominantly in a "fluid" state under physiological conditions (reviews, Singer, 1971; Bangham, 1972; Singer and Nicolson, 1972; Träuble, 1972; Edidin, 1974; Fox, 1975a; Nicolson, 1976a; and Kimelberg, this volume, p. 205).

TABLE I
SOME RECENT REVIEWS ON CELL SURFACE STRUCTURE AND DYNAMICS

(1) *Physical methods for studying cell membranes*
Weiss (1969)
McConnell and McFarland (1970)
Gitler (1972)
Horwitz (1972)
Wallach (1972)
Chapman (1973)
Oseroff et al. (1973)
Edidin (1974)
Wallach and Winzler (1974)

(2) *Lipids of cell membranes*
Bangham (1972)
Law and Snyder (1972)
Träuble (1972)
Bretscher (1973)
Papahadjopoulos and Kimelberg (1973)
Zwaal et al. (1973)
Fox (1975a)

(3) *Proteins of cell membranes*
Branton and Deamer (1972)
Guidotti (1972)
Razin (1972)
Singer and Nicolson (1972)
Steck and Fox (1972)
Juliano (1973)
Singer (1974a,b)
Steck (1974)
Vanderkooi (1974)

(4) *Oligosaccharides of cell membranes*
Roseman (1970)
Winzler (1970)
Kraemer (1971)
Nigam and Cantero (1972)
Parsons and Subjeck (1972)
Kefalides (1973)
Nicolson (1974)
Hughes (1975)

(5) *Dynamics of cell surface receptors*
Unanue and Karnovsky (1973)
Cuatrecasas (1974)
Edidin (1974)
Nicolson (1974, 1976a,b)
Singer (1974b)

(6) *Specialized structures of cell surfaces*
Branton (1971)
McNutt and Weinstein (1973)
Turner and Burger (1973)
Gilula (1974)
Staehelin (1974)
Weinstein et al. (1976)

(7) *Cell surfaces, transformation and morphogenesis*
Curtis (1967)
Weiss (1967)
Wallach (1972, 1975)
Burger (1973)
Oseroff et al. (1973)
Hynes (1974, 1976)
Robbins and Nicolson (1975)
Nicolson (1976b)
Nicolson and Poste (1976a)

The lipid bilayer is not a continuous structure, however, and is interrupted by numerous proteins which are inserted to varying degrees into the bilayer (Branton, 1969; Pinto da Silva and Branton, 1972; and chapters in this volume by McNutt [p. 75] and Gahmberg, [p. 371]. In addition, in at least some membranes the bilayer is asymmetric with respect to the distribution of specific phospholipid classes in the inner and outer halves of the bilayer (reviews, Bretscher, 1973; Zwaal et al., 1973; and Kimelberg, this volume [p. 205]).

The proteins of cell membranes are quite heterogeneous (reviews, Guidotti, 1972; Steck and Fox, 1972; Singer, 1974a; Wallach and Winzler, 1974) and can be operationally divided into two main classes: integral and peripheral (Singer and Nicolson, 1972; Singer, 1974a) or intrinsic and extrinsic (Capaldi and Green, 1972).

The integral membrane proteins (and glycoproteins) are globular, bimodal proteins which interact with both the hydrophobic (hydrocarbon) and hydrophilic regions of the membrane. This class of membrane proteins is characterized by its interactions with membrane lipids driven by the favorable entropy gained through sequestering integral hydrophobic structures away from the bulk aqueous phase. Consequently, these proteins are intercalated into the hydrophobic core of the lipid bilayer to varying depths dependent on their amino acid sequences and three-dimensional folding, while the hydrophilic portion of the molecules protrude from the bilayer into the external and/or internal aqueous phase. Some integral proteins span the entire bilayer and thus have hydrophilic regions protruding at both the external and cytoplasmic surfaces of the membrane (see Bretscher, 1971a,b; Segrest et al., 1973; Morrison et al., 1974; Hunt and Brown, 1975; and Gahmberg, this volume, p. 371). Certain integral proteins are also thought to exist as oligomeric complexes (reviews, Guidotti, 1972; Pinto da Silva and Nicolson, 1974; Wang and Richards, 1974; Hynes and Pearlstein, 1976; Nicolson, 1976a), and there is evidence indicating that a number of these integral proteins interact with peripheral membrane proteins (Nicolson and Painter, 1973; Elgsaeter and Branton, 1974; Ji and Nicolson, 1974; Elgsaeter et al., 1976).

In contrast to the integral membrane proteins, peripheral proteins are only weakly bound to the surfaces of membranes, and their binding is not dependent upon lipid–protein interactions for stability. Peripheral proteins can associate with integral membrane proteins, and perhaps also glycolipids, by ionic interactions that are usually disrupted by high salt concentrations or treatment with chelating agents.

The plasma membrane in its most simple form can therefore be viewed as a two-dimensional solution of a mosaic of integral membrane proteins embedded in a fluid lipid bilayer, with peripheral proteins bound loosely to either surface (Singer and Nicolson, 1972). This concept of membrane architecture has two important implications for membrane function.

First, this arrangement permits membrane components to be organized asymmetrically, thus allowing specific membrane components to be localized predominantly or exclusively in the outer or inner half of the membrane. For example, membrane glycoproteins and glycolipids exhibit striking asymmetry in that their carbohydrate residues are exposed exclusively on the outer surface of the membrane where they function as specific receptors for antibodies, hormones, lectins, viruses and other agents (review, Nicolson, 1976a). Maintenance of membrane asymmetry dictates that components oriented to either face of the bilayer, no matter whether on the cytoplasmic or the external surface, should not be able to rotate at appreciable rates from one side of the membrane to the other (Singer and Nicolson, 1972). As discussed below (section 2.1.3), such restraints to trans-membrane rotation do exist, and this arrangement is well suited to the vectorial flow of information across the cell surface.

The second important feature of this type of structural arrangement is that components can diffuse laterally within the plane of the membrane (reviews, Edidin, 1974: Cherry, 1975, 1976; McConnell, 1975a,b; Nicolson, 1976a; Nicolson and Poste, 1976a), permitting rapid and reversible changes in the topography of specific surface components. Measurements of the lateral mobility of membrane proteins (section 2.2.1) has revealed that some diffuse freely while others move much more slowly and appear to be under restraint. Identification of a graded hierarchy of mobilities for different components within the plasma membrane, and even within specific regions of the membrane, suggests that the cell may possess a control mechanism that restricts the mobility of certain components in order to maintain a specific topographic arrangement or "pattern" of components on the cell surface (Nicolson and Poste, 1976a).

It is clear, however, that the topographic arrangement of cell surface components is not a static phenomenon, and rapid and reversible changes in the topography of surface macromolecules can occur in response to both intra- and extracellular stimuli. While a considerable degree of speculation must currently surround the interpretation of many aspects of this subject, sufficient experimental data are already available to support a functional relationship between the topography and dynamics of cell surface macromolecules and the control of cell surface properties.

1.2. Some basic methodology

1.2.1. Physical techniques

A wide variety of physical techniques have been used to study the structure and dynamics of biological membranes. These fall into two general categories: those that can directly probe membrane structure with minimal perturbation of the membrane (X-ray diffraction, nuclear magnetic resonance [NMR] spectroscopy, Raman spectroscopy); and those that require the introduction of an extrinsic probe molecule into the membrane to monitor events at specific sites (electron spin resonance [ESR] spectroscopy; optical spectroscopy of chromophore-labeled probe molecules and fluorescence microscopy of labeled membrane molecules). Operationally, these techniques can be subdivided further on the basis of their ability to examine lipid or protein environments, or both. These various techniques have been reviewed fully by Oseroff et al. (1973), Wallach and Winzler (1974) and Podo and Blasie (1975) and only an outline of their general value and limitations will be given here.

X-Ray diffraction techniques have provided useful information on membrane structure, particularly in the examination of lipid organization in lipid-water dispersions (Luzzati et al., 1966; Luzzati, 1968, 1974; Overath and Träuble, 1973; Shipley, 1973; Gulik-Kryzwicki, 1975). Such studies have confirmed the existence of a bilayer structure for hydrated membrane lipids. In more complicated biological membrane systems such as myelin (Blaurock, 1971; Caspar and Kirschner, 1971; Worthington and King, 1971; Worthington, 1972), red blood cell membranes (Engelman, 1969; Wilkins et al., 1971); *Mycoplasma laidlawii* cell membranes (Engelman, 1971) retinal disc membranes (Blasie and Worthington, 1969; Blasie et al., 1969); Sindbis virus (Harrison et al., 1971b) and purple membranes from *Halobacterium halobium* (Blaurock and Stoeckenius, 1971), X-ray diffraction profiles are consistent with the predominant arrangement of lipids in a bilayer configuration. In addition, X-ray diffraction profiles of biological membranes (Blaurock, 1971, 1973; Engelman, 1971) virtually eliminates the proposal that the predominant location for membrane proteins is in a continuous coat at each surface of the phospholipid bilayer as suggested in the original Danielli–Davson model of membrane structure (Danielli and Davson, 1935).

In general, diffraction techniques cannot be used to gain information about the dynamics of molecular motion within a structure, and it is therefore necessary to correlate structure defined by diffraction techniques with information obtained from structural probes sensitive to molecular motion such as the various types of spectroscopy described below. Molecular movement can, of course, be inferred from diffraction data in which different patterns are detected in an initial membrane state and another (final) membrane state (see Blasie et al., 1969).

Podo and Blasie (1975) have pointed out the potential value of inelastic and quasi-elastic neutron scattering techniques as potentially useful tools for detecting the vibrational and diffusive motions, respectively, of membrane molecules, but their uses have so far been confined to lipid bilayer model membranes.

Insight into the molecular structure and conformation of phospholipid molecules within bilayers has also been obtained by NMR spectroscopy (reviews, Chapman and Dodd, 1971; Horwitz, 1972; Levine, 1972; Metcalfe, 1975; Podo and Blasie, 1975). Most information has been obtained from the spectra of ^{1}H, ^{13}C and ^{31}P environments in phospholipid bilayers. Collectively, such studies indicate that while the fatty acid chains within the hydrocarbon core of the bilayer are in a generally fluid state, the rate and anisotropy of intramolecular motion varies considerably within different regions of the lipid molecules. Data have been obtained which indicate that there is a flexibility gradient along the acyl chain of phospholipid molecules such that the anisotropic motion of lipid methylene groups increases toward the center of the bilayer with a large increase occurring near the terminal methyl ends of the hydrocarbon chains (see section 2.1.1 and Kimelberg, this volume, p. 205). Similar conclusions have been reached using Raman spectroscopy (Lippert and Peticolas, 1972). NMR techniques have also proved useful in examining the factors that affect both phase transitions and phase separations in lipid bilayers (section 2.1.4).

The more instructive applications of NMR have been in studies on the organization of pure phospholipids and mixtures of phospholipids in simple model membranes such as lipid vesicles. Attempts to use NMR techniques on unmodified biological membranes have encountered considerable technical difficulties in resolving ambiguities in the interpretation of proton resonance line widths (see Metcalfe, 1975).

The general conclusions obtained from NMR spectroscopy on the structure and dynamics of phospholipid molecules within bilayers and in the vicinity of proteins have been confirmed and, in some cases, extended by ESR studies using paramagnetic probes introduced into lipid vesicles and certain natural membranes (reviews, McConnell and McFarland, 1970; Jost et al., 1971; Levine, 1972). The most popular spin label probes have been nitroxide-containing amphipathic lipid structures (Fig. 1) which report on the local membrane

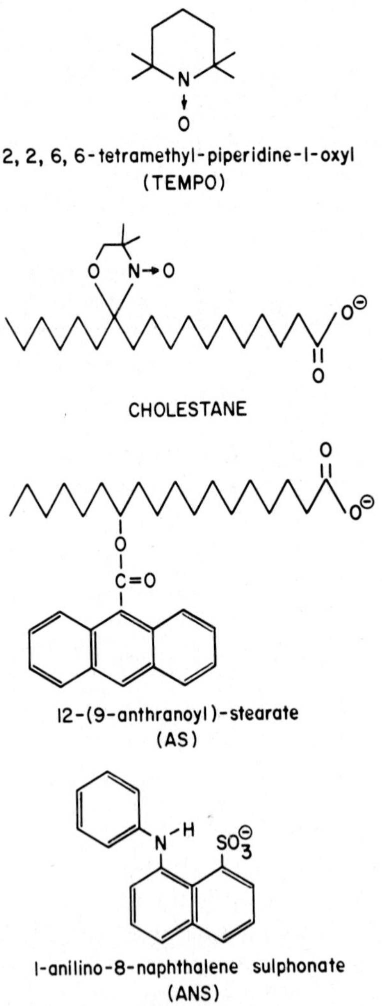

2, 2, 6, 6-tetramethyl-piperidine-1-oxyl
(TEMPO)

CHOLESTANE

12-(9-anthranoyl)-stearate
(AS)

1-anilino-8-naphthalene sulphonate
(ANS)

Fig. 1. Some common fluorescent and spin-label probes used in membrane research.

environment in the region of the probe (McConnell and McFarland, 1970; Jost et al., 1971). Paramagnetic probes of this type show very similar resonance spectra in both simple model membranes, such as lipid vesicles, and several natural membranes. The data consist of two partially overlapping signals originating from the probe fractions dissolved in the non-polar regions of the membrane and the external aqueous solution. The structure and mobility of (hydrocarbon regions) phospholipid molecules in model bilayers and natural membranes can be examined by analysis of the resonance spectra from membrane-intercalated phospholipid and fatty acid probes spin-labeled at selected positions. Experiments of this kind have provided useful information on lipid and protein packing, lateral diffusion of membrane lipids, lipid–protein interactions, the fluidity of membrane environments and the rate of transmembrane rotation (so called flip-flop) of molecules across the bilayer.

Dynamic measurements using spin-labeled probes will be discussed again in section 2.1, and it suffices to say here that ESR data have confirmed the existence of a lipid bilayer in cell membranes of erythrocytes (Hubbell and McConnell, 1968, 1969; Hubbell et al., 1970; Landsberger et al., 1971; Simpkins et al., 1971b), mycoplasma (Rottem et al., 1970; Tourtellotte et al., 1970) mouse fibroblasts (Wisnieski et al., 1974), nerves (Hubbell and McConnell, 1968, 1969; Simpkins et al., 1971a) and influenza (Landsberger et al., 1972), Newcastle disease (Wisnieski et al., 1974), equine encephalomyelitis (Hughes and Pederson, 1975), and murine leukemia viruses (Landsberger et al., 1971).

In common with all techniques in which a probe molecule is introduced into the membrane, ESR techniques suffer from the potential problem that the probe molecules themselves may cause a perturbation in the membrane because they introduce impurities which can create local molecular interactions not previously present in the membrane. The extent of the perturbation will, of course, depend on the physicochemical nature of the probe and its environment within the membrane. Since probes only "see" an approximation of the natural membrane condition, the less the perturbation the better the approximation. Additional problems are created by the possibilities that some probes may localize preferentially within more "fluid" regions of the membrane, and/or the probe may not be able to interact with certain membrane molecules due to alterations in their native structure (Oseroff et al., 1973; Wallach and Winzler, 1974). For example, Cadenhead et al. (1975) found that 12-nitroxide stearic acid, an ESR probe for fatty acid environments, does not behave like stearic acid in membranes and appears to "see" a viscosity environment approximately 10°C higher compared to the untagged molecule.

A further problem that arises in some ESR studies concerns the question of probe location. In simple systems with mainly one type of membrane, cellular penetration of the probe will probably not affect the interpretation of experimental data. However, in complicated multicompartment cells which possess a variety of intracellular membranes (nuclear, mitochondrial, endoplasmic reticulum, etc.), as well as hydrophobic non-membrane compartments, there is no reason to assume that the probe will localize exclusively in the plasma membrane. The interpretation of ESR data as providing information solely on the plasma membrane environments can only be justified if accompanied by information from subcellular fractionation studies showing the exact localization of the probe. Yet another problem, which has not received adequate recognition, is that nitroxide spin-labels are inactivated (reduced) to different degrees by different cells (Kaplan et al., 1973). Thus, differential uptake of spin probes by different cells, together with differing reduction of free radicals (e.g., due to variation in the rates of cellular aerobic glycolysis) may well frustrate comparative studies. This issue is particularly pertinent to the use of ESR techniques to define differences in membrane properties in tumor cells and their normal counterparts, since many tumor cells exhibit enhanced aerobic glycolysis.

Another useful tool for studying the lipids of biological membranes has been differential scanning calorimetry (DSC). Liquid crystal-gel transitions in lipid bilayers are endothermic processes which occur at specific transition temperatures, and they can be carefully monitored by this technique (reviews, Papahadjopoulos and Kimelberg, 1973; Kimelberg, this volume, p. 205). Although DSC measurements have been made on a limited number of natural membranes and/or mixtures of lipids extracted from natural membranes (see Steim et al., 1969; Overath et al., 1970), this technique is better suited for the investigation of the thermotropic behavior of pure phospholipids or phospholipid mixtures in model membranes.

Fluorescence probes can be used to study both membrane lipid and protein environments (Waggoner and Stryer, 1970). Fluorescence polarization techniques have been used extensively to monitor the fluidity and microviscosity parameters of the lipid bilayer in both model membranes and complex cellular membranes (Waggoner and Stryer, 1970; Shinitzky et al., 1971; Rudy and Gitler, 1972; Cogan et al., 1973; Vanderkooi and Callis, 1974; Papahadjopoulos et al., 1975a). Fluorescence probe molecules such as 1-anilino-8-naphthalenesulfonate (ANS) (Fig. 1) have been incorporated into plasma membranes in model lipid systems (Lesslaver et al., 1971) as well as in erythrocyte (Weiderkamm et al., 1970; Wahl et al., 1971; Fortes and Hoffman, 1971) electroplax (Kasai and Changeux, 1969; Patrick et al., 1971), and other surface membranes (Kita and Van der Kloot, 1972). The fluorescent quantum yield of ANS is enhanced nearly 100 times when incorporated from solution into phospholipid model membranes. The aromatic fluorophore 1,6,-diphenyl-1,3,5,-hexatriene has also been used by a number of investigators to probe lipid environments in model (Shinitzky and Barenholz, 1974; Papahadjopoulos et al., 1975a) and natural membranes (Shinitzky and Inbar, 1974; Poste et al., 1975a).

Bashford et al. (1976) have discussed the three major requirements affecting the validity of the fluorescent probe approach in studies of natural and artificial membranes: "Firstly, the location of the probe in the membrane must be known in order to relate the observed properties of the probe to the nature of its specific environment. Secondly, the introduction of the probe must not induce major perturbations in the system under investigation. Thirdly, the fluorescence properties of the probe must reflect its behavior in the membrane phase." Unfortunately, while the probes mentioned above meet these criteria in so far as studies with simple model membranes are concerned, they have shortcomings in the study of natural membrane environments. Like the spin-label probes, they suffer from the drawback that they can perturb membrane environments and localize preferentially in "fluid" regions of the membrane, and are also able to occupy membrane sites in both lipids and proteins (Feinstein et al., 1975). In addition, they localize to both the plasma membrane and various intracellular membranes (Poste et al., 1975a). An interesting example of the capacity of fluorescence probes to perturb membranes is the finding that increasing concentrations of ANS induce bovine erythrocytes to undergo a biconcave disc to sphere to crenated sphere transformation (Yoshida and Ikegami, 1974).

Fluorescent reporter groups attached to membrane proteins have enabled investigators to study the shape, mobility and disposition of membrane proteins using fluorescent depolarization and triplet-triplet decay (see Cherry, 1975). Using energy transfer between groups on the same molecule and calculating their distance of separation, Wu and Stryer (1972) were able to demonstrate that bovine rhodopsin, a membrane photoreceptor glycoprotein, is an oblong molecule. The rotational nature of the rhodopsin chromophore from retinal membrane in its natural membrane environment was measured by Brown (1972) and Cone (1972) using dichroism techniques. They found that under physiological conditions rhodopsin rotates so fast that polarized light reveals no dichroism unless the membranes are first fixed with glutaraldehyde to stop protein molecular rotations. However, by use of flash photolysis, Cone (1972) was able to measure the fast dichroic decay of a labile intermediate, lumirhodopsin, and hence calculate the rotational diffusion rate of rhodopsin.

Galla and Sackmann (1974) have developed an optical method of measuring lateral diffusion based on the formation of excited aromatic dimers (excimers). These excited states can only form after molecular collision, so excimer formation is a diffusion controlled process affected by the physical state of the membrane.

A number of optical techniques have been used to probe the conformation of membrane proteins (reviews, Wallach and Winzler, 1974; Wallach, 1975). Optically active substances such as proteins exhibit different refractive indices for left- and right-circularly polarized light, producing a change in optical rotation with the wavelength of light. This phenomenon is referred to as optical rotatory dispersion (ORD). A related technique for analysis of molecular conformations, circular dichroism (CD), is based on the different absorption of left- and right-circularly polarized light as a function of wavelength. The optical activity of peptide bonds arises from $\pi^\circ \to \pi^-$ electronic transitions near 190 nm and $n \to m$ transitions near 220 nm. The overall optical properties of polypeptides depends upon the spatial relationships of the individual peptide bonds, and α-helical, β- and "unordered" conformations yield distinct ORD and CD spectra.

Analysis of the secondary structure of membrane proteins has also been attempted using infrared (IR) spectroscopy (review, Wallach, 1975). IR measurements in some regions of the amide absorption band provide information on the folding of the peptide backbone and hydrogen bonding with other peptide linkages. Polypeptides in α-helical and/or unordered conformations exhibit IR spectra very different from those of β-structures, particularly in the Amide I region. Thus, the Amide I band of α-helical polypeptides lies at $1652\,\text{cm}^{-1}$ to $1656\,\text{cm}^{-1}$, but occurs at $1630\,\text{cm}^{-1}$ in β-structured polypeptides with an additional band near $1690\,\text{cm}^{-1}$ being found in antiparallel-β-structures (see Wallach, 1975). This technique does not, however, permit easy distinction between "unordered" and α-helical conformations, though this can be obtained by dichroism measurements.

Attempts to identify protein conformations in membranes by IR spectroscopy and optical activity measurements are based on two assumptions. The first is that the spectral properties of membrane proteins match the properties of the diverse synthetic polypeptides of known conformation which are used for comparison. The second is that the right-handed α-helix, the β-pleated sheet and "random" (unordered) conformations are the only ones present in the peptide arrays of membrane proteins. These assumptions, together with technical problems associated with the light-scattering properties of isolated membranes, have generated considerable controversy over the interpretation of data obtained with these methods, particularly optical activity measurements (review, Wallach, 1975). However, the extensive work done in this area has at least defined some important general patterns. Thus, despite technical obstacles that hinder exact calculations, these methods have shown that the proteins of most biomembranes contain more than 40% of their peptide linkages in a right-handed α-helical conformation (see, Lenard and Singer, 1966; Wallach and Zahler, 1966; Glaser et al., 1970; Avruch and Wallach, 1971; Glaser and Singer, 1971; Graham and Wallach, 1971; Urry, 1972; Wallach, 1975). These observations therefore indicate that globular structure predominates in membrane proteins and that there is little evidence to support earlier proposals for extensive β-structure as suggested in the original Davson–Danielli model of membrane structure in which the proteins were envisaged as extended β-structures situated on either side of the lipid bilayer with no intercalation of the proteins into the bilayer.

It should be remembered, however, that while many membranes (including the plasma membrane) contain proteins of more than average helicity, certain membranes (such as the inner mitochondrial membrane) appear to contain proteins with appreciable proportions of β-structure. Furthermore, in keeping with the dynamic properties of membranes, the secondary structure of proteins may well vary with changes in metabolic and functional activity (see Wallach, 1975).

1.2.2. Biological techniques
The main biochemical and biological techniques that have been useful in determining membrane dynamics are conventional and UV-fluorescence mi-

croscopy, electron microscopy (including freeze-cleavage electron microscopy) and radioisotope labeling methods.

One of the simplest demonstrations of the dynamic nature of cell surface organization is the movement and redistribution of large particles when attached to cell surface which can be seen by conventional light microscopy. Gold (Albrecht-Bühler, 1973; Albrecht-Bühler and Solomon, 1974) and carbon particles (Ingram, 1969; Abercrombie et al., 1970) can bind to the cell surface and undergo substantial movement on the cell surface. In many cases, it appears that particle movement is non-random and may follow membrane flow during cell motility (review, Trinkaus, 1976).

Much of our current understanding on the ability of membrane proteins and glycoproteins to move and undergo topographic rearrangements within membranes has come from studies on the interaction of multivalent ligands such as antibodies and plant lectins with the appropriate cell surface receptors. An extensive literature has developed over the past five years showing that ligands of this type can induce varying degrees of redistribution of cell surface receptors (reviews, Unanue and Karnovsky, 1973; Edidin, 1974; Nicolson, 1974; 1976a,b; and chapter by Sundqvist in this volume [p. 551]). Ligand-induced receptor redistribution can be easily monitored by UV light microscopy using fluorescein- and/or rhodamine-labeled antibodies or lectins. The general characteristics of the various kinds of receptor redistribution phenomena will be discussed in section 2.2.2.

By using fluorescent ligands directed against different surface receptors, information can also be obtained on the differing mobilities of specific receptors (Edidin and Weiss, 1972, 1974; Stackpole et al., 1974b,c; Mintz and Sachs, 1975). For example, in their now classic studies of receptor mobility on virus-induced interspecific heterokaryons, Frye and Edidin (1970) found that the intermixing of mouse and human cell surface antigens on hybrid cells occurred at different rates. These experiments were performed using Sendai virus to induce fusion of mouse and human cells followed by labeling with the appropriate alloantisera after which fluorescein-labeled anti-mouse and rhodamine-labeled anti-human antibodies were added (indirect immunofluorescent techniques) to demonstrate the mouse and human antigenic sites at different times after fusion.

The use of two different fluorescent surface labels that bind to unlike receptors can also be used to define the topographic relationship(s) of different surface components. For example, Neauport-Sautes et al. (1973a,b) showed that ligand-induced redistribution of H-$2D$ alloantigens on the surfaces of mouse lymphocytes was not accompanied by movement of H-$2K$ alloantigens, suggesting that these two antigenic specificities reside on separate surface molecules. Using this strategy, Poulik et al. (1973) and Solheim and Thorsby (1974) demonstrated co-migration of β_2-microglobulin and HL-A histocompatibility antigens on human lymphocytes, indicating a close topographic relationship between these components. In experiments of this type, however, the amount of ligand added is often crucial. For example, de Petris (1975) has shown that at low concentrations of concanavalin A (usually <10 μg/ml), certain high affinity lectin receptors on mouse lymphocytes can be labeled and redistributed independently from cell surface immunoglobulins, but at higher concentrations (>25 μg/ml) concanavalin A binds to both types of receptors, resulting in co-capping of the two determinants.

Accurate measurements of the mobilities of fluorescence-labeled ligands on the cell surface can also be used to obtain lateral diffusion coefficients for plasma membrane receptor molecules. This approach, referred to as fluorescence recovery after photobleaching, involves labeling a specific cell surface receptor with an appropriate fluorescent ligand followed by laser-induced bleaching of the fluorescence in a small region of the cell surface. The time required for the recovery of the original amount of fluorescence in the photobleached region is then measured. Assuming that spontaneous recovery of fluorescence has been excluded and that the exciting beam does not produce any photobleaching, the kinetics of fluorescence recovery must reflect the rate of lateral diffusion of neighboring unbleached fluorophores into the bleached region. These data can then be used to calculate a diffusion coefficient for the surface receptor in question (see section 2.2.1). While this method has so far been used only with membrane proteins and glycoproteins, there is no a priori reason why the same approach could not be used to monitor the translational mobility of membrane glycolipids or even lipids. A limiting factor in the case of membrane lipids, however, would be the problem of obtaining a suitable ligand.

Ligand-induced movement and redistribution of membrane components can also be monitored at the light microscope level by immunoautoradiography using radiolabeled antibodies. The autoradiograms obtained are generally of low resolution, but certain gross movements of antigenic sites such as ligand-induced receptor capping, can be detected (Perkins et al., 1972; Antoine et al., 1974). In general terms, approaches such as these are of little advantage since antibody-induced redistribution of receptors can be detected much more easily (and more quickly) using labeled antibodies conjugated to fluorescent molecules.

Electron microscopy offers considerably greater resolution than light microscope techniques, but, with the exception of freeze-cleavage electron microscopy, observations cannot be made on unfixed material. In addition, dynamic measurements of the mobility and redistribution of membrane components cannot be made on the same cell, and inferences concerning these processes are made by comparing static images of membranes in one state with those of similar membranes after a natural or experimental perturbation of the membrane.

There are a number of electron histochemical techniques that are suitable for the localization of surface antigens, lectin receptors, anionic sites and other membrane components (reviews, Hsu, 1967; Avrameas, 1970; Nicolson, 1976c). These techniques are based on two general strategies. The first involves direct labeling of surface components by covalent attachment of electron dense markers such as cationized ferritin or colloidal iron directly to the membrane component(s) in question, or using electron-dense ligands such as ferritin-conjugated antibodies, etc. to label specific receptors. In the second approach an electron-dense marker is produced by a chemical reaction on the cell surface involving directly or indirectly the membrane component of interest. A typical method is to introduce an external reagent such as an enzyme or an enzyme-conjugated antibody which can interact with the component under study to produce an electron-dense product or a product which subsequently can be made electron-dense.

With either approach, dynamic experiments are usually conducted by labeling cells (one-step labeling for direct, two-step labeling for indirect procedures), removing free unbound reagent by washing, etc. and then incubating the labeled cells for various times followed by fixation. Prefixed controls or unfixed cells held at low temperature (usually 0–4°C) are used for comparison of receptor movement or ligand-induced clustering.

Ultrastructural studies have also been done on membranes modified by specific enzymes or differential extraction procedures to test whether selective alteration or removal of particular membrane components is accompanied by changes in membrane structure. While having proven useful for monitoring gross changes in membrane organization, this approach in its present stage of development is not generally capable of demonstrating subtle changes in the molecular orientation and arrangement of individual membrane components.

The ultrastructural technique of freeze-cleavage electron microscopy (this term includes both freeze-fracture and freeze-etching) is now established as a valuable tool for analyzing the structural organization of the hydrophobic regions of cell membranes (see Branton, 1966, 1971; Pinto da Silva and Branton, 1970; Meyer and Winkelmann, 1972; Stolinski and Breathnach, 1976; and McNutt, this volume, p. 75). When membranes are rapidly frozen and fractured, they cleave down a plane formed by the middle of the lipid bilayer (Pinto da Silva and Branton, 1970; Tillack and Marchesi, 1970). The two complementary fracture faces produced by this technique thus provide information about the structural organization of the apolar center of membranes. The true inner and outer membrane surfaces are not revealed by fracturing but can be visualized by freeze-etching or sublimation of some of the ice surrounding the membrane. By combining freeze-fracturing and -etching, it is possible to observe the core of the membrane, as represented by the fracture faces, and the membrane surfaces, as revealed by freeze-etching.

The fracture faces of most membranes are populated with intramembranous particles which can be revealed by replica and heavy metal shadowing techniques. There is now convincing evidence that the intramembranous particles represent proteins or oligomeric protein complexes penetrating into and/or through the bilayer (MacLennan et al., 1971; Hong and Hubbell, 1972; Chen and Hubbell, 1973; Segrest et al., 1974; Vail et al., 1974; Kleemann and McConnell, 1976; and McNutt, this volume, p. 75). These particles are capable of rapid lateral movement under conditions where the membrane has been perturbed (see section 2.2.1). Particle movements can be detected by comparison to untreated control membranes that are rapidly frozen before the perturbation occurs.

The study of membrane dynamics is not limited to consideration of the complex structural and topographic rearrangements exhibited by different membrane components. An equally important aspect of this subject is the synthesis, assembly and turnover of different membrane components. An impressive array of analytical techniques are now available for monitoring the turnover of most membrane components. These have been discussed fully elsewhere (Wallach, 1975; Poste and Nicolson, 1977).

Turnover rates of membrane proteins and glycoproteins have been examined generally in two ways. One method is to label proteins and glycoproteins in cells with their radioactive precursor amino acids and carbohydrates. Leucine has been popular for protein labels, and glycoproteins have been labeled with fucose or glucosamine. Fucose has been demonstrated to be mainly incorporated into glycoproteins, and to a lesser extent into glycolipids, rather than broken down or converted to other metabolites. This is an important requirement for labeling studies of this type (Bosmann et al., 1969; Herscovics, 1970; Sakiyama and Burge, 1972). An alternative to the nonspecific labeling of proteins and glycoproteins is to identify marker molecules by enzyme assays and surface labels (see chapter by Gahmberg, this volume, p. 371). Surface-specific labels eliminate the need for the often difficult membrane isolation procedures. Another serious problem in studying the turnover of membrane components is the reutilization of the catabolized pieces of labeled molecules.

Turnover rates of membrane constituents are determined by their rates of synthesis, incorporation into membrane and degradation. The rates of synthesis and degradation can be studied independently by examining the labeling rate of precursors and the loss of prelabeled molecules, respectively. The net difference between synthesis and degradation may be used for estimating turnover rates. Arias et al. (1969) introduced a double-labeling method to measure turnover rates directly. Cells are labeled with a precursor of one isotope and then the cells are introduced to the same precursor of another isotope. The ratio of the two different isotopes is counted as the turnover rate of a molecule.

2. Dynamics of cell membrane components

2.1. Lipid motion

The motion and phase separations of lipids in phospholipid dispersions and cell membranes have been examined by X-ray diffraction, NMR, ESR, calorimetry and other physical techniques. Information has been obtained on lipid chain flexibility, rotational and lateral diffusion, viscosity, phase separation, and planar associations with other lipids and with membrane proteins.

2.1.1. Lipid viscosity

Hydrophobic interactions stabilize most artificial and biological membranes. The membranes are formed mainly from a matrix of "fluid" lipid which endows them with an average viscosity not unlike that of light machine oil (reviews, Edidin, 1974; Nicolson, 1976a). The bending and flexing of fatty acid hydrocarbon chains in the lipid matrix has been estimated using NMR and ESR. These techniques have demonstrated that there is a considerable degree of lipid acyl chain anisotropic motion in membranes which determine their bulk fluidity. NMR and ESR spectral parameters indicate that fatty acid acyl chain bending increases down the methylene chains from the polar head groups (Seelig, 1971; Hubbell and McConnell, 1971; Horwitz et al., 1972; Levine et al., 1972a,b; McFarland, 1972; Metcalfe, 1975; Kimelberg, this volume, p. 205). Interestingly, Levine et al. (1972a) found that molecular motion increased down the hydrocarbon chain to the methyl groups and also from the glycerol carbons toward the polar head groups of dipalmitoyl phosphatidylcholine, and Horwitz et al. (1973) determined with ^{31}P resonance that phosphatidylcholine and phosphatidylglycerol asymmetrically distribute between inner and outer surfaces of mixed lipid vesicles. However, the ESR data suggesting a gradual change in chain flexing and bending down the acyl chains is based on differences in order parameters which reflect different orientations of the chains relative to the nitroxide spin-labels. More recent NMR data obtained with unperturbed lipid bilayers indicates that the chain flexing and bending is relatively constant down the acyl chains with a rapid change occurring only near the methyl end (Horwitz et al., 1972; Seelig and Niederberger, 1974). Lee et al. (1973) believe that intermolecular interactions also contribute to these measurements. Under physiological conditions the acyl chains undergo rapid rotation and kinking (*trans-gauche* isomerization about C-C bonds) (Lippert and Peticolas, 1972; Träuble, 1972) where the kinks probably fluctuate rapidly up and down the acyl chains; thus, phospholipid acyl chain motion could be considered as being in a "defective ordered" rather than a "disordered" state (Träuble and Eibl, 1975).

Lipid composition and temperature play important roles in determining membrane fluidity. There is a growing body of evidence, reviewed in this volume by Kimelberg (p. 205), which suggests that the fluidity of the lipid matrix of membranes may be of considerable importance in regulating a number of membrane functions, and transport activity in particular. We will return to this topic in the next section when discussing membrane phase transitions.

In most animal cell membranes cholesterol plays an important role in modulating the bulk fluidity. Addition of cholesterol to artificial membranes made from extracted brain or erythrocyte lipids results in restriction of the mobility of spin-labeled fatty acids (Schreier-Muccillo et al., 1973) and steroids

(Butler et al., 1970b). Cholesterol interacts with certain fatty acids and sphingolipids (Brockerhoff, 1974) and appears to decrease their lateral molecular spacing and reduce flexibility of the carboxyl half of the phospholipid acyl chains (Hubbell and McConnell, 1971; Oldfield and Chapman, 1972). Hence it rigidifies membranes at physiological temperatures, but it does not solidify them (Ladbrooke et al., 1968; Oldfield and Chapman, 1971; Rottem et al., 1973; Vanderkooi et al., 1974). Oldfield and Chapman (1972) have concluded that cholesterol creates an "intermediate fluid condition" with either the fluid or solid lipid phases (see section 2.1.2) resulting in decreased phospholipid acyl chain mobility and flexing in most membranes at physiological temperatures.

Biological membranes are not simple phospholipid-cholesterol bilayers. They contain a variety of proteins and glycoproteins, and it is necessary to ask if the viscosities of biological membranes are determined mainly by their lipids or by lipid-protein complexes. Although the evidence is not complete, the bulk fluidity characteristics of biological membranes are certainly determined by lipids. Experimental evidence has been obtained in a variety of systems indicating that artificial lipid bilayers made from extracted membrane lipids possess similar viscosities to their native membrane precursors. Hubbell and McConnell (1968) used ESR techniques to study the rotational motion of a nitroxide spin label, TEMPO, (2,2,6,6-tetramethylpiperidine-1-oxyl; see Fig. 1) in lipid dispersions. TEMPO will partition between aqueous and non-polar environments, and the high field nitroxide line of the first derivative paramagnetic spectrum exhibits both apolar (hydrocarbon) and polar components, allowing estimates of the degree of partitioning in each environment. TEMPO is much more soluble in fluid versus solid phases, so the paramagnetic spectrum of TEMPO and similar paramagnetic lipid probes can be used to determine the degree of fluid apolar environment in membranes.

The most popular approach for measuring membrane viscosities has been polarization fluorescence measurements on the rotational diffusion of small organic fluorescent probes in membrane hydrophobic environments. Edidin (1974) has termed this type of measurement an estimate of lipid-in-lipid diffusion, as opposed to protein-in-lipid diffusion, which can also be measured by fluorescence polarization techniques in certain systems when a fluorescent chromophore is either present or attached to a membrane protein. Since the membrane properties being measured near the probe are local and do not reflect the bulk properties of membranes, some investigators have favored the use of the term "microviscosity" when discussing results of fluorescence polarization experiments (Shinitzky et al., 1971; Rudy and Gitler, 1972; Cogan et al., 1973). The available data (Table II), obtained in studies using either intrinsic chromophores or extrinsic fluorescent probes, suggest that diverse membranes have similar microviscosities ($\eta \cong 1\text{--}10$ poise).

Fluorescence polarization techniques have also been used to measure the diffusional rotation of molecules larger than fluorescent organic molecules. The pigment protein rhodopsin has attracted considerable attention in this respect due to its natural association with membranes and its possession of a natural chromophore, 11-*cis*-retinal. Rhodopsin is intercalated (but not completely submerged) into the lipid bilayer matrix of outer rod disc membranes and appears to undergo enough rapid rotational diffusion in native rods to eliminate the retinal dichroism seen in partially bleached, glutaraldehyde-fixed rods (Brown, 1972). Using flash photolysis to measure the half-time of dichroic decay of rhodopsin in frog disc membranes, Cone (1972) calculated that the chromophore had a high degree of rotational mobility about an axis parallel to the plane of the membrane with a rotational relaxation time of approximately 20 nsec. From these measurements, the viscosity of the medium in which rhodopsin rotates appears to be about $\eta = 2$ poise.

2.1.2. Lipid lateral motion

The lateral diffusion of lipid in lipid or in natural membranes has been estimated by a variety of techniques, but most studies have utilized ESR or

TABLE II
PLASMA MEMBRANE VISCOSITY DETERMINATIONS[a]

Plasma membrane	Label	Method	Viscosity (poise)	Reference
Lobster walking nerve	TEMPO	ESR-rotational relaxation	0.25–2.5	Ladbrooke et al., 1968
Human erythrocyte ghosts	Perylene	Fluorescence polarization	1–2	Rudy and Gitler, 1972
Human erythrocyte ghosts	Perylene	Fluorescence polarization	1.8	Feinstein et al., 1975
Human erythrocyte ghosts	Retinol	Fluorescence polarization	1–10	Radda and Smith, 1970
Human erythrocyte ghosts	Methanol	Diffusion partitioning	2 ± 1.7	Soloman, 1974
Kidney plasma membranes	Fatty acid, steroid, spin label	ESR-spin exchange	1–10	Grishman and Barnett, 1973
Rabbit polymorphonuclear leukocyte membrane	Perylene	Fluorescence polarization	3.35	Feinstein et al., 1975
Bovine brain myelin	Perylene	Fluorescence polarization	2.7	Feinstein et al., 1975
Human peripheral lymphocytes	Perylene	Fluorescence polarization	1.1	Rudy and Gitler, 1972
Mouse-human heterokaryons	Fluorescent antibody	Lateral diffusion of antigens	8–10	Frye and Edidin, 1970

[a] Modified from Edidin (1974).

NMR. Kornberg and McConnell (1971a,b) introduced spin-labeled phospholipids into lipid bilayer dispersions and monitored proton-nitroxide collisions between adjacent spin-labeled and unlabeled phospholipid molecules. By this method they were able to calculate a lower limit diffusion constant for spin-labeled lecithin in lecithin bilayers of approximately $D > 2.5 \cdot 10^{-11} \text{cm}^2 \text{sec}^{-1}$. Later measurements used ESR spectral parameters such as spin-exchange resonance line broadening to measure the diffusion constant of spin-labeled lecithin inserted in sarcoplasmic reticulum membranes yielding values of $D \cong 6 \pm 2 \cdot 10^{-8} \text{cm}^2 \text{sec}^{-1}$ at 40°C (Scandella et al., 1972). This can be compared to a value of $D = 12 \pm 2 \cdot 10^{-8} \text{cm}^2 \text{sec}^{-1}$ obtained with spin-labeled lecithin diffusion in lecithin multilayers (Devaux and McConnell, 1972) and $D = 1 \cdot 10^{-8} \text{cm}^2 \text{sec}^{-1}$ for spin-labeled androstane in dipalmitoyl lecithin at 40°C (Träuble and Sackmann, 1974). Other values of lateral diffusion constants obtained by ESR range from $0.9 \cdot 10^{-8}$ to $10 \cdot 10^{-8} \text{cm}^2 \text{sec}^{-1}$ (see Table 2 of Edidin, 1974). These values are dependent on the type of probe used, lipid composition (acyl chain length, degree of unsaturation, nature of the polar head group, etc.), divalent cation concentration and the methods of measurement and calculation, where, in many cases, different assumptions must be made to calculate diffusion constants (for example, Israelachvili et al., 1975).

Proton relaxation times (spin lattice relaxation times or T_1 and spin-spin relaxation or T_2) obtained from spin-echo measurements have been used to estimate rotation and lateral diffusion of native phospholipids in bilayer membranes, since these parameters cannot be entirely accounted for by intramolecular chain motion (see Lee et al., 1973). By diluting the bilayers with deuterated lipids, Lee et al. (1973) were able to measure the transverse T_2 relaxation due to rotational, perpendicular and lateral motions and found the latter to be the most reasonable parameter for estimating intermolecular motion which should dominate T_2 relaxation. With this assumption a value of $D = 0.9–1.8 \cdot 10^{-8} \text{cm}^2 \text{sec}^{-1}$ was calculated for the diffusion of dipalmitoyl lecithin at 20°C, and a lower limit of $D > 1–6 \cdot 10^{-9} \text{cm}^2 \text{sec}^{-1}$ was determined from proton line widths of lipid chain resonance in membranes of rabbit sciatic nerve, electroplax and sarcoplasmic reticulum.

These measurements indicate that lateral diffusion rates of phospholipids are quite high, such that a single phospholipid molecule can move a distance of a few μm per second when the membrane is in the fluid state (McConnell, 1975a,b). For comparison, the diffusion constant of ribonuclease (molecular weight 14 000 daltons) at 20°C is $D \cong 1 \cdot 10^{-6} \text{cm}^2 \text{sec}^{-1}$ (Tanford, 1968).

Divalent cations such as Ca^{2+} tend to condense phospholipid bilayers and restrict the mobility of anionic phospholipids as reflected by spin-labeled stearic acid analogs (Schnepel et al., 1974) or cholestane (Butler et al., 1970a). Calcium binding also decreases the mobility of ESR probes in *Bacillus subtilis* (Ehrström et al., 1973) and rat liver plasma membranes (Sauerheber and Gordon, 1975). In certain cases Ca^{2+} appears to enhance the lateral mobility of membrane lipids. Adams et al. (1976) found that introduction of Ca^{2+} into erythrocyte membranes during hemolysis results in increased lipid mobility measured with three fatty acid spin label probes. However, Ca^{2+}-induced membrane protein aggregation occurs under the same conditions which suggests that increasing Ca^{2+} causes decreased protein–lipid interactions (Adams et al., 1976) and increased protein–protein aggregations (Carraway et al., 1975; Elgsaeter et al., 1976).

Calcium appears to play an important role in determining the properties of biological membranes (see Changeux et al., 1967; Papahadjopoulos, 1972; Weiss, 1973; Smellie, 1974; Baker and Reuter, 1975), and as a regulator of a number of important cellular processes, including stimulus-secretion coupling, excitation-contraction coupling, and membrane fusion (reviews, Rasmussen, 1970; Poste and Allison, 1973; Douglas, 1974; Rubin, 1974; Carafoli et al., 1975). There is also an increasing amount of evidence to suggest that Ca^{2+} influx into cells may serve as an important initial transmembrane signal in triggering the response of cells to hormones, transmitters and perhaps also mitogens (reviews, Rasmussen, 1970; Poste and Allison, 1974; Douglas, 1974; Rubin, 1974; Berridge, 1975; Carafoli et al., 1975; McManus et al., 1975; Raff et al., 1976). Calcium is bound to high and low affinity sites on most membranes, and it is not unreasonable to assume that changes in Ca^{2+} binding to membranes would produce structural reorganization(s), including topographic rearrangement of membrane components and accompanying changes in membrane properties. The mechanism(s) by which Ca^{2+} is mobilized from membranes thus seems likely to emerge as a major area of research interest over the next few years.

From the evidence already available, it is clear that changes in Ca^{2+} binding to membranes can be triggered by a variety of biologically active molecules including cyclic nucleotides, ATP, neurotransmitters and certain hormones (see Shlatz and Marinetti, 1972; Triggle, 1972; Rubin, 1974; Berridge, 1975; Carafoli et al., 1975). Equally important, alterations in Ca^{2+} binding and mobilization usually occur as rapid and reversible events, a situation that is well suited to a possible regulatory role in the modulation of membrane properties.

2.1.3. Lipid perpendicular motion
It is now apparent that some, and probably all, biological membranes are asymmetric with specific components being oriented to different surfaces of the membrane. The techniques that have been developed to study asymmetry of membrane components include: asymmetric quenching of spin labels (discussed below) or fluorescent probes (Tsong, 1975a,b); the use of penetrating and non-penetrating site-reactive reagents (Berg, 1969; Bretscher, 1971a,b; De Pierre and Karnovsky, 1972; Gordesky et al., 1973; Kant and Steck, 1973; Bernacki, 1974; Whiteley and Berg, 1974; Yogeeswaran et al., 1974; Juliano and Behar-Bannelier, 1975); selective enzymatic modification of membrane components by proteases (Bender et al., 1971; Phillips and Morrison, 1971a; Steck, 1972; Triplett and Carraway, 1974), phospholipases (Zwaal et al., 1973, 1975), glycosidases (Cook et al., 1961; Eylar et al., 1962; Cook and Eylar, 1965; Steck, 1972; Steck and Dawson, 1974), protein kinases (Kinzel and Mueller, 1973), or lactoperoxidase (Phillips and Morrison, 1971a,b; Hubbard and Cohn, 1972;

Reichstein and Blostein, 1973; Morrison et al., 1974). Asymmetry of membrane proteins and glycoproteins has also been demonstrated by ultrastructural techniques using colloidal metals (Benedetti and Emmelot, 1967; Nicolson and Painter, 1973), ferritin-conjugated antibodies (Nicolson et al., 1971a; Painter et al., 1975), ferritin-conjugated lectins (Nicolson and Singer, 1971, 1974; Nicolson, 1976c), ferritin-avidin complexes (Heitzmann and Richards, 1974) and, in the case of enzymes, histochemical localization (Marchesi and Palade, 1967; Porter and Bernacki, 1975).

Singer and Nicolson (1972) proposed that the asymmetry of glycoproteins found in a variety of cells was probably due to their amphipathic structure which should make trans-membrane rotations or "flip-flop" energetically unfavorable. The free energies of activation required to send the hydrophilic and ionic portions of their structures through the hydrophobic core of the bilayer would be exceptionally large, making such flip-flop unlikely. This concept is supported by the finding that the glycoside portions of glycoproteins can only be detected at the exterior membrane surface (Nicolson and Singer, 1971, 1974; Steck, 1972; Steck and Dawson, 1974).

Membrane lipids are also amphipathic structures, and their asymmetric distribution in the plasma membrane has been postulated by Bretscher (1973). Lipid asymmetry can occur if lipids are inserted asymmetrically into the nascent bilayers or, alternatively, the bilayer may be first synthesized as a symmetric structure and rendered asymmetric later by selective enzymatic activity (Singer, 1974a). In addition, exchange of lipid molecules from the extracellular (Sakagami et al., 1965; Shohet and Nathan, 1970) and intracellular compartments (Zilversmit, 1971; Ehnholm and Zilversmit, 1973) could also modify membrane composition (for fuller discussion of this possibility see Kader, this volume, p. 127).

Maintenance of membrane asymmetry demands that the rotation of molecules through the membrane from one side to the other (flip-flop) must only occur very slowly. Kornberg and McConnell (1971b) measured the rates of "flip-flop" of a phosphatidylcholine analogue across phosphatidylcholine bilayers by spin-labeling techniques and found that the half-time for complete randomization was very slow, approx. 6.5 h at 30°C. Using radioactively labeled phosphatidylcholine vesicles and a beef heart phospholipid exchange protein, Johnson et al. (1975) found that about 60% of the label was rapidly transferred to the unlabeled vesicles by the exchange protein; the other 40% being transferred with a half-time of approx. 40 h at 20°C. Similarly, Sherwood and Montal (1975) estimated the half-time for flip-flop of oleyl acid phosphate to be 15–19 h at 22°C.

When mixed lipid dispersions are sonicated, or when cholesterol is added to vesicular lipid dispersions, preferential partitioning of specific lipids to inner or outer bilayer halves can occur. When Huang et al. (1974) examined the partitioning of added cholesterol to inner or outer bilayer halves in phosphatidylcholine membrane vesicles, they found a homogeneous distribution up to 30% cholesterol in egg lecithin, but above 30 mole percent cholesterol addition to the bilayer vesicles results in preferential cholesterol partitioning into the inner bilayer half. Equimolar mixtures of phosphatidylcholine (PC) with phosphatidylglycerol (PG) results in an outer bilayer concentration ratio of 2PG:PC (Michaelson et al., 1973).

Lipid asymmetry in biological membranes has been studied mainly in the human erythrocyte membrane. Use of lipid chemical labeling techniques (Bretscher, 1972; Gordesky and Marinetti, 1973) and enzymatic hydrolysis of phospholipids (Verkleij et al.; 1973; Zwaal et al., 1973; Kahlenberg et al., 1974) have demonstrated that sphingomyelin, phosphatidylethanolamine (PE) and phosphatidylserine (PS) are predominantly in the inner cytoplasmic half of the bilayer, while PC is present at higher concentrations in the outer half. Using a phospholipid exchange protein, Bloj and Zilversmit (1976) found that rat erythrocyte ghosts exchange approx. 75% PC rapidly and approx. 25% slowly to PC:cholesterol vesicles, while in inside-out ghost vesicle preparations 37% PC was exchanged rapidly and 63% at a slower rate. This yields flip-flop half-times of 2.3 h and 5.3 h for right-side-out ghosts and inside-out ghost vesicles, respectively. One note of caution concerning the use of inside-out vesicles of erythrocyte ghosts is the possibility that structural rearrangements can occur during membrane inversion. This may occur even when ghost membranes are made from intact cells. For example, differences in enzymatic degradation of phospholipids between intact cells and resealed ghosts have been found (Woodward and Zwaal, 1972). Higher rates of lipid flip-flop have been found in vesicles of nerve membranes and mycoplasma using spin-labeled phospholipids (Grant and McConnell, 1973; McNamee and McConnell, 1973). However, these fast flip-flop times have not been seen by other investigators (Rousselet et al., 1976). These latter authors used two different spin-labeled PC analogs and examined their flip-flop rate in human erythrocytes after incorporation by transfer of PC molecules to the erythrocytes during incubation with lipid vesicles. Spin-labeled PC in the outer bilayer could be selectively reduced by ascorbate treatment if the spin-label was on the α-chain, and on the inner surface PC could be selectively reduced by reductants in the cytosol if the spin-label was on the PC polar head group. The newly created anisotropic distribution of spin-labeled PC was stable for 4 h at 37°C. Rousselet et al. (1976) also found that the TEMPO-PC spin-labels of McNamee and McConnell (1973) and Grant and McConnell (1973) were also reduced by cytoplasmic reductants, probably explaining the very rapid rates of flip-flop obtained previously.

2.1.4. Lipid phase separation
When hydrated pure phospholipid dispersions are warmed from low temperatures where their acyl chains are in a solid or rigid state, they undergo endothermic phase transitions at characteristic temperatures which are indicative of acyl chain melting (reviews, Träuble, 1972; Chapman, 1973; Fox, 1975a; McConnell, 1975a,b; Kimelberg, this volume, p. 205). Below their phase transition temperatures phospholipid hydrocarbon chains are relatively rigid, parallel packed, and extended in the all-*trans* conformation. At temperatures

above the phase transition, the phospholipid acyl chains become more "fluid" or "liquid-crystalline" in nature and undergo cooperative chain motions such as flexing, bending, "kink" or *gauche* conformational changes (highly mobile C-C rotational isomers) and lateral motions (Träuble, 1972; Träuble and Eibl, 1974; McConnell, 1975b). These changes result in a reduction in bilayer thickness and an increase in the area occupied per molecule and in total volume (Phillips and Chapman, 1968; White, 1970; Träuble and Haynes, 1971; Sheetz and Chan, 1972; Nagle, 1973). For example, the main phase transition of dipalmitoylphosphatidylcholine (DPPC) occurs at about 41°C (Phillips et al., 1970; Shimshik and McConnell, 1973a). Above this transition the bilayer thickness is reduced by 6–7 Å, and the area occupied per DPPC molecule is increased from 48 Å^2 to 65–70 Å^2 (Phillips and Chapman, 1968). Also, lipid isothermal lateral compressibility increases markedly at the transition point (Linden et al., 1973). This transition results in an enthalpy change $\Delta H =$ 8.66 kcal mole^{-1} and an entropy change $\Delta S = 27.6$ cal deg^{-1} mole^{-1} (Phillips et al., 1969). Detailed thermotropic data for other phospholipids of biological importance are given in the chapter by Kimelberg in this volume (see Table I, p. 213).

The transition temperature range where a pure phospholipid undergoes phase change from a solid to liquid state is usually small. Thus, pure lipids are characterized by sharp melting points and large endothermic changes. These endothermic melting points can be measured accurately by differential scanning calorimetry (DSC) (Chapman et al., 1967; Papahadjopoulos, 1968; Steim et al., 1969; Phillips et al., 1970; Hinz and Sturtevant, 1972) where the transfer of heat to a sample of lipid can be accurately measured when the sample is heated or cooled at a constant rate. Phase changes can also be monitored by ESR probes such as TEMPO, because this probe shows enhanced partitioning into fluid lipids (Hubbell and McConnell, 1968). When ESR spectral parameters are calculated from the high field nitroxide resonance spectra and plotted as a function of temperature, a sharp (1–2°C) discontinuity occurs at the characteristic transition (Fig. 2). Phase transitions such as these occur at specific temperatures (both the onset and end of the transition), but these transitions are strongly affected by lipid composition, acyl chain length(s), the nature of the polar head group(s), the degree(s) of unsaturation and the presence of divalent cations (Gaffney and McConnell, 1974; Jacobson and Papahadjopoulos, 1975; Papahadjopoulos et al., 1976a).

Binary mixtures of bilayer phospholipids exhibit phase transitions of a more complicated type. Heating an all-solid, randomly dispersed mixture of two lipids results first in a broad onset of melting (lower transition temperature) where an equilibrium situation occurs and fluid and solid phases of differing lipid composition coexist. This is due mainly to lipid lateral separations in the plane of the bilayer. Lateral segregation must occur instead of perpendicular segregation, because the rates of lipid flip-flop across a bilayer do not occur at a fast enough rate to account for rapid equilibration of different phases (Kornberg and McConnell, 1971b; Shimshik and McConnell, 1973a). At the ter-

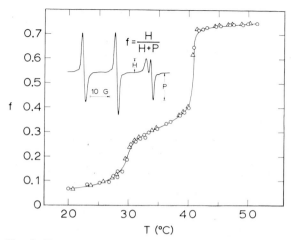

Fig. 2. Temperature-induced phase transition for dipalmitoylphosphatidylcholine and related phospholipids, determined by the change in order parameter (f) obtained from the paramagnetic resonance spectra of the spin label TEMPO (see Fig. 1). TEMPO lipid solubility increases when lipid is in the fluid state, and this is reflected by a sharp increase in order parameter at approximately 41°C. Reproduced with permission from McConnell (1975a).

mination of melting (upper transition temperature) all of the lipid is in a fluid state.

Lipid lateral phase separations require long-range molecular motion in the bilayer plane, and these motions result in lipid compositional segregation. Lateral phase separations are not characterized by sharp transitions; they usually occur over fairly broad temperature ranges (compared to pure lipids) where different phases coexist. McConnell and his collaborators (Shimshik and McConnell, 1973a,b; Grant et al., 1974; Wu and McConnell, 1974; McConnell, 1975b) have constructed schematic phase diagrams to explain the melting behavior of binary mixtures of lipids with complete solid and fluid phase miscibility (Fig. 3a, lipids of similar molecular structure) and diagrams with complete fluid phase miscibility but solid phase immiscibility (Fig. 3b, lipids of dissimilar molecular structure) (McConnell, 1975b). When any particular randomly mixed solid binary lipid mixture is warmed, a phase change occurs at the lower transition temperature (t_l) which marks the beginning of unequal melting and lateral phase separation (plotted as *solidus* curve). With continued heating, eventually the upper transition temperature (t_h) is reached which delineates the final melting of fluid + solid phase (plotted as *fluidus* curve) (Fig. 3).

Freeze-cleavage electron microscopy can be used to visualize the solid and fluid phases of at least some binary lipid mixtures (Pinto da Silva and Branton, 1972; Ververgaert et al., 1972; Chen and Hubbell, 1973; Shimshik et al., 1973; Grant et al., 1974; Kleemann et al., 1974; Verkleij et al., 1974). When heavy metal replicas of the fracture faces of freeze-cleaved phospholipid bilayers originally in solid phase are observed by electron microscopy, characteristic "banded" linear repeating ridges are seen indicating solid lipid phases (Ver-

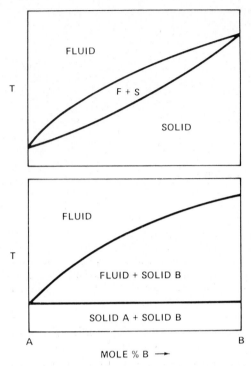

Fig. 3. Schematic phase diagrams for lateral phase separations in binary phospholipid mixtures (components A and B). In both cases, the components A and B are complete miscible in the fluid phase. In one case (upper), the components are miscible in the solid phase (forming a solid solution), and the other case (lower), the components are completely immiscible in the solid phase. Reproduced with permission from McConnell (1975b).

vergaert et al., 1972, 1973; Shimshik et al., 1973). When a binary mixture of lipids is examined after rapid quenching from a temperature where both solid and fluid phases are known to exist (F + S region in Fig. 3), smooth (indicating fluid phase lipid) and "banded" (indicating solid phase lipid) regions are obtained (see Fig. 4), and the relative areas of these regions correlates with the degree of fluid-versus-solid phase lipid domains estimated by ESR spectral analysis of spin labels such as TEMPO (Grant et al., 1974). Unfortunately, biological plasma membranes which contain appreciable concentrations of cholesterol do not show these banded regions (Verkleij et al., 1974; Kleemann and McConnell, 1976).

Certain biological membranes obtained from cells or bacteria grown on different fatty acids and from fatty acid bacterial auxotrophs have been observed to undergo lateral phase separations similar to binary lipid mixtures (James et al., 1972; Verkleij et al., 1972; Speth and Wunderlich, 1973; Haest et al., 1974; Kleemann and McConnell, 1974). Studies have been performed using freeze-cleavage electron microscopy to identify fluid and solid phase membrane regions in these biological systems, but in contrast to the

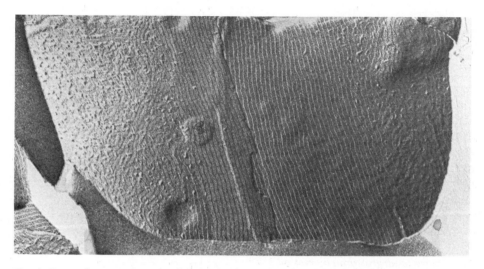

Fig. 4. Freeze-fracture electron micrograph of a 50:50 mixture of dielaidoylphosphatidylcholine and dipalmitoylphosphatidylcholine quenched from 23°C, a temperature corresponding to the midpoint in the phase diagram of Fig. 3a, where equal proportions of solid and fluid phase lipids are predicted to coexist. Notice the absence of particles in the banded regions which are thought to be solid lipid domains. Reproduced with permission from McConnell (1975b).

smooth (fluid) and banded (solid) lipid surfaces seen in model membranes containing only pure lipids, numerous intercalations or particles are visualized which are thought to be protein in nature (MacLennan et al., 1971; Hong and Hubbell, 1972; Segrest et al., 1974; Vail et al., 1974). These particles are excluded from certain membrane regions during slow cooling, and it is thought that the excluded regions represent solid phase lipid. To demonstrate that this was in fact the case, Kleemann and McConnell (1976) reconstituted model membrane vesicles with mixtures of lipid and the ATPase protein from rabbit sarcoplasmic reticulum membranes. In dimyristroyl phosphatidylcholine (DMPC) vesicles containing the ATPase (and in some cases, additionally, the spin label TEMPO) at temperatures above the phase transition temperature membrane particles were randomly dispersed, but below the phase transition the particles were aggregated and sequestered into regions identified as fluid, and no particles were found in the solid phase "banded" lipid regions. Similar results have been obtained with vesicles where the erythrocyte plasma membrane glycoprotein, glycophorin, was intercalated instead of ATPase (Grant and McConnell, 1974). In DMPC vesicles containing 10 mole percent cholesterol the characteristic bands in solid phase regions were diminished, and cholesterol concentrations equal to or greater than 20 mole percent abolished the bands seen by freeze-cleavage electron microscopy. Different phases do exist in these DMPC-cholesterol mixtures, as shown by ESR, and also by the non-uniform partitioning of membrane particles in vesicles containing 10 or 20 mole percent cholesterol. Below the known phase transition for a 10 mole percent cholesterol mixture (15°C), two protein particle phases have been shown to exist, a particle-rich phase and a particle-poor phase. Increasing the cholesterol to 20 mole percent caused particles to form string-like linear arrays when below the phase transition temperature. The lengths of the linear particle arrays decreased (reversibly) as the temperature approached 23°C, indicating a rather high particle lateral mobility even below the transition temperature. This appeared to be much higher than expected when compared to the lack of particle mobility in pure DMPC vesicles (Kleemann and McConnell, 1976). Calorimetric data indicate that phase transitions occur in lipid vesicles which have up to 33 mole percent cholesterol content (Darke et al., 1972; Engelman and Rothman, 1972; Hinz and Sturtevant, 1972; deKruyff et al., 1974), so the lack of visualization by freeze-cleavage techniques probably indicates that cholesterol modifies the packing of solid phase lipids, a conclusion supported by independent X-ray scattering experiments (Engelman, 1971; Shechter et al., 1972; Tardieu et al., 1973; Haest et al., 1974).

2.2. Protein and glycoprotein motion

2.2.1. Lateral mobility of membrane proteins and glycoproteins

It is now well established that integral membrane proteins are not static entities and can undergo dynamic change with respect to other membrane components. Although the lateral mobility of integral membrane proteins is generally lower than membrane lipids such as phospholipids, they can still have quite rapid rates of motion. Poo and Cone (1974) measured the rates of lateral motion of rhodopsin in rod outer segment disc membranes by photobleaching in one membrane region and measuring the time required for lateral diffusion across to the opposite end of the disc membrane and intermixing of the bleached rhodopsin molecule with unbleached rhodopsin. They were able to calculate a diffusion constant of $D \cong 4\text{–}5 \cdot 10^{-9} \, cm^2 \, sec^{-1}$ for rhodopsin. Photobleaching methods have also been used recently to monitor the mobility of specific plasma membrane proteins in single living cells cultured in vitro. Data obtained from measurements of fluorescence recovery after photobleaching have yielded lateral diffusion coefficients (D) within the range $3 \cdot 10^{-11} \, cm^2 \, sec^{-1}$ to $8 \cdot 10^{-12} \, cm^2 \, sec^{-1}$ for fluorescein-labeled concanavalin A and succinylated-concanavalin A bound to the surface of mouse fibroblasts (Jacobson et al., 1976; Zagyanksy and Edidin, 1976) and rat myoblasts (Schlessinger et al., 1976). In similar experiments using a fluorescent lipid probe (3,3'-dioctadecyl-indocarbocyanine iodide) instead of fluorescent-lectins, Schlessinger et al. (1976) noted a large difference between lipid-in-lipid and presumably protein-in-lipid lateral diffusion, the former being estimated at $D \cong 8 \pm 3 \cdot 10^{-9} \, cm^2 \, sec^{-1}$.

These experiments have also revealed that a minor, but nonetheless significant, class of lectin receptors appears to be relatively immobile. The lower mobility of concanavalin A receptors compared to lipid probes cannot be explained solely on the basis of lectin-induced cross-linking and receptor aggregation into clusters of lower lateral mobility, because non-specific labeling of surface proteins with fluorescein isothiocyanate (FITC) also yielded an immobilized fraction of surface proteins, Moreover, the lateral mobility of fluorescent lectins measured soon after lectin binding was not dependent on valence, and visible fluorescent patches were only observed some time after the data were taken. The lateral mobility of the fluorescent concanavalin A-receptor complexes was time- and temperature-dependent and was affected by sodium azide and cytochalasin B. This latter effect suggests the involvement of a cytoskeletal system in restraining mobility (see section 3). Interestingly, comparable photobleaching studies on the mobility of fluorescein-labeled wheat germ agglutinin (WGA)-receptor complexes have revealed that the percentage of mobile WGA-receptor complexes and their diffusion coefficient ($D = 2 \cdot 10^{-11}$ to $2 \cdot 10^{-10}$ $cm^2 sec^{-1}$) are higher than for concanavalin A-receptor complexes (Jacobson et al., 1976). In addition, the mobility of WGA-receptor complexes is unaffected by either colchicine or cytochalasin B suggesting that this receptor does not interact with membrane-associated cytoskeletal elements. This conclusion is consistent with earlier studies showing that while the mobility of concanavalin A receptors can be modulated by drugs acting on membrane-associated cytoskeletal elements, WGA receptors do not appear to be linked to these trans-membrane control elements (see Poste et al. 1975a; Edelman, 1976; Nicolson, 1976a,b; Jacobson et al., 1977).

Low rates of lateral mobility of what are most likely membrane glycoproteins have also been seen in photobleaching studies on intact human erythrocytes. Peters et al. (1974) non-specifically labeled the membrane surface proteins of human erythrocytes (mainly glycophorin and glycoprotein components [Band III] with molecular weights of approximately 90 000–100 000 daltons [Morrison et al., 1974; Hubbard and Cohn, 1972]) with fluorescein isothiocyanate and found that these components were not freely diffusing in the membrane plane ($D < 3 \cdot 10^{-12}$ $cm^2 sec^{-1}$). The existence of peripheral and trans-membrane controlling mechanisms (section 3.2) are also probably responsible for impeding lateral mobility in this system.

Diffusion of fluorescent antigen-antibody complexes was first used by Frye and Edidin (1970) in their classic experiments on histocompatibility antigen-antibody complex intermixing on mouse-human heterokaryons after Sendai-virus-induced fusion. As mentioned earlier, in these experiments complete intermixing of antigen-antibody complexes occurred within 40 min after cell fusion using a double fluorescent marker technique to visualize mouse H-2 and human HL-A antigens. A diffusion constant of approximately $D \cong 2 \cdot 10^{-10}$ $cm^2 sec^{-1}$ for the antibody-histocompatibility antigen complexes can be calculated from these data. In a more recent study Edidin and Fambrough (1973) examined the rate of lateral diffusion of fluorescent-Fab' monovalent antibodies after spot application to muscle fibers. By measuring the extent of patch spreading with time, Edidin and Fambrough were able to calculate a diffusion constant of $D \cong 1$–$2 \cdot 10^{-9}$ $cm\ sec^{-1}$ for Fab antibodies bound to their antigenic sites. Intact bivalent immunoglobulin molecules diffused within the membrane at about one-hundredth the rate of monovalent Fab'.

The movement of integral membrane proteins or protein complexes can also be detected by freeze-cleavage electron microscopy. A limitation here, as with other ultrastructural techniques, is that the same sample cannot be followed dynamically. A wide variety of treatments have been shown to induce redistribution of intramembranous particles, including: temperature changes (Pinto da Silva et al., 1971; James and Branton, 1973; Speth and Wunderlich, 1973; Kleemann and McConnell, 1974, 1976); alterations in membrane lipid composition (James et al., 1972; Chen and Hubbell, 1973; James and Branton, 1973; Kleemann and McConnell, 1976); pH changes (Pinto da Silva, 1972; Elgsaeter and Branton, 1974); enzymes (Wallach, 1969; Tillack et al., 1972); and membrane-active drugs (McIntyre et al., 1974; Poste et al., 1975a). Changes in particle distribution have also been seen when anionic sites on the outer membrane surface are labeled with colloidal iron particles (Nicolson, 1973a) and when external glycoproteins are labeled with ferritin-conjugated lectins (Tillack et al., 1972; Pinto da Silva and Nicolson, 1974).

Aggregation of the major glycoprotein of the human erythrocyte membrane (glycophorin) induced by a variety of techniques is accompanied by a corresponding clustering of intramembranous particles in freeze-fracture electron micrographs, suggesting that the particles represent the hydrophobic portion of the glycophorin molecules (Marchesi et al., 1972; Tillack et al., 1972), probably in association with the so called band III erythrocyte membrane glycoproteins (see Pinto da Silva and Nicolson, 1974). Similarly, aggregation of particles within the plasma membranes of *Entamoeba histolytica* trophozoites is accompanied by aggregation of anionic sites on the outer membrane surface (Pinto da Silva et al., 1975). However, other studies in different membrane systems have failed to identify any relationship between ligand-induced redistribution of determinants on the outer surface of the membrane and redistribution of intramembranous particles (see Karnovsky et al., 1972; Matter and Bonnet, 1974; Pinto da Silva et al., 1975).

2.2.2. Ligand-induced redistribution of cell surface components
The ability of multivalent ligands such as antibodies (reviews, Unanue and Karnovsky, 1973; Edidin, 1974; Loor, 1976) and plant lectins (Nicolson, 1974; 1976a,b,c) to induce lateral movement and topographic rearrangement of cell surface receptors has been demonstrated in numerous laboratories using a wide range of cell types.

For the purposes of present discussion, the vast literature on this subject can be reduced to the following general conclusions:

(1) The inherent topography of most, but not all, cell surface receptors that have been examined is uniform (so called random distribution) over the entire cell surface. Early immunoelectron microscopy studies in which cell surface antigens were thought to exist in topographically discrete "patches" (Aoki et al., 1969; Aoki, 1971; Aoki and Takahashi, 1972) are now recognized as ligand-induced redistribution artefacts (see Davis, 1972; de Petris and Raff, 1974; Stackpole et al., 1974c; Nicolson, 1974). This was demonstrated by the finding that the topography of ligand-receptor complexes on cells which had

been fixed before exposure to the ligand was uniform rather than patchy, indicating that the ligand was responsible for redistributing the receptors. In addition, studies on unfixed cells incubated with ligands for a short time at low temperatures (0–4°C) revealed a similar random, uniform distribution of receptors, but on warming to 37°C, the ligand-receptor complexes rapidly underwent redistribution to form patches;

(2) The binding of multivalent ligands to the cell surface and cross-linking of adjacent ligand-receptor complexes leads to redistribution of the previously random receptors to form aggregates, "clusters", larger "patches" and, in certain situations, the patches eventually coalesce to form a single large polar "cap" or aggregate of cross-linked ligand-receptor complexes. Following redistribution, the patched or capped ligand-receptor complexes are often internalized by endocytosis or, in some cases, shed from the cell surface (see reviews by Nicolson, 1976b and Sundqvist (p. 551) and de Petris (p. 645) in this volume);

(3) The rate and extent of ligand-induced redistribution of surface receptors is determined both by the cell type and by the nature of the ligand (see Edidin, 1974; Nicolson, 1976a,b). For example, treatment of mammalian lymphocytes with the plant lectin concanavalin A induced redistribution of concanavalin A receptors into a polar cap, yet the same ligand redistributes the surface receptors on mammalian fibroblasts into discrete patches rather than a cap. However, treatment of both lymphocytes and fibroblasts with antibodies to the same surface antigen results in capping of the antigen-antibody complexes.

Studies in several laboratories on the dynamics of cell surface lectin receptors have shown that they are more readily redistributed on certain cells, particularly transformed cells, while on other cell types, notably many untransformed cells, these receptors are relatively less mobile and undergo redistribution at much lower rates (see reviews by Nicolson, 1974, 1976b).

While the question of whether a particular ligand can induce capping of surface receptors appears to be influenced both by the nature of the ligand and by the cell type, there is also substantial variation between cell types in the final distribution of the cap. In lymphocytes, where the phenomenon was first recognized (Taylor et al., 1971), and for which the term "capping" was coined, the capped receptors are situated over one pole of the cell on the same side or opposite the Golgi apparatus depending on the ligand used. For example, Stackpole et al. (1974d) found that ~93% and ~83% of surface-immunoglobulin and concanavalin A caps, respectively, formed over the Golgi on splenic lymphocytes, but approx. 72% and approx. 75% of Thy-1 and TL caps were opposite the Golgi on thymic lymphocytes. On fibroblasts the caps are more usually found directly overlying the Golgi apparatus in the center of the cell (Ukena et al., 1974; Poste et al., 1975a,b).

Although antibody-induced redistribution often results in the formation of clusters, patches and caps (Edidin and Weiss, 1972; Unanue et al., 1973; Stackpole et al., 1974d; Yefenof and Klein, 1974; Cohen and Gilbertsen, 1975), it can also result in what Phillips and Perdue (1974, 1976a,b) have termed "marginal redistribution" and "coalescence of clusters". These authors examined several types of avian virus-transformed and leukosis virus-infected chick fibroblasts and found in certain cases that antibody-induced antigen redistribution results in the ligand-receptor complexes concentrating at the edges of cells that are attached to the substrate (so called "marginal redistribution"). Clusters sometimes formed on these same cells, and also on different cells in the population, and the clusters either coalesced into regions where endocytosis eventually occurred ("coalescence of clusters") or formed larger patches and eventually caps.

In many instances binding of a single ligand to the cell surface is sufficient to induce receptor redistribution. For some classes of receptors, however, addition of a second ligand (usually antibodies to the first ligand) is required to induce receptor redistribution (for detailed discussion see Nicolson, 1976a,b and Sundqvist, this volume, p. 551). It is of interest to note, however, that receptor redistribution has been induced by hybrid antibodies monovalent for two separate specificities (Stackpole et al., 1974a) and, in certain instances, even monovalent antibodies are effective (Stackpole et al., 1974c) suggesting that multivalency may not be an absolute requirement in order for a ligand to induce redistribution of surface components;

(4) As mentioned in section 1.2, different receptors situated on discrete surface molecules can be redistributed separately by the appropriate ligands. This approach thus offers a useful method for mapping the topographic relationships of specific classes of receptors on the cell surface and for defining whether different receptors reside on the same or different molecules or supramolecular complexes.

Despite the general comment in (1) above that most surface receptors appear to be distributed randomly over the cell surface, there are examples of receptors showing apparent non-random distributions. Wartiovaara et al. (1974) found that surface-fibroblast (SF) antigens on mouse fibroblasts attached to glass were nonrandomly distributed, apparently in association with fibrillar structures such as surface ridges and microvilli. Similarly, recent studies on mouse lymphocytes have shown that surface-immunoglobulin molecules appear to exist in interconnected networks rather than as random structures (Abbas et al., 1975), and non-random distributions of several neuronal surface antigens have been identified on chick neural cells (Rostas and Jeffrey, 1975). Using antisera against purified nerve synaptosomal membranes, Rostas and Jeffrey (1975) were also able to detect certain antigens only on the nerve presynaptic or axolemmal membranes.

A striking example of molecular segregation and regional specialization of membrane structure has been noted in the distribution of surface receptors on mammalian spermatozoa. Sperm from different species have distinct, discontinuous distributions of surface anionic sites on their surfaces (Bedford et al., 1972; Yanagimachi et al., 1972) which argues against free lateral mobility of the components bearing these residues. Nicolson and Yanagimachi (1974) have also shown that the mobility of lectin receptors differs significantly in different regions of the sperm plasma membrane, with greater receptor mobility being found in the past-acrosomal membrane region. In addition, certain antigens are localized to specific regions of the mammalian sperm head indicating their lack of free lateral diffusion (Koo et al., 1973; Koehler and Perkins, 1974; Koehler, 1975a,b). Freeze-cleavage electron microscopic studies on various mammalian spermatozoa have also revealed striking regional differences in the topography of intramembranous particles in these cells (Friend and Fawcett, 1975), though the question of whether the nonrandom distribution of intramembranous particles is accompanied by a nonrandom topography of components on the outer membrane surface has yet to be resolved.

Freeze-cleavage electron microscopic studies have also revealed nonrandom distributions of intramembranous particles in a wide variety of cell types in regions of the plasma membrane involved in the formation of different types of intercellular junctions (see McNutt, this volume p. 75).

Nonrandom clustering of plasma membrane receptors has long been recognized in specialized structures such as synaptic and neuromuscular junctions (Fambrough and Hartzell, 1972; Sytkowski et al., 1973; Fertuck and Salpeter, 1974). It seems likely that similar nonrandom clustering of receptors and other surface determinants will prove to be a fundamental feature of the molecular differentiation of plasma membranes in other cell types and that with the availability of more sophisticated ultrastructural probes, many more examples will be identified. For a detailed list of other examples of apparent regional differentiation of plasma membrane components in prokaryotic and eukaryotic cells, the reader is referred to the recent review by Stackpole (1977).

The final topic that is pertinent to this section is the possibility of preferential association(s) and topographic interrelationships between separate cell surface components. For example, there is some evidence which suggests that there may be a preferential topographic association of TL and H-$2D$ antigens on the surface of normal mouse thymocytes and leukemia cells (see Old and Boyse, 1973; Stackpole, 1977).

2.2.3. Protein and glycoprotein turnover

The rate of membrane synthesis described in most reports in fact represents a combination of synthesis and incorporation which are directly related to the biogenesis of plasma membranes. Among the many interesting and important questions to be answered in the area, we will be concerned only with turnover (for a more complete review, see Poste and Nicolson, 1977). One important question is whether individual molecules are inserted into the membrane independently in a manner something similar to the direct insertion of lipid molecules by phospholipid exchange proteins (Bloj and Zilversmit, 1976; Kader, this volume, p. 127). Alternatively, presynthesized lipoprotein complexes or vesicles could be inserted into the pre-existing plasma membrane (Fox, 1975b). Palade (1959) has proposed on the basis of extensive electron microscopic evidence on secretory mechanisms that plasma membranes are formed via an assembly line process. Polypeptides are synthesized initially on the rough endoplasmic reticulum and transferred to the smooth endoplasmic reticulum and then to the Golgi apparatus, where they are formed into vesicles. These vesicles migrate to the cell cortex and eventually fuse with the pre-existing plasma membrane. After fusion the original cisternal side of the vesicle has now become the outer side of the plasma membrane. The available evidence, while limited, is compatible with this hypothesis (see Hirano et al., 1972; Vitetta and Uhr, 1973; Melchers and Andersson, 1974; Morré et al., 1974; Schachter, 1974; Uhr et al., 1974; Palade, 1975). In addition, vesicles of unknown function have often been observed in the cytoplasm near to the plasma membrane in electron micrographs of a variety of cell types (Bennett and Leblond, 1970).

The most convincing evidence for the assembly line mechanism comes from data on the addition of saccharides during glycoprotein synthesis. In many cell types the predominant carbohydrate residues attached to polypeptide backbones such as the core saccharide mannose are incorporated into nascent molecules at the level of the rough endoplasmic reticulum (Melchers and Knopf, 1967; Molnar and Sy, 1967; Choi et al., 1971), though other carbohydrates can be incorporated into oligosaccharide chains in the smooth endoplasmic reticulum and the Golgi apparatus (Bennett and Leblond, 1970; Zagury et al., 1970; Choi et al., 1971; Melchers, 1971). When nascent proteins are labeled with radioactive precursors, the labels appear first in the rough endoplasmic reticulum and later at the cell surface (Siekevitz et al., 1967; Leblond and Bennett, 1974). Cycloheximide treatment stops protein synthesis but does not affect significantly the continued appearance of labeled proteins in plasma membranes for up to several hours (Ray et al., 1968). Other evidence suggests that nascent polypeptides are inserted through the rough endoplasmic reticulum membrane as they are synthesized with the amino terminal end penetrating into the membranes first, eventually protruding out to the cisternal side (Sabatini and Blobel, 1970). This asymmetric orientation of the polypeptide moiety in the membranes is compatible with the arrangement of glycoproteins in plasma membranes. For example, glycophorin, the major glycoprotein of human erythrocyte membranes, has its amino end at the outer membrane surface while the carboxyl terminal is located at the inner surface (Marchesi et al., 1972; Segrest et al., 1973). Other aspects of the asymmetry of the membranes are also in accord with the assembly line mechanism. Membrane-bound saccharides are located exclusively at the outer surface of plasma membranes (Nicolson and Singer, 1971, 1974; Hunt and Brown, 1974), but at the cisternal side of the endoplasmic reticulum (Hirano et al., 1972). The asymmetric distribution of phospholipids found in different membranes is also in accord with the assembly line hypothesis. For example, PC is enriched in the outer half of many plasma membranes and in the cisternal half of the endoplasmic reticulum membranes, whereas PS and PE are mostly found in the other respective half of these two membranes (Bretscher, 1972; Zwaal et al., 1973; Bloj and Zilversmit, 1976; Renooiz et al., 1976).

There are, however, data suggesting that the transfer of polypeptides from the rough endoplasmic reticulum to the Golgi apparatus is not unidirectional.

Some glycoproteins released from the Golgi apparatus are incorporated into endoplasmic reticulum instead of plasma membranes (Elhammer et al., 1975). Therefore, complex mechanisms appear to be involved in directing glycoproteins to plasma membranes, endoplasmic reticulum and probably other membranes of the cell.

The degradation rates of proteins and glycoproteins in plasma membranes of a variety of animal cells appears to follow first order kinetics, and half-lives can vary from a few hours to 18 days (reviews, Wallach, 1976; Tweto and Doyle, 1977). The degradation rate per cell generation, however, is less divergent, lying between 10% and 30% (Warren and Glick, 1968; Bock et al., 1971; Roberts and Yuan, 1974; Kaplan and Moskowitz, 1975a; Tweto and Doyle, 1976). Non-dividing cells (density-inhibited) and dividing, logarithmic growing cells show similar degradation rates, indicating that cell division is not directly related to degradation rates, and that membrane proteins and glycoproteins are constantly degraded regardless of the stage of growth or cell cycle (Warren and Glick, 1969; Kaplan and Moskowitz, 1975a; Tweto and Doyle, 1976). The major proportion of newly synthesized and incorporated membrane proteins and glycoproteins appears to contribute to the net increase in membrane proteins needed in growing cells, while they are degraded in nongrowing cells. In the latter case, net turnover is observed.

In most cells, with a few exceptions, glycoproteins are degraded noticeably faster than proteins. Mathews et al. (1976) suggested that both carbohydrate and polypeptide moieties were degraded at similar rates in neuroblastoma cells. However, others have proposed that the carbohydrate moieties are degraded independently of the conjugate polypeptide (Kaplan and Moskowitz, 1975b; Tweto and Doyle, 1976). Furthermore, partially degraded portions of glycoproteins are often required. For example, neuraminidase-treated CHO cells preferentially incorporate neuraminic acid into the positions of the cleaved neuraminic acids (Kraemer, 1967). The carbohydrates on many glycoproteins can also be modified directly at the cell surface by glycosyltransferases (Roth and White, 1972; Shur and Roth, 1975).

There is some disagreement on the turnover rates of membrane proteins. In L cells, cultured hepatoma and neuroblastoma cells the bulk of membrane proteins is degraded synchronously (Warren and Glick, 1969; Mathews et al., 1976; Tweto and Doyle, 1976), whereas noticeably heterogeneous rates are observed in rat liver cells and monkey kidney cells in culture (Dehlinger and Schimke, 1971; Kaplan and Moskowitz, 1975b). Heterogeneous turnover rates of proteins have also been observed in microsomal membrane proteins (Arias et al., 1969; Kiehn and Holland, 1970). The most extensive studies in this latter area have been performed by Palade's group. They examined variety of enzyme activities in rat liver microsome and in chloroplast membranes of green algae and found distinctively different turnover rates for individual enzymes (Schor et al., 1970; Bock et al., 1971). It has been suggested that there is a general correlation between the degradation rates of membrane proteins and their subunit size; i.e., the larger the size of subunits, the higher the degradation rate (Dehlinger and Schimke, 1971). Rigorous examination of a series of proteins and glycoproteins to test this premise in plasma membranes have thus far not been attempted.

While acknowledging the obvious importance of membrane assembly and turnover processes influencing membrane dynamics, we will not discuss these topics further, since the next volume of the Cell Surface Reviews series is devoted entirely to these phenomena. The remainder of this article, as well as the other chapters in this volume will therefore focus on the dynamics of membrane reorganization found in specific metabolic and physiologic states.

3. Mechanisms of receptor control

The findings discussed in section 2 that different cell surface receptors can have unique surface distributions and/or mobilities within the membrane suggests that controlling mechanisms exist which regulate cell surface dynamics and control the topographic distribution of surface components. In this section the evidence for such control and some possible mechanisms will be considered.

3.1. Planar (cis) control

It is now clear that plasma membrane components are mobile in the membrane plane, yet their movements can be restricted by planar associations or aggregations with other molecules, or by sequestration or exclusion into specific membrane domains (Nicolson, 1976a) (Fig. 5). The association of components in the membrane plane can be considered a form of cis-control (Singer and Nicolson, 1972; Nicolson, 1973b) where events occurring on one side of the membrane or in the plane of the membrane control the display, mobility, distribution, association, etc. of membrane components.

3.1.1. Planar or lateral associations

The non-covalent lateral association of cell membrane components into supramolecular aggregates or even paracrystalline arrays is common in some lower species (Oesterhelt and Stoeckenius, 1971), and it occurs under certain conditions in mammalian cells. One of the best examples of aggregated and even paracrystalline membrane structure is the cell junctions formed between contacting, adjacent cells (McNutt and Weinstein, 1973; Staehelin, 1974; Weinstein et al., 1976; McNutt, this volume, p. 75). Weinstein and McNutt (1972) have grouped cell junctions into two major categories: junctions that join cells in close or direct contact (zonulae occludentes, gap or nexus junctions and septate desmosomes); and junctions which join cells separated by 15–30 nm distances (zonulae adherentes, macula adherens, desmosomes). The morphology and ultrastructure of the various types of junctions are reviewed in detail elsewhere (Staehelin, 1974; Weinstein et al., 1976; McNutt, this volume, p. 75), but their structures appear to share the common property of developing from "subunits" which form supramolecular aggregates both in the membrane

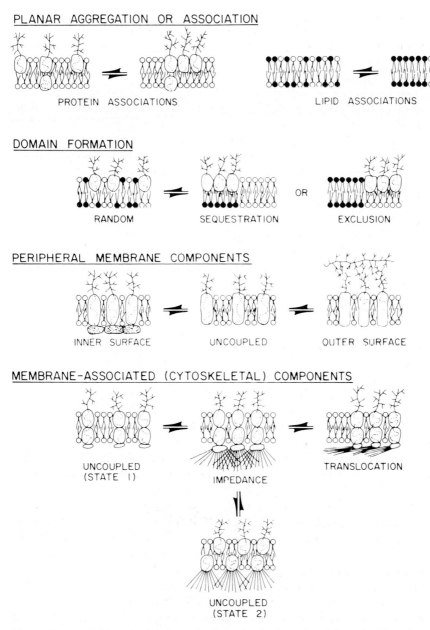

Fig. 5. Some possible restraining mechanisms on the lateral mobility of cell surface receptors. Reproduced with permission from Nicolson (1976a).

plane and across to adjacent plasma membranes. The forces holding junctional complexes between cells are formidable (Berry and Fried, 1969). Detergent treatment of membrane preparations has been utilized to solubilize plasma membranes in nonjunctional regions leaving the junctional areas intact (Benedetti and Emmelot, 1968; Goodenough and Stoeckenius, 1972). Once isolated, junctions appear to be as complex in molecular nature as they seem to be ultrastructurally. Analysis of protease-treated hepatocyte gap junctional complexes by Goodenough (1974) revealed that the junction is comprised mainly of two polypeptides with molecular weights of around 10 000 daltons, although these polypeptides are cleavage products of higher molecular weight (at least 34 000 daltons) components.

The nature of the linkage between gap junctional components on adjacent cells remains a mystery. However, gap functions can be disassociated in hypertonic disaccharide solutions suggesting that saccharide–saccharide interactions, and also possibly ionic bridges, may be involved in stabilizing these intercellar bridges (Goodenough and Gilula, 1974). Lateral associations between junctional subunits in the same cell membrane may be stabilized by disulfide bonds and possibly other noncovalent forces that stabilize their supramolecular organization. Cell junctions are probably capable of some (albeit restricted) lateral motion, probably as "frozen" protein-rich domains in a fluid-lipid matrix (Nicolson, 1976a). Since cell junctions can be broken down as well as assembled, their stability is not vital to the cell. Eventual breakdown of junctions probably occurs when cells are separated. This process could leave the half-junction free to move, ultimately resulting in their endocytosis and removal from the cell surface (Staehelin, 1973).

Freeze-fracture localization of nerve post-synaptic membrane components in tight, somewhat ordered, arrays (Sandri et al., 1972) may be another example of membrane lateral associations. The post-synaptic components (transmitter receptors and transmitter degradative enzymes) remain associated or complexed even after efferent presynaptic nerve endings have been stripped away (Mathews et al., 1976). However, this stabilization could also be due to another class of membrane control ("trans-membrane control by peripheral membrane components"; see section 3.2.1 below), because of the existence of a structure called the post-synaptic density observable by electron microscopy due to its heavy metal staining characteristics (Cotman et al., 1974).

Another possible example of membrane lateral associations that lead to the formation of complex supramolecular structures occurs during the plasma membrane budding of lipid-containing animal viruses. When intact viruses are produced during the later stages of viral infection, they are assembled at the host plasma membrane by a process called budding. An important step in the assembly and morphogenesis of viruses of this type is that the virus envelope is created by the clustering of virus-specified envelope proteins within the lipid bilayer of a region of the host plasma membrane with concomitant exclusion of host cell proteins from this region. There are several possible routes that could be followed to achieve this step (reviews, Lenard and Compans, 1974; Rott and

Klenk, 1977), but most interest has focused on two possibilities. The first involves random insertion of virus-specified proteins into the host cell plasma membrane followed by lateral aggregation into clusters to form the prospective budding site. The second possibility is that intracellular vesicles containing virus envelope proteins fuse with the plasma membrane. This would, of course, provide a means for achieving direct clustering of virus proteins within localized regions of the membrane, thereby reducing the need for substantial lateral aggregation of virus proteins within the plasma membrane. However, irrespective of which mechanism is involved, it seems likely that stable lateral association of the viral components is necessary to maintain the integrity of the virus budding site within a fluid membrane environment. However, an alternative, and perhaps additional, controlling mechanism may involve transmembrane restriction of the lateral mobility of the clustered virus envelope components via peripheral or membrane-associated structures on the inner surface of the prospective virus envelope region of the plasma membrane (cf., Garoff and Simons, 1974).

The forces involved in lateral (cis) membrane control and in stabilizing lateral associations could be of various types (Nicolson, 1976a). Integral membrane components could associate within the membrane (hydrophobic interaction) if these interactions are favored over protein–lipid interactions. This does not seem nearly as likely as associations driven by forces outside the membrane hydrophobic zone such as ionic bridge and hydrogen bond formation and possibly other short range forces (e.g., Van der Waals and London dispersion forces). Associations between oligosaccharides can be quite strong and are cooperative in many systems (Rees, 1969). Disulfide bond formation was mentioned previously, and it may be important in the maintenance of some supramolecular aggregates (Yu and Steck, 1974; Hynes et al., 1976).

3.1.2. Domain formation
The associations of lipids to form specific lateral phase separations has been discussed (section 2.1.4). Also, it appears that membrane components that organize to form intramembranous particles do not partition into solid phase lipid regions (Chen and Hubbell, 1973; Kleemann and McConnell, 1974; Grant et al., 1974). The preferential association of many membrane proteins within fluid lipid regions could be dependent on requirements of lipid bilayer thickness, degree of lipid acyl chain distortion, bending or rotation and lipid lateral compressibility (see section 2.1.4). If proteins are sequestered into fluid lipid domains, their dynamic display would be limited to only those regions. Alternatively, if they associate preferentially with solid phase lipids, their mobility would be impeded because of a more viscous lipid matrix, and they would also be limited to solid phase lipid regions. Thus, the dynamics and topographic display of membrane proteins may well be controlled, in part, by their associations with other proteins or with specific lipids in fluid or solid phases. In addition, differential fluidity from one lipid leaflet to the other could cause changes in partitioning *across* the bilayer and limit the number of

trans-membrane interactions which could occur (see next section). Since plasma membrane lipid composition can change during the cell cycle and during cell differentiation, it is conceivable that the display and dynamics of membrane proteins might be influenced by modifications in the membrane lipid matrix (Nicolson, 1976a).

3.2. Trans-membrane control

One of the more interesting ways in which the display and dynamics of proteins and glycoproteins situated on the outer surface of the plasma membrane can be controlled is through trans-membrane linkages to membrane peripheral components or membrane-associated components at the inner membrane surface (Nicolson, 1976a). The functional distinction between the terms peripheral and membrane-associated components has been made previously (Nicolson, 1974; 1976a), and it stems from the latter's similarity to cytoskeletal elements. Peripheral membrane components are proteins, glycoproteins, glycosylaminoglycans or other components that are held *to* the membrane by forces other than hydrophobic interactions within the lipid bilayer matrix. Membrane-associated components, on the other hand, are not membrane components in the true sense because of their transient nature, their dependency on cell energy systems for structural integrity, their sensitivity to drugs which disrupt cytoskeletal elements, and their structural linkage to other organelles, membranes and structures within the cell (Nicolson, 1976a).

3.2.1. Peripheral membrane components
There is substantial circumstantial evidence to suggest that in some membrane systems peripheral trans-membrane restraints are involved in controlling the display and mobility of cell surface receptors. In the human erythrocyte membrane the lateral mobilities of the major (outer surface) glycoproteins appear to be under restraint, and these molecules are prevented from undergoing free lateral diffusion (Nicolson et al., 1971b; Loor et al., 1972; Peters et al., 1974). This seems to be due to an inner surface peripheral protein network involving at least spectrin, a fibrous protein with a molecular weight of approximately 460 000 daltons (Marchesi et al., 1969; Clarke, 1971; Fuller et al., 1974) and erythrocyte membrane actin (band V on SDS-polyacrylamide gel electrophoresis [Fairbanks et al., 1971]). Tilney and Detmers (1975) have found that these two components can form high molecular weight associations in vitro, and they probably form a dense network on the inner erythrocyte membrane surface in situ (Nicolson et al., 1971a; Nicolson and Painter, 1973; Elgsaeter and Branton, 1974) affording the erythrocyte its characteristic biconcave shape, resistance to deformation and other physical properties (Jacob et al., 1972).

Evidence suggesting that a spectrin-actin network exerts peripheral transmembrane control is the following. In intact erythrocytes the movement of integral antigens and other proteins on the outer membrane surface and intramembranous particles in the membranes are normally impeded. However, in membrane ghosts in which some spectrin and actin is released (or reorganized), the lateral mobility of these integral components is enhanced dramatically (Pinto da Silva, 1972; Tillack et al., 1972; Nicolson, 1973a; Pinto da Silva and Nicolson, 1974; Victoria et al., 1975). Nicolson and Painter (1973) showed that perturbation of (inner surface) spectrin distribution in hypotonically lysed erythrocyte ghosts by sequestration of anti-spectrin IgG into ghosts resulted in concomitant perturbation (aggregation) of outer surface anionic sites known to be carried on the surface glycoproteins – mainly glycophorin, the major MN-sialoglycoprotein of the erythrocyte (Morawiecki, 1964; Marchesi et al., 1972) which traverses the membrane bilayer (Segrest et al., 1973; Morrison et al., 1974). The trans-membrane effects of anti-spectrin IgG on sialoprotein receptor distribution were found to be concentration-, time-, temperature- and valency-dependent (Fab would not substitute) (Nicolson and Painter, 1973).

Other experiments strongly suggest direct or indirect trans-membrane control over erythrocyte surface components by a spectrin-containing inner surface network. Ji and Nicolson (1974) used bifunctional imidate crosslinking agents to demonstrate this association by another experimental approach. Outer surface glycoproteins were first aggregated on erythrocyte ghosts (this is not possible on intact cells) by lectins such as *Phaseolus vulgaris* phytohemagglutinin (PHA) and wheat germ agglutinin (these lectins bind to glycophorin) or concanavalin A and *Ricinus communis* agglutinin (these bind to a band III component) (Adair and Kornfeld, 1974; Findlay, 1974). Then the lectin-labeled erythrocyte ghosts were crosslinked with dimethylmalonimidate which reacts with ϵ-amino groups on adjacent proteins. The spectrin molecules were found to be crosslinked under these conditions indicating that they had aggregated on the inner membrane surface following aggregation of the membrane glycoproteins at the outer surface induced by lectins. Spectrin was not crosslinked when the lectins were absent or removed by specific sugar haptens before addition of the bifunctional imidates. Direct or indirect linkage of spectrin, and probably other inner surface proteins (Wang and Richards, 1974), to the membrane-spanning surface glycoproteins (which in this system are organized into intramembranous particles [Pinto da Silva and Nicolson, 1974]) is also supported by experiments in which spectrin was shown to be aggregated after pH treatments known to produce aggregation of the intramembranous particles (Pinto da Silva, 1972; Elgsaeter et al., 1976). Elgsaeter et al. (1976) incubated freshly prepared erythrocyte ghosts in hypotonic solutions where the pH range was set to 3.8–5.3 and observed a slight contraction of ghosts which resulted in the formation of small membrane blebs. When these blebs were released from erythrocyte ghosts, collected and analyzed, they were found to be free of intramembranous particles and almost free of membrane proteins suggesting the contraction of a spectrin-actin meshwork resulted in lipid bilayer extrusion. If spectrin were first removed by pretreatment conditions known to elute spectrin and actin, the extrusion did not occur.

Other examples of trans-membrane peripheral control probably exist, but lack of detailed information does not permit us to speculate much further. One possibility is the dense array of relatively immobile intramembranous particles (Landis and Reese, 1974) and lectin receptors (Bittiger and Schnebli, 1974; Cotman and Taylor, 1974; Kelly et al., 1976) found in post-synaptic membranes. Even when pre-synaptic membranes are peeled away, the relative mobility of post-synaptic lectin receptors is low (Cotman and Taylor, 1974; Kelly et al., 1976). This lateral restraint could be due to proteins in the post-synaptic electron density which reside just at the inner surface of the post-synaptic membrane (Banker et al., 1974; Cotman et al., 1974).

3.2.2. Membrane-associated (cytoskeletal) components
Trans-membrane control over the distribution and dynamics of certain integral membrane proteins by membrane-associated cytoskeletal assemblies such as microtubules and microfilaments is one of the most interesting and complex restraining systems that cells probably use to control their surfaces. Microtubules are large rather rigid tubular structures with an outside diameter of approximately 25 nm and an inner core of 15 nm diameter. These structures are composed predominantly of a 55 000 daltons molecular weight protein, tubulin, which associates into dimers and higher order polymers to form the microtubules (Kirschner et al., 1974; Wilson and Bryan, 1974). These structures have been found in the nucleus, the cytoplasm and also often in close association with the plasma membrane (Weber et al., 1974; Brinkley et al., 1975) (Fig. 6) in a variety of cells (see Wilson, 1975). Cytoplasmic microtubules undergo rapid, reversible assembly and disassembly, and their state of polymerization is sensitive to temperature, Ca^{2+} concentration, cyclic nucleotide concentrations, pH, and pressure (Kirschner and Williams, 1974; Kirschner et al., 1974). Several drugs such as the alkaloids colchicine, colcemid, vinblastine sulfate and vincristine are effective in causing depolymerization of microtubules into subunit aggregates, as well as low temperature and high Ca^{2+} concentrations (Olmsted and Borisy, 1973; Wilson and Bryan, 1974).

Microtubules are often found in association with another class of membrane-associated cytoskeletal components, the cytoplasmic microfilaments. Microfilaments are thin protein polymers arranged in double helical filaments of diameter 6–8 nm and variable length. They occur in a variety of cellular organization states, the most common being the so called sheath or α-filaments lying parallel to and immediately beneath the plasma membrane and the microfilament "bundles" that penetrate deep into the cell from the inner surface of the plasma membrane. Alternatively, the microfilaments can form lattice or network filamentous structures that are present throughout the cytoplasm but are also common immediately adjacent to the plasma membrane (Goldman, 1972, 1975; Weber et al., 1974; Goldman et al., 1975; Lazarides, 1975a). Recent biochemical and immunofluorescence studies (Fig. 7) have greatly increased our knowledge of the properties of these cytoplasmic microfilament networks. Microfilaments have been shown to contain actin

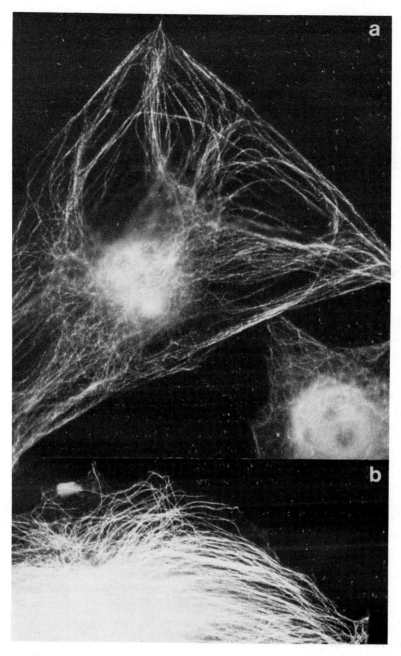

Fig. 6. Localization of tubulin in cytoplasmic microtubules by indirect immunofluorescence. a, low magnification showing cellular microtubular distribution and perinuclear organization. b, high magnification showing plasma membrane microtubule association. Reproduced with permission from Brinkley et al. (1975).

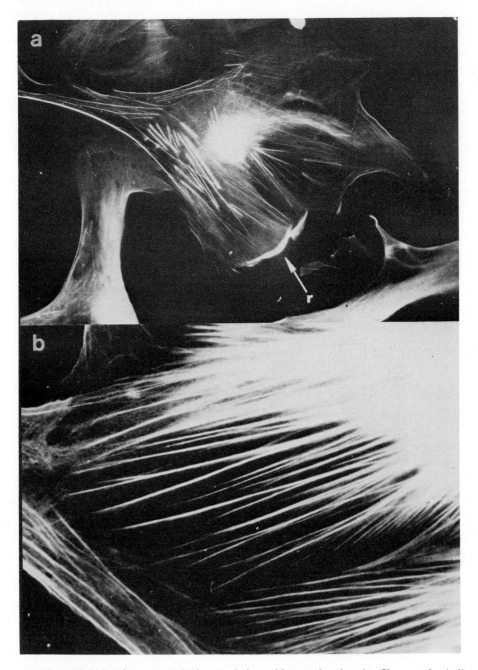

Fig. 7. Localization of tropomyosin in association with cytoplasmic microfilaments by indirect immunofluorescence techniques. a, low magnification showing tropomyosin localization in microfilament bundles and membrane ruffles (r) of human skin fibroblasts. b, high magnification of anti-tropomyosin labeled microfilaments in a rat embryo fibroblast. Reproduced with permission from Lazarides (1975a).

(Ishikawa et al., 1969; Goldman, 1972; Lazarides and Weber, 1974; Pollard and Weihing, 1974; Weber et al., 1974; Goldman et al., 1975; Lazarides 1975a) and tropomyosin (Lazarides, 1975a,b). These assemblies may be joined to the plasma membrane at regions containing dense arrays of α-actinin (Fig. 8) (Lazarides and Burridge, 1975; Tilney and Detmers, 1975) or by cross-bridges (Tilney and Detmers, 1975). Myosin has been found in association with microfilaments (Behnke et al., 1971; Weber and Groeschel-Stewart, 1974), but the former seems to be predominantly localized outside the microfilament regions (Weber and Groeschel-Stewart, 1974; Pollack et al., 1975), possibly organized into thicker cytoplasmic filaments (8–12 nm diameter). Microfilament assemblies are sensitive to cellular cyclic nucleotide concentration (Hsie et al., 1971; Willingham and Pastan, 1974), and they can be disrupted to varying degrees by a class of mold metabolites collectively called the cytochalasins (Schroeder, 1968; Wessels et al., 1971; Goldman, 1972). Microfilaments are thought to perform important muscle-like contractile functions in the cell such as locomotion, cytoplasmic streaming, cell shape maintenance, and movements or impedance of movement of certain plasma membrane components (see Poste et al., 1975b; Goldman et al., 1976; Nicolson, 1976a; Nicolson and Poste,

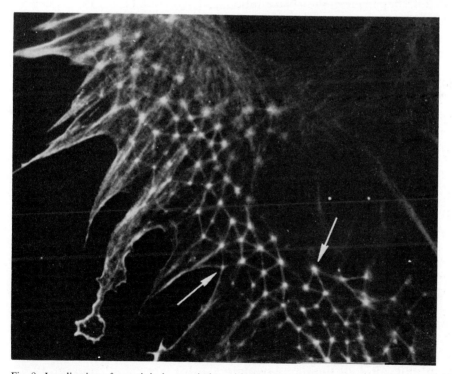

Fig. 8. Localization of α-actinin in association with the plasma membranes of rat embryo fibroblasts by indirect immunofluorescence. Arrows indicate tips of fluorescent nets which could be the membrane attachment points of the microfilaments. Reproduced with permission from Lazarides (1975a).

1976a). Conversely, the microtubules are thought to perform important anchoring functions that restrict the mobility of those plasma membrane components to which they are "linked" (Yahara and Edelman, 1972, 1973, 1975a; Berlin et al., 1974; Edelman, 1976).

The distribution and dynamics of surface receptors on several cell types seems to be under regulatory control by *both* of these classes of membrane-associated cytoskeletal systems (reviews, Unanue and Karnovsky, 1973; Edelman, 1976; Loor, 1976; Nicolson, 1976a,b). One of the most studied systems is the B-lymphocyte where the binding of multivalent ligands such as anti-immunoglobulin (anti-Ig) can stimulate dispersed → cluster → patch → cap receptor rearrangements, resulting eventually in endocytosis or shedding of receptor-ligand complexes from the cap region. There are a variety of drug and environmental conditions which prevent cap formation (Nicolson, 1976a), a process known to require cellular energy (Taylor et al., 1971; Loor et al., 1972; Sällström and Alm, 1972; Yahara and Edelman, 1972, 1975a; Unanue et al., 1973). Drugs that disrupt microfilament organization such as the cytochalasins prevent cap formation to variable degrees (Yahara and Edelman, 1972; Unanue et al., 1973; de Petris, 1974, 1975; Poste et al., 1975a,b), while drugs causing microtubule depolymerization have little effect (Unanue et al., 1973; Poste et al., 1975a) or may actually enhance cap formation (Yahara and Edelman, 1972; de Petris, 1975; Poste et al., 1975a,b). The fact that cytochalasin B *plus* colchicine or vinblastine administered simultaneously to lymphocytes (de Petris, 1974, 1975; Poste et al., 1975a,b) or polymorphonuclear leukocytes completely, or almost completely (Ryan et al., 1974a), blocks cap formation suggests that *both* microfilaments and microtubules are involved in the capping process. The role of the microfilament system is thought to be active (contractile), and these elements probably provide the stress necessary to translocate ligand-receptor clusters and patches into a cap at one end of the cell.

The binding of high concentrations (>25 μg/ml) of lectins such as concanavalin A to the surface of B-lymphocytes at 37°C results in inhibition of anti-Ig-induced capping (Loor et al., 1972; Yahara and Edelman, 1972, 1973, 1975a; de Petris, 1975). However, the concanavalin A block is not effective if cells are pretreated at low temperature (≤5°C) or labeled with low concentrations (≤5 μg/ml) of concanavalin A. Under these conditions concanavalin A stimulates capping of concanavalin A-receptors (Yahara and Edelman, 1972, 1973; de Petris, 1975). Low temperatures are known to disrupt microtubules (Wilson and Bryan, 1974) and also cause concanavalin A to disassociate from a tetrameric to dimeric form (Huet and Bernadac, 1975). (Dimeric succinyl-concanavalin A is unable to inhibit anti-Ig-induced capping [Edelman et al., 1973]). Interestingly, concanavalin A inhibition of anti-Ig-induced capping can be blocked by pretreatment with colchicine (Loor et al., 1972; Yahara and Edelman, 1972, 1973) or the Ca^{2+}-ionophore A23187 (Poste and Nicolson, 1976). Similarly, Oliver et al. (1975) have shown that colchicine treatment enhanced concanavalin A-induced capping of its own receptors on mouse leukocytes. Experiments such as these led Edelman et al. (1973, 1976) to propose that microtubules play an important role in "anchoring" or immobilizing cell surface receptors and that colchicine treatment breaks anchoring, presumably by disrupting microtubule "links" to distant receptors, resulting in the observed enhancement of cap formation induced by anti-Ig as well as concanavalin A at low concentrations.

Since crosslinking of adjacent receptors by direct concanavalin A linkages instead of indirect trans-membrane microtubule-linkages could also explain the immobilizing effect of concanavalin A (Loor, 1974; de Petris, 1975), concanavalin A was derivatized to nylon fibers, latex spheres and platelets so that the interaction of the lectin would be limited to a small percentage of the lymphocyte cell surface receptors. Lymphocytes that were interacted with the concanavalin A-derivatized nylon fibers did not form caps when treated with anti-Ig, even though only a small fraction of the total concanavalin A receptors on the cell surface were bound to the fibers (Rutishauser et al., 1974). As before, colchicine reversed the concanavalin A inhibition of anti-Ig-induced cap formation, suggesting that locally immobilized lectin receptors can cause anchoring of other receptors at long range. Similar results were obtained with concanavalin A-derivatized latex beads and platelets. When >10 beads or platelets were bound per lymphocyte, the ligand-induced redistribution of Ig, Fc and H-2 determinants by their respective antibodies was prevented, unless the cells were first pretreated with colchicine (Yahara and Edelman, 1975b).

Additional insight into the role of membrane-associated cytoskeletal assemblies in regulating the distribution and mobility of cell surface receptors has come from recent studies on the cellular effects of tertiary amine local anesthetics. This class of drugs was formerly thought to interact exclusively with membrane lipids (Papahadjopoulos, 1972; Seeman, 1972). In particular, they appear to partition into fluid membranes and bind through hydrophobic and ionic interactions to acidic phospholipids (Papahadjopoulos, 1970, 1972; Giotta et al., 1974). At high concentrations local anesthetics modify membrane lipid fluidity (Hubbell et al., 1970; Papahadjopoulos et al., 1975a). At low concentrations where membrane fluidity is unchanged (Papahadjopoulos et al., 1975a), they produce a variety of cellular effects including: inhibition of cell spreading (Rabinovitch and De Stefano, 1974), movement (Gail and Boone, 1972; Poste and Reeve, 1972), adhesion (O'Brien, 1962; Rabinovitch and De Stefano, 1973) and fusion (Poste and Reeve, 1972).

The site of action of local anesthetics in modifying the dynamics of ligand-induced receptor redistribution is now thought to be at the membrane-associated cytoskeletal system because: (a) agglutination of untransformed mouse fibroblastic cells by lectins does not occur at low lectin concentrations where a relatively low rate of lectin-induced redistribution occurs (Nicolson, 1974; 1976b), unless the cells are first treated with local anesthetics which enhance lectin-induced agglutinability and receptor redistribution (Poste et al., 1975a,b); (b) local anesthetics reverse the inhibition of concanavalin A-induced agglutination of transformed mouse fibroblastic cells by colchicine and vinblastine (Poste et al., 1975b); (c) lymphoid cell capping is prevented by local anesthetics (Ryan et al., 1974b; Poste et al., 1975b); and (d) local anesthetics induce breakdown of preformed caps on lymphocytes (Poste et al., 1975b).

The fact that the above effects of local anesthetic-treated cells can be duplicated by treatment with colchicine or vinblastine *together* with cytochalasin B (colchicine *or* cytochalasin B alone will not suffice) strongly suggests that colchicine-sensitive microtubules and cytochalasin B-sensitive microfilaments are *both* involved and, in fact, play opposing roles in the maintenance of surface receptor distribution and dynamics (Poste et al., 1975a,b; Nicolson, 1976a). This concept proposes that microtubules, as suggested earlier by Edelman (1974) (see above), serve to "anchor" receptors and restrict their lateral mobility within the membrane. The microfilaments are seen as also being linked to the same receptors (either directly or indirectly) and functioning in opposition to the microtubules via a contractile activity that redistributes receptors within the membrane to form caps. In this scheme the topography of membrane receptors at any time reflects the interplay between these two opposing, but coordinated, systems. This scheme accounts fully for the diverse effects of different drug treatments on receptor mobility. For example, functional dislocation of receptors from "anchoring" microtubules by such drugs as colchicine or vinblastine will favor ligand-induced receptor redistribution and, assuming that the contractile network of microfilaments remains "linked" to the receptors, capping of receptors will result as indeed observed (see Ukena et al., 1974; Poste et al., 1975b). Conversely, selective inhibition of microfilament function by cytochalasin B would be expected to prevent capping, as reported in numerous laboratories, but since the receptors remain linked to their microtubule "anchors", they will remain randomly distributed. Finally, in situations where both microtubules and microfilaments are perturbed, as in cells treated with local anesthetics or a combination of colchicine *plus* cytochalasin B, the loss of the microtubule anchors would facilitate ligand-induced receptor redistribution, but since the contractile microfilament system is also inhibited, the capping of ligand-receptor complexes cannot occur, and the ligand will instead induce the formation of multiple clusters of patches of receptors as is in fact seen (Poste et al., 1975a,b).

Evidence that local anesthetics may directly modify cytoskeletal elements or their plasma membrane attachment points has been obtained using electron microscopy. Low concentrations of local anesthetics such as dibucaine, tetracaine or procaine which modify surface receptor dynamics result in cell rounding and disappearance of cell membrane-associated microfilaments and microtubules of cultured cells. Within five minutes of exposure to local anesthetics, the plasma membranes on BALB/c 3T3 cells lift and in many places form surface "blebs" (Nicolson et al., 1976) resembling the surface zeiotic blebs induced by cytochalasin B (Carter, 1970). In other cell types it has been documented that both general (Allison and Nunn, 1968) and local anesthetics (Haschke et al., 1974; Hinkley and Telse, 1974) cause ultrastructural alterations in cytoskeletal assemblies. Local anesthetics such as lidocaine have also been reported to impair microtubule assembly in vitro (Haschke et al., 1974), and halothane, a gaseous general anesthetic, is known to cause microfilament disruption in mouse nerve cells (Hinkley and Telse, 1974). The exact mechanism by which local anesthetics disrupt cytoskeletal systems is presently unclear. One possibility is that they could increase intracellular Ca^{2+} concentrations to levels sufficient to depolymerize microtubules by displacing membrane Ca^{2+} (cf., Papahadjopoulos, 1972). The effects of local anesthetics on microfilament integrity is even less certain, but these drugs might affect Ca^{2+}-sensitive components such as actomyosin complexes, adenylcyclase or guanylcyclase, Ca^{2+}-requiring ATPases, or other Ca^{2+}-dependent processes (see discussion in Poste et al., 1975b; Poste and Nicolson, 1976; Nicolson and Poste, 1976a,b).

Calcium ionophores, such as A23187 and X537A have been used to examine the possible role of changes in cytoplasmic Ca^{2+} in regulating the cellular polymerization of microtubules (Poste and Nicolson, 1976; Nicolson and Poste, 1976b; Schreiner and Unanue, 1976). A23187 or X537A reduce the agglutinability of transformed 3T3 cells by concanavalin A (Poste and Nicolson, 1976) similar to the action of colchicine on these cells (Yin et al., 1972; Poste et al., 1975b). However, these drugs have no effect on the agglutinability of untransformed 3T3 cells unless cytochalasin B was also present. In this latter case, the concanavalin A-induced agglutination of untransformed 3T3 cells was enhanced dramatically (Poste and Nicolson, 1976; Nicolson and Poste, 1976b). In addition, A23187 or X537A was found to block concanavalin A-inhibition of anti-Ig-induced capping (Poste and Nicolson, 1976) or directly block capping (Schreiner and Unanue, 1976).

The effects of local anesthetics and Ca^{2+} ionophores on surface receptor dynamics and membrane-associated cytoskeletal functions suggests that Ca^{2+} sensitive functions are involved in these processes. Local anesthetics possess high affinity for Ca^{2+}-binding sites and can displace Ca^{2+} from membrane sites (Papahadjopoulos, 1970, 1972; Seeman, 1972). Modifications in Ca^{2+} binding to plasma membranes and to membrane-associated cytoskeletal assemblies induced by local anesthetics could functionally disturb the linkages of cell surface receptors to the cytoskeletal elements similar to the displacement of spectrin from the inner surface of human erythrocyte ghosts. For example, spectrin, which is involved in the trans-membrane control of receptor mobility in erythrocytes (Nicolson and Painter, 1973; Ji and Nicolson, 1974; Elgsaeter et al., 1976), is released from its association with the plasma membrane by Ca^{2+}-chelating agents (Marchesi and Steers, 1968) and will reassociate to spectrin-depleted ghost membranes only in the presence of Ca^{2+} (Juliano et al., 1971). Interestingly, local anesthetics cause dramatic changes in erythrocyte morphology (Sheetz and Singer, 1974; Sheetz et al., 1976), which could well be due to anesthetic displacement of Ca^{2+} from the plasma membrane and/or to modifications in Ca^{2+}-sensitive inner surface components involved in stabilizing the spectrin-actin network.

In other cell types such as 3T3 where extensive membrane-associated cytoskeletal assemblies exist, local anesthetics probably displace Ca^{2+} from cellular membranes which could produce a dual effect. First, the free Ca^{2+} concentration would be expected to increase, perhaps to levels capable of directly depolymerizing microtubules (Kirschner et al., 1974; Olmsted et al., 1974) (Fig. 9). The effects of local anesthetics on microfilaments are more elusive, but these components are known to possess Ca^{2+}-sensitive functions. Alternatively, Ca^{2+} could stabilize and link actin-containing microfilament systems to the plasma membrane via Ca^{2+} bridges similar to spectrin in the human erythrocyte membrane, and local anesthetics might break such linkages (Fig. 9). In either case an uncoupling of microfilaments and microtubules to the plasma membrane would ensue, an event that can be duplicated by treating cells with colchicine (or vinblastine) plus cytochalasin B.

4. Trans-membrane architecture

This final, brief section will attempt to synthesize the information from the previous sections into a scheme for membrane structure. It should be mentioned that any schematic representation of the cell surface and associated structures (such as Fig. 10) will necessarily be obsolete by press time. However, the main usefulness for such exercises is to stimulate further discussion and experimentation. We do not feel at this stage that such exercises are "premature" (Wallach and Winzler, 1974), since this hypothetical scheme is based on a large volume of constantly evolving experimental and theoretical data. A very similar scheme has been proposed recently and independently by Loor (1977).

The basic structure of the plasma membrane in our scheme (Fig. 10) is an elaboration of the Fluid Mosaic Model of membrane structure (Singer and Nicolson, 1972) with an added cellular structure (microvillus) (similar to Nicolson, 1976a and Loor, 1977). On the extracellular side of the membrane, glycosyl-

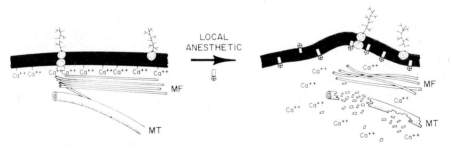

Fig. 9. Proposal to explain the action of tertiary amine local anesthetics on cell membranes and membrane-associated cytoskeletal assemblies. The partitioning of drug molecules into cellular membranes results in displacement of membrane-bound Ca^{2+} which could be involved in stabilizing microfilaments (MF) to the plasma membrane inner surface. In addition, the resulting increase in free cellular Ca^{2+} could be responsible for microtubule (MT) depolymerization (not drawn to scale). Reproduced with permission from Nicolson and Poste (1976b).

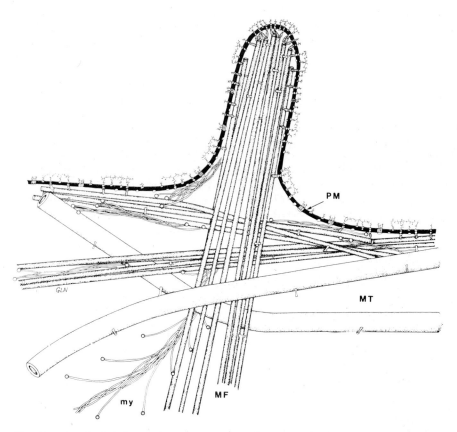

Fig. 10. Hypothetical interactions between membrane-associated microtubule (MT) and microfilament (MF) systems involved in trans-membrane control over cell surface receptor mobility and distribution. This model envisages opposite, but coordinated, roles for microfilaments (contractile) and microtubules (skeletal) and suggests that they are linked to one another or to the same plasma membrane (PM) inner surface components. This linkage may occur through myosin molecules (either in small bundles or the larger filaments [my] cf., Young et al., 1973) or through cross-bridging molecules such as α-actinin. In addition, peripheral membrane components linked at the inner or outer plasma membrane surface may extend this control over specific membrane domains.

aminoglycans or mucopolysacchardies are shown as binding to integral membrane glycoproteins, although certain parts of their structures may anchor hydrophobically in the membrane lipid matrix (Kraemer, 1975). Some of the integral membrane proteins or their supramolecular complexes are drawn as asymmetric trans-membrane molecules, consistent with the available evidence from at least certain membranes (reviewed in Nicolson, 1976a). Peripheral membrane proteins have been placed at the inner plasma membrane surface, and some of these are shown to interact with membrane-associated cytoskeletal elements. For example, α-actinin has been localized at the tips of microvilli (Schollmeyer et al., 1974; Mooseker and Tilney, 1975) and at various other regions on the plasma membrane inner surface in close association with actin-containing microfilaments (Lazarides, 1975a,b) (see Fig. 8). The α-actinin molecules appear to crosslink actin-containing microfilaments to the plasma membrane (Mooseker and Tilney, 1975). Parallel arrays of microfilaments are shown extending from the microvillus tip to deep within the cell cytoplasm and also running parallel to the plasma membrane (McNutt et al., 1971; Perdue, 1973; Lazarides and Weber, 1974; Lazarides, 1975a,b). It is likely that adjacent microfilament arrays are linked to myosin polymers similar to the attachment of microvilli microfilaments to the terminal web of intestinal epithelial cells by thick filaments (Mooseker and Tilney, 1975). Similar actin–myosin interactions probably occur immediately adjacent to the plasma membrane as well. In an earlier version of this diagram (Fig. 13 of Nicolson, 1976a) myosin molecules were also drawn on the extracellular side of the plasma membrane in concert with recent evidence suggesting its distribution in association with the cell surface (Willingham et al., 1974; Willingham and Pastan, 1975b). However, the antigens localized at the cell surface in these studies could have been nonspecifically absorbed or simply portions of myosin molecules since other studies have failed to detect myosin at the cell surface (cf., Weber and Groeschel-Stewart, 1974; Painter et al., 1975) and have instead localized it within the cytoplasm or on the cytoplasmic surface of the plasma membrane (Allison, 1973; Weber and Groeschel-Stewart, 1974).

A variety of "bridge" molecules are illustrated as being present in the cytoskeletal system (Fig. 10). In microvilli, bridge-like molecules (apparently different from α-actinin) link microfilaments to the plasma membrane. Their biochemical identity remains to be established, but these cross-bridges are distinctly identifiable in microvilli by electron microscopy (Mooseker and Tilney, 1975). Microtubules also possess cross-bridge molecules that link adjacent microtubules or link microtubules to cellular membranes (reviewed in Heath, 1974 and Hepler and Palevitz, 1974). Localization of microtubules by indirect immunofluorescence has revealed a close association of these structures with the plasma membrane (Brinkley et al., 1975; Weber et al., 1975a,b) (Fig. 6), consistent with several physiological studies using microtubule-disrupting drugs (section 3.2.2).

We have proposed that the membrane-associated microfilaments and microtubules play opposing and coordinating roles in the regulation of the movement and distribution of cell surface receptors (Poste et al., 1975a,b; Nicolson, 1976a; Nicolson and Poste, 1976a). To satisfy this proposal, microfilaments and microtubules must be connected to one another or to similar inner surface plasma membrane components, and in our scheme both possibilities are shown (Fig. 10). The trans-membrane components linked to the cytoskeletal assemblies are thought not to be the large macromolecular freeze-fracture intramembranous particles (Karnovsky et al., 1972; Matter and Bonnet, 1974; Pinto da Silva et al., 1975), but they could be smaller trans-membrane structures (<20 Å across) or larger structures linked together at the interface between the halves of the lipid bilayer (Nicolson, 1976a). In these latter cases the trans-membrane linkage components would not be visualized by freeze-fracture techniques. Edelman (1974, 1976) has advanced the idea that the cytoskeletal system exerts modulatory control over the cell surface by existing in "free" or "membrane-attached" equilibrium states, so that at any one time certain classes of trans-membrane linked surface receptors are under cytoplasmic control. It is also a possibility that only a few trans-membrane linkage components control the distribution and mobility of a large number of cell surface receptors through peripheral interactions as suggested by Hynes (1974). In Fig. 10 this possibility is illustrated in the form of glycosylaminoglycan interaction with a variety of surface components in "membrane domains". Finally, it is clear that not all surface components are coupled to one of the restraint mechanisms discussed in section 3, and these probably form a family of highly mobile, largely unregulated surface molecules.

Much more information will have to be gathered to fill in the many gaps in our knowledge which still exist. For example, what are the molecules which make up the trans-membrane linkages in most cells? In the human erythrocyte the surface glycoproteins (glycophorin and component[s] III) appear to be the major trans-membrane vehicles involved in cell surface receptor control, but the molecular nature of receptors and trans-membrane structures from other cell types have not been determined. What controls the assembly, disassembly, attachment and cross-bridging of membrane-associated cytoskeletal elements? The interesting effects of ATP, cyclic nucleotides and Ca^{2+} on the organizational status of microfilaments and microtubules have led to some interesting theories (for example, see Willingham and Pastan, 1975a), but their synthesis into a unifying hypothesis on cell surface control is still hampered by problems in the localization and compartmentalization of these and other cell regulatory molecules and the existence of exceptions to such theories (for example, see van Veen et al., 1976). The most obvious conclusion to be drawn here is that the cell surface can no longer be considered as a simple interfacial boundary for the cell and we must also consider the role of a variety of membrane-associated extra- and intracellular structures in regulating both the properties of the cell surface and a wide range of cellular activities.

Acknowledgments

We thank A. Brodginski for assistance. The authors' studies were supported by N.I.H.-N.C.I. contract CB-33879 from the Tumor Immunology Program (to G.L.N.), U.S.P.H.S. grants CA-15122-A1 (to G.L.N.), CA-13393 and CA-18260 (to G.P.) and NSF grants PCM76-18528 (to G.L.N.) and BMS75-09230 (to T.J.).

References

Abbas, A.K., Ault, K.A., Karnovsky, M.J. and Unanue, E.R. (1975) Non-random distribution of surface immunoglobulin on murine B lymphocytes. J. Immunol. 114, 1197–1204.

Abercrombie, M., Heaysman, J.E.M. and Pegrum, S.M. (1970) The locomotion of fibroblasts in culture, III. Movements of particles on the dorsal surface of the leading lamella. Exp. Cell Res. 62, 389–398.

Adair, W.L. and Kornfeld, S. (1974) Isolation of the receptors for wheat germ agglutinin and the *Ricinus communis* lectins from human erythrocytes using affinity chromatography. J. Biol. Chem. 249, 4696–4704.

Adams, D., Markes, M.E., Leivo, W.J. and Carraway, K.L. (1976) Electron spin resonance analysis of irreversible changes induced by calcium perturbation of erythrocyte membranes. Biochim. Biophys. Acta 426, 38–45.

Albrecht-Bühler, G. (1973) A quantitative difference in the movement of marker particles in the plasma membrane of 3T3 mouse fibroblasts and their polyoma transformants. Exp. Cell Res. 78, 67–70.

Albrecht-Bühler, G. and Solomon, F. (1974) Properties of particle movement in the plasma membrane of 3T3 mouse fibroblasts. Exp. Cell Res. 85, 225–233.

Allison, A.C. (1973) The role of microfilaments and microtubules in cell movement, endocytosis and exocytosis. In: Locomotion of Tissue Cells, Ciba Found. Symp. 14, 109–148.

Allison, A.C. and Nunn, J.F. (1968) Effects of general anaesthetics on microtubules. A possible mechanism of anaesthesia. Lancet 2, 1326–1329.

Antoine, J-C., Avrameas, S., Gonatas, N.K., Stieber, A. and Gonatas, J.O. (1974) Plasma membrane and internalized immunoglobulins of lymph node cells studied with conjugates of antibody or its Fab fragments with horseradish peroxidase. J. Cell. Biol. 63, 12–23.

Aoki, T. (1971) Surface antigens of murine leukemia cells and murine leukemia viruses. Transplant. Proc. 3, 1195–1198.

Aoki, T. and Takahashi, T. (1972) Viral and cellular surface antigens of murine leukemias and myelomas. Serological analysis by immunoelectron microscopy. J. Exp. Med. 135, 443–457.

Aoki, T., Hämmerling, U., de Harven, E., Boyse, E.A. and Old, L.J. (1969) Antigenic structure of cell surfaces: An immunoferritin study of the occurrence and topography of H-2, θ, and TL alloantigens on mouse cells. J. Exp. Med. 130, 979–1001.

Arias, I.M., Doyle, D. and Schimke, R.T. (1969) Studies on the synthesis and degradation of proteins of the endoplasmic reticulum of rat liver. J. Biol. Chem. 244, 3303–3315.

Avrameas, S. (1970) Emploi de la concanavaline-A pour l'isolement, la détection et la mesure des glycoprotéines et glucides extra- ou endocellulaires. C.R. Acad. Sci. 270, 2205–2208.

Avruch, J. and Wallach, D.F.H. (1971) Spectroscopic evidence of β structure in rat adipocyte plasma membrane protein. Biochim. Biophys. Acta 241, 249–253.

Baker, P.F. and Reuter, H. (1975) Calcium Movement in Excitable Cells, Pergamon Press, Oxford, 102 pp.

Bangham, A.D. (1972) Lipid bilayers and biomembranes. Ann. Rev. Biochem. 41, 753–776.

Banker, G., Churchill, L. and Cotman, C.W. (1974) Proteins of the post-synaptic density. J. Cell Biol. 63, 456–465.

Bashford, C.L., Morgan, C.G. and Radda, G.K. (1976) Measurement and interpretation of fluorescent polarizations in phospholipid dispersions. Biochim. Biophys. Acta 426, 157–172.

Bedford, J.M., Cooper, G.W. and Calvin, H.I. (1972) Post-meiotic changes in the nucleus and membranes of mammalian spermatozoa. In: R.A. Beatty and S. Glueecksohn-Waelsch (Eds.), The Genetics of the Spermatozoa, Bogtrykeriet Forum, Copenhagen, pp. 69–89.

Behnke, O., Kristensen, B.I. and Nielsen, L.E. (1971) Electron microscopical observation of actinoid and myosinoid filaments in blood platelets. J. Ultrastruct. Res. 87, 351–369.

Bender, W.W., Garan, H. and Berg, H.C. (1971) Proteins of the human erythrocyte membrane as modified by pronase. J. Mol. Biol. 58, 783–797.

Benedetti, E.L. and Emmelot, P. (1967) Studies on plasma membranes, IV. The ultrastructural localization and content of sialic acid in plasma membranes isolated from rat liver and hepatoma. J. Cell Sci. 2, 499–512.

Benedetti, E.L. and Emmelot, P. (1968) Hexagonal array of subunits in tight junctions separated from isolated rat liver plasma membranes. J. Cell Biol. 38, 15–24.

Bennett, G. and Leblond, C.P. (1970) Formation of cell coat material for the whole surface of columnar cells in the rat small intestine as visualized by radioautography with L-fucose-^{3}H. J. Cell Biol. 46, 409–416.

Berg, H.C. (1969) Sulfanilic acid diazonium salt: A label for the outside of the human erythrocyte membrane. Biochim. Biophys. Acta 183, 65–78.

Berlin, R.D., Oliver, J.M., Ukena, T.E. and Yin, H.H. (1974) Control of cell surface topography. Nature (London) 247, 45–46.

Bernacki, R.J. (1974) Plasma membrane ectoglycosyltransferase activity of L1210 murine leukemic cells. J. Cell. Physiol. 83, 457–466.

Berridge, M.J. (1975) The interaction of cyclic nucleotides and calcium in the control of cellular activity. In: P. Greengard and G.A. Robison (Eds.), Advances in Cyclic Nucleotide Research, Vol. 6, Raven Press, New York, pp. 1–98.

Berry, M.N. and Friend, D.S. (1969) High-yield preparation of isolated rat liver parenchymal cells. A biochemical and fine structural study. J. Cell Biol. 43, 506–520.

Bittiger, H. and Schnebli, H-P. (1974) Binding of concanavalin A and ricin to synaptic junctions of rat brain. Nature (London) 249, 370–371.

Blasie, J.K. and Worthington, C.R. (1969) Planar liquid-like arrangement of photopigment molecules in frog retinal receptor disk membrane. J. Mol. Biol. 39, 417–439.

Blasie, J.K., Worthington, C.R. and Dewey, M.M. (1969) Molecular localization of frog retinal receptor photopigment by electron microscopy and low-angle X-ray diffraction. J. Mol. Biol. 39, 407–416.

Blaurock, A.E. (1971) Structure of the nerve myelin membrane: Proof of the low-resolution profile. J. Mol. Biol. 56, 35–52.

Blaurock, A.E. (1973) The structure of a lipid-cytochrome C membrane. Biophys. J. 13, 290–298.

Blaurock, A.E. and Stoeckenius, W. (1971) Structure of the purple membrane. Nature (London) New Biol. 233, 152–155.

Bloj, B. and Zilversmit, D.B. (1976) Asymmetry and transposition rates of phosphatidylcholine in rat erythrocyte ghosts. Biochemistry 15, 1277–1283.

Bock, K.W., Siekevitz, P. and Palade, G.E. (1971) Localization and turnover studies of membrane nicotinamide adenine dinucleotide glycohydrolase in rat liver. J. Biol. Chem. 246, 188–195.

Bosmann, H.B., Hagopian, A. and Eylar, E.H. (1969) Glycoprotein biosynthesis: The characterization of two glycoproteins: fucosyltransferases in HeLa cells. Arch. Biochem. Biophys. 128, 470–481.

Branton, D. (1966) Fracture faces of frozen membranes. Proc. Natl. Acad. Sci. USA 55, 1048–1062.

Branton, D. (1969) Membrane structure. Ann. Rev. Plant Physiol. 20, 209–238.

Branton, D. (1971) Freeze-etching studies of membrane structure. Phil. Trans. Roy. Soc. Lond. B. 261, 133–138.

Branton, D. and Deamer, D. (1972) In: M. Alfert, H. Bauer, W. Sandritter, and P. Sitte (Eds.), Membrane Structure, Springer-Verlag, Vienna, 70 pp.

Bretscher, M.S. (1971a) Human erythrocyte membranes: Specific labeling of surface proteins. J. Mol. Biol. 58, 775–781.

Bretscher, M.S. (1971b) Major protein which spans the human erythrocyte membrane. J. Mol. Biol. 59, 351–357.

Bretscher, M.S. (1972) Phosphatidyl-ethanolamine: Differential labeling in intact cells of human erythrocytes by a membrane impermeable reagent. J. Mol. Biol. 71, 523–528.

Bretscher, M.S. (1973) Membrane structure: Some general principles. Membranes are asymmetric lipid bilayers in which cytoplasmically synthesized proteins are dissolved. Science 181, 622–629.

Brinkley, B.R., Fuller, G.M. and Highfield, D.P. (1975) Cytoplasmic microtubules in normal and transformed cells in culture: Analysis by tubulin antibody immunofluorescence. Proc. Natl. Acad. Sci. USA 72, 4981–4985.

Brockerhoff, H. (1974) Model of interaction of polar lipids, cholesterol, and proteins in biological membranes. Lipids 9, 645–650.

Brown, P.K. (1972) Rhodopsin rotates in the visual receptor membrane. Nature (London) New Biol. 236, 35–38.

Burger, M.M. (1973) Surface changes in transformed cells detected by lectins. Fed. Proc. 32, 91–101.

Butler, K.W., Dugas, H., Smith, I.C.P. and Schneider, H. (1970a) Cation-induced organization changes in a lipid bilayer model membrane. Biochem. Biophys. Res. Commun. 40, 770–776.

Butler, K.W., Smith, I.C.P. and Schneider, H. (1970b) Sterol structure and ordering effects in spin-labelled phospholipid multibilayer structures. Biochim. Biophys. Acta 219, 514–517.

Cadenhead, D.A., Kellner, B.M.J. and Müller-Landau, F. (1975) A comparison of a spin label and a fluorescent cell membrane probe using pure and mixed monomolecular films. Biochim. Biophys. Acta 382, 253–259.

Capaldi, R.A. and Green, D.E. (1972) Membrane proteins and membrane structure. FEBS Lett. 25, 205–209.

Carafoli, E., Clementi, F., Drabikowski, W. and Margreth, A. (1975) (Eds.) Calcium Transport in Contraction and Secretion, North-Holland, Amsterdam.

Carraway, K.L., Triplett, R.B. and Anderson, D.R. (1975) Calcium-promoted aggregation of erythrocyte membrane proteins. Biochim. Biophys. Acta 379, 571–581.

Carter, S.B. (1970) Cell movement and cell spreading: a passive or an active process? Nature (London) 225, 858–859.

Caspar, D.L.D. and Kirschner, D.A. (1971) Myelin membrane structure at 10 Å resolution. Nature (London) New Biol. 231, 46–52.

Changeux, J-P., Thiery, J., Tung, Y. and Kittel, C. (1967) On the cooperativity of biological membranes. Proc. Natl. Acad. Sci. USA 57, 335–341.

Chapman, D. (1973) Some recent studies of lipids, lipid-cholesterol and membrane systems. In: D. Chapman and D.F.H. Wallach (Eds.), Biological Membranes, Academic Press, New York, pp. 91–144.

Chapman, D., Williams, R.M. and Ladbrooke, B.D. (1967) Physical studies of phospholipids, VI. Thermotropic and lyotropic mesomorphism of some 1,2-diacyl-phosphatidylcholine (lecithins). Chem. Phys. Lipids 1, 445–475.

Chapman, P. and Dodd, G.H. (1971) Physicochemical probes of membrane structure, In: L.I. Rothfield (Ed.), Structure and Function of Biological Membranes, Academic Press, New York, pp. 13–81.

Chen, Y.S. and Hubbell, W.L. (1973) Temperature- and light-dependent structural changes in rhodopsin-lipid membranes. Exp. Eye Res. 17, 517–532.

Cherry, R.J. (1975) Molecular structure and protein mobility in cellular membranes, In: E.F. Walborg (Ed.), Cellular Membranes and Tumor Cell Behavior, Williams and Wilkins, Baltimore, pp. 41–60.

Cherry, R.J. (1976) Protein and lipid mobility in biological and model membranes, In: D. Chapman and D.F.H. Wallach (Eds.), Biological Membranes, Vol. 3, Academic Press, London, pp. 47–102.

Choi, Y.S., Knopf, P.M. and Lennox, E.S. (1971) Intracellular transport and secretion of an immunoglobulin light chain. Biochemistry 10, 668–679.

Clarke, M. (1971) Isolation and characterization of a water-soluble protein from bovine erythrocyte membranes. Biochem. Biophys. Res. Commun. 45, 1063–1070.

Cogan, U., Shinitzky, M., Weber, G. and Nishida, T. (1973) Microviscosity and order in the hydrocarbon region of phospholipid-cholesterol dispersions determined with fluorescent probes. Biochemistry 12, 521–528.

Cohen, H.J. and Gilbertsen, B.B. (1975) Human lymphocyte surface immunoglobulin capping. Normal characteristics and anomalous behavior of chronic lymphocytic leukemic lymphocytes. J. Clin. Invest. 55, 84–93.

Cone, R.A. (1972) Rotational diffusion of rhodopsin in the visual receptor membrane. Nature (London) New Biol. 236, 39–43.

Cook, G.M.W. and Eylar, E.H. (1965) Separation of the M and N blood-group antigens of the human erythrocyte. Biochim. Biophys. Acta 101, 57–66.

Cook, G.M.W., Heard, D.H. and Seaman, G.V.F. (1961) Sialic acids and the electrokinetic charge of the human erythrocyte. Nature (London) 191, 44–47.

Cotman, C.W. and Taylor, D. (1974) Localization and characterization of concanavalin A receptors in the synaptic cleft. J. Cell Biol. 62, 236–242.

Cotman, C.W., Banker, G., Churchill, L. and Taylor, D. (1974) Isolation of postsynaptic densities from rat brain. J. Cell Biol. 63, 441–455.

Cuatrecasas, P. (1974) Membrane receptors. Annu. Rev. Biochem. 43, 169–214.

Curtis, A.S.G. (1967) The Cell Surface: Its Molecular Role in Morphogenesis. Academic Press, New York.

Danielli, J.F. and Davson, H. (1935) A contribution to the theory of permeability of thin films. J. Cell. Comp. Physiol. 5, 495–508.

Darke, A., Finer, E.G., Flook, A.G. and Phillips, M.C. (1972) Nuclear magnetic resonance study of lecithin-cholesterol interactions. J. Mol. Biol. 63, 265–279.

Davis, W.C. (1972) *H-2* antigen on cell membranes: An explanation for the alteration of distribution by indirect labeling techniques. Science 175, 1006–1008.

Dehlinger, P.J. and Schimke, R.T. (1971) Size distribution of membrane proteins of rat liver and their relative rates of degradation. J. Biol. Chem. 246, 2574–2583.

De Kruyff, B., Van Dijck, P.W.M., Demel, R.A., Schuijff, A., Brants, F. and Van Deenen, L.L.M. (1974) Non-random distribution of cholesterol in phosphatidylcholine bilayers. Biochim. Biophys. Acta 356, 1–7.

De Pierre, J. and Karnovsky, M.L. (1972) Ectoenzymes, sialic acid and the internalization of cell membrane during phagocytosis. In: I.H. Lepow and P.D. Ward (Eds.), Inflammation, Mechanisms and Control, Academic Press, New York, pp. 55–70.

de Petris, S. (1974) Inhibition and reversal of surface capping by cytochalasin B, vinblastine and colchicine. Nature (London) 250, 54–56.

de Petris, S. (1975) Concanavalin A receptors, immunoglobulins and θ antigens of the lymphocyte surface. Interactions with concanavalin A and cytoplasmic structures. J. Cell Biol. 65, 123–146.

de Petris, S. and Raff, M.C. (1974) Ultrastructural distribution and redistribution of alloantigens and concanavalin A receptors on the surface of mouse lymphocytes. Eur. J. Immunol. 4, 130–137.

Devaux, P. and McConnell, H.M. (1972) Lateral diffusion in spin-labeled phosphatidylcholine multilayers. J. Am. Chem. Soc. 94, 4475–4481.

Douglas, W.W. (1974) Mechanism of release of neurophypophysial hormones: stimulus secretion-coupling. In: Handbook of Physiology, Vol. IV, Section 7, American Physiological Society, Washington, D.C., pp. 191–224.

Edelman, G.M. (1974) Surface alteration and mitogenesis in lymphocytes, In: B. Clarkson and R. Baserga (Eds.), Control of Proliferation in Animal Cells, Cold Spring Harbor Laboratory, New York, pp. 357–377.

Edelman, G.M. (1976) Surface modulation in cell recognition and cell growth. Science 192, 218–226.

Edelman, G.M., Yahara, I. and Wang, J.L. (1973) Receptor mobility and receptor-cytoplasmic interactions in lymphocytes. Proc. Natl. Acad. Sci. USA 70, 1442–1446.

Edidin, M. (1974) Rotational and translational diffusion in membranes. Annu. Rev. Biophys. Bioeng. 3, 179–201.

Edidin, M. and Fambrough, D. (1973) Fluidity of the surface of cultured cell muscle fibers. Rapid lateral diffusion of marked surface antigens. J. Cell Biol. 57, 27–37.

Edidin, M. and Weiss, A. (1972) Antigen cap formation in cultured fibroblasts: A reflection of membrane fluidity and of cell motility. Proc. Natl. Acad. Sci. USA 69, 2456–2459.

Edidin, M. and Weiss, A. (1974) Restriction of antigen mobility in the plasma membranes of some cultured fibroblasts, In: B. Clarkson and R. Baserga (Eds.), Control of Proliferation in Animal Cells, Cold Spring Harbor Laboratory, New York, pp. 213–220.

Ehrström, M., Eriksson, L.E.G., Israelachvili, J. and Ehrenberg, A. (1973) The effects of some cations and anions on spin labeled cytoplasmic membranes of *Bacillus subtilis*. Biochem. Biophys. Res. Commun. 55, 396–402.

Elgsaeter, A. and Branton, D. (1974) Intramembrane particle aggregation in erythrocyte ghosts, I. The effect of protein removal. J. Cell Biol. 63, 1018–1036.

Elgsaeter, A., Shotton, D.M. and Branton, D. (1976) Intramembrane particle aggregation in erythrocyte ghosts, II. The influence of spectrin aggregation. Biochim. Biophys. Acta 426, 101–122.

Elhammer, A., Svensson, H., Autuori, F. and Dallner, G. (1975) Biogenesis of microsomal membrane glycoproteins in rat liver, III. Release of glycoproteins from the Golgi fraction and their transfer to microsomal membranes. J. Cell Biol. 67, 715–725.

Engelman, D.M. (1969) Surface area per lipid molecule in the intact membrane of the human red cell. Nature (London) 223, 1279–1280.

Engelman, D.M. (1971) Lipid bilayer structure in the membrane of *Mycoplasma laidlawii*. J. Mol. Biol. 58, 153–165.

Engelman, D.M. and Rothman, J.E. (1972) The planar organization of lecithin-cholesterol bilayers. J. Biol. Chem. 247, 3694–3597.

Ehnholm, C. and Zilversmit, D.B. (1973) Exchange of various phospholipids and of cholesterol between liposomes in the presence of highly purified phospholipid exchange protein. J. Biol. Chem. 248, 1719–1724.

Eylar, E.H., Madoff, M.A., Brody, O.V. and Oncley, J.L. (1962) The contribution of sialic acid to the surface charge of the erythrocyte. J. Biol. Chem. 237, 1992–2000.

Fairbanks, G., Steck, T.L. and Wallach, D.F.H. (1971) Electrophoretic analysis of the major polypeptides of the human erythrocyte membrane. Biochemistry 10, 2606–2617.

Fambrough, D.M. and Hartzell, H.C. (1972) Acetylcholine receptors: Number and distribution at neuromuscular junctions in rat diaphragm. Science 176, 189–191.

Feinstein, M.B., Fernandez, S.M. and Sha'afi, R.I. (1975) Fluidity of natural membranes and phosphatidylserine and ganglioside dispersions. Effect of anesthetics, cholesterol and protein. Biochim. Biophys. Acta 413, 354–370.

Fertuck, H.C. and Salpeter, M.M. (1974) Localization of acetylcholine receptor by ^{125}I-labeled α-bungarotoxin binding at mouse motor end plates. Proc. Natl. Acad. Sci. USA 71, 1376–1378.

Findlay, J.B.C. (1974) The receptor proteins for concanavalin A and *Lens culinaris* phytohemagglutinin in the membrane of the human erythrocyte. J. Biol. Chem. 249, 4398–4403.

Fortes, P.A.G. and Hoffman, J.F. (1971) Interactions of the fluorescent anion 1-anilino-8-naphthalene sulfonate with membrane charges in human red cell ghosts. J. Memb. Biol. 5, 154–168.

Fox, C.F. (1975a) Phase transitions in model systems and membranes, In: C.F. Fox (Ed.), Biochemistry of Cell Walls and Membranes, University Park Press, Baltimore, pp. 279–306.

Fox, C.F. (1975b) Assembly of the *E. coli* membrane, In: F.O. Schmitt, D.M. Schneider and D.M. Crothers (Eds.), Functional Linkage of Biomolecular Systems, Raven Press, New York, pp. 131–137.

Friend, D.W. and Fawcett, D.W. (1974) Membrane differentiations in freeze-fractured mammalian spermatozoa. J. Cell Biol. 63, 641–652.

Frye, L.D. and Edidin, M. (1970) The rapid inter-mixing of cell surface antigens after formation of mouse-human heterokaryons. J. Cell Sci. 7, 319–333.

Fuller, G.M., Boughter, J.M. and Morazzani, M. (1974) Evidence for multiple polypeptide chains in the membrane protein spectrin. Biochemistry 13, 3036–3041.

Gaffney, B.J. and McConnell, H.M. (1974) The paramagnetic resonance spectra of spin labels in phospholipid membranes. J. Magn. Reson. 16, 1–28.

Gail, M.H. and Boone, C.W. (1972) Procaine inhibition of fibroblast motility and proliferation. Exp. Cell Res. 63, 252–255.

Galla, H-J. and Sackmann, E. (1974) Lateral diffusion in the hydrophobic region of membranes: Use of pyrene excimers as optical probes. Biochim. Biophys. Acta 339, 103–115.

Garoff, H. and Simons, K. (1974) Location of the spike glycoproteins in the Semliki Forest virus membrane. Proc. Natl. Acad. Sci. USA 71, 3988–3992.

Gilula, N.B. (1974) Junctions between cells, In: R.P. Cox (Ed.), Cell Communication, Wiley, New York, pp. 1–29.

Giotta, G.J., Chan, D.S. and Wang, H.H. (1974) Binding of spin-labeled local anesthetics to phosphatidylcholine and phosphatidylserine liposomes. Arch. Biochem. Biophys. 163, 453–458.

Gitler, C. (1972) Plasticity of biological membranes. Annu. Rev. Biophys. Bioeng. 1, 51–92.

Glaser, M. and Singer, S.J. (1971) Circular dichroism and the conformations of membrane proteins. Studies with RBC membranes. Biochemistry 10, 1780–1787.

Glaser, M., Simpkins, H., Singer, S.J., Sheetz, M. and Chan, S.I. (1970) On the interaction of lipids and proteins in the red blood cell membrane. Proc. Natl. Acad. Sci. USA 65, 721–728.

Goldman, R.D. (1972) The effects of cytochalasin B on the microfilaments of baby hamster kidney (BHK-21) cells. J. Cell Biol. 52, 246–254.

Goldman, R.D. (1975) The use of heavy meromyosin binding as an ultrastructural cytochemical method for localizing and determining the possible functions of actin-like microfilaments in non-muscle cells. J. Histochem. Cytochem. 23, 529–542.

Goldman, R.D., Lazarides, E., Pollack, R. and Weber, K. (1975) The distribution of actin in non-muscle cells. The use of actin antibody in the localization of actin within the microfilament bundles of mouse 3T3 cells. Exp. Cell Res. 90, 333–344.

Goldman, R., Pollard, T. and Rosenbaum, J. (1976) (Eds.) Cell Motility. Cold Spring Harbor Conferences on Cell Proliferation, Vol. 3, Cold Spring Harbor Laboratory, New York.

Goldup, A., Ohki, S. and Danielli, J.F. (1970) Black lipid films, In: J.F. Danielli, A.C. Ridderford and M.D. Rosenberg (Eds.), Recent Progress in Surface Science, Vol. 3, Academic Press, New York, pp. 193–260.

Goodenough, D.A. (1974) Bulk isolation of mouse hepatocyte gap junctions. Characterization of the principal protein, connexin. J. Cell Biol. 61, 557–563.

Goodenough, D.A. and Gilula, N.B. (1974) The splitting of hepatocyte gap junctions and zonulae occludentes with hypertonic disaccharides. J. Cell Biol. 61, 575–590.

Goodenough, D.A. and Stoeckenius, W. (1972) The isolation of mouse hepatocyte gap junctions. Preliminary chemical characterization and X-ray diffraction J. Cell Biol. 54, 646–656.

Gordesky, S.E. and Marinetti, G.V. (1973) The asymmetric arrangement of phospholipids in the human erythrocyte membrane. Biochem. Biophys. Res. Commun. 50, 1027–1031.

Gordesky, S.E., Marinetti, G.V. and Segel, G.B. (1973) The interaction of 1-fluoro-2,4-dinitrobenzene with amino-phospholipids in membranes of intact erythrocytes, modified erythrocytes, and erythrocyte ghosts. J. Memb. Biol. 14, 229–242.

Graham, J.M. and Wallach, D.F.H. (1971) Protein conformational transitions in the erythrocyte membrane. Biochim. Biophys. Acta 241, 180–194.

Grant, C.W.M. and McConnell, H.M. (1973) Fusion of phospholipid vesicles with viable Acholeplasma laidlawii. Proc. Natl. Acad. Sci. USA 70, 1238–1240.

Grant, C.W.M. and McConnell, H.M. (1974) Glycophorin in lipid bilayers. Proc. Natl. Acad. Sci. USA 71, 4653–4657.

Grant, C.W.M., Wu, S.H-W. and McConnell, H.M. (1974) Lateral phase separations in binary lipid mixtures: Correlation between spin label and freeze-fracture electron microscopic studies. Biochim. Biophys. Acta 363, 151–158.

Grisham, C.M. and Barnett, R.E. (1973) The role of lipid-phase transitions in the regulation of the (sodium + potassium) adenosine triphosphatase. Biochemistry 12, 2635–2637.

Guidotti, G. (1972) Membrane proteins. Annu. Rev. Biochem. 41, 731–752.

Gulik-Krzywicki, T. (1975) Structural studies of the associations between biological membrane components. Biochim. Biophys. Acta 415, 1–29.

Haest, C.W.M., Verkleij, A.J., De Gier, J. Scheek, R., Ververgaert, P.H.J. and Van Deenen, L.L.M. (1974) The effect of lipid phase transitions on the architecture of bacterial membranes. Biochim. Biophys. Acta 356, 17–26.

Harrison, S.C., Caspar, D.L.D., Camerini-Otero, R.D. and Franklin, R.M. (1971a) Lipid and protein arrangement in bacteriophase PM2. Nature (London) New Biol. 229, 197–201.

Harrison, S.C., David, A., Jumblatt, J. and Darnell, J.E. (1971b) Lipid and protein organization in Sindbis virus. J. Mol. Biol. 60, 523–528.

Haschke, R.H., Byers, M.R. and Fink, B.R. (1974) Effects of lidocaine on rabbit brain microtubular protein. J. Neurochem. 22, 837–843.

Heath, I.B. (1974) A unified hypothesis for the role of membrane bound enzyme complexes and microtubules in plant cell wall synthesis. J. Theoret. Biol. 48, 445–449.

Heitzmann, H. and Richards, F.M. (1974) Use of the avidin-biotin complex for specific staining of biological membranes in electron microscopy. Proc. Natl. Acad. Sci. USA 71: 3537–3541.

Hepler, P.K. and Palevitz, B.A. (1974) Microtubules and microfilaments. Annu. Rev. Plant Physiol. 25, 309–362.

Herscovics, A. (1970) Biosynthesis of thyroglobulin: Incorporation of ^{3}H-fucose into proteins by rat thyroids in vitro. Biochem. J. 117, 411–413.

Hinkley, R.E. and Telse, A.G. (1974) The effects of halothane on cultured mouse neuroblastoma cells, I. Inhibition of morphological differentiation. J. Cell Biol. 63, 531–540.

Hinz, H-J. and Sturtevant, J.M. (1972) Calorimetric investigation of the influence of cholesterol on the transition properties of bilayers formed from synthetic L-α-lecithins in aqueous suspension. J. Biol. Chem. 247, 3697–3701.

Hirano, H., Parkhouse, B., Nicolson, G.L., Lennox, E.S. and Singer, S.J. (1972) Distribution of saccharide residues on membrane fragments from a myeloma-cell homogenate: Its implications for membrane biogenesis. Proc. Natl. Acad. Sci. USA 69, 2945–2949.

Hong, K. and Hubbell, W.L. (1972) Preparation and properties of phospholipid bilayers containing rhodopsin. Proc. Natl. Acad. Sci. USA 69, 2617–2621.

Horwitz, A.F. (1972) Nuclear magnetic resonance studies on phospholipids and membranes, In: C.F. Fox and A.D. Keith (Eds.), Membrane Molecular Biology, Sinauer Associates, Stamford, Connecticut, pp. 164–191.

Horwitz, A.F., Horsley, W.J. and Klein, M.P. (1972) Magnetic resonance studies on membrane and model membrane systems: Proton magnetic relaxation rates in sonicated lecithin dispersions. Proc. Natl. Acad. Sci. USA 69, 590–593.

Horwitz, A.F., Michaelson, D.M. and Klein, M.P. (1973) Magnetic resonance studies of membrane and model membrane systems. Biochim. Biophys. Acta 298, 1–7.

Hsie, A.W., Jones, C. and Puck, T.T. (1971) Further changes in differentiation state accompanying the conversion of Chinese hamster cells of fibroblastic form by dibutyryl adenosine cyclic 3':5'-monophosphate and hormones. Proc. Natl. Acad. Sci. USA 68, 1648–1652.

Hsu, K.C. (1967) Ferritin labeled antigens and antibodies. Methods Immunol. Immunochem. 1, 397–400.

Huang, C-H., Sipe, J.P., Chow, S.T. and Martin, R.B. (1974) Differential interaction of cholesterol with phosphatidylcholine on the inner and outer surfaces of lipid bilayer vesicles. Proc. Natl. Acad. Sci. USA 71, 359–362.

Hubbard, A.L. and Cohn, Z. (1972) Enzymatic iodination of the red cell membrane. J. Cell Biol. 55, 390–405.

Hubbell, W.L. and McConnell, H.M. (1968) Spin-label studies of the excitable membranes of nerve and muscle. Proc. Natl. Acad. Sci. USA 61, 12–16.

Hubbell, W.L. and McConnell, H.M. (1969) Orientation and motion of amphiphilic spin labels in membranes. Proc. Natl. Acad. Sci. USA 64, 20–27.

Hubbell, W.L. and McConnell, H.M. (1971) Molecular motion in spin-labeled phospholipids and membranes. J. Am. Chem. Soc. 93, 314–326.

Hubbell, W.L., Metcalfe, J.C., Metcalfe, S.M. and McConnell, H.M. (1970) The interaction of small molecules with spin-labelled erythrocyte membranes. Biochim. Biophys. Acta 219, 415–427.

Huet, Ch. and Bernadac, A. (1975) Dynamic state of concanavalin A: receptor interactions on fibroblast surfaces. Biochim. Biophys. Acta 394, 605–619.

Hughes, F. and Pedersen, C.E., Jr. (1975) Paramagnetic spin label interactions with the envelope of a group A arbovirus: Lipid organization. Biochim. Biophys. Acta 394, 102–110.

Hughes, R.C. (1975) The complex carbohydrates of mammalian cell surfaces and their biological roles, In: P.N. Campbell and W.N. Aldridge (Eds.), Essays in Biochemistry, Vol. 11, Academic Press, New York, pp. 1–36.

Hunt, R.C. and Brown, J.C. (1974) Surface glycoproteins of mouse L cells. Biochemistry 13, 22–28.

Hunt, R.C. and Brown, J.C. (1975) Identification of a high molecular weight trans-membrane protein in mouse L cells. J. Mol. Biol. 97, 413–422.

Hynes, R.O. (1974) Role of surface alterations in cell transformation: The importance of proteases and surface proteins. Cell 1, 147–156.

Hynes, R.O. (1976) Cell surface proteins and malignant transformation. Biochim. Biophys. Acta 458, 73–107.

Hynes, R.O. and Pearlstein, E.S. (1976) Investigations of the possible role of proteases in altering surface proteins of virally transformed hamster fibroblasts. J. Supramol. Struct. 4, 1–14.

Hynes, R.O., Destree, A.T. and Mautner, V. (1976) Spatial organization at the cell surface. In: V.T. Marchesi (Ed.), Membranes and Neoplasia, Alan R. Liss, New York, pp. 189–202.

Ingram, V.M.H. (1969) A side view of moving fibroblasts. Nature (London) 222, 641–644.

Ishikawa, H., Bischoff, R. and Holtzer, H. (1969) Formation of arrowhead complexes with heavy meromyosin in a variety of cell types. J. Cell Biol. 43, 312–328.

Israelachvili, J., Sjösten, J., Eriksson, L.E.G., Ehrström, M., Gröslund, A. and Ehrenberg, A. (1975) ESR spectral analysis of the molecular motion of spin labels in lipid bilayers and membranes based on a model in terms of two angular motional parameters and rotational correlation times. Biochim. Biophys. Acta 382, 125–141.

Jacob, H., Amsden, T. and White, J. (1972) Membrane microfilaments of erythrocytes: Alternation in intact cells reproduces the hereditary spherocytosis syndrome. Proc. Natl. Acad. Sci. USA 69, 471–474.

Jacobson, K. and Papahadjopoulos, D. (1975) Phase transitions and phase separations in phospholipid membranes induced by changes in temperature, pH, and concentration of bivalent cations. Biochemistry 14, 152–161.

Jacobson, K., Wu, E-S. and Poste, G. (1976) Measurement of the translational mobility of concanavalin A in glycerol-saline solutions and on the cell surface by fluorescence recovery after photobleaching. Biochim. Biophys. Acta 433, 215–222.

Jacobson, K., Derzko, Z., Wu, E-S., Hou, Y. and Poste, G. (1977) Measurement of the lateral mobility of cell surface components in single, living cells by fluorescence recovery after photobleaching. J. Supramol. Struct., in press.

James, R. and Branton, D. (1973) Lipid- and temperature-dependent structural changes in Acholeplasma laidlawii cell membranes. Biochim. Biophys. Acta 323, 378–390.

James, R., Branton, D., Wisnieski, B. and Keith, A. (1972) Composition, structure and phase transition in yeast fatty acid auxotroph membranes. J. Supramol. Struct. 1, 38–49.

Ji, T.H. and Nicolson, G.L. (1974) Lectin binding and perturbation of the cell membrane outer surface induces a transmembrane organizational alteration at the inner surface. Proc. Natl. Acad. Sci. USA 71, 2212–2216.

Johnson, L.W., Hughes, M.E. and Zilversmit, D.B. (1975) Use of phospholipid exchange protein to measure inside-outside transportation in phosphatidylcholine liposomes. Biochim. Biophys. Acta 375, 176–185.

Jost, P., Waggoner, A. and Griffith, O.H. (1971) Spin labeling and membrane structure, In: L.I. Rothfield (Ed.), Structure and Function of Biological Membranes, Academic Press, New York, pp. 83–144.

Juliano, R.L. (1973) The proteins of the erythrocyte membrane. Biochim. Biophys. Acta 300, 341–378.

Juliano, R.L. and Behar-Bannelier, B. (1975) An evaluation of techniques for labelling the surface proteins of cultured mammalian cells. Biochim. Biophys. Acta 375, 249–267.

Juliano, R.L., Kimelberg, H.K. and Papahadjopoulos, D. (1971) Synergistic effects of membrane protein (spectrin) and Ca^{2+} on the Na^+ permeability of phospholipid vesicles. Biochim. Biophys. Acta 241, 894–905.

Kahlenberg, A., Walker, C. and Rohrlick, R. (1974) Evidence for an asymmetric distribution of phospholipids in the human erythrocyte membrane. Canad. J. Biochem. 52, 803–806.

Kant, J.A. and Steck, T.L. (1973) Specificity in the association of glyceraldehyde-3-phosphate dehydrogenase with isolated human erythrocyte membranes. J. Biol. Chem. 248, 8457–8464.

Kaplan, J. and Moskowitz, M. (1975a) Studies on the turnover of plasma membranes in cultured mammalian cells, I. Rates of synthesis and degradation of plasma membrane proteins and carbohydrates. Biochim. Biophys. Acta 389, 290–305.

Kaplan, J. and Moskowitz, M. (1975b) Studies on the turnover of plasma membranes in cultured mammalian cells, II. Demonstration of heterogeneous rates of turnover for plasma membrane protein and glycoproteins. Biochim. Biophys. Acta 389, 306–313.

Kaplan, J., Canonico, P.G. and Caspary, W.J. (1973) Electron spin resonance studies of spin-labeled mammalian cells by detection of surface-membrane signals. Proc. Natl. Acad. Sci. USA 70, 66–70.

Karnovsky, M.J., Unanue, E.R. and Leventhal, M. (1972) Ligand-induced movement of lymphocyte membrane macromolecules, II. Mapping of surface moieties. J. Exp. Med. 136, 907–917.

Kasai, M. and Changeux, J-P. (1969) In vitro interaction of 1-anilino-8-naphthalene sulfonate with excitable membranes isolated from the electrical organ of *Electrophorus electricus*. Biochem. Biophys. Res. Commun. 35, 420–427.

Kefalides, N.A. (1973) Structure and biosynthesis of basement membranes. Int. Rev. Connect. Tissue Res. 6, 63–104.

Kelly, P., Cotman, C.W., Gentry, C. and Nicolson, G.L. (1976) Distribution and mobility of lectin receptors on synaptic membranes of identified neurons in the central nervous system. J. Cell Biol. 71, 487–496.

Kiehn, E.D. and Holland, J.J. (1970) Membrane and nonmembrane proteins of mammalian cells. Synthesis, turnover and size distribution. Biochemistry 9, 1716–1728.

Kinzel, V. and Mueller, G.C. (1973) Phosphorylation of surface protein of HeLa cells using an exogenous protein kinase and [γ^{32}P]ATP. Biochim. Biophys. Acta 322, 337–351.

Kirschner, M.W. and Williams, R.C. (1974) The mechanism of microtubule assembly in vitro. J. Supramol. Struct. 2, 412–428.

Kirschner, M.W., Williams, R.C., Weingarten, M. and Gerhart, J.C. (1974) Microtubules from mammalian brain: Some properties of their depolymerization products and a proposed mechanism of assembly and disassembly. Proc. Natl. Acad. Sci. USA 71, 1159–1163.

Kita, H. and Van Der Kloot, W. (1972) Membrane probe. 1-anilino-8-naphthalene sulphonate promotes acetylcholine and catecholamine release. Nature (London) New Biol. 235, 250–252.

Kleemann, W. and McConnell, H.M. (1974) Lateral phase separations in *Escherichia coli* membranes. Biochim. Biophys. Acta 345, 220–230.

Kleemann, W. and McConnell, H.M. (1976) Interactions of proteins and cholesterol with lipids in bilayer membranes. Biochim. Biophys. Acta 219, 206–222.

Kleemann, W., Grant, C.W.M. and McConnell, H.M. (1974) Lipid phase separations and protein distribution in membranes. J. Supramol. Struct. 2, 609–616.

Koehler, J.K. (1975a) Studies on the distribution of antigenic sites on the surface of rabbit spermatozoa. J. Cell Biol. 67, 647–659.

Koehler, J.K. (1975b) Periodicities in the acrosome of acrosomal membrane: Some observations on mammalian spermatozoa, In: J.G. Duckett and P.A. Racey (Eds.), The Biology of the Male Gamete, Academic Press, London, pp. 337–342.

Koehler, J.K. and Perkins, W.D. (1974) Fine structure observations on the distribution of antigenic sites on guinea pig spermatozoa. J. Cell Biol. 60, 789–795.

Koo, G.C., Stackpole, C.W., Boyse, E.A., Hämmerling, U. and Lardis, M.P. (1973) Topographical location of H-Y antigen on mouse spermatozoa by immunoelectronmicroscopy. Proc. Natl. Acad. Sci. USA 70, 1502–1505.

Korn, E.D. (1969a) Current concepts of membrane structure and function. Fed. Proc. 28, 6–11.

Korn, E.D. (1969b) Cell Membranes. Structure and synthesis. Annu. Rev. Biochem. 38, 263–288.

Kornberg, R.D. and McConnell, H.M. (1971a) Lateral diffusion of phospholipids in a vesicle membrane. Proc. Natl. Acad. Sci. USA 68, 2564–2568.

Kornberg, R.D. and McConnell, H.M. (1971b) Inside-outside transitions of phospholipids in vesicle membranes. Biochemistry 10, 1111–1120.

Kraemer, P.M. (1967) Regeneration of sialic acid on the surface of Chinese hamster cells in culture, II. Incorporation of radioactivity from glucosamine-1-^{14}C. J. Cell. Physiol. 69, 199–208.

Kraemer, P.M. (1971) Complex carbohydrates of animal cells. Biochemistry and physiology of the cell periphery, In: L.A. Manson (Ed.), Biomembranes, Vol. I, Plenum Press, New York, pp. 67–190.

Kraemer, P.M. (1975) The cell surface: Opening remarks, In: C.R. Richmond, D.F. Petersen, P.F.

Mullaney and E.C. Anderson (Eds.), Mammalian Cells: Probes and Problems, USERDA Technical Information Center, Oak Ridge, Tennessee, pp. 242–245.

Ladbrooke, B.D., Williams, R.M. and Chapman, D. (1968) Studies on lecithin-cholesterol–water interactions by differential scanning calorimetry and X-ray diffraction. Biochim. Biophys. Acta 150, 333–340.

Landis, D.M.D. and Reese, T.S. (1974) Membrane structure in rapidly frozen, freeze-fractured cerebellar cortex. J. Cell Biol. 63, 184a.

Landsberger, F.R., Lenard, J., Paxton, J. and Compans, R.W. (1971) Spin-label electron spin resonance study on the lipid-containing membrane of influenza virus. Proc. Natl. Acad. Sci. USA 68, 2579–2583.

Landsberger, F.R., Compans, R.W., Paxton, J. and Lenard, J. (1972) Structure of the lipid phase of Rauscher murine leukemia virus. J. Supramol. Struct. 1, 50–59.

Law, J.H. and Snyder, W.R. (1972) Membrane lipids, In: C.F. Fox and A.D. Keith (Eds.), Membrane Molecular Biology, Sinauer Associates, Stamford, Connecticut, pp. 3–26.

Lazarides, E. (1975a) Immunofluorescence studies on the structure of actin filaments in tissue culture cells. J. Histochem. Cytochem. 23, 507–528.

Lazarides, E. (1975b) Tropomyosin antibody: the specific localization of tropomyosin in non-muscle cells. J. Cell Biol. 65, 549–561.

Lazarides, E. and Burridge, K. (1975) α-Actinin: Immunofluorescent localization of a muscle structural protein in nonmuscle cells. Cell 6, 289–298.

Lazarides, E. and Weber, K. (1974) Actin antibody: The specific visualization of actin filaments in non-muscle cells. Proc. Natl. Acad. Sci. USA 71, 2268–2272.

Leblond, C.P. and Bennett, G. (1974) Elaboration and turnover of cell coat glycoproteins, In: A.A. Moscona (Ed.), The Cell Surface in Development, Wiley, New York, pp. 29–49.

Lee, A.G., Birdsall, N.J.M. and Metcalfe, J.E. (1973) Measurement of fast lateral diffusion of lipids in vesicles and in biological membranes by ^{1}H nuclear magnetic resonance. Biochemistry 12, 1650–1658.

Lenard, J. and Compans, R.W. (1974) The membrane structure of lipid-containing viruses. Biochim. Biophys. Acta 344, 51–94.

Lenard, J. and Singer, S.J. (1966) Protein conformation in cell membrane preparations as studied by optical rotatory dispersion and circular dichroism. Proc. Natl. Acad. Sci. USA 56, 1828–1835.

Lesslaver, W., Cain, J. and Blasie, J.K. (1971) On the location of 1-anilino-8-naphthalene sulfonate in lipid model systems. Biochim. Biophys. Acta 241, 547–566.

Levine, Y.K. (1972) Physical studies of membrane structure. Prog. Biophys. Mol. Biol. 24, 1–74.

Levine, Y.K., Birdsall, N.J.M., Lee, A.G. and Metcalfe, J.G. (1972a) ^{13}C-nuclear magnetic resonance relaxation measurements of synthetic lecithins and the effect of spin-labeled lipids. Biochemistry 11, 1416–1421.

Levine, Y.K., Partington, P., Roberts, G.C.K., Birdsall, N.J.M., Lee, A.G. and Metcalfe, J.C. (1972b) ^{13}C nuclear magnetic relaxation times and models for chain motion in lecithin vesicles. FEBS Lett. 23, 203–207.

Linden, C.D., Wright, K.L., McConnell, H.M. and Fox, C.F. (1973) Lateral phase separations in membrane lipids and the mechanism of sugar transport in Escherichia coli. Proc. Natl. Acad. Sci. USA 70, 2271–2275.

Lippert, J.L. and Peticolas, W.L. (1972) Raman active vibrations in long chain fatty acids and phospholipid sonicates. Biochim. Biophys. Acta 282, 8–17.

Loor, F. (1974) Binding and redistribution of lectins on lymphocyte membrane. Eur. J. Immunol. 4, 210–220.

Loor, F. (1977) Structure and dynamics of the lymphocyte surface, In: F. Loor and G.E. Roelants (Eds.), B and T cells in Immune Recognition, Wiley, Chichester, England, in press.

Loor, F., Forni, L. and Pernis, B. (1972) The dynamic state of the lymphocyte membrane. Factors affecting the distribution and turnover of surface immunoglobulins. Eur. J. Immunol. 2, 203–212.

Luzzati, V. (1968) X-Ray diffraction studies of lipid water systems, In: D. Chapman (Ed.), Biological Membranes, Vol. 1, Academic Press, New York, pp. 71–121.

Luzzati, V. (1974) X-Ray diffraction approach to the structure of biological membranes, In: S.E.C. Gitler (Ed.), Perspectives in Membrane Biology, Academic Press, New York, pp. 25–44.

Luzzati, V., Reiss-Husson, F., Rivas, E. and Gulik-Krzywicki, T. (1966) Structure and polymorphism in lipid-water systems, and their possible biological implications. Ann. N.Y. Acad. Sci. 137, 409–413.

MacLennan, D.H., Seeman, P., Iles, G.H. and Yip, C.C. (1971) Membrane formation by the adenosine triphosphatase of sarcoplasmic reticulum. J. Biol. Chem. 246, 2702–2710.

Marchesi, V.T. and Palade, G.E. (1967) The localization of Mg-Na-K-activated adenosine triphosphatase on red cell ghost membranes. J. Cell Biol. 35, 385–404.

Marchesi, V.T. and Steers Jr., E. (1968) Selective solubilization of a protein component of the red cell membrane. Science 159, 203–204.

Marchesi, V.T., Steers Jr., E., Tillack, T.W. and Marchesi, S.L. (1969) Some properties of spectrin. A fibrous protein isolated from red cell membranes, In: G.A. Jamieson and T.J. Greenwalt (Eds.), The Red Cell Membrane, Structure and Function, Lippincott, Philadelphia, pp. 117–130.

Marchesi, V.T., Tillack, T.W., Jackson, R.L., Segrest, J.P. and Scott, R.E. (1972) Chemical characterization and surface orientation of the major glycoprotein of the human erythrocyte membrane. Proc. Natl. Acad. Sci. USA 69, 1445–1449.

Mathews, R.A., Johnson, T.C. and Hudson, J.E. (1976) Synthesis and turnover of plasma-membrane proteins and glycoproteins in a neuroblastoma cell line. Biochem. J. 154, 57–64.

Matter, A. and Bonnet, C. (1974) Effect of capping on the distribution of membrane particles in thymocyte membranes. Eur. J. Immunol. 4, 704–707.

McConnell, H.M. (1975a) Coupling between lateral and perpendicular motion in biological membranes. In: F.O. Schmitt, D.M. Schneider and D.M. Crothers (Eds.), Functional Linkage in Biomolecular Systems, Raven Press, New York, pp. 123–131.

McConnell, H.M. (1975b) Role of lipid in membrane structure and function, In: E.F. Walborg (Ed.), Cellular Membranes and Tumor Cell Behavior, Williams and Wilkins, Baltimore, pp. 61–80.

McConnell, H.M. and McFarland, B.G. (1970) Physics and chemistry of spin labels. Quart. Rev. Biophys. 3, 91–136.

McFarland, B.G. (1972) The molecular basis of fluidity in membranes. Chem. Phys. Lipids 8, 303–313.

McIntyre, J.A., Gilula, N.B. and Karnovsky, M.J. (1974) Cryoprotectant-induced redistribution of intramembranous particles in mouse lymphocytes. J. Cell Biol. 60, 192–203.

McManus, J.P., Whitfield, J.F., Boynton, A.L. and Rixon, R.H. (1975) In: G.I. Drummond, P. Greengard and G.A. Robison (Eds.), Advances in Cyclic Nucleotide Research, Vol. 5, Raven Press, New York, pp. 719–734.

McNamee, M.G. and McConnell, H.M. (1973) Transmembrane potentials and phospholipid flip-flop in excitable membrane vesicles. Biochemistry 12, 2951–2958.

McNutt, N.S. and Weinstein, R.S. (1973) Membrane ultrastructure at mammalian intercellular junctions. Prog. Biophys. Mol. Biol. 26, 45–101.

McNutt, N.S., Culp, L.A. and Black, P.H. (1971) Contact-inhibited revertant cell lines isolated from SV40-transformed cells, II. Ultrastructural study. J. Cell Biol. 50, 691–708.

Melchers, F. M. (1971) Biosynthesis of the carbohydrate portion of immuno-globulin. Radiochemical and chemical analysis of the carbohydrate moieties of two myeloma proteins purified from different subcellular fractions of plasma cells. Biochemistry 10, 653–659.

Melchers, F. and Andersson, J. (1974) Secretion of immunoglobulins, In: B. Ceccarelli, F. Clementi and F. Meldolesi (Eds.), Advances in Cytopharmacology. Vol. 2, Raven Press, New York, pp. 225–235.

Melchers, F. and Knopf, P.M. (1967) Biosynthesis of the carbohydrate portion of immunoglobulin chains. Possible relation to secretion. Cold Spring Harbor Symp. Quant. Biol. 32, 255–262.

Metcalfe, J.C. (1975) The dynamic properties of lipid molecules, In: F.O. Schmitt, D.M. Schneider and D.M. Crothers (Eds.), Functional Linkage in Biomolecular Systems, Raven Press, New York, pp. 90–101.

Meyer, H.W. and Winkelmann, H. (1972) Über die Einordnung der Membranproteine nach

Untersuchungen mit der Gefrierätzung an isolierten Erythrozytenmembranen. Protoplasma 75, 255–284.

Michaelson, D.M., Horwitz, A.F. and Klein, M.P. (1973) Transbilayer asymmetry and surface homogeneity of mixed phospholipids in cosonicated vesicles. Biochemistry 14, 2637–2645.

Mintz, U. and Sachs, L. (1975) Changes in the surface membrane of lymphocytes from patients with chronic lymphocytic leukemia and Hodgkin's disease. Int. J. Cancer 15, 253–259.

Molnar, J. and Sy, D. (1967) Attachment of a glucosamine to protein at the ribosomal site of rat liver. Biochemistry 6, 1941–1947.

Mooseker, M.S. and Tilney, L.G. (1975) Organization of an actin filament-membrane complex. Filament polarity and membrane attachment in the microvilli of intestinal epithelial cells. J. Cell Biol. 67, 725–743.

Morawiecki, A. (1964) Dissociation of M- and N-group mucoproteins into subunits in detergent solution. Biochim. Biophys. Acta 83, 339–347.

Morré, D.J., Keenan, T.W. and Huang, C.M. (1974) Membrane flow and differentiation: Origin of Golgi apparatus membranes from endoplasmic reticulum, In: G. Ceccarelli, F. Clementi and J. Meldolesi (Eds.), Advances in Cytopharmacology, Vol. 2, Raven Press, New York, pp. 107–125.

Morrison, M., Mueller, T.J. and Huber, C.T. (1974) Transmembrane orientation of the glycoproteins in normal human erythrocytes. J. Biol. Chem. 249, 2658–2660.

Nagle, J.F. (1973) Lipid bilayer phase transition: Density measurements and theory. Proc. Natl. Acad. Sci. USA 70, 3443–3444.

Neauport-Sautes, C., Lilly, F., Silvestre, E. and Kourilsky, F.M. (1973a) Independence of H-2K and H-2D antigenic determinants on the surface of mouse lymphocytes. J. Exp. Med. 137, 511–526.

Neauport-Sautes, C., Silvestre, D. and Lilly, F. (1973b) Independence of H-2K and H-2D antigenic determinants on the surface of mouse lymphocytes. Transplant. Proc. 5, 443–446.

Nicolson, G.L. (1973a) Anionic sites of human erythrocyte membranes, I. Effects of trypsin, phospholipase C, and pH on the topography of bound positively charged colloidal particles. J. Cell Biol. 57, 373–387.

Nicolson, G.L. (1973b) Cis- and trans-membrane control of cell surface topography. J. Supramol. Struct. 1, 410–416.

Nicolson, G.L. (1974) The interactions of lectins with animal cell surfaces. Int. Rev. Cytol. 39, 89–190.

Nicolson, G.L. (1976a) Transmembrane control of the receptors on normal and tumor cells, I. Cytoplasmic influence over cell surface components. Biochim. Biophys. Acta 457, 57–108.

Nicolson, G.L. (1976b) Transmembrane control of the receptors on normal and tumor cells, II. Surface changes associated with transformation and malignancy. Biochim. Biophys. Acta 458, 1–72.

Nicolson, G.L. (1976c) Ultrastructural localization of lectin receptors, In: J.K. Koehler (Ed.), Advanced Techniques in Biological Electron Microscopy, Vol. II, Springer, New York, in press.

Nicolson, G.L. and Painter, R.G. (1973) Anionic sites of human erythrocyte membranes, II. Transmembrane effects of anti-spectrin on the topography of bound positively charged colloidal particles. J. Cell Biol. 59, 395–406.

Nicolson, G.L. and Poste, G. (1976a) The cancer cell: Dynamic aspects and modifications in cell-surface organization. New Eng. J. Med. Part I: 295, 197–203; Part II: 295, 253–258.

Nicolson, G.L. and Poste, G. (1976b) Cell shape changes and transmembrane receptor uncoupling induced by tertiary amine local anesthetics. J. Supramol. Struct. 5, 65–72.

Nicolson, G.L. and Singer, S.J. (1971) Ferritin-conjugated plant agglutinins as specific saccharide stains for electron microscopy: Application to saccharides bound to cell membranes. Proc. Natl. Acad. Sci. USA 68, 942–945.

Nicolson, G.L. and Singer, S.J. (1974) The distribution and asymmetry of saccharides on mammalian cell membrane surfaces utilizing ferritin-conjugated plant agglutinins as specific saccharide stains. J. Cell Biol. 60, 236–248.

Nicolson, G.L. and Yanagimachi, R. (1974) Mobility and the restriction of mobility of plasma membrane lectin-binding components. Science 184, 1294–1296.

Nicolson, G.L., Marchesi, V.T. and Singer, S.J. (1971a) The localization of spectrin on the inner surface of human red blood cell membranes with ferritin-conjugated antibodies. J. Cell Biol. 51, 265–272.

Nicolson, G.L., Masouredis, S.P. and Singer, S.J. (1971b) Quantitative two-dimensional ultrastructural distribution of $Rh_0(D)$ antigenic sites on human erythrocyte membranes. Proc. Natl. Acad. Sci. USA 68, 1416–1420.

Nicolson, G.L., Smith, J.R. and Poste, G. (1976) Effect of local anesthetics on cell morphology and membrane-associated cytoskeletal organization in BALB/3T3 cells. J. Cell Biol. 68, 395–402.

Nigam, V.N. and Cantero, A. (1972) Polysaccharides in cancer. Adv. Cancer Res. 16, 1–96.

O'Brien, J.R. (1962) Platelet aggregation, Part I. Some effects on the adenosine phosphates, thrombin, and cocaine upon platelet adhesiveness. J. Clin. Pathol. 15, 446–455.

Oesterhelt, D. and Stoeckenius, W. (1971) Rhodopsin-like protein from the purple membrane of *Halobacterium halobium*. Nature (London) New Biol. 233, 149–152.

Old, L.J. and Boyse, E.A. (1973) Current enigmas in cancer research. The Harvey Lectures 67, 273–315.

Oldfield, E. and Chapman, D. (1971) Effects of cholesterol and cholesterol derivatives on hydrocarbon chain mobility in lipids. Biochem. Biophys. Res. Commun. 43, 610–616.

Oldfield, E. and Chapman, D. (1972) Dynamics of lipids in membranes: heterogeneity and the role of cholesterol. FEBS Lett. 23, 285–297.

Oliver, J.M., Zurer, R.B. and Berlin, R.D. (1975) Concanavalin A cap formation on polymorphonuclear leukocytes of normal and beige (Chediak-Higashi) mice. Nature (London) 253, 471–473.

Olmsted, J.B. and Borisy, G.G. (1973) Microtubules. Ann. Rev. Biochem. 42, 507–540.

Olmsted, J.B., Marcum, J.M., Johnson, K.A., Allen, C. and Borisy, G.G. (1974) Microtubule assembly: Some possible regulatory mechanisms. J. Supramol. Struct. 2, 429–450.

Oseroff, A.R., Robbins, P.W. and Burger, M.M. (1973) The cell surface membrane: Biochemical aspects and biophysical probes. Ann. Rev. Biochem. 42, 647–682.

Overath, P. and Träuble, H. (1973) Phase transitions in cells, membranes, and lipids of *Escherichia coli*. Detection by fluorescent probes, light scattering, and dilatometry. Biochemistry 12, 2625–2634.

Overath, P., Schairer, H. and Stoffel, W. (1970) Correlation of in vivo and in vitro phase transitions of membrane lipids in *Escherichia coli*. Proc. Natl. Acad. Sci. USA 67, 606–612.

Painter, R.G., Sheetz, M. and Singer, S.J. (1975) Detection and ultrastructural localization of human smooth muscle myosin-like molecules in human non-muscle cells by specific antibodies. Proc. Natl. Acad. Sci. USA 72, 1359–1363.

Palade, G. (1959) Functional changes in the structure of cell components, In: T. Hayashi (Ed.), Subcellular Particles, Ronald Press, New York, pp. 64–83.

Palade, G. (1975) Intracellular aspects of the process of protein synthesis. Science 189, 347–358.

Papahadjopoulos, D. (1968) Surface properties of acidic phospholipids: Interaction of monolayers and hydrated liquid crystals with uni- and bivalent metal ions. Biochim. Biophys. Acta 163, 240–254.

Papahadjopoulos, D. (1970) Phospholipid model membranes, III. Antagonistic effects of Ca^{2+} and local anesthetics on the permeability of phosphatidylserine vesicles. Biochim. Biophys. Acta 211, 467–477.

Papahadjopoulos, D. (1972) Studies on the mechanism of action of local anesthetics with phospholipid model membranes. Biochim. Biophys. Acta 265, 169–186.

Papahadjopoulos, D. and Kimelberg, H.K. (1973) Phospholipid vesicles (liposomes) as models for biological membranes: Their properties and interactions with cholesterol and proteins, In: S.G. Davison (Ed.), Progress in Surface Science, Vol. 3, Pergamon Press, Oxford, pp. 141–232.

Papahadjopoulos, D., Jacobson, K., Poste, G. and Sheperd, G. (1975a) Effect of local anesthetics on membrane properties, I. Changes in the fluidity of phospholipid bilayers. Biochim. Biophys. Acta 394, 504–519.

Papahadjopoulos, D., Vail, W.J., Jacobson, K. and Poste, G. (1975b) Cochleate cylinders: Formation by fusion of unilamellar lipid vesicles. Biochim. Biophys. Acta 394, 483–491.

Papahadjopoulos, D., Hui, S., Vail, W.J. and Poste, G. (1976a) Studies on membrane fusion, I.

Interactions of pure phospholipid membranes and the effect of fatty acids (myristic acid), lysolecithin, proteins and dimethylsulfoxide. Biochim. Biophys. Acta 448, 245–264.

Papahadjopoulos, D., Vail, W.J., Pangborn, W.A. and Poste, G. (1976b) Studies on membrane fusion, II. Induction of fusion in pure phospholipid membranes by Ca^{2+} and other divalent metals. Biochim. Biophys. Acta 448, 265–283.

Parsons, D.F. and Subjeck, J.R. (1972) The morphology of the polysaccharide coat of mammalian cells. Biochim. Biophys. Acta 256, 85–113.

Patrick, J., Valeur, B., Monnerie, L. and Changeux, J-P. (1971) Changes in extrinsic fluorescence intensity of the electroplax membrane during electrical excitation J. Memb. Biol. 5, 102–120.

Perdue, J.F. (1973) The distribution, ultrastructure, and chemistry of microfilaments in cultured chick embryo fibroblasts. J. Cell Biol. 58, 265–283.

Perkins, W.D., Karnovsky, M.J. and Unanue, E.R. (1972) An ultrastructural study of lymphocytes with surface-bound immunoglobulin. J. Exp. Med. 135, 267–276.

Peters, R., Peters, J., Tews, K.H. and Bähr, W. (1974) A microfluorimetric study of translational diffusion in erythrocyte membranes. Biochim. Biophys. Acta 367, 282–294.

Phillips, D.R. and Morrison, M. (1971a) Exposed protein on the intact human erythrocyte. Biochemistry 10, 1766–1771.

Phillips, D.R. and Morrison, M. (1971b) Position of glycoprotein polypeptide chain in the human erythrocyte membrane. FEBS Lett. 18, 95–97.

Phillips, E.R. and Perdue, J.F. (1974) Ultrastructural distribution of cell surface antigens in avian tumor virus-infected chick embryo fibroblasts. J. Cell Biol. 61, 743–756.

Phillips, E.R. and Perdue, J.F. (1976a) The dynamics of antibody-induced redistribution of viral envelope antigens in the plasma membranes of avian tumour virus-infected chick embryo fibroblasts. J. Cell Sci. 20, 459–477.

Phillips, E.R and Perdue, J.F. (1976b) The expression and localization of surface neoantigens in transformed and untransformed cultured cells infected with avian tumor viruses. J. Supramol. Struct. 4, 27–44.

Phillips, M.C. and Chapman, D. (1968) Monolayer characteristics of 1,2-diacylphosphatidylcholines (lecithins) and phosphatidylethanalamines at the air-water interface. Biochim. Biophys. Acta 163, 301–313.

Phillips, M.C., Williams, R.M. and Chapman, D. (1969) On the nature of hydrocarbon chain motions in lipid liquid crystals. Chem. Phys. Lipids 3, 234–244.

Phillips, M.C., Ladbrooke, B.D. and Chapman, D. (1970) Molecular interactions in mixed lecithin systems. Biochim. Biophys. Acta 196, 35–44.

Pinto da Silva, P. (1972) Translational mobility of the membrane intercalated particles of human erythrocyte ghosts. pH-dependent, reversible aggregation. J. Cell Biol. 53, 777–787.

Pinto da Silva, P. and Branton, D. (1970) Membrane splitting in freeze-etching: covalently bound ferritin as a membrane marker. J. Cell Biol. 45, 598–605.

Pinto da Silva, P. and Branton, D. (1972) Membrane intercalated particles: The plasma membrane as a planar fluid domain. Chem. Phys. Lipids 8, 265–278.

Pinto da Silva, P. and Nicolson, G.L. (1974) Freeze-etch localization of concanavalin A receptors to the membrane intercalated particles on human erythrocyte membranes. Biochim. Biophys. Acta 363, 311–319.

Pinto da Silva, P., Douglas, S.D. and Branton, D. (1971) Localization of A antigen sites on human erythrocyte ghosts. Nature 232, 194–196.

Pinto da Silva, P., Martinez-Palomo, A. and Gonzalez-Robles, A. (1975) Membrane structure and surface coat of *Entamoeba histolytica*. Topochemistry and dynamics of the cell surface: cap formation and microexudate. J. Cell Biol. 64, 538–550.

Podo, F. and Blasie, J.K. (1975) Mobility of membrane components, In: C.F. Fox (Ed.), Biochemistry of Cell Walls and Membranes, University Park Press, Baltimore, pp. 97–121.

Pollack, R., Osborn, M. and Weber, K. (1975) Patterns of organization of actin and myosin in normal and transformed cultured cells. Proc. Natl. Acad. Sci. USA 72, 994–998.

Pollard, T.D. and Weihing, R.R. (1974) Actin and myosin and cell movement. CRC Critical Rev. Biochem. 2, 1–65.

Poo, M. and Cone, R.A. (1974) Lateral diffusion of rhodopsin in the photoreceptor membrane. Nature (London) 247, 438–440.

Porter, C.W. and Bernacki, R.J. (1975) Ultrastructural evidence for ectoglycosyltansferase systems. Nature (London) 256, 648–650.

Poste, G. and Allison, A.C. (1973) Membrane fusion. Biochim. Biophys. Acta 300, 421–465.

Poste, G. and Nicolson, G.L. (1976) Calcium ionophores A23187 and X537A affect cell agglutination by lectins and capping of lymphocyte surface immunoglobulins. Biochim. Biophys. Acta 426, 148–155.

Poste, G. and Nicolson, G.L. (1977) (Eds.) The Synthesis, Assembly and Turnover of Cell Surface Components, Cell Surface Reviews, Vol. 4, North-Holland, Amsterdam. In press.

Poste, G. and Reeve, P. (1972) Inhibition of cell fusion by local anesthetics and tranquillizers. Exp. Cell Res. 72, 556–560.

Poste, G., Papahadjopoulos, D., Jacobson, K. and Vail, W.J. (1975a) Effects of local anesthetics on membrane properties, II. Enhancement of the susceptibility of mammalian cells to agglutination by plant lectins. Biochim. Biophys. Acta 394, 520–539.

Poste, G., Papahadjopoulos, D. and Nicolson, G.L. (1975b) The effect of local anesthetics on transmembrane cytoskeletal control of cell surface mobility and distribution. Proc. Natl. Acad. Sci. USA 72, 4430–4434.

Poulik, M.D., Bernoco, M., Bernoco, D. and Ceppellini, R. (1973) Aggregation of HL-A antigens at the lymphocyte surface by antiserum to β_2-microglobulin. Science 182, 1352–1355.

Rabinovitch, M. and De Stefano, M.J. (1973) Manganese stimulates adhesion and spreading of mouse sarcoma I ascites cells. J. Cell Biol. 59, 165–176.

Rabinovitch, M. and De Stefano, M.J. (1974) Macrophage spreading in vitro, III. The effect of metabolic inhibitors, anesthetics and other drugs on spreading induced by subtilisin. Exp. Cell Res. 88, 153–162.

Radda, G.K. and Smith, D.S. (1970) Retinol: A fluorescent probe for membrane lipids. FEBS Lett. 9, 287–289.

Raff, M.C., Freedman, M. and Gomperts, B. (1976) Immunological receptors as transducers, In: M. Seligman, J.L. Preud'homme and F.M. Kourilsky (Eds.), Membrane Receptors of Lymphocytes, North-Holland, Amsterdam, pp. 393–398.

Rasmussen, H. (1970) Cell communication, calcium ions, and cyclic adenosine monophosphate. Science 170, 404–412.

Ray, T.K., Lieberman, I. and Lansing, A.I. (1968) Synthesis of the plasma membrane of the liver cell. Biochem. Biophys. Res. Commun. 31, 54–65.

Razin, S. (1972) Reconstitution of biological membranes. Biochim. Biophys. Acta 265, 241–291.

Rees, D.A. (1969) Structure, conformation and mechanism in the formation of polysaccharide gels and networks. Adv. Carbohyd. Chem. Biochem. 24, 267–332.

Reichstein, E. and Blostein, R. (1973) Asymmetric iodination of the human erythrocyte membrane. Biochem. Biophys. Res. Commun. 54, 494–500.

Renooiz, W., Van Golde, L.M.G., Zwaal, R.F.A. and Van Deenen, L.L.M. (1976) Topological asymmetry of phospholipid metabolism in rat erythrocyte membranes. Evidence for flip-flop of lecithin. Eur. J. Biochem. 61, 53–58.

Robbins, J.C. and Nicolson, G.L. (1975) Surfaces of normal and transformed cells, In: F.F. Becker (Ed.), Cancer: A Comprehensive Treatise, Vol. 4, Plenum Press, New York, pp. 3–54.

Roberts, R.M. and Yuan, B.O. (1974) Chemical modification of the plasma membrane polypeptides of cultured mammalian cells as an aid to studying protein turnover. Biochemistry 13, 4846–4855.

Roseman, S. (1970) The synthesis of complex carbohydrates by multiglycosyltransferase systems and their potential function in intercellular adhesion. Chem. Phys. Lipids 5, 270–300.

Rostas, J.A.P. and Jeffrey, P.L. (1975) Restricted mobility of neuronal membrane antigens. Neurosci. Lett. 1, 47–53.

Roth, S. and White, D. (1972) Intercellular contact and cell-surface galactosyltransferase activity. Proc. Natl. Acad. Sci. USA 69, 485–489.

Rott, R. and Klenk, H.-D. (1977) Structure and assembly of viral envelopes, In: G. Poste and G.L.

Nicolson (Eds.), Virus Infection and The Cell Surface. Cell Surface Reviews, Vol. 2, North-Holland, Amsterdam, pp. 47–82.

Rottem, S., Hubbell, W.L., Hayflick, L. and McConnell, H.M. (1970) Motion of fatty acid spin labels in the plasma membrane of *Mycoplasma*. Biochim. Biophys. Acta 219, 104–113.

Rottem, S.J., Yashouv, J., Ne'eman, Z. and Razin, S. (1973) Cholesterol in mycoplasma membranes. Composition, ultrastructure and biological properties of membranes from *Mycoplasma myocoides* var. *capri* cells adapted to grow with low cholesterol concentrations. Biochim. Biophys. Acta 323, 496–508.

Rousselet, A., Guthmann, C., Matricon, J., Bienvenue, A. and Devaux, P.F. (1976) Study of the transverse diffusion of spin-labeled phospholipids in biological membranes, I. Human red blood cells. Biochim. Biophys. Acta 426, 357–371.

Rubin, R.P. (1974) Calcium and the Secretory Process. Plenum Press, New York, 320 pp.

Rudy, B. and Gitler, C. (1972) Microviscosity of the cell membrane. Biochim. Biophys. Acta 288, 231–236.

Rutishauser, U., Yahara, I. and Edelman, G.M. (1974) Morphology, motility, and surface behavior of lymphocytes bound to nylon fibers. Proc. Natl. Acad. Sci. USA 71, 1149–1153.

Ryan, G.B., Borysenko, J.S. and Karnovsky, M.J. (1974a) Factors affecting the redistribution of surface-bound concanavalin A on human polymorphonuclear leukocytes. J. Cell Biol. 62, 351–365.

Ryan, G.B., Unanue, E.R. and Karnovsky, M.J. (1974b) Inhibition of surface capping of macromolecules by local anaesthetics and tranquillisers. Nature (London) 250, 56–57.

Sabatini, D.D. and Blobel, G. (1970) Controlled proteolysis of nascent polypeptides in rat liver cell fractions. II. Location of the polypeptides in rough microsomes. J. Cell Biol. 45, 146–157.

Sakagami, T., Minari, O. and Orii, T. (1965) Behavior of plasma lipoproteins during exchange of phospholipids between plasma and erythrocytes. Biochim. Biophys. Acta 98, 111–116.

Sakiyama, H. and Burge, B.W. (1972) Comparative studies of the carbohydrate-containing components of 3T3 and Simian Virus-40 transformed 3T3 mouse fibroblasts. Biochemistry 11, 1366–1377.

Sällström, J.F. and Alm, G.V. (1972) Binding of concanavalin A to thymic and bursal chicken lymphoid cells. Exp. Cell Res. 75, 63–72.

Sandri, C., Akert, K., Livingston, R.E. and Moor, H. (1972) Particle aggregations at specialized sites in freeze-etched postsynaptic membranes. Brain Res. 41, 1–16.

Sauerheber, R.D. and Gordon, L.M. (1975) Spin label studies on rat liver plasma membrane: Calcium effects on membrane fluidity. Proc. Soc. Exp. Biol. Med. 150, 28–31.

Scandella, C.J., Devaux, P. and McConnell, H.M. (1972) Rapid lateral diffusion of phospholipids in rabbit sarcoplasmic reticulum. Proc. Natl. Acad. Sci. USA 69, 2056–2060.

Schachter, H. (1974) Glycosylation of glycoproteins during intracellular transport of secretory products, In: B. Ceccarelli, F. Clementi and J. Meldolesi (Eds.), Advances in Cytopharmacology, Vol. 2, Raven Press, New York, pp. 207–218.

Schlessinger, J., Koppel, D.E., Axelrod, D., Jacobson, K., Webb, W.W. and Elson, E.L. (1976) Lateral transport on cell membranes, I. The mobility of concanavalin A receptors on myoblasts. Proc. Natl. Acad. Sci. USA 73, 2409–2413.

Schnepel, G.H., Hegner, D. and Schummer, U. (1974) The influence of calcium on the molecular mobility of fatty acid spin labels in phosphatidylserine and phosphatidylinositol structures. Biochim. Biophys. Acta 367, 67–74.

Schollmeyer, J.V., Goll, D.E., Tilney, L., Mooseker, M., Robson, R. and Stromer, M. (1974) Localization of α-actinin in non-muscle material. J. Cell Biol. 63, 304a.

Schor, S., Siekevitz, P. and Palade, G.E. (1970) Cyclic changes in thylakoid membranes of synchronized *Chlamydomonas reinhardi*. Proc. Natl. Acad. Sci. USA 66, 174–189.

Schreier-Muccillo, S., Marsh, D., Dugas, H., Schneider, H. and Smith, I.C.P. (1973) A spin probe study of the influence of cholesterol on motion and orientation of phospholipids in oriented multibilayers and vesicles. Chem. Phys. Lipids 10, 11–27.

Schreiner, G.F. and Unanue, E.R. (1976) Calcium-sensitive modulation of Ig capping: Evidence supporting a cytoplasmic control of ligand-receptor complexes. J. Exp. Med. 143, 15–31.

Schroeder, T.E. (1968) Cytokinesis: Filaments in the cleavage furrow. Exp. Cell Res. 53, 272–278.
Seelig, J. (1971) On the flexibility of hydrocarbon chains in lipid bilayers. J. Am. Chem. Soc. 93, 5017–5022.
Seelig, J. and Niederberger, W. (1974) Two pictures of a lipid bilayer. A comparison between deuterium label and spin-label experiments. Biochemistry 13, 1585–1588.
Seeman, P. (1972) The membrane actions of anesthetics and tranquilizers. Pharmacol. Rev. 24, 583–655.
Segrest, J.P., Kahne, I., Jackson, R.L. and Marchesi, V.T. (1973) Major glycoprotein of the human erythrocyte membrane: Evidence for an amphiphatic molecular structure. Arch. Biochem. Biophys. 155, 167–183.
Segrest, J.P., Gulik-Krzywicki, T. and Sardet, C. (1974) Association of the membrane-penetrating polypeptide segment of the human erythrocyte MN-glycoprotein with phospholipid bilayers, I. Formation of freeze-etch intramembranous particles. Proc. Natl. Acad. Sci. USA 71, 3294–3298.
Shechter, E., Gulik-Krzywicki, T. and Kaback, H.R. (1972) Correlations between fluorescence, X-ray diffraction and physiological properties in cytoplasmic membrane vesicles isolated from *Escherichia coli*. Biochim. Biophys. Acta 274, 466–477.
Sheetz, M.P. and Chan, S.I. (1972) Effect of sonication on the structure of lecithin bilayers. Biochemistry 11, 4573–4581.
Sheetz, M.P. and Singer, S.J. (1974) Biological membranes as bilayer couples. A molecular mechanism of drug-erythrocyte interactions. Proc. Natl. Acad. Sci. USA 71, 4457–4461.
Sheetz, M.P., Painter, R.G. and Singer, S.J. (1976) Biological membranes as bilayer couples, III. Compensatory shape changes induced in membranes. J. Cell Biol. 70, 193–203.
Sherwood, D. and Montal, M. (1975) Transmembrane lipid migration in planar asymmetric bilayer membranes. Biophys. J. 15, 417–434.
Shimshik, E.J. and McConnell, H.M. (1973a) Lateral phase separation in phospholipid membranes. Biochemistry 12, 2351–2360.
Shimshik, E.J. and McConnell, H.M. (1973b) Lateral phase separations in binary mixtures of cholesterol and phospholipids. Biochem. Biophys. Res. Commun. 53, 446–451.
Shimshik, E.J., Kleemann, W., Hubbell, W.L. and McConnell, H.M. (1973) Lateral phase separations in membranes. J. Supramol. Struct. 1, 283–294.
Shinitzky, M. and Barenholz, Y. (1974) Dynamics of the hydrocarbon layer in liposomes of lecithin and sphingomyelin containing dicetylphosphate. J. Biol. Chem. 249, 2652–2656.
Shinitzky, M. and Inbar, M. (1974) Difference in microviscosity induced by different cholesterol levels in the surface membrane lipid layer of normal lymphocytes and malignant lymphoma cells. J. Mol. Biol. 85, 603–615.
Shinitzky, M., Dianoux, A-C., Gitler, C. and Weber, G. (1971) Microviscosity and order in the hydrocarbon region of micelles and membranes determined with fluorescent probes, I. Synthetic micelles. Biochemistry 10, 2106–2113.
Shipley, G.G. (1973) Recent X-ray diffraction studies of biological membranes and membrane components, In: D. Chapman and D.F.H. Wallach (Eds.), Biological Membranes, Vol. 2, Academic Press, New York, pp. 9–89.
Shlatz, L. and Marinetti, G.V. (1972) Calcium binding to the rat liver plasma membrane. Biochim. Biophys. Acta 290, 70–83.
Shohet, S.B. and Nathan, D.G. (1970) Incorporation of phosphatide precursors from serum into erythrocytes. Biochim. Biophys. Acta 202, 202–205.
Shur, B.D. and Roth, S. (1975) Cell surface glycosyltransferases. Biochim. Biophys. Acta 415, 473–512.
Siekevitz, P., Palade, G.E., Dallner, G., Ohad, I. and Omura, T. (1967) The biogenesis of intracellular membranes, In: H. Vogel, J.D. Lampen and V. Bryson (Eds.), Organizational Biosynthesis, Academic Press, New York, pp. 331–362.
Simpkins, H., Panko, E. and Tay, S. (1971a) Structural changes in the phospholipid regions of the axonal membrane produced by phospholipase C action. Biochemistry 10, 3851–3854.
Simpkins, H., Tay, S. and Panko, E. (1971b) Changes in the molecular structure of axonal and red blood cell membranes following treatment with phospholipase A_2. Biochemistry 10, 3579–3585.

Singer, S.J. (1971) The molecular organization of biological membranes, In: L.I. Rothfield (Ed.), Structure and Function of Biological Membranes, Academic Press, New York, pp. 145–222.
Singer, S.J. (1974a) The molecular organization of membranes. Ann. Rev. Biochem. 43, 805–833.
Singer, S.J. (1974b) Molecular biology of membranes with applications to immunology. Adv. Immunol. 19, 1–68.
Singer, S.J. and Nicolson, G.L. (1972) The fluid mosaic model of the structure of cell membranes. Science 175, 720–731.
Smellie, R.M.S. (1974) (Ed.) Calcium and Cell Regulation. Biochemical Society Symposium 39. The Biochemical Society, London.
Solheim, B.G. and Thorsby, E. (1974) β_2-Microglobulin is part of the HL-A molecule in the lymphocyte membrane. Nature (London) 249, 36–38.
Solomon, A.K. (1974) Apparent viscosity of human red cell membranes. Biochim. Biophys. Acta 373, 145–149.
Speth, V. and Wunderlich, F. (1973) Membrane of *Tetrahymena*, II. Direct visualization of reversible transitions in biomembrane structure induced by temperature. Biochim. Biophys. Acta 291, 621–628.
Stackpole, C.W. (1977) Topographical differentiation of the cell surface, In: J.F. Danielli, M.D. Rosenberg and D.A. Cadenhead (Eds.), Progress in Membrane and Surface Science, Vol. 14, Academic Press, New York, in press.
Stackpole, C.W., De Milio, L.T., Hämmerling, U., Jacobson, J.B. and Lardis, M.P. (1974a) Hybrid antibody-induced topographical redistribution of surface immunoglobulins, alloantigens and concanavalin A receptors on mouse lymphoid cells. Proc. Natl. Acad. Sci. USA 71, 932–936.
Stackpole, C.W., De Milio, L.T., Jacobson, J.B., Hämmerling, U. and Lardis, M.P. (1974b) A comparison of ligand-induced redistribution of surface immunoglobulins, alloantigens, and concanavalin A receptors on mouse lymphoid cells. J. Cell. Physiol. 83, 441–448.
Stackpole, C.W., Jacobson, J.B. and Lardis, M.P. (1974c) Antigenic modulation in vitro, I. Fate of thymus-leukemia (TL) antigen-antibody complexes following modulation of TL antigenicity from the surfaces of mouse leukemia cells and thymocytes. J. Exp. Med. 140, 939–953.
Stackpole, C.W., Jacobson, J.B. and Lardis, M.P. (1974d) Two distinct types of capping of surface receptors on mouse lymphoid cells. Nature (London) 248, 232–234.
Staehelin, L.A. (1973) Further observations on the fine structure of freeze-cleaved tight junctions. J. Cell Sci. 13, 763–785.
Staehelin, L.A. (1974) Structure and function of intercellular junctions. Int. Rev. Cytol. 39, 191–283.
Steck, T.L. (1972) The organization of proteins in human erythrocyte membranes, In: C.F. Fox (Ed.), Membrane Research, Academic Press, New York, pp. 71–93.
Steck, T.L. (1974) The organization of proteins in the human red blood cell membrane. A review. J. Cell. Biol. 62, 1–19.
Steck, T.L. and Dawson, G. (1974) Topographical distribution of complex carbohydrates in the erythrocyte membrane. J. Biol. Chem. 249, 2135–2142.
Steck, T.L. and Fox, C.F. (1972) Membrane proteins, In: C.F. Fox and A.D. Keith (Eds.), Membrane Molecular Biology, Sinauer Associates, Stamford, Connecticut, pp. 27–75.
Steim, J.M., Tourtellotte, M.E., Reinert, J.C., McElhaney, R.N. and Rader, R.L. (1969) Calorimetric evidence for the liquid-crystalline state of lipids in a biomembrane. Proc. Natl. Acad. Sci. USA 63, 104–109.
Stolinski, C. and Breathnach, A.S. (1976) Freeze-Fracture Replication of Biological Tissues, Academic Press, New York.
Sytkowski, A.J., Vogel, Z. and Nirenberg, M.W. (1973) Development of acetylcholine receptor clusters on cultured muscle cells. Proc. Natl. Acad. Sci. USA 70, 270–274.
Tanford, C. (1968) Physiological Chemistry of Macromolecules. Wiley, New York.
Tardieu, A., Luzzati, V. and Reman, F.C. (1973) Structure and polymorphism of the hydrocarbon chains of lipids: A study of lecithin-water phases. J. Mol. Biol. 75, 711–733.
Taylor, R.B., Duffus, W.P.H., Raff, M.C. and de Petris, S. (1971) Redistribution and pinocytosis of lymphocyte surface immunoglobulin molecules induced by anti-immunoglobulin antibody. Nature (London) New Biol. 233, 225–229.

Tillack, T.W. and Marchesi, V.T. (1970) Demonstration of the outer surface of freeze-etched red blood cell membranes. J. Cell Biol. 45, 649–653.

Tillack, T.W., Scott, R.E. and Marchesi, V.T. (1972) The structure of erythrocyte membranes studied by freeze-etching, II. Localization of receptors for phytohemagglutinin and influenza virus to the intramembranous particles. J. Exp. Med. 135, 1209–1227.

Tilney, L.G. (1976a) The polymerization of actin, II. How nonfilamentous actin becomes non-randomly distributed in sperm: Evidence for the association of this actin with membranes. J. Cell Biol. 69, 51–72.

Tilney, L.G. (1976b) The polymerization of actin, III. Aggregates of nonfilamentous actin and its associated proteins: A storage form of actin. J. Cell Biol. 69, 73–89.

Tilney, L.G. and Detmers, P. (1975) Actin in erythrocyte ghosts and its association with spectrin. Evidence for a nonfilamentous form of these two molecules in situ. J. Cell Biol. 66, 508–520.

Tourtellotte, M.E., Branton, D. and Keith, A. (1970) Membrane structure: Spin labeling and freeze etching of *Mycoplasma laidlawii*. Proc. Natl. Acad. Sci. USA 66, 909–916.

Träuble, H. (1972) Phase transitions in lipids. Biomembranes 3, 197–227.

Träuble, H. and Eibl, H. (1975) Cooperative structural changes in lipid bilayers, In: F.O. Schmitt, D.M. Schneider and D.M. Crothers (Eds.), Functional Linkage in Biomolecular Systems, Raven Press, New York, pp. 59–90.

Träuble, H. and Haynes, D.H. (1971) The volume change in lipid bilayer lamellae at the crystalline-liquid crystalline phase transition. Chem. Phys. Lipids 7, 324–335.

Träuble, H. and Sackmann, E. (1972) Studies of the crystalline-liquid crystalline phase transition of lipid model membranes, III. Structure of a steroid-lecithin system below and above the lipid-phase transition. J. Am. Chem. Soc. 94, 4499–4510.

Triggle, D.J. (1972) Effects of calcium on excitable membranes and neurotransmitter action, In: J.F. Danielli, M.D. Rosenberg and D.A. Cadenhead (Eds.), Progress in Surface and Membrane Science, Vol. 5, Academic Press, New York, pp. 267–331.

Trinkaus, J.P. (1976) On the mechanism of metazoan cell movements, In: G. Poste and G.L. Nicolson (Eds.), The Cell Surface in Animal Embryogenesis and Development, Cell Surface Reviews, Vol. 1, North-Holland, Amsterdam, pp. 225–330.

Triplett, R.B. and Carraway, K.L. (1972) Proteolytic digestion of erythrocytes, resealed ghosts, and isolated membranes. Biochemistry 11, 2897–2903.

Tsong, T.Y. (1975a) Effect of phase transition on the kinetics of dye transport in phospholipid bilayer structures. Biochemistry 14, 5409–5414.

Tsong, T.Y. (1975b) Transport of 8-anilino-1-naphthalenesulfonate as a probe of the effect of cholesterol on the phospholipid bilayer structures. Biochemistry 14, 5415–5417.

Turner, R.S. and Burger, M.M. (1973) The cell surface in cell interactions. Rev. Physiol. Biochem. Exp. Pharmacol. 68, 121–155.

Tweto, J. and Doyle, D. (1976) Turnover of the plasma membrane proteins of hepatoma tissue culture cells. J. Biol. Chem. 251, 872–882.

Tweto, J. and Doyle, D. (1977) Turnover of proteins of the eukaryotic cell surface, In: G. Poste and G.L. Nicolson (Eds.), Synthesis, Assembly and Turnover, Cell Surface Reviews, Vol. 4, North-Holland, Amsterdam, in press.

Uhr, J.W., Vitetta, E.S. and Melcher, U.K. (1974) Regulation of cell surface immunoglobulin and alloantigens on lymphocytes, In: G.M. Edelman (Ed.), Cellular Selection and Regulation in the Immune Response, Raven Press, New York, pp. 133–141.

Ukena, T.E., Borysenko, J.Z. and Karnovsky, M.J. (1974) Effects of colchicine, cytochalasin B and 2-deoxyglucose on the topographical organization of surface-bound concanavalin A in normal and transformed fibroblasts. J. Cell Biol. 61, 70–82.

Unanue, E.R. and Karnovsky, M.J. (1973) Redistribution and fate of Ig complexes on surface B lymphocytes: functional implications and mechanisms. Transplant. Rev. 14, 184–210.

Unanue, E.R., Karnovsky, M.J. and Engers, H.D. (1973) Ligand-induced movement of lymphocyte membrane macromolecules, III. Relationship between the formation and fate of anti-Ig surface Ig complexes and cell metabolism. J. Exp. Med. 137, 675–689.

Urry, D.W. (1972) Protein conformation in biomembranes. Optical rotation and absorption of membrane suspensions. Biochim. Biophys. Acta 265, 115–168.
Vail, W.J., Papahadjopoulos, D. and Moscarello, M.A. (1974) Interaction of a hydrophobic protein with liposomes. Evidence for particles seen in freeze fracture as being proteins. Biochim. Biophys. Acta 345, 463–467.
Vanderkooi, G. (1974) Organization of proteins in membranes with special reference to the cytochrome oxidase system. Biochim. Biophys. Acta 344, 307–344.
Vanderkooi, J.M. and Callis, J.B. (1974) Pyrene. A probe of lateral diffusion in the hydrophobic region of membranes. Biochemistry 13, 4000–4006.
Vanderkooi, J., Fischkoff, S., Chance, B. and Cooper, R.A. (1974) Fluorescent probe analysis of the lipid architecture of natural and experimental cholesterol-rich membranes. Biochemistry 13, 1589–1595.
Verkleij, A.J., Ververgaert, P.H.J., Van Deenen, L.L.M. and Elbers, P.F. (1972) Phase transitions of phospholipid bilayers and membranes of *Acholeplasma laidlawii* B visualized by freeze-fracture electron microscopy. Biochim. Biophys. Acta 288, 326–332.
Verkleij, A.J., Zwaal, R.F.A., Roelofsen, B., Comfurius, P., Kastelijn, D. and Van Deenen, L.L.M. (1973) The asymmetric distribution of phospholipids in the human red cell membrane. A combined study using phospholipases and freeze-etch electron microscopy. Biochim. Biophys. Acta 323, 178–193.
Verkleij, A.J., Ververgaert, P.H.J., De Kruyff, B. and Van Deenen, L.L.M. (1974) The distribution of cholesterol in bilayers of phosphatidylcholines as visualized by freeze-fracturing. Biochim. Biophys. Acta 373, 495–501.
Ververgaert, P.H.J., Elbers, P.F., Luitingh, A.J. and Van Den Berg, H.J. (1972) Surface patterns of freeze-fractured liposomes. Cytobiologie 6, 86–96.
Ververgaert, P.H.J., Verkleij, J.J., Verhoeven, J.J. and Elbers, P.F. (1973) Spray-freezing of liposomes. Biochim. Biophys. Acta 311, 651–654.
Victoria, E.J., Muchmore, E.A., Sudora, E.J. and Masouredis, S.P. (1975) The role of antigen mobility in anti-Rh_0(D)-induced agglutination. J. Clin. Invest. 56, 292–301.
Vitetta, E.S. and Uhr, J.W. (1973) Synthesis, transport, dynamics and fate of cell surface Ig and alloantigens in murine lymphocytes. Transplant. Rev. 14, 50–75.
Waggoner, A.S. and Stryer, L. (1970) Fluorescent probes of biological membranes. Proc. Natl. Acad. Sci. USA 67, 579–589.
Wahl, P., Kasai, M. and Changeux, J-P. (1971) A study on the motion of proteins in excitable membrane fragments by nanosecond fluorescence polarization spectroscopy. Eur. J. Biochem. 18, 332–341.
Wallach, D.F.H. (1969) Membrane lipids and the conformations of membrane proteins. J. Gen. Physiol. 54, 3s–26s.
Wallach, D.F.H. (1972) The Plasma Membrane: Dynamic Perspectives, Genetics and Pathology. Springer, New York.
Wallach, D.F.H. (1975) Membrane Molecular Biology of Neoplastic Cells. North-Holland, Amsterdam.
Wallach, D.F.H. and Winzler, R.J. (1974) Evolving Strategies and Tactics in Membrane Research. Springer, New York.
Wallach, D.F.H. and Zahler, P.H. (1966) Protein conformation in cellular membranes. Proc. Natl. Acad. Sci. USA 56, 1552–1559.
Wang, K. and Richards, F.M. (1974) An approach to nearest neighbor analysis of membrane proteins. Application to the human erythrocyte membrane of a method employing cleavable cross-linkages. J. Biol. Chem. 249, 8005–8018.
Warren, L. and Glick, M.C. (1968) Membranes of animal cells, II. The metabolism and turnover of the surface membrane. J. Cell Biol. 37, 729–746.
Warren, L. and Glick, M.C. (1969) Isolation of surface membranes of tissue culture cells, In: K. Habel and N.P. Salzman (Eds.), Fundamental Techniques in Virology, Academic Press, New York, pp. 66–71.
Wartiovaara, J., Linder, E., Ruoslahti, E. and Vaheri, A. (1974) Distribution of fibroblast surface

antigen. Association with fibrillar structures of normal cells and loss upon viral transformation. J. Exp. Med. 140, 1522–1533.

Weber, K. and Groeschel-Stewart, U. (1974) Antibody to myosin: the specific visualization of myosin-containing filaments in nonmuscle cells. Proc. Natl. Acad. Sci. USA 71, 4561–4564.

Weber, K., Lazarides, E., Goldman, R.D., Vogel, A. and Pollack, R. (1974) Localization and distribution of actin fibers in normal, transformed and revertant cells. Cold Spring Harbor Symp. Quant. Biol. 39, 363–369.

Weber, K., Bibring, Th. and Osborn, M. (1975a) Specific visualization of tubulin-containing structures in tissue culture cells by immunofluorescence. Exp. Cell Res. 95, 111–120.

Weber, K., Pollack, R. and Bibring, Th. (1975b) Antibody against tubulin: The specific visualization of cytoplasmic microtubules in tissue culture cells. Proc. Natl. Acad. Sci. USA 72, 459–463.

Weiderkamm, E., Wallach, D.F.H. and Fischer, H. (1970) The effect of phospholipase A upon the interaction of L-anilinonaphthalene-8-sulfonate with erythrocyte membranes. Biochim. Biophys. Acta 241, 770–778.

Weinstein, R.S. and McNutt, N.S. (1972) Cell junctions. New Eng. J. Med. 286, 521–524.

Weinstein, R.S., Mark, F.B. and Alroy, J. (1976) The structure and function of intercellular junctions in cancer. Adv. Cancer Res. 23, 23–89.

Weiss, D.E. (1973) The role of lipid in energy transmission and conservation in functional biological membranes. Subcell. Biochem. 2, 201–235.

Weiss, L. (1967) A. Neuberger and E.L. Tatum (Eds.), The Cell Periphery, Metastasis and Other Contact Phenomena, Frontiers of Biology, Vol. 7, North-Holland, Amsterdam.

Weiss, L. (1969) The cell periphery. Int. Rev. Cytol. 26, 63–105.

Wessells, N.K., Spooner, B.S., Ash, J.F., Bradly, M.O., Luduena, M.A., Taylor, E.L., Wrenn, J.T. and Yamada, K.M. (1971) Microfilaments in cellular and developmental processes. Science 171, 135–143.

White, S.H. (1970) A study of lipid bilayer membrane stability using precise measurements of specific capacitance. Biophys. J. 10, 1127–1148.

Whiteley, N.M. and Berg, H.C. (1974) Amidination of the outer and inner surfaces of the human erythrocyte membrane. J. Mol. Biol. 87, 541–561.

Wilkins, M.H.F., Blaurock, A.E. and Engelman, D.M. (1971) Bilayer structure in membranes. Nature (London) New Biol. 230, 72–76.

Willingham, M.C. and Pastan, I. (1974) Cyclic AMP mediates the concanavalin A agglutinability of mouse fibroblasts. J. Cell Biol. 63, 288–294.

Willingham, M.C. and Pastan, I. (1975a) Cyclic AMP and cell morphology in cultured fibroblasts. Effects on cell shape, microfilament and microtubule distribution, and orientation to substratum. J. Cell Biol. 67, 146–159.

Willingham, M.C. and Pastan, I. (1975b) Cyclic AMP modulates microvillus formation and agglutinability in transformed and normal mouse fibroblasts. Proc. Natl. Acad. Sci. USA 72, 1263–1267.

Willingham, M.C., Ostlund, R.E. and Pastan, I. (1974) Myosin is a component of the cell surface of cultured cells. Proc. Natl. Acad. Sci. USA 71, 4144–4148.

Wilson, L. (1975) Minireview. Action of drugs on microtubules. Life Sci. 17, 303–310.

Wilson, L. and Bryan, J. (1974) Biochemical and pharmacological properties of microtubules. Adv. Cell. Mol. Biol. 3, 21–72.

Winzler, R.J. (1970) Carbohydrates in cell surfaces. Int. Rev. Cytol. 29, 77–125.

Wisnieski, B.J., Parkes, J.G., Huang, Y.O. and Fox, C.F. (1974) Physical and physiological evidence for two phase transitions in cytoplasmic membranes of animal cells. Proc. Natl. Acad. Sci. USA 71, 4381–4385.

Woodward, C.B. and Zwaal, R.F.A. (1972) The lytic behavior of pure phospholipases A_2 and C towards osmotically swollen erythrocytes and resealed ghosts. Biochim. Biophys. Acta 274, 272–278.

Worthington, C.R. (1972) X-Ray studies on nerve and photoreceptors. Ann. N.Y. Acad. Sci. 195, 293–308.

Worthington, C.R. and King, G.I. (1971) Electron density profiles of nerve myelin. Nature (London) 234, 143–145.
Wu, C-W. and Stryer, L. (1972) Proximity relationships in rhodopsin. Proc. Natl. Acad. Sci. USA 69, 1104–1108.
Wu, S.H-W. and McConnell, H.M. (1975) Phase separations in phospholipid membranes. Biochemistry 14, 847–854.
Yahara, I. and Edelman, G.M. (1972) Restriction of the mobility of lymphocyte immunoglobulin receptors by concanavalin A. Proc. Natl. Acad. Sci. USA 69, 608–612.
Yahara, I. and Edelman, G.M. (1973) The effects of concanavalin A on the mobility of lymphocyte surface receptors. Exp. Cell Res. 81, 143–155.
Yahara, I. and Edelman, G.M. (1975a) Electron microscopic analysis of the modulation of lymphocyte receptor mobility. Exp. Cell Res. 91, 125–142.
Yahara, I. and Edelman, G.M. (1975b) Modulation of receptor mobility by locally bound concanavalin A. Proc. Natl. Acad. Sci. USA 72, 1579–1583.
Yanagimachi, R., Noda, Y.D., Fujimoto, M. and Nicolson, G.L. (1972) The distribution of negative surface charges on mammalian spermatozoa. Am. J. Anat. 135, 497–420.
Yefenof, E. and Klein, G. (1974) Antibody-induced redistribution of normal and tumor associated surface antigens. Exp. Cell Res. 88, 217–224.
Yin, H.H., Ukena, T.E. and Berlin, R.D. (1972) Effect of colchicine, colcemid and vinblastine on the agglutination by concanavalin A, or transformed cells. Science 178, 867–868.
Yogeeswaran, G., Laine, R. and Hakomori, S-I. (1974) Mechanism of cell contact-dependent glycolipid synthesis: Further studies with glycolipid-glass complex. Biochem. Biophys. Res. Commun. 59, 591–599.
Yoshida, S. and Ikegami, A. (1974) Disc to sphere transformation of erythrocytes, induced by 1-anilino-8-naphthalenesulfonate. Biochim. Biophys. Acta 367, 39–46.
Young, M., King, M.V., O'Hara, D.S. and Molberg, P.J. (1973) Studies on the structure and assembly pattern of the light meromyosin section of the myosin rod. Cold Spring Harbor Symp. Quant. Biol. 37, 65–76.
Yu, J. and Steck, T.L. (1974) Associations between the major red cell membrane penetrating protein and two inner surface proteins. Fed. Proc. 33, 1532.
Zagury, D., Uhr, J.W., Jamieson, J.D. and Palade, G.E. (1970) Immunoglobulin synthesis and secretion, II. Radioautographic studies of sites of addition of carbohydrate moieties and intracellular transport. J. Cell Biol. 46, 52–63.
Zagyansky, Y. and Edidin, M. (1976) Lateral diffusion of concanavalin A receptors in the plasma membrane of mouse fibroblasts. Biochim. Biophys. Acta 433, 209–214.
Zilversmit, D.B. (1971) Exchange of phospholipid classes between liver microsomes and plasma: comparison of rat, rabbit and guinea pig. J. Lipid Res. 12, 36–42.
Zwaal, R.F.A., Roelofsen, B. and Colley, C.M. (1973) Localization of red cell membrane constituents. Biochim. Biophys. Acta 300, 159–182.
Zwaal, R.F.A., Roelofsen, B., Comfurius, P. and Van Deenen, L.L.M. (1975) Organization of phospholipids in human red cell membranes as detected by the action of various purified phospholipases. Biochim. Biophys. Acta 406, 83–96.

Freeze-fracture techniques and applications to the structural analysis of the mammalian plasma membrane

N. Scott McNUTT

1. Introduction

Freeze-fracture is one of several techniques currently used for the electron microscopic study of cellular architecture at moderately high resolution. In combination with standard thin-section and negative staining methods, freeze-fracture has had its greatest impact on the analysis of the structure of membranes in cells. It is the aim of this review to present certain basic principles which are common to all of the various methods for freeze-fracturing membranes and which are necessary for an understanding of both the strengths and weaknesses of data obtained with this technique. Detailed discussion of the design and technical manipulation of freeze-fracture equipment will not be attempted since such considerations are not appropriate to this review. The primary purpose of this chapter will be to survey the contribution which freeze-fracture electron microscopy has made to our current understanding of the structural organization of the plasma membrane in cells from vertebrates, particularly mammals.

One of the earliest contributions of electron microscopy was to provide definitive evidence that the cytoplasm of a cell is enclosed by a membrane, named the plasma membrane, and that the cytoplasm was subdivided into a number of compartments by other membranes. In thin sections of cells, Robertson (1959, 1960) found that several types of membranes had a trilaminar profile composed of two electron-dense laminae, each approximately 25 Å in thickness, separated by an electron-lucent lamina, approximately 30 Å in thickness. Robertson put forward the "unit" membrane hypothesis that all membranes have a trilaminar structure which can be interpreted according to the Davson–Danielli model of membrane structure. In this hypothesis, the electron-lucent lamina in the center of the unit profile is interpreted as being a phospholipid bilayer which creates a hydrophobic interior in the membrane. As proposed initially, the membrane proteins were thought to reside in the electron-dense laminae (Robertson, 1959, 1960; Stoeckenius and Engelman, 1969).

The unit membrane hypothesis has served an important function in providing a definition for those structures which should be called membranes in electron micrographs. Also the trilaminar image has been a reliable indicator of the position of the hydrophobic permeability barrier of membranes. However, as more has been learned about membrane proteins, the unit membrane hypothesis has not been able to account either for the location of these proteins or for the diversity of transport functions of membranes. Also some membranes, particularly mitochondrial membranes, did not always have a trilaminar profile and instead had a globular appearance when fixed in potassium permanganate; however the interpretation of such images has been controversial (Sjöstrand, 1968). Other evidence for the presence of globular structures in membranes has been obtained in studies on erythrocyte membrane proteins in situ using circular dichroism which indicate that a considerable portion (about 40%) of the proteins are in an alpha helical form and exist as globular structures. These globular proteins have a diameter close to the thickness of the whole membrane and consequently could not fit into only one electron-dense lamina (reviewed by Singer and Nicolson, 1972).

The freeze-fracture technique has provided additional strong evidence on the localization of some membrane proteins. Images of freeze-fractured plasma membranes have been the subject of considerable controversy in the past but it is now generally agreed that the process of fracturing frozen membranes exposes structures within the hydrophobic interior of membranes (Branton, 1966; Branton et al., 1975). The images obtained are consistent with a fluid mosaic model of membrane structure in which some membrane proteins are inserted into the plasma membrane so that they lie partially within the hydrophobic lipid interior of the membrane and have a lateral mobility within the plane of the membrane (Singer and Nicolson, 1972).

This review will deal first with the freeze-fracture technique and how images are obtained, secondly with the interpretation of those images in terms of general membrane structure, and thirdly will discuss evidence from studies of certain specialized membrane structures.

2. Freeze-fracture technique

The freeze-fracture method has undergone a dramatic evolution from initial relatively crude techniques (Meryman and Kafig, 1954, 1955; Steere, 1957) to highly developed approaches (reviewed by Bullivant, 1973; Mühlethaler, 1973). The predominant method for carrying out the freeze-fracture process at the present time is the method of Moor et al. (1961) which utilizes a cryomicrotome inside a high vacuum chamber (Moor, 1966, 1971, 1973). Recently, a leading laboratory (Fisher and Branton, 1974) has published an up-to-date and detailed account of methods for performing the freeze-fracture technique using the

Balzers apparatus manufactured on the design of Moor (Balzers High Vacuum Corp., Balzers AG, Liechtenstein). Other types of freeze-fracture equipment employ a variety of designs and these have been reviewed recently (Bullivant, 1973).

The various technical approaches to freeze-fracture are based on the relatively simple principle that a high resolution replica can be cast on the surface of hydrated biological material at very low temperature (less than $-100°C$) in a vacuum (less than $1 \cdot 10^{-5}$ mm Hg). If this principle is kept in mind, the sequence of preparative steps in freeze-fracture is easily understood. The initial steps of the technique are involved with freezing of the specimen. In general, cells must be protected during freezing in order to avoid severe distortion of cellular architecture by large ice crystals (Bank and Mazur, 1972; Mazur et al., 1972). However, specimens frozen without any prior preparative procedures have been utilized and can give important information concerning the effects of cryoprotection procedures on membranes (Gilula et al., 1975). The size of ice crystals is inversely related to the freezing rate and is directly related to the amount of free water in the specimen. The freezing rate can be maximized by using, as quenching agents, fluids that have a high heat capacity and a low freezing point. The fluorocarbons dichlorodifluoromethane and monochlorodifluoromethane are popular quenching agents, since they have freezing points near $-150°C$, and have reasonable heat capacities. In addition, the amount of free water in the specimen can be reduced by replacing and binding water with substances such as glycerol, ethylene glycol, sucrose, dimethyl sulfoxide and many others. Glycerol is relatively non-toxic and is used often for cryoprotection as a 10–40% glycerol solution (Bank and Mazur, 1972). Since cellular membranes may differ in their glycerol permeability and glycerination may itself induce artifacts (Kirk and Tosteson, 1973; McIntyre et al., 1974; Martínez-Palomo et al., 1976), it is common practice to use a brief chemical treatment with aldehyde, such as glutaraldehyde, to increase membrane permeability to glycerol and to immobilize cellular architecture during glycerination. This aldehyde treatment is generally called "fixation", although this term can be misleading, since the stabilization of cellular structure for glycerination usually is not equivalent to the fixation employed for other electron microscopy preparative techniques, such as plastic embedding for thin sectioning. Typically weaker concentrations of glutaraldehyde (0.5 to 1%) are used for shorter intervals (5–30 min) to stabilize structures for freeze-fracture, whereas classical fixation involves higher glutaraldehyde concentrations (2–6%) for longer times (1–4 h). This can be an important consideration in studies correlating freeze-fracture and thin section images of membranes.

With the use of glycerination and quenching in fluorocarbon at $-150°C$, rapid freezing can be achieved in tissues up to 0.5 mm in thickness so that ice crystal formation is inapparent. With thicker specimens, the freezing rate progressively slows toward the center of the specimen due to the limited heat conduction through the shell of frozen aqueous matrix formed on the surface of the specimen during immersion cooling. Slow freezing produces several types of artifacts. For example, large extracellular ice crystals form and tend to withdraw any free supercooled water from the surrounding phase. This dehydration may be so marked as to shrink cells. In addition, the growth of ice crystals exposes the biological structures in the adjacent fluid phase to progressively higher solute concentrations which are potentially damaging (Mazur, 1970). Also the dendritic branches of the ice crystals may abut against and deform biological structures. Slow freezing would allow greater opportunity for phase transitions to occur in the phospholipids of the membranes, which could lead to lateral phase separations of lipids (Verkleij et al., 1972; Pagano et al., 1973). Deamer et al. (1970) have shown that rapid freezing can preserve lamellar and hexagonal phases of lipids for replication.

For studies of the true outer surface of cells, freezing in distilled water by conventional methods has been employed and the surface has appeared remarkably smooth (Tillack and Marchesi, 1970; Goodenough and Gilula, 1974). Although these studies have given valuable data, they must be interpreted with caution since freezing in distilled water produces the most extreme freezing artifacts, e.g., deformation by pressure from ice crystals. New freezing techniques are promising improvements for studies of the surface of cells. By suspending cells in droplets, approximately 10 μm in diameter, which are sprayed into liquid propane at $-180°C$, extremely rapid freezing of cells has been achieved (Bachmann and Schmitt, 1971; Bachmann and Schmitt-Fumian, 1973). It is interesting that the spray method preserves the extended state of delicate molecules on the surface of cells much better than conventional freezing (Plattner et al., 1973) due to the lack of deformation by ice crystals.

Once the specimen is frozen, it is fractured usually with a cold knife (at $-196°C$) using one of two basic technical approaches, either fracturing in a vacuum chamber (Moor et al., 1961; Steere, 1969, 1973) or fracturing at atmospheric pressure and then placing the specimen into a vacuum chamber (Steere, 1957; Bullivant and Ames, 1966). Both approaches can give comparable results and each approach has both advantages and limitations which have been reviewed fully elsewhere (Bullivant, 1973). Certain modifications of the fracturing procedure have been devised for fracturing special types of specimens, such as monolayers of cultured cells (Collins et al., 1975; Pfenninger and Rinderer, 1975). Also the specimen can be frozen between two specimen carriers and fractured by a free break which is produced when the carriers are separated or snapped apart. This technique allows the topography above and below the fracture plane to be retrieved and replicated (Steere and Moseley, 1969; Chalcroft and Bullivant, 1970; Wehrli et al., 1970; Weinstein et al., 1970a). The pair of replicas, one above and one below the fracture plane, are called complementary replicas. The study of complementary replicas has been important in the interpretation of membrane images and can be used to detect distortion of structures produced by the fracture process.

When the fractured surface is at high vacuum, a high resolution replica is cast on its topography. The replica is prepared by evaporation of a heavy metal, such as platinum, from an electrode and allowing this metal vapor to condense on the specimen. When a metal is evaporated from a source at an acute angle to the fractured surface, the resulting replica shows the presence of metal deposits on the structures on that side facing the source ("windward side") and an absence of deposit on the side facing away from the source ("leeward side"). In relation to the position of the shadowing source, the pattern of metal deposition appears opposite for elevations and depressions (Fig. 1). Structures which are represented as elevations receive the deposit of metal on the side closest to the source and no deposit on the side farthest from the source. On the other hand, depressions have metal deposited on the side farthest from the source and not on the side closest to the source. This pattern gives the so-called "shadow cast" effect since this is similar to the play of light and shade on projections and depressions. The metal may be evaporated either by resistance heating of an electrode or by bombardment of the metal in an electron gun (Moor, 1973). In standard shadow-cast replicas, the area free of metal deposit represent holes or weaknesses in the replica. To provide greater stability, a subsequent, more uniform coating layer of carbon is condensed onto the surface from a separate electrode at 90°C to the specimen. The carbon film is relatively electron-lucent compared to the metal film and ordinarily does not obscure specimen detail.

Replicas can be examined in a scanning electron microscope without removing them from the fractured surface. However, to take advantage of the high resolution of the transmission electron microscope, the replica must be thin and must be removed from the fractured specimen. These two requirements oppose each other since the thinnest and highest resolution replicas are

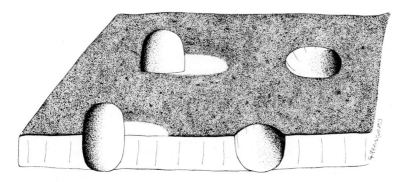

Fig. 1. Schematic representation of a replica lying on a freeze-fractured membrane face. Platinum metal (black stipples) has been condensed onto the upper surface of the membrane following evaporation from a platinum electrode (not shown) at an angle from the upper left of the figure. This drawing illustrates that, in the case of elevations, platinum grains are deposited on the near side ("windward side") and not on the far side ("leeward side"). In terms of distance from the shadowing source, platinum deposits are on the near side of elevations but are on the far side of depressions. This drawing also shows a positive image of a replica, i.e. platinum grains are black and areas free of platinum appear white, since this is the conventional method for printing pictures of freeze-fracture replicas. The various membrane components represented in this figure are not drawn to scale with respect to each other.

the most easily damaged during separation from the specimen. To remove the replica from the surface, a variety of agents have been used to dissolve the specimen, including sulphuric acid, chromic acid, sodium hydroxide, sodium hypochlorite, enzymes and organic solvents. The nature of the specimen dictates the type of agent used. Some of the agents will dissolve the carbon support film and weaken the replica but otherwise do not distort the replica. Consequently, in most instances the replica accurately represents the fractured topography. After the replica is isolated, it is washed, placed on a grid, and examined in a transmission electron microscope. When viewed on the screen of the electron microscope, the regions of dense metal deposit have deflected electrons and appear dark whereas areas free of metal transmit electrons and appear light (Fig. 1). This is called a positive image. From the early work of Moor et al. (1961) and Branton (1966), a convention has developed in which freeze-fracture pictures are published as positive images. This convention presents images that are the opposite of how light would strike a similar surface, i.e., dense metal appears black and "shadows" are white whereas light striking the objects would reflect as white light and shadows would be dark. However, the convention is thoroughly ingrained in the field and has the advantage that it minimizes the amount of photographic manipulation of the images. With a small amount of visual training, most observers readily adapt to the "white shadows" convention and can easily interpret images of depressions and elevations, particularly if the images are oriented so that the direction of shadow-casting is from the approximate bottom of the field of view.

The technique called freeze-etching is the freeze-fracture technique with the addition of a so-called "etching" step after fracturing but before replication. Etching is jargon for freeze-drying of the fractured surface, which occurs if the specimen temperature is maintained near −100°C for a short time (seconds to minutes) in a high vacuum. At this temperature, water vapor sublimes from the surface of ice crystals causing hydrated compartments to recede below the level of the fractured surface. Etching is useful for demonstrating ice crystal size in the specimen, for outlining structures in the solute compartment adjacent to the ice (when the ice crystals are small), and for exposing the true surface of structures which have been frozen in distilled water or very dilute solutes. Etching has disadvantages in that the water vapor released from ice crystals may partially recondense onto the fractured topography and obscure detail.

3. Interpretation of freeze-fracture images

Proper interpretation of freeze-fracture images is dependent on being given enough information about the methodology. For example, the following information should be specified: specimen fixation and cryoprotection protocol; temperature and conditions of fracturing; whether or not an etching step has been included; the direction of shadowing; and the type of photographic printing procedure and magnification. Unfortunately, the literature contains many manuscripts which fail to give adequate data of this type. When all these data are given, further interpretation depends on how faithfully the replica represents the fractured topography and any additional topography revealed by etching.

Good replication technique requires that the evaporated metal condenses directly onto the fractured surface of the specimen. Ideally, once condensed, the metal atoms should not move on the surface. This condition is rarely (if ever) met. One of the standard metals used for shadow casting is platinum. Platinum atoms condense onto the frozen, fractured specimen but migrate slightly and form crystallites on the surface. The size of the crystallites limits the resolution of the shadowing technique and depends on specimen temperature, vacuum level during shadowing, precise composition of the metal vapor and the amount of radiant heating of the surface during shadowing (Moor, 1973). Pure platinum produces large crystallites (Fig. 2), whereas alloys, or the co-condensation of a slight amount of carbon vapor, evidently disrupts crystallite growth so that very small crystallites form, thus giving a very fine grain replica. In practice, crystallite size can be reduced to approximately 10 to 15 Å in diameter and this forms the present resolution limit of the standard shadowing method with platinum-carbon mixtures (Moor, 1973).

A far greater limitation on resolution is the inability of any of the freeze-fracture techniques to eliminate the condensation of extraneous volatile substances onto the very cold, fractured topography. Kreutziger (1968a) called attention to the problem of specimen surface contamination and illustrated how contamination obscures very fine details present in the fractured surface. Contamination is minimized if replication is performed as soon as possible after fracture of the specimen in vacuo (Staehelin and Bertaud, 1971). Contamination of a surface fractured at atmospheric pressure can be reduced by protecting the specimen with liquid nitrogen or cold metal blocks prior to replication (Bullivant and Ames, 1966) or by attempting to evaporate off contaminants (Steere, 1957). All of the freeze-fracture methods fail to eliminate contamination completely. Contamination of the cold fractured surfaces may arise from sources outside the specimen or from the specimen itself. External sources of contamination include condensation of atmospheric water entering through small leaks in the vacuum system and condensation of hydrocarbons from a hot oil diffusion pump or from an oil-type mechanical pump. Also extraneous volatile materials may be released from the metal evaporation source during replication.

Endogenous sources of contamination often arise from the partial recondensation of water vapor released from the specimen during fracturing (due to the inadvertent warming of small fragments of the specimen) or during etching (when relatively large amounts of water vapor may be discharged from the ice present in the specimen). The presence of contamination has caused some confusion in the past regarding the interpretation of freeze-etch images (Staehelin, 1968; Staehelin and Bertaud, 1971). However, contamination of a gross sort is readily recognized and can be avoided (Staehelin and Bertaud,

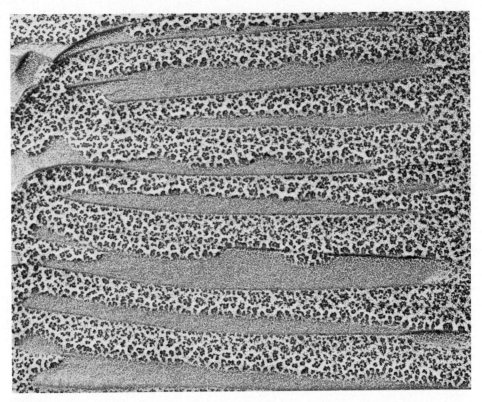

Fig. 2. Electronmicrograph of the surface of a 20% glycerol solution which has been frozen, fractured, and replicated with a film of relatively pure platinum. The heavy black spots are large platinum crystallites that have formed by migration of platinum on the surface. Intervening, more homogenous, stripes are regions of less platinum migration. The exact reasons for the differences in platinum crystallite size are unknown but may be related to the presence of ice crystals and glycerol-water eutectic mixture in the frozen solution. The large size of platinum crystallites shown here can be reduced to 10–15 Å by sufficient codeposition of carbon with the platinum to disrupt crystallite growth (replicated at $-150°C$; ×64 000).

1971; Bullivant, 1973). The possibility of a very fine coating of contaminants renders some of the smallest details in a replica of doubtful value. If contaminants are deposited evenly over the fractured surface, then correction factors might be calculated for measurements. In fact, contaminants are rarely distributed evenly and seem to have preferential sites for nucleation on specimens (Fig. 3). This may limit the reliability of comparison of sizes of small structures (100 Å or less) in freeze-etch replicas with the sizes of similar structures as seen by thin section or negative-staining techniques.

A third problem is that the fracturing of a frozen object is a crude process which appears to distort some structures in that they are plastically deformed. The most striking example of this is the severe deformation of polystyrene spheres which occurs during fracturing (Clark and Branton, 1968; Dunlop and

Fig. 3. Freeze-fracture replica of a human red blood cell membrane suspended in a 40% glycerol solution. This electronmicrograph demonstrates an outer (or "E") half-membrane with its split face (E face) appearing as a sheet on which small elevations, or "particles", are present. This E face illustrates a subtle form of contamination, which is numerous globules, 100–200 Å in diameter (at C), selectively deposited on membrane faces. Note that the surrounding glycerol-water eutectic (at G) appears free of particles. The particulate contamination is most concentrated on the far right side of the E face, whereas relatively uncontaminated E face, is present at the far left. This contamination may be condensed water vapor or oil vapor and can be minimized by proper positioning of the cold knife and rapid replication in the Balzers freeze-etch apparatus (Staehelin and Bertaud, 1971). This type of contamination mimicking particles may produce artifactually high particle counts for membranes (see text) (replicated at −100°C; ×28 000).

This electronmicrograph, and all of the other electronmicrographs in this chapter, are printed as positive images. Freeze-fracture replicas are oriented so that the direction of metal shadow-casting is approximately from the bottom of the field of view. Unless otherwise specified, specimens were frozen in monochlorodifluoromethane at −150°C.

Robarts, 1972). More pertinent to the study of biological membranes are the demonstrations by Bullivant et al. (1972) and by Bullivant (1974) that myosin and membrane proteins are deformed during fracture, although the deformation is subtle and may involve slippage of the proteins relative to the ice and frozen lipid. Another possible source of plastic deformation is intramolecular slippage, such as the unravelling of the tertiary structure of helical proteins. Plastic deformation of biological structures during fracturing appears limited by the very low temperature since complementary replicas show a close-to-perfect match of the major structures but not of some small details of the fracture face. Plastic deformation places a severe limitation on the meaning of fine measurements of sizes of structures in freeze-fracture replicas.

4. Freeze-fracture images of plasma membranes

4.1. The erythrocyte

Studies on human erythrocyte membranes have contributed greatly to our understanding of freeze-fracture images of membranes, and freeze-fracture has made very significant contributions to the understanding of erythrocyte membrane structure. When a suspension of erythrocytes is frozen and fractured, the fracture propagates through the specimen by following planes of least resistance. When a replica is prepared on such a specimen and examined, four types of fractured surfaces can be seen. The extracellular space can be clearly identified and often appears smooth. The intracellular space appears finely granular in whole cells but appears smooth in hemoglobin-free erythrocyte ghosts. Therefore, the small granules comprising the fractured surface of the erythrocyte cytoplasm have been interpreted as being hemoglobin molecules (Weinstein and Bullivant, 1967). Hemoglobin is a globular protein with an effective diameter of 65 Å (Sears et al., 1964) which, if enlarged by 20 Å by platinum shadowing, would be the same size as many of the cytoplasmic granules. Smaller granules in the erythrocyte cytoplasm are interpreted as being hemoglobin molecules that are only partially uncovered by the fracture.

At the interface between extracellular space and cytoplasm, the fracture reveals the membrane as a relatively smooth sheet on which there are numerous globular granules or "particles", that are approximately 80 Å in diameter. On close inspection, the fractured membrane has two appearances, one with numerous particles and the other with only 1/4 to 1/5 the number of particles (Fig. 4).

The particles are present in approximately the same numbers in whole cells and in hemoglobin-free ghosts (Weinstein, 1969). Therefore, the particles are associated with the sheet-like membrane, although their precise relationship to the membrane has been the subject of some controversy in the past. Initially such membrane particles were considered to reside either on the true outer and inner surfaces of the membrane (Moor and Mühlethaler, 1963; Branton

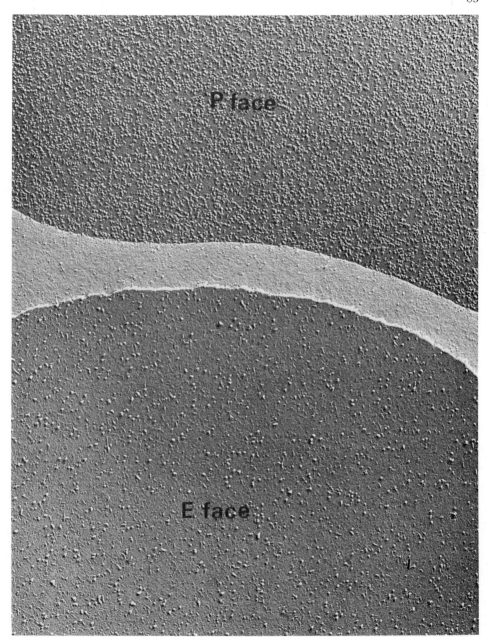

Fig. 4. Electronmicrograph of a freeze-fracture replica of human erythrocyte membranes. At the upper portion of the figure, one membrane has been split to reveal the inner half-membrane ("P" half-membrane) and its P face. At the lower portion, another membrane has its outer half-membrane (E half-membrane) exposed along its E face, according to terminology suggested by Branton et al. (1975). The P face has more intramembrane particles, 80–100 Å in diameter, than the E face. Also P face particles appear to belong to a relatively homogeneous population whereas the E face particles are heterogeneous in size and shape (40% glycerol; replicated at $-115°C$; ×60 000).

and Moor, 1964; Mühlethaler et al., 1965; Weinstein, 1969) or within the membrane (Branton, 1966). Now the particles are thought to reside within the interior of the membrane and to be exposed when the fracture splits the membrane into two lamellae or "half-membranes". Acceptance of this interpretation is based on complementary replicas which demonstrate that, wherever there is a half-membrane appearance above the fracture plane, a complementary half-membrane appears below the fracture plane (Steere and Moseley, 1969; Chalcroft and Bullivant, 1970; Wehrli et al., 1970; Weinstein et al., 1970a). The exact position of the fracture plane within membranes is probably down the center of a lipid bilayer, as shown by the experiments of Deamer and Branton (1967) on artificial phospholipid bilayers in ice and by the fact that plasma membranes appear to split into two lamellae of approximately equal thickness in most regions (Branton, 1966, 1969; McNutt and Weinstein, 1970). A membrane may be split into two lamellae but the thickness of each lamella will depend on the amount of additional material associated with either the inner or outer lamellae. The occasional suggestions that the membrane may be split asymmetrically probably should be interpreted as reflecting an asymmetry in material associated with a lipid bilayer rather than as a fracture passing outside a lipid bilayer (Chalcroft and Bullivant, 1970; McNutt and Weinstein, 1970). Regardless of whether or not such asymmetry is present, it is common for authors to refer to the two lamellae as half-membranes (Fisher, 1976).

The terminology for the lamellae has varied widely and recently fourteen authors have joined efforts to suggest a uniform terminology (Branton et al., 1975). The terms are relatively simple for plasma membrane and are based on giving a letter designation to each half-membrane (Fig. 5). The inner half-membrane remains attached to the frozen cytoplasm during fracture so that this inner half-membrane is called the protoplasmic half-membrane or the "P" half. The outer half-membrane remains attached to the extracellular space and is called the exoplasmic or "E" half-membrane, following the concept of an

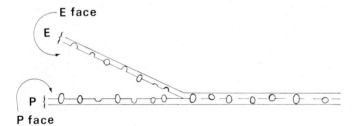

Fig. 5. Idealized schematic drawing of one interpretation of membrane splitting during freeze-fracture. The membrane is split into two halves, a P half (inner half) and an E half (outer half). Two surfaces, called faces, are generated by the fracture, i.e. the P face and E face. It is customary to consider that fracturing produces complementary P and E faces as shown here. However, an exact matching of particles on one face with depressions on the other face is not evident in most replicas of nonjunctional plasma membrane (see text).

exoplasmic space as proposed by de Duve (Branton et al., 1975). The true inner or juxtacytoplasmic surface of the membrane is called the P surface and is labelled "PS". The true outer or extracellular surface is called the E surface and is labelled "ES". The artificial surfaces generated by membrane splitting are called "faces" to distinguish them from true membrane surfaces. The split face on the inner half-membrane is called the P face and labelled "PF". The split face on the outer half-membrane is called the E face and labelled "EF". Earlier terminology varied, but a widely used system referred to the "A" and the "B" faces of the membrane (McNutt and Weinstein, 1970). The A face is equivalent to the P face and the B face is equivalent to the E face in the new terminology.

The particles present on the P and E faces of the membrane have been the subject of numerous studies. Before discussing these studies, the evidence must be examined which indicates that the particles represent real structures, especially in the light of the foregoing discussion of shadow-casting metal nucleation, contamination and plastic deformation artifacts of freeze-fracture. One basic test of the reality of particles is whether the number of particles is consistent for a given membrane fracture face. Independently, different laboratories using different freeze-etch techniques to study human erythrocyte membranes have published results on particle counts that are in close agreement for P face particles but not in agreement for E face particles (Table I). Most authors agree on a mean of 2600 to 2800 particles per square μm of human erythrocyte P face (Branton, 1969; Weinstein, 1969; Lessin et al., 1972; Kirk and Tosteson, 1973; Parish, 1975). In contrast the same authors obtained mean values ranging from 540 to 1400 particles per square μm for the E faces of human erythrocytes. The exact reasons for the rather wide discrepancies in these quantitative data are not clear, but the disparity suggests that the number of particles observed depends in part on the specimen preparation procedure. From the publications cited, one cannot be certain what aspect of the preparation procedure might be responsible for the differences. In studies of endothelial cell plasma membranes chemical fixation with glutaraldehyde has

TABLE I

MEAN PARTICLE COUNTS ON HUMAN ERYTHROCYTE MEMBRANES

Particles (per square μm)			Reference
P face	E face	Total	
2600	575	3175	Weinstein (1969)
2700	540	3240	Lessin et al. (1972)
2600	1200	3800	Parish (1975)
2658	1218	3876	Kirk and Tosteson (1973)
2800	1400	4200	Branton (1969)
3800	1400	5200	Meyer and Winkelmann (1969)
...	...	4500 (?)	Tillack et al. (1972)

been shown to alter the number of particles on some fracture faces apparently by affecting their "polarity", that is, by affecting whether they remain attached to the P or the E face (Dempsey et al., 1973). However, in a recent study of glutaraldehyde-fixed human red cells, Parish (1975) found that the number of particles per square μm of P and E faces was the same in buffer-washed and glutaraldehyde-fixed red cells. Therefore glutaraldehyde fixation is unlikely to be the cause of the diverse particle counts found on human red cell membranes. The glycerination protocols might introduce some of the variability since Kirk and Tosteson (1973) found a marked reduction in E face particle density as a consequence of glycerination. However, the highest E face particle counts without glycerol (approximately 1200 particles per square μm) reported by Kirk and Tosteson (1973) are equal to the counts of others using glycerination as cryoprotection. Consequently glycerination seems unlikely to account for all of the observed variation.

As noted by Weinstein (1974a) the discrepancies in particle counts raise the problem of defining of what constitutes a particle. In the published reports of particle counts the investigators have not defined exactly what sort of elevation is included as a particle and what elevations are excluded from the count. This apparently elementary consideration is a more difficult one than it seems because the exact dimensions of particles depend on replica thickness, angle of shadowing, and viewing angle. With high angle shadowing, very small elevations are lost in the replica thickness, whereas at low angles of shadowing, small elevations are visible (Pinto da Silva and Miller, 1975). Obviously, if the angle of viewing duplicates the angle of shadowing, the appearance of small elevations is lost. For the erythrocyte P face, the particles are small globular elevations above the background plane of the membrane. The particles measure approximately 80–100 Å in diameter depending on replica thickness. In paired photographs for stereoscopic views, the P face particles generally appear to be about as tall as they are wide. There is a striking uniformity of the particles on the P face of the erythrocyte membrane (Fig. 4) and most authors are in agreement on the number of particles per square μm. However there is a remarkable heterogeneity of the particles on the E face (Fig. 4). The heterogeneity in size and shape of E face particles presents a major problem in the definition of a particle. It is possible that shadowing and viewing angle considerations account for some of the variability in E face particle counts. It is this author's opinion that more attention should be paid to the observations of Bullivant (1969, 1973) that the number of particles is reduced by increasing the protection of the fracture face between the fracturing and replication steps. Also Deamer et al. (1970) and Staehelin and Bertaud (1971) clearly demonstrated that contamination of fracture faces can mimic particles (Fig. 3). Consequently, excessively high counts of particles on membranes are suspect of being high due to additional particles that represent contamination.

Despite these problems, most investigators consider the particles to reflect real structures and, in some cases, this is assuredly true. Another test of the reality of particles is to determine whether a globular elevation on one membrane face has a matching depression on the complementary fracture face or whether the particles undergo plastic deformation during fracturing (Fig. 6). In practice, this is an extremely difficult technical feat due to the high magnification analysis that is required. Steere (1973) and Bullivant (1974) have published complementary replicas that demonstrate variability in whether the particles on the P face have matching structures on the E face. Bullivant (1974) has shown clearly that a particle on the P face of the erythrocyte membrane may have another particle, a depression, or a smooth surface in the matching location on the E face. Bullivant (1974) interprets his data as being strong evidence that some of the particles are plastically deformed during fracturing but also admits that some of the particles may represent contamination. Flower (1973) has shown that certain invertebrate plasma membranes fracture so that E faces have enough depressions to be complementary to the P face elevations while other plasma membranes do not show such complementarity, even when processed with the same techniques. He attributed the differences to be due to different degrees of plastic deformation of the outer half membrane and its E face (Flower, 1973). What comes through clearly from these studies is that even the most careful studies have difficulty in proving the reality of those particles that lack complementary structures but they also cannot prove them to be total artifacts. It is encouraging, however, that with improvements in freeze-etch techniques greater complementarity is found between P and E faces of membranes.

A considerable amount of evidence indicates that the number and distribution of particles reflects the number and distribution of a portion of the membrane proteins. Branton (1969) noted that myelin has a low protein content and has no particles, whereas other membranes, such as those of chloroplasts, have a high protein content and abundant particles. This general correlation has been confirmed in numerous investigations. Recently, with improved replication techniques, small particles have been observed on myelin fracture faces indicating that one of the myelin proteins may have a structural representation in the center of the membrane (cf. Bischoff and Moor, 1967a,

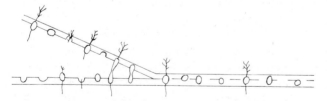

Fig. 6. Schematic drawing of membrane splitting accompanied by plastic deformation. During fracturing, several types of appearance may be generated depending on the amount of slippage of membrane proteins relative to the surrounding matrix of lipid, ice, glycerol-water eutectic, and other proteins. A given intramembrane protein may remain with either the E half or P half-membrane. Under some conditions, a complementary depression may remain or may be filled partially to completely with a portion of the protein. Also a degree of uncoiling of the tertiary structure of the membrane protein may occur, so that a particle on one face has another particle on the matching location of the opposite face.

1967b; Vail et al., 1974; Pinto da Silva and Miller, 1975). Only a portion of the total membrane protein appears to be responsible for the appearance of membrane particles. On incubation with pronase, up to 70% of membrane protein was removed from erythrocytes before the number of particles began to decrease (Branton, 1971). Similar results have been obtained with trypsin (Speth et al., 1972). Treatment of erythrocyte ghost membranes with the detergent, lithium diodosalicylate, extracts the major glycoprotein of the membrane, called glycophorin (Marchesi et al., 1972), which is one of the constituents of the intramembrane particles. Glycophorin has a molecular weight of approximately 55 000 daltons, is 40% protein and 60% carbohydrate, and has a considerable amount of sialic acid that would contribute negatively charged groups to the cell surface. Glycophorin exhibits MN blood group moieties and also has receptors for influenza virus, phytohemagglutinin, and wheat germ agglutinin. Cationic ferritin bound to anionic sites (at pH 5.5) and ferritin-labelled antibody to blood group A antigens are localized on the surface of erythrocytes in a distribution resembling that of the intramembrane particles (Pinto da Silva et al., 1971, 1973). In trypsin-treated erythrocyte ghosts, the glycophorin particles migrate into clusters on the P face and leave behind large smooth areas with few particles. When influenza virus or phytohemagglutinin are bound to the true surface of trypsin-treated erythrocytes, the binding sites for these markers are clustered and overlie the intramembrane particle clusters (Tillack et al., 1972). When erythrocyte ghost membranes with clustered particles are sonicated, particle-rich and particle-poor vesicles are formed and the particle-rich vesicles contain more glycoprotein than particle-poor vesicles (Di Pauli and Brdiczka, 1974). Furthermore, surface labelling studies have shown that the glycoprotein receptor for concanavalin A (Con A) follows the distribution of intramembrane particles (Pinto da Silva and Nicolson, 1974), although some workers have not confirmed these results (Bächi and Schnebli, 1975). The receptor for Con A is found not on glycophorin but instead on a minor glycoprotein present in Band III in the standard nomenclature for the bands produced by electrophoresis of erythrocyte membrane proteins in sodium dodecyl sulfate gels (Findlay, 1974). Thus the particles on the P face may correspond to at least two separate membrane proteins, a Band III component and glycophorin (Pinto da Silva and Nicolson, 1974). A working hypothesis is that each individual intramembrane particle contains both glycophorin and the Band III component, since: (1) two populations of particles are not apparent; (2) the number of Band III component molecules and glycophorin molecules are each approximately equal to the number of particles seen by freeze-etching; and (3) the size of the hydrophobic segment of glycophorin is not large enough to account for a particle 85 Å in diameter (Pinto da Silva and Nicolson, 1974). Attempting to relate the size of a particle to the size of the hydrophobic portion of a molecule not only requires knowledge of variables such as replica thickness, but also assumes that fracturing occurs without slippage of the molecules relative to the membrane and ice. Slippage of glycophorin during fracturing could cause the hydrophilic portion of glycophorin to protrude from the membrane face along with the hydrophobic portion so that both would contribute to particle size. Evidence of plastic deformation of erythrocyte particles during fracturing (Bullivant, 1974) indicates that the fracture faces may show more of the molecules than only the portion originally contained within the lipid bilayer.

Although Con A binding sites localize with intramembrane particles in erythrocytes, there is variability in whether Con A distribution follows intramembrane particle distribution in other cells. Multivalent concanavalin A induces clustering of its binding sites on many cells (Rosenblith et al., 1973). In mouse plasmacytoma cells, multivalent Con A also induces clustering of intramembrane particles and this correlates with agglutinability with Con A (Guérin et al., 1974). However in other cells, e.g. 3T3 fibroblasts and *Entamoeba histolytica*, clustering of Con A binding sites is not accompanied by clustering of intramembrane particles (Pinto da Silva and Martínez-Palomo, 1975; Pinto da Silva et al., 1975). Similarly, the phytohemagglutinin receptor is associated with the intramembrane particles in erythrocytes (Tillack et al., 1972) but in platelets there is no correlation between phytohemagglutinin binding and intramembrane particle distribution (Feagler et al., 1974). Unfortunately, the morphologic appearance of intramembrane particles is not sufficiently distinctive to allow an accurate identification of a molecular species. Consequently, studies on the distribution of specific molecules in one cell type cannot be applied readily to other cell types. Even glycophorin is not recognizable on the basis of its morphology in the red blood cell when compared to purified glycophorin incorporated into purified phospholipid vesicle membranes (Grant and McConnell, 1974). The reason for the dissimilar appearance of glycophorin in red cells and in phospholipid vesicles will require further experimentation but seems likely to be due to the association of glycophorin with other membrane components in intact membranes (Grant and McConnell, 1974). Segrest et al., (1974) studied a hydrophobic fragment of glycophorin, prepared by tryptic digestion, whose sequence of 35 amino acid residues is known. It is this segment of the glycophorin molecule which is most certain to contribute to intramembrane particles and, when added to lecithin liposomes, forms particles approximately 70–80 Å in diameter (Segrest et al., 1974). The number of such particles is dependent on the concentration of hydrophobic peptide. In contrast to the particles of purified glycophorin visualized by Grant and McConnell (1974), the particles formed by only the hydrophobic fragment of glycophorin in lecithin do resemble in situ intramembrane particles (Segrest et al., 1974). The 70–80 Å particles formed by the hydrophobic fragment were multimers of approximately 25 monomers per particle. It is not known how this number of monomers relates to intramembrane particles in situ. Similarly, Vail et al. (1974) found that a hydrophobic protein isolated from myelin produces particles when inserted into liposome membranes, but the particles appear rather large (85–100 Å) compared to the somewhat smaller particles observed on fracture faces of myelin by Pinto da Silva and Miller (1975), who did not state sizes for what they describe as particles and "subparticles". Such size differences probably reflect different states of aggregation of the monomeric units making up the individual particles.

Relatively little attention has been given to structures visible on the E face of fractured erythrocyte membranes. These structures may represent separate entities from those visible on the P face or they may simply be complementary or plastically deformed portions of the particles visualized on the P face. Several observations suggest that some E face structures are not simply portions of P face structures. This is particularly evident in studies of plasma membranes other than erythrocytes (McNutt, 1970, 1975), but is also true for erythrocytes. Weinstein (1969) and others (McNutt, 1976) noted a filamentous appearance to the E faces of human erythrocyte membranes in some preparations (Fig. 7). Since the P faces lack such filaments and lack complementary grooves, these E face filaments apparently arise by plastic deformation of cytoplasmic filaments which are firmly attached to the plasma membrane. The filaments on the E face are approximately 80 Å in diameter and of variable length up to 500–600 Å. Since the cytoplasmic fibrous protein "spectrin" (Marchesi et al., 1969, 1970) is firmly attached directly or indirectly to glycophorin and another very closely associated membrane protein (Nicolson and Painter, 1973; Ji and Nicolson, 1974), the obvious question is raised: is spectrin plastically deformed during fracture to produce E face filaments? One of the difficulties in answering this question is that a filamentous appearance is uncommon for erythrocyte E faces and great variation is found within a single blood sample as well as from sample to sample. Willison (1975) observed that the amount of plastic deformation of a protein is dependent on the orientation of its bonding forces relative to the fracture plane. He was able to determine this on cytoplasmic crystals of "fraction 1" protein in chloroplasts since if the fracture occurred in a square packed plane of the crystal, plastic deformation was inapparent. If, however, the fracture passed through a hexagonally packed plane of the crystal, the protein was stretched into strands, often over 1000 Å in length. Presumably this stretching caused unravelling of the tertiary coiled structure of the protein. Alternatively, since the strands were close to the diameter of the protein particles themselves, the proteins could have slipped relative to each other in the crystal without major unravelling of the polypeptide chains. This evidence suggests that the precise angle of fracture may be one important determinant that generates the great variation observed in filamentous structures on E faces of human erythrocytes. Further evidence in favor of this interpretation is that E face filaments are most commonly observed in areas where the fracture plane changes abruptly from the cytoplasm to the intramembranous location so that the fracture plane passes almost parallel to filaments attaching to the membrane. In such regions, the filamentous strands appear to originate from the cytoplasm and possibly represent plastically deformed intracellular cytoskeletal elements, e.g., spectrin or actin (McNutt, 1975).

Although freeze fracture can demonstrate attachments of filaments to membranes, it does not give any direct evidence on the identity of the

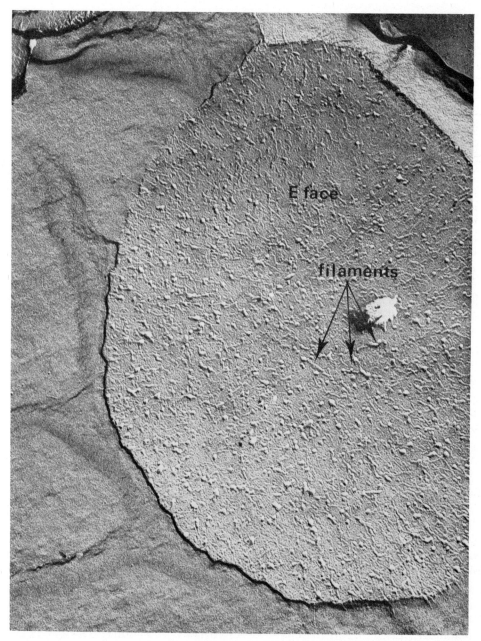

Fig. 7. Freeze-fracture replica of an erythrocyte E face. In occasional preparations of human erythrocytes, a small proportion of E faces have filamentous strands of varying diameter (approx. 60–120 Å) and of varying length. These filaments do not have complementary grooves on P faces and appear to arise by plastic deformation of intramembrane proteins or cytoplasmic proteins bound to the membrane (20% glycerol; replication at −150°C; ×73 800).

molecules attaching to the membrane or how these attachments relate to the distribution of the intramembrane particles seen on the erythrocyte P face. However, interesting freeze-fracture experiments have been devised to give indirect evidence that spectrin and actin attach to the intramembrane particles and stabilize their position within the plane of the membrane (Elgsaeter and Branton, 1974; Elgsaeter et al., 1976). These freeze-fracture experiments support the earlier work, using other techniques, which indicated that spectrin is attached to the inner surface of the membrane (Nicolson et al., 1971) and that clumping of spectrin with antispectrin antibody causes a clustering of anionic sites on the outer surface of the erythrocyte membrane (Nicolson and Painter, 1973). From these studies it was considered very likely that spectrin is attached to intramembrane particles since the anionic sites studied were sialic acid, a component of glycophorin, and therefore of the erythrocyte intramembrane particles (Nicolson and Painter, 1973). Recently freeze-fracture has shown this interpretation to be correct (Elgsaeter et al., 1976). For example, freeze-fracture studies have shown that the intramembrane particles in washed erythrocyte ghosts exhibit a reversible clumping which is pH dependent and shows an optimum between pH 4.5 and 5.0 (Pinto da Silva, 1972; Elgsaeter and Branton, 1974). Elgsaeter and Branton (1974) demonstrated that a substantial amount of spectrin (and actin) had to be removed from the membrane during ghost washing before the intramembrane particle aggregation at pH 5 could be observed. Recently Elgsaeter et al. (1976) have observed that, at pH 5, an isoelectric precipitation of any residual spectrin and actin would be expected to occur on the cytoplasmic surface of the erythrocyte ghost. They have demonstrated that other conditions which would be expected to aggregate spectrin and actin result in aggregation of intramembrane particles but they could not exclude a direct effect of the experimental conditions on the intramembrane particles themselves (Elgsaeter et al., 1976). The results of both freeze-fracture and other studies indicate that the binding and distribution of spectrin on the inner surface of the membrane exerts an important influence on the stability and distribution of intramembrane particles within the plane of the membrane.

A different application of the freeze-fracture procedure has been to obtain half-membrane fragments for microchemical analysis of the distribution of some membrane components (Fisher, 1976). Outer half-membrane fragments can be harvested after binding human erythrocytes to polylysine-coated coverslips, freezing, and fracturing off the coverslip. During fracturing, those membranes closest to the coverslip are split, leaving outer half-membranes attached to the coverslip. Fisher (1976) has reported, with this technique, that more cholesterol is present in the outer half-membrane than the inner half-membrane. These results shed light on the earlier observations of Seeman et al. (1973) that saponin-treated erythrocytes show holes in the outer half-membrane (visible by deep etching) but not in the inner half-membrane (as seen by fracturing). Seeman et al. (1973) speculated that the holes or pits were produced by removal of cholesterol from the membrane by saponin treatment. Also other lipid components may be distributed asymmetrically in the membrane (Verkleij et al., 1973; Zwaal et al., 1973). These freeze-fracture results point out the usefulness of the half-membrane harvesting technique for the further analysis of membrane asymmetry.

4.2. Other cells

The plasma membranes of many mammalian cell types have been freeze-fractured and their P and E faces described. They all have intramembrane particles that bear a general resemblance to those found in erythrocyte membranes but the numbers of particles vary with the type of cell. In addition, the junctions between cells in tissues have a specialized appearance and will be discussed separately. The studies of non-junctional membrane in cells primarily have been descriptive but hold much promise in the future for increasing our knowledge of plasma membrane functions as experiments are devised to study the particles in each cell type. A selected list of recent examples of some of the mammalian cell types studied includes: skeletal muscle (Rayns et al., 1975; Ellisman et al., 1976); cardiac muscle (Rayns et al., 1968; Sommer et al., 1972; McNutt, 1975); smooth muscle (Devine and Rayns, 1975); lymphocytes (McIntyre et al., 1974; Wunderlich et al., 1974); platelets (Ruska and Schulz, 1968; Hoak 1972; Feagler, et al., 1974); pancreatic islet cells (Orci et al., 1974); intestinal epithelial cells (Staehelin, 1974); kidney epithelial cells (Kuhn and Reale, 1975; Kuhn et al., 1975; Pricam et al., 1975; Ryan et al., 1975; Caulfield et al., 1976); spermatozoa (Friend and Fawcett, 1974; Stackpole and Devorkin, 1974); skin (Breathnach et al., 1972, 1973b; Orwin and Thomson, 1973); vascular endothelium (Simionescu et al., 1974); retina (Clark and Branton, 1968; Raviola and Gilula, 1975). In addition, studies on these cell types in tissues have yielded data of general interest concerning pinocytosis vesicles and specialized orthogonal arrays of particles.

Pinocytosis vesicles (Fig. 8) are flask-shaped invaginations of the plasma membrane (Palade, 1961) which consist of a spherical-shaped region approximately 600 to 1000 Å in diameter connected to the plasma membrane by a short neck approximately 200 Å in length and 400 to 600 Å in diameter. They are widely accepted as being involved in the transport of large molecular weight substances from one side to the other side of endothelial cells in capillaries of the type found in skeletal muscle (Bruns and Palade, 1968a, 1968b). They are also found in small numbers in many cells, e.g., at the basal surface of keratinocytes in skin, on the sarcolemma of skeletal, cardiac and smooth muscle, and the blood front of hepatocytes. They are readily observed in freeze-etch replicas. Most frequently the fracture plane breaks through the necks of vesicles rather than passing around vesicles (Fig. 9). Occasionally the fracture splits entire vesicle membranes, leaving vesicles projecting from the E half membrane (Fig. 9). Also, vesicle membranes may be cleaved to show longitudinal profiles. In their involvement in transport, pinocytosis vesicles present interesting problems related to the fusion of membranes, e.g., what are the determinants of vesicle formation? How do cytoplasmic vesicles ap-

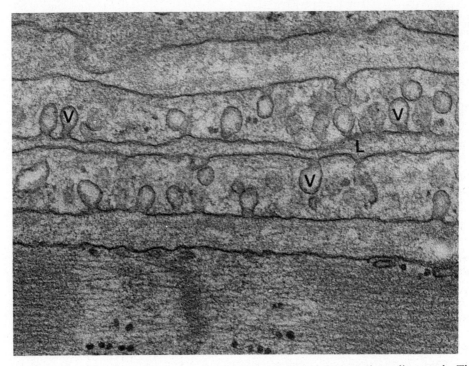

Fig. 8. Electronmicrograph of a thin section through a capillary in guinea pig cardiac muscle. The plasma membrane of the endothelial cells have numerous invaginations, named pinocytosis vesicles (at V). Each vesicle has an approximately spherical portion which is connected by a short neck to the planar region of the membrane. Immersion fixation in 3% glutaraldehyde, allowed the lumen of the capillary to be reduced to a narrow slit (at L) (×60 000).

proaching the membrane fuse with the membrane? What receptors are involved? Evidence relating to these questions was obtained by Satir et al. (1973) who observed that intramembrane particles formed a 'rosette' or cluster in the plasma membrane prior to fusion with the membrane of mucocysts in the protozoan *Tetrahymena pyriformis* during mucocyst discharge. These data suggest that particle clusters may be important in membrane fusion. Orci and Perrelet (1973) found that intramembrane particles were approximately three times more abundant in regions of pinocytotic vesicles than in regions without such vesicles in the plasma membrane (sarcolemma) from smooth muscle cells in intestine. While this study of Orci and Perrelet (1973) may indicate that intramembrane particles are involved in pinocytosis vesicle activity, an alternative interpretation must be considered. The particle density observed by Orci and Perrelet (1973) is not very high, approximately 1800 particles per square μm of P face of membrane with vesicles. The area called nonpinocytotic region of the smooth muscle sarcolemma corresponds in thin sections to the region of attachment of the actomyosin system to the membrane. It has been noted in cardiac muscle that regions of attachment of actin to the plasma

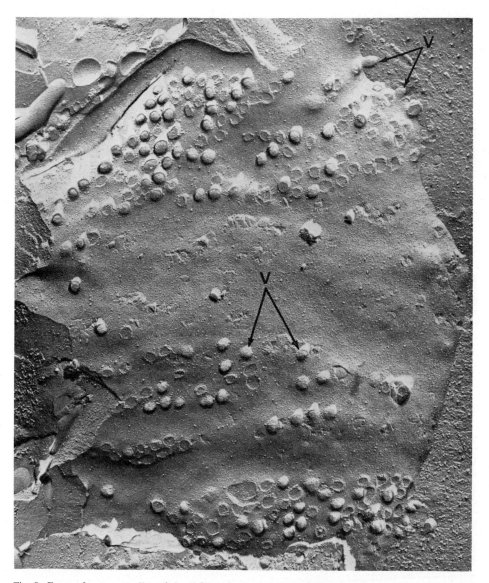

Fig. 9. Freeze-fracture replica of the E face of a human smooth muscle cell membrane. Pinocytosis vesicles (at V) are numerous and project from the membrane face. In many areas, the vesicles have been fractured off the membrane where their necks connect to the planar portion of the membrane, thus producing a circular break in the E face (40% glycerol; replication at −100°C; ×45 800).

membrane have relatively few of the ordinary intramembrane particles (McNutt, 1970; McNutt and Weinstein, 1973). Consequently, the relative increase of particles observed in pinocytotic areas by Orci and Perrelet (1973) could be either a reflection of the sparsity of intramembrane particles in regions of actin insertion or an indication of particle involvement in pinocytosis vesicle activity. In mammalian cells that form secretory vesicles, the fusion of these vesicles with the plasma membrane has not been associated with rosette arrays on the plasma membrane (Orci et al., 1973b; Smith et al., 1973; Chi et al., 1975). These contrasting data suggest that the rosette formation preceding mucocyst discharge in *Tetrahymena* may be a phenomenon which does not occur in mammalian cells.

A different membrane specialization visible in high resolution replicas, is the presence of small orthogonal arrays of particles on membrane P faces and of depressions on E faces (Fig. 10). These arrays are formed by particles (approximately 40 Å in diameter) with approximately 70 Å center-to-center spac-

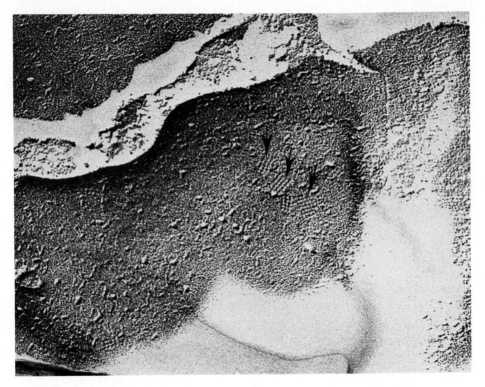

Fig. 10. Freeze-fracture replica of the E face of a plasma membrane of a mouse cardiac myocyte at high magnification. Orthogonal arrays of E face depressions (at arrows) are found on membranes of both electrically excitable and inexcitable membranes. Similar P face particle arrays can be found (but are not shown). The center-to-center spacing in the array is approx. 60 Å. Such arrays are rare in cardiac muscle membranes and have an unknown function (20% glycerol; replication at −150°C; ×160 000).

ing. Orthogonal arrays were observed first by Kreutziger (1968b) in mouse liver and have been found subsequently in plasma membranes of intestinal epithelial cells (Staehelin, 1972; Rash et al., 1974), skeletal muscle cells (Rash et al., 1974; Rash and Ellisman, 1974; Ellisman et al., 1976), astrocytes (Landis and Reese, 1974) and cardiac muscles cells (McNutt, 1975). Rash and Ellisman (1974) have noted that the orthogonal arrays are distributed on skeletal muscle plasma membrane so that the arrays are not present at the neuromuscular junction but are found in other regions. This distribution correlates with the zones of electrical excitability of the membrane. However, Heuser et al., (1974) have found similar orthogonal arrays at the neuromuscular junction in frog sartorius muscles. Ellisman et al. (1976) have reported that orthogonal arrays are more abundant in the sarcolemma of fast twitch than in the sarcolemma of slow twitch skeletal muscle. Also, similar orthogonal arrays have been seen on the nonexcitable membranes of glial cells (Landis and Reese, 1974; Hanna et al., 1976). The function of these orthogonal arrays remains undetermined.

Another interesting specialization is found only in the plasma membrane of those epithelial cells facing onto the lumen of the urinary bladder. Unique rigid plaques, composed of hexagonally arrayed large particles, form approximately 73% of the plasma membrane surface (Staehelin et al., 1972). Although these plaques had been described previously in thin section and negative stain preparations (Porter et al., 1967; Vergara et al., 1969; Hicks and Ketterer, 1970), freeze-etch preparations have demonstrated clearly that the plaques not only are at the membrane surface but also extend through the lipid bilayer and are anchoring sites for cytoplasmic filaments (Staehelin et al., 1972). It is speculated that these plaques and their attached filaments are structural reinforcements that protect the cell against tears in the membrane during maximal distention of the bladder (Staehelin et al., 1972).

The numbers and distributions of intramembrane particles on the general plasma membrane surfaces of cancer cells (Weinstein, 1974b) and virus-transformed cells in culture have attracted considerable attention (Scott et al., 1973; Torpier et al., 1975). Initially it was reported that 3T3 fibroblasts had randomly distributed intramembrane particles in cultures of low cell density but, with increasing cell density and cell-to-cell contact, the general intramembrane particles became aggregated (Scott et al., 1973). In contrast, SV40 virus-transformed 3T3 fibroblasts did not show this response to cell contact and had randomly distributed intramembrane particles in high density cultures, with numerous cell-to-cell contacts. The implication was made that these results were not altered by chemical fixation with 1% glutaraldehyde (Barnett et al., 1974). However, these results were not in agreement with previous freeze-fracture studies of normal and transformed cells in culture (Revel et al., 1971; Pinto da Silva and Gilula, 1972). Recently, Pinto da Silva and Martínez-Palomo (1975) have published that they were unable to confirm the results of Scott et al. (1973) and that particle aggregation could be produced by incubation of either 3T3 or SV40 transformed 3T3 cells in saline. Gilula et al. (1975) have demonstrated that results similar to those of Scott et al. (1973) could be obtained by glycerination of unfixed 3T3 cells but not if the cells were adequately prefixed with glutaraldehyde or if the cells were frozen without glycerol or glutaraldehyde. Consequently, the results of Scott et al. (1973) appear not to show a real distribution of intramembrane particles but instead to reflect the ability of glycerol to induce intramembrane particle aggregation in untransformed cells (Gilula et al., 1975). The basis of this glycerol-induced particle aggregation requires further experimentation but appears likely to be related to the presence of an extensive microfilament system beneath the plasma membranes of untransformed 3T3 cells in confluent culture and less extensive microfilaments in transformed cells (McNutt et al., 1971, 1973). This microfilament system might influence intramembrane particle distribution in 3T3 cells (cf. Poste et al., 1975) in a fashion similar to that shown for the spectrin-actin system of erythrocytes (Elgsaeter et al., 1976).

5. Intercellular junctions

The study of the structure of intercellular junctions has advanced dramatically from applications of freeze-fracture techniques. Perhaps it is because intercellular junctions generally represent only a small fraction of the total cell surface area that freeze-fracture has contributed a great amount of information. The reader is referred to several recent reviews that deal with structure-function correlations in intercellular junctions (McNutt and Weinstein, 1973; Gilula, 1974; Staehelin, 1974) and discuss data from techniques other than freeze-fracture.

In brief, the category of intercellular junctions can be subdivided into symmetrical junctions and asymmetrical junctions. The symmetrical junction has apparently equal contributions from each cell of the pair and the junction has a mirror-image symmetry about a central plane. In an asymmetrical junction, both cells participate but their contributions differ and mirror-image symmetry is lacking in the fully formed junction. Examples of the symmetrical types are desmosomes (macula adherens), tight junctions (macula or zonula occludens), gap junctions (macula communicans or nexus) and "intermediate" junctions (similar to fascia adherens). Examples of asymmetrical junctions are the chemical synapses between neurons and neuromuscular junctions.

Some types of intercellular junctions, visible in thin sections, do not have a recognizable counterpart in freeze-fracture preparations, e.g., the "septate-like junction" described in thin sections of adrenal epithelial cells (Friend and Gilula, 1972a). The name spacing junction has been suggested for this structure since it lacks an intramembrane specialization whereas the septate junctions found in invertebrates have prominent linear arrays of particles (Staehelin, 1974).

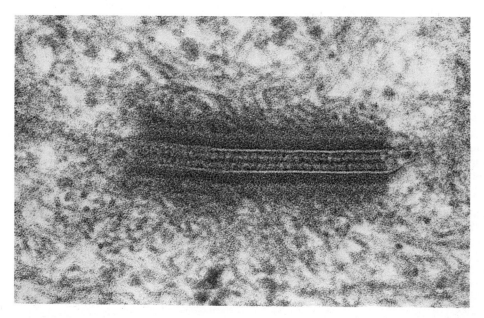

Fig. 11. Electron micrograph of a thin section through a desmosome in the stratified squamous epithelium of mouse esophagus at high magnification. At a desmosome the adjacent plasma membranes are separated by a 250 Å cleft filled with proteinaceous material, which has a central dense stratum. Cytoplasmic tonofilaments attach to dense plaques on the cytoplasmic surfaces of the membranes (× 190 000).

5.1. Symmetrical junctions

5.1.1. Adherens junctions (desmosome, macula adherens)

The desmosome is a round to oval region of attachment of two plasma membranes which are 25–35 nm apart and have electron-dense material attached to both cytoplasmic and extracellular surfaces (reviewed by McNutt and Weinstein, 1973; Staehelin, 1974) (Fig. 11). The dense material within the extracellular cleft between the membranes usually has a central plane of maximum condensation called the central dense stratum. In some thin-section preparations, fine filaments have been visualized connecting the central dense stratum to each of the membranes (Kelly, 1966; Rayns et al., 1969). On the cytoplasmic surface of the membrane is an amorphous dense plaque which is the site of attachment for tonofilaments (Kelly, 1966; Skerrow and Matoltsy, 1974a). The proteins in desmosomes isolated from bovine epidermis have approximately 50% of their structure as nonpolar amino acid residues (Skerrow and Matoltsy, 1974b), implying that these proteins may be anchored in the hydrophobic lipid interior of the membrane.

Freeze-fracture preparations show that the internal region of the membrane is specialized at desmosomes (McNutt and Weinstein, 1973; Orwin et al., 1973b). In glutaraldehyde-fixed tissue, both the P and E faces of desmosome membranes are round to oval areas, 0.1 to 0.5 µm in diameter, having a granular to fibrillar texture (Fig. 12). The granules and fibrils are approximately 80–100 Å in diameter and the fibrils are irregular in length, some up to 400 Å. The granules and fibrils probably represent short segments of fibrils penetrating into the interior of the membrane from either the cytoplasmic or extracellular surfaces of the membrane, or both. In regions of such irregular texturing, it is difficult to know whether individual granules and filaments have complementary depressions; however, the general appearance suggests that the filaments may be plastically deformed during fracturing. When the membranes are fractured obliquely, filamentous strands can be seen stretching from the

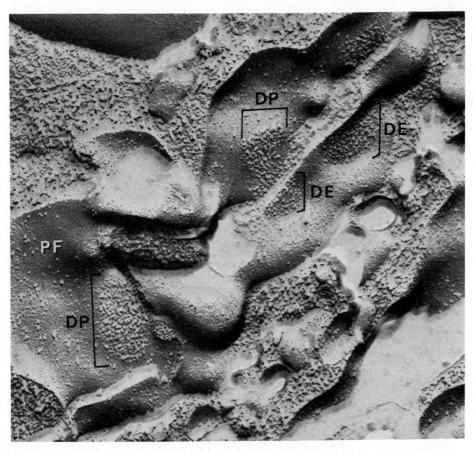

Fig. 12. Freeze-fracture replica of the plasma membranes of cells in the stratified squamous epithelium of mouse esophagus, at a lower magnification than Fig. 11. On the P face of one cell membrane (at PF), desmosomes are evident as round to oval patches of granules and short filaments that are approximately 100 Å in diameter (at DP). In chemically fixed tissue, the E faces of desmosomes (at DE) are similar to the P faces in the lower layers of this epithelium. Proteins within the interspaces are not defined by freeze-fracturing by standard methods (×71 400).

cytoplasm to the E face. In cross-fractured desmosomes, the central dense stratum occasionally is visible in etched preparations (Staehelin, 1974), but attempts to visualize any filaments connecting the central dense stratum to membranes have been disappointing.

In some epithelia, desmosomes are very prominent on the membrane fracture faces (McNutt and Weinstein, 1973) while in other epithelia the fracture faces of desmosomes are almost inapparent (Friend and Gilula, 1972b; Staehelin, 1974).

The desmosome fracture faces are changed by chemical fixation with glutaraldehyde. In unfixed desmosomes, the particles and granules generally are more prominent on the E face than on the P face. In fixed desmosomes, the number of particles and granules is approximately equal in the two faces (Breathnach et al., 1972; McNutt and Weinstein, 1973; Staehelin, 1974). The polarity of desmosome particles changes as the cells move from the basal layers of the epidermis into the stratum corneum (Breathnach et al., 1973a). In unfixed human skin, the majority of particles and granules are on the E face in basal layer but, in the stratum corneum, are distributed on both faces with a preference for the particles to appear on the P face (Breathnach et al., 1973a). This change of polarity has not been related to any changes in cell adhesion and its significance is unknown.

5.1.2. Occludens junctions (zonulae and maculae occludentes; tight junctions)
At an occludens junction in thin sections, the adjacent cell membranes are in such intimate contact that extracellular space is excluded (Fig. 13). The region of occlusion generally is a linear domain of varying length. If the linear domains form a continuous belt completely encircling the cell, then the resulting junction is called a zonula occludens and represents a major barrier to diffusion of substances through the extracellular space (Farquhar and Palade, 1963). When the linear domains do not form a continuous belt around the cell, the individual linear domains are often named macula occludens. Owing to their shape, maculae occludens do not prevent extracellular tracers from diffusing around or bypassing the junctional domain (Karnovsky, 1967, 1968).

Farquhar and Palade (1963) found that high resolution images of occludens junctions showed that membrane structure is modified by the loss of the outer leaflet of the trilaminar "unit" membrane at the sites of contact. Freeze-fracture reveals unique and dramatic specialization of membrane structure which is similar at both maculae and zonulae occludentes (Fig. 14). When the membranes are split during freeze-fracture, the linear domain is found to contain a fibril approximately 80 Å in diameter. In the occludens junctions of most epithelia, several fibrils form a branching and anastomosing network within the plane of the membrane (Kreutziger, 1968b; Staehelin et al., 1969). During fracture, the fibril usually remains adherent to the P face and leaves a complementary groove in the E face of the membrane (Chalcroft and Bullivant, 1970). Occasional exceptions to this pattern have been noted, particularly

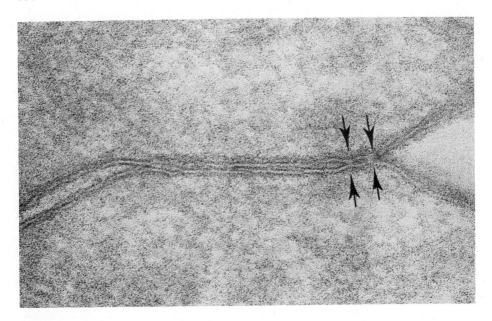

Fig. 13. Thin section of an occludens junction in mouse intestinal epithelium. At these junctions, the adjacent plasma membranes are in contact but have a substructure that is difficult to resolve. In favorable orientations, points of membrane-to-membrane contact show loss of the outer leaflets of the unit membranes (at arrows) (×270 000).

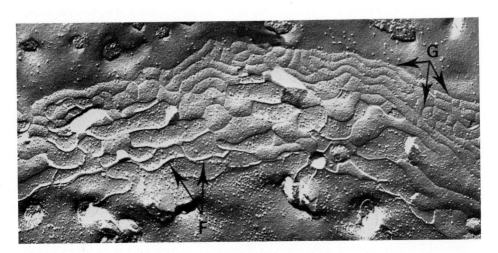

Fig. 14. Freeze-fracture replica of an occludens junction attaching the plasma membranes of adjacent epithelial cells in the human prostate. On the P face of one cell membrane, the occludens junction contains fibrils (at F) which form a network within the plane of the membrane. The E face of the other cell membrane has a similar network of grooves (at G). From the freeze-fracture replica, it can be seen that a thin section would show points of membrane contact where there are fibrils within the membrane (40% glycerol; replication at −100°C; ×41 000).

in tissue that is not fixed in glutaraldehyde. In unfixed preparations, the fibrils may break into a series of granules and short strands on both P and E faces and leave corresponding grooves in the complementary faces (Staehelin et al., 1969; Weinstein et al., 1970b). The effect of glutaraldehyde fixation in stabilizing the fibrils suggests that they contain protein (Staehelin et al., 1969).

It is somewhat controversial whether each membrane participating in an occludens junction contributes a fibril to the junction or whether a single fibril is shared between the two membranes (Chalcroft and Bullivant, 1970; Staehelin, 1973, 1974; Wade and Karnovsky, 1974). In essence, the dispute rests on whether the fibrils on the P face are sufficient in height to equal the thickness of two membranes or only a single membrane. Wade and Karnovsky (1974) have observed that the fibril on the P face appears to have a height consistent with a single fibril model. However, this does not rule out the possibility that each fibril is formed by two filaments, one contributed by each membrane. The observation of occasional fibrils with approximately half the height of the usual fibrils (Staehelin, 1973) is consistent with the "two filament" hypothesis. The single fibril data and the two filament data are not mutually exclusive if the two filaments are so strongly bonded to each other that the usual fracture leaves them attached as a single fibril.

In some tissues, the fibrils characteristically fracture so that they project from the E face rather than the P face. Examples are found in arterial endothelium (Simionescu et al., 1975, 1976) and in cells in Henle's layer of the wool follicle (Orwin et al., 1973c). In arterial endothelium, a few strands of occludens junctions must provide an effective permeability barrier, whereas many strands are necessary in columnar epithelia to produce a "very tight" zonula occludens (Claude and Goodenough, 1973). In epithelia with only a few P face strands, the trans-epithelial resistance is relatively low and the epithelium is considered to be "leaky" (Claude and Goodenough, 1973; Humbert et al., 1976). It is possible that occludens junction fibrils are more firmly bound into the membrane when the fibril appears on the E face than when it remains on the P face since the fracture would have to split less of the fibril-lipid interface when the fibril appears on the E face than when it remains on the P face (see Wade and Karnovsky, 1974). Such a hypothesis remains to be tested. Also, it has been noted that the complexity of the network of fibrils varies with the stage of secretion of some epithelia (Pitelka et al., 1973) and with the degree of mechanical distention of the lumen (Hull and Staehelin, 1976).

Recently Martínez-Palomo and Erlij (1975) have challenged the concept that the functional status of the occludens junction permeability barrier can be assessed by examination of the continuity and complexity of the fibril network in freeze-fracture replicas. Their work must be considered in relation to the studies of Wade et al. (1973) on toad bladder and to the general correlation of occludens junction complexity with "tightness" to paracellular diffusion of ions (Claude and Goodenough, 1973).

Wade et al. (1973) observed in thin sections that when toad bladder epithelium is placed in an osmotic gradient with the greater tonicity on the lumenal side large "blisters" or dilations occur in the lenticular clefts of extracellular space trapped within the meshes of the occludens junction. The formation of these blisters in the occludens junctions coincided with a drop in electrical resistance of the bladder epithelium. This effect was dependent on the nature of the solute and occurred if the gradient was produced with urea, thiourea, mannitol, or sodium chloride, but was not produced by raffinose, glycerol, or ethylene glycol. Raffinose and glycerol gradients also were stated to produce minimal effects on water flow and electrical potential difference across the toad bladder. The urea gradient was found not only to swell the interstices of the occludens junction by hydrostatic pressure but also to increase the permeability of the junction to horseradish peroxidase and barium ions. As a further extension of this work, Martínez-Palomo and Erlij (1975) examined toad bladder after treatment with hypertonic urea using freeze-fracture and confirmed that the blisters were located within the interstices of the fibrillar network of the zonula occludens. However, if bladder epithelium was treated with a lysine gradient, the transepithelial resistance fell and the occludens junctions became permeable to lanthanum, but there were no "blisters" and the array of fibrils appeared unaltered by freeze-fracture. Martínez-Palomo and Erlij (1975) also noted that the array of fibrils was approximately equal in depth and complexity in frog urinary bladder and rabbit ileum even though there is a 100-fold greater transepithelial resistance for the frog bladder. Perhaps a unifying idea is that, if the fibrillar array of an occludens junction is disrupted or consists of only a few P face fibrils, the junction is likely to be "leaky". Whereas, if the junction has a complex and intact fibrillar array, it is difficult to judge its degree of permeability to small solutes but in many instances is relatively "tight" (Claude and Goodenough, 1973).

Studies of occludens junction formation and turnover have been possible only by freeze-fracture. Orci et al. (1973a) found that in vitro proteolytic digestion with low concentrations of the crude enzyme preparation "pronase" induced the formation of extensive occludens junctions in cells of pancreatic islets. The significance of this response to proteolysis is unknown. In regard to the morphology of forming junctions, Montesano et al. (1975) have studied developing occludens junctions in fetal rat liver and have observed junction formation from the addition of particles and short fibrils onto preexisting fibril networks. Another mechanism for occludens junction formation is the de novo assembly of fibrils by fusion of particles, often in close association with gap junctions (Elias and Friend, 1976). The close association of occludens junctions and gap junctions (Friend and Gilula, 1972b) has led to speculation that the two types of junctions may have a common precursor particle (Elias and Friend, 1976). Occasionally, occludens junctions and desmosomes appear to be closely associated although the significance of this association is unknown (Orwin et al., 1973c; Alroy and Weinstein, 1976). One variant of the occludens junction resembles a primitive or partially formed occludens junction and is found between endothelial cells in both capillaries and venules. These occludens junctions consist of only a few closely approximated linear domains of specialized membrane without fibrils and with relatively few particles (Simionescu et al., 1975; Staehelin, 1975; Yee and Revel, 1975).

The disassembly of occludens junctions has been described during amphibian neurulation (Decker and Friend, 1974) where initially zonulae occludentes appear complete at 42 h of development (*Rana pipiens*, stage 12) but are lost over the succeeding 30 h as neurulation proceeds. In this tissue, the zonula occludens fragments into maculae and then the linear ridges break into a number of short segments and large particles. A comparable rapid disassembly of zonulae occludentes has not been studied in mammalian tissues. A different mode of removal of occludens junctions appears to be their interiorization on a vesicle of plasma membrane and subsequent degradation in lysosomes (Staehelin, 1973, 1974).

5.1.3. Gap junctions (macula communicans)

5.1.3.1. Small subunit gap junction (type I gap junctions; nexus). Gap junctions are regions of membrane-to-membrane contact (Fig. 15) that have been implicated in low resistance ionic coupling between cells (reviews, Revel et al., 1971; McNutt and Weinstein, 1973; Staehelin, 1974). On the basis of electrophysiological studies, Loewstein (1966) clearly stated that the membranes at the sites of coupling must be specialized in that they are much more permeable to small ions carrying electric current than non-junctional plasma membrane. Also, the membranes of adjacent cells had to be in sufficient contact to prevent leakage of current from intracellular to extracellular space at the coupling site. On the basis of thin-section electron microscopy, there was some indication that the membranes at the sites of electrical coupling had a specialized array of globular subunits, closely packed and often with hexagonal packing in many areas (Robertson, 1963; Revel and Karnovsky, 1967). However, the extent of specialization of the membrane was not determined by thin section data alone. Freeze-fracture contributed striking evidence that the internal structure of the plasma membrane is highly specialized at gap junctions and that the globular subunit structure seen at the surface of membranes continues through the hydrophobic portion of the membrane (Kreutziger, 1968b; McNutt and Weinstein, 1969, 1970, 1973; Goodenough and Revel, 1970; Goodenough and Gilula, 1974) (Fig. 16).

In thin-sectioned and negative-stained gap junctions, the globular subunits are cylindrical in shape, having a diameter of approximately 70 Å and an undetermined height. When the cylindrical subunits are packed in hexagonal arrays, the center-to-center spacing is 90–100 Å (Revel and Karnovsky, 1967; Benedetti and Emmelot, 1968). In freeze-fracture preparations, the gap junction fractures so that numerous globular particles appear on the P face and discrete depressions or "pits" remain in the E face of the membrane. The particles are 60–80 Å in diameter (depending on replica quality). The particles are usually closely packed and often packed in hexagonal arrays with a 90–110 Å center-to-center spacing. E face depressions are complementary to P

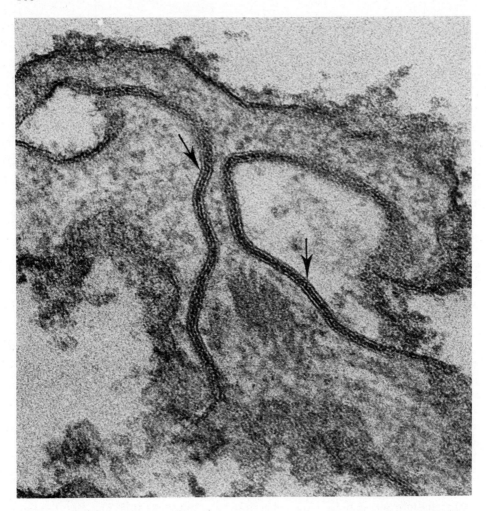

Fig. 15. Thin section of the processes from several cells in the stratified squamous epithelial cells in human oral mucosa. The processes are attached to each other by gap junctions (at arrows). At a gap function, adjacent membranes are parallel and separated by a 20 to 30 Å cleft or "gap". Electron-dense tracers which fill the extracellular space outline electron-lucent subunits spanning the gap and connecting the two membranes (not shown) (×170 000).

face particles (Chalcroft and Bullivant, 1970; Steere and Sommer, 1972). The depressions measure 25–40 Å in diameter depending on replica thickness and shadowing angle (McNutt and Weinstein, 1970, 1973; Staehelin, 1974). One might ask how a 40 Å depression can be complementary to an 80 Å particle. It is inherent in the replication process that platinum deposits will tend to fill depressions and enlarge elevations. Thus a 20 Å layer of platinum crystallites may enlarge a 60 Å particle up to 80 Å and reduce a 60 Å depression to 40 Å. Another factor reducing the complementarity in P and E faces is the fact that extraneous contaminants tend to obscure E face depressions at the gap junc-

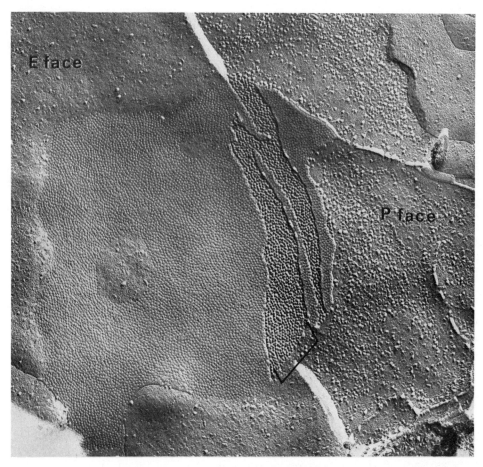

Fig. 16. Freeze-fracture replica of mouse hepatocyte plasma membranes attached at a gap junction, small subunit type. On the P face of one cell membrane, there are numerous, closely packed particles, 60–80 Å in diameter (at GP). The E face of the apposed cell membrane contains a similar array of closely packed depressions that would be complementary to P face particles (20% glycerol; replication at $-100°C$; ×78 000).

tion before obscuring P face particles (Kreutziger, 1968a). The center-to-center spacing of the globular subunits is 90–100 Å but the real dimensions of each subunit are very difficult to determine since the subunit dimensions would have to be corrected by subtraction of replica thickness and platinum grain size (McNutt and Weinstein, 1970). Also, since Bullivant (1974) has demonstrated that plastic deformation of erythrocyte intra-membrane particles can occur, distortion of the gap junction particles themselves must be taken into account. These considerations greatly limit the ability to correlate fine details of freeze-fracture images with thin section or negative stain images of the small subunit gap junction. A more productive approach is that of Goodenough and Stoeckenius (1972) to analyze the X-ray diffraction patterns of isolated gap junctions and to isolate the proteins from gap junctions (Goodenough, 1974, 1976). The fact that Goodenough (1976) has isolated a single polypeptide with a molecular weight of 18 000 daltons from gap junctions makes it very likely that the structures seen by freeze-fracture and by thin sections are identical and slight differences in dimensions are the result of preparative techniques. Consequently, the major contribution of freeze-fracture to the study of the small subunit gap junction has been the demonstration that the specialized structure of the junction continues through the hydrophobic interior of the membrane.

Morphological studies have raised the possibility that each gap junction subunit is a hollow cylinder (Fig. 17). In thin section preparations, a central dot stains with potassium permanganate in each of the gap junction subunits (Robertson, 1963). When gap junctions are exposed to a very fine dispersion of colloidal lanthanum hydroxide, some of the subunits have a central dot of lanthanum, approximately 10–15 Å in diameter (Revel and Karnovsky, 1967). Also, in negatively stained isolated gap junctions, a 20 Å region in the center of each subunit fills with the stain solution (Goodenough, 1976). The importance of the central dot is that it may well correspond to a channel that connects the cytoplasm of adjacent cells and accounts for low resistance coupling (Loewenstein, 1966) and may allow metabolites to pass directly from cell to cell (Payton et al., 1969; Gilula et al., 1972). Freeze-fracture confirms that the center of each subunit in gap junctions is specialized. With low-angle shadowing and high resolution replication techniques, there is a depression visible in the center of the gap junction particles (McNutt and Weinstein, 1969, 1970, 1973). This depression is at the limit of resolution of the freeze-fracture technique and is approximately 20–25 Å in diameter and of unknown depth. The presence of this central depression reinforces the concept that the subunits may be hollow cylinders which serve as passive conduits for transport of small ions that couple adjacent cells (Fig. 17). Central depressions are occasionally found on nonjunctional particles and it is possible to speculate that these particles also may have some other type of transmembrane transport function. Since central depressions are not visible in all particles, they probably are not a platinum grain nucleation artifact.

5.1.3.2. Large subunit gap junction (type II gap junction). Freeze-fracture has demonstrated that there are two types of gap junctions formed between mammalian cells. By far the most common is the small subunit gap junction, but in the intestinal epithelium of the rat, Staehelin (1972, 1974) has observed a gap junction with a larger intramembrane particle on the P face and a greater center-to-center spacing of the hexagonal array. The particles are 100–110 Å in diameter and are spaced at 190–200 Å (Fig. 18). An array of depressions is present on E faces of the membranes of the intestinal epithelial cells that has the same center-to-center spacing and is presumably complementary to the large particle arrays (Staehelin, 1972, 1974). It is interesting that the large subunit gap junctions are always seen in close association with the small subunit

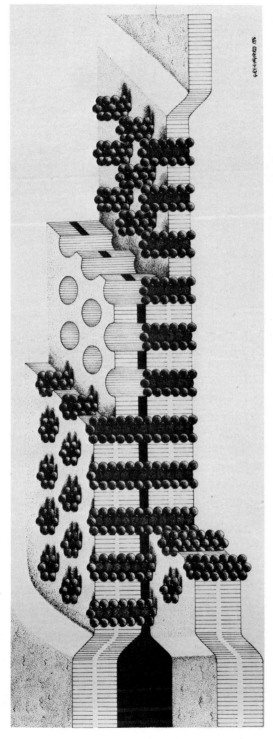

Fig. 17. Interpretative drawing of small subunit gap junction structure. The left half of the figure shows two plasma membranes coupled by hollow cylindrical assemblies of monomeric units. As the cylinders span the narrow cleft of extracellular space between the two membranes, the cylinders can be outlined by electron-dense extracellular tracer substances. In the right half of the figure, the stepwise splitting of this junction is depicted much like it appears in freeze-fracture replicas. As the frozen membranes are split, the gap junction cylinders are broken off so that they remain attached to the juxtacytoplasmic half (P half) of each plasma membrane leaving holes or "pits" in the outer half (E half) of each membrane. Each cylinder is shown as a simple stack of monomeric units arranged around a central pore which permits small molecules to pass across the junction. The exact shape of the monomers and their number per cylinder have yet to be determined.

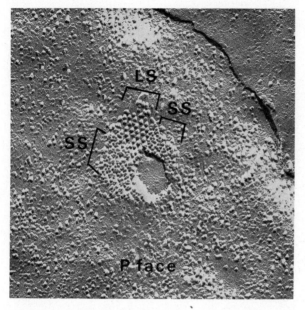

Fig. 18. Freeze-fracture replica of mouse intestinal plasma membrane. Closely packed subunits of a small subunit gap junction (SS) contrast with the large subunit gap junction (LS). The particles of the large subunit gap junction are approximately 100 Å in diameter and have a 190 Å center-to-center spacing, whereas the small subunit gap junction particles are 60–80 Å in diameter, with a 90–100 Å center-to-center spacing in the arrays (40% glycerol; replication at −100°C; ×100 000).

gap junctions. The large subunit junctions cover less area and are generally at the edge of the small subunit junctions. This appearance suggests some type of transitional state either in gap junction formation or breakdown. However, hexagonally-packed arrays of large subunits have not been observed near small subunit gap junctions, including those found in developing systems (Albertini and Anderson, 1975; Elias and Friend, 1976) and during gap junction interiorization presumably for degradation (Albertini et al., 1975).

In invertebrates, a large subunit gap junction has been described and is formed by globular subunits, 125 Å in diameter with a 200 Å center-to-center spacing (Peracchia, 1973). In contrast to the large subunit gap junction in mammals, this invertebrate junction tends to have the subunits on the E face of the membrane rather than the P face. This characteristic has led most authors to place it in a category distinct from the large subunit gap junction of mammals (reviewed by Staehelin, 1974).

5.1.3.3. Gap junction formation and degradation. Freeze-fracture replicas suggest that, during the formation of gap junctions, there is a sequence of membrane specialization which results in the typical small subunit gap junction. Decker and Friend (1974), Johnson et al. (1974) and Albertini and Anderson (1975) observed: (1) that developing gap junctions have an associated region of smooth P face membrane (with few particles) which either precedes junction formation or surrounds enlarging junctions; (2) that a population of large particles (100–120 Å in diameter) appears on P faces very early in gap junction formation or is seen near the margin of enlarging junctions; and (3) that small clusters of 60–80 Å particles assemble in this P face region and are distinctive in their close packing (often hexagonal packing) and complementary E face pits. The large particles on the P face leave large pits on the E face (Decker and Friend, 1974). The presence of complementary pits on the E face suggests that these large particles may be identical to those found in the large subunit gap junctions of mammals (see Type II gap junctions in Staehelin, 1972, 1974) but they lack the hexagonal packing seen in this junction.

In some tissues, particularly those which are forming very large gap junctions, each gap junction appears subdivided into regions or domains with very close packing of subunits partially separated from each other by particle-free membrane or "aisles" (Albertini et al., 1975; Elias and Friend, 1976). Each particle domain consists of approximately 3 to 5 rows of particles, with 12 to 25 particles per row (Albertini et al., 1975), or they may be somewhat more irregular. Superficially, the particles in the domains may appear to be packed orthogonally but close inspection reveals that particles in adjacent rows are offset relative to each other by half the center-to-center spacing within the row. This type of close packing is obviously hexagonal when the domains coalesce into larger aggregates.

In some studies, the observation has been made that P face particle packing in gap junctions is less orderly than packing of the depressions in the E face (Albertini et al., 1975). If this were true, this would be strong evidence against the complementarity of P and E faces of the junction. A more likely interpretation of the images is that P face particle packing only appears less orderly due to replication technique. More specifically, in some laboratories, gap junction particles are estimated to be individually 80–100 Å in diameter but are stated to be packed at a 90–100 Å center-to-center spacing (reviewed by Staehelin, 1974). These dimensions leave only a very slender 10 Å cleft between particles which could easily be lost due to contamination, large platinum grains, or inadequacy of focussing on the replica. Inability to resolve this slender cleft would cause loss of the order in the hexagonal packing. In contrast, greater contamination or loss of resolution would be required to obscure a 30 to 40 Å depression in the E face. Consequently, statements about differences in packing order on P and E faces of gap junctions would have to be supported by high resolution and quantitative data before they can be accepted as evidence against P and E face complementarity.

Two interesting variants of small-subunit gap junctions have been described which are relevant to the formation and degradation of these junctions. The first variant is the gap junction that has been observed within the interstices of the fibrillar array of an occludens junction (Friend and Gilula, 1972b) (Fig. 19). In such a position gap junctions are relatively inaccessible to extracellular tracers and would therefore be extremely difficult to identify in thin sectioned

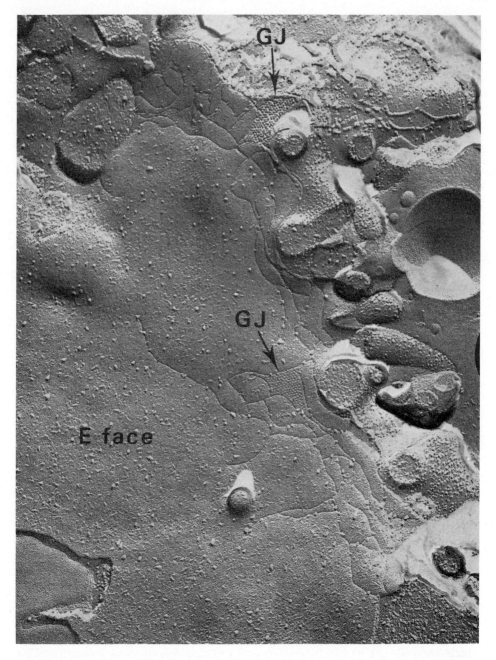

Fig. 19. Freeze-fracture replica of mouse liver plasma membranes near a bile canaliculus. Within the interstices of the occludens junction network are small gap junctions. These gap junctions would be identifiable only by freeze-fracture, since they could not be outlined by tracers added to the extracellular space. The functional significance of this close association between gap junctions and occludens junctions is unknown (20% glycerol; replication at −100°C; ×83 000).

material. Such areas require a freeze-fracture identification. It is not known whether this spatial proximity between occludens and gap junctions has any particular functional significance. Usually cells having gap junctions within the interstices of occludens junctions also have gap junctions outside these regions. Perhaps the significance of this proximity of gap and occludens junctions is only that it is an efficient use of available membrane area. Since the membrane trapped within an occludens junction cannot be used easily for cell-to-extracellular fluid transport, it is efficient to use the area for cell-to-cell communication. A close association between gap junctions and occludens junctions also occurs during gap junction formation in some tissues (Decker and Friend, 1974; Albertini and Anderson, 1975; Albertini et al., 1975; Elias and Friend, 1976). This close association of tight and gap junctions during development in some epithelia has led to speculation that the tight junctions may aid as a scaffolding in the assembly of gap junctions (Decker and Friend, 1974) or that they share a common precursor particle (Elias and Friend, 1976). During the development of the cells in the lens of the calf eye, gap junctions are formed from particles which are packed in prominent linear rows as well as the typical polygonally packed particles in small domains (Benedetti et al., 1974). The striking linear rows of single particle width resemble developing and regressing occludens junctions (Decker and Friend, 1974; Montesano et al., 1975). However, occludens junctions have not been demonstrated between developing lens fiber cells. Also rows of single particle width have been identified in the P face images of gap junctions between photoreceptor cells in vertebrate retina (Raviola and Gilula, 1973). The presence of linear particle arrays reinforces the overlap in the appearance of developing occludens and gap junctions. Another feature shared by the two developing systems is an early zone of relatively smooth membrane before particle or fibril assembly (Montesano et al., 1975). Plasma membrane growth in general may begin with the addition of smooth membrane and be followed by the insertion of intramembrane particles, as shown for the growth cone membrane of rat neurons by Pfenninger and Bunge (1974). Further elucidation of the relationships between these junctions awaits isolation of occludens junction protein and its chemical characterization.

The other variant gap junctions mentioned, i.e. the spherical cytoplasmic inclusions (Fig. 20), are likely to be part of a process of gap junction turnover and degradation. When thin sections initially demonstrated gap junctions with a ring-shaped profile, two possible interpretations were evident: either that some cells had long finger-like processes which deeply indented the adjacent cell membrane at gap junctions so that cross-sectioned processes gave a ring (gap junction) profile or that actual spherical cytoplasmic inclusions were present that were derived from gap junction membrane. Serial thin section and lanthanum tracer studies have shown that both interpretations are correct for the ring-shaped gap junctions or "annular nexuses" found in the granulosa cells of the rat ovarian follicle (Merk et al., 1973). Similar lanthanum tracer observations have been made on gap junctions in renal adenocarcinoma (Letourneau et al., 1975). Freeze-fracture studies have shown that the spherical

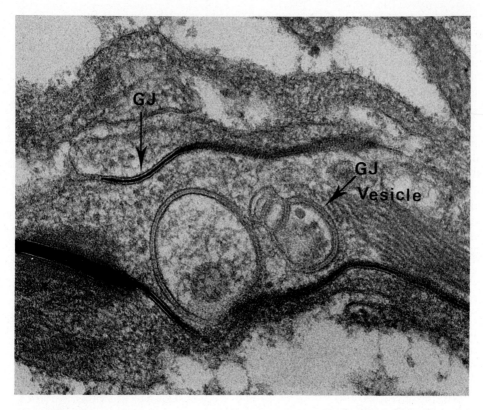

Fig. 20. Thin section of a gap junction in the upper layers of stratified squamous epithelium of the human cervix. An electron opaque tracer (lanthanum hydroxide) has been added to the extracellular space and penetrates the interior of the gap junctions (GJ) which are in continuity with the cell surface. Several vesicles (GJ vesicle) derived from gap junctions are evident within the cytoplasm of one cell and are not penetrated by the tracer. Freeze-fracture has shown gap junction vesicles to be typical small subunit gap junctions in terms of particle distribution. However freeze-fracture rarely allows unequivocal identification of these interiorized gap junctions ($\times 72\,000$).

inclusions have a typical appearance of small subunit gap junctions, i.e. hexagonally packed P face particles and E face depressions in the apposed membranes (Albertini and Anderson, 1974). The spherical inclusions of gap junction membrane are interesting in that their formation involves the transfer of cytoplasm from one cell to another, although the functional significance and ultimate fate of these inclusions are unknown. The recent study by Goodenough (1976) on gap junction vesicle formation may clarify the mechanism of formation of the spherical inclusions. Goodenough (1976) found that, after treatment of a hepatocyte plasma membrane pellet with trypsin, collagenase, and hyaluronidase, there was conversion of sheet-like gap junctions to vesicular gap junctions, 20% of which appeared to be completely closed spheres. Although one cannot extrapolate directly from these studies in vitro to the formation of spherical junctions in vivo, the relationship between these phenomena deserves further study. A role for freeze-fracture in such studies is clear since this technique is necessary for showing the membrane specialization on the spheres, which are difficult to study by thin sections alone and are not accessible to extracellular tracers (Orwin et al., 1973a; Albertini and Anderson, 1974).

5.2. Asymmetrical junctions

Recent freeze-fracture studies show that the internal structure of membranes contains specialized assemblies of particles at the chemical synapses in the brain and retina (Moor et al., 1969; Sandri et al., 1972; Akert et al., 1974; Raviola and Gilula, 1975; Hanna et al., 1976), at neuromuscular junctions (Heuser et al., 1974; Rash and Ellisman 1974; Ellisman et al., 1976) and at glial-axonal junctions (Livingston et al., 1973). The membranes at these sites have a variety of specializations which are complex and have been studied primarily only at a descriptive level. At neuromuscular junctions, large intramembrane particles (110–140 Å in diameter) have been visualized on the crests of the junctional folds in postsynaptic membrane (Ellisman et al., 1976). A similar distribution has been demonstrated for acetylcholine receptor sites using ^{125}I-alpha bungarotoxin and high resolution autoradiography (Fertuck and Salpeter, 1974). These data suggest (but do not prove) that the large particles at these sites are acetylcholine receptors (Ellisman et al., 1976). Future freeze-fracture studies of these asymmetrical junctions hold promise for elucidating their functions.

6. Conclusions

In this review, an attempt has been made to outline the principles underlying the major approaches to freeze-fracture and how they apply to the study of the plasma membrane of vertebrate cells, particularly mammals. Emphasis has been placed on the types of information produced by the technique as well as on its limitations and seldom-discussed artifacts. Hopefully this will provoke careful descriptions of pertinent experimental conditions and greater, in-depth analysis of freeze-fracture images. Although freeze-fracture has distinct limitations, some of these may be turned to advantages by particular types of studies. For example, plastic deformation is severely damaging to attempts to correlate intramembrane particle size with the sizes of proteins measured by other methods. However, future studies might show that the degree of plastic deformation and the conditions under which it occurs provide unique data on intramolecular and intermolecular bonding forces of membrane macromolecules. Also the microchemical analysis of membrane halves (Fisher, 1976) appears very promising for determining the effects of fracturing on frozen membrane proteins and should aid in further interpretation of membrane fracture faces.

Acknowledgement

Any previously unpublished work presented was supported in part by Grant No. CA 16482 from the National Cancer Institute, National Institutes of Health, U.S. Public Health Service and by Veterans Administration funds for project No. 01.

References

Akert, K., Livingston, R.B., Moor, H. and Streit, P. (1974) Proceedings: Ultrastructure of synapses in the waking state. A laboratory report on recent advances. J. Neurol. Transm. Suppl. 11, 1–11.

Albertini, D.F. and Anderson, E. (1974) The appearance and structure of intercellular connections during the ontogeny of the rabbit ovarian follicle with particular reference to gap junctions. J. Cell Biol. 63, 234–250.

Albertini, D.F. and Anderson, E. (1975) Structural modifications of lutein gap junctions during pregnancy in the rat and the mouse. Anat. Rec. 181, 171–194.

Albertini, D.F., Fawcett, D.W., and Olds, P.J. (1975) Morphological variations in gap junctions of ovarian granulosa cells. Tissue and Cell 7, 389–405.

Alroy, J. and Weinstein, R.S. (1976) Unusual cell junction complexes in canine mammary adenoacanthomas. J. Natl. Cancer Inst. 56, 667–670.

Bächi, T. and Schnebli, H.P. (1975) Reaction of lectins with human erythrocytes, II. Mapping of Con A receptors by freeze-etching electron microscopy. Exp. Cell Res. 91, 285–295.

Bachmann, L. and Schmitt, W.W. (1971) Improved cryofixation applicable to freeze etching. Proc. Natl. Acad. Sci. USA 68, 2149–2152.

Bachmann, L. and Schmitt-Fumian, W.W. (1973) Spray-freeze-etching of dissolved macromolecules, emulsions and subcellular components and spray-freezing and freeze-etching. In: E.L. Benedetti and P. Favard (Eds.), Freeze-Etching: Techniques and Applications, Chapters 6 and 7, Societé Française de Microscopie Electronique, Paris, pp. 63–79.

Bank, H. and Mazur, P. (1972) Relation between ultrastructure and viability of frozen-thawed Chinese hamster tissue-cultured cells. Exp. Cell Res. 71, 441–454.

Barnett, R.E., Furcht, L.T. and Scott, R.E. (1974) Differences in membrane fluidity and structure in contact-inhibited and transformed cells. Proc. Natl. Acad. Sci. USA 71, 1992–1994.

Benedetti, E.L. and Emmelot, P. (1968) Structure and function of plasma membranes isolated from liver, In: A.J. Dalton and F. Haguenau (Eds.), Ultrastructure in Biological Systems. Vol. 4, The Membranes, Academic Press, New York, pp. 33–120.

Benedetti, E.L., Dunia, I. and Bloemendal, H. (1974) Development of junctions during differentiation of lens fibers. Proc. Natl. Acad. Sci. USA 71, 5073–5077.

Bischoff, A. and Moor, H. (1967a) Ultrastructural differences between the myelin sheaths of peripheral nerve fibers and CNS white matter. Z. Zellforsch. 81, 303–310.

Bischoff, A. and Moor, H. (1967b) The ultrastructure of the "difference factor" in myelin. Z. Zellforsch. 81, 571–580.

Branton, D. (1966) Fracture faces of frozen membranes. Proc. Natl. Acad. Sci. USA 55, 1048–1056.

Branton, D. (1969) Membrane structure. Ann. Rev. Plant Physiol. 20, 209–238.

Branton, D. (1971) Freeze-etching studies of membrane structure. Phil. Trans. R. Soc. Lond. Ser. B. 261, 133–138.

Branton, D. and Moor, H. (1964) Fine structure in freeze-etched *Allium cepa* L. root tips. J. Ultrastruct. Res. 11, 401–411.

Branton, D., Bullivant, S., Gilula, N.B., Karnovsky, M.J., Moor, H., Mühlethaler, K., Northcote, D.H., Packer, L., Satir, B., Satir, P., Speth, V., Staehelin, L.A., Steere, R.L., Weinstein, R.S. (1975) Freeze-etching nomenclature. Science 190, 54–56.

Breathnach, A.S., Stolinski, C. and Gross, M. (1972) Ultrastructure of foetal and post-natal human skin as revealed by the freeze-fracture replication technique. Micron 3, 287–304.

Breathnach, A.S., Goodman, T., Stolinski, C. and Gross, M. (1973a) Freeze-fracture replication of cells of stratum corneum of human epidermis. J. Anat. 114, 65–81.

Breathnach, A.S., Gross, M. and Martin, B. (1973b) Freeze-fracture replication of melanocytes and melanosomes. J. Anat. 116, 303–320.

Bruns, R.R. and Palade, G.E. (1968a) Studies on blood capillaries, I. General organization of blood capillaries in muscle. J. Cell Biol. 37, 244–276.

Bruns, R.R. and Palade, G.E. (1968b) Studies on blood capillaries, II. Transport of ferritin molecules across the wall of muscle capillaries. J. Cell Biol. 37, 277–299.

Bullivant, S. (1969) Freeze-fracturing of biological materials. Micron 1, 46–51.

Bullivant, S. (1973) Freeze-etching and freeze-fracturing, In: J.K. Koehler (Ed.), Advanced Techniques in Biological Electron Microscopy. Springer, New York, pp. 67–112.

Bullivant, S. (1974) Freeze-etching techniques applied to biological membranes. Phil. Trans. Roy. Soc. Lond. Ser. B 268, 5–14.

Bullivant, S. and Ames, A. III (1966) A simple freeze fracture replication method for electron microscopy. J. Cell Biol. 29, 435–447.

Bullivant, S., Rayns, D.G., Bertaud, W.S., Chalcroft, J.P. and Grayston, G.F. (1972) Freeze-fractured myosin filaments. J. Cell Biol. 55, 520–524.

Caulfield, J.P., Reid, J.J. and Farquhar, M.G. (1976) Alterations of the glomerular epithelium in acute aminonucleoside nephrosis. Evidence for formation of occluding junctions and epithelial cell detachment. Lab. Invest. 34, 43–59.

Chalcroft, J.P. and Bullivant, S. (1970) An interpretation of liver cell membrane and junction structure based on observation of freeze-fracture replicas of both sides of the fracture. J. Cell Biol. 47, 49–60.

Chi, E.Y., Lagunoff, D. and Koehler, J.K. (1975) Electron microscopy of freeze fractured rat peritoneal mast cells. J. Ultrastruct. Res. 51, 46–54.

Clark, A.W. and Branton, D. (1968) Fracture faces in frozen outer segments from guinea pig retina. Z. Zellforsch. Mikrosk. Anat. 91, 586–603.

Claude, P. and Goodenough, D.A. (1973) Fracture faces of zonulae occludentes from "tight" and "leaky" epithelia. J. Cell Biol. 58, 390–400.

Collins, T.R., Bartholomew, J.C. and Calvin, M. (1975) A simple method for freeze-fracture of monolayer cultures. J. Cell Biol. 67, 904–911.

Deamer, D.W. and Branton, D. (1967) Fracture planes in an ice-bilayer model membrane system. Science 158, 655–657.

Deamer, D.W., Leonard, R., Tardieu, A. and Branton, D. (1970) Lamellar and hexagonal lipid phases visualized by freeze-etching. Biochim. Biophys. Acta 219, 47–60.

Decker, R.S. and Friend, D.S. (1974) Assembly of gap junctions during amphibian neurulation. J. Cell Biol. 62, 32–47.

Dempsey, G.P., Bullivant, S. and Watkins, W.B. (1973) Endothelial cell membranes: polarity of particles as seen by freeze-fracturing. Science 179, 190–192.

Devine, C.E. and Rayns, D.G. (1975) Freeze-fracture studies of membrane systems in vertebrate muscle, II. Smooth muscle. J. Ultrastruct. Res. 51, 293–306.

DiPauli, G. and Brdiczka, D. (1974) Localization of glycoproteins within erythrocyte membranes of sheep. A freeze-etching and biochemical study. Biochim. Biophys. Acta 352, 252–259.

Dunlop, W.F. and Robards, A.W. (1972) Some artifacts of the freeze-etching technique. J. Ultrastruct. Res. 40, 391–400.

Elgsaeter, A. and Branton, D. (1974) Intramembrane particle aggregation in erythrocyte ghosts, I. The effects of protein removal. J. Cell Biol. 63, 1018–1030.

Elgsaeter, A., Shotten, D.M. and Branton, D. (1976) Intramembrane particle aggregation in erythrocyte ghosts, II. The influence of spectrin aggregation. Biochim. Biophys. Acta 426, 101–122.

Elias, P.M. and Friend, D.S. (1976) Vitamin-A-induced mucous metaplasia. An in vitro system for modulating tight and gap junction differentiation. J. Cell Biol. 68, 173–188.

Ellisman, M.H., Rash, J.E., Staehelin, L.A., and Porter, K.R. (1976) Studies of excitable membranes, II. A comparison of specializations at neuromuscular junctions and nonjunctional sarcolemmas of mammalian fast and slow twitch muscle fibers. J. Cell Biol. 68, 752–774.

Farquhar, M.G., and Palade, G.E. (1963) Junctional complexes in various epithelia. J. Cell Biol. 17, 375–412.
Feagler, J.R., Tillack, T.W., Chaplin, D.D. and Majerus, P.W. (1974) The effects of thrombin on phytohemagglutinin receptor sites in human platelets. J. Cell Biol. 60, 541–553.
Fertuck, H.C. and Salpeter, M.M. (1974) Localization of acetylcholine receptor by I^{125}-labelled alpha-bungarotoxin binding at mouse motor endplates. Proc. Natl. Acad. Sci. USA 71, 1376–1378.
Findlay, J.B.C. (1974) The receptor proteins for Concanavalin A and *Lens culinaris* phytohemagglutinin in the membrane of the human erythrocyte. J. Biol. Chem. 249, 4398–4403.
Fisher, K.A. (1976) Analysis of membrane halves: Cholesterol. Proc. Natl. Acad. Sci. USA 73, 173–177.
Fisher, K. and Branton, D. (1974) Application of the freeze-fracture technique to natural membranes, In: S. Fleischer and L. Packer (Eds.), Methods in Enzymology, Vol. XXXII, Biomembranes, Part B. Academic Press, New York, pp. 35–44.
Flower, N.E. (1973) Complementary plasma membrane fracture faces in freeze-etch replicas. J. Cell Sci. 12, 445–452.
Friend, D.S. and Fawcett, D.W. (1974) Membrane differentiations in freeze-fractured mammalian sperm. J. Cell Biol. 63, 641–664.
Friend, D.S. and Gilula, N.B. (1972a) A distinctive cell contact in the rat adrenal cortex. J. Cell Biol. 53, 148–163.
Friend, D.S. and Gilula, N.B. (1972b) Variations in tight and gap junctions in mammalian tissues. J. Cell Biol. 53, 758–776.
Gilula, N.B. (1974) Junctions between cells, In: R.P. Cox (Ed.), Cell Communication, Wiley, New York, pp. 1–29.
Gilula, N.B., Reeves, O.R. and Steinbach, A. (1972) Metabolic coupling, ionic coupling, and cell contacts. Nature (London) 235, 262–265.
Gilula, N.B., Eger, R.R. and Rifkin, D.B. (1975) Plasma membrane alteration associated with malignant transformation in culture. Proc. Natl. Acad. Sci. USA 72, 3594–3598.
Goodenough, D.A. (1974) Bulk isolation of mouse hepatocyte gap junctions. Characterization of the principal protein, connexin. J. Cell Biol. 61, 557–563.
Goodenough, D.A. (1976) In vitro formation of gap junction vesicles. J. Cell Biol. 68, 220–231.
Goodenough, D.A. and Gilula, N.B. (1974) The splitting of hepatocyte gap junctions and zonulae occludentes with hypertonic disaccharides. J. Cell Biol. 61, 575–590.
Goodenough, D.A. and Revel, J.P. (1970) A fine structural analysis of intercellular junctions in the mouse liver. J. Cell Biol. 45, 272–290.
Goodenough, D.A. and Stoeckenius, W. (1972) The isolation of mouse hepatocyte gap junctions. Preliminary chemical characterization and X-ray diffraction. J. Cell Biol. 54, 646–656.
Grant, C.W.M. and McConnell, H.M. (1974) Glycophorin in lipid bilayers. Proc. Natl. Acad. Sci. USA 71, 4653–4657.
Guérin, C., Zachowski, A., Prigent, B., Paraf, A., Dunia, I., Diawara, M.A. and Benedetti, E.L. (1974) Correlation between the mobility of inner plasma membrane structure and agglutination by Concanavalin A in two cell lines of MOPC 173 plasmacytoma cells. Proc. Natl. Acad. Sci. USA 71, 114–117.
Hanna, R.B., Hirano, A. and Pappas, G.D. (1976) Membrane specializations of dendritic spines and glia in the weaver mouse cerebellum: A freeze-fracture study. J. Cell Biol. 68, 403–410.
Heuser, J.E., Reese, T.S. and Landis, D.M.D. (1974) Functional changes in frog neuromuscular junctions studied with freeze-fracture. J. Neurocytology 3, 109–131.
Hicks, R.M. and Ketterer, B. (1970) Isolation of the plasma membrane of the lumenal surface of rat bladder epithelium, and the occurrence of a hexagonal lattice of subunits both in negatively stained whole mounts and in sectioned membranes. J. Cell Biol. 45, 542–553.
Hoak, J.C. (1972) Freeze-etching studies of human platelets. Blood 40, 514–522.
Hull, B.E. and Staehelin, L.A. (1976) Functional significance of the variations in the geometrical organization of tight junction networks. J. Cell Biol. 68, 688–704.
Humbert, F., Grandchamp, A., Pricam, C., Perrelet, A. and Orci, L. (1976) Morphological changes in tight junctions of *Necturus maculosus* proximal tubules undergoing saline diuresis. J. Cell Biol. 69, 90–96.

Ji, T.H. and Nicolson, G.L. (1974) Lectin binding and perturbation of the outer surface of the cell membrane induces a transmembrane organizational alteration at the inner surface. Proc. Natl. Acad. Sci. USA 71, 2212–2216.

Johnson, R.G., Hammer, M., Sheridan, J.D. and Revel, J.P. (1974) Gap junction formation between reaggregated Novikoff hepatoma cells. Proc. Natl. Acad. Sci. USA 71, 4536–4540.

Karnovsky, M.J. (1967) The ultrastructural basis of capillary permeability studied with peroxidase as a tracer. J. Cell Biol. 35, 213–236.

Karnovsky, M.J. (1968) The ultrastructural basis of transcapillary exchanges. J. Gen Physiol. (Suppl.) 52, 64–95.

Karnovsky, M.J. and Ryan, G.B. (1975) Substructure of the glomerular slit diaphragm in freeze-fractured normal rat kidney. J. Cell Biol. 65, 233–236.

Kelly, D.E. (1966) Fine structure of desmosomes, and an adepidermal globular layer in developing newt epidermis. J. Cell Biol. 28, 51–72.

Kirk, R.G. and Tosteson, D.C. (1973) Cation transport and membrane morphology. J. Membrane Biol. 12, 273–285.

Kreutziger, G.O. (1968a) Specimen surface contamination and the loss of structural detail in freeze-fracture and freeze-etch preparations, In: Proceedings of the 26th Annual Meeting of the Electron Microscopy Society of America, Claitor's Publishing Division, Baton Rouge, Louisiana, pp. 138–139.

Kreutziger, G.O. (1968b) Freeze-etching of intercellular junctions of mouse liver, In: Proceedings of the 26th Annual Meeting of the Electron Microscopy Society of America, Claitor's Publishing Division, Baton Rouge, Louisiana, pp. 234–235.

Kuhn, K. and Reale, E. (1975) Junctional complexes of the tubular cells in the human kidney as revealed with freeze-fracture. Cell Tissue Res. 160, 193–205.

Kuhn, K., Reale, E. and Wermbter, G. (1975) The glomeruli of the human and the rat kidney studied by freeze-fracturing. Cell Tissue Res. 160, 177–191.

Landis, D.M.D. and Reese, T.S. (1974) Arrays of particles in freeze-fractured astrocytic membranes. J. Cell Biol. 60, 316–320.

Lessin, L.S., Jensen, W.N. and Klug, P. (1972) Ultrastructure of the normal and hemoglobinopathic red blood cell membrane. Freeze-etching and stereoscan electron microscopic studies. Arch. Int. Med. 129, 306–319.

Letourneau, R.J., Li, J.J., Rosen, S. and Villee, C.A. (1975) Junctional specialization in estrogen-induced renal adenocarcinomas of the golden hamster. Cancer Res. 35, 6–10.

Livingston, R.B., Pfenninger, K., Moor, H. and Akert, K. (1973) Specialized paranodal and interparanodal glial-axonal junctions in the peripheral and central nervous system: A freeze-etching study. Brain Res. 58, 1–24.

Loewenstein, W.R. (1966) Permeability of membrane junctions. Ann N.Y. Acad. Sci. 137, 441–472.

Marchesi, S.L., Steers, E., Marchesi, V.T. and Tillack, T.W. (1970) Physical and chemical properties of a protein isolated from red cell membranes. Biochemistry 9, 50–57.

Marchesi, V.T., Steers, E., Tillack, T.W. and Marchesi, S.L. (1969) Some properties of spectrin. A fibrous protein isolated from red cell membranes, In: G.A. Jamieson and T.J. Greenwalt (Eds.), Red Cell Membrane, Structure and Function, Lippincott, Philadelphia, pp. 117–130.

Marchesi, V.T., Tillack, T.W., Jackson, R.L., Segrest, J.P. and Scott, R.E. (1972) Chemical characterization and surface orientation of the major glycoprotein of the human erythrocyte membrane. Proc. Natl. Acad. Sci. USA 69, 1445–1449.

Martínez-Palomo, A. and Erlij, D. (1975) Structure of tight junctions in epithelia with different permeability. Proc. Natl. Acad. Sci. USA 72, 4487–4491.

Martínez-Palomo, A., Pinto da Silva, P. and Chavez, B. (1976) Membrane structure of *Entamoeba histolytica*: Fine structure of freeze-fractured membranes. J. Ultrastruct. Res. 54, 148–158.

Mazur, P. (1970) Cryobiology: The freezing of biological systems. Science 168, 939–949.

Mazur, P., Leibo, S.P. and Chu, E.H.Y. (1972) A two-factor hypothesis of freezing injury. Evidence from Chinese hamster tissue-culture cells. Exp. Cell Res. 71, 345–355.

McIntyre, J.A., Gilula, N.B. and Karnovsky, M.J. (1974) Cryoprotectant-induced redistribution of intramembranous particles in mouse lymphocytes. J. Cell Biol. 60, 192–203.

McNutt, N.S. (1970) Ultrastructure of intercellular junctions in adult and developing cardiac muscle. Am. J. Cardiol. 25, 169–183.

McNutt, N.S. (1975) Ultrastructure of the myocardial sarcolemma. Circ. Res. 37, 1–13.

McNutt, N.S. (1976) Plastic deformation of actin in the production of filaments on the E faces of freeze-fractured plasma membranes (Abstr.). J. Cell Biol. 70, 188a.

McNutt, N.S. and Weinstein, R.S. (1969) Interlocking subunit arrays forming nexus membranes, In: Proceedings of the 27th Annual Meeting of the Electron Microscopy Society of America, Claitor's Publishing Division, Baton Rouge, Louisiana, pp. 330–331.

McNutt, N.S. and Weinstein, R.S. (1970) The ultrastructure of the nexus: a correlated thin-section and freeze-cleave study. J. Cell Biol. 47, 666–688.

McNutt, N.S. and Weinstein, R.S. (1973) Membrane ultrastructure at mammalian intercellular junctions, In: J.A.V. Butler and D. Noble (Eds.), Progress in Biophysics and Molecular Biology, Volume 26, Pergamon, Oxford, pp. 45–101.

McNutt, N.S., Culp, L.A. and Black, P.H. (1971) Contact-inhibited revertant cell lines isolated from SV40-transformed cells, II. Ultrastructural study. J. Cell Biol. 50, 691–708.

McNutt, N.S., Culp, L.A. and Black, P.H. (1973) Contact-inhibited revertant cell lines isolated from SV40-transformed cells, IV. Microfilament distribution and cell shape in untransformed, transformed, and revertant Balb/c 3T3 cells. J. Cell Biol. 56, 412–428.

Merk, F.B., Albright, J.T. and Botticelli, C.R. (1973) The fine structure of granulosa cell nexuses in rat ovarian follicles. Anat. Rec. 175, 107–125.

Meryman, H.T. and Kafig, E. (1954) Replication of frozen solutions for electron microscopy, In: Proceedings of the Third International Conference on Electron Microscopy, London, pp. 486–488.

Meryman, H.T. and Kafig, E. (1955) The study of frozen specimens, ice crystals, and ice crystal growth by electron microscopy. Research Report of Naval Med. Res. Inst., Natl. Naval Med. Ctr. 13, 527–544.

Meyer, H.W. and Winkelmann, H. (1969) Die Gefrierätzung und die Struktur biologischer Membranen. Protoplasma 68, 253–270.

Montesano, R., Friend, D.S., Perrelet, A. and Orci, L. (1975) In vivo assembly of tight junctions in fetal rat liver. J. Cell Biol. 67, 310–319.

Moor, H. (1966) Use of freeze-etching in the study of biological ultrastructure, In: G.W. Richter and M.A. Epstein (Eds.), International Review of Experimental Pathology, Vol. 5, Academic Press, New York, pp. 179–216.

Moor, H. (1971) Recent progress in the freeze-etching technique. Phil. Trans. Roy. Soc. Lond. Ser. B. 261, 121–131.

Moor, H. (1973) Cryotechnology for the structural analysis of biological material, Etching and related problems, and Evaporation and electron guns, In: E.L. Benedetti and P. Favard (Eds.), Freeze-Etching: Techniques and Applications, Société Française de Microscopie Electronique, Paris, pp. 11–30.

Moor, H. and Mühlethaler, K. (1963) Fine structure in frozen-etched yeast cells. J. Cell Biol. 17, 609–623.

Moor, H., Mühlethaler, K., Waldner, H. and Frey-Wyssling, A. (1961) A new freezing ultramicrotome. J. Biophys. Biochem. Cytol. 10, 1–13.

Moor, H., Pfenninger, K., Sandri, C. and Akert, K. (1969) Freeze-etching of synapses. Science 164, 1405–1407.

Mühlethaler, K. (1973) History of freeze-etching, In: E.L. Benedetti and P. Favard (Eds.), Freeze-Etching: Techniques and Applications, Société Française de Microscopie Electronique, Paris, pp. 1–10.

Mühlethaler, K., Moor, H. and Szarkowski, J.W. (1965) The ultrastructure of the chloroplast lamellae. Planta 67, 305–323.

Nicolson, G.L. and Painter, R.G. (1973) Anionic sites of human erythrocyte membranes, II. Antispectrin-induced transmembrane aggregation of the binding sites for positively charged colloidal particles. J. Cell Biol. 59, 395–406.

Nicolson, G.L., Marchesi, V.T. and Singer, S.J. (1971) The localization of spectrin on the inner surface of human red blood cell membranes by ferritin conjugated antibodies. J. Cell Biol. 59, 395–406.

Orci, L. and Perrelet, A. (1973) Membrane-associated particles: increase at sites of pinocytosis demonstrated by freeze-etching. Science 181, 868–869.

Orci, L., Amherdt, M., Henquin, J.C., Lambert, A.E., Unger, R.H. and Renold, A.E. (1973a) Pronase effect on pancreatic beta cells secretion and morphology. Science 180, 647–649.

Orci, L., Amherdt, M., Malaisse-Lagae, F., Rouiller, C. and Reynold, A.E. (1973b) Insulin release by emicytosis: demonstration with freeze-etching technique. Science 179, 82–84.

Orci, L., Amherdt, M., Malaisse-Lagae, F., Perrelet, A., Dulin, W.E., Gerritsen, G.C., Malaisse, W.J. and Renold, A.E. (1974) Morphological characterization of membrane systems in A- and B-cells of the chinese hamster. Diabetologia 10 (Suppl.), 529–539.

Orwin, D.F.G. and Thomson, R.W. (1973) Plasma membrane differentiations of keratinizing cells of the wool follicle, IV. Further membrane differentiations. J. Ultrastruct. Res. 45, 41–49.

Orwin, D.F.G., Thomson, R.W. and Flower, N.E. (1973a) Plasma membrane differentiations of keratinizing cells of the wool follicle, I. Gap junctions. J. Ultrastruct. Res. 45, 1–14.

Orwin, D.F.G., Thomson, R.W. and Flower, N.E. (1973b) Plasma membrane differentiations of keratinizing cells of the wool follicle, II. Desmosomes. J. Ultrastruct. Res. 45, 15–29.

Orwin, D.F.G., Thomson, R.W. and Flower, N.E. (1973c) Plasma membrane differentiations of keratinizing cells of the wool follicle, III. Tight junctions. J. Ultrastruct. Res. 45, 30–40.

Pagano, R.E., Cherry, R.J. and Chapman, D. (1973) Phase transitions and heterogenity in lipid bilayers. Science 181, 557–559.

Palade, G.E. (1961) Blood capillaries of the heart and other organs. Circulation 24, 368–388.

Parish, G.R. (1975) Changes of particle frequency in freeze-etched erythrocyte membranes after fixation. J. Microsc. 104, 245–256.

Payton, B.W., Bennett, M.V.L. and Pappas, G.D. (1969) Permeability and structure of junctional membranes at an electrotonic synapse. Science 166, 1641–1643.

Peracchia, C. (1973) Low resistance junctions in crayfish. II. Structural details and further evidence for intercellular channels by freeze-fracture and negative staining. J. Cell. Biol. 57, 66–76.

Pfenninger, K.H. and Bunge, R.P. (1974) Freeze-fracturing of nerve growth cones and young fibers. A study of developing plasma membrane. J. Cell Biol. 63, 180–196.

Pfenninger, K.H. and Rinderer, E.R. (1975) Methods for the freeze-fracturing of nerve tissue cultures and cell monolayers. J. Cell Biol. 65, 15–28.

Pinto da Silva, P. (1972) Translational mobility of the membrane intercalated particles of human erythrocyte ghosts. pH-dependent, reversible aggregation. J. Cell Biol. 53, 777–787.

Pinto da Silva, P. and Branton, D. (1970) Membrane splitting in freeze-etching. Covalently bound ferritin as a membrane marker. J. Cell Biol. 45, 598–605.

Pinto da Silva, P. and Gilula, N.B. (1972) Gap junctions in normal and transformed fibroblasts in culture. Exp. Cell Res. 71, 393–401.

Pinto da Silva, P. and Martínez-Palomo, A. (1975) Distribution of membrane particles and gap junctions in normal and transformed 3T3 cells studied in situ, in suspension, and treated with concanavalin A. Proc. Natl. Acad. Sci. USA 72, 572–576.

Pinto da Silva, P. and Miller, R.G. (1975) Membrane particles on fracture faces of frozen myelin. Proc. Natl. Acad. Sci. USA 72, 4046–4050.

Pinto da Silva, P. and Nicolson, G. (1974) Freeze-etch localization of Concanavalin A receptors to the membrane intercalated particles of human erythrocyte ghost membranes. Biochim. Biophys. Acta 363, 311–319.

Pinto da Silva, P., Douglas, S.D. and Branton, D. (1971) Localization of A antigenic sites on human erythrocyte ghosts. Nature (London) 232, 194–196.

Pinto da Silva, P., Moss, P.S., and Fudenberg, H.H. (1973) Anionic sites on the membrane intercalated particles of human erythrocyte ghost membranes. Freeze-etch localization. Exp. Cell Res. 81, 127–138.

Pinto da Silva, P., Martínez-Palomo, A. and Gonzales-Robles, A. (1975) Membrane structure and surface coat of *Entamoeba histolytica*. Topochemistry and dynamics of the cell surface: cap formation and microexudate. J. Cell Biol. 64, 538–550.

Pitelka, D.R., Hamamoto, S.T., Duafala, J.G. and Nemanic, M.K. (1973) Cell contacts in the mouse mammary gland. Normal gland in postnatal development and secretory cycle. J. Cell Biol. 56, 797–818.

Plattner, H., Schmitt-Fumian, W.W. and Bachmann, L. (1973) Cryofixation of single cells by spray-freezing, In: E.L. Benedetti and P. Favard (Eds.), Freeze Etching: Techniques and Applications, Société Française de Microscopie Electronique, Paris, pp. 81–100.

Porter, K. R., Kenyon, K. and Badenhausen, S. (1967) Specializations of the unit membrane. Protoplasma 63, 262–274.

Poste, G., Papahadjopoulos, D., Jacobson, K. and Vail, W.J. (1975) Effects of local anesthetics on membrane properties, II. Enhancement of the susceptibility of mammalian cells to agglutination by plant lectins. Biochim. Biophys. Acta 394, 520–539.

Pricam, C., Humbert, F., Perrelet, A., Amherdt, M. and Orci, L. (1975) Intercellular junctions in podocytes of the nephrotic glomerulus as seen with freez-fracture. Lab. Invest. 33, 209–218.

Rash, J.E. and Ellisman, M.H. (1974) Studies on excitable membranes, I. Macromolecular specializations of the neuromuscular junction and nonjunctional sarcolemma. J. Cell Biol. 63, 567–586.

Rash, J.E., Staehelin, L.A. and Ellisman,M.H. (1974) Rectangular arrays of particles on freeze-cleaved plasma membranes are not gap junctions. Exp. Cell Res. 86, 187–190.

Raviola, E. and Gilula, N.B. (1973) Gap junctions between photoreceptor cells in the vertebrate retina. Proc. Natl. Acad. Sci. USA 70, 1677–1681.

Raviola, E. and Gilula, N.B. (1975) Intramembrane organization of specialized contacts in the outer plexiform layer of the retina. A freeze-fracture study in monkeys and rabbits. J. Cell Biol. 65, 192–222.

Rayns, D.G., Simpson, F.O. and Bertaud, W.S. (1968) Surface features of straited muscle, I. Guinea pig cardiac muscle. J. Cell Sci. 3, 467–474.

Rayns, D.G., Simpson, F.O. and Ledingham, J.M. (1969) Ultrastructure of desmosomes in mammalian intercalated disc; appearances after lanthanum treatment. J. Cell Biol. 42, 322–326.

Rayns, D.G., Devine, C.E. and Sutherland, C.L. (1975) Freeze-fracture studies of membrane systems in vertebrate muscle, I. Striated muscle. J. Ultrastruct. Res. 50, 306–321.

Revel, J.P. and Karnovsky, M.J. (1967) Hexagonal array of subunits in intercellular junctions of the mouse heart and liver. J. Cell Biol. 33, C7–C12.

Revel, J.P., Yee, A.G. and Hudspeth, A.J. (1971) Gap junctions between electrotonically coupled cells in tissue culture and in brown fat. Proc. Natl. Acad. Sci. USA 68, 2924–2927.

Robertson, J.D. (1959) The ultrastructure of cell membranes and their derivatives. Biochem. Soc. Symp. 16, 3–43.

Robertson, J.D. (1960) The molecular structure and contact relationships of cell membranes, In: J.A.V. Butler and B. Katz (Eds.), Progress in Biophysics and Biophysical Chemistry, Vol. 10, Pergamon, Oxford, pp. 343–418.

Robertson, J.D. (1963) The occurrence of a subunit pattern in the unit membranes of club endings in Mauthner cell synapses in goldfish brains. J. Cell Biol. 19, 201–221.

Rosenblith, J.Z., Ukena, T.E., Yin, H.H., Berlin, R.D. and Karnovsky, M.J. (1973) A comparative evaluation of the distribution of concanavalin A-binding sites on the surfaces of normal, virally-transformed, and protease-treated fibroblasts. Proc. Natl. Acad. Sci. USA 70, 1625–1629.

Ruska, C. and Schulz, H. (1968) Elektronenmikroskopische Darstellung von Thrombocyten mit der gefrierätztechnik. Klin. Wochenschr. 46, 689–696.

Ryan, G.B., Leventhal, M. and Karnovsky, M.J. (1975) A freeze-fracture study of the junctions between glomerular epithelial cells in aminonucleoside nephrosis. Lab. Invest. 32, 397–403.

Sandri, C., Akert, K., Livingston, R.B. and Moor, H. (1972) Particle aggregations at specialized sites in freeze-etched postsynaptic membranes. Brain Res. 41, 1–16.

Satir, B., Schooley, C. and Satir, P. (1973) Membrane fusion in a model system: mucocyst secretion in *Tetrahymena*. J. Cell Biol. 56, 153–176.

Scott, R.E., Furcht, L.T. and Kersey, J.H. (1973) Changes in membrane structure associated with cell contact. Proc. Natl. Acad. Sci. USA 70, 3631–3635.

Sears, D.A., Weed, R.I. and Swisher, S.N. (1964) Differences in the mechanism of in vitro immune hemolysis related to antibody specificity. J. Clin. Invest. 43, 975–985.

Seeman, P., Cheng, D. and Iles, G.H. (1973) Structure of membrane holes in osmotic and saponin hemolysis. J. Cell Biol. 56, 519–527.

Segrest, J.P., Gulik-Krzywicki, T. and Sardet, C. (1974) Association of the membrane-penetrating polypeptide segment of the human erythrocyte MN-glycoprotein with phospholipid bilayers, I. Formation of freeze-etch intramembranous particles. Proc. Natl. Acad. Sci. USA 71, 3294–3298.

Simionescu, M., Simionescu, N. and Palade, G.E. (1974) Morphometric data on the endothelium of blood capillaries. J. Cell Biol. 60, 128–152.

Simionescu, M., Simionescu, N. and Palade, G.E. (1975) Segmental differentiations of cell junctions in the vascular endothelium. The microvasculature. J. Cell Biol. 67, 863–885.

Simionescu, M., Simionescu, N. and Palade, G.E. (1976) Segmental differentiations of cell junctions in the vascular endothelium. Arteries and veins. J. Cell Biol. 68, 705–723.

Singer, S.J. and Nicolson, G.L. (1972) The fluid mosaic model of the structure of cell membranes. Science 175, 720–731.

Sjöstrand, F.S. (1968) Ultrastructure and function of cellular membranes, In: A.J. Dalton and F. Haguenau (Eds). Ultrastructure in Biological Systems, Vol. 4, The Membranes, Academic Press, New York, pp. 151–210.

Skerrow, C.J. and Matoltsy, A.G. (1974a) Isolation of epidermal desmosomes. J. Cell Biol. 63, 515–523.

Skerrow, C.J. and Matoltsy, A.G. (1974b) Chemical characterization of isolated epidermal desmosomes. J. Cell Biol. 63, 524–530.

Smith, U., Smith, D.S., Winkler, H. and Ryan, J.W. (1973) Exocytosis in the adrenal medulla demonstrated by freeze-etching. Science 179, 79–82.

Sommer, J.R., Steere, R.L., Johnson, E.A. and Jewett, P.H. (1972) Ultrastructure of cardiac muscle: Comparative review with emphasis on the muscle fibers of the ventricles, In: F.E. South, J.P. Hannon, J.R. Willis, E.T. Pengelley and N.R. Alpert (Eds.), Hibernation and Hypothermia: Perspectives and Challenges, Elsevier, New York, pp. 291–355.

Speth, V., Wallach, D.F.H., Weidekamm, E. and Knüfermann, H. (1972) Micromorphologic consequences following perturbation of erythrocyte membranes by trypsin, phospholipase A, lysolecithin, sodium dodecyl sulfate, and saponin. A correlated freeze-etching and biochemical study. Biochim. Biophys. Acta 255, 386–394.

Stackpole, C.W. and Devorkin, D. (1974) Membrane organization in mouse spermatozoa revealed by freeze-etching. J. Ultrastruct. Res. 49, 167–187.

Staehelin, L.A. (1968) The interpretation of freeze-etched artificial and biological membranes. J. Ultrastruct. Res. 22, 326–347.

Staehelin, L.A. (1972) Three types of gap junctions interconnecting intestinal epithelial cells visualized by freeze-etching. Proc. Natl. Acad. Sci. USA 69, 1318–1321.

Staehelin, L.A. (1973) Further observations on the fine structure of freeze-cleaved tight junctions. J. Cell Sci. 13, 763–786.

Staehelin, L.A. (1974) Structure and function of intercellular junctions. Int. Rev. Cytol. 39, 191–283.

Staehelin, L.A. (1975) A new occludens-like junction linking endothelial cells of small capillaries (probably venules) of rat jejunum. J. Cell Sci. 18, 545–551.

Staehelin, L.A. and Bertaud, W.S. (1971) Temperature and contamination dependent freeze-etch images of frozen water and glycerol solutions. J. Ultrastruct. Res. 37, 146–168.

Staehelin, L.A., Mukherjee, T.M. and Williams, A.W. (1969) Freeze-etch appearance of the tight junctions in the epithelium of small and large intestine of mice. Protoplasma 67, 165–184.

Staehelin, L.A., Chlapowski, F.J. and Bonneville, M.A. (1972) Lumenal plasma membrane of the urinary bladder, I. Three-dimensional reconstruction from freeze-etch images. J. Cell Biol. 53, 73–91.

Steere, R.L. (1957) Electron microscopy of structural detail in frozen biological specimens. J. Biophys. Biochem. Cytol. 3, 45–60.

Steere, R.L. (1969) Freeze-etching simplified. Cryobiology 5, 306–323.

Steere, R.L. (1973) Preparation of high-resolution freeze-etch, freeze-fracture, frozen-surface, and freeze-dried replicas in a single freeze-etch module, and the use of stereo electron microscopy to obtain maximum information from them, In: E.L. Benedetti and P. Favard (Eds.), Freeze-Etching: Techniques and Applications, Société Française de Microscopie Electronique, Paris, pp. 223–255.

Steere, R.L. and Moseley, M. (1969) New dimensions in freeze-etching, In: Proceedings of the 27th Annual Meeting of the Electron Microscopy Society of America, Claitor's Publishing Division, Baton Rouge, Louisiana, pp. 202–203.

Steere, R.L. and Sommer, J.R. (1972) Stereo ultrastructure of nexus faces exposed by freeze-fracturing. J. Microsc. (Paris) 15, 205–218.

Stoeckenius, W. and Engelman, D.M. (1969) Current models for the structure of biological membranes. J. Cell Biol. 42, 613–646.

Tillack, T.W. and Marchesi, V.T. (1970) Demonstration of the outer surface of freeze-etched red blood cell membranes. J. Cell Biol. 45, 649–653.

Tillack, T.W., Scott, R.E. and Marchesi, V.T. (1972) The structure of erythrocyte membranes studied by freeze-etching, II. Localization of receptors for phytohemagglutinin and influenza virus to the intramembranous particles. J. Exp. Med. 135, 1209–1227.

Torpier, G., Montagnier, L., Biquard, J-M. and Vigier, P. (1975) A structural change of the plasma membrane induced by oncogenic viruses: Quantitative studies with the freeze-fracture technique. Proc. Natl. Acad. Sci. USA 72, 1695–1698.

Vail, W.J., Papahadjopoulos, D. and Moscarello, M.A. (1974) Interaction of a hydrophobic protein with liposomes. Evidence for particles seen in freeze fracture as being proteins. Biochim. Biophys. Acta 345, 463–467.

Vergara, J., Longley, W. and Robertson, J.D. (1969) A hexagonal arrangement of subunits in membrane of mouse urinary bladder. J. Mol. Biol. 46, 593–596.

Verkleij, A.J., Ververgaert, P.H.J., Van Deenen, L.L.M. and Elbers, P.F. (1972) Phase transitions of phospholipid bilayers and membranes of *Acholeplasma laidlawii B* visualized by freeze fracturing electron microscopy. Biochim. Biophys. Acta 288, 326–332.

Verkleij, A.J., Zwaal, R.F.A., Roelofsen, B., Comfurius, P., Kastelijn, D. and Van Deenen, L.L.M. (1973) The asymmetric distribution of phospholipids in the human red cell membrane. A combined study using phospholipases and freeze-etch electron microscopy. Biochim. Biophys. Acta 323, 178–193.

Wade, J.B. and Karnovsky, M.J. (1974) The structure of the zonula occludens. A single fibril model based on freeze-fracture. J. Cell Biol. 60, 168–180.

Wade, J.B., Revel, J.P. and Di Scala, V. (1973) Effect of osmotic gradients on intercellular junctions of the toad bladder. Am. J. Physiol. 224, 407–415.

Wehrli, E., Mühlethaler, K. and Moor, H. (1970) Membrane structure as seen with a double replica method for freeze-fracturing. Exp. Cell Res. 59, 336–339.

Weinstein, R.S. (1969) Electron microscopy of surfaces of red cell membranes, In: G.A. Jamieson and T.J. Greenwalt (Eds.), Red Cell Membrane Structure and Function, Lippincott, Philadelphia, pp. 36–76.

Weinstein, R.S. (1974a) The morphology of adult red cells, In: D. Mac N. Surgenor (Ed.), The Red Blood Cell, 2nd ed., Academic Press, New York, pp. 213–268.

Weinstein, R.S. (1974b) Changes in cell membrane structure associated with neoplastic transformation in human urinary bladder epithelium (Abstr.). J. Cell Biol. 63, 367a.

Weinstein, R.S. and Bullivant, S. (1967) The application of freeze-cleaving techniques to studies on red blood cell fine structure. Blood 29, 780–789.

Weinstein, R.S., Clowes, A.W. and McNutt, N.S. (1970a) Unique cleavage plane in frozen red cell membranes. Proc. Soc. Exp. Biol. Med. 134, 1195–1198.

Weinstein, R.S., McNutt, N.S., Nielsen, S.L. and Pinn, V.W. (1970b) Intramembranous fibrils at tight junctions, In: Proceedings of the 28th Annual Meeting of the Electron Microscopy Society of America, Claitor's Publishing Division, Baton Rouge, Louisiana, pp. 108–109.

Willison, J.H.M. (1975) The relationship of plastic deformation in freeze-etching to the orientation of a protein particle. J. Microsc. 105, 81–85.

Wunderlich, F., Wallach, D.F., Speth, V. and Fischer, H. (1974) Differential effects of temperature on the nuclear and plasma membranes of lymphoid cells. A study by freeze-etch electron microscopy. Biochim. Biophys. Acta 373, 34–43.

Yee, A.G. and Revel, J.P. (1975) Endothelial cell junctions. J. Cell Biol. 66, 200–204.

Zwaal, R.F., Roelofsen, B. and Colley, C. (1973) Localization of red cell membrane constituents. Biochim. Biophys. Acta 300, 159–182.

Exchange of phospholipids between membranes

3

J.C. KADER

1. Introduction

The discovery of the exchange of phospholipids between biological membranes came in a large part from unsuccessful attempts to study mitochondrial phospholipid synthesis in vitro. Carefully purified, isolated mitochondria from various cells were found to be unable to synthesize the nitrogen-containing phospholipids, phosphatidylcholine or phosphatidylethanolamine, even though they represent the major mitochondrial phospholipids. The idea of a cooperation between the microsomes, which are able to synthesize these lipids, and the mitochondria, began to emerge in the late 1960s (Dawson, 1966; Kadenbach, 1968; Mazliak et al., 1968; Wirtz and Zilversmit, 1968; McMurray and Dawson, 1969), with the possibility that this cooperation might involve exchange of phospholipids between the membranes of these organelles, since exchange processes had already been demonstrated between lipoproteins (Dawson, 1966).

The experimental demonstration of phospholipid exchange between these organelles came within months of each other from the independent investigations of Wirtz and Zilversmit (1968), Kadenbach (1968), McMurray and Dawson (1969) and Akıyama and Sakagami (1969). These experiments involved the incubation of a microsomal fraction containing radiolabeled phospholipids with a nonradioactive mitochondrial fraction. The cell fractions were then carefully separated from each other, and it was found that the initially unlabeled fraction had not only acquired the radiolabel, but also the specific radioactivity increased with the time of incubation, while the specific radioactivity of the phospholipids of the "donor" fraction decreased concomitantly. These results were interpreted as due to exchange of phospholipids between membranes.

These investigations had the merit of opening a new field of research which concerns many areas of cell biology and especially the structure of the biological membranes. According to the most recent and generally accepted models of membrane structure (Singer, 1971; Singer and Nicolson, 1972; Bretscher, 1972, 1973, 1974; Zwaal et al., 1973), biological membranes consist of a phospholipid bilayer with partially embedded integral (intrinsic) proteins and a variety of peripheral (extrinsic) proteins that are associated with the membrane by non-hydrophobic interactions. It has been shown that movements of phospholipids within the membranes are of two types (Fig. 1): (1) within each monolayer of the bilayer there is a lateral diffusion of the phospholipids. This movement is rapid, being in the range of 1 to $2 \cdot 10^{-8}$ cm^2 sec^{-1} according to data from electron spin resonance measurements on spin-labeled phospholipids (Devaux and McConnell, 1972; Scandella et al., 1972; Träuble and Sackmann, 1972); and nuclear magnetic resonance studies (A.G. Lee et al., 1973; Brûlet and McConnell, 1975; Lee, 1975). These studies were all done with artificial lipid bilayers; (2) phospholipids can also move between each monolayer of the bilayer. However, this "flip-flop" movement is much slower as determined by studying the distribution of a spin-labeled phosphatidylcholine between the two monolayers of a phosphatidylcholine vesicle (half-time of 6.5 h at 30°C) (Kornberg and McConnell, 1971b). Other measurements of "flip-flop" times will be discussed in section 11.

The exchange of phospholipids between membranes (Fig. 1) represents a third type of movement that reflects the fluidity of the membrane constituents. Many questions arose with the discovery of this exchange process in vitro. Is it possible to show the existence of phospholipid exchange in vivo? Is exchange of phospholipids between intracellular membranes, different from the exchange of these constituents between cell surface membranes and the external medium? Is phospholipid exchange active between all the membranes of a living cell? What part of the phospholipid pool is concerned in the exchange? Are there factors controlling the exchange of phospholipids? What is the mechanism of the transfer of a phospholipid from one membrane to another? It is the aim of this review to present the most recent answers to these

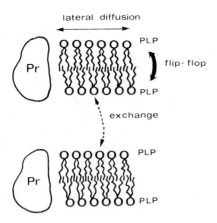

Fig. 1. Schematic representation of phospholipid movements in membranes. Pr, integral membrane proteins; PLP, phospholipids (see text for details).

questions. The excellent reviews written by Dawson (1973) and Wirtz (1974) and the thesis of Kamp (1975) have helped greatly in this task.

2. The origin of the hypothesis for exchange of phospholipids within the cell

As mentioned briefly in the introduction, the idea of an exchange of phospholipids between membranes originated from a comparison of the phospholipid metabolism of microsomes and mitochondria.

2.1. In vivo experiments

Observations in several laboratories on the incorporation of radiolabeled phospholipid precursors into microsomal and mitochondrial lipids following injection of the precursors into animals or incubation of various animal and plant tissues in aqueous solutions containing the precursors revealed that both the microsomal and mitochondrial fraction became rapidly labeled. Incorporation of precursors into lipids in both fractions has been demonstrated in rat liver (Ada, 1949; Gurr et al., 1965; Miller and Cornatzer, 1966; Kadenbach, 1968; Bygrave, 1969; McMurray and Dawson, 1969; Wirtz and Zilversmit, 1969a; Blok et al., 1971), in isolated rat liver cells (Jungalwala and Dawson, 1970), in rat liver slices (Wojtczak et al., 1971) and in plant tissues (Abdelkader and Mazliak, 1970; Mazliak et al., 1975). Kinetic measurements revealed, however, that the microsomal phospholipids displayed a higher specific radioactivity from the outset, and a sequence of decreasing radioactivity could be defined in which microsomes incorporated more radioactivity than the outer mitochondrial membrane which, in turn, had a higher specific activity than the inner mitochondrial membranes (Bygrave, 1969; McMurray and Dawson, 1969; Blok et al., 1971; Wojtczak et al., 1971). These results were confirmed by Jungalwala and Dawson (1970) and Wojtczak et al. (1971) who performed pulse-chase experiments by labeling rat liver cells and rat liver slices, respectively, with different precursors, including [^{14}C]choline, for one hour before transferring the labeled cells or slices into a nonradioactive medium. They observed a fall in the specific radioactivity of the microsomal phospholipids, and an increase in the specific radioactivity of the mitochondrial phospholipids (Fig. 2). The three major phospholipids each behaved in the same manner (Jungalwala and Dawson, 1970). In addition, the specific labeling of phosphatidylcholine in the outer mitochondrial membrane increased more rapidly than phosphatidylcholine in the inner membrane (Wojtczak et al., 1971). According to Jungalwala and Dawson (1970) the fall in specific radioactivity of the microsomal pool was too large to be explained by dilution of this pool by freshly synthesized phospholipids and could not be attributed to a breakdown of the phospholipid pool. The easiest explanation of these results was that transfer of phospholipids had occurred from the rapidly labeled endoplasmic reticulum to the mito-

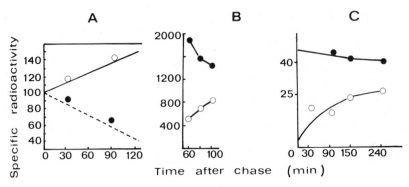

Fig. 2. In vivo kinetics of phospholipid labeling of subcellular fractions: A, incorporation of [Me-^{14}C]choline into mitochondria and microsomes from rat liver slices during pulse-chase labeling. The chase was applied after 60 min. The slices were then washed to remove the precursor and then reincubated in the absence of radioactive precursor for the various times indicated on the abscissa. The specific radioactivity of the phospholipids present in microsomal (●) or mitochondrial (○) fractions at the onset of the chase was taken as 100 (reproduced from Wojtczak et al., 1971); B, similar experiments as in A, but starting from isolated rat liver cells, labeled with [^{14}C]ethanolamine. The specific radioactivity of phosphatidylethanolamine was expressed in cpm/μg of phospholipid phosphorus (reproduced from Jungalwala and Dawson, 1970); C, potato tuber slices incubated for 30 min with [^{14}C]choline and then treated as indicated in A. The specific radioactivity of phosphatidylcholine was expressed in cpm/μg phospholipid (reproduced from Mazliak et al., 1975). Reproduced with the permission of Biochimica Biophysica Acta for (A) and (B) and Academic Press for (C).

chondria. Kinetics of labeling of phosphatidylcholine and phosphatidylethanolamine in rat liver also strongly suggested a similar intermembrane transfer of these phospholipids (Landriscina and Marra, 1973; Parkes and Thompson, 1973, 1975).

Analogous experiments have been performed with plant tissues, viz. potato tuber (Abdelkader and Mazliak, 1970; Mazliak et al., 1975) (Fig. 2). Also, in vivo incorporation of [^{14}C]choline in castor bean endosperm has been followed in three different cell fractions: a light membrane fraction (mainly endoplasmic reticulum) which was rapidly labeled; and two heavier ones, mitochondria and glyoxysomes, which were slowly labeled. The changes in the distribution of the [^{14}C]-label between these fractions were again consistent with a transfer of [^{14}C]choline labeled phospholipids from the light fraction to the heavier ones (T. Kagawa et al., 1973).

Other evidence in favor of the existence of the phospholipid exchange in vivo was provided by the autoradiographic studies of Stein and Stein (1967, 1969) using [^{3}H]choline or [^{3}H]palmitate and the controlled modifications of microsomal pools studied by Wirtz and Zilversmit (1969a). In these latter experiments, treatment of rats with phenobarbital was used to induce expansion of the microsomal phospholipid pool, and a parallel decrease in the specific radioactivity of [^{32}P]-labeled microsomal and mitochondrial phospholipids was observed. In contrast, carbon tetrachloride treatment produced an increase in the specific radioactivities of the mitochondrial and microsomal pools. The parallel nature of the changes in the specific radioactivities of both compartments is also illustrated by the fact that the ratio of radiolabeled phosphatidylcholine in the mitochondrial and microsomal fractions remained constant (approximately 0.55) with the different treatments. A similar ratio was also found in an earlier study by Bjornstad and Bremer (1966). Additional data on the effects of carbon tetrachloride and phenobarbital on phospholipid metabolism have been presented by Kamath and Rubin (1974b,c).

Beattie (1969a) showed that inhibition of protein synthesis in rat liver by cycloheximide results (after a lag period) in a reduced incorporation of [^{14}C]glycerol into lecithin in both the microsomal and mitochondrial fractions. A similar reduction in the incorporation of [^{14}C]acetate and [^{14}C]choline precursors in the presence of cycloheximide has been observed in experiments with potato tuber organelles (Abdelkader, 1972; Abdelkader and Mazliak, 1972).

All of these results are compatible with the existence of a phospholipid exchange process in which the microsomal phospholipids are in equilibrium with part of the mitochondrial phospholipid pool. This exchange process explains why mitochondria can incorporate radiolabeled phospholipid precursors in the intact cell, yet do not show any incorporation when studied as isolated organelles in vitro as described in the next section.

2.2. In vitro experiments

In spite of many attempts, definitive evidence for the synthesis of nitrogen-containing phospholipids (phosphatidylcholine and phosphatidylethanolamine) or phosphatidylinositol by purified mitochondria in vitro has yet to be obtained (reviews, Van den Bosch et al., 1972; Dawson, 1973; Mazliak, 1973; McMurray, 1973; Bjerve, 1974; Kates and Marshall, 1975), even though these organelles are rich in these phospholipids (see Rouser et al., 1968; Van den Bosch et al., 1972; Dawson, 1973; Mazliak, 1973; McMurray, 1973). This inability of mitochondria to synthesize lecithin does not seem to be due to limiting amounts of diglycerides (McMurray, 1975). The possibility of phospholipid biosynthesis in isolated mitochondria raised in the studies of McMurray and Dawson (1969) and McMurray (1973) may reflect base-exchange phenomena and/or contamination of the mitochondrial fractions.

It is established, however, that animal and plant mitochondria can synthesize phosphatidic acid de novo by acylation of sn glycero-3-phosphate (Stoffel and Schiefer, 1968; Shephard and Hübscher, 1969; Zborowski and Wojtczak, 1969; Daae and Bremer, 1970; Sarzala et al., 1970) or from dihydroxyacetone phosphate (Hajra and Agranoff, 1968a,b) and diphosphatidylglycerol (Davidson and Stanacev, 1971; Hostetler et al., 1971, 1972; Dennis and Kennedy, 1972; Hostetler and Van den Bosch, 1972). In animal cells phosphatidylglycerol is mainly synthesized in mitochondria (Kiyasu et al., 1963; Possmayer et al., 1968; Stanacev et al., 1968, 1969; Davidson and Stanacev, 1970) while in plant cells, the biosynthesis seems to be located both in mitochondrial (Douce, 1968; Douce and Dupont, 1969; Moore, 1974) and microsomal (Marshall and Kates, 1972; Moore, 1974) fractions. Nevertheless, differences in lipid biosynthetic capacity do exist between organelles isolated from different animal tissues (see Davidson and Stanacev, 1974).

It is now accepted that the endoplasmic reticulum is the main site of biosynthesis of nitrogen-containing phospholipids. This cell fraction contains CDPcholine:1,2-diglyceride phosphotransferase (EC 2.7.8.2) which catalyzes the final step of phosphatidylcholine biosynthesis. This has been shown in both animal cells (Schneider, 1963; Wilgram and Kennedy, 1963; Gurr et al., 1965; McCaman and Cook, 1966; Stoffel and Schiefer, 1968; McMurray and Dawson, 1969; Sarzala et al., 1970; Swartz and Mitchell, 1970; Williams and Bygrave, 1970) and in plant cells (Devor and Mudd, 1971; Lord et al., 1972). Similarly, phosphatidylinositol is also synthesized by the endoplasmic reticulum in animal (Jungalwala et al., 1971; Strunecka and Zborowski, 1975) and plant cells (Moore et al., 1973; Bowden and Lord, 1975).

Since mitochondria are unable to synthesize all of their own phospholipids, a cooperation between this organelle and the endoplasmic reticulum seems necessary to explain the origin of the mitochondrial phospholipids. The experimental evidence in support of this concept is presented in the next section.

3. Exchange of phospholipids between membranes

3.1. Identification of phospholipid exchange processes

The interpretation of the data obtained in many early studies of phospholipid metabolism in subcellular organelles is hindered by the questionable purity of the subcellular fractions used. The study of the phospholipid biosynthetic capacity of different organelles, as well as the demonstration of the phospholipid exchange process, obviously requires that the subcellular fraction(s) used are pure and free from contaminants from other cellular material.

Assuming that these important prerequisites can be met, phospholipid exchange is typically monitored by incubating microsomes containing [^{32}P]-labeled phospholipids with unlabeled mitochondria. After incubation of the two fractions together, the fractions are separated and their respective specific radioactivities measured. Since it can be shown that no significant change in the concentration of total phospholipids occurs during the incubation (Fig. 3), it is clear that the observed decrease in radioactivity in the microsomal lipids, and the appearance of radiolabeled phospholipids in the mitochondrial fraction, represents a measure of the extent of phospholipid exchange between these fractions. A key feature of most exchange experiments of this kind is that the fractions are incubated together in the presence of a post-microsomal supernatant which is obtained by centrifugation at 105 000 g for one hour. As

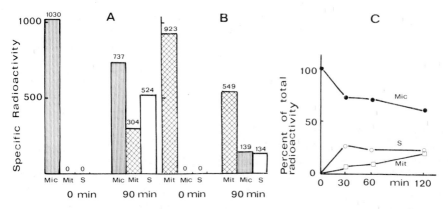

Fig. 3. Exchange of phospholipids between microsomes and mitochondria from rat liver (A and B) or potato tuber (C). A, [^{32}P]-labeled microsomes from rat liver were incubated with unlabeled mitochondria and 105 000 g supernatant for 0 or 90 min at 37°C. The fractions were then separated. The specific radioactivity of the phospholipids, expressed in cpm/μg of phospholipid phosphorus, is indicated at the top of each column; B, similar experiments as in A but starting with labeled mitochondria; C, potato tuber microsomes labeled with [1-^{14}C], incubated with non-radioactive mitochondria and 105 000 g supernatant. The distribution of the label (as % of total radioactivity) between the subcellular fractions was determined at various times of incubation. Mic, microsomes; Mit, mitochondria; S, supernatant. Reproduced with permission from Wirtz and Zilversmit (1968) (A and B) and from Abdelkader and Mazliak (1972) (C).

discussed below, this 105 000 g supernatant significantly enhances phospholipid exchange.

For exchange experiments of the kind outlined above, a number of steps have been devised to restrict cross-contamination of the fractions and/or accurately quantitate the exchange process. For example, to limit cross-contamination of the unlabeled fraction by the labeled one, the incubated fractions are washed many times and purified on sucrose density gradients. In addition, the purity of the fractions, particularly the mitochondrial fraction, can be checked by electron microscopy (Blok et al., 1971). The degree of contamination of one fraction by the other can also be estimated by monitoring the activities of appropriate marker enzymes; i.e., mitochondrial succinate-cytochrome c reductase or cytochrome oxidase and microsomal NADPH cytochrome c reductase (McMurray and Dawson, 1969; Blok et al., 1971) or glucose-6-phosphatase (Akiyama and Sakagami, 1969; Kamath and Rubin, 1973). A level of 1 to 2% of contamination of mitochondria by microsomes has been found in many experiments. Similar exchange assays, but involving membranes containing proteins labeled with [^{3}H]leucine again indicate that the maximum degree of contamination is 1 to 2%, if it is assumed that this contamination is due to coprecipitation of labeled membranes with unlabeled ones (Wirtz and Zilversmit, 1968). A further method involves addition of known amounts of [^{3}H]phosphatidylcholine-labeled microsomes at the end of the incubation to a mixture of [^{14}C]phosphatidylcholine-labeled microsomes and unlabeled mitochondria. Determination of [^{14}C]/[^{3}H] ratio then provides an indicator of the extent of the contamination (Wirtz et al., 1972). Finally, since the exchange processes are in general very slow at low temperatures, the extent of transfer of radiolabeled phospholipid between the labeled and unlabeled fractions at 0°C can be used as a simple control to monitor the extent of cross-contamination between the fractions (Kamath and Rubin, 1973).

In addition, the labeling pattern of the phospholipids of the initially nonradioactive fraction has been found to differ from the pattern of the phospholipids in the initially labeled fraction (Wirtz and Zilversmit, 1968; Akiyama and Sakagami, 1969). Further metabolism of the transferred phospholipid has also been observed (Butler and Thompson, 1975).

By careful checking to distinguish transfer of radiolabel between fractions due to simple cross-contamination from that resulting from the inter-membrane exchange of phospholipid molecules, a large body of evidence has accumulated which indicates that the observed exchange is clearly not due to cross-contamination.

3.2. Distribution of phospholipid exchange processes

Exchange of phospholipids has been shown to occur between membranes isolated from a wide variety of eukaryotic cells (Table I), but studies on prokaryotic cells are lacking. However, exchange of branched fatty acids has been demonstrated between isolated mesosomes and protoplasts of *Bacillus subtilis* (Bureau et al., 1976), but it is not known if this exchange of fatty acids involves phospholipid exchange.

Exchange processes occurring in the rat liver cell have attracted the most attention (Table I). Exchange of phospholipids has been demonstrated between various membranes, including total microsomes, rough or smooth microsomes, calcium-loaded microsomes ("low speed" microsomes of Kamath and Rubin, 1973, 1974a), whole mitochondria, inner or outer membranes of the mitochondria and plasma membranes. Except for the dictyosomal, nuclear and lysosomal membranes, which were not studied, all the membranes of the liver cell can exchange their phospholipids with other subcellular fractions. Phospholipid exchange is not, however, a feature of all mammalian cell membranes. For example, Miller and Dawson (1972) found that the rate of phospholipid exchange between microsomes and synaptosomes in guinea pig brain was very slow, and no exchange could be detected between microsomes and myelin from the same tissue. Similar findings have been obtained by Pasquini et al. (1975) in studies on exchange phenomena between rat brain microsomes and myelin.

Phospholipid exchange capacity has also been found in plant membranes (Table I), though evidence for phospholipid exchange between chloroplast and other cellular fractions is lacking. However, a cooperation between plastids and microsomes was suggested for the biosynthesis of linolenic acid in young pea leaves (Trémolières and Mazliak, 1974).

Exchange of phospholipids has also been studied in certain systems not listed in Table I. Nozawa and Thompson (1972) obtained evidence from in vivo labeling experiments which suggested that exchange of phospholipids occurs in *Tetrahymena pyriformis* and the possibility of intracellular exchange of phospholipids was also raised in Chlapowski and Band (1971a,b) in their studies on lipid metabolism in *Acanthamoeba palestinensis*.

3.3. Properties of phospholipid exchange between membranes

From the experiments listed in Table I, the following general characteristics of the phospholipid exchange process can be defined.

3.3.1. Phospholipid exchange is time and temperature dependent

This dependence was revealed in the first exchange experiments (Wirtz and Zilversmit, 1968; Akiyama and Sakagami, 1969; McMurray and Dawson, 1969) in which it was shown that the intensity of the exchange could be stimulated by incubating organelles at higher temperatures or for longer times (Fig. 4). At 0°C or 4°C the amount of exchange is limited (Wirtz and Zilversmit, 1968) or absent (McMurray and Dawson, 1969). However, appreciable exchange was found to occur at 4°C by Akiyama and Sakagami (1969), Dyatlovitskaya et al. (1972) and Kader (1975). Generally, exchange assays are performed at incubation temperatures varying from 20°C to 37°C using incubation times of 20 to 120 min.

3.3.2. Phospholipid exchange involves intact phospholipid molecules

The use of different radiolabeled phospholipid precursors (phosphate, choline, ethanolamine, serine, inositol, linoleate, acetate and glycerol) has revealed that intact phospholipid molecules are exchanged between membranes (Wirtz and

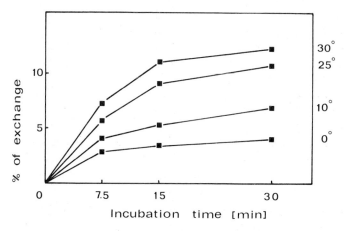

Fig. 4. Time and temperature dependence of intracellular phospholipid exchange. [^{32}P]-labeled microsomes from potato tuber were incubated with unlabeled mitochondria at various temperatures. The percent exchange corresponds to the percentage of the initial microsomal phospholipid radioactivity recovered in mitochondria at the end of incubation. Reproduced with permission from Kader (1975).

TABLE I
EXCHANGE OF PHOSPHOLIPIDS BETWEEN MEMBRANES

Origin	Labeled membrane	Unlabeled membrane	Precursor	Reference
Rat liver	Microsomes	Mitochondria	[^{32}P]phosphate and [1-3^{14}C]glycerol	Wirtz and Zilversmit, 1968
Rat liver	Mitochondria	Microsomes	[^{32}P]phosphate	Wirtz and Zilversmit, 1968
Rat liver	Microsomes	Mitochondria	[^{32}P]phosphate	Kadenbach, 1968
Rat liver	Microsomes	Mitochondria	CDP [1,2-^{14}C]choline and [^{32}P]phosphate	McMurray and Dawson, 1969
Rat liver	Microsomes	Mitochondria	[^{32}P]phosphate	Akiyama and Sakagami, 1969
Rat liver	Microsomes + 105 000 g S^a	Mitochondria	[^{32}P]phosphate	Akiyama and Sakagami, 1969
Rat liver	105 000 g S^a	Mitochondria	[^{32}P]phosphate	Akiyama and Sakagami, 1969
Rat liver	Microsomes	Mitochondria	[^{14}C]choline [^{14}C]glycerol	Beattie, 1969
Rat liver	Microsomes	Mitochondria	[^{32}P]phosphate	Blok et al., 1971
Rat liver	Microsomes	Inner mitochondrial membrane	[^{32}P]phosphate	Blok et al., 1971 Sauner and Lévy, 1971
Rat liver	Microsomes	Outer mitochondrial membrane	[^{32}P]phosphate	Wojtczak et al., 1971
Rat liver	Outer mitochondrial membrane	Inner mitochondrial membrane	[^{32}P]phosphate	Blok et al., 1971 Wojtczak et al., 1971
Rat liver	Microsomes	Mitochondria	[^{32}P]phosphate [^{14}C]choline	Wirtz and Kamp, 1972
Rat liver	Microsomes	Calcium-loaded microsomes	[^{32}P]phosphate	Kamath and Rubin, 1973
Rat liver	Microsomes	Mitochondria	[^{32}P]phosphate	Kamath and Rubin, 1973
Rat liver	Microsomes	Plasma membrane	[^{32}P]phosphate	Kamath and Rubin, 1973
Rat liver	Mitochondria	Microsomes	[^{32}P]phosphate	Kamath and Rubin, 1973
Rat liver	Calcium-loaded microsomes	Microsomes	[^{32}P]phosphate	Kamath and Rubin, 1973
Rat liver	Rough microsomes	Smooth calcium-loaded microsomes	[^{32}P]phosphate	Kamath and Rubin, 1973

Tissue	Fraction	Label	Reference
Rat liver	Smooth microsomes	Rough calcium-loaded microsomes	[^{32}P]phosphate	Kamath and Rubin, 1973
Rat liver	Microsomes	Mitochondria	[^{3}H]inositol	Strunecka and Zborowski, 1975
Rat liver	Rough microsomes	Mitochondria	[^{32}P]phosphate	Taniguchi and Sakagami, 1975
Rat liver	Smooth microsomes	Mitochondria	[^{32}P]phosphate	Taniguchi and Sakagami, 1975
Rat liver	Microsomes	Mitochondria	[L-3^{14}C]serine	Butler and Thompson, 1975
Rat small intestinal mucosa	Microsomes	Mitochondria	[^{32}P]phosphate	Horiuchi, 1973
Rat small intestinal smooth muscle	Microsomes	Mitochondria	[^{32}P]phosphate	Minakawa, 1974
Rat hepatoma	Microsomes	Mitochondria	[^{32}P]phosphate	Dyatlovitskaya et al., 1972
Rat brain	Microsomes	Mitochondria	[^{3}H]phosphatidyl-inositol, [^{14}C]choline	Possmayer, 1974
Beef heart	Mitochondria	Microsomes	[^{32}P]phosphate	Wirtz and Zilversmit, 1970
Beef liver	Microsomes	Mitochondria	[^{32}P]phosphate	Wirtz et al., 1972
Beef liver	Mitochondria	Microsomes	[^{14}C]choline	Wirtz et al., 1972
Pig thyroid	Microsomes	Mitochondria	[^{32}P]phosphate	Jungalwala et al., 1971
Guinea pig liver	Microsomes	Mitochondria	[^{32}P]phosphate	Miller and Dawson, 1972
Guinea pig liver	Mitochondria	Microsomes	[^{32}P]phosphate	Miller and Dawson, 1972
Guinea pig liver	Microsomes	Synaptosomal fractions	[^{32}P]phosphate	Miller and Dawson, 1972
Potato tuber	Microsomes	Mitochondria	[^{32}P]phosphate	Abdelkader and Mazliak, 1970
Potato tuber	Mitochondria	Microsomes	[1-3^{14}C]glycerol	Abdelkader and Mazliak, 1970
Cauliflower florets	105 000 g S[a]	Mitochondria Microsomes	[^{14}C]acetate	Abdelkader and Mazliak, 1970
	105 000 g S[a]			
Potato tuber	Microsomes	Mitochondria	[^{32}P]phosphate	Kader, 1975

[a] 105 000 g post-microsomal supernatant.

Zilversmit, 1968; Abdelkader and Mazliak, 1970). When microsomes containing double-labeled phosphatidylcholine ([^{3}H]choline to label the base moiety and [^{14}C]linoleate to label the acyl moiety) were incubated with unlabeled mitochondria, the phosphatidylcholine molecules recovered in mitochondria possessed similar ^{3}H/^{14}C ratios to the original microsomes (Wirtz and Kamp, 1972). However, as explained in section 4, lysophosphatidylcholine is exchanged between plasma and erythrocytes more rapidly than phosphatidylcholine (Sakagami et al., 1965b).

3.3.3. Phospholipid exchange is bidirectional
The decrease in specific radioactivity of the phospholipids in the initially radiolabeled subcellular fraction (A) and the increase in this parameter in the unlabeled fraction (B) are easily explained by a transfer of labeled phospholipids from (A) to (B) and a concomitant transfer of unlabeled phospholipids from (B) to (A), since no change in the concentration of the total phospholipid pool is observed during the incubation. The reverse experiment, starting from (B) with a radiolabeled preparation and unlabeled (A) has confirmed this assertion.

3.3.4. Phospholipid exchange is energy-independent and does not require biologically active membranes
McMurray and Dawson (1969) showed that the exchange of phospholipids from microsomes to mitochondria in rat liver did not require provision of an energy substrate. Similarly, Kamath and Rubin (1973) found that addition of inhibitors of energy metabolism such as KCN, dinitrophenol or rotenone did not alter the exchange process. Moreover, mitochondria denatured by trichloracetic acid or by boiling can exchange their phospholipids to the same extent as untreated control fractions (Kamath and Rubin, 1973; Butler and Thompson, 1975). Thus, biological activity does not seem necessary in either the donor or the acceptor membranes for exchange of phospholipids to occur.

3.3.5. Phospholipid exchange involves a major proportion of the phospholipid pool
Up to 35% (Wirtz and Zilversmit, 1968) or 40% (McMurray and Dawson, 1969) or 10 to 40% of the total radioactivity of microsomal phospholipids (Abdelkader and Mazliak, 1970) have been recovered in mitochondria in exchange experiments. These values clearly indicate the considerable scale of the exchange process.

It is easy to calculate the amount of phospholipids transferred between two membranes. The percentage of decrease in specific radioactivity of phospholipids in the initially radioactive membrane indicates the extent of exchange of phospholipids (e.g., 28.5% in Fig. 3). An alternative calculation uses the previously unlabeled fraction. For example, if A (cpm) is the *total* radioactivity of phospholipids in this fraction after incubation (A is, of course, corrected for eventual loss of fraction and for cross-contamination) and if B is the initial *specific* radioactivity of phospholipids in the labeled fraction (in cpm/µg of phospholipid), then the amount of phospholipids transferred between these fractions is given by the ratio, A/B.

Despite the large amount of phospholipid exchange that can occur, total depletion of labeled phospholipids from the initially radiolabeled fraction has never been observed, and the unlabeled fraction can thus never incorporate 100% of the total, initial radioactivity at the end of the incubation. This may reflect reverse transfer of nonradioactive phospholipids originating from the initially unlabeled compounds. However, another possibility is that exchange of phospholipids concerns only a part of the phospholipid pool of the membrane, and this will be examined in section 11.

3.3.6. Phospholipid exchange does not involve all the phospholipid classes

The distribution of the radiolabel in specific phospholipids has been found to be different in the initially labeled microsomes from rat liver and in the unlabeled mitochondria preincubated with this fraction (Wirtz and Zilversmit, 1968; McMurray and Dawson, 1969). For instance, 62% of the [^{32}P]-label in phospholipids was associated with mitochondrial phosphatidylcholine when this organelle was preincubated with microsomes in which phosphatidylcholine represented only 40.9% of the [^{32}P] radioactivity (Wirtz and Zilversmit, 1968). This finding suggested that phosphatidylcholine exchanged more rapidly than the other phospholipids, and it was shown later that phosphatidylinositol and phosphatidylethanolamine exchanged at a slower rate than phosphatidylcholine, at least in rat liver membranes, while diphosphatidylglycerol and phosphatidic acid were almost inactive (Wirtz and Zilversmit, 1968; McMurray and Dawson, 1969). However, except for diphosphatidylglycerol, which was not exchanged, no such differences have been detected in the exchange capacity of different phospholipids in plant organelles (Abdelkader and Mazliak, 1970) or in guinea pig brain membranes (Miller and Dawson, 1972).

The striking differences in the exchange kinetics of individual phospholipids observed in rat liver membranes may reflect the location of particular phospholipids within the membrane which, in turn, may influence their availability for exchange. For instance, it is suggested that the polar head groups of the diphosphatidylglycerol molecules are buried within the membrane or tightly bound to other membrane constituents (Guarnieri et al., 1971) and this may serve as sufficient restraint to limit exchange of this particular phospholipid. However, another factor, the 105 000 g supernatant, plays a role in the different behavior of individual phospholipids.

3.3.7. Phospholipid exchange is stimulated by the 105 000 g supernatant

Phospholipid exchange between microsomes and mitochondria is more intense in presence of the 105 000 g supernatant than with buffered sucrose solution alone (Wirtz and Zilversmit, 1968; Akiyama and Sakagami, 1969; McMurray and Dawson, 1969). Almost all the experiments described in this chapter were performed in the presence of a 105 000 g supernatant. This supernatant enhances the phospholipid exchange between membranes, but the 105 000 g supernatant from rat liver is more effective in stimulating the exchange of phosphatidylcholine than that of phosphatidylethanolamine (Wirtz and Zilversmit, 1968; McMurray and Dawson, 1969; Blok et al., 1971; Wirtz et al., 1972; Taniguchi and Sakagami, 1975). The exchange of phosphatidylinositol (Strunecka and Zborowski, 1975) and phosphatidylserine (Butler and Thompson, 1975) is also stimulated by the addition of 105 000 g supernatant. The postmicrosomal supernatant also stimulated the exchange of phospholipids in guinea pig brain except for diphosphatidylglycerol (Miller and Dawson, 1972). The same results have been obtained in studies with plant tissues such as potato tuber and cauliflower florets (Abdelkader and Mazliak, 1970).

All these supernatants contain phospholipids. For example, it was determined in rat liver that $0.2 \mu g$ of phospholipid per mg of protein were present in the 105 000 g supernatant. These phospholipids are probably associated to proteins in the form of various lipid-protein complexes, and it is logical to suppose that these lipid-protein complexes play a role in the exchange of phospholipids between membranes. For example, the phospholipids bound in the lipid-protein complexes may be exchanged with membrane phospholipids and transferred to another membrane. A collision complex may precede this exchange. However, in contradiction of this suggestion, concentration of assay mixtures to small volumes did not stimulate the exchange (Miller and Dawson, 1972). In spite of this result, the lipid-protein complex hypothesis is presently considered the most logical potential mechanism to account for the observed exchange, and this concept will be discussed in more detail later in this article.

4. Phospholipid exchange involving lipoproteins

The intracellular exchange of phospholipids is presumed to involve the transfer to a membrane of phospholipid molecules which have not been synthesized in this structure. At the level of the cell or of entire organism, transfer of lipids can be viewed as being of considerable physiologic importance. Two examples will suffice: (1) The erythrocyte cannot synthesize its phospholipids de novo. Only acylation of lysophospholipids is possible, and transfer of phospholipids or lysophospholipids from the plasma lipoproteins to the erythrocyte membrane may influence the phospholipid composition of the erythrocyte membrane; and (2) linoleic acid is an "essential" fatty acid which is not synthesized by the animal cell but which is formed by plant cells (Vijay and Stumpf, 1972; Abdelkader et al., 1973). The linoleic acid molecules, present in the diet, are taken by the animal and are incorporated into the intracellular membranes, and it seems likely that a transfer mechanism from outside to inside the cell is necessary for achieving this.

Phospholipid exchange between lipoproteins, between erythrocytes and plasma lipoproteins and between non-erythroid cells and the external medium has been detected in numerous studies and these processes will be discussed in the following sections.

4.1. Lipoprotein–lipoprotein exchange

As emphasized by Dawson (1973) and Wirtz (1974), experiments involving human (Kunkel and Bearn, 1954; Eder, 1957; Florsheim and Morton, 1957) or rabbit lipoproteins (Eder et al., 1954) were the first to show movements of phospholipids between lipid-protein complexes. These investigations were based on incubation of [^{32}P]-labeled plasma lipoproteins with nonradioactive lipoproteins of differing density such as high density (Scanu, 1969) or low or very low density lipoproteins (Margolis, 1969). Separation of these molecular species by ultracentrifugation after incubation together revealed that rapid exchange of phospholipids occurred. Illingworth and Portman (1972b) determined the rates of exchange of phospholipids between low and high density lipoproteins of the plasma of squirrel monkeys. Phosphatidylcholine and sphingomyelin exchanged rapidly between lipoproteins, and complete equilibration of the phospholipid pools was attained after 4 or 5 h. Phospholipid exchange was also observed between very low and high density lipoproteins from rats (Rubenstein and Rubinstein, 1972).

4.2. Lipoproteins and cellular fractions

Plasma lipoproteins can exchange phospholipids with subcellular structures such as mitochondria (Tarlov, 1968; Wirtz and Zilversmit, 1969b) and microsomes (Zilversmit, 1971a). A net transfer of microsomal phosphatidylcholine and phosphatidylethanolamine to plasma was noted when rat or guinea pig liver microsomes were incubated with plasma. By using [^{32}P]-labeled microsomes, an active exchange of phospholipids, mainly phosphatidylcholine and, to a lesser extent, sphingomyelin, was demonstrated. Furthermore, the addition of the 105 000 g supernatant strongly increased this exchange (Zilversmit, 1971a). A similar stimulatory effect with this supernatant was observed with the mitochondria-plasma assay performed with rat liver (Wirtz and Zilversmit, 1969b) and also between plasma membrane of squirrel monkey liver cells and high or low density lipoproteins (Illingworth et al., 1973). However, no stimulation by rat liver supernatant was noted for the exchange of phospholipids between human erythrocytes and rat liver microsomes (Wirtz, 1974).

4.3. Chylomicrons and serum lipoproteins

McCandless and Zilversmit (1958) and Havel and Clarke (1958) observed phospholipid exchange between the membranes of "chylomicrons" (see Zilversmit, 1969 for definition) and serum. Incubation of dog lymph chylomicrons with serum produced a loss of chylomicron phospholipids, accompanied by an exchange of phosphatidylcholine and, to a lesser extent, sphingomyelin, but no net transfer of sphingomyelin was observed. The phospholipids lost from the chylomicrons appeared in the high-density lipoproteins. The exchange of phospholipids could not be accounted for by intact lipoproteins since, while phospholipids were transferred from chylomicrons to serum, free cholesterol moved in the opposite direction, and the cholesterol esters did not move at all (Minari and Zilversmit, 1963).

4.4. Erythrocytes and plasma lipoproteins

Since the ability of erythrocytes to synthesize phospholipids (Percy et al., 1973) or fatty acids (Mulder and Van Deenen, 1965a; Pittman and Martin, 1968) is limited, the renewal of phospholipids in this cell depends to a major extent on phospholipid exchange with plasma (reviews, Van Deenen and De Gier, 1964, 1974). This exchange has been demonstrated in a variety of mammalian erythrocytes (Hahn and Hevesy, 1939; Reed, 1959; Lovelock et al., 1960; Rowe, 1960; Polonovski and Paysant, 1963; Mulder and Van Deenen, 1965b; Sakagami et al., 1965a,b; Shohet and Nathan, 1970). According to Soula et al. (1972), the exchange of phospholipids between chicken erythrocytes and plasma was slow, but subsequent investigations from the same laboratory (Dousset and Douste-Blazy, 1974) have shown that gamma irradiation (40 krad) of rat erythrocytes stimulates the exchange of phospholipids, particularly lysophosphatidylcholine, between erythrocytes and plasma.

The exchange involves not only intact phospholipids such as phosphatidylcholine and sphingomyelin, but also lysophosphatidylcholine (Sakagami et al., 1965b; Soula et al., 1967; Dousset and Douste-Blazy, 1974). This lysophospholipid can be acylated by the erythrocyte (Oliveira and Vaughan, 1964; Van Deenen and De Gier, 1964; Robertson and Lands, 1964; Mulder et al., 1965; Waku and Lands, 1968; Shohet and Nathan, 1970). Attention must be paid to this conversion for the determination of exchange rates. Not all of the membrane phospholipids are involved in the exchange between erythrocytes and plasma. Phosphatidylserine and phosphatidylethanolamine are not exchanged, at least in human and dog erythrocytes (Reed, 1968), and it is interesting to note that these latter phospholipids have not been found in the plasma of several mammalian species (Nelson, 1967).

An interesting question concerns the importance of the phospholipid pool available for the exchange (see Dawson, 1973). Reed (1968) calculated that about 60% of the total membrane phosphatidylcholine and 30% of the sphingomyelin were available for exchange in dog and human erythrocytes. Smith and Rubinstein (1974) found that the phosphatidylcholine of the erythrocyte labeled in vivo with [^{32}P] exchanged more rapidly than phosphatidylcholine labeled in vitro with [^{14}C]linoleate. The phosphatidylcholine pool labeled in vivo corresponds to the portion of cellular phosphatidylcholine available for exchange (about 55%). This pool can be labeled in vitro with [^{14}C]linoleate, while [^{14}C]linoleate labels the entire phosphatidylcholine pool. It is assumed that the phosphatidylcholine pool unavailable for the exchange is located in the inner surface of the membrane, since disruption of the erythrocyte membrane stimulates the exchange. According to Bretscher (1972, 1973, 1974), Verkleij et al. (1973) and Zwaal et al. (1973), the phosphatidylcholine molecules are mainly located on the outside of the erythrocyte membrane. Since phosphatidylcholine is the major phospholipid of the plasma, it is tempting to assume that the exchange of phosphatidylcholine and sphingomyelin between the plasma (which is rich in these phospholipids) and the outer monolayer of the membrane is responsible for this asymmetry. However, phospholipid asymmetry could also originate in the initial biogenesis of the membrane (see Nicolson et al., this volume, p. 20).

4.5. Tissues and the external medium

Exchange of phospholipids between cells and the medium is not restricted to the erythrocyte-plasma system. Phospholipid exchange has been observed to occur between cultured chick embryo fibroblasts and their growth medium (Peterson and Rubin, 1969, 1970), between rat liver slices and serum (Kook and Rubinstein, 1969, 1970) and between rapidly dividing human cells and plasma lipoproteins (Illingworth et al., 1973). In these experiments, attention was given to the rupture of intact cells as a source of phospholipids by following the release of cellular DNA into the medium (Peterson and Rubin, 1969). These investigations showed a release of phospholipids (labeled with $[^{32}P]$phosphate, $[^{3}H]$palmitate or $[^{14}C]$choline) into the medium. This release was accelerated by the addition of plasma lipoproteins to the medium (Peterson and Rubin, 1969; Illingworth et al., 1973).

Intact cells are also able to take up labeled phospholipids present in lipoproteins. These data suggest that two processes occur, a net transfer of phospholipids from the cells to the external medium and an exchange of cellular phospholipids with phospholipids associated with the lipoproteins.

The movement of phospholipids between lipoproteins and intact cells probably plays a role in the renewal of the phospholipid pool of tissue cells. It is not known if this exchange concerns mainly the plasma membrane or whether intracellular membranes also participate. If the former is the case, one might suggest that the phospholipid composition of the plasma membrane partially reflects the composition of the external medium. Other evidence to support the occurrence of exchange processes in vivo can be found in the distribution in different animal tissues of double labeled phosphatidylcholine after intravenous injection (Lekim et al., 1972).

Other means for the transport of phospholipids as lipid-protein complexes may occur within the cell via the activity of the endoplasmic reticulum or dictyosomes. For instance, Glauman et al. (1975) showed that lipoproteins assembled in the rough and smooth endoplasmic reticulum of rat liver were transferred to the Golgi apparatus and were then finally released into the circulation.

All of the exchange processes described so far in this section have involved lipoproteins. It is thus reasonable to assume that the apolipoproteins function as carriers of phospholipids and that the phospholipids associated with lipoproteins can exchange with the phospholipids of other lipoproteins or cell membranes by formation of collision-complexes. To explain the different rates of exchange of different phospholipid species, it can be suggested that these are related to variation in the steric availability within the donor structure.

It is tempting to conclude that similar mechanisms are involved in the phospholipid exchange between subcellular fractions, since a soluble fraction stimulates the exchange. However, a careful study of this soluble fraction has revealed additional properties, and these are discussed in the next section.

Incubation of cultured cells with artificial lipid vesicles (liposomes) prepared from a variety of natural and synthetic phospholipids has also been shown to result in extensive cellular incorporation of phospholipids (up to $1 \cdot 10^9$ molecules per cell) (Papahadjopoulos et al., 1973; Huang and Pagano, 1975; Pagano and Huang, 1975; Poste and Papahadjopoulos, 1976). While the bulk of the incorporated lipid appears to arise via uptake of intact vesicles, either by their fusion with the plasma membrane or by endocytosis, the exchange of phospholipids also occurs and accounts for about 10% of the total uptake of radiolabeled lipid.

5. Involvement of proteins in the exchange of phospholipids between membranes

Since the post-microsomal 105 000 g supernatant strongly increases phospholipid exchange between membranes, considerable attention has been given to the properties and mode of action of this fraction.

5.1. Role of the pH 5.1 supernatant

The 105 000 g supernatant involved in stimulating phospholipid exchange contains phospholipids and proteins. Wirtz and Zilversmit (1969b) showed that about 95% of the phospholipids associated with the supernatant from rat liver could be removed by simple adjustment of the pH of the supernatant to 5.1. This procedure also produces loss of about 40% of the protein originally associated with the supernatant. Importantly, this pH adjustment is not accompanied by any loss in its phospholipid exchange stimulation activity.

The extent of the exchange induced by the pH 5.1 supernatant is related directly to the amount introduced into the incubation medium. A typical experiment is presented in Fig. 5 which shows that exchange in the absence of pH 5.1 supernatant is very low.

5.1.1. Generalization of the pH 5.1 supernatant stimulated exchange
The stimulation of phospholipid exchange by the pH 5.1 supernatant first observed with rat liver has since been found with similar material isolated from a wide variety of sources. Table II indicates that pH 5.1 supernatants or 105 000 g supernatants isolated from animal or plant cells are able to stimulate the exchange of phospholipids between many membranous structures, in-

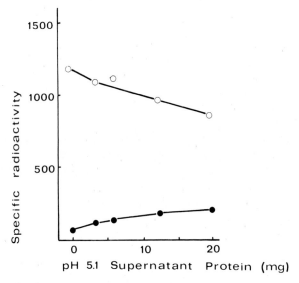

Fig. 5. Exchange of phospholipids labeled with [^{32}P]phosphate between rat liver mitochondria (●) and microsomes (○) incubated with increasing amounts of pH 5.1 supernatant from rat liver. The specific radioactivity of the phospholipids is expressed in cpm/μg phospholipid phosphorus. Reproduced with permission from Wirtz and Zilversmit (1969b).

cluding lipoproteins. Rat kidney or heart cytosols also contain stimulatory factors (Wirtz, 1971). Also of interest, however, is the finding that plasma from the rat (Zilversmit, 1971b) or the squirrel monkey (Illingworth and Portman, 1972a) cannot stimulate exchange of phospholipids, respectively, between liposomes and rat liver mitochondria or between high-density and low-density lipoproteins. In some cases, a net transfer of phospholipids from microsomes to plasma was catalyzed by pH 5.1 supernatant. However, rat liver pH 5.1 supernatant did not stimulate phospholipid exchange between human erythrocytes and rat liver microsomes (Wirtz, 1974). The same supernatant enhanced cholesterol exchange between rat liver microsomes and mitochondria and between erythrocytes and microsomes or mitochondria (Bell, 1975).

5.1.2. Which phospholipids are preferentially transferred?
The pH 5.1 and 105 000 g supernatants of rat liver preferentially enhance the exchange of phosphatidylcholine between membranes and between microsomes and plasma (Table II). Phosphatidylethanolamine was exchanged to a smaller extent. The same result was obtained for a beef liver pH 5.1 supernatant, while plant and guinea pig brain 105 000 g supernatants do not appear to possess particular specificities.

What determines this exchange specificity? Some insight into this question was provided by the experiments of Wirtz (1974) and Harvey et al. (1974). They examined pH 5.1 supernatants from rat liver and beef brain, which differed markedly in their ability to stimulate phosphatidylcholine and phosphatidyl-

TABLE II
ENHANCEMENT OF PHOSPHOLIPID EXCHANGE BETWEEN MEMBRANES BY pH 5.1 SUPERNATANT OR 105 000 g SUPERNATANT

Origin	Exchange assay	Molecular specificity[c]	References
Rat liver	Microsomes–mitochondria	PC > PE	Wirtz and Zilversmit, 1969b
Rat liver	Mitochondria–plasma	—	Wirtz and Zilversmit, 1969b
Rat liver	Microsomes–mitochondria[a]	PC and FI > PE	McMurray and Dawson, 1969
Rat liver	Microsomes–mitochondria[a]	PC > PE	Akiyama and Sakagami, 1969
Rat liver	Microsomes–plasma	PC > SM > PE	Zilversmit, 1971b
Rat liver	Microsomes–inner membrane	PC > PE	Blok et al., 1971
Rat liver	Microsomes–outer membrane	PC > PE	Blok et al., 1971
Rat liver	Different microsomal fractions–mitochondria–plasma membranes	—	Kamath and Rubin, 1973
Rat liver	Rough microsomes–mitochondria Smooth microsomes–mitochondria	PC > PE	Taniguchi and Sakagami, 1975
Rat liver	Microsomes–mitochondria	PI[b]	Strunecka and Zborowski, 1975
Rat liver	Microsomes–mitochondria	PS[b]	Butler and Thompson, 1975
Rat brain	Microsomes–mitochondria[a]	PI	Possmayer, 1972
Guinea pig brain	Microsomes–mitochondria[a]	All phospholipids except DPG	Miller and Dawson, 1972
Pig thyroid	Microsomes–mitochondria	PI	Jungalwala et al., 1971
Beef liver	Microsomes–mitochondria	PC > PI > PE	Wirtz et al., 1972
Beef brain	Microsomes–mitochondria	PI > PC > PE	Harvey et al., 1974
Squirrel monkey	High density lipoproteins–low density lipoproteins	PC, SM	Illingworth and Portman, 1972b
Squirrel monkey	Plasma membrane–high or low density lipoproteins	PC, SM	Illingworth et al., 1973
Potato tuber, cauliflower	Microsomes–mitochondria	All phospholipids except DPG	Abdelkader and Mazliak, 1970

[a]105 000 g post-microsomal supernatant was used.
[b]The other phospholipids were not studied.
[c]PC, phosphatidylcholine; PE, phosphatidylethanolamine; PI, phosphatidylinositol; SM, sphingomyelin; DPG, diphosphatidylglycerol.

inositol exchange, respectively (Table II). Using rat liver microsomes and mitochondria to follow this exchange, these investigators found that when rat liver pH 5.1 supernatant was introduced in the exchange mixture, phosphatidylcholine was the most rapidly exchanged, whereas when the beef brain pH 5.1 supernatant was added, phosphatidylinositol was preferentially exchanged (Fig. 6). This experiment clearly indicates that the molecular specificity of the phospholipid exchange is imposed by a factor(s) in the supernatant.

In connection with this study, it can be mentioned that exchange of phospholipids also occurs between microsomes and mitochondria of different plant tissues incubated with 105 000 g supernatants obtained from the same species and from different species. Interestingly, more labeled phospholipids accumulated in the supernatant when fractions from two species were mixed than when all fractions were from the same species. This would suggest that a certain degree of genetic specificity is also involved in the exchange process (Abdelkader and Mazliak, 1970).

5.1.3. Evidence for the protein nature of the active fraction
The stimulatory effect of pH 5.1 supernatants isolated from various sources (rat liver, guinea pig brain, beef liver and brain) is diminished by heat treatment and by trypsin digestion (Wirtz and Zilversmit, 1968; McMurray and Dawson, 1969; Miller and Dawson, 1972; Helmkamp et al., 1974; Kamath and Rubin, 1974a; Lutton and Zilversmit, 1976). When the pH 5.1 supernatant from rat liver was heated for 10 min at 60°C, a 50% reduction in phospholipid exchange between labeled mitochondria and unlabeled microsomes was noted (Wirtz and Zilversmit, 1968). Another important observation is that the phospholipid exchange stimulatory activity can be recovered in an ammonium sulfate precipitate obtained from the pH 5.1 supernatant, and activity is also retained after dialysis (Wirtz and Zilversmit, 1969b; Miller and Dawson, 1972; Lutton

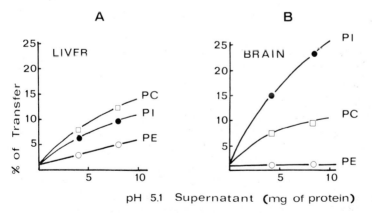

Fig. 6. Exchange of phospholipids between [^{32}P]-labeled microsomes and unlabeled mitochondria from rat liver in the presence of pH 5.1 supernatant prepared from rat liver (A) or rat brain (B). The extent of the exchange was expressed as a percentage of the radioactivity of the microsomal phospholipids recovered in mitochondria. Reproduced with permission from Wirtz (1974).

and Zilversmit, 1976). All these results strongly suggest that the phospholipid exchange activity of pH 5.1 supernatants is associated with a protein(s). This protein must be specific, since other proteins such as serum albumin are unable to promote this exchange (Wirtz and Zilversmit, 1968; Wirtz and Kamp, 1972). However, McMurray (1973) has shown that high concentrations of cytochrome c could stimulate phospholipid exchange between microsomes and mitochondria, but no physiological role can be reasonably attributed to this effect.

The obvious next step was, therefore, to attempt to isolate and purify these soluble proteins, and this was first done by Wirtz and Zilversmit (1970).

5.2. Discovery of phospholipid exchange proteins (PLEP)

Using ammonium sulfate precipitation, hydroxylapatite adsorption-desorption and Sephadex G-100 chromatography, Wirtz and Zilversmit (1970) achieved an 82-fold purification of a protein from beef heart postmicrosomal supernatant (Fig. 7) which was able to stimulate the exchange of [^{32}P]-labeled phospholipids between mitochondria and microsomes. This protein fraction, designated as *phospholipid exchange protein* (PLEP), stimulated only the exchange of phosphatidylcholine and phosphatidylinositol and not the exchange of phosphatidylethanolamine, while the unpurified pH 5.1 supernatant was able to stimulate the exchange of all three phospholipids.

Many other PLEP have now been isolated from various sources (see section 7). The discovery of PLEP accounted for many properties of the stimulatory effect of soluble supernatant on this exchange and introduced new concepts concerning the phospholipid exchange mechanism, since soluble proteins

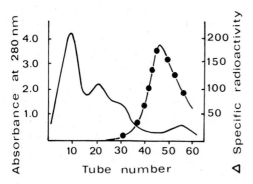

Fig. 7. Isolation of a phospholipid exchange protein from an animal cytosol (beef heart). After adjustment to pH 5.1, followed by ammonium sulfate precipitation and hydroxylapatite adsorption-desorption, the active proteins from the pH 5.1 supernatant were chromatographed on a Sephadex G-100 column. The elution pattern (continuous line) was determined at 280 nm. The different fractions were assayed for phospholipid exchange activity by incubating microsomes containing [^{32}P]-labeled phospholipids and unlabeled mitochondria with various amounts of these fractions and the activity expressed as the increase in specific radioactivity of the phospholipids (△) produced by the addition of 1 mg of protein of each fraction (●). Reproduced with permission from Wirtz and Zilversmit (1970).

seemed able to control the preferential exchange of particular phospholipids between membranes. How do these proteins act? Do they carry phospholipids, or do they act on phospholipid–protein interactions in the membranes to increase membrane fluidity and thus facilitate movement of lipids?

To answer these questions it was necessary to simplify the exchange assay performed between two types of membranes by replacing one of the membranes, or both, by bilayers or monolayers of phospholipids of defined composition. The use of artificial phospholipid vesicles (liposomes) has been of great value in this respect.

6. Phospholipid exchange in artificial (model) membranes

6.1. Phospholipid exchange between liposomes and cell fractions

Zilversmit (1971b) studied phospholipid exchange between unlabeled mitochondria and liposomes prepared from [^{32}P]phosphatidylcholine. The liposomes were made by vigorous shaking of an aqueous solution into which the labeled phospholipid was introduced (in later experiments ultrasonic irradiation replaced the shaking) and a trace of [^{14}C]triolein was also added as a non-exchangeable marker. Rat liver mitochondria were then incubated at 37°C for 40 min with the labeled liposomes. After centrifugation and washing to separate the liposomes and the mitochondria, the phospholipids of the mitochondrial pellets were found to contain from 2.8 to 10.3% of the initial radioactivity of the liposomes, while contamination of mitochondria by liposomes was less than 1%. Moreover, when a pH 5.1 post-microsomal supernatant fraction was introduced in the incubation mixture, a considerable increase in the percentage of exchange was observed with 24.2% to 39% of the original radioactivity appearing in mitochondrial pellets. The stimulation of the exchange by a rat liver pH 5.1 supernatant depended on the amount of protein introduced (Fig. 8A). This finding probably explains why Fleischer and Brierley (1961) failed to observe exchange of phospholipids between liposomes and mitochondria. Dobiasova and Linhart (1970) observed uptake of emulsified radiolabeled phosphatidylcholine, phosphatidylethanolamine or cholesterol by rat heart mitochondria. In this system divalent cations were found to stimulate uptake, which is due partly to lipid exchange and partly to lipid aggregation. A similar uptake was detected with rat brain mitochondria (Dobiasova and Radin, 1968).

Phospholipid exchange has also been observed between unlabeled mitochondria and radiolabeled chylomicrons or artificial fat emulsions (Zilversmit, 1971b). Again, the pH 5.1 supernatant enhanced the phospholipid exchange or, more correctly, induced exchange, since little movement of phospholipids was detectable in the absence of the pH 5.1 supernatant. It is interesting to note that plasma lipoproteins did not stimulate this exchange at all (Zilversmit, 1971b).

Beef liver pH 5.1 supernatant has also been shown to strongly stimulate

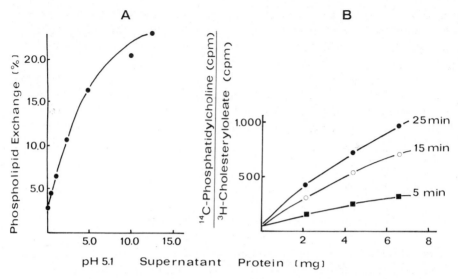

Fig. 8. Stimulation of phospholipid exchange between liposomes and organelles by pH 5.1 supernatant. A, effect of addition of pH 5.1 supernatant from rat liver (in mg of protein) on phospholipid exchange between [^{32}P]-labeled phosphatidylcholine liposomes (also containing [^{14}C]triolein) and unlabeled mitochondria. The percentage exchange was calculated as the corrected percentage of radioactivity recovered in the mitochondrial phospholipids after incubation. Reproduced with permission from Zilversmit (1971b); B, beef liver microsomes containing [^{14}C]phosphatidylcholine were incubated with phosphatidylcholine liposomes containing [^{3}H]-cholesteryloleate, in the presence of various amounts of beef liver pH 5.1 supernatant for various times as indicated. The microsomes were then precipitated by adjustment of the pH to 5.1. The increase in ^{14}C/^{3}H ratio of the liposomal phosphatidylcholine indicated the extent of the exchange. Reproduced with permission from Kamp et al. (1973).

phosphatidylcholine exchange between liposomes and microsomes (Kamp et al., 1973). Microsomes containing [^{14}C]phosphatidylcholine were incubated with liposomes prepared from this phospholipid and containing [7α-^{3}H]cholesteryl-oleate as a non-exchangeable marker. The exchange of phosphatidylcholine, indicated by the increase in ^{14}C/^{3}H ratio of the liposomal phospholipids, was enhanced by the addition of pH 5.1 supernatant (Fig. 8B).

Plant mitochondria are also able to accept lipids from labeled liposomes (Douady and Mazliak, 1975). [^{14}C]Acetate-labeled liposomes, prepared from total cell lipids, were incubated with cauliflower floret mitochondria for various times. Depending on experimental conditions, up to 30% of the initial radioactivity was recovered in mitochondrial lipids (Fig. 9). However, it is not known if phospholipid molecules were actually transferred in these experiments, since only total lipids, labeled from [^{14}C]acetate were examined.

Table III summarizes the various exchange assays between [^{3}H]- or [^{14}C]phosphatidylcholine liposomes of [^{3}H]phosphatidylinositol-containing liposomes and mitochondria or erythrocyte ghosts. Enhancement of the exchange of phospholipid was obtained by introduction in the incubation medium of pH 5.1 supernatant or purified PLEP isolated from beef, rat or calf

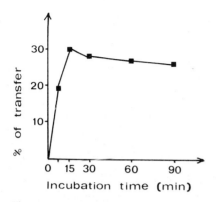

Fig. 9. Transfer of radioactive lipids from [^{14}C]acetate labeled liposomes to cauliflower floret mitochondria. Unlabeled mitochondria were incubated with radioactive liposomes for the indicated times and the percentage transfer was determined as a percentage of the original liposomal radioactivity recovered in the mitochondrial pellets at the end of the incubation. Reproduced with permission from Douady and Mazliak (1975).

livers, beef heart or brain, potato tuber or cauliflower floret (Table III). These proteins presumably also exist in photosynthetic cells, and it was recently found that a pH 5.1 post-microsomal supernatant prepared from spinach leaves was able to transfer phosphatidylcholine from [^{3}H]phosphatidylcholine liposomes containing [^{14}C]cholesteryloleate (as non-exchangeable tracer) to a spinach plastid fraction (Dubacq and Kader, unpublished observations).

It is remarkable that the capacity of the pH 5.1 supernatant isolated from various tissues to specifically enhance the movement of particular phospholipids between subcellular fractions is also manifest in studies with liposomes and subcellular fractions (Wirtz et al., 1972). Moreover, as is discussed in section 6, purified PLEP are able to discriminate between the different phospholipids of liposomes with, in some cases, exchange of only a single phospholipid.

In experiments of the type outlined in Table III, double labeling of liposomes with both an exchangeable radioactively labeled phospholipid and a non-exchangeable radiolabel offers a convenient inbuilt control for assessing the extent of cross-contamination, since the amount of non-exchangeable liposome label recovered in the cell fraction provides an accurate index of the level of contamination by liposomes. A variety of non-exchangeable radioactive markers have been used for this purpose, including [^{14}C]triolein (Zilversmit, 1971; Ehnholm and Zilversmit, 1973; Johnson and Zilversmit, 1975), [7α-^{3}H]cholesteryloleate, or [1-^{14}C]cholesteryloleate (Kamp et al., 1973; Hellings et al., 1974; Helmkamp et al., 1974; Rothman and Dawidowicz, 1975; Wirtz et al., 1976); and [^{3}H]estradiol (Kader and Douady, unpublished observations).

Recently, Lutton and Zilversmit (1976) have begun to study the anatomical distribution of phospholipid exchange activities in different cells and tissues. Their approach involves isolation of 105 000 g post-microsomal supernatants

TABLE III
EXCHANGE OF PHOSPHOLIPIDS BETWEEN ARTIFICIAL MEMBRANES

Origin	Exchange assay	Simulatory soluble proteins	References
(A) *Exchange between liposomes and membranes*			
Rat liver	[^{32}P]Phosphatidylcholine liposomes or [^{32}P]chylomicrons or [^{32}P]fat emulsions + unlabeled mitochondria	pH 5.1 supernatant	Zilversmit, 1971
Beef liver	[^{32}P]Mitochondria or [^{32}P]microsomes + liposomes	pH 5.1 supernatant or purified PLEP*	Wirtz et al., 1972
Beef liver	Liposomes + [^{14}C]choline microsomes	purified PLEP	Kamp et al., 1973
Beef heart	[^{32}P]Phosphatidylcholine liposomes + mitochondria	purified PLEP	Ehnholm and Zilversmit, 1973
Beef brain	[^{3}H]Phosphatidylinositol microsomes + liposomes	purified PLEP	Helmkamp et al., 1974; Harvey et al., 1974
Beef liver	[^{14}C]Phosphatidylcholine liposomes + rat liver mitochondria	purified PLEP	Wirtz et al., 1976
Rat liver	[^{3}H]Phosphatidylserine microsomes + mitochondria	pH 5.1 supernatant	Butler and Thompson, 1975
Calf liver	[^{3}H]Phosphatidylcholine liposomes + human erythrocyte ghosts	purified PLEP	Rothman and Dawidowicz, 1975
Rat liver	[^{32}P]Phosphatidic acid containing liposomes + unlabeled mitochondria or inner mitochondrial membrane–matrix	cytoplasmic supernatant	Wojtczak and Zborowski, 1975
Potato tuber	[^{32}P]Liposomes + mitochondria	purified PLEP	Kader and Douady, unpublished observations
Rat intestine	[^{32}P]Liposomes + beef heart mitochondria	purified PLEP	Lutton and Zilversmit, 1976
Rat liver	Spin-labeled phosphatidylcholine liposomes + mitochondria	pH 5.1 supernatant beef liver PLEP	Rousselet et al., 1976
(B) *Exchange between different liposomes*			
	[^{32}P]Liposomes "sensitized" or non-sensitized	PLEP from beef heart	Ehnholm and Zilversmit, 1973
	[^{32}P]Phosphatidylcholine liposomes + vesicles lacking phosphatidylcholine	PLEP from beef heart	Kagawa Y. et al., 1973
	"Donor" liposome–"acceptor" liposome	PLEP from beef liver	Hellings et al., 1974; Van den Besselaar et al., 1975

*PLEP = phospholipid exchange protein.

from specific tissues which are then assayed for their ability to enhance exchange of radiolabeled phosphatidylcholine from liposomes to unlabeled mitochondria. In an initial study on rat intestine, these investigators have shown that when expressed on the basis of activity per gram of tissue or per mg protein, the phospholipid exchange activity of intestinal villi or crypts is greater than that of the intestinal wall, but both are weaker than the corresponding fraction from liver. In addition, the phospholipid exchange activity was found to be higher in villi from the jejunum than from the ileum.

6.2. Exchange of phospholipids between liposomes

Within the last few years a number of attempts have been made to study phospholipid exchange processes occurring between populations of liposomes. While this approach has the obvious advantage that the purity of the interacting membrane systems can be controlled, the final separation of two liposome preparations after incubation can prove troublesome. Nonetheless, four approaches have been successful.

Ehnholm and Zilversmit (1972, 1973) have studied the exchange of [^{32}P]-labeled phospholipids between labeled and unlabeled vesicles using immunological methods to separate the donor and acceptor vesicle populations. These investigators incorporated Forssman antigen into the membrane of liposomes prepared from phosphatidylcholine, phosphatidylethanolamine, sphingomyelin and cholesterol. These antigen-containing liposomes (A) were then incubated with liposomes without antigen (B) for 2 h at room temperature, after which the mixed liposome population was treated with anti-Forssman antibodies to induce aggregation of liposome population A which could then be separated from the B liposome population by simple centrifugation. The [^{32}P]-labeled phospholipids can be incorporated into either A or B type liposomes in this system. Ehnholm and Zilversmit (1973) found that phospholipids exchange weakly between the two populations of liposomes in the absence of a pH 5.1 supernatant. However, when a beef heart pH 5.1 supernatant was added to the incubation mixture, a spectacular increase in the exchange rate was observed, and up to 50% transfer was attained. This result suggests that phospholipids do not exchange spontaneously between liposomes, and soluble proteins from the pH 5.1 supernatant are required to promote such movement. Cholesterol, however, exchanged readily between the two liposome populations; no stimulation by the pH 5.1 supernatant was noted.

Another elegant method developed by Hellings et al. (1974) and Van den Besselaar et al. (1975) involves separation of vesicles on the basis of differences in surface charge. This method involves preparation of "donor" liposomes containing [^{14}C]phosphatidylcholine and a sufficient amount of phosphatidic acid (9 moles/100) to assure the binding of these liposomes to a DEAE cellulose column. The "acceptor" liposomes containing [^{3}H]cholesteryloleate, [^{14}C]phosphatidylcholine and a weaker amount of phosphatidic acid (2 moles/100) are not retained by the column. After incubation of "donor" and "acceptor" liposomes together in the presence of purified PLEP, the mixture is chromatographed on DEAE cellulose to separate the two liposome populations after which their respective content of radiolabeled material is measured. In this system, as in the experiments described immediately above, little spontaneous phospholipid exchange was detectable, and significant exchange was demonstrable only in the presence of an exchange protein (Hellings et al., 1974).

Demel et al. (1973a) have described a novel approach in which phospholipid exchange between monolayers and liposomes can be monitored by changes in the surface radioactivity of a phosphatidylcholine monolayer resulting from exchange of molecules from the monolayer with liposomes in the subphase. This approach will be discussed in more detail in section 8.1.

Exchange of phospholipid molecules between liposome membranes can also be followed by NMR spectroscopy (Barsukov et al., 1974, 1975). This approach has the advantage that separation of the two interacting liposome populations is not required, since exchange phenomena can be detected directly by changes in proton line width spectra.

It must be emphasized that in all the exchange experiments listed in Table III, crude supernatants or purified PLEP were present. These soluble proteins were required, since no phospholipid exchange between artificial liposomes was observed (Kornberg and McConnell, 1971b; Ehnholm and Zilversmit, 1972, 1973). Simple collisions occurring between liposomes do not produce significant spontaneous phospholipid exchange, but exchange can be detected with certain liposome systems after prolonged incubation (>24 h), even in the absence of exchange-promoting supernatants or proteins (see Papahadjopoulos et al., 1976a). Studies on exchange phenomena between liposomes may also be complicated by fusion of the interacting liposomes (Maeda and Ohnishi, 1974; Papahadjopoulos et al., 1974, 1976a,b).

Phospholipid exchange between liposome membranes represents a highly simplified assay system, since membrane proteins are eliminated. Experiments completed to date with these systems have clearly demonstrated that phospholipid exchange must be explained in terms of lipid–protein interactions. In short, these studies have established that exchange is highly dependent on specific exchange proteins. These data, together with the findings discussed earlier on phospholipid exchange occurring between subcellular fractions, have thus served to focus attention on the properties of the various PLEP found in different tissues and their mode of action.

7. Purification and properties of phospholipid exchange proteins

The ability to follow phospholipid movements between liposomes and subcellular fractions, or between two types of liposomes, has greatly facilitated the isolation and the study of the properties of the PLEP. Various PLEP have been isolated, but the best purification to date has been achieved by Wirtz et al. (1972) and Kamp et al. (1973) who isolated a highly purified protein from beef liver cytosol that is specific for phosphatidylcholine exchange and by Ehnholm and Zilversmit (1973) and Johnson and Zilversmit (1975) who have isolated a PLEP from beef heart cytosol that is specific for phosphatidylcholine and sphingomyelin.

7.1. Purification

In the experimental investigations described here, the unit of PLEP activity is defined as the amount (in nmoles or in percentage of the initially labeled pool) of phospholipid transferred between membrane fractions in an interval of time in the presence of a defined amount of PLEP.

7.1.1. Phosphatidylcholine exchange protein from beef liver (PC-PLEP)

Wirtz et al. (1972), starting from the pH 5.1 supernatant of beef liver (28 g of protein) isolated 14 mg of a 310-fold purified PLEP which was active in the following exchange assays: [^{32}P]-labeled microsomes and unlabeled mitochondria; or [^{14}C]choline labeled microsomes and unlabeled mitochondria. Ammonium sulfate precipitation, DEAE Sephadex A-50 separation and gel filtration on Sephadex G-50 and G-75 were used at the different steps in the purification. This protein fraction gives one to four bands on polyacrylamide gel electrophoresis, has a molecular weight around 22 000 daltons and shows a high specificity for phosphatidylcholine exchange. Other fractions eluted in the void volume of DEAE Sephadex A-50 column exchanged phosphatidylethanolamine and phosphatidylinositol more actively than phosphatidylcholine.

A 2680-fold purification of this monospecific PLEP (called PC-PLEP) was achieved by Kamp et al. (1973). The phospholipid exchange activity was assayed using labeled liposomes (containing phosphatidylcholine and phosphatidic acid in the molar proportions of 98/2) and [7α-^{3}H]cholesteryloleate and unlabeled microsomes containing [^{14}C]phosphatidylcholine. The ^{14}C/^{3}H ratio found in the liposomal phospholipids after incubation indicated the intensity of exchange (see Fig. 8B).

The different steps of purification of the active protein, indicated in Table IV, involve ion-exchange chromatography on DEAE-cellulose and carboxymethylcellulose followed by gel filtration on Sephadex G-50 (Kamp and Wirtz, 1974). Kamp et al. (1973) also introduced an additional pH adjustment of pH 3.0 in their method of purification. A surprising result of this treatment was an increase in the recovery of the exchange activity. This increase was tentatively attributed to removal of a hypothetical inhibitor or to changes in surface charge produced by transfer of acidic phospholipids in the liposomes. Evidence to support this latter suggestion was given by the change in transfer specificity of the pH 3.0 supernatant (i.e., the pH 5.1 supernatant adjusted to pH 3.0) in comparison with the original pH 5.1 supernatant. While the latter was able to exchange phosphatidylcholine, phosphatidylinositol and phosphatidylethanolamine, the pH 3.0 supernatant stimulated phosphatidylcholine exchange only.

The final preparation obtained by Kamp et al. (1973) contained a highly

TABLE IV
PURIFICATION OF PHOSPHATIDYLCHOLINE EXCHANGE PROTEIN FROM BEEF LIVER[a]

Purification steps	Protein (mg)	Purification factor	Recovery (%)	Specificity[b]
PH 5.1 supernatant	407 000	—	100	PC, PI, PE
pH 3.0 supernatant	131 000	4	135	PC
$(NH_4)_2SO_4$ precipitation	66 000	6	104	PC
DEAE cellulose	4 600	56	63	PC
CM cellulose	235	860	50	PC
Sephadex G-50	39	2680	26	PC

[a]From Kamp et al., 1973.
[b]PC, phosphatidylcholine; PI, phosphatidylinositol; and PE, phosphatidylethanolamine.

purified protein which transferred only phosphatidylcholine between liposomes and microsomes or between microsomes and mitochondria (Table V). The molecular specificity of this PC-PLEP was confirmed by other techniques (see section 8). No pH optimum for the PC-PLEP activity was found between 3.5 and 8.0 (Wirtz, 1974). Only a few μg of the PC-PLEP (instead of mg quantities of the pH 5.1 supernatant protein) were sufficient to actively transfer phosphatidylcholine under the same experimental conditions. However, this purified protein was unable to transfer phosphatidylcholine between human erythrocytes and liposomes (Hellings et al., 1974; Rothman and Dawidowicz, 1975).

The PC-PLEP from beef liver was found to be homogeneous by gel filtration, sodium dodecyl sulfate polyacrylamide gel electrophoresis, immunoelectrophoresis and isoelectrophoresis focusing (Fig. 10). Sodium dodecyl sulfate electrophoresis gave only one band and microimmunoelectrophoresis performed with purified PC-PLEP and an anti-PLEP immune serum also revealed

TABLE V
SPECIFICITY OF BEEF LIVER PHOSPHATIDYLCHOLINE EXCHANGE PROTEIN (PC-PLEP)[a]

Additions	% of transfer		
	Phosphatidylcholine	Phosphatidylinositol	Phosphatidylethanolamine
pH 5.1 supernatant	29.2	20.5	6.0
PC-PLEP	23.4	0	0

[a][^{32}P]-Labeled microsomes isolated from beef liver were incubated with unlabeled mitochondria in presence of pH 5.1 supernatant or 2680-fold purified PC-PLEP. The percentage of microsomal radioactivity recovered in the mitochondrial phospholipids at the end of the incubation, corrected for a standard without the exchange protein, indicated the intensity of the transfer. From Kamp et al., 1973.

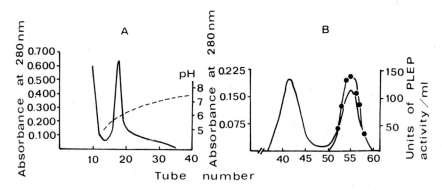

Fig. 10. Homogeneity of the beef liver phosphatidylcholine exchange protein (PC-PLEP) determined by isoelectric focusing and Sephadex G-50 gel filtration. A, isoelectric focusing of purified PC-PLEP between pH 5.0 and 7.0 indicating the presence of only one active protein; B, purification of PC-PLEP on Sephadex G-50. The exchange activity (●), expressed as units of phospholipid exchange protein activity, was recovered in only one of the two protein peaks determined by absorbance at 280 nm. Reproduced with permission from Kamp et al. (1973).

only a single precipitin line. This fraction can be stored at −20°C in 50% glycerol without loss of activity for several months (Kamp, 1975). At concentrations as low at 100 μg/ml purified PC-PLEP has a tendency to form aggregates and is very resistant to trypsin treatment. The molecular properties of the protein might explain these observations (see section 9).

7.1.2. PLEP from beef heart

Purification of the beef heart PLEP described by Wirtz and Zilversmit (1970) was achieved by Ehnholm and Zilversmit (1973) using Sephadex G-75 gel filtration and isoelectric focusing. Two proteins, 9 and 13, differing by their isoelectric points (5.5 and 4.7, respectively) were able to transfer phosphatidylcholine and, to a lesser extent, sphingomyelin between liposomes, but phosphatidylethanolamine was hardly exchanged, and cholesterol exchange was unaffected by these proteins. Protein 13 showed only one band after polyacrylamide gel electrophoresis, while protein 9 gave one major band of lower mobility (in agreement with the lower isoelectric point) and some minor bands.

It must be noted that the specificity of these proteins toward phosphatidylcholine and sphingomyelin was also found in the crude pH 5.1 supernatant, and it seems plausible to assume that a single protein is able to simultaneously exchange these two compounds, since they both contain choline as terminal moiety.

A 10-fold purified PLEP, also from beef heart cytosol, was isolated recently by Zilversmit and Johnson (1975). This protein facilitated phosphatidylcholine exchange between labeled phosphatidylcholine liposomes and unlabeled mitochondria, but was unable to stimulate the exchange of phosphatidylcholine between phosphatidylcholine/cholesterol liposomes and intact blood cells. However, when sealed erythrocyte ghosts were used, exchange of phosphatidylcholine and, to a lesser extent, of sphingomyelin, phosphatidylinositol and lysophosphatidylcholine was enhanced (Bloj and Zilversmit, 1976).

7.1.3. PLEP from rat intestine

Recently Lutton and Zilversmit (1976) have isolated two PLEP differing in their isoelectric points from rat intestine supernatant. The first protein (containing about 65% of the exchange activity) has an isoelectric point between 4.5 and 5.3, similar to PLEP from other tissues, but the second one (responsible for 30% of the exchange activity) had a basic isoelectric point between 8 and 9. The discovery of this basic protein, which is able to enhance the exchange of phosphatidylcholine between liposomes and mitochondria serves to emphasize the multiplicity of PLEP in animal tissues.

7.1.4. PLEP from beef brain (PI-PLEP)

The pH 5.1 supernatant from beef brain differs markedly from similar supernatants from other tissues in its specificity for phosphatidylinositol (PI) (Helmkamp et al., 1974). This specificity suggested the presence of a PLEP capable of preferentially transferring phosphatidylinositol. The isolation of such proteins has been achieved successfully using DEAE cellulose chromatography, isoelectric focusing and Sephadex G-100 gel filtration (Helmkamp et al., 1974). Two proteins, I and II (isoelectric points 5.2 and 5.5, respectively) (Fig. 11), have been isolated which can stimulate phosphatidylinositol exchange between [^{3}H]phosphatidylinositol-containing microsomes and liposomes containing phosphatidylcholine and phosphatidylinositol in the molar ratio of 98/2. The purification factor was as high as 508 (protein I) and 426 (protein II). Although these two purified proteins were able to exchange phosphatidylcholine between [^{14}C]phosphatidylcholine-containing microsomes and phosphatidylcholine-phosphatidylinositol liposomes, they preferentially stimulated

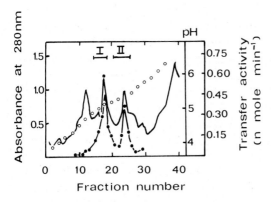

Fig. 11. Isoelectric focusing of beef brain PLEP. The pH gradient varies from 4 to 6 (○); (−), absorbance at 280 nm, (●), exchange activity expressed as transfer activity (nmoles. min^{-1}) as determined by exchange between liposomes and microsomes. Reproduced with permission from Helmkamp et al. (1974).

the exchange of phosphatidylinositol. No movement of phosphatidylethanolamine was detected between microsomes and mitochondria incubated with these proteins.

The same protein probably catalyzes the exchange of both phosphatidylinositol and phosphatidylcholine, since DEAE cellulose chromatography, tryptic digestion and heat treatment did not modify the activity of the protein I relative to protein II. The difference in the isoelectric properties of the two proteins may reflect differences in their molecular conformation (Helmkamp et al., 1974). Also, because of the heterogeneity of the brain tissue, one cannot exclude the possibility that each protein originated from different classes of cells. A third exchange protein, probably PC-PLEP, was also presumed to exist in these cytosols (Helmkamp et al., 1974). It must be noted that PI-PLEP activity has also been detected in beef liver cytosol (Wirtz et al., 1972) and purified from this source (unpublished experiments cited by Kamp, 1975).

7.1.5. PLEP from plant cytosols
The presence of active lipid-degrading enzymes in the potato tuber cell (Galliard, 1973, 1975) hinders the direct study of the pH 5.1 supernatant isolated from this tissue. As shown in Fig. 12, the pH 5.1 supernatant from this tissue contains proteins which can be separated into two major peaks by Sephadex G-75 purification (Kader, 1975). When fractions from the peak corresponding to high molecular weight proteins were tested in a phospholipid exchange assay using [^{32}P]-labeled microsomes and unlabeled mitochondria, a decrease in the exchange of the phospholipids was noted. This was interpreted as resulting from a degradation of the phospholipid pools by acyl-hydrolase(s) present in this fraction (the enzymes from potato tuber have molecular weights of about 100 000 daltons [see Galliard, 1971]). When fractions from the second peak were checked for their phospholipid exchange activity in the same assay system, a stimulation was noted. The stimulatory effect was diminished by preheating and also depended on the amount of protein added. The fraction was able to stimulate the exchange of phospholipids between mitochondria and microsomes but also between double-labeled liposomes containing [^{32}P]phospholipids and [^{3}H]estradiol (Kader and Douady, unpublished observations) (Fig. 12). Similar experiments performed on cauliflower floret and Jerusalem artichoke have revealed the presence of PLEP activity in the cytosols from these plants (Kader and Douady, 1975).

Unlike PLEP from animal tissues, plant tissue PLEP show no specificity toward any particular phospholipid (Kader, 1975). However, nitrogen-containing phospholipids were exchanged slightly more actively by potato tuber PLEP, while diphosphatidylglycerol was not exchanged at all. It is possible, however, that this low specificity of potato PLEP is due to the presence of many specific PLEP, since the preparations used in this work were only partially purified. The fragility of the potato protein, which gradually loses its activity during the purification steps, probably explains the low purification (×4.0) obtained to date. The low specificity of the potato tuber PLEP might, however,

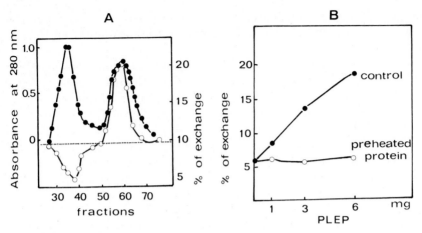

Fig. 12. Detection of PLEP activity in potato tuber tissue. A, the pH 5.1 supernatant proteins were separated on a Sephadex G-75 column and assayed for exchange activity in a microsome–mitochondria exchange assay. ●, percent exchange as a percentage of the original radioactivity in the microsomal fraction recovered in the mitochondria at the end of incubation; ○, absorbance at 280 nm; ·—·—·—·, control exchange assay performed without the protein fraction. Reproduced with permission from Kader (1975); B, stimulation of phospholipid exchange between [^{32}P]-labeled liposomes and unlabeled mitochondria by increasing amounts of potato tuber PLEP purified as in A. The exchange protein was introduced in an exchange mixture comprising [^{32}P]-labeled liposomes (also containing a trace of [^{3}H]estradiol) and unlabeled mitochondria. The protein fraction was untreated (●) or preheated at 70°C for 30 min (○). After incubation for 30 min at 30°C, the mitochondria were separated and their phospholipids were analyzed. The percent of exchange was determined as the percentage of the initial radioactivity recovered in the mitochondrial phospholipids after correction for loss of organelles and cross contamination. From Kader and Douady, unpublished observations.

be a peculiar feature of PLEP in plant tissues. Further purification of plant PLEP is clearly needed to resolve this question.

7.2. *Common properties of the phospholipid exchange proteins*

If one summarizes the properties of the PLEP isolated from various cells (Table VI), many common points emerge. The molecular weights are of the same order, varying from 21 000 to 32 800 daltons. The PLEP are acidic with isoelectric points between 4.7 and 5.8, except for the basic rat intestine PLEP. The specificity of different exchange proteins toward phospholipids varies strongly. This variation does not reflect the presence of membrane proteins, since this specificity is also found in exchange between liposomes (Ehnholm and Zilversmit, 1973). In addition to the data given in Table VI, it has been shown that beef liver PC-PLEP can specifically exchange phosphatidylcholine between liposomes and mitochondria (Wirtz et al., 1976) or between liposomes of differing surface charge (Hellings et al., 1974; Van den Besselaar et al., 1975), and PLEP from beef brain (protein I) can produce marked transfer of phosphatidylinositol between liposomes and microsomes (Harvey et al., 1974).

TABLE VI
PROPERTIES OF PHOSPHOLIPID EXCHANGE PROTEINS ISOLATED FROM VARIOUS CYTOSOLS

	Assay system	Molecular weight	Isoelectric point	Molecular specificity	References
Beef liver	Liposome–microsome Liposome–mitochondria Liposome–liposome Liposome–monolayer	21 320[a] 22 000[b] 23 000[c]	5.8	PC	Kamp et al., 1973
Beef liver	Liposome–microsome	22 000[b]	—	PC	Wirtz et al., 1972
Beef heart	Liposome–liposome	Protein 9: 21 000[c] Protein 13: 25 900[c]	4.7 5.5	PC, SM PC, SM	Ehnholm and Zilversmit, 1973 Ehnholm and Zilversmit, 1973
Beef brain	Liposome–microsome	Protein I: 26 500[b] 32 500[c] 29 000[d] Protein II: 27 500[b] 32 800[c] 30 000[d]	5.2 5.5	PI, PC PI, PC	Helmkamp et al., 1974 Helmkamp et al., 1974
Rat intestine	Liposome–mitochondria	—	Protein 1: 4.5–5.3 Protein 2: 7–9.2	Only PC was studied	Lutton and Zilversmit, 1976
Potato tuber	Microsome–mitochondria	22 000[b] 22 000[c]	?	PI, PC, PE	Kader, 1975

[a] Determined by amino acid analysis.
[b] Determined by gel filtration.
[c] Determined by sodium dodecyl sulfate polyacrylamide gel electrophoresis.
[d] Average.

It seems unlikely that the specificity of PC-PLEP is caused by eventual alteration of exchange proteins specific for other phospholipids during the isolation procedure because: (a) the pH 5.1 supernatant itself shows a relative specificity toward phosphatidylcholine; and (b) PLEP, which preferentially stimulates the exchange of phosphatidylethanolamine or phosphatidylinositol, can be detected in the same cytosol (Wirtz et al., 1972).

Another interesting feature of these proteins is their immunological properties. Harvey et al. (1973) succeeded in preparing an antiserum against PC-PLEP from beef liver. Below the equivalence point, the exchange of phosphatidylcholine induced by this protein was totally inhibited in the presence of antiserum (Fig. 13A), whereas a residual level of exchange activity still remained when pH 5.1 supernatant was tested in the same manner (Fig. 13B). This residual activity was probably due to the presence in the beef brain of phosphatidylinositol exchange proteins, which are also able to exchange phosphatidylcholine (Harvey et al., 1974). This explanation is supported by the finding that the antiserum against the beef liver PC-PLEP did not inhibit the exchange of phosphatidylinositol or phosphatidylethanolamine mediated by the pH 5.1 supernatant of beef liver, nor did it alter the exchange activity of proteins I and II from beef brain (Harvey et al., 1974).

These investigations have effectively demonstrated the immunological specificity of PLEP. Nevertheless, these experiments also indicate that exchange proteins from one animal are able to exchange phospholipids between mem-

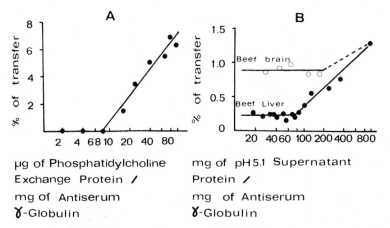

Fig. 13. Effect of antiserum against phosphatidylcholine exchange protein of beef liver (PC-PLEP) on phosphatidylcholine exchange mediated by beef liver PC-PLEP (A) or beef liver (●) or beef brain (○) pH 5.1 supernatant (B). After incubation of the purified protein or pH 5.1 supernatants in the presence of various amounts of antiserum and centrifugation to remove antigen-antibody precipitate, the exchange activity of the liquid phase was assayed in the microsome-liposome system of Kamp et al. (1973). The exchange activity is expressed as the percent transfer of microsomal [^{14}C]phosphatidylcholine, mediated by either μg (A) or mg (B) amounts of protein. The various amounts of purified protein (A) or of pH 5.1 supernatant (B) preincubated with a fixed amount of antiserum are expressed in μg (A) or in mg (B) of protein per 1 mg of serum. Reproduced with permission from Harvey et al. (1973).

branes of others. Moreover, it has been shown recently that phospholipid exchange between rat liver mitochondria and liposomes can also be mediated by PLEP from potato tuber (Kader and Douady, unpublished observations). It is not known to what extent genetic requirements play a role in these "hybridization" experiments.

Another type of exchange protein has been discovered recently in rabbit plasma. This protein, which has an isoelectric point of 5.2, a high molecular weight and the solubility characteristics of a globulin, stimulates the exchange of cholesterol ester between very low and low-density lipoproteins from hypercholesterolemic rabbit plasma (Zilversmit et al., 1975).

8. Do phospholipid exchange proteins act as phospholipid carriers?

Highly purified PC-PLEP from beef liver and PLEP from beef heart have proven very useful in obtaining answers to the question posed in the title of this section. The suggestion that exchange proteins might act as phospholipid carriers arose from the discovery that one mole of the 2680-fold purified PC-PLEP contained one mole of bound phosphatidylcholine (Wirtz, 1974; Kamp, 1975; Kamp et al., 1975b). The earlier data of Kamp et al. (1973), indicating that only 0.4 and 0.6 mole of phosphatidylcholine were bound per one mole of purified PC-PLEP, can be explained by an incomplete extraction of the bound phospholipid. It is tempting to suppose that it is this bound phosphatidylcholine that is exchangeable with phosphatidylcholine molecules present in biological or artificial membranes. To check the validity of this hypothesis, three different experimental approaches have been followed.

8.1. Studies with phospholipid monolayers

In a series of elegant experiments, Demel et al. (1973a) studied the interactions of PC-PLEP from beef liver with a radioactively labeled phospholipid monolayer made from 1-palmitoyl-2-oleoyl-sn-glycero-3-[^{14}C]methylphosphorylcholine which was set up in such a manner that the surface pressure was constantly maintained at 30 dynes cm^{-1}. Under these conditions, the exchange protein did not penetrate the monolayer. However, when the exchange protein was injected in the subphase medium in equal concentrations (1 nmole of exchange protein for 1 nmole of phosphatidylcholine in the monolayer), the surface radioactivity decreased rapidly, and the [^{14}C]phospholipid label was recovered in the exchange protein after gel filtration of the subphase (Fig. 14A). An equilibrium was attained when the surface radioactivity was 50% of the initial value. This is consistent with the presence of one mole of phosphatidylcholine per mole of exchange protein. Moreover, the exchange protein was able to exchange phosphatidylcholine between two separate monolayers, consisting of labeled (monolayer 1) and unlabeled (monolayer 2) phosphatidylcholine. After addition of exchange protein, the surface radioactivity in

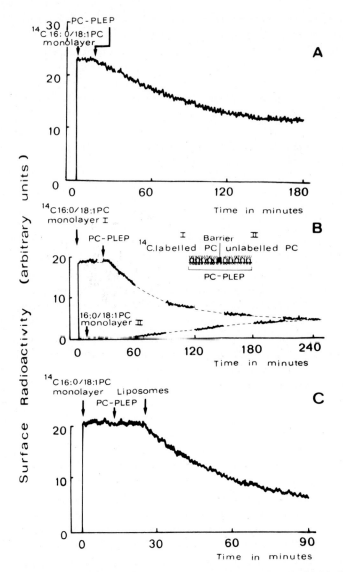

Fig. 14. Phosphatidylcholine exchange between a phosphatidylcholine monolayer and beef liver PC-PLEP containing equal quantities of phosphatidylcholine molecules (A) and between 2 monolayers separated by a barrier (B), or between a monolayer and phosphatidylcholine liposomes in the presence of beef liver PC-PLEP (C).

The monolayers were obtained as described in the text. In (A), equal quantities of phosphatidylcholine were present in the monolayer and in PC-PLEP. In (B), the sum of the amounts of phosphatidylcholine present in the two monolayers equaled the amount of this phospholipid bound to PC-PLEP molecules. In (C), the exchange protein was introduced in amounts insufficient to affect surface radioactivity (the concentration was reduced ten times), but sufficient to mediate phospholipid exchange with liposomes. In all these experiments, the surface pressure was maintained at 30 dynes cm^{-1}. The movement of phosphatidylcholine molecules was followed by measuring the change in the surface radioactivity (expressed in arbitrary units). Injections are indicated by arrows. [^{14}C]$_{16:0}$/C$_{18:1}$ PC = 1-palmitoyl-2-oleyl-sn-glycero-3-[^{14}C]methylphosphorylcholine. Reproduced with permission from Demel et al. (1973a).

monolayer 1 decreased, while [^{14}C]phosphatidylcholine appeared in monolayer 2 (Fig. 14B). Since no lateral diffusion of phosphatidylcholine between the two monolayers was possible, this result again indicated that the PC-PLEP was able to act as a phosphatidylcholine carrier.

The introduction of non-radioactive liposomes into the subphase in the presence of limiting amounts of exchange protein (0.1 nmole of protein/nmole of phosphatidylcholine in the monolayer) also rapidly diminished the surface radioactivity. The [^{14}C]-labeled phosphatidylcholine of the monolayer was replaced by unlabeled phosphatidylcholine of the liposomes (Fig. 14C). An equilibrium was reached when 35% of the initial label had been exchanged. The exchange protein was required to mediate the movement of phosphatidylcholine between the monolayer and liposomes, and this probably explains why no exchange of phospholipids was noted between monolayers and liposomes in earlier studies (Dawson, 1966). When radioactive phosphatidylethanolamine or phosphatidylinositol monolayers were used, no change in the surface radioactivity was observed after addition of liposomes and PC-PLEP into the subphase, confirming again the high molecular specificity of this particular exchange protein. As mentioned, because of the experimental conditions, penetration of the exchange protein in the monolayers was not possible, and no perturbation of the monolayer was produced. Indeed, Barsukov et al. (1974) have shown in NMR spectroscopic investigations that the exchange proteins present in rat liver 105 000 g supernatant do not disturb the integrity of the liposomes and do not alter the fluidity of the bilayer.

All these observations are consistent with the concept that the PC-PLEP acts as a phosphatidylcholine carrier.

8.2. Binding experiments

The availability of radiolabeled phosphatidylcholine of high specificity and highly purified PLEP from beef liver or heart has allowed the binding of this phospholipid to the exchange proteins to be studied in some detail. Two independent investigations have successfully shown this binding. Kamp et al. (1975a) incubated phosphatidylcholine liposomes (labeled on a [^{14}C]methyl group) with the exchange protein for 30 min at 37°C, after which the mixture was chromatographed on Sephadex G-75 or electrophoresed on polyacrylamide gels. In both cases [^{14}C]-radioactivity was recovered in association with the exchange protein (Fig. 15). The second study, by Johnson and Zilversmit (1975), examined the binding of phosphatidylcholine to beef heart PLEP. These authors first checked, however, that the exchange protein did not bind to the liposomes. This was established by showing that when liposomes and PLEP that had been incubated together were separated on a Sepharose 4B column, the exchange activity remained totally associated with the protein peak. Further evidence that liposomes were not bound to the exchange protein was obtained by isoelectric focusing of a preincubated mixture of exchange protein and double-labeled phosphatidylcholine liposomes containing [^{32}P]-

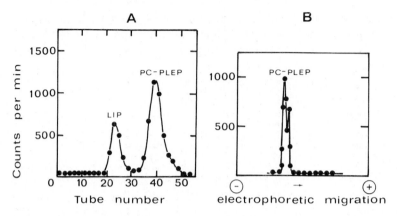

Fig. 15. Binding of radiolabeled phosphatidylcholine to beef liver PC-PLEP. A, the preincubated mixture of labeled liposomes and PLEP was separated on a Sephadex G-75 column and the distribution of radioactivity (in cpm) in two peaks corresponding to liposomes (Lip) and to exchange protein (PC-PLEP) measured. B, polyacrylamide gel electrophoresis profile of labeled exchange protein showing that the radioactive peak corresponds to the stained band. Reproduced with permission from Kamp et al. (1975b).

labeled phosphatidylcholine and [^{14}C]triolein. In this system, the two protein peaks of exchange activity contained no [^{14}C]-label, indicating that liposomes or liposome fragments were not bound to these proteins, but these two peaks coincided with the two peaks of [^{32}P]-labeled phosphatidylcholine, clearly indicating transfer of radioactive phosphatidylcholine from the liposomes to the exchange protein.

Phospholipid binding has yet to be described for other PLEP. For instance, no evidence for phosphatidylinositol binding has been obtained for PI-PLEP. In some experiments (Akiyama and Sakagami, 1969; Abdelkader and Mazliak, 1970), radioactive lipids were recovered in the supernatant after incubation of unlabeled supernatant containing PLEP with radioactive organelles. To what extent this radioactivity can be accounted for by binding of phospholipids to the exchange protein is unclear, since membrane fragments were probably also present in the supernatants.

In connection with these observations, it can be noted that phosphatidylcholine binds to the major protein of human plasma, the high-density lipoprotein. The cleavage of this protein into two fractions, III and IV, has revealed that the binding of phosphatidylcholine seems to be located in the carboxyl-terminal portion of the molecule (Lux et al., 1972). However, binding of phosphatidylcholine liposomes to serum albumin was not detected unless phosphatidylcholine was dispersed on Celite (Jonas, 1975).

8.3. Net transfer

Additional evidence of a phospholipid carrier function for PLEP comes from the transfer of phospholipid from one structure to another. A clear demonstration of transfer of phospholipids was made by Y. Kagawa et al. (1973) who showed that beef heart PLEP was able to introduce phosphatidylcholine into vesicles lacking phosphatidylcholine. These vesicles contained all the constituents needed for an efficient partial reaction of the oxidative phosphorylation, the ATP-[^{32}P]-inorganic phosphate exchange, except for phosphatidylcholine and were therefore inactive. The incorporation of phosphatidylcholine into these vesicles, mediated by the exchange protein, restored the activity. The restoration of function paralleled the transfer of phosphatidylcholine as monitored by [^{32}P]-labeled lipid (Fig. 16). This work underlines the potential importance of PLEP as experimental tools for manipulating membrane phospholipid composition and membrane function.

A net transfer of sphingomyelin has also been shown to occur from Forssman-antigen containing liposomes to liposomes lacking this antigen (Ehnholm and Zilversmit, 1973). However, no exchange in the opposite direction occurred, possibly because of the slight difference in membrane surface properties of the two liposomal populations.

A systematic study of the net transfer of [^{3}H]phosphatidylinositol during incubation of labeled microsomes with liposomes lacking phosphatidylinositol in the presence of purified PLEP from beef brain revealed that [^{3}H]phosphatidylinositol was recovered in the liposomes and that a decrease in the phosphatidylinositol content of the microsomes (about 25%, calculated relative to phosphatidylethanolamine, which is not exchanged) occurred during incubation (Harvey et al., 1974). A similar net transfer of phosphatidylinositol

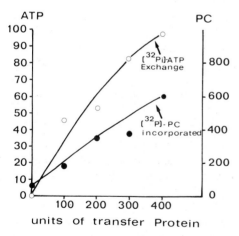

Fig. 16. Relationship between the transfer of [^{32}P]-labeled phosphatidylcholine, mediated by various amounts of beef heart phospholipid exchange protein and [^{32}P]–ATP exchange reaction of phosphatidylcholine-less vesicles. The ATP–[^{32}P]inorganic phosphate exchange activity is expressed in nmoles of ATP mg^{-1} · 10 min^{-1}, while the incorporation of [^{32}P]phosphatidylcholine is indicated in nmoles mg^{-1} h^{-1}. A unit of phospholipid exchange protein is defined as the amount required to transfer 80 pmoles of phosphatidylcholine from [^{32}P]-labeled liposomes containing 300 nmoles of phosphatidylcholine to 12.5 mg of mitochondria. Reproduced with permission from Y. Kagawa et al. (1973).

from microsomes to liposomes has been shown by Zborowski and Wojtczak (1975) using rat liver pH 5.1 supernatant. All the data presented in this section are consistent with the hypothesis that PLEP act as phospholipid carriers, at least in the case of phosphatidylcholine and phosphatidylinositol transfer. In none of these studies, however, has it been clearly demonstrated that a net transfer of phospholipid occurred, i.e., that a transfer of a phospholipid from membrane A to membrane B lacking this phospholipid, was not accompanied by a concomitant transfer of a biochemically different phospholipid from B to A. Are phospholipid exchange movements between membranes bidirectional or is one-way transfer possible? This important question will be examined in section 10.

Before concluding this section, it should be noted that the existence of lipid-binding proteins is not without precedent. For example, the apoproteins of certain lipoproteins or bovine serum albumin avidly bind fatty acids, while fatty-acid binding proteins, designated as Z proteins (Ockner et al., 1972; Mishkin et al., 1973; Suzue and Marcel, 1975) have been isolated from the cytosol of a variety of animal cells. However, they differ from the phospholipid exchange proteins, since they do not react with membrane structures to exchange their bound lipids. The phospholipid binding capacity of PLEP is not their primary feature, which is capacity to exchange bound phospholipids with membrane phospholipids. A distinction must also be made between the PLEP and other soluble proteins involved in lipid metabolism, such as the sterol-carrier proteins (review, Scallen et al., 1975). Thus, to understand the mode of action of these proteins, two areas need to be explored: (1) the interactions between specific phospholipids and the appropriate exchange proteins; and (2) the interactions between the exchange protein and membranes. These interactions will be examined in the following sections.

9. *Nature of the interactions between phospholipid exchange proteins and bound phospholipids*

The interaction of PLEP with membrane phospholipids would appear to involve conflicting requirements, since both hydrophobic and electrostatic interactions are probably required for effective binding between these molecules.

9.1. Hydrophobic interactions

The amino acid sequence of highly purified PC-PLEP from beef liver has been examined by Kamp et al. (1973) (Table VII). Its composition is characterized by a higher amount of acidic amino acids than basic ones, which is reflected in the isoelectric point of the protein at pH 5.8. The protein is comprised of 190 amino acid residues and a low content of half-cystine residues participating in disulfide bonds. This no doubt accounts for the finding that N-ethyl maleimide

TABLE VII
AMINO ACID SEQUENCE OF HIGHLY PURIFIED PHOSPHATIDYLCHOLINE
EXCHANGE PROTEIN ISOLATED FROM BEEF LIVER[a]

Amino acids (moles/mole protein)	Asp	Thr	Ser	Glu	Pro	Gly	Ala	Val	Cyt (1/2)
	16	5	11	29	10	16	14	16	2
	Met	Ile	Leu	Tyr	Phe	Lys	His	Arg	Trp
	4	6	15	9	8	16	3	8	2

[a]From Kamp et al., 1973.

inhibited beef heart PLEP (Johnson and Zilversmit, 1975) and rat liver PLEP (Illingworth and Portman, 1972a). From this amino acid composition, the method of Bigelow (1967) was used to calculate the average hydrophobicity of the exchange protein. A high value (1109 calories per amino acid residue) was found, relative to the molecular weight of the exchange protein (Kamp et al., 1973). This high hydrophobicity, leading to intermolecular bonding, probably explains the tendency of highly purified PC-PLEP to form aggregates.

Similar high hydrophobicity characterizes two proteins of human high-density lipoproteins which are involved in extracellular exchange of phosphatidylcholine (Jackson et al., 1972; Lux et al., 1972). These two proteins have respectively a molecular weight and a hydrophobicity of 27 000 daltons and 1106 calories/mole (for fraction III) and 17 000 daltons and 1165 calories/mole (for fraction IV) (Edelstein et al., 1972). However, it seems unlikely that these proteins affect phospholipid exchange between membranes (Kamp, 1975), since plasma lipoproteins do not stimulate this exchange (Zilversmit, 1971a).

As indicated above, the beef liver PC-PLEP contains one mole of phosphatidylcholine per mole of protein. Because of the high hydrophobicity of the exchange protein, it is not unreasonable to suggest that hydrophobic binding is involved in the interaction between exchange protein(s) and phospholipid(s). Since detergents and organic solvents can compete with lipids for hydrophobic sites on proteins (Helenius and Simons, 1975), Kamp et al. (1975a) studied the effects of these compounds on phosphatidylcholine binding to beef liver PC-PLEP. At concentrations where delipidation did not occur, deoxycholate, at concentrations lower than critical micelle concentration (0.1%, w/v), and isobutanol (7.6%, v/v) were able to displace the phosphatidylcholine molecules bound to the PC-PLEP, rendering the displaced molecules susceptible to phospholipase A_2 attack. Hydrolysis of [^{14}C]phosphatidylcholine bound to PLEP was followed by a decrease in the radioactivity associated with PLEP (Fig. 17A). A similar result was obtained with dioxane which lowers the dielectric constant of the medium.

By increasing the concentration of deoxycholate up to 0.42% (v/v), [^{14}C]phosphatidylcholine could be extracted from the exchange protein (Fig. 17B). Other ionic and non-ionic detergents (e.g., Triton X-100) and sodium dodecyl sulfate or lysophospholipids (e.g., lysophosphatidylcholine) had the

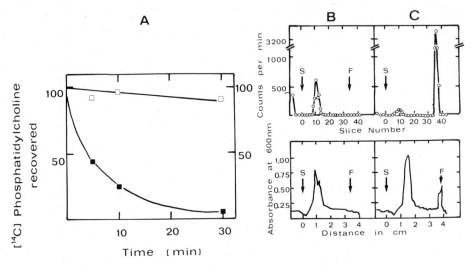

Fig. 17. Binding of phosphatidylcholine to beef liver PC-PLEP. A, The effect of low concentrations of deoxycholate on the rate of hydrolysis of [^{14}C]-labeled phosphatidylcholine bound to beef liver PC-PLEP by phospholipase A_2 of bee venom. □, untreated PLEP; ■, PLEP treated with 0.05% deoxycholate (w/v) before hydrolysis. The decrease in the % of initial radioactivity (in ordinate) indicates the extent of the hydrolysis; B and C, the effect of high concentrations of deoxycholate on binding of [^{14}C]phosphatidylcholine to the exchange protein. B, untreated protein and C, protein treated with 0.42% deoxycholate. The mixtures were electrophoresed on polyacrylamide gels. S and F indicate the start and the front of the electrophoresis profile. The stained bands were read at 600 nm and the radioactivity was expressed in c.p.m. Reproduced with permission from Kamp et al. (1975a) (experiment A) and Kamp et al. (1975b) (experiment B).

same effect (Kamp et al., 1975b). The detergent molecules were thus able to interact with the hydrophobic sites of the exchange protein and to compete with the phosphatidylcholine molecules bound to the exchange protein. This interaction was also indicated by an increase in the solubility of the PC-PLEP in the presence of these detergents, and by changes in electrophoretic mobility and immunological properties of the exchange protein to which detergent molecules were irreversibly bound (particularly sodium dodecyl sulfate and cetyltrimethylammonium bromide).

The fact that the phosphatidylcholine molecules bound to the PLEP were extractable by organic solvents suggests that the phospholipid molecules are bound to the protein by non-covalent bonds, thus indicating that PLEP does not work as an enzyme.

Hydrophobic interactions therefore play an important role in the lipid–PLEP interactions. Tanford (1972, 1973) calculated the free energy changes associated with binding of hydrocarbon chains or of amphiphiles such as phospholipids to proteins. He studied different types of association between proteins and amphiphiles. Spontaneous lipid-protein aggregations are possible, and at free amphiphile concentrations below the critical micelle concentration, a competition for amphiphiles between proteins and membranes or micelles was shown. A protein such as bovine serum albumin readily combines with molecules containing long hydrocarbon chains but, when amphiphiles are considered, specific affinity for the polar head groups is also involved.

Evidence against the crucial role of hydrophobic binding in the phospholipid exchange function was obtained by Wirtz et al. (1970). They studied the transfer of various molecular species of phosphatidylcholine between rat liver microsomes and mitochondria. However, no significant change in the rate of transfer was observed with changes in the nature of the acyl moieties of the phospholipid. Taniguchi et al. (1973) confirmed these results in the same tissue, and Illingworth and Portman (1972b) could not detect differences in the rate of exchange of molecular species of phosphatidylcholine between lipoproteins. Studies on the kinetics of labeling in vivo in guinea pig liver also suggest an independence of the movement of phospholipids and the degree of unsaturation (Parkes and Thompson, 1973).

From these observations, it must be concluded that hydrophobic binding is not the only type of interaction between the exchange protein and the phospholipid.

9.2. Electrostatic interactions

The existence of different forms of PLEP with varying specificities toward different phospholipids indicates that electrostatic interactions may play an important role in the binding of phospholipid to the exchange protein. The following evidence also supports this hypothesis: (a) changes in ionic strength of the medium inhibit phospholipid exchange activity (Johnson and Zilversmit, 1975) (Fig. 18); and (b) the introduction of cations (La^{3+}, Ca^{2+}, Mg^{2+} and, to a lesser extent, Na^+ and K^+) strongly reduces protein-mediated exchange of phosphatidylcholine between liposomes and a monolayer (Wirtz et al., 1976)

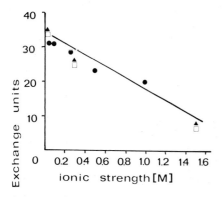

Fig. 18. Effects of salts on phospholipid exchange. NaCl (●), $CaCl_2$ (□) or $MgCl_2$ (▲) were introduced in exchange assays performed with [^{32}P]-labeled liposomes, unlabeled mitochondria and PLEP from beef heart. One unit of exchange activity is defined as the transfer of 1% of the labeled phosphatidylcholine between liposomes and mitochondria incubated for 40 min. Reproduced with permission from Johnson and Zilversmit (1975).

(Fig. 19). According to these authors, the extent of inhibition depends on the valency of the cations rather than of the ionic strength. However, it should be noted that Kamath and Rubin (1973) found that La^{3+} at 0.1 M did not inhibit phospholipid exchange between membranes. These observations are consistent with the view that ionic interactions occur between the exchange protein and its associated lipid.

However, some findings are more difficult to explain. For instance, the exchange of sphingomyelin, which possesses a similar head group to phosphatidylcholine, is very slow, both between membranes and between liposomes (Ehnholm and Zilversmit, 1973). One possible interpretation of this difference is that the hydrophobic interaction of phospholipids with the exchange protein is weaker in the case of sphingomyelin than for phosphatidylcholine. Another possibility might be that sphingomyelin is extracted from the membrane less readily than phosphatidylcholine. Similarly, the fact that phosphatidylethanolamine is not exchanged by PLEP may be due either to more tight embedding of this phospholipid within the membrane or to a particular disposition of phosphatidylethanolamine molecules in the membranes. For example, Phillips et al. (1972) indicated that the polar head groups of phosphatidylethanolamine are disposed tangentially to the plane of a vesicle bilayer, thus producing a net neutralization of the charges. In contrast, the zwitterionic polar groups of phosphatidylcholine molecules are extended normal to the bilayer, leading to dipolar repulsion between adjacent bilayers. These differences in zwitterionic conformation produced differences in the stabilities of phosphatidylcholine and phosphatidylethanolamine emulsions.

What part of the phosphatidylcholine molecule governs the interactions with

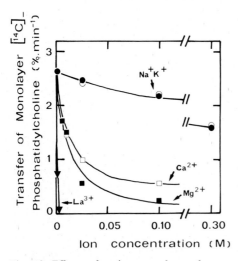

Fig. 19. Effects of cations on the exchange of phosphatidylcholine between unlabeled phosphatidylcholine liposomes and a monolayer containing [^{14}C]phosphatidylcholine in the presence of beef liver PC-PLEP. The percent transfer of [^{14}C]phosphatidylcholine was calculated from changes in the surface radioactivity of the monolayer. Reproduced with permission from Wirtz et al., 1976.

the exchange protein? The lack of effect of beef liver PC-PLEP on lysophosphatidylcholine exchange between microsomes and mitochondria (Kamp, 1975) can be explained by the need for hydrophobic interactions which depend on the two acyl chains, and not on only one. Johnson and Zilversmit (1975) observed that analogues of phosphatidylcholine (choline chloride, phosphorylcholine, phosphate, sn-glycerophosphorylcholine and lysophosphatidylcholine) did not inhibit the phospholipid exchange between liposomes and mitochondria. This suggests that these compounds are unable to displace the phosphatidylcholine from the exchange protein or to compete with this phospholipid in the binding to the protein. Other analogues of phosphatidylcholine were introduced by Kamp (1975) as *trace amounts* into "donor" or "acceptor" liposomes prepared from phosphatidylcholine and phosphatidic acid according to the method of Hellings et al. (1974) and Van den Besselaar et al. (1975). The intensity of phospholipid exchange mediated by beef liver PC-PLEP depended on the distance between phosphorus and nitrogen in the molecule, since (Table VIII) the addition of methylene groups inhibited the exchange (compounds 2 and 3). The number of methyl groups linked to the nitrogen was also important, since dimethylphosphatidylcholine (compound 4) was not exchanged. Steric requirements were also suggested by a partial restoration of the exchange on addition of an ethyl group to the dimethylphosphatidylcholine. As a final confirmation, binding of compounds 1 and 2, but not 3 and 4 to the PC-PLEP was observed. Kamp (1975) suggested that a zwitterionic amino acid pair on the PC-PLEP permits an electrostatic interaction with the polar head group of phosphatidylcholine.

It is not possible, however, in our present state of knowledge, to propose that electrostatic interactions are of exclusive importance. Using the technique of introducing labeled analogues as trace amounts in liposome assays, Kamp (1975) investigated the effect of changing the nature of the acyl chains of the phosphatidylcholine (Table IX). It was found that the optimal composition of phosphatidylcholine acyl residues was in R_1, palmitoyl and in R_2, oleyl chains for the activity of beef liver PC-PLEP (the main molecular species present in egg phosphatidylcholine is palmitoyl/oleyl sn-glycerophosphorylcholine). If palmitoyl chains were introduced in R_1 and R_2, the exchange of this compound was slower than that of egg lecithin (5.4% of transfer instead of 14.8%). Table IX also shows that changes in the bonds between the acyl chains and the glycerol backbone diminished the transfer of the following analogues: D-enantiomer of palmitoyl/oleyl-glycerophosphorylcholine, 1-acyl, 2-alkyl deoxyglycerophosphorylcholine and 1-acyl, 2-alkyl glycerophosphorylcholine.

The formation of a binding complex between the exchange protein and phosphatidylcholine is thus governed both by hydrophobic and electrostatic interactions. Since hydrolysis of phosphatidylcholine bound to beef liver PC-PLEP is only catalyzed by phospholipase A_2 on deoxycholate-pretreated exchange protein, it was suggested (Kamp et al., 1975a, 1975c) that phosphatidylcholine was embedded in the exchange protein (Fig. 20). However, these conclusions are presently valid only for PC-PLEP and it is not known

TABLE VIII
EFFECT OF MODIFICATION OF THE POLAR HEAD GROUP OF PHOSPHATIDYLCHOLINE ON THE PLEP-MEDIATED PHOSPHOLIPID TRANSFER OF THESE ANALOGUES[a]

Compound	Polar head group	% of transfer
1	---P-O-CH$_2$-CH$_2$-$\overset{+}{N}$(CH$_3$)$_3$	14.8
2	---P-O-CH$_2$-CH$_2$-CH$_2$-$\overset{+}{N}$(CH$_3$)$_3$	8
3	----P-O-CH$_2$-CH$_2$-CH$_2$-CH$_2$-$\overset{+}{N}$(CH$_3$)$_3$	0
4	---P-O-CH$_2$-CH$_2$-$\overset{+}{N}$H(CH$_3$)$_2$	0
5	---P-O-CH$_2$-CH$_2$-$\overset{+}{N}$(CH$_3$)$_2$-C$_2$H$_5$	6
6	--P-OCH$_2$-CH$_2$-$\overset{+}{N}$-(CH$_3$)$_2$-C$_3$H$_7$	0

[a]Different radiolabeled phosphatidylcholine analogues were introduced as traces into "donor" liposomes containing phosphatidylcholine and phosphatidic acid (91/9 molar ratio). These donor liposomes were incubated with "acceptor" liposomes and beef liver PC-PLEP, then were treated according to the techniques of Hellings et al. (1974) and Van den Besselaar et al. (1975). The % of transfer was corrected for control incubations.

P represents:
$$-O-\underset{\underset{O}{\|}}{\overset{\overset{O}{|}}{P}}-O$$

The acyl chains were palmitoyl and oleyl in the *sn* 1 and 2 positions of the lipid.
From Kamp, 1975.

TABLE IX
EFFECT OF MODIFICATION OF THE APOLAR PART ON THE TRANSFER OF PHOSPHORYLCHOLINE-CONTAINING PHOSPHOLIPIDS BY BEEF LIVER PC-PLEP[a]

Phospholipid	Apolar chain composition	% of transfer
[^{14}C]1,2 diacylglycerophosphorylcholine	C$_{16:0}$/C$_{18:1}$	15
D-[^{14}C]diacylglycerophosphorylcholine	C$_{16:0}$/C$_{18:1}$	5
rac[^{14}C]1-acyl, 2-alkyl, 2-deoxyglycerophosphorylcholine	C$_{18:1}$/C$_{16:0}$	7.5
rac[^{14}C]1-acyl, 2-alkyl glycerophosphorylcholine	C$_{18:1}$/C$_{16:0}$	4.8

[a]The radioactive lipids were incorporated as traces in donor phosphatidylcholine/phosphatidic acid (91/9, mole/mole) liposomes and then incubated as indicated in Table VIII.
From Kamp, 1975.

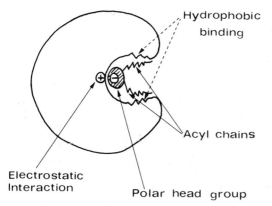

Fig. 20. Schematic representation of possible interactions in the complex between beef liver PC-PLEP and a phosphatidylcholine molecule.

how other phospholipids bind to PLEP. In particular, it would be of interest to study whether a simultaneous binding of two phospholipids (for instance, phosphatidylcholine and phosphatidylinositol) to the PLEP of beef brain or of phosphatidylcholine and sphingomyelin on beef heart PLEP is possible.

The binding of phospholipids to exchange proteins is only one step in the exchange process. To achieve actual transfer, the PLEP must form a collision-complex with the membrane. Since electrostatic interactions play a role in phospholipid binding to PLEP, it does not seem unreasonable to suggest that the surface properties of the membrane will also play a crucial role in the exchange.

10. Control of phospholipid exchange by membrane surface charge and the mode of action of phospholipid exchange proteins

Since the phospholipid exchange reaction must involve interaction of the exchange protein–phospholipid "complex" with both "donor" and "acceptor" membranes, several studies have been made to examine the influence of membrane surface properties on exchange processes.

10.1. Liposome–liposome interactions

Hellings et al. (1974) introduced increasing amounts of phosphatidic acid or phosphatidylinositol into "donor" phosphatidylcholine liposomes and studied the effect of beef liver PC-PLEP on the exchange of [^{14}C]phosphatidylcholine between labeled donor liposomes and unlabeled phosphatidylcholine "acceptor" liposomes. The results showed that exchange decreased as the content of phosphatidic acid in the donor liposomes increased, and exchange was completely inhibited at 20 mole % phosphatidic acid. Similar inhibitory effects were observed when phosphatidylinositol was included in the donor liposomes (Fig. 21). In contrast, uncharged sphingomyelin-phospholipid mixtures became inhibitory only above 20 mole % sphingomyelin. The transfer of phospholipids between liposomes in these systems follows the kinetics of an ideal isotope reaction, since a linear relationship between $\ln(1 - R_t/R_\infty)$ and time (R_t and R_∞) is the percent of labeled phosphatidylcholine transferred by the exchange protein from the "donor" to "acceptor" liposomes at time t and at infinity (when equilibrium was attained), respectively (Hellings et al., 1974).

Van den Besselaar et al. (1975) have proposed a model to describe how PLEP carries phosphatidylcholine from one liposome to another. They assumed that an initial collision complex between the PC-PLEP and membrane is necessary, after which the PC-PLEP leaves the membrane with a phospholipid bound to it. Steady-state equations were presented for this reaction. According to these equations, the initial rate of transfer of phosphatidylcholine from one liposomal population to another depends on the total concentration of phosphatidylcholine in the two populations of liposomes and on the total amount of exchange protein in the medium. The experimental data on the rate of transfer of phosphatidylcholine between two populations of liposomes containing various amounts of phosphatidic acid (which diminished the intensity of transfer) were fitted to the theoretical equation by computer analysis. The calculated values of the initial rates of transfer showed a very close fit to the observed experimental rates. This mathematical treatment of the results also allowed calculation of the association constant of the exchange protein with the liposomes. It was found that the association of the exchange protein with the donor liposome increased when the phosphatidic acid content of the

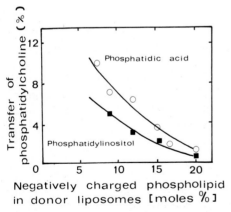

Fig. 21. Effects of acidic phospholipids on the transfer of [^{14}C]phosphatidylcholine from "donor" to "acceptor" liposome. "Donor" liposomes contained [^{14}C]phosphatidylcholine and various amounts of phosphatidylinositol or phosphatidic acid while the "acceptor" liposomes contained a fixed phosphatidylcholine/phosphatidic acid molar ratio (98/2). The incubations were performed at 25°C for 20 min in the presence of beef liver PC-PLEP. Reproduced with permission from Hellings et al. (1974).

liposomes increased. The decrease in the dissociation constant, which is implied, thus leaves less exchange protein available for the exchange reaction. The PLEP associates more easily with highly negative charged interfaces and dissociates with more difficulty. The protein may penetrate the liposomal surface, since the negative charges of the polar head groups of phosphatidylinositol (or phosphatidic acid) molecules repel each other. Under these conditions, less exchange protein will be free in the medium and able to carry phosphatidylcholine. It should be noted, however, that the investigations of Demel et al. (1973a) on phosphatidylcholine monolayers did not reveal any penetration of the PLEP when the surface pressure exceeded 20 dynes cm^{-1}.

Another way of modifying membrane surface properties is to change the acyl chains of the phospholipids. Kamp (1975) compared phospholipid exchange from donor liposomes containing phosphatidylcholine in which the two acyl chains were saturated fatty acids and from liposomes containing palmitoyl and oleyl as acyl residues (the main molecular species of egg phosphatidylcholine). Two main conclusions emerged from these experiments (Fig. 22). First,

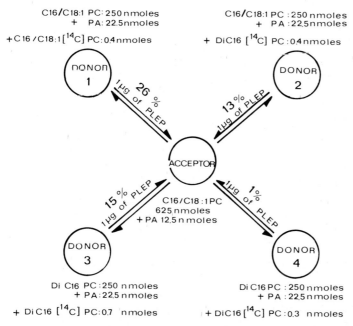

Fig. 22. The effects of acyl chain composition on phosphatidylcholine exchange between liposomes. The exchange reaction was followed between "acceptor" liposomes which have a fixed composition and different "donor" liposomes using the technique of Hellings et al. (1974) and Van den Besselaar et al. (1975). "Donor" liposomes contained a fixed amount of phosphatidic acid and unsaturated ($C_{16}/C_{18:1}$ PC = 1-palmitoyl 2-oleyl-sn glycero-3-phosphorylcholine) or saturated (Di-C_{16}PC = dipalmitoyl-sn glycero-3-phosphorylcholine) depending on the experiments. The values on the arrows indicate the percent of the initial radioactivity of the "donor" labeled phosphatidylcholine recovered in "acceptor" liposome after incubation for 40 min at 37°C in the presence of 1 µg of beef liver PC-PLEP. Figure constructed from the data of Kamp (1975).

saturated phosphatidylcholine is transferred more slowly than the unsaturated species (in Fig. 22; 13% of transfer from donor 2 to acceptor instead of 20% from donor 1 to acceptor). This may be due to a weak interaction between saturated phospholipids and the PLEP. However, this does not appear to be the case, since unsaturated phosphatidylcholine is also slowly transferred if the donor liposome is made from saturated phosphatidylcholine (donor 3 to acceptor in Fig. 22). Lack of transfer of saturated phosphatidylcholine is also observed when the liposomes are prepared from the saturated phospholipid (donor 4 to acceptor in Fig. 22).

These data strongly suggest that interactions between the liposomes and the PLEP play an important part in the transfer process (Kamp, 1975). As shown in Fig. 22, it is clear that the replacement of unsaturated phosphatidylcholine with the saturated species in the donor liposomes produced a decrease in rates of transfer, since all the other conditions were similar. The low fluidity of the artificial membranes made from saturated phosphatidylcholine may explain the weak interaction of the exchange protein with the membranes. A decrease in the mobility of the acyl chains can also be produced by cholesterol which has a condensing effect on membranes (Chapman, 1973; Van Deenen, 1975), but addition of cholesterol to phosphatidylcholine liposomes does not alter the exchange mediated by beef heart PLEP at a cholesterol/phosphatidylcholine molar ratio up to 0.4:1.0 (Johnson and Zilversmit, 1975).

It is not easy to discriminate between the importance of the interactions between liposomes and PLEP and between lipid and PLEP in the overall exchange reaction. However, the method used by Kamp (1975) in the experiments shown in Fig. 22 involved introduction of *trace amounts* of the compound to be studied into phosphatidylcholine/phosphatidic acid liposomes, which are able to form a proper complex with the exchange protein. Since these traces did not modify protein–liposome interactions, it was thus possible to examine whether the complex can be exchanged with other membrane constituents.

In connection with these observations, Wirtz (1974) demonstrated that the beef liver PC-PLEP was unable to transfer phospholipids from rat liver microsomes to high-density lipoprotein apoprotein isolated from human and rat serum. A lack of interaction between the apoprotein and the PLEP may explain this inactivity.

10.2. Liposome–mitochondria and liposome–microsome interactions

Similar inhibition of transfer of labeled phosphatidylcholine from beef liver mitochondria to liposomes has been observed when phosphatidic acid was introduced into liposomes (Wirtz, 1972; Wirtz et al., 1975, 1976). Introduction of stearylamine (which is positively charged) at concentrations higher than 10 mole % also produced a decrease in the transfer rate. Beef liver PC-PLEP was used in both these experiments (Fig. 23). Other negatively charged phospholipids also inhibited the exchange reaction, including phosphatidyl-

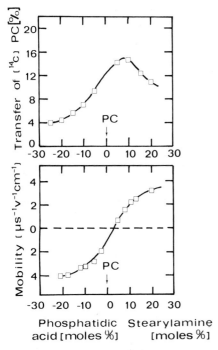

Fig. 23. The relationship between liposome surface charge (as reflected by the electrophoretic mobility) and the rate of phospholipid exchange. Various amounts of phosphatidic acid or stearylamine were added to phosphatidylcholine liposomes (PC) which were incubated with [^{14}C]choline-labeled mitochondria in the presence of PC-PLEP. The rate of transfer (expressed as corrected percentage of radioactivity recovered in the liposomal phosphatidylcholine) was determined between each type of liposome and mitochondrion. The electrophoretic mobility of the liposomes is expressed as $\mu\sec^{-1} V^{-1} cm^{-1}$. Reproduced with permission from Wirtz et al., 1976.

serine, phosphatidylinositol, phosphatidylglycerol and to a lesser extent phosphatidylethanolamine. These authors measured the surface charge of the various liposomes by microelectrophoresis and found an inverse relationship between electrophoretic mobility and the rate of transfer with all the phospholipids tested, except for phosphatidylethanolamine (Wirtz et al., 1976). It was noted, however, that liposomes with a slight net positive surface charge allowed an optimal transfer, probably by facilitating the interactions of the PLEP with the membrane.

Harvey et al. (1974) examined the transfer of [^{3}H]phosphatidylinositol between microsomes containing labeled phosphatidylinositol and phosphatidylcholine and unlabeled liposomes containing various amounts of phosphatidylinositol in the presence of the protein I PLEP from beef brain. The transfer of phosphatidylinositol from microsomes to liposomes was reduced by increasing the phosphatidylinositol content of the liposome (Fig. 24). The fact that a higher relationship was found between the amount of phosphatidylinositol present in the liposome and the transfer of this phospholipid in the

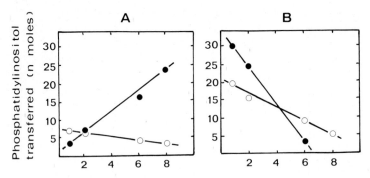

Fig. 24. Effects of liposomal phosphatidylinositol content on transfer of phosphatidylinositol (A) or phosphatidylcholine (B) between beef brain microsomes and liposomes in the presence of protein I from beef brain.

The transfer was determined from [^{3}H]phosphatidylinositol or [^{14}C]phosphatidylcholine-labeled microsomes to liposomes (○) or from liposomes to microsomes (●). The phosphatidylinositol content of the phosphatidylcholine liposomes is expressed in mole percentages. Reproduced with permission from Harvey et al., 1974.

reverse direction (liposome → microsomes) may be explained by the equilibration of the phosphatidylinositol pools. The transfer of phosphatidylcholine in both directions was also diminished by increasing the phosphatidylinositol content (Fig. 24).

From these data it appears that acidic phospholipids can significantly influence the efficiency of PLEP activity. This effect is probably mediated via the influence of acidic phospholipids on membrane surface charge which presumably affects the interaction of the PLEP with the membrane. The inhibitory effect of increasing concentrations of membrane acidic phospholipids on PLEP activity appears to be a general phenomenon for PLEP from animal tissues, but the situation concerning PLEP from plant tissue is unclear.

Since the increasing negative surface charge reduces phospholipid transfer, it should be possible to restore the initial transfer rate by neutralizing the additional charges. This was demonstrated effectively by Wirtz et al. (1976) who stimulated exchange activity that had been previously inhibited by a high liposomal phosphatidic acid content by introducing Mg^{2+} into the incubation mixture. The electrophoretic mobilities of the liposomes were diminished in parallel. This stimulation by Mg^{2+} was limited to concentrations sufficient to neutralize the surface charge of the liposome (2 mM).

However, the effects of cations in the medium on the protein-mediated exchange cannot be explained totally in terms of effects on membrane surface charge because: (a) introduction of phosphatidylethanolamine inhibits the exchange of phosphatidylcholine, although to a lesser extent than the acidic phospholipids, but does not affect the surface charge on the membranes (Wirtz et al., 1976); (b) the addition of high concentrations of Mg^{2+} to the medium decreases exchange, but has no effect on membrane surface charge (Wirtz et al., 1976); and (c) polyvalent cations, added in the subphase, inhibit the exchange of phosphatidylcholine between liposomes and a [^{14}C]phosphatidylcholine monolayer (Demel et al., 1973a) (Fig. 19), but the surface pressure was unaffected under these conditions. All these experiments were performed with beef liver PC-PLEP. It was suggested in section 9 that cations can disturb the electrostatic interactions between the polar head groups of the phosphatidylcholine and the PLEP. This hypothesis is also consistent with the observations presented above.

10.3. Mode of action of PLEP

According to the data obtained by the Utrecht group, by Zilversmit and coworkers at Cornell and by other laboratories, the transfer of phospholipids by PLEP may proceed as follows (Fig. 25):

(*1*) *Binding of phospholipid molecules.* Binding has been studied in most detail with the phosphatidylcholine exchange protein. The binding of phosphatidylcholine by this protein involves both hydrophobic interactions with the acyl chains and electrostatic interactions involving the polar head group. The phospholipid molecule is bound non-covalently to the PLEP, since the phospholipid can be extracted by organic solvents.

(*2*) *Formation of a collision complex between the PLEP and the membrane.* The formation of this "complex" is essential, and the surface properties of the membrane appear to be of considerable importance in regulating this step, and both acidic phospholipids and saturated acyl chains decrease complex formation. The chemical composition of the membrane thus determines the "apparent" specificity of the PLEP (Kamp, 1975).

(*3*) *Release of phospholipid from the PLEP to the interface.* This step can be revealed by the presence of labeled phospholipids in membranes using exchange assays performed with different structures, including monolayers.

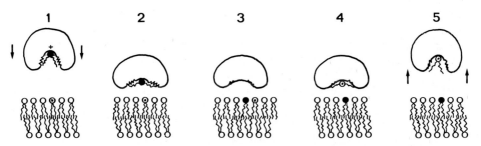

Fig. 25. Hypothetical scheme indicating possible steps in phosphatidylcholine exchange mediated by PC-PLEP. Identical polar head groups in the phosphatidylcholine bilayer have been represented differently in order to permit the schematic sequence of phosphatidylcholine exchange to be illustrated more easily. (1) PLEP, to which phospholipid is bound, approaches a membrane; (2) Formation of a "complex" between PLEP and membrane. A conformational change of the protein occurs; (3) Release of the bound phospholipid from PLEP to the membrane; (4) Extraction of a phospholipid molecule from the membrane to PLEP; (5) Detachment of PLEP from the membrane. If step 4 is not operating, PLEP will not contain a phospholipid.

(4) *Extraction of a phospholipid molecule from the membrane by the PLEP.* The extraction step is not necessarily linked to the preceding step, and steps (3) and (4) do not obligatorily follow step (2).

It is not known precisely how the release and extraction of phospholipid takes place. Recent observations of Wirtz et al. (1976) on the intrinsic fluorescence of the PLEP suggest that it undergoes a conformational change within the liposome membrane to form a zwitterionic amino acid pair with which phosphatidylcholine from the membrane could interact. It is logical that conformation changes would occur, since the PLEP will pass from a hydrophilic to a hydrophobic phase in its interaction with the membrane.

(5) *Detachment of the PLEP with or without a bound phospholipid.* It is assumed that the most frequent sequence is for a detached PLEP molecule, together with its bound phospholipid, to exchange the phospholipid by interacting with another "acceptor" membrane. It is not known if protein–protein interactions can also result in exchange of phospholipids between exchange proteins and membrane proteins.

The mode of action of the PLEP involves interactions between membrane surfaces and the exchange protein. One might expect that phospholipid exchange essentially involves the external monolayers of the membranes and that final insertion of the transferred phospholipid into the new acceptor membrane is essentially the reverse of steps (2)–(4), above.

11. Membrane asymmetry: implications for phospholipid exchange

To what extent are the different phospholipids found in membranes involved in the exchange process? This question has been examined in experiments with both natural and model membrane systems.

11.1. Natural membranes

In vivo labeling of phospholipids in microsomes and mitochondria from rat liver suggested to McMurray and Dawson (1969) and to Wirtz and Zilversmit (1969b) that only a small part of the phospholipid pool in the microsomes were involved in the exchange process. The latter authors found that the ratio of mitochondrial to microsomal phosphatidylcholine radioactivity was about 0.5, a result that could be explained if only 50% of the microsomal phospholipid pool was available for exchange. Results for the specific radioactivity of the phospholipids during incubation of labeled cell supernatant and microsomes are also consistent with the involvement of only a portion of the phospholipid pool in the exchange (Wirtz and Zilversmit, 1969b).

As indicated in section 4, only between 55% and 63% of the membrane phosphatidylcholine in erythrocytes is available for exchange; this available pool of phosphatidylcholine is probably located in the outer monolayer of the membrane. Asymmetry in the distribution of phospholipid in the erythrocyte plasma membrane was established by Zwaal et al. (1973) and Verkleij et al. (1973) who showed that only 66% of the phosphatidylcholine in intact human erythrocytes was degradable by phospholipase. This result suggested that this proportion of phosphatidylcholine was located in the outer monolayer of the membrane and that the remaining 33% was present on the inner monolayer and thus not accessible to the enzyme. Using phospholipase A_2, Renooij et al. (1974) have shown that the in vitro incorporation of fatty acids into erythrocyte phosphatidylcholine involves mainly the *inner* monolayer of the membrane, but the exchange process involves intact phospholipids in the *outer* monolayer. Other evidence that the exchange of certain phospholipids may be restricted is provided by observations on the movements of phospholipids between the inner and the outer membranes of mitochondria (Beattie, 1969b; Blok et al., 1971; Sauner and Lévy, 1971; Wojtczak et al., 1971). These investigators showed that the outer membrane was predominantly engaged in phospholipid exchange between microsomes and intact rat liver mitochondria (Fig. 26), and significantly lower amounts of the exchanged phospholipids were found in the inner membrane. Indeed, the radioactive phospholipids, recovered in the inner membrane in the experiments of Blok et al. (1971) and Sauner and Lévy (1971), were considered by Wojtczak et al. (1971) to have resulted from partial destruction of the outer membrane which then permitted direct exchange between microsomes and inner mitochondrial membranes. However, Blok et al. (1971) provided good evidence for the integrity of 90% of the mitochondria used in their experiments. The controversy still remains. Nevertheless, all these authors agree that phospholipid exchange between the outer and inner membrane is slower than between microsomes and the outer membrane. Blok et al. (1971) suggested that the intermembrane space lacked a phospholipid exchange

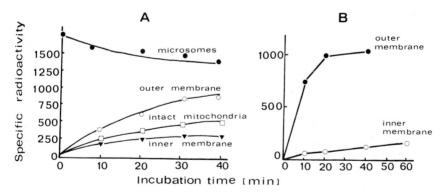

Fig. 26. Exchange experiments showing that the outer mitochondrial membrane of rat liver is predominantly involved in the exchange of phospholipids. Unlabeled mitochondria were incubated with [^{32}P]-labeled microsomes and unlabeled 105 000 g supernatant after which the inner and outer mitochondrial membranes were separated by centrifugation. The curves show the specific radioactivity of the phospholipids recovered in the membranes at the end of the incubation expressed in cpm/μg phospholipid phosphorus for (A) or cpm/mg protein for (B). Reproduced with permission from Blok et al. (1971) for A and from Wojtczak et al. (1971b) for B.

protein of the kind found in the cytoplasm. Wojtczak et al. (1971) suggested that the outer membrane was mainly concerned in the exchange, but because of the slow transverse movement of phospholipids within the membranes, the phospholipids were only slowly transferred to the inner membranes. Rousselet et al. (1976) have shown that beef liver PC-PLEP promoted selective incorporation of spin-labeled phosphatidylcholine into the outer layer of the mitochondrial inner membrane of rat liver, while in the absence of PLEP, the labeled molecules were incorporated on both sides of the membrane. Recent observations by Wojtczak and Zborowski (1975) have also revealed that "mitoplasts" (mitochondria without their outer membranes) are better acceptors than intact mitochondria of phosphatidic acid molecules transferred from liposomes in the presence of cytoplasmic supernatant. Also, a soluble fraction was isolated from sonicated mitochondria in this study which was able to mediate this transfer, suggesting that a PLEP may also be partly located within this organelle.

If phospholipids are slowly transferred to the inner membrane, what is the origin of the phospholipids of this membrane in vivo? Are they synthesized only during the early biogenesis of the whole mitochondrion, and then remain stable during the life of a mitochondrion, as suggested by Wojtczak et al. (1971)? Ruigrok et al. (1972) have proposed that direct contact occurs between the inner and outer membranes and is responsible for the observed transfer. These authors observed that mitochondrial swelling in situ produced by calcium acetate perfusion of rat liver was accompanied by increased transfer of phosphatidylcholine from the outer to the inner mitochondrial membrane.

It is not known if selective phospholipid exchange or localized exchange reactions are responsible for the differences in the lipid composition between the outer and inner mitochondrial membranes. Variations in phospholipid and fatty acid composition have been reported in these membranes in mitochondria from certain animal (Huet et al., 1968; Colbeau et al., 1971) and plant cells (Meunier and Mazliak, 1972; Moreau et al., 1974), but no significant difference has been detected in mitochondria from guinea pig liver (Parkes and Thompson, 1970).

It is obvious from these observations that phospholipid and protein movement within the double membrane of intracellular organelles (mitochondria or chloroplasts) will have to be investigated systematically with attention also being given to the exchange of other constituents.

11.2. Model membranes

By studying phosphatidylcholine exchange between liposomes containing [^{32}P]phosphatidylcholine and unlabeled mitochondria in the presence of beef heart PLEP, and by repeatedly replacing mitochondria by fresh ones at 10–20 min intervals, Johnson et al. (1975) observed a loss of [^{32}P]-label from the liposomes. This loss occurs very rapidly for about 10 min, after which a slower decrease was observed, with an equilibrium being reached at about 50%

of the initial radioactivity. The PLEP remained fully active during these two stages of loss. Since the liposomes used in these experiments were true single bilayer liposomes, this result suggests that about 50% of the phospholipid in the liposomes was not available for exchange (Fig. 27). Long-term experiments were also performed in which radiolabeled liposomes, unlabeled mitochondria and beef heart PLEP were incubated for 40 min, after which the mitochondria were removed, and the liposomes were again incubated with the PLEP. Surprisingly, a very slow decrease in the label was observed. The conclusion drawn from these experiments was that only a portion of phosphatidylcholine pool, situated in the outer monolayer of the bilayer vesicle, is accessible for exchange. This pool in sonicated liposomes represents about 60 to 63% of the total phosphatidylcholine. This value is also in agreement with data on the proportion of phosphatidylcholine available for hydrolysis by phospholipase D (Johnson et al., 1975).

In independent studies, Rothman and Dawidowicz (1975) showed that only 70% of the phosphatidylcholine of "sized" liposomes was available for transfer to human erythrocyte ghosts mediated by calf liver PLEP. This phosphatidylcholine pool was again located in the outer layer of the vesicle. Similar experiments, performed by Lenard and Rothman (1976) have revealed that only the phospholipids of the outer surface of a viral bilayer are available for rapid exchange mediated by beef heart PLEP. This study also showed that while two approximately equal pools of cholesterol exist in the viral bilayer, only the pool situated in the outer surface is rapidly exchangeable. Similarly, only cholesterol in the outer monolayer of phosphatidylcholine/cholesterol vesicles is available for exchange with erythrocyte ghosts. Barsukov et al. (1975) have shown that the PLEP from rat liver introduced phospholipid molecules mainly into the outer monolayer of phosphatidylcholine or phosphatidylinositol bilayer vesicles. Bloj and Zilversmit (1976) reported that about 75% of phosphatidylcholine of sealed erythrocyte ghosts was available for exchange

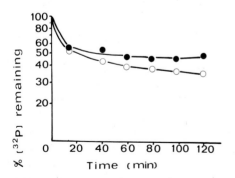

Fig. 27. Loss of [^{32}P]phosphatidylcholine from liposomes to unlabeled beef heart mitochondria in the presence of beef heart PLEP. The loss of [^{32}P]phosphatidylcholine was checked after various incubation times as indicated in the text. The curves correspond to two separate experiments. Reproduced with permission from Johnson et al., 1975.

with liposomes, mediated by beef heart PLEP, whereas when "inside-out" vesicles were used, only 37% of this phospholipid was exchangeable. The partial availability of phosphatidylcholine in liposomes was also postulated by Hellings et al. (1974) and Van den Besselaar et al. (1975) who concluded that 30 to 35% of this phospholipid in single bilayer liposomes did not participate in the exchange. Mathematical treatment of the experimental results showed that the data fitted with the hypothesis of an ideal isotope exchange discussed earlier (Hellings et al., 1974) and with the model for the formation of a collision complex between liposomes and PLEP (Van den Besselaar et al., 1975).

Several other techniques have been used to demonstrate asymmetry of phospholipids in single bilayer liposomes. By measuring choline methyl proton resonance, Michaelson et al. (1973) found that 63% of the choline signal was affected by Eu^{3+} which interacts only with the outer layer. Similar techniques gave comparable results (70%: Finer et al., 1972; Hauser and Barratt, 1973), while the distribution of spin-labeled phosphatidylcholine indicated that 65% of this phospholipid is present in the outer monolayer (Kornberg and McConnell, 1971b).

In all these experiments the liposomes were submitted to prolonged ultrasonic irradiation to produce single bilayer liposomes. Wirtz et al. (1976) have shown that phospholipid exchange is more active in sonicated liposomes than in hand-shaken multilamellar liposomes and suggested that this might be due to a lower amount of phosphatidylcholine (9%) present in the outer monolayer of the latter liposomes.

Since the exchange protein interacts mainly with the external monolayer, it is important to know the extent of natural phospholipid exchange occurring within the membrane, i.e., "flip-flop" movement. Kornberg and McConnell (1971b), using a spin-labeled nitroxide of phosphatidylcholine, reported a half-time for "flip-flop" of 6.5 h at 30°C in egg phosphatidylcholine liposomes. The use of the same technique in biological membranes has revealed much shorter half-times for "flip-flop" movement with values as low as 3.8–7 min at 15°C being reported for the electroplax of *Electrophorus electricus* (McNamee and McConnell, 1973). Half-times of the same order have also been found in the cytoplasmic membranes of *Acholeplasma laidlawii* (Grant and McConnell, 1973).

As shown in Fig. 27, the use of PLEP offers an original way of studying "flip-flop" movement in artificial or natural membranes. Completely different values were obtained when this approach was compared with the data discussed above. Rothman and Dawidowicz (1975) found a half-time of 11 days for the transposition of phospholipid in phosphatidylcholine liposomes, in agreement with Johnson et al. (1975) who found half-times for the "flip-flop" as long as 25 days. Even more recent studies by Lenard and Rothman (1976) have indicated half-times for the transposition of phosphatidylcholine or sphingomyelin in the lipid bilayer of virus envelopes of higher than 10 and 30 days at 30°C, respectively. Similarly, Roseman et al. (1975) have determined the half-time for equilibration of phosphatidylethanolamine across the bilayer of phosphatidylcholine/phosphatidylethanolamine (molar ratio of 9/1) vesicles to be at least 80 days at 22°C.

The large discrepancy between the results obtained by spin-labeling experiments and the availability of phospholipids for exchange, might be explained by the fact that spin-labeled phosphatidylcholine is not exactly similar to the authentic phospholipid. However, Rousselet et al. (1976) have calculated that the rate of "flip-flop" of a spin-labeled phosphatidylcholine in the inner membrane of rat liver mitochondria was slow (half-time for transposition longer than 24 h), while Bloj and Zilversmit (1976), using beef heart PLEP as a tool for the determination, obtained much faster rates of transposition of phosphatidylcholine in sealed rat erythrocyte ghosts with a half-time of 2.3 h for equilibration of this phospholipid in the two monolayers.

These recent results indicate that the situation is still controversial. This controversy is of considerable interest, since phospholipid asymmetry is a characteristic of many artificial and biological membranes, and a very rapid "flip-flop" movement would certainly not be consistent with the maintenance of asymmetry.

Half-lives of about 11 days have been found for the renewal of phospholipids in subcellular organelles from animals or plants (Fletcher and Sanadi, 1961; Bailey et al., 1967; Gross et al., 1969; Abdelkader and Mazliak, 1971). Half-lives of phospholipids in inner and outer mitochondrial membranes (McMurray and Dawson, 1969) and microsomes are similar, while plasmalemmal sphingomyelin has a slower turnover than microsomal sphingomyelin (T. Lee et al., 1973).

The other components of the membranes certainly influence the movement of phospholipids. The degree of unsaturation affects the fluidity of membranes (Chapman, 1973; Emmelot and Van Hoeven, 1975), and interactions of lipids with integral membrane proteins may create lipid domains of differing mobility. For instance, Warren et al. (1974a,b,c; 1975) have suggested that each Ca^{2+}-Mg^{2+} ATPase molecule in sarcoplasmic reticulum membranes is surrounded by an annulus of 30 or more phospholipid molecules which appear to be essential to its activity. Cholesterol cannot replace these phospholipids. A similar enzyme-lipid complex has also been postulated for the structural organization of cytochrome oxidase (Jost et al., 1973). It may be that the constituents of these lipid microenvironments are more stable than the bulk lipids (see Kimelberg, this volume, p. 205). If this concept can be extended to all the membranes, it would be of interest to examine whether the lack of exchangeability of some lipid pools is related to their involvement in similar "lipid annuli" linked to membrane enzymes.

12. The physiological significance of phospholipid exchange

Discussion of the possible physiological relevance of phospholipid exchange must inevitably involve substantial speculation, since many unanswered questions remain. However, researchers working on this subject have not been timid in offering speculative proposals concerning the functional significance of this phenomenon. Most attention has focused on the possible contribution of phospholipid exchange processes in renewal of membrane lipids and the related question of membrane biogenesis.

Dawson (1973) has suggested that the phospholipid exchange process participates in the maintenance of the phospholipid composition of membranes by introducing phospholipids from the site of biosynthesis to repair membrane damage produced by phospholipases or other lipid-degrading enzymes.

In view of the growing recognition that many membrane-bound enzymes appear to require phospholipids for this activity in both animal (Lester and Fleischer, 1961; Coleman, 1973; Jost et al., 1973; Roelofsen and Van Deenen, 1973; Warren et al., 1975) and plant cells (Jolliot and Mazliak, 1973), it is tempting to speculate that exchange proteins may be able to influence the activity of such enzymes by modulating the lipid environment in the immediate vicinity of the enzyme molecule. For instance, it has been shown that phosphatidylcholine is specifically involved in microsomal hydroxylation (Lu and Levin, 1974), microsomal acyl-coenzyme desaturate (Gurr et al., 1969; Abdelkader et al., 1973) and in ATP-[^{32}P]phosphate exchange (Y. Kagawa et al., 1973). In this latter system, a correlation was established between the amount of phosphatidylcholine introduced into the vesicles by a transfer process, mediated by the PC-PLEP, and the activity of the reconstituted system. Exchange proteins could thus be a potential mechanism for regulating enzymatic activities by supplying phospholipids needed in a reaction. There are, therefore, a number of interesting problems concerning interactions between the membrane-bound enzymes and exchange proteins which remain to be examined.

Renewal of membrane phospholipids need not, however, depend exclusively on phospholipid exchange phenomena. For example, the proposals concerning possible structural continuity between various cellular membranes, if correct, would mean that phospholipids synthesized at one site in the cell could diffuse laterally within a continuous membrane network to other sites (Guidoni and Thomas, 1969; Ruby et al., 1969; Bracker and Grove, 1971; Franke and Kartenbeck, 1971; Morré et al., 1971a,b).

In addition to a possible role in the renewal of membrane phospholipids, the exchange of phospholipids may also play an important role in the biogenesis of membranes. By this process, phospholipids could be distributed from the sites of biosynthesis to other membranes of the cell which themselves lack the metabolic machinery required for de novo biosynthesis of phospholipids. Most attention has been paid to phospholipid movement between microsomes and mitochondria. The phospholipids transferred from microsomes to mitochondria are not only integrated with the membranes in an intact form, but can also be metabolized further. For example, phosphatidylserine transferred from microsomes can be decarboxylated to phosphatidylethanolamine by mitochondrial enzymes (Butler and Thompson, 1975). However, many other membranes in the cell are involved in phospholipid exchange (Fig. 28), including the

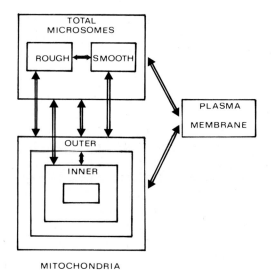

Fig. 28. Schematic representation of possible exchange processes occurring between the various membrane systems of the cell. The data on which this figure is based have been obtained largely from experiments with rat liver cells.

plasma membrane. Since the latter is unable to synthesize phospholipids (Victoria et al., 1971; Victoria and Korn, 1975), it is tempting to suggest that at least some of the phospholipids of the plasma membrane may arise via exchange processes, though other mechanisms may well contribute to an equal or greater extent (see Chapter 1, this volume, pp. 20 and 31).

Exchange of phospholipids represents a kind of cooperation between the different membranous structures of the cell. A symbiotic relationship between mitochondria, which are thought to be of bacterial origin (Cohen, 1970), and other cell membranes is necessary to account for the composition of the membranes of this organelle. Mitochondria also have only a limited capacity to form proteins, and few hydrophobic proteins were synthesized in this organelle (Beattie, 1968; Beattie et al., 1970; Kadenbach, 1971; Tzagoloff, 1971; Tzagoloff and Akai, 1972; Tzagoloff and Meagher, 1972; Hadvary and Kadenbach, 1973; Schatz and Mason, 1974; Wheeldon et al., 1974). This therefore prompts the question of whether a simultaneous assembly of mitochondrial proteins and lipids occurs in other membranes with subsequent transfer of both components to the mitochondrial membrane(s). Indeed, according to Beattie (1969a,b,c), the soluble proteins of the mitochondrial matrix and the major part of the proteins of the inner and outer membranes of rat liver mitochondria are not synthesized in these organelles, but transferred from rough endoplasmic reticulum. Kadenbach (1968) and Beattie (1969a,b,c) have both proposed from labeling kinetics that a simultaneous transfer of both constituents occurs. However, Kadenbach (1970) has shown that only the apoprotein of cytochrome c is synthesized in the endoplasmic reticulum of rat liver and is transported through the outer mitochondrial membrane to the inner one while the attachment of the haem group, which is synthesized in the mitochondria, to the apoprotein takes place within mitochondria. Similar transfer of protein and lipid components partially synthesized on rat liver microsomes with subsequent transfer to mitochondria, has also been reported for cytochrome oxidase (Schiefer, 1969).

Despite this evidence which supports simultaneous transfer of both protein and lipid components, the discovery of phospholipid exchange proteins would seem to make this process unlikely. According to Dawson (1973), it is reasonable to suppose that the transfer of protein and phospholipids are independent processes; nevertheless, membrane proteins, interacting with phospholipids, probably control the phospholipid composition of the membrane. On the other hand, Godinot (1973) and Godinot and Lardy (1973) have suggested that glutamate dehydrogenase synthesized on the microsomes of rat liver is transferred to mitochondria; interactions between the enzyme and diphosphatidylglycerol control this transfer of the dehydrogenase to the mitochondrial matrix.

However, the concept that the mitochondrion is completely dependent on the subcellular environment is difficult to reconcile with the longevity of mitochondrial suspensions in vitro, which suggests considerable autonomy from biosynthetic and renewal processes (Romani and Ozelkok, 1973).

It is also important to point out that there is no direct evidence for the participation of phospholipid exchange in the biogenesis of membranes in vivo. Only indirect evidence exists to support this concept. Although some support for the view that phospholipid exchange may be needed to supplement the limited biosynthetic capacity of organelles such as mitochondria, all the experimental evidence has been obtained using in vitro systems. The determination of the capacity of subcellular fraction to synthesize phospholipid(s) in vitro requires destruction of the cell, and the negative results obtained to date may reflect the fragility of the enzymes involved in lipid synthesis and may not accurately reflect their absence in a subcellular fraction. Furthermore, active phospholipid exchange is clearly not a general feature of all cellular membranes (Miller and Dawson, 1972).

The view that PLEP play a fundamental role in membrane biogenesis, while no doubt attractive to those working with exchange proteins, must be tempered by the fact that equally compelling evidence has been presented for other mechanisms of membrane biogenesis. There is now considerable circumstantial evidence to support the "membrane flow" concept of membrane biogenesis (see Morré et al., 1971a; and Chapter 1, this volume, p. 32) in which newly synthesized vesicles of Golgi-derived membrane are transferred to other intracellular membranes where they fuse with the existing membrane. In this way intact segments of new membrane are introduced directly into the old membrane. There is no obvious reason to conclude, however, why this process and phospholipid exchange do not occur in parallel. Indeed, phospholipid exchange may be required to generate asymmetry in newly inserted membrane segments. For example, if the lipid bilayer in the new membrane synthesized in the Golgi apparatus exhibits symmetry, then asymmetry must be generated later, either within the Golgi or when the new membrane segment has been inserted at its final location. Lipid asymmetry might therefore be imposed as a post-synthetic event via the selective exchange of phospholipids by PLEP acting on different sides of the bilayer.

13. Unanswered questions

In the eight years since the first discovery of the phospholipid exchange process, an enormous amount of work has been done on this phenomenon. Many unanswered questions still remain, however, and these will no doubt receive considerable attention over the next few years. More information is required on the molecular specificity of PLEP from different cells and tissues. For example, are these PLEP monospecific for phospholipids other than phosphatidylcholine? Apart from the intrinsic relevance of this question to exchange phenomena, the availability of a range of PLEP with monospecific exchange activity for most of the phospholipids found in biological membranes would provide a useful tool for the experimental manipulation of membrane composition. Also of interest is the location of various PLEP within the cell. More information is also needed on the mode of action of PLEP and the related issue of whether their activity is subject to physiologic modulation. The study of these questions will require detailed analysis of the biosynthesis of PLEP and the events controlling the induction of different amounts of these proteins within the cell. Finally, it will be of interest to see if PLEP deficiency can contribute to altered cell function and disease.

Acknowledgments

I am indebted to Dr. T. Galliard (Food Research Institute, Norwich, Great Britain) and to Professor P. Mazliak (Director of the Laboratory of Physiologie Cellulaire) for critical reading of the manuscript and for helpful suggestions, and I thank Professor L.L.M. Van Deenen and Dr. K.W.A. Wirtz (Laboratory of Biochemistry, Rijksuniversiteit of Utrecht, The Netherlands) for stimulating discussions. I am grateful to Dr. D.B. Zilversmit, Dr. J.E. Rothman and Dr. P. Devaux for providing preprints of their work and to Dr. H.H. Kamp, Dr. R.M.C. Dawson, Professor T. Sakagami, Professor L. Wojtczak, Dr. L.W. Johnson and Professor L. Douste-Blazy for supplying useful material. I also thank the publishing companies and the authors for their kind permission to reproduce previously published material.

References

Abdelkader, A.B. (1972) Biogénèse des lipides membranaires pendant la survie de tranches de tubercules de pomme de terre (*Solanum tuberosum* L.). Thesis, University of Paris VI, pp. 153–170.

Abdelkader, A.B. and Mazliak, P. (1970) Echanges de lipides entre mitochondries, microsomes et surnageant cytoplasmique de cellules de pomme de terre ou de chou-fleur. Eur. J. Biochem. 15, 250–262.

Abdelkader, A.B. and Mazliak, P. (1971) Renouvellement des lipides dans diverses fractions cellulaires de parenchyme de pomme de terre ou d'inflorescence de chou-fleur. Physiol. Vég. 9, 227–240.

Abdelkader, A.B. and Mazliak, P. (1972) Biogenesis of mitochondria, endoplasmic reticulum membranes and polyribosomes in aged potato tuber slices. 8th FEBS Meeting, Abstract 665, Amsterdam.

Abdelkader, A.B., Chérif, A., Demandre, C. and Mazliak, P. (1973) The oleyl-coenzyme A desaturase of potato tubers. Enzymatic properties, intracellular localization and induction during "aging" of tuber slices. Eur. J. Biochem. 32, 155–163.

Ada, G.L. (1949) Phospholipid metabolism in rabbit liver cytoplasm. Biochem. J. 45, 422–428.

Akiyama, M. and Sakagami, T. (1969) Exchange of mitochondrial lecithin and cephalin with those in rat liver microsomes. Biochim. Biophys. Acta 187, 105–112.

Bailey, E., Taylor, C.B. and Bartley, W. (1967) Turnover of mitochondrial components of normal and essential fatty acids-deficient rats. Biochem. J. 104, 1026–1032.

Barsukov, L.I., Shapiro, Yu.E., Viktorov, V.I., Volkova, V.I., Bystrow, V.F. and Bergelson, L.D. (1974) Study of intervesicular phospholipid exchange by NMR. Biochim. Biophys. Res. Commun. 60, 196–203.

Barsukov, L.I., Shapiro, Yu.E., Viktorov, A.V., Volkova, V.I., Bystrow, V.F. and Bergelson, L.D. (1975) Intervesicular phospholipid exchange. An NMR study. Chem. Phys. Lipids 14, 211–226.

Beattie, D.S. (1968) Studies on the biogenesis of mitochondrial protein components in rat liver slices. J. Biol. Chem. 243, 4027–4033.

Beattie, D.S. (1969a) The relationship of protein and lipid synthesis during the biogenesis of mitochondrial membranes. J. Membrane Biol. 1, 383–401.

Beattie, D.S. (1969b) The biosynthesis of the protein and lipid components of the inner and outer membranes of rat liver mitochondria. Biochem. Biophys. Res. Commun. 35, 67–74.

Beattie, D.S. (1969c) The turnover of the protein components of the inner and outer membrane fractions of rat liver mitochondria. Biochem. Biophys. Res. Commun. 35, 721–727.

Beattie, D.S., Patton, G.M. and Stuchell, R.N. (1970) Studies in vitro on amino acid incorporation into purified components of rat liver mitochondria. J. Biol. Chem. 245, 2177–2184.

Bell, F.P. (1975) Cholesterol exchange between microsomal, mitochondrial and erythrocyte membranes and its enhancement by cytosol. Biochim. Biophys. Acta 398, 18–27.

Bigelow, C.C. (1967) On the average hydrophobicity of proteins and the relation between it and protein structure. J. Theoret. Biol. 16, 187–211.

Bjerve, K.S. (1974) Biosynthesis of phospholipids. Thesis, Universitets Forlagetstrykningssentral, Oslo, pp. 8–24.

Bjornstad, P. and Bremer, J. (1966) In vivo studies on pathways for the biosynthesis of lecithin in the rat. J. Lipid Res. 7, 38–45.

Bloj, B. and Zilversmit, D.B. (1976) Asymmetry and transposition rates of phosphatidylcholine in rat erythrocyte ghosts. Biochemistry (in press).

Blok, M.C., Wirtz, K.W.A. and Scherphof, G.L. (1971) Exchange of phospholipids between microsomes and inner and outer mitochondrial membranes of rat liver. Biochim. Biophys. Acta 233, 61–75.

Bowden, L. and Lord, J.M. (1975) Development of phospholipid synthesizing enzymes in castor bean endosperm. FEBS Lett. 49, 369–372.

Bracker, C.E. and Grove, S.N. (1971) Continuity between cytoplasmic endomembranes and outer mitochondrial membranes in fungi. Protoplasma 73, 15–34.

Bretscher, M.S. (1972) Asymmetrical lipid bilayer structure for biological membranes. Nature (London) New Biol. 236, 11–12.
Bretscher, M.S. (1973) Membrane structure: some general principles. Science 181, 622–629.
Bretscher, M.S. (1974) Some aspects of membrane structure, In: O.S. Estrada and C. Gitler (Eds.), Perspectives in Membrane Biology, Academic Press, New York, pp. 3–24.
Brûlet, P. and McConnell, H.M. (1975) Magnetic resonance spectra of membranes. Proc. Natl. Acad. Sci. USA 72, 1451–1455.
Bureau, G., Kader, J.C. and Mazliak, P. (1976) Isolement d'une fraction protéique stimulant les échanges d'acides gras ramifiés entre les mésosomes et les protoplastes de Bacillus subtilis var niger, C.R. Acad. Sci. Paris, Sér. D 282, 119–122.
Butler, M.M. and Thompson, W. (1975) Transfer of phosphatidylserine from liposomes or microsomes to mitochondria. Stimulation by a cell supernatant factor. Biochim. Biophys. Acta 388, 52–57.
Bygrave, F.L. (1969) Studies on the biosynthesis and turnover of the phospholipid components of the inner and outer membranes of rat liver mitochondria. J. Biol. Chem. 244, 4768–4772.
Chapman, D. (1973) Some recent studies of lipids, lipid-cholesterol and membrane systems, In: D. Chapman and D.F.H. Wallach (Eds.), Biological Membranes, Vol. 2, Academic Press, London, pp. 91–144.
Chlapowski, F.J. and Band, R.N. (1971a) Assembly of lipids into membranes in Acanthamoeba palestinensis, I. Observations on the specificity and stability of choline-[^{14}C] and glycerol-[^{3}H] as labels for membrane phospholipids. J. Cell Biol. 50, 625–633.
Chlapowski, F.J. and Band, R.N. (1971b) Assembly of lipids into membranes in Acanthamoeba palestinensis, II. The origin and fate of glycerol-[^{3}H]-labeled phospholipids of cellular membranes. J. Cell Biol. 50, 634–651.
Cohen, S.S. (1970) Are/were mitochondria and chloroplasts microorganisms? Am. Scientist, 58, 281–289.
Colbeau, A., Nachbaur, J. and Vignais, P.M. (1971) Enzymic characterization and lipid composition of rat liver subcellular membranes. Biochim. Biophys. Acta 249, 462–492.
Coleman, R. (1973) Membrane-bound enzymes and membrane ultrastructure. Biochim. Biophys. Acta 300, 1–30.
Daae, L.N.W. and Bremer, J. (1970) The acylation of glycerophosphate in rat liver. A new assay procedure for glycerophosphate acylation, studies on its subcellular and submitochondrial localization and determination of the reaction products. Biochim. Biophys. Acta 210, 92–104.
Davidson, J.B. and Stanacev, N.Z. (1970) Biochemistry of polyglycerophosphatides in central nervous tissues, I. On the biosynthesis, structure and enzymatic degradation of phosphatidylglycerophosphate and phosphatidylglycerol in isolated sheep brain mitochondria. Can. J. Biochem. 48, 634–642.
Davidson, J.B. and Stanacev, N.Z. (1971) Biosynthesis of cardiolipin in mitochondria isolated from guinea-pig liver. Biochem. Biophys. Res. Commun. 42, 1191–1199.
Davidson, J.B. and Stanacev, N.Z. (1974) Evidence for the biosynthetic differences between isolated mitochondria and microsomes. Can. J. Biochem. 52, 936–939.
Dawson, R.M.C. (1966) The metabolism of animal phospholipids and their turnover in cell membranes, In: P.N. Campbell and G.D. Greville (Eds.), Essays in Biochemistry, Academic Press, London, pp. 69–115.
Dawson, R.M.C. (1973) The exchange of phospholipids between cell membranes. Sub-Cell. Biochem. 2, 69–89.
Demel, R.A., Wirtz, K.W.A., Kamp, H.H., Geurts Van Kessel, W.S.M. and Van Deenen, L.L.M. (1973a) Phosphatidylcholine exchange protein from beef liver. Nature (London) New Biol. 246, 102–105.
Demel, R.A., London, Y., Geurts Van Kessel, W.S.M., Vossenberg, F.G.A. and Van Deenen, L.L.M. (1973b) The specific interaction of myelin basic proteins with lipids at the air-water interface. Biochim. Biophys. Acta 311, 507–519.
Dennis, E.A. and Kennedy, E.P. (1972) Intracellular sites of lipid synthesis and the biogenesis of mitochondria. J. Lipid Res. 13, 263–267.

Devaux, P. and McConnell, H.M. (1972) Lateral diffusion in spin-labeled phosphatidylcholine multilayers. J. Am. Chem. Soc. 94, 4475–4481.

Devor, K.A. and Mudd, J.B. (1971) Biosynthesis of phosphatidylcholine by enzyme preparations from spinach leaves. J. Lipid Res. 12, 403–411.

Dobiasova, M. and Linhart, J. (1970) Association of phospholipid-cholesterol micelles with rat heart mitochondria: stimulators and inhibitors. Lipids 5, 445–451.

Dobiasova, M. and Radin, N.S. (1968) Uptake of cerebroside, cholesterol and lecithin by brain myelin and mitochondria. Lipids 3, 439–448.

Douady, D. and Mazliak, P. (1975) Etude des transferts d'acides gras entre liposomes et mitochondries isolées de Chou-fleur. Physiol. Vég. 13, 383–392.

Douce, R. (1968) Mise en évidence du cytidine diphosphate diglycéride dans les mitochondries végétales isolées. C.R. Acad. Sci. Paris, Sér. D 267, 534–537.

Douce, R. and Dupont, J. (1969) Biosynthèse du phosphatidylglycérol dans les mitochondries végétales isolées: mise en évidence du phosphatidylglycérophosphate. C.R. Acad. Sci. Paris, Sér. D 268, 1657–1660.

Dousset, N. and Douste-Blazy, L. (1974) Influence de l'irradiation sur les échanges in vitro de phospholipides entre plasma et globules rouges de rat. Int. J. Radiat. Biol. 26, 505–510.

Dyatlovitskaya, E.V., Trusova, V.M., Greshnickh, K.P., Gorkova, N.P. and Bergelson, L.D. (1972) Tumor lipids. In vitro exchange of phospholipids between mitochondria and microsomes from the Zajdela rat hepatoma. Biokhimiya (USSR) 37, 607–613.

Edelstein, C., Lin, C.T. and Scanu, A.M. (1972) On the subunit structure of human serum high density lipoprotein, I. A study of its major polypeptide component (Sephadex, fraction III). J. Biol. Chem. 247, 5842–5849.

Eder, H.A. (1957) The lipoproteins of human serum. Am. J. Med. 23, 269–282.

Eder, H.A., Bragdon, J.I. and Boyle, E. (1954) The in vitro exchange of phospholipid phosphorus between lipoproteins. Circulation 10, 603.

Ehnholm, C. and Zilversmit, D.B. (1972) Use of Forssman antigen in the study of phosphatidylcholine exchange between liposomes. Biochim. Biophys. Acta 274, 652–657.

Ehnholm, C. and Zilversmit, D.B. (1973) Exchange of various phospholipids and of cholesterol between liposomes in the presence of highly purified phospholipid exchange proteins. J. Biol. Chem. 248, 1719–1724.

Emmelot, P. and Van Hoeven, R.P. (1975) Phospholipid unsaturation and plasma membrane organization. Chem. Phys. Lipids 14, 236–246.

Finer, E.G., Flook, A.G. and Hauser, H. (1972) Mechanism of sonication of aqueous egg yolk lecithin dispersions and nature of the resultant particles. Biochim. Biophys. Acta 260, 49–58.

Fleischer, S. and Brierley, G. (1961) The equilibration between soluble micelles of phospholipids and the bound lipid of mitochondrial particles. Biochim. Biophys. Acta 53, 609–612.

Fletcher, M.J. and Sanadi, D.R. (1961) Turnover of rat liver mitochondria. Biochim. Biophys. Acta 51, 356–360.

Florsheim, W.H. and Morton, M.E. (1957) Stability of phospholipid binding in human serum lipoproteins. J. Appl. Physiol. 10, 301–304.

Fowler, S. and DeDuve, C. (1969) Digestive activity of lysosomes, III. The digestion of lipids by extracts of rat liver lysosomes. J. Biol. Chem. 244, 471–481.

Franke, W.W. and Kartenbeck, J. (1971) Outer mitochondrial membrane continuous with endoplasmic reticulum. Protoplasma 73, 35–41.

Galliard, T. (1971) The enzymic deacylation of phospholipids and galactolipids in plants. Purification and properties of a lipolytic-acyl hydrolase from potato tubers. Biochem. J. 121, 379–390.

Galliard, T. (1973) Phospholipid metabolism in photosynthetic plants, In: G.B. Ansell, J.N. Hawthorne and R.M.C. Dawson (Eds.), Form and Function of Phospholipids, Elsevier, Amsterdam, pp. 253–288.

Galliard, T. (1975) Degradation of plant lipids by hydrolytic and oxidative enzymes, In: T. Galliard and E.I. Mercer (Eds.), Recent Advances in the Chemistry and Biochemistry of Plant Lipids, Academic Press, London, pp. 319–357.

Glaser, M., Ferguson, K.A. and Vagelos, P.R. (1974) Manipulation of the phospholipid composition of tissue culture cells. Proc. Natl. Acad. Sci. USA 71 4072–4076.

Glaumann, H., Bergstrand, A. and Ericsson, J.L.E. (1975) Studies on the synthesis and intracellular transport of lipoprotein particles in rat liver. J. Cell Biol. 64, 356–377.

Godinot, C. (1973) Nature and possible functions of association between glutamate dehydrogenase and cardiolipin. Biochemistry 12, 4029–4034.

Godinot, C. and Lardy, H.A. (1973) Biosynthesis of glutamate dehydrogenase in rat liver. Demonstration of its microsomal localization and hypothetical mechanism of transfer to mitochondria. Biochemistry 12, 2051–2060.

Gordesky, S.E. and Marinetti, G.V. (1973) The asymmetric arrangement of phospholipids in the human erythrocyte membrane. Biochem. Biophys. Res. Commun. 50, 1027–1031.

Grant, C.W.M. and McConnell, H.M. (1973) Fusion of phospholipid vesicles with viable Acholeplasma laidlawii. Proc. Natl. Acad. Sci. USA 70, 1238–1240.

Gross, N.J., Getz, G.S. and Rabinowitz, M. (1969) Apparent turnover of mitochondrial deoxyribonucleic acid and mitochondrial phospholipids in the tissues of the rat. J. Biol. Chem. 244, 1552–1562.

Guarnieri, M., Stechmiller, B. and Lehninger, A.L. (1971) Use of antibody to study the location of cardiolipin in mitochondrial membranes. J. Biol. Chem. 246, 7526–7532.

Guidoni, J.J. and Thomas, H. (1969) Connection between a mitochondrion and endoplasmic reticulum in liver. Experientia 25, 632–633.

Gurr, M.I., Robinson, M.P. and James, A.T. (1969) The mechanism of formation of polyunsaturated fatty acids by photosynthetic tissue. The tight coupling of oleate desaturation with phospholipid synthesis in Chlorella vulgaris. Eur. J. Biochem. 9, 70–78.

Gurr, M.J., Prottey, C. and Hawthorne, J.N. (1965) The phospholipids of liver-cell fractions, II. Incorporation of [^{32}P]orthophosphate in vivo in normal and regenerating rat liver. Biochim. Biophys. Acta 106, 357–370.

Hadvary, P. and Kadenbach, B. (1973) Isolation and characterization of chloroform-soluble proteins from rat liver mitochondria and other fractions. Eur. J. Biochem. 39, 11–20.

Hahn, L. and Hevesy, G. (1939) Interaction between the phosphatides of the plasma and corpuscles. Nature (London) 144, 204–207.

Hajra, A.K. and Agranoff, B.W. (1968a) Acyl dihydroxyacetone phosphate. Characterization of a [^{32}P]-labeled lipid from guinea-pig liver mitochondria. J. Biol. Chem. 243, 1617–1622.

Hajra, A.K. and Agranoff, B.W. (1968b) Reduction of palmitoyl dihydroxyacetone phosphate by mitochondria. J. Biol. Chem. 243, 3542–3543.

Hallinan, T. (1974) Lipid effects of enzyme activities: phospholipid dependence and phospholipid constraint. Biochem. Soc. Trans. 2, 817–820.

Hannig, K. and Heidrich, H.G. (1974) The use of continuous preparative free-flow electrophoresis for dissociating cell fractions and isolation of membranous components, In: S. Fleischer and L. Packer (Eds.), Methods in Enzymology, Vol. XXXI, Academic Press, New York, pp. 746–761.

Harvey, M.S., Wirtz, K.W.A., Kamp, H.H., Zegers, B.J.M. and Van Deenen, L.L.M. (1973) A study of phospholipid exchange proteins present in the soluble fractions of beef liver and brain. Biochim. Biophys. Acta 323, 234–239.

Harvey, M.S., Helmkamp, G.M., Wirtz, K.W.A. and Van Deenen, L.L.M. (1974) Influence of membrane phosphatidylinositol content on the activity of bovine brain phospholipid transfer protein. FEBS Lett. 46, 260–262.

Hauser, H. and Barratt, M.D. (1973) Effect of chain length on the stability of lecithin bilayers. Biochem. Biophys. Res. Commun. 53, 399–405.

Havel, R.J. and Clarke, J.C. (1958) Metabolism of chylomicron phospholipids. Clin. Res. 6, 254–265.

Heidrich, H.G., Stahn, R. and Hannig, K. (1970) The surface charge of rat liver mitochondria and their membranes. Clarification of some controversies concerning mitochondrial structure. J. Cell Biol. 46, 137–150.

Helenius, A. and Simons, K. (1975) Solubilization of membranes by detergents. Biochim. Biophys. Acta 415, 29–79.

Hellings, J.A., Kamp, H.H., Wirtz, K.W.A. and Van Deenen, L.L.M. (1974) Transfer of phosphatidylcholine between liposomes. Eur. J. Biochem. 47, 601–605.

Helmkamp Jr., G.M., Harvey, M.S., Wirtz, K.W.A. and Van Deenen, L.L.M. (1974) Phospholipid exchange between membranes. Purification of bovine brain proteins that preferentially catalyze the transfer of phosphatidylcholine. J. Biol. Chem. 249, 6382–6389.

Horiuchi, I. (1973) Studies on phospholipids in small intestinal mucosa, 2. On exchange of phospholipids between mitochondria and microsomes in rat small intestinal mucosa. Sapporo Med. J. 42, 457–464.

Hostetler, K.Y. and Van den Bosch, H. (1972) Subcellular and submitochondrial localization of the biosynthesis of cardiolipin and related phospholipids in rat liver. Biochim. Biophys. Acta 260, 380–386.

Hostetler, K.Y., Van den Bosch, H. and Van Deenen, L.L.M. (1971) Biosynthesis of cardiolipin in liver mitochondria. Biochim. Biophys. Acta 239, 113–119.

Hostetler, K.Y., Van den Bosch, H. and Van Deenen, L.L.M. (1972) The mechanism of cardiolipin biosynthesis in liver mitochondria. Biochim. Biophys. Acta 260, 507–513.

Huang, L. and Pagano, R.E. (1975) Interaction of phospholipid vesicles with cultured mammalian cells, I. Characteristics of uptake. J. Cell Biol. 67, 38–48.

Huet, C., Lévy, M. and Pascaud, M. (1968) Spécificité de constitution en acides gras des phospholipides des membranes mitochondriales. Biochim. Biophys. Acta 150, 521–524.

Illingworth, D.R. and Portman, O.W. (1972a) Independence of phospholipid and protein exchange between plasma lipoproteins in vivo and in vitro. Biochim. Biophys. Acta 280, 281–289.

Illingworth, D.R. and Portman, O.W. (1972b) Exchange of phospholipids between low and high density lipoproteins of squirrel monkeys. J. Lipid Res. 13, 220–227.

Illingworth, D.R., Portman, O.W., Robertson, A.L. and Mayar, W.A. (1973) The exchange of phospholipids between plasma lipoproteins and rapidly dividing human cells grown in tissue culture. Biochim. Biophys. Acta 306, 422–436.

Jackson, R.L., Baker, H.N., David, J.S.K. and Gotto, A.M. (1972) Isolation of a helical, lipid-binding fragment from the human plasma high density lipoprotein, apolp-Gln I. Biochem. Biophys. Res. Commun. 49, 1444–1451.

Johnson, L.W. and Zilversmit, D.B. (1975) Catalytic properties of phospholipid exchange protein from bovine heart. Biochim. Biophys. Acta 375, 165–175.

Johnson, L.W., Hughes, M.E. and Zilversmit, D.B. (1975) Use of phospholipid exchange proteins to measure inside-outside transposition in phosphatidyl liposomes. Biochim. Biophys. Acta 375, 176–185.

Jolliot, A. and Mazliak, P. (1973) Rôle des lipides dans diverses activités enzymatiques de la chaine de transport des électrons d'une fraction mitochondriale isolée d'inflorescences de Chou-fleur. Plant Sci. Lett. 1, 21–29.

Jonas, A. (1975) Phosphatidylcholine interaction with bovine serum albumin: effect of the physical state of the lipid on protein-lipid complex formation. Biochem. Biophys. Res. Commun. 4, 1003–1008.

Jost, P.C., Griffiths, O.H., Capaldi, R.A. and Van der Kooi, G. (1973) Evidence for boundary lipid in membranes. Proc. Natl. Acad. Sci. USA 70, 480–484.

Jungalwala, F.B. and Dawson, R.M.C. (1970) Phospholipid synthesis and exchange in isolated liver cells. Biochem. J. 117, 481–490.

Jungalwala, F.B., Freinkel, N. and Dawson, R.M.C. (1971) The metabolism of phosphatidylinositol in the thyroid gland of the pig. Biochem. J. 123, 19–33.

Kadenbach, B. (1968) Transfer of proteins from microsomes into mitochondria. Biosynthesis of cytochrome c, In: E.C. Slater, J.M. Tager, S. Papa and E. Quagliariello (Eds.), Biochemical Aspects of the Biogenesis of Mitochondria, Adriatica, Bari, pp. 415–429.

Kadenbach, B. (1970) Biosynthesis of cytochrome c. The sites of synthesis of apoprotein and holoenzyme. Eur. J. Biochem. 12, 392–398.

Kadenbach, B. (1971) Isolation and characterization of a peptide synthesized in mitochondria. Biochem. Biophys. Res. Commun. 44, 724–730.

Kader, J.C. (1975) Proteins and the intracellular exchange of lipids, I. The stimulation of phospholipid exchange between mitochondria and microsomal fractions by proteins isolated from potato tuber. Biochim. Biophys. Acta 380, 31–44.

Kader, J.C. and Douady, D. (1975) A new group of plant proteins: the phospholipid exchange proteins, which catalyze transfers of phospholipid between membranes. 10th FEBS Meeting, Abstract 1039, Paris.

Kagawa, T., Lord, J.M. and Beevers, H. (1973) The origin and turnover of organelle membranes in castor bean endosperm. Plant Physiol. 51, 61–65.

Kagawa, Y., Johnson, L.W. and Racker, E. (1973) Activation of phosphorylating vesicles by net transfer of phosphatidylcholine by phospholipid transfer protein. Biochem. Biophys. Res. Commun. 50, 245–251.

Kamath, S.A. and Rubin, E. (1973) The exchange of phospholipids between subcellular organelles of the liver. Arch. Biochem. Biophys. 158, 312–322.

Kamath, S.A. and Rubin, E. (1974a) A short procedure to measure phospholipid exchange in mitochondria and microsomes. Anal. Biochem. 59, 172–177.

Kamath, S.A. and Rubin, E. (1974b) Effect of carbon tetrachloride and phenobarbital on plasma membranes: enzymes and phospholipid transfer. Lab. Invest. 30, 494–499.

Kamath, S.A. and Rubin, E. (1974c) Alterations in the rate of phospholipid exchange between cell membranes. Lab. Invest. 30, 500–504.

Kamp, H.H. (1975) Purification and properties of phosphatidylcholine exchange protein from bovine liver, Thesis, Rijksuniversiteit of Utrecht, pp. 9–86.

Kamp, H.H. and Wirtz, K.W.A. (1974) Phosphatidylcholine exchange protein from beef liver, In: S.P. Colowick and N.O. Kaplan (Eds.), Methods in Enzymology, Vol. 32, Academic Press, New York, pp. 140–146.

Kamp, H.H., Wirtz, K.W.A. and Van Deenen, L.L.M. (1973) Some properties of phosphatidylcholine exchange proteins purified from beef liver. Biochim. Biophys. Acta 318, 313–325.

Kamp, H.H., Sprengers, E.D., Westerman, J., Wirtz, K.W.A. and Van Deenen, L.L.M. (1975a) Action of phospholipases on the phosphatidylcholine exchange protein from beef liver. Biochim. Biophys. Acta 398, 415–423.

Kamp, H.H., Wirtz, K.W.A. and Van Deenen, L.L.M. (1975b) Delipidation of the phosphatidylcholine exchange protein from beef liver by detergents. Biochim. Biophys. Acta 398, 401–414.

Kamp, H.H., Wirtz, K.W.A. and Van Deenen, L.L.M. (1975c) Lipid–protein interactions within the phosphatidylcholine exchange protein from beef liver. 18th International Conference on the Biochemistry of Lipids, Abstract B12, Graz.

Kates, M. and Marshall, M.O. (1975) Biosynthesis of phosphoglycerides in plants, In: T. Galliard and E.I. Mercer (Eds.), Recent Advances in the Chemistry and Biochemistry of Plant Lipids, Academic Press, London, pp. 115–159.

Kiyasu, T.Y., Pieringer, R.A., Paulus, H. and Kennedy, E.P. (1963) The biosynthesis of phosphatidylglycerol. J. Biol. Chem. 238, 2293–2298.

Kook, A.I. and Rubinstein, D.R. (1969) The release of lipoproteins by rat liver slices. Can. J. Biochem. 47, 65–69.

Kook, A.I. and Rubinstein, D. (1970) The role of serum lipoproteins in the release of phospholipids by rat liver slices. Biochim. Biophys. Acta 202, 396–398.

Kornberg, R.D. and McConnell, H.M. (1971a) Lateral diffusion of phospholipids in a vesicle membrane. Proc. Natl. Acad. Sci. USA 68, 2564–2568.

Kornberg, R.D. and McConnell, H.M. (1971b) Inside-outside transitions of phospholipids in vesicle membranes. Biochemistry 10, 1111–1120.

Kramer, R., Schatter, C. and Zahler, P. (1972) Preferential binding of sphingomyelin by membrane proteins of the sheep red cell. Biochim. Biophys. Acta 282, 146–156.

Kunkel, H.G. and Bearn, A.G. (1954) Phospholipid studies of different serum lipoproteins employing [^{32}P]. Proc. Soc. Exp. Biol. Med. 86, 887–891.

Landriscina, C. and Marra, E. (1973) Fatty acid synthesis in vivo. Transfer of lipids from microsomes to other subcellular fractions of rat liver. Life Sci. 13, 1373–1381.

Lee, A.G. (1975) Lipid on the run. Nature (London) 256, 370–371.

Lee, A.G., Birdsall, N.J.M. and Metcalfe, J.C. (1973) Measurement of fast lateral diffusion of lipids in vesicles and in biological membranes by ^{1}H nuclear magnetic resonance. Biochemistry 12, 1650–1659.

Lee, T., Stephens, N., Moehl, A. and Snyder, F. (1973) Turnover of rat liver plasma membrane phospholipids. Comparison with microsomal membranes. Biochim. Biophys. Acta 291, 86–92.

Lekim, D., Betzing, H. and Stoffel, W. (1972) Incorporation of complete phospholipid molecules in cellular membranes of rat liver after uptake from blood serum. Z. Physiol. Chem. 353, 949–964.

Lenard, J. and Rothman, J.E. (1976) Transbilayer distribution and movement of cholesterol and phospholipid in the membrane of influenza virus. Proc. Natl. Acad. Sci. USA 73, 391–395.

Lester, R.L. and Fleischer, S. (1961) Studies on the electron-transport system, XXVII. The respiratory activity of acetone-extracted beef heart mitochondria. Role of coenzyme Q and other lipids. Biochim. Biophys. Acta 47, 358–377.

Lord, J.M., Kagawa, T. and Beevers, H. (1972) Intracellular distribution of enzymes of the cytidine diphosphate choline pathway in castor bean endosperm. Proc. Natl. Acad. Sci. USA 69, 2429–2432.

Lovelock, J., James, A. and Rowe, C. (1960) The lipids of whole blood, 2. The exchange of lipids between the cellular constituents and the lipoproteins of human blood. Biochem. J. 74, 137–140.

Lu, A.Y.H. and Levin, W. (1974) The resolution and reconstitution of the liver microsomal hydroxylation system. Biochim. Biophys. Acta 344, 205–240.

Lutton, C. and Zilversmit, D.B. (1976) Phospholipid exchange proteins in rat intestine. Lipids (in press).

Lux, S.E., John, K.M., Fleischer, S., Jackson, R.L. and Gotto, A.M. (1972) Identification of the lipid-binding cyanogen bromide fragment from the cystine-containing high density apolipoprotein, apoLP-GLN, II. Biochem. Biophys. Res. Commun. 49, 23–29.

Maeda, T. and Ohnishi, S. (1974) Membrane fusion. Transfer of phospholipid molecules between phospholipid bilayer membranes. Biochem. Biophys. Res. Commun. 60, 1509–1516.

Margolis, S. (1969) Structure of very low and low density lipoproteins, In: E. Tria and A.M. Scanu (Eds.), Structural and Functional Aspects of Lipoproteins in Living Systems, Academic Press, London, pp. 369–424.

Marshall, M.O. and Kates, M. (1972) Biosynthesis of phosphatidylglycerol by cell-free preparations from spinach leaves. Biochim. Biophys. Acta 260, 558–570.

Mazliak, P. (1973) Lipid metabolism in plants. Ann. Rev. Plant Physiol. 24, 287–310.

Mazliak, P., Stoll, U. and Abdelkader, A.B. (1968) Coopération entre les mitochondries et le reste de la cellule pour le renouvellement des lipides mitochondriaux du parenchyme de pomme. Biochim. Biophys. Acta 152, 414–417.

Mazliak, P., Douady, D., Demandre, C. and Kader, J.C. (1975) Exchange processes between organelles involved in membrane lipid biosynthesis, In: T. Galliard and E.I. Mercer (Eds.), Recent Advances in the Chemistry and Biochemistry of Plant Lipids, Academic Press, London, pp. 301–318.

McCaman, R.E. and Cook, K. (1966) Intermediary metabolism of phospholipids in brain tissue, III. Phosphocholine-glyceride transferase. J. Biol. Chem. 241, 3390–3394.

McCandless, E.L. and Zilversmit, D.B. (1958) Fate of triglycerides and phospholipids and artificial fat emulsions: disappearance from the circulation. Am. J. Physiol. 193, 294–300.

McMurray, W.C. (1973) Phospholipids in subcellular organelles and membranes, In: G.B. Ansell, J.N. Hawthorne and R.M.C. Dawson (Eds.), Form and Function of Phospholipids, Elsevier, Amsterdam, pp. 205–251.

McMurray, W.C. (1975) Biosynthesis of mitochondrial phospholipids using endogenously generated diglycerides. Can. J. Biochem. 53, 784–795.

McMurray, W.C. and Dawson, R.M.C. (1969) Phospholipid exchange reactions within the liver cell. Biochem. J. 12, 91–108.

McNamee, M.G. and McConnell, H.M. (1973) Transmembrane potentials and phospholipid flip-flop in excitable membrane vesicles. Biochemistry 12, 2951–2958.

Meunier, D. and Mazliak, P. (1972) Différence de composition lipidique entre les deux membranes des mitochondries de pomme de terre. C.R. Acad. Sci. Paris, Sér. D 275, 213–216.

Michaelson, D.M., Horwitz, A.F. and Klein, M.P. (1973) Transbilayer asymmetry and surface homogeneity of mixed phospholipids in co-sonicated vesicles. Biochemistry 12, 2637–2645.

Miller, E.K. and Dawson, R.M.C. (1972) Exchange of phospholipids between brain membranes in vitro. Biochem. J. 126, 823–835.

Miller, J.E. and Cornatzer, W.E. (1966) Phospholipid metabolism in mitochondria and microsomes of rabbit liver during development. Biochim. Biophys. Acta 125, 534–541.

Minakawa, K. (1974) Studies on phospholipids in small intestine smooth muscle, II. On exchange of phospholipids between mitochondria and microsomes in rat small intestinal smooth muscle. Sapporo Med. J. 43, 98–103.

Minari, O. and Zilversmit, D.B. (1963) Behavior of dog lymph chylomicron lipid constituents during incubation with serum. J. Lipid Res. 4, 424–436.

Mishkin, S., Stein, L., Gatmaitan, Z. and Arias, I.M. (1973) The binding of fatty acids to cytoplasmic proteins: binding to Z proteins in lipid and other tissues of the rat. Biochem. Biophys. Res. Commun. 47, 997–1003.

Moore, Jr., T.S. (1974) Phosphatidylglycerol synthesis in castor bean endosperm. Kinetics, requirements and intracellular localization. Plant Physiol. 54, 164–168.

Moore, Jr., T.S., Lord, J.M., Kagawa, T. and Beevers, H. (1973) Enzymes of phospholipid metabolism in the endoplasmic reticulum of castor bean endosperm. Plant Physiol. 52, 50–53.

Moreau, F., Dupont, J. and Lance, C. (1974) Phospholipid and fatty acid composition of outer and inner membranes of plant mitochondria. Biochim. Biophys. Acta 345, 294–304.

Morré, J.D., Merrit, W.D. and Lambi, C.A. (1971a) Connections between mitochondria and endoplasmic reticulum in rat liver and onion stem. Protoplasma 73, 43–49.

Morré, D.J., Mollenhauer, H.H. and Bracker, C.E. (1971b) Origin and continuity of Golgi apparatus, In: J. Reinert and H. Ursprung (Eds.), Origin and Continuity of Cell Organelles, Springer, Berlin, pp. 82–126.

Mulder, E. and Van Deenen, L.L.M. (1965a) Metabolism of red-cell lipids, I. Incorporation in vitro of fatty acids into phospholipids from mature erythrocytes. Biochim. Biophys. Acta 106, 106–117.

Mulder, E. and Van Deenen, L.L.M. (1965b) Metabolism of red-cell lipids, III. Pathways for phospholipid renewal. Biochim. Biophys. Acta 106, 348–365.

Mulder, E., Van den Berg, J.W.D. and Van Deenen, L.L.M. (1965) Metabolism of red cell lipids, II. Conversions of lysophosphoglycerides. Biochim. Biophys. Acta 106, 118–127.

Nelson, G.J. (1967) The phospholipid composition of plasma in various mammalian species. Lipids 2, 323–328.

Nozawa, Y. and Thompson Jr., G.A. (1972) Studies on membrane formation in *Tetrahymena pyriformis*, V. Lipid incorporation into various cellular membranes of stationary phase cells, starving cells, and cells treated with metabolic inhibitors. Biochim. Biophys. Acta 282, 93–104.

Ockner, R.K., Manning, J.A., Poppenhausen, R.B. and Lo, W.K.L. (1972) A binding protein for fatty acids in cytosol of intestinal mucosa, liver, myocardium and other tissues. Science 177, 56–58.

Oliveira, M.M. and Vaughan, M. (1964) Incorporation of fatty acids into phospholipids of erythrocyte membranes. J. Lipid Res. 5, 156–162.

Oursel, A., Lamant, A., Salsac, L. and Mazliak, P. (1973) Etude comparée des lipides et de la fixation passive du calcium dans les racines et les fractions subcellulaires du *Lupinus luteus* et de la *Vicia faba*. Phytochemistry 12, 1865–1874.

Pagano, R.E. and Huang, L. (1975) Interaction of phospholipid vesicles with cultured mammalian cells, II. Studies of mechanism. J. Cell Biol. 67, 49–60.

Papahadjopoulos, D., Poste, G. and Schaeffer, B.E. (1973) Fusion of mammalian cells by unilamellar lipid vesicles. Influence of lipid charge, fluidity and cholesterol. Biochim. Biophys. Acta 323, 23–42.

Papahadjopoulos, D., Poste, G., Schaeffer, B.E. and Vail, W.J. (1974) Membrane fusion and molecular segregation in phospholipid vesicles. Biochim. Biophys. Acta 352, 10–28.

Papahadjopoulos, D., Hui, S., Vail, W.J. and Poste, G. (1976a) Studies on membrane fusion, I. Interactions of pure phospholipid membranes and the effect of fatty acids (myristic acid), lysolecithin, proteins and dimethylsulfoxide. Biochim. Biophys. Acta 448, 245–264.

Papahadjopoulos, D., Vail, W.J., Pangborn, W.A. and Poste, G. (1976b) Studies on membrane fusion, II. Induction of fusion in pure phospholipid membranes by Ca^{2+} and other divalent metals. Biochim. Biophys. Acta 448, 265–283.

Parkes, J.G. and Thompson, W. (1970) The composition of phospholipids in outer and inner mitochondrial membranes from guinea-pig liver. Biochim. Biophys. Acta 196, 162–169.

Parkes, J.G. and Thompson, W. (1973) Structural and metabolic relation between molecular classes of phosphatidylcholine in mitochondria and endoplasmic reticulum of guinea-pig liver. J. Biol. Chem. 248, 6655–6662.

Parkes, J.G. and Thompson, W. (1975) Phosphatidylethanolamine in liver mitochondria and endoplasmic reticulum: molecular species, distribution and turnover. Can. J. Biochem. 53, 698–705.

Pasquini, J.M., Gomez, C.J., Najle, R. and Soto, E.F. (1975) Lack of phospholipid transport mechanisms in cell membranes of the central nervous system. J. Neurochem. 24, 439–443.

Percy, A.K., Schmell, E., Earles, B.J. and Lennarz, W.J. (1973) Phospholipid biosynthesis in the membranes of immature and mature red blood cells. Biochemistry, 12, 2456–2461.

Peterson, J.A. and Rubin, H. (1969) The exchange of phospholipids between cultured chick embryo fibroblasts and their growth medium. Exp. Cell Res. 58, 365–378.

Peterson, J.A. and Rubin, H. (1970) The exchange of phospholipids between chick embryo fibroblasts as observed by autoradiography. Exp. Cell Res. 60, 382–392.

Phillips, M.C., Finer, E.G. and Hauser, H. (1972) Difference between conformations of lecithin and phosphatidylethanolamine polar groups and their effects on interactions of phospholipid bilayer membranes. Biochim. Biophys. Acta 290, 397–402.

Pittman, J.G. and Martin, D.B. (1968) Fatty acid biosynthesis in human erythrocytes: evidence in mature erythrocytes for an incomplete long chain fatty acid synthesizing system. J. Clin. Invest. 45, 165–172.

Polonovski, J. and Paysant, M. (1963) Métabolisme phospholipidique du sang, VIII. Echange des phospholipides marqués entre globules et plasma sanguin in vitro. Bull. Soc. Chim. Biol. 45, 339–348.

Possmayer, F. (1974) Evidence for a specific phosphatidylinositol transferring protein in rat brain. Brain Res. 74, 167–174.

Possmayer, F., Balakrishnan, G. and Strickland, K.P. (1968) The incorporation of labeled glycerophosphoric acid into the lipids of rat brain preparations, III. On the biosynthesis of phosphatidylglycerol. Biochim. Biophys. Acta 164, 79–87.

Poste, G. and Papahadjopoulos, D. (1976) Lipid vesicles as carriers for introducing materials into cultured cells: influence of vesicle lipid composition on mechanism(s) of vesicle incorporation into cells. Proc. Natl. Acad. Sci. USA 73, 1603–1607.

Reed, C.F. (1959) Studies of in vivo and in vitro exchange of erythrocyte and plasma phospholipids. J. Clin. Invest. 38, 1032–1033.

Reed, C.F. (1968) Phospholipid exchange between plasma and erythrocytes in man and in the dog. J. Clin. Invest. 47, 749–760.

Renooij, W., Van Golde, L.M.G., Zwaal, R.F.A., Roelofsen, B. and Van Deenen, L.L.M. (1974) Preferential incorporation of fatty acids at the inside of human erythrocyte membranes. Biochim. Biophys. Acta 363, 287–292.

Robertson, A.F. and Lands, W.E.M. (1964) Metabolism of phospholipids in normal and spherocytic human erythrocytes. J. Lipid Res. 5, 88–93.

Roelofsen, B. and Van Deenen, L.L.M. (1973) Lipid requirement of membrane-bound ATPase. Studies on human erythrocyte ghosts. Eur. J. Biochem. 40, 245–257.

Romani, R.J. and Ozelkok, S. (1973) "Survival" of mitochondria in vivo. Plant Physiol. 51, 702–707.

Roseman, M., Litman, B.J. and Thompson, T.E. (1975) Transbilayer exchange of phosphatidylethanolamine for phosphatidylcholine and N-acetamidoylphosphatidylethanolamine in single-walled bilayer vesicles. Biochemistry 14, 4826–4830.

Rothman, J.E. and Dawidowicz, E.A. (1975) Asymmetric exchange of vesicle phospholipids catalyzed by the phosphatidylcholine exchange protein. Measurement of inside-outside transition. Biochemistry 14, 2809–2816.

Rouser, G., Nelson, G.J., Fleischer, S. and Simon, G. (1968) Lipid composition of animal cell membranes, organelles and organs, In: D. Chapman (Ed.), Biological Membranes, Vol. 1, Academic Press, London, pp. 5–69.

Rousselet, A., Colbeau, A., Vignais, P.M. and Devaux, P.F. (1976) Study of the transverse diffusion of spin-labelled phospholipids in biological membranes, II. Inner mitochondrial membranes of rat liver. Use of phosphatidylcholine exchange protein. Biochim. Biophys. Acta, in press.

Rowe, C.E. (1960) The phospholipids of human blood plasma and their exchange with the cells. Biochem. J. 76, 471–475.

Rubenstein, B. and Rubinstein, D. (1972) Interrelationship between rat serum very low density and high density lipoproteins. J. Lipid Res. 13, 317–324.

Ruby, J.R., Dyer, R.F. and Skalks, R.G. (1969) Continuities between mitochondria and endoplasmic reticulum in the mammalian ovary. Z. Zellforsch. 97, 30–37.

Ruigrok, Th.J.C., Van Zaane, D., Wirtz, K.W.A. and Scherphof, G.L. (1972) The effects of calcium acetate on mitochondria in the perfused rat liver, II. Enhanced transfer of phosphatidylcholine from outer to inner mitochondrial membranes. Cytobiologie 5, 412–421.

Sakagami, T., Minari, O. and Orii, T. (1965a) Behavior of plasma lipoproteins during exchange of phospholipids between plasma and erythrocytes. Biochim. Biophys. Acta 98, 111–116.

Sakagami, T., Minari, O. and Orii, T. (1965b) Interaction of individual phospholipids between rat plasma and erythrocytes in vitro. Biochim. Biophys. Acta 98, 356–364.

Sarzala, M.G., Van Golde, L.M.G., De Kruyff, B. and Van Deenen, L.L.M. (1970) The intramitochondrial distribution of some enzymes involved in the biosynthesis of rat liver phospholipids. Biochim. Biophys. Acta 202, 106–119.

Sauner, M.T. and Lévy, M. (1971) Study of the transfer of phospholipids from the endoplasmic reticulum to the outer and inner mitochondrial membranes. J. Lipid Res. 12, 71–75.

Scallen, T.J., Seetharam, B., Srikantaiah, M.V., Hansbury, E. and Lewis, M.K. (1975) Sterol carrier protein hypothesis: requirement for three substrate-specific soluble proteins in liver cholesterol biosynthesis. Life Sci. 16, 853–874.

Scandella, C.J., Devaux, P. and McConnell, H.M. (1972) Rapid lateral diffusion of phospholipids in rabbit sarcoplasmic reticulum. Proc. Natl. Acad. Sci. USA 69, 2056–2060.

Scanu, A.M. (1969) Serum high density lipoproteins, In: E. Tria and A.M. Scanu (Eds.), Structural and Functional Aspects of Lipoproteins in Living Systems, Academic Press, London, pp. 425–445.

Schatz, G. and Mason, T.L. (1974) The biosynthesis of mitochondrial proteins. Ann. Rev. Biochem. 43, 51–87.

Schiefer, H.G. (1969) Studies on the biosynthesis of the lipid and protein components of a membranous cytochrome oxidase. Preparation of rat liver mitochondria in vivo. Z. Physiol. Chem. 350, 235–244.

Schneider, W.C. (1963) Intracellular distribution of enzymes, XIII. Enzymatic synthesis of deoxycytidine diphosphate choline and lecithin in rat liver. J. Biol. Chem. 238, 3572–3578.

Seimiya, T. and Ohki, S. (1973) Ionic structure of phospholipid membranes and binding of calcium ions. Biochim. Biophys. Acta 298, 546–561.

Shephard, E.H. and Hübscher, G. (1969) Phosphatidate biosynthesis in mitochondrial subfractions of rat liver. Biochem. J. 113, 429–440.

Shohet, S.B. and Nathan, D.G. (1970) Incorporation of phosphatidate precursors from serum into erythrocytes. Biochim. Biophys. Acta 202, 202–205.

Singer, S.J. (1971) The molecular organization of biological membranes, In: L.I. Rothfield (Ed.), Structure and Function of Biological Membranes, Academic Press, New York, pp. 145–222.

Singer, S.J. and Nicolson, G.L. (1972) The fluid mosaic model of the structure of cell membranes. Science 175, 720–741.

Smith, A.D. and Winkler, H. (1968) Lysosomal phospholipases A_1 and A_2 of bovine adrenal medulla. Biochem. J. 108, 867–874.

Smith, N. and Rubinstein, D. (1974) Some characteristics of the exchange of lecithin between rabbit erythrocytes and serum, Can. J. Biochem. 52, 706–717.

Soula, G., Valdiguié, P. and Douste-Blazy, L. (1967) Métabolisme lipidique dans les globules rouges de lapin, II. Echanges in vitro de phospholipides marqués au ^{32}P entre globules rouges et plasma. Bull. Soc. Chim. Biol. 49, 1317–1330.

Soula, G., Souillard, C. and Douste-Blazy, L. (1972) Métabolisme phospholipidique des hématies nuclées. Incorporation de [^{32}P]orthophosphate dans les phospholipides plasmatiques et globulaires de poulet in vivo. Eur. J. Biochem. 30, 93–99.

Stanacev, N.Z., Isaac, D.C. and Brookes, K.B. (1968) The enzymatic synthesis of phosphatidylglycerol in sheep brain. Biochim. Biophys. Acta 152, 806–808.

Stanacev, N.Z., Stuhne-Sekalec, L., Brookes, K.B. and Davidson, J.B. (1969) Intermediary metabolism of phospholipids. The biosynthesis of phosphatidylglycerophosphate and phosphatidylglycerol in heart mitochondria. Biochim. Biophys. Acta 176, 650–653.

Stein, O. and Stein, Y. (1967) Lipid synthesis intracellular transport, storage and secretion, I. Electron microscopic radioautographic study of liver after injection of tritiated palmitate or glycerol in fasted and ethanol-treated rats. J. Cell Biol. 33, 319–340.

Stein, O. and Stein, Y. (1969) Lecithin synthesis, intracellular transport and secretion in rat liver, IV. A radioautographic and biochemical study of choline-deficient rats injected with choline-[^{3}H]. J. Cell Biol. 40, 461–483.

Stoffel, W. and Schiefer, H.G. (1968) Biosynthesis and composition of phosphatides in outer and inner mitochondrial membranes. Z. Physiol. Chem. 349, 1017–1026.

Strunecka, A. and Zborowski, J. (1975) Microsomal synthesis of phosphatidylinositol and its exchange between subcellular structures of rat liver. Comp. Biochem. Physiol. 50, 55–60.

Suzue, G. and Marcel, Y.L. (1975) Studies on the fatty acid binding proteins in cytosol of rat liver. Can. J. Biochem. 53, 804–809.

Swartz, J.G. and Mitchell, J.E. (1970) Biosynthesis of retinal phospholipids: incorporation of radioactivity from labeled phosphorylcholine and cytidine diphosphate choline. J. Lipid Res. 11, 544–560.

Tanford, C. (1972) Hydrophobic free energy, micelle formation and the association of proteins with amphiphiles. J. Mol. Biol. 67, 59–74.

Tanford, C. (1973) The Hydrophobic Effect, Wiley, New York, pp. 120–142.

Taniguchi, M. and Sakagami, T. (1975) Exchange of phospholipids between mitochondria and rough or smooth microsomes in vitro. J. Biochem. (Tokyo). 77, 1245–1248.

Taniguchi, M., Hirayama, H. and Sakagami, T. (1973) Exchange of molecular species of phosphatidylcholines and phosphatidylethanolamines between rat liver mitochondria and microsomes in vitro. Biochim. Biophys. Acta 296, 65–70.

Tarlov, A. (1968) Turnover of mitochondrial phospholipids by exchange with soluble lipoproteins in vitro. Fed. Proc. 27, 458.

Träuble, H. and Sackmann, E. (1972) Studies on the crystalline-liquid crystalline phase transition of lipid model membranes, III. Structure of a steroid-lecithin system below and above the lipid-phase transition. J. Am. Chem. Soc. 94, 4499–4510.

Trémolières, A. and Mazliak, P. (1974) Biosynthetic pathway of α-linolenic acid in developing pea-leaves. In vivo and in vitro study. Plant Sci. Lett., 2, 193–201.

Tzagoloff, A. (1971) Assembly of the mitochondrial membrane system, IV. Role of mitochondrial and cytoplasmic protein synthesis in the biosynthesis of the rutamycin-sensitive adenosine triphosphatase. J. Biol. Chem. 246, 3050–3056.

Tzagoloff, A. and Akai, A. (1972) Assembly of the mitochondrial membrane system, VIII. Properties of the products of mitochondrial protein synthesis in yeast. J. Biol. Chem. 247, 6517–6523.

Tzagoloff, A. and Meagher, P. (1972) Assembly of the mitochondrial membrane system, VI. Mitochondrial synthesis of subunit proteins of the rutamycin-sensitive adenosine triphosphatase. J. Biol. Chem. 247, 594–603.

Tzagoloff, A., Akai, A., Needleman, R.B. and Zulch, G. (1975) Assembly of mitochondrial membrane system. Cytoplasmic mutants of *Saccharomyces cerevisiae* with lesions in enzymes of respiratory chain and in mitochondrial ATPase. J. Biol. Chem. 250, 8236–8242.

Van Deenen, L.L.M. (1975) Lipid-protein interactions in model systems and biomembranes, In: G. Gardos and I. Szasz (Eds.), Biomembranes: Structure and Function, North-Holland, Amsterdam, pp. 3–16.

Van Deenen, L.L.M. and De Gier, J. (1964) Chemical composition and metabolism of lipids in red cells of various animal species, In: C. Bishop and D.M. Surgenor (Eds.), The Red Blood Cell, Academic Press, New York, pp. 243–302.

Van Deenen, L.L.M. and De Gier, J. (1974) Lipids of the red cell membrane, In: D.M. Surgenor (Ed.), The Red Blood Cell, Vol. I, Academic Press, New York, pp. 147–211.

Van den Besselaar, A.M.H.P., Helmkamp Jr., G.M. and Wirtz, K.W.A. (1975) Kinetic model of the protein-mediated phosphatidylcholine exchange between single bilayer liposomes. Biochemistry 14, 1852–1858.

Van den Bosch, H., Van Golde, L.M.G. and Van Deenen, L.L.M. (1972) Dynamics of phosphoglycerides. Rev. Physiol. (Ergeb. Physiol.) 66, 13–145.

Verkleij, A.J., Zwaal, R.F.A., Roelofsen, B., Comfurius, P., Kastelijn, O. and Van Deenen, L.L.M. (1973) The asymmetric distribution of phospholipids in the human red cell membrane. A combined study using phospholipases and freeze-etching electron microscopy. Biochim. Biophys. Acta 323, 178–193.

Victoria, E.J. and Korn, E.D. (1975) Enzymes of phospholipid metabolism in the plasma membrane of *Acanthamoeba castellanii*. J. Lipid Res. 16, 54–60.

Victoria, E.J., Van Golde, L.M.G., Hostetler, K.Y., Scherphof, C.L. and Van Deenen, L.L.M. (1971) Some studies on the metabolism of phospholipids in plasma membranes from rat liver. Biochim. Biophys. Acta 239, 443–457.

Vijay, I.K. and Stumpf, P.K. (1972) Fat metabolism in higher plants, XLVIII. Properties of oleyl-coenzyme A desaturase of *Carthamus tinctorius*. J. Biol. Chem. 247, 360–366.

Waku, K. and Lands, W.E.M. (1968) Control of lecithin in erythrocyte membranes. J. Lipid Res. 9, 12–18.

Warren, G.B., Birdsall, N.J.M., Lee, A.G. and Metcalfe, J.C. (1974a) Lipid substitution: the investigation of functional complexes of single species of phospholipid and a purified calcium transport protein, In: G.F. Azzone, M.E. Klingenberg, E. Quagliariello and N. Siliprandi (Eds.), Membrane Proteins in Transport and Phosphorylation, North-Holland, Amsterdam, pp. 1–12.

Warren, G.B., Toon, P.A., Birdsall, N.J.M., Lee, A.G. and Metcalfe, J.C. (1974b) Reversible lipid titrations of the activity of pure adenosine triphosphatase-lipid complexes. Biochemistry 13, 5501–5507.

Warren, G.B., Toon, P.A., Birdsall, N.J.M., Lee, A.G. and Metcalfe, J.C. (1974c) Complete control of the lipid environment of membrane-bound protein: application to a calcium transport system. FEBS Lett. 41, 122–124.

Warren, G.B., Houslay, M.D. and Metcalfe, J.G. (1975) Cholesterol is excluded from the phospholipid annulus surrounding an active calcium transport protein. Nature (London) 255, 684–687.

Wheeldon, L.W., Dianoux, A.C., Bof, M. and Vignais, P.W. (1974) Stable and labile products of mitochondrial protein synthesis in vitro. Eur. J. Biochem. 46, 189–199.

White, D.A. (1973) The phospholipid composition of mammalian tissues, In: G.B. Ansell, R.M.C. Dawson and J.N. Hawthorne (Eds.), Form and Function of Phospholipids, Elsevier, Amsterdam, pp. 441–482.

Wilgram, G.F. and Kennedy, E.P. (1963) Intracellular distribution of some enzyme catalyzing reactions in the biosynthesis of complex lipids. J. Biol. Chem. 238, 2615–2619.

Williams, M.L. and Bygrave, F.L. (1970) Incorporation of inorganic phosphate into phospholipids by the homogenate and by sub-cellular fractions of rat liver. Eur. J. Biochem. 17, 32–38.

Wirtz, K.W.A. (1971) Intracellular transfer of membrane phospholipids. Thesis. State University of Utrecht, pp. 10–53.

Wirtz, K.W.A. (1972) Exchange of phosphatidylcholine between rat liver mitochondria and liposomes: effect of zeta-potential. 8th FEBS Meeting, Abstract 69, Amsterdam.

Wirtz, K.W.A. (1974) Transfer of phospholipids between membranes. Biochim. Biophys. Acta 344, 95–117.

Wirtz, K.W.A. and Kamp, H.H. (1972) Exchange of phosphatidylcholine between mitochondria and microsomes: stimulation by a specific protein, In: L. Bolis, R.D. Keynes and W. Wilbrandt (Eds.), Role of Membranes in Secretory Processes, North-Holland, Amsterdam, pp. 52–61.

Wirtz, K.W.A. and Zilversmit, D.B. (1968) Exchange of phospholipids between liver mitochondria and microsomes in vitro. J. Biol. Chem. 243, 3596–3602.

Wirtz, K.W.A. and Zilversmit, D.B. (1969a) The use of phenobarbital and carbon tetrachloride to examine liver phospholipid exchange in intact rats. Biochim. Biophys. Acta 187, 468–476.

Wirtz, K.W.A. and Zilversmit, D.B. (1969b) Participation of soluble liver proteins in the exchange of membrane phospholipids. Biochim. Biophys. Acta 193, 105–116.

Wirtz, K.W.A. and Zilversmit, D.B. (1970) Partial purification of phospholipid exchange protein from beef heart. FEBS Lett. 7, 44–46.

Wirtz, K.W.A., Van Golde, L.M.G. and Van Deenen, L.L.M. (1970) The exchange of molecular species of phosphatidylcholine between mitochondria and microsomes of rat liver. Biochim. Biophys. Acta 218, 176–179.

Wirtz, K.W.A., Kamp, H.H. and Van Deenen, L.L.M. (1972) Isolation of a protein from beef liver which specifically stimulates the exchange of phosphatidylcholine. Biochim. Biophys. Acta 274, 606–617.

Wirtz, K.W.A., Kamp, H.H. and Van Deenen, L.L.M. (1975) Protein-mediated transfer of phosphatidylcholine between membranes and liposomes. Biochem. Soc. Trans. 3, 608–611.

Wirtz, K.W.A., Geurts Van Kessel, W.S.M., Kamp, H.H. and Demel, R.A. (1976) The protein mediated transfer of phosphatidylcholine between membranes. The effect of membrane lipid composition and ionic composition of the medium. Eur. J. Biochem. 61, 513–523.

Wojtczak, L. and Zborowski, J. (1975) Transport of phosphatidic acid between membranes. 10th FEBS Meeting, Abstract 1042, Paris.

Wojtczak, L., Baranska J., Zborowski, J. and Drahota, Z. (1971) Exchange of phospholipids between microsomes and mitochondrial outer and inner membranes. Biochim. Biophys. Acta 249, 41–52.

Zborowski, J. and Wojtczak, L. (1969) Phospholipid synthesis in rat liver mitochondria. Biochim. Biophys. Acta 187, 73–84.

Zborowski, J. and Wojtczak, L. (1975) Net transfer of phosphatidylinositol from microsomes and mitochondria to liposomes catalyzed by the exchange protein from rat liver. FEBS Lett. 51, 317–320.

Zilversmit, D.B. (1969) Chylomicrons, In: E. Tria and A.M. Scanu (Eds.), Structural and Functional Aspects of Lipoproteins in Living Systems, Academic Press, London, pp. 329–368.

Zilversmit, D.B. (1971a) Exchange of phospholipid classes between liver microsomes and plasma: comparison of rat, rabbit and guinea-pig. J. Lipid Res. 12, 36–42.

Zilversmit, D.B. (1971b) Stimulation of phospholipid exchange between mitochondria and artificially prepared phospholipid aggregates by a soluble fraction from liver. J. Biol. Chem. 246, 2645–2649.

Zilversmit, D.B. and Johnson, L.W. (1975) Purification of phospholipid exchange proteins from beef heart, In: J.M. Lowenstein (Ed.), Methods in Enzymology, vol. XXXV, Academic Press, New York, pp. 262–269.

Zilversmit, D.B., Hughes, L.B. and Balmer, J. (1975) Stimulation of cholesterol ester exchange by lipoprotein-free rabbit plasma. Biochim. Biophys. Acta 409, 393–398.

Zwaal, R.F.A., Roelofsen, B. and Colley, C.M. (1973) Localization of red cell membrane constituents. Biochim. Biophys. Acta 300, 159–182.

The influence of membrane fluidity on the activity of membrane-bound enzymes

H.K. KIMELBERG

1. Introduction

The important role of the lipid component in the functioning of cell membranes has been recognized since the work of Overton in the 1890s (Overton, 1899), which showed that the ability of compounds to permeate cell membranes was related directly to their lipid solubility. The structural basis for this role was first indicated by the work of Gorter and Grendel (Gorter and Grendel, 1925), who showed that the amount of lipids extracted from red blood cell membranes was sufficient to form a continuous bilayer. The basic involvement of lipids in membrane structure and function has thus been recognized for some time. Subsequently, the details of protein–lipid interactions in membranes and their implications for the structure and function of biomembranes, have attracted considerable attention (for reviews of the various proposed models of membrane organization, see Hendler, 1971, 1974; Singer, 1971, 1974).

At the present time, there is general acceptance, as a working model, of the fluid mosaic model of membrane structure proposed by Singer (1971) and by Singer and Nicolson (1972). The principal features of this model are that most, but not all, of the lipids in membranes are present in the form of a bilayer, with the membrane proteins either bound to the charged surface of the bilayer (peripheral proteins), or inserted to varying degrees (integral proteins) into the bilayer (Fig. 1). The experimental evidence and background for this model are summarized in the review articles given above.

Seen from above, the membrane may be viewed as a mosaic of protein and lipid, but with the relative amounts varying from membrane to membrane (Fig. 2). Thus, Fig. 2A is representative of a membrane with a relatively low ratio of protein to lipid such as myelin (approximately 20%), while Fig. 2B illustrates a membrane with a relatively high protein content such as mammalian mitochondria (approximately 75%). The fluid aspect of the mosaic model envisages that the proteins have considerable lateral translational mobility which, in the absence of other processes, should be diffusion-limited. The rate of such diffusion will, therefore, depend both on the viscosity of the lipid bilayer and to a lesser degree on the size of the protein or aggregate of several proteins

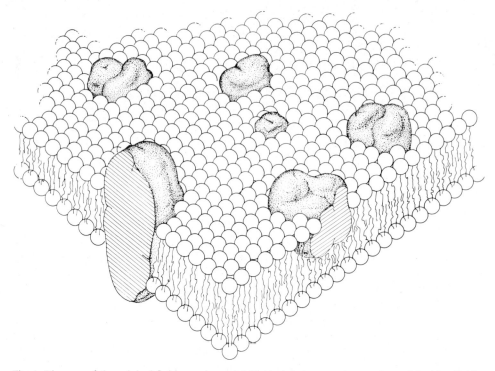

Fig. 1. Diagram of the original fluid mosaic model (lipid-globular protein mosaic model with a lipid matrix). The small open spheres represent the polar head groups of the phospholipids, with the attached fatty acyl chains shown in the fluid state as wiggly lines. The solid bodies with the stippled surfaces are the integral membrane proteins which penetrate the bilayer to varying degrees (reproduced with permission from Singer and Nicolson, 1972).

(Scandella et al., 1972). Measurements of the microviscosity of various biological membranes have given values of 0.25–10 poise at 25°C (Edidin, 1974; Feinstein et al., 1976), and diffusion coefficients for proteins of 10^{-9} to 10^{-12} cm^2 sec^{-1} have been found for several membranes (Poo and Cone, 1974; Cherry, 1975; Jacobson et al., 1975; Zagyansky and Edidin, 1976). The lateral diffusion of lipid molecules such as lecithin and steroids is faster and has been estimated at from 0.10 to $10 \cdot 10^{-8}$ cm^2 sec^{-1} for both lipid bilayers (Devaux and McConnell, 1972; Träuble and Sackmann, 1972; Edidin, 1974), and biological membranes (Scandella et al., 1972; Lee et al., 1973; Sackmann et al., 1973). In practice, it appears that the movement of some membrane proteins may also be controlled by cytoplasmic cytoskeletal elements (microfilaments and microtubules) that are closely associated with the plasma membrane (Berlin, 1975; Bretscher and Raff, 1975; Poste et al., 1975; Edelman, 1976; also see Chapter 1 in this volume). It is not clear at the present time to what degree and under what conditions lateral movement of membrane proteins is determined by the viscosity of the membrane and/or by membrane-associated cytoskeletal elements. The ability of integral plasma membrane proteins to move laterally within the membrane and

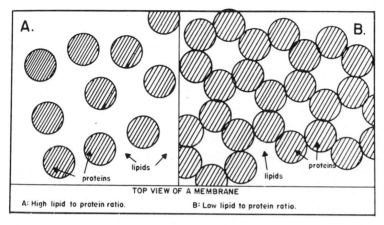

Fig. 2. Top views of the fluid mosaic membrane shown in three-dimensional view in Fig. 1. The hatched circles represent both integral and peripheral proteins and the open area is the lipid bilayer. The view shown in A represents a low protein to lipid ratio membrane such as myelin, and B represents a high protein to lipid ratio membrane such as mitochondria (reproduced with permission from Papahadjopoulos, 1973).

thus produce topographic rearrangement of cell surface is now viewed as a potentially very important mechanism in regulating the expression of specific cell surface properties.

Another aspect of molecular motion in membranes is the possibility of transmembrane inversion of membrane components, a process often suggested as a basis for transmembrane transport. The rate at which lipid molecules can invert from one monolayer to another of the bilayer, a process which has been termed "flip-flop" (Kornberg and McConnell, 1971), is very much slower than the lateral diffusion of either lipids or proteins, occurring with a half time of approximately 6.5 h at 37°C. This appears to be able to occur at least an order of magnitude faster in some natural membranes (McNamee and McConnell, 1973). It is interesting to note that Langmuir (1938) had earlier speculated that such movements might exist and could provide the molecular basis for such membrane properties as nerve excitation. The slowness of the flip-flop process is now interpreted as providing a mechanism for maintaining the phospholipid assymetry of the bilayer in natural membranes (Bretscher, 1973). It is still possible, however, that local modification of membranes might accelerate this process to rates which could be functionally significant, as suggested above. To date, however, there are no experimental data showing inversion or rotation of membrane protein components. Finally, membrane fluidity also involves the molecular movement of the fatty acyl chains of membrane phospholipids. This forms a major aspect of the work to be discussed in this review and will be dealt with in detail later.

After this highly abbreviated survey of membrane structure it is necessary to state that the real objective of this article is to review the rapidly growing literature which suggests that various aspects of membrane fluidity can influence the activity of membrane-bound enzymes, both in the plasma membrane and in the membranes of intracellular organelles. Although lateral motion of the enzyme itself may be required for the function of certain enzymes (see section 3.7) more work has been done on the effect of the intrinsic membrane fluidity or microviscosity on enzyme activity. The fact that most, and perhaps all, hydrophobic membrane enzymes are lipid-dependent (section 3.1), has now focused a great deal of attention on the role of lipid fluidity in the functioning of membrane enzymes, both in situ and in reconstituted systems. A complicating factor in the control of enzyme activity by the lipid environment, is that the enzyme may well be responding not to the bulk fluidity of the membrane, but to specific changes in its own microenvironment which can be restricted to a single monolayer of lipid molecules surrounding the protein (Jost et al., 1973a,b; Warren et al., 1975a). In turn, it is clear that the protein itself can influence the properties of such tightly bound lipids (Hong and Hubbell, 1972; Jost et al., 1973a,b). This microheterogeneity offers, however, a very effective way of independently altering the activities of a number of enzymes bound within the same membrane.

Much of the conceptual basis for the functional importance of lipid fluidity in natural membranes is derived from an intensive period of work on the properties of model membranes formed from phospholipids, and mixtures of phospholipids and other lipids. This review will therefore first briefly describe some of the basic properties of model lipid bilayers in the form of liposomes in relation to their fluid and non-fluid properties (sections 2.1 and 2.2). It is assumed that the reader at least tentatively accepts that the available evidence supports the basic premise that the lipids of biological membranes are in the form of a bilayer and the evidence for this will be reviewed in section 2.5. The next step is to consider how the properties of the lipid bilayer are altered when major membrane components such as cholesterol and proteins are incorporated (sections 2.3 and 2.4). With this background of work with model membranes it is then possible to enquire to what degree similar effects occur in biological membranes and how they might influence the enzyme activity of membrane proteins. This question has been studied simultaneously in both model systems by reconstitution studies and in intact membranes of animal cells or microorganisms where the lipid composition can be altered by dietary changes (see section 3). All these studies have, in a sense, shown the *potential importance* of lipid fluidity as factor in controlling membrane enzyme activity. In the final section I will consider those studies that indicate that lipid fluidity may indeed play a regulatory role in cell function both in normal and pathological states (section 4).

Reflecting the growing and recent interest in the structure and function of cellular membranes, there is now a large number of reviews dealing with various aspects of this subject, and some of these articles have discussed, to varying degrees, the involvement of membrane fluidity in enzyme activity (Raison, 1973; Papahadjopoulos and Kimelberg, 1973; Fourcans and Jain, 1974; Singer, 1974; Farias et al., 1975; Fox, 1975).

2. Lipid fluidity in model and biological membranes

2.1. Structure and formation of liposomes

The recent explosive increase in research devoted to membranes has partially obscured the long period of less conspicuous work, often on seemingly unrelated systems, which, by an unpredicted catalysis, has led over the past decade to the present significant commitment to this particular field. A significant part of this "catalytic" effect was due to work in the 1960s which showed that spontaneously forming vesicular phospholipid bilayer membranes have many of the permeability properties of cellular membranes (Bangham et al., 1965; Bangham, 1968). The existence of such spontaneously forming membranes which "exploit the water-oil interfacial activity of certain lipids to define anatomical boundaries" (Bangham et al., 1974), had been known since the last century and their identification as consisting of concentric lamellae of bimolecular leaflets was suggested as early as 1940 on the basis of X-ray diffraction and electron microscope studies. This background has been reviewed briefly in the introduction to an experimental paper by Stoeckenius (1959). The aptly descriptive term "myelin figures" formerly given to these membrane structures has now been replaced by the term liposome (Bangham et al., 1974) or, more simply, phospholipid vesicles. The practical importance of these preparations can be appreciated by considering a brief section on biological implications in Gaines' monograph on "Insoluble Monolayers at the Liquid-Gas Interface" written in 1966 (Gaines, 1966) in which it was stated that although planar bilayers (Mueller et al., 1962) were closer to biological membranes, their technical difficulties made lipid monolayers a useful alternative technique. It is now clear that liposomes very successfully bridge the gap between these techniques both in terms of experimental simplicity and biological relevance, and liposomes have rapidly become the dominant experimental system for work on model membranes.

Liposomes are formed simply by shaking dried films of phospholipid or mixtures of phospholipids and other lipid-soluble compounds with aqueous buffers. Such preparations are, however, quite heterogeneous with respect to size as well as being multilamellar (Fig. 3A). Each lamella constitutes a single bilayer and the reviews by Bangham (1968) and Papahadjopoulos (1973) should be consulted for the experimental evidence for bilayer structure. In these preparations therefore a large proportion of the lipid bilayers are not exposed to the external medium. It was soon found (Papahadjopoulos and Miller, 1967; Huang, 1969) that ultrasonication of such suspensions transformed most of the multilamellar liposomes into much smaller liposomes bounded by a single bilayer (Fig. 3B). These unilamellar liposomes could be further purified by Sepharose chromatography (Huang, 1969; Bangham et al., 1974). The small size of these liposomes (approx. 250–500 Å) impose packing constraints on the molecules such that in these preparations about two thirds of the lipids are in the external monolayer (Kornberg and McConnell, 1971). With these pre-

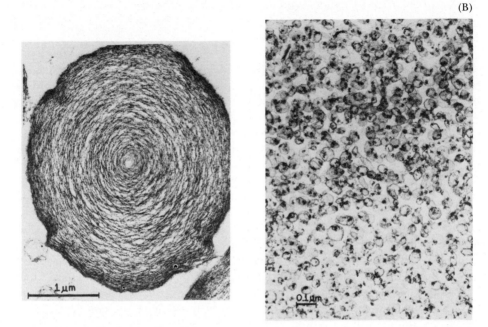

Fig. 3. Left figure shows a non-sonicated multilamellar liposome of bovine heart cardiolipin prepared by shaking 5 mg lipid in 2 ml of 10 mM NaCl buffered to pH 7.4. The suspension was then fixed in 1.0% OsO_4 in veronal acetate buffer, dehydrated in acetone and embedded in Epon A12. The right figure shows the same preparation after ultrasonication in a bath-type sonicator under N_2 for 30 min at approx. 22°C showing small unilamellar vesicles. Note difference in magnification; the bar represents 1 μm for the non-sonicated liposome and 0.1 μm for the sonicated liposomes (reproduced with permission from Papahadjopoulos, 1973).

parations permeability studies can be done on a single compartment system with a known surface area. Such preparations are also very useful for enzyme reconstitution studies.

2.2. Lipid fluidity and phase transitions in liposomes

Phospholipids in the liposome bilayer structures described above show considerable motion of their fatty acyl chains at temperatures greater than their gel-to-liquid crystalline transition temperature and also in the absence of "ordering" agents such as cholesterol. This fluidity is generally considered to be due to rotational isomerization around the single C-C bonds of the fatty acid chains (see Hubbell and McConnell, 1971; Träuble, 1971; Seelig and Niederberger, 1974). Thus, a *trans* conformation can be converted to a *gauche* conformation by rotating a C-C bond by 120°. The continuous and rapid occurrence of these motions, the life time of any one conformation may be approx. 10^{-6} sec (Seelig and Niederberger, 1974), will result in overall motion of the chain and a fluid condition or disordered state of the lipids. Below the phase transition the fatty acyl chains are in the all-*trans* conformation and are thus tightly and regularly packed in a crystalline array. This state represents the solid or ordered state. The effect of a change of a single C-C bond from a *trans* to *gauche* conformation for a simple ten-carbon fatty acid is shown in Fig. 4 in the model on the far right (Seelig and Niederberger, 1974). The effect of two *gauche* rotations in opposite directions separated by an unaltered *trans* bond (i.e. g^+tg^- or g^-tg^+) is shown in the second model from the left in Fig. 4 and results in a kink. The first and third models from the left are shown in the all-*trans* condition for comparison. Such changes can occur for the C-C bonds of the fatty acyl groups of phospholipids (Hubbell and McConnell, 1971; Träuble, 1971).

Phospholipids are relatively complicated molecules in that there are two fatty acyl groups of usually 16–18, but also up to 26, carbons long. These are esterified to the 1 and 2 positions of a glycerol molecule, with a phosphate group esterified to the 3 position. The head group specificity of the phospholipid is determined by the nature of the group (e.g. choline, ethanolamine or serine) which is esterified in turn to this phosphate group. The major phase transition of a phospholipid in the form of a liposome in excess water represents the transition from the solid all-*trans* state of its fatty acyl group, to one with considerable *trans-gauche* rotational isomerism. This transition occurs

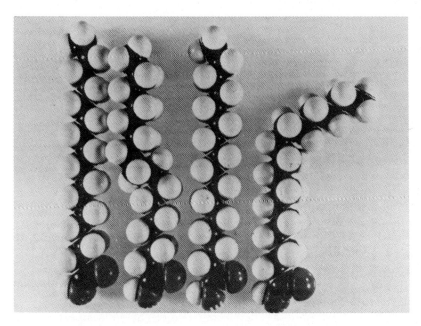

Fig. 4. Molecular models of decanoic acid showing the effect of *gauche* conformations. The 1st and 3rd chains from the left show the all-*trans* conformation. The 2nd chain shows the effect of two opposite *gauche* rotations, separated by one carbon (g^+tg^- or g^-tg^+). It results in a kink involving a perpendicular shift of the part of the molecule distal to the kink. This remains parallel to the segment before the kink but is displaced by 1.5 Å and the chain becomes shortened by one CH_2 unit. The effect of a single *gauche* conformation is shown in the 4th chain on the right (reproduced with permission from Seelig and Niederberger, 1974).

within a characteristic temperature range. The enthalpy of this transition is, in part, the energy required for the total *gauche* isomerizations, the disruption of Van der Waals forces between neighboring alkyl chains and for the structure-breaking of ordered solvent around phospholipid head groups. This transition can be quite abrupt for some phospholipids suggesting a highly cooperative change between the two states. Thus, for dipalmitoylphosphatidylcholine, in which the 16 carbon palmitic acid is esterified to both the 1 and 2 position of the glycerol backbone and a choline group is esterified to the phosphate group, the mid-point (T_m) of the transition has been found to occur at values from 42° to 45°C, within a narrow temperature spread of several degrees (Ladbrooke and Chapman, 1969; Hinz and Sturtevant, 1972; Chapman et al., 1974; Jacobson and Papahadjopoulos, 1975). It should be noted that most studies have been done on non-sonicated dispersions. For dipalmitoylphosphatidylcholine it has been reported, however, that sonication broadens the transition and lowers the T_m by 3–5°C (Jacobson and Papahadjopoulos, 1975).

Values for transition temperatures obtained from differential scanning calorimetry studies for a variety of phospholipids are shown in Table I. It can be seen that alterations in the structure of phospholipid molecules alters the T_m and temperature range of these transitions. Thus, changing the choline of the head group in dimyristoyl-phosphatidylcholine (DMPC) to dimyristoyl-phosphatidylethanolamine (DMPE) increases the T_m from 23°C to 48°C (Oldfield and Chapman, 1972; Chapman et al., 1974). Dipalmitoylphosphatidylcholine melts at 42.4°C, while dipalmitoylphosphatidic acid melts at 67°C at pH 6.5 (Jacobson and Papahadjopoulos, 1975). This indicates that specific head group interactions can markedly alter the melting behavior of the hydrocarbon chains. However, it should be noted that the replacement of the phosphatidylcholine with the phosphatidylglycerol headgroup has little effect on the phase transition (see Table I). An even more important determinant of the melting behavior of phospholipids are alterations in the fatty acyl chains themselves. With saturated chains the T_m shifts to higher temperatures with increasing chain length. Phospholipids in biological membranes, however, usually contain fatty acyl chains with one or more unsaturated *cis* C-C bonds. This has the extremely important effect of markedly lowering the T_m by putting a permanent kink in the chain and interfering with the regular all-*trans* packing of the chains since the bond angle between double-bonded carbon is 16° greater than that for single bonded carbons (109°28′). Fig. 5A is a diagram of a typical tracing of a heating curve for an endothermic type phospholipid transition, with the temperature points identified. Chapman and his co-workers (e.g. Phillips et al., 1970) usually quote T_1 (beginning of melt) as the temperature of the transition, since they feel that the mid-point simply represents the maximal rate of change while the slope representing the completion of the change on the high temperature side is partly a reflection of instrument performance. Other authors either quote the midpoint, T_m, or quote the entire temperature range occupied by the transition.

Naturally occurring phospholipids generally have one unsaturated and one

TABLE I

PHASE TRANSITION DATA OF VARIOUS PHOSPHOLIPID DISPERSIONS MEASURED BY DIFFERENTIAL SCANNING CALORIMETRY

Phospholipid	T_1(°C)	T_m(°C)	T_2(°C)	ΔH(kcal/mole)	Reference
Dimyristoyl (C14) PC	23	25.5	28	6.7	Oldfield and Chapman, 1972
		23.7		6.2	Hinz and Sturtevant, 1972
Dipalmitoyl (C16) PC	41			8.7	Ladbrooke and Chapman, 1969
	41.4	42.4	45	8.9	Papahadjopoulos et al., 1975
		41.8		9.7	Hinz and Sturtevant, 1972
Distearyl (C18) PC	58			10.7	Ladbrooke and Chapman, 1969
		58.8		10.8	Hinz and Sturtevant, 1972
Dioleoyl (C18:1) PC	−22			7.6	Ladbrooke and Chapman, 1969
Egg PC	−5				Ladbrooke and Chapman, 1969
Dimyristoyl (C14) PE	48	51	55		Oldfield and Chapman, 1972
Dimyristoyl (C14) PG	21.4	23.1	27		Kimelberg and Papahadjopoulos, 1974
Dipalmitoyl (C16) PG	39.9	41.0	43.9	7.9	Jacobson and Papahadjopoulos, 1975
Distearyl (C18) PG	52	53.7	56		Kimelberg and Papahadjopoulos, 1974
Dipalmitoyl (C16) PA (pH 6.5)	63	67	70.0		Jacobson and Papahadjopoulos, 1975

T_1, temperature at which slope of arm of main endothermic peak intercepts the base line (beginning of melt). T_2, corresponding point on high temperature side (end of melt). T_m is temperature of maximum peak height. The peak may or may not be symmetric (see diagram in Fig. 5). These values are either those quoted by the authors or obtained by inspection of their data. ΔH is the total enthalpy of the transition (major + pre-transition) and is equivalent to the area under the two peaks and is given if the value is quoted by the authors. All at pH 7.4 unless indicated otherwise. PC, phosphatidylcholine; PE, phosphatidylethanolamine; PG, phosphatidylglycerol; PA, phosphatidic acid.

saturated chain, predominantly with the unsaturated chain on the 2 position of the glycerol backbone. The melting behavior of phospholipid bilayers is also affected when simple mixtures of phospholipids or phospholipids with different fatty acyl chains on the same glycerol backbone (intramolecular mixing) are measured. This is analogous to the situation actually found in biological membranes which consist of different phospholipids with heterogeneous fatty acyl chains. With intramolecular mixtures, sharp transitions intermediate between those that would be found for phospholipids consisting of either fatty acid alone are seen. When phospholipids with different head groups are mixed, an intermediate behavior is found, mixtures of DMPC and DMPE showing very broad intermediate transitions (Phillips et al., 1970; Phillips et al., 1972; Chapman et al., 1974). Thus, biological membranes may

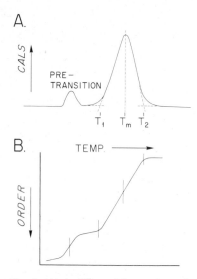

Fig. 5. (A) A differential scanning calorimeter heating curve of a phospholipid liposome suspension. The absorption of heat is indicated in the upward direction and represents calories/unit time. The pretransition sometimes occurring on the low temperature side of the main endothermic transition is shown. The positions of the usually quoted temperatures of the main transition are also indicated. T_m represents the mid-point of the transition and T_1 and T_2 the lower and upper limits respectively. (B) Shows the comparable behavior of a fluorescence or spin labelled probe in a phospholipid liposome. The degree of order is arbitrarily represented on the ordinate, increasing with decreasing temperature. It can be seen that the features of the main endothermic melt and the pretransition are easily detectable as sharp changes in the slope of the curve or as mid-points of the sharp rate of change in the order parameter between such changes in slope.

consist of both fluid and solid patches of lipid bilayers. However, the influence of other major components of biological membranes, such as cholesterol and proteins further complicates the type of states that will occur. This will be discussed later. In addition to differential scanning calorimetry, membrane phase transitions are measured by a variety of other sensitive physical techniques. These techniques are able to quantitatively measure the degree of order or fluidity of the membrane. Such techniques include: electron spin resonance (Hubbell and McConnell, 1971) and nuclear magnetic resonance (Seelig and Niederberger, 1974; Godici and Landsberger, 1974); X-ray diffraction (Engelman, 1971); fluorescence (Radda and Vanderkooi, 1972); light scattering (Träuble and Eibl, 1974); dilatometry (Overath and Träuble, 1973); and viscometry (Zimmer and Schirmer, 1974). These references are simply illustrative rather than comprehensive and references to earlier work can be found in these papers. These techniques can measure the overall bulk fluidity (X-ray diffraction, dilatometry, light scattering, proton magnetic resonance and viscometry) or the fluidity in the environment of the probe (electron spin resonance and fluorescence) and may thus be limited in accurately analyzing the heterogeneity likely to be found in biological membranes. These techniques, however, when coupled with work on simple model systems do at least allow one to predict the basic type of behavior to be expected and to set certain limits for experiments with natural membranes.

As mentioned above, the presence of unsaturated bonds will lower the phospholipid transition temperature markedly. A single *cis* double bond disturbs the all-*trans* packing, and the solid to fluid transition will occur at a much lower temperature. Thus, the T_m for dioleoylphosphatidylcholine (two 18 carbon fatty acyl chains with a single *cis* double bond) is $-22°C$, whereas for the corresponding phospholipid with no unsaturated bonds, distearyl-phosphatidylcholine, the T_m is 58°C (Phillips et al., 1972), both values having been measured by differential scanning calorimetry. It is of interest that for phosphatidylserine from beef brain a broad transition from 5–20°C (Kimelberg and Papahadjopoulos, 1974) and 0–12°C (Jacobson and Papahadjopoulos, 1975) has been found. This phospholipid has been reported to consist of 49% stearic (18:0), 37% oleic (18:1) and about 4% each of (20:1), (20:4) and (22:6) fatty acids (Papahadjopoulos and Miller, 1967). The first figure in the brackets represents the number of carbons in the fatty acid and the second after the colon refers to the number of C=C bonds. Egg phosphatidylcholine with 34% (16:0), 20% (18:1) and 11.2% (18:2) (Papahadjopoulos and Miller, 1967), shows a transition at $-5°C$ (Ladbrooke and Chapman, 1969). Thus, naturally occurring phospholipids show low melting temperatures as well as broad endothermic melts, as predicted from studies using synthetic phospholipids.

In model bilayers electron spin resonance techniques have been useful in detecting variations in fluidity along the fatty acyl chain progressing from the polar head group towards the interior of the bilayer. Using N-oxyloxazolidine labelled fatty acids as ESR probes (e.g. Hubbell and McConnell, 1971) (see Fig. 6B), a profile of increasing motion or fluidity of the hydrocarbon chain has been detected from the polar head group region towards the terminal methylene group. It has been suggested that the gradual increase in fluidity detected by the spin label is in fact a result of the large ESR probe perturbing the packing of the bilayer and reporting an average fluidity. Deuterium magnetic resonance studies have found that in decanoic acid there is a constant order parameter until the last 3 carbons, where there was an abrupt increase in fluidity (Seelig and Niederberger, 1974). A spin label probe in the same molecule showed a gradual non-discontinuous increase in fluidity. A similar profile was also seen for naturally occurring egg lecithin (Godici and Landsberger, 1974) and, in addition, some motion of the quaternary nitrogen of the choline group was also detected. These papers should be consulted for further references on this subject.

It has been mentioned above that mixtures of phospholipids with widely different T_m values can undergo two separate solid to fluid melts thus producing a bilayer consisting of separate solid and fluid areas (Phillips et al., 1972). With mixtures of different head groups or binary mixtures of the same phospholipid whose saturated chains only differ by 2 carbon atoms, broad intermediate transitions indicating varying proportions of solid and fluid states producing an overall varying intermediate fluidity are seen. These various

A.

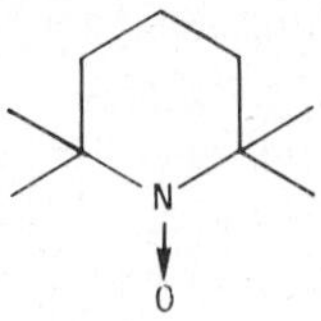

TEMPO

B.

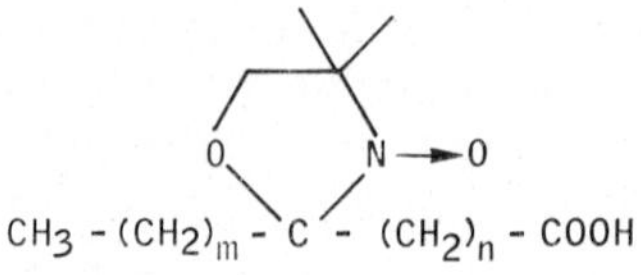

$CH_3-(CH_2)_m-C-(CH_2)_n-COOH$

STEARIC ACID (C18)
+
N-OXYLOXAZOLIDINE RING

Fig. 6. Formulae of commonly used spin label probes. (A) TEMPO (2,2,6,6-tetramethylpiperidine-1-oxy). This probe preferentially partitions into fluid bilayer regions from water, with characteristic changes in the ESR spectrum from which a partition ratio (f) can be derived. (B) Stearic acid with an N-oxyloxazolidine ring attached to one of the methylene groups. Values for m and n designate the position of the ring on the fatty acid. Also referred to as nitroxide-labelled fatty acid.

phenomena have been referred to as phase separations (Shimshick and McConnell, 1973a,b). Such broad transitions are also seen by differential scanning calorimetry for phospholipids with homogeneous saturated fatty acyl chains but possessing head groups with a net negative charge as mentioned above (see Table I) and for natural phospholipids with heterogeneous fatty acyl chains and charged or neutral head groups (Ladbrooke and Chapman, 1969; Chapman et al., 1974; Kimelberg and Papahadjopoulos, 1974; Papahadjopoulos et al., 1974; Jacobson and Papahadjopoulos, 1975).

The limits of such broad transitions can also be detected by the various probe techniques mentioned above by abrupt changes in the slope of the curve (discontinuities) when the probe behavior is plotted as a function of temperature. This is shown in Fig. 5B where it is compared to a diagram of an equivalent differential scanning calorimeter tracing, and also shows a pretransition melt. The two discontinuities seen for the probe indicate the beginning and end of a broad transition, with a mixture of solid and fluid in between. Such techniques permit the estimation of the relative proportion of solid and fluid bilayer as a function of temperature or composition, but not the molecular details. However, freeze fracture electron micrographs of liposomes consisting of 1:1 dielaidoyl-phosphatidylcholine and dipalmitoyl-phosphatidylcholine, which were quick-frozen from a temperature intermediate between the T_m of the components show two large areas that have been identified as fluid and solid (Grant et al., 1974), as shown in Fig. 7. The area showing a light jumbled pattern has been identified as the fluid membrane, and the area with a distinct, ordered or banded pattern is the solid membrane. Similar results have

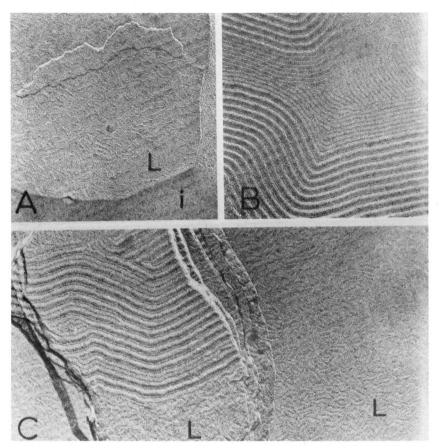

Fig. 7. Freeze-fracture electronmicrographs of non-sonicated liposomes formed from 1:1 mole % mixtures of dielaidoyl-(C18 *trans* 1) and dipalmitoyl (C16) phosphatidylcholine and quenched from different temperatures. The shadowing is from the bottom to top of page. (A) Quenched from 36°C, all fluid. L, lipid; i, ice. (B) Quenched from 10°C–11°C. All solid, showing uniform, banded pattern. (C) Quenched from 30°C. Shows separate domains of solid areas with a banded pattern and fluid areas with a jumbled pattern (reproduced with permission from Grant, Wu and McConnell, 1974).

been reported by Ververgaert et al. (1973) for different phosphatidylcholine liposomes and would imply exclusive segregation of the two phases. For such segregation to occur, the cohesive forces of each phase must be sufficient to overcome the tendency of phospholipids to migrate rapidly in the plane of the membrane. Biological membranes may be too heterogenous for such discrete phase separations, seen in model systems under certain defined conditions, to occur by themselves. There is evidence, however, that divalent cations such as Ca^{2+}, pH and proteins can promote such separations in model systems (Papahadjopoulos et al. 1974, 1976a,b,c) and these agents may provide the increased cohesive forces required in biological membranes.

The ability of phospholipids to form bilayers spontaneously in aqueous solutions is now well established, as is the fact that the fatty acyl chains can exist in a fluid or solid state, or in some intermediate mixed state. Further details on the behavior of model phospholipid systems at the molecular level will undoubtedly continue to enlarge our understanding of membrane structure. For instance, it has been suggested recently that the pretransition seen in the differential scanning calorimeter for dipalmitoyl-phosphatidylcholine is associated with a transition from a tilted solid form to a perpendicular solid at around 35°C. The main transition at 42°C is, as usual, assigned to the melting of the fatty acyl groups (Rand et al., 1975). This is shown in Fig. 8 and can be compared with the melting behavior detected by differential scanning calorimetry or probe studies shown in the diagram in Fig. 5. For the purposes of determining plausible mechanisms for the effects of lipid fluidity on membrane enzyme activity, however, the major problem at the present time is to determine to what degree lipids existing in highly complex biological membranes show fluidity changes corresponding to those seen in model membranes and what effects these might have on the conformation and activity of membrane proteins. To help answer these questions the model membrane systems can be modified by investigating the effects of adding major components from natural membranes such as proteins and cholesterol.

2.3. Effects of cholesterol, divalent cations and pH on lipid fluidity

The effects of cholesterol on the properties of model membranes have attracted considerable interest in recent years and have been reviewed extensively (Oldfield and Chapman, 1972; Phillips, 1972; Chapman, 1973; Papahadjopoulos and Kimelberg, 1973; Jain, 1975). These reviews can be consulted for the details of the experimental approaches and results, and only the main conclusions will be mentioned here. Cholesterol consists of a rigid sterol ring region approx. 11 Å long and about 25 Å^2 cross-sectional area, with a 3β–OH group at one end. Attached to the other end is an approx. 8 Å long 8 carbon hydrocarbon chain with about half the cross-sectional area of the sterol ring (Rothman and Engelman, 1972). This molecule inserts into phospholipid bilayers with the –OH group in the region of the polar head groups and the remainder of the molecule parallel to the phospholipids (Chapman, 1968; Rand

Fig. 8. Diagram of a dipalmitoylphosphatidylcholine bilayer showing structural changes at 35°C and 42°C. Circles represent polar head groups, straight lines the solid, fully extended hydrocarbon chains and the wiggly lines represent melted hydrocarbon chains. The change from a tilted to straight orientation is thought to represent the pre-transition at 35°C (see Fig. 5), and the solid to fluid hydrocarbon chains the main endothermic melt (from Rand, Chapman and Larsson, 1975).

and Luzzati, 1968). The effect of cholesterol is to condense the average area per molecule and reduce the motion of the hydrocarbon chains of *fluid* phospholipid bilayers. On the other hand, by disrupting the all-*trans* crystalline packing of *solid* phospholipid bilayers it increases the fluidity of solid bilayers. This effect results in what has been termed an "intermediate fluid state" (Chapman, 1968; Ladbrooke et al., 1968; Oldfield and Chapman, 1972). As seen by differential scanning calorimetry increasing amounts of cholesterol produce a progressive decrease in the height and some broadening of the transition peak, resulting in no effect on the T_m but a gradual decrease in the ΔH of the transition and a decreasing T_1 and increasing T_2 (see Fig. 5). Finally, a state of intermediate fluidity is obtained at 30 to 40 mole per cent cholesterol, at which point no transition can be detected by differential scanning calorimetry (Ladbrooke et al., 1968). Associated with the condensation effect on fluid phospholipid bilayers is a small increase in thickness, which in egg lecithin multilayers is from 39 Å to 42 Å (Levine and Wilkins, 1971).

The effect of cholesterol on the fluidity profile of the phospholipids going from the lipid-water interface towards the terminal methylene group in the interior of the bilayer is not continuous. Thus, at a 2:1 egg lecithin to cholesterol mole ratio the first eight carbons from the fatty acid-glycerol linkage are relatively rigid (Hubbell and McConnell, 1971). There is, however, rapidly increasing motion, or increased probability for *gauche* states, for the remaining carbons towards the terminal methyl group. It has been suggested that this is explicable by considering the space filling properties of cholesterol and palmitic acid (Rothman and Engelman, 1972). Thus, the first seven or so methylenes are immobilized by the relatively bulky planar ring of cholesterol while the remaining methylenes are less restricted by the narrower hydrocarbon chain of the cholesterol. Thus, no requirements for a specific interaction of cholesterol with phospholipids are required and it can act, as is in fact found (Oldfield and Chapman, 1972), as a general modifier of the hydrocarbon chains of phospholipids. Recently Kroon et al. (1975) have studied the molecular motion of cholesterol in liposomes of deuterated dipalmitoyl-phosphatidylcholine. Deuteration suppressed interference from the dipalmitoyl-phosphatidylcholine protons and these workers were able to detect considerable motion in the cholesterol aliphatic chain by proton magnetic resonance, consistent with the Rothman–Engelman model. As will be seen later, modifying the cholesterol level in both biological membranes and in reconstituted systems markedly affects their permeability properties and the activity of membrane-bound enzymes. Recent work has indicated that above approx. 30% mole ratio, cholesterol may preferentially displace egg phosphatidylcholine molecules from the inner monolayer of single unilamellar liposomes of approx. 250 Å diameter (Huang et al., 1974). Differential scanning calorimetry studies have also indicated that cholesterol may associate and bind in a binary mixture to the phospholipid with the lower melting temperature (De Kruyff et al., 1974), implying that like other substances, e.g. proteins and spin-labelled probes, it may localize in the more fluid areas of membranes.

Divalent cations, such as Ca^{2+} or Mg^{2+} are even more ubiquitous components of biological systems than cholesterol, and are more likely to undergo rapid and reversible changes in their binding to membranes and membrane-associated macromolecules. Possible localized changes in pH at the membrane interface might also fit into this category. It is therefore of considerable interest that both divalent cations and pH can have significant effects on the fluidity of phospholipid bilayers.

Because the effects of Ca^{2+} and Mg^{2+} on phospholipids are likely to be due to their binding to polar head groups, marked effects are usually only seen with phospholipids possessing a net charge (always negative for naturally occurring phospholipids). Thus, no effects are generally seen with neutral, zwitterionic phospholipids such as phosphatidylcholine. Ca^{2+} has been reported to increase the cation permeability of liposomes made from the negatively charged phospholipid, phosphatidylserine (Papahadjopoulos and Bangham, 1966; Papahadjopoulos et al., 1976b,c). It has been found that Ca^{2+} added to one side of a membrane destabilizes it, thus explaining the increase in membrane permeability in liposomes, whereas Ca^{2+} added to both sides of the bilayer produces a highly stable membrane (Papahadjopoulos and Ohki, 1969; Papahadjopoulos et al., 1976b, 1976c). Consistent with this effect it has been found that Ca^{2+} concentrations >1 mM abolish the phase transition of phosphatidylserine and dipalmitoyl-phosphatidylglycerol (Jacobson and Papahadjopoulos, 1975) due to the formation, as identified by X-ray diffraction studies, of solid fatty acyl chains. Ca^{2+} at lower concentrations of 0.5 mM, and Mg^{2+} at concentrations of 5–10 mM cause an increase in the T_m of 10–20°C with negatively charged phospholipids such as dipalmitoyl-phosphatidylglycerol and brain phosphatidylserine (Träuble and Eibl, 1974; Kimelberg and Papahadjopoulos, 1974; Verkleij et al., 1974; Jacobson and Papahadjopoulos, 1975; Papahadjopoulos et al., 1976b). Mg^{2+} is only effective at $10 \times$ higher concentrations that Ca^{2+} and appears to form less solid membranes. Recently, Ca^{2+} at concentrations of 120 mM, has been reported to greatly increase the T_m of dimyristoylphosphatidylglycerol to a temperature of 89°C (Van Dijck et al., 1975).

Intuitively an increased charge on the polar head group might be thought to lead to increased fluidity due to increased repulsion between neighboring molecules (Träuble and Eibl, 1974), but this is not always the case. Thus, some degree of specific interactions have to be considered to explain the differences in the effects of Ca^{2+} and Mg^{2+} on the phase transitions of acidic lipids. Also a specific head group effect is suggested in the case of DMPE, which has a T_m value some 25°C higher than DMPC. This latter effect may be due to a specific interaction of the head groups, namely intermolecular pair formation between the phosphate and amine group of the phosphatidylethanolamine head group (Papahadjopoulos and Weiss, 1969), leading to increased cohesive forces between neighboring molecules.

These effects of Ca^{2+} and Mg^{2+} have the fundamentally important implication for the subject matter of this review of enabling phase transitions or fluidity changes to occur isothermally. Another aspect of potentially equal biological importance is the phenomenon of divalent cation-induced lateral phase separation of acidic lipids when present in a mixture with neutral lipids. Thus, 10 mM Ca^{2+} can induce the formation of solid patches of Ca^{2+}-phosphatidylserine complexes in a mixture of fluid phosphatidylserine-phosphatidylcholine mixtures as determined by ESR measurements (Ohnishi and Ito, 1973), or differential scanning calorimetry (Papahadjopoulos et al., 1974; Papahadjopoulos and Poste, 1975). These effects were not produced by Mg^{2+}. In addition Ca^{2+} and to a lesser extent Mg^{2+} induces a similar lateral separation in egg phosphatidylcholine-phosphatidic acid mixtures but not with other charged phospholipids such as phosphatidylethanolamine and dipalmitoylphosphatidylglycerol (Ito and Ohnishi, 1974). The phosphatidic acid was prepared by hydrolysis of egg phosphatidylcholine with phospholipase D.

Consistent with divalent metal effects being partly mediated through charge neutralization of the polar head groups is the finding that alteration in pH has a marked effect on the phase transitions of charged phospholipids when it occurs in the ionization range of their charged groups (Träuble and Eibl, 1974; Jacobson and Papahadjopoulos, 1975). Thus, the T_m of dipalmitoyl phosphatidic acid is altered from 67°C at pH 6.5 when the free phosphate has only one negative charge to 58°C at pH 9.1 when the secondary phosphate is completely ionized (Jacobson and Papahadjopoulos, 1975). Fluorescence polarization data also show that phase transitions can be induced to occur isothermally by altering the pH (Träuble and Eibl, 1974; Jacobson and Papahadjopoulos, 1975).

2.4. Effects of proteins on lipid fluidity

The protein content of biological membranes can vary from 20% (myelin) to 76% (mitochondria) of their dry weight (review, Papahadjopoulos, 1973). Consequently, it has been recognized for some time that the effects of proteins on the behavior of model bilayer membranes, as well as the effect of such interactions on the properties of the proteins themselves, is of considerable interest (see reviews by Papahadjopoulos and Kimelberg, 1973; Gulik-Krzywicki, 1975; Kimelberg, 1976a). It has been found that a number of soluble proteins such as albumin (Sweet and Zull, 1969) and cytochrome *c* and lysozyme (Kimelberg and Papahadjopoulos, 1971a,b) can increase the permeability of liposomes to glucose and Na^+. Although this effect was correlated with increased monolayer penetration (Kimelberg and Papahadjopoulos, 1971b), it was not clear to what degree this resulted in an ordering or disordering of the bilayer structure. More recent studies using differential scanning calorimetry have shown that those proteins which showed electrostatic binding to negative liposomes, monolayer penetration and increased permeability effects, caused a marked decrease in T_m. Thus, the myelin basic protein and cytochrome *c*, which have been classified as peripheral membrane proteins (Singer and Nicolson, 1972), decreased the T_m of dipalmitoyl-phosphatidylglycerol membranes by up to 10°C, as well as decreasing the enthalpy (Papahadjopoulos et al., 1975). These results suggest that the increased permeability and monolayer penetration caused by both cytochrome c (Quinn and Dawson, 1969; Kimelberg and Papahadjopoulos, 1971b) and myelin basic protein (Demel et al., 1973) might be accompanied by fluidization of the bilayer. As discussed previously, however, the differential scanning calorimeter only strictly measures the temperature at which the transition occurs. These effects were not seen with neutral dipalmitoyl-phosphatidylcholine membranes suggesting an electrostatic requirement. Similar effects are also apparently not detected by spin-labelled fatty acids with cytochrome c-cardiolipin-phosphatidylcholine mixtures (Van and Griffith, 1975). It is of interest that binding at neutral pH of the positively charged proteins ribonuclease and polylysine (which cause a much smaller permeability increase) produced either no effect or an increase in the T_m (Papahadjopoulos et al., 1975b).

Thus, both non-membrane and membrane associated proteins such as lysozyme, cytochrome c and myelin basic protein can show some degree of non-polar interactions with the hydrocarbon portion of the fatty acyl chains (Gulik-Krzywicki et al., 1969; Kimelberg and Papahadjopoulos, 1971a,b). These effects are in contrast to the effects of divalent cations, which increased the T_m of the same phospholipid. The effect of these positively charged proteins cannot be due simply to charge neutralization. While the interactions of cytochrome c and lysozyme occur only at low ionic strength (10 mM NaCl), the basic myelin protein is even more effective at 100 mM NaCl (Papahadjopoulos et al., 1973a, 1975b). Some studies on the interaction of soluble proteins with liposomes have also been done with spin labels and fluorescent probes (reviews, Papahadjopoulos and Kimelberg, 1973; Gulik-Krzywicki, 1975). These studies indicate both ordering and disordering of the phospholipid bilayer dependent on the experimental conditions. Systematic studies comparing calorimetric data with results obtained with these other techniques would be very useful. It is of interest that Galla and Sackmann (1975) have recently shown that polylysine, like Ca^{2+}, can cause strong segregation of phosphatidic acid from phosphatidylcholine, and the polylysine-bound phosphatidic acid also shows an increased melting temperature.

Studies using less polar "integral" membrane proteins show a different behavior pattern. Myelin proteolipid did not alter the T_m but did progressively lower the enthalpy of the transition of dipalmitoylphosphatidylcholine liposomes, thus showing no requirement for an initial charge interaction (Papahadjopoulos et al., 1975b), a situation resembling the behavior of cholesterol. This is probably due to the progressive immobilization of fatty acyl chains adjacent to the protein, since such a class of immobilized boundary lipid had been previously identified for cytochrome oxidase membranes using spin-labelled stearic acid (Jost et al., 1973a,b; Vanderkooi, 1974). By progressive extraction of phospholipid it was determined that a limiting amount of about 0.2 mg lipid/mg protein was immobilized, while any additional phospholipid was fluid and mobile. By making certain reasonable assumptions about the size of the lipid molecules and proteins, it was determined that this amount of lipid could form a single layer of boundary lipid around the protein. A similar layer of immobilized boundary lipid has been identified for the Ca^{2+} ATPase of muscle sarcoplasmic reticulum (Warren et al., 1975a). Reconstituted membranes of rhodopsin and pure egg phosphatidylcholine also show immobilization of phospholipid (Hong and Hubbell, 1972). The Ca^{2+} ATPase has recently been shown, using TEMPO (Fig. 6A), to broaden the phase transition of DMPC by increasing its upper temperature limit. Freeze-etch studies indicated that, at low temperatures, the protein was excluded from the solid, banded area (Kleeman and McConnell, 1976).

The work to date on the interaction of various types of proteins with model and natural membranes can be conveniently summarized in the diagram shown in Fig. 9. The protein shown in (1) is an "ideal" peripheral protein showing only charge interactions with the polar surface of bilayers. Proteins showing both polar and non-polar interactions could be classified as "intermediate" proteins which could interact with lipids in several different ways as shown in (2) and (3). This category would appear to include a large number of membrane proteins since "peripheral" proteins such as cytochrome c and myelin basic protein alter the properties of liposomes in a way which indicates that some interaction with the hydrocarbon interior of the bilayer is occurring. Spectrin has also been reported to increase the permeability of liposomes and to penetrate phospholipid monolayers (Juliano et al., 1971). It is not known whether similar interactions occur with these proteins in situ. These proteins are classified as peripheral because they can be removed from membranes by manipulations of ionic strength, without using detergents. Thus, the definition is primarily an operational one. However, they do require a combination of

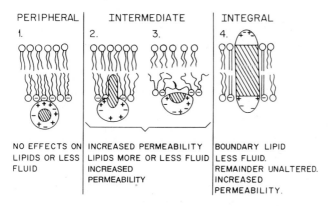

Fig. 9. Schematic representation of the possible modes of interaction of proteins with a phospholipid bilayer. The small open circles are phospholipid polar head groups with two fluid (wiggly lines), or solid (straight lines in 4) fatty acyl chains. Larger bodies represent proteins, with the hatched area indicating their hydrophobic portions. The open areas are the more hydrophilic surfaces with + or − indicating surface charges. Possible effects on the phospholipid bilayer properties are indicated below each diagram. See text for further details of these interactions.

procedures such as hypotonic plus hypertonic treatment in the case of cytochrome c (Jacobs and Sanadi, 1960) and EDTA plus alkaline pH in the case of spectrin (Marchesi et al., 1969), and these treatments could indirectly weaken hydrophobic associations in the membrane.

The degree of hydrophobic interaction of a protein with the membrane interior is likely to be determined by the amount of its non-polar surface area. A membrane protein such as adenylate cyclase which is classified as integral because it requires the detergent Lubrol to solubilize it (Levey, 1971, 1973), was found to bind <0.2 mg Triton X-100/mg protein, indicating that about 5% of its surface is hydrophobic (Neer, 1974). Other membrane proteins and lipoproteins bind 0.3–0.6 mg Triton X-100/mg protein, and soluble proteins such as ovalbumin and cytochrome c bind <0.03 mg Triton X-100/mg protein (Helenius and Simons, 1972). These results indicate a surprisingly small amount of hydrophobic surface area for some detergent-solubilized membrane proteins but the possibility of conformational changes in the protein upon binding to the membrane or after extraction complicates such analysis of membrane–protein interaction.

Recently a class of amphipathic proteins containing a distinct soluble catalytic part and a hydrophobic binding part has been recognized (Ito and Sato, 1968; Spatz and Strittmatter, 1971). Examples of this are cytochrome b_5 and NADH cytochrome b reductase (Spatz and Strittmatter, 1973). This would also correspond to types (2) and (3) in Fig. 9. Thus, categories (2) and (3) could involve a wide spectrum of membrane proteins, ranging from those showing minimal hydrophobic association to those showing considerable hydrophobic association. The term "integral" protein might then be restricted to those membrane proteins such as the (Na + K) ATPase, cytochrome oxidase and the red blood cell membrane protein, glycophorin, which have been shown to penetrate through the entire membrane. The various membrane enzymes which will be discussed in the remainder of this review fall into categories 2, 3 and 4 shown in Fig. 9. Proteins of type 2 and 3 are well-suited to reactions taking place on only one side of the membrane, with the binding portion inserted into the membrane allowing modulation of enzyme activity by the membrane. The integral protein enzymes are well-suited to transmembrane transport functions and might thus be expected to show the greatest change in activity in response to variations in membrane fluidity.

2.5. Lipid fluidity in biological membranes

This section will provide a brief discussion of lipid fluidity in biological membranes, emphasizing to what degree departures from the behavior of pure phospholipid bilayers is a reflection of the known effects of membrane components, such as cholesterol and proteins. The general subject of the fluid properties of lipids in natural membranes has been reviewed extensively by Gitler (1972), Siekevitz (1972), Edidin (1974) Bretscher and Raff (1975) and is also discussed by Nicolson et al. in Chapter 1 of this volume (p. 1).

The first indication that the lipids of biological membranes might have considerable molecular motion was the study by Steim et al. (1969) of *Mycoplasma laidlawii* membranes using differential scanning calorimetry. This study utilized *M. laidlawii* because it normally contains no membrane cholesterol which, at sufficiently high levels, abolishes distinct phase transitions in phospholipids (see section 2.3). It was found that the intact membranes showed a broad, reversible endothermic phase transition whose T_m corresponded to that of the solid to fluid phase transitions shown by a total membrane lipid extract, and varied with the saturation of the membrane phospholipids. The T_m was distinct from the irreversible transition due to protein denaturation, which occurred at higher temperatures. Based on phospholipid content the enthalpy of the transition shown by the intact membranes was about 75% of that shown by extracted lipids. The authors interpreted these results as essentially confirming the Danielli–Davson model of biological membranes (see reviews by Hendler, 1971, 1974). In the same year, studies with the hydrophobic spin label probe, TEMPO (Fig. 6A) showed that the hydrophobic regions of some mammalian membranes were also relatively fluid (Hubbell and McConnell, 1968).

Following the initial studies on *Mycoplasma*, other calorimetric studies established that the membranes of related microorganisms, such as *Escherichia coli* as well as non- or low-cholesterol mammalian membranes such as mitochondria and microsomes also showed broad endothermic transitions (see Blazyck and Steim, 1972). These data also indicated, however, that while thermal denaturation of proteins did not affect the position of the endothermic peaks, the peak was shifted by up to 10°C to lower temperatures in the lipids extracted from mitochondria and microsomes. This suggested a partial immobilization of the lipids by the protein in the original membranes. On the basis of lipid phosphorus it appeared that 75–80% of the lipids in the original mitochondrial and microsomal membranes participated in the thermotropic melt. The contribution of the mitochondrial outer membrane, which contains considerable cholesterol (11% mole/mole phospholipid) compared to the inner mitochondrial membrane (<2%, see Munn, 1974) was apparently not significant in this study.

Spin label studies (see review by Mehlhorn and Keith, 1972) have confirmed the early work of Hubbell and McConnell (1968) using the TEMPO label (see Fig. 6A). These techniques have revealed considerable membrane fluidity as measured by the partition or motion of such labels in the membrane. Thus, using a nitroxide-labelled stearic acid probe Hubbell and McConnell (1971) found that a lobster nerve fibre showed increasing fluidity toward the terminal methyl groups of the phospholipid fatty acids. This increase was much less than that seen in phospholipid liposomes above their phase transition temperature and the overall order parameter (S), obtained from the relative positions of the electron spin resonance lines (see Hubbell and McConnell, 1971) was greater. The nerve membrane profile actually resembled that from a 2:1 egg lecithin–cholesterol mixture.

Using the ratio A/B obtained from the TEMPO electron spin resonance spectra, and which represents the partition of the probe between a fluid hydrophobic region and water, McConnell et al. (1972) calculated that the proportion of the lipid in sarcoplasmic reticulum membranes that was in a fluid state was in the range 79–86%. It was assumed that the TEMPO partitions proportionally into the fluid lipid phase. This value for the amount of lipid in a fluid state agrees surprisingly well with the proportion of lipids that should be in a fluid state in mitochondria and microsomal membranes as determined by differential scanning calorimetry.

In model membrane systems protein tends to decrease the motion of phospholipids. This has also been found to be the case in natural membranes. Thus, increasing the amount of protein in reaggregates of *Mycoplasma* membranes leads to decreasing freedom of motion of nitroxide-labelled fatty acids (Rottem et al., 1970). Esser and Lanyi (1973) examined cell envelope vesicles of the bacteria *Halobacterium cutirubrum*, using labelled fatty acids of the general type shown in Fig. 6B. These membranes have a lipid/protein ratio of <0.2 and the stearic acid label showed a value for the order parameter (S_n) of 0.79 where n = 3 (see Fig. 6B) and a very similar value of 0.73 where n = 10. By comparison, for sarcoplasmic reticulum vesicles, the relative values for S_3 were 0.63 and S_{10} was 0.35 (McConnell et al., 1972). Similarly for lobster walking nerve $S_3 = 0.63$ and $S_{10} - 0.45$. A lower value indicates decreased order. The high protein levels also abolished all appearance of a phase transition for the fatty acid label in *Halobacterium cutirubrum* except for S_{14}. The influence of other membrane components on membrane fluidity is also emphasized by the study of Sefton and Gaffney (1974), who showed that the order parameter (S) is greater for the Sindbis viral membrane compared with the membranes of the chick embryo cells on which they were grown, for labels with a number of different values for n. However, the order at the center of the bilayer, measured with S_{14}, was indistinguishable. These effects were ascribed to higher protein in the viral membrane (2/1 protein to lipid), more cholesterol and interactions with RNA nucleoprotein. Although specific localization of spin label in the plasma membrane is often assumed, it should be noted that for studies on mammalian cells which possess a large amount of internal membrane structure, considerable amounts of spin label have been found in all subcellular membrane fractions (Kaplan et al., 1973). For experiments involving brief exposure to the probe specific localization within the plasma membrane might possibly be obtained (Gaffney, 1975), though definitive proof requires assay of cell fractions and this has rarely been done in the spin label work reported in the literature.

Fluorescence polarization studies (see review by Radda and Vanderkooi, 1972) have been used recently to obtain values for the microviscosity of a number of membranes (Feinstein et al., 1975). Values from 0.90 to 0.95 poise for rat liver mitochondria and microsomes respectively were obtained. A value of 1.8 poise was found for human erythrocyte membrane, 2.7 poise for bovine brain myelin and 3.35 poise for rabbit polymorphonuclear leukocytes, using the fluorescent probe perylene. This is within the range of values quoted for a variety of membranes using fluorescent polarization and other techniques by Edidin (1974). Feinstein et al. (1975) also quoted a value of 1.73 for phosphatidylserine liposomes that was slightly increased to 1.93 by the addition of membrane basic protein and 2.22 by the more hydrophobic proteolipid, while cholesterol at 1:1 mole ratio increased it to 4.06. As a comparison white oil had a microviscosity of 1.15. Thus, the phospholipids in biological membranes have considerable molecular motion, and in some cases show definite phase transitions. In one case in *E. coli*, an excellent correlation has been found between the phase transition behavior of the intact membranes and the isolated membrane lipids (Overath and Träuble, 1973). Other membranes containing increased amounts of protein or cholesterol show a decreased motion but this is entirely consistent with the behavior seen in model systems and suggests that lipid fluidity in natural membranes may be of the same type found in pure phospholipid bilayers.

Another important type of mobility seen for phospholipids in model membranes is a rapid translational movement mentioned briefly in the introduction. Thus, diffusion constants of 1–$1.8 \cdot 10^{-8}$ cm^2/sec in pure phosphatidylcholine liposomes have been reported for spin-labelled phosphatidylcholine (Devaux and McConnell, 1972) or spin-labelled androstan molecules (Träuble and Sackmann, 1972). Scandella et al. (1972), using a spin-labelled phosphatidylcholine found a diffusion constant of $12 \cdot 10^{-8}$ cm^2/sec for 4:1 phosphatidylcholine : cholesterol mixtures at 40°C which is actually faster than the constant for phosphatidylcholine alone using the same label (Devaux and McConnell, 1972). Other workers have found cholesterol decreases the diffusion constant from $1 \cdot 10^{-8}$ cm^2/sec to $1 \cdot 10^{-9}$ cm^2/sec (Lee et al., 1973). For a more complete discussion of such studies see the review by Edidin (1974).

These rates found in pure phospholipid membranes seem to be similar to the rates in biological membranes, with constants from 0.5 to $13.7 \cdot 10^{-8}$ cm^2/sec having been reported (see Edidin, 1974). Using a spin-labelled phosphatidylcholine a rate of $6 \cdot 10^{-8}$ cm^2/sec has been reported for rabbit sarcoplasmic reticulum (Scandella et al., 1972). This value only varied from 6 to $12 \cdot 10^{-8}$ cm^2/sec for a temperature increase from 40–70°C. These results indicate that the lipid molecules are moving very fast in the plane of the membrane. With diffusion constants of $\sim 10^8$ Å^2/sec, each molecule will exchange with its neighbor about 10^6 times a second. It is clear that such rapid motion of lipids would tend to obliterate any specific lipid effects on proteins ("that field and roof lie level and the same so fast I move defying time" Thomas, 1957). This point emphasizes the importance of segregation by proteins of the particular lipid(s) required in their activity and/or control.

In contrast to the rapid rate of lateral diffusion, phospholipids in model bilayers show a very slow rate of inversion from one monolayer to the other. Such "flip-flop" exchange has been determined to occur with a half time of 6.5 h at 30°C (Kornberg and McConnell, 1971). In contrast, the flip-flop of labelled dioleoylphosphatidyl TEMPOcholine in excitable membrane vesicles from *Electrophorus* was in the range of 3.8–7 min at 15°C (McNamee and

McConnell, 1973), which is considerably faster than that found in pure phosphatidylcholine liposomes. These workers also gave a tentative value of 20–30 min for the half-time of the flip-flop of the same label in red blood cells. These results may be somewhat at variance with experiments described later (section 3.5.3) showing that there is considerable asymmetry of the phospholipids in the two monolayers of the red blood cell membrane. From the available evidence it does not appear that phospholipid flip-flop is likely to be fast enough to be significant in most transport processes or in the rapid control of enzyme activity.

Another important aspect of the mobility of membrane components for enzyme–enzyme interactions is the ability of protein molecules to diffuse laterally in the plane of the membrane. The fast lateral motion of lipids indicates that the diffusion of lipid-soluble substrates is not likely to be rate-limiting, while the diffusion of water-soluble substrates would occur in the aqueous phase (Stier and Sackmann, 1973). As mentioned in the introductory section, proteins diffuse in the plane of the membrane with diffusion constants of 10^{-9} to 10^{-12} cm^2/sec (see Edidin, 1974). The values of 10^{-9} cm^2/sec leads to collision rates in the millisecond range for proteins separated by around 100 Å, which will usually be sufficient to account for most enzyme reactions. Such lateral mobility of membrane proteins was detected originally by the mixing of antigens combined with different fluorescent-labelled antibodies, in mouse–human heterokaryons (Frye and Edidin, 1970). Assuming the antigen molecules were 100 Å radius, a viscosity of 1–2 poise was originally calculated, well within the 1–10 poise range seen for a variety of membranes by different techniques (Edidin, 1974), and that of 1.7 poise for pure phosphatidylserine and 4.1 poise for 1:1 phosphatidylserine/cholesterol membranes (Feinstein et al., 1975). Lateral motion of proteins has also been seen in freeze-etch pictures of membranes, since the proteins tend to aggregate when the membranes are quenched from low temperatures (see chapter in this volume by McNutt, p. 75). It has been suggested recently that proteins can move out of the solid phase into remaining fluid areas of membrane (Kleeman and McConnell, 1976). Thus, when the membrane is completely fluid, a random distribution of protein is found, whereas when rapidly quenched from a temperature when the membrane is in a relatively solid state, aggregates of protein particles are seen, as shown in Fig. 10 for *E. coli* (Verkleij et al., 1972; Haest et al., 1974).

3. Lipid fluidity and membrane enzymes

3.1. Lipid dependence of membrane enzymes

Well before the current interest in the effects of the dynamic properties of membrane lipids on membrane enzyme activity, it was known that most membrane enzymes required phospholipid for activity. Some of the earliest work in this area involved studies on mitochondrial enzymes. The observation that

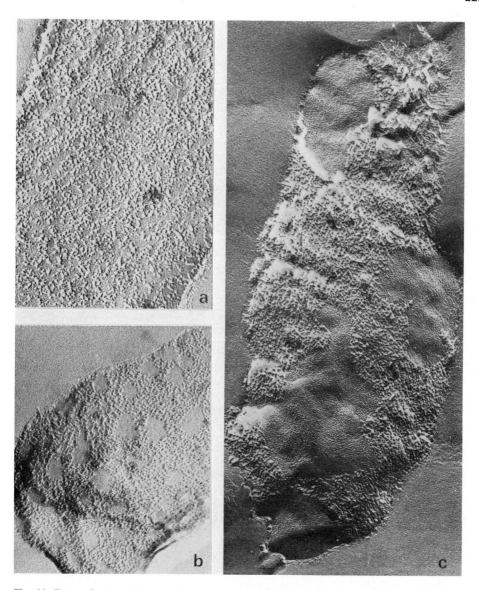

Fig. 10. Freeze fracture electronmicrographs of unsaturated fatty acid requiring *E. coli* mutants grown on different fatty acids and quenched from varying temperatures. (a) Grown on oleate (18: *cis* 1) and quenched from 37°C. (b) Grown on oleate quenched from 22°C or grown on elaidate (18: *trans* 1) and quenched from 37°C. (c) Grown on oleate or elaidate, quenched from 0°C. Magnification ×120 000 (reproduced with permission from Haest et al., 1974).

β-hydroxybutyrate dehydrogenase had a specific requirement for lecithin (Jurtshuk et al., 1961) is among the first reports of a lipid-dependent enzyme. It was also found that synthetic lecithins containing saturated (C 16) fatty acyl chains were barely active, implying a requirement for fluid lipid membranes, since the assays were done below the melting temperature (approx. 42°C) of the dipalmitoyl-phosphatidylcholine used (Jurtshuk et al., 1961). More recent work has indicated, however, that at relatively higher concentrations saturated lecithins below their T_m can activate (Grover et al., 1975). As the authors themselves recognized, the lack of activation by dipalmitoyl-phosphatidylcholine and DMPC (Green and Fleischer, 1963) could also have been attributed to the inability of the phospholipids to form liposomes at room temperature (they were incorrectly referred to as micelles). It is now well-known that liposomes are only formed at temperatures above the phase transition temperature of the phospholipid used, i.e. when the lipids are in a fluid state. The requirement for water-clear "micellar" solutions made by sonication or by rapid dispersion of an ethanolic solution of lecithin into water (Jurtshuk et al., 1961), a technique later shown by Batzri and Korn (1973) to form single bilayer liposomes, indicates that small single-walled liposomes were actually used in reactivating the lipid-depleted enzyme in this early study. Electron micrographs of the lipid dispersions used show, however, a typical multilamellar myelin-type figure, which appears to have been correctly interpreted as consisting of multiple bimolecular leaflets (Green and Fleischer, 1963). They were, however, represented as solid cylinders consisting of a continuous spiral bilayer, and not as vesicular multilamellar structures with considerable free internal volume. Subsequent to this work, a large number of enzymes of both mitochondrial and non-mitochondrial origin have been found to be lipid-dependent after extraction of phospholipids by organic solvents, detergents or phospholipases. This subject has been reviewed extensively in the past few years and the reader should refer to these articles for fuller details (e.g. Triggle, 1970; Rothfield and Romeo, 1971; Razin, 1972; Coleman, 1973; Fourcans and Jain, 1974).

The following sections of this review will deal principally with the effects of the fluidity of the phospholipid hydrocarbon chains on enzyme activity, but it seems necessary to at least mention that an equally important aspect of such interactions is whether there is any specificity for the polar head group, as described above for the phosphatidylcholine requirement of β-hydroxybutyrate dehydrogenase. Specific interactions of this type, coupled with the ability of phospholipids to show lateral phase separation, could provide a means of localizing a unique phospholipid microenvironment around each enzyme. It is clear, however, that while some enzymes show a specific requirement, others do not. In some cases such as adenylate cyclase, different functional aspects of the same enzyme can involve different phospholipids, thereby increasing the potential for membrane modulation of enzyme activity (Levey, 1971, 1973; Rethy et al., 1972).

Reports of requirements for different phospholipids by the same enzyme after delipidation presumably reflect differences in experimental techniques.

This could arise from: (1) different procedures of varying efficiency used for extracting the lipid; (2) variable activity remaining, which may be due to variations in the lipid remaining; and (3) the purity of the phospholipids used for reactivation. The divergence of results that can be obtained with the same enzyme from the same tissue, is well illustrated by the results for (Na + K) ATPase reactivation, although a majority of recent studies on this enzyme are in agreement that of the naturally occurring phospholipids phosphatidylserine is the most efficient activator of enzyme activity (Roelofsen and Van Deenen, 1973). Examples of different extraction procedures used and the activating lipids found for specific enzymes are shown in Table II. This table is derived from a more comprehensive table given in Kimelberg (1976a) plus the inclusion of some more recent work. Fourcans and Jain (1974) have also published a similar compendium summarizing the same type of data for a number of lipid-dependent enzymes.

3.2. Membrane fluidity and Arrhenius plots of enzyme activity

To study the behavior of membrane enzymes as a function of membrane fluidity it is clearly necessary to be able to alter fluidity in a predictable manner. One way of achieving this isothermally is by altering the phospholipid composition and saturation, and/or the cholesterol content of the membrane either in situ or in reconstituted systems. Alternatively, if the composition of the membrane is such that it will undergo phase transitions within a temperature range which does not inactivate the enzyme, the fluidity of both model and natural membranes can be altered by changing the temperature. The resulting changes can then be conveniently represented as Arrhenius plots (Dixon and Webb, 1964).

The Arrhenius equation relates temperature and reaction velocity as follows:

$$\frac{d \ln k}{dT} = \frac{E_a}{RT^2} \tag{1}$$

where, k is the rate constant, R the gas constant, T the absolute temperature in degrees Kelvin and E_a is known as the activation energy. See also Dixon and Webb (1964).

Equation (1) can be integrated to a more useful form for graphical representation:

$$\ln k = -\frac{E_a}{RT} + \text{constant} \tag{2}$$

Thus, if 1/T is plotted against the natural logarithm of the rate constant, a straight line with a negative slope with respect to 1/T is usually obtained, since reaction rates generally increase with increasing temperature. The slope of this line thus gives the activation energy E_a, of the reaction. E_a is related to the

TABLE II
LIPID REQUIREMENTS FOR (Na$^+$ + K$^+$) ATPase

Source	Method of delipidation	Percent activity remaining	Lipid effective in reactivation (<20% in parentheses)[a]	Reference
Horse erythrocyte	phospholipase A	5	PS (PE, PI)	Ohnishi and Kawamura, 1964
Beef brain	deoxycholate	10	PC > lyso PC (di PG)	Tanaka and Strickland, 1965
Beef brain	deoxycholate	—	Dodecylphosphate, lyso PE (decylalcohol monoctodecyl phosphate)	Tanaka and Sakamoto, 1969
Beef brain	deoxycholate	—	lyso PC, PS	Tanaka et al., 1971
Beef brain	phospholipase A	10	PS = PA > PI > PE = inosithin. Albumin activates	Hokin and Hexum, 1972
Rabbit kidney	deoxycholate	0	PS > PG (PI, PA, di PG, PE, PC)	Kimelberg and Papahadjopoulos, 1972
Rabbit kidney	lubrol	18	PS > PG (PE, PI di PG, PC)	Wheeler et al., 1975
Ox brain	deoxycholate	21	PS (PA, PI)	Wheeler and Whittam, 1970
Human erythrocytes	phospholipase A$_2$ and C. Ether extraction + PS decarboxylase	0	PS > PA (PC, PE)	Roelofsen and Van Deenen, 1973

[a] PI, phosphatidylinositol; PS, phosphatidylserine; di PG, ciphosphatidylglycerol. The abbreviations for the other phospholipids are as in the footnote to Table I.

enthalpy of the activated complex by:

$$Ea = \Delta H^* + RT \qquad (3)$$

where ΔH^* is the enthalpy of the activated state according to the Eyring rate theory.

For many membrane enzymes, and some soluble enzymes as well, a plot of the logarithm of the reaction velocity against 1/T does not give a straight line but shows a sharp change in slope. The departures from linearity can take the form of actual discontinuities or breaks in the curve, or two or more linear lines with points of intersection (Raison, 1973). Since even with different intersecting lines the behavior is a sharp or discontinuous change in slope, all these different effects will be referred to simply as "discontinuities". For enzymes the lower temperature lines usually, but not always, have larger slopes than the upper temperature lines (see Figs. 11, 13, 14). The two most likely reasons for a sharp, rather than a curvilinear change in slope are: (1) a sharp phase change in the solvent (i.e. water or the membrane environment); or (2) a change from one conformation of the enzyme to another with differing activation energies (for a discussion on this point see Dixon and Webb, 1964, pp. 158–166). It has been argued that at the point of intersection, where the reaction velocity is only marginally different, both phases are coexisting with about the same free energy of activation (Kumamoto et al., 1971; Raison, 1973). Since ΔH^* for the two phases, derived from the two slopes, is clearly different, a compensating change in the entropy of activation (ΔS^*) must occur to give the same ΔF^*. If this does not occur, two non-intersecting lines of different slopes with a gap between them indicating a significant change in ΔF^* should be obtained, which is in fact sometimes found (Kumamoto et al., 1971; Raison, 1973). Thermodynamic calculations for both the Ca^{2+} ATPase from sarcoplasmic reticulum (Inesi et al., 1973; Madeira and Antunes-Madeira, 1975) and phospholipid reconstituted (Na + K) ATPase (Kimelberg, 1976a) show large changes in ΔS^* for the differing values of ΔH^*. ΔH^* can be obtained simply from the measured Ea according to equation (3). ΔF^* can be obtained from the measured reaction rate (k) at any temperature from:

$$\Delta F^* = - RT \ln \frac{(k \cdot h)}{(k_B T)} \qquad (4)$$

where k_B is Boltzmann's constant, and h is Planck's constant.

ΔS^* is then simply calculated from the usual relationship

$$\Delta F^* = \Delta H^* - T \cdot \Delta S^*. \qquad (5)$$

The various values of these quantities for the Ca^{2+} ATPase and (Na + K) ATPase are shown in Table III. It is clear that the decrease in Ea and therefore ΔH^* in the higher temperature range is compensated by a decrease in ΔS^* thus keeping the free energy of activation (ΔF^*) relatively constant with changing temperature. A positive ΔS^* value indicates a gain of entropy by the

TABLE III
FREE ENERGY AND ENTROPY OF ACTIVATION FOR (Na + K) ATPase AND Ca^{2+} ATPase ACTIVITY[a]

	Transition temp. in Arrhenius plot of activity (°C)	Temp. °C	ΔH^* kcal $mole^{-1}$	ΔS^* cal. deg^{-1} $mole^{-1}$	ΔF^* kcal $mole^{-1}$
Ca^{2+} ATPase	20	5–20	27 to 29	+34 to +36	17 to 19
		20–35	16 to 18	+3 to +4	15 to 17
(Na + K) ATPase reconstituted with dioleoyl phosphatidylglycerol	—	20	16	−18.5	21.5
(Na + K) ATPase reconstituted with dimyristoyl phosphatidylglycerol	20	12.7	29	+25.9	22.0
		30	13	−28.1	21.9

[a]Data from Inesi et al. (1973) for the Ca^{2+} ATPase data, and Kimelberg (1976a) for the (Na + K) ATPase.

activated complex, and implies that the activated complex is less ordered than the reactants. The values shown for ΔS^* in Table III are what one might expect if the transition temperature corresponds to a solid to fluid melt, since the ΔS^* values are greater (more positive) below the transition temperature suggesting that the reactants are more ordered than the activated complex. It is assumed that the activated complex has a constant degree of order throughout the temperature range. It is of interest that the (Na + K) ATPase reconstituted with dioleoyl-phosphatidylglycerol which is fluid above 0°C shows a single Ea and negative ΔS^* value, comparable to dimyristoyl-phosphatidylglycerol above its transition temperature.

In most studies in which Arrhenius plots of enzyme activity have been used, the logarithm of the reaction velocity is usually plotted against 1/T. It has been pointed out, however, that the measured velocity of an enzyme is usually the result of a number of different reaction rates, all of which may have a different temperature dependence (Dixon and Webb, 1964). The simplest procedure is to ensure that one is always measuring the maximal velocity of the enzyme reaction (V_{max}). Under these conditions $V = k_{+2}e$ (Dixon and Webb, 1964), so that one might assume that one was measuring the effect of temperature on the rate constant k_{+2} alone. k_{+2} is the rate at which the final ES complex breaks down to give product, and e is the total enzyme concentration. This would be true only if this remained the rate-limiting step throughout the temperature range studied which need not necessarily be the case. The V_{max} should be accurately determined at each temperature value at infinite substrate concentration by Lineweaver-Burk plots, since one critical kinetic parameter, the substrate affinity or K_m, might vary markedly with temperature and the measured rate may no longer approximate V_{max} (Dixon and Webb, 1964). Current practice has been, however, to determine initial rates at substrate levels which are saturating at the highest temperature measured. This is a reasonable operational technique as long as one is only asking the basic question of whether enzyme activity is affected by membrane fluidity.

A recent study (Sullivan et al., 1974) has shown that the K_m for O-nitrophenyl-β-D-galactopyranoside (ONPG) transport by *E. coli* increases 4–6-fold from 0–28°C, while V_{max} increases 100- to 180-fold. Arrhenius plots of these two parameters indicated that while the change in V_{max} was gradual and curvilinear, the change in K_m was sharply sigmoid. On this basis the authors suggested that the changes in K_m may be responsible for discontinuities in Arrhenius plots of overall ONPG transport. Future work will undoubtedly focus far more on the influence of temperature and membrane fluidity in general on different kinetic parameters and regulation of membrane enzyme activity. A number of studies have dealt with the effects of dietary changes on the cooperativity of a number of enzymes as shown by Hill plots, and these studies have been reviewed recently (Farias et al., 1975). Zakim and Vessey (1975) have recently reported that the activation and inhibition of UDPglucuronyltransferase by other UDP-sugars was markedly affected by microsomal membrane fluidity. A recent report has also shown that cholesterol inhibits the V_{max} rather than the K_m in a reconstituted (Na + K) ATPase (Kimelberg, 1975).

Theoretical and experimental considerations (see section 3.5) indicate that the discontinuities in Arrhenius plots do fit the concept of a phase change occurring in either the enzyme itself or in its environment. It is well-known, however, that many soluble enzymes such as fumarase (Massey, 1953) and D-amino acid oxidase (Massey et al., 1966) also show discontinuities when their temperature dependence is plotted as Arrhenius plots. This has been explained in terms of two stable conformation states of the enzymes with different Ea values, which undergo a sharp transition from one form to the other at the discontinuity temperature (t_d). Indeed, early studies which showed definite discontinuities in Arrhenius plots of the enzyme activity of membrane-bound enzymes such as the (Na + K) ATPase (Gruener and Avi-Dor, 1966; Bowler and Duncan, 1968) were interpreted in terms of changes in protein conformation. The view that temperature-dependent alterations in protein conformation are responsible for such behavior, and even for effects of temperature on more general membrane properties has been reviewed recently (Konev et al., 1975). Clearly, effects on enzyme activity basically involve changes in the active sites via local or general changes in the conformation of the protein molecule. The following sections will therefore deal with the evidence that in membrane enzymes such changes are fundamentally influenced by the state of the membrane lipids. It can be envisaged that when the lipids are in a relatively "solid" state, conformational changes of the enzyme protein or even translational movements of the enzyme molecule necessary for its activity are restricted. Thus, the enzyme shows a lower activity and greater Ea values in this state. In contrast, in a fluid membrane, such changes should occur more easily with a lower ΔH^* and a correspondingly smaller entropy increase. The term "viscotropic" has been suggested as a descriptive term for these effects of membrane fluidity on enzyme activity (Kimelberg and Papahadjopoulos, 1972).

3.3. Alterations in membrane lipid composition and enzyme activity

3.3.1. In Escherichia coli

The concept that alterations in membrane lipid fluidity can alter the function of membrane enzymes has extremely important implications in view of the fundamental role of membranes in cellular function. The first experimental studies suggesting the influence of membrane lipid fluidity involved studies using lipid mutants of *Escherichia coli* whose membrane lipids could be predictably altered. Other early evidence also came from studies on chilling sensitive plants and poikilotherms and will be discussed later (3.3.4). A brief review of the biochemistry and genetics of *E. coli* mutants which are unable to complete certain steps in fatty acid biosynthesis, together with the original references to this work will be found in the review by Silbert et al. (1974). Most of the studies have used mutants, first isolated by Silbert and Vagelos (1967), which cannot synthesize unsaturated fatty acids. Therefore the fatty acids have to be supplied in the medium. When such mutations are combined with a mutation which also prevents degradation of fatty acids the supplemental fatty acid will be the only one incorporated into the cell membranes.

Between 1969 and 1972 a number of workers using mutants found that the temperature-dependence of several enzyme and/or associated transport activities varied with the type of fatty acid supplied in the growth medium. Specifically, the temperature at which discontinuities in Arrhenius plots occurred depended on the nature of the supplemental fatty acid. Chemical analysis showed significant incorporation of the fatty acid into the cell membrane. Schairer and Overath (1969) found that the uptake of thiomethyl-β-galactoside (TMG) by *E. coli* cells containing *trans*-unsaturated fatty acids showed a discontinuity around 30°C. Although induction of the *lac* operon in the absence of unsaturated fatty acids resulted in synthesis of β-galactosidase and M-protein, a functional β-galactosidase transport system required the simultaneous presence of unsaturated fatty acids. Continuing their earlier studies, Overath et al. (1970), using the Langmuir monolayer technique, correlated Arrhenius plots of TMG efflux, respiration and growth of *E. coli* on different fatty acid supplements, with the temperature at which phosphatidylethanolamines containing these same fatty acids showed a transition from a liquid-expanded to liquid-condensed force-area curve (i.e. from a collapse area of $\sim$50 Å/molecule to $\sim$30 Å/molecule). *E. coli* membranes contain about 69% phosphatidylethanolamine (Silbert et al., 1974). While a good correspondence was found for TMG efflux and the temperature of the phase transition as measured with the monolayer technique, the other activities showed less correspondence or even the absence of a definite discontinuity in the Arrhenius plot. The accurate use of monolayers requires knowledge of the actual area per molecule for the phospholipids in the membrane. The correspondence of the phosphatidylethanolamine phase transition temperature and TMG transport might suggest that phosphatidylethanolamine influences TMG transport and has an area of around 50 $Å^2$/molecule in the membrane. The presumably cyclopropane (19:0) fatty acid chain phosphatidylethanolamine melts some 5°C lower than (*cis* 18:1) phosphatidylethanolamine using this technique. Wilson et al. (1970) measured the uptake of *O*-nitrophenyl-β-galactoside and *p*-nitrophenyl-β-glucoside in cells grown on oleic (18:1) and linoleic (18:2) acid. They obtained discontinuities in Arrhenius plots at 13° and 7°C respectively, the lower temperature being found for the more unsaturated acid and therefore consistent with a transition occurring at a lower temperature. Further studies on the independent β-glucoside transport systems showed that the discontinuities in the Arrhenius plot varied with the supplemental fatty acid for a wide range of different fatty acids (Wilson and Fox, 1971). Furthermore, in cells grown consecutively on different fatty acids the temperature profile corresponded closely to that of the supplemental fatty acid present when the transport system was induced, except when induction occurred in the presence of the less fluid fatty acid. Under these conditions, the temperature profile corresponded to that of the more fluid fatty acid, in agreement with other data suggesting a preferential association of membrane enzymes with the more fluid regions of the membrane, either at the time of synthesis or subsequently by rapid lateral diffusion. Work by Overath et al. (1971) indicated significant changes in the temperature profiles of β-galactoside hydrolysis after shifting to a different medium, which suggested considerable lateral motion and randomization of the lipids. Tsukagoshi and Fox (1973) found that this randomization was most complete when the temperature was maintained above the T_c of both fatty acids, obscuring the characteristic temperature profile of the fatty acid associated with the β-galactosidase during induction.

The general implication of these studies was that alteration of the saturation of the fatty acyl side-chains of membrane phospholipids influences membrane transport and enzyme activities. This has been substantiated by a number of subsequent studies using *E. coli*, *Mycoplasma* and related microorganisms, yeast mutants, and chilling-sensitive plants and poikilotherms which are known to respond to temperature alterations by changing the fatty acid composition of their membranes. It also suggested a heterogeneity or non-uniform relationship between membrane phospholipids and membrane enzymes, since not all activities showed the same temperature dependence. Other work has emphasized this heterogeneity and differences in the responses of enzymes localized in the same membrane to temperature.

Mavis and Vagelos (1972) studied the effects of different fatty acid supplements in *E. coli* mutants on the activity of the two membrane-bound transferase enzymes responsible for acylating glycerol 3-phosphate, as well as the membrane-bound glycerol 3-phosphate dehydrogenase. The transferase enzymes showed a curvilinear or linear Arrhenius plot whose slope was affected by the degree of saturation of the fatty acid supplement. The slope of the glycerol 3-phosphate dehydrogenase was unaffected, however, by the saturation of the fatty acid, though the actual activity did vary. The authors suggested that these diverse effects indicated microheterogeneity of membrane lipids, especially in the environment of membrane enzymes, which could in part be influenced by the properties of the enzyme proteins themselves. Studies by Esfahani et al. (1971) have shown that the succinic dehydrogenase activity and proline transport of *E. coli* mutants is also influenced by the nature of the fatty acid present in the growth medium. Studies combining X-ray diffraction and activity measurements indicated that the discontinuities in the Arrhenius plots occurred at different temperatures and below the beginning of the phase transition of the membrane lipids. Thus, these data also suggested some heterogeneity in the distribution of lipids in the plane of the membrane. It was also found that after delipidation and reconstitution the behavior of succinic dehydrogenase activity reflected the properties of the lipids used for the reconstitution rather than the enzyme protein (Esfahani et al., 1972). Sineriz et al. (1973) found no significant differences in the effect of an oleic acid or linoleic acid supplement on Arrhenius plots of *E. coli* Ca^{2+}-ATPase activity. When the effects of Na^+ inhibition were plotted as Hill plots, however, n values of 1.60 and 2.20 were found for the oleic acid- and linoleic acid-supplemented cells, respectively. These authors suggest that for certain cases such types of Hill plots may be more sensitive indicators of the interactions of proteins with membrane lipids (see also Farias et al., 1975, for a comprehensive review of this approach).

3.3.2. *Mycoplasma and Acholeplasma*

Studies on *Acholeplasma laidlawii* (De Kruyff et al., 1973) and *Mycoplasma mycoides* (Rottem et al., 1973), microorganisms that do not possess a cell wall, have resulted in similar conclusions to those derived from bacterial studies. The membrane fatty acid composition of the cell membranes of these organisms can be varied by altering the fatty acids of the growth medium, in a similar way to the *E. coli* mutants. In addition, the cholesterol content can be varied (0–25%) by growing them in the presence or absence of cholesterol. De Kruyff et al., (1973) showed that the discontinuity in the Arrhenius plot of the Mg^{2+} ATPase activity of *Acholeplasma laidlawii* can be varied from 15°C to 18°C depending on the fatty acid supplement. Measurements of the phase transition using differential scanning calorimetry indicated that the discontinuity in the Arrhenius plot of enzyme activity corresponded with the beginning of the lipid phase transition. Cholesterol somewhat inhibited the ATPase activity and shifted the discontinuity to lower temperatures. Rottem et al., (1973) using *Mycoplasma mycoides* incorporated considerably more cholesterol into their membranes (25%), and also found an inhibition of enzyme activity. However, in this case the discontinuity was abolished.

3.3.3. Yeast
Studies utilizing mutant and wild-type forms of yeast have indicated that changes in the saturation of membrane lipids or the amount of cholesterol alters the temperature dependence of mitochondrial enzymes. Thus, enrichment of cells with oleic acid resulted in a discontinuity (t_d) in Arrhenius plots at 10.2°C for cytochrome oxidase (Ainsworth et al., 1972) and 12°C for Mg^{2+}-ATPase and kyneurine hydroxylase, an outer membrane enzyme (Janki et al., 1974). The comparable values for membranes enriched with linoleic acid were 7.4°C, 8°C and 9°C, showing that increasing unsaturation decreased the t_d value as also seen in *E. coli*. The effect of increasing the sterol (ergosterol) content was seen as a decrease of some 4–5°C in the t_d for several mitochondrial enzyme activities (Cobon and Haslam, 1973; Thompson and Parks, 1974). Wild-type yeast cells grown anaerobically were found to have about one tenth of the percentage of unsaturated fatty acids found in aerobically grown cells (Watson et al., 1973). Consistent with this the single discontinuity found in Arrhenius plots of activity occurred 6°C higher for the ATPase and 11°C higher for the succinic dehydrogenase activity, in anaerobically grown cells. In addition, the activities of both enzymes were inhibited throughout the temperature range studied in anaerobically grown cells.

3.3.4. Effects of temperature acclimation in plants and poikilotherms
Poikilotherms are known to alter the saturation of their membrane lipids according to the ambient temperature, with increased unsaturation occurring at lower temperatures (see review by Hazel and Prosser, 1974, pp. 663–664). Poikilotherms and chilling-resistant plants generally contain more unsaturated fatty acids than homeotherms and chilling-sensitive plants. This is reflected in differences in Arrhenius plots which generally show no breaks for poikilotherms and chilling-resistant plants. Thus, studies of mitochondrial succinoxidase activity in mitochondria from a poikilotherm (rainbow trout) showed a linear Arrhenius plot from 4°–30°C, but a homeotherm (rat) showed a discontinuity at approx. 23°C (Lyons and Raison, 1970). Subsequent studies (Raison et al., 1971) showed that a nitroxide-labelled stearic acid spin label probe (see Fig. 6B) also showed discontinuities when its correlation time was plotted in the form of an Arrhenius plot. This occurred at 23°C for rat liver mitochondria and at the same temperature in sonicated liposomes of phospholipids extracted from the mitochondria. Mitochondria and liposomes from a chilling-sensitive plant (sweet potato), showed discontinuities at 12°C and the chilling-resistant potato plant showed no discontinuities. It should be recalled that Blazyk and Steim (1972), using differential scanning calorimetry, found a broad endothermic phase transition between about −20°C to 20°C for rat liver mitochondria, showing a good correspondence for the t_d of succinoxidase activity and the upper limit of the broad phase transition (see also section 3.5.1 for a fuller discussion of Arrhenius plots and phase transitions in mitochondria). Recently, Raison and McMurchie (1974) detected a lower discontinuity at 8°C in addition to the previously described one at 24°C in rat liver mitochondria

for both a spin-labelled probe and succinoxidase activity. The consequences of the effects of temperature on the activity of mitochondrial enzymes in both animals and plants has been reviewed by Lyons (1972).

3.3.5. Diet-induced alterations

A number of studies have demonstrated the influence of dietary changes on selected enzyme activities and, in many cases, these have been correlated with changes in membrane lipids. Replacement of cholesterol by desmosterol in vitro leads to a 2-fold increase in the (Na + K) ATPase activity of rat erythrocytes (Fiehn and Seiler, 1975). It seems, as the authors suggest, that the presence of the double bond in the desmosterol side-chain is sufficient to alter the packing properties of cholesterol, and possibly lead to an increased fluidity. Rats fed an essential fatty acid deficient diet, show a decreased bile-salt-independent flow of bile (Sarfeh et al., 1974). Since it has been suggested that the bile-salt-independent flow is dependent on (Na + K) ATPase activity in the canalicular membrane, the authors suggested that the essential fatty acid-deficient rats might have a reduced (Na + K) ATPase activity due to membrane lipid alterations. Other authors (Sun and Sun, 1974) have found a modest (10–20%) increase in synaptosomal (Na + K) ATPase activity after feeding mice a fatty acid-deficient diet for six months. This was correlated with increased levels of some unsaturated fatty acids and decreased levels in others. Apart from the species difference this might indicate a different response of the brain and liver to dietary deprivation of essential fatty acids (Farias et al., 1975). Lipid deprivation in rats led, after four months, to decreased activity of brain monoamine oxidase and 5' mononucleotidase (Bernsohn and Spitz, 1974). There have also been a number of studies in which the cooperativity of a number of enzymes from different organs of the rat, including the (Na + K) ATPase and acetylcholinesterase, were affected by changes in the lipid content of the diet. This suggests a dependence of the cooperative behavior of some membrane enzymes on lipid fluidity, and has recently been comprehensively reviewed (Farias et al., 1975).

3.4. Lipid phase separation and enzyme activity

Suggestions as to the mechanism by which lipid fluidity affects enzyme activity usually favor the possibility of greater freedom for protein conformational changes and/or translational motion that might be expected in a fluid as opposed to a solid lipid environment. Recently, a more precise suggestion, at least for protein-mediated transport processes, based on the greater compressibility of lipids during a broad phase transition or phase separation has been advanced (Linden et al., 1973). Consistent with such broad separations at least two discontinuities were detected in Arrhenius plots of β-glucoside and β-galactoside transport in *E. coli* mutants. Furthermore, these correlated directly with the limits of the phase separation as detected by a spin label, 2,2,6,6-tetramethylpiperidine-1-oxy, TEMPO (see Fig. 6A), which preferentially distributes into a fluid rather than a solid bilayer (Shimshick and

McConnell, 1973a). The transitions found for p-nitrophenyl-β-glucoside transport and TEMPO partitioning (f) are shown in Figs. 11 and 12 and summarized in Table IV for both *E. coli* cells and liposomes of phospholipids extracted from their inner membranes.

It is clear from Table IV that the discontinuities in the Arrhenius plot of glucoside transport (see Fig. 11) approximate closely to those of the lateral phase separation as determined by TEMPO partitioning (see Fig. 12). The t_h values for TEMPO partitioning in liposomes also agree quite well. There is, however, a marked decrease for t_l in liposomes for the oleic and linoleic cells. As Linden et al. (1973) suggest, this might be due to preferential binding and immobilization of the more fluid lipids by membrane enzymes. Linden et al. (1973) interpreted the complex discontinuities in the Arrhenius plots of the oleic acid supplemented cells at t_h as due to a large increase in the lateral compressibility due to the coexistence of solid and fluid phases which would presumably increase protein penetration into the membrane, thereby facilitating transport or enzymic processes. These effects appeared to occur with a zero or even negative Ea (see Fig. 11). Some evidence for increased lateral compressibility promoting the penetration of dipalmitoyl PC monolayers by a hydrophobic protein ([1-^{14}C]acetyl-β casein A) has recently been reported (Phillips et al., 1975).

Consistent, in part, with the above explanation is a recent study showing that the binding of cytochrome b_5 to phosphatidylcholine liposomes results in a unique effect of temperature on the b_5 tryptophan fluorescence. Maximum fluorescence is found at the mid-point of the transition of either dipalmitoylphosphatidylcholine or DMPC liposomes (Dufourcq et al., 1975). This effect could be mediated via protein penetration of the liposome membrane bilayer resulting in alterations of protein conformation. This penetration would therefore be maximal at the mid-point of the transition. A similar explanation can explain the recent observation that phospholipase A_2 shows maximum hydrolysis of liposomes of saturated lecithins at the mid-point of their phase transitions (Op den Kamp et al., 1975). Papahadjopoulos et al. (1973b) found a maximum for $^{22}Na^+$ exchange diffusion at the mid-point of the phase transition of dipalmitoylphosphatidylcholine liposomes. All these effects therefore show a maximum at the mid-point of the phase transition when the ratio of solid to fluid phase is likely to be essentially equal, resulting in a maximum amount of solid to fluid interface.

Recently Morrisett et al. (1975) instead of using TEMPO have studied the phase transitions of elaidic acid-enriched *E. coli* (66–70% of membrane lipids) using a stearic acid labelled at different positions with a N-oxyloxazolidine ring (Fig. 6B). Using the probe labelled in the fifth, twelfth and sixteenth carbon they have found two separate phase separations from 25.5°C–35.5°C and from 9.5°C–21°C. Similar phase separations were found in the extracted lipids, plus a third covering the 3.0°C–9.0°C range. Based both on its temperature and magnitude, the 25.5°C–35.5°C separation was assigned to dielaidoylphosphatidylethanolamine which constitutes the major phospholipid. NADH oxidase showed two discontinuities in an Arrhenius plot at 32°C and 27°C and D-lactate

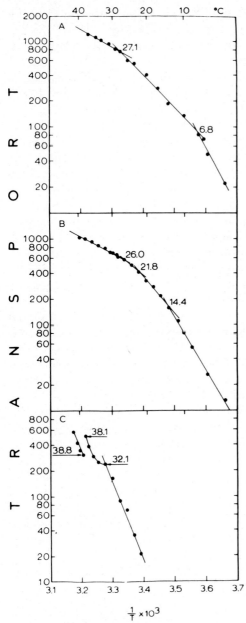

Fig. 11. Arrhenius plots of the transport of p-nitrophenyl-β-glucoside by unsaturated fatty acid requiring *E. coli* mutants. Grown at 37°C and growth medium supplemented with: (A) linoleic acid (C18: *cis* 2), (B) oleic acid (C18: *cis* 1), and (C) elaidic acid (C18: *trans* 1). Units of transport are nmole/20 min per $2 \cdot 10^9$ cells (reproduced with permission from Linden et al., 1973).

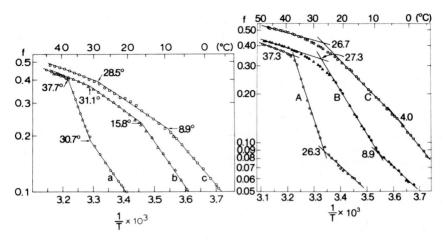

Fig. 12. Left figure: Arrhenius plot of TEMPO partition parameter for membranes of *E. coli* cells used in Fig. 11 with varying fatty acid supplements in growth medium; (a) elaidic acid (b) oleic acid and (c) linoleic acid. Right figure: Arrhenius plot of TEMPO partition parameter for liposomes of phospholipids extracted from the membranes of same *E. coli* mutants as used in Figs. 11 and 12 (left figure). Fatty acid supplement (A) elaidic acid, (B) oleic acid, (C) linoleic acid (reproduced with permission from Linden et al., 1973).

TABLE IV
CORRELATION OF DISCONTINUITY TEMPERATURES OF GLUCOSIDE TRANSPORT AND TEMPO PARTITIONING IN *E. coli* AND PHOSPHOLIPID LIPOSOMES OF *E. coli* MEMBRANE PHOSPHOLIPIDS[a]

	Transport (p-nitrophenyl-β-glucoside)		TEMPO partitioning (f) in cells		TEMPO partitioning (f) in phospholipid liposomes	
E. coli grown on	t_h	t_l	t_h	t_l	t_h	t_l
Elaidic acid	38.7	32.1	37.7	30.7	37.3	26.3
Oleic acid	26.0, 21.8	14.4	31	15.8	27.3	8.9
Linoleic acid	27.1	6.8	28.5	8.9	26.7	4.0

[a]Data taken from Linden et al. (1973).

oxidase at 31°C and 36°C. Further discontinuities may not have been seen due to the difficulty of detecting activities at the lower temperatures. By comparison, Linden et al. (1973) using TEMPO only detected the higher temperature phase separation, which also correlated with glucoside and galactoside transport (see Table IV). On the basis of these results, Morrisett et al. (1975) suggest that there is a high degree of lipid and protein heterogeneity even in the relatively simple bacterial membrane. Thus, the lack of correlation between enzyme activities and phase transitions found in earlier *E. coli* studies (Overath et al., 1970; Esfahani et al., 1971) did indeed indicate heterogeneity, but later work using more refined techniques has identified further phase separations to which these activities do correspond. The data are, however, still circumstantial and direct analysis of responses of enzymes to particular phase separations by use of labelled enzymes in situ, has been attempted infrequently (Inesi et al., 1973). The demonstration of multiple phase separations begins to complicate the exact relation between membrane fluidity and enzyme activity, and more direct proof of such associations is now required. The correlation of phase separations and membrane activities has been reviewed recently by Fox (1975).

A further suggestion as to the mechanism responsible for the appearance of discontinuities in Arrhenius plots of membrane enzyme activity has been put forward by Lee et al. (1974). They found that a discontinuity occurred in dioleoyl-phosphatidylcholine substituted Ca^{2+} ATPase at about 26°C, close to a break in the partitioning of TEMPO into dioleoyl-phosphatidylcholine liposomes and well above the T_c at −22°C found by differential scanning calorimetry. They proposed the existence of short-lived above average density "clusters" of dioleoyl-phosphatidylcholine molecules to explain these results.

3.5. Correlation of Arrhenius plots of enzyme activity and lipid fluidity in natural membranes

It has already been pointed out that prior to, or contemporary with, the illuminating work on the *E. coli* mutants, several workers had published Arrhenius plots of membrane enzymes and non-membrane enzymes, which showed discontinuities. The interpretation of such data at the time was, however, unclear. Publication of the *E. coli* work was followed quickly by a rapid increase in the number of studies, as the significance of such behavior for membrane-bound enzymes became evident. Prior to the paper of Linden et al. (1973), most of these studies showed a single discontinuity in their Arrhenius plots (but see Gruener and Avi Dor, 1966; Smith, 1967; Bowler and Duncan, 1968). Following this work, however, many workers extended the temperature range of their studies and indeed found two discontinuities, consistent with the occurrence of a broad lateral phase separation rather than sharp phase transitions (Fox, 1975). This section will attempt to survey the many studies of enzymes in different membranes for which such discontinuities have been found, and which have, in some cases, been correlated with independent estimations of membrane lipid fluidity using spin-labelled probes or other techniques. The presentation of this material can conveniently be organized according to the types of membranes studied. These mainly include enzymes from mitochondrial (both outer and inner membranes), microsomal (endoplasmic reticulum) and sarcoplasmic reticulum membranes and plasma membranes.

3.5.1. Mitochondria
Table V summarizes the discontinuity temperatures for a variety of mitochondrial enzymes. As can be seen from the selected data shown in this table, there

is a discontinuity found at 17°C–27°C for most activities, and a lower one also found in a few cases at around 12°C. The discontinuities for cytochrome oxidase at 30°C and 20°C were only found in one study (Wilschut and Scherphof, 1974), although a discontinuity at 10°C was reported for oleic acid-enriched yeast cells (Ainsworth et al., 1972). Lee and Gear (1974) did not find discontinuities, but they measured their cytochrome oxidase activity in the presence of the detergent Lubrol, which is known to disrupt membranes and convert discontinuous Arrhenius plots to linear ones (see below). The discontinuities for enzyme activity in rat liver mitochondria can be compared with a single spin label transition at 23°C (Raison et al., 1971), and broad transition from about −20°C to 20°C seen by differential scanning calorimetry (Blazyck and Steim, 1972). On this basis, it is tempting to assign the discontinuities seen at 17–27°C to the transition detected as the upper limit of the very broad phase separation by differential scanning calorimetry and the single transition in spin label studies. The low discontinuities seen at around 12°C might represent local solidification around the enzymes, possibly due to the ability of proteins to partially immobilize adjacent lipids and this is only seen for a limited number of activities. Thus, the approx. 20°C transition is a general response, presumably representing the completely fluid membrane. Lee and Gear (1974) ascribe the discontinuity that they find at 27°C for a number of activities, to a general membrane conformation change. It is of interest that whereas the early work of Kemp et al. (1969) identified the lower discontinuity for succinoxidase activity in rat liver mitochondria seen by Lee and Gear (1974), Lenaz et al. (1972) in later work identified only the upper discontinuity. No significant differences are seen between the state 3 (+ADP) and state 4 (−ADP) succinoxidase respiration. Both the succinoxidase activity studies of Lyons and Raison (1970) and the spin label studies of Raison et al. (1971) identify a somewhat lower single transition at 23°C. It is also possible that these different groups of temperatures may represent transitions in the two monolayers of the membrane (see section 3.5.3).

It has been mentioned already that Lee and Gear (1974) measured their cytochrome oxidase activity in the presence of a detergent. Since detergents disrupt protein-lipid associations it is not unexpected that it produces a linear Arrhenius plot. Bertoli et al. (1973) found a single discontinuity at approx. 17°C for the ATPase activity of either fresh or frozen beef heart mitochondria. Treatment of frozen beef heart mitochondria with the detergent Triton X-100 resulted in a considerable straightening of the plot. The linear plot for ATPase activity found by Lee and Gear (1974) in sonicated rat liver mitochondria is of interest. They suggested that the formation of "inside-out" small, inner membrane vesicles by ultrasonication might have disrupted the membrane and/or ATPase sufficiently to abolish temperature-induced conformational changes. On the other hand, Kemp et al. (1969) found a discontinuity for rat liver mitochondrial ATPase at 19°C in the presence of uncoupler. After sonication, uncoupler independent ATPase showed two discontinuities at 9°C and 18°C. Thus, it is important to establish to what degree preparative

TABLE V
DISCONTINUITY TEMPERATURES AND ACTIVATION ENERGIES FOR SOME MITOCHONDRIAL ENZYMES

Enzyme	Discontinuity temperature(s)		Activation energies (kcal mole^{-1})			References
	T_h or single discontinuity	T_l	$>T_h$	$(T_l - T_h)$	$<T_l$	
Cytochrome oxidase	—			9.3		Lenaz et al., 1972
Cytochrome oxidase	—			13.2		Lee and Gear, 1974
Cytochrome oxidase	30	20	7	9	12	Wilschut and Scherphof, 1974
Succinate cytochrome c reductase	20		15.7		32.8	Lenaz et al., 1972
	27.5		4.1		15.2	Lee and Gear, 1974
	23	12	12	17	23	Wilschut and Scherphof, 1974
Succinoxidase	17.9 (state 3)		9.2		24.4	Kemp et al., 1969
	16.8 (state 4)		8.9		18.8	Kemp et al., 1969
	27.0		9.2		17.0	Lenaz et al., 1972
	27.5	17.5 (state 3)	2.5	10.6	26.8	Lee and Gear, 1974
	30.0	16.0 (state 4)	3.9	9.9	31.4	Lee and Gear, 1974
ATPase (F_1)	17.7 (frozen)		9.8		27.1	Bertoli et al., 1973
	16.8 (fresh)		12.8		23.4	Bertoli et al., 1973

						Reference
	23 (sonicated)		8		29	Quoted in Raison, 1973
	19 (+uncoupler)	—	16.8			Lee and Gear, 1974
	18 (sonicated)			9		Kemp et al., 1969
	18.6–24.6					Kemp et al., 1969
	14.2 (sonicated)					Watson et al., 1973
						Bruni et al., 1975
NADH malate dehydrogenase		—	13.0			Lenaz et al., 1972
			25.3			Lee and Gear, 1974
Succinoxidase spin label	23	2.5			23.4	Lyons and Raison, 1970
Lactate oxidase spin label	23	3.2			7.2	Raison et al., 1971
	16					Eletr et al., 1974
	21					Eletr et al., 1974
Respiration-dependent Ca^{2+} uptake	26.5	12.5	4.5	15.7	25.3	Lee and Gear (1974)
Valinomycin-induced K^+ uptake	28.5	12.0	7.3	12.6	24.6	Lee and Gear (1974)

Activation energies are only given where they are quoted in the paper. The type of mitochondria used in the studies in this table were as follows: beef heart mitochondria, Lenaz et al. (1972), Bertoli et al. (1973), Bruni et al. (1975); rat liver mitochondria, Kemp et al. (1969), Lyons and Raison (1970), Raison et al. (1971), Wilschut and Scherphof (1974), Lee and Gear (1974); and yeast mitochondria, Eletr et al. (1974), Watson et al. (1973).

procedures such as ultrasonication, as well as different detergent treatments, result in significant alteration of protein-lipid associations in the membrane for each enzyme studied. Different results seem to greatly depend on small variations in such procedures.

Lee and Gear (1974) suggested that a lower transition at approx. 17°C which they found for some activities was due to a change in the adenine nucleotide translocase responsible for transporting ADP or ATP across the inner mitochondrial membrane, and which is the rate-limiting step for ADP or ATP requiring reactions. The discontinuity at 16°C for state 4 succinoxidase activity was suggested as being due to the rate of succinate oxidation in state 4 being dependent on slow hydrolysis of endogenous ATP by mitochondrial ATPase. A t_d of approx. 17°C had been found by Bertoli et al. (1973) for frozen beef heart mitochondria. It was speculated that an even lower discontinuity at approx. 12°C for ion uptake (see Table V) was due to some undefined ion translocase in the inner mitochondrial membrane. It is of interest that outer mitochondrial membrane-bound enzymes such as monoamine oxidase and rotenone-insensitive NADH cytochrome c reductase showed no discontinuity (Lee and Gear, 1974), as well as soluble Kreb's cycle enzymes such as malate dehydrogenase located in the inner mitochondrial matrix space (Lenaz et al., 1972; Lee and Gear, 1974).

3.5.2. Sarcoplasmic reticulum and microsomal enzymes
Inesi et al. (1973) studied the temperature dependence of Ca^{2+} accumulation by rabbit sarcoplasmic reticulum membranes and the activity of the Ca^{2+} activated ATPase, and found a sharp discontinuity for both at 20°C. Spin-labelled stearic acid probes of the type shown in Fig. 6B also showed a discontinuity at 20°C. Spin-labelling the ATPase protein in sarcoplasmic reticulum is relatively easy since it is ≥50% of the protein present, and such a label also showed a change in its order parameter at approx. 20°C. A higher change at 42°C was ascribed to protein denaturation. As shown in Table III the high temperature range of the Arrhenius plot of activity had a lower ΔS^* value than the lower temperature range. As discussed previously, this could be due to there being a smaller difference between the order of the reactants and activated complex in the higher temperature range when the membrane is fluid, which is intuitively what one might expect if the activated state requires a certain degree of molecular motion. These results have been confirmed recently by Madeira and Antunes-Maderia (1975) for the Ca^{2+} ATPase. They found a t_d for Ca^{2+} ATPase and perylene fluorescence polarization at 17–18°C. Phospholipase A_2 digestion of the membrane was also inhibited below 20°C. It is of interest that no transition for perylene fluorescence or phospholipase digestion was seen for liposomes made from lipids extracted from sarcoplasmic reticulum. Ca^{2+} ATPase in lobster sarcoplasmic reticulum showed a lower t_d at 11.5°C which was correlated with increased unsaturation of PC.

Membrane fluidity has also been reported to affect the activity of several microsomal enzymes. Eletr et al. (1973) have reported that Arrhenius plots of

UDPglucuronyltransferase activity and glucose 6-phosphatase activity from guinea pig liver show a discontinuity at 19°C, and glucuronyltransferase also had a second discontinuity at 32°C. This was correlated with the behavior of lipophilic nitroxide radicals which also showed discontinuities at the same temperature. Treatment with phospholipase A abolished the 19°C break for both enzymes and the breaks at both temperatures for the nitroxide radicals. Sonicated microsomes showed only the 19°C break, which was also seen in sonicated liposomes prepared from extracted lipids, implying that the 19°C transition is a general property of the lipid component of the membrane. The transition at 32°C could then be due to a transition of the glucuronyltransferase enzyme protein itself, or a local lipid environment around the enzyme, as has been found for cytochrome P450 reductase from rabbit liver (Stier and Sackmann, 1973) which also happens to show a transition at 32°C. Later work has shown (Zakim and Vessey, 1975) that UDPglucose inhibition of UDPglucuronyltransferase activity is only seen at temperatures below 16°C. Conversely the activation of UDPglucuronyltransferase activity by UDP-N-Acetylglucosamine is only found at 16°C and above. The effect at 16°C is likely to be a reflection of the lipid transition previously reported at 19°C. In this case, the membrane lipid fluidity alters not only the enzyme activity but also its regulation by other ligands.

3.5.3. Plasma membrane enzymes

Because of the critical role of the plasma membrane in the control of cellular activity and its likely involvement in many disease states (see Wallach, 1972 and section 4) the possible control of plasma membrane enzyme activity by membrane fluidity is of particular interest. Since the plasma membrane has several critical transport properties there have also been studies dealing with the influence of membrane fluidity on this parameter. The reader is referred to sections in recent reviews on this latter subject by Plagemann and Richey (1974, pp. 288–291) and Berlin and Oliver (1975, pp. 311–313) as well as the large amount of work on bacterial membrane transport as described in section 3.3.1.

The most widely studied mammalian plasma membrane enzyme with regard to membrane fluidity has been the (Na + K) ATPase which is responsible for maintaining the Na^+ and K^+ concentration gradients seen in virtually all cells. As mentioned above, several early studies had observed discontinuities in Arrhenius plots of this enzyme from a variety of species (Gruener and Avi Dor, 1966; Smith, 1967; Bowler and Duncan, 1968; Charnock et al., 1971a,b). The first indirect evidence that such discontinuities reflected changes in the fatty acid saturation and therefore the fluidity of membrane lipids was that of Smith (1967), who showed that the (Na + K) ATPase activity of goldfish intestinal mucosa acclimated at 8°C had a t_d at 12°C, whereas goldfish acclimated at 30°C had a t_d at 21°C. It is well-known that poikilotherms, including the goldfish, respond to colder acclimation temperatures by increasing the unsaturation of the fatty acids of their membrane phospholipids (Hazel and Prosser, 1974). Kemp and Smith (1970) later found that increasing the adaptation temperature increased the fatty acid saturation, especially (18:0) phosphatidylethanolamine, of goldfish intestinal lipids.

The known phospholipid requirement of (Na+K) ATPase, as well as its functional importance, make this enzyme an excellent choice for directly comparing alterations in phospholipid fluidity with changes in enzyme activity and/or kinetics. As shown by the *E. coli* work, enzymes respond to such alterations by shifts in the discontinuity temperatures of their Arrhenius plots which therefore is a useful diagnostic procedure for such changes. Since bacteria do not contain the (Na+K) ATPase this enzyme was not studied in the *E. coli* work. Several papers published in 1972 showed that the discontinuous Arrhenius plots already reported for the (Na+K) ATPase could be directly altered by changes in the lipid composition of the membrane preparation. Taniguchi and Iida (1972) showed that the discontinuity for the (Na+K) ATPase activity of ox brain microsomes at 20°C was abolished by phospholipase A treatment, resulting in a linear Arrhenius plot. Most significantly, addition of phosphatidylserine and phosphatidylinositol to such a preparation restored the original discontinuous plot. These authors suggested that the t_d at 20°C is a reflection of the phase transition of the phospholipids essential for (Na+K) ATPase activity, although these experiments do not relate the effect directly to the phase transitions of the lipids. Priestland and Whittam (1972) found that a rabbit brain homogenate gave, in this case, a linear Arrhenius plot of (Na+K) ATPase activity, but this activity was completely inhibited by deoxycholate treatment, which is known to remove lipids. This preparation was reactivated by phosphatidylserine and showed a nonlinear Arrhenius plot with a discontinuity at 15°, which the authors suggested was a reflection of the physical state of the phosphatidylserine, although again, this was not directly established. A study by Kimelberg and Papahadjopoulos (1972) also used deoxycholate to delipidize a microsomal preparation from rabbit kidney cortex. This resulted in a preparation that was markedly stimulated by phosphatidylserine or phosphatidylglycerol. Phosphatidylglycerol obtained as the direct hydrolysis product of egg phosphatidylcholine by phospholipase D action in the presence of excess glycerol (Papahadjopoulos et al., 1973b) showed a linear Arrhenius plot. Dipalmitoyl-phosphatidylglycerol, however, gave a discontinuity at 32°C, which was close to the phase transition of dipalmitoylphosphatidylcholine at approx. 40°C. Furthermore, it was found that the presence of cholesterol inhibited reactivation of dipalmitoyl-phosphatidylglycerol above 32°C. Although all these results were certainly consistent with a fluidity effect, a direct correspondence between the t_d and the phase transition of the lipids was lacking. Further studies on membrane fluidity in the reconstituted (Na+K) ATPase system and other enzymes will be dealt with more fully in section 3.6.

Grisham and Barnett (1972), using an oxyloxazolidine ring-labelled steroid probe, found that the lipids of a highly purified (Na+K) ATPase were more ordered than the less pure microsomal preparation. In contrast, a probe of the type shown in Fig. 6B with the oxyloxazolidine ring on the twelfth carbon from the α carbon, showed the purified preparation to be more fluid. Thus, the interior of the bilayer adjacent to the ATPase appeared to be more fluid. Later work (Grisham and Barnett, 1973) showed that using stearic acid which was labelled at different carbons, there was a marked increase in fluidity in the purified preparation at carbon 5 to 6 of the spin-labelled stearic acid. A crude plasma membrane fraction was equally ordered up to carbon 5 but was markedly less fluid from 5 to 8. This preparation then showed the same fluidity as the purified preparation at carbon 12. Assuming equal orientation of the label in the two preparations these results indicate that the lipid environment of the (Na + K) ATPase was more fluid than average. This is to be contrasted with work on cytochrome c oxidase which shows increasing immobilization of spin-labelled probes in the lipid adjacent to the enzyme (Jost et al., 1973a,b). On the basis of their activities and binding of ligands, both these enzymes are thought to span the entire membrane. Grisham and Barnett (1973) also showed an exact relationship between the t_d of an Arrhenius plot of activity and a t_d for a spin-labelled probe of the general type shown in Fig. 6B (stearic acid with n = 6) in the purified (Na + K) ATPase. Liposomes made from the lipids extracted from the (Na + K) ATPase preparation also showed a discontinuity at the same temperature (see Table VI and Fig. 13). These results constitute evidence for a direct relationship between lipid fluidity (as directly measured by the spin-labelled probe) and enzyme activity. It suggests either that it is relatively easy to isolate the (Na + K) ATPase with only its immediate microenvironment or else it responds to a greater degree than other enzymes to bulk membrane lipid fluidity. Indeed, it can be seen from Table VI that the t_d for (Na + K) ATPase activity is remarkably constant at about 20°C for all the different mammalian species and preparations measured. As might be expected, there is a difference when other poikilothermic vertebrate classes are examined. Even in these cases, however, the goldfish gave the characteristic discontinuity at 21°C when it was acclimated at 30°C rather than 8°C (Smith, 1967). Also, when the frog and bovine enzymes were delipidated and reconstituted with the same crude mammalian phospholipid mixture (containing phosphatidylcholine, phosphatidylethanolamine and phosphatidylserine), they showed an identical t_d value at 17°C (Tanaka and Teruya, 1973).

The K^+-dependent phosphatase reaction has been reported to have a linear Arrhenius plot, which is also seen for the 50% of this activity retained after lipid depletion (Walker and Wheeler, 1975a). These data indicate that the K^+-dependent phosphatase, which is considered to represent the K^+-dependent hydrolysis of the E_2P complex (e.g. see Goldman and Albers, 1973) may not respond to changes in phospholipid fluidity, and that some other partial reaction of the (Na + K) ATPase is involved. Goldman and Albers (1975), however, have reported a curvilinear plot possibly indicating a discontinuity at 15°C–18°C, and Goldman and Albers (1973) concluded that both the Na^+-dependent phosphorylation to form E_1P and the K^+-dependent dephosphorylation of E_2P were dependent on phosphatidylserine.

A study using spin-labelled probes and X-ray diffraction has shown that

TABLE VI
DISCONTINUITY TEMPERATURES (t_d) FROM ARRHENIUS PLOTS OF ENZYME ACTIVITY AND MEMBRANE FLUIDITY OF PLASMA MEMBRANES

Enzyme activity	Species and tissue	t_d (Activity) °C	t_d (Fluidity) °C	References
(Na + K) ATPase	Lamb kidney outer medulla	20°	20°	Grisham and Barnett, 1973
	Rabbit kidney	20°	—	Charnock et al., 1971a
		20°	—	Walker and Wheeler, 1975a
	Rabbit kidney cortex	17°	—	Charnock et al., 1975
	Sheep kidney cortex and outer medulla	22°	18–22°	Charnock and Bashford, 1975
	Rat brain	20°; 6°	—	Gruener and Avi-Dor, 1966
	Rat brain	20°; 8°	—	Bowler and Duncan, 1968
	Rabbit skeletal muscle	20° (no t_d at Mg/ATP ratios >1)	—	Boldyrev et al., 1974

	Goldfish intestinal mucosa	21° or 12° (dependent on acclimation temp.)	—	Smith, 1967
	Cultured mouse SV40T3 cells	24°	—	Kimelberg and Mayhew, 1975
	Cultured mouse LM cells	37°; 31°; 23°; 15°	37°; 31°; 21°; 15°	Wisnieski et al., 1975
	Bovine brain	17°	—	Tanaka and Teruya, 1973
	Frog kidney	10°	—	Tanaka and Teruya, 1973
	Ox brain	20°	—	Taniguchi and Iida, 1972
Adenyl cyclase	Rat liver	32°	—	Kreiner et al., 1973
	Mouse Ehrlich cells	24°	—	Bar, 1974
	Rat brain	24°	—	Bar, 1974
Acetylcholinesterase	Rat erythrocytes	20°	—	Bloj et al., 1974

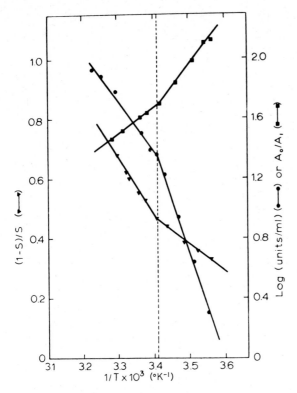

Fig. 13. Arrhenius plot comparing (Na + K) ATPase activity of purified lamb kidney outer medulla enzyme with behavior of spin-labelled probes in enzyme preparation and extracted lipids. ●—●, (Na + K) ATPase activity in units/ml; ▼—▼, order parameter S in form (1-S)/S for spin-labelled fatty acid of general type shown in Fig. 6B with m = 10 and n = 4; ■—■, ratio (A°/A) of intensities of middle and low-field lines for label in liposomes made from extracted lipids (reproduced with permission from Grisham and Barnett, 1973).

phospholipase A treatment results in inhibition of (Na + K) ATPase activity and leads to alterations in both bilayer structure and protein conformation (Simpkins and Hokin, 1973). Charnock and Bashford (1975) have recently shown that both the activity and fluorescence polarization of an (Na + K) ATPase preparation from sheep kidney cortex show a discontinuity at 22°C. A similar discontinuity was found for right angle light scattering (18°C–22°C), which suggests that the presence of fluorescent probes did not result in significant perturbations. It was of interest that light scattering by liposomes made from lipids extracted from the (Na + K) ATPase preparation also showed a discontinuity at 21°C.

Cholesterol is known to decrease the fluidity of membrane lipids and has been reported to inhibit reactivation of lipid-depleted (Na + K) ATPase by added lipid (Kimelberg and Papahadjopoulos, 1972, 1974; Roelofsen and Van Deenen, 1973). It is therefore of interest that progressive replacement of rat red blood cell cholesterol by its precursor desmosterol by inhibition of desmosterol reductase, results in an up to 2-fold increase in (Na + K) ATPase activity (Fiehn and Seiler, 1975). Increased cholesterol content of guinea pig red blood cells is associated with a 50% decrease in ouabain-sensitive Na^+ efflux, but no decrease in ^3H ouabain binding (Kroes and Ostwald, 1971). By contrast, in human red blood cells, depletion of cholesterol leads to inhibition of ouabain-sensitive K^+ uptake (Posnansky et al., 1973).

Farias et al. (1975) have argued that for some enzymes, including the (Na + K) ATPase, analysis of the Hill coefficients for the action of activators or inhibitors is a more sensitive measure of changes in membrane lipid fluidity than Arrhenius plots. Also it enables the interaction to be measured isothermally. In these studies the membrane lipid composition is altered by feeding animals a fat-free diet alone or supplemented with hydrogenated fat or cholesterol, and is compared to animals fed an essential fatty acid sufficient diet. A fat-free diet caused a decrease in the Hill coefficient n for Na^+ and K^+ activation of (Na + K) ATPase (Farias et al., 1975). In contrast, increased cholesterol (Bloj et al., 1974), or increased saturation of membrane lipids (Farias et al., 1975), led to an increase in n.

Adenylate cyclase is known to be localized in the cell plasma membrane (Solyom and Trams, 1972; Kimelberg, 1976b) and has important regulatory properties. Kreiner et al. (1973) found a discontinuity at 32°C in the Arrhenius plot of hormone-stimulated adenylate cyclase activity from rat liver, but obtained a linear curve for the basal activity. In this case the E_a was higher in the temperature range above the discontinuity. In contrast, Bar (1974) reported that the basal adenylate cyclase activity of rat brain cerebral cortex has a discontinuity in its Arrhenius plot at 24°C, with activation energies of 2.2 and 9 kcal/mole above and below this temperature, respectively. The fluoride-stimulated enzyme showed a completely linear Arrhenius plot. A similar discontinuity at 24°C was also seen for the Arrhenius plot of the basal and norepinephrine-stimulated adenyl cyclase activity of *Ehrlich ascites* tumor cells. Again, the Arrhenius plot of the fluoride-stimulated activity was linear (see Table VI).

Acetylcholinesterase in rat erythrocytes has also been reported to have a discontinuity at 20°C for animals fed a fatty acid sufficient diet (Bloj et al., 1974). In animals fed an essential fatty acid deficient diet, or one supplemented with saturated fatty acid, the t_d was found at 28°C. When the enzyme was solubilized with Triton X-100 the t_d moved to lower temperatures and the activation energies increased. Upon rebinding to the depleted red cell membranes, however, the original Arrhenius plot profile was restored, showing that its characteristics are determined partly by the membrane.

As can be seen from Table VI, the plasma membrane enzymes of homeothermic animals, with few exceptions, show a discontinuity in the range 17°C–24°C. A second discontinuity is sometimes seen at 6°C–8°C. Recently, an interesting paper by Wisnieski et al. (1975) has attempted to correlate multiple transitions for the (Na + K) ATPase of cultured mouse LM cells with transitions seen for partitioning of a nitroxide labelled fatty acid in purified plasma membranes of these cells. Discontinuities in this parameter at 37°C, 30°C, 21°C and 15°C showed good correspondence with the (Na + K) ATPase transitions (see Table VI), and those for the transport of α-aminoisobutyrate (37°C, 29°C, 20°C, and 16°C). The authors consider that these multiple transitions might represent at least two separate phase separations. Since it is known that there is an asymmetric distribution of phospholipids between the inner and outer monolayer of the cell membrane of the human erythrocyte (Bretscher, 1973; Verkleij et al., 1973) and possibly other plasma membranes (Emmelot and Van Hoeven, 1975), this could provide the chemical basis for two distinct lateral phase separations. Based on certain assumptions Wisnieski et al. (1975) assigned 14°C–32°C as the limits of the phase separation for the outer monolayer and 22°C–38°C for the inner monolayer. The (Na + K) ATPase and the amino acid transport appeared to respond to all the transitions, which the authors suggested reflected the fact that the proteins responsible for these reactions were likely to span the entire membrane. On this basis other authors appear to have identified only the beginning of the inner monolayer transitions. The approx. 38°C transition is not normally reported since it is near the temperature where many enzymes begin to be inhibited. This work, however, certainly indicates that mammalian cells, like microorganisms (Oldfield and Chapman, 1972), may function at temperatures within a broad phase separation. Future work will determine whether these multiple transitions are a general property of cell membranes.

In this connection it is of interest to note that Rottem (1975) has suggested that the outer monolayer of the membrane of *Mycoplasma hominis* was more fluid than the inner monolayer on the basis of spin-label experiments, which agrees with the findings of Wisnieski et al. (1975) on LM cells. However, a more fluid outer monolayer is difficult to reconcile with a higher concentration of cholesterol in this monolayer (Emmelot and Van Hoeven, 1975). Eletr et al. (1973) reported a discontinuity at 32°C for glucuronyltransferase and Stier and Sackmann (1973) also found a discontinuity at 32°C for the reduction of nitroxide-labelled stearic acid by cytochrome P-450 reductase. These are both microsomal endoplasmic reticulum enzymes from guinea pig and rabbit liver, respectively. These enzymes might thus be responding to the upper limit of the endoplasmic reticulum phase separation equivalent to the outer monolayer of the plasma membrane, assuming the transitions observed by Wisnieski et al. (1975) are of general applicability.

3.6. Enzyme activity and lipid fluidity in reconstituted systems

Faced with complex, interdependent biological systems, biochemists have traditionally attempted to simplify matters by purifying the components before analyzing their properties. The interactions of the individual components can then be examined until a reconstituted system is obtained that mimics the natural state. Since membranes are clearly complex systems this technique has naturally been applied to the problem of the lipid involvement in membrane enzyme activity, including the question of lipid fluidity. The single most important step in the use of this technique for membrane-bound enzymes is the method(s) used for removing the lipid as well as maintaining the enzyme in an active state and in its native conformation. The lipid can be removed with organic solvents, phospholipases or detergents. For further information on these methods the reader is referred to several recent reviews (Razin, 1972; Gulik-Krzywicki, 1975; Helenius and Simons, 1975). Briefly, it appears that removal of lipids by non-ionic or weakly anionic detergents like deoxycholate provides the best compromise between removing the lipids but not denaturing the enzyme. Denaturation of proteins is a problem with organic solvents and the extraction is best done at low temperatures. Phospholipases have the problem of a Ca^{2+} requirement for phospholipases C and D, the production of inhibitory products such as fatty acids and some preparations give incomplete hydrolysis of the lipids. As can be gathered from previous sections of this review, however, these latter two techniques can provide useful information when used under appropriate conditions. In this section I will consider several enzymes for which the fluidity aspects of the reactivation of the delipidated enzyme have been studied.

As mentioned previously (section 3.1), the solubilized mitochondrial enzyme β-hydroxybutyrate dehydrogenase was shown to have a specific requirement for lecithin, and a saturated dipalmitoyl-phosphatidylcholine did not activate the enzyme at 37°C (Jurtshuk et al., 1961). This question has recently been re-examined in some detail (Grover et al., 1975). The specific requirement for the phosphoryl choline head-group was confirmed, indicating that a negative and a positive charge was required. Also, a single hydrophobic chain as in stearyl-phosphorycholine or lysolecithin was necessary as a minimum requirement. Such compounds however, form micelles rather than bilayer liposomes, and result in unstable complexes which lead to inhibition upon incubation. Grover et al. (1975) concluded that formation of a bilayer-type liposome was required for fully effective activation. Their reactivation studies were done at 25°C and the complex was also assayed at 25°C. The most effective activators were the di C(9:0) to di C(14:0), and the di C(18) containing at least one double bond. These are the short chain saturated fatty acids and unsaturated fatty acids which are fluid below 25°C. Phosphatidylcholine molecules having acyl chains shorter than C(9:0) do not form bilayer liposomes, whereas those having saturated chains of di C(16:0) and di C(18:0) are solid until temperatures of approx. 40°C and approx. 60°C, respectively, are reached. This data indicates that, as discussed previously, a fluid phospholipid bilayer, plus in this case a specific requirement for the lecithin head group, is most effective in restoring enzyme activity. It was also found that the concentration of L α di C(14:0) lecithin, which has a T_m of 23°C (see Table I), required for 50% activation at 25°C was 2.3 μM but 60 μM was required for a similar activation at 10°C. Nielsen et al. (1973) have recently shown that pure beef heart lecithin gives a tighter binding of NAD^+ and NADH than seen in the intact membrane or with total mitochondrial phospholipids, suggesting a possible function for the phosphorylcholine head group. The requirement for fluid phospholipid fatty acyl side-chains could be a result of the protein having to penetrate to some degree into the non-polar interior of the bilayer (see section 2.4), and this is more effective when the lipids are in a fluid state (Papahadjopoulos et al., 1973a). On this basis it is difficult to understand why much higher concentrations of phospholipids in the solid state can also reconstitute the enzyme. There may be some concentration dependent change in the properties of liposomes, but the exact mechanism of these effects awaits further study.

Houslay et al. (1975) also found a specific requirement for lecithin for β-hydroxybutyrate dehydrogenase. Other lipids, including cholesterol, were inhibitory. They also concluded that fluid phospholipid acyl chains are essential for activity of the enzyme, and showed that it may preferentially associate with dioleoyl lecithin in the presence of other mitochondrial lipids. These authors used a lipid exchange technique involving cholate in which excess exogenous lipid replaces endogenous lipid (Racker, 1972; Hinkle et al., 1972; Warren et al., 1974b), and was done on whole mitochondria without further purification. The cholate can then be removed by slow dialysis, or more rapidly by gradient centrifugation. The latter process results in the formation of smaller enzyme-lipid complexes which can have a single monolayer of phospholipid around the protein, while the dialysis technique usually results in the formation of membranous vesicles. Apparently, after equilibration, simple addition of the cholate mixture to the reaction medium resulting in a very high dilution, is sufficient to retain the effect of the added lipids, which show normal thermal transition effects (Warren et al., 1974b). Such techniques are likely to result in optimal hydrophobic interactions between lipid and protein. It was found that β-hydroxybutyrate dehydrogenase was completely inactive below 24°C for DMPC, 29°C for dipalmitoyl-phosphatidylcholine and 50°C for distearyl-phosphatidylcholine. Although this temperature is very similar to the T_m for pure DMPC, the temperatures for the others are 10°C–15°C below their transition temperatures. It was suggested that this could be due to a fluidizing effect of the protein on adjacent lipids (Houslay et al., 1975). It would not appear to be due to residual cholate because, as mentioned above, activity in the presence of DMPC began near the T_m for the phospholipid (see Table I). The lower temperatures for the other phospholipids seems superficially at variance with the immobilization found for cytochrome oxidase, but is similar to the results for the (Na+K) ATPase both in its natural membrane (Grisham and Barnett, 1973) or after interaction with liposomes (Kimelberg and Papahadjopoulos, 1972, 1974).

Using the above exchange technique both Racker (1972) and Warren et al. (1974a) have reconstituted a functional Ca^{2+} pump and ATPase activity. In a separate paper Warren et al. (1974b) also showed that the activity of the reconstituted ATPase was affected by the saturation of the fatty acyl groups of the phosphatidylcholine used. Thus, with dioleoyl-phosphatidylcholine the complex was active down to 10°C while with DMPC the enzyme was inactivated at 25°C and with dipalmitoyl-phosphatidylcholine at 29°C. This is very similar to the results found with β-hydroxybutyrate dehydrogenase (Houslay et al., 1975) and the same arguments apply. It is of interest that in this system, unlike the one described by Grover et al. (1975), the di C (12:0) phosphatidylcholine is completely inactive. Later work on the Ca^{2+} ATPase (Warren et al., 1975a) has suggested that the "annulus" or monolayer of phospholipid around the ATPase (approx. 30 molecules phospholipid/ATPase) was able to exclude cholesterol present in the added lipid, until quite high concentrations of cholate were used. The "annulus" is equivalent to the "boundary lipid" concept for cytochrome oxidase (Jost et al., 1973a,b). Green (1975) has reported similar results for the Ca^{2+} ATPase delipidated with Triton X-100. A minimum amount of phospholipid of 40–45 moles/mole of enzyme was found to be required for maximum activity. Phosphatidylcholine was found to be the most effective phospholipid in this regard, and was only effective above its T_m. The and Hasselbach (1973) have also found that Ca^{2+} ATPase delipidated with phospholipase A could be most effectively reactivated with unsaturated fatty acids. Recently an inhibitory effect of 2–10 mM Mg^{2+} on reconstituted Ca^{2+} ATPase activity has been described by Warren et al. (1975b). Mg^{2+} inhibited the activity only when the enzyme was reconstituted with dioleoylphosphatidic acid and had no effect with dioleoylphosphatidylcholine. This strongly suggests that the inhibitory effect of Mg^{2+} was due to its interaction with the head groups of the negatively charged phospholipid.

Glucose-6-phosphatase from liver microsomes was partially inactivated by removal of >90% of its endogenous lipids by 0.1–0.4% deoxycholate (Garland and Cori, 1972). Sonicated dioleoylphosphatidylcholine liposomes fully reactivated the enzyme at 30°C, while dipalmitoylphosphatidylcholine liposomes were ineffective.

As discussed in section 3.5.3, some of the earliest reports of discontinuous Arrhenius plots for enzyme activity were for the (Na + K) ATPase. Since this enzyme has been known to be phospholipid-dependent for some time (Skou, 1961; Schatzmann, 1962; Ohnishi and Kawamura, 1964), it is an attractive system in which to examine the effects of lipid fluidity on enzyme activity. As described in the previous section, studies in cells showed a good correspondence between alterations in membrane lipid fatty acyl chain saturation or measurements of membrane fluidity by probes, and the temperature at which discontinuities in Arrhenius plots occurred for several eukaryotic and prokaryotic membrane enzymes. Since the (Na + K) ATPase is not present in bacteria it was not studied in the *E. coli* fatty acid auxotroph system. This section will discuss a number of studies in which the endogenous lipid has been removed from the (Na + K) ATPase, thereby inactivating or partially inactivating it. The effects of lipid fluidity can then be directly assessed by reconstituting the enzyme with phospholipids of differing fatty acyl side chain composition, as well as adding other components capable of altering phospholipid fluidity in a mixture with the phospholipids. It is also, of course, the most convenient method of examining phospholipid head-group specificity.

An early study (Kimelberg and Papahadjopoulos, 1972) utilized the deoxycholate phospholipid extraction techniques of Tanaka and Strickland (1965). This technique was found to remove ⩾93% of the endogenous phospholipid of rabbit kidney microsomes as determined by organic phosphorus. (Na + K) ATPase activity could then be restored by adding sonicated phospholipid liposomes made from phosphatidylglycerol or phosphatidylserine. Phosphatidylglycerol, prepared from egg phosphatidylcholine, was nearly as effective as brain phosphatidylserine, which had previously been shown to be most effective in reactivating such a preparation (Fenster and Copenhaver, 1967). Egg phosphatidylglycerol, retaining the fatty acyl chain composition of the parent egg phosphatidylcholine, showed a linear Arrhenius plot from 20°C–39°C for the reactivated enzyme with an activation energy of 10 kcal/mole. In contrast, dipalmitoyl-phosphatidylglycerol, obtained from egg phosphatidylglycerol reacylation, show a non-linear Arrhenius plot with a discontinuity or intercept at 32°C and activation energies of 20 and 42 kcal/mole above and below this temperature, respectively. Although cholesterol at a 1:1 mole ratio had little effect on phosphatidylserine-activated activity, it decreased the reactivation by dipalmitoyl-phosphatidylglycerol five-fold (Kimelberg and Papahadjopoulos, 1972).

A subsequent study (Kimelberg and Papahadjopoulos, 1974) directly compared the discontinuities obtained for Arrhenius plots of enzyme activity reactivated with different disaturated phosphatidylglycerols, and direct determinations of their melting characteristics by differential scanning calorimetry. These results are shown in Figs. 14 and 15. It is clear from Fig. 14 that increasing the chain length of the saturated di-phosphatidylglycerol increases the temperature at which the discontinuity in the Arrhenius plot occurs. Correlating this with the endothermic phase transitions shown in Fig. 15, it can be seen that they occur at temperatures lower than the actual onset of melting. The dioleoyl-phosphatidylglycerol should have a T_m well below 0°C by analogy with dioleoyl-phosphatidylcholine ($T_m = -22$°C, see Table I) and indeed shows a linear Arrhenius plot from 10°C–40°C, with an Ea of 14.7 kcal/mole. If this represented the Ea of a completely fluid membrane it was argued that the discontinuity for dimyristoyl-phosphatidylglycerol (20.2°C) and bovine brain phosphatidylserine (15.0°C) represented the upper limits of a broad phase separation, by analogy with the findings of Linden et al. (1973). The discontinuities of dipalmitoyl- and distearylphosphatidylglycerol would then represent the lower limits of the transition, since their Ea values were greater. Only in the case of phosphatidylserine, however, did the position of the Arrhenius plot discontinuity approach the end of melt temperature of the broad phase transition seen for this phospholipid (5°C–20°C). It was noted however, and other studies have shown (Papahadjopoulos et al., 1975b) that sonicated liposomes can melt approx. 5°C lower than non-sonicated, which would alter the beginning of the melt for dipalmitoyl-phosphatidylglycerol so that it coincided more nearly with the discontinuity for the Arrhenius plot at 31.4°C. It has been noted that in the reconstitution of both Ca^{2+} ATPase (Warren et al., 1974) and β-hydroxybutyrate dehydrogenase (Houslay et al., 1975) by disaturated phosphatidylcholines of differing chain length, the inhibition of activity is coincident with the mid-point

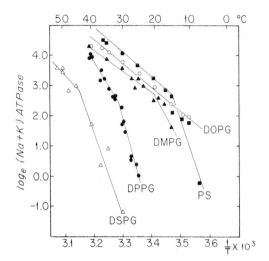

Fig. 14. Arrhenius plots of phosphatidylserine and phosphatidylglycerol-activated (Na + K) ATPase from rabbit kidney outer medulla, depleted of endogenous phospholipid with deoxycholate. Sonicated liposomes made from different phosphatidylglycerol preparations with varying fatty acyl chain substitutions were used as follows: DOPG, dioleoyl-phosphatidylglycerol, 0.18 μ moles; DMPG, dimyristoylphosphatidylglycerol, 0.24 μmoles; DPPG, dipalmitoylphosphatidylglycerol, 0.59 μmoles; DSPG, distearoyl-phosphatidylglycerol, 0.64 μmoles; and PS, bovine phosphatidylserine, 0.64 μmoles. These concentrations gave optimal activation. The amount of enzyme protein was 0.028–0.056 mg in a total volume of 1.5 ml. The maximal specific activity was 100–120 μmoles P_i/mg protein/h with PS or DOPG at pH 7.2 and 40°C for (Na + K) ATPase, which was the activity sensitive to 1 mM ouabain (reproduced with permission from Kimelberg and Papahadjopoulos, 1974).

of the transition for DMPC but about 10°C lower for dipalmitoyl-phosphatidylcholine. Similarly De Kruyff et al. (1973) have noted that the discontinuity in the Arrhenius plot of Mg^{2+} ATPase activity from *Acholeplasma laidlawii* occurs several degrees lower than the onset of melting as determined by differential scanning calorimetry. Since these studies used different methods of delipidation or no delipidation at all, it may reflect the property of some proteins of being able to increase the fluidity of their adjacent "boundary" or "annulus" lipids (see previous sections).

By examining activities down to a lower temperature range for delipidated (Na + K) ATPase reactivated by bovine brain phosphatidylserine, two discontinuities can now be identified at 8°C and 17°C (Kimelberg, 1975) (Fig. 16). This was reasonably close to the limits of the broad transition previously observed for phosphatidylserine. The transition seen at 15°C–17°C is in agreement with the transition at 15°C reported by Priestland and Whittam (1972) for deoxycholate-treated (Na + K) ATPase reactivated by bovine brain phosphatidylserine, and the numerous studies which report transitions at approx. 20°C for (Na + K) ATPase from brain and other tissues (see Table VI). This cannot, of course, be taken as evidence by itself that phosphatidylserine specifically activates (Na + K) ATPase in brain but is consistent enough with it for the question to be more

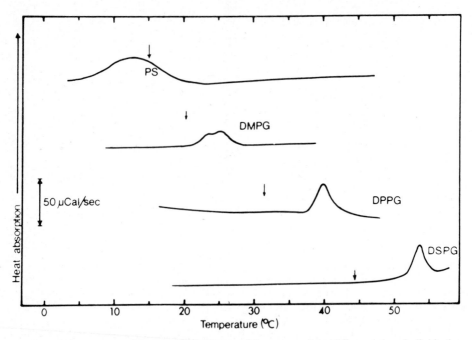

Fig. 15. Differential scanning calorimetry curves of liposomes of the different phospholipids shown for the Arrhenius plot in Fig. 14. Sonicated liposomes containing 1 μmole phospholipid in 15 μl sample volume at pH 7.0 were scanned at 2.5°C/min. See Fig. 14 for definitions of abbreviations. The vertical arrows indicate the temperatures at which the discontinuities of the Arrhenius plots of (Na + K) ATPase activity in Fig. 14 occurred. Because of the high concentrations used the dispersions remained turbid after normal sonication procedures. Subsequent experiments showed that when sonicated to clarity at lower concentrations and then concentrated by ultrafiltration, the onset of the melt is lowered from 38°C to 33°C for DPPG, which is very close to the t_d from the Arrhenius plot (reproduced with permission from Kimelberg and Papahadjopoulos, 1974).

directly examined. It also suggests that a similar enzyme-phospholipid complex may exist in a variety of tissues. It was also found that cholesterol increases the Ea values and appears to shift the temperature of the discontinuity about 3.5°C higher (see Fig. 16). This is in contrast to other work on *Acholeplasma* (De Kruyff et al., 1973) and yeast (Cobon and Haslam, 1973), where increasing levels of cholesterol have been reported to decrease the temperature of the discontinuity for several enzyme activities. In *Mycoplasma mycoides* (Rottem et al., 1973), increasing levels of cholesterol abolished the discontinuity, and the resulting linear Arrhenius plot showed a low Ea value. Since, as previously discussed, cholesterol progressively broadens and then abolishes the transition, these differing results may reflect differing amounts of cholesterol in the enzyme environment.

The results shown for (Na + K) ATPase activity in Fig. 16 were for a 1:1 mole ratio of phosphatidylserine: cholesterol which generally abolishes the phase transition (see section 2.3). The results are consistent with cholesterol broaden-

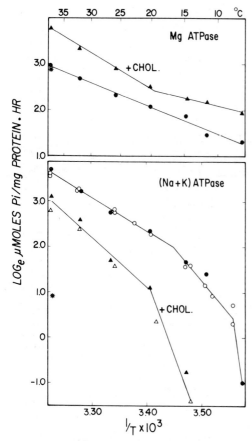

Fig. 16. Effects of cholesterol on Arrhenius plots of (Na + K) ATPase and Mg^{2+} ATPase activity of phospholipid depleted rabbit kidney ATPase (see Fig. 14). The activities were assayed in the presence of 0.5 μmoles sonicated phosphatidylserine or of sonicated liposomes consisting of 0.5 μmoles phosphatidylserine plus 0.5 μmoles cholesterol. For the results shown as closed symbols 0.076 mg enzyme protein was present and for the open symbols 0.132 mg enzyme protein was present in a total volume of 1.5 ml. The (Na + K) ATPase activity was the total activity sensitive to 1 mM ouabain and the pH was maintained at 6.91–7.07. The Mg ATPase was the residual activity insensitive to ouabain. The asterisks indicate the activities in the absence of added lipid (reproduced with permission from Kimelberg, 1975).

ing the phospholipid phase transition such that the discontinuity remains the upper limit of this transition, but also suggests that the cholesterol level in the enzyme environment must have a lower than average level. It has been suggested that cholesterol is excluded from the boundary lipid of proteins (Warren et al., 1975a). It is of interest that cholesterol increases the activity of the ouabain-insensitive Mg^{2+} ATPase and converts a linear Arrhenius plot to a discontinuous one, which now shows a higher E_a value in the upper temperature range (Fig. 16). Cholesterol at a 1:1 mole ratio with phosphatidylserine was also found to have no significant effect on the K_m of the ouabain-sensitive or insensitive lipid-depleted Mg^{2+} ATPase compared to phosphatidylserine alone. It did, however, decrease the V_{max} of the ouabain-sensitive (Na + K) ATPase and increase the V_{max} of the ouabain-insensitive Mg^{2+} ATPase. Mg^{2+} at concentrations in excess of the ATP concentration inhibited the enzyme and abolished discontinuities in the Arrhenius plot (Kimelberg, 1975). Inhibition by Mg^{2+} has also been described for the Ca^{2+} ATPase (Warren et al., 1975b). It had previously been shown, and was discussed in section 2.3, that free Mg^{2+} increases the T_m of acidic lipids.

Walker and Wheeler (1975b) have studied the effects of varying phospholipid fatty acyl composition on the reactivation of an NaI-treated microsomal fraction from rabbit kidney, which had been delipidated with the detergent Lubrol WX. Maximal reactivation by phosphatidic acid derived from phosphatidylserine was about two-fold greater than phosphatidic acid derived from egg phosphatidylcholine, which the authors suggested could be due to the parent phosphatidylserine containing more unsaturated fatty acyl side chains, and specifically more di-unsaturated molecules than egg phosphatidylcholine. Further work is required to substantiate this interesting suggestion. Phosphatidylglycerol prepared from phosphatidylserine was only more effective than phosphatidylglycerol from egg phosphatidylcholine at low lipid concentrations. In addition, the reactivation by the phospholipids derived from egg phosphatidylcholine showed a sigmoid relationship with concentration.

Marked effects of the degree of saturation of the fatty acyl side chains of phospholipids have also been reported for the mitochondrial ATPase activity of beef heart submitochondrial particles (Bruni et al., 1975) depleted of phospholipid by cholate treatment (Kagawa and Racker, 1966). While micelle-forming disaturated short chain (8:0 or less) phosphatidylcholine molecules or bilayer-forming longer chain phosphatidylcholine molecules above their T_m did reactivate, most effective reactivation was obtained with dioleoyl-phosphatidylcholine. Thus a fluid bilayer formed from unsaturated phospholipids provided the most effective reactivation, and also showed greater inhibition by oligomycin, a potent inhibitor of the enzyme in the mitochondria. This requirement of oleoyl side chains was apparently not so critical when phospholipids with negative head groups, such as dimyristoyl-phosphatidylglycerol, were used. Consistent with the requirement for lipid bilayer fluidity it was found that the addition of cholesterol at a 1:1 mole ratio with dioleoyl-phosphatidylcholine inhibited the reaction by 32–35%. Arrhenius plots of enzyme activity in submitochondrial particles showed a discontinuity at 14.2°C (See Table V). The enzyme reactivated with DMPC plus dicetyl phosphate or dimyristoyl-phosphatidylglycerol showed discontinuities at 23.9°C and 19.7°C, respectively, which are reasonably close to the measured transitions of these phospholipids. Surprisingly, however, reactivation by dioleoyl-phosphatidylglycerol gave a discontinuity in the Arrhenius plot at 16°C. This phospholipid by analogy with dioleoylphosphatidylcholine (Table I), should have a transition well below 0°C but this has not been measured directly. Curvilinear rather than discontinuous Arrhenius plots were obtained for the dioleoyl- and dipalmitoyl-phosphatidylcholine reactivated enzymes. As the authors suggest, this could be due to the formation of quasi-crystalline clusters, presumably around the enzyme protein (also see Lee et al., 1974, and section 3.4). However, independent identification of areas with separate fluidity characteristics would be needed to confirm this.

The studies on enzyme lipid dependency covered in this section, apart from the Ca^{2+} ATPase, (Racker, 1972; Warren et al., 1974a) and the work of Houslay et al. (1975) on β-hydroxybutyrate dehydrogenase have generally added the lipid-depleted enzyme to sonicated dispersions of phospholipids. Thus, it is not clear whether an actual *trans*-membrane orientation has been achieved with this method. There may simply be a partial penetration of the membrane by the enzymes. Indeed, for the (Na+K) ATPase an increased $^{86}Rb^+$ efflux from sonicated liposomes was found consistent with such a penetration, but this was not ATP-dependent and therefore no evidence of a reconstitution of the pump requiring a *trans*-membrane orientation as a minimal condition was obtained. An ATP hydrolysis uncoupled from cation transport was consistent with the ATP hydrolysis rate and amounts of K^+ inside the liposomes (Kimelberg and Papahadjopoulos, 1974). These conditions also gave very low levels of recombination when measured directly after being run on 0.5 M NaCl sucrose gradients compared to the detergent dialysis technique (Slack et al., 1973). Thus, the reactivation that occurs when the enzyme is added to pre-formed liposomes only shows some of the characteristics of hydrophobic interaction, such as dependency on the state of the fatty acyl chains.

The (Na+K) ATPase and Ca^{2+} ATPase discussed in this section not only function enzymatically as ATP phosphohydrolases, but also utilize the energy from this reaction to actively transport ions. The transport properties of these enzymes have been successfully reconstituted with added phospholipids, using the cholate dialysis technique, to form intact vesicles with the enzyme presumably penetrating through the entire membrane. This technique has been used successfully for the Ca^{2+} ATPase (Racker, 1972; Warren et al., 1974a) and the (Na+K) ATPase (Goldin and Tong, 1974; Hilden et al., 1974; Sweadner and Goldin, 1975; Hilden and Hokin, 1975; Hokin, 1975). Reconstitution of ATP-dependent pumps has also been achieved by prolonged sonication of the enzyme with phospholipid in the absence of detergent for both the Ca^{2+} ATPase (Racker and Eytan, 1973) and the (Na+K) ATPase (Racker and Fisher, 1975).

Utilizing these reconstituted systems the next logical step would now appear to involve varying the composition of the fatty acyl side chains or adding other components such as cholesterol, to systematically examine the effects of lipid fluidity on the transport processes. These could then be compared directly with effects on enzymatic activity, thereby providing a more complete description of the functioning of these transport ATPases. This approach has been attempted with bacterial rhodopsin which when reconstituted with DMPC catalyzes a light-driven proton uptake by the vesicles (Racker and Hinkle, 1974). Since there was no marked effect of temperature on the rate of uptake between 5°C and 25°C, and since transport by mobile carriers has been shown to be inhibited when the bilayer is in a solid state (Krasne et al., 1971), the authors concluded that transport in this case was due to a channel mechanism.

3.7. Membrane microenvironments and lateral translational mobility

A corollary of the fluid mosaic model of cell membranes explicitly mentioned in the original proposal (Singer and Nicolson, 1972), was that a proportion of the lipid was not in the bilayer form but might be tightly and specifically bound to proteins. Thus, about 25% of the total lipids in *Mycoplasma laidlawii* did not contribute to the endothermic transition seen by differential scanning calorimetry (Steim et al., 1969). Certain proteins might also be arranged in relatively permanent arrays, unable to undergo independent lateral motion. This has been illustrated in diagrammatic form in Fig. 2 (see section 1) from Papahadjopoulos (1973), where the circles represent proteins spaced widely apart in protein-poor, and as contiguous structures in protein-rich membranes. Indeed, the existence of sequences of interacting enzymes in membranes such as mitochondria and endoplasmic reticulum suggest a requirement for multi-enzyme arrays. However, if lateral diffusion is rapid enough, specific interactions alone, based on an adequate number of successful collisions without specific permanent enzyme arrays, might be sufficient. This section will put together some of the evidence relating to the existence of modified lipid areas around proteins distinct from the bulk lipid. It will also review the still somewhat limited studies on whether lateral diffusion occurs rapidly enough for specific functional protein-protein interactions to occur, or whether permanent enzyme "chains" are required. Some of this material has also been referred to in sections 1, 2.4, and 2.5.

The concept of specific lipid microenvironments distinct from the bulk lipid originated with early observations that enzyme and/or transport activities in bacterial (Overath et al., 1970; Esfahani et al., 1971; Mavis and Vagelos, 1972) and mammalian mitochondrial membranes (Lenaz et al., 1972) showed discontinuities which occurred at temperatures that did not correspond to independently measured transitions, or occurred at different temperatures for different enzymes from the same membrane. Later work showed the existence of multiple transitions in bacterial (Linden et al., 1973; Morrisett et al., 1975) and mammalian cells (Wisnieski et al., 1975), for both membrane fluidity and enzyme activity, which may correspond to at least some of these specific microenvironments. Support for this concept also came from the observation that in a highly purified enzyme preparation of the (Na + K) ATPase the discontinuity in the Arrhenius plot of enzyme activity and also that for a spin-labelled probe reporting lipid bilayer fluidity coincided, and was different from the behavior of the probe in a less pure plasma membrane preparation (Grisham and Barnett, 1973).

Possible mechanisms for these effects were based on the numerous observations of a lipid polar head-group specificity for membrane-bound enzymes (see section 3.1) which indicated that such enzymes had specific lipid affinities, and therefore might sequester the appropriate lipids. By analogy with studies on the interaction of phospholipids with soluble proteins (see section 2.4) this could be based on specific polar head-group interactions alone. Peripheral proteins could also cause localization of lipids, since it has been recently shown that polylysine, like Ca^{2+}, can segregate negatively charged lipids in a binary mixture with a neutral phospholipid (Galla and Sackmann, 1975).

Using spin-labelled probes Jost et al. (1973a,b) have directly demonstrated a monolayer of immobilised lipid surrounding cytochrome oxidase by the simple expedient of extracting increasing amounts of phospholipid from membranous cytochrome oxidase preparations and noting that the spectrum of the spin-labelled stearic acid localised within the lipid membrane indicated increasing immobilisation. Maximum immobilisation occurred at a concentration of 0.2 μ moles phospholipid/mg protein, which was calculated, on the basis of reasonable assumptions regarding the size of the protein and phospholipid molecules, to represent a single monolayer of phospholipid around the protein. This was termed "boundary" lipid. Warren et al. (1974b) also removed increasing amounts of phospholipid from the Ca^{2+} ATPase with cholate, and observed that a minimum of thirty molecules of phospholipid per enzyme molecule was required for activity, which also represented a monolayer of phospholipid around the protein. This layer, which is equivalent to that described by Jost et al. (1973a) was termed the "annulus". A subsequent study indicated that cholesterol was restricted from entering the intact annulus since no inhibition of enzyme activity was obtained until the enzyme plus annulus was equilibrated with cholesterol-containing mixtures in the presence of sufficiently high concentrations of detergent that were known to strip off the annulus lipid (Warren et al., 1975a). By analogy with a spin-labelled steroid which shows a lateral diffusion coefficient of 10^{-8} cm^2/sec (Träuble and Sackmann, 1972), cholesterol should be able to rapidly diffuse laterally in the plane of the membrane. Warren et al. (1975a) suggested that the basis of cholesterol exclusion might be that the phospholipids in the annulus are relatively immobilised. Since it has been shown that cholesterol preferentially associates with the more fluid lipids of a binary mixture (De Kruyff et al., 1974), this could be the basis of cholesterol exclusion from the annulus. This is in contrast, however, to their earlier data (Warren et al., 1974b) which showed that inhibition of Ca^{2+} ATPase activity, reconstituted with dipalmitoyl-phosphatidylcholine occurred some 11°C lower than the T_c for pure dipalmitoyl-phosphatidylcholine, which presumably reflects the behavior of the immediate lipid environment of the enzyme.

Stier and Sackmann (1973) took advantage of the ability of the NADH-dependent cytochrome P-450 reductase system to reduce nitroxide spin labels and abolish their ESR signal, to study the effect of temperature on the lateral diffusion of a spin-labelled stearic acid (Fig. 6B) interacting with the reductase molecule. Plotting the initial rate of this reduction in an Arrhenius plot gave a discontinuity at 32°C, with Ea values of 8.7 and 30.8 kcal/mole. The order parameter measured for the same spin label in the absence of NADH showed no evidence of a fluidity transition down to 5°C. Furthermore, NADH-dependent reduction of a water soluble phosphate derivative of the TEMPO label (similar to label shown in Fig. 6A), which would remain in the aqueous phase, showed a linear Arrhenius plot. Thus, the discontinuity in the Arrhenius plot was not due to a change in state of the enzyme molecule itself but was likely to be due to a property of the phospholipid around the reductase molecule. It was suggested that this existed in a relatively solid or quasi-crystalline state below 32°C thereby inhibiting the diffusion of the lipophilic spin label up to this temperature. It represented less than 20% of the total phospholipid.

The concept of a specific lipid domain for membrane proteins suggested by such work is undoubtedly an important one in membrane organization. It enables lipids to independently influence specific proteins, permitting a much more precise control of membrane function. Also, although the rapid lateral translational mobility of lipids is undoubtedly important for many membrane functions, the segregation of lipids around membrane proteins enables specific functions to be retained in an overall fluid membrane.

Since the lateral diffusion of proteins in the membrane should be directly proportional to the fluidity of the membrane, some effect of fluidity on enzyme activity which involves the actual collision of separate enzyme molecules should be observed. This has recently been found for cytochrome b_5 and cytochrome b_5 reductase of the microsomal P-450 system (Strittmatter and Rogers, 1975). These proteins had been shown to consist of a hydrophilic catalytic part and a smaller hydrophobic part required for binding to the membrane (Spatz and Strittmatter, 1971, 1973). They were sequentially added to pre-formed DMPC liposomes. These were of uniform 250–400 Å diameter after sonication and subsequent centrifugation of remaining multilamellar liposomes at 100 000 g for 30 min. Prolonged incubation resulted in the binding of b_5 and then the reductase to these liposomes. The direct reduction of ferricyanide by the NADH-dependent reductase gave a linear Arrhenius plot. However, the reduction of added cytochrome c according to the known sequence, NADH reductase → cytochrome b_5 → cytochrome c, gave a sharp increase in activity between 19°C–23°C and much smaller increase between 10°C–12°C. This corresponded quite well to the main endothermic phase transition for DMPC at 23.7°C–23.8°C and the pre-transition at 12.5°C–13.5°C (Hinz and Sturtevant, 1972; see also Table I), allowing for some perturbation of the phospholipid by the protein. These effects of temperature were greatest at the lowest amounts of bound enzymes, when the requirement for lateral diffusion is greater. Binding and reduction of excess cytochrome b_5 by microsomes, since no electron transfer occurs between cytochrome b_5 molecules themselves, indicates considerable lateral diffusion of b_5 and/or reductase in the endoplasmic reticulum membrane as well (Rogers and Strittmatter, 1974). Thus, the multi-enzyme sequences found from studies of enzymatic activity may partly reflect functional interactions and rapid lateral diffusion of a spin-labelled stearic acid (Fig. 6B) interacting with the reductase structural sequences in some membranes, however, is indicated by actual isolation of multi-enzyme complexes (Green and Tzagoloff, 1966).

It would be of interest to perform similar experiments on other systems where enzyme-enzyme collision reactions are thought to occur, such as the chloroplast and mitochondrial electron transfer chains. Indeed, certain components of the latter, such as the lipid soluble co-enzyme Q, have been considered to have had the characteristics of "mobile" carriers for some time (e.g. Green and Tzagoloff, 1966). It has recently been pointed out that the asymmetrical arrangement of phospholipids between the inner and outer leaflets of the plasma membrane, with phosphatidylcholine, sphingomyelin and cholesterol predominantly in the outer and phosphatidylserine, phosphatidylinositol and phosphatidylethanolamine in the inner leaflet, would lead to a more fluid inner leaflet (Emmelot and Van Hoeven, 1975; Bretscher and Raff, 1975). If this proposal is verified, it suggests that proteins or other components localized in separate monolayers can diffuse at different rates. This could be an effective regulatory step for *trans*-membrane permeability channels composed of two components present in each monolayer.

4. A regulatory role for membrane fluidity?

The preceding sections of this review have shown how experimental modification of the lipid composition of biological membranes or reconstituted model membranes resulting in decreased membrane lipid fluidity, can alter the properties of associated enzymes. In this section I will discuss the somewhat limited evidence on whether changes in membrane lipid fluidity actually have functional significance in vivo. The experimental approach to this question generally involves studying alterations in lipid composition, especially with regard to fatty acyl side-chain composition and cholesterol content, and correlating this with changes in enzyme activity and/or permeability rates. This type of analysis cannot directly determine a causal relationship and would have to be corroborated by other studies, such as with model and reconstituted systems resembling the putative in vivo condition. This section will consider studies in which at least some part of this experimental sequence has been established and which suggest that alterations in the membrane lipid composition may play an important functional role. They will be considered under two headings: those involving membrane changes in normal conditions, such as growth and development; and studies of abnormal and disease conditions.

4.1. Alterations in normal states

The clearest case in which changes in membrane lipid composition have been shown to have a normal functional role is the effect of altered environmental temperatures in poikilotherms or hibernators. Aspects of this subject have already been discussed in section 3.3.4, in which it was pointed out that studies on the temperature dependence of enzymes in poikilotherms and chilling-resistant plants were among the earliest evidence demonstrating the effects of lipid membrane fluidity on membrane enzyme activity. This early work has been reviewed by Lyons (1972) and Raison (1973). Section 3.5.3 also refers to changes in (Na + K) ATPase activity in poikilotherms. Some of this work and other studies will be briefly discussed together in this section.

There has been an extensive amount of work on changes in lipid saturation as a function of acclimation temperature in poikilotherms (see Hazel and Prosser, 1974, pp. 636–639, 657 and 663). Generally, but not always, a drop in acclimation temperature is correlated with increasing unsaturation of the fatty acids, but this has usually been measured as total fatty acids in whole animals or organs rather than the fatty acids from membrane phospholipids. In the goldfish (*Carassius auratus*), decreasing temperature is associated with increasing unsaturation and a decrease in the cholesterol–phospholipid ratio of total body fat (Hoar and Cottle, 1951). An increased unsaturation with decreasing temperature was also found for the fatty acids of goldfish brain (Johnston and Roots, 1964). In the case of goldfish intestinal lipids, increasing temperature was associated with increased saturation of the fatty acids obtained from the phospholipid fraction (Kemp and Smith, 1970). This finding is of interest because it had previously been shown (Smith, 1967) that discontinuities in Arrhenius plots of (Na + K) ATPase activity from this tissue showed an increase from 8°C to 21°C for an 18°C rise in acclimation temperature. Since these alterations in (Na + K) ATPase were not accompanied by alterations in [^{3}H]ouabain binding (Smith and Ellory, 1971), the enzyme level was apparently unaffected. Thus, it seems reasonable to ascribe such alterations in enzyme activity to changes in the fluidity of the membrane lipid environment, in view of the known effects of lipid fluidity on this enzyme. Studying succinic dehydrogenase from goldfish muscle, Hazel (1972) found that the temperature activity profiles of goldfish adapted at 5°C or 25°C were different. The enzyme from both types was then delipidated and reactivated with lipids extracted from either a 5°C or 25°C adapted fish. The temperature profile of the reactivated enzyme corresponded to the acclimation temperature of the animal from which the lipids were obtained rather than the animal from which the enzyme was obtained, showing a clear dependence of the temperature effect on lipid properties.

Marked changes in lipid composition have also been found in hibernating mammals, which can lower their body temperatures from 37°C to as low as 1°C during hibernation. Although this results in a marked reduction in metabolic rates, the animal has to maintain a minimal degree of metabolism as well as avoid irreversible cellular damage. Hibernation has also been shown to be associated with increased unsaturation of fatty acids, such as the lipids of hibernating hamster brain (Goldman, 1975) and the phospholipids of the heart of hibernating ground squirrels (Aloia et al., 1974). These latter workers showed a 3.5-fold increase in the amount of lysoglycerophosphatides. This is of interest since incubation of (Na + K) ATPase preparations with phospholipase A, which removes the fatty acid on the two position of the glycerol resulting in the formation of lysoglycerophosphatides, converts discontinuous Arrhenius plots of (Na + K) ATPase activity to linear plots (Taniguchi and Iida, 1972; Charnock et al., 1973). The control Arrhenius plot can be restored by adding intact phospholipids back to the phospholipase A-treated preparation. As discussed in section 3.3.4 many poikilotherms show linear Arrhenius plots for several enzyme activities, whereas homeotherms show discontinuous plots. The hibernating and warm-adapted hamsters, however, both show discontinuous Arrhenius plots, but with reduced activation energies for the hibernator (Goldman and Albers, 1975). Aloia et al. (1974) mentioned that a corollary study using spin labels showed a linear Arrhenius plot for the hibernating ground squirrel, whereas the active ground squirrel showed a discontinuity at approx. 24°C.

Differences in K^+ and Na^+ flux have also been reported for the erythrocytes of ground squirrels as compared to a non-hibernating mammal, the guinea pig (Kimzey and Willis, 1971). A sharp drop in the percentage of total ouabain-sensitive K^+ influx was found for the guinea pig at approx. 20°C, while this fraction was constant for the ground squirrel down to 5°C. Differences in the temperature profile of passive K^+ efflux and active influx suggest that the effect on influx was specific to the (Na + K) ATPase, and not a general permeability response of the membrane. Similarly, the K^+ uptake and (Na + K) ATPase activity of a hibernator (the hamster) was less affected by a temperature decrease than that of a non-hibernator (the rat). The (Na + K) ATPase activity was more sensitive to temperature changes than K^+ uptake (Willis and Li, 1969). In unadapted erythrocytes from the dog a response of specific transport processes to temperature rather than a general membrane response has been suggested by Elford (1975), on the basis of a different effect of temperature on Na^+ and K^+ transport.

Growth and cell division should potentially involve cell membrane changes, both in terms of increasing cell transport processes and the membrane changes involved in the cell cycle. This is clearly a difficult problem to approach experimentally but rapid increases in cell permeability to various substances concomitant with increased growth has often been observed (Plagemann and Richey, 1974; Holley, 1975). This suggests a potential role for membrane fluidity changes, perhaps in addition to effects on the transport proteins themselves since it has been shown that transport processes are markedly affected by membrane fluidity changes (section 3.3.1). Many of these transport processes show discontinuous Arrhenius plots (Plagemann and Richey, 1974; Berlin and Oliver, 1975). Stimulation of growth in lymphocytes by phytohemagglutination (Quastel and Kaplan, 1970) and quiescent 3T3 cells by addition of serum (Rozengurt and Heppel, 1975) leads to a rapid increase in ouabain-sensitive K^+ uptake, whereas decreased growth in both 3T3 and SV403T3 cells leads to a reduction in ouabain-sensitive K^+ uptake (Kimelberg and Mayhew, 1975). Since ouabain inhibits cell growth in lymphocytes (Quastel and Kaplan, 1970), Ehrlich ascites cells (Mayhew, 1972) and DNA synthesis in 3T3 cells (Rozengurt and Heppel, 1975) there would appear to be a definite positive correlation between cation transport and cell growth. The fact that stimulation of K^+ uptake by a number of mitogens was correlated with 2–4 fold increases in (Na + K) ATPase activity, but had no appreciable effect on [^{3}H]ouabain binding, suggests an increase in the activity of the (Na + K) ATPase (cf. studies on goldfish intestinal (Na + K) ATPase above), rather than increases in enzyme levels (Averdunck and Lauf, 1975). This might result from a change in membrane state, and indeed, an increased lymphocyte membrane fluidity following concanavalin A treatment has recently been detected by spin label studies (Barnett et al., 1974b), though there is some controversy concerning this study (Dodd, 1975). A further argument for such increased transport processes being a result of changes in membrane fluidity is that the transport of a number of compounds with different carrier mechanisms is usually stimulated by a single stimulatory event, such as lectin binding in lymphocytes (Averdunck and Lauf, 1975) or growth in cultured cells (Plagemann and Richey, 1974). Temperature studies of these activities, together with studies of the temperature dependence of the physical state of the membrane should help to clarify this possibility. However, Stein and Rozengurt (1976) have recently reported that although the V_{max} for uridine transport shows a 10-fold increase for growing serum-stimulated 3T3 cells as compared to quiescent 3T3 cells, Arrhenius plots show no change in activation energies. They therefore concluded that the stimulation was due to an increase in the number of carriers, rather than an increase in the rate of turnover of the carriers.

Cholesterol and dipalmitoyl-phosphatidylcholine as well as the addition of Mg^{2+} and relatively high concentrations of Ca^{2+}, have been reported to inhibit the fusion of cultured muscle cells (Van der Bosch et al., 1973). Arrhenius plots of the fusion rate showed a discontinuity at 35°C, with Ea values of 22 and 73 kcal/mole. In the presence of cholesterol this discontinuity was abolished and a much higher activation energy of 140 kcal/mole was found. It would also be of interest to find a correlation between aging and membrane lipid properties. Intuitively we might expect increased peroxidation and breakdown of membrane lipids as a result of aging. Inhibition of several membrane, lipid-dependent enzymes has been reported in the rat as a function of age at 6 and 24 months, but no corresponding alterations in total phospholipid content were found (Grinna and Barber, 1972). Possible differences in fatty acid saturation, however, were not investigated. Bowler and Tirri (1974) concluded that although young rats tolerate lower body temperatures than adults and show a marked rise in (Na + K) ATPase during the 10–20-day postnatal period associated with myelination and proliferation of glial cells, there was no qualitative difference in the temperature dependence of the (Na + K) ATPase activity of a synaptic membrane preparation. Curvilinear Arrhenius plots rather than ones showing a sharp intersection were found for (Na + K) ATPase, but the Mg^{2+} ATPase showed an abrupt change of slope. The Ea over the 10°C–30°C range was 25 kcal/mole for the adult and 36 kcal/mole for the preparation from the immature rat brain. The authors concluded that on the basis of their experiments there was only evidence for an increase in (Na + K) ATPase level, in contrast to the studies on the response of hibernators and poikilotherms to lowered body temperatures previously mentioned. A single paper showing that the binding of ribosomes to endoplasmic reticulum results in discontinuous

Arrhenius plots for [^{14}C]leucine incorporation into proteins (Towers et al., 1972), suggests the intriguing possibility of membrane lipid fluidity having pleiotropic effects by influencing protein synthesis.

4.2. Alterations in disease

The current active interest in membranes has been accompanied, not unexpectedly, by an interest in the involvement of membranes in disease. Several authors have dealt with this topic recently (see Wallach 1972, and several chapters in Weissman and Claiborne, 1976). This section will concern itself with those abnormal conditions for which there is some evidence of changes in membrane fluidity resulting in the altered function of membrane enzymes.

It has long been appreciated that alterations in the cell surface must play a critical role in the metastasis and invasiveness of cancer cells (see Weiss, 1976). The possibility of some of these alterations in cancer cells being due to increased membrane fluidity seems to have first been suggested in relation to the action of lectins, which were found to agglutinate transformed but not untransformed cells (see review by Burger, 1973). This was not related to changes in the amounts of lectin bound, and was considered to possibly reflect changes in the translational mobility of their glycoprotein receptors (see review by Nicolson, 1974). Studies on the rotational relaxation time of fluorescein conjugated concanavalin A (Con A) showed that the mobility of these sites in lymphomas was lower than in lymphocytes, but were more mobile in transformed fibroblasts as compared to normal fibroblasts (Inbar et al., 1973). It was assumed that the rotation of the fluorescein labelled concanavalin A was directly related to the movement in the membrane of the glycoprotein to which it was bound, as indicated by cluster and cap formation of fluorescein-labelled Con A (Inbar and Sachs, 1973). Other studies involving freeze-fracture electronmicroscope techniques have shown that density-inhibited 3T3 cells show an aggregated pattern of intramembraneous particles, usually thought to represent membrane proteins, whereas SV40 transformed 3T3 cells show a random pattern (Furcht and Scott, 1974). This resembles the pattern of behavior discussed for model and cell membranes in sections 2.4 and 2.5, where solidification of membrane lipids led to aggregation of membrane proteins. In cultured mammalian cells aggregation of such particles by Con A can also be promoted by treatment with local anesthetics (Poste et al., 1975). It is not clear, however, to what degree such behavior does in fact depend on small increases in membrane lipid fluidity due to the anesthetics, or the interplay of such effects and the activity of the sub-membrane microfilament and microtubular systems and peripheral membrane proteins (Berlin, 1975; Poste et al., 1975; Edelman, 1976).

Direct measurements of membrane fluidity in normal and transformed cells has led to conflicting results. Barnett et al., (1974a) labelled cells with a spin-labelled fatty acid of the type shown in Fig. 6B, but with the position of the label on the chain not clearly specified, and found an increased order parameter for confluent 3T3 cells compared with SV40 or polyoma virus transformed 3T3 cells. Using the same label with m = 10 (see Fig. 6B), Gaffney (1975) found no difference for confluent 3T3 and transformed 3T3 cells. However, as discussed in previous sections, the fatty acyl chain near the polar head groups of the phospholipids is relatively immobilized and it would be of interest to compare labels located nearer to the terminal methyl group where the chains have greater motion. It is not clear in these studies, as well as others, whether the label is distributed exclusively in the surface membrane. However, on the basis of cellular destruction of the label and on the fact that the measurements were performed within a few minutes of adding the label, Gaffney (1975) argues that it probably is. Proton and ^{13}C NMR studies have indicated that transformed hamster embryonic fibroblasts are more fluid than their normal counterparts (Nicolau et al., 1975). In contrast, recent fluorescence polarization studies have indicated that SV40 or polyoma virus transformed 3T3 cells show 50% increased microviscosity (Fuchs et al., 1975).

Studies on the total lipid composition of normal and transformed cells showed a small decrease in the polyunsaturated fatty acids of zwitterionic phospholipids after transformation of chick embryo fibroblasts by Rous sarcoma virus (Yau and Weber, 1972). It is noteworthy, however, that small consistent increases were found in many of the unsaturated fatty acyl groups of the phosphatidylinositol plus phosphatidylserine fraction, and these minor acid lipids have been found to specifically activate several membrane enzymes (see section 3.1). Also studies on the phospholipid composition of purified plasma membranes, rather than total lipids, might yield greater differences. There have been both positive (Bergelson et al., 1970) and negative reports (Feo et al., 1973) of differences in the lipid composition of isolated plasma membranes from normal liver and hepatoma cells, of which one of the most recent indicates an increased cholesterol content and decreased fatty acyl saturation in rat but not mouse hepatoma membranes (Van Hoeven et al., 1975). These authors concluded on the basis of this and other work that neoplastic cells did not have a uniformly altered phospholipid saturation. A recent study has indicated that the cholesterol content per cell surface is 50% higher in SV101 and PY transformed 3T3 cells compared to normal 3T3 cells (Adam et al., 1975). This was simply due, however, to the cell surface of 3T3 cells being 50% higher and the total cell cholesterol to total cell phospholipid mole ratio, which is the important relationship, was constant at 0.30:1. If this is representative of the plasma membrane it indicates that the bulk of the membrane lipid is in an intermediate fluid state (see section 2.3), but since these were total cell values their precise meaning is unclear.

All these measurements are bulk measurements and, as has been argued before, functionally important membrane changes may well be localized to specific membranes. Thus, Inbar and Shinitsky (1974a) have noted that treating lymphoma cells with 1:1 cholesterol:lecithin liposomes, which resulted in a 55% increase in free cholesterol per cell lipid, resulted in a decreased killing of mice by ascites tumor cell development after intraperitoneal injection. Similarly, they also found that leukemia in mice and in man is associated with a decrease of free cholesterol in the plasma membrane of leukemic cells as compared to normal leukocytes (Inbar and Shinitsky, 1974b). The marked increases in transport of sugars (Hatanaka, 1974), amino acids (Foster and Pardee, 1969; Isselbacher, 1972) and K^+ (as $^{86}Rb^+$) (Kimelberg and Mayhew, 1975) by transformed cells in comparison to normal cells could be attributable to changes in the membrane lipid environment. In the case of increased K^+ uptake this possibility was suggested on the basis of Arrhenius plots of the increased (Na + K) ATPase activity of SV403T3 as compared to 3T3 cells (Kimelberg and Mayhew, 1975). In an analogous situation, the rapid stimulation of ouabain-sensitive K^+ (as $^{86}Rb^+$) uptake in quiescent 3T3 cells concomitant with the initiation of rapid growth due to serum stimulation, was insensitive to cycloheximide. This indicated that the process involved increased activity of the transport sites, rather than synthesis of new enzyme molecules (Rozengurt and Heppel, 1975). These increased transport effects are usually associated with an increase in V_{max} for the process with no effect on K_m (Isselbacher, 1972; Weber, 1973; Rozengurt and Heppel, 1975). The elucidation of such effects, especially if due to localized changes, is clearly a formidable experimental problem. Thus, in spite of considerable effort, definite conclusions as to how the surface membranes of cancer cells differ from normal cells remains elusive. The present concepts of membrane structure and its dynamic aspects are of relatively recent vintage and if they prove to be basically sound they will undoubtedly lead to some consensus on this subject in the future. At present however, the remarks of Abercrombie and Ambrose in 1962 still seem the most pertinent: "it is wiser for the present merely to say that the cell surface, at whatever remove from the primary transformation it is altered, seems to play an important part in the manifestation of malignancy".

A disease state in which it might be intuitively surmised that membrane fluidity plays an important role is atherosclerosis. This is because this disease is usually associated with high plasma levels of cholesterol, or can be induced experimentally in many animals by feeding them a high cholesterol diet. As has been discussed in previous sections of this review, increased cholesterol or increased saturation of phospholipids leads to marked decreases in the activity of many membrane enzymes, as well as decreased passive permeability to numerous substances. A paper proposing these effects as important early steps in the etiology of atherosclerosis and which briefly summarizes the relevant evidence, has been published recently (Papahadjopoulos, 1974).

However, there have, as yet, been few studies aimed directly at relating the high cholesterol levels in atherosclerosis to effects on membrane enzyme activity. Cholesterol can rapidly exchange from circulating β lipoproteins into red blood cell membranes, and possibly other membranes including aortic intima cells (reviewed in Papahadjopoulos, 1974). Thus, cholesterol can exchange from 1:1 cholesterol-lecithin mixtures into red blood cell membranes to produce cells that resemble the high cholesterol "spur" and red cells resulting from severe liver disease (Cooper et al., 1975). These cells also have decreased

Na$^+$ and K$^+$ passive and active transport, and also show decreased fluidity of the hydrophobic region of the membrane as measured by a fluorescent probe (Vanderkooi et al., 1974). Such general exchange mechanisms would result in a number of tissue cellular membranes showing increased cholesterol levels on a high cholesterol diet, and presumably certain tissues such as aortic intima cells are more sensitive and/or functionally important. Rats fed a butter-cholesterol diet showed a 200% increase in free cholesterol levels in mitochondria, lysosome and microsome subcellular fractions of liver. This was associated with a 60% decrease in the glucose-6-phosphatase levels of the microsomal fraction (Morrison et al., 1970). As described earlier, glucose-6-phosphatase is an endoplasmic reticulum marker which requires unsaturated lipids for activity (Garland and Cori, 1972). In the same study there was only a 10% decrease in the cytochrome oxidase activity of the mitochondrial fraction perhaps reflecting the efficient exclusion of cholesterol from the cytochrome oxidase "annulus" or "boundary" lipid and/or a relatively low cholesterol content of the mitochondrial inner membrane. Paradoxically, another study showed an increase in the activity of brain (Na + K) ATPase of rabbits fed an atherogenic diet which was associated with a decreased unsaturated/saturated fatty acid ratio for brain membrane fractions (Toffano et al., 1974). This result implies an influence of high plasma cholesterol levels on the central nervous system in spite of the blood-brain barrier. Studies on enzyme activity and associated transport properties are clearly needed in arterial wall tissue, such as the intima plus media cells of the aorta, in both normal and atherosclerotic animals.

Membrane abnormalities have been detected in the hereditary cardiomyopathy of the Syrian hamster. Ventricular tissue of the myopathic hamster shows increased cholesterol levels and a significant decrease in Ca^{2+} binding but not in Ca^{2+} uptake. There was no effect on Ca^{2+} or Mg^{2+} ATPase activity, and the apparent decrease in (Na + K) ATPase activity was unreliable because of the very low specific activities (Owens et al., 1973). Duchenne muscular dystrophy involves a lipid metabolism defect, and increased cholesterol and sphingomyelin levels are found in tissues from experimental animals in which a similar disease has been inherited (Owens and Hughes, 1970). These changes may involve the plasma membrane, and the finding of an increased intracellular Na$^+$/K$^+$ ratio and a decreased muscle resting membrane suggests an inhibition of the (Na + K) ATPase as one possible mechanism (see refs. in Bray, 1973). Indeed, Bray (1973) has found that the (Na + K)-stimulated fraction of the ATPase activity of the skeletal muscle of dystrophic white mice was 71% less than in unaffected litter mates. In the presence of ouabain the decrease was somewhat less, 37%. In contrast, the basal Mg^{2+} ATPase was 158% greater. This resembles the effect of cholesterol on phospholipid-activated (Na + K) ATPase activity (see Fig. 16), in inhibiting (Na + K) ATPase activity but activating the ouabain-insensitive Mg^{2+} ATPase associated with it. Rodan et al. (1974) investigated ATPase and adenylate cyclase activity in dystrophic chicks and found a complex pattern of results. Thus, the ouabain-sensitive ATPase was less in dystrophic liver plasma membrane than in the control membrane but greater in dystrophic plasma membrane from muscle. Amphotericin B, which interacts with cholesterol, resulted in stimulation by ouabain of ATPase activity in normal muscle and dystrophic muscle and liver but increased inhibition in normal liver. Tissue specific effects were also found for adenylate cyclase. Both these tissues, however, as well as erythrocytes, showed up to a 2-fold increase in plasma membrane viscosity in dystrophic animals as measured by fluorescence polarization using perylene as a probe (Sha'afi et al., 1975). However, using electron spin resonance, Roses et al. (1975) have found an increase in membrane fluidity in red blood cells from patients with myotonic muscular dystrophy. Thus, while the various types of muscular dystrophy seem to involve some effect on membrane fluidity it is not yet clear whether there is a single consistent effect. An answer to the question of whether atherosclerosis or other diseases with increased membrane lipid saturation and/or cholesterol also involve an early decrease of membrane enzyme activity due to decreased membrane fluidity clearly depends on further study.

Marked increases in both saturated and unsaturated free fatty acids have been observed in the brains of homeotherms, but not poikilotherms, starting a few minutes after postdecapitation ischemia (Aveldano and Bazan, 1975). Although a small part of this may be due to deacylation of triglycerides, the authors suggest that the major part derives from membrane phospholipids by the action of endogenous brain phospholipase A. Such changes, however, do not seem to be reflected in changes in membrane enzyme activity since we have found no effect on the activity or Arrhenius plot profile of the (Na + K) ATPase activity of total rat brain homogenates an hour after decapitation (Kimelberg and Biddlecome, unpublished observations). Hamsters fed an essential fatty acid-deficient diet show decreased bile-salt independent bile flow, which is thought to depend on the (Na + K) ATPase activity located in the canalicular membrane (Sarfeh et al., 1974). It was therefore suggested that the essential fatty acid deficiency caused a decrease of this (Na + K) ATPase activity by some effect on its membrane lipid environment.

Finally, it is now being suggested that a number of pharmacologically active agents may exert their effect via an effect on membrane fluidity. Thus, it has been recently reported that active Δ'-*trans*-tetrahydrocannabinol, $(-)$ Δ'-THC, inhibits both the Mg^{2+} and (Na + K) ATPase of rat ileum, although the (Na + K) ATPase is more sensitive. Complete inhibition occurs at 10^{-5} M although 10^{-7} M is the normal pharmacological blood level. However, there could be local membrane concentration of this drug (Laurent et al., 1974). The inhibition of (Na + K) ATPase does not correlate in any simple way with effects on membrane fluidity since it has been recently reported that the active $(-)$ Δ'-THC fluidizes cholesterol/lecithin liposomes. The less active $(+)$ Δ'-THC stereoisomer has a smaller effect. In contrast, the non-psychoactive cannabinol and cannabidiols decrease the fluidity of the liposomes (Lawrence and Gill, 1975). It is possible, however, that these agents could also cause disruption of the interaction of the enzyme with specific phospholipids essential for its activity, which would mask any stimulatory effect due to increased fluidity. Recent work is now indicating that both general (Seeman, 1972) and local anesthetics (Papahadjopoulos et al., 1975a) may also act partly through a general fluidizing effect on the lipid component of the membrane.

5. Concluding remarks

This review has discussed the evidence which has led many workers to the viewpoint that changes in membrane lipid fluidity can influence the activity of membrane enzymes. Objectively, it does appear, however, that all that can be justifiably said at present is that a membrane with a fluid lipid component is required for the activity of membrane enzymes, and probably for cellular health and vigor in general. The basic mechanism of these effects still remains unclear. The possible complications of membrane structure and function suggest a mosaic of almost limitless complexity in which some membranes may well rival the diversity we can presently perceive for the whole cell itself.

Nonetheless, the principle properties of the lipid bilayer as the basic structural fabric for membranes seem well-established and its modification in biological membranes is amenable to investigation. In this regard the concepts of lateral phase separation together with separate protein microenvironments with differing properties and functions, give an indication of the nature of the complexities of biological membranes. Indeed, it is the detailed properties of the membrane proteins and their specific interactions with lipids which looms as the most formidable obstacle to understanding membrane function. Yet even here a few principles are beginning to be established. Suffice it to say, however, that we are still a long way from knowing the molecular details of the interaction of any membrane protein with its neighboring lipid molecules in the membrane and the way in which changes in the lipid component influence the function of such proteins. These data will have to be determined by the precise types of physical techniques discussed in this review to reconstituted systems known to exactly mimic the membrane protein in situ. The details of how lipid fluidity changes affect protein function also need to be clarified and may well differ from protein to protein. Thus, although increasing fatty acyl chain fluidity generally decreases the Van der Waals and other cohesive forces between neighboring phospholipid molecules, it is not known whether similar changes increase or decrease the interactions of phospholipids with proteins. An important question is to what degree are the phospholipid microenvironments of enzymes fluid, and how do they respond to alterations in the fluidity of the bulk membrane lipids? Thus, not unexpectedly, it is still true that "our model systems have shed very little light on active and/or facilitated diffusion and none at all on the coordinated systems of membrane activity" (Bangham, 1972). The assumption now seems to be, however, that some day they will.

This review covers the experimental work of just over half a decade from the inception of its subject matter as an established area of investigation. This has made the compilation effort somewhat less onerous but definite conclusions proportionally scantier. For someone attempting the same task five years hence the literature to be reviewed will have undoubtedly increased at the usual geometric rate. Hopefully the increased labor will be compensated for by clearer insights into the problem.

Acknowledgements

I should like to thank Dr. D.S. Berns of the New York State Division of Labs. and Research and Dr. S.H.G. Allen, Dr. E. Morrison, Dr. D. Treble and Steven Ullrich of Albany Medical College for reading the manuscript and for their valuable suggestions. I should also like to acknowledge former colleagues at Roswell Park Memorial Institute, Drs. D. Papahadjopoulos, K. Jacobson, E. Mayhew, R. Juliano and G. Poste among others for stimulating and helpful debates on many of the concepts and work discussed in this review. Finally, I should like to express my deep appreciation to Mrs. Elvira Graham who willingly spent so many hours typing and proofreading the manuscript.

I should like to thank both authors and publishers for their kind permission to reproduce the following figures: Fig. 1 (copyright 1972 by the American Association for the Advancement of Science); Figs. 2, 3 (Academic Press); Figs. 4, 13 (copyright by the American Chemical Society); Figs. 7, 10, 16 (Elsevier/North-Holland Biomedical Press); Fig. 8 (The Rockefeller University Press); Fig. 14 (Pergamon Press); Fig. 15 (American Society of Biological Chemists).

References

Abercrombie, M. and Ambrose, E.F. (1962) The surface properties of cancer cells; a review. Cancer Res. 22, 525–548.
Adam, G., Alpes, H., Blaser, K. and Neubert, B. (1975) Cholesterol and phospholipid content of 3T3 cells and transformed derivatives. Z. Naturforsch. 30, 638–642.
Ainsworth, P.J., Tustanoff, E.R. and Ball, A.J.S. (1972) Membrane phase transitions as a diagnostic tool for studying mitochondriogenesis. Biochem. Biophys. Res. Commun. 47, 1299–1305.
Aloia, R.C., Pengelley, E.T., Bolen, J.L. and Rouser, G. (1974) Changes in phospholipid composition in hibernating ground squirrel, *Citellis lateralis*, and their relationships to membrane function at reduced temperatures. Lipids 9, 993–999.
Aveldano, M.I. and Bazan, N.G. (1975) Differential lipid deacylation during brain ischemia in a homeotherm and a poikilotherm. Content and composition of free fatty acids and triacylglycerols. Brain Res. 100, 99–110.
Averdunck, R. and Lauf, P.K. (1975) Effects of mitogens on sodium-potassium transport, ^{3}H-ouabain binding and adenosine triphosphatase activity in lymphocytes. Exp. Cell Res. 93, 331–342.
Bangham, A.D. (1968) Physical structure and behavior of lipids and lipid enzymes, In: J.A.V. Butler and D. Noble (Eds.), Progress in Biophysics and Molecular Biology, Vol. 18, Pergamon, Oxford, pp. 29–95.
Bangham, A.D. (1972) Lipid bilayers and biomembranes. Annu. Rev. Biochem. 41, 753–776.
Bangham, A.D., Hill, M.W. and Miller, N.G.A. (1974) Preparation and use of liposomes as models of

biological membranes. In: E.D. Korn (Ed.), Methods in Membrane Biology, Vol. 1, Plenum, New York, pp. 1–68.

Bangham, A.D., Standish, M.M. and Watkins, J.C. (1965) Diffusion of univalent ions across the lamellae of swollen phospholipids. J. Mol. Biol. 13, 238–252.

Bar, H-P. (1974) On the kinetics and temperature dependence of adrenaline–adenylate cyclase interactions. Mol. Pharmacol. 10, 597–604.

Barnett, R.E., Furcht, L.T. and Scott, R.E. (1974a) Differences in membrane fluidity and structure in contact-inhibited and transformed cells. Proc. Natl. Acad. Sci. USA 71, 1992–1994.

Barnett, R.E., Scott, R.E., Furcht, L.T. and Kersey, J.H. (1974b) Evidence that mitogenic lectins induce changes in lymphocyte membrane fluidity. Nature (London) 249, 465–466.

Batzri, S. and Korn, E.D. (1973) Single bilayer liposomes prepared without sonication. Biochim. Biophys. Acta 298, 1015–1019.

Bergelson, L.D., Dyatlovitskaya, E.V., Torkhovskaya, T.I., Sorokina, I.B. and Gorkova, N.P. (1970) Phospholipid composition of membranes in the tumor cell. Biochim. Biophys. Acta 210, 287–298.

Berlin, R.D. (1975) Microtubules and the fluidity of the cell surface. Ann. N.Y. Acad. Sci. 253, 445–454.

Berlin, R.D. and Oliver, J.M. (1975) Membrane transport of purine and pyrimidine bases and nucleosides in animal cells. Int. Rev. Cytol. 42, 287–336.

Bernsohn, J. and Spitz, F.J. (1974) Linoleic- and linolenic acid dependency of some brain membrane-bound enzymes after lipid deprivation in rats. Biochem. Biophys. Res. Commun. 57, 293–298.

Bertoli, E., Parenti-Castelli, G., Landi, L., Sechi, A.M. and Lenaz, G. (1973) Activation energies of mitochondrial adenosine triphosphatase under different conditions. Bioenergetics 4, 591–598.

Blazyk, J.F. and Steim, J.M. (1972) Phase transitions in mammalian membranes. Biochim. Biophys. Acta 266, 737–741.

Bloj, B., Morero, R.D. and Farias, R.N. (1974) Effect of essential fatty acid deficiency on the Arrhenius plot of acetylcholinesterase from rat erythrocytes. J. Nutr. 104, 1265–1272.

Boldyrev, A.A., Tkachuk, V.A. and Titanji, P.V.R. (1974) Activation energy of skeletal muscle sarcolemmal Na^+, K^+-adenosine triphosphatase. Biochim. Biophys. Acta 357, 319–324.

Bowler, K. and Duncan, C.J. (1968) The effect of temperature on the Mg^{2+}-dependent and Na^+-K^+ ATPases of a rat brain microsomal preparation. Comp. Biochem. Physiol. 24, 1043–1054.

Bowler, K. and Tirri, R. (1974) The temperature characteristics of synaptic membrane ATPases from immature and adult rat brain. J. Neurochem. 23, 611–613.

Bray, G.M. (1973) A comparison of the ouabain-sensitive $(Na^+ + K^+)$-ATPase of normal and dystrophic skeletal muscle. Biochim. Biophys. Acta 298, 239–245.

Bretscher, M.S. (1973) Membrane structure: some general principles. Science 181, 622–629.

Bretscher, M.S. and Raff, M.C. (1975) Mammalian plasma membranes. Nature (London) 258, 43–49.

Bruni, A., Van Dijck, P.W.M. and De Gier, J. (1975) The role of phospholipid acyl chains in the activation of mitochondrial ATPase complex. Biochim. Biophys. Acta 406, 315–328.

Burger, M.M. (1973) Surface changes in transformed cells detected by lectins. Fed. Proc. 32, 91–101.

Chapman, D. (1968) Recent physical studies of phospholipids and natural membranes, In: D. Chapman (Ed.), Biological Membranes, Vol. 1, Academic Press, London, pp. 125–202.

Chapman, D. (1973) Some recent studies of lipids, lipid-cholesterol and membrane systems, In: D. Chapman and D.F.H. Wallach (Eds.), Biological Membranes, Vol. 2, Academic Press, London, pp. 91–144.

Chapman, D., Urbina, J. and Keough, K.M. (1974) Biomembrane phase transitions. Studies of lipid-water systems using differential scanning calorimetry. J. Biol. Chem. 249, 2512–2521.

Charnock, J.S. and Bashford, C.L. (1975) A fluorescent probe study of the lipid mobility of membranes containing sodium- and potassium-dependent adenosine triphosphatase. Mol. Pharmacol. 11, 766–774.

Charnock, J.S., Cook, D.A., Almeida, A.F. and To, R. (1973) Activation energy and phospholipid requirements of membrane-bound adenosine triphosphatase. Arch Biochem. Biophys. 159, 393–399.

Charnock, J.S., Cook, D.A. and Casey, R. (1971a) The role of cations and other factors on the apparent energy of activation of (Na + K) ATPase. Arch. Biochem. Biophys. 147, 323–329.

Charnock, J.S., Doty, D.M. and Russell, J.C. (1971b) The effect of temperature on the activity of ($Na^+ + K^+$) ATPase. Arch. Biochem. Biophys. 142, 633–637.

Charnock, J.S., Almeida, A.F. and To, R. (1975) Temperature-activity relationships of cation activation and ouabain inhibition of (Na + K) ATPase. Arch. Biochem. Biophys. 167, 480–487.

Cherry, R.J. (1975) Protein mobility in membranes. FEBS Letters 55, 1–7.

Cobon, G.S. and Haslam, J.M. (1973) The effect of altered membrane sterol composition on the temperature dependence of yeast mitochondrial ATPase. Biochem. Biophys. Res. Commun. 52, 320–326.

Coleman, R. (1973) Membrane-bound enzymes and membrane ultrastructure. Biochim. Biophys. Acta, 300, 1–30.

Cooper, R.A., Arner, E.C., Wiley, J.S. and Shattil, S.J. (1975) Modification of red cell membrane structure by cholesterol-rich lipid dispersions. J. Clin. Invest. 55, 115–126.

De Kruyff, B., Van Dijck, P.W.M., Godbach, R.W., Denez, R.A. and Van Deenen, L.L.M. (1973) Influence of fatty acid and sterol composition on the lipid phase transition and activity of membrane-bound enzymes in *Acholeplasma laidlawii*. Biochim. Biophys. Acta 330, 269–282.

De Kruyff, B., Van Dijck, P.W.M., Demel, R.A., Schuijff, A., Brants, F. and Van Deenen, L.L.M. (1974) Non random distribution of cholesterol in phosphatidylcholine bilayers. Biochim. Biophys. Acta 356, 1–7.

Demel, R.A., London, Y., Geurts van Kessel, W.S.M., Vossenberg, F.G.A. and Van Deenen, L.L.M. (1973) The specific interaction of myelin basic protein with lipids at the air–water interface. Biochim. Biophys. Acta 311, 507–519.

Devaux, P. and McConnell, H.M. (1972) Lateral diffusion in spin-labeled phosphatidylcholine multilayers. J. Am. Chem. Soc. 94, 4475–4481.

Dixon, M. and Webb, E.C. (1964) Enzymes, Academic Press, New York, pp. 145–166.

Dodd, N.J.F. (1975) PHA and lymphocyte membrane fluidity. Nature (London) 257, 827–828.

Dufourcq, J., Faucon, J.F., Lussan, C. and Bernon, R. (1975) Study of lipid-protein interactions in membrane models: Intrinsic fluorescence of cytochrome b_5-phospholipid complexes. FEBS Lett. 57, 112–116.

Edelman, G.M. (1976) Surface modulation in cell recognition and cell growth. Science 192, 218–226.

Edidin, M. (1974) Rotational and translational diffusion in membranes. Ann. Rev. Biophys. Bioeng. 3, 179–201.

Eletr, S., Zakim, D. and Vessey, D.A. (1973) A spin-label study of the role of phospholipids in the regulation of membrane-bound microsomal enzymes. J. Mol. Biol. 78, 351–362.

Eletr, S., Williams, M.A., Watkins, T., and Keith, A.D. (1974) Perturbations of the dynamics of lipid alkyl chains in membrane systems: Effect on the activity of membrane-bound enzymes. Biochim. Biophys. Acta 339, 190–201.

Elford, B.C. (1975) Interactions between temperature and tonicity on cation transport in dog red cells. J. Physiol. 246, 371–397.

Emmelot, P. and Van Hoeven, R.P. (1975) Phospholipid unsaturation and plasma membrane organization. Chem. Phys. Lipids 14, 236–246.

Engelman, D.M. (1971) Lipid bilayer structure in the membrane of *Mycoplasma laidlawii*. J. Mol. Biol. 58, 153–165.

Esfahani, M., Limbrick, A.R., Knutton, S., Oka, T. and Wakil, S.J. (1971) The molecular organization of lipids in the membrane of *E. coli*: Phase transitions. Proc. Natl. Acad. Sci. USA 68, 3180–3184.

Esfahani, M., Crowfoot, P.D. and Wakil, S.J. (1972) Molecular organization of lipids in *E. coli* membranes. II. Effect of phospholipids on succinic-ubiquinone reductase activity. J. Biol. Chem. 247, 7251–7256.

Esser, A.F. and Lanyi, J.K. (1973) Structure of the lipid phase in cell envelope vesicles from *Halobacterium cutirubrum*. Biochemistry 12, 1933–1938.

Farias, R.N., Bloj, B., Morero, R.D., Sineriz, F. and Trucco, R.E. (1975) Regulation of allosteric membrane-bound enzymes through changes in membrane lipid composition. Biochim. Biophys. Acta 415, 231–251.

Feinstein, M.B., Fernandez, S.M. and Sha'afi, R.E. (1975) Fluidity of natural membranes and phosphatidylserine and ganglioside dispersions. Biochim. Biophys. Acta 413, 354–370.

Fenster, L.J. and Copenhaver Jr., J.H. (1967) Phosphatidylserine requirement of ($Na^+ + K^+$)-activated adenosine triphosphatase from rat kidney and brain. Biochim. Biophys. Acta 137, 406–408.

Feo, F., Canuto, R.A., Bertone, G., Garcea, R. and Pani, P. (1973) Cholesterol and phospholipid composition of mitochondria and microsomes isolated from Morris Hepatoma 5123 and rat liver. FEBS Lett. 33, 229–232.

Fiehn, W. and Seiler, D. (1975) Alteration of erythrocyte ($Na^+ + K^+$)-ATPase by replacement of cholesterol by desmosterol in the membrane. Experientia 31, 773–775.

Foster, D.O. and Pardee, A.B. (1969) Transport of amino acids by confluent and nonconfluent 3T3 and polyoma virus-transformed 3T3 cells growing on glass cover slips. J. Biol. Chem. 2675–2681.

Fourcans, B. and Jain, M.K. (1974) Role of phospholipids in transport and enzymic reactions, In: R. Paoletti and D. Kritchevsky (Eds.), Advances in Lipid Research, Vol. 12, Academic Press, New York, pp. 147–226.

Fox, C.F. (1975) Transitions in model systems and membranes, In: C.F. Fox (Ed.), MTP International Review of Science, Biochemistry of Cell Walls and Membranes, Biochemistry Series 1, Vol. 2, University Park Press, Baltimore, pp. 279–306.

Frye, L.D. and Edidin, M. (1970) The rapid intermixing of cell surface antigens after formation of mouse-human heterokaryons. J. Cell Sci. 7, 319–335.

Fuchs, P., Parola, A., Robbins, P.W. and Blout, E.R. (1975) Fluorescence polarization and viscosities of membrane lipids of 3T3 cells. Proc. Natl. Acad. Sci. USA 72, 3351–3354.

Furcht, L.T. and Scott, R.E. (1974) Influence of cell cycle and cell movement on the distribution of intramembranous (IMP) particles in contact-inhibited and transformed cells. Exp. Cell Res. 88, 311–318.

Gaffney, B.J. (1975) Fatty acid chain flexibility in the membranes of normal and transformed fibroblasts. Proc. Natl. Acad. Sci. USA 72, 664–668.

Gaines, G.L. (1966) Insoluble monolayers at liquid-gas interfaces. Interscience, New York, pp. 347–361.

Galla, H.J. and Sackmann, E. (1975) Chemically induced lipid phase separation in model membranes containing charged lipids: A spin label study. Biochim. Biophys. Acta 401, 509–529.

Garland, R.C. and Cori, C.F. (1972) Separation of phospholipids from glucose-6-phosphatase by gel chromatography. Specificity of phospholipid reactivation. Biochemistry 11, 4712–4718.

Gitler, C. (1972) Plasticity of biological membranes. Ann. Rev. Biophys. Bioeng. 1, 51–92.

Godici, P.E. and Landsberger, F.R. (1974) The dynamic structure of lipid membranes. A ^{13}C nuclear magnetic resonance study using spin labels. Biochemistry 13, 362–368.

Goldin, S.M. and Tong, S.W. (1974) Reconstitution of active transport catalyzed by the purified sodium and potassium ion-stimulated adenosine triphosphatase from canine renal medulla. J. Biol. Chem. 249, 5907–5915.

Goldman, S.S. (1975) Cold resistance of the brain during hibernation, III. Evidence of a lipid adaptation. Am. J. Physiol. 228, 834–838.

Goldman, S.S. and Albers, R.W. (1973) (Na + K) activated ATPase, IX. Role of phospholipids. J. Biol. Chem. 243, 867–874.

Goldman, S.S. and Albers, R.W. (1975) Cold resistance of the brain during hibernation. Temperature sensitivity of the partial reactions of the Na^+, K^+-ATPase. Arch. Biochem. Biophys. 169, 540–544.

Gorter, E. and Grendel, F. (1925) On bimolecular layers of lipoids on the chromocytes of the blood. J. Exp. Med. 41, 439–443.

Grant, C.W.M., Wu, S. H-W., and McConnell, H.M. (1974) Lateral phase separations in binary lipid mixtures. Correlation between spin label and freeze-fracture electron microscopic studies. Biochim. Biophys. Acta 363, 151–158.

Green, D.E. and Fleischer, S. (1963) The role of lipids in mitochondrial electron transfer and oxidative phosphorylation. Biochim. Biophys. Acta 70, 554–582.

Green, D.E. and Tzagoloff, A. (1966) The mitochondrial electron transfer chain. Arch. Biochem. Biophys. 116, 293–304.
Green, N.M. (1975) The influence of lipids on the conformation of the adenosine triphosphatase of sacroplasmic reticulum and of other membrane proteins. Biochem. Soc. Trans. 3, 604–605.
Grinna, L. and Barber, A. (1972) Age-related changes in membrane lipid content and enzyme activities. Biochim. Biophys. Acta 288, 347–353.
Grisham, C.M. and Barnett, R.E. (1972) The interrelationship of membrane and protein structure in the functioning of the $(Na^+ + K^+)$ activated ATPase. Biochim. Biophys. Acta 266, 613–624.
Grisham, C.M. and Barnett, R.E. (1973) The role of lipid-phase transitions in the regulation of the (sodium + potassium) adenosine triphosphatase. Biochemistry 12, 2635–2637.
Grover, A.K., Slotboom, A.J. De Haas, G.H. and Hammes, G.G. (1975) Lipid specificity of β-hydroxybutyrate dehydrogenase activation. J. Biol. Chem. 250, 31–38.
Gruener, N. and Avi-Dor, Y. (1966) Temperature-dependence of activation and inhibition of rat-brain adenosine triphosphatase activated by sodium and potassium ions. Biochem. J. 100, 762–767.
Gulik-Krzywicki, T. (1975) Structural studies of the associations between biological membrane compounds. Biochim. Biophys. Acta 415, 1–28.
Gulik-Krzywicki, T., Shechter, E., Luzzati, V. and Faure, M. (1969) Interactions of proteins and lipids: structure and polymorphism of protein-lipid water phases. Nature (London) 223, 1116–1121.
Haest, C.W.M., Verkleij, A.J., De Gier, J., Scheek, R., Ververgaert, P.H.J. and Van Deenen, L.L.M. (1974) The effect of lipid phase transitions on the architecture of bacterial membranes. Biochim. Biophys. Acta 356, 17–26.
Hatanaka, A.M. (1974) Transport of sugars in tumor cell membranes. Biochim. Biophys. Acta 355, 77–104.
Hazel, J.R. (1972) The effect of temperature acclimation upon succinic dehydrogenase activity from the epaxial muscle of the common goldfish (*Carassius auratus*), II. Lipid reactivation of the soluble enzyme. Comp. Biochem. Physiol. 43B, 863–882.
Hazel, J.R. and Prosser, D.L. (1974) Molecular mechanisms of temperature compensation in poikilotherms. Physiol. Rev. 54, 620–677.
Helenius, A. and Simons, K. (1972) The binding of detergents to lipophilic and hydrophilic proteins. J. Biol. Chem. 247, 3656–3661.
Helenius A. and Simons, K. (1975) Solubilization of membranes by detergents. Biochim. Biophys. Acta 415, 29–79.
Hendler, R.W. (1971) Biological membrane ultrastructure. Physiol. Rev. 51, 66–97.
Hendler, R.W. (1974) Protein disposition in biological membranes. In: L.A. Manson (Ed.), Biomembranes, Vol. 5, Plenum, New York, pp. 251–273.
Hilden, S. and Hokin, L.E. (1975) Active potassium transport coupled to active sodium transport in vesicles reconstituted from purified sodium and potassium ion-activated adenosine triphosphatase from the rectal gland of *Squalus acanthias*. J. Biol. Chem. 250, 6296–6303.
Hilden, S., Rhee, H.M. and Hokin, L.E. (1974) Sodium transport by phospholipid vesicles containing purified sodium and potassium ion activated adenosine triphosphatase. J. Biol. Chem. 249, 7432–7440.
Hinkle, P.C., Kim, J.J. and Racker, E. (1972) Ion transport and respiratory control in vesicles formed from cytochrome oxidase and phospholipids. J. Biol. Chem. 247, 1338–1339.
Hinz, H-J. and Sturtevant, J.M. (1972) Calorimetric studies of dilute aqueous suspensions of bilayers formed from synthetic L-α-lecithins. J. Biol. Chem. 247, 6071–6075.
Hoar, W.S. and Cottle, M.K. (1951) Some effects of temperature acclimatization on the chemical constituents of goldfish tissues. Can. J. Zool. 30, 49–54.
Hokin, L.E. (1975) Purification and molecular properties of the (sodium + potassium)-adenosinetriphosphatase and reconstitution of coupled sodium and potassium transport in phospholipid vesicles containing purified enzyme. J. Exp. Zool. 194, 197–206.
Hokin, L.E. and Hexum, T.C. (1972) Studies on the characterization of the (Na + K) transport ATPase, IX. On the role of phospholipids in the enzyme. Arch. Biochem. Biophys. 151, 453–463.

Holley, R.F. (1975) Control of growth of mammalian cells in cell culture. Nature (London) 258, 487–490.

Hong, K. and Hubbell, W.L. (1972) Preparation and properties of phospholipid bilayers containing rhodopsin. Proc. Natl. Acad. Sci. USA 69, 2617–2621.

Houslay, M.D., Warren, G.B., Birdsall, N.J.M. and Metcalfe, J.C. (1975) Lipid phase transitions control β-hydroxybutyrate dehydrogenase activity in defined lipid protein complexes. FEBS Lett. 51, 146–151.

Huang, C-H. (1969) Studies on phosphatidylcholine vesicles. Formation and physical characteristics. Biochemistry 8, 344–352.

Huang, C-H., Sipe, J.P., Chow, S.T. and Martin, R.B. (1974) Differential interaction of cholesterol with phosphatidylcholine on the inner and outer surfaces of lipid bilayer vesicles. Proc. Natl. Acad. Sci. USA 71, 359–362.

Hubbell, W.L. and McConnell, H.M. (1968) Spin-label studies of the excitable membranes of nerve and muscle. Proc. Natl. Acad. Sci. USA 61, 12–16.

Hubbell, W.L. and McConnell, H.M. (1971) Molecular motion in spin-labeled phospholipids and membranes. J. Am. Chem. Soc. 93, 314–326.

Inbar, M. and Sachs, L. (1973) Mobility of carbohydrate containing sites on the surface membrane in relation to the control of cell growth. FEBS Lett. 32, 124–128.

Inbar, M. and Shinitzky, M. (1974a) Increase of cholesterol level in the surface membrane of lymphoma cells and its inhibitory effect on ascites tumor development. Proc. Natl. Acad. Sci. USA 71, 2128–2130.

Inbar, M. and Shinitzky, M. (1974b) Cholesterol as a bioregulator in the development and inhibition of leukemia. Proc. Natl. Acad. Sci. USA 71, 4229–4231.

Inbar, M., Shinitzky, M. and Sachs, L. (1973) Rotational relaxation time of concanavalin A bound to the surface membrane of normal and malignant transformed cells. J. Mol. Biol. 81, 245–253.

Inesi, G., Millman, M. and Eletr, S. (1973) Temperature-induced transitions of function and structure in sarcoplasmic reticulum membranes. J. Mol. Biol. 81, 483–504.

Isselbacher, K.J. (1972) Increased uptake of amino acids and 2-deoxy-D-glucose by virus-transformed cells in culture. Proc. Natl. Acad. Sci. USA 69, 585–589.

Ito, T. and Ohnishi, S.-I. (1974) Ca^{2+}-induced lateral phase separations in phosphatidic acid-phosphatidylcholine membranes. Biochim. Biophys. Acta 352, 29–37.

Ito, A. and Sato, R. (1968) Purification by means of detergents and properties of cytochrome b_5 from liver microsomes. J. Biol. Chem. 243, 4922–4930.

Jacobs, E.E. and Sanadi, D.R. (1960) The reversible removal of cytochrome c from mitochondria. J. Biol. Chem. 235, 531–534.

Jacobson, K. and Papahadjopoulos, D. (1975) Phase transitions and phase separations in phospholipid membranes induced by changes in temperature, pH, and concentration of bivalent cations. Biochemistry 14, 152–162.

Jacobson, K., Wu, E. and Poste, G. (1976) Measurement of the translational mobility of concanavalin A in glycerol-saline solutions and on the cell surface by fluorescence recovery after photobleaching. Biochim. Biophys. Acta 433, 215–227.

Jain, M.K. (1975) Role of cholesterol in biomembranes and related systems, In: F. Bronner and A. Kleinzeller (Eds.), Current Topics in Membranes and Transport, vol. 7, Academic Press, New York, pp. 1–57.

Janki, R.M., Aithal, H.N., McMurray, W.C. and Tustanoff, E.R. (1974) The effect of altered membrane-lipid composition on enzyme activities of outer and inner mitochondrial membranes of *Saccharomyces cerevisiae*. Biochem. Biophys. Res. Commun. 56, 1078–1085.

Johnston, P.V. and Roots, B.I. (1964) Brain lipid fatty acids and temperature acclimation. Comp. Biochem. Physiol. 11, 303–309.

Jost, P., Griffith, O.H., Capaldi, R.A. and Vanderkooi, G. (1973a) Evidence for boundary lipid in membranes. Proc. Natl. Acad. Sci. USA 70, 480–484.

Jost, P., Griffith, O.H., Capaldi, R.A. and Vanderkooi, G. (1973b) Identification and extent of fluid bilayer regions in membraneous cytochrome oxidase. Biochim. Biophys. Acta 311, 141–152.

Juliano, R.L., Kimelberg, H.K. and Papahadjopoulos, D. (1971) Synergistic effects of a membrane protein (spectrin) and Ca^{2+} on the Na$^+$ permeability of phospholipid vesicles. Biochim. Biophys. Acta 241, 894–905.

Jurtshuk, P., Sekuzu, I., and Green, D.E. (1961) The interaction of the D(-)β-hydroxybutyric apoenzyme with lecithin. Biochem. Biophys. Res. Commun. 6, 76–80.

Kagawa, Y. and Racker, E. (1966) Partial resolution of the enzymes catalyzing oxidative phosphorylation, IX. Reconstruction of oligomycin-sensitive adenosine triphosphatase. J. Biol. Chem. 241, 2467–2474.

Kaplan, J., Canonico, P.G. and Caspary, W.J. (1973) Electron spin resonance studies of spin-labeled mammalian cells by detection of surface-membrane signals. Proc. Natl. Acad. Sci. USA 70, 66–70.

Kemp, A., Groot, G.S.P. and Reitsma, H.J. (1969) Oxidative phosphorylation as a function of temperature. Biochim. Biophys. Acta 180, 28–34.

Kemp, P. and Smith, M.W. (1970) Effect of temperature acclimatization on the fatty acid composition of goldfish intestinal lipids. Biochem. J. 117, 9–15.

Kimelberg, H.K. (1975) Alterations in phospholipid-dependent (Na + K) ATPase activity due to lipid fluidity. Effects of cholesterol and Mg^{2+}. Biochim. Biophys. Acta 413, 143–156.

Kimelberg, H.K. (1976a) Protein-liposome interactions and their relevance to the structure and function of cell membranes. Mol. Cell Biochem. 10, 171–190.

Kimelberg, H.K. (1976b) Enzyme composition of plasma membranes: Mammalian tissues and cultured cells. Cell Biology Data Book, FASEB Biological Handbooks, Bethesda, Md. pp. 95–98.

Kimelberg, H.K. and Mayhew, E. (1975) Increased ouabain-sensitive ^{86}Rb$^+$ uptake and sodium and potassium ion-activated adenosine triphosphatase activity in transformed cell lines. J. Biol. Chem. 250, 100–104.

Kimelberg, H.K. and Papahadjopoulos, D. (1971a) Interactions of basic proteins with phospholipid membranes. Binding and changes in the Na$^+$ permeability of PS vesicles. J. Biol. Chem. 246, 1142–1148.

Kimelberg, H.K. and Papahadjopoulos, D. (1971b) Phospholipid-protein interactions: membrane permeability correlated with monolayer penetration. Biochim. Biophys. Acta 233, 805–809.

Kimelberg, H.K. and Papahadjopoulos, D. (1972) Phospholipid requirements for (Na$^+$ + K$^+$)-ATPase activity: Head-group specificity and fatty acid fluidity. Biochim. Biophys. Acta 282, 277–292.

Kimelberg, H.K. and Papahadjopoulos, D. (1974) Effects of phospholipid acyl chain fluidity, phase transitions and cholesterol on (Na$^+$ + K$^+$)-stimulated adenosine triphosphatase. J. Biol. Chem. 249, 1071–1080.

Kimzey, S.L. and Willis, J.S. (1971) Temperature adaptation of active sodium-potassium transport and of passive permeability in erythrocytes of ground squirrels. J. Gen. Physiol. 58, 634–649.

Kleeman, W. and McConnell, H.M. (1976) Interactions of proteins and cholesterol with lipids in bilayer membranes. Biochim. Biophys. Acta 419, 206–222.

Konev, S.V., Chernitski, E.A., Aksentisev, S.L., Mazhul, V.M., Volotovski, I.D. and Nisenbaum, G.D. (1975) Nondenaturational structural transitions of proteins and biological membranes. Mol. Cell Biochem. 7, 5–17.

Kornberg, R.D. and McConnell, H.M. (1971) Inside-outside transitions of phospholipids in vesicle membranes. Biochemistry 10, 1111–1120.

Krasne, S., Eisenman, G., Szabo, G. (1971) Freezing and melting of lipid bilayers and the mode of action of nonactin, valinomycin and gramicidin. Science 174, 412–415.

Kreiner, P.W., Keirns, J.J. and Bitensky, M.W. (1973) A temperature-sensitive change in the energy of activation of hormone-stimulated hepatic adenylyl cyclase. Proc. Natl. Acad. Sci. USA 70, 1785–1789.

Kroes, J. and Ostwald, R. (1971) Erythrocyte membranes—effect of increased cholesterol content on permeability. Biochim. Biophys. Acta 249, 647–650.

Kroon, P.A., Kainosho, M. and Chan, S.I. (1975) State of molecular motion of cholesterol in lecithin bilayers. Nature (London) 256, 582–584.

Kumamoto, J., Raison, J.K. and Lyons, J.M. (1971) Temperature "breaks" in Arrhenius plots: A thermodynamic consequence of a phase change. J. Theoret. Biol. 31, 47–51.

Ladbrooke, B.D. and Chapman, D. (1969) Thermal analysis of lipids, proteins and biological membranes. A review and summary of some recent studies. Chem. Phys. Lipids 3, 304–367.

Ladbrooke, B.D., Williams, R.M. and Chapman, D. (1968) Studies on lecithin-cholesterol-water interactions by differential scanning calorimetry and X-ray diffraction. Biochim. Biophys. Acta 150, 333–340.

Langmuir, I. (1938) Overturning and anchoring of monolayers. Science 87, 493–500.

Laurent, B., Roy, P.E. and Gailis, L. (1974) Inhibition by Δ^1-tetrahydrocannabinol of Na^+-K^+ transport ATPase from rat ileum. Can. J. Physiol. Pharmacol. 52, 1110–1113.

Lawrence, D.K. and Gill, E.W. (1975) The effects of Δ^1-tetrahydrocannabinol and other cannabinoids on spin-labeled liposomes and their relationship to mechanisms of general anesthesia. Mol. Pharmacol. 11, 595–692.

Lee, A.G., Birdsall, N.J.M., Metcalfe, J.C. (1973) Measurement of fast lateral diffusion of lipids in vesicles and in biological membranes by ^{1}H nuclear magnetic resonance. Biochemistry 12, 1650–1659.

Lee, A.G., Birdsall, N.J.M., Metcalfe, J.C., Toon, P.A. and Warren, G.B. (1974) Clusters in lipid bilayers and the interpretation of thermal effects in biological membranes. Biochemistry 13, 3699–3705.

Lee, M.P. and Gear, A.R.L. (1974) The effect of temperature on mitochondrial membrane-linked reactions. J. Biol. Chem. 249, 7541–7549.

Lenaz, G., Sechi, A.M., Parenti-Castelli, G., Landi, L. and Bertoli, E. (1972) Activation energies of different mitochondrial enzymes: Breaks in Arrhenius plots of membrane-bound enzymes occur at different temperatures. Biochem. Biophys. Res. Commun. 49, 536–542.

Levey, G.S. (1971) Restoration of glucagon responsiveness of solubilized myocardial adenyl cyclase by phosphatidylserine. Biochem. Biophys. Res. Commun. 43, 108–113.

Levey, G.S. (1973) The role of phospholipids in hormone activation of adenylate cyclase, In: R.O. Greep (Ed.), Recent Progress in Hormone Research, vol. 29, Academic Press, New York, pp. 361–382.

Levine, Y.K. and Wilkins, M.H.F. (1971) Structure of oriented lipid bilayers. Nature (London) New Biol., 230, 69–72.

Linden, C.D., Wright, K.L., McConnell, H.M. and Fox, C.F. (1973) Lateral phase separations in membrane lipids and the mechanisms of sugar transport in *Escherichia coli*. Proc. Natl. Acad. Sci. USA 70, 2271–2275.

Lyons, J.M. (1972) Phase transitions and control of cellular metabolism at low temperatures. Cryobiology 9, 341–350.

Lyons, J.M. and Raison, J.K. (1970) A temperature-induced transition in mitochondrial oxidation: Contrasts between cold and warm blooded animals. Comp. Biochem. Physiol. 37, 405–411.

Madeira, V.M.C. and Antunes-Madeira, M.C. (1975) Thermotropic transitions in sarcoplasmic reticulum. Biochem. Biophys. Res. Commun. 65, 997–1003.

Marchesi, S.L., Steers, E., Marchesi, V.T. and Tillack, T.W. (1969) Physical and chemical properties of a protein isolated from red cell membranes. Biochemistry 9, 50–57.

Massey, V. (1953) Studies on Fumarase, 3. The Effect of Temperature. Biochem. J. 53, 72–79.

Massey, V., Curti, B. and Ganther, H. (1966) A temperature-dependent conformational change in D-amino acid oxidase and its effect on catalysis. J. Biol. Chem. 241, 2347–2357.

Mavis, R.D. and Vagelos, P.R. (1972) The effect of phospholipid fatty acid composition on membranous enzymes in *E. coli*. J. Biol. Chem. 247, 652–659.

Mayhew, E. (1972) Ion transport by ouabain resistant and sensitive Ehrlich ascites carcinoma cells. J. Cell Physiol. 79, 441–451.

McConnell, H.M., Wright, K.L. and McFarland, B.G. (1972) The fraction of the lipid in a biological membrane that is in a fluid state: A spin label assay. Biochem. Biophys. Res. Commun. 47, 273–281.

McNamee, M.G. and McConnell, H.M. (1973) Transmembrane potentials and phospholipid flip-flop in excitable membrane vesicles. Biochemistry 12, 2951–2958.

Mehlhorn, R.J. and Keith, A.D. (1972) Spin labeling of biological membranes. In: C.F. Fox and A.D. Keith (Eds.), Membrane Molecular Biology, Sinauer, Stamford, Connecticut, pp. 192–227.

Morrisett, J.D., Pownall, H.J., Plumlee, R.T., Smith, L.C., Zehner, Z.E., Esfahani, M. and Wakil, S.J. (1975) Multiple thermotropic phase transitions in *Escherichia coli* membranes and membrane lipids. J. Biol. Chem. 250, 6969–6976.

Morrison, E.S., Scott, R.F., Imai, H., Kroms, M., Nour, B.A. and Briggs, R.G. (1970) Effect of thrombogenic and atherogenic diets on aspects of hepatic energy metabolism in rats. Atherosclerosis 12, 139–158.

Mueller, P., Rudin, D.O., Tien, H.T. and Wescott, W.C. (1962) Reconstitution of cell membrane structure in vitro and its transformation into an excitable system. Nature (London) 194, 979–980.

Munn, E.A. (1974) The Structure of Mitochondria, Academic Press, London, pp. 218–232.

Neer, E.J. (1974) The size of adenylate cyclase. J. Biol. Chem. 249, 6527–6531.

Nicolau, Cl., Dietrich, W., Steiner, M.E., Steiner, S. and Melnick, J.L. (1975) ^{1}H and ^{13}C nuclear magnetic resonance spectra of the lipids in normal and SV40 virus-transformed hamster embryo fibroblast membranes. Biochim. Biophys. Acta 382, 311–321.

Nicolson, G.L. (1974) The interactions of lectins with animal cell surfaces. Int. Rev. Cytol. 39, 89–190.

Nielsen, N.C., Zahler, W.L. and Fleischer, S. (1973) Mitochondrial D-β-hydroxybutyrate dehydrogenase, IV. Kinetic analysis of reaction mechanism. J. Biol. Chem. 248, 2556–2562.

Ohnishi, S. and Ito, T. (1973) Clustering of lecithin molecules in phosphatidylserine membranes induced by calcium ion binding to phosphatidylserine. Biochem. Biophys. Res. Commun. 51, 132–138.

Ohnishi, T. and Kawamura, H., (1964) Rôle des phosphatides dans l'adénosine triphosphatase sensitive à l'ouabaine localiseé dans les membranes d'erythrocyte. J. Biochem. 56, 377–378.

Oldfield, E. and Chapman, D. (1972) Dynamics of lipids in membranes: Heterogeneity and the role of cholesterol. FEBS Lett. 23, 285–297.

Op den Kamp, J.A.F., Kauerz, M. Th., Van Deenen, L.L.M. (1975) Action of pancreatic phospholipase A_2 on phosphatidylcholine bilayers in different physical states. Biochim. Biophys. Acta 406, 169–177.

Overath, P. and Träuble, H. (1973) Phase transitions in cells, membranes, and lipids of *Escherichia coli*. Detection by fluorescent probes, light scattering and dilatometry. Biochemistry, 12, 2625–2634.

Overath, P., Hill, F.F. and Lamnek-Hirsch, I. (1971) Biogenesis of *E. coli* membrane: Evidence for randomization of lipid phase. Nature (London) New Biol. 234, 264–267.

Overath, P., Schairer, H.U. and Stoffel, W. (1970) Correlation of in vivo and in vitro phase transitions of membrane lipids in *Escherichia coli*. Proc. Natl. Acad. Sci. USA 67, 606–612.

Overton, E., (1899) Über die allgemein osmotischen Eigenschaften der Zelle, ihre vermutlichen Ursachen und ihre Bedeutung für die Vierteljahresschr. Naturforsch. Ges., Zürich 44, 88.

Owens, K. and Hughes, B.P. (1970) Lipids of dystrophic and normal mouse muscle: Whole tissue and particulate fractions. J. Lipid Res. 11, 486–495.

Owens, K., Weglicki, W.B., Ruth, R.C., Stam, A.C. and Sonnenblick, E.H. (1973) Lipid compositions, Ca^{2+} uptake, and Ca^{2+}-stimulated ATPase activity of sarcoplasmic reticulum of the cardiomyopathic hamster. Biochim. Biophys. Acta 296, 71–78.

Papahadjopoulos, D. (1973) Phospholipid membranes as experimental models for biological membranes, In: L. Prince and D.F. Sears (Eds.), Biological Horizons in Surface Science, Academic Press, New York, pp. 160–220.

Papahadjopoulos, D. (1974) Cholesterol and cell membrane function. A hypothesis concerning the etiology of atherosclerosis. J. Theoret. Biol. 43, 329–337.

Papahadjopoulos, D. and Bangham, A.D. (1966) Biophysical properties of phospholipids, II. Permeability of phosphatidylserine liquid crystals to univalent ions. Biochim. Biophys. Acta 126, 185–188.

Papahadjopoulos, D. and Kimelberg, H.K. (1973) Phospholipid vesicles (liposomes) as models for biological membranes: Their properties and interactions with cholesterol and proteins, In: S.G. Davison (Ed.), Progress in Surface Science, Pergamon, Oxford, pp. 141–232.

Papahadjopoulos, D. and Miller, N. (1967) Phospholipid model membranes, I. Structural characteristics of hydrated liquid crystals. Biochim. Biophys. Acta 135, 624–638.

Papahadjopoulos, D., Moscarello, M., Eylar, E.H. and Isac, T. (1975b) Effects of proteins on thermotropic phase transitions of phospholipid membranes. Biochim. Biophys. Acta 401, 317–335.

Papahadjopoulos, D. and Ohki, S. (1969) Stability of asymmetric phospholipid membranes. Science 164, 1075–1077.

Papahadjopoulos, D. and Poste, G. (1975) Calcium-induced phase separation and fusion in phospholipid membranes. Biophys. J. 15, 945–948.

Papahadjopoulos, D. and Weiss, L. (1969) Amino groups at the surfaces of phospholipid vesicles. Biochim. Biophys. Acta 183, 417–426.

Papahadjopoulos, D., Cowden, M. and Kimelberg, H.K. (1973a) Role of cholesterol in membranes. Effects on phospholipid-protein interactions, membrane permeability and enzymatic activity. Biochim. Biophys. Acta 330, 8–26.

Papahadjopoulos, D., Jacobson, K., Nir, S. and Isac, T. (1973b) Phase transitions in phospholipid vesicles. Fluorescence polarization and permeability measurements concerning the effect of temperature and cholesterol. Biochim. Biophys. Acta 311, 330–348.

Papahadjopoulos, D., Poste, G., Schaeffer, B.E. and Vail, W.J. (1974) Membrane fusion and molecular segregation in phospholipid vesicles. Biochim. Biophys. Acta 352, 10–28.

Papahadjopoulos, D., Jacobson, K., Poste, G. and Shepherd, G. (1975a) Effects of local anesthetics on membrane properties, I. Changes in the fluidity of phospholipid bilayers. Biochim. Biophys. Acta 394, 504–520.

Papahadjopoulos, D., Hui, S., Vail, W.J. and Poste, G. (1976a) Studies on membrane fusion, I. Interactions of pure phospholipid membranes and the effect of fatty acids, lysolecithin, proteins and dimethylsulfoxide. Biochim. Biophys. Acta 448, 245–264.

Papahadjopoulos, D., Vail, W.J., Pangborn, W.A. and Poste, G. (1976b) Studies on membrane fusion, II. Induction of fusion in pure phospholipid membranes by Ca^{2+} and other divalent metals. Biochim. Biophys. Acta 448, 265–283.

Papahadjopoulos, D., Vail, W.J., Newton, C., Nir, S., Jacobson, K., Poste, G. and Lazo, R. (1976c) Studies on membrane fusion, III. The role of calcium-induced phase changes. Biochim. Biophys. Acta 465, 579–598.

Phillips, M.C. (1972) The physical state of phospholipids and cholesterol in monolayers, bilayers and membranes, In: J.F. Danielli, M.D. Rosenberg and D.A. Cadenhead (Eds.), Progress in Surface and Membrane Science, vol. 5, Academic Press, New York, pp. 139–221.

Phillips, M.C., Ladbrooke, D.B. and Chapman, D. (1970) Molecular interactions in mixed lecithin systems. Biochim. Biophys. Acta 196, 35–44.

Phillips, M.C., Hauser, H. and Paltauf, F. (1972) The inter- and intra-molecular mixing of hydrocarbon chains in lecithin/water systems. Chem. Phys. Lipids. 8, 127–133.

Phillips, M.C., Graham, D.E. and Hauser, H. (1975) Lateral compressibility and penetration into phospholipid monolayers and bilayer membranes. Nature (London) 254, 154–156.

Plagemann, P.G.W. and Richey, D.P. (1974) Transport of nucleosides, nucleic acid bases, choline and glucose by animal cells in culture. Biochim. Biophys. Acta 344, 263–305.

Poo, M.M. and Cone, R.A. (1974) Lateral diffusion of rhodopsin in the photoreceptor membrane. Nature (London) 247, 438–441.

Posnansky, M., Kirkwood, D. and Solomon, A.K. (1973) Modulation of red cell K^+ transport by membrane lipids. Biochim. Biophys. Acta 330, 351–355.

Poste, G., Papahadjopoulos, D., Jacobson, K. and Vail, W.J. (1975) Effects of local anesthetics on membrane properties. II. Enhancement of the susceptibility of mammalian cells to agglutination by plant lectins. Biochim. Biophys. Acta 394, 520–540.

Priestland, R.N. and Whittam, R. (1972) The temperature dependence of activation by phosphatidylserine of the sodium pump adenosine triphosphatase. J. Physiol. 220, 353–361.

Quastel, M.R. and Kaplan, J.G. (1970) Early stimulation of potassium uptake in lymphocytes treated with PHA. Exp. Cell Res. 63, 230–233.

Quinn, P.J. and Dawson, R.M.C. (1969) Interactions of cytochrome c and ^{14}C carboxymethylated cytochrome c with monolayers of phosphatidylcholine, phosphatidic acid and cardiolipin. Biochem. J. 115, 65–75.

Racker, E. (1972) Reconstitution of a calcium pump with phospholipids and a purified Ca^{2+} adenosine triphosphatase from sacroplasmic reticulum. J. Biol. Chem. 247, 8198–8200.

Racker, E. and Eytan, E. (1973) Reconstitution of an efficient calcium pump without detergents. Biochem. Biophys. Res. Commun. 55, 174–178.

Racker, E. and Fisher, L.W. (1975) Reconstitution of an ATP-dependent sodium pump with an

ATPase from electric eel and pure phospholipids. Biochem. Biophys. Res. Commun. 67, 1144–1150.

Racker, E. and Hinkle, P.C. (1974) Effect of temperature on the function of a proton pump. J. Membrane Biol. 17, 181–188.

Radda, G.K. and Vanderkooi, J. (1972) Can fluorescent probes tell us anything about membranes? Biochim. Biophys. Acta 265, 509–549.

Raison, J.K. (1973) The influence of temperature-induced phase changes on the kinetics of respiratory and other membrane-associated enzyme systems. Bioenergetics 4, 285–309.

Raison, J.K. and McMurchie, E.J. (1974) Two temperature-induced changes in mitochondrial membranes detected by spin labelling and enzyme kinetics. Biochim. Biophys. Acta 363, 135–140.

Raison, J.K., Lyons, J.M., Melhorn, R.J., and Keith, A.D. (1971) Temperature-induced phase changes in mitochondrial membranes detected by spin labeling. J. Biol. Chem. 246, 4036–4040.

Rand, R.P. and Luzzati, V. (1968) X-Ray diffraction study in water of lipids extracted from human erythrocytes. The position of cholesterol in the lipid lamellae. Biophys. J. 8, 125–137.

Rand, R.P., Chapman, D. and Larsson, K. (1975) Tilted hydrocarbon chains of dipalmitoyl lecithin become perpendicular to the bilayer before melting. Biophys. J. 15, 1117–1124.

Razin, S. (1972) Reconstitution of biological membranes. Biochim. Biophys. Acta 265, 241–296.

Rethy, A., Tomasi, V., Trevisani, A. and Barnabei, O. (1972) The role of phosphatidylserine in the hormonal control of adenylate cyclase of rat liver plasma membranes. Biochim. Biophys. Acta 290, 58–69.

Rodan, S.B., Hintz, R.L., Sha'afi, R.I. and Rodan, G.A. (1974) The activity of membrane-bound enzymes in muscular dystrophic chicks. Nature (London) 252, 589–590.

Roelofsen, B. and van Deenen, L.L.M. (1973) Lipid requirement of membrane bound ATPase. Studies on human erythrocyte ghosts. Eur. J. Biochem. 40, 245–257.

Rogers, M.J. and Strittmatter, P. (1974) Evidence for random distribution and translational movement of cytochrome b_5 in endoplasmic reticulum. J. Biol. Chem. 249, 895–900.

Roses, A.D., Butterfield, D.A., Appel, S.H. and Chestnut, D.B. (1975) Phenytoin and membrane fluidity in myotonic dystrophy. Arch. Neurol. 32, 535–538.

Rothman, J.E. and Engelman, D.M. (1972) Molecular mechanism for the interaction of phospholipid with cholesterol. Nature (London) New Biol. 237, 42–44.

Rothfield, L.I. and Romeo, D. (1971) Enzyme reactions in biological membranes, In: L.I. Rothfield (Ed.), Structure and Function of Biological Membranes, Academic Press, New York, pp. 251–284.

Rottem, S. (1975) Heterogeneity in the physical state of the exterior and interior regions of Mycoplasma membrane lipids. Biochem. Biophys. Res. Commun. 64, 7–12.

Rottem, S., Cirillo, V.P., De Kruyff, B., Shinitzky, M. and Razin, S. (1973) Cholesterol in mycoplasma membranes. Correlation of enzymic and transport activities with physical state of lipid in membranes of *Mycoplasma mycoides* var. cipri adapted to grow with low cholesterol concentrations. Biochim. Biophys. Acta 323, 509–519.

Rottem, S., Hubbell, W., Hayflick, L. and McConnell, H. (1970) Motion of fatty acid spin labels in the plasma membrane of *Mycoplasma*. Biochim. Biophys. Acta 219, 104–113.

Rozengurt, E. and Heppel, L.A. (1975) Serum rapidly stimulates ouabain-sensitive $^{86}Rb^+$ influx in quiescent 3T3 cells. Proc. Natl. Acad. Sci. USA 72, 4492–4495.

Sackmann, E., Träuble, H., Galla, H-J. and Overath, P. (1973) Lateral diffusion, protein mobility, and phase transitions in *Escherichia coli* membranes. A spin label study. Biochemistry 12, 5360–5368.

Sarfeh, I.J., Beeler, D.A., Treble, D.H. and Balint, J.A. (1974) Studies of the hepatic excretory defects in essential fatty acid deficiency. J. Clin. Invest. 53, 423–430.

Scandella, D.J., Devaux, P. and McConnell, H.M. (1972) Rapid lateral diffusion of phospholipids in rabbit sarcoplasmic reticulum. Proc. Natl. Acad. Sci. USA 69, 2056–2060.

Schairer, H.V. and Overath, P. (1969) Lipids containing trans-unsaturated fatty acids change the temperature characteristic of thiomethylgalactoside accumulation in *Escherichia coli*. J. Mol. Biol. 44, 209–214.

Schatzmann, H.J. (1962) Lipoprotein nature of red cell adenosine triphosphatase. Nature (London) 196, 677.

Seelig, J. and Niederberger, W. (1974) Two pictures of a lipid bilayer. A comparison between deuterium label and spin-label experiments. Biochemistry 13, 1585–1588.

Seeman, P. (1972) The membrane actions of anesthetics and tranquilizers. Pharmacol. Rev. 24, 583–655.

Sefton, B.M. and Gaffney, B.J. (1974) Effect of the viral proteins on the fluidity of the membrane lipids in Sindbis virus. J. Mol. Biol. 90, 343–359.

Sha'afi, R.I., Rodan, S.B., Hintz, R.L., Fernandez, S.M. and Rodan, G.A. (1975) Abnormalities in membrane microviscosity and ion transport in genetic muscular dystrophy. Nature (London) 254, 525–526.

Shimshick, E.J. and McConnell, H.M. (1973a) Lateral phase separations in binary mixtures of cholesterol and phospholipids. Biochem. Biophys. Res. Commun. 53, 446–451.

Shimshick, E.J. and McConnell, H.M. (1973b) Lateral phase separation in phospholipid membranes. Biochemistry 12, 2351–2360.

Siekevitz, P. (1972) Biological membranes: The dynamics of their organisation. Annu. Rev. Physiol. 34, 112–140.

Silbert, D.F. and Vagelos, P.R. (1967) Fatty acid mutant of $E.\ coli$ lacking β-hydroxydecanoyl thioester dehydrase. Proc. Natl. Acad. Sci. USA 58, 1579–1586.

Silbert, D.F., Cronan Jr., J.E., Beacham, I.R. and Harder, M.E. (1974) Genetic engineering of membrane lipid. Fed. Proc. 33, 1725–1732.

Simpkins, H. and Hokin, L.E. (1973) Studies on the characterization of the sodium-potassium transport adenosine triphosphatase, XIII. On the organisation and role of phospholipids in the purified enzyme. Arch. Biochem. Biophys. 159, 897–902.

Sineriz, F., Farias, R.N. and Trucco, R.E. (1973) Lipid–protein interactions in membranes: Arrhenius plots and Hill plots in membrane bound (Ca^{2+})-ATPase of $Escherichia\ coli$. FEBS Lett. 32, 30–32.

Singer, S.J. (1971) The molecular organization of biological membranes, In: L.I. Rothfield (Ed.), Structure and Function of Biological Membranes, Academic Press, New York, pp. 145–222.

Singer, S.J. (1974) The molecular organization of membranes. Ann. Rev. Biochem. 43, 805–833.

Singer, S.J. and Nicolson, G.L. (1972) The fluid mosaic model of the structure of cell membranes. Science 175, 720–731.

Skou, J.C. (1961) The relationship of a ($Mg^{2+} + Na^+$)-activated, K^+-stimulated enzyme or enzyme system to the active, linked transport of Na^+ and K^+ across the cell membrane, In: A. Kleinzeller and A. Kotyk (Eds.), Membrane Transport and Metabolism, Academic Press, New York, pp. 228–236.

Slack, J.R., Anderton, B.H. and Day, W.A. (1973) A new method for making phospholipid vesicles and the partial reconstitution of the ($Na^+ + K^+$)-activated ATPase. Biochim. Biophys. Acta 323, 547–559.

Smith, M.W. (1967) Influence of temperature acclimatization on the temperature-dependence and ouabain-sensitivity of goldfish intestinal adenosine triphosphatase. Biochem. J. 105, 65–71.

Smith, M.W. and Ellory, J.C. (1971) Temperature-induced changes in sodium transport and Na^+/K^+-adenosine triphosphatase activity in the intestine of goldfish ($Carassius\ auratus$ L.) Comp. Biochem. Physiol. 39A, 209–218.

Solyom, A. and Trams, E.G. (1972) Enzyme markers in characterization of isolated plasma membranes. Enzyme 13, 329–372.

Spatz, L. and Strittmatter, P. (1971) A form of cytochrome b_5 that contains an additional hydrophobic sequence of 40 amino acid residues. Proc. Natl. Acad. Sci. USA 68, 1042–1046.

Spatz, L. and Strittmatter, P. (1973) A form of reduced nicotinamide adenine dinucleotide-cytochrome b_5 reductase containing both the catalytic site and an additional hydrophobic membrane-binding segment. J. Biol. Chem. 248, 793–799.

Steim, J.M., Tourtelliotte, M.E., Reinert, J.C., McElhaney, R.N. and Rader, J.C. (1969) Calorimetric evidence for the liquid-crystalline state of lipids in a biomembrane. Proc. Natl. Acad. Sci. USA 63, 104–109.

Stein, W.D. and Rozengurt, E. (1976) Temperature dependence of uridine transport in quiescent and serum-stimulated 3T3 cells. Biochim. Biophys. Acta 419, 112–118.

Stier, A. and Sackmann, E. (1973) Spin labels as enzyme substrates. Heterogeneous lipid distribution in liver microsomal membranes. Biochim. Biophys. Acta 311, 400–408.

Stoeckenius, W. (1959) An electron microscope study of myelin figures. J. Biophys. Biochem. Cytol. 5, 491–500.

Strittmatter, P. and Rogers, M.J. (1975) Apparent dependence of interactions between cytochrome b_5 and cytochrome b_5 reductase upon translational diffusion in dimyristoyl lecithin liposomes. Proc. Natl. Acad. Sci. USA 72, 2658–2661.

Sullivan, K.H., Jain, M.K. and Koch, A.L. (1974) Activation of the β-galactoside transport system in Escherichia coli ML-308 by n-alkanols. Biochim. Biophys. Acta 352, 287–297.

Sun, G.Y. and Sun, A.Y. (1974) Synaptosomal plasma membranes: Acyl group composition of phosphoglycerides and $(Na^+ + K^+)$-ATPase activity during fatty acid deficiency. J. Neurochem. 22, 15–18.

Sweadner, K.J. and Goldin, S.M. (1975) Reconstitution of active ion transport by the sodium and potassium ion-stimulated adenosine triphosphatase from canine brain. J. Biol. Chem. 250, 4022–4024.

Sweet, C. and Zull, J.E. (1969) Activation of glucose diffusion from egg lecithin liquid crystals by serum albumin. Biochim. Biophys. Acta 173, 94–103.

Tanaka, R. and Sakamoto, T. (1969) Molecular structure in phospholipid essential to activate $(Na^+ + K^+ - Mg^{2+})$ dependent ATPase and $(K^+ - Mg^{2+})$ dependent phosphatase of bovine cerebral cortex. Biochim. Biophys. Acta 193, 384.

Tanaka, R. and Strickland, K.P. (1965) Role of phospholipid in the activation of Na^+, K^+-activated ATPase of beef brain. Arch. Biochem. Biophys. 111, 583–592.

Tanaka, R. and Teruya, A. (1973) Lipid dependence of activity-temperature relationship of $(Na^+ + K^+)$-activated ATPase. Biochim. Biophys. Acta 323, 584–591.

Tanaka, R., Sakamoto, T. and Sakamoto, Y. (1971) Mechanism of lipid activation of Na, K, Mg-activated ATPase and K, Mg-activated phosphatase of bovine cerebral cortex. J. Membrane Biol. 4, 42–51.

Taniguchi, K. and Iida, S. (1972) The effect of phospholipids on the apparent activation energy of $(Na^+ + K^+)$-ATPase. Biochim. Biophys. Acta 274, 536–541.

The, R. and Hasselbach, W. (1973) Unsaturated fatty acids as reactivators of the calcium-dependent ATPase of delipidated sarcoplasmic membranes. Eur. J. Biochem. 39, 63–68.

Thomas, D. (1957) Should lanterns shine, In: Collected Poems, New Directions Publishing Corp., New York, p. 72.

Thompson, E.D. and Parks, L.W. (1974) The effect of altered sterol composition on cytochrome oxidase and adenosylmethionine: Δ 24 sterol methyltransferase enzymes of yeast mitochondria. Biochem. Biophys. Res. Commun. 57, 1207–1213.

Toffano, G., Gonzato, P., Aporti, F. and Castellani, A. (1974) Variations of brain enzymatic activities in experimental atherosclerosis Atherosclerosis 20, 427–436.

Towers, N.R., Raison, J.K., Kellerman, G.M. and Linnane, W. (1972) Effects of temperature-induced phase changes in membranes on protein synthesis by bound ribosomes. Biochim. Biophys. Acta 287, 301–311.

Träuble, H. (1971) The movement of molecules across lipid membranes: A molecular theory. J. Membrane Biol. 4, 193–208.

Träuble, H. and Eibl, H. (1974) Electrostatic effects on lipid phase transitions: Membrane structure and ionic environment. Proc. Natl. Acad. Sci. USA 71, 214–219.

Träuble, H. and Sackmann, E. (1972) Studies of crystalline-liquid crystalline phase transition of lipid model membranes, III. Structure of a steroid-lecithin system below and above the lipid-phase transition. J. Am. Chem. Soc. 94, 4499–4510.

Triggle, D.J. (1970) Some aspects of the role of lipids in lipid-protein interactions and cell membrane structure and function, In: J.F. Danielli, A.C. Riddiford and M.D. Rosenberg (Eds.), Recent Progress in Surface Science, Vol. 3, Academic Press, New York, pp. 273–290.

Tsukagoshi, N. and Fox, C.F. (1973) Transport system assembly and the mobility of membrane lipids in Escherichia coli. Biochemistry 12, 2822–2829.

Van, S.P., and Griffith, O.H. (1975) Bilayer structure in phospholipid-cytochrome c model membranes. J. Membrane Biol. 20, 155–170.

Van der Bosch, J., Schudt, Chr. and Pette, D. (1973) Influence of temperature, cholesterol, dipalmitoyllecithin and Ca^{2+} on the rate of muscle cell fusion. Exp. Cell Res. 82, 433–438.

Vanderkooi, G. (1974) Organization of proteins in membranes with special reference to the cytochrome oxidase system. Biochim. Biophys. Acta 344, 307–345.

Vanderkooi, J., Fischkoff, S., Chance, B. and Cooper, R.A. (1974) Fluorescent probe analysis of the lipid architecture of natural and experimental cholesterol-rich membranes. Biochemistry 13, 1589–1595.

Van Dijck, P.W.M., Ververgaert, P.H.J.Th., Verkleij, A.J., Van Deenen, L.L.M. and De Gier, J. (1975) Influence of Ca^{2+} and Mg^{2+} on the thermotropic behaviour and permeability properties of liposomes prepared from dimyristoyl phosphatidylglycerol and mixtures of dimyristoyl phosphatidylglycerol and dimyristoyl phosphatidylcholine. Biochim. Biophys. Acta 406, 465–478.

Van Hoeven, R.P., Emmelot, P., Krol, J.H. and Oomen-Meulemans, E.P.M. (1975) Studies on plasma membranes, XXII. Fatty acid profiles of lipid classes in plasma membranes of rat and mouse livers and hepatomas. Biochim. Biophys. Acta 380, 1–11.

Verkleij, A.J., Ververgaert, P.H.J. Van Deenen, L.L.M. and Elbers, P.F. (1972) Phase transitions of phospholipid bilayers and membranes of *Acholeplasma laidlawii* B visualized by freeze fracturing electron microscopy. Biochim. Biophys. Acta 288, 326–332.

Verkleij, A.J., Zwaal, R.F.A., Roelofsen, B., Comfurius, P., Kastelijn, D. and Van Deenen, L.L.M. (1973) The assymetric distribution of phospholipids in the human red cell membrane. Biochim. Biophys. Acta 323, 178–193.

Verkleij, A.J., De Kruyff, B., Ververgaert, P.H.J.Th., Tocanne, J.F. and Van Deenen, L.L.M. (1974) The influence of pH, Ca^{2+} and protein on the thermotropic behaviour of the negatively charged phospholipid, phosphatidylglycerol. Biochim. Biophys. Acta 339, 432–437.

Ververgaert, P.H.J.Th., Verkleij, A.J., Elbers, P.F. and Van Deenen, L.L.M. (1973) Analysis of the crystallization process in lecithin liposomes: A freeze-etch study. Biochim. Biophys. Acta 311, 320–329.

Walker, J.A. and Wheeler, K.P. (1975a) Differential effects of temperature on a membrane adenosine triphosphatase and associated phosphatase. Biochem. J. 151, 439–442.

Walker, J.A. and Wheeler, K.P. (1975b) Polar head-group and acyl-side chain requirements for phospholipid-dependent $(Na^+ + K^+)$ ATPase. Biochim. Biophys. Acta 394, 135–144.

Wallach, D.F.H. (1972) The Plasma Membrane: Dynamic Perspectives, Genetics and Pathology. Springer, New York.

Warren, G.B., Toon, P.A., Birdsall, N.J.M., Lee, A.G. and Metcalfe, J.C. (1974a) Reconstitution of a calcium pump using defined membrane components. Proc. Natl. Acad. Sci. USA 71, 622–626.

Warren, G.B., Toon, P.A., Birdsall, N.J.M., Lee, A.G. and Metcalfe, J.C. (1974b) Reversible lipid titrations of the activity of pure adenosine triphosphatase-lipid complexes. Biochemistry 13, 5501–5507.

Warren, G.B., Houslay, M.D., Metcalfe, J.C. and Birdsall, N.J.M. (1975a) Cholesterol is excluded from the phospholipid annulus surrounding an active calcium transport protein. Nature 255, 684–687.

Warren, G.B., Metcalfe, J.C., Lee, A.G. and Birdsall, N.J.M. (1975b) Mg^{2+} regulates the ATPase activity of a calcium transport protein by interacting with bound phosphatidic acid. FEBS Lett. 50, 261–264.

Watson, K., Bertoli, E. and Griffiths, D.E. (1973) Phase transitions in yeast mitochondrial membranes. The transition temperatures of succinate dehydrogenase and F_1-ATPase in mitochondria of aerobic and anaerobic cells. FEBS Lett. 30, 120–124.

Weber, M.J. (1973) Hexose transport in normal and in Rous sarcoma virus-transformed cells. J. Biol. Chem. 248, 2978.

Weiss, L. (1976) (Ed.) Fundamental Aspects of Metastasis. North-Holland, Amsterdam.

Weissmann, G. and Claiborne, R. (1976) Cell Membranes, Biochemistry, Cell Biology and Pathology, H.P. Publishing Co., New York.

Wheeler, K.P. and Whittam, R. (1970) The involvement of PS in ATPase activity of the sodium pump. J. Physiol. 207, 303–328.
Wheeler, K.P., Walker, J.A. and Barker, D.M. (1975) Lipid requirement of the membrane sodium-plus-potassium ion-dependent adenosine triphosphatase system. Biochem. J. 146, 713–722.
Willis, J.S. and Li, N.M. (1969) Cold resistance of Na-K-ATPase of renal cortex of the hamster, a hibernating mammal. Am. J. Physiol. 217, 321–326.
Wilschut, J.C. and Scherphof, G.L. (1974) The effect of partial degradation of mitochondrial phospholipids by phospholipase A on the temperature dependence of succinate-cytochrome c reductase and cytochrome c oxidase. Biochim. Biophys. Acta 356, 91–99.
Wilson, G. and Fox, C.F. (1971) Biogenesis of microbial transport systems: Evidence for coupled incorporation of newly synthesized lipids and proteins into membranes. J. Mol. Biol. 55, 49–60.
Wilson, G., Rose, S.P. and Fox, C.F. (1970) The effect of membrane lipid unsaturation on glycoside transport. Biochem. Biophys. Res. Commun. 38, 617–623.
Wisnieski, B.J., Parkes, J.G., Huang, Y.O. and Fox, C.F. (1975) Physical and physiological evidence for two phase transitions in cytoplasmic membranes of animal cells. Proc. Natl. Acad. Sci. USA 71, 4381–4385.
Yau, T.M. and Weber, M.J. (1972) Changes in acyl group composition of phospholipids from chicken embryonic fibroblasts after transformation by Rous sarcoma virus. Biochem. Biophys. Res. Commun. 49, 112–120.
Zagyansky, Y. and Edidin, M. (1976) Lateral diffusion of concanavalin A receptors in the plasma membrane of mouse fibroblasts. Biochim. Biophys. Acta 433, 209–214.
Zakim, D. and Vessey, D.A. (1975) The effect of a temperature-induced phase change within membrane lipids on the regulatory properties of microsomal uridine diphosphate glucuronyltransferase. J. Biol. Chem. 250, 342–343.
Zimmer, G. and Schirmer, H. (1974) Viscosity changes of erythrocyte membrane and membrane lipids at transition temperature. Biochim. Biophys. Acta 345, 314–320.

Man
culti

Alar

1.

C
p
e

are associated with serum lipoproteins and the free
albumin.
In the presence of serum, the cellular pathway
several lipid components of serum, the cellular pathway
exogenous supply. Under these conditions,
cellular membranes comes to reflect that
which cannot convert demosterol to chol
as the principle sterol in cells grown i
1972). The synthesis of some o
phatidylethanolamine, phosphatid
not inhibited by the presence of
sera.
Since the cellular needs fo
be a net transfer of some
when cells are grown i
large amounts of lipi
absence of serum
therefore, that t
of the cell an
1969, 1970,
catalyzed
underst
of the
nea

and the subject ...
therefore present a brief over...
membrane composition in mammalian and a...
possible ways in which this experimental approach can be
membrane organization and the biosynthesis and assembly of memb...
components.

2. Lipid metabolism

Most animal cell lines grown in vitro derive the bulk of their lipid from the serum that supplements their growth medium (Rothblat and Kritchevsky, 1967; Rothblat, 1972; Bailey and Dunbar, 1973; Howard and Howard, 1974; Spector, 1972). A few cell lines do not require serum but these are presently the exception rather than the rule. Animal sera contain large amounts of lipid in the form of phospholipids, cholesterol, cholesteryl esters, triglycerides, and free fatty acids. The free and esterified fatty acids display a variety of chain lengths and degrees of unsaturation. The major phospholipids represented are lecithin, sphingomyelin, and lysolecithin. Most of the lipid in sera is not found free but is associated with serum proteins. The phospholipids, cholesterol and triglycerides

G. Poste & G.L. Nicolson (eds.) *Dynamic Aspects of Cell Surface Organization*
© *Elsevier/North-Holland Biomedical Press, 1977*

fatty acids are bound to
 ... s for the de novo synthesis of
 ... the cells rely primarily on an
 ... the fatty acyl composition of the
 ... of the serum. In the mouse L cell,
 ... esterol, the former replaces demosterol
 ... the presence of serum lipid (Rothblat,
 ... the phospholipid classes, e.g. phos-
 ... ylserine and phosphatidylinositol, is probably
 ... serum lipid since they are not present in most

 ... lipid are satisfied, in part, by the serum, there must
 ... lipid from the serum to the cell. On the other hand,
 ... the presence of delipidized serum proteins they efflux
 ... into the medium. This latter observation is not seen in the
 ... protein and is presumably mediated by it. It appears,
 ... here are net fluxes of lipid catalyzed by serum both into and out
 ... this has been confirmed experimentally (Peterson and Rubin,
 ... llingworth et al., 1973). The mechanisms by which these fluxes are
 ... and controlled are of considerable interest but are not completely
 ... ood. A related problem is the mechanisms by which the relative amounts
 ... different lipids are determined and controlled since the lipid ratios remain
 ... rly constant in cells grown with and without serum lipid.

3. Lipid alterations

Two general procedures are available for altering the lipids of animal cells: (1) addition of exogenous lipid(s); and (2) inhibition of the lipid biosynthetic or catabolic pathways. The addition of exogenous phospholipid has been exploited to incorporate phospholipids with desired fatty acyl chains (Papahadjopoulos et al., 1973, 1974; Pagano et al., 1974; Huang and Pagano, 1975; Pagano and Huang, 1975; Poste and Papahadjopoulos, 1976) and to change the cellular levels of cholesterol (Shinitzky and Inbar, 1974; Alderson and Green, 1975) and glycolipid (Dawson et al., 1972; Bara et al., 1973; Brailovsky et al., 1973; Cuatrecasas, 1973; Clayton et al., 1974; Keenan et al., 1975; Révész and Greaves, 1975). The general method involves removing the growth medium and incubating the cells with the appropriate lipid dispersion. Alternatively, the glycolipids can be altered simply by growing the cells in their presence.

To date, the extent of incorporation of exogenous phospholipids into the plasma membrane fraction has been relatively small. Only 5 to 10% of the total plasma membrane phospholipid can be substituted for by exogenous lipids. The incorporated lipids are, however, largely undergraded. The levels of cholesterol can be lowered by incubating the cells with egg yolk lecithin (EYL) liposomes and raised by incubation with EYL-cholesterol liposomes. There appear to be limits, however, on the extent to which membrane cholesterol levels can be altered. The maximum and minimum levels differ from each other by a factor of 2–3. There is a report, yet unsubstantiated, that alterations of cholesterol levels may regulate growth control in lymphocytes (Inbar and Shinitzky, 1974). Finally, several fold increases in natural glycolipids and the incorporation of unnatural glycolipids have been reported. Changes in the cell's physiology often accompany the incorporation of glycolipids. The most striking of these is the demonstration that incorporation of the ganglioside GM_1, the receptor for cholera toxin in isolated fat cells, results in an enhanced binding of toxin and increased sensitivity to its lipolytic effects (Cuatrecasas, 1973). Other physiologic alterations induced by exogenous glycolipids include induction of erythropoiesis (Clayton et al., 1974) and changes in cell growth control (Bara et al., 1973; Brailovsky et al., 1973; Laine and Hakomori, 1973; Keenan et al., 1975).

The principal merits of procedures utilizing the uptake of exogenous lipid are that it is rapid, occurs under defined conditions, and allows incorporation of foreign molecules not readily introduced by other means. The major disadvantages are that the mechanism of uptake is not well understood and that the introduced molecules may or may not mix with the bulk membrane lipid or those associated with proteins. Another consideration is that the incubation of cells with phospholipids containing defined fatty acyl chains results in changes in the relative amounts of the lipid classes as well as in the fatty acyl composition.

The second class of methods for altering animal cell lipids involves inhibition of the endogenous biosynthetic and catabolic pathways. This approach relies on the observations that cultured cells are able to take up lipids from the growth medium and incorporate them into membrane components and that delipidized sera still support appreciable proliferation of serum requiring cells (Geyer, 1967; Bailey and Dunbar, 1973; Howard and Howard, 1974; Horwitz, 1976). The proliferation of cells in such a lipid depleted medium (LDM) requires de novo fatty acid synthesis, and removal of biotin, a vitamin required for de novo fatty acid synthesis, inhibits this endogenous synthesis and forces the cell to rely on fatty acids supplied exogenously either free, bound to albumin or esterified to Tween or phospholipids (Wisnieski et al., 1973; Horwitz et al., 1974; Ferguson et al., 1975). Using these systems, a large variety of natural fatty acids, fatty acid analogues and mixtures of fatty acids have been screened for their ability to support growth. The extent of proliferation depends on the cell line as well as the particular fatty acid supplemented (Horwitz et al., 1974; 1976; Williams et al., 1974). The dependence on the cell line is evident from comparing the growth of mouse 3T3 cells and their SV101 viral transformant (SV101-3T3). The saturation density and growth rate of 3T3 cells approaches that of the control only when their lipid-deficient growth medium is supplemented with a mixture of fatty acids representing those initially present in the serum. In contrast, the proliferation of SV101-3T3 cells supplemented with either oleate or biotin is nearly identical to that of the control, i.e. cells grown in normal serum. Several fatty acids are toxic to some cell lines. Cells grown in their presence generally exhibit distinctive morphologies and time courses of death suggesting different underlying mechanisms (Horwitz et al., 1974).

Analyses of the lipids from both the total cellular membranes and plasma membrane fraction derived from cells grown in LDM supplemented with exogeneous fatty acids show substantial alterations in their fatty acyl composition (Horwitz et al., 1974, 1976; Williams et al., 1974). For example, over 40% of the phospholipid fatty acyl positions are occupied by nonadecanoate (19:0), an unnatural analogue, in L-cells grown in LDM supplemented with nonadecanoate. There is little, if any, desaturation or degradation to heptadecanoate (17:0) in the mouse L cell although there is appreciable metabolism in some other cells. As another example, elaidate $(18:1_t)$, another unnatural analogue, occupies well over 50% of the acyl positions of L_6 cells, a myogenic cell line, and in chick pectoral myoblasts and myotubes grown in LDM supplemented with elaidate. Finally, cells grown in the presence of regular serum show about 30 to 40% of their fatty acyl positions occupied by oleate and when grown in LDM supplemented with oleate, this value approaches 60%. This value did not increase significantly in L_6 cells grown in oleate supplemented LDM carried through several passages, and proliferation was greatly inhibited unless another fatty acid like stearate was added along with the oleate. To date, no one has reported total replacement with a single fatty acid. One unusual observation is that whereas cells grown in LDM containing linoleate and biotin have approx. 35% of their fatty acyl positions occupied by 18:2 and 20:4, there is little enrichment in the plasma membrane fraction.

The levels of cholesterol in L cell and L_6 cells have been lowered using inhibitors of 3-hydroxy-3-methyl glutaryl-CoA(HMG-CoA) reductase (Chen et al., 1974; Horwitz, 1977). This enzyme catalyzes the conversion of HMG to mevalonate, a highly regulated reaction in their de novo synthesis that is inhibited by cholesterol and its oxygenated derivatives. Cells grown in LDM in the presence of one of these inhibitors, e.g. 25-OH cholesterol, possess cholesterol levels that are two to three-fold less than normal. This decrease is evident also in the sterol to phospholipid ratio of the plasma membrane fraction. Growth in the presence of the inhibitor and exogenous cholesterol or mevalonate restores the levels of sterol to values between those of the inhibited and the controlled cells. One study reports a replacement of approx. 90% of the cellular sterol by one of the inhibitors itself (Chen et al., 1974). In some cell lines the product of endogenous synthesis is desmosterol, the immediate precursor of cholesterol. In these cells, the enzyme responsible for this conversion is either missing or possesses a low activity (Rothblat, 1972). The subcellular distribution of sterol in inhibited and control cells has not been reported. Addition of exogenous cholesterol to L_6 cells growing in media containing regular serum produces only a small increase in the level of sterol and a large increase in the amount of cholesterol esters.

Removal of choline from the growth medium inhibits cell proliferation and presumably the biosynthesis of lecithin (and possibly sphingomyelin), a principal phosphatide in most animal cells. Under normal conditions choline is available both as the lecithin in the serum and as a synthetic component of the growth medium. Most cells, in culture at least, do not efficiently convert phosphatidylethanolamine to lecithin and do not synthesize appreciable amounts of free choline. When L_6 or L cells are grown in the absence of exogenous choline, proliferation ceases and the ratio of phosphatidylethanolamine to phosphatidylcholine, the two principal phosphatides, increases by about a factor of two. The addition of exogenous ethanolamine to LDM results in little or no net proliferation but increases the phosphatidylethanolamine:phosphatidylcholine ratio to about a factor of three greater than that of the control. In these altered cells there is little change in the relative amounts of the other phospholipids (Horwitz, 1977). Several other alcohols and choline analogues have been similarly screened primarily in L-cells. Of these, dimethylethanolamine supports appreciable growth, and the phosphatide derived from it becomes a principal membrane component primarily replacing lecithin. Most of the other molecules assayed do not support appreciable proliferation but occupy a significant proportion of the membrane phospholipids (Glaser et al., 1974; Blank et al., 1975). No one has reported total replacement of a phospholipid class.

N-(n-Hexyl)-o-β-glycosylsphingosine (HGls) is an inhibitor of glycolipid catabolism at the level of glycosylceramide. A rat astrocytoma and mouse neuroblastoma grown in its presence show 5-fold increases in the levels of this lipid. However, there is no detectable effect on proliferation or morphology accompanying this increase (Dawson et al., 1974).

In summary, it now appears possible to control, within limits, the composition of several of the lipid components in animal cells grown in vitro. There is evidence that each of the lipid alterations reported are reflected in plasma membrane fractions as well as in whole cell extracts. While there do appear to be changes in other lipid classes consequent to changes in any particular class, in general they are relatively small and often in a direction opposite to that expected for compensation. Systematic studies of possible changes in other lipid classes accompanying the alteration of any particular class and the subcellular distribution of the alterations remain to be reported.

4. Some observations and generalizations

Though studies of the effects of lipid alterations on membrane phenomena are only just beginning, some observations have been made that appear consistent. The evidence providing their foundation has been discussed elsewhere and will only be outlined briefly here (Horwitz, 1976). It may seem premature and potentially misleading to try to draw conclusions and to generalize at this early time, but it may also be beneficial since it serves to focus thought and effort.

A striking observation is the variety of lipid compositions that allow growth and expression of differentiated cell properties. Though limits do exist, significant variations in the degree of unsaturation, proportion of different chain lengths, and relative amounts of cholesterol and glycolipid appear tolerated. This result is not surprising and occurs under normal growth conditions as well, since the fatty acyl composition of the cell reflects that of the serum in which it is grown (Geyer, 1967). The fatty acyl composition of the serum in turn is somewhat variable. Though not reported in detail to date, it is possible that changes in any one lipid may be compensated for by offsetting changes in other membrane components. On the other hand, one should note that alterations in the temperature dependence of some membrane activities and characteristic morphologies often accompany the lipid alterations.

There appear to be well defined limits on the variability of the lipid components. One example is that of cellular sterol. Using either biosynthetic inhibition or incubation with liposomes of appropriate composition, the level of cholesterol can be decreased maximally by about a factor of between two and three. Attempts to decrease the levels further result in cell death. It also appears that there is maximum level of sterol. Attempts to raise the level of free cholesterol in cellular membranes appear limited at a characteristic value and additional cholesterol is converted into cholesterol esters which are not found in the plasma membrane (Rothblat, 1972; Rothblat et al., 1976). The degree of variation of the fatty acyl composition, like cholesterol, also appears to be limited. To date, no one has reported total replacement with a single fatty acid in animal cells though it is possible in some microorganisms. In elaidate-supplemented mammalian cells the fraction of acyl positions occupied by elaidate can become relatively high, however, and cell proliferation ceases prematurely. Increases in the ratio of PE/PC by over a factor of 2–3 relative to that of cells grown normally result in decreases in attachment and/or viability.

The strategy of the cell in designing its lipid composition remains a mystery. If the major goal were only to provide a fluid environment, then the cell would need only two principal fatty acids and a single phospholipid head group. The bacterium *Escherichia coli* approximates this situation. The strategy of animal cells may be more complex since they contain several principal lipids – including cholesterol – and several different fatty acyl chains. It could be, for example, that the cells aspire toward a fixed viscosity or fluidity which would result in optimal rates of membrane activities yet provide an adequate permeability barrier. Alternatively, they may seek to maintain a solid-liquid equilibrium to regulate membrane topography or to optimize or regulate membrane activities. If these were the strategies, it appears that they are not highly tuned since the lipid composition and hence membrane properties do not appear precisely regulated.

There has been considerable interest and speculation concerning the role of the physical state of the lipids in modulating cell growth control, agglutinability and tumorigenicity – properties that change on neoplastic transformation. Although a systematic and detailed analysis of the lipid composition of several normal and transformed cells has not been presented, inspection of the published analyses indicate similarities rather than large differences (Howard and Howard, 1975; Yau et al., 1976). Physical studies using spin-labels and a fluorescent probe also show small differences, if any, between transformed and untransformed cells (Gaffney, 1975; Fuchs et al., 1975; Hatten et al., 1976a; Horwitz, 1977; Yau et al., 1976). Although not definitive, these studies of the cell surface do not indicate the large differences one may have expected if changes in the physical properties of the lipids were directly relevant to transformation. Some direct studies on cells grown with altered fatty acyl compositions or at different temperatures support this position as well (Horwitz et al., 1974; Hatten et al., 1976a; Williams et al., 1976). These studies demonstrate that although the physical properties of the lipid can modulate saturation density and agglutinability, they most likely do not do so in unaltered cells grown at 37°C.

A noteworthy exception to this discussion is the report that lymphomas have lowered levels of cholesterol and hence a lowered microviscosity. On restoring the cholesterol level of lymphomas to that of lymphocytes their tumorigenicity declines (Inbar and Shinitzky, 1974; Shinitzky and Inbar, 1974). This very interesting result remains to be repeated elsewhere, and the proliferative capacity of the altered lymphoma remains to be demonstrated.

Little work has focused on the physical properties of animal cell membranes. The effective membrane microviscosity as estimated by a fluorescent probe decreases as the cholesterol level is lowered and increases as it is raised (Shinitzky and Inbar, 1974). The melting, as determined by spin labels, of the membrane lipids is complex showing several deviations from linearity which vary with temperature and with changes in the fatty acyl composition (Wisnieski et al., 1974a,b; Hatten et al., 1976a,b; Horwitz, 1977). Some of these inflections probably correspond to the temperatures of onset and completion of lateral phase separations. Recent electron diffraction studies have provided direct evidence for their existence (Hui and Parsons, 1976). Several plasma membrane activities, e.g. the response to ligands and cell locomotion and adhesion, respond dramatically to changes in temperature and possibly to the physical state of the membrane, while others, like transport, change very little (Horwitz et al., 1974; Rittenhouse et al., 1974; Wisnieski et al., 1974a; Horwitz, 1977).

The role(s) of the lipids undoubtedly extend beyond those of determining the physical properties of the membrane. Many cells can be grown in the absence of linoleate and linolenate – two "essential fatty acids" (Bailey and Dunbar, 1973). A likely role for them is as precursors of the prostaglandins which serve as hormones in some cell types. The fatty acid palmitate is used in the synthesis of the sphingosine base. Phosphatidylinositol is rapidly turned over and metabolized during the stimulation of cells with a variety of different agents (Mitchell, 1975). Cholesterol, though it plays a major role in determining membrane structure, can be metabolized further to form steroid hormones. It is likely that some lipids, like phosphatidylcholine and phosphatidylethanolamine, function primarily in structural roles, while others, like the polyunsaturated fatty acids and phosphatidylinositol, serve more specialized roles divorced from that of determining bilayer fluidity.

5. Conclusion

The ability to control the lipid composition of animal cells should prove useful in studies of membrane structure. The determination of the fatty acyl chain dynamics and ordering, lipid asymmetry, headgroup conformations, and phase equilibria and their changes in response to different effectors will be greatly facilitated by controlled alterations in the membrane composition. Also the incorporation of ^{13}C and deuterium-enriched analogues will facilitate the use of magnetic resonance and diffraction techniques. The recent description of a class of photoactivatible fatty acid analogues, if they can be incorporated into animal cells, may help elucidate several features of membrane structure including the nature of the lipid environment around selected proteins (Chakrabarti and Khorana, 1975).

The role of the different lipid components and the cell's strategy in designing its lipids remain largely speculative. Lipid alterations can provide methods for deciphering these roles and mechanisms and provide tools for other studies. This latter point is illustrated by a recent description of a method, based on the physical properties of membrane lipids, for synchronizing 3T6 cells out of G2 (Shodell, 1975). In addition, the effects of some fatty acids and lipids on such cellular properties as growth, morphology, and adhesion provide selective systems for the isolation of mutants and their subsequent genetic analysis.

Finally, the ability to control the synthesis of different membrane lipids may help answer questions about the coordination of the synthesis of various membrane components, the relation of lipid synthesis to the cell cycle and expression of differentiated membrane properties, and the assembly of membrane components into functional entities (Cornell et al., 1977).

Acknowledgements

I thank Allie Wight and Pat Ludwig for their contributions and dedication and Mr. J. Cannon for reading the manuscript. My own research was supported by the Cystic Fibrosis Foundation and NIH Grant GM 23244 and benefited from facilities made available through NIH grant GM 20138, and was done during the tenureship of the Dr. William Daniel Stroud Established Investigatorship of the American Heart Association.

References

Alderson, J.C.E. and Green, C. (1975) Enrichment of lymphocytes with cholesterol and its effect on lymphocyte activation. FEBS Lett. 52, 208–211.

Bailey, J.M. and Dunbar, L.M. (1973) Essential fatty acid requirement of cells in tissue culture: A review. Exp. Mol. Pathol. 18, 142–161.

Bara, J., Lallier, R., Brailovsky, C. and Nigam, U.N. (1973) Fixation of a *Salmonella minnesota* R-form glycolipid on the membrane of normal and transformed rat embryo fibroblasts. Eur. J. Biochem. 35, 489–494.

Blank, M.L., Piantadosi, C., Ishag, K.S. and Snyder, F. (1975) Modification of glycerolipid metabolism in LM fibroblasts by an unnatural amino alcohol, N-isopropylethanolamine. Biochem. Biophys. Res. Commun. 62, 983–988.

Brailovsky, C., Trudel, M., Lallier, R. and Nigam, V.N. (1973) Growth of normal and transformed rat embryo fibroblasts. Effects of glycolipids from *Salmonella minnesota* R mutants. J. Cell Biol. 57, 124–132.

Chakrabarti, P. and Khorana, H.G. (1975) A new approach to phospholipid–protein interactions in biological membranes. Synthesis of fatty acids and phospholipids containing photosensitive groups. Biochemistry 14, 5021–5033.

Chen, A.W., Kandutsch, A.A. and Waymouth, C.A. (1974) Inhibition of cell growth by oxygenated derivatives of cholesterol. Nature (London) 251, 419–421.

Clayton, R.B., Cooper, J.M., Curstedt, T., Sjovall, J., Brosook, H., Chin, J. and Schwarz, A. (1974) Stimulation of erythroblast maturation in vitro by sphingolipids. J. Lipid Res. 15, 557–562.

Cornell, R., Grove, G., Rothblat, G. and Horwitz, A. (1977) Lipid requirement for cell cycling: The effect of selective inhibition of lipid synthesis. Exp. Cell Res. (in press).

Cronan, J.R. and Gelmann, E.P. (1975) Physical properties of membrane lipids: Biological relevance and regulation. Bacteriol. Rev. 39, 232–256.

Cronan, J. and Vagelos, P.R. (1972) Metabolism and function of the membrane phospholipids of *E. coli*. Biochim. Biophys. Acta 265, 25–60.

Cuatrecasas, P. (1973) Gangliosides and membrane receptors for cholera toxin. Biochemistry 12, 3358–3566.

Dawson, G., Matalon, R. and Dorfman, H. (1972) Glycosphingolipids in cultured human fibroblasts. J. Biol. Chem. 247, 5951–5958.

Dawson, G., Stoolmiller, A.C. and Radin, N.S. (1974) Inhibition of β-glucosidase by N-(n-hexyl)-o-glycosylsphingosine in cell strains of neurological origin. J. Biol. Chem. 249, 4638–4646.

Ferguson, K.A., Glaser, M., Bayer, W.A. and Vagelos, P.R. (1975) Alteration of fatty acid composition of LM cells by lipid supplementation and temperature. Biochemistry 14, 146–151.

Fox, C.F. (1975) Phase transition in model systems and membranes. In: C.F. Fox (Ed.), MTP International Review of Science. Biochemistry Series One, Vol. 2, University Park Press, Baltimore, pp. 279–306.

Fuchs, P., Pavola, A., Robbins, P.W. and Blout, E.R. (1975) Fluorescence polarization and viscosities of membrane lipids of 3T3 cells. Proc. Natl. Acad. Sci. USA 72, 3351–3354.

Gaffney, B.J. (1975) Fatty acid chain flexibility in the membranes of normal and transformed fibroblasts. Proc. Natl. Acad. Sci. USA 72, 664–668.

Geyer, R.P. (1967) Uptake and retention of fatty acids by tissue culture cells. In: G. Rothblat and D. Kritchevsky (Eds.), Lipid Metabolism in Tissue Culture Cells. The Wistar Institute Press, Philadelphia, pp. 33–47.

Glaser, M., Ferguson, K.A. and Vagelos, R.P. (1974) Manipulation of the phospholipid composition of tissue culture cells. Proc. Natl. Acad. Sci. USA 71, 4072–4076.

Hatten, M.E., Scandella, C.J., Horwitz, A.F. and Burger, M.M. (1976) Similarities in the membrane fluidity of 3T3 and SV101-3T3 cells and its relation to ConA and WGA induced agglutination. Submitted.

Hatten, M.E., Horwitz, A.F. and Burger, M.M. (1977) The influence of membrane lipids on proliferation in transformed and untransformed cell lines. Exp. Cell Res., Submitted.

Horwitz, A.F. (1976) The structural and functional roles of lipids in the surfaces of animal cells grown in vitro. In: G. Rothblat and V. Cristafalo (Eds.), Growth, Nutrition and Metabolism of Cells in Culture, Vol. 3. Academic Press, New York.

Horwitz, A.F., Hatten, M.E. and Burger, M.M. (1974) Membrane fatty acid replacement and their effect on growth and lectin-induced agglutinability. Proc. Natl. Acad. Sci. USA 71, 3115–3119.

Howard, B.V. and Howard, W.J. (1974) Lipid metabolism in cultured cells. Adv. Lipid Res. 12, 51–96.

Howard, B.V. and Howard, W.J. (1975) Lipids in normal and tumor cells in culture. Prog. Biochem. Pharmacol. 10, 135–166.

Huang, L. and Pagano, R.E. (1975) Interaction of phospholipid vesicles with culture mammalian cells, I. Characteristics of uptake. J. Cell Biol. 67, 38–48.

Hui, S.W. and Parsons, D.F. (1976) Phase transition of plasma membranes of rat hepatocytes and hepatoma cells identified by electron diffraction. Cancer Res. 36, 1918–1922.

Illingworth, D.R., Portmann, O.W., Robertson, A.L. and Magyor, W.A. (1973) The exchange of phospholipids between plasma lipoproteins and rapidly dividing cells grown in tissue culture. Biochim. Biophys. Acta 306, 422–436.

Inbar, M. and Shinitzky, M. (1974) Increase of cholesterol level in the surface membrane of lymphoma cells and its inhibitory effect on ascites tumor development. Proc. Natl. Acad. Sci. USA 71, 2128–2130.

Keenan, T.W., Schmid, E., Franke, E.E. and Weigandi, H. (1975) Exogenous glycosphingolipids suppress growth rate of transformed and untransformed 3T3 mouse cells. Exp. Cell Res. 92, 259–268.

Keith, A.D., Wisnieski, B.J., Henry, S. and Williams, J.C. (1973) Membranes of yeast and Neurospora: Lipid mutants and physical studies. In: J.A. Erwin (Ed.), Lipids and Biomembranes of Eukaryotic Microorganisms. Academic Press, New York, pp. 259–322.

Laine, R.A. and Hakomori, S-I. (1973) Incorporation of exogenous glycosphingolipids into plasma membranes of cultured hamster cells and concurrent changes of growth behavior. Biochem. Biophys. Res. Commun. 54, 1039–1045.

Linden, C.D. and Fox, C.F. (1975) Membrane physical state and function. Accts. Chem. Res. 8, 321–327.

Mitchell, R.H. (1975) Inositol phospholipids and cell surface receptor function. Biochim. Biophys. Acta 415, 81–147.

Pagano, R.E. and Huang, L. (1975) Interaction of phospholipid vesicles with cultured mammalian cells, II. Studies of mechanism. J. Cell Biol. 67, 49–60.

Pagano, R.E., Huang, L. and Wey, C. (1974) Interaction of phospholipid vesicles with cultured mammalian cells. Nature (London) 252, 166–167.

Papahadjopoulos, D., Poste, G. and Schaeffer, B.E. (1973) Fusion of mammalian cells by unilammelar lipid vesicles: Influence of lipid surface change, fluidity, and cholesterol. Biochim. Biophys. Acta 323, 23–42.

Papahadjopoulos, D., Mayhew, E., Poste, G. and Smith, S. (1974) Incorporation of lipid vesicles by mammalian cells provides a potential method for modifying cell behavior. Nature (London) 252, 163–166.

Peterson, J.A. and Rubin, H. (1969) The exchange of phospholipids between cultured chick embryo fibroblasts and their growth medium. Exp. Cell Res. 58, 365–378.

Peterson, J.A. and Rubin, H. (1970) The exchange of phospholipids between cultured chick embryo fibroblasts as observed by autoradiography. Exp. Cell Res. 60, 383–392.

Poste, G. and Papahadjopoulos, D. (1976) Lipid vesicles as carriers for introducing materials into cultured cells: influence of vesicle lipid composition on mechanism(s) of vesicle incorporation into cells. Proc. Natl. Acad. Sci. USA 73, 1603–1607.

Révész, T. and Greaves, M. (1975) Ligand-induced redistribution of lymphocyte membrane ganglioside GM1. Nature (London) 257, 103–106.

Rittenhouse, H.G., Williams, R.E., Wisnieski, B. and Fox, C.F. (1974) Alterations of characteristic temperatures for lectin interactions in LM cells with altered lipid composition. Biochem. Biophys. Res. Commun. 58, 222–228.

Rothblat, G.H. (1972) Cellular sterol metabolism. In: G. Rothblat and V. Cristafalo (Eds.), Growth, Nutrition and Metabolism in Cells in Culture, Vol. 1. Academic Press, New York, pp. 297–325.

Rothblat, G.H. and Kritchevsky, D. (1967) (Eds). Lipid Metabolism in Tissue Culture Cells. The Wistar Institute Press, Philadelphia.

Rothblat, G.H., Arbogast, L., Kritchevsky, D. and Naftulin, M. (1976) Cholesteryl ester metabolism in tissue culture cells, II. Lipids 11, 97–108.

Shinitzky, M. and Inbar, M. (1974) Difference in microviscosity induced by cholesterol levels in the surface membrane lipid layer of normal lymphocytes and malignant lymphoma cells. J. Biol. Chem. 85, 613–615.

Shodell, M. (1975) Reversible arrest of mouse 3T6 cells in the G2 phase of growth: A membrane mediated G2 function. Nature (London) 256, 578–580.

Spector, A. (1972). Fatty acid, glyceride, and phospholipid metabolism. In: G. Rothblat and V. Cristafalo (Eds.), Growth, Nutrition and Metabolism of Cells in Culture, Vol. 1, Academic Press, New York, pp. 257–296.

Williams, R.E., Wisnieski, B.J., Rittenhouse, H.G. and Fox, C.F. (1974) Utilization of fatty acid supplements by cultured animal cells. Biochemistry 13, 1969–1977.

Williams, R.E., Iwata, K., Rittenhouse, H. and Fox, C.F. (1977) The effect of alteration in the physical state of mammalian cell lipids on the ability of an oncogenic mammalian cell line to proliferate at low temperature. Exp. Cell. Res. (in press).

Wisnieski, B.J., Williams, R.E. and Fox, C.F. (1973) Manipulation of fatty acid composition in animal cells grown in culture. Proc. Natl. Acad. Sci. USA 10, 3369–3673.

Wisnieski, B.J., Parkes, J.G., Huang, V.O. and Fox, C.F. (1974a) Physical and physiological evidence for two phase transitions in cytoplasmic membranes of animal cells. Proc. Natl. Acad. Sci. USA 71, 4381–4385.

Wisnieski, B.J., Huang, V.O. and Fox, C.F. (1974b) Physical properties of lipid phase of membranes from cultured animal cells. J. Supramol. Struct. 2, 593–603.

Yau, T.M., Buckman, T., Hale, A.H. and Weber, M.J. (1976) Alterations in lipid acyl group composition and membrane structure in cells transformed by Rous sarcoma virus. Biochemistry 15, 3212–3219.

Glycolipids as membrane receptors important in growth regulation and cell–cell interactions

D.R. CRITCHLEY and M.G. VICKER

1. Introduction

The cell surface is a logical place to look for possible causes of the breakdown in cellular interaction characteristics of malignant cells. Evidence from a number of approaches recently reviewed by Pardee (1975) and Nicolson (1976a,b) has justified this line of research, and it is now generally accepted that the organisation of the surface of malignant cells is different from that of their normal counterpart, although the significance of many of the observations is still far from clear.

Indications that changes in cellular glycolipids were related to the malignant process stemmed in part from the work of Rapport and associates who isolated a pure lipid hapten from human epidermoid carcinoma which reacted with rabbit antisera directed against many different types of human tumour (Rapport et al., 1958). The lipid hapten was characterised as lactosyl ceramide (cytolipin H) and more recently a ceramide tetrahexoside (cytolipin R) has been identified as a lipid hapten in rat lymphosarcoma (Rapport et al., 1967; Rapport, 1969; Inoue et al., 1972; Laine et al., 1972). The possible importance of lactosyl ceramide as an antigenic determinant in malignancy was also highlighted by the work of Tal et al. (1964) and Tal and Halperin (1970). In addition, alterations in metabolism of more complex glycolipids related to blood group activities were found in human adenocarcinomas by Hakomori et al. (1967) and changes in glycolipid patterns have also been found in a number of other tumours (see section 6). Glycolipids are thought to be relatively tissue-specific compared to say phospholipids (Martensson, 1969) and are subject to variation depending on species (Svennerholm, 1970), genetic strain (Coles et al., 1970), sex (Gray, 1971), age (Coles et al., 1970), and pathological state (Adams and Gray, 1967; Brady, 1973) of the animal. The sensitive metabolic control and their antigenic properties apparently expressed at the cell surface made glycolipids of considerable interest to those studying changes in the cell surface associated with malignancy.

Although considerable emphasis has been given in recent years to changes in cell surface glycoproteins and proteoglycans our knowledge of changes in glycolipids is perhaps more detailed partly because the precise structure of many of the molecules has been known for some time (see Hakomori, 1973, 1975a,b; Brady and Fishman, 1974; for recent reviews). It is the purpose of this review to consider our knowledge about glycolipid changes in malignancy in the light of exciting new data which implicates them as receptors for a wide variety of molecules which can influence intracellular events.

2. Glycolipid structure, biosynthesis, and nomenclature

The glycolipids found in mammalian cells are predominantly glycosphingolipids, i.e. molecules based on sphingosine or a derivative, although glycolipids containing a glycerol (Shaw, 1970), polyisoprenoid alcohol (Lennarz and Scher, 1972; Lennarz, 1975) or a fatty acid glucosamine backbone (Osborn, 1971) are found elsewhere in biological systems. Typically a long chain fatty acid is linked to the sphingosine base through an amide bond, to give ceramide, and the primary hydroxyl of the sphingosine is substituted with a sugar residue, frequently glucose, to give the simplest glycosphingolipid, glucosyl ceramide. More complex molecules are formed by a sequential addition of sugars, reactions catalysed by membrane bound glycolipid glycosyltransferases. The essential features of glycospingolipids are shown in Fig. 1.

The likely biosynthetic routes for some of the glycolipids commonly found in mammalian organs are shown in Fig. 2. Most of the enzymes involved have been demonstrated in extraneural tissue and details of the reactions catalysed are discussed in reviews by Stoffel (1971), and Morrell and Braun (1972). From Fig. 2 it can be seen that lactosyl ceramide is a key intermediate in the synthesis of more complex glycolipids, the ones that mainly concern us here being the neutral glycolipids and sialic acid containing glycolipids or gangliosides. Neutral glycolipids are labelled according to the number of sugar residues present, from GL1, a monohexosyl ceramide, through to GL5, a pentahexosyl ceramide. However, not all neutral glycolipids containing the same number of sugar residues need have the same structure and these are distinguished by referring to say GL4a (β GalNac 1→3 α Gal 1→4 β Gal 1→4 Glc 1→1 Cer) or GL4b (α GalNac 1→3 β GalNac 1→3 α Gal 1→4 β Gal→Cer) (see Fig. 3 and Table I). Gangliosides are labelled according to the nomenclature of Svennerholm (1970), where GM, GD and GT refer to mono-, di- and tri-sialogangliosides, and the numeral represents $5-n$, n being the number of neutral sugar residues.

3. Glycolipids in normal and transformed cells

One of the first observations on the effect of transformation on glycolipid composition of cultured cells was made by Hakomori and Murakami (1968).

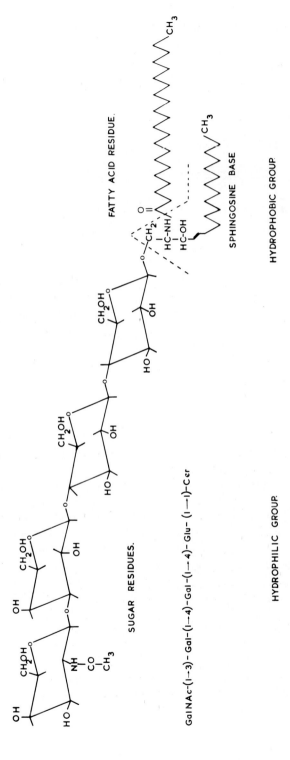

Fig. 1. Structural features of a typical glycosphingolipid.

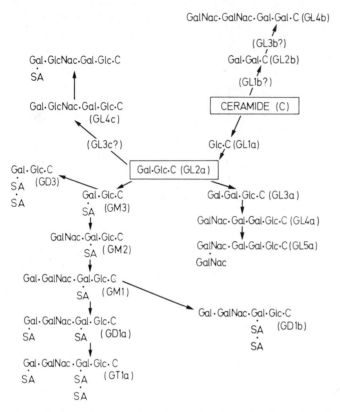

Fig. 2. Probable biosynthetic pathways for the main glycosphingolipids found in animal cells in culture. The nomenclature for sialoglycosphingolipids (gangliosides) is that of Svennerholm (1970). Further details of the structures of neutral glycolipids can be found in Table I.

They found that whereas normal poorly tumorigenic BHK 21 cells contained predominantly GM3, virally transformed cells which were highly tumorigenic showed a four-fold reduction in levels of GM3 and a ten-fold increase in the levels of its precursor GL2, (see Fig. 2 for structure and nomenclature of glycolipids). Spontaneously transformed cells were intermediate both in their tumorigenic potential and glycolipid pattern. The results were similar in cells transformed by a DNA tumour virus, polyoma virus and an RNA tumour virus, Rous sarcoma virus (Schmidt Rupin or Bryan strain), although accumulation of GL2 was somewhat variable (Hakomori et al., 1968). This exciting result led to the suggestion that the carbohydrate chains of glycolipids from tumour cells are often less complete than those from non-tumorigenic cells.

These initial observations were soon followed by similar studies on a variety of mouse cell lines, Swiss 3T3 and BALBc/3T3 fibroblasts, and AL/N mouse embryo epithelial cells, transformed by Simian virus (SV40) or polyoma virus (Py) (Mora et al., 1969; Brady and Mora, 1970). Whereas all of the normal cell

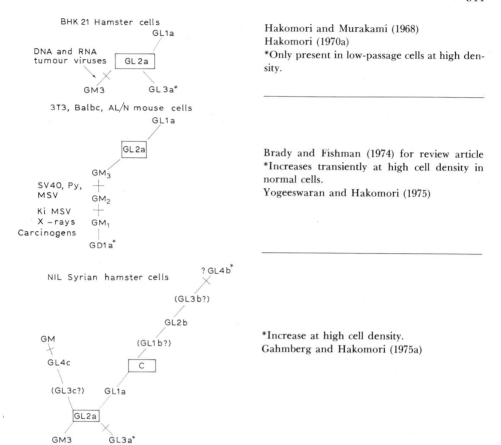

Fig. 3. The effect of transformation on glycolipid metabolism in cultured cells. ⊁— Steps inhibited in cells transformed by DNA or RNA tumour viruses.

lines were found to contain a spectrum of complex gangliosides (GM3, GM2, GM1, GD1a), the transformed derivatives showed a striking loss of gangliosides more complex than GM3. It is of interest to note that the modification in glycolipid pattern is apparently cell-specific and not virus-specific. For example, transformation of BHK21 cells by polyoma virus leads to reduced levels of GM3 whereas in mouse cell lines transformed by the same virus, GM3 is unaffected and the levels of GM2, GM1 and GD1a are reduced. The above interpretation is complicated by the recent observation that different strains of the same virus interfere with ganglioside metabolism at different points. BALBc/3T3 cells transformed by the Kirsten strain of Murine Sarcoma Virus (KiMSV) have reduced levels of GM1 and GD1a but normal amounts of GM2 (Fishman et al., 1974a) whereas the same cell line transformed by the Moloney

TABLE I
NIL 2 HAMSTER CELL GLYCOLIPIDS: STRUCTURE AND NOMENCLATURE[a]

			Induction of levels	
			NIL	NIL Py[b]
GL1a	Glc→Cer		±	±
GL2a	β Gal 1→4 Glc→Cer		+	+
Glycolipids synthesised from lactosyl ceramide				
GL3a	α Gal 1→4 β Gal 1→4 Glc→Cer		++	+
GL4a	β GalNac 1→3 α Gal 1→4 β Gal 1→4 Glc→Cer	Globoside	++	±
GL5a	α GalNac 1→3 β GalNac 1→3 α Gal 1→4 β Gal 1→4 Glc→Cer	Classical structure for Forssman antigen	++	+
(GL3c?)	β GlcNac 1→3 β Gal 1→4 Glc→Cer	Paragloboside	−	+
GL4c	β Gal 1→4 β GlcNac 1→3 β Gal 1→4 Glc→Cer	Sialylparagloboside	+	−
GM	Sial→β Gal 1→4 β Glc Nac 1→3 β Gal 1→4 Glc→Cer	Hematoside	++	+
GM$_3$	Sial 2→3 β Gal 1→4 Glc→Cer			
Glycolipids synthesised from digalactosyl ceramide				
(GL1b?)	Gal Cer			
GL2b	α Gal 1→4 β Gal→Cer		±	±
(GL3b?)	β Gal Nac 1→3 α Gal 1→4 β Gal→Cer			
GL4b	α Gal Nac 1→3 β Gal Nac 1→3 α Gal 1→4 β Gal→Cer	New type Forssman antigen	++	±

[a]From Gahmberg and Hakomori (1975a,b).
[b]Polyoma virus-transformed NIL cells.

isolate of MSV has reduced levels of all gangliosides more complicated than GM3, (Mora et al., 1973). Nevertheless, the results are in general agreement with the statement that virus-transformed cells contain reduced levels of the more complex glycolipids. Such findings have now also been made on BALBc/3T3 cells transformed by the chemical carcinogens, methylcholanthrene and benzopyrene, and also by X-irradiation (Coleman et al., 1975).

The above observations are now substantiated by a wealth of literature not only on changes in gangliosides accompanying transformation, but also on alterations in neutral glycolipids (see section 4.1.1 and Fig. 3), and fucolipids (see section 3.3.2). Additional support for the idea that changes in glycolipids also occur in chemically transformed cell lines grown in culture (Coleman et al., 1975) has been scarce. Methylcholanthrene-transformed rat embryo fibroblasts were reported to have reduced levels of GM2 compared to control cells, as were the cells transformed by Rauscher leukemia virus (Langenbach, 1975). In this case there were also some additional differences between the chemically and virally transformed cells. These studies are supported by the original observation that hepatoma cell lines in culture also had a simplified glycolipid pattern (Brady et al., 1969; see also section 6).

3.1. Enzymatic basis for change in glycolipid pattern on transformation

The observed decrease in levels of gangliosides more complex than GM3 in mouse cells transformed by DNA (Py or SV40) or RNA (MSV) tumour viruses is apparently explained by a marked reduction in the activity of the N-acetylgalactosaminyl transferase, which converts GM3 to GM2 (Cumar et al., 1970; Fishman et al., 1971; Mora et al., 1973). The activity of the enzyme was not reduced in Swiss 3T3 cells lytically infected with polyoma virus (Mora et al., 1971), or infected with murine leukemia virus (MLV) alone, the helper virus function necessary for transformation with MSV (Mora et al., 1973). This suggests that stable insertion of the viral genome accompanied by transformation is necessary to alter the cell pattern of ganglioside synthesis. The activity of other enzymes in the biosynthetic pathway were not modified in any consistent pattern by transformation, the activities of some being markedly elevated, others being reduced or unaffected. These results have been interpreted as meaning that transformation results in modification of the activity of a specific glycosyl transferase in the biosynthetic chain, resulting in a failure to accumulate gangliosides beyond the enzymatic block (Brady and Fishman, 1974). Support for this concept has come from studies with BALBc/3T3 cells transformed by KiMSV, chemical carcinogens or X-rays, which contain near normal levels of GM2 but reduced amounts of GM1 (Fishman et al., 1974a, Coleman et al., 1975). In this case the activity of a galactosyl transferase (GM2–GM1) was reduced, although the galactosaminyl transferase (GM3–GM2) and sialyl transferase (GM1–GD1a) were present at near normal levels. The kinetic properties of the residual galactosyl transferase have been studied and found to be similar to the normal enzyme (Coleman et al., 1975). It has therefore been inferred that transformation alters ganglioside biosynthesis by blocking enzyme synthesis either at transcriptional or translational levels. The possibility that many of the enzyme molecules produced by transformed cells are inactive for some reason has not been excluded.

That the decreased activity of the biosynthetic pathway is a key event in determining the glycolipid pattern in transformed cells is also supported by evidence that the activities of degradative enzymes are about the same in both normal and transformed cells (Cumar et al., 1970). However, these results may bear re-examination in the light of the results of Schengrund et al. (1973) who found increased sialidase activity toward added extracellular di- and trisialogangliosides associated with a number of transformed cell lines. Variations in the levels of sialidase have also been implicated as an important factor modulating the levels of GD1a in 3T3 cells (Yogeeswaran and Hakomori, 1975; see also section 4.1.2). A reduction in the activity of specific glycosyl transferase is also thought to explain low levels of GM3 in Py–BHK 21 cells (Den et al., 1971), and low levels of GL3 in transformed NIL Syrian hamster fibroblasts (Kijimoto and Hakomori, 1971; Critchley et al., 1974; Chandrabose and Macpherson, 1976) and a subclone of BHK 21 transformed by polyoma virus (Kijimoto and Hakomori, 1971). A cell line (DMN4B) selected from BHK cells following mutagenesis with dimethylnitrosamine showed a transformed phenotype at 38°C but appeared normal at 32°C (Di Mayorca et al., 1973). [^{14}C]Palmitate incorporation into GL1, 2 and 3 was markedly reduced at 38°C and the activity of the UDPgal lactosyl ceramide α galactosyl transferase was found to be heat-labile (Buehler and Moolten, 1975).

The possibility that the activity of certain specific glycosyl transferases is reduced in chemically, X-irradiated and virus (DNA and RNA)-transformed cells by a class of inhibitory molecule has been considered. Experiments in which normal and transformed cells have been co-cultivated (Mora et al., 1971; Critchley and Macpherson, 1973) or homogenates of the two cell types mixed (Mora et al., 1971, 1973; Den et al., 1974; Coleman et al., 1975; Chandrabose and Macpherson, 1976) have failed to demonstrate such an inhibitor. However, evidence has been presented that such a molecule is involved in controlling the activity of a lactosyl ceramide sialyl transferase during development of rat brain (Duffard and Caputto, 1972). The molecule (molecular weight 70–80 000 daltons), which was found in cytosol, mitochondrial and microsomal fractions increased the K_m of the enzyme for lactosyl ceramide. It is possible that such a competitive inhibitor has been overlooked in the cell culture mixed homogenate experiments perhaps because of the use of high concentrations of glycolipid substrate.

That these experiments truly establish the molecular basis of the altered glycolipid pattern in transformed cells is brought into question by reports that glycosyltransferase activities are low or absent in cells known to contain the predicted product of the reactions (Critchley et al., 1974; Den et al., 1974; Hakomori, 1975a; Chandrabose and Macpherson, 1976). For example, Chandrabose and Macpherson (1976) were unable to demonstrate the synthesis of GL4 or GL5 in NIL cells when their supposed precursors GL3 or GL4 were used as substrate with UDP N-acetylgalactosamine as the sugar nucleotide donor, although the cells were known to contain GL4 and 5. In view of the heterogeneity of the carbohydrate sequences in these molecules in hamster NIL cells (Gahmberg and Hakomori, 1975a, see Fig. 3) it is possible that the GL3/4 used in the assay (isolated from horse spleen) had a different structure from the naturally occurring NIL cell substrates, a problem previously encountered by Stoffyn et al. (1973) when assaying an N-acetylgalactosaminyltransferase in rat tissue. However, even when the GL3 or GL4 was isolated from NIL cells no N-acetylgalactosaminyltransferase activity was detected (Chandrabose and Macpherson, 1976). A further possibility is that there may be more than one pool of, say, GL3 in the cell, and the N-acetylgalactosaminyl transferase may be specific for one with a particular sphingosine base or fatty acid composition. The idea is supported by studies on pig erythrocytes which showed that although palmitic, stearic, and oleic acids are the principal fatty acids in glycolipids with short carbohydrate moieties (cerebrosides), the proportion of these fatty acids in glycolipids with more complex carbohydrate moieties was markedly reduced. The difference was made up by increased amounts of behenic, lignoceric and nervonic acids which were present only in trace amounts in the cerebrosides (Sweeley and Dawson, 1969). Similar variations in the fatty acid composition of glycolipids with increasing length of the carbohydrate residue have been found in a number of systems (Siddiqui and Hakomori, 1970; Weinstein et al., 1970; Yogeeswaran et al., 1970) although the possibility that acyl chain rearrangement takes place via a ceramidase cannot be discounted.

Glycolipid glycosyltransferases are notoriously difficult to assay because: (1) they are membrane-bound enzymes; (2) they are difficult to solubilize and the detergents commonly used may cause partial inactivation; (3) substrates are frequently in micellar form; and (4) glycosylhydrolases are often present in the fraction assayed for glycosyl transferase and nucleotide pyrophosphatase activities can make sugar nucleotide concentrations rate-limiting (Geren and Ebner, 1974; Coleman et al., 1975). In an attempt to avoid some of these difficulties Maccioni et al. (1972) have described an assay in which the membranes are not disrupted and incorporation of labelled sugars from sugar nucleotides into endogenous substrates is followed. As the spatial relationship between glycosyltransferase and glycolipid intermediate may be strictly ordered within the membrane, the method offers a more physiological approach to the study of these enzymes. An additional problem relating specifically to assay of these enzymes in cultured cells stems from the presence of a wide variety of lipids including glycolipids (Vance and Sweeley, 1967; Slomiany and Horowitz, 1970; Tao and Sweeley, 1970; Yogeeswaran et al., 1970; Yu and Ledeen, 1972) in serum. De novo synthesis of lipids by cultured cells is known to be markedly suppressed where serum lipids are available (Bailey et al., 1972; Howard and Howard, 1975). It is therefore possible that the activity of the glycosyltransferases is modulated by serum glycolipids, especially as glycolipids added to the culture medium are known to be rapidly accumulated by cells (Cuatrecasas, 1973b; Laine and Hakomori, 1973; Keenan et al., 1974). Notwithstanding these various considerations it is generally believed that a decrease in the activity of specific glycolipid glycosyltransferase is responsible for the altered glycolipid metabolism of transformed cells.

3.2. Subcellular distribution of glycolipids

The concept that the glyclolipid changes discussed so far are indeed relevant to cell surface properties rested until fairly recently on immunological evidence (Rapport et al., 1958; Tal, 1965; Marcus and Janis, 1970) and observations that glycolipids were important components of the erythrocyte (Sweeley and Dawson, 1969), the synaptosomal (Wiegandt, 1967) and rat liver plasma membranes (Dod and Gray, 1968a,b). Partly because of improved membrane fractionation techniques the results have now been extended to a number of other systems, e.g. rat liver and bovine mammary gland plasma membranes (Keenan et al., 1972a,b), the microvillous membrane of rat (Forstner and Wherret, 1973), mouse small intestine (Kawai et al., 1974), the plasma membrane of human (Levis et al., 1976a) and pig lymphocytes (Levis et al., 1976b). However, there are still relatively few reports on glycolipid distribution in cell culture systems. In a comparison of enzymatically and antigenically characterised subcellular fractions isolated from BHK21-C13 cells, Renkonen and co-workers (1972) reported a ten-fold enrichment of ganglioside (mainly GM3) in plasma membrane fractions compared with whole cells on a protein basis. The levels of gangliosides in the plasma membrane were also five to seven-fold greater than in an endoplasmic reticulum fraction. Importantly, the molar ratios of total glycolipid to phospholipid were 0.045 for the whole cell, 0.076 for the plasma membrane and 0.023 for the endoplasmic reticulum, although no data were given for mitochondrial or nuclear fractions. If one assumes, for example, that of the total cellular protein 5% is in the plasma membrane then calculations show that 50% of the ganglioside of BHK21/C13 is located at this level. Comparable data are as yet unavailable for the mouse cell lines although in Swiss 3T3 cells, Yogeeswaran et al. (1972) found a three to five-fold enrichment of gangliosides in the plasma membrane compared to whole cells. This was considered to be only a modest enrichment when compared to marker enzymes whose specific activity increased ten to twenty-fold, a problem also encountered in other systems (Keenan et al., 1972a,b).

From these studies alone it would seem likely that the marked reduction in levels of gangliosides that occur on transformation of BHK21 cells or mouse cell lines (see section 3) will be reflected at the level of the cell surface. This conclusion is supported by a number of other cell fractionation studies on the subcellular distribution of gangliosides and neutral glycolipids (Klenk and Choppin, 1970; Weinstein et al., 1970; Critchley et al., 1973), and by the availability of glycolipids to enzymes (Yogeeswaran et al., 1972; Gahmberg and Hakomori, 1975a,b; and see section 5), antibodies (Hakomori and Kijimoto,

1972 also see section 5) and other proteins in intact cells in culture (Hollenberg et al., 1974; also see section 7).

3.3. Correlation between glycolipid changes and transformation

3.3.1. Generality of the effect of transformation on glycolipid pattern

The number of exceptions to the rule that the glycolipid pattern of cells is simplified on transformation are small and not well documented. Yogeeswaran et al. (1972) described two clones of SV40 transformed 3T3 fibroblasts which showed an actual elevation of the levels of the more complex gangliosides, although other clones transformed by SV40 or polyoma virus showed the expected simplification in ganglioside pattern. Lack of absolute correlation between transformation and simplification of glycolipid pattern was also noted in a STU mouse cell line transformed by SV40 or Friend leukemia virus (Diringer et al., 1972). In a recent review article, Brady and Fishman (1974) reported that out of 26 individually transformed (DNA viruses) mouse cell lines that were available to them from various sources, only two failed to show altered ganglioside patterns. One of the exceptions was an SV40-transformed mouse cell line selected for resistance to 5-bromo-2-deoxyuridine (Dubbs et al., 1967). The second exception was the SV40 transformed Swiss 3T3 cell line described by Renger and Basilico (1972) which is temperature-sensitive for transformation in a host cell function. None of the above exceptions have been intensively studied and it is therefore difficult to judge the significance of the results. It would be interesting, for example, to know more about the other properties frequently associated with expression of the transformed phenotype in these cells e.g. morphology, cyclic nucleotide levels (Pastan et al., 1975; Willingham, 1976), transport activities (Hatanaka, 1974), surface protein profiles, (Hynes, 1976), secretion of plasminogen activator (Ossowski et al., 1973; Unkeless et al., 1973) and so on. In addition, studies of the effect of transformation on glycolipid patterns are complicated by the possibility of clonal variation. For example, although the overall pattern of gangliosides in a population may show GM3, GM2, GM1, GD1a as the only components, a small percentage of cells may also have GD1b and GT gangliosides. Transformation of this minority population may produce clones in which accumulation of the more complex gangliosides is indeed blocked, but of course they would not be recognised as different from the bulk of the starting normal population. The fact that the clones of SV40 transformed 3T3 cells described by Yogeeswaran et al., (1972) showed an unusual elevation of the levels of disialogangliosides may have been due to the effect of transformation on the unrecognised ability of the parent cells to synthesize trisialogangliosides.

The above problem of clonal variation affecting glycolipid pattern was first highlighted in a study of NIL hamster fibroblasts (Sakiyama et al., 1972). From initial experiments the cells were reported to contain GL1, GL2 and GM3 and a series of neutral glycolipids GL3, GL4 and GL5 (Robbins and Macpherson, 1971). Subsequently, several clones were isolated which varied in morphology, saturation density, and glycolipid composition (Sakiyama et al., 1972). Contrary to the then current ideas, they found no correlation between saturation density and glycolipid pattern. In fact the clone with the highest saturation density was the only one to show a complete set of glycolipids previously found in NIL cells. A similar lack of correlation between saturation density and levels of complex gangliosides was also reported in mouse cell lines (Yogeeswaran et al., 1972).

3.3.2. Studies with viral mutants temperature-sensitive for transformation

The degree of correlation between glycolipid changes and transformation has also been studied in cells infected with mutant viruses temperature-sensitive for the maintenance of transformation. A switch in temperature can convert the cells from the transformed to the normal phenotype or vice versa, although the viral genome still remains integrated. Changes which relate to transformation can therefore be separated from those associated with insertion of the viral genome. The results from such studies have so far lacked the desired reproducibility but nevertheless certain trends have emerged.

Initial studies by Hakomori et al. (1971) showed that chick embryo fibroblasts (CEF) transformed by Rous sarcoma virus (RSV) had reduced levels of GM3, GD3 and more complex gangliosides not yet fully characterised. This was accompanied by accumulation of their presumed precursors, ceramide and GL1. The GM3/GL1 ratio decreased from about 58 to around 5 on transformation, although significantly it was altered little in cells infected with the non-transforming Rous associated virus (RAV-1) used as a control for possible effects of virus release in this system. However, Warren et al. (1972) were unable to show this dramatic shift in ratio using [^{14}C]-palmitate labelling of chick embryo fibroblast transformed with a temperature-sensitive mutant (Martin, 1970) of Rous sarcoma virus. In a recent review article, Hakomori (1975a) made no mention of elevated levels of ceramide and GL1*. Levels of GM3 were, however, found to be temperature-sensitive in cells infected with a mutant of RSV (Wyke, 1973), although there was no correspondence between the activity of the lactosyl ceramide sialyltransferase and GM3 levels. Similar discrepancies between the activity of this enzyme and transformation in chick embryo fibroblasts were reported in a review article by Brady and Fishman (1974). An additional problem was that levels of higher gangliosides reduced by transformation failed to revert to normal under non-permissive conditions (Hakomori, 1975a). At least one other transformation induced change in CEF, the tumour specific surface antigen (Bauer et al., 1974), does not consistently revert at non-permissive temperature, although agglutinability (Burger and Martin, 1972), sugar transport and the LETS glycoprotein (Hynes and Wyke, 1975; Critchley et al., 1976) are all temperature-sensitive.

Similar studies have been made on BHK 21 cells transformed with a temperature-sensitive mutant of polyoma virus (ts3) (Dulbecco and Eckhart, 1970). At permissive temperature the cells showed a dramatic reduction in levels of GM3 and increased GL2 (Gahmberg et al., 1974) in agreement with the predicted results (Hakomori and Murakami, 1968; see section 3), the pattern

---

*See also Hakomori et al. (1977) Virology 76, 485–493.

being reversed at the non-permissive temperature. Unfortunately, control experiments comparing cells transformed by wild-type virus with normal BHK21 cells failed to show decreased GM3 although GL2 was somewhat elevated. In analogy with results on CEF, it was noted that levels of a certain glycolipid GL3 reduced by transformation did not revert to normal in mutant-infected cells grown under non-permissive conditions (Hammerstrom and Bjursell, 1973; Gahmberg et al., 1974). Growth in agar, and serum requirements, have also been reported to lack temperature dependence in this system although density-dependent inhibition of DNA synthesis, agglutinability by lectins, and the level of LETS glycoprotein were temperature-sensitive (Dulbecco and Eckhart, 1970; Eckhart et al., 1971; Gahmberg et al., 1974).

The results imply that not all virus-induced changes (including some glycolipid modifications) are related directly to loss of growth control, but may be associated with permanent changes related to transformation.

More encouragingly, the simplification of the pattern of fucose-containing glycolipids induced in rat embryo cells by MSV (Steiner et al., 1973) was temperature-sensitive in cells infected with mutant virus (Steiner et al., 1974). The structure of these glycolipids, which also change in human cancer cells (Steiner and Melnick, 1974), is novel in that the amino group of the sphingosine base is not substituted with a long chain fatty acid (Skelly et al., 1976). Levels of GM3 and GL3 were also found to relate to the transformed phenotype in normal rat kidney cells (NRK) infected with a temperature sensitive mutant of the KiMSV (Brady and Fishman, 1974).

Because of the considerable technical difficulties in quantitation, no one has yet looked at the rate of change of glycolipid pattern when cells infected with a temperature sensitive viral mutant are shifted from non-permissive to permissive temperature. It would be interesting to know whether the change in glycolipid pattern accompanied the rapid changes in morphology, transport activities (Hatanaka and Hasafusa, 1970; Weber, 1973; Eckhart and Weber, 1974) and secretion of plasminogen activator (Rifkin et al., 1975), or lagged behind, as does the change in iodination of LETS glycoprotein (Hynes and Wyke, 1975). Hakomori et al. (1971) have studied the change in glycolipid pattern as a function of time in cells primarily infected with RSV. The most rapid change, which approximately paralleled the number of transformed cells in the population, was in GD3, subsequently found to be more closely associated with insertion of the viral genome than transformation. Reduction in levels of GM3, which has been suggested to be related directly to transformation in other studies (Hakomori, 1975a) was said to follow, rather than accompany, transformation in this primary infection system. The kinetics of change in glycolipid pattern have also been studied during mass transformation of Swiss 3T3 cells with MSV(MLV) (Mora et al., 1973). Two days after infection morphological transformation was virtually complete but there was no change in the activity of any of the glycolipid glycosyltransferases until 4 days after infection.

These studies suggest that altered glycolipid metabolism is a late event in the transformation process.

3.3.3. Studies with revertant cell lines

The correlation between glycolipid pattern and growth characteristics has also been examined in revertants selected from a number of transformed cell lines by a variety of techniques. Revertants selected from SV40 transformed Swiss 3T3 mouse fibroblast populations using FUdR showed a flattened morphology, exhibited contact inhibition of growth and were poorly tumorigenic when compared to the original starting parent cell population, (Pollack et al., 1968; Mora et al., 1971). Whereas SV40 Swiss 3T3 cells contained much reduced amounts of gangliosides more complex than GM3, the levels in the revertants were nearly identical to the untransformed cells (Mora et al., 1971). This was also reflected in the increased activity of the N-acetylgalactosaminyl transferase responsible for GM2 synthesis, which was specifically reduced in the transformed cells. Revertants of polyoma virus transformed Swiss 3T3 cells with growth properties intermediate between the normal and fully transformed cells, showed an intermediate ganglioside pattern.

In a similar study, revertants of a cloned SV40 induced hamster tumour cell line, selected for resistance to actinomycin D, contained higher levels of gangliosides more complex than GM3 compared to the parent cell line (Nigam et al., 1973). The revertants, which also had a higher N-acetylgalactosaminyl transferase activity responsible for synthesis of GM2 from GM3, showed a lower saturation density, reduced agglutinability with wheat germ agglutinin, reduced tumorigenicity and a failure to grow in agar. Back revertants were isolated, intermediate in these various properties and with an intermediate ganglioside pattern. These observations suggest a close parallel between glycolipid pattern and other expressions of the normal phenotype. It is unclear whether revertant cells repress the expression of most of the integrated viral genes. Revertants of SV40 transformed 3T3 cells are known to carry a complete rescuable viral genome and express certain viral functions such as the nuclear T antigen (Pollack et al., 1968, 1970; Mora et al., 1971). It may be relevant that certain revertants show an increased chromosome number (Pollack et al., 1970). Experiments with hybrids between SV40 transformed BHK21 and normal 3T3 mouse fibroblasts showed that synthesis of complex gangliosides characteristic of the mouse partner was suppressed (Wiblin and Macpherson, 1973). Partial but selective depletion of the chromosomes of the SV40-BHK component resulted in the reappearance of the more complex gangliosides. Perhaps reversion depends on the number of chromosomes lacking viral genes which are present in the cell.

In contrast to these observations revertants of hamster cells originally transformed by SV40, polyoma virus or chemical carcinogens, failed to regain normal ganglioside patterns and the activity of a key N-acetylgalactosaminyl transferase remained low (Den et al., 1974). The revertants which had a low saturation density and tumorigenicity also had an altered chromosome number. So called "flat" revertants selected from transformed hamster fibroblasts with FUdR were also found to retain the glycolipid pattern characteristic of the parent transformed cells (Critchley and Macpherson, 1973).

In summary, not all changes in glycolipid pattern relate either to transformation or loss of growth control, although the correlation is surprisingly good.

3.4. Molecular basis of the viral effect of glycolipid synthesis

Precisely how a tumour virus can produce a modification in the glycolipid biosynthetic machinery of the host cell is an open question. The fact that the block in general is cell-specific, rather than virus-specific, rules out a direct analogy with modifications of the carbohydrate structure of the *Salmonella* O-antigen glycolipid by lysogenic phages. In this case immediately after infection the phage: (1) represses synthesis of a key host enzyme; (2) produces an inhibitor of another; and (3) directs the synthesis of a virus-coded enzyme. The overall result is a change in the detailed structure of the carbohydrate residues of the O-antigen (Losick and Robbins, 1969).

Possible mechanisms to explain the effect of transformation on glycosyl transferase levels in cultured cells have recently been discussed by Brady and Fishman (1974). These included: (1) an integration site model in which viral DNA would be inserted at the cistron coding for a specific glycosyl transferase; (2) synthesis of a virus-coded repressor molecule which would inhibit translation of a specific mRNA coding for a glycosyl transferase; (3) a direct effect of a viral gene product on the enzyme. None of the three possibilities appears tenable in the light of experiments discussed so far. For example, it seems unlikely that the integration site model is correct because glycolipid synthesis is temperature sensitive in cells infected with a ts viral mutant, yet the viral genome is thought to remain integrated at both permissive and non-permissive temperatures. A similar argument can be put forward based on the results of lytically infected cells and revertants. The idea of a virus-coded repressor molecule can apparently be discounted because it should be virus-specific yet one can clearly see that the precise step at which glycolipid synthesis is inhibited is governed predominantly by the cell type rather than the virus. Similarly a direct effect of a viral gene product on the glycosyl transferase would also demand virus, not host cell specificity. Our increased understanding of host cell–virus interactions may lead to modification of some of these criticisms.

All these considerations assume that the effects of transformation on glycolipid pattern is a primary effect of the virus. Other possibilities will be discussed elsewhere in this article (see sections 4.3.1, 4.3.2 and 4.3.3).

3.5. Effects of exogenous glycolipids on cell growth

The idea that the altered growth characteristics of transformed cells are in some way related to the loss of specific cell surface glycolipids has led to attempts to re-insert the deleted molecules back into the transformed cell membrane. Thus, addition of glycolipid GL4 to the growth medium of virus-transformed NIL hamster fibroblasts slightly reduced their growth rate although the effect was more apparent with the untransformed cells (Laine and Hakomori, 1973). Glycolipid GL4 also prolonged the G_0/G_1 phase when quiescent NIL cells were stimulated to divide by trypsinization. The glycolipid was taken up most rapidly by freshly trypsinized cells, doubling the cellular content of the molecule. The bulk of the GL4 was reported to be recovered in an isolated plasma membrane fraction. However, the specificity of the effect of the glycolipid on cell growth was not determined, and we have found (Critchley et al., unpublished observations) that GM3, GL2 or GL1 molecules, which are relatively unaffected after transformation of NIL cells, also inhibited growth when used at the same high concentrations ($8 \cdot 10^{-4}$ M) as those used by Laine and Hakomori (1973). Indeed certain lipids, e.g. palmitate, can be toxic at relatively low concentrations ($6 \cdot 10^{-5}$ M) with the accumulation of lipid droplets leading eventually to cell lysis (Geyer, 1967).

Similar experiments have also been performed with 3T3 cells (Keenan et al., 1974, 1975), which specifically lose gangliosides more complex than GM3 on transformation (Brady and Fishman, 1974; also see section 3). Uptake of ganglioside GM1 (approximately $1 \cdot 10^{-4}$ M) was biphasic; an early saturable component, possibly representing binding, being followed by a gradual accumulation over 4–5 h (Keenan et al., 1974). Interestingly, gangliosides competed against each other for the initial binding process. Gangliosides reduced the growth rate and saturation density of both SV40-transformed and untransformed 3T3 cells, monosialogangliosides being more effective than disialogangliosides (Keenan et al., 1975). The isolated carbohydrate moieties or ceramide alone were ineffective as growth inhibitors. However, neutral glycolipids were not markedly affected by transformation in these cells (GL1, GL2 and GL3) and trisialogangliosides were equally as effective as monosialoganglioside, again bringing into question the specificity of these growth inhibitory effects. Evidence was also presented that accumulated ganglioside is arranged differently in the cell membrane from endogenous ganglioside (Keenan et al., 1975) and may therefore not be in the correct position to play its possible physiological role.

Stemming from the observation that glycolipids from *Salmonella minnesota* R mutants cause tumour necrosis in vivo, Brailovsky et al. (1973) studied the effect of these molecules on normal and SV40 transformed rat embryo fibroblasts in vitro. The glycolipids bound preferentially to the transformed cells, elevating levels of cAMP and causing cessation of growth when the cells reached confluency. The glycolipids, which were localised at the cell surface by immunofluorescence techniques (Bara et al., 1973a), may mimic the endogenous glycolipids deleted by transformation (Bara et al., 1973b).

4. Variations in glycolipid pattern in normal cells

4.1. Growth-dependent variation

4.1.1. Cell density-dependent glycolipids

Comparison of the glycolipid pattern of sparse and dense normal cells in culture has shown that levels of certain glycolipids increase at high cell densities. The best studied example is the accumulation of neutral glycolipids GL3, GL4 and

GL5* in dense NIL Syrian hamster fibroblasts (Robbins and Macpherson, 1971; Kijimoto and Hakomori, 1972; Sakiyama et al., 1972; Critchley and Macpherson, 1973). Similar observations have been made for levels of GL3 in low passage BHK21/C13 (Hakomori, 1970a), GD3 in human fibroblasts (Hakomori, 1970a), GM3 and more complex gangliosides in chicken embryo fibroblasts (Hakomori et al., 1971) and GM3 in some clones of NIL hamster cells (Sakiyama et al., 1972, Hirschberg et al., 1975). Indeed, in certain NIL cell clones the levels of both gangliosides (GM3) and neutral glycolipids (GL5) increase at high cell density (Kijimoto and Hakomori, 1972; Sakiyama et al., 1972). Perhaps significantly, levels of these density-dependent glycolipids are all dramatically reduced in cells transformed by a variety of RNA or DNA tumour viruses.

Simplification of glycolipid pattern has also been reported to accompany the change from growth in monolayer to suspension in KB cells, a human epidermoid carcinoma strain (Chatterjee et al., 1975a). Levels of GL3, GL4 and GD1a were reduced by approximately two-fold in suspension cultures with a corresponding increase in the levels of GL1, GL2 and GM3. However, monolayer cells were allowed to become confluent before harvesting, whereas suspension cultures were maintained at a fixed concentration by daily additions of fresh medium. The possibility that the results were due to cell density effects was not fully discussed. Studies with revertants, some of which retain their transformed glycolipid pattern (Critchley and Macpherson, 1973; Den et al., 1974; see section 3.3.3), suggest that the effect was not related simply to a change from a rounded to flattened morphology.

That the density-dependent glycolipids are in a position to influence cell surface properties has been confirmed in NIL cells by subcellular fractionation (Critchley et al., 1973) and by the availability of these molecules in intact cells to bind antibodies (Hakomori and Kijimoto, 1972) and react with galactose oxidase (Gahmberg and Hakomori, 1973b, 1975a,b; see section 5.1).

The density dependence of GL3 in NIL and BHK 21 cells is probably explained by the demonstration that the activity of a UDPgal: lactosyl ceramide α-galactosyl transferase is markedly increased in dense cells (see also section 3.1). The enzyme was absent or much reduced in transformed derivatives (Kijimoto and Hakomori, 1971; Critchley et al., 1974; Chandrabose et al., 1976). In contrast the β-galactosyltransferase involved in synthesis of GL2 was much less affected by cell density or viral transformation. The results again point to the importance of synthetic enzymes in controlling the extent of accumulation of certain glycolipids, although the activities of the degradative enzymes have been little studied. In NIL cells the activity of an α-galactosidase utilizing GL3 as substrate was not affected by cell density although there was an increase on transformation (Kijimoto and Hakomori, 1971; Chandrabose et al., 1976). It has therefore not been excluded that transformed NIL cells have reduced amounts of GL3 etc., because of decreased synthesis coupled to increased degradation.

---

*For a fuller description of the structures of these molecules see Gahmberg and Hakomori (1975a); also see Fig. 3 and Table I.

Possibly related to this discussion is the observation that the density-dependent effect in normal cells is frequently larger if assessed using isotope incorporation rather than chemical quantitation. This suggests that turnover may also increase at high cell density (Hakomori, 1970a; Kijimoto and Hakomori, 1971; Sakiyama et al., 1972; Gahmberg and Hakomori, 1975a). The situation is complex because whereas the steady-state level of GL3 is influenced by the rate of synthesis, utilization to form GL4 and degradation by glycosidases, that of GL5 is coupled to synthesis and degradation only. Increased turnover of GL3 would therefore be predicted at high cell density if indeed it is the precursor of GL4. Elucidation of turnover rates and precursor product relationships is made more difficult because of the presence of at least two different carbohydrate sequences in GL4 (Gahmberg and Hakomori, 1975a; see also Fig. 3 and Table I). Residual GL4 in transformed cells has yet a different structure from the two present in the normal cell. These considerations point to our poor understanding of the mechanisms involved in the control of glycolipid levels in cultured cells especially as they relate to density-dependent glycolipids.

4.1.2. Kinetics of the change in glycolipid pattern as related to cell density
Increased incorporation of $[1-^{14}C]$galactose or $[1-^{14}C]$palmitate into density-dependent glycolipids of various NIL cell clones occurred just prior to confluency and significantly before the cells had ceased to divide, according to the results of Kijimoto and Hakomori (1972), Critchley and Macpherson (1973) and Critchley et al., (1974). This conclusion was also supported by direct assay of an α-galactosyltransferase responsible for GL3 synthesis (Critchley et al., 1974; Chandrabose et al., 1976). On the other hand, Hirschberg et al. (1975) found that incorporation of $[1-^{14}C]$palmitate into a density-dependent GM3 in a NIL subclone only increased as the cells became confluent and stopped growing. Unfortunately, they failed to discuss this aspect of their results in terms of other literature on the subject. The question of whether the increase in synthesis of density-dependent glycolipids is an early event occurring prior to quiescence is also complicated by the fact that accumulation of isotope into the various molecules occurs at different rates (Kijimoto and Hakomori, 1972; Critchley and Macpherson, 1973; Sakiyama and Robbins, 1973a; Critchley et al., 1974). For example, Sakiyama and Robbins (1973a) found that whereas incorporation of $[1-^{14}C]$palmitate into GL5 increased with increasing cell density, incorporation into GL4 was only slightly elevated until the cell number had plateaued. At this stage, when the cells were presumably non-dividing, there was a more dramatic increase in incorporation into GL4. The result may be explained by the presumed precursor product relationship between GL4 and GL5. Label may not accumulate in GL4 until GL5 has increased to a certain critical level.

Incorporation of $[1-^{14}C]$palmitate into the density-dependent glycolipids persists for at least 12 h when dense non-dividing NIL cells are trypsinized and replated at low density (Critchley and Macpherson, 1973; Critchley et al., 1974).

A marked change in incorporation occurred between 12–19 h by which time many cells had initiated DNA synthesis. After 26 h a cell doubling had occurred and the pattern characteristic of sparse cells was established. Dense cells trypsinized and reseeded dense also showed some reduction in incorporation into density-dependent glycolipids around the onset of DNA synthesis, but the effect was transient. Because trypsinization of dense quiescent cells produces a partially synchronous culture, the possible influence of cell cycle on incorporation patterns has to be remembered (see section 4.2). Preliminary experiments indicate that the existing density-dependent glycolipids are not rapidly degraded before the onset of DNA synthesis but are probably diluted out by cell division (Critchley and Macpherson, 1973). In summary, in at least some NIL subclones the change in glycolipid metabolism is an early event when cells become crowded, but occurs slowly when cells are re-stimulated to grow.

One of the major questions about the significance of the density-dependent glycolipid effect is the generality of the phenomenon. Until recently, many publications paid little attention to the importance of growth densities before analysing the glycolipid composition. Considerable doubt was cast by the apparent absence of the effect in mouse cell lines which exhibit strong density dependent inhibition of growth (Fishman et al., 1971; Yogeeswaran et al., 1972; Brady and Fishman, 1974). More recently, Yogeeswaran and Hakomori (1975) have reported a 1.5-fold increase in levels of GD1a at an early stage of cell contact which was, however, transient, i.e. sparse and dense cells had identical glycolipid patterns. The increase paralleled an elevated biosynthetic activity (assayed using endogenous acceptors) and a decreased sialidase activity. Predictably, these transient effects were not found in transformed cells. The results emphasize the possible importance of a fine balance between biosynthesis and degradation in influencing glycolipid pattern, but certain apparent inconsistencies within the study make it important to substantiate the data. For example, although biosynthetic activities were reported to increase and sialidase activities decrease at the early stage of confluence, overall activities in normal and transformed cells were very similar although we know the ganglioside patterns to be quite different (Mora et al., 1969; Brady and Mora, 1970; Yogeeswaran et al., 1972; Yogeeswaran and Hakomori, 1975; see also section 3). Such considerations lead one to query whether the activities measured were related to ganglioside pattern. Interestingly, a transient three-fold increase in incorporation $D[1-^{14}C]$galactose has also been found to occur in rat myoblasts just prior to fusion (Whatley et al., 1976). Mutant cells which were unable to fuse failed to show the effect and had an overall simplification of glycolipid pattern.

In conclusion, although a number of points need clarification, and examination of more cell lines is needed, the overall picture relating to density-dependent glycolipids is relatively clear. That is, certain specific glycolipids increase in dense cells and the effect is lost on transformation with tumour viruses.

4.2. Glycolipid metabolism as a function of cell cycle

To check if the density-dependent effect was in any way related to comparison of dividing versus non-dividing cells, Critchley and Macpherson (1973) studied the glycolipids of NIL cells made quiescent in sparse culture. Cells blocked with excess thymidine or by glutamine deprivation retained the sparse glycolipid pattern, but cells blocked by serum deprivation incorporated significantly increased amounts of [1-^{14}C]palmitate into GL3 although the other density-dependent glycolipids (GL4, GL5) were little affected. This result has subsequently been confirmed by assaying directly for the specific α-galactosyltransferase responsible for GL3 synthesis (Chandrabose et al., 1976). Sparse cells are thus capable of synthesizing at least one of the density-dependent glycolipids. One possible explanation is that GL3, at least, is synthesized as a function of cell cycle. Cells in G_0-phase of the cell cycle (serum deprivation) might synthesize increased levels of GL3 whereas cells in S-phase (thymidine block) might synthesize little of this component.

These studies were subsequently extended by looking at incorporation of [1-^{14}C]palmitate into glycolipids in NIL cells synchronised by release from low-serum block. Whereas GL3 was preferentially labelled in G_1-phase in sparse non-contacted or dense cells, incorporation into GL4, GL5 and GM3 was not so markedly influenced by the cell-cycle (Critchley et al., 1974). Similar, but not identical, results have been reported for other NIL cell subclones synchronised by a double thymidine block, mitotic collection method (Wolf and Robbins, 1974; Hirschberg et al., 1975). In agreement with the isotope incorporation data, Gahmberg and Hakomori (1975a) found that whereas chemically determined levels of GL3 increased in NIL cells in G_1 phase, GL4 and GL5 did not show marked cell cycle changes. A marked increase in the levels of glycolipids during G_2/M/early G_1-phase was also found in KB cells synchronised by release from thymidine block (Chatterjee et al., 1973, 1975a). Perhaps significantly, the activity of a ceramide glucosidase and trihexosidase peaked in S phase (Chatterjee et al., 1975b). These results thus offer an explanation for increased labelling of GL3 in sparse cultures of NIL cells blocked in G_0/G_1-phase of the cell cycle by serum deprivation. This attractive interpretation ignores the finding that glutamine and isoleucine deprivation, both of which have been reported to block cells at a similar point in the cell cycle to serum deprivation (Pardee, 1974), fail to increase incorporation of [1-^{14}C]palmitate into GL3. One possibility is that deprivation of essential amino acids inhibits protein synthesis in general, and therefore affects the synthesis of the glycosyltransferase involved in GL3 formation. In fact, dense non-dividing cells subsequently deprived of glutamine for 48 h also showed reduced incorporation into GL3, although reducing the serum concentration had no such effect. An additional consideration is that whereas serum starved cells have high levels of cAMP comparable with those in dense quiescent cells (Seifert and Paul, 1972; Kram et al., 1973; Rozengurt and De Asua, 1973), isoleucine- (Sheppard and Prescott, 1972) or glutamine and histidine-deprived cells (Rudland et al., 1974) have significantly lower levels of this cyclic nucleotide. If such generalizations are true for NIL cells, it is possible that cAMP levels influence the synthesis of density-dependent glycolipids. Levels of cAMP have been reported to increase with cell density in both 3T3 (Otten et al., 1972; Bannai and Sheppard, 1974) and NRK cells (Anderson et al., 1973; Carchman et al., 1974), and in myoblasts just prior to fusion (Zalin and Montague, 1974). However, Sakiyama and Robbins (1973a) found that cultivation of sparse NIL cells in the presence of dibutyryl cAMP failed to increase incorporation of [1-^{14}C]palmitate into cell density-dependent GL4 and GL5. Similar treatments were reported to be without effect on the ganglioside pattern of normal and transformed Swiss 3T3 cells (Sheinin et al., 1974).

4.3. Possible mechanisms to explain density-dependent glycolipids

At present, it is difficult to present a tenable theory to explain the mechanism of the density-dependent effect which encompasses all of the foregoing results. A discussion of three of the most attractive explanations that have received some consideration is nevertheless worthwhile.

4.3.1. Cell cycle effects
As GL3 is probably synthesized in G_1-phase in at least some NIL subclones, it is possible that GL3 is strongly labelled in dense populations because of the accumulation of cells in the G_0/G_1 phase of the cell cycle. The density dependence of GL4 and GL5 is not explained on this basis unless one makes the not unreasonable assumption that some of the GL3 is subsequently converted to GL4 and GL5. Extending the argument, transformed cells may always show a sparse glycolipid pattern because they never accumulate in the G_0/G_1 phase of the cell cycle.

There are several points against this proposal. The idea that cells arrested in G_0 phase convert much of their GL3 to GL4 and GL5 is not borne out by the data on G_0 arrested sparse cells (Critchley and Macpherson, 1973). As a theory it also fails to account for the change in glycolipid pattern being an early event occurring at least in some NIL cell clones as the cells become crowded, but before cessation of division. In addition, Hirschberg et al. (1975) studying a NIL subclone with a density-dependent GM3, found that although GM3 was not synthesised predominantly in any one phase of the cell cycle, its concentration only increased when the cell growth curve had plateaued. In this case it is difficult to explain the density dependence either as an accumulation of G_0/G_1 phase cells or as an early event in cell crowding. In contrast, another subclone of NIL cells was reported to synthesize GL4 predominantly in the G_1-phase and the density dependence was only seen when cell number was no longer increasing (Sakiyama and Robbins, 1973a). In this case the lack of effect in transformed cells could be explained by their failure to stop dividing, although transformed derivatives, where growth was inhibited by addition of dibutyryl cAMP, failed to regain density-dependent GL4. It is possible, however, that the transformed cells failed to arrest in G_0 phase of the cell cycle (Smets, 1972).

That the effect of viral transformation on glycolipid metabolism is related simply to failure of the transformed cells to stop dividing seems unlikely from other considerations. For example, Yogeeswaran and Hakomori (1975) found no difference between normal growing *non-contacted* and quiescent Swiss 3T3 cells although there remains a major difference between the normal and transformed cells (Mora et al., 1971; Brady and Fishman, 1974). Similarly, Fishman et al. (1971) found a marked difference between the activities of certain key glycosyltransferases in *growing* normal and transformed cells. On balance these studies suggest that although glycolipid metabolism does vary considerably during the cell cycle, accumulation of cells in the G_0/G_1 phase cannot solely explain the phenomenon of density-dependent glycolipids and the lack of these molecules in transformed cells.

4.3.2. Synthesis of density-dependent glycolipids by transglycosylation

It has been suggested that increased synthesis of density-dependent glycolipids at an early stage of cell crowding is related to cell contact (Hakomori 1975a), although there is no direct evidence for this. One attractive possibility which assumes a *primary* role for cell contact in the synthesis of density-dependent glycolipids is based on the work of Roth and co-workers (Roth et al., 1971a; Roth and White, 1972; Webb and Roth, 1974; Shur and Roth, 1975). They have suggested that cellular recognition and adhesion may be mediated by interaction of surface glycosyl transferases with substrate molecules on the surface of adjacent cells. Under certain conditions, the interaction may lead to addition of a sugar residue to the substrate, i.e. transglycosylation. They have reported the presence of glycosyl transferases on the surface of a number of cell types including mouse 3T3 fibroblasts which supposedly incorporated extracellular UDPgal into endogenous glycoprotein and glycolipid only when there was appreciable cell contact. Although these data have been criticised on a number of grounds (Deppert et al., 1974; Evans, 1974; Keenan and Morré, 1975), the suggestion that glycosyl transferases do exist on the cell surface and may be involved in interaction with other cells or soluble proteins has received considerable support (reviewed recently by Shur and Roth, 1975). Without entering the controversy, it is interesting to extend the theory to synthesis of density-dependent glycolipids. In NIL cells for example, a UDPgal: lactosyl ceramide α-galactosyl transferase on the cell surface would interact with GL2 on a neighbouring cell and depending on the availability of UDPgal* result in synthesis of GL3. An attempt to test this hypothesis was made by Yogeeswaran et al. (1974a) who incubated intact NIL or BHK 21 cells with glass particles (2–10 μm in diameter) or glass coverslips to which glycolipids were covalently

---

*Or a lipid-sugar intermediate (see, Lennarz, 1975).

attached. Exogenous radioactive sugar nucleotide was added, or the intracellular pool was made radioactive by pre-incubating the cells with [^{14}C]galactose. In either case glycosylation of the glycolipid bound to the glass occurred. Reactions catalysed by a galactosyl transferase (GL2→GL3), a galactosaminyl transferase (GL4→GL5) and a sialyl transferase (GL2→GM3) were demonstrated. Perhaps significantly, the extent of the glycosylation reactions was less with the transformed cells. Although these elegant experiments suggest that there are glycolipid glycosyl transferases associated with the outside of the plasma membrane, it is impossible to exclude that a few leaky cells were responsible for the low levels of glycosylation observed. Glycolipid galactosyl transferases have been detected in plasma membrane fractions isolated from NIL cells (Critchley et al., 1974; Chandrabose et al., 1976) but the fate of the Golgi apparatus in the fractionation scheme was unclear.

At present there is little additional evidence that trans-glycosylation is involved in synthesis of density-dependent glycolipids. Indeed there is some doubt as to whether the galactose residue of GL2 is in fact available at the cell surface. The glycolipid is not labelled when either intact erythrocytes or NIL cells are exposed to galactose oxidase followed by borotritiide (Gahmberg and Hakomori, 1973b, 1975a). Indeed space filling models suggest that the lactose residue of GL2 would barely project beyond the base head groups of phospholipids (Gray, 1974). However, in an interesting study Chatterjee and Sweeley (1973) found that within 10 min of thrombin-induced platelet aggregation, there was a marked reduction in levels of GL2 and a corresponding increase in GM3. A protease inhibitor, phenylmethylsulphonyl fluoride, which inhibited aggregation also inhibited conversion of GL2 to GM3. It is tempting to speculate that proteolysis of platelet surface proteins produces aggregation and leads to exposure of GL2, which subsequently interacts with a sialyl transferase on an adjacent platelet within the aggregate, producing GM3.

If transglycosylation is indeed a mechanism by which GL3, GL4 and GL5 are synthesized in NIL cells, it is not the only one. Thus, cells seeded from dense to sparse culture continued to incorporate high levels of [1-^{14}C]palmitate into these glycolipids for at least 12 h after cell contacts had been disrupted by trypsinization (Critchley and Macpherson, 1973; Critchley et al., 1974, also see section 4.1.2). Also the levels of incorporation of [1-^{14}C]palmitate into GL3 and the activity of the specific α-galactosyl transferase were high when sparse cells were blocked in the G_1-phase of the cell cycle (Critchley and Macpherson, 1973; Chandrabose et al., 1976; also see section 4.2). In low serum, NIL cells show an extremely flat morphology with long processes, and it cannot be totally excluded that just sufficient cell contact was made to trigger GL3 synthesis.

Further studies are needed to clarify the role of cell contact in the synthesis of density-dependent glycolipids, and to unequivocally demonstrate a surface location for *specific* glycolipid glycosyl transferases. It will also be important to know more about the availability of GL2 at the surface and to establish whether the products of glycosylation GL3, GL4 and GL5 are located solely in the plasma membrane (Critchley et al., 1973).

4.3.3. Enzyme induction

The third and perhaps most obvious mechanism to explain density-dependent glycolipids synthesis is based on enzyme induction. The activity of the α-galactosyl transferase involved in synthesis of GL3 has been shown to increase 10–20-fold when NIL cells grow from very sparse to dense culture (Critchley et al., 1974; Chandrabose et al., 1976). It is possible that cell contact acts as the inducer, but the activity of the enzyme was near maximal when the cells were still quite sparse and cell contacts apparently minimal. A further possibility is that diffusable factors are present in the growth medium which might induce the enzyme. Sparse and dense NIL cells were therefore grown on different areas of the same dish whilst sharing the same growth medium (Robbins and Macpherson, 1971; Critchley and Macpherson, 1973). However, the glycolipid pattern of sparse cells was not affected by co-cultivation. Also, transformed NIL-HSV cells did not influence the pattern in dense normal cells grown in the same dish. This tends to exclude the possibilities: (a) that dense cells secrete a molecule into the medium which directs GL3 production in other cells; (b) that normal cells in dense culture deplete the medium of an inhibitor of GL3 synthesis; (c) that a reduction of pH of the medium typical of dense normal or transformed cells is a critical factor; and (d) that transformed cells produce a medium transported factor which inhibits GL3 synthesis in neighbouring cells. There is therefore a total lack of evidence that diffusible molecules are important in the present system. If enzyme induction by unknown factors is the basis for synthesis of density-dependent glycolipids, then it is not surprising that synthesis of the glycolipids would continue for sometime after the inducer was removed. For example, the fact that dense cells seeded as sparse cultures continue to synthesize GL3 for 12–19 h may be a reflection of the half life of the enzyme (see section 4.1.2).

A fuller understanding of the mechanism of induction may come from the chance observation that addition of butyrate to HeLa cells *specifically* increases the activity of a CMP-sialic lactosyl ceramide sialyl transferase 7–20-fold (Fishman et al., 1974b, 1976; Simmons et al., 1975). Although induction showed a lag period of 4 h the activity was maximal by 24 h. The process required RNA and protein synthesis, but not DNA synthesis, and the continued presence of the inducer. The half life of the enzyme was 16–24 h. The relationship between induction of the enzyme and a subsequent change in morphology of HeLa cells involving extensive process formation is unclear.

5. *Exposure of glycolipids at the cell surface*

Much of the information on this subject has been obtained by looking at the ability of glycolipids of intact cells to react with antisera to specific glycolipids or enzymes such as neuraminidase or galactose oxidase. For example antisera to the major human erythrocyte glycolipid (GL4) reacts poorly with adult cells but strongly with foetal erythrocytes, although both have about the same concentration of the glycolipid (Hakomori, 1971). This interesting result suggests that the reactive groups become masked as the cells mature. Indeed, GL4 of adult erythrocytes reacts with antisera if the cells are first treated with neuraminidase or trypsin. However, GL4 is labelled to about the same extent in either cell type by the galactose oxidase/borotritiide technique (Gahmberg and Hakomori, 1973a) possibly because of the smaller molecular weight of the enzyme (mol. wt. 75 000 daltons) compared with an immunoglobulin (mol. wt. 180 000 daltons). If this interpretation is correct, then the significance of the results from these methods will be limited until we know something about the size of molecule that might interact with glycolipids. Nevertheless, it is tempting to speculate from the foregoing discussion that alterations in the exposure of glycolipids might be an important way of controlling their influence on cell surface properties.

5.1. In cultured cells

Initial results which showed that gangliosides of some culture cell lines were not susceptible to neuraminidase lead to the not unreasonable concept that the short carbohydrate chains of glycolipids in tissue culture cells were buried, possibly underneath a coat of surface glycoprotein. For example, Weinstein et al. (1970) found that GM3 and a disialoganglioside of intact mouse L cells were barely attacked by neuraminidase even though they were shown to be localised in the plasma membrane by subcellular fractionation techniques and that neuraminidase removed 45% of total cell sialic acid, presumably from glycoproteins. Levels of GM3 in a B-16 melanoma, Morris hepatoma, and Green Monkey kidney cells, and a disialoganglioside in a human neuroblastoma, were also reported to be unchanged after exposure of intact cells to neuraminidase (Barton and Rosenberg, 1973). Similarly Keenan et al. (1975) found that little of the lipid bound sialic acid of SV40 transformed 3T3 cells (mainly GM3) was released by neuraminidase. Their conclusion that gangliosides in membranes are generally masked was supported by evidence that GD3 present in intact bovine erythrocytes or milk fat globules was inaccessible to the enzyme (Tomich et al., 1976). Significantly, pre-treatment with trypsin or EDTA rendered the ganglioside susceptible to attack. There is, however, considerable evidence that not all gangliosides are buried under an umbrella of cell surface glycoproteins. For example, although GM3 of intact erythrocytes was not susceptible to neuraminidase attack, a glucosamine-containing ganglioside and GD3 were both susceptible (Wintzer and Uhlenbruck, 1967). In contrast to the study of Keenan et al. (1975) on Swiss 3T3 cells, Yogeeswaran et al. (1972) found that 60% of the lipid bound sialic acid of Swiss 3T3 and Py-3T3 cells was removed by the enzyme, much of the GM3 being converted to GL2, and GD1a to GM1. We have also found that the GD1a of Swiss 3T3 cells is available to neuraminidase (see Fig. 4). The result is supported by the finding that neuraminidase treatment of a transformed mouse epithelial cell line SVS AL/N increased its capacity to bind cholera toxin, which is thought to specifically interact with GM1 (Hollenberg et al., 1974; see section 7.1.1). The possibility that these latter

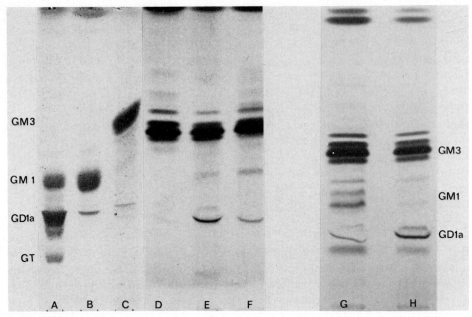

Fig. 4. Metabolic labelling of the gangliosides of cultured cells, and the susceptibility of GD1a to neuraminidase. Chinese hamster ovary (D), Balb or Swiss 3T3 (E) and NIL hamster cells (F) were grown to confluency in the presence of 1 μCi/ml [1-^{14}C]palmitate. The gangliosides were extracted, run on Silica gel G plates using the solvent $CHCl_3:CH_3OH:H_2O$, 60:35:8, and detected by autoradiography. Individual gangliosides were identified by comparison with standards; A, mixed gangliosides (Sigma); B, GM1 prepared by neuraminidase treatment of mixed gangliosides; C, GM3 prepared from horse spleen. A faint band in the position of GM1 is present in Balbc or Swiss 3T3 cells (E), but not CHO cells (D). The label in this position in NIL cells is probably sialylparagloboside not GM1, see Table I. That the GD1a in 3T3 cells is exposed to the extracellular environment was shown by incubating prelabelled cells in monolayer with 1.5 ml PBS ($Ca^{2+}Mg^{2+}$) plus or minus 50 units of neuraminidase (Behringwerke) for 60 min at room temperature. In cells incubated with neuraminidase (G) there was a marked reduction in label in GD1a and a corresponding increase in GM1, compared to control cells (H).

results are explained by a small population of leaky or dead cells in which the availability of gangliosides is altered cannot be excluded. It is also possible that the neuraminidase preparations used in this work contained sufficient proteolytic enzymes to result in exposure of ganglioside by removing cell surface proteins. We have looked for proteolytic contamination of neuraminidase (Behringwerke; isolated from *Vibrio cholera*) by studying the effect of the enzyme on the major iodinatable surface protein of NIL cells (LETS glycoprotein; see Critchley et al., 1976). There was no significant reduction in incorporation of label into LETS glycoprotein after 60 min to 50 units/ml of neuraminidase although 100% of the protein was removed by 5 μg/ml trypsin for 5–10 min.

The exposure of the cell density-dependent glycolipids of NIL hamster fibroblasts (GL3, GL4, GL5) have been studied extensively by Hakomori and co-workers. For example, although the concentration of GL5 (Forssman-reactive glycolipid) increased with increasing cell density in NIL cells, its reactivity to antisera specific to the glycolipid decreased (Hakomori and Kijimoto, 1972). The reactivity of dense cells could be increased by taking them into suspension with EDTA, a finding of great potential importance. Unfortunately, whether this was a specific effect, perhaps representing interaction of a density-dependent glycolipid with molecules on adjacent cells, (see section 7.5), or simply represented general masking of glycolipid, was not determined.

Density-dependent glycolipids of NIL cells are also labelled by the galactose oxidase/borotritiide method, with much of the ^{3}H label being introduced into GL4, although GL3 and GL5 are also labelled (Gahmberg and Hakomori, 1973b). The reasons for these variations in extent of labelling are unclear. Glycolipid GL3 may be less well labelled than GL4 because the shorter carbohydrate residue means it extends less far from the lipid bilayer and may therefore be masked by surrounding glycoproteins. In fact GL2 was barely labelled. This does not of course explain the paucity of label in GL5 since both GL4 and GL5 have similar terminal N-acetylgalactosamine residues. Unfortunately no study has been made to see if labelling of these glycolipids varies as a function of cell density, in a similar manner to their reactivity to Forssman antisera (Hakomori and Kijimoto, 1972). Gahmberg and Hakomori (1974) have also reported decreased labelling of these glycolipids in transformed NIL cells but not to the extent predicted by the much reduced levels of GL3, GL4 and GL5 in these cells. A more detailed study showed that the specific activities of transformed cell glycolipids were 10–20-fold greater than in normal cells (Gahmberg and Hakomori, 1975a). A similar increase in reactivity of glycolipid in transformed BHK21 or 3T3 cells has been found using an antisera against GM3 (Hakomori et al., 1968). Possible reasons for this apparent increased exposure will be discussed later (see section 5.2).

Exposure of the density-dependent glycolipids as detected by the galactose oxidase method has also been found to vary as a function of the phase of the cell cycle (Gahmberg and Hakomori, 1974, 1975a). NIL cells synchronised by trypsinization of confluent cultures showed a marked increase in glycolipid labelling in G_1-phase, reduced labelling in S/G_2 phase, and a further but somewhat smaller increase in the following G_1-phase. The fact that cells were synchronised by trypsinization places a considerable limitation on the interpretation of the results in terms of exposure. The results are complicated by: (1) possible altered exposure due to trypsinization per se; and (2) the possible regeneration of surface components, stripped off by proteolysis, during the experiment. However, Gahmberg and Hakomori (1974, 1975a) reported that exposure of glycolipids was very low after trypsinization in contrast to previous results on NIL cells (Gahmberg and Hakomori, 1973b) and the observation that trypsinization frequently leads to exposure of previously masked glycolipids (Hakomori, 1971; Tomich et al., 1976). Although such considerations are clearly important, Gahmberg and Hakomori (1974, 1975a) found the major increase in labelling occurred about 8 h after trypsinization suggesting that it was not due simply to removal of surface proteins. The results were partially confirmed using a double thymidine block synchronization method, although in this system the increased reactivity occurred in G_2/M phase. The interpretation of the results is also complicated by the possibility that synthesis of glycolipids may be maximal in G_1-phase and increased labelling may simply represent increased levels of these molecules in the surface membrane. However, on a per cell basis only GL2 and GL3 increased in G_1-phase and the most marked increase in labelling was in GL4 and GL5 (Gahmberg and Hakomori, 1974, 1975a). It would seem therefore that the increased labelling of glycolipids in G_1-phase represents a genuine increase in exposure of the molecules. Perhaps significantly, although the reactivity of glycolipids in transformed cells was higher than in normal cells, there was no phasic exposure during the cell cycle.

A further point of great interest was that pre-treatment of NIL cells with low concentration of *Ricinus communis* lectin (RCA I) increased labelling of glycolipids by the galactose oxidase/borotritiide method. Low concentrations of RCA I and concanavalin A (Con A) (10 μg/ml) also led to increased labelling of the LETS glycoprotein, although labelling was reduced at high concentrations presumably because the interaction of the lectins with carbohydrate chains sterically inhibits galactose oxidase (Gahmberg and Hakomori, 1975b). However, whereas both lectins had a similar effect on the surface glycoproteins, only RCA I enhanced labelling of glycolipid (mainly GL4a). Labelling was suppressed to a small extent at higher concentrations of RCA I (100 μg/ml) although it is unclear whether lectins do interact with glycolipids in membranes (Allan et al., 1972; Yogeeswaran et al., 1974b; Surolia et al., 1975).

5.2. Possible mechanisms to explain alterations in exposure of glycolipids

The results on exposure of glycolipids during the cell cycle in normal and transformed cells have been related to changes in the amount and distribution of cell surface glycoproteins and to membrane fluidity (Hakomori, 1975a). For example the glycolipids of transformed cells might be more exposed because the cells are known to lack the LETS glycoprotein (Gahmberg and Hakomori, 1973b; Hynes, 1973, 1976). This may result in more space between the remaining surface molecules, exposing larger areas of lipid bilayer and therefore glycolipid. There are at least two points against this attractive idea. Firstly, Gahmberg and Hakomori (1974, 1975a) claim that trypsinization of normal NIL cells, which is known to remove a number of surface glycoproteins including LETS (Critchley, 1974), failed to markedly increase labelling of glycolipids by galactose oxidase. It is difficult to judge the severity of this criticism because a number of other studies, admittedly in other systems, suggest that trypsin does indeed lead to exposure of glycolipids (Hakomori, 1971; Tomich et al., 1976). Secondly, the greater exposure of glycolipids of NIL cells in G_1-phase does not coincide with a reduction in labelling of the LETS glycoprotein which is itself maximally exposed in G_1-phase of the cell cycle (Hynes and Bye, 1974). The fact that the lectin RCA I enhanced exposure of glycolipids to galactose oxidase may be explained by the observation that lectins can cause clustering of cell surface glycoproteins (review, Nicolson, 1974). This in turn may be related to their ability to agglutinate cells. It was therefore significant that normal NIL cells were more readily agglutinated by RCA I than Con A, and only RCA I enhanced labelling of glycolipid by galactose oxidase (Gahmberg and Hakomori, 1975b). However, the lack of effect of lectins in enhancing exposure of glycolipids in transformed cells is odd as the cells are more readily agglutinated and therefore presumably also exhibit clustering of surface molecules. The argument that the glycolipids of transformed cells are already maximally exposed because of a reduced complement of surface glycoproteins is not consistent with the lack of effect of trypsin in enhancing labelling in normal cells, as previously discussed. This suggests that other factors must be involved.

One such factor might be membrane fluidity (Hakomori, 1975a). An increase in fluidity in G_1-phase cells and transformed cells throughout the cycle could conceivably lead to increased collision frequencies between galactose oxidase and glycolipids leading to increased labelling. A similar logic might also explain the increased labelling of the LETS glycoprotein in G_1-phase cells (Hynes and Bye, 1974). It has been suggested that loss of this protein may be in part responsible for increased fluidity of lipids in transformed cells (Hakomori, 1975a). This idea is not consistent with the continued presence of LETS glycoprotein in G_1-phase cells unless one postulates that its conformation and/or association with other membrane proteins is broken during this period. Direct evidence for an increase in membrane fluidity in cells stimulated from G_0 into G_1 phase (Barnett et al., 1974a) and in transformed cells (Barnett et al., 1974b) has been presented although doubts have been raised about the validity of the data (Gaffney, 1975). The topic has been discussed in a recent review by Nicolson (1976b) and the net conclusion is that there is presently no compelling evidence to support the view that the membrane lipids in transformed cells are more "fluid" than their untransformed counterparts.

6. Correlation between altered glycolipid pattern and malignancy

Very few definitive studies have been made to establish whether the glycolipid composition of tumours excised from animals is different from that of the parent tissue. Such comparisons are complicated anyway by the cellular heterogeneity within the tumour and parent tissue. However, results from studies both on spontaneously formed human tumours and those induced in animals, are generally in agreement with the results from in vitro studies using cell cultures transformed by tumour viruses i.e. tumour cells have a simplified glycolipid pattern.

Human malignancies that have been investigated include a number of brain tumours (Seifert and Uhlenbruck, 1965; Kostic and Buchheit, 1970), kidney carcinoma (Karlsson et al., 1974) and leukemic leukocytes (Hildebrand et al., 1971, 1972). One of the best documented is that of human adenocarcinomata which, although lacking glycolipids with blood group A or B activity, accumulated less complex precursor fucose-containing glycolipids (Hakomori et al., 1967). The activities of the glycosyltransferases responsible for synthesis of blood group A or B active glycolipids were found to be present at much reduced levels in tissues from these patients (Stellner et al., 1973). The data on fucolipids and blood group glycolipids in normal and tumour tissue have been reviewed recently by Hakomori (1975b). In contrast to these results at least one situation has been found in which human cancer tissue, a hepatoma, had an apparently normal neutral glycolipid composition although ganglioside and blood group-type glycolipids were not extensively investigated (Kawanami and Tsuji, 1968).

Experimental tumours in animals in which glycolipid composition has been found to differ significantly from control material include: avian transplantable lymphoid tumours (Keenan and Doak, 1973); a mammary carcinoma induced with 7,12-dimethylbenz[α]anthracene (Keenan and Morré, 1973); and a Novikoff ascites hepatoma in rats (Leblond Larouche et al., 1975). Perhaps the most detailed study is that of Siddiqui and Hakomori (1970) on the Morris hepatoma system in rats. A comparison of normal rat liver, rapidly dividing neonatal liver, and three hepatomas gave a complete spectrum of rates of cell division, with one of the hepatomas in fact growing more slowly than neonatal liver. Adult liver contained GM3, GM1, GD1a and GT1. Interestingly the glycolipid pattern of neonatal liver showed a greater preponderance of GT ganglioside showing that rapid cell division per se is not the limiting factor in the synthesis of more complex glycolipids. In contrast, all the hepatomas consistently lacked significant amounts of GT ganglioside, although there was a marked accumulation of GD1a. The tumour lines also contained more total ganglioside/mg protein and elevated levels of GL2 and ceramide. Essentially similar observations were made by Cheema et al. (1970) and by Dinistrian et al. (1975). Gangliosides of an hepatic epithelial cell line have also been compared with those in hepatoma cells grown in culture (Brady et al., 1969). Normal hepatocytes contained GM3, GM1 and GD1a. In this case the level of GD1a was markedly decreased in the tumour cell line, although as previously observed by Siddiqui and Hakomori (1970) in the in vivo studies, the total ganglioside content/mg protein increased. The results can be interpreted to indicate a simplification in glycolipid pattern in tumour cells, but the elevation of total ganglioside levels is in marked contrast to virus-transformed cells where the reduction in complexity of the glycolipids is generally accompanied by an overall decrease in glycolipid concentration.

The glycolipid composition of tumorigenic cells derived from cell lines grown in culture (without the use of oncogenic viruses) has also been investigated. Although the glycolipid pattern of spontaneously tumorigenic BHK21 cells tended toward that seen after viral transformation (Hakomori and Murakami, 1968), spontaneously tumorigenic lines derived from BALBc/3T3 and AL/N cells had a normal glycolipid pattern (Brady and Mora, 1970). Levels of the density-dependent, transformation-sensitive glycolipids of NIL hamster fibroblasts have also been studied in a number of derived malignant cell lines. Tumours developed by injecting large numbers (10^7) of untransformed NIL cells into hamsters were subsequently taken into culture and the glycolipid pattern characterised (Critchley and Macpherson, 1973). Of eleven tumour lines studied, ten had a significantly modified glycolipid pattern in that they no longer accumulated [1-^{14}C]palmitate into GL3 and GL4. Incorporation into GL5 remained near normal and was still dependent on cell density. One of the tumour cell lines retained all three glycolipids and they were still density-dependent. The pattern was only slightly modified on re-passage through the animal, although viral transformation led to a marked reduction and incorporation into all three glycolipids. Interestingly this same tumour cell line also retained the LETS glycoprotein although the other tumour lines with modified glycolipid pattern had little of this protein (Hynes, 1976). Sakiyama and Robbins (1973b) also found that although some tumour cell lines derived from NIL sub-clones had a simplified glycolipid pattern, one retained the density-dependent glycolipid of the parent cell line. In addition, they found no clear correlation between tumorigenicity of the clones and agglutinability by Con A or wheat germ agglutinin, although these two properties had been reported previously to run parallel (Inbar et al., 1972). A lack of correlation between presence of normal gangliosides and agglutinability has also been noted in a transformed variant of 3T3 cells (Yogeeswaran et al., 1972).

The idea that glycolipids might be important tumour-associated antigens (Rapport, 1969) has recently received support from studies of the glycolipids of Py-NIL induced tumours (Sundsmo and Hakomori, 1976). In addition to a marked reduction in levels of GL3, GL4a, GL4b and GL5a in Py-NIL cells, a further component, a sialyl paragloboside, is also depleted on transformation and its precursor paragloboside (GL4c) accumulates (see Fig. 3, Table I). The sera of Py-NIL tumour bearing hamsters had significant levels of complement fixing antibody to GL4c but showed no reaction to GL4a or GL5a. Significantly, the serum titre of antibody to GL4c increased with the size of the tumour. Using an Ouchterlony double diffusion technique the sera produced a precipitin band with GL4c but not with GL3, GL4a, GL5a or glycoprotein fractions. However, monospecific antisera to GL4c prepared in rabbits cross-reacted with a glycoprotein fraction from the Py-NIL tumour cells. The possibility that the effective tumour-associated surface antigen of Py-NIL is a glycoprotein containing a similar carbohydrate sequence to GL4c cannot be excluded.

Glycolipid antigens have also been implicated in graft rejection (Esselman et al., 1973). Membranes prepared from canine kidneys, which had been transplanted but rejected by the recipient, had elevated levels of glycolipid including Forssman glycolipid (GL5a or GL4b). One possibility discussed was that the increased levels of Forssman antigen generated cytolytic antibodies which subsequently lead to graft rejection. In conclusion, it would appear that: (1) simplification of the glycolipid pattern *frequently* accompanies tumour formation both in naturally occurring and experimental tumour systems; and (2) glycolipids may also be important tumour-associated cell surface antigens. Unfortunately, the exceptions to the first statement lead to a marked reduction in interest in glycolipids. In retrospect it must be remembered that our understanding of the factors which influence growth regulation and the tumorigenic potential of cells is in its infancy. If, for example, we propose that there are 20 determinants of these properties, then a change in a few may be enough to alter the cell's characteristics. Many other determinants may remain unaltered and glycolipids might occasionally fall into this category.

7. Glycolipids as cell surface receptors

A marked renewal of interest in glycolipids has been stimulated by the finding that they can act as membrane receptors for proteins which in turn are known to influence intracellular events. Thus the receptors for a number of bacterial toxins e.g. tetanus toxin (Clowes et al., 1972; Van Heyningen, 1974b), botulinum toxin (Simpson and Rapport, 1971), cholera toxin (Cuatrecasas, 1973a,b; Van Heyningen, 1974a) and *E. coli* enterotoxin (Holmgren, 1973; Donta and Viner, 1974; Zenser and Metzger, 1974) can all be glycolipid in nature. The mammalian proteins interferon (Besançon and Ankel, 1974a; Besançon et al., 1976), macrophage migration inhibitory factory (Higgins et al., 1976) and thyrotropin (Mullin et al., 1976) can also recognise specific sequences in the carbohydrate moiety of glycolipids. Bacterial glycolipids, the lipopolysaccharides, are important in host phage interactions (Losick and Robbins, 1969) and gangliosides may act as receptors for Sendai virus (Haywood, 1974) and Rubella virus (Shortridge and Biddle, 1972). In addition, glycolipids are also known to be important blood group antigens (Hakomori, 1970b, 1975b) and both Forssman antigen (Siddiqui and Hakomori, 1971; Gahmberg and Hakomori, 1975b) and θ antigen in mouse lymphocytes (Esselman and Miller, 1974; Miller and Esselman, 1975) can be represented in the carbohydrate sequences of glycolipids. It is clear from this brief survey that glycolipids can no longer be considered unimportant components of the lipid bilayer.

The field which has generated much of the current excitement about glycolipids as possible membrane receptors is that of cholera toxin. Although some of the literature is only of direct relevance to work on the molecular basis of the disease, much is pertinent to those interested in the general significance of glycolipids in membranes. Because it is the only example of protein–cell surface glycolipid interaction which has been extensively studied, the work will be reviewed in some detail. Inferences drawn from these studies will be discussed with respect to the possible significance of loss of certain glycolipids in cells transformed by tumour viruses (sections 7.4 and 7.5).

7.1. Studies with cholera toxin

7.1.1. Binding specificity
In an initial study, Cuatrecasas (1973a) found that binding of ^{125}I-cholera toxin to rat liver membranes or fat cells was an extremely rapid saturable process, essentially irreversible, with dissociation constants of $1.1 \cdot 10^{-9}$ M and $4.6 \cdot 10^{-10}$ M respectively. Binding was partially inhibited (20%) by high concentrations of D-galactose (0.2 M) suggesting that the toxin might interact with a saccharide

component of the cell surface. Of a range of glycoproteins tested, fetuin (150 μg/ml) was the most potent inhibitor of binding (58% inhibition), and exposure of the subterminal galactose residues with neuraminidase increased its effectiveness by 30%. However, certain sialoglycolipids were much more effective, 20 ng/ml GM1 resulting in 50% inhibition of binding. Molecules lacking the free terminal galactose, e.g. GM2, GD1a, GM3 were 35, 50 and 6500-fold less effective than GM1, respectively. Binding was not reduced by treating membranes with high concentrations (1 mg/ml) of pronase or trypsin (37°C, 90 min), but was almost completely abolished by extraction with chloroform methanol. Pre-incubating cells or membranes with GM1 markedly enhanced binding of toxin, however. These results strongly suggest that cholera toxin interacts with a high degree of specificity with the cell surface glycolipid GM1, in agreement with the results of Van Heyningen et al. (1971), Holmgren et al. (1973), King and Van Heyningen (1973), and Van Heyningen (1974b).

The diverse actions of cholera toxin (Table II) are a result of stimulation of adenyl cyclase with a subsequent increase in levels of intracellular cAMP (Finkelstein, 1973; Sharp, 1973). In contrast, the toxin was reported to have no effect on cAMP phosphodiesterase activity (Kimberg et al., 1971).

On present evidence, it can not be totally excluded that some other molecule with a similar carbohydrate structure to that contained in GM1 is the natural receptor for cholera toxin. In a series of experiments with transformed derivatives of a mouse cell line AL/N, Hollenberg et al. (1974) concluded there was a good correlation between binding of ^{125}I-cholera toxin, elevation of levels of cAMP, inhibition of DNA synthesis and cellular levels of GM1. In contrast with this conclusion, they found that the toxin inhibited DNA synthesis in an SV40 transformed derivative of AL/N in which GM1 was undetectable either by chemical analysis or more sensitive radiolabelling methods. However, these cells were 10-fold less sensitive to toxin than cells containing detectable quantities of

TABLE II
SOME OF THE DIVERSE ACTIONS OF CHOLERA TOXIN

(1) Salt and fluid imbalance in intestine (Pierce et al., 1971; Hynie and Sharp, 1972).
(2) Increased secretion by rat pancreas slices (Kempen et al., 1975).
(3) Stimulation of lipolysis in fat cells (Cuatrecasas, 1973b,c).
(4) Increased steroidgenesis in adrenal cells (Donta et al., 1973; Wolf et al., 1973; Palfreyman and Schulster, 1975).
(5) Glycogenolysis in liver and platelets (Zieve et al., 1971).
(6) Increased glucose oxidation in thyroid tissue (Mashiter et al., 1973).
(7) Inhibition of cytolysis by sensitised lymphocytes (Strom et al., 1972).
(8) Antigen and IgE-induced histamine release from lymphocytes (Lichtenstein et al., 1973; Sharp, 1973).
(9) Inhibition of DNA synthesis in stimulated lymphocytes (Sulzer and Craig, 1973; Holmgren et al., 1974; Révész and Greaves, 1975).
(10) Inhibition of DNA synthesis in fibroblasts (Hollenberg and Cuatrecasas, 1973; Hollenberg et al., 1974).
(11) Induces differentiation in melanoma cells (O'Keefe and Cuatrecasas, 1974).

GM1. Pre-incubation of SV40 AL/N cells with neuraminidase increased their toxin binding capacity 4-fold. It is difficult to explain this result by postulating conversion of GD1a to GM1, as GD1a was also undetectable in these cells. It should be noted that AL/N cells contain a large number of toxin receptors (Hollenberg et al., 1974), and since it has been calculated that about 8000 molecules of bound toxin are enough to produce maximal inhibition of DNA synthesis, the SV40 AL/N cells may have just sufficient residual GM1, possibly derived from serum (Yogeeswaran et al., 1970; Yu and Ledeen, 1972), to give a response. Certain clones of chemically transformed mouse fibroblasts (NCTC 2071) grown in defined medium failed to respond to cholera toxin and contained no detectable GM1 (Moss et al., 1976). Responsiveness to toxin was demonstrated with binding of as few as 17 000 molecules of [^{3}H]GM1 per cell. Such low levels of GM1 would not have been detected by the methods used to quantitate the gangliosides of SV40 AL/N cells in the previous study (Hollenberg et al., 1974).

Good correlation between the cellular content of GM1 and responsiveness to cholera toxin has also been found in intestinal cells from several species (Holmgren et al., 1975). In our laboratory we have found reasonable agreement between GM1 content and cholera toxin binding when comparing 3T3 cells, NIL Syrian hamster fibroblasts and Chinese hamster ovary cells (Table III and Fig. 4).

TABLE III
BINDING OF ^{125}I-LABELLED CHOLERA TOXIN TO CELL LINES IN CULTURE[a]

Experimental series[b]	Cell line	Method of removing cells from monolayer	Relative binding efficiency (%)[c]
(a)	3T3	Trypsin (Difco)	100
		EDTA	96
		Scraped	110
(b)	3T3	Trypsin	100
	3T3 SV40	Trypsin	93
	NIL	Trypsin	9
	NIL HSV	Trypsin	9
	CHO	Trypsin	8
(c)	3T3	Trypsin, pretreatment none	100
		Trypsin, pretreatment neuraminidase	202
		Trypsin, pretreatment GM1	156
		Trypsin, pretreatment unlabelled toxin	2

[a]Binding was determined on cells suspended at $1.5 \cdot 10^5$ cells/ml in phosphate buffered saline pH 7.4 containing 0.1% w/v bovine serum albumin. After incubation for 15 min with 250 ng/ml of ^{125}I-toxin at 20°C, the cells were collected on EAWP Millipore filters which were subsequently washed 4× with 1 ml aliquots of the buffer prior to counting.

[b]In experiment (c) cells were pre-incubated in monolayer with neuraminidase (50 units/ml in PBS; Behringwerke), 10 μg/ml GM_1 or PBS for 60 min at 37°C prior to trypsinization. Non-specific binding of ^{125}I-toxin was determined by pre-incubating 3T3 cells with 1 μg/ml of unlabelled toxin for 15 min.

[c]All results are expressed as a percentage of the counts bound to 3T3 cells removed by trypsin.
(Critchley, D.R. and Dilks, S., unpublished observations.)

In addition the vast majority of cell surface glycoproteins with terminal or subterminal galactose residues are removed by treating these cells with 250 μg/ml trypsin for 10 min at 37°C (Critchley, 1974) yet this caused no reduction in the amount of ^{125}I-cholera toxin bound (Table III).

Although there is still some slight uncertainty about the identity of the natural receptor for cholera toxin, there is no doubt that it will bind to GM1 covalently linked to Sepharose (Cuatrecasas et al., 1973) and to GM1 incorporated either into micelles, (Van Heyningen, 1974a) or cell membranes (Cuatrecasas, 1973b; Révész and Greaves, 1975; Moss et al., 1976; Sedlacek et al., 1976).

7.1.2. Cholera toxin: interaction with adenyl cyclase in intact cells

In contrast to the rapid binding of cholera toxin to intact cells, the onset of the biological effect exhibits a variable lag phase of the order of 20–60 min (Cuatrecasas, 1973b, and references in Table II). During the lag phase adenyl cyclase activity remains unaffected for some time after which activity begins to increase (Bennett and Cuatrecasas, 1975b) elevating levels of cAMP which in turn produce a particular biological response depending on the cell type (see Table II). For example, lipolysis in fat cells is evident 60–90 min after the initial exposure to toxin. Interestingly lipolysis induced by the hormones epinephrine or glucagon is not characterised by a lag phase (Cuatrecasas, 1973b,c). The reason behind the lag in toxin-induced activation of adenyl cyclase is not entirely clear. In most systems the process is temperature-dependent but does not require DNA, RNA, or protein synthesis, (Cuatrecasas, 1973c; Kantor, 1975; Kimberg et al., 1973) although there is at least one report that protein synthesis is required (Palfreyman and Schulster, 1975). A further possibility was that the toxin first stimulated synthesis of prostaglandins which in turn activated adenyl cyclase (Bennett, 1971) similar to proposals made for the action of certain hormones (Kuehl et al., 1970; Sato et al., 1972). However, inhibitors of prostaglandin biosynthesis such as indomethacin and acetylsalicylic acid failed to inhibit activation of adenyl cyclase by cholera toxin (Cuatrecasas, 1973c; Kimberg et al., 1974).

A more likely explanation for the lag phase may lie in the structure of the toxin molecule, approximate molecular weight 84 000 daltons (Finkelstein et al., 1974; Sattler et al., 1975; Gill, 1976). It consists of two types of subunit, the larger one, (molecular weight 54 000 daltons, approximately; the B subunit), itself composed of a number of smaller molecular weight subunits (approximately 10 600 daltons; Gill, 1976), is responsible for interaction of the toxin with the membrane receptor. The smaller subunit (molecular weight approximately 29 000 daltons; the A subunit), a single polypeptide with an intrachain disulphide bridge, effects the biological actions of the protein through activation of adenyl cyclase. It has been suggested that five B subunits form a closed ring with the A subunit located on its axis (Gill, 1976). Such "doughnut shaped" structures have been visualised in the electron microscope by negative staining (Ohtomo et al., 1976). The lag phase is most likely explained by the time the A or toxic subunit takes to reach adenyl cyclase molecules on the inner face of the lipid bilayer. In fact A subunit-cyclase associations have been demonstrated (Bennett et al., 1975; Sayhoun and Cuatrecasas, 1975). Cyclase solubilised in Lubrol PX co-chromatographed on Sepharose 6B with some of the bound ^{125}I-toxin if the extract was made from cells in which the lag phase was completed. In contrast, much less radioactivity was present in the cyclase peak when the extract was made only 10 min after binding of toxin. Antisera to the whole toxin or the toxic A subunit was also found to selectively precipitate adenyl cyclase extracted from cells after completion of the lag phase (Bennett et al., 1975; Sayhoun and Cuatrecasas, 1975). Agarose affinity columns containing the A subunit also specifically absorbed solubilised adenyl cyclase activity (Bennett et al., 1975).

What precise steps are involved in bringing the toxic or A-subunit into a position where it can interact with the cyclase is the subject of some controversy. It has been shown that during the lag phase, interaction of cholera toxin with lymphocytes leads to a marked redistribution of the toxin and its receptor, i.e. "patching" and "capping" (Holmgren et al., 1974; Craig and Cuatrecasas, 1975; Révész and Greaves, 1975). Redistribution by itself was not sufficient to activate the cyclase as a derivative of toxin lacking the A subunit (toxoid, choleragenoid) also induced "patching" and "capping", but failed to activate the enzyme. Lymphocytes coated with cholera toxin will still bind to Sepharose particles coated with GM1 suggesting that the toxin is at least bivalent (Craig and Cuatrecasas, 1975). Interestingly microfilament, microtubule and metabolic inhibitors which block the capping of immunoglobulin molecules in lymphocytes (De Petris, 1974; Unanue and Karnovsky, 1974) were also found to inhibit redistribution of bound toxin in lymphocytes (Craig and Cuatrecasas, 1975; Révész and Greaves, 1975). As the dimensions of glycolipids suggest they can not span the membrane, it is possible that they associate with specific membrane proteins which in turn are linked to a submembranous cytoskeleton of microfilaments and microtubules (Nicolson, 1976a,b). In fact there is some evidence using cross-linking reagents that erythrocyte membrane glycolipids are closely associated with the specific membrane proteins (Ji, 1974), although which groups on the glycolipid would react with the reagent used was unclear. The recent finding that some erythrocyte glycosphingolipids have a free amino group may be relevant in this context (Skelly et al., 1976). GM1 added to cells normally lacking the molecule could also be made to redistribute in response to toxin (Révész and Greaves, 1975), possibly arguing against specific association with a membrane protein. However, it is possible that GM1 was sufficiently like the endogenous glycolipid to interact with such putative proteins. By incorporating a fluorescent derivative of GM1 into lymphocytes Sedlacek et al. (1976) showed that redistribution of the freshly incorporated GM1 by toxin was also sensitive to microtubule inhibitors.

A possible relationship between capping of a GM1-toxin complex and activation of adenyl cyclase was examined by Craig and Cuatrecasas (1975). They tested the effect of inhibitors of toxin induced capping of GM1 in lymphocytes on the subsequent activation of cyclase. Colchicine, vinblastine and vincristine, which inhibited capping by approximately 60%, had little effect on cyclase activation by toxin, and it was argued that sufficient redistribution may have taken place to maximally stimulate the enzyme. Sodium azide which produced a 96% inhibition of toxin induced capping in lymphocytes resulted in a 64% inhibition of cyclase activation. However, Bennett and Cuatrecasas (1975b) found that azide had no effect on toxin activation of cyclase in red cells. Craig and Cuatrecasas (1975) also found that undiluted anti-toxin sera which produced a 90% inhibition of capping in lymphocytes drastically reduced cyclase activation. At present these results are very difficult to interpret. Sayhoun and Cuatrecasas (1975) suggest that as a result of progressive multivalent binding of cholera toxin to membrane-bound GM1, the toxin molecule is destabilized, and the A subunit of the toxin (which is reported to be hydrophobic – Bennett and Cuatrecasas 1975a; unpublished observations) is released into the lipid phase of the membrane. The A subunit would then diffuse in the plane of the membrane where collisions with adenyl cyclase molecules would lead to the formation of stable complexes and activation of adenyl cyclase. The fact that activation of adenyl cyclase occurs very slowly below certain critical temperatures (e.g. rat fat cells exhibit a discontinuity at 26–30°C, toad erythrocytes at 15–17°C) suggests that "fluidity" of membrane phospholipids may be required throughout the process of toxin induced enzyme activation (Bennett and Cuatrecasas, 1975b).

7.1.3. Cholera toxin: interaction with adenyl cyclase in isolated membranes

The adenyl cyclase of lysed cells or isolated membranes can be activated by the A subunit directly, without involvement of the GM1 receptor and without a lag phase (Gill and King, 1975; King and Van Heyningen, 1975; Sayhoun and Cuatrecasas, 1975; Moss et al., 1976). There are some discrepancies between the results of Sayhoun and Cuatrecasas (1975) and the others which remain to be resolved, the most important of which is whether intracellular factors are involved in the process of cyclase activation. Activation of the enzyme by toxin in intact cells appears to be irreversible since the elevated activity is still apparent in membranes isolated from the exposed cells. However, it has proved difficult to use toxin to activate the enzyme in lysed cells or isolated membranes. Gill (1975) has reported that the lysates of rabbit erythrocytes which hydrolyse their endogenous source of NAD, fail to respond to toxin. Addition of exogenous NAD considerably increased the response of the lysates to toxin, although other nucleotides including ATP were ineffective. Further, the response of erythrocyte ghosts to toxin was greater if NAD plus a non-dialysable cytoplasmic factor was present. In apparent contrast, Sayhoun and Cuatrecases (1975) found that homogenates or membranes isolated from fat cells responded to toxin only if Mg-ATP plus an ATP regenerating system was present. NAD or NADH were without effect and no consideration was given to the need for a macromolecular cytoplasmic factor.

Gill (1975) and Gill and King (1975) have proposed that the A subunit of the toxin may itself be an enzyme which catalyses a reaction between NAD and a target molecule on the membrane, leading to a permanent change in the conformation of adenyl cyclase. They have suggested that the target molecule might be a catalytic or regulatory component of the cyclase or some other membrane component which once modified interacts with the cyclase leading to its irreversible activation. In this context, Holmgren and Lonnroth (1975) reported that exposure of lymphocytes to certain carboxyl group modifying reagents specifically inhibited subsequent toxin-induced activation of cyclase although binding was unchanged, and the enzyme was still stimulated by epinephrine and prostaglandins. They propose to have interfered with a cell component required to translate the initial GM1-toxin complex into a cyclase-activating signal. The idea of the toxin itself having enzymatic activity comes from direct analogy with the mechanism of action of diphtheria toxin on protein synthesis (Gill et al., 1973). Diphtheria toxin is also composed of two subunits one involved in binding to the membrane, the other a toxic subunit which catalyses the transfer of ADP-ribose from NAD to the polypeptidyl elongation factor 2 of eukaryotic cells, thus inhibiting protein synthesis. Bennett and Cuatrecasas (1975b) have argued against the idea that cholera toxin contains a catalytic activity after analysis of the kinetics of activation of adenyl cyclase.

Gill (1976) has recently proposed a model to account for the events occurring up to the activation of adenyl cyclase which incorporates much of the available data (Fig. 5). Initial binding of toxin would involve interaction of one of the five B polypeptides with GM1. By diffusion of the GM1-toxin complex in the plane of

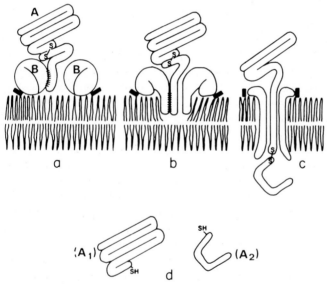

Fig. 5. One possible mode of entry of peptide A_1. (a) Hatching represents noncovalent bonds between A_2 and B. It is suggested that the initial binding of toxin to surface receptor may be followed (b) by a change in the conformation of the B subunits which enter the membrane, bind to lipids, and create a hydrophilic channel. (c) The modified B no longer binds to A_2. Component A, now free to diffuse, may sometimes diffuse inward, unfolding on the outside and spontaneously refolding inside the cell. B remains as a membrane protein. (d) The disulfide bond between A_1 and A_2 is reduced by intracellular glutathione. Reproduced with permission from D.M. Gill (1976).

the membrane the remaining four B polypeptides might also interact with 4 additional GM1 molecules such that the "doughnut shaped" toxin molecule would now be held in a flat position against the membrane. Such interaction may weaken the non-covalent binding between A and B subunits releasing the A or toxic subunit in free form. Indeed Bitensky et al. (1975) have shown that membrane preparations from mouse ascites cells will release a 2.7S fragment from cholera toxin (5.5S) as long as the toxin is first exposed to NAD, NADH, or dithiothreitol, a procedure which in itself did not lead to generation of the 2.7S fragment. Sayhoun and Cuatrecasas (1975) have also presented evidence that the two subunits of the toxin dissociate after binding to fat cells. If the A subunit is hydrophobic (Bennett and Cuatrecasas, 1975a) it may then enter the lipid phase of the membrane and subsequently interact with adenyl cyclase. Alternatively, Gill (1976) has suggested that hydrophobic sites involved in interaction of the B peptides become unmasked allowing the B subunit to penetrate the lipid bilayer. The central core of the toxin then forms a hydrophilic channel through which component A diffuses into the cytosol. The full cyclase-stimulating activity of the A subunit is apparently only expressed when the intrachain disulphide bond is reduced releasing peptides A_1 (molecular weight 23 000 daltons) and A_2 (molecular weight approximately 5500 daltons). This may occur when the A subunit encounters intracellular glutathione, the A_1 peptide generated subsequently activating adenyl cyclase (Fig. 5).

7.2. Protein–glycolipid interactions in other systems

It is now becoming evident that the interaction of cholera toxin with glycolipids is not just an interesting but isolated example of protein–glycolipid receptor interaction. Very preliminary evidence that pre-incubation of macrophages with glycolipids enhances their response to migration inhibition factor has been presented (Higgins et al., 1976). A better documented example stems from the observation that gangliosides inhibit the binding of ^{125}I-labelled thyrotropin to bovine thyroid plasma membranes (Mullin et al., 1976). The inhibitory potencies were GD1b > GT > GM1 > GM2 = GD1a. Fluorescence studies show that the hormone–ganglioside interaction induces a conformational change in the thyrotropin molecule, the change induced by the best inhibitor of binding (GD1b) being distinct from that induced by the minimal inhibitor (GD1a). The interaction appears to be hormonally specific and insulin, glucagon, prolactin, growth hormone, follicle-stimulating hormone and corticotropin did not prevent GD1b inhibition of thyrotropin binding. Initially these results appear surprising in that a putative receptor fragment for thyrotropin had previously been isolated by trypsinization of thyroid plasma membranes (Tate et al., 1975). The molecule, with a molecular weight of 25–30 000 daltons was characterised as a glycoprotein, with sialic acid residues playing a key role in its receptor function. The role played by sialic acid is also clearly reflected when considering the hormone–ganglioside interaction; GD1b with an additional internally linked sialic acid residue was much more effective than GD1a in inhibiting $[I^{125}]$thyrotropin binding to thyroid plasma membranes (Mullin et al., 1976). Which of the two molecular species of receptor, the glycoprotein or glycolipid, is the physiological receptor remains unclear, although the thyroid plasma membrane was found to have unusually high levels of the more complex gangliosides for an extraneural tissue.

One of the results of the interaction of thyrotropin with its receptor is activation of the adenyl cyclase system (Wolf et al., 1974; Marshall et al., 1976), although precisely how this is achieved is unknown. Similar problems arise when considering the mechanism by which peptide hormones such as glucagon, ACTH, growth hormone etc., activate the adenyl cyclase system, (Birnbaumer, 1973; Cuatrecasas, 1974; Hammes and Rodbell, 1976). Although the molecular basis by which cholera toxin results in adenyl cyclase activation is still far from clear, the structure of the molecule provides the basis for the simple model previously discussed, i.e. a binding subunit (B) responsible for interaction of the molecule with the receptor and a cyclase activating subunit (A) which is reponsible for the biological effect of the toxin. On initial inspection it seemed unlikely that molecules with similar structure to cholera toxin would be involved in cyclase activation in mammalian systems. This impression was confirmed by examination of the structure of glucagon, (for a review see Tager and Steiner, 1974). At present the molecular features of this molecule give no such obvious clue as to how interaction with its receptor leads to cyclase activation. However, the glycoprotein hormones thyrotropin, luteinizing hormone, human chorionic gonadotropin, and follicle-stimulating hormone have two subunits, one of which, the α, is common to each hormone, and the other, the β subunit, contains the information which specifies the particular hormonal activity to be expressed. Interestingly the β-subunits from different hormones have marked sequence homologies and may have evolved from a common gene product (reviews, Tager and Steiner, 1974; Pierce et al., 1976). Additional information can be found in section 7 of the *Handbook of Physiology*, published by the American Physiology Society). These observations were made even more exciting by the finding that the B subunit of cholera toxin which interacts with the glycolipid receptor has a peptide sequence in common with the β subunit of thyrotropin, luteinizing hormone, human chorionic gonadotrophin and follicle-stimulating hormone (Ledley et al., 1976). A possible sequence homology was also found between the A_1 peptide of the toxin and the α subunit of the above hormones. The results support the idea that the β subunit of thyrotropin is responsible for binding of the hormone to its glycolipid or glycoprotein acceptor, and as at least some of the effects of the hormones on tissues are mediated through cAMP it is not unreasonable to suggest that the α subunit might bind and subsequently activate adenyl cyclase. It remains to be seen if glycolipids also act as receptors for the other glycoprotein hormones.

Although the finding that physiological molecules have similar structures to cholera toxin makes it easier to see how the interaction of hormone with receptor leads to activation of adenyl cyclase, there is still the problem of how the A or α subunit comes into contact with the enzyme. The idea that there is direct structural contiguity between the receptor and adenyl cyclase is not borne out by the data on hormonal stimulation of fat cells. Although at least six hormones can activate adenyl cyclase in fat cells, the effect is not additive as would be predicted if each hormone receptor was associated with a specific adenyl cyclase (Cuatrecasas, 1974). An alternative view, proposed by Cuatrecasas (1974), is that interaction of a hormone (H) with its receptor (R) forms a complex (HR) with a specific affinity for adenyl cyclase (AC). Collisions between HR and AC would be dependent on diffusion in the lipid phase of the membrane and would thus be strongly influenced by factors that modulate the translational mobility of membrane components (see Chapter 1 of this volume). The theory predicts that if diffusion of HR in the plane of the membrane is slow, discrepancies will exist between the kinetics of binding of H to R and the onset of cyclase activation. Such a lag phase has indeed been demonstrated for cholera toxin activation of adenyl cyclase although apparently, activation of adenyl cyclase by thyrotropin does not show a similar phenomenon (Van Sande and Dumont (1973) Biochim. Biophys. Acta 313, 320–328).

7.3. Glycolipids as receptors for interferon

There is now some evidence to suggest that a heteroglycan structure, particularly that of the ganglioside GM2, functions as the specific cell surface receptor for interferon. This glycoprotein (Dorner et al., 1973) is synthesized in virus-infected cells, but only when it binds to the surface of other cells does it induce intracellular changes which inhibit viral replication (Metz, 1975) and possibly the growth of normal (Tan, 1976) and tumorigenic cells (Gressor et al., 1973).

The effect of interferon occurs without its entry into the cell. Ankel et al. (1973) have reported that interferon which is covalently bound to Sepharose beads is also able to induce the anti-viral state in mouse cells. Besançon and Ankel (1974a) found that a mixed ganglioside preparation blocked the effect of Sepharose-interferon and that ganglioside was bound to the interferon. Several lines of evidence suggest that GM2 is the specific receptor for interferon. Firstly, phytohemagglutinin (PHA) and Con A pre-treated cells are nearly unaffected by Sepharose-interferon (Besançon and Ankel, 1974b) and PHA inhibits ganglioside-interferon binding directly (Besançon and Ankel, 1974a). Presumably the lectins compete with interferon for a cell surface binding site. Secondly, when Sepharose-interferon is pre-incubated with GM2 it fails to inhibit virus reproduction. The effect of GM2 is more than 5-fold greater than that seen when GM3 is used (Besançon et al., 1976). Thirdly, the binding to interferon depends specifically on the terminal saccharide, and part of the subterminal structure of GM2. Sialylactose releases GM2 from interferon; sialic acid plus lactose is much less effective, and either sugar alone or a number of heterologous glycoproteins, including fetuin, have little or no effect (Besançon et al., 1976).

Besançon et al. (1976) proposed that the loss of sensitivity to interferon by some virus transformed cells (Brailovsky et al., 1969) is due to the effects of transformation in decreasing cell surface GM2 levels. Revel et al. (1976) have shown that antibodies against a human chromosome 21 product inhibit the binding of interferon to human cells and they have suggested that this product is the cell surface receptor for interferon. However, Tan (1976) has considered the possibility that more than one chromosome 21 product is involved in the binding and response to interferon. It would be valuable to learn if hybrid cells which lack human chromosome 21, or cells which do not respond to interferon, contain GM2 or the glycosyltransferases necessary for its synthesis. However, the effect of interferon is generally species-specific (Lockart, 1973) and it is therefore possible that a different glycolipid acts as the receptor in species other than mouse. Interestingly interaction of interferon with its receptor subsequently leads to activation of adenyl cyclase (Weber and Stewart, 1975). It is tempting to speculate that the glycoprotein interferon has a structure similar to that of the glycoprotein hormones and cholera toxin which also bind to glycolipid receptors and stimulate adenyl cyclase. Finally, how a bacterial product, cholera toxin, came to possess a similar structure and sequence homology with mammalian hormones is a fascinating problem for protein evolutionists. Other examples of a partial similarity of structure between bacterial and mammalian proteins have been described (see, Hartley, 1974). Although the structure of hormones such as thyrotropin has a similarity to that of cholera toxin, the latter has not been found to contain carbohydrate (Lospalluto and Finkelstein, 1972). Interestingly, the terminal sialic acid and galactose residues of interferon can be removed or chemically modified, without significant loss of its activity (Dorner et al., 1973; Bose et al., 1976).

7.4. Possible role of glycolipids as receptors for molecules involved in growth regulation

The fact that interaction of cholera toxin with GM1 leads to elevation of intracellular levels of cAMP, a molecule strongly implicated in growth regulation and differentiation (Chlapowski et al., 1975; Pastan et al., 1975; Willingham, 1976) suggests that other glycolipids might act as receptors for the physiological equivalent of the toxin, i.e. molecules normally present in the cellular milieu which can inhibit growth or stimulate the expression of differentiated functions. The finding that many transformed cells, which apparently fail to respond to such putative growth-regulatory signals, have reduced levels of complex glycolipids leads one to suggest that deletion of these molecules might be a key event in the cells loss of growth regulation.

Attempts to re-insert deleted glycolipids into transformed cells are therefore of considerable importance. The results from such experiments have given encouraging but equivocal results (reviewed in section 3.5). Nevertheless, it may be highly significant that elevation of the levels of GM1 in SV40 AL/N cells restored their ability to respond to normal levels of cholera toxin by cessation of growth (Hollenberg et al., 1974). Firstly, this indicates that GM1 re-inserted into cell membranes is orientated in much the same way as endogenous GM1 such that it is able to bind cholera toxin in a manner which leads to adenyl cyclase activation. This suggests that studying the effect of glycolipids on growth and morphology of transformed cells is a meaningful approach. Secondly, it implies that intracellular mechanisms involved in control of cell division are in part functional in SV40 AL/N cells, and therefore re-emphasizes the possible importance of loss of the more complex cell surface glycolipids which might act as receptors for growth regulatory signals. However, it is unclear whether toxin-treated SV40 AL/N cells were arrested in G_0/G_1 phase of the cell cycle (Hollenberg et al., 1974). Transformed cells blocked with dibutyryl cAMP were found to accumulate in G_2-phase (Smets, 1972), and it has been proposed that transformed cells lack a restriction point in G_1-phase where a commitment to a new round of cell division is normally made (Pardee, 1974; Pardee and Rozengurt, 1975).

Perhaps the significance of the cell density-dependent increase in certain glycolipids is that it provides more receptors capable of interacting with growth regulatory molecules. Such molecules may be structures on the surface of neighbouring cells which specifically recognize the density-dependent glycolipids (see section 4.3.2 and 7.5) or they may be serum components or factors released into the medium. A number of inhibitory factors have been identified in vivo (Houck and Hennings, 1973) and in vitro. Thus growth of a highly malignant melanocyte cell line was reported to be inhibited by a macromolecule released from density-inhibited melanocytes (Lipkin and Knecht, 1974). Lymphocytes and macrophages were also found to produce molecules that inhibit fibroblast growth (Calderon et al., 1974; Uhlrich 1974) and serum from some animals contains inhibitory as well as stimulatory factors (Otsuka, 1972; Leffert, 1974; Zimmerman and Kern, 1976). The possibility that thyrotropin interacts with a glycolipid in thyroid plasma membrane (Mullin et al., 1976) probably resulting in adenyl cyclase activation (Wolf et al., 1974; Marshall et al., 1976) also suggests that the whole family of glycoprotein hormones could influence growth behaviour and/or the expression of differentiated functions in cells containing the correct receptor in vivo. Applying the ideas of Cuatrecasas (1974) on hormone-receptor interaction (section 7.2) to regulation of cell division in vitro, binding of a growth-inhibitory molecule (I) to a glycolipid receptor (R) might produce a complex (IR) with a high affinity for AC. The second step formation of an IR-AC complex would be dependent on the concentration of the individual components and their diffusion properties within the plane of the membrane. Clearly, a large excess of the glycolipid receptor (R) *could* increase the formation of IR complexes and increase the collision frequency between IR and AC. Hence the rationale behind cell density-dependent glycolipids. Interaction of IR with AC would stimulate adenyl cyclase, increase the levels of cAMP and, through a sequence of undefined events, lead to growth inhibition. Other complexes between growth-stimulatory molecules and their receptors may also have a high affinity for AC possibly resulting in dissociation of IR-AC complexes, decreased AC activity and resumption of cell growth. Growth regulation would therefore be a fine balance between inhibitory and stimulatory signals.

A receptor role for glycolipids would explain several points that were considered initially to limit their likely importance in cell growth control. Thus the fact that levels of glycolipids do not strictly correlate with saturation density or tumorigenicity (sections 3.3.1 and 6) might be explained by altered characteristics elsewhere in the chain of reactions which lead to growth inhibition. To illustrate this point, we have recently found that although certain clones of Swiss 3T3 cells have the receptor for, and do bind cholera toxin, the interaction fails to inhibit serum-stimulated uridine transport (Critchley, unpublished observations), a process known to be inhibited by elevated levels of cAMP (Rozengurt and Jimenez de Asua, 1973; Jiminez de Asua et al., 1974). Of a number of possible explanations for this result the following are relevant to this discussion: (1) coupling between the toxin receptor complex and AC may be defective; (2) the enzyme may be activated generating cAMP which then fails to accumulate because of an active phosphodiesterase; and (3) cAMP may accumulate but fail to activate, for example, a protein kinase involved in modulating the activity of the transport system. It is therefore clear that in a complex process such as growth regulation a defect at any one of a number of key points may lead to loss of control although much of the system remains operative. Considered in this light it is not surprising that there is no absolute correlation between levels of complex glycolipids and saturation density (see section 3.3.1). The lack of effect of dibutyryl cAMP in restoring the normal glycolipid pattern to transformed cells (Sakiyama and Robbins, 1973a; Sheinin et al., 1974; also see section 4.2) would also be predicted from the foregoing discussion as the need for the glycolipid receptor is bypassed. Similarly the slow change in glycolipid patterns that occurs when dense quiescent cells are stimulated to grow (Critchley and Macpherson, 1973; section 4.1.2) can be rationalised by postulating a dissociation of an IR-AC complex. The effect of the glycolipid is then negated without the need to expect a rapid change in glycolipid pattern. Such changes in interaction between membrane components could lead to a rapid reduction in cyclase activity and decreased levels of intracellular cAMP which, in turn, might direct some of the early events thought to prepare the cell for a round of DNA replication (Pardee and Rozengurt, 1975). The later changes in membrane glycolipid composition, occurring once the cell is committed to divide would seem to establish the surface membrane in a configuration characteristic of dividing cells. Synthesis and increased exposure of glycolipids in the G_1 phase of subsequent cell cycles (Gahmberg and Hakomori, 1975a; also see section 5.1) would allow the cell to interact with available regulatory molecules prior to a further commitment to yet another round of cell division.

In vivo, cells with altered membrane glycolipids may arise spontaneously, or as a result of virus transformation or chemical carcinogenesis. In view of the results discussed previously, it is not unreasonable to speculate that loss of certain key cell surface receptors, conceivably glycolipid in nature, would result in a cell population outside the sphere of influence of the mechanisms which must be important in the control and integration of cell and tissue functions. Quite clearly such a loss of a membrane receptor could be vital in subsequent tumour formation whether as a primary or secondary consequence of the initial "transformation" event. Identification of molecules that interact with cell surface glycolipids, whether cell membrane components, hormones, or other as yet uncharacterised products secreted by cells, may be made easier now that methods for preparation of glycolipid-sepharose affinity columns are available (Cuatrecasas et al., 1973; Laine et al., 1974).

7.5. Glycolipids as cell surface receptors involved in intercellular recognition

The function of glycolipids may not be limited to the binding of soluble regulatory molecules at the cell surface. They could also share with glycoproteins an intercellular role in the regulation of cell social behaviour which has been postulated to depend on the cell-cell adhesion of complementary plasma membrane components (Tyler, 1946; Weiss, 1947; Balsamo and Lilien, 1974; Burger, 1974). Cell social interactions are important, for example, in many aspects of cellular morphogenetic movements, pattern formation and differentiation in development (see Trinkaus, 1969, 1976) which are at least partially manifest through a number of phenomena observable in cell culture. These include contact inhibition of locomotion (Abercrombie and Heaysman, 1954), growth or topoinhibition (Dulbecco, 1970), formation of low resistance junctions (Furshpan and Potter, 1968), cell fusion (Okasaki and Holtzer, 1965; Poste and Allison, 1973) and expression of differentiated functions (Morris and Moscona, 1970). Each of these interactions probably requires or is initiated following cohesive contact between cells (Abercrombie, 1970; Gregg, 1971).

Glycolipids and glycoproteins have at least two features in common from which a potential role in cell interaction might be inferred: (1) the carbohydrate residues of both classes of molecule (which are basically very similar and can be identical with respect to the terminal saccharide sequence (Roseman, 1970; Hughes, 1975) are exposed to the extracellular environment at the cell surface; and (2) the carbohydrate residues have an enormous potential for structural heterogeneity. Hughes (1975) has calculated that a trihexoside of three different sugars may exist in several hundred configurations. Such diversity may generate the specificity in cell surface molecules which some authors think could govern selective aspects of cell interactions (Roseman, 1970, 1974; Roth et al., 1971b; Kemp et al., 1973; McGuire and Burdick, 1975). Some of the clearest examples of regulated cell surface carbohydrate changes controlling cellular interactions are found in bacteriophage adhesion (Robbins and Uchida, 1962), mating reactions in bacteria (Sneath and Lederberg, 1961), yeast (Crandall et al., 1974), *Chlamydomonas* (Weise and Hayward, 1972) and possibly cell interactions in tissue pattern formation (Barbera et al., 1973). Recent work on the molecular basis of cell aggregation in the cellular slime moulds may hold especially close analogies for developing systems generally and will therefore be considered in some detail.

The myxamoebae of *Dictyostelium discoideum* live as independent cells until they experience starvation. Each cell then ceases growth and within a few hours begins a cAMP-modulated chemotactic migration which results in streams of cohering cells coalescing into the multicellular form of the organism, the slug or grex (Bonner, 1971). The cohesion of growing and of non-growing differentiating cells can be distinguished on morphological, immunochemical and molecular levels. Cells of either type will aggregate when agitated in single cell suspensions, but while EDTA inhibits the adhesion of growing cells it does not prevent aggregation of non-growing ones. This insensitivity to EDTA arises 2 h (Garrod, 1972) or 6 h (Rosen et al., 1973) after starvation, and is also expressed morphologically. Thus, in the presence of EDTA, non-growing cells adhere only at their terminal ends while side-to-side contact, a characteristic which growing and non-growing cells share, is suppressed (Gerisch, 1968). New antigens appear on the cell surface as growth stops (Takeuchi, 1963; Gregg, 1971; Beug et al., 1973a, Wilhelms et al., 1974). Univalent antibody fragments (Fab) directed against growth phase cells inhibited only the side-to-side contacts on growing or non-growing cells, whereas Fab against non-growing cells (which was first adsorbed with fragments of growing cells) specifically inhibited only the terminal cell contacts (Beug et al., 1973b). Wilhelms et al., (1974) studied aggregation mutants and suggested that at least two cell surface antigens are required for the specific adhesion of non-growing cells. In one such pleiotropic mutant the authors found that terminal fucose residues were lacking from the antigens which included a glycosphingolipid, but this molecule is probably not involved in cell aggregation. Weeks (1975) has demonstrated that while growing cells bind the lectin Con A to one type of site, the non-growing cells display two distinct types of binding site.

Candidates for the adhesive molecules of non-growing cells have recently been isolated and characterized (Huesgen and Gerisch, 1975, Reitherman et al., 1975). These include two specific carbohydrate-binding proteins, lectins, which have been visualized on the surface of *D. discoideum* (Chang et al., 1975) and on five other species of slime mould (Rosen et al., 1975). These lectins, as well as their specific, high affinity, carbohydrate binding sites appear on the cell surface within 3–6 h after starvation (Reitherman et al., 1975), i.e. at the same time as the cells become cohesive. This suggests that the regulated appearance of these molecules and their receptors might indeed be responsible for the newly acquired adhesive behaviour of differentiating slime mould cells. The lectins themselves probably lack carbohydrate (Simpson et al., 1975) and have an overall molecular weight of 250 000 daltons (*Polysphondylium pallidum*) or 100 000 daltons (*D. discoideum*) made up of subunits with molecular weights of approximately 25 000 daltons (Simpson et al., 1974, 1975). Purified lectin from *D. discoideum* is able to agglutinate formalinized sheep erythrocytes, and glutaraldehyde-fixed differentiated, but not vegetative, homologous and heterologous myxamoebae (Rosen et al., 1973; Reitherman et al., 1975). The agglutination is inhibited by simple sugars, the sugar specificity depending on the lectin species, but the most potent inhibitor of lectin-induced *D. discoideum* aggregation is an extract containing a glycopeptide derived from pronase-treated cell homogenates (Rosen et al., 1973). Presumably each lectin must contain at least two binding sites; both could be specific for carbohydrate or alternatively there could be two types of site one for carbohydrate and one which is hydrophobic and is anchored in the lipid phase of the plasma membrane. Studies of lectin binding to myxamoebae show that while the association constant markedly increases upon differentiation and reflects relative species specificity the number of binding sites per cell changes very little (Reitherman et al., 1975). Freeze fracture observations of myxamoebae, which show that membrane (glycoprotein) particles are dispersed in growing cells, but become aggregated in the differentiating cells (Aldrich and Gregg, 1973), may account for this apparent discrepancy.

Interestingly, agglutinins and/or lectins similar to those of the slime moulds have now been detected in a variety of vertebrate tissues and cell lines. Glycoproteins which enhance homologous cell aggregation have been extracted from chick (Merrell et al., 1975) and rat cells (Lloyd and Cook, 1975), and lectins specific for terminal β-D-galactose residues have been found in chick muscle, a myoblast cell line (Nowak et al., 1976) and in the electric organ of *Electrophorus electricus* (Teichberg et al., 1975). Their function is generally unknown, but in muscle cells they arise just before cell fusion occurs (Nowak et al., 1975). In liver tissue one carbohydrate binding protein present in the plasma membrane may have an important role in the clearance of desialized glycoproteins from the blood (Hudgin et al., 1974). Yamada et al (1975, 1976) have purified a fraction from chick embryo fibroblasts which agglutinates formalised sheep erythrocytes and homologous or heterologous cell lines, increases cell substrate adhesion and alignment of cells in culture. Agglutination was inhibited by EDTA, and somewhat by amino sugars (as well as other amines e.g. tris, arginine, lysine etc.), but not by other monosaccharides. The agglutinin co-purified and may be identical with the LETS glycoprotein. Neither the agglutinin nor LETS glycoprotein were found in virus transformed cells and Yamada et al. (1976) have suggested the molecule(s) play(s) an important role in normal cell adhesion, but not growth control.

In comparison with the slime mould system, it is interesting that levels of LETS glycoprotein expressed at the cell surface increase in situations where formation of stable cell contacts might be important e.g. cell crowding (Hynes and Bye, 1974). Which molecule might interact with LETS glycoprotein if it possesses an adhesive function is unknown. However, changes in levels of density-dependent glycolipids (section 4) and LETS glycoprotein often occur in parallel, and are coincident, with variations in cell density, cell cycle and viral transformation. These results again show a superficial similarity with the slime mould system where a lectin and its specific receptor appear as the cells become cohesive. It is therefore tempting to speculate that LETS glycoprotein might interact with the cell density-dependent glycolipids. Alternatively, it is possible that density-dependent glycolipids and glycoproteins are part of a similar, but separate, system of cell cohesion and do not themselves interact. The idea of glycolipids interacting with cell surface proteins, specifically glycosyl transferases has already been discussed elsewhere in this article (4.3.2). It was exciting therefore that a fraction purified from the surface of normal NIL hamster fibroblasts and enriched in LETS glycoprotein contained high levels of galactosyl transferase activity although this was assayed using a glycoprotein substrate (J.M. Graham, personal communication).

There have been few attempts so far to correlate the levels of surface glycoproteins and glycolipids with cell cohesion using in vitro assay systems. Treatment of BHK21 cells with relatively low concentrations (100 μg/ml for 2 min) of pure trypsin produces suspension of single cells which rapidly aggregate into small clumps when freed of trypsin and agitated at 37°C in simple (glucose and salts) media. High levels of trypsin, (1 mg/ml) inhibit aggregation completely (Edwards and Campbell, 1971; Edwards et al., 1975). In analogy with the slime moulds, aggregation is greatest in cells harvested from dense, non-growing cultures and is virtually undetectable in normal sparse-growing (Edwards and Campbell, 1971), virus-transformed (Edwards et al., 1971), or virus-infected cultures (O'Neill, 1973). Sparse or dense cells blocked in G_0 by serum deprivation were also able to aggregate but within 2 h of the addition of fresh serum to the cultures adhesion was markedly reduced (O'Neill, 1973). These results show that cohesion is at its greatest under conditions where the cell might be expected to contain elevated levels of certain glycolipids (see section 4.2), LETS glycoprotein (see Hynes, 1976) and a decreased amount of one group of large glycopeptides derived from cell surface glycoproteins (Buck et al., 1970, 1971). However, LETS glycoprotein, which is extremely sensitive to removal by trypsin (5 μg trypsin for 2 min; Hynes, 1976) would be expected to vanish completely from the cell surface under the harvesting conditions. Thus it is unlikely to be involved in the aggregation of BHK21 cells unless significant amounts are regenerated during the aggregation assay. In addition, the rapid reduction in cohesiveness of serum-starved cells after the addition of serum is probably not due to a rapid loss of the accumulated glycolipids or LETS glycoproteins, levels of which decline at a lower rate (Critchley and Macpherson, 1973; Hynes and Bye, 1974). It is possible that although the putative cohesive molecules are present, aggregation is inhibited due to the effect of serum on some other component required in the adhesion process. For example, an organised microtubule system is known to be required for aggregation of BHK21 cells (Waddell et al., 1974; Edwards et al., 1975) and serum has been shown to disrupt the submembranous cytoskeleton (Pollack and Rifkin, 1975).

Recently Vicker (1976) reported that a specific glycopeptide fraction derived from a pronase-treated digest of the cell surface of BHK21 cells (method of Buck et al., 1971) could inhibit cell aggregation. Analogous glycopeptides from various transformed cells and other cell surface glycopeptides from homologous cells had little or no effect on aggregation. The active molecule(s) bind to the cell surface (Allen and Minnikin, 1975) and their activity increases following pretreatment with neuraminidase. However, periodate or galactose oxidase destroys all inhibitory activity (Vicker, 1976). Presumably the inhibiting glycopeptide(s) compete with intact parent cell surface molecules for binding to specific cell surface receptors, suggesting that adhesion measurable in the shaken-suspension system is indeed accounted for by this interaction. The outline features of these molecules are now known (Ogata et al., 1976) and suggest a role for cell surface α or β galactosyl terminal residues in cell cohesion.

In the foregoing sections we have emphasized the relationship, coincidental as it might be, between changes in cell cohesion and cell surface glycolipids and glycoproteins. We wish to point out the evidence that in some cases binding molecules arise in coordination with their heteroglycan receptors (Reitherman et al., 1975). The similar behaviour in this regard of glycolipids and glycoproteins, together with the similarity of their structures, leads us to consider both molecular classes as intercellular receptors (see Roseman, 1974). Although there is no *direct* evidence that glycolipids are involved in cell–cell interaction, it should be remembered that it has long been recognised that they act as receptors for certain bacteriophages (Robbins and Uchida, 1962; Losick and Robbins, 1969), and circumstantial evidence points to a role for glycolipids as receptors for viruses in mammalian cells (Shortridge and Biddle, 1972; Haywood, 1974). A possible role for glycolipids in adhesion and cell social interactions should not be excluded.

References

Abercrombie, M. (1970) Control mechanisms in cancer. Eur. J. Cancer 6, 7–13.

Abercrombie, M. and Heaysman, J.E.M. (1954) Observations on the social behaviour of cells in tissue culture, I. "Monolayering" of fibroblasts. Exp. Cell Res. 6, 293–306.

Adams, E. P. and Gray, G.M. (1967) Effect of BP8 ascites (sarcoma) tumours on glycolipid composition in kidneys of mice. Nature (London) 216, 277–278.

Aldrich, H.C. and Gregg, J.H. (1973) Unit membrane structural changes following cell association in Dictyostelium. Exp. Cell Res. 81, 407–412.

Allan, D., Auger, J. and Crumpton, M.J. (1972) Glycoprotein receptors for Con A isolated from pig lymphocyte plasma membrane by affinity chromatography in sodium deoxycholate. Nature (London) New Biol. 236, 23–25.

Allen, A. and Minnikin, S.M. (1975) The binding of the mucoprotein from gastric mucus to cells in tissue culture and the inhibition of cell adhesion. J. Cell Sci. 17, 617–631.

Anderson, W.B., Russell, T.R., Carchman, R.A. and Pastan, I. (1973) Interrelationships between adenylate cyclase activity, adenosine 3',5' cyclic monophosphate, phosphodiesterase activity, adenosine 3',5' cyclic monophosphate levels, and growth of cells in culture. Proc. Natl. Acad. Sci. USA 70, 3802–3805.

Ankel, H., Chany, C., Galliot, B., Chevalier, M.J. and Robert, M. (1973) Antiviral effects of interferon covalently bound to Sepharose. Proc. Natl. Acad. Sci. USA 70, 2360–2363.

Bailey, J.M., Howard, B.V., Dunbar, L.M. and Tillman, S.F. (1972) Control of lipid metabolism in cultured cells. Lipids 7, 125–134.

Balsamo, J. and Lilien, J. (1974) Functional identification of three components which mediate tissue-type specific embryonic cell adhesion. Nature (London) 251, 522–524.

Bannai, S. and Sheppard, J.R. (1974) Cyclic AMP, ATP and cell contact. Nature (London) 250, 62–64.

Bara, J., Lallier, R., Brailovsky, C. and Nigam, V.N. (1973a) Fixation of *Salmonella minnesota* R. form glycolipid on the membrane of normal and transformed rat embryo fibroblasts. Eur. J. Biochem. 35, 489–494.

Bara, J., Lallier, R., Brailovsky, C. and Nigam, V.N. (1973b) Molecular models on the insertion of a Salmonella R form glycolipid onto the cell membrane of normal and transformed cells. Eur. J. Biochem. 35, 495–498.

Barbera, A.J., Marchase, R.B., and Roth, S. (1973) Adhesive recognition and retinotectal specificity. Proc. Natl. Acad. Sci. USA 70, 2482–2486.

Barnett, R.E., Furcht, L.T. and Scott, R.E. (1974a) Differences in membrane fluidity and structure in contact inhibited and transformed cells. Proc. Natl. Acad. Sci. USA 71, 1992–1994.

Barnett, R.E., Scott, R.E., Furcht, L.T. and Kersey, J.H. (1974b) Evidence that mitogenic lectins induce changes in lymphocyte membrane fluidity. Nature (London) 249, 465–466.

Barton, N.W. and Rosenberg, A. (1973) Action of *Vibrio cholerae* neuraminidase (sialidase) upon the surface of intact cells and their isolated sialolipid components. J. Biol. Chem. 248, 7353–7358.

Bauer, H., Kurth, R., Rohrschneider, L., Pauli, G., Friis, R.R. and Gelderblom, H. (1974) The role of cell surface changes in RNA tumour virus-transformed cells. Cold Spring Harbor Symp. Quant. Biol. 39, part 2, 1181–1185.

Bennett, A. (1971) Cholera and prostaglandins. Nature (London) 231, 536–538.

Bennett, V. and Cuatrecasas, P. (1975a) Mechanism of action of *Vibrio cholerae* enterotoxin. Effects on adenylate cyclase of toad and rat erythrocyte plasma membranes. J. Membrane Biol. 22, 1–28.

Bennett, V. and Cuatrecasas, P. (1975b) Mechanism of activation of adenylate cyclase by *Vibrio cholerae* enterotoxin. J. Membrane Biol. 22, 29–52.

Bennett, V., O'Keefe, E. and Cuatrecasas, P. (1975) Mechanism of action of cholera toxin and the mobile receptor theory of hormone receptor–adenylate cyclase interactions. Proc. Natl. Acad. Sci. USA 72, 33–37.

Besançon, F. and Ankel, H. (1974a) Binding of interferon to gangliosides. Nature (London) 252, 478–480.

Besançon, F. and Ankel, H. (1974b) Inhibition of interferon action by plant lectins. Nature (London) 250, 784–786.

Besançon, F., Ankel, H. and Basu, S. (1976) Specificity and reversibility of interferon ganglioside interaction. Nature (London) 259, 576–578.

Beug, H., Katz, F.E. and Gersch, G. (1973a) Dynamics of antigenic membrane sites relating to cell aggregation in *Dictyostelium discoideum*. J. Cell. Biol. 56, 647–658.

Beug, H., Katz, F.E., Stein, A. and Gerisch, G. (1973b) Quantitation of membrane sites in aggregating *Dictyostelium* cells by use of tritiated univalent antibody. Proc. Natl. Acad. Sci. USA 70, 3150–3154.

Birnbaumer, L. (1973) Hormone-sensitive adenyl cyclases useful models for studying hormone receptor functions in cell-free systems. Biochim. Biophys. Acta 300, 129–158.

Bitensky, M.W., Wheeler, M.A., Mehta, H. and Miki, N. (1975) Cholera toxin activation of adenylate cyclase in cancer cell membrane fragments. Proc. Natl. Acad. Sci. USA 72, 2572–2576.

Bonner, J.T. (1971) Aggregation and differentiation in the cellular slime molds. Ann. Rev. Microbiol. 25, 75–92.

Bose, S., Gurari-Rotman, D., Ruegg, U., Corley, L. and Afinsen, C.B. (1976) Apparent dispensibility of the carbohydrate moiety of human interferon for antiviral activity. J. Biol. Chem. 251, 1659–1662.

Brady, R.O. (1973) The abnormal biochemistry of inherited disorders of lipid metabolism. Fed. Proc. 32, 1660–1667.

Brady, R.O. and Fishman, P.H. (1974) Biosynthesis of glycolipids in virus transformed cells. Biochim. Biophys. Acta 355, 121–148.

Brady, R.O. and Mora, P.T. (1970) Alteration in ganglioside pattern and synthesis in SV 40 and polyoma virus-transformed mouse cell lines. Biochim. Biophys. Acta 218, 308–319.

Brady, R.O., Borek, C. and Bradley, R.M. (1969) Composition and synthesis of gangliosides in rat hepatocyte and hepatoma cell lines. J. Biol. Chem. 244, 6552–6554.

Brailovsky, C.A., Berman, L.D. and Chany, C. (1969) Decreased interferon sensitivity and production in cells transformed by SV40 and other oncogenic agents. Int. J. Cancer 4, 194–203.

Brailovsky, C.A., Trudel, M., Lallier, R. and Nigam, V.N. (1973) Growth of normal and transformed rat embryo fibroblasts. Effects of glycolipids from *Salmonella minnesota* R mutants. J. Cell Biol. 57, 124–132.

Buck, C.A., Glick, M.C. and Warren, L. (1970) A comparative study of glycoproteins from the surface of control and Rous Sarcoma virus transformed hamster cells. Biochemistry 9, 4567–4575.

Buck, C.A., Glick, M.C. and Warren, L. (1971) Effect of growth on the glycoproteins from the surface of control and Rous sarcoma virus transformed hamster cells. Biochemistry 10, 2176–2180.

Buehler, R. and Moolten, F. (1975) Abnormal response of glycolipid synthesis to temperature in transformed BHK cells thermosensitive for growth control. Biochem. Biophys. Res. Commun. 67, 91–96.

Burger, M.M. (1974) The isolation of surface components involved in specific cell-cell adhesion and cellular recognition. In: S. Estrada-O and C. Gitler (Eds.), Perspectives in Membrane Biology. Academic Press, New York, pp. 509–528.

Burger, M.M. and Martin, G.S. (1972) Agglutination of cells transformed by Rous sarcoma virus by wheat germ agglutinin and concanavalin. Nature (London) New Biol. 237, 9–12.

Calderon, J., Williams, R.T. and Unanue, E.R. (1974) An inhibitor of cell proliferation released by cultures of macrophages. Proc. Natl. Acad. Sci. USA 71, 4273–4277.

Carchman, R.A., Johnson, G.S. and Pastan, I. (1974) Studies on the levels of cyclic AMP in cells transformed by wild-type and temperature-sensitive Kirsten sarcoma virus. Cell 1, 59–64.

Chandrabose, K.A. and Macpherson, I.A. (1976) Glycolipid glycosyltransferases of a hamster cell line in culture, I. Kinetic constants, substrate, and donor nucleotide sugar specificities. Biochim. Biophys. Acta 429, 96–111.

Chandrabose, K.A., Graham, J.M. and Macpherson, I.A. (1976) Glycolipid glycosyltransferases of a hamster cell line, II. Subcellular distribution and effect of culture age and density. Biochim. Biophys. Acta 429, 112–122.

Chang, C-W., Reitherman, R.W., Rosen, S.D. and Barondes, S.H. (1975) Cell surface location of discoidin, a developmentally regulated protein from *Dictyostelium discoideum*. Exp. Cell Res. 95, 136–142.

Chatterjee, S. and Sweeley, C.C. (1973) The effect of thrombin induced aggregation on human platelet glycosphingolipids. Biochem. Biophys. Res. Commun. 53, 1310–1316.

Chatterjee, S., Sweeley, C.C. and Velicier, L.F. (1973) Biosynthesis of proteins, nucleic acids and glycosphingolipids by synchronised KB cells. Biochem. Biophys. Res. Commun. 54, 585–592.

Chatterjee, S., Sweeley, C.C. and Velicier, L.F. (1975a) Glycosphingolipids of human KB cells grown in monolayer, suspension, and synchronized cultures. J. Biol. Chem. 250, 61–66.

Chatterjee, S., Velicier, L.F. and Sweeley, C.C. (1975b) Glycosphingolipid glycosyl hydrolases and glycosidases of synchronised human KB cells. J. Biol. Chem. 250, 4972–4979.

Cheema, P., Yogeeswaran, G., Morris, M.P. and Murray, R.K. (1970) Ganglioside patterns of three Morris minimal deviation hepatomas. FEBS Lett. 11, 181–184.

Chlapowski, F.J., Kelly, L.A. and Butcher, R.W. (1975) Cyclic nucleotides in cultured cells. Adv. Cyclic Nucleotide Res. 6, 245–338.

Clowes, A.W., Cherry, R.J. and Chapman, D. (1972) Physical effects of tetanus toxin on model membranes containing gangliosides. J. Mol. Biol. 67, 49–57.

Coleman, P.L., Fishman, P.H., Brady, R.O. and Todaro, G.J. (1975) Altered ganglioside biosynthesis in mouse cell cultures following transformation with chemical carcinogens and X-irradiation. J. Biol. Chem. 250, 55–60.

Coles, L., Hay, J.B. and Gray, G.M. (1970) Factors affecting the glycosphingolipid composition of mouse tissues. J. Lipid Res. 11, 158–163.

Craig, S.W. and Cuatrecasas, P. (1975) Mobility of Cholera toxin receptors on rat lymphocyte membranes. Proc. Natl. Acad. Sci. USA 72, 3844–3848.

Crandall, M., Lawrence, L.M. and Saunders, R.M. (1974) Molecular complementarity of yeast glycoprotein mating factors. Proc. Natl. Acad. Sci. USA 71, 26–29.

Critchley, D.R. (1974) Cell surface proteins of NIL 1 hamster fibroblasts labelled by a galactose oxidase, tritiated borohydride method. Cell 3, 121–125.

Critchley, D.R. and Macpherson, I.A. (1973) Cell density dependent glycolipids in NIL 2 hamster cells, derived malignant and transformed cell lines. Biochim. Biophys. Acta 296, 145–159.

Critchley, D.R., Graham, J.M. and Macpherson, I. (1973) Subcellular distribution of glycolipids in a hamster cell line. FEBS. Lett. 32, 37–40.

Critchley, D.R., Chandrabose, K.A., Graham, J.M. and Macpherson, I. (1974) Glycolipids of NIL hamster cells as a function of cell density and cell cycle. In: B. Clarkson and R. Baserga (Eds.), Control of Proliferation in Animal Cells. Cold Spring Harbor Laboratory Press, New York, pp. 481–493.

Critchley, D.R., Wyke, J.A. and Hynes, R.O. (1976) Cell surface and metabolic labelling of the proteins of normal and transformed chicken cells. Biochim. Biophys. Acta 436, 335–352.

Cuatrecasas, P. (1973a) Interaction of *Vibrio cholerae* enterotoxin with cell membranes. Biochemistry 12, 3547–3557.

Cuatrecasas, P. (1973b) Gangliosides and membrane receptors for cholera toxin. Biochemistry 12, 3558–3566.

Cuatrecasas, P. (1973c) Cholera toxin–fat cell interaction and the mechanism of activation of the lipolytic response. Biochemistry 12, 3567–3577.

Cuatrecasas, P. (1974) Membrane receptors. Ann. Rev. Biochem. 43, 169–214.

Cuatrecasas, P., Parikh, I. and Hollesberg, M.D. (1973) Affinity chromatography and structural analysis of *Vibrio cholerae* enterotoxin–ganglioside agarose and the biological effects of ganglioside-containing soluble polymers. Biochemistry 12, 4253–4264.

Cumar, F.A., Brady, R.O., Kolodny, E.H., McFarland, V.W. and Mora, P.T. (1970) Enzymatic block in the synthesis of gangliosides in DNA virus-transformed tumorigenic mouse cell lines. Proc. Natl. Acad. Sci. USA 67, 757–764.

Den, H., Shultz, A.M., Bals, M. and Roseman, S. (1971) Glycosyltransferase activities in normal and polyoma-transformed BHK cells. J. Biol. Chem. 246, 2721–2723.

Den, H., Sela, B., Roseman, S. and Sachs, L. (1974) Blocks in ganglioside synthesis in transformed hamster cells and their revertants. J. Biol. Chem. 249, 659–661.

De Petris, S. (1974) Inhibition and reversal of capping by cytochalasin B vinblastine and colchicine. Nature (London) 250, 54–56.

Deppert, W., Werchau, H. and Walter, G. (1974) Differentiation between intracellular and cell surface glycosyltransferases: Galactosyl transferase activity in intact cells and in cell homogenates. Proc. Natl. Acad. Sci. USA 71, 3068–3072.

Di Mayorca, G., Greenblatt, M., Trauthen, T., Soller, A. and Giordano, R. (1973) Malignant transformation of BHK_{21} Clone 13 cells in vitro by nitrosamines – a conditional state. Proc. Natl. Acad. Sci. USA 70, 46–49.

Dinistrian, A.M., Skipski, V.P., Barclay, M., Essner, E.S. and Stock, C.C. (1975) Gangliosides of plasma membranes from normal rat liver and Morris hepatoma. Biochem. Biophys. Res. Commun. 64, 367–375.

Diringer, H., Ströbel, G. and Koch, M.A. (1972) Glycolipids of mouse fibroblasts and virus transformed mouse cell lines. Z. Physiol. Chem. 353, 1769–1774.

Dod, B.J. and Gray, G.M. (1968a) The localization of the natural glycosphingolipids on rat liver cells. Biochem. J. 110, 50p.

Dod, B.J. and Gray, G.M. (1968b) The lipid composition of rat liver plasma membrane. Biochim. Biophys. Acta 150, 397–404.

Donta, S.T. and Viner, J.P. (1974) Inhibition of the steroidgenic effects of cholera and heat-labile *E. coli* enterotoxins by ganglioside. J. Clin. Invest. 53, 20a.

Donta, S.T., King, M. and Sloper, K. (1973) Induction of steroidgenesis in tissue culture by cholera enterotoxin. Nature (London) New Biol. 243, 246–247.

Dorner, F., Scriba, M. and Weil, R. (1973) Interferon: evidence for its glycoprotein nature. Proc. Natl. Acad. Sci. USA 70, 1981–1985.

Dubbs, D.R., Kitt, S., de Torres, R.A. and Anken, M. (1967) Virogenic properties of bromodeoxyuridine-sensitive and bromodeoxyuridine-resistant Simian virus 40-transformed mouse kidney cells. J. Virol. 1, 968–979.

Duffard, R.O. and Capputto, R. (1972) A natural inhibitor of sialyl transferase and its possible influence on this enzyme activity during brain development. Biochemistry 11, 1396–1400.

Dulbecco, R. (1970) Topoinhibition and serum requirement of transformed and untransformed cells. Nature (London) 227, 802–806.

Dulbecco, R. and Eckhart, W. (1970) Temperature-dependent properties of cells transformed by a thermosensitive mutant of polyoma virus. Proc. Natl. Acad. Sci. USA 67, 1775–1781

Eckhart, W. and Weber, M.J. (1974) Uptake of 2-deoxyglucose by BALB/3T3 cells: changes after polyoma infection. Virology 61, 223–228.

Eckhart, W., Dulbecco, R. and Burger, M. (1971) Temperature-dependent surface changes in cells infected or transformed by a thermosensitive mutant of polyoma virus. Proc. Natl. Acad. Sci. USA 68, 283–286.

Edwards, J.G. and Campbell, J.A. (1971) The aggregation of trypsinized BHK 21 cells. J. Cell. Sci. 8, 53–71.
Edwards, J.G., Campbell, J.A. and Williams, J.T. (1971) Transformation by polyoma virus affects adhesion of fibroblasts. Nature (London) New Biol. 231, 147–148.
Edwards, J.G., Campbell, J.A., Robson, R.T. and Vicker, M.G. (1975) Trypsinized BHK 21 cells aggregate in the presence of metabolic inhibitors and in the absence of divalent cations. J. Cell. Sci. 19, 653–667.
Esselman, W.J. and Miller, H.C. (1974) Brain and thymus lipid inhibition of antibrain-associated θ cytotoxicity. J. Exp. Med. 139, 445–450.
Esselman, W.J., Ackerman, J.R. and Sweeley, C.C. (1973) Glycosphingolipids of membrane fractions from normal and transplanted canine kidney. J. Biol. Chem. 248, 7310–7317.
Evans, W.H. (1974) Nucleotide pyrophosphatase, a sialoglycoprotein located on the hepatocyte surface. Nature (London) 250, 391–394.
Finkelstein, R.A. (1973) Cholera. CRC Crit. Rev. Microbiol. 2, 533–623.
Finkelstein, R.A., Boesman, M., Neoh, S.H., LaRue, M.K. and Delaney, R. (1974) Dissociation and recombination of the subunits of the cholera enterotoxin (choleragen). J. Immunol. 113, 145–150.
Fishman, P.H., McFarland, V.W., Mora, P.T. and Brady, R.O. (1971) Ganglioside biosynthesis in mouse cells: glycosyltransferase activities in normal and virally-transformed lines. Biochem. Biophys. Res. Commun. 48, 48–57.
Fishman, P.H., Brady, R.O., Bradley, R.M., Aaronson, S.A. and Todaro, G.J. (1974a) Absence of a specific ganglioside galactosyltransferase in mouse cells transformed by murine sarcoma virus. Proc. Natl. Acad. Sci. USA 71, 298–301.
Fishman, P.H., Simmons, J.L., Brady, R.O. and Freese, E. (1974b) Induction of glycolipid biosynthesis by sodium butyrate in HeLa cells. Biochem. Biophys. Res. Commun. 59, 292–299.
Fishman, P.H., Bradley, R.M. and Henneberry, R.C. (1976) Butyrate-induced glycolipid biosynthesis in HeLa cells: properties of the induced sialyltransferase. Arch. Biochem. Biophys. 172, 618–626.
Forstner, G.G. and Wherrett, J.R. (1973) Plasma membrane and mucosal glycosphingolipids in the rat intestine. Biochim. Biophys. Acta 306, 446–457.
Furshpan, E.J. and Potter, D.D. (1968) Low-resistance junctions between cells in embryos and tissue culture. In: A.A. Moscona and A. Monroy (Eds.), Current Topics in Developmental Biology, Vol. 3, Academic Press, New York, pp. 95–127.
Gaffney, B.J. (1975) Fatty acid chain flexibility in the membranes of normal and transformed fibroblasts. Proc. Natl. Acad. Sci. USA 72, 664–668.
Gahmberg, C.G. and Hakomori, S. (1973a) External labelling of cell surface galactose and galactosamine in glycolipid and glycoprotein of human erythrocytes. J. Biol. Chem. 248, 4311–4317.
Gahmberg, C.G. and Hakomori, S. (1973b) Altered growth behaviour of malignant cells associated with changes in externally labelled glycoprotein and glycolipid. Proc. Natl. Acad. Sci. USA 70, 3329–3333.
Gahmberg, C.G. and Hakomori, S. (1974) Organisation of glycolipids and glycoproteins in surface membranes: Dependency on cell cycle and transformation. Biochem. Biophys. Res. Commun. 59, 283–291.
Gahmberg, C.G. and Hakomori, S. (1975a) Surface carbohydrates of hamster fibroblasts, I. Chemical characterization of surface-labelled glycosphingolipids and a ceramide tetrasaccharide specific for transformants. J. Biol. Chem. 250, 2438–2446.
Gahmberg, C.G. and Hakomori, S. (1975b) Surface carbohydrates of hamster fibroblasts, II. Interaction of hamster NIL cell surfaces with *Ricinus communis* lectin and concanavalin A as revealed by surface galactosyl label. J. Biol. Chem. 250, 2447–2451.
Gahmberg, C.G., Kiehn, D. and Hakomori, S. (1974) Changes in a surface-labelled galactoprotein and in glycolipid concentrations in cells transformed by a temperature-sensitive polyoma virus mutant. Nature (London) 248, 413–415.
Garrod, D.R. (1972) Acquisition of cohesiveness by slime mould cells prior to morphogenesis. Exp. Cell Res. 72, 588–591.

Geren, L.M. and Ebner, K.E. (1974) Folic acid effects on glycoprotein-galactosyltransferase: a re-assessment. Biochem. Biophys. Res. Commun. 59, 14–21.
Gerisch, G. (1968) Cell aggregation and differentiation in *Dictyostelium*. In: A.A. Moscona and A. Monroy (Eds.), Current Topics in Developmental Biology, Vol. 3, Academic Press, New York, pp. 157–197.
Geyer, R.P. (1967) Uptake and retention of fatty acids by tissue culture cells. In: G. Rothblat and D. Kritchevsky (Eds.), Lipid Metabolism in Tissue Culture Cells. Wistar Press Monograph, Philadelphia, pp. 33–47.
Gill, D.M. (1975) Involvement of nicotinamide adenine dinucleotide in the action of cholera toxin in vitro. Proc. Natl. Acad. Sci. USA 72, 2064–2068.
Gill, D.M. (1976) The arrangement of subunits in Cholera toxin. Biochemistry, 15, 1242–1248.
Gill, D.M. and King, C.A. (1975) The mechanism of action of cholera toxin in pigeon erythrocyte lysates. J. Biol. Chem. 250, 6424–6432.
Gill, D.M., Poppenheimer, A.M. and Uchida, T. (1973) Diphtheria toxin, protein synthesis, and the cell. Fed. Proc. 32, 1508–1515.
Gray, G.M. (1971) The effect of testosterone on the biosynthesis of the neutral glycosphingolipids in the C57/BL mouse kidney. Biochim. Biophys. Acta 239, 494–500.
Gray, G.M. (1974) Glycosphingolipids in biological membranes. In: S. Estrada-O. and C. Gitler (Eds.), Perspectives in Membrane Biology. Academic Press, New York, pp. 85–106.
Gregg, J.H. (1971) Developmental potential of isolated *Dictyostelium* myxamoebae. Dev. Biol. 26, 478–85.
Gressor, I., Bandu, M.T., Tovey, M., Bodo, G., Paucker, K. and Stewart, W. II (1973) Interferon and cell division, VII. Inhibitory effect of highly purified interferon preparations on the multiplication of leukemia L1210 cells. Proc. Soc. Exp. Biol. Med. 142, 7–10.
Hakomori, S. (1970a) Cell density-dependent changes of glycolipid concentrations of fibroblasts, and loss of this response in virus-transformed cells. Proc. Natl. Acad. Sci. USA 67, 1741–1747.
Hakomori, S. (1970b) Glycosphingolipids having blood group ABH and Lewis specificities. Chem. Phys. Lipids 5, 96–115.
Hakomori, S. (1971) Glycolipid changes associated with malignant transformation. In: D.F.H. Wallach and H. Fisher (Eds.), Dynamic Structure of Cell Membranes. Springer, Berlin, pp. 65–96.
Hakomori, S. (1973) Glycolipids of tumour cell membranes. Adv. Cancer Res. 18, 265–315.
Hakomori, S. (1975a) Structures and organization of cell surface glycolipids: dependency on cell growth and malignant transformation. Biochim. Biophys. Acta 417, 53–80.
Hakomori, S. (1975b) Fucolipids and glood group glycolipids in normal and tumor tissue. Progr. Biochem. Pharmacol. 10, 167–196.
Hakomori, S. and Kijimoto, S. (1972) Forssman reactivity and cell contact in cultured hamster cells. Nature (London) New Biol. 239, 87–88.
Hakomori, S. and Murakami, W.T. (1968) Glycolipids of hamster fibroblasts and derived malignant-transformed cell lines. Proc. Natl. Acad. Sci. USA 59, 254–261.
Hakomori, S., Koscielak, J., Bloch, H. and Jeanloz, R. (1967) Immunologic relationship between blood group substances and a fucose-containing glycolipid of human adenocarcinoma. J. Immunol. 98, 31–38.
Hakomori, S., Teather, C. and Andrews, H. (1968) Organizational difference of cell surface "Hematoside" in normal and virally transformed cells. Biochem. Biophys. Res. Commun. 33, 563–568.
Hakomori, S., Saito, T. and Vogt, P.K. (1971) Transformation by Rous sarcoma virus: effects on cellular glycolipids. Virology 44, 609–621.
Hammerstrom, S. and Bjursell, G. (1973) Glycolipid synthesis in baby hamster-kidney fibroblasts transformed by a thermosensitive mutant of polyoma virus. FEBS Lett. 32, 69–72.
Hammes, G.G. and Rodbell, M. (1976) Simple model for hormone-activated adenylate cyclase systems. Proc. Natl. Acad. Sci. USA 73, 1189–1192.
Hartley, B.S. (1974) Enzyme families. In: M.J. Carlile and J.J. Skehel (Eds.), Evolution in the Microbial World. The Society for General Microbiology Symposium 24. Cambridge University Press, pp. 151–182.

Hatanaka, M. (1974) Transport of sugars in tumor cell membranes. Biochim. Biophys. Acta 355, 77–104.

Hatanaka, M. and Hanafusa, H. (1970) Analysis of a functional change in membranes in the process of cell transformation by Rous sarcoma virus: Alteration in the characteristics of sugar transport. Virology 41, 647–652.

Haywood, A.M. (1974) Characteristics of Sendai virus receptors in a model membrane. J. Mol. Biol. 83, 427–436.

Higgins, T., Sabatino, A., Remold, H. and David, J. (1976) Enhancement of migration inhibitory factor activity by preincubating macrophages with macrophage glycolipids. Fed. Proc. 35, Abstr. 1015, p. 389.

Hildebrand, J., Stryckmans, P. and Stoffyn, P. (1971) Neutral glycolipids in leukemic and non-leukemic leukocytes. J. Lipid Res. 12, 361–366.

Hildebrand, J., Stryckmans, P. and Vahoud, J. (1972) Gangliosides in leukemic and non-leukemic human leukocytes. Biochim. Biophys. Acta 260, 272–278.

Hirschberg, C.B., Wolf, B.A. and Robbins, P.W. (1975) Synthesis of glycolipids and phospholipids in hamster cells: Dependence on cell density and the cell cycle. J. Cell Physiol. 85, 31–39.

Hollenberg, M.D. and Cuatrecasas, P. (1973) Epidermal growth factor: receptors in human fibroblasts and modulation of action by cholera toxin. Proc. Natl. Acad. Sci. USA 70, 2964–2968.

Hollenberg, M.D., Fishman, P.H., Bennett, V. and Cuatrecasas, P. (1974) Cholera toxin and cell growth: Role of membrane gangliosides. Proc. Natl. Acad. Sci. USA 71, 4224–4228.

Holmgren, J. (1973) Comparison of the tissue receptors for *Vibrio cholerae* and *Escherichia coli* enterotoxins by means of ganglioside and natural cholera toxoid. Infect. Immun. 8, 851–859.

Holmgren, J. and Lonnroth, I. (1975) Mechanism of action of cholera toxin. Specific inhibition of toxin-induced activation of adenylate cyclase. FEBS Lett. 55, 138–142.

Holmgren, J., Lonnroth, I. and Svennerholm, L. (1973) Fixation and inactivation of cholera toxin by GM_1 ganglioside. Scand. J. Infect. Dis. 5, 77–78.

Holmgren, J., Lindholm, L. and Lonnroth, I. (1974) Interaction of cholera toxin and toxin derivatives with lymphocytes, 1. Binding properties and interference with lectin-induced cellular stimulation. J. Exp. Med. 139, 801–819.

Holmgren, J., Lonnroth, I., Mansson, J.E. and Svennerholm, L. (1975) Interaction of cholera toxin and membrane GM_1 ganglioside of small intestine. Proc. Natl. Acad. Sci. USA 72, 2520–2524.

Houck, J.C. and Hennings, H. (1973) Chalones. Specific endogenous mitotic inhibitors. FEBS Lett. 32, 1–8.

Howard, B.V. and Howard, W.J. (1975) Lipids in normal and tumour cells in culture. In: Progress in Biochemical Pharmacology, Vol. 19, Karger, Basel, pp. 135–166.

Hudgin, R.L., Price, Jr., W.E., Ashwell, G., Stockert, R.J. and Morell, A.G. (1974) The isolation and properties of a rabbit liver binding protein specific for asialoglycoproteins. J. Biol. Chem. 249, 5536–5543.

Huesgen A. and Gerisch G. (1975) Solubilized contact sites A from cell membranes of *Dictyostelium discoideum*. FEBS Lett. 56, 46–49.

Hughes, R.C. (1975) The complex carbohydrates of mammalian cell surfaces and their biological roles. In: P.N. Campbell and W.N. Aldridge (Eds.), Essays in Biochemistry, Vol. 11, Academic Press, London, pp. 1–36.

Hynes, R.O. (1973) Alteration of cell-surface proteins by viral transformation and by proteolysis. Proc. Natl. Acad. Sci. USA 70, 3170–3174.

Hynes, R.O. (1976) Cell surface proteins and malignant transformation. Biochim. Biophys. Acta 458, 73–107.

Hynes, R.O. and Bye, J.M. (1974) Density and cell cycle dependence of cell surface proteins in hamster fibroblasts. Cell 3, 113–120.

Hynes, R.O. and Wyke, J.A. (1975) Alterations in surface proteins in chicken cells transformed by temperature-sensitive mutants of Rous sarcoma virus. Virology 64, 492–504.

Hynie, S. and Sharp, G.W.G. (1972) The effect of cholera toxin on intestinal adenyl cyclase. Adv. Cyclic Nucleotide Res. 1, 163–174.

Inbar, M., Ben-Bassat, H. and Sachs, L. (1972) Membrane changes associated with malignancy. Nature (London) New Biol. 236, 3–4.

Inoue K., Graf L. and Rapport, M.M. (1972) Immunochemical studies of organ and tumour lipids, XIX. Cytolytic action of antibodies directed against cytolipin R. J. Lipid Res. 13, 119–127.

Ji, T.H. (1974) Cross linking of glycolipids in erythrocyte ghost membrane. J. Biol. Chem. 249, 7841–7847.

Jimenez de Asua, L., Rozengurt, E. and Dulbecco, R. (1974) Kinetics of early change in phosphate and uridine transport and cyclic AMP levels stimulated by serum in density-inhibited 3T3 cells. Proc. Natl. Acad. Sci. USA 71, 96–98.

Kantor, H.S. (1975) Enterotoxins of *Escherichia coli* and *Vibrio cholerae*: Tools for the molecular biologist. J. Infect. Dis. 131, S22–S32.

Karlsson, K.A., Samuelsson, B.E., Schersten, T. and Steen, G.O. (1974) Sphingolipid composition of human renal carcinoma. Biochim. Biophys. Acta 337, 349–355.

Kawai, K., Fiujita, M. and Nakao, M. (1974) Lipid composition of two different regions of an intestinal epithelial cell membrane of mouse. Biochim. Biophys. Acta 369, 222–233.

Kawanami, J. and Tsuji, T. (1968) Lipids of cancer tissues, III. Glycolipids of human hepatoma tissue. Jap. J. Exp. Med. 38, 11–18.

Keenan, T.W. and Doak, R.L. (1973) Enzymatic block in higher ganglioside biosynthesis in avian transplantable lymphoid tumor. FEBS Lett. 37, 124–128.

Keenan, T.W. and Morré, D.J. (1973) Mammary Carcinoma: Enzymatic block in disialoganglioside biosynthesis. Science 182, 935–937.

Keenan, T.W. and Morré, D.J. (1975) Glycosyltransferases: do they exist on the surface membrane of mammalian cells? FEBS Lett. 55, 8–13.

Keenan, T.W., Morré, D.J. and Huang, C.M. (1972a) Distribution of gangliosides among subcellular fractions from rat liver and bovine mammary gland. FEBS Lett. 24, 204–208.

Keenan, T.W., Huang, C.M. and Morré, D.J. (1972b) Gangliosides: nonspecific localization on the surface membranes of bovine mammary gland and rat liver. Biochem. Biophys. Res. Commun. 47, 1277–1283.

Keenan, T.W., Franke, W.W. and Wiegandt, H. (1974) Ganglioside accumulation by transformed murine fibroblasts (3T3) cells and canine erythrocytes. Z. Physiol. Chem. 355, 1543–1550.

Keenan, T.W., Schmidt, E., Franke, W.W. and Wiegandt, H. (1975) Exogenous glycosphingolipids suppress growth rate of transformed and untransformed 3T3 mouse cells. Exp. Cell Res. 92, 259–270.

Kemp, R.B., Lloyd, C.W. and Cook, G.M.W. (1973) Glycoproteins in cell adhesion. In: J.F. Danielli (Ed.), Progress in Surface and Membrane Science, Vol. 7, Academic Press, London, pp. 271–318.

Kempen, H.J.M., de Pont, J.J.H.H.M. and Bonting, S.L. (1975) Rat pancreas adenylate cyclase, III. Its role in pancreatic secretion assessed by means of cholera toxin. Biochim. Biophys. Acta 392, 276–287.

Kijimoto, S. and Hakomori, S. (1971) Enhanced glycolipid: α-galactosyltransferase activity in contact inhibited hamster cells, and loss of this response in polyoma transformants. Biochem. Biophys. Res. Commun. 44, 557–563.

Kijimoto, S. and Hakomori, S. (1972) Contact-dependent enhancement of net synthesis of Forssman glycolipid antigen and hematoside in NIL cells at the early stage of cell-to-cell contact. FEBS Lett. 25, 38–42.

Kimberg, D.V., Field, M., Johnson, J., Henderson, A. and Gershon, E. (1971) Stimulation of intestinal mucosal adenyl cyclase by cholera enterotoxin and prostaglandins. J. Clin. Invest. 50, 1218–1230.

Kimberg, D.V., Field, M., Gershon, E., Schooley, R.T. and Henderson, A. (1973) Effects of cycloheximide on the response of intestinal mucosa to cholera enterotoxin. J. Clin. Invest. 52, 1376–1383.

Kimberg, D.V., Field, M., Gershon, E. and Henderson, A. (1974) Effect of prostaglandins and cholera enterotoxin on intestinal mucosal cAMP accumulation. Evidence against an essential role for prostaglandins in the action of toxin. J. Clin. Invest. 53, 941–949.

King, C.A. and Van Heyningen, W.E. (1973) Deactivation of cholera toxin by sialidase-resistant monosialosyl ganglioside. J. Infect. Dis. 127, 631–647.

King, C.A. and van Heyningen, S. (1975) Subunit A from cholera toxin is an activator of adenylate cyclase in pigeon erythrocytes. Biochem. J. 146, 269–271.

Klenk, H.D. and Choppin, P.W. (1970) Glycosphingolipids of plasma membranes of cultured cells and an enveloped virus (SV5) grown in these cells. Proc. Natl. Acad. Sci. USA, 66, 57–64.

Kostic, D. and Buchleit, F. (1970) Gangliosides in human brain tumours. Life Sci. 9, 589–596.

Kram, R., Mamont, P. and Tomkins, G.M. (1973) Pleiotypic control by adenosine 3',5'-cyclic monophosphate: A model for growth control in animal cells. Proc. Natl. Acad. Sci. USA 70, 1432–1436.

Kuehl, F.A., Humes, J.L., Tarnoff, J., Cirillo, V.J. and Ham, E.A. (1970) Prostaglandin receptor site: Evidence for an essential role in the action of luteinizing hormone. Science 169, 883–886.

Laine, R.A. and Hakomori, S. (1973) Incorporation of exogenous glycosphingolipids in plasma membranes of cultured hamster cells and concurrent change of growth behaviour. Biochem. Biophys. Res. Commun. 54, 1039–1045.

Laine, R.A., Yogeeswaran, G. and Hakomori, S. (1974) Glycosphingolipids covalently linked to agarose gel or glass beads. Use of the compounds for purification of antibodies directed against globoside and hematoside. J. Biol. Chem. 249, 4460–4466.

Laine, R., Sweeley, C.C., Li, Y-T, Kisic, A., and Rapport, M.M. (1972) On the structure of cytolipin R; a ceramide tetrahexoside from rat lymphosarcoma. J. Lipid Res. 13, 519–530.

Langenbach, R. (1975) Gangliosides of chemically and virally transformed rat embryo cells. Biochim. Biophys. Acta 388, 231–242.

Leblond Larouche, L., Morais, R., Nigam, V.N. and Karasaki, S. (1975) A comparative study of carbohydrate content, protein, glycoprotein, and ganglioside pattern of cell membranes isolated from Novikoff ascites hepatoma and normal liver cells. Arch. Biochem. Biophys. 167, 1–12.

Ledley, F.O., Mullin, B.R., Lee, G., Aloj, S.M., Fishman, P.H., Hunt, L.T., Dayhoff, M.O. and Kohn, L.D. (1976) Sequence similarity between cholera toxin and glycoprotein hormones. Implications for structure activity relationship and mechanism of action. Biochem. Biophys. Res. Commun. 69, 852–859.

Leffert, A.L. (1974) Growth control of differentiated foetal rat hepatocytes in primary monolayer culture. J. Cell Biol. 62, 767–779.

Lennarz, W.J. (1975) Lipid linked sugars in glycoprotein synthesis. Science 188, 986–991.

Lennarz, W.J. and Scher, M.G. (1972) Metabolism and functions of polyisoprenol sugar intermediates in membrane associated reactions. Biochim. Biophys. Acta 205, 417–441.

Levis, G.M., Evangeltos, G.P., and Crumpton, M.J. (1976a) Lipid compositions of lymphocyte plasma membrane from pig mesenteric lymph node. Biochem. J. 156, 103–110.

Levis, G.M., Karli, J.N. and Crumpton, M.J. (1976b) Plasma membrane glycosphingolipids (GSLS) of the human lymphoblastoid cell-line BRI8 and differences between the GSLS of BRI8 cells and those of peripheral lymphocytes. Biochem. Biophys. Res. Commun., 68, 336–342.

Lichtenstein, L.M., Henney, C.S., Bourne, H.R. and Greenough, W.B. (1973) Effects of cholera toxin on in vitro models of immediate and delayed hypersensitivity. Further evidence for the role of cyclic adenosine 3',5'-monophosphate. J. Clin. Invest. 52, 691–697.

Lipkin, G. and Knecht, M.E. (1974) A diffusable factor restoring contact inhibition of growth to malignant melanocytes. Proc. Natl. Acad. Sci. USA 71, 849–853.

Lloyd, C.W. and Cook, G.M.W. (1975) A membrane glycoprotein-containing fraction which promotes cell aggregation. Biochem. Biophys. Res. Commun. 67, 696–700.

Lockart Jr., R.Z. (1973) Criteria for the acceptance of a viral inhibitor as an interferon and a general description of the biological properties of known interferons. In: N.B. Finter (Ed.), Interferon and Interferon Inducers, Frontiers of Biology, Vol. 2. North-Holland, Amsterdam, pp. 11–27.

Losick, R.M. and Robbins, P.W. (1969) The receptor site for a bacterial virus. Sci. Am. 221, 121–124.

Lospalluto, J.J. and Finkelstein, R.A. (1972) Chemical and physical properties of cholera exoenterotoxin (choleragen) and its spontaneously formed toxoid (choleragenoid). Biochim. Biophys. Acta 257, 158–166.

Maccioni, H.J.F., Arce, A. and Capputto, R. (1972) A method of determining the sequence of incorporation of monosaccharides in the synthesis of the branched oligosaccharide chain of a structural compound. FEBS Lett. 23, 136–138.

Marcus, D.M. and Janis, R. (1970) Localization of glycosphingolipids on human tissues by immunofluorescence. J. Immunol. 104, 1530–1539.

Marshall, N.J., von Borcki, S. and Eakins, R.P. (1976) Independence of β-adrenergic and thyrotropin receptors linked to adenylate cyclase in thyroid. Nature (London) 261, 603–604.

Martennson, E. (1969) Glycosphingolipids of animal tissue. In: R.T. Holman (Ed.), Progress in the Chemistry of Fats and Other Lipids, Vol. 10. Pergamon, New York, pp. 367–407.

Martin, G.S. (1970) Rous sarcoma virus: a function required for the maintenance of the transformed state. Nature (London) 227, 1021–1023.

Mashiter, K., Mashiter, G.D., Haugher, R.L. and Field, J.B. (1973) Effects of cholera and *E. coli* enterotoxins on cyclic adenosine 3'5'-monophosphate levels and intermediary metabolism in the thyroid. Endocrinology 92, 541–549.

McGuire, E.J. and Burdick, C.L. (1975) Intercellular adhesive selectivity, I. An improved assay for the measurement of embryonic chick intercellular adhesion (liver and other tissues). J. Cell Biol. 68, 80–89.

Merrell, R., Gottlieb, D.I. and Glaser, L. (1975) Embryonal cell surface recognition, extraction of an active plasma membrane component. J. Biol. Chem. 250, 5655–5659.

Metz, D.H. (1975) The mechanism of interferon action. Cell 6, 429–439.

Miller, H.C. and Esselman, W.J. (1975) Modulation of immune response by antigen-reactive lymphocytes after cultivation with gangliosides. J. Immunol. 115, 839–843.

Mora, P.T., Brady, R.O., Bradley, R.M. and McFarland, V.W. (1969) Gangliosides in DNA virus-transformed and spontaneously transformed tumorigenic mouse cell lines. Proc. Natl. Acad. Sci. USA 63, 1290–1296.

Mora, P.T., Cumar, F.A. and Brady, R.O. (1971) A common biochemical change in SV40 and polyoma virus transformed mouse cells coupled to control of cell growth in culture. Virology 46, 60–72.

Mora, P.T., Fishman, P.H., Bassin, R.H., Brady, R.O. and McFarland, V.W. (1973) Transformation of Swiss 3T3 cells by murine sarcoma virus is followed by decrease in a glycolipid glycosyltransferase. Nature (London) 245, 226–228.

Morrell, P. and Braun, P. (1972) Biosynthesis and metabolic degradation of sphingolipids not containing sialic acid. J. Lipid Res. 13, 293–310.

Morris, J.E. and Moscona, A.A. (1970) Induction of glutamine synthetase in embryonic retinae: its dependence on cell interactions. Science 167, 1736–1737.

Moss, J., Fishman, P.H., Manganiello, V.C., Vaughan, M. and Brady, R.O. (1976) Functional incorporation of ganglioside into intact cells: Induction of choleragen responsiveness. Proc. Natl. Acad. Sci. USA 73, 1034–1037.

Mullin, B.J., Fishman, P.H., Lee, G., Aloj, S.M., Ledley, F.D., Winand, R.J., Kohn, L.D. and Brady, R.O. (1976) Thyrotropin-ganglioside interactions and their relationship to the structure and function of thyrotropin receptors. Proc. Natl. Acad. Sci. USA, 73, 842–846.

Nicolson, G.L. (1974) The interactions of lectins with animal cell surfaces. Int. Rev. Cytol. 39, 89–190.

Nicolson, G.L. (1976a) Transmembrane control of the receptors on normal and tumour cells, I. Cytoplasmic influence over cell surface components. Biochim. Biophys. Acta 457, 57–108.

Nicolson, G.L. (1976b) Transmembrane control of the receptors on normal and tumor cells, II. Surface changes associated with transformation and malignancy. Biochim. Biophys. Acta 458, 1–72.

Nigam, V.N., Lallier, R. and Brailovsky, C. (1973) Ganglioside patterns and phenotypic characteristics in a normal variant and a transformed back variant of a Simian virus 40-induced hamster tumor cell line. J. Cell Biol. 58, 307–316.

Nowak, T.P., Haywood, P.L. and Barondes, S.H. (1976) Developmentally regulated lectin in embryonic chick muscle and a myogenic cell line. Biochem. Biophys. Res. Commun. 68, 650–657.

Ogata, S.I., Muramatsu, T. and Kobata, D. (1976) New structural characteristics of the large glycopeptides from transformed cells. Nature (London) 259, 580–582.

Ohtomo, N., Muraoka, T., Tashiro, A., Zinnaka, Y and Amako, K. (1976) Size and structure of the cholera toxin molecule and its subunits. J. Infect. Dis. 133 suppl., S31–S40.

Okasaki, K. and Holtzer, H. (1965) An analysis of myogenesis in vitro using fluorescein-labelled antimyosin. J. Histochem. Cytochem. 13, 726–739.

O'Keefe, E. and Cuatrecasas, P. (1974) Cholera toxin mimics melanocyte stimulating hormone in inducing differentiation in melanoma cells. Proc. Natl. Acad. Sci. USA 71, 2500–2504.

O'Neill, C.H. (1973) Growth induction by serum or polyoma virus inhibits the aggregation of trypsinized suspensions of BHK21 tissue culture fibroblasts. Exp. Cell Res. 81, 31–39.

Osborn, M.J. (1971) The role of membranes in the synthesis of macromolecules. In: L.I. Rothfield (Ed.), Structure and Function of Biological Membranes. Academic Press, New York, pp. 343–400.

Ossowski, L., Unkeless, J.C., Tobia, A., Quigley, J.P., Rifkin, D.B. and Reich, E. (1973) An enzymatic function associated with transformation of fibroblasts by oncogenic viruses, II. Mammalian fibroblast cultures transformed by DNA and RNA tumour viruses. J. Exp. Med. 137, 112–126.

Otsuka, H. (1972) An inhibitor present in calf serum which prevents growth of BHK 21 cells in suspension culture. J. Cell Sci. 10, 137–152.

Otten, J., Johnson, G.S. and Pastan, I. (1972) Regulation of cell growth by cyclic adenosine 3',5'-monophosphate. Effect of cell density and agents which alter cell growth or cyclic adenosine 3',5'-monophosphate levels in fibroblasts. J. Biol. Chem. 247, 7082–7087.

Palfreyman, J.W. and Schulster, D. (1975) On the mechanism of action of cholera toxin on isolated rat adrenocortical cells. Biochim. Biophys. Acta 404, 221–229.

Pardee, A.B. (1974) A restriction point for control of normal animal cell proliferation. Proc. Natl. Acad. Sci. USA 71, 1286–1290.

Pardee, A.B. (1975) The cell surface and fibroblast proliferation some current research trends. Biochim. Biophys. Acta 417, 153–172.

Pardee, A.B. and Rozengurt, E. (1975) Role of the surface in production of new cells. In: C.F. Fox (Ed.), Biochemistry of Cell Walls and Membranes. Medical and Technical Publications, London, pp. 155–185.

Pastan, I.H., Johnson, G.S. and Anderson, W.B. (1975) Role of cyclic nucleotides in growth control. Ann. Rev. Biochem. 44, 491–522.

Pierce, J.G., Faith, M.R., Giudice, L.C. and Reeve, J.R. (1976) Structure and structure function relationships in glycoprotein hormones, In: Polypeptide Hormones: Molecular and Cellular Aspects. CIBA Foundation Symposium 41 (new series). Elsevier, Amsterdam, pp. 225–250.

Pierce, N.F., Greenough, W.B. and Carpenter, C.C.J. (1971) *Vibrio cholerae* enterotoxin and its mode of action. Bacteriol. Rev. 35, 1–13.

Pollack, R. and Rifkin, D. (1975) Actin-containing cables within anchorage-dependent rat embryo cells are dissociated by plasmin and trypsin. Cell 6, 495–506.

Pollack, R.E., Green, H. and Todaro, G.J. (1968) Growth control in cultured cells: Selection of sublines with increased sensitivity in contact inhibition and decreased tumor-producing ability. Proc. Natl. Acad. Sci. USA 60, 126–133.

Pollack, R.E., Wolman, S. and Vogel, A. (1970) Reversion of virus-transformed cell lines: Hyperploidy accompanies retention of viral genes. Nature (London) 228, 967–970.

Poste, G. and Allison, A.C. (1973) Membrane fusion. Biochim. Biophys. Acta 300, 421–465.

Rapport, M.M. (1969) Immunological properties of lipids and their relation to the tumor cell. Ann. N.Y. Acad. Sci. 159, 446–450.

Rapport, M.M., Graf, L., Skipski, V.P. and Alonzo, N.F. (1958) Cytolipin H, a pure lipid hapten isolated from human carcinoma. Nature (London) 181, 1803–1804.

Rapport, M.M., Schneider, H. and Graf, L. (1967) Cytolipin R: A pure lipid hapten isolated from rat lymphosarcoma. Biochim. Biophys. Acta 137, 409–411.

Reitherman, R.W., Rosen, S.D., Frazier, W.A. and Barondes, S.H. (1975) Cell surface species-specific high affinity receptors for discoidin: developmental regulation in *Dictyostelium discoideum*. Proc. Natl. Acad. Sci. USA 72, 3541–3545.

Renger, H.C. and Basilico, C. (1972) Mutation causing temperature-sensitive expression of cell transformation by a tumour virus. Proc. Natl. Acad. Sci. USA, 69, 109–114.

Renkonen, O., Gahmberg, C.G., Simons, K. and Kääriäinen, L. (1972) The lipids of the plasma membranes and endoplasmic reticulum from cultured baby hamster kidney cells (BHK 21). Biochim. Biophys. Acta 255, 66–78.

Revel, M., Bash, D. and Ruddle, F.H. (1976) Antibodies to a cell-surface component coded by human chromosome 21 inhibit action of interferon. Nature (London) 260, 139–141.

Révész, T. and Greaves, M. (1975) Ligand induced redistribution of lymphocyte membrane ganglioside—GM_1. Nature (London) 257, 103–106.

Rifkin, D.B., Beal, L.P. and Reich, E. (1975) Macromolecular determinants of plasminogen activator synthesis. In: E. Reich, D.B. Rifkin and E. Shaw (Eds.), Proteases and Biological Control. Cold Spring Harbor Laboratory, Cold Spring Harbor, N.Y., pp. 841–847.

Robbins, P.W. and Macpherson, I. (1971) Glycolipid synthesis in normal and transformed animal cells. Proc. Roy. Soc. Lond. Ser. B 177, 49–58.

Robbins, P.W. and Uchida, T. (1962) Studies on the chemical basis of the phage conversion of O-antigens in the E-group Salmonellae. Biochemistry 1, 323–335.

Roseman, S. (1970) The synthesis of complex carbohydrates by multiglycosyltransferases and their potential function in intercellular adhesion. Chem. Phys. Lipids 5, 270–297.

Roseman, S. (1974) Complex carbohydrates and intercellular adhesion. In: E.Y.C. Lee and E.E. Smith (Eds.), Biology and Chemistry of Eucaryotic Cell Surfaces, Miami Winter Symposia, Vol. 7, Academic Press, New York, pp. 317–354.

Rosen, S.D., Kafka, J.A., Simpson, D.L. and Barondes, S.H. (1973) Developmentally regulated, carbohydrate-binding protein in *Dictyostelium discoideum*. Proc. Natl. Acad. Sci. USA 70, 2554–2557.

Rosen, S.D., Reitherman, R.W. and Barondes, S.H. (1975) Distinct lectin activities from six species of cellular slime moulds. Exp. Cell Res. 95, 159–166.

Roth, S. and White, D. (1972) Intercellular contact and cell-surface galactosyl transferase activity. Proc. Natl. Acad. Sci. USA 69, 485–489.

Roth, S., McGuire, E. and Roseman, S. (1971a) Evidence for cell-surface glycosyltransferases: their potential role in cellular recognition. J. Cell Biol. 51, 536–547.

Roth, S.A., McGuire, E.J. and Roseman, S. (1971b) An assay for intercellular adhesive specificity. J. Cell Biol. 51, 974–980.

Rozengurt, E. and Jimenez De Asua, L. (1973) Role of cyclic 3'5'-adenosine monophosphate in the early transport changes induced by serum and insulin in quiescent fibroblasts. Proc. Natl. Acad. Sci. USA 70, 3609–3612.

Rudland, P.S., Seeley, M. and Siefert, W. (1974) Cyclic GMP and cyclic AMP levels in normal and transformed fibroblasts. Nature (London) 251, 417–419.

Sakiyama, H. and Robbins, P.W. (1973a) The effect of dibutyryl adenosine 3'5'-cyclic monophosphate on the synthesis of glycolipids by normal and transformed cells. Arch. Biochem. Biophys. 154, 407–414.

Sakiyama, H. and Robbins, P.W. (1973b) Glycolipid synthesis and tumorigenicity of clones isolated from the NIL 2 line of hamster embryo fibroblasts. Fed. Proc. 32, 86–90.

Sakiyama, H., Gross, S.K. and Robbins, P.W. (1972) Glycolipid synthesis in normal and virus-transformed hamster cell lines. Proc. Natl. Acad. Sci. USA 69, 872–876.

Sato, S., Szako, M., Kowalski, K. and Burke, G. (1972) Role of prostaglandin in thyrotropin action on thyroid. Endocrinology 90, 343–356.

Sattler, J., Wiegandt, H., Staerk, J., Kranz, T., Ronneberger, H.J., Schmidtberger, R. and Zilg, H. (1975) Studies of the subunit structure of choleragen. Eur. J. Biochem. 57, 309–316.

Sayhoun, N. and Cuatrecasas, P. (1975) Mechanism of activation of adenylate cyclase by cholera toxin. Proc. Natl. Acad. Sci. USA, 72, 3438–3442.

Schengrund, C.L., Lausch, R.N. and Rosenberg, A. (1973) Sialidase activity in transformed cells. J. Biol. Chem. 248, 4424–4428.

Sedlacek, H.H., Stärk, J., Seiler, F.R., Ziegler, W. and Wiegandt, H. (1976) Cholera toxin induced redistribution of sialoglycolipid receptor at the lymphocyte membrane. FEBS Lett. 61, 272–276.

Seifert, W.E. and Paul, D. (1972) Levels of cyclic AMP on sparse and dense cultures of growing and quiescent 3T3 cells. Nature (London) New Biol. 240, 281–283.

Sharp, G.W.G. (1973) Action of cholera toxin on fluid and electrocyte movement in the small intestine. Ann. Rev. Med. 24, 19–28.

Shaw, N. (1970) Bacterial glycolipids. Bacteriol. Rev. 34, 365–377.

Sheinin, R., Yogeeswaran, G. and Murray, R.K. (1974) Synthesis of surface glycoproteins and glycosphingolipids in db-cAMP treated normal and virus transformed cells. Exp. Cell Res. 89, 95–104.

Sheppard, J.R. and Prescott, D.M. (1972) Cyclic AMP levels in synchronised mammalian cells. Exp. Cell Res. 75, 293–296.

Shortridge, K.F. and Biddle, F. (1972) Rubella virus non-specific haemagglutination inhibitor: Evidence for the role of glycolipid bound to low density (β) lipoprotein. Clin. Chim. Acta 42, 285–294.

Shur, B.D. and Roth, S. (1975) Cell surface glycosyltransferases. Biochim. Biophys. Acta 415, 473–512.

Siddiqui, B. and Hakomori, S. (1970) Change of glycolipid pattern in Morris hepatomas 5123 and 7800. Cancer Res. 30, 2930–2937.

Siddiqui, B. and Hakomori, S. (1971) A revised structure for the Forssman glycolipid hapten. J. Biol. Chem. 246, 5766–5769.

Siefert, H. and Ühlenbruck, G. (1965) About gangliosides in human tumours. Naturwissenschaften 32, 190.

Simmons, J.L., Fishman, P.H., Freese, E. and Brady, R.O. (1975) Morphological alterations and ganglioside sialyltransferase activity induced by small fatty acids in HeLa cells. J. Cell Biol. 66, 414–424.

Simpson, L.L. and Rapport, M.M. (1971) Binding of botulinum toxin to membrane lipids, sphingolipids, steroids and fatty acids. J. Neurochem. 18, 1751–1759.

Simpson, D.L., Rosen, S.D. and Barondes, S.H. (1974) Discoidin, a developmentally regulated carbohydrate-binding protein from *Dictyostelium discoideum*: Purification and characterization. Biochemistry 13, 3487–3493.

Simpson, D.L., Rosen, S.D. and Barondes, S.H. (1975) Pallidin. Purification and characterization of a carbohydrate-binding protein from *Polysphondylium pallidum* implicated in intercellular adhesion. Biochim. Biophys. Acta 412, 109–114.

Skelly, J., Gacto, M., Steiner, M.R. and Steiner, S. (1976) Preliminary evidence for novel fucose containing lipids in normal and murine sarcoma virus transformed rat cells. Biochem. Biophys. Res. Commun. 68, 442–449.

Slomiany, B.L. and Horowitz, M.I. (1970) Glycolipids of bovine serum. Biochim. Biophys. Acta 218, 278–287.

Smets, L.A. (1972) Contact inhibition of transformed cells incompletely restored by dibutryl cyclic AMP. Nature (London) New Biol. 239, 123–124.

Sneath, P.H.A. and Lederberg, J. (1961) Inhibition by periodate of mating in *Escherichia coli* K-12. Proc. Natl. Acad. Sci. USA 77, 86–90.

Steiner, S. and Melnick, J.L. (1974) Altered fucolipid patterns in cultured tumour cancer cells. Nature (London) 251, 717–718.

Steiner, S., Brennan, P.J. and Melnick, J.L. (1973) Fucosylglycolipid metabolism in oncornavirus transformed cell lines. Nature (London) 245, 19–21.

Steiner, S.M., Melnick, J.L., Kit, S. and Somers, K.D. (1974) Fucosylglycolipids in cells transformed by a temperature sensitive mutant of murine sarcoma virus. Nature (London) 248, 682–684.

Stellner, K., Hakomori, S. and Warner, G. (1973) Enzymic conversion of "H_1-glycolipid" to A or B-glycolipid and deficiency of these enzyme activities in adenocarcinoma. Biochem. Biophys. Res. Commun. 55, 439–445.

Stoffel, W. (1971) Sphingolipids. Ann. Rev. Biochem. 40, 57–82.

Stoffyn, P., Stoffyn, A. and Hauser, G. (1973) Structure of trihexosylceramide biosynthesised in vivo. J. Biol. Chem. 248, 1920–1923.

Strom, T.B., Deisseroth, A., Morganroth, J., Carpenter, C.B. and Merrill, J.P. (1972) Alteration of the cytotoxic action of sensitized lymphocytes by cholinergic agents and activators of adenylate cyclase. Proc. Natl. Acad. Sci. USA 69, 2995–2999.

Sulzer, B.M. and Craig, J.P. (1973) Cholera toxin inhibits macromolecular synthesis in mouse spleen cells. Nature (London) New Biol. 244, 178–180.

Sundsmo, J.S. and Hakomori, S. (1976) Lacto-N-neotetraosylceramide ("Paragloboside") as a possible tumour associated surface antigen of hamster NIL-Py tumor. Biochem. Biophys. Res. Commun. 68, 799–806.

Surolia, A., Bachhawat and Podder, S.K. (1975) Interaction between lectin from *Ricinus communis* and liposomes containing gangliosides. Nature (London) 257, 802–804.

Svennerholm, L. (1970) Ganglioside metabolism. In: M. Florkin and E.H. Stotz (Eds.), Comprehensive Biochemistry, Vol. 18, Elsevier, Amsterdam, pp. 201–225.

Sweeley, C.C. and Dawson, G. (1969) Lipids of the erythrocyte. In: G.A. Jamieson and T.J. Greenwalt (Eds.), Red Cell Membrane Structure and Function. Lippincott, Philadelphia, pp. 172–227.

Tager, H.S. and Steiner, D.F. (1974) Peptide hormones. Ann. Rev. Biochem. 43, 509–538.

Takeuchi, I. (1963) Immunochemical and immunohistochemical studies on the development of the cellular slime mold *Dictyostelium discoideum* Dev. Biol. 8, 1–26.

Tal, C. (1965) The nature of the cell membrane receptor for the agglutination factor present in the serum of tumor patients and pregnant women. Proc. Natl. Acad. Sci. USA 54, 1318–1321.

Tal, C. and Halperin, M. (1970) Presence of serologically distinct protein in serum of cancer patients and pregnant women. An attempt to develop a diagnostic cancer test. Israel J. Med Sci. 6, 708–716.

Tal, C. Dishon, T. and Gross, J. (1964) The agglutination of tumour cells in vitro by sera from tumour patients and pregnant women. Brit. J. Cancer 18, 111–119.

Tan, Y.H. (1976) Chromosome 21 and the cell growth inhibitory effect of human interferon preparations. Nature (London) 260, 141–143.

Tao, R.V.P. and Sweeley, C.C. (1970) Occurrence of hematoside in human plasma. Biochim. Biophys. Acta 218, 372–375.

Tate, R.L., Holmes, J.M., Kohn, L.D. and Winand, R.J. (1975) Characteristics of a solubilised thyrotropin receptor from bovine thyroid plasma membrane. J. Biol. Chem. 250, 6527–6533.

Teichberg, V.I., Silman, I., Beitsch, D.D. and Resheff, G. (1975) A β-D-galactoside binding protein from electric organ tissue of *Electrophorus electricus*. Proc. Natl. Acad. Sci. USA 72, 1383–1387.

Tomich, J.M., Mather, I.H. and Keenan, T.W. (1976) Proteins mask gangliosides in milk fat globule and erythrocyte membranes. Biochim. Biophys. Acta 433, 357–364.

Trinkaus, J.P. (1969) Cells into Organs. Prentice Hall, New Jersey.

Trinkaus, J.P. (1976) Mechanisms of metazoan cell movements. In: G. Poste and G.L. Nicolson (Eds.), The Cell Surface in Animal Embryogenesis and Development, Cell Surface Reviews, Vol. 1, North-Holland, Amsterdam, pp. 225–329.

Tyler, A. (1946) An autoantibody concept of cell structure, growth and differentiation. Growth 10, 7–19.

Ulrich, F. (1974) A dialyzable protein synthesis inhibitor released by mammalian cells in vitro. Biochem. Biophys. Res. Commun. 60, 1453–1459.

Unanue, E.R. and Karnovsky, M.J. (1974) Ligand-induced movement of lymphocyte membrane macromolecules, V. Capping, cell movement, and microtubular function in normal and lectin-treated lymphocytes. J. Exp. Med. 140, 1207–1220.

Unkeless, J.C., Tobia, A., Ossowski, L., Quigley, J.B., Rifkin, D.B. and Reich, E. (1973) An enzymatic function associated with transformation of fibroblasts by oncogenic viruses, I. Chick embryo fibroblast cultures transformed by avian RNA tumour viruses. J. Exp. Med. 137, 85–111.

Vance, D.E. and Sweeley, C.C. (1967) Quantitative determination of neutral glycosyl ceramides in human blood. J. Lipid. Res. 8, 621–630.

Van Heyningen, S. (1974a) Cholera toxin: Interaction of subunits with ganglioside GM_1. Science 183, 656–657.
Van Heyningen, W.E. (1974b) Gangliosides as membrane receptors for tetanus toxin, cholera toxin and serotonin. Nature (London) 415–417.
Van Heyningen, W.E., Carpenter, C.C.J., Pierce, N.F. and Greenough, W.B. (1971) Deactivation of cholera toxin by ganglioside. J. Infect. Dis. 124, 415–418.
Vicker, M.G. (1976) BHK21 fibroblast aggregation inhibited by glycopeptides from the cell surface. J. Cell Sci. 21, 161–173.
Waddell, A.W., Robson, R.T., Edwards, J.G. (1974) Colchicine and vinblastine inhibit fibroblast aggregation. Nature (London) 248, 239–241.
Warren, L., Critchley, D.R. and Macpherson, I. (1972) Surface glycoproteins and glycolipids of chicken embryo cells transformed by a temperature-sensitive mutant of Rous sarcoma virus. Nature (London) 235, 275–278.
Webb, G.C. and Roth, S. (1974) Cell contact dependence of surface galactosyltransferase activity as a function of cell cycle. J. Cell Biol. 63, 796–805.
Weber, M.J. (1973) Hexose transport in normal and Rous sarcoma virus-transformed cells. J. Biol. Chem. 248, 2978–2983.
Weber, J.M. and Stewart, R.B. (1975) Cyclic AMP potentiation of interferon antiviral activity and effects of interferon on cyclic AMP levels. J. Gen. Virol. 28, 363–372.
Weeks, G. (1975) Studies of the cell surface of *Dictyostelium discoideum* during differentiation. The binding of ^{125}I-concanavalin A to the cell surface. J. Biol. Chem. 250, 6706–6710.
Weinsten, D.B., Marsh, J.B., Glick, M.C. and Warren, L. (1970) Membranes of animal cells, VI. The glycolipids of the L cell and its surface membrane. J. Biol. Chem. 245, 3928–3937.
Weise, L. and Hayward, P.C. (1972) On sexual agglutination and mating-type substances in isogamous dioecious *Chlamydomonads*, III. The sensitivity of sex cell contact to various enzymes. Am. J. Bot. 59, 530–536.
Weiss, P. (1947) The problems of specificity in growth and development. Yale J. Biol. Med. 19, 235–278.
Whatley, R., Ng, S.K.C., Rogers, J., McMurray, W.C. and Sanwal, B.D. (1976) Developmental changes in gangliosides during myogenesis of a rat myoblast cell line and its drug resistant variants. Biochem. Biophys. Res. Commun. 70, 180–185.
Wiblin, C.N. and Macpherson, I. (1973) Reversion in hybrids between SV40 transformed hamster and mouse cells. Int. J. Cancer 12, 148–161.
Wiegandt, H. (1967) The subcellular localization of gangliosides in the brain. J. Neurochem. 14, 671–674.
Wilhelms, O.H., Lüderlitz, O., Westphal, O. and Gerisch, G. (1974) Glycosphingolipids and glycoproteins in the wild type and in a non-aggregating mutant of *Dictyostelium discoideum*. Eur. J. Biochem. 78, 89–101.
Willingham, M.C. (1976) Cyclic AMP and cell behaviour in cultured cells. Int. Rev. Cytol. 44, 319–363.
Wintzer, G. and Uhlenbruck, G. (1967) Topischem Anordnung von Gangliosiden in der Erythrozytenmembran. Z. Immunitätsforsch. Allerg. Klin. Immunol. 133, 60–67.
Wolf, B.A. and Robbins, P.W. (1974) Cell mitotic cycle synthesis of NIL hamster glycolipids including the Forssman antigen. J. Cell Biol. 61, 676–687.
Wolf, J., Winand, R.J. and Kohn, L.D. (1974) The contribution of subunits of thyroid stimulating hormone to the binding and biological activity of thyrotropin. Proc. Natl. Acad. Sci. USA 71, 3460–3464.
Wolff, J., Temple, R. and Cook, G.H. (1973) Stimulation of steroid secretion in adrenal tumor cells by choleragen. Proc. Natl. Acad. Sci. USA 70, 2741–2744.
Wyke, J.A. (1973) The selective isolation of temperature-sensitive mutants of Rous sarcoma virus. Virology 52, 587–590.
Yamada, K.M., Yamada, S.S. and Pastan, I. (1975) The major cell surface glycoprotein of chick embryo fibroblasts is an agglutinin. Proc. Natl. Acad. Sci. USA 72, 3158–3162.
Yamada, K.M., Yamada, S.S. and Pastan, I. (1976) Cell surface protein partially restores mor-

phology, adhesiveness, and contact inhibition of movement to transformed fibroblasts. Proc. Natl. Acad. Sci. USA 73, 1217–1221.

Yogeeswaran, G. and Hakomori, S. (1975) Cell contact-dependent ganglioside changes in mouse 3T3 fibroblasts and a suppressed sialidase activity on cell contact. Biochemistry 14, 2151–2156.

Yogeeswaran, G., Wherrett, J.R., Chatterjee, S. and Murray, R.K. (1970) Partial characterization and demonstration of ^{14}C-glucosamine incorporation. J. Biol. Chem. 245, 6718–6725.

Yogeeswaran, G., Sheinin, R., Wherett, J.R. and Murray, R.K. (1972) Studies on the glycosphingolipids of normal and virally transformed 3T3 mouse fibroblasts. J. Biol. Chem. 247, 5146–5158.

Yogeeswaran, G., Laine, R. and Hakomori, S. (1974a) Mechanism of cell contact-dependent glycolipid synthesis: further studies with glycolipid glass complex. Biochem. Biophys. Res. Commun. 59, 591–599.

Yogeeswaran, G., Murray, R.K. and Wright, J.A. (1974b) Glycosphingolipids of wild-type and mutant lectin-resistant Chinese hamster ovarian cells. Biochem. Biophys. Res. Commun. 56, 1010–1016.

Yu, R.K. and Ledeen, R.W. (1972) Gangliosides of human, bovine, and rabbit plasma. J. Lipid Res. 13, 680–686.

Zalin, R.J. and Montague, W. (1974) Changes in adenylate cyclase cyclic AMP and protein kinase levels in chick myoblasts and their relationship to differentiation. Cell 2, 103–108.

Zenser, T.V. and Metzger, J.F. (1974) Comparison of the action of *Escherichia coli* enterotoxin on the thymocyte adenylate cyclase–cyclic adenosine monophosphate system to that of cholera toxin and prostaglandin EI. Infect. Immun. 10, 503–509.

Zieve, P.D., Pierce, N.F. and Greenough, W.B. (1971) Stimulation of glycogenolysis by purified cholera exotoxin in disrupted cells. Johns Hopkins Med. J. 129, 299–303.

Zimmerman, D.H. and Kern, M. (1976) Differentiation of lymphoid cells: A selective anti-mitogenic component of normal serum which inhibits the induction of immunoglobulin production. J. Biol. Chem. 251, 2469–2474.

Cell surface proteins: changes during cell growth and malignant transformation

Carl G. GAHMBERG

1. Introduction

The purpose of this review is to provide an introductory outline to the current methods used in characterizing cell membrane components and to describe recent major findings concerning plasma membrane organization in normal and malignant cells. Special emphasis will be given to the glycoproteins of the cell surface, especially their structure and organization within the membrane, their role in cellular function and the possible contribution of alterations in these components in determining the altered surface properties and abnormal social behavior displayed by malignant cells.

During the past several years it has become increasingly apparent that the cell surface is involved in the control of cell growth, but the actual mechanism(s) remain unclear. Numerous studies have shown that chemical changes occur in the plasma membrane during normal cell growth and also after malignant transformation. Differences between normal and malignant cells have been found for most components of the plasma membrane that have been studied but few changes are consistent for the malignant phenotype.

Recently, interest has focused on the cell surface glycoproteins and glycolipids of normal and malignant cells. This interest is due partially to the fact that changes in the composition of these components are known to accompany malignant transformation, and also to an increasing awareness of their functional role(s) in determining the properties of the cell surface. For example, it is known that carbohydrate sequences may have strong immunologic specificity and that relatively small chemical changes may result in dramatic differences in antigenic specificities (Hakomori and Kobata, 1975). Cell surface glycoproteins are also involved in a variety of surface-mediated processes some of which may be intimately connected with the control of cell growth. These include: cell adhesion and recognition (Roseman, 1970; Roth et al., 1971; Roth and White, 1972; Merrell et al., 1975); cell homing patterns (Gesner and Ginsburg, 1964; Jancik and Schauer, 1974; Durocher et al., 1975) and the transport of nutrients and other compounds into the cell (Cabantchik and Rothstein, 1974a,b; Holley and Kiernan, 1974; Kyte, 1974; Ho and Guidotti, 1975; Gahmberg et al., 1976c).

Recent studies of the human erythrocyte membrane have shown that the major proteins exposed on the external surface of the cell are glycoproteins (Thomas and Winzler, 1969; Bretscher, 1971a; Marchesi et al., 1972; Tanner and Boxer, 1972). Some, and possibly all, of these proteins penetrate through the membrane to the cytoplasmic side (Bretscher, 1971a; Morrison et al., 1974). Interestingly, the carbohydrate residues found on plasma membrane glycoproteins and glycosphingolipids are exposed only to the external milieu and the plasma membrane has no carbohydrate on the inner half of the membrane (Gahmberg and Hakomori, 1973a; Nicolson and Singer, 1974; Steck and Dawson, 1974; Gahmberg, 1976). On the other hand, it is possible that there are no cell surface proteins which are not glycoproteins (Gahmberg, 1976). If this is true also in nucleated cells, as there is good reason to believe, it means that all of the proteins exposed on the cell surface are glycoproteins with their carbohydrate exclusively on the outer surface. The external carbohydrates could thus function as a hydrophilic "lock" to keep the glycoproteins properly oriented. The complex carbohydrate structures may also discharge additional functions on the cell surface and the elucidation of these activities represents a major focal point for current research in cell biology.

2. The human erythrocyte membrane as a model for the structure of mammalian cell surface membranes

Most of our current knowledge of membrane structure has been obtained from observations on the human erythrocyte membrane. The plasma membrane of nucleated mammalian cells is much more difficult to study because of isolation problems and difficulties in obtaining enough cells for chemical studies. Therefore I will first focus on the structure of the erythrocyte membrane to give a general description of membrane structure.

2.1. Lipids

About half of the mass of the erythrocyte membrane is protein and the other half lipid (Sweeley and Dawson, 1969). A few per cent is carbohydrate, which is found bound both to lipids, as glycosphingolipids, and to proteins as glycoproteins. The most important lipids quantitatively are the different phospholipids and cholesterol. The phospholipids include phosphatidylcholine(lecithin), phosphatidylethanolamine, phosphatidylserine and sphingomyelin (Sweeley and Dawson, 1969). Phosphatidylinositol is found in very low concentrations in the erythrocyte membrane and is usually not a component of plasma membranes but is enriched in the endoplasmic reticulum (Renkonen et al., 1972). Cholesterol is a characteristic constituent of plasma membranes and is found here in higher concentrations than in other membranes (Renkonen et al., 1972). Also glycolipids are enriched in the cell surface membrane (Dod and Gray, 1968, Klenk and Choppin, 1970, Renkonen et al., 1970). Low angle X-ray diffraction studies have shown that at least most of the membrane is composed of a lipid bilayer (Oseroff et al., 1973). In this bilayer the polar lipid head groups are oriented both to the external milieu and to the cytoplasm, while the hydrophobic hydrocarbon chains of the phospholipids and glycolipids, as well as the backbone of the cholesterol molecule, are oriented to the interior of the membrane.

A characteristic feature of the lipids is their ability to diffuse laterally in the membrane with appreciable speed, but there appears to be very little exchange of lipid between the two halves of the lipid bilayer (Kornberg and McConnell, 1971a; Kornberg and McConnell, 1971b). Another important feature of the lipids is their assymmetrical distribution between the two halves of bilayer (Bretscher, 1972a,b; Verkleij et al., 1973). Some phospholipids like phosphatidylcholine and sphingomyelin are located mainly in the outer leaflet of the human erythrocyte membrane, whereas phosphatidylethanolamine and phosphatidylserine are enriched in the inner half. Glycolipids probably exist exclusively in the outer part of the bilayer (Gahmberg and Hakomori, 1973a; Steck and Dawson, 1974), but only a relatively small proportion are actually accessible at the surface (Gahmberg and Hakomori, 1975a). Whether cholesterol is oriented assymmetrically is not clear, but recent results indicate a preferential localization in the external half of the bilayer (Fisher, 1976). How and why this lipid assymmetry is generated is not known.

The physiological functions of the plasma membrane lipids are not well understood. The main function of the phospholipids is of course in the formation of the bilayer and they thus form the basis for this structure. Some phospholipids are known to specifically activate membrane enzymes in vitro (Emmelot and Bos, 1968; Widnell and Unkeless, 1968; also see chapter in this volume by Kimelberg, p. 205) and it is possible that these lipids have other important functions that we are not yet aware of. Cholesterol is supposed to have a stabilizing role in the membrane by intercalating between the phospholipid hydrocarbon chains (Oseroff et al., 1973), but more specific functions are unknown. The physiological functions of the glycolipids are not known, but they are known to play an important pathophysiological role as receptors for viruses (Haywood, 1974), interferon (Besançon et al., 1976) and bacterial toxins, like cholera (Cuatrecasas, 1973; Van Heyningen, 1974; Holmgren et al., 1975) and tetanus (Van Heyningen, 1974).

2.2. Proteins

The frame-work of the plasma membrane is formed by the lipid bilayer but the proteins embedded in it are responsible for most of the specific functions of the plasma membrane. As first pointed out clearly by Singer and Nicolson in their fluid mosaic model of membrane structure (Singer and Nicolson, 1972) (Fig. 1) there are two major different groups of membrane proteins, the so called integral (intrinsic) and peripheral (extrinsic) proteins. The integral proteins are strongly associated with the membrane and are not released by hypotonic or

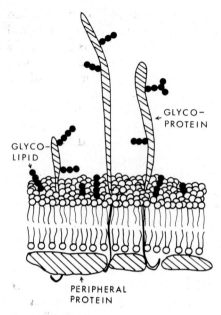

Fig. 1. The fluid mosaic model of membrane structure.
The lipid bilayer is formed by the phospholipids (♠) with their polar head groups pointing to the outside and the cytoplasm and their hydrocarbon chains within the hydrophobic core of the membrane. Glycolipids (♠) are only present in the outer part of the bilayer with their saccharide chains oriented externally between the membrane glycoproteins. The glycoproteins, which are integral, membrane-penetrating proteins, also have their hydrophilic carbohydrate chains oriented only to the outside. In addition, they interact with peripheral membrane proteins on the inside of the membrane.

hypertonic media, chelating agents or extreme pH. The peripheral proteins can be removed from the membrane by changing pH, salt concentration or by chelating agents and once in solution they resemble "ordinary" soluble proteins, but under certain conditions they aggregate to form net-works (Steck, 1974). Detergents, chaotropic agents and lipid solvents are required to solubilize integral proteins. Once in solution they commonly need the presence of detergents or other membrane solubilizers, but sometimes they remain soluble, as is the case with the major erythrocyte glycoprotein, PAS 1 (for nomenclature see Fairbanks et al., 1971). However, such proteins then probably exist as macromolecular aggregates with their hydrophobic segments protected from water.

The integral proteins are mostly, if not always, glycoproteins and they penetrate through the membrane. The most extensively studied protein of this type is the major erythrocyte glycoprotein, variously referred to as PAS 1, MN protein or glycophorin (Marchesi et al., 1972). The primary structure of this protein has been determined (Tomita and Marchesi, 1975) (Fig. 2). It is composed of three structurally different domains (Fig. 3). On the external side of the lipid bilayer is a hydrophilic portion containing the carbohydrate. This

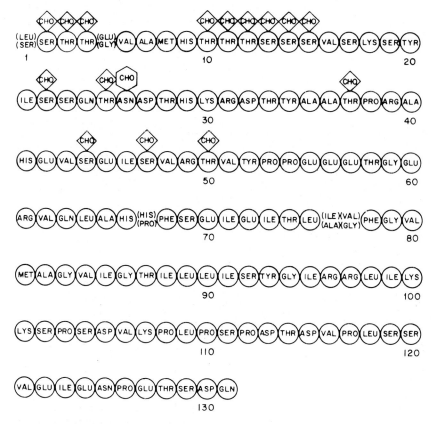

Fig. 2. The primary structure of glycophorin.
The polypeptide chain is composed of 131 amino acids. All carbohydrate is located between the 2nd and 50th amino acid and most of it is clustered in four regions in this part of the polypeptide chain. There are no charged amino acids between residues 68 and 95. The COOH-terminal region contains numerous charged amino acids. There is no cysteine in the molecule.
(Reproduced with permission from Tomita and Marchesi, 1975, and the National Academy of Sciences, USA.)

domain is responsible for the influenza virus receptor and MN blood group activities of the molecule. Protease treatment of intact cells cleaves the external part of the protein. The fragment of the protein which penetrates the membrane is enriched in hydrophobic amino acids and no charged amino acid exists for a sequence of 32 residues. The COOH-terminal cytoplasmic end is also hydrophilic. The PAS 1 protein is especially rich in carbohydrate, containing about 60%. There is also a very high content of sialic acid (Tomita and Marchesi, 1975). The oligosaccharide side chains are linked to 15 serine and threonine residues, each containing four monosaccharide residues, and, in addition, there is one complex asparagine-linked oligosaccharide chain. The carbohydrate gives it a rather hydrophilic character once freed from the

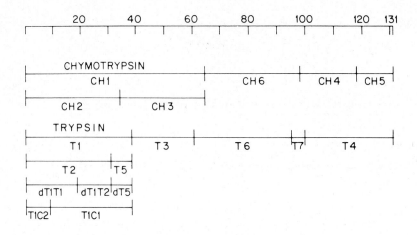

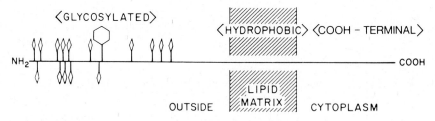

Fig. 3. The different domains of the glycophorin molecule.
The cleavage points of chymotrypsin and trypsin are shown. The lower part shows the three different domains of the molecule. Outside the membrane is the hydrophilic carbohydrate-containing region, within the membrane the hydrophobic region and inside the membrane the COOH-terminal end which probably is in the cytoplasm.
(Reproduced with permission from Tomita and Marchesi, 1975, and the National Academy of Sciences, USA.)

membrane. This property has greatly facilitated its isolation and further characterization. It probably exists as a dimer in the membrane (Marton and Garvin, 1973, Slutzky and Ji, 1974), as is the case for the most predominant erythrocyte membrane glycoprotein, band 3 (Steck, 1972). Band 3 protein also penetrates the membrane and its carbohydrate is located externally (Tanner and Boxer, 1972; Steck and Dawson, 1974; Gahmberg, 1976). In contrast to the PAS 1 glycoprotein with its numerous relatively simple oligosaccharide chains, band 3 contains two types of oligosaccharide chains, one very complex, which probably exists only as one copy per polypeptide chain, and one smaller oligosaccharide (Gahmberg et al., 1976b). When the membrane is cleaved along the internal region of the lipid bilayer by freeze-etching, the fracture faces contain many intramembranous particles (Pinto da Silva and Branton, 1970; Pinto da Silva et al., 1971; Tillack et al., 1972; Pinto da Silva and Nicolson, 1974). There is reason to believe that these particles are formed by integral membrane proteins, because lipid vesicles lacking protein do not show such particles. According to the present view at least both PAS 1 and band 3 polypeptides together form these particles. Trypsin treatment of intact cells, which cleaves only the sialoglycoproteins, aggregates the intramembranous particles (Marchesi et al., 1972), and interaction with carbohydrate-binding ligands like lectins or anti-blood group antibodies also aggregates these structures (Tillack et al., 1972; Pinto da Silva and Nicolson, 1974). This indicates that the integral proteins can diffuse within the plane of the membrane.

Another important feature of the erythrocyte integral proteins is their connection with peripheral proteins on the cytoplasmic side of the membrane. This has been shown by using antibodies against spectrin, a high molecular weight polypeptide present on the cytoplasmic side of the membrane. Such antibodies redistribute lectin binding sites on the outer surface of the membrane (Nicolson and Painter, 1973). Spectrin itself is a peripheral protein and does not penetrate the membrane. Also perturbation of glycoproteins on the outer surface can result in secondary changes at the inner surface (Ji and Nicolson, 1974). This may be important for transmission of signals through the membrane, or so called trans-membrane coupling.

The peripheral proteins, which seem to be located exclusively on the cytoplasmic side of the bilayer, are fundamentally different from the integral proteins. Although associated with the membrane, they are bound more loosely than the integral proteins. The actual mechanism of binding to the membrane is not clear but ionic bonds are known to be partially involved. Some of these proteins may show specific binding to the cytoplasmic side of the bilayer (Kant and Steck, 1973), which could mean that specific binding sites are involved. The predominant polypeptide on the inner side is spectrin, which on SDS-gel electrophoresis forms two closely spaced bands, with molecular weights of 200 000 and 220 000 daltons (Trayer et al., 1971). In addition band 5 is associated with the spectrin polypeptides (Steck, 1974).

2.3. Carbohydrates

The oligosaccharide side chains in glycoproteins are formed by relatively few monosaccharides. Sialic acid, N-acetyl D-glucosamine, N-acetyl D-galactosamine, D-galactose, D-mannose, L-fucose and sometimes D-glucose are found. Glycosphingolipids from mammalian cells lack mannose, but often contain glucose as the most internal sugar linked to the ceramide. The oligosaccharides are formed by glycosidic bonds between the reducing end of a monosaccharide and a hydroxyl group of another in α or β configuration. This gives a large number of possible linkages and small changes may result in antigenically different structures. Most internal oligosaccharides are linked either to asparagine (and form an alkali-stable linkage) or to serine or threonine, (which results in an alkali labile bond). The oligosaccharides are often very complex and there is only limited information available on the structure of plasma membrane glycoprotein oligosaccharides. In some cases they seem to resemble the oligosaccharides of serum glycoproteins (Baenziger

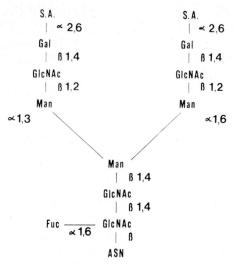

Fig. 4. Structure of a complex asparagine-linked glycoprotein oligosaccharide. The structure is that of immunoglobulin E complex oligosaccharide (Baenziger and Kornfeld, 1974). Recent results indicate that some membrane glycoproteins contain similar saccharide chains.

and Kornfeld, 1974) (Fig. 4). A characteristic feature is microheterogeneity within the oligosaccharide chains, which could be due to a limited capacity of the sugar transferases. The carbohydrate structure of the PAS 1 glycoprotein has been determined (Thomas and Winzler, 1969) but the glycopeptides of other erythrocyte glycoproteins are still poorly known. The major polypeptide of erythrocyte membranes, band 3, has a very complex oligosaccharide chain and this is even bigger in the erythrocyte variant En(a-) which lacks the PAS 1 glycoprotein (Gahmberg et al., 1976b).

3. Methods for studying cell surface proteins

One of the most important goals in membrane biochemistry is to isolate specific membrane proteins in a pure and active form so that experiments can be done to study their structure and function. In many instances this requires isolation of the plasma membrane.

3.1. Isolation of plasma membranes

The erythrocyte membrane can simply be isolated after hypotonic lysis by centrifugation, but the isolation of plasma membranes from nucleated cells presents a much more difficult problem. The main obstacle here is the presence of many different membranes which cross-contaminate the various fractions. Although it is still necessary to isolate the plasma membrane to answer certain questions, modern surface labeling and immunological methods have radically altered the situation. Cell surface molecules can now be analyzed directly and identified in intact cells and can often be isolated without first isolating the plasma membrane. This saves much time and effort. In addition, the yield of plasma membranes is often low, many proteins may be lost or modified during the isolation procedure, and the final preparation is always more or less contaminated with other cell organelles. Surface labeling techniques circumvent these difficulties.

It must also be remembered that many solid tissues like liver, are composed of different types of cells, and thus the isolated membrane fractions may contain plasma membranes from various cells. Also, different regions of the membrane within a single cell may differ and this again may result in several plasma membrane fractions (Evans, 1970).

Plasma membrane isolation methods have been reviewed recently by DePierre and Karnovsky (1973). The isolation of plasma membranes from solid tissues or from isolated cells often needs a different strategy. When cells from solid tissues are homogenized gently in hypotonic buffers, the surface membrane tends to form large sheets. On differential centrifugation the membrane sediments in the crude nuclear fraction at low $g \times min$ values. The surface membrane can then be separated from the nuclei by density gradient centrifugation. However many types of cells, especially those that mostly have been widely used in model studies of cancer, are obtained by cell culture or directly as suspension cells from animals like ascitic and leukemic cells. When such cells are homogenized the surface membrane often forms small vesicles, which on differential centrifugation are recovered with the microsomal fraction (DePierre and Karnovsky, 1973). Warren and co-workers have circumvented this problem by more or less prefixing the surface membrane in intact cells by fluorescein mercuric acetate or Zn^{2+}. The membrane is stiffened after such treatments and is not easily fragmented and larger membrane sheets are formed which can be isolated relatively easily (Warren et al., 1966). A major drawback, however, is that many proteins are denatured and the modified membrane loses important biological characteristics.

Wallach and co-workers (Wallach and Kamat, 1966) first introduced a method to homogenize cells by subjecting them to high pressure under nitrogen in a pressure chamber and then rapidly releasing the cells to atmospheric pressure (Hunter and Commerford, 1961). This treatment fragments the surface membrane to small vesicles and the homogenization step can readily be monitored. A significant draw-back in this method is the rather laborious isolation procedure, but this is offset by the very reproducible results. This method has been used successfully with Ehrlich ascites cells and hamster fibroblasts (Gahmberg and Simons, 1970). The two-phase polymer system introduced by Albertsson for separating proteins (Albertsson, 1971), has also proved to be a very convenient and rapid method for isolating surface membranes (Brunette and Till, 1971; Laine and Hakomori, 1973), especially when membranes are isolated from many different cell samples and the amount of cells is limited.

After isolation of the membrane it is important to determine the purity of the preparation and this problem is often neglected. Both positive and negative markers should be used. There are a number of markers for the plasma membrane. Enzymes such as $Na^+ - K^+$ activated ATPase (Gahmberg and Simons, 1970), 5'-nucleotidase (Widnell and Unkeless, 1968) and adenyl cyclase (Marinetti et al., 1969) are used widely and appear to be specific for this structure. Also some enzymes located on the external surface of the plasma membrane may be used as plasma membrane markers (DePierre and Karnovsky, 1974), but [^{125}I]-labelled lectins (Chang et al., 1975), antiplasma membrane antibodies (Gahmberg and Simons, 1970), radioactive fucose (Gahmberg, 1971) or glycosphingolipids (Klenk and Choppin, 1970; Renkonen et al., 1970) are, in general, more convenient external markers. Electron microscopy should also be used to characterize preparations, but the different types of cellular membranes are difficult to differentiate on morphological criteria alone and this method gives only a qualitative estimation of the compositions of the fractions. In addition, markers for intracellular structures should be assayed: e.g. DNA for nuclei; acid hydrolases for lysosomes; uricase for peroxisomes; RNA for ribosomes; succinate dehydrogenase for mitochondria; and glycolytic enzymes for the soluble fraction of the cell. In general it is more difficult to find good markers for intracellular organelles than for the plasma membrane. The adsorption of soluble serum proteins to the cell surface may also be a problem. It should be emphasized that the simultaneous use of several markers is required to give sufficiently reliable results. Also because cells vary enormously, plasma membrane markers used for one cell type may not be suitable in some other cell. In addition, it is often difficult to compare plasma membrane preparations from two different cells. When one compares plasma membranes from normal cells and their malignant counterparts, the same isolation procedure may give preparations with different degrees of contamination. The exact level may be difficult to determine because the markers used to identify cell fractions in normal cells may also have changed following malignant transformation.

3.2. Isolation of membrane proteins

Classical methods for protein isolation which were developed for soluble proteins have proved unsatisfactory for the isolation of plasma membrane proteins. This is especially true for the isolation of integral membrane proteins which must be solubilized by special treatments.

The hydrophobic nature of membrane proteins often results in aggregation in aqueous media. To avoid this, different types of membrane-solubilizing reagents have been used. The most important are the detergents. The solubilization of membranes by these compounds has been reviewed recently (Helenius and Simons, 1975). Detergents have an amphipathic nature and form micelles in aqueous media above the critical micellar concentration (CMC). For membrane solubilization a concentration above the CMC is needed (Helenius and Simons, 1975). Detergents are often classified as nonionic and ionic. Among the nonionic detergents Triton X-100 and Nonidet P40 are widely used. These detergents usually do not denature membrane proteins and the biological activities are preserved (Helenius and Simons, 1975). This is also valuable for subsequent purification of the isolated protein by affinity chromatography. These mild detergents can also be used successfully to selectively extract integral membrane proteins from peripheral membrane proteins (Steck, 1974). Nonionic detergents evidently bind to hydrophobic areas in membrane proteins and thus substitute for lipids. In this way the hydrophobic areas are protected from the aqueous environment and the proteins stay in solution.

Sodium deoxycholate, though an anionic detergent, resembles the nonionic detergents in many respects. It solubilizes membranes more effectively than nonionic detergents, but many biological activities are preserved. In addition this detergent can be dialyzed more easily than the nonionic detergents.

Sodium dodecyl sulfate (SDS) is the most widely used detergent in membrane research. It strongly denatures proteins and, in the presence of a reducing agent, the protein usually dissociates into individual polypeptide chains. SDS binds more extensively than nonionic detergents to various parts of the membrane proteins (Reynolds and Tanford, 1970; Helenius and Simons, 1972) and these form rodlike particles (Reynolds and Tanford, 1970). An interesting exception is provided by various proteases, which often remain active in SDS if the sample is not boiled. This is important to bear in mind when membrane proteins are isolated.

Membranes can also be solubilized with varying efficiency by many other types of solvents. Phenol and pyridine extraction has been used successfully to isolate the sialoglycoproteins from erythrocyte membranes (Blumenfeld and Zvilichovsky, 1972; Howe et al., 1972), but these proteins are exceptional because of their high content of carbohydrate, which makes them hydrophilic. The chaotropic agent lithium diiodosalicylate has been used by Marchesi and co-workers to isolate the major sialoglycoprotein of erythrocyte membranes (Marchesi, 1972), by Hunt et al. (1975) for the isolation of a concanavalin A receptor from mouse fibroblasts, and by Merrell et al. (1975) to isolate a retinal recognition protein. Butanol has been used to isolate erythrocyte membrane proteins (Maddy, 1966) and hydrophobic bacterial proteins (Sandermann and Strominger, 1972). In addition buffers containing high salt concentrations have been used with appreciable success in certain cases. Guanidine hydrochloride and urea tend to keep membrane proteins in solution. Recently the high molecular weight surface glycoprotein of fibroblasts was isolated by extracting cell monolayers with urea (Yamada and Weston, 1974), leaving the cell intact. Here it is possible that proteases were activated during the urea treatment and the isolated protein represents a proteolytic fragment.

If it is possible, it is advantageous to avoid SDS because its use usually results in loss of biological activity. Therefore wherever possible the membranes are first extracted with nonionic detergents or salts. However, the dissociation of membrane proteins into individual polypeptides is often incomplete using nonionic detergents or sodium deoxycholate. Mixed micelles may be formed containing different proteins and lipids, and the resolution in subsequent steps may be poor.

Affinity columns containing Sepharose bound to lectins like *Lens culinaris* or concanavalin A can be used to separate glycoproteins from non-glycoproteins (Allan and Crumpton, 1972; Adair and Kornfeld, 1974). Sometimes the glycoproteins bind very strongly to the lectin columns and elution using the inhibitory monosaccharide gives poor yields. Lentil lectin (*Lens culinaris*) columns have a lower affinity for glycoproteins than concanavalin A and the yields are higher. Indirect precipitation after use of specific antibodies has worked very well especially in the isolation of *H-2* and *HLA* antigens (Nathenson and Cullen, 1974), but for the isolation of most membrane proteins the immunological approach, though attractive, is often limited by the lack of specific antibodies.

Gel filtration in the presence of SDS often gives a good resolution (Ho and Guidotti, 1975) and similar results may be obtained by using preparative polyacrylamide gel electrophoresis. Hydroxyl apatite column chromatography in the presence of SDS has also proved useful because here the separation is not based on molecular weight but on adsorption to the column. This system has been used for isolation of viral membrane proteins (Garoff et al., 1974; Moss and Rosenblum, 1974) and band 3 from erythrocyte membranes (Gahmberg et al., 1976b).

Another popular approach for isolating cell surface proteins has been to use proteases acting on living cells. The external portions of surface proteins are cleaved, depending on which protease is used and the resistance of the surface polypeptides. However, in this case only a modified protein is obtained which lacks at least the intramembranous portion and, when present, also the cytoplasmic end. Since the antigenic and biological activities often reside in the outer carbohydrate containing part, the examination of such fragments is not without value. Papain has extensively been used to solubilize *H-2*, *HL-A* and thymus-leukemia (TL) antigens from lymphocytes (Nathenson and Cullen, 1974; Anundi et al., 1975) and these have been purified and characterized. By immunizing with the papain-solubilized fragment of the large external surface glycoprotein of fibroblasts, a specific antiserum against protein was obtained. This was then used to isolate the complete molecule (Kuusela et al., 1975).

3.3. Polyacrylamide gel electrophoresis

The use of polyacrylamide gel electrophoresis in the presence of SDS to separate membrane proteins was a major break-through in membrane biochemistry. In this system the membranes are solubilized, preferably at 100°C in the presence of SDS and a reducing agent such as 2-mercaptoethanol and electrophoresed in a polyacrylamide matrix containing SDS. All polypeptides bind SDS and acquire a strong negative charge and thus move against the anode. Usually proteins migrate according to their molecular weights (Weber and Osborn, 1969), but carbohydrate-rich glycoproteins may behave anomalously (Bretscher, 1971a; Segrest et al., 1971) and give apparent molecular weights which are too high. However, glycoproteins containing only a few per cent carbohydrate migrate like standard proteins and present no special problems. Another possibility which should be remembered, however, is that hydrophobic proteins may bind excess detergent, acquire a high negative charge and thus move faster than expected.

Polyacrylamide gels containing 5–10% acrylamide are most commonly used, either as cylindrical or slab gels. For a quantitative estimation of radioactivity cylindrical gels are prefered, especially if the protein patterns are relatively simple. These can be stained for protein, usually with Coomassie blue (Weber and Osborn, 1969) or for sialic acid with the periodic acid-Schiff (PAS) stain (Fairbanks et al., 1971). Glycoproteins lacking sialic acid do not stain well with the PAS procedure. Radioactive gels can be sliced to 1–2 mm slices, which are then treated for example with NCS solubilizer (Amersham/Searle) containing 10% water. After incubation at 50°C overnight, the radioactivities are determined using a toluene-based scintillation fluid. 30% hydrogen peroxide also solubilizes proteins well.

In comparison with cylindrical gels, the use of slab gels offers a number of advantages. Especially using discontinuous electrophoresis systems (Laemmli,

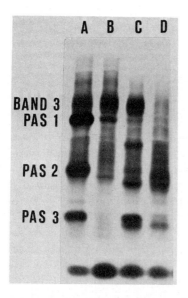

Fig. 5. Human erythrocyte membrane glycoproteins labeled by the galactose oxidase method. Cells were reduced with [^{3}H]NaBH$_4$ after treatment with: A, neuraminidase plus galactose oxidase; B, galactose oxidase; C, trypsin plus neuraminidase plus galactose oxidase; and D, pronase plus neuraminidase plus galactose oxidase. After labeling the membranes were isolated and electrophoresed on slab gels and the gels subjected to fluorography.

1970), the resolution is better and many samples can be compared in the same slab. In addition, staining for protein is more sensitive and faster and there is a lower need of material. The greatest advantage, however, is the use of autoradiography or fluorography to identify radioactive proteins in the gel. In this approach, gels are dried under vacuum to a thin sheet and by applying an X-ray film in close contact with the gel a good resolution of radioactive bands is obtained. It was not possible previously to visualize a tritium label in dried gels by autoradiography because of the low energy of the β-emission. This problem has been solved recently, however, by introducing a scintillator into the gel (Bonner and Laskey, 1974). Routinely, the gels are fixed after electrophoresis in 20% sulfosalicylic acid overnight, stained with Coomassie blue and destained (Weber and Osborn, 1969) and then treated twice with dimethyl sulfoxide in a glass or metal container for 30 min at room temperature. The gels are then transferred into 200–300 ml of a solution of 22% 2,5-diphenyl oxazole (PPO) in dimethyl sulfoxide and kept there for 3 h. The dimethyl sulfoxide solutions can be reused about 10 times. The gels are then transferred to water, the PPO precipitates immediately and after 1 h the gels are dried for fluorography. The fluorography is performed at −70°C. This treatment is necessary for visualization of tritium but it is also much more sensitive for [^{14}C] and [^{35}S] and thus the incubation times are much shorter than without a scintillator in the gel. Recently, the procedure has been modified slightly and made even more sensitive using X-ray film (Kodak RP Royal X-Omat) which has been pre-exposed briefly to a weak light source (Laskey and Mills, 1975). Radioactive standard proteins are of great value and can easily be obtained for example by labeling with [^{14}C]formaldehyde (Rice and Means, 1971).

3.4. Specific labeling of the outer surface of the plasma membrane

A number of methods to label external molecules on the cell surface have been developed recently. The methods are often relatively simple, a small amount of cells can be used and when combined with sensitive analytical techniques like SDS-polyacrylamide gel electrophoresis considerable information can be obtained. Much of what we presently know about membrane structure and the changes accompanying malignant transformation has been obtained using such techniques. These methods also offer the advantage of direct visualization of surface molecules. In contrast, labeling cell surface components with lectins or antibodies is usually complicated by uncertainties about the molecular nature of the receptors that we are studying.

The external labeling methods can be divided conveniently into techniques that specifically label cell surface carbohydrate and those that label cell surface polypeptides. No specific labeling method for lipids exists, though some lipids are labeled by the carbohydrate and protein methods. Many of these methods can also be combined with each other or with protease or glycosidase treatments to get additional information.

3.4.1. Labeling of cell surface carbohydrates

3.4.1.1. The galactose oxidase method for labeling galactose and N-acetyl galactosamine. The galactose oxidase method of labeling cell surface carbohydrate and its applications have been reviewed recently (Gahmberg and Hakomori, 1976; Gahmberg et al., 1976d). Galactose oxidase is an enzyme from *Dactylium dendroides* with a molecular weight of about 80 000 daltons (Amaral et al., 1963). The enzyme acts on nonreducing terminal D-galactosyl and N-acetyl D-galactosaminyl residues and forms the corresponding carbon-6 aldehydes. The enzyme has been used by Ashwell and co-workers for labeling serum proteins for metabolic studies (Ashwell and Morell, 1974). Because of its size the enzyme does not easily penetrate the plasma membrane. Therefore only carbohydrate located relatively far out on the external surface is oxidized by the enzyme. Because sialic acids are often linked to penultimate galactosyl groups, more efficient labeling can often be achieved by treating the cells with neuraminidase to cleave off sialic acids before or simultaneously with the galactose oxidase treatment (Gahmberg and Hakomori, 1973a; Gahmberg et al., 1976a). After oxidation the cells are reduced with tritiated sodium borohydride ([^{3}H]NaBH$_4$). The borohydride ion itself can probably penetrate the membrane easily. The use of [^{3}H]NaBH$_4$ of high specific activity results in better incorporation of label. In this way the oxidized sugars are reduced back to the original molecules and very little modification of the cells has occurred.

The procedure is as follows: $1-10 \cdot 10^7$ cells are washed in PBS and divided into three equal lots in 1 ml of Dulbecco's PBS, containing Ca^{2+}. To one lot is added 12 units of *Vibrio cholerae* neuraminidase (Behringwerke, Marburg-Lahn, Germany) and 5 units galactose oxidase (Kabi AB, Stockholm, Sweden). The commercially available enzymes are usually free of proteolytic activity, but the galactose oxidase can easily be purified on unsubstituted Sepharose by affinity chromatography (Gahmberg and Hakomori, 1975a). To another tube is added only galactose oxidase and the third tube is left without enzymes. All three tubes are incubated at 37°C with gentle shaking for 30 min and then the cells are washed twice with PBS. When labeling cell monolayers, three parallel dishes are used and treated with enzymes as described above. After enzyme treatment the cells are taken off the dishes by EDTA treatment or by scraping with a rubber policeman, and washed in PBS. Then 0.5 ml PBS is added and about 0.5 mCi [^{3}H]NaBH$_4$ per tube in 0.05 ml 0.01 N NaOH. After incubation for 30 min at room temperature, the cells are washed four times with PBS by centrifugation. If the samples are to be used for SDS-polyacrylamide gel electrophoresis, the cells are dissolved in SDS after labeling, boiled and kept frozen until analyzed. The [^{3}H]NaBH$_4$ preparation is prepared by dissolving the dry powder (usually 0.25 Ci) in 0.5 ml 0.01 N NaOH and quickly transferred to tubes containing 0.1 ml each. These are immediately frozen in dry ice/methanol and one tube at a time is dissolved in 2.5 ml 0.01 N NaOH and divided into 25 tubes containing 2 mCi each. After freezing these are preserved at −70°C where they remain relatively stable for at least a year.

The above procedure, combined with SDS polyacrylamide gel electrophoresis, can be used to establish the following points: (1) which proteins are labeled and exposed on the cell surface; (2) that the labeled proteins are glycoproteins containing galactosyl/N-acetyl galactosaminyl terminal groups; (3) by comparing samples treated with neuraminidase plus galactose oxidase with those exposed to galactose oxidase alone an estimate can be obtained of the presence of sialic acid; (4) the apparent molecular weights of the labeled glycoproteins can be determined; (5) an approximate quantitation of the labeled glycoproteins can be performed; and (6) are there proteins present which are labeled by [^{3}H]NaBH$_4$ alone without the use of enzymes. Such proteins are found in fibroblasts and in leukemic cells (Gahmberg et al., 1974; Andersson et al., 1976).

The reagents needed for labeling are relatively inexpensive and especially when combined with fluorography of slab gels the resolution and information provided is appreciable. Some modifications of the galactose oxidase technique have been used. We have also tried labeling the oxidized groups with [^{35}S]methionine sulfone hydrazide (Itaya et al., 1975). The main purpose here was to obtain a strong label which could be detected by autoradiography, because at that time the fluorography method was not available. Another possible advantage was lower nonspecific labeling, that is without using enzymes, because the negatively charged methionyl sulfone hydrazide should only penetrate the membrane poorly. However, the advantages over [^{3}H]NaBH$_4$, if any, are small and the method is much more expensive and laborious. Weber and Hof (1975) have used dansyl hydrazine after the galactose oxidase treatment and the fluorescently labeled glycoproteins could quickly be visualized from polyacrylamide gels. This may facilitate isolation of membrane proteins.

3.4.1.2. The periodate–[^{3}H]NaBH$_4$ method for labeling sialic acids. Periodate in low concentration specifically oxidizes sialic acids and 5-acetamido-3,5-dideoxy-L-arabino-2-heptulosonic acid is formed (Liao et al., 1973). Because most cellular sialic acid is localized at the cell surface (Eylar et al., 1962), the method can be used to label mainly cell surface proteins. However, it does not provide information about the actual exposure on the surface and its most important use is therefore for obtaining radioactive glycoproteins and for a semiquantitative estimation of the amount of sialic acid present in the membrane (Gahmberg et al., 1976a). The labeling can be done in the following way: cells are washed in PBS, and sodium metaperiodate is added to a final concentration of 2 mM in PBS. After incubation for 10 min in the dark at room temperature, the excess periodate is removed by four washes in PBS. Then 0.5 ml PBS is added and 0.5 mCi [^{3}H]NaBH$_4$. After incubation for 30 min at room temperature, the cells are washed four times in PBS.

3.4.1.3. The cytidine monophosphate (CMP)-[^{14}C]sialic acid method for labeling cell surface glycoproteins. Incubation of cells with CMP-[^{14}C]sialic acid in PBS results in incorporation of [^{14}C]sialic acid primarily into cell surface glycoproteins (Datta, 1974). Phosphate esters do not easily penetrate the plasma membrane and therefore only the surface-exposed carbohydrates are predominantly labeled. Pretreatment with neuraminidase enhances the incorporation of label. Labeling by this procedure depends on the presence of cell surface sialyl transferase, and the labeling pattern depends on the activity of the enzyme rather than the absolute chemical quantities of acceptor molecules at the cell surface. Nevertheless, the method can detect changes in cell surface properties, which may be sensitively adapted to the physiological state of the cell. CMP-[^{14}C]sialic acid is expensive, however, and there are also problems in obtaining reasonable levels of incorporated radioactivity.

3.4.2. Labeling of the polypeptide portion of cell surface glycoproteins

There are numerous methods available for labeling the polypeptide portion of cell surface proteins. Lactoperoxidase catalyzed iodination has been the most widely used method because of its simplicity and the high levels of radioactivity that can be incorporated into cells. If used properly, this technique should be specific for cell surface proteins. For more detailed analysis of the labeled proteins other labeling methods such as [^{35}S]formyl methionyl sulfone methyl phosphate or isethionyl acetimidate are of great value, especially when combined with peptide mapping techniques.

3.4.2.1. The lactoperoxidase method for labeling cell surface proteins.

Lactoperoxidase, like galactose oxidase, has a molecular weight of about 80 000 daltons and therefore does not easily penetrate the plasma membrane. It catalyzes the oxidative iodination of exposed tyrosine(histidine) residues in the presence of hydrogen peroxide. The hydrogen peroxide can be supplied either by repeated additions (Phillips and Morrison, 1971) but a more gentle method for cells involves constant generation of a low amount of hydrogen peroxide by using simultaneously glucose oxidase plus glucose (Hubbard and Cohn, 1972). Either [^{125}I] or [^{131}I] can be used and the radioactivities determined separately. [^{131}I] gives a better resolution of closely spaced bands from polyacrylamide gels (Teng and Chen, 1976).

3.4.2.2. The pyridoxal phosphate-[^{3}H]NaBH$_4$ method for labeling cell surface proteins.

Like most phosphate esters, pyridoxal phosphate, though a low molecular weight compound, penetrates the cell membrane only poorly. At alkaline pH it reacts with exposed amino groups and forms a Schiff-base. This can then readily be reduced with [^{3}H]NaBH$_4$, which stabilizes the bond and introduces a radioactive label (Rifkin et al., 1972; Cabantchik et al., 1975). This method is simple and easy to use because of commercially available reagents. [^{32}P] Pyridoxal phosphate could be advantageous because of the background activity when using [^{3}H]NaBH$_4$, but the short half-life of [^{32}P] and the problems with the preparation of the reagent have limited its use. Also amino group containing phospholipids may be labeled.

3.4.2.3. The [^{35}S] formyl methionyl sulfone methyl phosphate method for labeling cell surface proteins. This strongly negative reagent, which can be obtained with high specific activity, has been used successfully to label cell surface proteins and amino phospholipids in erythrocytes (Bretscher, 1971b) and viruses (Gahmberg et al., 1972a). At alkaline pH it reacts with amino groups and forms an amide bond. Because of the negative charge it does not penetrate the membrane. One clear advantage is the combination with peptide mapping techniques which can be used for determining which parts of the protein are exposed on either surface of the membrane. Disadvantages are the high price of radioactive methionine, the rather complicated method of synthesis and the instability of the compound. These factors have limited its use.

3.4.2.4. The [^{3}H]/[^{14}C] isothionyl acetimidate method for labeling cell surface proteins. Whiteley and Berg (1974) have introduced a very promising method using a nonpenetrating imidoester, isothionyl acetimidate, which can be labeled with [^{3}H] or [^{14}C]. This compound reacts with exposed amino groups, and is evidently very gentle to cells. Even when most amino groups are blocked, the cells remain viable. By combining it with use of [^{3}H] or [^{14}C]-labeled ethyl acetimidate, which has the same specificity, but readily penetrates the membrane, important information can be obtained concerning the location of different parts of membrane proteins. The synthesis and stability of the compounds may represent problems.

3.4.2.5. Photochemical labeling of cell surface proteins. A method of considerable potential value is photochemical labeling. Staros et al. (1974) have used N-(4-azido-2-nitrophenyl)-2-amino ethyl sulfonate, a photolabile nitrene precursor to label erythrocyte membrane proteins. At 0°C this reagent does not penetrate the membrane, but the membrane is permeable to it at 37°C. In the dark the reagent is unreactive but when exposed to light it becomes highly reactive. By using appropriate conditions either the external or the internal side of the membrane can be labeled (Staros et al., 1975). Theoretically, one should also be able to use reagents of this type to specifically label certain parts of the cells by using high-energy beams of light.

3.4.2.6. Labeling of cell surface proteins using transglutaminase. Transglutaminases are present in various tissues and catalyze the formation of ε(γ-glutamyl) lysyl crosslinks or the substitution of various primary amines in glutamine residues of proteins. Therefore these enzymes are potentially useful for either labeling cell surface proteins with radioactive amines like [^{14}C]putrescine and fluorescent amines like dansylcadaverine and for the formation of protein-protein crosslinks (Dutton and Singer, 1975; Lorand et al., 1975; Mosher, 1975a,b). Intact erythrocytes are not significantly crosslinked or labeled, but leaky ghosts interact much more efficiently. As pointed out by Dutton and Singer (1975), this method offers some advantages over other methods. The labeling conditions are very mild and many different types of amines can be used for labeling.

3.5. Chemical analysis of membranes

3.5.1. Proteins

Determination of the protein content of membranes and preparations of isolated membrane proteins is usully done according to the method of Lowry et al. (1951). Because mainly tyrosines are measured by this method and hydrophobic membrane proteins may contain unusual amounts of this amino acid compared to standard proteins erroneous results may be obtained. More reliable protein measurement can be obtained by the ninhydrin reaction (Spies, 1957) or amino acid analysis. The use of p-toluene sulfonic acid for hydrolysis (Liu and Chang, 1971) may be advantageous compared to classical hydrolysis in 6 N HCl.

Peptide mapping of isolated proteins is also a powerful technique for various problems. Rarely is there enough protein for peptide mapping followed by staining and radioactive techniques are used instead. Proteins from surface-labeled membranes can be analyzed, and the labeled (exposed) peptides determined. A useful technique is radioiodination by the chloramine T method of separated proteins from SDS polyacrylamide gels (Bray and Brownlee, 1973), requiring only a few micrograms of protein. This method was successfully used to study peptides of the band 3 polypeptide of normal erythrocytes and erythrocyte variants (Gahmberg et al., 1976b). The labeled protein is digested with proteases, for example a mixture of trypsin and chymotrypsin, or thermolysin, and chromatographed in n-buthanol–acetic acid–water (17:5:25 by vol. on Whatman 3 MM paper, followed by electrophoresis at pH 3.5 (Bray and Brownlee, 1973).

Important information about the structure of membrane proteins can also be obtained by column chromatography of peptides from enzyme-digested or cyanogen bromide cleaved proteins (Roy and Konigsberg, 1972). In addition, purification of peptides may be achieved in this way. The hydrophobic segment of the glycophorin molecule in human erythrocytes was isolated by these techniques (Marchesi et al., 1972).

Membrane proteins labeled by a mixture of [^{3}H]amino acids can also be sequenced by microtechniques (Silver and Hood, 1976) and this offers a powerful tool for future use.

3.5.2. Carbohydrates

For analysis of carbohydrate structure, the first important step after isolation of the glycopeptides is to determine the monosaccharide composition. Sialic acid is determined by the thiobarbituric acid method (Warren, 1959) or by the resorcinol method (Svennerholm, 1957). Neutral sugars can be determined by different methods. A total estimate may be done by the anthrone reaction (Hewitt, 1958). For determination of D-galactose, D-mannose, L-fucose and D-glucose, trimethyl silyl derivatives can conveniently be used (Laine et al., 1972) using mannitol or inositol as internal standard. Another possibility is the use of alditol acetate derivatives (Niedermeyer, 1971). This method is more laborious but more exact, because only single monosaccharide peaks are formed and often peaks from contaminating substances present in the trimethyl silyl samples are eliminated. Total amino sugars are determined by the Elson-Morgan reaction (Gatt and Berman, 1966). The amino sugars can also be determined as trimethyl silyl or alditol acetate derivatives, but often these give unreliable low results. An alternative method is to use the amino acid analyzer for determination of amino sugars, which gives both the total amino sugar content and separately N-acetyl D-glucosamine and N-acetyl D-galactosamine (Bahl, 1969).

For more detailed analysis of glycoproteins, the individual glycopeptides must be isolated. This must be preceded by digestion of the glycoprotein, with pronase being used widely for this purpose. We have used 1 mg pronase per mg membrane protein in 0.1% SDS at 60°C for 24 h in PBS. The pronase solution should be autodigested at 37°C for 1 h to destroy contaminating glycosidases. Another, recent method, which would appear to hold some promise, is to digest the glycoprotein with hydrazine (Bayard and Roux, 1975). After lyophilization 0.3 ml anhydrous hydrazine (for less than 2 mg protein) is added, the tubes capped under nitrogen and the digestion performed at 100°C for 30 h. Sialic acids should first be removed with neuraminidase or weak acid. The intact oligosaccharide is released by this treatment and the polypeptide portion is digested forming amino acid hydrazides.

It is also possible to use certain endoglycosidases to liberate part of the oligosaccharides (Koide and Muramatsu, 1974). After digestion, the oligosaccharides are isolated by gel filtration. Further characterization of the glycopeptides is achieved by determining the anomeric linkages using specific glycosidases. A major problem is to obtain specific and pure glycosidases. Commercial preparations are often not good enough. After each digestion the glycopeptide must usually be reisolated and then digested with the next glycosidase and so on. This is often difficult and needs much material, either radioactively labeled or not. The final structure of the glycopeptides is obtained by methylation studies to get the linkage positions. The Hakomori method or modifications of it are generally used (Lindberg, 1972; Stellner et al., 1973). The methylated oligosaccharides are cleaved by acid hydrolysis, reduced with borohydride and acetylated. Identification is then performed by combined gas chromatography-mass spectrometry.

3.6. Fluorescence microscopy and electron microscopy

Fluorescence and electron microscopy of intact cells and isolated membranes have provided considerable useful information about cell surface proteins and membrane structure, but the interpretation of the results is often complicated by the limited knowledge of the reactive molecules on the cell surface.

To visualize the location of cell surface glycoproteins (and glycolipids) lectins coupled to a fluorescent label, to peroxidase or to ferritin have been much used (Nicolson, 1971; Martinez-Palomo et al., 1972; Edelman et al., 1973). Some lectins like concanavalin A and wheat germ agglutinin are commercially available and many other lectins are easily purified by specific affinity chromatography using Sephadex or agarose columns (Nicolson and Blaustein, 1971; Agrawal and Goldstein, 1972; Allan and Crumpton, 1972; Olsnes et al., 1974). Covalent coupling to ferritin can for example be done using glutaraldehyde (Siess et al., 1971; Kishida et al., 1975) or carbodiimides (Kishida et al., 1975).

Detailed information about the location of surface antigens can be obtained by use of specific antibodies. The main problem with this approach is to obtain a good antiserum, which usually, but not always, requires purification of the antigen.

For ultrastructural visualization of carbohydrate-containing cell surface molecules, oxidation of galactosyl/N-acetyl galactosaminyl residues can be performed with galactose oxidase followed by reaction with biotin hydrazide and avidin coupled to ferritin (Heitzmann and Richards, 1974). Instead of galactose oxidase, periodate can be used to oxidize sialic acids, followed by biotin hydrazide and avidin-ferritin. Sialic acids can also be visualized in electronmicrographs with cationized ferritin (Nicolson, 1973a).

4. Studies of surface glycoproteins of normal and transformed cells

4.1. External labeling of normal and malignant cells

4.1.1. Fibronectin

Using the galactose oxidase and lactoperoxidase labeling techniques and immunological methods, several groups of investigators have found that normal fibroblasts have a high molecular weight cell surface protein which is absent in transformed cells. This protein, which has been variously called galactoprotein a (Gahmberg and Hakomori, 1973b) (it is known to contain galactose/N-acetyl galactosamine), large external transformation sensitive (LETS) protein (Hynes, 1973), Z protein (Blumberg and Robbins, 1975), fibroblast surface (SF) antigen (Vaheri and Ruoslahti, 1974) and now fibronectin (fibroblast connecting or binding protein) (Kuusela et al., 1975) has generated substantial interest because of its possible role in cell growth regulation. This protein is a major component of the cell surface in hamster (Gahmberg and Hakomori, 1973b; Hynes, 1973), rat (Stone et al., 1974), mouse (Hogg, 1974), chicken (Stone et al., 1974; Blumberg and Robbins, 1975) and human (Vaheri et al., 1976) fibroblasts. It is a glycoprotein which is readily labeled by the galactose oxidase method and by metabolic labeling using radioactive monosaccharides (Pearlstein and Waterfield, 1974). The tentative molecular weight of the protein on SDS polyacrylamide gels is 200 000–250 000 daltons (Fig. 6), but the uncertainty concerning molecular weight determinations of membrane glycoproteins must again be emphasized. Fibronectin differs depending on the source from which it is obtained. This is illustrated by the galactose oxidase technique, where the protein is readily labeled in hamster fibroblasts by galactose only (Gahmberg and Hakomori, 1973b), while in mouse 3T3 cells successful labeling requires

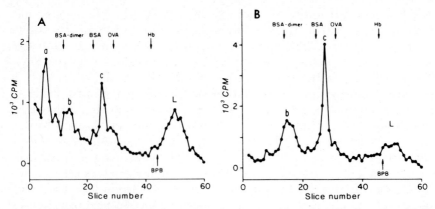

Fig. 6. SDS-polyacrylamide gel electrophoresis patterns of hamster Nil cells surface-labelled by the galactose oxidase method: A, normal Nil cells; a, fibronectin; b, galactoprotein b; c, nonspecifically labelled protein by [^{3}H]NaBH$_4$ alone; L, lipid peak; and B, polyoma virus-transformed Nil cells. Note the absence of fibronectin peak and the prominent b and c peaks. The positions of the [^{14}C]labelled marker proteins are indicated: BSA-dimer, bovine serum albumin dimer; BSA, bovine serum albumin; OVA, ovalbumin; and Hb, hemoglobin.
(Reproduced with permission from Gahmberg and Hakomori, 1973a, and the National Academy of Sciences, USA.)

pretreatment with neuraminidase (Gahmberg and Hakomori, unpublished observations). Differences between fibronectins of different species have also been observed (Kuusela et al., 1976).

Fibronectin is an integral protein which is not soluble in simple aqueous buffers and requires solubilization by detergents or high salt. The external part of the molecule is easily cleaved by treating intact cells with low levels of proteases (Hynes, 1973; Gahmberg and Hakomori, 1974) and immobilized papain has been used to isolate this part of the molecule (Ruoslahti et al., 1973). The protein also appears to penetrate the plasma membrane. This was shown by Hunt and Brown (1975) using mouse L cell fibroblasts and was demonstrated by the following method. Intact cells were first labeled with [^{125}I] using the lactoperoxidase technique. The cells were then incubated with polystyrene latex beads, which were ingested by the cells forming inverted plasma membrane vesicles around the beads. These inverted vesicles were isolated after rupture of the cells and treated with trypsin. In this way it was shown that trypsin could degrade fibronectin from the original inner surface of the plasma membrane to a 65 000 molecular weight fragment, indicating that the protein penetrated the membrane. However, when intact cells are treated with proteases, the cleaved fragment of fibronectin has a molecular weight of about 200 000 daltons, and this makes the results more difficult to interpret. As mentioned earlier, it is well known that the major glycoproteins of erythrocytes penetrate the plasma membrane, but whether this is also so in nucleated cells remains to be confirmed experimentally.

Fibronectin does not occur uniformly on the cell surface but appears to be localized to fibrillar structures (Wartiovaara et al., 1974). The distribution of these fibrils resembles that of actin (Pollack et al., 1975) or tubulin (Weber et al., 1975), and it is possible that fibronectin is associated with these intracellular proteins. When fibronectin is isolated by immunological methods, it co-purifies with a protein with a molecular weight of 45 000 daltons (Kuusela et al., 1975) which is similar to actin. An interesting possibility is that fibronectin has some receptor function, and by its interaction with submembrane cytoskeletal structures, transductive signals through the membrane can be regulated.

Fibronectin has also been purified by extraction of intact cell monolayers with urea (Yamada and Weston, 1974). Surprisingly, fibronectin is one of the major proteins released into the medium by cultured cells. However, it is possible that the released protein is a result of proteolytic degradation, because it is known to be very sensitive to proteases. The intramembranous portion of surface membrane proteins is small in those proteins studied to date, having molecular weights of 3000–5000 daltons (Gahmberg et al., 1972b; Nathenson and Cullen, 1974; Tomita and Marhesi, 1975). The size of the cytoplasmic fragment on the inside of the membrane is not known, and a small change in molecular weight would not be easily observed.

It has been proposed repeatedly that the absence of fibronectin from transformed cells results from proteolytic activity at the cell surface. It is known that transformed cells synthesize the protein, but it is not retained in the membrane (Vaheri and Ruoslahti, 1975). It is found as a soluble protein in the culture medium and in sera of different species (Keski-Oja et al., 1976). This molecule most probably represents a fragment of the membrane bound form. It is also found in relatively high concentrations in plasma (0.3–0.5 mg/ml) and corresponds to a previously characterized plasma protein, cold insoluble globulin (Edsall et al., 1955; Mossesson and Umfleet, 1970; Ruoslahti and Vaheri, 1975). In the circulation it exists as a dimer and binds strongly to fibrin and insolubilized fibrinogen. Crude fibrin is a good source of the protein (Mosher, 1975a; Ruoslahti and Vaheri, 1975).

In normal fibroblasts, fibronectin appears at the cell surface in the G_1 phase of the cell cycle (Gahmberg and Hakomori, 1974; Hynes and Bye, 1974) and is absent or present only in very low concentrations in mitotic cells (Gahmberg and Hakomori, 1974; Hynes and Bye, 1974). Little is expressed during the S and G_2 phases of the cell cycle, but during the next G_1 phase synthesis is again accelerated. It occurs in much higher concentrations in confluent normal cells than in actively growing cells (Gahmberg and Hakomori, 1974) (Fig. 7). In transformed cells, however, it is not demonstrable on the surface during any phase of the cell cycle. In a temperature-sensitive line of polyoma virus-transformed fibroblasts (i.e., the cells show a transformed phenotype when grown at 32°C, but a normal phenotype when grown at 39°C), fibronectin is present only at the nonpermissive temperature. Normal fibroblasts, however, contain the protein at both temperatures and cells transformed by a wild-type polyoma virus do not show the protein at either temperature (Gahmberg et al., 1974) (Fig. 8). Similar results have been obtained in chick cells transformed with

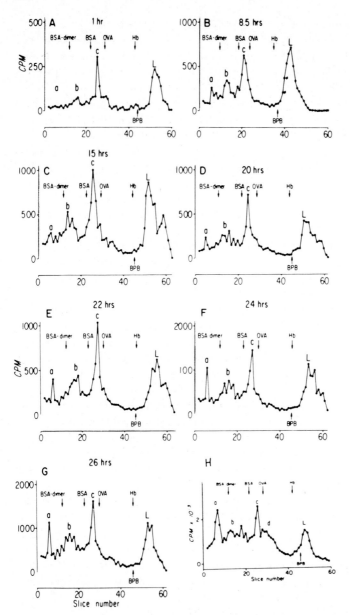

Fig. 7. Cell cycle and cell density dependency of cell surface labeling with galactose oxidase. Confluent contact-inhibited Nil hamster cells were trypsinized and seeded at low density on plastic dishes: A, 1 h after seeding, note the low levels of fibronectin(a) and galactoprotein b; B, 8.5 h after seeding, (G_1 phase), fibronectin and galactoprotein b have increased; C to D, 15–20 h after seeding (S to G_2 phases), no obvious changes in the labelling patterns; E, 22 h after seeding (mitosis), relatively little label in fibronectin and galactoprotein b; F to G, 24–26 h after seeding (beginning of next G_1 phase), fibronectin is increasing; and H, pattern of confluent Nil cells with a large fibronectin peak. The positions of the [^{14}C]labelled marker proteins are indicated (abbreviations are as in the legend to Fig. 6). (Reproduced with permission from Gahmberg and Hakomori, 1974.)

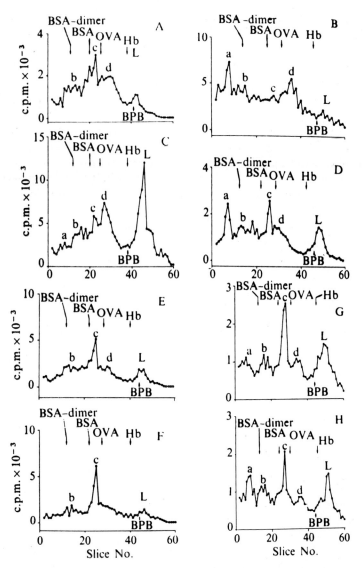

Fig. 8. Expression of fibronectin in normal and virus-transformed cells.
Cells were labelled by the galactose oxidase method and electrophoresed on polyacrylamide gels in the presence of SDS: A, normal BHK cells, sparsely grown at 32°C; B, normal confluent BHK cells grown at 32°C, note the presence of fibronectin; C, normal BHK cells sparsely grown at 39°C; D, normal confluent BHK cells grown at 39°C, fibronectin is present; E, confluent wild-type polyoma virus-transformed BHK cells grown at 32°C; F, confluent wild-type polyoma virus-transformed BHK cells grown at 39°C; G, polyoma virus ts-3 transformed BHK cells grown at 32°C showing the transformed phenotype, note the absence of clear fibronectin peak; H, polyoma virus ts-3 transformed BHK cells grown at 39°C showing a normal phenotype. Note the presence of fibronectin peak. The positions of the [^{14}C]labelled marker proteins are indicated (abbreviations are as in the legend to Fig. 6).
(Reproduced with permission from Gahmberg et al., 1974.)

temperature-sensitive mutants of Rous sarcoma virus (Hynes and Wyke, 1975).

In normal cells fibronectin functions as the major receptor for concanavalin A and *Ricinus communis* agglutinins (Gahmberg and Hakomori, 1975b). When monolayers of normal cells were incubated with 10 μg/ml of these lectins, fibronectin was more strongly labeled by the galactose oxidase method than in control cells without lectins. At lectin concentrations of 100 μg/ml (which correspond to agglutinating concentrations) the oxidation of fibronectin by galactose oxidase was strongly depressed (Fig. 9). The lectins themselves did not influence the activity of the galactose oxidase. Fibronectin, isolated by the urea extraction method, can agglutinate formalinized sheep erythrocytes at a concentration of 2 μg/ml, (Yamada et al., 1975). The agglutination is inhibited by antiserum against the protein by chelating agents like EDTA and also by amino sugars.

Although most studies on fibronectin have been done with fibroblasts, fibronectin, or a similar protein, is probably present in glial cells (Vaheri et al.,

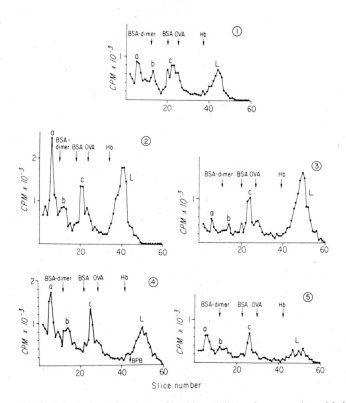

Fig. 9. Interaction of hamster fibroblast (Nil) surface proteins with lectins: 1, surface proteins of control Nil cells; 2, 3, 4 and 5, label of Nil cells pretreated with 10 μg and 100 μg of *Ricinus communis* agglutinin/ml, and 10 μg and 100 μg of Con A/ml, respectively; 6, control Nilpy cells; 7, 8, 9 and 10, polyoma virus-transformed Nil cells pretreated with 10 μg and 100 μg of *Ricinus communis* agglutinin/ml, and 10 μg and 100 μg Con A/ml, respectively; 11, control Nil cells treated

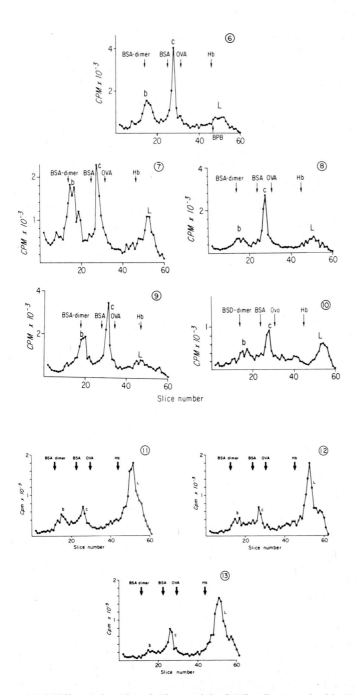

with 0.05% trypsin; 12 and 13, trypsinized Nil cells treated with 10 μg and 100 μg of *Ricinus communis* agglutinin/ml, followed by labelling, respectively. The positions of [^{14}C]-labelled marker proteins are indicated (abbreviations as in the legend to Fig. 6).
(Reproduced with permission from Gahmberg and Hakomori, 1975b.)

1976) and possibly also in lymphocytes (Trowbridge et al., 1975; Andersson et al., 1976; Gahmberg et al., 1976a). Mouse and human B lymphocytes have a similar high molecular weight protein on the surface. This protein is also very sensitive to proteases (Gahmberg et al., 1975a) and, as in the case of fibroblasts, proteases stimulate B lymphocytes to divide (Vischer, 1974). In at least some instances this high molecular weight B cell protein is lost from malignant lymphocytes (Trowbridge et al., 1975; Andersson et al., 1976). A schematic representation of possible differences in the architecture of the cell surface of normal and malignant cells is shown in Fig. 10.

4.1.2. Other cell surface proteins

In addition to fibronectin there are numerous cell surface proteins in fibroblasts that are labeled by external labeling techniques. However, many of these are relatively weakly labeled and this is due presumably to the presence of only

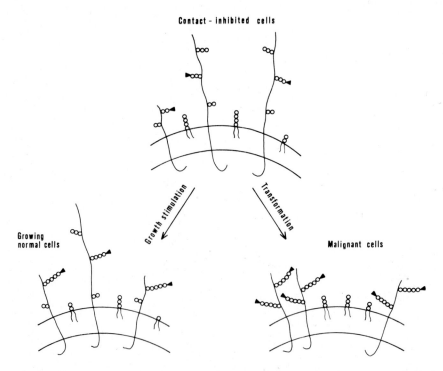

Fig. 10. A schematic representation of the cell surface glycoproteins and glycolipids in normal confluent cells, actively growing normal cells, and virus-transformed cells.
The contact-inhibited cells contain high levels of high molecular weight glycoproteins. The glycolipids are deeply embedded in the bilayer and not much exposed to the outside. Growth stimulation results in loss of some of the high molecular weight surface proteins and a relative increase in proteins of lower molecular weights. Malignant cells lack the high molecular weight glycoprotein, the oligosaccharide side chains of the glycoproteins are efficiently substituted with sialic acids and the glycolipids are more exposed to the outside than in normal cells.

a few copies per cell. With increasing sensitivity of detection methods, additional proteins will no doubt be revealed. Excluding fibronectin, few differences are consistently found in the surface glycoproteins between normal and transformed cells and differences are usually of a quantitative nature. Numerous laboratories have tried to find surface glycoproteins that are specific for transformed cells but, in general, such proteins have not been found. It should be emphasized that small changes, including alterations in cell surface antigens, may not easily be seen. Of possible functional importance is a group of surface glycoproteins found in transformed cells using external labeling techniques which we have called galactoprotein b (Gahmberg and Hakomori, 1975b). This protein(s) has an apparent molecular weight of 130 000 daltons and is also present in normal fibroblasts but is not readily exposed in these cells. Trypsin treatment exposes the protein. As mentioned below, this protein is a major receptor for various lectins.

A discussion of surface proteins which are present in minor quantities and are not detected by direct analysis of the cell surface by external labeling techniques is given on p. 401. Combined with sensitive immunological methods, such proteins may be isolated and further characterized.

4.2. Cell surface proteins and regulation of cell growth

An attractive role for fibronectin is in the regulation of cell growth. Transformed cells, which do not show contact-inhibition of growth, have lost this protein, and actively growing normal cells contain low levels. Trypsin treatment of the type used to disperse cell monolayers and to initiate cell division rapidly removes this protein from the cell surface. In addition, this protein is synthesized during the G_1 phase of the cell cycle. All of the proteases which have been studied so far can stimulate cell growth but not all remove fibronectin from the cell surface (Blumberg and Robbins, 1975; Teng and Chen, 1975). For example, bovine thrombin in low concentrations stimulates chick cells to divide but it does not remove fibronectin from the cell surface. Similarly, fibronectin is not removed by neuraminidase or insulin, both of which stimulate cells to divide. It seems more plausible, therefore, that conformational changes of the cell surface induced by these agents are enough to stimulate cell growth. One possibility is that by changing the three-dimensional structure of the membrane, transport sites for various nutrients could be activated. Similarly, the stimulation of T lymphocytes by lectins (Nowell, 1960; Janossy and Greaves, 1971), anti-lymphocyte antibodies (Gräsbeck et al., 1964) and periodate or neuraminidase plus galactose oxidase (Thurman et al., 1973; Novogrodsky, 1975) is probably due not to cleavage of surface proteins but to reorganization of the membrane. However, Teng and Chen (1976) have reported recently that thrombin specifically removes another high-molecular weight surface protein from chick cells. This protein is also present in low concentrations in actively growing cells and is sensitive to trypsin. Future experiments will no doubt reveal the significance of this finding.

Reich and co-workers (Ossowski et al., 1973; Unkeless et al., 1973a,b) have shown that transformed cells release a protease and show higher levels of proteolytic activity than untransformed cells. This protease, which has a molecular weight of 38 000 daltons (Unkeless et al., 1973b), activates serum plasminogen of plasma to plasmin which, in turn, hydrolyzes fibrin. The production of this fibrinolytic activity is temperature-sensitive in chick cells transformed by temperature-sensitive mutants of Rous sarcoma virus and is expressed within as little as one hour after the shift to permissive temperature (Unkeless et al., 1973a). Similar fibrinolytic activity has also been found in a variety of different tumors including: hepatomas, mammary carcinomas, and a range of animal and human tumor cell lines. Plasmin removes fibronectin from the cell surface and this might offer a possible correlation between this cell surface protein and Reich et al.'s findings. However, as noted earlier, many other proteases can remove fibronectin from the cell surface. Indeed, transformed cells, when mixed with normal cells in the absence of plasma, are still able to remove the fibronectin from normal cells (Hynes et al., 1975). Thus, plasmin may be involved in some cases in the regulation of cell growth but other mechanisms must also exist. Obviously, it is possible that the specific protease studied by Reich's group may still be involved in removal of fibronectin from transformed cells and thus may influence cell growth. It should also be remembered that cancer cells develop from a wide variety of normal cells and show very different characteristics from one cell line to another, even with the same histologic type of tumor. Therefore, it may be fruitless to search for a unifying rule for most or all types of cancer and exceptions to the rule(s) may not be unexpected.

Cyclic AMP and cyclic GMP have repeatedly been implicated in regulation of cell growth (review, Pardee, 1975). Cyclic AMP is formed from ATP by adenylate cyclase and destroyed by phosphodiesterase (Robison et al., 1972). Adenyl cyclase is localized to the plasma membrane, and there is no doubt of its importance in the regulation of hormone activity (Robison et al., 1972). Transformed cells show lower levels of cyclic AMP than normal cells (Bürk, 1968) and normal cells show a decreased cyclic AMP level during mitosis (Burger et al., 1972). Also, agents that stimulate cell division decrease the level of cyclic AMP. Cyclic GMP seems to behave in the opposite way and is often increased in cells stimulated to divide (Goldberg et al., 1974). Phorbol myristate acetate, a tumour promoter, increases the level of cyclic GMP very quickly (Estensen et al., 1974). Exogenous butyryl cyclic AMP changes the morphology of transformed cells toward that of normal cells (Hsie and Puck, 1971; Johnson et al., 1971; Sheppard, 1971) and decreases cell growth. Interestingly, surface glycoproteins (and glycolipids) in the plasma membrane of transformed fibroblasts show a changed organization after treatment with cyclic AMP (Gahmberg and Hakomori, 1973b). The surface profile resembles that of normal fibroblasts, though it is not identical. This finding further suggests that there is a correlation between cell growth control and expression of particular surface glycoproteins.

4.3. Cell surface antigens of normal and transformed cells

Most experimentally induced neoplasms contain tumor-specific transplantation antigens which can be detected both in vivo and in vitro (Klein, 1966). Most of these antigens are associated with the cell surface. The chemistry of these antigens is beginning to be elucidated, but detailed data are still very limited. It may be expected, however, that these antigens are surface glycoproteins or glycolipids, and there is already considerable evidence in support of this conclusion.

In general, virus-induced tumors contain antigens specific for the virus, whereas chemically induced tumors do not show such specificity and instead display a variety of different antigens. There is evidence that carbohydrate-containing surface molecules represent some tumor antigens. Human blood group substances have been well characterized chemically, and in many neoplasms these antigens are changed. Often this involves a loss of blood group specificity and this may be due to incomplete synthesis of carbohydrate chains in cell surface glycolipids (Hakomori and Kobata, 1975). Neuraminidase treatment of tumor cells sometimes results in accelerated rejection by the host, possibly by exposing carbohydrate containing antigens (Sanford and Codington, 1971; Simmons and Rios, 1971). Sialic acid is usually linked to galactosyl residues in glycoproteins, and galactosyl residues may be important for tumor rejection. When normal lymphocytes are treated with neuraminidase, they react with antibodies naturally present in serum (Rogentine, 1975), and the increased sialylation of glycopeptides found in many tumor cells might thus contribute to their escape from detection by the host's immune defense mechanisms.

There are also some recent reports which describe the presence of oncornavirus-specific glycoproteins on the surface of virus-induced lymphoma cells (Del Villano et al., 1975; Obata et al., 1975; Tung et al., 1975). These glycoproteins, probably specified by the virus, have molecular weights of 69 000 and 71 000 daltons. They are also components of the virus and present on its surface. These antigens appear identical to the G_{IX} thymocyte antigen. The glycoproteins in question are not restricted to malignant cells but are also present in normal thymocytes, though their presence is always correlated with the presence of virus components. Another potentially important antigen is the tumor-specific antigen found in chick cells transformed by Rous sarcoma virus (Rohrschneider et al., 1975). This surface glycoprotein, with an apparent molecular weight of 100 000 daltons, was isolated by indirect immune precipitation of [^{3}H]fucose-labeled cells. Whether this protein is coded for by the virus or by the host genome is presently not known.

Another interesting question concerns the relationship between tumor antigens and transplantation antigens. The thymus leukemia antigen, for example, is present on normal thymocytes of some strains of mice and also on leukemic cells. This antigen is composed of four polypeptides, two of which are identical to β_2-microglobulin and the remaining two heavy chains contain the antigenic determinants (Davies et al., 1969; Muramatsu et al., 1973b; Anundi et al., 1975; also see chapter in this volume by Hyman, p. 513). The mouse transplantation antigens (H-2), which are also cell surface glycoproteins, have a similar structure but differ in their amino acid and carbohydrate composition (Muramatsu et al., 1973a). There is also other evidence for a close connection between tumor and transplantation antigens. Invernizzi and Parmiani (1975) tested tumor immunity in mice by immunizing with tissues of mice of different H-2 specificities and showed that there was a very clear immune response against challenge with chemically induced tumors, indicating a cross reactivity between the H-2 antigens of the host and the tumor antigens. Virus infection may also result in new H-2 specificities (Garrido et al., 1976) suggesting that these antigens are easily changed. Recently, microsequencing techniques applied to the N-terminal ends of H-2 antigens have shown an enormous variability in the H-2 heavy chains, similar to that seen in immunoglobulins (Silver and Hood, 1976). This is rather difficult to explain as alleles of the same gene locus and a hypermutation mechanism must exist.

4.4. Interaction of normal and transformed cells with lectins: insights into plasma membrane glycoprotein organization

In 1963 Aub et al. observed that crude wheat germ lipase agglutinated transformed fibroblasts at a much lower concentration than that needed to agglutinate normal cells. This interesting phenomenon was followed up by Burger (Burger and Goldberg, 1967) using purified wheat germ agglutinin and by Inbar and Sachs (1969) using concanavalin A. Agglutination of transformed cells by lectins has since been studied in numerous laboratories, but the results obtained have been difficult to interpret. One reason for this is the relative lack of specificity of the lectins which results in binding to a variety of surface receptors, the nature of which is often poorly known. However, when combined with other techniques, such as surface labeling, the results may become more meaningful. Binding of wheat germ agglutinin to cell receptors is specifically inhibited by glucosamine and most lectins studied are inhibited by different monosaccharides and even better by cell membrane oligosaccharides (Adair and Kornfeld, 1974).

It was first thought that transformed cells possessed more lectin receptors than normal cells and that this was responsible for the increased agglutination of transformed cells. In addition, it was found that when normal cells were treated with trypsin they also became readily agglutinable by lectins (Burger, 1969). This was proposed as resulting from the presence of cryptic receptors in normal cells, which normally were covered by trypsin-sensitive structures. At this stage the story seemed relatively simple and clear. However, it was soon found that both normal and transformed cells bound radioactively labeled lectins to approximately the same extent (Ozanne and Sambrook, 1971). The binding of lectins to the cell surface varies, however, depending on the specificity of the lectin and the treatment of cells. For example, *Ricinus communis* agglutinin, which is specific for D-galactosyl residues, binds more to neuraminidase-treated hamster cells than to untreated cells (Nicolson, 1973b). Sialic acids are known to be linked to penultimate galactosyl residues, and thus it might be expected that neuraminidase treatment would result in exposure of new terminal galactosyl residues. This has also been shown with the same cells using the galactose oxidase surface labeling technique in combination with neuraminidase treatment (Gahmberg et al., 1974). However, the correlation between lectin binding and agglutination is also rather poor in this instance.

The next stage in the story was initiated when Nicolson observed that the topographic distribution of lectin receptors on the surfaces of normal and transformed cells was different (Nicolson, 1971, 1972). On normal cells the distribution was random, but in transformed and trypsin-treated cells the receptors were clustered. Later studies by Rosenblith et al. (1973) and Nicolson (1973c) showed that if cells were fixed before incubation with lectins, the distribution was random in all cells. The clustering of receptor sites observed in unfixed cells was thus induced by the lectins themselves. The clustering depends, in part, on the fluidity of the membrane and the ability of receptors to move within the membrane. At +4°C the lipid bilayer is "frozen" and no clustering occurs. If, however, cells are incubated with lectins at this temperature, washed free of soluble lectin and then transferred to 37°C, the lectin receptors cluster and the cells agglutinate. Normal cells in mitosis are more easily agglutinated than cells in interphase, and thus in this respect resemble transformed cells (Fox et al., 1971).

There is increasing evidence that microtubular-like structures are involved in lectin-mediated cell agglutination, because agents such as vinblastine, colchicine or colcemid, which bind to these structures, decrease the agglutination of cells (Yin et al., 1972). Microtubular proteins are probably not present on the cell surface but on the cytoplasmic side of the plasma membrane (Weber et al., 1975) and transmembrane communication may thus exist (review, Edelman, 1976). This resembles the situation in erythrocytes, where it is known that membrane spanning glycoproteins interact with peripheral proteins on the inner surface of the membrane (Nicolson and Painter, 1973; Ji and Nicolson, 1974).

Inbar et al. (1973) have used fluorescence polarization to measure the rotational relaxation time of concanavalin A bound to receptors on normal and SV40 transformed hamster fibroblasts and on the other hand normal lymphocytes and malignant lymphoma cells. In spite of some problems associated with the technique, where the movement of receptors is measured rather far from the marker that is monitored, the results appear interesting. They show that the mobility is increased in transformed fibroblasts but in contrast decreased in lymphoma cells. Lymphoma cells also contain decreased cholesterol in the membrane, which should result in increased mobility (Inbar and Shinitzky, 1974). The same group has also found that the cap formation by fluorescently labeled concanavalin A which occurs in normal lymphocytes, is decreased in lymphoma cells (Mintz and Sachs, 1975). Thus also the lateral movement of lectin receptors in malignant lymphocytes is slower than in normal cells.

Why then do transformed cells and trypsinized normal fibroblasts agglutinate more easily than normal control cells? The exact answer is not known, but in surveying the available experimental data I favor the following hypothesis. The major cell surface glycoprotein difference between transformed cells and their untransformed counterparts seems to be the presence of fibronectin in the untransformed cells. By saturating the lectin-binding sites of fibroblast monolayers and then labeling with galactose oxidase, we have shown that the major lectin receptor in normal non-trypsinized cells is fibronectin (Gahmberg and Hakomori, 1975b). This protein is absent from transformed and trypsin-treated cells. However, both transformed and trypsinized normal cells have on their surface a group of similar proteins which interact with the lectins (Gahmberg and Hakomori, 1975b). These proteins are present in normal cells but since fibronectin is present it competes more effectively for lectin binding. It is reasonable to expect that fibronectin is rather immobile on the cell surface under these conditions. There is some evidence for this because it is associated with fibrillar structures (Wartiovaara et al., 1974). On the other hand, the smaller molecular weight lectin-binding proteins found on transformed and trypsinized cells would be more mobile in the membrane and would thus be able to form clusters more easily, which, in turn, would facilitate the agglutination process.

4.5. Glycopeptides and oligosaccharides from the surface of normal and transformed cells

Warren, Glick and Buck have studied the glycopeptides from the surface and intracellular membranes of hamster, mouse and chicken fibroblasts (Buck et al., 1970, 1971, 1974) (Fig. 11). Cells were prelabeled with radioactive fucose or glucosamine and then treated with trypsin to obtain glycoprotein fragments. These were then digested extensively with pronase to obtain glycopeptides. On Sephadex G50 gel filtration the radioactive glycopeptides from transformed cells had higher apparent molecular weights than those from normal cells. Also, glycopeptides from normal cells in the logarithmic phase of growth had higher molecular weights than those from confluent stationary phase cells. If the glycopeptides were treated with neuraminidase, the differences were abolished. This indicates that there is more extensive sialylation of the oligosaccharides in transformed and actively growing cells. Very recently Ogata et al. (1976) have provided evidence that not only sialic acid residues contribute to these differences. Neuraminidase-treated glycopeptides from untransformed and transformed cells also differed in their affinity for concanavalin A, and differences in molecular weights were detected by column chromatography on Sephadex G-50.

A disturbing fact concerning studies of this kind is that we do not know from which protein(s) these glycopeptides originate and this has made interpretation

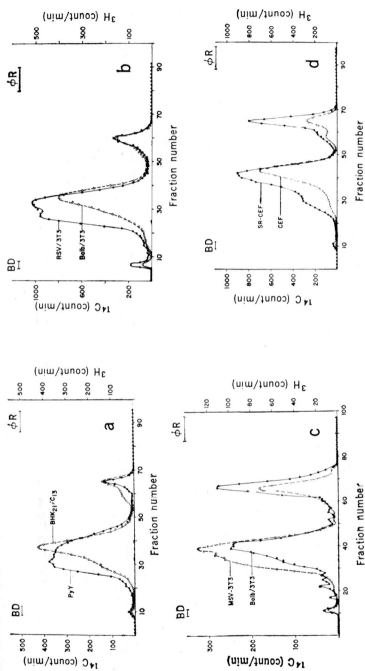

Fig. 11. Comparison of pronase-digested glycopeptides from the surfaces of normal and transformed cells removed from the surface of control and transformed cells by trypsin digestion.

Cells were grown as monolayers for 72 h in the presence of 50 μCi L-[^{3}H]fucose (●——●), or L-[^{14}C] fucose (○---○). Monolayers were washed and the cells were digested with purified trypsin (1 mg per ml of tris-buffered saline, pH 7.5) for 15 min at 37°C. Trypsin inhibitor was added and the cells were removed from the tryptic digest by centrifugation. The tryptic digests ("trypsinates") to be compared were pooled and digested with pronase (0.1 mg/ml of digest) for 5 days at 37°C in the presence of toluene. The digests were then cochromatographed on Sephadex G-50 columns (0.8 cm × 100 cm). Columns were developed in buffer containing tris-acetate (pH 9.0), 0.1% SDS, 0.01 M methylene-diamine tetraacetate, and 0.1% 2-mercaptoethanol. Fractions of 0.7 ml were collected, and radioactivity was determined by scintillation counting in Aquasol (New England Nuclear): (a) "trypsinates from BHK$_{21}$/C$_{13}$ cells and BHK$_{21}$/C$_{13}$ cells transformed with polyoma virus (PyY); (b) "trypsinates" from Balb/3T3 cells and Balb/3T3 cells transformed with RSV (RSV-3T3); (c) "trypsinates" from Balb/3T3 cells and Balb/3T3 cells transformed by MSV (MSV-3T3); and (d) "trypsinates" from chick embryo fibroblasts (CEF) and CEF transformed by SR-RSV (SR-CEF). BD, fractions in which blue dextran 2000 was eluted; φR, fractions in which phenol red was eluted. (Reproduced with permission from Buck et al. (1971).)

more difficult. Recent experiments, however, show that fibronectin is the major protein removed by proteases from the surface of normal confluent cells and that this protein can be readily labeled by galactose oxidase in hamster cells without using neuraminidase. The low molecular weight glycopeptides released by proteases may therefore originate mainly from this protein in normal cells, but in transformed and trypsinized cells the glycopeptides certainly come from quite different glycoproteins. There is also evidence that these differences are not restricted to the plasma membrane but are also present on intracellular membranes (Buck et al., 1974). Meezan et al. (1969) have also studied the glycopeptides from different cellular fractions of untransformed mouse 3T3 cells and SV40 transformed 3T3 cells. Gel filtration of pronase digested samples on Sephadex G-50 revealed that the elution patterns were different with a prominent high molecular weight peak being obtained from the normal cells. A second broad peak was more prominent from transformed cells.

Muramatsu et al. (1973a) have compared the glycopeptides obtained by pronase digestion of human diploid cells in growing and non-growing states. Again the glycopeptides were of a larger size when isolated from growing cells. These investigators used endoglycosidases (Koide and Muramatsu, 1974) to obtain a glycopeptide fragment of the protein-carbohydrate linkage region containing fucose and glucosamine plus a few amino acids. These fragments were different when isolated from growing or non-growing cells. In non-growing cells two fragments were obtained with molecular weights of 700 and 900 daltons, but growing cells contained only one fragment with a molecular weight of 800 daltons. Whether the differences are in the peptide or carbohydrate portions is not known.

Van Beek et al. (1973) have studied the glycopeptides from normal and malignant lymphoblasts, normal and transformed fibroblasts and normal liver cells and Novikoff hepatoma cells. The glycopeptides were obtained by pronase digestion of fucose labeled cells. In all cases the malignant cells contained glycopeptides of higher molecular weights than the normal cells. This would thus seem to be a general property of malignant cells. The reason for the extension of the carbohydrate chains in growing and malignant cells is not known. One interesting possibility is the following. We have observed recently that a human erythrocyte variant, En(a-), lacks the major surface glycoprotein, PAS 1. In these cells the other major surface glycoprotein, band 3, has longer oligosaccharide chains than the band 3 of normal cells (Gahmberg et al., 1976b). Both transformed fibroblasts and En(a-) erythrocytes thus lack a major surface glycoprotein. In these cells the sugar transferases may then direct their activity toward other acceptors resulting in more complex oligosaccharide chains on the glycoproteins which remain.

4.6. Transport of glucose and glucose analogs in normal and malignant cells

One of the earliest changes associated with malignant transformation is an increased uptake of glucose and glucose analogs (Hatanaka and Hanafusa,

1970). This was first shown using chick embryo fibroblasts, which were massively transformed with Rous sarcoma virus. Such changes have since been confirmed in a number of laboratories (Martin et al., 1971; Weber, 1973; Kletzien and Perdue, 1974a; Kletzien and Perdue, 1974b). There is good reason to believe that the sugar transport proteins are surface glycoproteins and thus changes in these may be anticipated. Transformed cells show the highest rate of transport, growing normal cells a slower rate of transport and contact-inhibited cells an even slower transport (Weber, 1973; Kletzien and Perdue, 1974a; Kletzien and Perdue, 1974b). The K_m values for transport were similar, but the V_{max} for hexose uptake was appreciably higher in transformed cells. This indicates that the molecular change is not a modification of preexisting transport sites but an increase in the number of these. Cells temperature-sensitive for transformation transport glucose at a higher rate at the permissive temperature than at the non-permissive (Weber, 1973; Kletzien and Perdue, 1974b). Enhanced glucose transport can be induced in resting contact-inhibited cells by very different types of reagents. Protease treatment results in an increase in transport (Vaheri et al., 1973), but similar enhancement is seen using neuraminidase (Vaheri et al., 1972) or insulin (Vaheri et al., 1973). Proteases and neuraminidase probably induce a conformational change in the plasma membrane resulting in an increased number of transport sites. The physiological function of insulin is to increase glucose transport, but the molecular mechanism is still rather unclear. Insulin probably binds to its receptor at the cell surface (Hollenberg and Cuatrecasas, 1975) and somehow activates the transport process. The actual glucose transport molecule(s) in the plasma membrane of nucleated cells is not known. In erythrocytes there is evidence that glucose transport is mediated through band 3, a membrane-penetrating glycoprotein, and membrane glycoproteins may be similarly responsible for glucose transport in other cells.

Holley and Kiernan (1974) and others have put forward the hypothesis that the fundamental change occurring in transformed cells is an increased transport of various nutrients. Because plasma membrane glycoproteins seem to be surface glycoproteins (Cabantchik and Rothstein, 1974a; Kyte, 1974; Ho and Guidotti, 1975) the observed chemical changes in surface glycoproteins may also be causally related to changes in transport.

4.7. Cell surface recognition and cellular adhesion

Altered adhesive behavior is a fundamental characteristic of transformed cells and may be partially responsible for metastatic spreading. It is important therefore to understand the molecular basis of cell–cell adhesive interactions. It is now well established from studies of lower organisms that cell surface glycoproteins are involved in cellular aggregation. Yen and Ballou (1973) have characterized a sexual agglutination factor from the yeast *Hansenula wingei* type 5. This unusual protein contains 85% carbohydrate, 10% protein and 5% phosphate. The carbohydrate is composed solely of mannose residues, forming oligosaccharide side chains varying in length from 1 to 15 residues. Most of the amino acids are serine and threonine and mannose is attached to every second of these. Humphreys and co-workers have characterized an aggregation factor from the marine sponge *Microciona parthena* which is composed of similar amounts of amino acids and carbohydrate (Henkart et al., 1973). The sugars include galactose, mannose, uronic acid, glucosamine and galactosamine. The aggregation factor has an unusual shape with radiating arms from a small core. Cell aggregation induced by this factor requires the presence of divalent cations.

Siu et al. (1976) have studied a surface protein from the slime mold *Dictyostelium discoideum*. This organism responds to depletion of nutrients by aggregation and mutual cohesiveness. This results in an aggregate which develops into a multicellular unit. The protein responsible for aggregation is also able to agglutinate sheep erythrocytes. It is easily purified by its affinity to Sepharose beads and has a molecular weight of 100 000 daltons and is composed of four subunits with molecular weights of 26 000 daltons. The carbohydrate-binding protein is synthesized during the times when the cells acquire the capacity for intercellular adhesion.

Similarly, in higher organisms there is increasing evidence that cell surface glycoproteins are involved in cell recognition and adhesion. Merrell et al. (1975) have isolated a plasma membrane protein fraction from neural retina and optic tectum. The factor shows specificity for the two cell surfaces and is capable of inhibiting the specific aggregation of homologous cells. The active component seems to be rather small with a nominal molecular weight below 10 000 daltons, because it passes an Amicon PM-10 filter. On SDS polyacrylamide gel electrophoresis the molecular weight is 60 000 daltons. It stains with the PAS method for sialic acid and would thus seem to be a glycoprotein.

Much attention has focused recently on the possible role of cell surface glycosyltransferases in cell recognition and adhesion. This concept, advanced by Roseman, Roth and others, proposes that a glycosyltransferase on the surface of one cell recognizes an uncompleted oligosaccharide sequence on a neighbouring cell and instead of transferring the appropriate sugar, binds to the acceptor resulting in cell adhesion. There is a large body of evidence that cell surface glycosyl transferases exist (Roth et al., 1971; Roth and White, 1972; LaMont et al., 1974a; Patt and Grimes, 1974), but whether these represent remnants of Golgi glycosyltransferases which were transported to the external surface of cells during the biosynthesis of the plasma membrane and do not have any physiological function is still unclear. Roth and White (1972) have presented evidence that normal, contact-inhibited 3T3 cells can only transfer galactose from UDP-galactose to galactosyl receptors on adjacent cells, whereas 3T12 cells, which do not show contact-inhibition of growth, do not show this requirement and can transfer sugars to the same cell as the enzyme is located. Isselbacher and his colleagues have found a close correlation between lectin agglutinability and cell surface galactosyltransferase (LaMont et al., 1974a; Podolsky et al., 1974). For example, human erythrocytes are not easily ag-

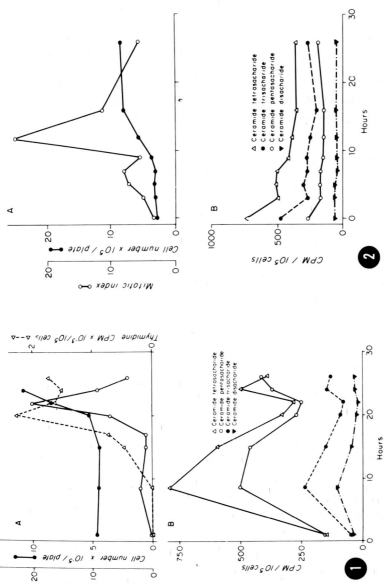

Fig. 12. Variation of surface-labelled glycolipids in hamster Nil and polyoma virus-transformed Nil cells (Nilpy). Synchronization started from confluent trypsinized cells for Nil cells and was obtained by a double thymidine block for Nilpy cells: (1) Nil cells; and (2) Nilpy cells. A shows the changes of mitotic index, cell number and thymidine incorporation and B shows the radioactivities in glycolipids at the corresponding time as in A. Note the pronounced peak in the G_1 phase in normal cells, which was absent in the transformed cells. (Reproduced with permission from Gahmberg and Hakomori, 1974.)

glutinated by concanavalin A but rabbit erythrocytes are. Rabbit erythrocytes display cell surface galactosyltransferase activity but similar activity is absent from human erythrocytes. Purified galactosyltransferase from rabbit erythrocyte membranes can be adsorbed to human erythrocytes, which results in sensitivity to agglutination by concanavalin A. Intestinal villus crypt cells and tumor cells from the small intestine which show a high surface galactosyltransferase activity are also easily agglutinated by concanavalin A (LaMont et al., 1974b) whereas cells from the villus tip show a decreased enzymatic activity and are less agglutinable. Thus it is possible that cell surface glycosyltransferases themselves function as receptors for cell surface recognition molecules.

Bosmann et al. (1973) have studied the surface properties of two melanoma cell lines with high and low metastatic potential. The high metastatic line has a much higher level of glycosyltransferase activity and also displays a high level of surface glycosyltransferase acceptor reactions. This again provides support for the hypothesis that cell surface glycosyltransferases may be involved in cell adhesion.

The very recent finding that the major fibroblast surface glycoprotein (fibronectin) has the ability to agglutinate cells (Yamada et al., 1975), strongly argues for a role of this protein in cell adhesion. This is the major surface protein removed by trypsin and EDTA, both of which are used for dispersing cell monolayers. This protein is synthesized during the G_1 phase of the cell cycle, which also represents that part of the cell cycle where contact-inhibited cells remain during the stationary growth phase. We also know that glycolipids, which are potential substrates for such a lectin, also are most strongly exposed in the G_1 phase (Gahmberg and Hakomori, 1975a) (Fig. 12). The finding that the elongation of glycolipid oligosaccharides, which normally occurs on cell contact (Hakomori, 1970) is absent from transformed cells, also fits the hypothesis that fibronectin and complex glycoconjugates may interact in the G_1 phase resulting in contact-inhibition of growth.

Acknowledgements

The personal research cited in this paper was supported by the Finnish Cancer Society, the Finska Läkaresällskapet and the Academy of Finland. The skilful technical assistance of Marja Wilkman is acknowledged.

References

Adair, W.L. and Kornfeld, S. (1974) Isolation of the receptors for wheat germ agglutinin and the *Ricinus communis* lectins from human erythrocytes using affinity chromatography. J. Biol. Chem. 249, 4696–4704.

Agrawal, B.B.L. and Goldstein, I.J. (1972) Concanavalin A, the jack bean agglutinin. Methods Enzymol. 28, part B, 313–318.

Albertsson, P.Å. (1971) Partition of Cell Particles and Macromolecules, 2nd ed., Wiley, New York.

Allan, D. and Crumpton, M.J. (1972) Glycoprotein receptors for concanavalin A isolated from pig lymphocyte plasma membrane by affinity chromatography in sodium deoxycholate, Nature (London) New Biol. 236, 23–25.

Amaral, D., Bernstein, L., Morse, D. and Horecker, B.L. (1963) Galactose oxidase of *Polyporus circinatus*: a copper enzyme. J. Biol. Chem. 238, 2281–2284.

Andersson, L.C., Wasastjerna, C. and Gahmberg, C.G. (1976) Different surface glycoprotein patterns on human T-, B- and leukemic-lymphocytes. Int. J. Cancer 17, 40–46.

Anundi, H., Rask, L., Östberg, L. and Peterson, P.A. (1975) The subunit structure of thymus leukemia antigens. Biochemistry 14, 5046–5054.

Ashwell, G. and Morell, A.G. (1974) The role of surface carbohydrates in hepatic recognition and transport of circulating glycoproteins. In: A. Meister (Ed.), Advances in Enzymology, Vol. 41, Wiley, New York, pp. 99–128.

Aub, J.C., Tieslau, C. and Lankester, A. (1963) Reactions of normal and tumor cell surfaces to enzymes, I. Wheat germ lipase and associated mucopolysaccharides. Proc. Natl. Acad. Sci. USA 50, 613–619.

Baenziger, J. and Kornfeld, S. (1974) Structure of the carbohydrate units of IgE immunoglobulins, II. Sequence of the sialic acid-containing glycopeptides. J. Biol. Chem. 249, 1897–1903.

Bahl, O.P. (1969) Human chorionic gonadotropin, I. Purification and physicochemical properties. J. Biol. Chem. 244, 567–574.

Bayard, B. and Roux, D. (1975) Hydrazinolysis and nitrous deamination of glycoproteins. Evidence for a common inner core in carbohydrate moiety. FEBS Lett. 55, 206–211.

Besançon, F., Ankel, H. and Basu, S. (1976) Specificity and reversibility of interferon–ganglioside interaction. Nature (London) 259, 576–578.

Blumberg, P.M. and Robbins, P.W. (1975) Effect of proteases on activation of resting chick embryo fibroblasts and on cell surface proteins. Cell 6, 137–147.

Blumenfeld, O.O. and Zvilichovsky, B. (1972) Isolation of glycoproteins from red cell membranes using pyridine. Methods Enzymol. 28, part B, 245–252.

Bonner, W.M. and Laskey, R.A. (1974) A film detection method for tritium-labelled proteins and nucleic acids in polyacrylamide gels. Eur. J. Biochem. 46, 83–88.

Bosmann, H.B., Bieber, G.F., Brown, A.E., Case, K.R., Gersten, D.M., Kimmerer, T.W. and Lione, A. (1973) Biochemical parameters correlated with tumor cell inplantation. Nature (London) 246, 487–489.

Bray, D. and Brownlee, S.M. (1973) Peptide mapping of proteins from acrylamide gels. Anal. Biochem. 55, 213–221.

Bretscher, M.S. (1971a) Major human erythrocyte glycoprotein spans the cell membrane. Nature (London) New Biol. 231, 229–232.

Bretscher, M.S. (1971b) Human erythrocyte membranes: specific labeling of surface proteins. J. Mol. Biol. 58, 775–781.

Bretscher, M.S. (1972a) Asymmetrical lipid bilayer structure for biological membranes. Nature (London) New Biol. 236, 11–12.

Bretscher, M.S. (1972b) Phosphatidyl-ethanolamine: Differential labelling in intact cells and cell ghosts of human erythrocytes by a membrane-impermeable reagent. J. Mol. Biol. 71, 523–528.

Brunette, D.M. and Till, J.E. (1971) A rapid method for the isolation of L-cell surface membranes using an aqueous two-phase polymer system. J. Membrane Biol. 5, 215–224.

Buck, C.A., Glick, M.C. and Warren, L. (1970) A comparative study of glycoproteins from the surface of control and Rous sarcoma virus transformed hamster cells. Biochemistry 9, 4567–4576.

Buck, C.A., Glick, M.C. and Warren, L. (1971) Glycopeptides from the surface of control and virus-transformed cells. Science 172, 169–171.

Buck, C.A., Fuhrer, J.P., Soslau, G. and Warren, L. (1974) Membrane glycopeptides from subcellular fractions of control and virus-transformed cells. J. Biol. Chem. 249, 1541–1550.

Burger, M.M. (1969) A difference in the architecture of the surface membrane of normal and virally transformed cells. Proc. Natl. Acad. Sci. USA 62, 994–1001.

Burger, M.M. and Goldberg, A.R. (1967) Identification of a tumor-specific determinant on neoplastic cell surfaces. Proc. Natl. Acad. Sci. USA 57, 359–366.

Burger, M.M., Bombik, B.M., Breckenridge, B.M. and Sheppard, J.R. (1972) Growth control and cyclic alterations of cyclic AMP in the cell cycle. Nature (London) New Biol. 239, 161–163.

Bürk, R.R. (1968) Reduced adenyl cyclase activity in a polyoma virus transformed cell line. Nature (London) 219, 1272–1275.

Cabantchik, Z.I. and Rothstein, A. (1974a) Membrane proteins related to anion permeability of human red blood cells, I. Localization of disulfonic stilbene binding sites in proteins involved in permeation. J. Membrane Biol. 15, 207–226.

Cabantchik, Z. I. and Rothstein, A. (1974b) Membrane proteins related to anion permeability of human red blood cells, II. Effects of proteolytic enzymes on disulfonic stilbene sites of surface proteins. J. Membrane Biol. 15, 227–248.

Cabantchik, Z.I., Balshin, M., Breuer, W. and Rothstein, A. (1975) Pyridoxal phosphate. An anionic probe for protein amino groups exposed on the outer and inner surfaces of intact human red blood cells. J. Biol. Chem. 250, 5130–5136.

Chang, K.-J., Bennett, V. and Cuatrecasas, P. (1975) Membrane receptors as general markers for plasma membrane isolation procedures. J. Biol. Chem. 250, 488–500.

Cuatrecasas, P. (1973) Interaction of *Vibrio cholerae* enterotoxin with cell membrane. Biochemistry 12, 3547–3558.

Datta, P. (1974) Labeling of the external surface of hamster and mouse fibroblasts with [^{14}C]sialic acid. Biochemistry 13, 3987–3991.

Davies, D.A.L., Alkins, B.J., Boyse, E.A., Old, L.J. and Stockert, E. (1969) Soluble TL and H-2 antigens prepared from a TL positive leukaemia of a TL negative mouse strain. Immunology 16, 669–676.

Del Villano, B.C., Nave, B., Croher, B.P., Lerner, R.A. and Dixon, F.J. (1975) The oncornavirus glycoprotein GP 69/71: a constituent of the surface of normal and malignant thymocytes. J. Exp. Med. 141, 172–187.

DePierre, J.W. and Karnovsky, M.L. (1973) Plasma membranes of mammalian cells. A review of methods for their characterization and isolation. J. Cell Biol. 56, 275–303.

DePierre, J.W. and Karnovsky, M.L. (1974) Ecto-enzymes of the guinea pig polymorphonuclear leukocyte II. Properties and suitability as markers for the plasma membrane. J. Biol. Chem. 249, 7121–7129.

Dod, B.J. and Gray, G.M. (1968) The lipid composition of rat liver plasma membranes. Biochim. Biophys. Acta 150, 397–404.

Durocher, J.R., Payne, R.C. and Conrad, M.E. (1975) Role of sialic acid in erythrocyte survival. Blood 45, 11–20.

Dutton, A. and Singer, S.J. (1975) Crosslinking and labeling of membrane proteins by transglutaminase-catalyzed reactions. Proc. Natl. Acad. Sci. USA 72, 2568–2571.

Edelman, G.M. (1976) Surface modulation in cell recognition and cell growth. Science 192, 218–226.

Edelman, G.M., Yahara, I. and Wang, J.L. (1973) Receptor mobility and receptor-cytoplasmic interactions in lymphocytes. Proc. Natl. Acad. Sci. USA 70, 1442–1446.

Edsall, J.T., Gilbert, G.A. and Scheraga, H.A. (1955) The non-clotting component of human plasma fraction I-1 ("cold – insoluble globulin"). J. Am. Chem. Soc. 77, 157–161.

Emmelot, P. and Bos, C.J. (1968) Studies on plasma membranes, V. On the lipid dependence of some phosphohydrolases of isolated rat-liver plasma membranes. Biochim. Biophys. Acta 150, 341–353.

Estensen, R.D., Hadden, J.W., Hadden, E.M., Touraine, F., Touraine, J.-L., Haddox, M.K. and Goldberg, N.D. (1974) Phorbol myristate acetate: effects of a tumor promoter on intracellular cyclic GMP in mouse fibroblasts and as a mitogen on human lymphocytes. In: B. Clarkson and R. Baserga (Eds.), Control of Proliferation in Animal Cells. Cold Spring Harbor Conferences on Cell Proliferation, Vol. 1, Cold Spring Harbor Laboratory, New York, pp. 627–634.

Evans, W.H. (1970) Fractionation of liver plasma membranes prepared by zonal centrifugation. Biochem. J. 166, 833–842.

Eylar, E.H., Madoff, M.A., Brody, O.V. and Oncley, J.L. (1962) The contribution of sialic acid to the surface charge of the erythrocyte. J. Biol. Chem. 237, 1992–2000.

Fairbanks, G., Steck, T.L. and Wallach, D.F.H. (1971) Electrophoretic analysis of the major polypeptides of the human erythrocyte membrane. Biochemistry 10, 2606–2617.

Fisher, K.A. (1976) Analysis of membrane halves: cholesterol. Proc. Natl. Acad. Sci. USA 73, 173–177.

Fox, T.O., Sheppard, J.R. and Burger, M.M. (1971) Cyclic membrane changes in animal cells: transformed cells permanently display a surface architecture detected in normal cells only during mitosis. Proc. Natl. Acad. Sci. USA 68, 244–247.

Gahmberg, C.G. (1976) External labeling of human erythrocyte glycoproteins. Studies with galactose oxidase and fluorography. J. Biol. Chem. 251, 510–515.

Gahmberg, C.G. and Simons, K. (1970) Isolation of plasma membrane fragments from BHK21 cells. Acta Pathol. Microbiol. Scand. 78, 176–182.

Gahmberg, C.G. (1971) Proteins and glycoproteins of hamster kidney fibroblast (BHK21) plasma membranes and endoplasmic reticulum. Biochim. Biophys. Acta 249, 81–85.

Gahmberg, C.G., Simons, K., Renkonen, O. and Kääriäinen, L. (1972a) Exposure of proteins and lipids in the Semliki Forest virus membrane. Virology 50, 259–262.

Gahmberg, C.G., Utermann, G. and Simons, K. (1972b) The membrane proteins of Semliki Forest virus have a hydrophobic part attached to the viral membrane. FEBS Lett. 28, 179–182.

Gahmberg, C.G. and Hakomori, S. (1973a) External labeling of cell surface galactose and galactosamine in glycolipid and glycoprotein of human erythrocytes. J. Biol. Chem. 248, 2135–2142.

Gahmberg, C.G. and Hakomori, S. (1973b) Altered growth behavior of malignant cells associated with change in externally labeled glycoprotein and glycolipid. Proc. Natl. Acad. Sci. USA 70, 3329–3333.

Gahmberg, C.G. and Hakomori, S. (1974) Organization of glycolipids and glycoproteins in surface membranes: dependency on cell cycle and on transformation. Biochem. Biophys. Res. Commun. 59, 283–291.

Gahmberg, C.G. and Hakomori, S. (1975a) Surface carbohydrates of hamster fibroblasts, I. Chemical characterization of surface-labeled glycosphingolipids and a specific ceramide tetrasaccharide for transformants. J. Biol. Chem. 250, 2438–2446.

Gahmberg, C.G. and Hakomori, S. (1975b) Surface carbohydrates of hamster fibroblasts, II. Interaction of hamster Nil cell surfaces with *Ricinus communis* lectin and concanavalin A as revealed by surface galactosyl label. J. Biol. Chem. 250, 2447–2455.

Gahmberg, C.G., Kiehn, D. and Hakomori, S. (1974) Changes in a surface-labelled galactoprotein and in glycolipid concentrations in cells transformed by a temperature-sensitive polyoma virus mutant. Nature (London) 248, 413–415.

Gahmberg, C.G. and Hakomori, S. (1976) Organization of glycoprotein and glycolipid in the plasma membrane of normal and transformed cells as revealed by galactose oxidase. In: L. Manson (Ed.), Biomembranes, Vol. 8, Wistar Press, Philadelphia, pp. 131–165.

Gahmberg, C.G., Häyry, P. and Andersson, L.C. (1976a) Characterization of surface glycoproteins of mouse lymphoid cells. J. Cell Biol. 68, 642–653.

Gahmberg, C.G., Myllylä, G., Leikola, A., Pirkola, A. and Nordling, S. (1976b) Absence of the major sialoglycoprotein in the membrane of human En(a-) erythrocytes and increased glycosylation of band 3. J. Biol. Chem. 251, 6108–6116.

Gahmberg, C.G., Myllylä, G., Leikola, A., Pirkola, A. and Nordling, S. (1976c) Decreased transport of anions in the human erythrocyte variant En(a-). (Submitted for publication).

Gahmberg, C.G., Itaya, K. and Hakomori, S. (1976d) External labeling of cell surface carbohydrates. In: E.D. Korn (Ed.), Methods in Membrane Biology, Vol. 17, Plenum Press, New York, pp. 175–206.

Garoff, H., Simons, K. and Renkonen, O. (1974) Isolation and characterization of the membrane polypeptides of Semliki Forest virus. Virology 61, 493–504.

Garrido, F., Schirrmacher, V. and Festenstein, H. (1976) H-2-like specificities of foreign haplotypes appearing on a mouse sarcoma after vaccinia virus infection. Nature (London) 259, 228–230.

Gatt, R. and Berman, E.R. (1966) A rapid procedure for the estimation of amino sugars on a micro scale. Anal. Biochem. 15, 167–171.

Gesner, B.M. and Ginsburg, V. (1964) Effect of glycosidases on the fate of transfused lymphocytes. Proc. Natl. Acad. Sci. USA 52, 750–755.

Goldberg, N.D., Haddox, M.K., Dunham, E., Lopez, C. and Hadden, J.W. (1974) The Yin Yang hypothesis of biological control: opposing influences of cyclic GMP and cyclic AMP in the regulation of cell proliferation and other biological processes. In: B. Clarkson and R. Baserga (Eds.), Control of Proliferation in Animal Cells. Cold Spring Harbor Conferences on Cell Proliferation, Vol. 1, Cold Spring Harbor Laboratory, New York, pp. 609–625.

Gräsbeck, R., Nordman, C.T. and de la Chapelle, A. (1964) Leukocyte-mitogenic effect of serum from rabbits immunized with human leukocytes. Acta Med. Scand. supp. 412, 39–47.

Hakomori, S. (1970) Cell density-dependent changes of glycolipid concentrations in fibroblasts, and loss of this response in virus-transformed cells. Proc. Natl. Acad. Sci. USA 67, 1741–1747.

Hakomori, S. and Kobata, A. (1975) Blood group antigens, In: M. Sela (Ed.), The Antigens, Vol. 2. Academic Press, New York, pp. 80–140.

Hatanaka, M. and Hanafusa, J. (1970) Analysis of a functional change in membrane in the process of cell transformation by Rous sarcoma virus; alteration in the characteristics of sugar transport. Virology 41, 647–652.

Haywood, A.M. (1974) Characteristics of Sendai virus receptors in a model membrane. J. Mol. Biol. 83, 427–436.

Heitzmann, H. and Richards, F.M. (1974) Use of the avidin-biotin complex for specific staining of biological membranes in electron microscopy. Proc. Natl. Acad. Sci. USA 71, 3537–3541.

Helenius, A. and Simons, K. (1972) The binding of detergents to lipophilic and hydrophilic proteins. J. Biol. Chem. 247, 3656–3661.

Helenius, A. and Simons, K. (1975) Solubilization of membranes by detergents. Biochim. Biophys. Acta 415, 29–79.

Henkart, P., Humphreys, S. and Humphreys, T. (1973) Characterization of sponge aggregation factor. A unique proteoglycan complex. Biochemistry 12, 3045–3050.

Hewitt, B.R. (1958) Spectrophotometric determination of total carbohydrate. Nature (London) 182, 246.

Ho, M.K. and Guidotti, G. (1975) A membrane protein from human erythrocytes involved in anion exchange. J. Biol. Chem. 250, 675–683.

Hogg, N.M. (1974) A comparison of membrane proteins of normal and transformed cells by lactoperoxidase labelling. Proc. Natl. Acad. Sci. USA 71, 489–492.

Hollenberg, M.D. and Cuatrecasas, P. (1975) Insulin: interaction with membrane receptors and relationship to cyclic purine nucleotides and cell growth. Fed. Proc. 34, 1556–1563.

Holley, R.W. and Kiernan, J.A. (1974) Control of the initiation of DNA synthesis in 3T3 cells: low-molecular-weight nutrients. Proc. Natl. Acad. Sci. USA 71, 2942–2945.

Holmgren, J., Lönnroth, I., Månsson, J.-E. and Svennerholm, L. (1975) Interaction of cholera toxin and membrane GMl ganglioside of small intestine. Proc. Natl. Acad. Sci. USA 72, 2520–2524.

Howe, C., Lloyd, K.O. and Lee, L.T. (1972) Isolation of glycoproteins from red cell membranes using phenol. Methods Enzymol., Vol. 28 part B, 236–245.

Hsie, A.W. and Puck, T.T. (1971) Morphological transformation of Chinese hamster cells by dibutyryl adenosine cyclic 3′:5′-monophosphate. Proc. Natl. Acad. Sci. USA 68, 358–361.

Hubbard, A.L. and Cohn, Z.A. (1972) The enzymatic iodination of the red cell membrane. J. Cell Biol. 55, 390–405.

Hunt, R.C. and Brown, J.C. (1975) Identification of a high molecular weight trans-membrane protein in mouse L cells. J. Mol. Biol. 97, 413–422.

Hunt, R.C., Bullis, C.M. and Brown, J.C. (1975) Isolation of a concanavalin A receptor from mouse L cells. Biochemistry 14, 109–115.

Hunter, M.J. and Commerford, S.L. (1961) Pressure homogenization of mammalian tissues. Biochim. Biophys. Acta 47, 580–586.

Hynes, R.O. (1973) Alteration of cell-surface proteins by viral transformation and proteolysis. Proc. Natl. Acad. Sci. USA 70, 3170–3174.

Hynes, R.O. and Bye, J.M. (1974) Density and cell cycle dependence of cell surface proteins in hamster fibroblasts. Cell 3, 113–120.

Hynes, R.O. and Wyke, J.A. (1975) Alterations in surface proteins in chicken cells transformed by temperature-sensitive mutants of Rous sarcoma virus. Virology 64, 492–504.

Hynes, R.O., Wyke, J.A., Bye, J.M., Humphreys, K.C. and Pearlstein, E.S. (1975) Are proteases involved in altering surface proteins during viral transformation? In: E. Reich, D.B. Rifkin and E. Shaw (Eds.), Proteases and Biological Control. Cold Spring Harbor Conferences on Cell Proliferation, Vol. 2. Cold Spring Harbor Laboratory, New York, pp. 931–944.

Inbar, M. and Sachs, L. (1969) Interaction of a carbohydrate binding protein concanavalin A with normal and transformed cells. Proc. Natl. Acad. Sci. USA 63, 1418–1424.

Inbar, M. and Shinitzky, M. (1974) Cholesterol as a bioregulator in the development and inhibition of leukemia. Proc. Natl. Acad. Sci. USA 71, 4229–4231.

Inbar, M., Shinitzky, M. and Sachs, L. (1973) Rotational relaxation time of concanavalin A bound to the surface membrane of normal and malignant transformed cells. J. Mol. Biol. 81, 245–253.

Invernizzi, G. and Parmiani, G. (1975) Tumour-associated transplantation antigens of chemically induced sarcomata cross reacting with allogeneic histocompatibility antigens. Nature (London) 254, 713–714.

Itaya, K., Gahmberg, C.G. and Hakomori, S. (1975) Cell surface labeling of erythrocyte glycoproteins by galactose oxidase and Mn^{2+}-catalyzed coupling reaction with methionine sulfone hydrazide. Biochem. Biophys. Res. Commun. 64, 1028–1035.

Jancik, J. and Schauer, R. (1974) Sialic acid – a determinant of the life-time of rabbit erythrocytes. Z. Physiol. Chem. 355, 395–400.

Janossy, G. and Greaves, M.F. (1971) Lymphocyte activation, I. Response of T and B lymphocytes to phytomitogens. Clin. Exp. Immunol. 9, 483–498.

Ji, T.H. and Nicolson, G.L. (1974) Lectin binding and perturbation of the outer surface of the cell membrane induces a transmembrane organizational alteration at the inner surface. Proc. Natl. Acad. Sci. USA 71, 2212–2216.

Johnson, G.S., Friedman, R.M. and Pastan, I. (1971) Restoration of several morphological characteristics of normal fibroblasts in sarcoma cells treated with adenosine-3':5'-cyclic monophosphate and its derivatives. Proc. Natl. Acad. Sci. USA 68, 425–429.

Kant, J.A. and Steck, T.L. (1973) Specificity in the association of glyceraldehyde 3-phosphate dehydrogenase with isolated human erythrocyte membranes. J. Biol. Chem. 248, 8457–8464.

Keski-Oja, J., Vaheri, A. and Ruoslahti, E. (1976) Fibroblast surface antigen (SF): the external glycoprotein lost in proteolytic stimulation and malignant transformation. Int. J. Cancer 17, 261–269.

Kimelberg, H.K. (1977) The influence of membrane fluidity on the activity of membrane-bound enzymes. In: G. Poste and G.L. Nicolson (Eds.), Cell Surface Reviews, Vol. 3, North-Holland, Amsterdam, pp. 205–293.

Kishida, Y., Olsen, B.R., Berg, R.A. and Prockop, D.J. (1975) Two improved methods for preparing ferritin-protein conjugates for electron microscopy. J. Cell Biol. 64, 331–339.

Klein, G. (1966) Tumor antigens. Annu. Rev. Microbiol. 20, 223–252.

Klenk, H.D. and Choppin, P.W. (1970) Glycosphingolipids of plasma membranes of cultured cells and an enveloped virus (SV5) grown in these cells. Proc. Natl. Acad. Sci. USA 66, 57–64.

Kletzien, R.F. and Perdue, J.F. (1974a) Sugar transport in chick embryo fibroblasts, I.A functional change in the plasma membrane associated with the rate of cell growth. J. Biol. Chem. 249, 3366–3374.

Kletzien, R.F. and Perdue, J.F. (1974b) Sugar transport in chick embryo fibroblasts, II. Alterations in transport following transformation by a temperature-sensitive mutant of the Rous sarcoma virus. J. Biol. Chem. 249, 3375–3382.

Koide, N. and Muramatsu, T. (1974) Endo-β-N-acetylglucosaminidase acting on carbohydrate moieties of glycoproteins. J. Biol. Chem. 249, 4897–4904.

Kornberg, R.D. and McConnell, H.M. (1971a) Lateral diffusion of phospholipids in a vesicle membrane. Proc. Natl. Acad. Sci. USA 68, 2564–2568.

Kornberg, R.D. and McConnell, H.M. (1971b) Inside-outside transitions of phospholipids in vesicle membranes. Biochemistry 10, 1111–1120.

Kuusela, P., Ruoslahti, E. and Vaheri, A. (1975) Polypeptides of a glycoprotein antigen present in serum and surface of normal but not of transformed chicken fibroblasts. Biochim. Biophys. Acta 379, 295–303.

Kuusela, P., Ruoslahti, E., Engvall, E. and Vaheri, A. (1976) Immunological interspecies cross-reactions of fibroblast surface antigen (fibronectin). Immunochemistry 13, 639–642.

Kyte, J. (1974) The reactions of sodium and potassium ion-activated adenosine triphosphatase with specific antibodies. Implications for the mechanism of active transport. J. Biol. Chem. 249, 3652–3660.

Laemmli, U.K. (1970) Cleavage of structural proteins during the assembly of the head of bacteriophage T4. Nature (London) 227, 680–685.

Laine, R.A. and Hakomori, S. (1973) Incorporation of exogenous glycosphingolipids in plasma-membranes of cultured hamster cells and concurrent change of growth behavior. Biochem. Biophys. Res. Commun. 54, 1039–1045.

Laine, R.A., Esselman, W.J. and Sweeley, C.C. (1972) Gas-liquid chromatography of carbohydrates. Methods Enzymol. 28, part B, 159–167.

LaMont, J.T., Perrotto, J.L., Weiser, M.M. and Isselbacher, K.J. (1974a) Cell surface galactosyltransferase and lectin agglutination of thymus and spleen lymphocytes. Proc. Natl. Acad. Sci. USA 71, 3726–3730.

LaMont, J.T., Weiser, M.M. and Isselbacher, K.J. (1974b) Cell surface glycosyltransferase activity in normal and neoplastic intestinal epithelium of the rat. Cancer Res. 34, 3225–3228.

Laskey, R.A. and Mills, A.D. (1975) Quantitative film detection of [^{3}H] and [^{14}C] in polyacrylamide gels by fluorography. Eur. J. Biochem. 56, 335–341.

Liao, T.-H., Gallop, P.M. and Blumenfeld, O.O. (1973) Modification of sialyl residues of sialoglycoprotein(s) of the human erythrocyte surface. J. Biol. Chem. 248, 8247–8253.

Lindberg, B. (1972) Methylation analysis of polysaccharides. Methods Enzymol. 28, part B, 178–195.

Liu, T.-Y. and Chang, Y.H. (1971) Hydrolysis of proteins with p-toluenesulfonic acid. Determination of tryptophan. J. Biol. Chem. 246, 2842–2848.

Lorand, L., Shishido, R., Parameswaran, K.N. and Steck, T.L. (1975) Modification of human erythrocyte ghosts with transglutaminase. Biochem. Biophys. Res. Commun. 67, 1158–1166.

Lowry, O.H., Rosebrough, N.J., Farr, A.L. and Randall, R.J. (1951) Protein measurement with the Folin phenol reagent. J. Biol. Chem. 193, 265–275.

Maddy, A.H. (1966) The properties of the protein of the plasma membrane of ox erythrocytes. Biochim. Biophys. Acta 117, 193–200.

Marchesi, V.T. (1972) Isolation of membrane bound glycoproteins with lithium diiodosalicylate. Methods Enzymol. 28, part B, 252–254.

Marchesi, V.T., Tillack, T.W., Jackson, R.L., Segrest, J.P. and Scott, R.E. (1972) Chemical characterization and surface orientation of the major glycoprotein of the human erythrocyte membrane. Proc. Natl. Acad. Sci. USA 69, 1445–1449.

Marinetti, G.V., Ray, T.K. and Tomasi, V. (1969) Glucagon and epinephrine stimulation of adenyl cyclase in isolated rat liver plasma membranes. Biochem. Biophys. Res. Commun. 36, 185–193.

Martin, G.S., Venuta, S., Weber, M. and Rubin, H. (1971) Temperature-dependent alterations in sugar transport in cells infected by a temperature-sensitive mutant of the Rous sarcoma virus. Proc. Natl. Acad. Sci. USA 68, 2739–2741.

Martinez-Palomo, A., Wicher, R. and Bernhard, W. (1972) Ultrastructural detection of concanavalin surface receptors in normal and in polyoma-transformed cells. Int. J. Cancer 9, 676–684.

Marton, L.S.G. and Garvin, J.S. (1973) Subunit structure of major human erythrocyte glycoprotein. Depolymerization by heating ghosts with sodium dodecyl sulfate. Biochem. Biophys. Res. Commun. 52, 1457–1462.

Meezan, E., Wu, H.C., Black, P.H. and Robbins, P.W. (1969) Comparative studies on the carbohydrate-containing membrane components of normal and virus-transformed mouse fibroblasts, II. Separation of glycoproteins and glycopeptides by Sephadex chromatography. Biochemistry 8, 2518–2524.

Merrell, R., Gottlieb, D.I. and Glaser, L. (1975) Embryonal cell recognition. Extraction of an active plasma membrane component. J. Biol. Chem. 250, 5655–5659.

Mintz, U. and Sachs, L. (1975) Changes in the surface membrane of lymphocytes from patients with chronic lymphocytic leukemia and Hodgkin's disease. Int. J. Cancer 15, 253–259.

Morrison, M., Mueller, T.J. and Huber, C.T. (1974) Transmembrane orientation of the glycoproteins in normal human erythrocytes. J. Biol. Chem. 249, 2658–2660.

Mosher, D.F. (1975a) Cross-linking of cold-insoluble globulin by fibrin-stabilizing factor. J. Biol. Chem. 250, 6614–6621.

Mosher, D.F. (1975b) Labeling of cold-insoluble globulin and a fibroblast protein by fibrin-stabilizing factor. Fed. Proc. 34, 1567.

Moss, B. and Rosenblum, E.N. (1974) Hydroxylapatite chromatography of protein-sodium dodecyl sulfate complexes. J. Biol. Chem. 247, 5194–5198.

Mossesson, M.W. and Umfleet, R.A. (1970) The cold insoluble globulin of human plasma, I. Purification, primary characterization and relationship to fibrinogen and other cold insoluble fraction components. J. Biol. Chem. 245, 5728–5736.

Muramatsu, T., Atkinson, P.H. and Nathenson, S.G. (1973a) Cell-surface glycopeptides: growth-dependent changes in the carbohydrate-peptide linkage region. J. Mol. Biol. 80, 781–799.

Muramatsu, T., Nathenson, S.G., Boyse, E.A. and Old, L.J. (1973b) Some biochemical properties of thymus leukemia antigens solubilized from cell membranes by papain digestion. J. Exp. Med. 137, 1256–1262.

Nathenson, S.G. and Cullen, S.E. (1974) Biochemical properties and immunochemical-genetic relationships of mouse H-2 alloantigens. Biochim. Biophys. Acta 344, 1–25.

Nicolson, G.L. (1971) Difference in the topology of normal and tumour cell membranes shown by different surface distributions of ferritin-conjugated concanavalin A. Nature (London) New Biol. 233, 244–246.

Nicolson, G.L. (1972) Topography of membrane concanavalin A sites modified by proteolysis. Nature (London) New Biol. 239, 193–197.

Nicolson, G.L. (1973a) Anionic sites of human erythrocyte membranes, I. Effect of trypsin, phospholipase C, and pH on the topography of bound positively charged colloidal particles. J. Cell. Biol. 57, 373–387.

Nicolson, G.L. (1973b) Neuraminidase "unmasking" and failure of trypsin to "unmask" β-D-galactose-like sites on erythrocyte, lymphoma, and normal and virus-transformed fibroblast cell membranes. J. Natl. Cancer Inst. 50, 1443–1451.

Nicolson, G.L. (1973c) Temperature-dependent mobility of concanavalin A sites on tumor cells. Nature (London) New Biol. 243, 218–220.

Nicolson, G.L. and Blaustein, J. (1971) The interaction of *Ricinus communis* agglutinin with normal and tumor cell surfaces. Biochim. Biophys. Acta 266, 543–547.

Nicolson, G.L. and Painter, R.G. (1973) Anionic sites of human erythrocyte membranes, II. Anti-spectrin induced transmembrane aggregation of the binding sites for positively charged colloidal particles. J. Cell Biol. 59, 395–406.

Nicolson, G.L. and Singer, S.J. (1974) Distribution and asymmetry of mammalian cell surface saccharides utilizing ferritin-conjugated plant agglutinins as specific saccharide stains. J. Cell Biol. 60, 236–248.

Niedermeyer, W. (1971) Gas chromatography of neutral and amino sugars in glycoproteins. Anal. Biochem. 40, 465–475.

Novogrodsky, A. (1975) Induction of lymphocyte cytotoxicity by modification of the effector or target cells with periodate or with neuraminidase and galactose oxidase. J. Immunol. 114, 1089–1093.

Nowell, P.C. (1960) Phytohemagglutinin: an initiator of mitosis in cultures of human leukocytes. Cancer Res. 20, 462–466.

Obata, Y., Ikeda, H., Stockert, E. and Boyse, E.A. (1975) Relation of G_{IX} antigen of thymocytes to envelope glycoprotein of murine leukemia virus. J. Exp. Med. 141, 188–197.

Ogata, S.-I., Muramatsu, T. and Kobata, A. (1976) New structural characteristic of the large glycopeptides from transformed cells. Nature (London) 259, 578–580.

Olsnes, S., Saltvedt, E. and Pihl, A. (1974) Isolation and comparison of galactose-binding lectins from *Abrus precatorius* and *Ricinus communis*. J. Biol. Chem. 249, 803–810.

Oseroff, A.R., Robbins, P.W. and Burger, M.M. (1973) The cell surface membrane: biochemical aspects and biophysical probes. Annu. Rev. Biochem. 42, 647–682.

Ossowski, L., Quigley, J.P., Kellerman, G.M. and Reich, E. (1973) Fibrinolysis associated with oncogenic transformation. Requirement of plasminogen for correlated changes in cellular morphology, colony formation in agar, and cell migration. J. Exp. Med. 138, 1056–1064.

Ozanne, B. and Sambrook, J. (1971) Binding of labelled con A and wheat germ agglutinin to normal and transformed cells. Nature (London) New Biol. 232, 156–160.

Pardee, A.B. (1975) The cell surface and fibroblast proliferation. Some current research trends. Biochim. Biophys. Acta 417, 153–174.

Patt, L.M. and Grimes, W.J. (1974) Cell surface glycolipid and glycoprotein glycosyltransferases of normal and transformed cells. J. Biol. Chem. 249, 4157–4165.

Pearlstein, E. and Waterfield, M.D. (1974) Metabolic studies on ^{125}I-labeled baby hamster kidney cell plasma membranes. Biochim. Biophys. Acta 362, 1–12.

Phillips, D.R. and Morrison, M. (1971) Exposed protein on the intact human erythrocyte. Biochemistry 10, 1766–1771.

Pinto da Silva, P. and Branton, D. (1970) Membrane splitting in freeze-etching. J. Cell Biol. 45, 598–605.

Pinto da Silva, P. and Nicolson, G.L. (1974) Freeze-etch localization of concanavalin-A receptors to membrane intercalated particles of human erythrocyte ghost membranes. Biochim. Biophys. Acta 363, 311–314.

Pinto da Silva, P., Douglas, S.D. and Branton, D. (1971) Localization of A antigen sites on human erythrocyte ghosts. Nature (London) 232, 194–196.

Podolsky, D.K., Weiser, M.M., LaMont, J.T. and Isselbacher, K.J. (1974) Galactosyltransferase and concanavalin A agglutination of cells. Proc. Natl. Acad. Sci. USA 71, 904–908.

Pollack, R., Osborn, M. and Weber, K. (1975) Patterns of organization of actin and myosin in normal and transformed cultured cells. Proc. Natl. Acad. Sci. USA 72, 994–998.

Renkonen, O., Gahmberg, C.G., Simons, K. and Kääriäinen, L. (1970) Enrichment of gangliosides in plasma membranes of hamster kidney fibroblasts. Acta Chem. Scand. 24, 733–735.

Renkonen, O., Gahmberg, C.G., Simons, K. and Kääriäinen, L. (1972) The lipids of the plasma membranes and endoplasmic reticulum from cultured baby hamster kidney cells (BHK21). Biochim. Biophys. Acta 255, 66–78.

Reynolds, J.A. and Tanford, C. (1970) The gross conformation of protein-sodium dodecyl sulfate complexes. J. Biol. Chem. 245, 5161–5165.

Rice, R.H. and Means, G.E. (1971) Radioactive labeling of proteins in vitro. J. Biol. Chem. 246, 831–832.

Rifkin, D.B., Compans, R.W. and Reich, E. (1972) A specific labeling procedure for proteins on the outer surface of membranes. J. Biol. Chem. 247, 6432–6437.

Robison, A.G., Butcher, R.W. and Sutherland, E.W. (1972) Cyclic AMP. Academic Press, New York.

Rogentine, G.N. (1975) Naturally occurring human antibody to neuraminidase-treated human lymphocytes. Antibody levels in normal subjects, cancer patients, and subjects with immunodeficiency. J. Natl. Cancer Inst. 54, 1307–1311.

Rohrschneider, L.R., Kurth, R. and Bauer, H. (1975) Biochemical characterization of tumor-specific cell surface antigens on avian oncornavirus transformed cells. Virology 66, 481–491.

Roseman, S. (1970) Synthesis of complex carbohydrates by multiglycosyltransferase systems and their potential functions in intercellular adhesion. Chem. Phys. Lipids 5, 270–297.

Rosenblith, J.Z., Ukena, T.E., Yin, H.H., Berlin, R.D. and Karnovsky, M.J. (1973) A comparative evaluation of the distribution of concanavalin-A binding sites on the surfaces of normal, virally-transformed and protease-treated fibroblasts. Proc. Natl. Acad. Sci. USA 70, 1625–1629.

Roth, S. and White, D. (1972) Intercellular contact and cell-surface galactosyl transferase activity. Proc. Natl. Acad. Sci. USA 69, 485–489.

Roth, S., McGuire, E.J. and Roseman, S. (1971) Evidence for cell-surface glycosyltransferases. Their potential role in cellular recognition. J. Cell Biol. 51, 536–547.

Roy, D. and Konigsberg, W. (1972) Chromatography of proteins and peptides on diethylaminoethyl cellulose. Methods Enzymol. 25, part B, 221–231.

Ruoslahti, E. and Vaheri, A. (1975) Interaction of soluble fibroblast surface antigen with fibrinogen and fibrin. Identity with cold insoluble globulin of human plasma. J. Exp. Med. 141, 497–501.

Ruoslahti, E., Vaheri, A., Kuusela, P. and Linder, E. (1973) Fibroblast surface antigen: a new serum protein. Biochim. Biophys. Acta 322, 352–358.

Sandermann, H. and Strominger, J.L. (1972) Purification and properties of C_{55}-isoprenoid alcohol phosphokinase from *Staphylococcus aureus*. J. Biol. Chem. 247, 5123–5131.

Sanford, B.H. and Codington, J.F. (1971) Further studies of the effect of neuraminidase on tumor cell transplantability. Tissue Antigens 1, 153–161.

Segrest, J.P., Jackson, R.L., Andrews, E.P. and Marchesi, V.T. (1971) Human erythrocyte membrane glycoprotein: a reevaluation of the molecular weight as determined by SDS polyacrylamide gel electrophoresis. Biochem. Biophys. Res. Commun. 44, 390–395.

Sheppard, J.R. (1971) Restoration of contact-inhibited growth to transformed cells by dibutyryl adenosine 3':5'-cyclic monophosphate. Proc. Natl. Acad. Sci. USA 68, 1316–1320.

Siess, E., Wieland, O. and Miller, F. (1971) A simple method for the preparation of pure and active γ-globulin-ferritin conjugates using glutaraldehyde. Immunology 20, 659–665.

Silver, J. and Hood, L. (1976) Preliminary amino acid sequences of transplantation antigens: genetic and evolutionary implications. Contemp. Topics Mol. Immunol., in press.

Simmons, R.L. and Rios, A. (1971) Immunotherapy of cancer: immunospecific rejection of tumors on recipients of neuraminidase-treated tumor cells plus BCG. Science 174, 591–593.

Singer, S.J. and Nicolson, G.L. (1972) The fluid mosaic model of the structure of cell membranes. Science 175, 720–731.

Siu, C.-H., Lerner, R.A., Ma, G., Firtel, R.A. and Loomis, W.F. (1976) Developmentally regulated proteins of the plasma membrane of Dictyostelium discoideum. The carbohydrate-binding protein. J. Mol. Biol. 100, 157–178.

Slutzky, G.M. and Ji, T.H. (1974) The dissimilar nature of two forms of the major human erythrocyte membrane glycoprotein. Biochim. Biophys. Acta 373, 337–346.

Spies, J.R. (1957) Colorimetric procedures for amino acids. Methods Enzymol. 3, 467–477.

Staros, J.V., Haley, B.E. and Richards, F.M. (1974) Human erythrocytes and resealed ghosts. A comparison of membrane topology. J. Biol. Chem. 249, 5004–5007.

Staros, J.V., Richards, F.M. and Haley, B.E. (1975) Photochemical labeling of the cytoplasmic surface of the membranes of intact human erythrocytes. J. Biol. Chem. 250, 8174–8178.

Steck, T.L. (1972) Cross-linking the major proteins of the isolated erythrocyte membrane. J. Mol. Biol. 66, 295–305.

Steck, T.L. (1974) The organization of proteins in the human red blood cell membrane. J. Cell Biol. 62, 1–19.

Steck, T.L. and Dawson, G. (1974) Topographical distribution of complex carbohydrates in the erythrocyte membrane. J. Biol. Chem. 249, 2135–2142.

Stellner, K., Saito, H. and Hakomori, S. (1973) Determination of amino sugar linkages in glycolipids by methylation. Amino sugar linkages of ceramide pentasaccharides of rabbit erythrocytes and of Forssman antigen. Arch. Biochem. Biophys. 155, 464–472.

Stone, K.R., Smith, R.E. and Joklik, W.K. (1974) Changes in membrane polypeptides that occur when chick embryo fibroblasts and NRK cells are transformed with avian sarcoma viruses. Virology 58, 86–100.

Svennerholm, L. (1957) Quantitative estimation of sialic acids, II. A colorimetric resorcinol hydrochloric acid method. Biochim. Biophys. Acta 24, 604–611.

Sweeley, C.C. and Dawson, G. (1969) Lipids of the erythrocyte. In: G.A. Jamieson and T.J. Greenwalt (Eds.), Red Cell Membrane Structure and Function. Lippincott, Philadelphia, pp. 172–227.

Tanner, M.J.A. and Boxer, D.H. (1972) Separation and some properties of the major proteins of the human erythrocyte membrane. Biochem. J. 129, 333–347.

Teng, N.N.H. and Chen, L.B. (1975) The role of surface proteins in cell proliferation as studied with thrombin and other proteases. Proc. Natl. Acad. Sci. USA 72, 413–417.

Teng, N.N.H. and Chen, L.B. (1976) Thrombin-sensitive surface protein of cultured chick embryo cells. Nature (London) 259, 578–580.

Thomas, D.B. and Winzler, R.J. (1969) Structural studies on human erythrocyte glycoproteins. J. Biol. Chem. 244, 5943–5946.

Thurman, G.B., Giovanella, B. and Goldstein, A.L. (1973) Evidence for the T cell specificity of sodium periodate-induced lymphocyte blastogenesis. J. Immunol. 113, 810–812.

Tillack, T.W., Scott, R.E. and Marchesi, V.T. (1972) The structure of erythrocyte membranes studied by freeze-etching, II. Localization of receptors for phytohemagglutinin and influenza virus to the intramembranous particles. J. Exp. Med. 135, 1209–1227.

Tomita, M. and Marchesi, V.T. (1975) Amino acid sequence and oligosaccharide attachment sites of human erythrocyte glycophorin. Proc. Natl. Acad. Sci. USA 72, 2964–2968.

Trayer, H.R., Nozaki, Y., Reynolds, J.A. and Tanford, C. (1971) Polypeptide chains from human red blood cell membranes. J. Biol. Chem. 246, 4485–4488.

Trowbridge, I.S., Ralph, P. and Bevan, M.J. (1975) Differences in the surface proteins of mouse B and T cells. Proc. Natl. Acad. Sci. USA 72, 157–161.

Tung, J.-S., Vitetta, E.S., Fleissner, E. and Boyse, E.A. (1975) Biochemical evidence linking the G_{IX} thymocyte surface antigen to the gp 69/71 envelope glycoprotein of murine leukemia virus. J. Exp. Med. 141, 198–205.

Unkeless, J.C., Tobia, A., Ossovski, L., Quigley, J.P., Rifkin, D.B. and Reich, E. (1973a) An enzymatic function associated with transformation of fibroblasts by oncogenic viruses, I. Chick embryo fibroblast cultures transformed by avian RNA tumor viruses. J. Exp. Med. 137, 85–111.

Unkeless, J.C., Kellerman, G.M., Danø, K. and Reich, E. (1973b) Fibrinolysis associated with oncogenic transformation. Partial purification and characterization of the cell factor, a plasminogen activator. J. Biol. Chem. 249, 4295–4305.

Vaheri, A. and Ruoslahti, E. (1974) Disappearance of a major cell type specific surface glycoprotein antigen (SF) after transformation of fibroblasts by Rous sarcoma virus. Int. J. Cancer 13, 579–586.

Vaheri, A. and Ruoslahti, E. (1975) Fibroblast surface antigen produced but not retained by virus-transformed human cells. J. Exp. Med. 142, 530–538.

Vaheri, A., Ruoslahti, E. and Nordling, S. (1972) Neuraminidase stimulates division and sugar uptake in density-inhibited cell cultures. Nature (London) New Biol. 85, 211–212.

Vaheri, A., Ruoslahti, E., Hovi, T. and Nordling, S. (1973) Stimulation of density-inhibited cell cultures by insulin. J. Cell. Physiol. 81, 355–364.

Vaheri, A., Ruoslahti, E., Westermark, B. and Pontén, J. (1976) A common cell-type specific surface antigen (SF) in cultured human glial cells and fibroblasts: loss in malignant cells. J. Exp. Med. 143, 64–72.

Van Beek, W.P., Smets, L.A. and Emmelot, P. (1973) Increased sialic acid density in surface glycoprotein of transformed and malignant cells – a general phenomenon? Cancer Res. 33, 2913–2922.

Van Heyningen, W.E. (1974) Gangliosides as membrane receptors for tetanus toxin, cholera toxin and serotonin. Nature (London) 249, 415–417.

Verkleij, A.J., Zwaal, R.F.A., Roelofsen, B., Confucius, P., Kastelijn, D. and Van Deenen, L.L.M. (1973) The asymmetric distribution of phospholipids in the human red cell membrane. Biochim. Biophys. Acta 323, 178–193.

Vischer, T.L. (1974) Stimulation of mouse B lymphocytes by trypsin. J. Immunol. 113, 58–62.

Wallach, D.F.H. and Kamat, V.B. (1966) Preparation of plasma-membrane fragments from mouse ascites tumor cells. Methods Enzymol. 8, 164–172.

Warren, L. (1959) The thiobarbituric acid assay of sialic acids. J. Biol. Chem. 234, 1971–1975.

Warren, L., Glick, M.C. and Nass, M.K. (1966) Membranes of animal cells, I. Methods of isolation of the surface membrane. J. Cell. Physiol. 68, 269–287.

Wartiovaara, J., Linder, E., Ruoslahti, E. and Vaheri, A. (1974) Distribution of fibroblast surface antigen. Association with fibrillar structures of normal cells and loss upon viral transformation. J. Exp. Med. 140, 1522–1533.

Weber, K. and Osborn, M. (1969) The reliability of molecular weight determination by dodecyl sulfate-polyacrylamide gel electrophoresis. J. Biol. Chem. 244, 4406–4412.

Weber, K., Pollack, R. and Bibring, T. (1975) Antibody against tubulin: the specific visualization of cytoplasmic microtubules in tissue culture cells. Proc. Natl. Acad. Sci. USA 72, 459–463.

Weber, M.J. (1973) Hexose transport in normal and in Rous sarcoma virus-transformed cells. J. Biol. Chem. 248, 2978–2983.

Weber, P. and Hof, L. (1975) Introduction of a fluorescent label into the carbohydrate moiety of glycoconjugates. Biochem. Biophys. Res. Commun. 65, 1298–1302.

Whiteley, N.M. and Berg, H.C. (1974) Amidination of the outer and inner surfaces of the human erythrocyte membrane. J. Mol. Biol. 87, 541–561.

Widnell, C.C. and Unkeless, J.C. (1968) Partial purification of a lipoprotein with 5'-nucleotidase activity from membranes of rat liver cells. Proc. Natl. Acad. Sci. USA 61, 1050–1057.

Yamada, K.M. and Weston, J.A. (1974) Isolation of a major cell surface glycoprotein from fibroblasts. Proc. Natl. Acad. Sci. USA 71, 3492–3496.

Yamada, K.M., Yamada, S.S. and Pastan, I. (1975) The major cell surface glycoproteins of chick embryo fibroblasts is an agglutinin. Proc. Natl. Acad. Sci. USA 72, 3158–3162.

Yen, P.H. and Ballou, C.E. (1973) Composition of a specific intercellular agglutination factor. J. Biol. Chem. 248, 8316–8318.

Yin, H.H., Ukena, T.E. and Berlin, R.D. (1972) Effect of colchicine, colcemid, and vinblastine on the agglutination by concanavalin A of transformed cells. Science 178, 867–868.

Shedding of tumor cell surface antigens   8

Michael R. PRICE and Robert W. BALDWIN

1. Introduction

The concept that the surface components of tumor cells are released into the extracellular environment of the neoplasm and may subsequently be found in the serum or other body fluids of the tumor-bearing individual has recently received much attention. Since many, if not all, neoplasms express tumor antigens which are not present on (or only minimally associated with) normal adult tissues, the possibility that these products may be identified in the circulation of the tumor-bearing host has several important implications. This is particularly relevant with tumors which express organ-related neoantigens, since the ability to monitor patients' sera for these products using simple assays would represent an important adjunct to the early diagnosis of cancer. Furthermore, with the development of quantitative assays for tumor antigens, sequential monitoring of circulating antigen levels has the potential for detection of tumor progression or regression following therapy and for early diagnosis of local recurrence or metastases. Obviously, similar arguments may be applied to advocate monitoring of patients for other malignancy-associated defects (e.g. elevated or depressed hormone levels and various functional tests) although identification of tumor antigens may provide a more specific marker for neoplasia.

A second and more fundamental consequence of the detection of circulating tumor antigens is that these components possibly interfere with the effective mediation of cellular immunity against the progressing neoplasm. The "shedding" of antigenic material may provide the tumor with "soluble antigenic camouflage" by which it escapes host immunological control. These materials may also participate in the formation of "blocking factors" such as specific immune complexes which are protective toward the tumor in the face of an active cell-mediated immune response. Finally, host responses to acellular forms of tumor antigen in body fluids may lead to a "deviation" of the immune response to these antigens as they are expressed upon tumor cells with the consequence that tumor rejection reactions are diminished. Before any account of the immunobiological effects of shed tumor antigens in the tumor-bearing host is given, it is pertinent to consider the expression and nature of these antigens so that their shedding from the tumor cell surface may be related to current views regarding the organization of components in the cell surface membrane.

2. *Nature of tumor antigens and molecular expression at the cell surface*

The very nature of the variety of in vitro immunological assays for tumor antigen detection suggests that the predominant expression of these components is at the cell surface. The bases of these tests frequently involve antibody binding to cell surface receptors or cytolysis or cytostasis occurring following interaction of sensitized lymphoid cells with tumor cells carrying the appropriate antigens. Indeed, it is difficult to conceive how tumor rejection responses could operate in vivo unless the antigenic determinants are exposed at the tumor cell surface so that interaction with antibody or sensitized lymphoid cells is possible. Nevertheless, the molecular organization of tumor antigens within the plasma membrane remains somewhat elusive. Taking the Fluid Mosaic Model of cell surface membrane organization (Singer and Nicolson, 1972) as being appropriate to the present discussion, it is possible that some tumor antigens may be integral components of the plasma membrane while others may show peripheral association with the cell surface. In this respect, the method of extraction of antigens from tumor tissue gives some indication as to the probable organization and association of tumor antigens with the cell surface membrane since the classification of membrane components (particularly membrane proteins and glycoproteins) as peripheral or integral components is an operational one and is defined by the conditions required to remove them in a soluble form from the lipid-bilayer (Singer and Nicolson, 1972; Robbins and Nicolson, 1975). Table I summarizes these criteria and also presents the problem with regard to consideration of tumor antigen expression and organization at the cell surface.

Examining firstly experimental animal tumors, particularly those induced by chemical carcinogens, many of these express a tumor specific antigen which is individually distinct for each transplanted tumor line as shown by the specificity of tumor rejection tests in appropriately immunized hosts or using in vitro assays of cell-mediated and humoral immunity (reviewed by Baldwin, 1973; Baldwin and Price, 1975). In several instances, especially with 4-dimethylaminoazobenzene (DAB)-induced hepatomas and 3-methylcholanthrene (MCA)-induced sarcomas, it was found necessary to resort to vigorous treatments of isolated cell membrane preparations to liberate soluble tumor-specific antigens and a reasonable conclusion from these findings is that these antigens are integral membrane components. In particular, with hepatomas and sarcomas in the rat, soluble antigens were released from both intact tumor cells (Harris et al., 1973) and from isolated cell membrane preparations (Baldwin and Glaves, 1972; Baldwin et al., 1973e; Thomson and Alexander, 1973) by

TABLE I

CLASSIFICATION OF MEMBRANE-ASSOCIATED TUMOR ANTIGENS AS INTEGRAL OR PERIPHERAL MEMBRANE COMPONENTS

(A) DEFINITIONS

(1) *Integral membrane components*:
    (a) Primary interactions with membrane: hydrophobic associations within lipid bilayer (e.g. strong interactions with acyl groups of membrane lipids).
    (b) Secondary interactions within membrane: cytoskeletal restraints (e.g. with proteins or glycoproteins spanning the membrane, further immobilization may result from interaction with microfilaments, microtubules or macromolecules associated with the inner membrane surface) and peripheral restraints (e.g. association with peripheral components or lattices, see below).
    (c) Solubilization requires membrane disruption (e.g. enzymes, detergents, organic solvents, chaotropic agents and physically disruptive methods, such as sonication).

(2) *Peripheral membrane components*:
    (a) Primary interaction with membrane: weak interactions with integral membrane components (e.g. protein, glycoprotein, glycolipid) by ionic and hydrogen bonding.
    (b) Secondary interactions with membrane: stabilization within peripheral lattices (e.g. "glycocalyx").
    (c) Solubilization does not require membrane disruption (e.g. peripheral components may be solubilized by low ionic strength, chelating agents).

(B) PROBLEM
    Can membrane-associated tumor antigens be classified as integral or peripheral components of the cell plasma membrane?

limited enzymatic digestion with papain. Also, purified rat hepatoma plasma membrane preparations retained an increased level of specific antigenic activity compared with homogenates, and no demonstrable activity was associated with intracytoplasmic protein, nuclei or nuclear membranes (Price and Baldwin, 1974a,b). Conversely with sarcomas induced in the guinea pig by MCA or 7,12-dimethylbenz[a]anthracene (DMBA), early studies established that tumor antigens capable of eliciting tumor rejection responses, delayed cutaneous hypersensitivity reactions and inhibiting the migration of sensitized macrophages, could be isolated from the soluble intracytoplasmic fraction of homogenates (Oettgen et al., 1968; Bloom et al., 1969). This would suggest that either this antigen is a peripheral component of the cell surface or that tumor antigen may be present as a soluble moiety in the cell cytoplasm, possibly as a product to be secreted, or as an antigenic precursor of the macromolecule to be inserted into the plasma membrane.

Recently, the 3 M KCl extraction procedure of Reisfeld and Kahan (1970, 1972) used in the isolation of histocompatibility antigens has been applied extensively to the problem of solubilizing tumor antigens from a variety of experimental animal tumors (e.g. Meltzer et al., 1971, 1972; Leonard et al., 1972, 1975; Thomson and Alexander, 1973; Brannen et al., 1974; Plata and Levy, 1974; Smith and Leonard, 1974; Steele et al., 1975; Zöller et al., 1976) and human tumors (e.g. Mavligit et al., 1973, 1974; Vanky et al., 1974; Powell et al.,

1975; Roth et al., 1976). In their original description of the technique, Reisfeld and Kahan (1970, 1972) proposed that antigen solubilization was accomplished by a reduction of the ordered structure of water molecules intimately associated with membrane protein, thus allowing the hydrophobic region of the proteins to become detached from their lipid environment and thus dispersing membrane macromolecules into the aqueous media. If this is the underlying mechanism for the release of tumor antigens following exposure to high salt concentrations, this would imply that the antigenic components associated with many experimental animal and human tumors are well integrated into the general structure of the tumor cell plasma membrane. However, an alternative mechanism for antigen release is suggested by the experiments of Mann (1972) who found that 3 M KCl solubilization of *HL-A* histocompatibility antigens was effected by the action of soluble intracellular and possibly membrane-bound proteolytic enzymes. This interpretation is supported in part by Fairbanks et al. (1971) who demonstrated that proteolysis of erythrocyte membranes occurs, particularly under conditions of high salt concentrations at neutral pH. Comparably, Tökés and Chambers (1975) showed that isolated red cell membranes or intact erythrocytes have the capacity to digest ^{125}I-labelled casein. Nevertheless, treatment of isolated membrane fractions, rather than cell homogenates containing high levels of autolytic enzymes, has in some instances proved successful in liberating soluble antigenic components. For example, Prat and Comoglio (1976) released a fraction from purified MOPC-460 plasmacytoma cell plasma membranes which was termed SMA (soluble membrane antigens) for which a radioimmunoassay was developed. Similarly, tumor-specific antigens on a diethylnitrosamine-induced guinea pig hepatoma could be extracted by 3 M KCl treatment of purified tumor cell plasma membranes and these preparations were capable of eliciting delayed cutaneous hypersensitivity reactions in immune animals (Leonard et al., 1975). Antigenic activity was, however, associated with a protein of molecular weight greater than 250 000 daltons as assessed by gel filtration on Sephadex G200, whereas when the same extraction procedure was applied to intact cells, antigenic activity was eluted predominantly in the included volumes of these columns. However, comparable tests using isolated membranes from an aminoazo dye-induced rat hepatoma failed to solubilize tumor specific antigen (M.R. Price, unpublished observations) whereas the same procedure applied to whole tissue was effective (Zöller et al., 1976).

An alternative explanation for the mode of action of antigen solubilization by 3 M KCl has been proposed by Pincus (1974) who, from studies on platelets, has suggested that solubilization may be dependent upon the presence of cyclic AMP. By increasing the concentration of cyclic AMP, membrane-bound protein kinase is activated and in the platelet membrane, there are four specific membrane proteins that are phosphorylated. Thus, phosphorylation of cell membrane components may cause some kind of dissociation by increasing the negative charge rendering some proteins extractable in hypertonic salt. Alternatively, another possibility may be that cyclic AMP combines directly with a class of proteins on the membrane and that this interaction somehow causes disaggregation (Pincus, 1974).

From these considerations, it is evident that the widely adopted 3 M KCl antigen solubilization procedure gives little information as to whether tumor antigens are integral components of the plasma membrane. A further difficulty associated with interpretation of data obtained with this procedure is that tumor antigens present within the cell cytoplasm as soluble entities are also extracted by this technique and unless information is available on the subcellular localization and distribution of antigen within a particular tumor system, it is not possible to state that the isolated soluble materials were originally associated with the plasma membrane.

As mentioned already, the 3 M KCl extraction procedure has found extensive application in the preparation of tumor antigens from human neoplastic tissues, although from the above considerations, this approach alone provides only very limited information about antigen expression at the cell surface. Nevertheless, several studies have indicated that some human tumor antigens are intimately associated with cell membranes and may be considered as integral components since vigorous treatments were required to render them water-soluble. In particular, with melanoma and colon carcinoma, tumor-associated antigens of apparent organ-related specificity have been solubilized by papain treatment of cell membranes. These soluble fractions retained the capacity to inhibit the cytotoxicity of patients' peripheral blood leukocytes (Baldwin et al., 1973c; Embleton, 1973; Embleton and Price, 1975) or could be used to prepare xenogeneic antisera which after appropriate absorption were monospecific for melanoma-associated antigens (Viza and Phillips, 1975; Viza et al., 1975). Similarly, surface antigens associated with human leukemic cells have been solubilized by papain treatment of cell membrane preparations (Harris et al., 1971; Mann et al., 1971, 1974).

Several other degradational methods used to isolate soluble tumor antigens support the view that they are integral membrane components. Exposure of tumor cells or their membranes to low intensity ultrasound for short periods has proved effective in the extraction of soluble antigens from both experimental animal and human tumors and, in some instances, partial purification of antigenically active materials from the sonicates has been achieved (Holmes et al., 1970; Hollinshead et al., 1974; Herberman et al., 1975). In particular, extracts of human breast carcinoma and melanoma have been fractionated by Sephadex G200 gel filtration chromatography to give two fractions, one of which (Sephadex Fraction II, molecular weight approximately 39 000 daltons) shows histological tumor type specificity in delayed hypersensitivity skin tests, while the other (Sephadex Fraction III, molecular weight approximately 10 000 daltons) gives more non-specific reactivity in these tests (Hollinshead et al., 1974; Herberman et al., 1975). Alternatively, attempts have been made to combine the physical disruption of sonication with salt extraction techniques for membrane antigen solubilization. This approach has been applied to the isolation of human and rat histocompatibility antigens using either KCl (Reisfeld and Kahan, 1970) or KI (Jones and Feldman, 1975).

Other procedures known to dissociate cell membranes such as treatment with detergents or organic solvents have received less attention, since frequently antigenic activity is lost or impaired. However, Mann (1975) has extracted human melanoma cell membranes with the non-ionic detergent Brij-98 and fractionated the material by chromatography on Bio-gel, isoelectric focusing, and separation on acrylamide gel by electrophoresis. Selected components were eluted from the gels and these materials were injected into rabbits in an attempt to prepare specific anti-melanoma antisera.

The two most characterized human tumor antigens are carcinoembryonic antigen and α-fetoprotein associated predominantly with colo-rectal and hepatocellular carcinomas respectively (reviewed by Laurence and Neville, 1972; Abelev, 1974; Fuks et al., 1974; Ruoslahti et al., 1974; Terry et al., 1974; Zamcheck, 1975), but these antigens most probably represent secretory products of the tumor cell rather than integral membrane components. With carcinoembryonic antigen (CEA) for example, colon carcinoma cells were easily agglutinated by specific anti-CEA antisera or cell surface immunofluorescence was observed using fluorescein conjugated anti-CEA antisera, indicating antigen localization at the cell surface (Gold et al., 1968; Gold, 1970). However, when tumor cells were treated with ferritin-conjugated goat anti-CEA antiserum and examined in the electron microscope, the ferritin label was located in the glycocalyx or "fuzzy coat" immediately adjacent to and strictly a part of the surface membrane (Gold et al., 1970). It was therefore concluded that CEA is not an integral structural component of the plasma membrane, but is localized at the extreme periphery of the cells. Whether CEA is an integral component of the glycocalyx itself is not known. Although CEA is not generally considered to be a target for immune reactions in the tumor-bearing host, specific humoral immune responses to CEA particularly in the IgM immunoglobulin class, have been demonstrated by Gold and his associates (Gold, 1967; Gold and Gold, 1973; Gold et al., 1973). This antigen has, however, been found to be distinguishable from the component isolated following papain digestion of colon carcinoma cell membranes and which inhibited the in vitro cytotoxicity of patients' blood leukocytes for cultured target cells (Baldwin et al., 1973c; Embleton, 1973). Also, Hollinshead et al. (1970, 1972) obtained a soluble fraction containing CEA by sonication of tumor cell membranes and this fraction was reactive in delayed hypersensitivity tests in colon carcinoma patients. However, the skin reactive antigen was separable from CEA by polyacrylamide gel electrophoresis.

With the exception of CEA and α-fetoprotein and possibly other human tumor antigens, which are abnormal tumor products recognized by xenogeneic antisera and therefore may not be immunogenic in the tumor-bearing host (e.g. the fetal sulphoglycoprotein antigen associated with gastric cancer, described by Häkkinen, 1974), little is known of the chemical nature of tumor-associated antigens. From the various reports already discussed, there is evidence in a number of instances that some tumor antigens are intimately associated with the plasma membrane and may be true integral membrane components. This poses the next question: what are the most likely macromolecular candidates representing tumor antigen? While present opinion favors membrane-associated proteins or glycoproteins as fulfilling this role, on purely theoretical grounds a carbohydrate moiety could display the required diversity in saccharide sequence and/or configuration to account for the multitude of individual specificities of tumor antigens. Furthermore, although changes in plasma membrane glycolipids may be common phenomena associated with neoplastic transformation, a functional correlation between these changes and antigenic alterations in tumor cells has yet to be established (Hakomori, 1975). Evidence has been obtained, particularly from studies with tumors in experimental animals, to support the view that some tumor antigens are membrane-associated proteins or glycoproteins. With one aminoazo dye-induced rat hepatoma, papain digestion of cell membranes liberated a soluble antigenic fraction which showed heterogeneity in molecular size of components carrying the antigenic determinant (Baldwin et al., 1973e). Heterogeneity in antigen molecular size and/or charge has been reported by other groups (Suter et al., 1972; Leonard et al., 1975). Although this may be due to aggregation, a problem frequently encountered when handling isolated membrane proteins, Leonard et al. (1975) have suggested that this is not necessarily the case. In the rat hepatoma experiments, it was possible to isolate one fraction with an approximate molecular weight of 55 000 daltons which gave a single but diffuse band of stained protein on analytical polyacrylamide electrophoretic gels. Compositional analysis of this material revealed marked similarities in the overall distribution of amino acids between this antigen and histocompatibility antigens isolated from cells of rat, mouse and human origin (Fig. 1). Several other comparisons between tumor antigens and histocompatibility antigens have been made. This is pertinent since the latter have been identified as membrane-associated glycoproteins (see, for example, Shimada and Nathenson, 1969; Reisfeld and Kahan, 1971). There has recently been further speculation as to whether specific antigens on experimental animal tumors, especially those induced by chemical carcinogens, are related to, or even represent modified histocompatibility antigens (Haywood and McKhann, 1971; Bowen and Baldwin, 1975; Germain et al., 1975; Invernizzi and Parmiani, 1975; Parmiani and Invernizzi, 1975). In contradiction to this postulate, however, Klein and Klein (1975) found that the genetic determinant for one MCA-induced murine sarcoma-specific antigen is not localized on the same chromosome as that determining the major histocompatibility (H-2) complex. Nevertheless, the hypothesis that tumor-specific antigens are related to histocompatibility antigens requires further evaluation since methods developed for the isolation of histocompatibility antigens have almost uniformly proved effective for extracting tumor antigens. Experiments performed in this laboratory have established that there may be points of immunological identity between purified rat hepatoma specific antigen and transplantation antigens expressed by nor-

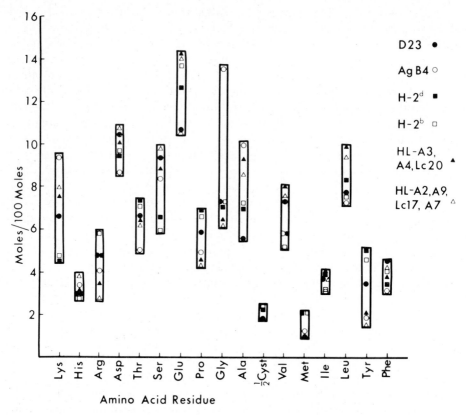

Fig. 1. Comparison of the amino acid composition of rat hepatoma D23 specific antigen and rat, murine and human histocompatibility antigen. Rat hepatoma D23 specific antigen was prepared according to Baldwin et al. (1973e) following papain digestion of tumor cell membranes. Values for individual amino acid residues were calculated from chromatograms given by duplicate samples of 24 and 72 h acid hydrolysates (M.R. Price, unpublished observations). The rat AgB4 histocompatibility antigens were prepared by specific immunoabsorption procedures applied to 3 M KCl extracts of rat lymphoid tissues (Callahan and Dewitt, 1975). Murine H-2 (H-2^d and H-2^b) and human HL-A (HL-A3, A4, Lc20 and HL-A2, A9, Lc17, A7) histocompatibility antigens were isolated following papain digestion of cell membranes as described by Mann et al. (1970).

mal rat liver (Bowen and Baldwin, 1975). Similarly, Germain et al. (1975) determined that exposure of murine P815 mastocytoma cells to alloantiserum protected them from the cytotoxic action of syngeneic immune lymphoid cells as assessed using a ^{51}Cr release test, suggesting that there may be a linkage between tumor and histocompatibility antigens. Also, in the case of murine histocompatibility antigens, it has been suggested that these can be physically associated on the cell surface with viral antigens and possibly other foreign antigens and this has led to the hypothesis that *H-2* molecules serve as adaptors that combine with viral antigens on the cell surface to form hybrid antigens containing self (*H-2*) and non-self (virus) elements (Schrader et al., 1975). The adaptor-antigen complex may then be recognized by a subclass of T-lymphocytes that possesses a repertoire of receptors directed against hybrids of foreign and *H-2* antigens. An as yet relatively unexplored area with regard to these considerations would be an evaluation of whether tumor-specific antigens possessed structural subunits such as β-2-microglobulin which represents a common portion of *HL-A* and other histocompatibility antigens (Poulik et al., 1974; Rask et al., 1974; Strominger et al., 1974; Tanigaki and Pressman, 1974). Also it should be recognized that progress has been rapid in defining the molecular nature of histocompatibility antigens and their association with the cell membrane. For example, studies on the subunit structure, cell surface orientation and partial amino acid sequences of murine *H-2* antigens are presently being conducted and models have been proposed for the molecular expression of these membrane associated antigens (Henning et al., 1976).

An alternative approach to determining the chemical nature of tumor-associated antigens has involved treatment of tumor cells or their membranes with various degradative agents and then assaying for antigenic activity. In this way, any loss in activity may reflect degradation of the antigenic site or residues in the environment of the determinant necessary for the full expression of antigenic activity, or the treatment may result in antigen solubilization. This type of procedure has been applied to the analysis of the rat hepatoma D23 specific antigen (Baldwin et al., 1974a) and also to a human acute lymphoblastic leukemia (ALL) associated antigen (Brown et al., 1975). In both cases, the tumor antigens were not susceptible to trypsin or neuraminidase, although the hepatoma antigen was solubilized by papain and ALL-associated antigenic activity was lost following pronase digestion. Again in both studies, evidence was presented indicating that oligosaccharide structures may be associated to some extent with the molecule bearing the antigenic determinant.

As a final comment on the nature and expression of tumor antigens at the cell surface, several investigators have established that antigen redistribution occurs following incubation of tumor cells at 37°C with specific antisera and an appropriate fluorescein conjugated heterologous anti-immunoglobulin (Leonard, 1973; Yefenof and Klein, 1974). The ability to "cap" tumor-associated antigens indicates that these components are free to undergo lateral redistribution within the lipid bilayer without appreciable transmembrane restraints anchoring them to the cell interior. In this respect, Yefenof and Klein (1974) observed that a murine leukemia virus-induced antigen (termed MCSA) showed very limited capping in contrast to other cell surface antigens examined, and this was related to the fact that this antigen may be determined by a budding virus, thus explaining the rather stable anchorage of MCSA sites. Other tumor-associated and histocompatibility antigens exhibited good capping, reflecting their mobility within the cell membrane. Another factor limiting the mobility of membrane antigens may be related to the cholesterol content of the plasma membrane, since the low cholesterol: phospholipid ratios found in chronic lymphocytic leukemia cells (Vlodavsky and Sachs, 1974) would be expected to increase membrane fluidity and thus possibly increase the lateral mobility of integral membrane proteins. Also, since the phenomenon of capping appears to be restricted to only a few cell types and/or receptors, other effects such as surface antigen density in the cell membrane, may influence their ability to participate in interactions resulting in cap formation.

3. Tumor antigen shedding in vitro

3.1. Background considerations

Although it is well recognized that cell surfaces are dynamic assemblies, constantly undergoing synthesis, degradation and regeneration, the mechanisms of loss of surface components from the plasma membrane are less defined (review, Nicolson, 1976). Clearly, tumor cell death due to inadequate nutrient supply or induced by cytotoxic agents, followed by fragmentation and autolysis of cell constituents, represents one pathway by which cell surface components may be liberated in a soluble form into the extracellular environment (Fig. 2). However, metabolic turnover of the cell surface is also implicated in these processes. Experiments on the incorporation of radiolabeled precursors into viable cells and tissues have demonstrated that plasma membrane proteins and glycoproteins are synthesized at similar rates (Warren and Glick, 1968; Evans and Gurd, 1971) although with mouse liver plasma membranes, synthesis of glycolipids occurred more rapidly, emphasizing their active metabolic state (Evans and Gurd, 1971). In mouse fibroblasts (L cells) both growing and non-growing cells synthesize approximately similar amounts of surface membrane and cell particulate matter, but in growing cells the material is incorporated with a net increase in substance (Warren and Glick, 1968). However, with non-growing cells, newly synthesized material is incorporated, but a corresponding amount of material is released without net increase in substance. That the synthesis and degradative aspects of turnover may be coupled is suggested by the finding that metabolic inhibitors or omission of amino acids in the culture media result in a decrease in synthesis of the surface membrane and cell particulates and cause an equivalent decrease in the rate of degradation of surface membranes and particulates (Warren and Glick, 1968). Similarly, Cone et al. (1971) determined that shedding of lactoperoxidase iodinated surface proteins of mouse lymphoid cells from both normal and neoplastic sources occurred at a rapid rate, and that this was dependent on cellular respiration and protein synthesis. Blocking of protein synthesis by puromycin treatment of human peripheral lymphocytes also prevented the regeneration of surface HL-A2 histocompatibility antigen determinants after their removal by papain, though the regeneration of these determinants was not fully blocked by inhibition of DNA-dependent RNA synthesis (Turner et al., 1972). These experiments indicated that HL-A2 determinants are synthesized de novo, and not from a pool of preformed precursors and that mRNA coding for antigen sites exists within the cell. In

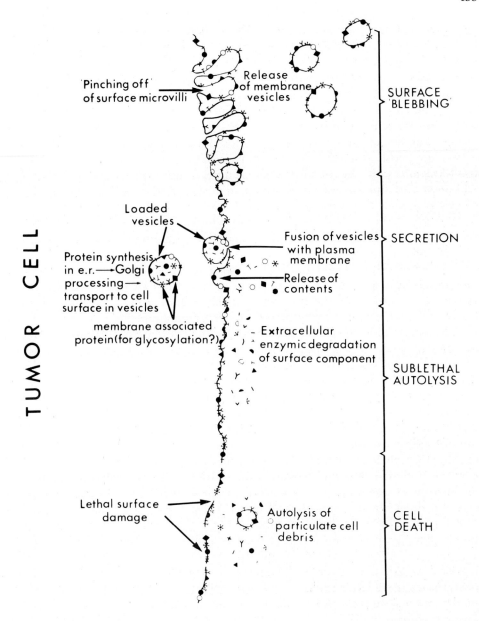

Fig. 2. Some possible pathways for the release of tumor cell surface components into the extracellular environment.

this case, therefore, antigen expression may be controlled at the translational level or the presence of the messenger may indicate continuous turnover of surface antigen (Turner et al., 1972). In other studies, the expression of histocompatibility antigens at the cell surface was found to be dependent upon the stage of the cell cycle, being maximal during early interphase, presumably part of the Gl period (Cikes, 1970; Karb and Goldstein, 1971; Cikes and Klein, 1972; also see chapter in this volume by Cikes and Friberg, p. 473).

The mobility of tumor cell surface components also displays a cell cycle dependency with a marked increase only at mitosis, this being the converse to the situation with the normal cells examined (Shoham and Sachs, 1974). Although this finding was interpreted in terms of the nuclear and surface events associated with various stages of the cell cycle being out of phase in the neoplastic cell, these observations emphasize possible variation in the stability of the surface during the cell cycle which may have consequences in determining the rate of antigen shedding.

The loss of cell surface constituents may result from the release of plasma membrane by a process of pinching off of microvilli or other surface projections (Nowotny et al., 1974) or alternatively shedding may take place at the molecular level rather than by membrane fragmentation.

Neoplastic cells have been characterized by the presence of increased surface proteolytic activity which has been implicated as providing a perpetual stimulus to divide and so increase their invasiveness to surrounding normal tissue (Burger, 1973; Nachbar et al., 1974; Poste and Weiss, 1976). A further consequence of the presence of surface proteases as well as that of released extracellular lysosomal enzymes (Poste, 1971) could well be the continuous digestion of the tumor cell surface leading to the liberation of active soluble tumor antigen determinants. In this context, Kapeller et al. (1973) noted that glucosamine-labeled glycopeptides in trypsin digests of chick embryo fibroblasts gave very similar elution profiles from DEAE cellulose columns as those obtained by chromatography of naturally shed materials. In the same study, it was also determined that with 3T3 mouse cells, culture supernatants contain *H-2* histocompatibility antigens. Furthermore, the spent culture medium of human lymphoid cell lines has been used as a source for the isolation of *HL-A* antigens (Pellegrino et al., 1973) and the "HL-A common portion fragment" identical to β-2-microglobulin (Tanigaki and Pressman, 1974). It was also observed by these latter workers and their colleagues that material was present in the plasma carrying *HL-A* determinants and gel filtration yielded three active fractions of molecular size to $8 \cdot 10^6$, 48 000 and 10 000 daltons, respectively (Miyakawa et al., 1973a). The high molecular size material appeared to represent a definite portion of the cell membrane and digestion of this fraction with papain yielded fragments identical to the second. It is of note that the molecules of 48 000 daltons were equivalent to the molecular fragments that have been derived from cultured lymphoid cells by digestion of membrane fractions with papain (Miyakawa et al., 1973b). In studies with a tumor-specific antigen associated with an aminoazo dye-induced rat hepatoma, tumor antigen has been isolated from the serum of tumor-bearers (Bowen and Baldwin, 1976a) and the molecular characteristics of this component are comparable to the antigen released from tumor membranes by proteolysis (Baldwin et al., 1973e; Baldwin and Price, 1976).

The evidence just described lends support to the concept that some surface components may be disassociated from the plasma membrane by proteolytic pathways. In this respect, proteolysis of viable rat hepatoma cells results in the loss of surface projections and microvilli (Harris et al., 1973) so that sublethal self-autolysis (Poste, 1971) may be contributory in inducing phenomena such as cell membrane "blebbing" as already described (Nowotny et al., 1974). Alternatively, tumor antigen and other cellular components may be released into the extracellular environment by secretion. However, with tumor antigens which are immunogenic in the host, this pathway of release presents certain logistic problems. If the antigens are to function as targets for lymphocytotoxic reactions, or for antibody binding, the question arises as to whether a secretory product showing possibly transient expression at the cell surface would be an appropriate target for these reactions. Investigations on the secretion of Ig from lymphoid cells suggest that it is possible for a secretory component of the cell also to be membrane-associated (Knopf, 1973; Melchers and Andersson, 1973; Parkhouse, 1973; Vitetta and Uhr, 1973, 1975). Following synthesis and then assembly of L and H chains into dimer and tetramer molecules in the endoplasmic reticulum, initial saccharides are added and the Ig molecules are transported to the Golgi for further glycosylation. These molecules are packaged into membrane vesicles for transport to the plasma membrane where they fuse with the plasma membrane with resulting release and secretion of the IgG. Some of the Ig is considered to be membrane-associated during this transportation process (to facilitate glycosylation) so that while most Ig is secreted, some remains bound to the vesicle which becomes fused into the plasma membrane. Furthermore, Melcher et al. (1975) have provided evidence that IgM and IgD-like molecules on mouse lymphocytes are integral membrane proteins.

Finally, it is worth mentioning that if tumor antigens are peripheral components attached to integral membrane protein and/or glycoprotein by noncovalent bonding, their release into the surrounding environment (as well as possible endocytosis) would occur with continued antigen synthesis.

3.2. Detection of antigens shed in vitro

The general conclusion by Kapeller et al. (1973) that renewal of surface constituents is rapid and occurs within 1 to 24 h depending on the surface marker and cell type suggests that the problem of studying antigen turnover in vitro is open to investigation. However, direct evidence which conclusively demonstrates that tumor antigens are continuously being shed from cells by normal metabolic processes is in most instances lacking. One reason for this is due partly to technical problems, in that present assays for antigen detection are not sufficiently sensitive to demonstrate released antigen in any quantitative manner. Experiments already performed that attempt to identify antigen in tissue culture supernatants are open to this criticism, although they do lend support to the view that macromolecules retaining antigenic activity are released into culture supernatants. For example, in order to demonstrate the shedding of soluble antigenic material by human melanoma cells, cultures were deprived of serum nutrients so that contamination with serum components could be avoided when assaying for antigenic activity (Grimm et al., 1976). This procedure, while leaving the majority of the cells intact, by no means guarantees that *all* cells remain in a viable state. Therefore, non-viable cells and their debris, both particulate and soluble, and material rendered soluble by autolysis, will be found in the supernatants and it is not possible to determine to what extent a particular product identified by its antigenic properties arose via cultivation artefacts or by actual membrane turnover.

In studies by Currie and Alexander on the detection of antigens in tissue culture supernatants of MCA-induced rat sarcomas, it was concluded that the rate of antigen shedding may actually determine the capacity of a tumor to metastasize (Alexander, 1974; Currie and Alexander, 1974). With one MCA-induced rat sarcoma, MC-3, antigenic activity was detected in tissue culture supernatants by the inhibition of tumor-bearer lymphoid cell cytotoxicity, whereas supernatants from another sarcoma, MC-1, did not inhibit lymphocytotoxic reactions for cultured tumor target cells. Also, since sarcoma MC-3 regularly metastasized by blood-borne and lymphatic spread, and sarcoma MC-1 was non-metastasizing, it was suggested that the metastasizing MC-3 tumor escaped host immune control by rapid antigen shedding providing a soluble antigen camouflage and allowing the deposition of metastases. One difficulty in the interpretation of these tests is that it was not possible to identify tumor-associated rejection antigens by conventional immunoprotection assays using the metastasizing sarcoma MC-3 even though specifically cytotoxic lymphoid cells are detectable in the lymph nodes of MC-3-tumor-bearing rats (Currie and Gage, 1973). Nevertheless, the findings of Doljanski (1973) and Ben-Sasson et al. (1974) showing that cultured Rous sarcoma virus (RSV)-transformed cells release surface macromolecules that bind specifically to lymphocytes of chickens bearing RSV-induced tumors supports the general conclusions of Currie and Alexander (1974). Also, Hayami et al. (1974) determined that culture supernatants harvested from Japanese quail RSV-induced tumors had "blocking activity" attributable to tumor antigen in in vitro tests. When cytotoxic spleen cells were pre-exposed to culture supernatants admixed with serum from quails in which tumors had regressed, their cytotoxicity for cultured tumor cells was abolished. This phenomenon was ascribed to the interaction of immune complexes with sensitized lymphoid cells although no identification of antigen and/or antibody was presented (Hayami et al., 1974). More direct evidence showing that tumor antigens released from cultured cells retain biological activity in vivo has been obtained using exhausted tumor cell culture media to immunize animals against challenge with viable tumor cells (Pellis and Kahan, 1975a). In agreement with other tests using 3 M KCl solubilized tumor extracts as the immunizing material (Pellis and Kahan, 1975b), significant protection to tumor cell challenge was demonstrable although antigenic activity was found only in the particulate fraction of tissue culture supernatants.

Tumor antigens have also been identified in the culture supernatants of murine L-1117 lymphoma cells. This material was defined by its capacity to inhibit the complement-dependent cytotoxicity of rabbit antisera directed against tumor antigens in a ^{51}Cr release test (Fujimoto et al., 1973). The criterion adopted for solubility of this material was the ability of the active fraction to survive sedimentation at 100 000 g for 60 min, although this may be inadequate in view of the findings of Rapaport et al. (1965) who demonstrated that material could be sedimented at 200 000 g from 100 000 g supernatants of homogenates of human leukocytes which consisted of fibrillar particulate components together with some small membrane fragments. Furthermore, this subcellular fraction also retained normal transplantation antigenic activity (Rapaport et al., 1965). Evidence was also presented by Fujimoto et al. (1973) to indicate that the lymphoma-associated antigen was closely linked with normal histocompatibility antigens, although both activities were located on a large macromolecular complex or covalently linked molecule which eluted in the excluded volume of a Sephadex G200 column.

An antigen(s) associated with a spontaneous murine melanoma was identified by Bystryn et al. (1974) in tissue culture supernatants using an antigen-binding radioimmunoassay following labelling of growing cultures of tumor cells with [^{3}H]leucine. In this case, the melanoma-associated antigen(s) had a molecular weight of 150 000 to 200 000 daltons and showed the characteristics of a glycoprotein. Acid precipitable activity was found to be higher in culture media than in cell lysates leading the authors to conclude that rapid shedding or secretion of antigen(s) was occurring. Further studies confirmed these findings (Bystryn, 1976), and suggested that antigen release was more rapid than the average release of other macromolecules. Although these studies illustrate the in vitro shedding of tumor-associated products, some of the antigen(s) detected was present to a lesser and variable extent in normal murine tissues (Bystryn et al., 1974) thus raising questions regarding the tumor specificity of the material assayed.

Roth et al. (1976) have identified a human melanoma-associated antigen in concentrated tissue culture supernatants of melanoma cells cultured in spinner flask using a specially formulated chemically defined serum-free medium. This was evaluated for reactivity in tests for delayed cutaneous hypersensitivity responses and inhibition of complement fixation. As in many studies analyzing immune reactions to human tumor antigens, a proportion of normal subjects reacted positively in this investigation although a significantly greater responsiveness was observed in melanoma patients.

To date, no single study on the detection of released antigen has given a completely satisfactory account of the synthesis and turnover of membrane-associated tumor antigens, though a valid general conclusion is that these products can be liberated by normal metabolic processes rather than cell death alone. To illustrate the actual quantities of antigen released by growing cells, Goldenberg et al. (1972) determined CEA levels in cultures of human colonic carcinoma. With cells cultured in 150 cm^2 glass bottles, an average of 3.75 ng of CEA were synthesized and shed into the medium per day. Since CEA is probably one of the more abundantly produced abnormal tumor-associated components, the amounts of other more weakly expressed antigens that are released into the media are likely to be much lower.

3.3. Induction of release of cell surface antigens by immune mechanisms

The only clearly understood mechanism by which immune responses in the tumor-bearing host may actually induce the release of cell surface antigens is by immune cytolysis leading to cell fragmentation and autolysis. Although there are several, still poorly understood, pathways of varying complexity by which immune cytolysis may be effected (Hellström and Hellström, 1974b; Cerottini and Brunner, 1974; Price and Baldwin, 1975), the end result is the same and tumor products may be liberated into the extracellular environment of the tumor. It is tempting to invoke the possibility that antibody-induced effects such as the redistribution, patching, capping of surface antigens might contribute to antigen turnover and release, possibly in the form of immune complexes which may interfere with cell-mediated immune responses (Fig. 3). The combination of antibody with cell surface antigens may also produce allosteric changes in the antigen which then modify its interaction with the protein-lipid bilayer. In this way the affinity of the antigen in its association with the cell membrane may be reduced sufficiently to allow the release of an antigen-antibody complex. Evidence showing that these phenomena are operative in vivo is generally lacking, however, and these effects may represent in vitro artefacts, even though they have enlarged our understanding of the expression and stability of antigens and other receptors at the cell surface. Capping phenomena have been induced in vivo by passaging various MCA-induced sarcoma ascites cells in irradiated mice and then administering relatively large quantities of syngeneic or allogeneic antiserum into the peritoneal cavity (Klein, 1975). After harvesting, the cells were exposed to human complement and an anti-complement (β-1 C) conjugate. The pattern of capping for several cell surface antigens was essentially the same or more impressive than that demonstrated previously in tests in vitro (Yefenof and Klein, 1974). Positive identification of "capping like" effects occurring under normal physiological conditions has recently been obtained by R.W. Stoddart, W. Jacobson and R.D. Collins (manuscripts in preparation) who found that in apparently unaffected areas of lymph node or spleen from patients with Hodgkin's disease, cells showing caps were present. These were revealed by staining of fixed tissue specimens with fluorescein-labeled concanavalin A. Interestingly,

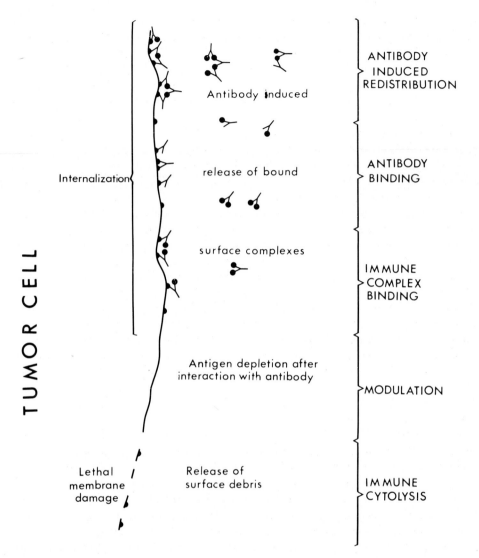

Fig. 3. Some possible contributions of the immune response in inducing the release of membrane-associated antigens into the extracellular environment.

the caps faced towards the affected area. Caps and patches of lectin receptors have also been identified by staining with a variety of fluorescein-labeled lectins on megakaryocytes in cases of idiopathic thrombocytopenic purpura and on histiocytes in sinus histiocytosis. In each situation, caps were observed after prefixation indicating that they were not a result of lectin-induced redistribution of surface binding sites (Stoddart et al., manuscript in preparation).

Studies employing radioiodinated specific alloantibodies against DBA/2J mastocytoma cells have provided some evidence that antibody binding to antigen may result in shedding from the cell surface which, in turn, may lead to a reduction in the susceptibility of target cells to sensitized lymphoid cells (Faanes and Choi, 1974). In this case 20 to 30% of the cell bound antibody was released into culture supernatants within 3 h after incubation at 37°C. This release of cell surface bound antibody, either free or as immune complexes, may also account for the finding that in this system isoantibody blocks lymphocyte-mediated cytotoxicity in the early phase of incubation, but that the effect is abolished following further incubation (Faanes et al., 1973). Similarly, Hellström (1974) reported that tumor cells blocked by serum regained their susceptibility to the cytotoxic action of sensitized lymphoid cells after incubation for about 5 h.

Measurement of the stability of cell surface-bound antibodies has also been employed to distinguish between two tumor antigens associated with aminoazo dye-induced rat hepatomas and MCA-induced rat sarcomas. These tumors express both a tumor-specific antigen, individually distinctive for each transplanted line, and a cross-reacting embryonic antigen which can be demonstrated by the development of membrane immunofluorescence staining of tumor cells after treatment with multiparous rat serum (Baldwin et al., 1974b; Baldwin and Price, 1975). As shown in Fig. 4, when tumor cells were treated with syngeneic tumor-immune serum or a diluted Slonaker anti-Wistar alloantiserum, cell-bound antibody was still detectable by membrane immunofluorescence staining following incubation at 37°C for 4 h. However, when cells were treated with multiparous rat serum, although the immunofluorescence staining was comparable initially to that given by tumor-immune or diluted alloantiserum, after further incubation at 37°C for 2 h cell-bound antibody was no longer detectable upon addition of the fluorescein-labeled rabbit anti-rat immunoglobulin conjugate (Fig. 4). These findings were interpreted as indicating differences in the expression of tumor-specific and embryonic antigen at the tumor cell surface, and that the tumor-specific component showed greater stability (at least in its association with antibody).

Antibody-induced loss of antigen-antibody complexes from the cell surface has been proposed to occur in the process of antigen modulation first described in studies upon the murine thymus-leukemia (TL) antigen whereby lymphoid cells treated with specific antisera gradually lose their susceptibility to attack (Boyse et al., 1967; Old et al., 1968). While this phenomenon may be mediated by endocytosis or shedding of complexes, or both, a further contribution to modulation may occur as a result of capping or patching requiring that surface components be relatively mobile and resulting in the immobilization or aggregation of surface-bound Ig. For example, capping or patching of TL antigen with antisera rendered cells insusceptible to complement-mediated lysis (Stackpole et al., 1974; Loor et al., 1975). The redistribution of surface-bound Ig involved in such processes may impede complement fixation by steric hinderance or by the inability of Ig molecules to conform to the correct orientation necessary to initiate the complement cascade. Also, in a study upon the effects of Clq on the redistribution of cell-bound Ig, Sundqvist et al. (1974)

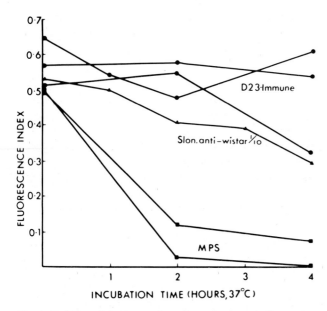

Fig. 4. Stability of the interaction of syngeneic and allogeneic antibodies with cell surface antigens on rat hepatoma D23 cells determined using the indirect membrane immunofluorescence test. Hepatoma D23 cells were reacted with syngeneic tumor-immune serum (reactive with the tumor specific antigen; ●—●, D23 immune) syngeneic multiparous rat serum (reactive with cross-reactive tumor-associated embryonic antigens; ■—■, MPS) and a multispecific Slonaker anti-Wistar liver antiserum (▲—▲, Slon anti-Wistar 1/10), and excess antibody was removed by washing. Cell-bound antibody remaining on tumor cells after various incubation times at 37°C was visualized by staining with fluorescein conjugated rabbit anti-rat immunoglobulin. Fluorescence indices were calculated by determining the percentage of cells unstained following incubation with normal rat serum samples minus the percentage of cells unstained by test serum samples, divided by the former figure. Fluorescence indices of greater than 0.3 represent significant membrane immunofluorescence staining of tumor cells.

determined that Clq promoted clustering and the maximum susceptibility to lysis occurred when there was a clustered or patchy distribution of Ig.

Davey et al. (1976) have reported data which indicate that there may be a correlation between the ability of surface membrane components to modulate and the capacity of a tumor to metastasize. A metastasizing murine lymphoma was found to be more susceptible to antibody-induced modulation of histocompatibility antigens than a non-metastasizing lymphoma. Also, since the metastasizing lymphoma was found to release histocompatibility antigens at a greater rate both in vitro and in vivo, and was more resistant to complement-dependent antibody lysis, it was concluded that the lability of surface antigens might well be related to the metastasizing properties of the tumor.

The ultimate fate of antigen lost by modulation after interaction with specific antibody may vary according to cell type and nature of receptor. With the TL antigen (Yu and Cohen, 1974) and *H-2* antigens (Lengerová et al., 1972) endocytosis of surface antigen-antibody complexes was the predominant fate in in vitro tests. When extending these considerations to tumor-associated antigens expressed upon a developing neoplasm several effects may be operating simultaneously as a consequence of host immune responses, some of which result in immune complex release. Immune complexes may be shed following interaction of cell surface antigens with antibody, while antigen released by turnover or cell degradation may lead to the production of immune complexes either in the local environment of the tumor or systematically in the circulation where antigen concentration gradients may exist. The relevance of such immune complexes in modifying cellular responses to the developing tumor will be discussed in a subsequent section (see page 447).

Although the possible contribution of antibody in inducing shedding of antigens is not known, there is considerable evidence that tumor cells in vivo may be coated with immunoglobulin (Witz, 1973). This has been established in experiments with a variety of tumors including carcinogen-induced hepatomas and sarcomas (Witz et al., 1967; Ran and Witz, 1970; Robins, 1975), spontaneous mammary carcinomas (Ran and Witz, 1970) and polyoma and SV40 virus-induced tumors (Sobczak and De Vaux St. Cyr, 1971; Ran et al., 1976), as well as several human tumors (Eilber and Morton, 1971; Philips and Lewis, 1971; Thunold et al., 1973; Gupta and Morton, 1975). Although in some examples, non-specific immunoglobulin binding may occur (Witz, 1973; Robins, 1975), in tests on melanoma patients it was found that tumor eluates contain specific antibodies detectable by complement fixation (Gupta and Morton, 1975).

Further evidence suggesting that tumor host responses may be involved in determining antigen release has been obtained by Thomson et al. (1973c) who observed a decrease in the release of tumor antigen into the circulation from a transplanted rat sarcoma when recipients received 500 rad whole-body radiation. This was interpreted as indicating that release of tumor antigen may be influenced by host responses as well as by metabolic turnover and cell degradation during tumor growth.

4. Detection of antigens shed in vivo

4.1. Identification of tumor-associated antigens in body fluids

The most conclusive evidence showing that tumor antigens are shed in vivo is provided by data on the identification and quantitation of antigenic activity in the serum of tumor-bearing hosts. By far the most attention has been applied to products such as α-fetoprotein (AFP) and carcinoembryonic antigen (CEA) which have been defined by their reaction with appropriate xenoantisera. The association of elevated serum levels of AFP and CEA with various neoplastic and non-neoplastic conditions has been the subject of several extensive reviews to which the reader is referred (for reviews on AFP see Abelev, 1971, 1974; Laurence and Neville, 1972; Ruoslahti et al., 1974 and on CEA see Fuks et al.,

1974; Terry et al., 1974; Shuster et al., 1975; Zamcheck, 1975). However, several comments are worth making in the light of experience gained from assays of AFP and CEA levels in cancer patients. Firstly, the clinical applications of monitoring CEA and AFP are limited by the finding that these materials can also be detected in the serum of normal individuals and of patients with various non-malignant diseases. Fuks et al. (1974) have noted that it has not proved possible as yet to purify CEA-active material from these sera in sufficient quantities to allow immunological and physicochemical comparison with colon carcinoma derived CEA. It is possible that the CEA-like activity measured by radioimmunoassay in the serum of non-colon carcinoma patients may not be identical to colon carcinoma CEA suggesting that the non-specificity of CEA assays for defining malignant disease may be due to cross-reacting antigenic protein. Furthermore, Vrba et al. (1976) have shown that circulating CEA-like activity in some sera of patients with carcinoma of the digestive tract may be immunologically different from the available CEA standards purified from human tumor extracts. Studies such as these illustrate the potential difficulties which may be encountered when assaying circulating antigens. Also, although the contribution of AFP or CEA in promoting immune responses in the tumor-bearing host is not known, there have been reports implicating their involvement in host reactions. Specific humoral immune responses against CEA, primarily in the IgM immunoglobulin class have been demonstrated in Gold's laboratory (Gold, 1967; Gold and Gold, 1973; Gold et al., 1973), and Dattwyler and Tomasi (1975) have presented data showing that AFP blocks the generation of specific immune cytotoxic T-cells directed against a mouse mastocytoma.

Tumor antigens which are immunogenic in the host have been identified in the serum of tumor-bearers in studies with a variety of experimental animal tumors. For example, with one aminoazo dye-induced rat hepatoma, D23, tumor-specific antigen was identified in tumor-bearer serum both in free form and as immune complexes in rats with large intraperitoneal growths (Baldwin et al., 1973a). In these studies, tumor antigen in free form was isolated in the low molecular weight included fraction of serum separated by Sephadex G150 gel filtration at neutral pH. Tumor antigen present in immune complexes was isolated from the Sephadex G150 exluded fraction of serum following dissociation at pH 3 and refractionation on G150 at this pH. Tumor-specific antigen in these separated serum fractions was defined firstly by their capacity to neutralize specific antibody in tumor-immune rat serum, as assayed by reduction of membrane immunofluorescence staining with hepatoma D23 cells. Secondly, these serum fractions were shown to induce specific antibody responses when used to immunize syngeneic rats, this again being assayed by membrane immunofluorescence methods.

These methods were employed subsequently in studies to monitor the sequential changes in the pattern of serum borne tumor products during the progressive growth of this hepatoma D23 when implanted subcutaneously (Bowen et al., 1975). Under these conditions, tumor-specific antigen was

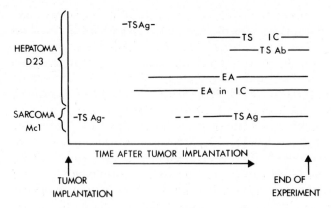

Fig. 5. Schematic representation of various serum factors detected in tumor-bearer serum during the growth of the rat hepatoma D23 and rat sarcoma MC-1. TSAg, tumor-specific antigen; TSAb, tumor-specific antibody; TS IC, tumor-specific immune complexes; EA, embryonic antigen. Data compiled from Bowen et al. (1975), Rees et al. (1975), Thomson et al. (1973b) and Thomson (1975).

released early into the circulation, being detected within 7 to 10 days of tumor implantation (Fig. 5). With further progressive growth of the tumor, free antigen became undetectable but concomitantly tumor-specific immune complexes were identified, suggesting that these changes partly reflected the induction and development of a tumor-immune response. Then in the terminal phase of tumor growth, when large subcutaneous masses were present, it was possible to detect free tumor-specific antibody in serum, but again in the presence of specific immune complexes. Therefore, apart from the initial phase of tumor growth, when it might be expected that host responses to the tumor were poorly developed, the tumor-bearer serum contained specific immune complexes.

There is, however, evidence from studies on the effect of these sera on cell-mediated immunity (see page 455) that the composition of immune complexes changes as the tumor progresses. This might be expected since at the early phase of tumor growth it appears that there is excess tumor antigen in the serum, whilst in the terminal stage excess tumor-specific antibody can be demonstrated (Fig. 5). Similarly, Smith and Leonard (1974) determined that guinea pigs bearing progressively growing transplants of a diethylnitrosamine-induced hepatoma, developed antibody after 4 to 5 weeks of tumor growth. Before antibody was detected, their serum inhibited the binding of antitumor antibody to target cells, this effect being attributed to free antigen, antigen-antibody complexes, low affinity antibodies or a non-specific factor.

As already commented upon, chemically induced rat tumors like the amino-azo dye-induced hepatomas express both tumor-specific and embryonic antigens (Baldwin et al., 1974b). It is of interest, therefore, to note that the pattern of release into the serum of tumor-associated embryonic antigens during subcutaneous growth of hepatoma D23 differed from that observed with tumor-specific antigen, in that free embryonic antigen together with immune complexes was present at all stages of tumor growth (Rees et al., 1975). This may reflect the relative instability of the tumor-associated embryonic antigen at the cell surface when compared with the tumor-specific product. Alternatively variations in serum levels may be ascribed to differences in the rate of synthesis and secretion of these tumor products whilst their persistence in the circulation will be governed to some degree by the efficacy of specific humoral immune responses elicited by these tumor products.

Another parameter influencing the levels of circulating serum factors is the site of tumor growth. For example when cells from the rat hepatoma D23 are injected intravenously, pulmonary tumor nodules develop and in this artificial metastasis model, free serum antigen was detected only during later stages of tumor growth (from day 14 to 24 after tumor induction) (Bowen and Baldwin, 1976b). Immune complexes of tumor specific antigen and antibody were identified slightly earlier (from day 10) and both complexes and free antigen persisted until tumor burden resulted in respiratory distress. This pattern of events markedly differs from the situation observed when hepatoma D23 is grown from a subcutaneous implant of tumor cells (see Fig. 5; Bowen et al., 1975).

The release of tumor-associated products into the circulation has similarly been demonstrated with a number of other experimental animal tumors, although there are quite marked variations between the pattern of events occurring during progressive growth of different types of tumor. Thus rapid release of tumor-specific antigen into the serum of rats bearing transplants of a 3-methylcholanthrene-induced sarcoma (MC-1) has been observed by Thomson and his colleagues using a solid phase radioimmunoassay technique (Thomson et al., 1973b; Thomson, 1975). Antigenic activity was detected initially in serum 1 to 3 days after tumor implantation, this being attributable primarily to cells in the tumor implant undergoing autolysis (Fig. 5). Progressive growth of this tumor was accompanied by a slow rise in total serum antigen levels and these only decreased following surgical removal of the implanted tumor (Thomson et al., 1973b). Comparably, with a BALB/c murine plasmacytoma, free tumor antigen and immune complexes have been detected in tumor-bearer serum by their specific inhibition of antibody-induced agglutination of erythrocytes coated with tumor extracts (Kolb et al., 1974) or using a spleen cell migration inhibition assay (Poupon et al., 1974). In other studies, tumor antigen-containing moieties in the serum of rats bearing grafts of a Rous virus-induced sarcoma were identified by demonstrating that immunization of normal rats with tumor-bearer serum induced protection to subsequent challenge with viable tumor cells (El Ridi and Bubeník, 1975). These findings extend those of earlier studies (El Ridi and Bubeník, 1973) where it was shown that tumor-bearer sera did not contain tumor-specific antibody levels demonstrable by membrane immunofluorescence techniques.

From these considerations, it is worth noting that tumor-bearer serum or other body fluids may represent a convenient source of tumor-specific antigen.

For example, the tumor-specific antigen associated with the aminoazo dye-induced hepatoma D23 has been isolated from the serum of tumor-bearing rats and this serum product has similar physicochemical characteristics (molecular weight 55 000 daltons; isoelectric point 4.5) to the material liberated by papain digestion of tumor membrane fractions (Baldwin et al., 1973e; Baldwin and Price, 1976; Bowen and Baldwin, 1976a). Similarly, tumor-specific antigen has been isolated from the ascitic fluid of mice bearing grafts of a spontaneous lymphoma and partial purification achieved by ammonium sulphate precipitation, gel filtration and ion exchange chromatography (Wolf and Steele, 1975).

Whilst the experimental tumor studies do indicate that tumor antigen either in free form or as immune complexes appears in the circulation of the tumor-bearing hosts, it is evident that the methods employed in these analyses are still completely inadequate. Clearly, if tumor antigen is to be monitored during tumor growth precise immunoassays similar to the radioimmunoassay of Thomson and colleagues (Thomson et al., 1973b; Thomson 1975) must be employed. Moreover, if a precise analysis of serum factors is to be obtained in terms of free antigen as well as immune complexes, the procedures already developed must be improved.

Clinical investigations on the release of tumor antigens into the serum and body fluids are hindered by the lack of suitable in vitro methods for monitoring tumor-immune reactions in patients. For example, Viza and Phillips (1975) have employed rabbit antisera prepared against soluble melanoma membrane material to monitor sera from melanoma patients. In these tests, 19% of patients' sera tested gave positive immunoprecipitation with the anti-melanoma antibody whereas no reactions were obtained with sera taken from control subjects consisting of healthy volunteers or patients with measles or leukaemia. Although the antiserum raised against melanoma extracts was absorbed with normal human serum and spleen extracts this does not establish conclusively that the reactions detected in melanoma patients' serum involved a melanoma-specific antigen. This point is emphasized by other studies where it has been conclusively established that soluble *HL-A* antigens can be identified in human serum (Miyakawa et al., 1973a).

Similar conclusions were drawn by Carrel and Theilkaes (1973) in their study on melanoma-associated antigens using an absorbed antiserum prepared in the rabbit against concentrated dialysed specimens of patients' urine. Although the experiments indicated the presence of a common antigen in the tumor which was excreted in patients' urine, the question remains as to whether this was specific for melanoma or whether the activity represented material of embryonic or viral origin (Carrel and Theilkaes, 1973).

Since it has been reported that tumor-associated antigens in solubilized extracts of human tumors including carcinoma of breast and colon and melanoma can be monitored by their capacity to inhibit leukocyte migration (Jones and Turnbull, 1974; McCoy et al., 1974, 1975) this assay has been used in studies designed to detect tumor-related products in the plasma from bladder carcinoma patients (Bowen, 1975). In this study, plasma fractions containing material in the molecular weight range 20 000 to 100 000 daltons was isolated by Sephadex G150 chromatography using essentially the procedures designed for the isolation of tumor antigen from serum of experimental tumor-bearing animals (Baldwin et al., 1973a; Bowen and Baldwin, 1976a). These fractions were then tested for their capacity to inhibit migration of leukocytes from urinary bladder patients. Whilst these studies did indicate that the techniques were able to detect leukocyte-inhibitory material in bladder cancer patients' plasma, it was not established conclusively that this factor was tumor-specific.

A similar uncertainty also exists in recent studies designed to monitor circulating immune complexes in patients with Burkitt's lymphoma and nasopharyngeal carcinoma (Mukojima et al., 1973; Oldstone et al., 1975; Heimer and Klein, 1976). For example, in the experiments of Heimer and Klein (1976) immune complexes were tested for by consumption of hemolytic complement and the results indicated that the levels were increased in patients when compared to healthy controls. These immune complexes were sedimented between 10 and 19 sec and were retained by concanavalin A-Sepharose linked columns. This observation, together with the finding that the bound material was eluted by α-methyl-D-mannoside, suggests that the complexes may contain glycoprotein, but this product has not been identified.

4.2. Blocking factors

Material in the serum of tumor-bearing individuals which has the capacity to abrogate the cytotoxicity of sensitized lymphoid cells for cultured tumor cells has been rather loosely described as "blocking factor" or in early studies as "blocking antibody". The association of blocking activity with tumor-bearer serum in both experimental and human cancer has been reviewed recently (Hellström and Hellström, 1974b) and the possible mechanisms now thought to be involved in these reactions are shown in Fig. 6. While the original proposal was that blocking effects were mediated by antibody (Hellström and Hellström, 1969), this interpretation has subsequently been found to be inadequate and there is evidence suggesting the involvement of specific immune complexes, although, as already discussed, the mechanism by which antigen in immune complexes arises in the serum is still ill defined.

The observation that serum blocking activity was rapidly lost in animals whose tumors spontaneously regressed (Hellström and Hellström, 1969, 1970) or where a developing tumor was surgically removed (Baldwin et al., 1973d) was inconsistent with the view that antibody was the active factor since antibody alone would be expected to persist much longer in the circulation. Similarly, in human studies, blocking activity was present in 85% of patients with clinical evidence of tumor, but in only 14% of patients where there was no clinical manifestation of disease (Hellström and Hellström, 1974a).

The first indication that the blocking factor in tumor-bearer serum might be tumor-specific antigen-antibody complexes was obtained by Sjögren and his

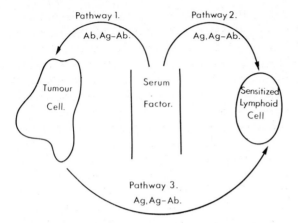

Fig. 6. Schematic representation of some possible interactions of humoral factors interfering with cell-mediated immune reactions.

Blocking reactions: these may be mediated by antibody (free or immune complexed) interacting with tumor cell surface antigens thereby protecting tumor cells from lymphocytotoxic attack (Pathway 1).

Inhibitory reactions: these may be mediated by antigen (free or immune complexed) interacting with specific receptors on sensitized lymphoid cells so diminishing their cytotoxic capabilities (Pathway 2). Alternatively, antibody initially bound to the tumor cell surface membrane may be released, as an immune complex which is then capable of interacting with receptors on sensitized lymphoid cells (Pathway 3).

At an operational level, blocking and inhibitory activities of serum factors are determined by pretreating the tumor or lymphoid cell preparations respectively with test or control sera, before lymphoid cells are added to the cultured tumor cells and cytotoxicity assessed.

colleagues (1971). In these studies, the blocking factor in the serum of mice bearing Moloney virus-induced sarcomas was absorbed onto intact tumor cells, eluted with low pH buffer and then separated by membrane ultrafiltration into high molecular weight (greater than 100 000 daltons) and low molecular weight (10 000 to 100 000 daltons) fractions. Neither of these individual fractions alone displayed blocking activity when added to cultured sarcoma cells, although activity was restored upon their recombination. This procedure was also applied to blocking factor eluted directly from surgical specimens of human tumors where it was found that acid eluates were blocking but that this activity was lost following fractionation into "low" and "high" molecular weight material (Sjögren et al., 1972). Again, blocking activity was restored upon recombination of the two fractions and in addition a similar effect was obtained when tumor cells were first exposed to the high molecular weight fraction (antibody?) and then the low molecular weight fraction (antigen?) but not vice versa. Neither antigen nor antibody were positively identified in these fractions, and it is quite possible that acid extraction of tumor tissue or cells releases membrane-associated antigens as well as surface-bound immune complexes.

More direct evidence supporting the concept that tumor-bearer serum blocking reactions are mediated by specific immune complexes comes from tests showing that complexes formed by the addition of papain-solubilized tumor-specific antigen prepared from a transplanted rat hepatoma to serum taken from rats following surgical removal of tumor grafts were effective in blocking tumor cells from the cytotoxic attack by immune lymph node cells (Baldwin et al., 1972). When increasing amounts of antigen were added to post-excision sera, the blocking effect was lost, presumably due to the formation of immune complexes in antigen excess which are no longer able to bind to the surface of tumor target cells. That these post-excision sera do in fact contain tumor-specific antibody was previously established by complement-dependent serum cytotoxicity reactions measured either by the colony inhibition test (Baldwin and Embleton, 1971) or using the microcytotoxicity assay (Baldwin et al., 1973d). Similarly, tumor-bearer serum from rats bearing transplants of an aminoazo dye-induced hepatoma contained immune complexes after approximately two weeks subcutaneous growth, these being identified following their dissociation by detection of antigen and antibody after fractionation by gel filtration at low pH, and serum taken at this stage was also found to be blocking in lymphocytotoxicity assays (Bowen et al., 1975).

That immune complexes may exert their maximum blocking activity when composed of antigen and antibody in equivalence has been reported by José and Seshadri (1974). In these tests, immune complexes together with either excess tumor antigen or antibody were prepared by the addition of radioiodinated tumor antigen (supernatant from tumor cell cultures) to serial dilutions of serum from patients in remission. These mixtures were then incubated with cultured neuroblastoma target cells and autochthonous blood leukocytes and the blocking activity was determined. Using a radioimmunocounterelectrophoresis assay it was found that the maximum blocking effect of these mixtures was obtained when there was antigen-antibody equivalence (José and Seshadri, 1974).

Removal of blocking factors from sera by passage through immunoabsorbent columns designed to combine either with antigen or antibody has been taken as evidence supporting the view that serum blocking activity resides in immune complexes. Anti-IgG_{2a} or anti-IgG_{2b} immunoabsorbent columns removed the blocking activity of multiparous mouse sera which prevented the killing of plated tumor cells by lymphocytes from such mice, although the blocking factor could be recovered in column eluates (Tamerius et al., 1975). Similarly, by passage of multiparous serum through an immunoabsorbent column prepared using an absorbed rabbit anti-mouse embryo antiserum, the blocking activity was retained on the column but was recovered following elution with 3 M NaSCN. The implication from these studies was that the blocking factor in multiparous serum consisted of immune complexes composed of antigen of possibly embryonic origin and antibody of the IgG_{2a} and IgG_{2b} class.

In a similar investigation, the nature of the blocking factor in the serum of mice bearing a transplanted MCA-induced sarcoma was analyzed (Tamerius et al., 1976). Serum from hyperimmune mice was initially coupled to Sepharose 4B, this then being used as the immunoabsorbent. After passage of tumor-bearer serum through this column, blocking activity was removed. Blocking activity was regained in the eluate following elution of the column. The eluate was further fractionated by gel filtration on Sephadex G200 and the blocking factor identified in the region of the column in which marker IgG was eluted.

An alternative approach for investigating the nature of blocking factors in the serum of tumor-bearing individuals has been to establish that this activity can be neutralized by the addition of tumor-specific antibody (review, Baldwin and Robins, 1975). The usual procedure employed to demonstrate the counteraction of blocking activity of tumor-bearer sera involves exposing plated tumor cells to a blocking serum in admixture with serum from a tumor-free or actively immunized donor, and with the appropriate ratio of sera, cultured tumor cells regain their susceptibility to the cytotoxicity of sensitized lymphoid cells. For example, in the initial studies with murine Moloney sarcoma virus-induced tumors, it was shown that pre-exposure of plated target cells to heat-inactivated tumor-bearer serum blocked their susceptibility to cytotoxic lymph node cells and that this blocking could be neutralized or "unblocked" by mixing this serum with serum taken from regressor mice (Hellström and Hellström, 1970). Similar unblocking has been detected with rat polyoma virus-induced tumors using serum taken following tumor resection (Bansal and Sjögren, 1971, 1972, 1973). With human tumors, serum from clinically free patients was found to be unblocking when mixed with serum from tumor-bearing individuals (Hellström et al., 1971). The conclusion from these tests is that unblocking serum containing tumor-specific antibody neutralizes the factor, presumably tumor antigen (most likely in the form of immune complexes), thus allowing sensitized lymphoid cells to exert their cytotoxic reactivity. This interpretation is further supported in studies using the rat hepatoma D23 in which unblocking sera were shown initially to positively contain tumor-specific antibody either by membrane immunofluorescence staining of tumor cells or complement-dependent cytotoxicity (Robins and Baldwin, 1974). In these tests, both syngeneic tumor-immune sera and an absorbed rabbit antiserum neutralized the blocking activity of hepatoma D23 tumor-bearer serum.

Investigations have been undertaken in an attempt to substantiate that the effects of "unblocking sera" observed in vitro may be relevant in vivo. For example, Bansal and Sjögren (1972) prepared unblocking sera by immunizing rats or rabbits with tumor cells and reported that regression of both primary and transplanted tumors induced by polyoma virus occurred after repeated administration of these antisera. Even though tumor regression was accompanied by a loss of serum-blocking activity, the results of these experiments remain equivocal since passively transferred antisera may themselves mediate a variety of responses (e.g. antibody killing in the presence of complement or cell-dependent antibody cytolysis).

A further aspect of studies upon blocking and related serum factors has been the demonstration of an "arming factor" which collaborates with non-sensitized lymphoid cells to produce specific killing of tumor cells (Pollack and Nelson, 1974). This appears in the serum within 24 to 48 h after initiation of the tumor by inoculation of Moloney sarcoma virus or transplants of syngeneic sarcoma cells and although cell-dependent antibody cytotoxicity may be positively identified in this system in donors bearing tumors for longer times, this factor is apparently neither IgG nor IgM (Pollack and Nelson, 1975). Comparably, when tumor-bearer spleen cells were cultured for 2 days, their supernatants were found to exhibit both cell-dependent antibody cytotoxicity and classical blocking activity whereby they protected tumor target cells from the cytotoxic attack by immune lymphocytes (Nelson et al., 1975a,b). Further studies involving the cultivation of tumor-bearer spleen cells with [^{14}C]leucine confirmed that the factors responsible for blocking and cell-dependent antibody killing were synthesized by these cells in vitro (Nelson et al., 1975c). These investigations emphasize that the host responds to exposure to tumor cells to produce factors of diverse properties although it is not known to what extent circulating tumor antigens released from the tumor contribute to the induction of these factors or whether this solely represents a reaction to antigens expressed on the intact tumor cell.

4.3. Inhibitory factors

The release of tumor antigens from the developing tumor into the circulation may have profound consequences upon the immune response other than the formation of immune complexes which interact with tumor cell surface antigens (Fig. 6). For example, it has been found that brief exposure of immune lymphoid cells to isolated tumor antigens inhibits their cytotoxicity for cultured tumor cells. This was established initially in tests with the rat hepatoma D23 in which partially purified, papain-solubilized, tumor-specific antigen inhibited the cytotoxicity of lymph node cells from syngeneic tumor-immune donors for hepatoma D23 target cells (Baldwin et al., 1973f). This effect could also be demonstrated using tumor-bearer serum, although in this case it was not known whether the inhibitory reaction observed was mediated by free circulating tumor antigen or immune complexes.

Other studies have supported the view that antigens solubilized from the tumor may modify the cytotoxicity of sensitized lymphoid cells for tumor target cells. With murine sarcoma virus-induced tumors, specific inhibition of cytotoxicity was obtained using 3 M KCl extracts of tumors and either immune spleen cells or purified populations of splenic T cells (Plata and Levy, 1974). Comparably, with human tumors, in particular colon carcinoma and melanoma, specific inhibition of peripheral blood effector cell cytotoxicity has been obtained after exposure of the effector cells to papain-solubilized tumor membrane fractions (Baldwin et al., 1973c; Embleton, 1973; Embleton and Price, 1975) and other crude tumor antigen preparations including 3 M KCl extracts of tumors (Nind et al., 1975).

In each of the tests described, cytotoxicity was evaluated by determining cell survival in the microcytotoxicity test by visual counting methods. Other attempts have been made to evaluate inhibition of cytotoxicity by tumor-associated antigens using the chromium release test. In experiments designed to detect inhibition of the cytotoxicity of sensitized spleen cells for MSV-induced tumors, no effect was observed with intact MSV (M) virus or the viral envelope antigen (VEA) (Gorczynski and Knight, 1975). Inhibitory activity was, however, demonstrable in MSV (M) virus preparations following their disruption by Triton X-100, the interpretation being that spleen cells recognize internal virus group-specific antigens on the cell surface. It was further possible to show that one of the major antigens of this group, P30, inhibited cytotoxicity. Conversely, Plata and Levy (1974) were unable to detect inhibition of cytotoxicity by 3 M KCl extracts of MSV-induced tumors using the chromium release assay whereas equivalent extracts showed inhibitory activity when assayed by the microcytotoxicity test.

The cytotoxicity inhibition test has proved of value in discriminating the antigenic targets against which tumor-bearer effector cell cytotoxicity is directed using transplanted MCA-induced sarcomas and DAB-induced hepatomas in the rat. As already mentioned, these tumors possess an individually distinct surface antigen as well as cross-reactive embryonic antigens and it was found that spleen and lymph node cells from tumor bearers showed cross-reactive cytotoxicity (Zöller et al., 1975). By choice of appropriate tumor and embryo cell extracts, this cytotoxicity was determined to be directed both against the tumor-specific and embryonic antigens, contrasting with the situation in the hyperimmune rat which shows individual specificity in these cell-mediated responses (Zöller et al., 1975, 1976). Similarly, tumor-associated embryonic antigen preparations inhibit the cytotoxicity of multiparous rat lymph node cells for tumor or embryonic target cells (Rees et al., 1974).

In other tests using a 1,2-dimethylhydrazine-induced rat colon carcinoma, similar approaches have been employed to identify both an organ-related embryonic antigen as well as an embryonic component of more widespread specificity. These products were initially separated from a soluble extract of tumor by a procedure involving immune complex formation and dissociation using an absorbed multiparous rat serum (Steele et al., 1975). Briefly, the experiments showed that the cytotoxicity of multiparous rat lymph node cells for fetal gut cells was inhibited by the embryonic antigen preparation of gut-related specificity whereas this material was without effect upon lymphoid cell reactivity for fetal lung cells, this reaction being inhibited by the widespread embryonic antigen preparation (Steele et al., 1975).

These studies illustrate the fact that the cytotoxicity of sensitized lymphoid cells for tumor target cells may be prevented following exposure of effector cells to tumor antigen preparations. They also emphasize that tumor antigens released from the developing tumor play an important role in modifying host responses since many of the inhibitory reactions referred to have also been demonstrated using tumor-bearer serum as a source of "inhibitory factor" rather than isolated tumor antigen preparations. Findings such as those of Currie and Basham (Currie and Basham, 1972; Currie, 1973) showing that repeated washing of peripheral blood lymphocytes from patients with disseminated melanoma, bladder carcinoma, hypernephroma and sarcoma increases their cytotoxicity may also be interpreted as reflecting the removal of bound tumor antigen. Alternatively, Laux and Lausch (1974) determined that spleen cells from tumor-bearing hamsters were not cytotoxic for PARA-7 target cells but after incubation in vitro overnight at 37°C, the cells became specifically reactive.

In many of the investigations on specific inhibition of cell-mediated immune responses by isolated tumor antigens, similar effects were also demonstrable using tumor-bearer serum as the "inhibitory factor", the implication being that circulating tumor antigens may mediate antagonistic reactions in the tumor-bearing host. The nature of this inhibitory reaction is not clear although the experiments of Currie and Basham (Currie and Basham, 1972; Currie, 1973) and Laux and Lausch (1974) just mentioned suggest that the inhibitory factor (putative tumor antigen either free or in immune complexes) is bound to the surface of the sensitized lymphoid cell or to a cell essential for the full expression of cytotoxicity and, furthermore, that this interaction is reversible by various manipulations. Bruce et al. (1976) have described similar findings in studies upon the in vitro stimulation of lymphocytes primed in vivo with Gross virus-induced lymphoma cells. In these tests, cultures of sensitized lymphocytes (responder cells) and mitomycin-treated leukemic cells (stimulator cells) generate effector cells which show high cytotoxic activity against leukemic cells. The effector cells emerge from culture in a "partially blocked" condition and their cytotoxicity was increased by subjecting them to a deblocking procedure involving incubation at 37°C for 3 h followed by washing (Bruce et al., 1976), this being designed to allow antigen to elute from the surface of effector cells (Kontiainen and Mitchison, 1975).

It is evident that further definition of the mechanism of inhibition of cytotoxicity by tumor antigens is required. Nevertheless, progress has been made in determining the nature of the effector cell involved in these reactions, the current view being that this belongs to the T cell population (Plata and Levy, 1974; Blair et al., 1975; Knight et al., 1975; Lane et al., 1975; Nind et al., 1975; Shellam and Knight, 1975).

Apart from these specific inhibitory responses thought to be mediated by tumor antigens (either free or as immune complexes) originating from the developing tumor, other studies have indicated that non-specific humoral factors may be involved in the tumor–host relationship. For example, non-specific inhibition of mitogen stimulation by serum from cancer patients has been demonstrated (Silk, 1967; Sample et al., 1971) and an immunosuppressive α-globulin has been implicated in processes of impairment of lymphocyte function (Glaser and Herberman, 1974; Glasgow et al., 1974a,b). Therefore, although these reactions have only been performed in in vitro tests, the possibility cannot be excluded that these humoral factors may play a role in non-specific immunosuppression caused by the tumor and in the immune responsiveness against the tumor itself.

5. Immunobiological effects of circulating antigens

5.1. Correlation of serum factors with tumor growth

The impetus for attempting to determine the levels of circulating factors (especially tumor antigens) during tumor growth stems from the proposal that such information would be invaluable for monitoring therapy and/or for designing effective immunotherapeutic protocols. Furthermore, with improvements in the sensitivity of immunological assays, these tests may represent an important immunodiagnostic aid for the identification of small primary tumors or for detecting the early recurrence of residual disease. It is as yet unknown as to whether the assay of circulating factors may be used as a measure of the extent of tumor burden although such a correlation would obviously provide important information to the clinician. In this context, despite the non-specificity for the diagnosis of colon carcinoma, the application of assays for CEA in the clinical management of patients with this disease has recently been summarized by Zamcheck (1975). While it was recognized that an intensified search for more specific markers of neoplasia should be encouraged, Zamcheck emphasized the usefulness of the CEA assay provided it is employed judiciously, with full awareness of its limitation.

From the considerations already mentioned, serum levels of tumor-associated products will be controlled by a number of phenomena including the rate of antigen release and rate of elimination from the circulation. Each of these components represents a composite series of events, one of which involves the effectiveness of the immune response against tumor neoantigens since this will be contributory in determining the serum levels of antigen in a free form or as immune complexes. This is exemplified in a model immune elimination experiment in which the levels of ^{125}I-labeled bovine serum albumin after intraperitoneal injection into rats immunized against this antigen were approximately five to ten times lower than in untreated rats (V. Preston, M.R. Price and R.A. Robins, unpublished observations). Antigen-antibody complexes are known to be eliminated from the circulation by phagocytosis effected by the reticuloendothelial system and by deposition in renal glomeruli and other tissues (Dixon et al., 1958; Benacerraf et al., 1959). These studies illustrate that serum antigen levels may be effectively reduced by immune complex formation and removal from the circulation. It is therefore of significance to note that immune complexes have been identified in the kidneys of mice bearing the B-16 melanoma (Poskitt et al., 1974) or the C1300 neuroblastoma (Oldstone, 1975). In this latter study, bound IgG was eluted and recovered from the glomeruli, and exhibited membrane immunofluorescence reactivity when added to C1300 tumor cells. A further possibility was raised by Prather and Lausch (1976) who suggested that the presence of abundant tumor antigen-antibody complexes may induce or enhance production of naturally occurring anti-immunoglobulins (Kunkel and Tan, 1964) with consequent clearance from the circulation. The tumor specificity of this response could be accounted for if the anti-gammaglobulin antibodies were reacting with idiotypic determinants (Prather and Lausch, 1976). Also, it may be noted that the levels of circulating antigens may be influenced by pathological changes such as the observed rise in permeability of the kidney in advanced cancer patients leading to increases in urinary protein excretion (Rudman et al., 1969). It may further be of significance that renal failure is a frequent complication in terminal disease.

From these considerations, it is certainly not surprising that the presence of serum antigen, antibody and immune complexes does not correlate with the development of the tumor when different animal tumor models are compared or different neoantigens analyzed. This is exemplified as shown earlier in Fig. 5, showing the results of sequentially monitoring the serum of rats bearing transplants of hepatoma D23 or sarcoma MC-1 for the individually distinct tumor-specific antigen and for cross reactive embryonic antigen activity. It may be noted that in the tests with sarcoma MC-1, free antigen was detected early after tumor implantation and this was attributed to autolytic degradation of a high proportion of tumor cells in the inoculum. Similar findings were obtained by Bray et al. (1975) using the murine B16 melanoma model and transient serum blocking activity was detected as early as 1 h after tumor inoculation, this being demonstrated both when homogenates or intact tissue-cultured tumor cells were injected. With progressive growth of sarcoma MC-1 the initial peak of antigenic activity fell and later continued to steadily rise (Thomson et al., 1973b; Thomson, 1975). Conversely, with hepatoma D23, tumor-specific antibody together with specific immune complexes was detected at the terminal stages of growth (Bowen et al., 1975).

Similar variations in the levels of serum factors have been observed when serum samples from tumor-bearers were monitored for biological activity (Bowen et al., 1975; Prather and Lausch, 1976). In these tests, serum factor-mediated responses occurring primarily at the level of the target cell (blocking reactions) may be differentiated from those operative at the effector cell level (inhibitory reactions) by pre-exposing the cultured tumor cells or sensitized lymphoid cells to the appropriate test or control normal serum. In this way, blocking reactions may be considered to be mediated by antibody (free or in immune complexes) and inhibitory responses may be effected by antigen (again free or in immune complexes). From the results of biological assays such as these, and also tests in which positive identification of antigen and antibody are made, several tentative conclusions regarding in vitro correlates and tumor growth may be drawn, these being illustrated in Table II. Perhaps the major feature of this Table relates to the fact that the presence of tumor antigen, antibody or immune complexes may not itself be indicative of a favorable or unfavorable prognosis, but rather the levels of such factors monitored on a sequential basis have greater significance. Unfortunately, in vitro correlates of the measurements of these factors have proved to be equivocal indicators of malignant disease. For example, while the Hellströms and their colleagues (Hellström et al., 1973; Hellström and Hellström, 1974a) found that the presence of serum-blocking activity was associated with active disease, there

TABLE II

SERUM FACTORS AS PROGNOSTIC INDICATORS DURING EARLY NEOPLASTIC DISEASE: GENERAL PROPOSALS FROM STUDIES WITH EXPERIMENTAL ANIMAL TUMORS

Indicators of progressive disease
Increasing levels of circulating antigenic activity (either in a free form or as immune complexes) or decreasing free antibody with increasing levels of immune complexes, i.e. persistence or increase of serum-blocking activity and/or persistence or increase in serum-inhibitory activity.

Indicators of favorable prognosis
Decreasing levels of circulating antigenic activity (either in a free form or as immune complexes) accompanied by increasing free antibody levels, i.e. decrease in serum-blocking activity and/or decrease in serum-inhibitory activity.

have been other reports in which the blocking reactivity of tumor-bearer or tumor-free serum samples did not relate to the clinical status of the donor (Heppner et al., 1973; Pierce and De Vald, 1975). It must be appreciated, however, that at the present time several criticisms have been levelled at the microcytotoxicity test as applied to the analysis of cell-mediated immune reactions in human cancer (reviews, Baldwin, 1975; Herberman and Oldham, 1975; Baldwin and Embleton, 1977) so that although caution is recommended in applying the proposals in Table II directly to human neoplasia, it is clear that more precise techniques for the identification of specific circulating tumor products are required.

Returning to studies performed with experimental animals, the assay of various circulating factors has proved useful in monitoring the effects of immunotherapeutic manipulations using the adjuvant bacillus Calmette Guérin (BCG). For example, in tests with an MCA-induced rat sarcoma, serum levels of tumor antigen increased with progressive tumor growth as assessed using a solid phase radioimmunoassay, but then rapidly fell following surgical resection of the tumor, this being replaced by tumor-specific antibody (Thomson et al., 1973a,b). When, however, the growth of this sarcoma was suppressed by a contralateral injection of tumor cells in admixture with BCG, treated rats developed a tumor-specific antibody response and serum-borne antigen was not detected (Thomson, 1975). Comparably, increases in the level of circulating tumor antigen was associated with the development of lung tumor nodules in rats following intravenous injection of hepatoma D23 cells, whereas when tumor growth was suppressed by BCG treatment serum antigen was no longer detectable (Bowen and Baldwin, 1976b). Similar effects were observed when monitoring the immune rejection of rat hepatoma D23 implants following treatment with a contralateral injection of a mixed cell inoculum of tumor cells and BCG (Embleton, 1976a). Effective treatment and regression of tumors was accompanied by a rapid loss of serum blocking activity whereas failure of treatment was associated with persistent serum blocking. This is in accord with the conclusions of sequential immunologic studies on patients with malignant melanoma since in these patients the presence of blocking activity was found to reflect an unfavorable prognosis (Hellström et al., 1973; Hellström and Hellström, 1974a). Also, in studies with melanoma patients undergoing active immunotherapy with irradiated allogeneic tumor cells mixed with BCG as well as cytotoxic drugs, the ability of patients' sera to modify peripheral blood effector cell cytotoxicity was correlated with the clinical outcome in these patients (Currie and McElwain, 1975). The implication from these latter tests was that the appearance of circulating inhibitory factor represented increasing levels of antigen in the serum originating from the progressing tumor.

5.2. Immune responses to circulating tumor antigens in the tumor host

In several instances, soluble antigens prepared from tumor tissues possess the capacity to elicit protection to challenge with viable tumor cells. These fractions have been obtained by various procedures including sonication (Holmes et al., 1970), hypertonic salt extraction (Meltzer et al., 1972; Pellis et al., 1974; Pellis and Kahan, 1975b; Price et al., 1976) enzymatic hydrolysis (Drapkin et al., 1974; Law and Appella, 1975) and by isolation of the soluble intracytoplasmic fraction of tumor homogenates using either chemically induced or virally induced tumors in the rat, mouse and guinea pig (Oettgen et al., 1968). It is thus clear that although soluble tumor antigens may be less immunogenic than the intact cell or cell particulates, they may provide limited immunoprotection to the treated animal and, in some investigations, this may only be revealed in animals immunized with antigen within restricted dose ranges (Pellis et al., 1974; Pellis and Kahan, 1975b; Price et al., 1976). Furthermore, tumor antigens isolated from the serum of tumor-bearing rats elicit the production of tumor-specific antibody in treated rats (Baldwin et al., 1973a). These studies serve to emphasize the possibility that the action of soluble antigens from a developing tumor may not be entirely antagonistic in the induction of protective responses to a developing neoplasm although as inferred in the previous section, their presence in the circulation may modify an established immune reaction. Similarly, if tumor antigens are released from the growing tumor associated with particulate membrane elements (see Fig. 2), it is relevant to note that membrane preparations in a variety of physical forms and from a variety of experimental animal tumors have been found to be effective in the induction of tumor rejection responses in treated animals (see, Baldwin and Price, 1975). A further feature has recently arisen in defining the conditions for promoting tumor rejection responses using acellular antigen preparations. In experimental studies with a variety of rat tumors, tumor-associated embryonic antigens do not appear to contribute to the induction of resistance to transplanted tumor cells (Baldwin et al., 1974c). However, using an artificial metastasis model system, it was found possible to limit the formation and development of tumor nodules in the lung after intravenous challenge with tumor cells in rats which have been immunized with rat embryo cell membranes (Shah et al., 1977). The significance of this finding requires further evaluation although it does emphasize that under limited circumstances, tumor-associated embryonic antigens may function as tumor rejection antigens.

Another consequence of the administration of acellular forms of tumor antigens to normal recipients is that this may lead to the development of an "immune deviation" response. This was originally demonstrated in experiments with aminoazo dye-induced rat hepatomas where injection of tumor membrane preparations containing tumor-specific antigen rendered the rats incapable of producing a tumor rejection response when subsequently immunized with γ-irradiated tumor cells (Baldwin et al., 1973b). This type of response has subsequently been evaluated in greater detail using transplanted MCA-induced rat sarcomas with which it was again shown that pre-immunization of rats with tumor antigen-containing fractions (in this case, either crude membrane preparations or 3 M KCl extracts) renders the recipients incapable of producing a tumor rejection reaction when subsequently immunized with irradiated tumor cells (Embleton, 1976b). The mechanisms involved in this type of "immune deviation" are largely unresolved, although the possibility that "suppressor cells" are produced by immunization with acellular forms of tumor antigen is supported by the finding that lymph node cells from antigen-treated rats interfere with the in vitro cytotoxicity of lymph node cells from tumor-immune rats for the appropriate target sarcoma (Embleton, 1976b). The relevance of suppressor or inhibitory cells in malignant disease remains largely to be explored, although cells capable of enhancing tumor growth in vivo have been detected in mice bearing an MCA-induced sarcoma as early as 24 h after inoculation of tumor cells (Fujimoto et al., 1975a,b). Similarly, in vitro tests have established that some lymphocyte functions of tumor-bearing animals can be inhibited by suppressor cells found in the spleens of such animals (Gorczynski, 1974; Kirchner et al., 1974a), although in some experimental tumors suppressor cells are not detected until the tumor has reached an appreciable size (Kirchner et al., 1974b; Glaser et al., 1975). These cells may act non-specifically since in analyses of lymphoproliferative reactions, they inhibited the responses to mitogens as well as tumor-associated antigens (Glaser et al., 1975). From these brief comments, and consideration of the wider roles of suppressor or inhibitory cells in the control of immune responses (Asherson and Zembala, 1975), it is evident therefore that promotion of this cell population may be one feature in the induction of immunity to tumor-associated antigens.

Another series of experiments in which acellular tumor antigens modify tumor-immune responses have been performed recently, these tests showing that administration of antigen preparations from an MCA-induced rat sarcoma could abrogate a developing tumor rejection response (Baldwin, 1976). An immunotherapy model was chosen for these studies in which sarcoma cells injected in admixture with BCG fail to grow out and the immune response developing from this treatment will produce rejection of a contralateral challenge of the same tumor either administered simultaneously or up to 4 days earlier (Baldwin and Pimm, 1973). This tumor rejection response which is mediated by lymphoid cells sensitive to 450 rad whole-body irradiation, could be abrogated when rats received additional treatment with large doses of soluble tumor antigen administered intraperitoneally (Baldwin, 1976). It is not known whether this type of response involves the induction of "immune deviation" mechanisms possibly including the promotion of suppressor cells as already described, since an alternative explanation for these findings is that large quantities of tumor antigen present in rats treated by daily injection of tumor extract for a period of 5 days, inhibits the cytotoxicity of sensitized host lymphoid cells. Previous studies have shown for example that cytotoxic lymph node and spleen cells are detectable as early as 2 days after tumor induction (Zöller et al., 1975) and that their in vitro cytotoxicity for tumor cells may be inhibited by exposure to tumor antigen preparations even when lymphoid cells are taken as early as this (M. Zöller and M.R. Price, unpublished observations).

The findings of similar tests using tumor-bearer serum to abrogate tumor immunity also appear to be open to several interpretations and indeed such tests have provided inconsistent results since both suppression and enhancement of tumor growth have been achieved (Bansal et al., 1972; El Ridi and Bubeník, 1973). Probably these effects are due to qualitative or quantitative differences in the levels of tumor antigen, antibody or immune complexes in the test sera employed and, in almost all cases, these factors have not been defined. In this respect, abrogation of tumor immunity in mice immunized by surgical excision of transplanted MCA-induced sarcomas has been achieved by administration of tumor-bearer serum taken from donors immunosuppressed by whole-body irradiation, in order to obtain high levels of circulating tumor antigen (Vaage, 1974).

Unfortunately, the results of immune manipulations designed to counteract serum factors (e.g. circulating antigens or immune complexes) which may be antagonistic to the effective mediation of host cell-mediated immunity remain somewhat equivocal, since the reagent used to produce the desired result, may itself elicit beneficial effects by alternative mechanisms. For example, in tests showing that passive transfer of antibody to tumor-bearers produces tumor regression or retardation of growth (Hellström et al., 1969; Bansal and Sjögren, 1972), it cannot be concluded that this occurred through immune neutralization of serum factors because transferred antibody itself may mediate a number of responses including complement and cell-dependent killing of tumor cells.

The potential impairment of host responses directed against the tumor by circulating factors derived from the tumor itself has been emphasized in this review and it is reasonable to propose that soluble antigens or other blocking factors (antibody, immune complexes) exert their maximal effect in the local microenvironment of the tumor itself by interfering with cell-mediated immunity. This may be particularly relevant with small progressing tumors where antigen concentration in the locality of the tumor itself, but not in the circulation, is sufficient to induce these inhibitory reactions and evade host control of growth although with a much larger tumor (which may be of a size beyond host control anyway) the concentrations of these factors is probably large enough to cause a systemic modification of the effector cell population. Indeed, this latter situation may correspond to the "eclipse" phase associated with advanced tumor growth when tumor-specific cell-mediated immunity is depressed (Le François et al., 1971; Barski et al., 1972, 1974; Howell et al., 1975; Youn et al., 1975).

6. Conclusion

The weight of the evidence presented here suggests that with a developing tumor, antigen shedding occurs to a greater or lesser extent, this being dependent upon the nature of the antigen and properties of the tumor itself as well as the involvement of host immune responses. It is perhaps unfortunate that the term "shedding" has been so widely adopted to describe these phenomena since it has led to an oversimplified view of a number of complex related and unrelated processes. Nevertheless, it is apparent that such processes may have profound effects in both modifying the induction of appropriate immune responses and interfering with the efficacy of mediation of established tumor immunity. One of the most important considerations, however, which must now, and in the immediate future, receive attention is the development of more sensitive immunoassays for antigen detection. The natural progression in this respect would be to use radioimmunoassays for this purpose although techniques offering similar sensitivity but which require less capital investment such as enzyme immunoassays, are possibly more readily available to a wider number of laboratories. The essential requirement of these assays is that antigenic activity is determined in a precise and quantitative manner since this would be invaluable in assessing the potential usefulness of monitoring circulating antigens (or other serum factors) in the cancer patient as a diagnostic and prognostic aid in the management of human cancer. Furthermore although progress has been made in the biochemistry of tumor associated antigens in neoplasia, these studies are also at a stage where they would be significantly advanced with the development of more reliable and sensitive antigen assays.

Acknowledgements

The personal research cited in this article was supported by a grant from the Cancer Research Campaign, and by a government equipment grant obtained through the Royal Society.

References

Abelev, G.I. (1971) Alpha-fetoprotein in oncogenesis and its association with malignant tumors. Adv. Cancer Res. 14, 295–358.

Abelev, G.I. (1974) α-Fetoprotein as a marker of embryo-specific differentiations in normal and tumor tissues. Transplant. Rev. 20, 3–37.

Alexander, P. (1974) Escape from immune destruction by the host through shedding of surface antigens: is this a characteristic shared by malignant and embryonic cells? Cancer Res. 34, 2077–2082.

Asherson, G.L. and Zembala, M. (1975) Inhibitory T cells. Curr. Topics Microbiol. Immunol. 72, 55–100.

Baldwin, R.W. (1973) Immunological aspects of chemical carcinogenesis. Adv. Cancer Res. 18, 1–75.

Baldwin, R.W. (1975) In vitro assays of cell-mediated immunity to human solid tumors: problems of quantitation, specificity, and interpretation. J. Natl. Cancer Inst. 55, 745–748.

Baldwin, R.W. (1976) Role of immunosurveillance against chemically induced rat tumors. Transplant. Rev. 28, 62–74.

Baldwin, R.W. and Embleton, M.J. (1971) Demonstration by colony inhibition methods of cellular and humoral immune reactions to tumour-specific antigens associated with aminoazo-dye-induced rat hepatomas. Int. J. Cancer 7, 17–25.

Baldwin, R.W. and Embleton, M.J. (1977) Assessment of cell-mediated immunity to human tumour associated antigens. Int. Rev. Exp. Pathol. 17, 49–96.

Baldwin, R.W. and Glaves, D. (1972) Solubilization of tumour-specific antigen from plasma membrane of an aminoazo-dye-induced rat hepatoma. Clin. Exp. Immunol. 11, 51–56.

Baldwin, R.W. and Pimm, M.V. (1973) BCG immunotherapy of a rat sarcoma. Brit. J. Cancer 28, 281–287.

Baldwin, R.W. and Price, M.R. (1975) Neoantigen expression in chemical carcinogenesis. In: F.F. Becker (Ed.), Cancer: A Comprehensive Treatise, Vol. 1, Plenum Press, New York, pp. 353–386.

Baldwin, R.W. and Price, M.R. (1976) Immunology of experimental liver cancer. In: H.M. Cameron, G.P. Warwick and C.A. Linsell (Eds.), Liver Cell Cancer, Elsevier, Amsterdam, pp. 203–242.

Baldwin, R.W. and Robins, R.A. (1975) Humoral factors abrogating cell-mediated immunity in the tumor-bearing host. Curr. Topics Microbiol. Immunol. 72, 21–53.

Baldwin, R.W., Price, M.R. and Robins, R.A. (1972) Blocking of lymphocyte-mediated cytotoxicity for rat hepatoma cells by tumour-specific antigen-antibody complexes. Nature (London) New Biol. 238, 185–187.

Baldwin, R.W., Bowen, J.G. and Price, M.R. (1973a) Detection of circulating hepatoma D23 antigen and immune complexes in tumour bearer serum. Brit. J. Cancer 28, 16–24.

Baldwin, R.W., Embleton, M.J. and Moore, M. (1973b) Immunogenicity of rat hepatoma membrane fractions. Brit. J. Cancer 28, 389–399.

Baldwin, R.W., Embleton, M.J. and Price, M.R. (1973c) Inhibition of lymphocyte cytotoxicity for human colon carcinoma by treatment with solubilized tumour membrane fractions. Int. J. Cancer 12, 84–92.

Baldwin, R.W., Embleton, M.J. and Robins, R.A. (1973d) Cellular and humoral immunity to rat hepatoma-specific antigens correlated with tumour status. Int. J. Cancer 11, 1–9.

Baldwin, R.W., Harris, J.R. and Price, M.R. (1973e) Fractionation of plasma membrane-associated tumour-specific antigen from an aminoazo dye-induced rat hepatoma. Int. J. Cancer 11, 385–397.

Baldwin, R.W., Price, M.R. and Robins, R.A. (1973f) Inhibition of hepatoma-immune lymph-node cell cytotoxicity by tumour-bearer serum, and solubilized hepatoma antigen. Int. J. Cancer 11, 527–535.

Baldwin, R.W., Bowen, J.G. and Price, M.R. (1974a) Solubilization of membrane-associated tumour-specific antigens by β-glucosidase. Biochim. Biophys. Acta 367, 47–58.

Baldwin, R.W., Embleton, M.J., Price, M.R. and Vose, B.M. (1974b) Embryonic antigen expression on experimental rat tumours. Transplant. Rev. 20, 77–99.

Baldwin, R.W., Glaves, D. and Vose, B.M. (1974c) Immunogenicity of embryonic antigens associated with chemically induced rat tumours. Int. J. Cancer 13, 135–142.

Bansal, S.C. and Sjögren, H.O. (1971) Unblocking serum activity in vitro in the polyoma system may correlate with antitumour effects of antiserum in vivo. Nature (London) New Biol. 233, 76–77.

Bansal, S.C. and Sjögren, H.O. (1972) Counteraction of the blocking of cell-mediated tumor immunity by inoculation of unblocking sera and splenectomy: immunotherapeutic effects on primary polyoma tumors in rats. Int. J. Cancer 9, 490–509.

Bansal, S.C. and Sjögren, H.O. (1973) Regression of polyoma tumor metastasis by combined unblocking and BCG treatment – correlation with induced alterations in tumor immunity status. Int. J. Cancer 12, 179–193.

Bansal, S.C., Hargreaves, R. and Sjögren, H.O. (1972) Facilitation of polyoma tumor growth in rats by blocking sera and tumor eluate. Int. J. Cancer 9, 97–108.

Barksi, G., Youn, J.K., Belehradek, J. and Le François, D. (1972) Variation des mécanismes de défense immunologique spécifique de type cellulaire en fonction de la croisance tumorale. Ann. Inst. Pasteur 122, 633–643.

Barski, G., Youn, J.K., Le François, D. and Belehradek, J. (1974) Evolution of specific cell-bound immunity in hosts bearing solid tumors as related to tumor growth and treatment. Israel J. Med. Sci. 10, 913–924.

Benacerraf, B., Sebestyen, M. and Cooper, N.S. (1959) The clearance of antigen antibody complexes from the blood by the reticulo-endothelial system. J. Immunol. 82, 131–137.

Ben-Sasson, Z., Weiss, D.W. and Doljanski, F. (1974) Specific binding of factor(s) released by Rous sarcoma virus-transformed cells to splenocytes of chickens with Rous sarcomas. J. Natl. Cancer Inst. 52, 405–412.

Blair, P.B., Lane, M.A. and Yagi, M.J. (1975) Blocking of spleen cell activity against target mammary tumor cells by viral antigens. J. Immunol. 115, 190–194.

Bloom, B.R., Bennett, B., Oettgen, H.F., McLean, E.P. and Old, L.J. (1969) Demonstration of delayed hypersensitivity to soluble antigens of chemically induced tumors by inhibition of macrophage migration. Proc. Natl. Acad. Sci. USA 64, 1176–1180.

Bowen, J.G. (1975) Tumour antigen in human cancer patients' sera. Brit. J. Cancer 32, 242.

Bowen, J.G. and Baldwin, R.W. (1975) Tumour-specific antigen related to rat histocompatibility antigens. Nature (London) 258, 75–76.

Bowen, J.G. and Baldwin, R.W. (1976a) Isolation and characterization of tumour-specific antigen from the serum of rats bearing transplanted aminoazo dye-induced hepatomas. Transplantation 21, 213–219.

Bowen, J.G. and Baldwin, R.W. (1976b) Serum factor levels during the growth of rat hepatoma nodules in the lungs. Int. J. Cancer 17, 254–260.

Bowen, J.G., Robins, R.A. and Baldwin, R.W. (1975) Serum factors modifying cell mediated immunity to rat hepatoma D23 correlated with tumour growth. Int. J. Cancer 15, 640–650.

Boyse, E.A., Stockert, E. and Old, L.J. (1967) Modification of the antigenic structure of the cell membrane by thymus-leukemia (TL) antibody. Proc. Natl. Acad. Sci. USA 58, 954–961.

Brannen, G.E., Adanis, J.S. and Santos, G.W. (1974) Tumor-specific immunity in 3-methylcholanthrene-induced murine fibrosarcomas, I. In vivo demonstration of immunity with three preparations of soluble antigens. J. Natl. Cancer Inst. 53, 165–175.

Bray, A.E., Holt, P.G., Roberts, L.M. and Keast, D. (1975) Early onset of serum blocking in a murine melanoma model. Int. J. Cancer 16, 607–615.

Brown, G., Hogg, N. and Greaves, M. (1975) Candidate leukaemia-specific antigen in man. Nature (London) 258, 454–456.

Bruce, J., Mitchison, N.A. and Shellam, G.R. (1976) Studies on a Gross-virus-induced lymphoma in the rat, III. Optimisation, specificity and applications of the in vitro immune response. Int. J. Cancer 17, 342–350.

Burger, M.M. (1973) Surface changes in transformed cells detected by lectins. Fed. Proc. 32, 91–101.

Bystryn, J.-C. (1976) Release of tumor associated antigens by murine melanoma cells. J. Immunol. 116, 1302–1305.

Bystryn, J.-C., Schenkein, I., Baur, S. and Uhr, J.W. (1974) Partial isolation and characterization of antigen(s) associated with murine melanoma. J. Natl. Cancer Inst. 52, 1263–1269.

Callahan, G.N. and Dewitt, C.W. (1975) Rat cell surface antigens, I. Isolation and partial characterization of an Ag-B antigen. J. Immunol. 114, 776–778.

Carrel, S. and Theilkaes, L. (1973) Evidence for a tumour-associated antigen in human malignant melanoma. Nature (London) 242, 609–610.
Cerottini, J.-C. and Brunner, K.T. (1974) Cell mediated cytotoxicity, allograft rejection and tumor immunity. Adv. Immunol. 18, 67–132.
Cikes, M. (1970) Relationship between growth rate, cell volume, cell cycle kinetics, and antigenic properties of cultured murine lymphoma cells. J. Natl. Cancer Inst. 45, 979–988.
Cikes, M. and Klein, G. (1972) Quantitative studies of antigen expression in cultured murine lymphoma cells. I. Cell-surface antigens in "asynchronous" cultures. J. Natl. Cancer Inst. 49, 1599–1606.
Cone, R.E., Marchalonis, J.J. and Rolley, R.T. (1971) Lymphocyte membrane dynamics: metabolic release of cell surface proteins. J. Exp. Med. 134, 1373–1384.
Currie, G.A. (1973) The role of circulating antigen as an inhibitor of tumour immunity in man. Brit. J. Cancer 28, Suppl. I, 153–161.
Currie, G.A. and Alexander, P. (1974) Spontaneous shedding of TSTA by viable sarcoma cells: its possible role in facilitating metastatic spread. Brit. J. Cancer 29, 72–75.
Currie, G.A. and Basham, C. (1972) Serum mediated inhibition of the immunological reactions of the patient to his own tumour: a possible role for circulating antigen. Brit. J. Cancer 26, 427–438.
Currie, G.A. and Gage, J.O. (1973) Influence of tumour growth on the evolution of cytotoxic lymphoid cells in rats bearing a spontaneously metastasizing syngeneic fibrosarcoma. Brit. J. Cancer 28, 136–146.
Currie, G.A. and McElwain, T.J. (1975) Active immunotherapy as an adjunct to chemotherapy in the treatment of disseminated malignant melanoma: a pilot study. Brit. J. Cancer 31, 143–156.
Dattwyler, R.J. and Tomasi, T.B. (1975) Inhibition of cytotoxic sensitization of T cells by alpha-fetoprotein. Int. J. Cancer 16, 942–945.
Davey, G.C., Currie, G.A. and Alexander, P. (1976) Spontaneous shedding and antibody induced modulation of histocompatibility antigens on murine lymphomata: correlation with metastatic capacity. Brit. J. Cancer 33, 9–14.
Dixon, F.J., Vazquez, J.J., Weigle, W.O. and Cochrane, C.G. (1958) Pathogenesis of serum sickness. Arch. Pathol. 65, 18–28.
Doljanski, F. (1973) A new look at the cell surface. Israel J. Med. Sci. 9, 251–257.
Drapkin, M.S., Appella, E. and Law, L.W. (1974) Immunogenic properties of a soluble tumor-specific transplantation antigen induced by Simian virus. J. Natl. Cancer Inst. 52, 259–264.
Eilber, F.R. and Morton, D.L. (1971) Immunologic response to human sarcomas: relation of antitumor antibody to the clinical course. In: B. Amos (Ed.), Progress in Immunology, Vol. 1, Academic Press, New York, pp. 951–957.
El Ridi, R. and Bubeník, J. (1973) Tumour associated antigen in the serum of rats with large Rous sarcoma virus-induced tumours. Folia Biol. (Prague) 19, 273–280.
El Ridi, R. and Bubeník, J. (1975) Tumour associated transplantation antigen in sera of rats with large RSV-induced sarcomas. Int. J. Cancer 16, 83–90.
Embleton, M.J. (1973) Significance of tumour associated antigens on human colonic carcinomata. Brit. J. Cancer 28, Suppl. I, 142–152.
Embleton, M.J. (1976a) Effect of BCG on cell-mediated cytotoxicity and serum blocking factor during growth of a rat hepatoma. Brit. J. Cancer 33, 584–595.
Embleton, M.J. (1976b) Inhibition of cell-mediated immunity to rat sarcomas following treatment with isolated tumour antigen preparations. Brit. J. Cancer 34, 316.
Embleton, M.J. and Price, M.R. (1975) Inhibition of in vitro lymphocytotoxic reactions against tumor cells by melanoma membrane extracts. Behring Inst. Mitt. 56, 157–160.
Evans, W.H. and Gurd, J.W. (1971) Biosynthesis of liver membranes. Incorporation of [^{3}H]leucine into proteins and of [^{14}C]glucosamine into proteins and lipids of liver microsomal and plasma-membrane fractions. Biochem. J. 125, 615–624.
Faanes, R.B. and Choi, Y.S. (1974) Interaction of isoantibody and cytotoxic lymphocytes with allogeneic tumor cells. J. Immunol. 113, 279–288.
Faanes, R.B., Choi, Y.S. and Good, R.A. (1973) Escape from isoantiserum inhibition of lymphocyte-

mediated cytotoxicity. J. Exp. Med. 137, 171–182.
Fairbanks, G., Steck, T.L. and Wallach, D.F.H. (1971) Electrophoretic analysis of the major polypeptides of the human erythrocyte. Biochemistry 10, 2606–2624.
Fujimoto, S., Chen, C.H., Sabbadini, E. and Sehon, A.H. (1973) Association of tumor and histocompatibility antigens in sera of lymphoma-bearing mice. J. Immunol. 111, 1093–1100.
Fujimoto, S., Greene, M.I. and Sehon, A.H. (1975a) Regulation of the immune response to tumor antigens, I. Immunosuppressor T cells in tumor bearing hosts. J. Immunol. 116, 791–799.
Fujimoto, S., Greene, M.I. and Sehon, A.H. (1975b) Regulation of the immune response to tumor antigens, II. The nature of immunosuppressor T cells in tumor bearing hosts. J. Immunol. 116, 800–806.
Fuks, A., Banjo, C., Shuster, J., Freedman, S.O. and Gold, P. (1974) Carcinoembryonic antigen (CEA): molecular biology and clinical significance. Biochim. Biophys. Acta 417, 123–152.
Germain, R.W., Dorf, M.E. and Benacerraf, B. (1975) Inhibition of T lymphocyte-mediated tumor-specific lysis by alloantisera directed against the H-2 serological specificities of the tumor. J. Exp. Med. 142, 1023–1028.
Glaser, M. and Herberman, R.B. (1974) Effect of immunoregulatory α-globulin on in vitro proliferative response of murine rat lymphocytes to syngeneic Gross virus-induced lymphoma. J. Natl. Cancer Inst. 53, 1767–1769.
Glaser, M., Kirchner, H. and Herberman, R.B. (1975) Inhibition of in vitro lymphoproliferative responses to tumor-associated antigens by suppressor cells from rats bearing progressively growing Gross leukemia virus-induced tumors. Int. J. Cancer 16, 384–393.
Glasgow, A.H., Menzoian, J.O., Nimberg, R.B., Cooperband, S.R., Schmid, K. and Mannick, J.A. (1974a) An immunosuppressive peptide fraction in the serum of cancer patients. Surgery 76, 35–42.
Glasgow, A.H., Nimberg, R.B., Menzoian, J.O., Saporoschetz, I., Cooperband, S.R., Schmid, K. and Mannick, J.A. (1974b) Association of anergy with an immunosuppressive peptide fraction in the serum of patients with cancer. New Engl. J. Med. 291, 1263–1267.
Gold, J.M. and Gold, P. (1973) The blood group A-like site on the carcinoembryonic antigen. Cancer Res. 33, 2821–2824.
Gold, J.M., Freedman, S.O. and Gold, P. (1973) Human anti-CEA antibodies detected by radioimmunoelectrophoresis. Nature (London) New Biol. 239, 60–62.
Gold, P. (1967) Circulating antibodies against carcinoembryonic antigens of the human digestive system. Cancer 20, 1663–1667.
Gold, P. (1970) The model of colonic cancer in the study of human tumor-specific antigen. In: W.J. Burdette (Ed.), Carcinoma of the Colon. Charles C. Thomas, Springfield, Illinois, pp. 131–142.
Gold, P., Gold, M. and Freedman, S.O. (1968) Cellular location of carcinoembryonic antigens of the human digestive system. Cancer Res. 28, 1331–1334.
Gold, P., Krupey, J. and Ansari, H. (1970) Position of the carcinoembryonic antigen of the human digestive system in ultrastructure of tumor cell surface. J. Natl. Cancer Inst. 45, 219–225.
Goldenberg, D.M., Pavia, R.A., Hansen, H.J. and Vandevoorde, J.P. (1972) Synthesis of carcinoembryonic antigen in vitro. Nature (London) New Biol. 239, 189–190.
Gorczynski, R.M. (1974) Immunity to murine sarcoma virus-induced tumors, II. Suppression of T cell-mediated immunity by cells from progressor animals. J. Immunol. 112, 1826–1838.
Gorczynski, R.M. and Knight, R.A. (1975) Immunity to murine sarcoma virus induced tumors. Direct cellular cytolysis of ^{51}Cr-labelled target cells in vitro and analysis of blocking factors which modulate cytotoxicity. Brit. J. Cancer 31, 387–404.
Grimm, E.A., Silver, H.K.B., Roth, J.A., Chee, D.O., Gupta, R.K. and Morton, D.L. (1976) Detection of tumor-associated antigen in human melanoma cell line supernatants. Int. J. Cancer 17, 559–564.
Gupta, R.K. and Morton, D.L. (1975) Suggestive evidence for in vivo binding of specific antitumor antibodies of human melanomas. Cancer Res. 35, 58–62.
Häkkinen, I.P.T. (1974) FSA – foetal sulphoglycoprotein antigen associated with gastric cancer. Transplant. Rev. 20, 61–76.
Hakomori, S. (1975) Structures and organization of cell surface glycolipids: dependency on cell

growth and malignant transformation. Biochim. Biophys. Acta 417, 55–89.
Harris, J.R., Price, M.R. and Baldwin, R.W. (1973) The purification of membrane-associated tumour antigens by preparative polyacrylamide gel electrophoresis. Biochim. Biophys. Acta 311, 600–614.
Harris, R., Viza, D., Todd, R., Phillips, J., Sugar, R., Jennison, R.F., Marriott, G. and Gleeson, M.H. (1971) Detection of human leukaemia associated antigens in leukaemic serum and normal embryos. Nature (London) 233, 556–557.
Hayami, M., Hellström, I., Hellström, K.E. and Lannin, D.R. (1974) Further studies on the ability of regressor sera to block cell-mediated destruction of Rous sarcomas. Int. J. Cancer 13, 43–53.
Haywood, G.R. and McKhann, C.F. (1971) Antigenic specificities on murine sarcoma cells. Reciprocal relationship between normal transplantation antigens (H-2) and tumor-specific immunogenicity. J. Exp. Med. 133, 1171–1187.
Heimer, R. and Klein, G. (1976) Circulating immune complexes in sera of patients with Burkitt's lymphoma and nasopharyngeal carcinoma. Int. J. Cancer 18, 310–316.
Hellström, I. and Hellström, K.E. (1969) Studies on cellular immunity and its serum-mediated inhibition in Moloney-virus-induced mouse sarcomas. Int. J. Cancer 4, 587–600.
Hellström, I. and Hellström, K.E. (1974a) Cell-mediated immune reactions to tumor antigens with particular emphasis on immunity to human neoplasms. Cancer 34, 1461–1468.
Hellström, I., Hellström, K.E., Pierce, G.E. and Fefer, A. (1969) Studies on immunity to autochthonous mouse tumors. Transplant. Proc. 1, 90–94.
Hellström, I., Hellström, K.E., Sjögren, H.O. and Warner, G.A. (1971) Serum factors in tumor-free patients cancelling the blocking of cell-mediated tumor immunity. Int. J. Cancer 8, 185–191.
Hellström, I., Warner, G.A., Hellström, K.E. and Sjögren, H.O. (1973) Sequential studies on cell-mediated tumor immunity and blocking serum activity in ten patients with malignant melanoma. Int. J. Cancer 11, 280–292.
Hellström, K.E. (1974) Discussion. In: Schering Symposium on Immunopathology. Adv. Biosci. 12, 551.
Hellström, K.E. and Hellström, I. (1970) Immunological enhancement as studied by cell culture techniques. Ann. Rev. Microbiol. 24, 373–398.
Hellström, K.E. and Hellström, I. (1974b) Lymphocyte-mediated cytotoxicity and blocking serum activity to tumor antigens. Adv. Immunol. 18, 209–277.
Henning, R., Milner, R.J., Reske, K., Cunningham, B.A. and Edelman, G. (1976) Subunit structure, cell surface orientation, and partial amino-acid sequences of murine histocompatibility antigens. Proc. Natl. Acad. Sci. USA 73, 118–122.
Heppner, G.H., Stolbach, L., Byrne, M., Cummings, F.J., McDonough, E. and Calabresi, P. (1973) Cell-mediated and serum blocking reactivity to tumor antigens in patients with malignant melanoma. Int. J. Cancer 11, 245–260.
Herberman, R.B. and Oldham, R.K. (1975) Problems associated with study of cell-mediated immunity to human tumors by microcytotoxicity assays. J. Natl. Cancer Inst. 55, 749–753.
Herberman, R.B., Hollinshead, A., Char, D., Oldham, R., McCoy, J. and Cohen, M. (1975) In vivo and in vitro studies of cell-mediated immune response to antigens associated with malignant melanoma. Behring Inst. Mitt. 56, 131–138.
Hollinshead, A., Glew, D., Bunnag, B., Gold, P. and Herberman, R. (1970) Skin-reactive soluble antigen from intestinal cancer-cell-membranes and relationship to carcinoembryonic antigens. Lancet 1, 1191–1195.
Hollinshead, A.C., McWright, C.G., Alford, T.C., Glew, D.H., Gold, P. and Herberman, R.B. (1972) Separation of skin reactive intestinal cancer antigen from the carcinoembryonic antigen of Gold. Science 177, 887–889.
Hollinshead, A.C., Herberman, R.B., Jaffurs, W.J., Alpert, L.K., Minton, J.P. and Harris, J.E. (1974) Soluble membrane antigens of human malignant melanoma cells. Cancer 34, 1235–1243.
Holmes, E.C., Kahan, B.D. and Morton, D.L. (1970) Soluble tumor-specific transplantation antigens from methylcholanthrene-induced guinea pig sarcomas. Cancer 25, 373–379.
Howell, S.B., Dean, J.H. and Law, L.W. (1975) Defects in cell-mediated immunity during growth of

a syngeneic Simian virus-induced tumor. Int. J. Cancer 15, 152–169.

Invernizzi, G. and Parmiani, G. (1975) Tumour-associated transplantation antigens of chemically induced sarcomata cross reacting with allogeneic histocompatibility antigens. Nature (London) 254, 713–714.

Jones, B.M. and Turnbull, A.R. (1974) In vitro cellular immunity in mammary carcinoma. Brit. J. Cancer 29, 337–339.

Jones, J.M. and Feldman, J.D. (1975) Soluble membrane antigen fractions that react with rat enhancing alloantibodies. Transplantation 19, 219–225.

José, D.G. and Seshadri, R. (1974) Circulating immune complexes in human neuroblastoma. Direct assay and role in blocking specific cellular immunity. Int. J. Cancer 13, 824–838.

Kapeller, M., Gal-Oz, R., Grover, N.B. and Doljanski, F. (1973) Natural shedding of carbohydrate-containing macromolecules from cell surfaces. Exp. Cell Res. 79, 152–158.

Karb, K. and Goldstein, G. (1971) Combination autoradiography and membrane fluorescence in studying cell cycle transplantation antigen relationships. Transplantation 11, 569–573.

Kirchner, H., Chused, T.M., Herberman, R.B., Holden, H.T. and Lavrin, D.H. (1974a) Evidence of suppressor cell activity in spleens of mice bearing primary tumors induced by Moloney sarcoma virus. J. Exp. Med. 139, 1473–1487.

Kirchner, H., Herberman, R.B., Glaser, M. and Lavrin, D.H. (1974b) Suppression of in vitro lymphocyte stimulation in mice bearing primary Moloney sarcoma virus-induced tumors. Cell. Immunol. 13, 32–40.

Klein, G. (1975) Discussion. In: R.T. Smith and M. Landy (Eds.), Immunobiology of the Tumor–Host Relationship. Academic Press, New York, p. 56.

Klein, G. and Klein, E. (1975) Are methylcholanthrene-induced sarcoma-associated, rejection inducing (TSTA) antigens, modified forms of H-2 or linked determinants? Int. J. Cancer 15, 879–887.

Knight, R.A., Mitchison, N.A. and Shellam, G.R. (1975) Studies on a Gross-virus-induced lymphoma in the rat, II. The role of cell-membrane-associated and serum P30 antigen in the antibody and cell-mediated response. Int. J. Cancer 15, 417–428.

Knopf, P.M. (1973) Pathways leading to expression of immunoglobulins. Transplant. Rev. 14, 145–162.

Kolb, J.-P., Poupon, M.-F. and Lespinats, G. (1974) Tumor-associated antigen (TAA) and anti-TAA antibodies in the serum of BALB/c mice with plasmacytomas. J. Natl. Cancer Inst. 52, 723–727.

Kontiainen, S. and Mitchison, N.A. (1975) Blocking antigen-antibody complexes on the T-lymphocyte surface identified with defined protein antigens. Immunology 28, 523–533.

Kunkel, H.G. and Tan, E.M. (1964) Autoantibodies and disease. Adv. Immunol. 4, 351–395.

Lane, M.A., Roubinian, J., Slomich, M., Trefts, P. and Blair, P.B. (1975) Characterization of cytotoxic effector cells in the mouse mammary tumor system. J. Immunol. 114, 24–29.

Laurence, D.J.R. and Neville, A.M. (1972) Foetal antigens and their role in the diagnosis and clinical management of human neoplasms: a review. Brit. J. Cancer 26, 335–355.

Laux, D. and Lausch, R.N. (1974) Reversal of tumor-mediated suppression of immune reactivity by in vitro incubation of spleen cells. J. Immunol. 112, 1900–1908.

Law, L.W. and Appella, E. (1975) Studies of soluble transplantation and tumor antigens. In: F.F. Becker (Ed.), Cancer: A Comprehensive Treatise, Vol. 4, Plenum, New York, pp. 135–157.

Le François, D., Youn, J.K., Belehradek, J. and Barski, G. (1971) Evolution of cell-mediated immunity in mice bearing tumors produced by a mammary carcinoma cell line. Influence of tumor growth, surgical removal and treatment with irradiated tumor cells. J. Natl. Cancer Inst. 46, 981–987.

Lengerová, A., Pokorná, Z., Viklický, V. and Zelney, V. (1972) Phenotypic suppression of H-2 antigens and topography of the cell surface. Tissue Antigens 2, 332–340.

Leonard, E.J. (1973) Cell surface antigen movement: induction in hepatoma cells by antitumor antibody. J. Immunol. 110, 1167–1169.

Leonard, E.J., Meltzer, M.S., Borsos, T. and Rapp, H.J. (1972) Properties of soluble tumor-specific antigen solubilized by hypertonic potassium chloride. Natl. Cancer Inst., Monogr. 35, 129–134.

Leonard, E.J., Richardson, A.K., Hardy, A.S. and Rapp, H.J. (1975) Extraction of tumor-specific antigen from cells and plasma membranes of line-10 hepatoma. J. Natl. Cancer Inst. 55, 73–79.

Loor, F., Block, N. and Little, J.R. (1975) Dynamics of the TL antigens on thymus and leukemia cells. Cell. Immunol. 17, 351–365.

Mann, D.L. (1972) The effect of enzyme inhibitors on the solubilization of HL-A antigens with 3 M KCl. Transplantation 14, 398–401.

Mann, D.L. (1975) An approach to the development of antisera to tumor associated antigens: experience with acute leukemia and melanoma. Behring Inst. Mitt. 56, 103–106.

Mann, D.L., Fahey, J.L. and Nathenson, S.G. (1970) Molecular comparisons of papain solubilized H-2 and HL-A alloantigens. In: P.I. Terasaki (Ed.), Histocompatibility Testing 1970. Munksgaard, Copenhagen, pp. 461–470.

Mann, D.L., Rogentine, G.N., Halterman, R. and Leventhal, B. (1971) Detection of an antigen associated with acute leukaemia. Science 174, 1136–1137.

Mann, D.L., Halterman, R. and Leventhal, B. (1974) Acute leukaemia-associated antigens. Cancer 34, 1446–1451.

Mavligit, G., Ambus, U., Gutterman, J.U., McBride, C.M. and Hersh, E.M. (1973) Antigen solubilized from human solid tumours: lymphocyte stimulation and cutaneous delayed hypersensitivity. Nature (London) New Biol. 243, 188–190.

Mavligit, G.M., Hersh, E.M. and McBride, C.M. (1974) Lymphocyte blastogenesis induced by autochthonous human solid tumor cells: relationship to stage of disease and serum factors. Cancer 34, 1712–1721.

McCoy, J.L., Jerome, L.F., Dean, J.H., Alford, T.C., Doering, T. and Herberman, R.B. (1974) Inhibition of leukocyte migration by tumor-associated antigens in soluble extracts of human breast carcinoma. J. Natl. Cancer Inst. 53, 11–17.

McCoy, J.L., Jerome, L.F., Dean, J.H., Perlin, E., Oldham, R.K., Char, D.H., Cohen, M.H., Felix, E.L. and Herberman, R.B. (1975) Inhibition of leukocyte migration by tumor-associated antigens in soluble extracts of human malignant melanoma. J. Natl. Cancer Inst. 55, 19–23.

Melcher, U., Eidels, L. and Uhr, J.W. (1975) Are immunoglobulins integral membrane proteins? Nature (London) 258, 434–435.

Melchers, F. and Andersson, J. (1973) Synthesis, surface deposition and secretion of immunoglobulin M in bone marrow-derived lymphocytes before and after mitogenic stimulation. Transplant. Rev. 14, 76–130.

Meltzer, M.S., Leonard, E.J., Rapp, H.J. and Borsos, T. (1971) Tumor-specific antigen solubilized by hypertonic potassium chloride. J. Natl. Cancer Inst. 47, 703–709.

Meltzer, M.S., Oppenheim, J.J., Littman, B.H., Leonard, E.J. and Rapp, H.J. (1972) Cell-mediated tumor immunity measured in vitro and in vivo with soluble tumor-specific antigens. J. Natl. Cancer Inst. 49, 727–734.

Miyakawa, Y., Tanigaki, N., Kreiter, V.P., Moore, G.E. and Pressman, D. (1973a) Characterization of soluble substances in the plasma carrying HL-A alloantigenic activity with HL-A common antigenic activity. Transplantation 15, 312–319.

Miyakawa, Y., Tanigaki, N., Yagi, Y. and Pressman, D. (1973b) Common antigen structures of HL-A antigens, I. Antigenic determinants recognizable by rabbits on papain-solubilized HL-A molecular fragments. Immunology 24, 67–76.

Mukojima, T., Gunvén, P. and Klein, G. (1973) Circulating antigen-antibody complex associated with Epstein-Barr virus in recurrent Burkitt's lymphoma. J. Natl. Cancer Inst. 51, 1319–1321.

Nachbar, M.S., Oppenheim, J.D. and Aull, F. (1974) Cell surface contributions to the malignant process. Am. J. Med. Sci. 268, 122–138.

Nelson, K., Pollack, S.B. and Hellström, K.E. (1975a) Specific antitumor responses by cultured immune spleen cells, I. In vitro culture method and initial characterization of factors which block immune cell-mediated cytotoxicity in vitro. Int. J. Cancer 15, 806–814.

Nelson, K., Pollack, S.B. and Hellström, K.E. (1975b) Specific antitumor responses by cultured immune spleen cells, II. Culture supernatants induce specific anti-tumor cytotoxicity by non-immune lymphoid cells in vitro. Int. J. Cancer 16, 292–300.

Nelson, K., Pollack, S.B. and Hellström, K.E. (1975c) In vitro synthesis of tumor-specific factors with blocking and antibody-dependent cellular cytotoxicity (ADC) activities. Int. J. Cancer 16, 932–941.

Nicolson, G.L. (1976) Trans-membrane control of the receptors on normal and tumor cells. II. Surface changes associated with transformation and malignancy. Biochim. Biophys. Acta 458, 57–108.

Nind, A.P.P., Matthews, N., Pihl, E.A.V., Rolland, J.M. and Nairn, R.C. (1975) Analysis of inhibition of lymphocyte cytotoxicity in human colon carcinoma. Brit. J. Cancer 31, 620–629.

Nowotny, A., Grohsman, J., Abdelnoor, A., Rote, N., Yang, C. and Waltersdorff, R. (1974) Escape of TA3 tumors from allogeneic immune rejection: theory and experiments. Eur. J. Immunol. 4, 73–78.

Oettgen, H.F., Old, L.J., McLean, E.P. and Carswell, E.A. (1968) Delayed hypersensitivity and transplantation immunity elicited by soluble antigens of chemically induced tumours in inbred guinea-pig. Nature (London) 220, 295–297.

Old, L.J., Stockert, E., Boyse, E.A. and Kim, J.H. (1968) Antigenic modulation: loss of TL antigen from cells exposed to TL antibody. Study of the phenomenon in vitro. J. Exp. Med. 127, 523–539.

Oldstone, M.B.A. (1975) Immune complexes in cancer: demonstration of complexes in mice bearing neuroblastomas. J. Natl. Cancer Inst. 54, 223–226.

Oldstone, M.B.A., Theofilopoulos, A.N., Gunvén, P. and Klein, G. (1975) Immune complexes associated with neoplasia: presence of Epstein-Barr virus antigen-antibody complexes in Burkitt's lymphoma. Intervirology 4, 292–302.

Parkhouse, R.M.E. (1973) Assembly and secretion of immunoglobulin M (IgM) by plasma cells and lymphocytes. Transplant. Rev. 14, 131–144.

Parmiani, G. and Invernizzi, G. (1975) Alien histocompatibility determinants on the cell surface of sarcomas induced by methylcholanthrene, I. In vivo studies. Int. J. Cancer 16, 756–767.

Pellegrino, M.A., Pellegrino, A., Ferrone, S., Kahan, B.D. and Reisfeld, R.A. (1973) Extraction and purification of soluble HL-A antigens from exhausted media of human lymphoid cell lines. J. Immunol. 111, 783–788.

Pellis, N.R. and Kahan, B.D. (1975a) Immunoprotection by fractions of tumor cell cultures: the immunogenicity of exhausted culture medium. Fed. Proc., 34, 1042.

Pellis, N.R. and Kahan, B.D. (1975b) Specific tumor immunity induced with soluble materials: restricted range of antigen dose and of challenge tumor load for immunoprotection. J. Immunol. 115, 1717–1722.

Pellis, N.R., Tom, B.H. and Kahan, B.D. (1974) Tumor-specific and allospecific immunogenicity of soluble extracts from chemically induced murine sarcomas. J. Immunol. 113, 708–711.

Philips, T.M. and Lewis, M.G. (1971) A method for the elution of immunoglobulin from the surface of living cells. Rev. Eur. Études Clin. Biol. 16, 1052–1053.

Pierce, G.E. and De Vald, B.L. (1975) Effects of human sera on reactivity of lymphocytes in microcytotoxicity assays. Cancer Res. 35, 2729–2737.

Pincus, J. (1974) Discussion. In: B.D. Kahan and R.A. Reisfeld (Eds.), The Cell Surface: Immunological and Chemical Approaches. Adv. Exp. Med. Biol. 51, 86.

Plata, F. and Levy, J.P. (1974) Blocking of syngeneic effector T cells by soluble tumour antigens. Nature (London) 249, 271–274.

Pollack, S.B. and Nelson, K. (1974) Early appearance of a lymphoid arming factor and cytotoxic lymph-node cells after tumor induction. Int. J. Cancer 14, 522–529.

Pollack, S.B. and Nelson, K. (1975) Evidence for two factors in sera of tumor-immunized mice which induce specific lymphoid cell-dependent cytotoxicity: IgG_2 and a rapidly appearing factor not associated with IgG or IgM. Int. J. Cancer 16, 339–346.

Poskitt, P.K.F., Poskitt, T.R. and Wallace, J.H. (1974) Renal deposition of soluble immune complexes in mice bearing B-16 melanoma. J. Exp. Med. 140, 410–425.

Poste, G. (1971) Sub-lethal autolysis. Modification of cell periphery by lysosomal enzymes. Exp. Cell Res. 67, 11–16.

Poste, G. and Weiss, L. (1976) Some considerations on cell surface alterations in malignancy. In: L. Weiss (Ed.), Fundamental Aspects of Metastasis. North-Holland, Amsterdam, pp. 25–47.

Poulik, M.D., Ferrone, S., Pellegrino, M.A., Sevier, D.E., Oh, S.K. and Reisfeld, R.A. (1974)

Association of HL-A antigens and β-2-microglobulin: concepts and questions. Transplant. Rev. 21, 106–125.

Poupon, M.-F., Lespinats, G. and Kolb, J.-P. (1974) Blocking effect of the migration-inhibition reaction by sera from immunized syngeneic mice and by sera from plasmacytoma-bearing BALB/c mice. Detection of free, circulating tumor antigen. J. Natl. Cancer Inst. 52, 1127–1134.

Powell, A.E., Sloss, A.M., Smith, R.N., Makley, J.T. and Hybay, C.A. (1975) Specific responsiveness of leukocytes to soluble extracts of human tumors. Int. J. Cancer 16, 905–913.

Prat, M. and Comoglio, P.M. (1976) A solid-state competitive binding radioimmunoassay for measurement of antigens solubilized from membranes. J. Immunol. Methods 9, 267–272.

Prather, S.O. and Lausch, R.N. (1976) Kinetics of serum factors mediating blocking, unblocking and antibody-dependent cellular cytotoxicity in hamsters given isografts of para-7 tumor cells. Int. J. Cancer 17, 380–388.

Price, M.R. and Baldwin, R.W. (1974a) Preparation of aminoazo dye induced rat hepatoma membrane fractions retaining tumour specific antigen. Brit. J. Cancer 30, 382–393.

Price, M.R. and Baldwin, R.W. (1974b) Immunogenic properties of rat hepatoma subcellular fractions. Brit. J. Cancer 30, 394–400.

Price, M.R. and Baldwin, R.W. (1975) Immunobiology of chemically induced tumors. In: F.F. Becker (Ed.), Cancer: A Comprehensive Treatise, Vol. 4, Plenum, New York, pp. 209–236.

Price, M.R., Preston, V.E. and Zöller, M. (1976) Immunisation with solubilized extracts of rat tumours and fractions of tumour bearer serum. Brit. J. Cancer 34, 316.

Ran, M. and Witz, I.P. (1970) Tumor-associated immunoglobulins. The elution of IgG_2 from mouse tumors. Int. J. Cancer 6, 361–372.

Ran, M., Klein, G. and Witz, I.P. (1976) Tumor bound immunoglobulin. Evidence for the in vivo coating of tumor cells by potentially cytotoxic antibodies. Int. J. Cancer 17, 90–97.

Rapaport, F.T., Dausset, J., Converse, J.M. and Lawrence, H.S. (1965) Biological and ultrastructural studies of leucocyte fractions as transplantation antigens in man. Transplantation 3, 490–500.

Rask, L., Östberg, L., Lindblom, B., Fernstedt, Y. and Peterson, P.A. (1974) The subunit structure of transplantation antigens. Transplant. Rev. 21, 85–105.

Rees, R.C., Price, M.R., Baldwin, R.W. and Shah, L.P. (1974) Inhibition of rat lymph node cell cytotoxicity by hepatoma-associated embryonic antigen. Nature (London) 252, 751–753.

Rees, R.C., Price, M.R., Shah, L.P. and Baldwin, R.W. (1975) Detection of hepatoma-associated embryonic antigen in tumour-bearer serum. Transplantation 19, 424–429.

Reisfeld, R.A. and Kahan, B.D. (1970) Biological and chemical characterization of human histocompatibility antigens. Fed. Proc., 29, 2034–2040.

Reisfeld, R.A. and Kahan, B.D. (1971) Extraction and purification of soluble histocompatibility antigens. Transplant. Rev. 6, 81–112.

Reisfeld, R.A. and Kahan, B.D. (1972) Markers of biological individuality. Sci. Am. 226, No. 6, 28–37.

Robbins, J.C. and Nicolson, G.L. (1975) Surfaces of normal and transformed cells. In: F.F. Becker (Ed.), Cancer: A Comprehensive Treatise, Vol. 4, Plenum, New York, pp. 3–54.

Robins, R.A. (1975) Serum antibody responses to an ascitic variant of rat hepatoma D23. Brit. J. Cancer 32, 21–27.

Robins, R.A. and Baldwin, R.W. (1974) Tumour-specific antibody neutralization of factors in rat hepatoma-bearer serum which abrogate lymph-node-cell cytotoxicity. Int. J. Cancer 14, 589–597.

Roth, J.A., Holmes, E.C., Reisfeld, R.A., Slocum, H.K. and Morton, D.L. (1976) Isolation of a soluble tumor-associated antigen from human melanoma. Cancer 37, 104–110.

Rudman, D., Del Rio, A., Akgun, S. and Frumin, E. (1969) Novel proteins and peptides in the urine of patients with advanced neoplastic disease. Am. J. Med. 46, 174–187.

Ruoslahti, E., Pihko, H. and Seppälä, M. (1974) Alpha-fetoprotein: immunochemical purification and chemical properties. Expression in normal state and in malignant and non-malignant liver disease. Transplant. Rev. 20, 38–60.

Sample, W.F., Gertner, H.R. and Chretien, P.B. (1971) Inhibition of phytohemagglutinin-induced in vitro lymphocyte transformation by serum from patients with carcinoma. J. Natl. Cancer Inst. 46, 1291–1297.

Schrader, J.W., Cunningham, B.A. and Edelman, G.M. (1975) Functional interactions of viral and histocompatibility antigens at tumor cell surfaces. Proc. Natl. Acad. Sci. USA 72, 5066–5069.

Shah, L.P., Rees, R.C., Price, M.R. and Baldwin, R.W. (1977) Tumour rejection responses in rats sensitized to rat embryonic tissue, III. Development of experimental lung metastases in rats immunized with cell membrane fractions prepared from rat embryonic and tumour tissue. Brit. J. Cancer (Submitted for publication).

Shellam, G.R. and Knight, R.A. (1974) Antigenic inhibition of cell-mediated cytotoxicity against tumour cells. Nature (London) 252, 330–332.

Shimada, A. and Nathensen, S.G. (1969) Murine histocompatibility-2 (H-2) alloantigens. Purification and some chemical properties of soluble products from $H-2^b$ and $H-2^d$ genotypes released by papain digestion of membrane fractions. Biochemistry 8, 4048–4062.

Shoham, J. and Sachs, L. (1974) Different cyclic changes in the surface membrane of normal and malignant transformed cells. Exp. Cell Res. 85, 8–14.

Shuster, J., Freedman, S.O. and Gold, P. (1975) Fetal antigens in clinical medicine. In: G.N. Vyas, D.P. Stites and G. Brecher (Eds.), Laboratory Diagnostics of Immunologic Disorders. Grune and Stratton, New York, pp. 239–258.

Silk, M. (1967) Effect of plasma from patients with carcinoma on in vitro lymphocyte transformation. Cancer 20, 2088–2089.

Singer, S.J. and Nicolson, G.L. (1972) The fluid mosaic model of the structure of cell membranes. Science 175, 720–731.

Sjögren, H.O., Hellström, I., Bansal, S.C. and Hellström, K.E. (1971) Suggestive evidence that the "blocking antibodies" of tumor bearing individuals may be antigen-antibody complexes. Proc. Natl. Acad. Sci. USA 68, 1372–1375.

Sjögren, H.O., Hellström, I., Bansal, S.C., Warner, G.A. and Hellström, K.E. (1972) Elution of "blocking antibodies" from human tumors, capable of abrogating tumor cell destruction by specifically immune lymphocytes. Int. J. Cancer 9, 274–283.

Smith, H.G. and Leonard, E.J. (1974) Humoral immune responses to tumor-specific antigens in strain-2 guinea pigs. J. Natl. Cancer Inst. 53, 187–194.

Sobczak, E. and De Vaux Saint Cyr, Ch. (1971) Study of the in vivo fixation of antibodies on tumors provoked in hamsters by injection of SV40-transformed cells (TSV_5Cl_2). Int. J. Cancer 8, 47–52.

Stackpole, C.W., Jacobsen, J. and Lardis, M.P. (1974) Antigenic modulation in vitro, I. Fate of thymus-leukemia (TL) antigen-antibody complexes following modulation of TL antigenicity from the surfaces of mouse leukemia cells and thymocytes. J. Exp. Med. 140, 939–953.

Steele, G., Sjögren, H.O. and Price, M.R. (1975) Tumor-associated and embryonic antigens in soluble fractions of a chemically-induced rat colon carcinoma. Int. J. Cancer 16, 33–51.

Strominger, J.L., Cresswell, P., Grey, H., Humphreys, R.E., Mann, D., McCune, J., Parham, P., Robb, R., Sanderson, A.R., Springer, T.A., Terhorst, C. and Turner, M.J. (1974) The immunoglobulin-like structure of human histocompatibility antigens. Transplant. Rev. 21, 126–143.

Sundqvist, K.G., Svehag, S.E. and Thorstensson, R.T. (1974) Dynamic aspects of the interaction between antibodies and complement at the cell surface. Scand. J. Immunol. 3, 237–250.

Suter, L., Bloom, B.R., Wadsworth, E.M. and Oettgen, H.F. (1972) Use of the macrophage migration inhibition test to minitor fractionation of soluble antigens of chemically induced sarcomas of inbred guinea pigs. J. Immunol. 109, 766–775.

Tamerius, J., Hellström, I. and Hellström, K.E. (1975) Evidence that blocking factors in the sera of multiparous mice are associated with immunoglobulins. Int. J. Cancer 16, 456–464.

Tamerius, J., Nepom, J., Hellström, I. and Hellström, K.E. (1976) Tumor-associated blocking factors: isolation from sera of tumor-bearing mice. J. Immunol. 116, 724–730.

Tanigaki, N. and Pressman, D. (1974) The basic structure and the antigenic characteristics of HL-A antigens. Transplant. Rev. 21, 15–34.

Terry, W.D., Henkart, P.A., Coligan, J.E. and Todd, C.W. (1974) Carcinoembryonic antigen: characterization and clinical applications. Transplant. Rev. 20, 100–129.

Thomson, D.M.P. (1975) Soluble tumour-specific antigen and its relationship to tumour growth. Int. J. Cancer 15, 1016–1029.

Thomson, D.M.P. and Alexander, P. (1973) A cross-reacting embryonic antigen in the membrane of rat sarcoma cells which is immunogenic in the syngeneic host. Br. J. Cancer 27, 35–47.

Thomson, D.M.P., Eccles, S. and Alexander, P. (1973a) Antibodies and soluble tumour-specific antigens in blood and lymph of rats with chemically induced sarcomata. Br. J. Cancer 28, 6–15.

Thomson, D.M.P., Sellens, V., Eccles, S. and Alexander, P. (1973b) Radioimmunoassay of tumor-specific transplantation antigen of a chemically-induced rat sarcoma: circulating soluble tumour antigen in tumour bearers. Br. J. Cancer 28, 377–388.

Thomson, D.M.P., Steele, K. and Alexander, P. (1973c) The presence of tumour-specific membrane antigen in the serum of rats with chemically induced sarcomata. Br. J. Cancer 27, 27–34.

Thunold, S., Tonder, O. and Larsen, O. (1973) Immunoglobulins in eluates of malignant human tumors. Acta Pathol. Microbiol. Scand. Suppl. 236, 97–100.

Tökés, Z. and Chambers, S. (1975) Proteolytic activity associated with human erythrocyte membranes. Self-digestion of isolated human erythrocyte membranes. Biochim. Biophys. Acta 389, 325–338.

Turner, M.J., Strominger, J.L. and Sanderson, A.R. (1972) Enzymic removal and re-expression of a histocompatibility antigen, HL-A2, at the surface of human peripheral lymphocytes. Proc. Natl. Acad. Sci. USA 69, 200–202.

Vaage, J. (1974) Circulating tumor antigens versus immune serum factors in depressed concomitant immunity. Cancer Res. 34, 2979–2983.

Vanky, F., Klein, E., Stjernswärd, J. and Nilsonne, U. (1974) Cellular immunity against tumor-associated antigens in humans: lymphocyte stimulation and skin reaction. Int. J. Cancer 14, 272–288.

Vitetta, E.S. and Uhr, J.W. (1973) Synthesis, transport, dynamics and fate of cell surface Ig and alloantigens in murine lymphocytes. Transplant. Rev. 14, 50–75.

Vitetta, E.S. and Uhr, J.W. (1975) Immunoglobulins and alloantigens on the surface of lymphoid cells. Biochim. Biophys. Acta 415, 253–271.

Viza, D. and Phillips, J. (1975) Identification of an antigen associated with malignant melanoma. Int. J. Cancer 16, 312–317.

Viza, D., Phillips, J. and Trejdosiewicz, L.K. (1975) Cell surface and serum melanoma associated antigens. Behring Inst. Mitt. 56, 83–86.

Vlodavsky, I. and Sachs, L. (1974) Difference in the cellular cholesterol to phospholipid ratio in normal lymphocytes and lymphocytic leukaemic cells. Nature (London) 250, 67–68.

Vrba, R., Alpert, E. and Isselbacher, K.J. (1976) Immunological heterogeneity of serum carcinoembryonic antigen (CEA). Immunochemistry 13, 87–89.

Warren, L. and Glick, M.C. (1968) Membranes of animal cells, II. The metabolism and turnover of the surface membrane. J. Cell Biol. 37, 729–746.

Witz, I.P. (1973) The biological significance of tumor-bound immunoglobulins. Curr. Topics Microbiol. Immunol. 61, 151–171.

Witz, I., Yagi, Y. and Pressman, D. (1967) IgG associated with microsomes from autochthonous hepatomas and normal liver of rats. Cancer Res. 27, 2295–2299.

Wolf, A. and Steele, K.A. (1975) Separation of a tumour specific transplantation-type antigen from the ascitic fluid of mice bearing a syngeneic lymphoma. Br. J. Cancer 31, 684–688.

Yefenof, E. and Klein, G. (1974) Antibody-induced redistribution of normal and tumor associated surface antigens. Exp. Cell Res. 88, 217–224.

Youn, J.K., Le François, D., Hue, G., Santillana, M. and Barski, G. (1975) Activation of "eclipsed" lymphoid cells from advanced tumor-bearing mice through adoptive transfer to sublethally irradiated syngeneic hosts. Int. J. Cancer 16, 629–638.

Yu, A. and Cohen, E.P. (1974) Studies on the effect of specific antisera on the metabolism of cellular antigens. J. Immunol. 112, 1296–1307.

Zamcheck, N. (1975) The present status of CEA in diagnosis, prognosis, and evaluation of therapy. Cancer 36, 2460–2468.

Zöller, M., Price, M.R. and Baldwin, R.W. (1975) Cell-mediated cytotoxicity to chemically-induced rat tumours. Int. J. Cancer 16, 593–606.

Zöller, M., Price, M.R. and Baldwin, R.W. (1976) Inhibition of cell-mediated cytotoxicity to chemically induced rat tumours by soluble tumour and embryo cell extracts. Int. J. Cancer 17, 129–137.

Expression of cell surface antigens on cultured tumor cells

M. CIKES and S. FRIBERG Jr.

1. Introduction

Early studies of the antigenic properties of cultured mammalian cells revealed that genetically homogeneous cell populations contained both antigen-positive and antigen-negative cells and that the proportion of antigen-positive cells changed when cells were tested at intervals (Franks, 1967). Clones derived from such cell lines contained the same proportion of antigen-positive cells as the parental cell populations (Franks and Dawson, 1966). Mutation was not regarded as a plausible explanation for these findings because the changes were too frequent. It was also noted that some cell surface antigens are stable during growth of cells in culture while others disappeared soon after explantation (Franks, 1968; Harris, 1968). These observations, now made on a variety of cell types in many different laboratories, suggest that the expression of certain cell surface antigens is a dynamic phenomenon, a notion in keeping with current concepts which view the plasma membrane of mammalian cells as a dynamic structure that is able to undergo rapid and reversible change in its functional properties in response to specific stimuli arising in both the intracellular and extracellular environments. In this review, we will examine the possible mechanisms underlying variations in the expression of cell surface antigens by cultured cells. Particular emphasis will be given to cell cycle dependent antigenic changes and to the alterations in antigenic composition found in tumor cell populations.

2. Cell surface antigens in non-synchronized cultures

2.1. Variations in cell surface antigen expression during a single growth cycle

Mouse lymphomas induced by Moloney leukemia virus (MLV) and converted to the ascitic form can readily be established as culture lines. These cultured cells are particularly suitable for studies of cell surface antigen expression because: (a) the cells carry on their surface both *H-2* and MLV-determined cell surface antigens (MV-CSA); (b) they are susceptible to complement-dependent immune lysis; (c) a large number of cells can easily be obtained for antibody absorption and antigen extraction studies; (d) the viability of cells harvested at any time point during a single growth cycle is excellent (>95% viable cells); and (e) their growth as suspension cultures greatly facilitates handling of the cells.

The sensitivity of MLV-induced mouse lymphoma cells (YCAB) to complement-dependent lysis with *H-2* and MV-CSA antisera changes considerably when examined at various times during a single growth cycle (Cikes, 1970a). The extent of these variations, expressed in terms of the cytotoxicity index (which denotes the proportion of cells specifically killed by antiserum and complement), ranges between 0.10 and 0.90. The cells exhibit maximal sensitivity to both *H-2* and MV-CSA antisera at the time they reach the stationary phase, whereas during the logarithmic phase they are almost completely resistant to the same antisera and complement. The sensitivity of these cells to immune lysis follows closely the population doubling time characteristic for the time periods between the two testings. This is a reproducible finding irrespective of the cell concentration in the initial inoculum, though the latter does, of course, influence the growth curves of the cell populations.

An inverse relationship between cytotoxic sensitivity and modal cell volume was observed in the initial study by Cikes (1970a). It was also shown that the antibody absorptive capacity of YCAB cells paralleled their sensitivity to the lytic potency of the respective antisera. This suggested that the number of antibody-binding sites per cell was reduced on logarithmic phase cells. Taking into account the larger volume of logarithmic phase cells (see below), the difference in the absorptive capacity between the stationary and logarithmic phase cells would appear more pronounced if it were expressed per unit of cell surface area.

These studies were subsequently extended and confirmed by using monospecific *H-2* antisera and by relating antigen expression to the cell cycle kinetics during a single growth cycle (Cikes, 1970b). In addition, a decrease in cell-surface antigen content was noted in YCAB cells which had been in the stationary phase for a prolonged period, most probably reflecting the changed growth properties of YCAB cells at the time these experiments were performed.

It is appropriate to discuss here the possible influence of cell volume on the cytotoxic sensitivity of cells. The resistance to specific immune lysis of logarithmic phase cells, which have an approximately 40 to 50% larger volume than stationary phase cells, could at first sight be explained by a "dilution" of antigenic determinants due to the larger surface area of these cells. It could be argued that the binding of bi- or multivalent antibody molecules is more efficient if each combining site of antibody could bind to one antigenic determinant than if the distance between antigenic determinants exceeds a certain critical level, as might be the case on the larger logarithmic phase cells. For example, in complement-dependent immune lysis of red blood cells it seems that two adjacent IgG molecules are required to activate the complement components required for cell lysis (Borsos, 1965; Humphrey, 1965) or that a single IgM molecule must combine with more than one antigenic determinant in order to activate the complement system (Ishizaka, 1968). Extrapolation of these findings to other cell types would mean that the resistance of logarithmic phase cells to immune lysis could be due to the paucity of cell surface antigenic determinants resulting from the increase in cell volume. However, two findings argue against this assumption. Firstly, cells from the late stationary phase are smaller than the logarithmic phase cells, yet the former are less sensitive to immune lysis than the latter (Cikes, 1970b). Secondly, the amounts of extractable *H-2* antigens are greater in stationary phase cells than in logarithmic phase cells (Cikes and Klein, 1972a). It can be concluded therefore that cell volume does not seem to be an important factor in determining the sensitivity of cells to complement dependent immune lysis.

Lerner et al. (1971) also found that the cytotoxic sensitivity of YCAB lymphoma cells was maximal during the stationary phase and practically absent during the logarithmic phase. These authors suggested, however, that the variation in cytotoxic sensitivity of YCAB cells was not due to changes in the extent of antigen expression and proposed that other factors were responsible, including such possibilities as: inefficient activation of complement system; a different susceptibility of the cell membrane to complement damage brought about by changes in membrane configuration, charge and structure; a different ability of cells to repair the membrane damage; and an inhibition of the complement system in certain cell physiological states. Lerner et al.'s dismissal of variation in the number of antigens on the cell surface as an explanation for differences in cytotoxic sensitivity was based on immunofluorescence studies which failed to reveal any significant variation in accessibility of "viral" antigen on YCAB cells to an antiserum obtained from Fischer rats with a transplanted Moloney sarcoma virus tumor. In addition, electron microscopic studies showed the presence of viral particles on almost all cells during both logarithmic and stationary phases. Lerner et al. also showed that the extent of complement activation was constant during the growth of YCAB cells and that at least the initial portion of the complement reaction was taking place on the surface of YCAB cells, whereas the terminal portion of this reaction, which is probably required for cytolysis to occur, might have proceeded at a distance from the cell surface. They therefore suggested that the binding sites for the latter complement components might be available on the cell surface only during a limited period of the cell life cycle.

The failure of Lerner et al. to demonstrate quantitative changes in antigen expression on YCAB cells by indirect membrane immunofluorescence was most likely due to a condition of antibody excess which can mask manifold differences in cell surface antigen concentration, though such changes could have readily been demonstrated by a suitable dilution of the antiserum. The electron microscopic observations of Lerner et al. are even less conclusive because the release of MLV was not quantitated.

Indirect membrane immunofluorescence has also been used to study the expression of *H-2* and MV-CSA antigens on cultured YCAB lymphoma cells (Cikes, 1970b) and JLS-V9 cells derived from bone marrow of a BABL/c mouse (Cikes, 1971a). The intensity of the immunofluorescence staining and the proportion of antigen-positive YCAB cells followed closely their sensitivity to complement-dependent immune lysis. A similar, but more pronounced, pattern of antigen expression was observed on JLS-V9 cells. In the late stationary phase the proportion of antigen-positive cells declined rapidly in the JLS-V9 cultures. Cell populations with overlapping volume distribution spectra displayed a markedly different proportion of antigen-positive cells, again indicating that cell volume was not a major factor in determining the expression of cell surface antigens. Thus, a similar pattern of antigen expression during a single growth cycle was observed in two cell lines differing in origin, growth and cultural properties.

Membrane immunofluorescence is a sensitive and technically simple procedure, but its evaluation is dependent on visual observation and it is not therefore suitable for quantitation of cell-surface antigens unless instrumental measurements of fluorescence are used. In order to obtain more quantitative data on the expression of cell surface antigens, Cikes snd Klein (1972a) used the quantitative antibody absorption method which is one of the most reliable methods for measurement of the relative content of cell surface antigens accessible to antibody. This method was used to follow the expression of *H-2* and MV-CSA antigens on YAC lymphoma cells during a single growth cycle. It was found that logarithmic phase cells had only about 1/2 to 1/3 of the surface antigen content of stationary phase cells. The amount of extractable antigens showed a similar variation to those exposed on the cell surface. This study demonstrated that the number of antibody binding sites changed inversely with the rate of cell multiplication and that a direct correlation existed between the susceptibility to specific immune lysis and the number of the available antibody binding sites on the cell surface. Reisfeld's group used a similar approach to study antigen expression on several cultured cell lines. L1210 murine leukemia cells showed maximal susceptibility to complement dependent *H-2* antibody-mediated lysis during the mid-logarithmic phase of the cell growth cycle. Antigen expression on these cells was, however, low during the first 10 h in culture. The yields of soluble *H-2* antigens and the antibody-absorbing capacity of L1210 cells for *H-2* antisera correlated with their cytotoxic sensitivity (Goetze et al., 1972)*. Pellegrino et al. (1973) studied *HL-A* antigens in the WIL_2 cultured human cell line derived from splenic lymphocytes of a normal donor. The cytotoxic sensitivity and antibody-absorbing capacity of these cells did not change appreciably during the growth cycle, but the yield of extractable *HL-A* antigens was much

---

*It should be mentioned here that the growth curves of cultured cells are influenced by the age (i.e. lag, logarithmic or stationary phase) of seeded cells (Cikes, unpublished observations). Consequently, the time after seeding does not necessarily determine the position of cells within their life cycle. This emphasizes the importance of measuring simultaneously the antigen content of cells and their position within the cell cycle in cell populations originating from the same culture.

greater from cells in the late logarithmic or stationary phase compared with that from early logarithmic phase cells.

The same group of investigators examined *HL-A* antigen expression in a cultured human lymphoid cell line (RPMI 8866) derived from a donor with myelogenous leukemia (Pellegrino et al., 1974). They found a low cytotoxic sensitivity of RPMI 8866 cells during early logarithmic phase (which was attributed to the G_1 period) but no change in the concentration of *HL-A* antigens on the cell surface was detectable using the quantitative antibody absorption assay, isotopic antiglobulin tests or the direct extraction of the *HL-A* antigens; nor was there a change in binding and activation of complement components during the growth cycle of RPMI 8866 cells. The authors attributed the changes in cytotoxic susceptibility of these cells to the changes in structural properties of the cell membrane and a cell cycle-dependent ability of RPMI 8866 cells to repair the membrane damage caused by antibody and complement. Everson et al. (1974) found maximal absorptive capacity of cultured human lymphoblastoid cell lines for *HL-A* antibody during the logarithmic phase of growth (see section 4). Zwerner and Acton (1975) have reported variations in anti-Thy.l (but not anti-TL) absorptive capacity in a series of mouse lymphoblastoid cell lines, with maximum absorption occurring during the logarithmic phase of growth.

Since the results cited above reveal discrepancies between cytotoxic sensitivity and the concentration of cell surface antigens, some studies which have attempted to tackle this problem will be mentioned. Möller and Möller (1962) found a direct correlation between concentration of *H-2* antigens and cytotoxic sensitivity in a number of mouse cell lines. Similar observations have been made in cultured murine lymphoma cells derived from a methylcholantrene-induced lymphoma (Cann and Herzenberg, 1963b), in two sublines of a spontaneous mouse mammary adenocarcinoma (Friberg, 1972), in mouse thymocytes and spleen cells (Basch, 1974) and in cultured mouse myeloma, lymphoma and mastocytoma cells (Lesley et al., 1974). It seems, however, that *different* cell lines may exhibit a susceptibility to complement-dependent immune lysis which is not a function of the concentration of their cell surface antigens (Lesley et al., 1974; Ohanian et al., 1974). Conflicting results have also been reported regarding the influence of the movement of cell surface antigens on complement-dependent immune cytolysis (Edidin and Henney, 1973; Boyle et al., 1975).

2.2. Cell cycle kinetics during a single growth cycle

The inverse relationship between the extent of antigen expression and cell volume described in the previous section suggests that cell surface antigens are maximally expressed during the G_1 period (Cikes, 1970a,b; 1971a). This notion was based initially on findings showing that the position of cells within the cell cycle is directly proportional to the cell mass and that DNA replication occurs when the cells reach a certain "critical" mass (Killander and Zetterberg, 1965;

Zetterberg, 1970). Although the latter finding could not be confirmed in all instances (Fox and Pardee, 1972), the mass of individual cells (Killander and Zetterberg, 1965) and the modal cell volume (Enger and Tobey, 1969; Cikes and Friberg, 1971) increased during interphase by a factor of approximately two. It appears therefore that the modal cell volume is a function of the position of cells within their life cycle. To test this hypothesis, YCAB lymphoma cells were pulsed with [^{3}H]thymidine for 30 min at different time intervals during a single growth cycle and the proportion of cell incorporating the isotope, together with the modal cell volume, were related to the growth rate during the corresponding time intervals. A direct relationship was found between the growth rate on the one hand, and the proportion of cells with incorporated [^{3}H]thymidine and the modal cell volume, on the other. For instance, about 80% of YCAB cells incorporated [^{3}H]thymidine when their population doubling time was about 25 hours, whereas only 10 to 20% of them incorporated isotope when cells reached the stationary phase (Cikes, 1971a). The modal cell volume of the stationary phase cells increased from 440 to 640 μ^3 when cells were diluted and incubated in fresh medium for 30 to 40 h (Cikes, 1970b). It was thus demonstrated that large YCAB cells were predominantly S phase cells.

To get more insight into the cell cycle kinetics during a single growth cycle, the duration of cell cycle periods was measured at different time intervals during the growth cycle of YCAB cells (Cikes, 1970b). The results showed that prolongation of the population doubling time was due mainly to the prolongation of the G_1 period whereas the relative durations of S, G_2 and mitosis were rather constant despite different growth rates. It should be noted, however, that the absolute duration of the S period was also prolonged in stationary phase cultures, but when the duration of S period was related to the population doubling time (i.e. the *relative* duration of S period), it did not change significantly. When the duration of the cell cycle periods was used to compute the proportion of cells positioned in the respective periods, it was found that deceleration of the growth rate was accompanied by accumulation of cells in the G_1 period (Cikes, 1970b).

Similar observations regarding cell cycle kinetics in non-synchronized cultures have been made by other authors. Hill et al. (1959) measured the DNA content of mouse L strain fibroblasts growing in agitated suspension cultures. When stationary phase cells were diluted and incubated in fresh medium, the DNA content of cells doubled during the first day after subculture while the cell count remained constant. These results suggested that dilution and incubation of the stationary phase cells in fresh medium caused a synchronous entry of cells into the S period. Todaro et al. (1965) reported that addition of fresh culture medium to contact inhibited 3T3 cells produced synchronous division of a small fraction of cells occurring 30 h after the medium change and that this was preceded by an increase in cellular protein and RNA synthesis and DNA replication. Thus, contact inhibited 3T3 cells were reversibly arrested in the G_1 period. A similar sequence of events following addition of fresh culture medium was observed in contact inhibited BSC-1 monkey kidney cells (Becker and Levitt, 1968). The cell number in contact inhibited BSC-1 cell cultures doubled 60 to 70 h after addition of fresh culture medium and cell proliferation was preceded by a gradual increase of [^{3}H]thymidine uptake which reached a maximum 30 h after medium change. Nilausen and Green (1965) also found that 3T3 cells were arrested in the G_1 period when the cultures reached the saturation density. This was ascertained by the measurement of DNA content per cell at different time points during a single growth cycle and by following the incorporation of [^{3}H]thymidine and the proportion of cells in mitosis after dilution and incubation of the stationary phase cells in fresh medium. Tobey and Ley (1970) described a technique of growing Chinese hamster cells to the stationary phase as a simple means for providing large quantities of cells synchronized in the G_1 period. Dilution or resuspension of these cells in fresh culture medium induced a synchronous traverse of the cell cycle. It was shown subsequently that the same effect could be achieved by addition of appropriate amounts of isoleucine and glutamine to the G_1-arrested cells and that the logarithmically phase cells accumulated in the G_1 period when cultured for 30 h in medium deficient in isoleucine and glutamine (Ley and Tobey, 1970). Enger and Tobey (1972) showed that isoleucine deficient F-10 medium was an efficient means to synchronize Chinese hamster cells in the G_1 period. Thus the two amino acids affected the redistribution of the cells within the cell cycle periods. Various inhibitors of macromolecular synthesis (Yeh and Fischer, 1969; Bellanger et al., 1970) and serum factors (Holley and Kiernan, 1968; Baker and Humphreys, 1971) have been postulated to have a similar effect. Todo et al. (1971) analyzed the cell cycle kinetics of the SK-L7 culture line established from peripheral blood of a patient with acute myeloblastic leukemia and found that the transition from the logarithmic to stationary phase entailed a 3.7-fold prolongation of the G_1 period whereas the duration of S,G_2 and mitosis were less affected. Glinos and Werrlein (1972) reported that medium renewal and dilution of stationary L-929 fibroblasts triggered these cells to enter synchronously into the S period and to divide, suggesting that a large proportion of cells in stationary phase cultures was arrested in the G_1 period.

It should be noted, however, that stationary phase cells, especially those held in the stationary phase for a longer time period, might differ physiologically from the cells merely arrested in that period. Levine et al. (1965) showed that the progressive inhibition of macromolecular synthesis occurring in cultures approaching the saturation density was accompanied by disappearance of free cytoplasmic polyribosomes. Becker et al. (1971) found that stationary phase hamster embryo fibroblasts had less than 70% of the ribosome complement of the early G_1-phase cells obtained by the selective detachment of mitotic cells. These changes, which would be expected to reduce the protein synthetic capacities of the cell, may explain the low antigen content found in the late stationary and "starved" cells (Cikes, 1970b; 1971a; Cikes and Klein, 1972b).

3. Cell surface antigens in synchronized cultures

3.1. Variations in cell surface antigen expression during the cell cycle

Direct evidence for the cell cycle-dependent expression of cell surface antigens has been obtained in several studies with synchronized cell cultures. Kuhns and Bramson (1968) synchronized S3 strain HeLa cells by the single thymidine treatment and followed the expression of H blood-group activity during the cell cycle by mixed agglutination. 7 to 8 h after thymidine release a mitotic peak appeared and the highest proportion of antigen-positive cells was observed during this time period. Control cultures showed neither the mitotic nor the antigenic activity noted in synchronized cultures. It does not necessarily follow, however, that H blood-group activity was maximal in mitotic cells since the antigen-positive cells observed during the prolonged mitotic wave might have belonged to the G_1 fraction which is short or absent in rapidly growing cells during the logarithmic phase (Robbins and Scharff, 1967).

Cikes and Friberg (1971) studied the expression of *H-2* and MV-CSA antigens in synchronized JLS-V9 cells infected in vitro with MLV (Wright et al., 1967). Cells were synchronized in mitosis by exposing subconfluent monolayer cultures to colcemid. Mitotic cells were collected, incubated in fresh culture medium and their subsequent traverse of the cell cycle followed by monitoring the incorporation of [^{3}H]thymidine, determination of modal cell volumes and direct counting of cells with mitotic figures. The expression of the cell surface antigens was monitored during synchronous growth by indirect membrane immunofluorescence. The highest proportion of antigen-positive cells was found during the G_1 period. During S and G_2 antigen expression was low but it increased again when the majority of cells divided and entered the G_1 period of the next cycle. The *H-2* and MV-CSA antigens were temporally coexpressed. In control "asynchronous", logarithmically growing cultures, the antigen expression was low throughout the observation period. Data on thymidine incorporation and cell volume suggested that the majority of cells in the control cultures were in the S period.

Recently, several groups of investigators have reported cell cycle-dependent antigen expression in cultured cells. Thomas (1971) studied the expression of blood group B and H activities in synchronized cultures of mouse mastocytoma cells (P815Y) and phytohemagglutinin (PHA)-treated mouse lymphocytes. P815Y cells were synchronized at the beginning of S period by the double thymidine method and indirect membrane immunofluorescence used to monitor the expression of B and H blood group activities. It was found that these two activities behaved reciprocally during synchronous growth of mastocytoma cells. In the G_1 period cells were B− H+, whereas in the S period they were B+ H−. To find out whether the changes in H blood-group activity occurred in mitosis or early G_1, P815Y cells were synchronized by the single thymidine treatment and then placed in medium with colcemid. The proportion of H+ cells was found to increase in parallel with the increase of cells arrested in metaphase, whereas the proportion of B+ cells remained constant during this time period. The interpretation in this study that the expression of H blood-group activity was maximal during mitosis is in contradiction with the results obtained in asynchronous cultures where H was maximally expressed during the stationary phase of growth when mitotic activity was practically absent. A more likely interpretation is that colcemid might have arrested merely the nuclear events in the cells while the cytoplasmic processes of the colcemid-arrested cells might have proceeded to the G_1 period of the next cycle. Thomas also showed that B− mouse lymphocytes, which are normally in G_0 phase, acquired B+ activity upon stimulation with PHA.

Karb and Goldstein (1971) used an interesting approach combining autoradiography and direct membrane immunofluorescence to study the expression of *HL-A* antigens at the single cell level in cultured Burkitt lymphoma cells. Using a combination of filters these investigators were able to simultaneously observe the immunofluorescent staining pattern and autoradiographic grains in the same cell. They found that a higher proportion of antigen-negative cells incorporated [^{3}H]thymidine than did the antigen-positive cells. This technique does not distinguish the cells in early and late S period from cells that are in the middle of S period, unless the number of grains per cell is counted.

Shipley (1971) synchronized V79-753B-3 Chinese hamster cells by incubating them with 2 mM hydroxyurea for 3.5 h which synchronizes the cells at the G_1/S boundary. Following the removal of hydroxyurea the cells synchronously traversed the cell cycle and reached the middle of the S and G_2 periods after 3 and 3.5 to 5 h, respectively. The sensitivity of the synchronized cells to complement-dependent immune damage was tested by their colony-forming ability following exposure to a heterologous antiserum (prepared in rabbits against whole Chinese hamster cells) and complement. Shipley found that the cells in the middle of the S period (1 to 3 h after removal of hydroxyurea) were maximally resistant to specific immune lysis. The cytotoxic sensitivity started to reappear 5 h after hydroxyurea release when the cells were beginning to enter the G_2 period. 7 h after hydroxyurea removal, cells were again resistant to the immune lysis. The reappearance of the resistant fraction of cells was interpreted as resulting from an increase in cell numbers subsequent to mitosis and re-entry of cells into the S phase. To exclude the possibility that the hydroxyurea treatment might have contributed to the observed behaviour of the synchronized cells, an asynchronous population of Chinese hamster cells was exposed to the antiserum and complement, and the surviving fraction tested for its sensitivity to a 540 R X-ray dose. The pattern of X-ray sensitivity of the surviving cell population was that expected from the cells synchronized in the middle of S period. These data provide additional evidence that cells in the middle of the S period were maximally resistant to the complement-dependent immune lysis.

Pasternak et al. (1971) used zonal centrifugation to synchronize P815Y mouse mastocytoma cells. This procedure separates cells according to their size which, in turn, is dependent on the position of cells within the cell cycle (Warmsley and Pasternak, 1970). The extent of complement-dependent immune cytolysis in cells harvested from different regions of the gradient was measured by the ^{51}Cr-release test using a suitable dilution of H-2 antiserum. The cytotoxic sensitivity was high during the early G_1 period, decreased progressively during the G_1/S transition, reached its lowest value in the middle of S, and was restored in G_2. The cytotoxic sensitivity of P815Y cells from different portions of the gradient paralleled their ability to inhibit the specific immune lysis suggesting that the extent of the specific immune lysis was measuring the cell surface antigen content.

Lerner et al. (1971) synchronized YCAB mouse lymphoma cells in the G_1 period by growing the cells to the final cell density in stationary suspension cultures and then followed their sensitivity to a rat anti-MLV serum and complement using both ^{51}Cr-release and the trypan blue dye exclusion test to monitor cell viability. The cytotoxic sensitivity of these cells was confined to the G_1 period (see also section 2.1). Cikes et al. (1972) used the quantitative antibody absorption technique to follow the expression of H-2 antigens in synchronized cultures. YAC mouse lymphoma cells were synchronized with hydroxyurea at the G_1/S boundary. Following removal of hydroxyurea the cells started to incorporate [^{3}H]thymidine, attaining maximum incorporation of the label after 28 hours. The number of cells did not change during this time period. The antibody absorptive capacity was low shortly after hydroxyurea removal (due presumably to "starvation" in the stationary phase), rose to maximal values during the G_1/S transition period, and attained the lowest values in the middle of S period. The re-expression of H-2 antigens after the first cell division following hydroxyurea removal was dependent on the final cell density characteristic for the culture conditions used in these experiments. If the cells had reached the final density, the antigens were fully re-expressed. If, however, the final cell density was higher than the density attained after the first cell division, antigen expression remained low. Control, logarithmically growing cells incorporated [^{3}H]thymidine at a high rate and had a low antibody absorptive capacity, comparable to that of the synchronized S-phase cells. Similar results were obtained in YAC cultures synchronized by growing the cells to the final cell density (stationary phase of growth) and resuspending them in fresh culture medium (see section 2).

These results indicated that considerable changes in number of antibody binding sites on the cell surface were taking place during the cell cycle. The temporal coincidence of the maximal expression of cell surface antigens and the maximal sensitivity to specific immune cytolysis can hardly be regarded as a fortuitous finding. In the absence of supportive data for other alternatives, changes in the extent of cell surface antigen expression presently seems to be the most convincing explanation to account for the changes in cytotoxic sensitivity during the cell cycle. Factors which might influence the union between antibody and antigen molecules on the cell surface will be discussed later in section 3.2.

Pellegrino et al. (1972) examined *HL-A* antigen expression in synchronized cultures of a long established human cell line (WIL_2) derived from splenic lymphocytes of a normal donor. The synchronization in G_1 was achieved by growing the cells to saturation density. The authors found maximal antibody absorptive capacity occurred 12 hours after release from the G_1 block. DNA synthesis had reached only 40% of its maximum at this time, maximum synthesis occurring 21 hours after the release from the G_1 block. The antibody absorptive capacity during the presynthetic period was 35 to 50% of the maximum. In addition, Pellegrino et al. found that 2-day old cells had a higher antibody absorptive capacity than 4-day old cells. Unfortunately, the complete growth curves and the antibody absorptive capacity of WIL_2 cells at the time of maximum DNA synthesis were not reported. The results of Pellegrino et al. (1972) are reminiscent of the re-expression of cell surface antigens following refeeding of "starved" cells observed by other investigators (Cikes et al., 1972; Cikes and Klein, 1972a,b).

Summer et al. (1973) assayed the expression of *H-2* antigens in Ficoll gradient fractions of P815Y mastocytoma cells using complement-dependent immune cytolysis, binding of radiolabeled antibody and the inhibition of specific immune cytolysis. In addition, the fragility of cells was assayed by their resistance to detergents, hypotonicity and freeze-thawing. Immune cytolysis and cell fragility reached minimal values during the S period. The remaining tests showed an increase of the amount of *H-2* antigens during the G_1 period; thereafter their accessibility to antibody remained constant. The authors concluded that *H-2* antigens were synthesized and inserted into the cell membrane during the G_1 period and that their concentration did not change during the S and G_2 periods. Ficoll gradient centrifugation was also used by Everson et al. (1974) to synchronize RPMI 8866 human lymphoblatoid cell line. *HL-A* antigen expression on these cells was monitored in Ficoll gradient fractions by quantitative antibody absorption and complement-dependent immune lysis. The antibody absorptive capacity of cells increased progressively during the cell cycle with a maximum in the G_2 period whereas the cytotoxicity tests gave unclear results due to the high level of lysis found in the complement and antiserum controls. Everson et al. (1974) also examined *HL-A* antigen expression in asynchronous RPMI 8866 cultures enriched for cells in G_0, G_1, and S periods. They found that 1-day old, S-enriched cells were more susceptible to complement-dependent immune lysis and had a greater antibody absorptive capacity than 2-, 3-, and 4-day-old cells (see also section 2). The method of producing these enriched cell populations (splitting of logarithmic phase cells in half every day to give 1-, 2-, 3- and 4-day-old cells) is, however, strongly suggestive of the starvation effect in overcrowded cultures (see section 2). Starvation effects might also have influenced the results obtained in the Ficoll gradient synchronization studies if cells from such overcrowded cultures had been used for separation.

Ferrone et al. (1973) found no growth-dependent changes in *HL-A* antigen expression in synchronized WIL_2 human lymphoid cell line as judged by the sensitivity of these cells to complement-dependent immune lysis, the extent of activation of the complement system, cellular binding of labeled complement components, and the antibody absorptive capacity of the cells. The discrepancy between these results and those showing cell cycle-dependent changes in the expression of cell surface antigens were attributed to the fact that a different source of the cells was used (i.e. normal cells versus neoplastic cells). This interpretation is not convincing, however, since Rosenfeld et al. (1973a,b) demonstrated cell cycle-dependent changes of cell surface antigens (detected by antilymphocytic globulin) in lymphoblastoid human cell lines derived from normal donors. The results reported by Thomas and Phillips (1973) and by Thomas (1974) introduce an additional possibility to explain the discrepancies encountered in studies on changes in cell surface antigen expression during the cell cycle. Thomas and Phillips (1973) found that dividing lymphoid cells possess membrane antigens which are not present on normal resting lymphocytes. Thomas (1974) also showed that erythroagglutinin, anti-i, sera recognize a cell surface antigen which is present in foetal tissues, transformed cell lines and mitogen stimulated lymphocytes and absent from normal adult tissues. This antigen, referred to as i^1 antigen, was maximally expressed in human lymphoblastoid cell lines during the S and G_2 periods.

Thus the existence of "cell cycle period-specific antigens" seems conceivable and a possible heterogeneity of antisera used in cell synchronization studies might explain the divergent results in different systems.

Killander et al. (1974) used cytophotometric techniques to measure the expression of *HL-A* and IgM in human lymphoblastoid cells on the single cell level. In addition, the authors measured the dry mass and DNA content of single cells and identified the cells in the S phase by [^{3}H]thymidine incorporation. The *HL-A* and IgM expression increased continuously during the interphase but the ratio of *HL-A* or IgM to total protein remained constant. More recently, Burk et al. (1976) demonstrated cell cycle-dependent expression of a sarcoma associated tumor antigen (SATA) on a human neurosarcoma cell line. SATA was defined by sarcoma patients' sera. Its expression in synchronized cultures was maximal during the mid-G_1 period and declined to minimal levels during S and G_2 periods.

The results of studies on cell surface antigen expression in synchronized cell cultures obtained in various laboratories are summarized in Table I.

3.2. Possible mechanisms of cell cycle-dependent antigen expression

Three mechanisms may be invoked to explain cell cycle-dependent expression of cell surface antigens:
   (1) changes in the topology of antigenic determinants on the cell surface;
   (2) alterations in the rate of degradation (or shedding); and
   (3) variation in the rate of synthesis of antigenic determinants.

The first mechanism finds some support from studies on the agglutination of normal and transformed cells by plant agglutinins (lectins). It has been shown by numerous investigators that normal cells are not agglutinated by concentrations of plant agglutinins that agglutinate transformed cells though mild trypsinization of normal cells renders them equally agglutinable as transformed cells (Burger and Goldberg, 1967; Burger, 1969; Inbar and Sachs, 1969a,b; Sela et al., 1970). Saturation binding experiments using purified ^{125}I-labeled wheat germ agglutinin (Ozanne and Sambrook, 1971), concanavalin A (Con A) (Cline and Livingston, 1971; Ozanne and Sambrook, 1971), and soy bean agglutinin (Sela et al. 1971) have shown that similar numbers of agglutinin receptors are present on both normal and transformed cells. Three mechanisms have been proposed to explain the greater susceptibility of transformed cells to agglutination by plant agglutinins: (a) exposure of cryptic agglutinin sites (Burger, 1969); (b) concentration of exposed sites due to a decrease of cell volume (Ben-Bassat et al., 1971); and (c) changes in plasma membrane organization which allow redistribution of the agglutinin sites with resulting clustering of the sites which is more favorable to cell agglutination (Singer and Nicolson, 1972). Most evidence has been obtained in support of the last mechanism in studies in which the topography of cell surface agglutinin receptors using ferritin-conjugated Con A has been monitored by immunoelectronmicroscopic techniques (review, Nicolson, 1974). These studies have shown that the agglutinin receptors on transformed cells can be redistributed by agglutinins more easily than comparable receptors on untransformed cells (review, Nicolson, 1974). No evidence has so far been obtained to support the other hypotheses of cryptic sites or changes in cell volume (for full discussion see Nicolson, 1974).

Agglutinability of normal and transformed cells has been reported to be cell cycle-dependent (Fox et al., 1971; Noonan et al., 1973; Smets, 1973; Shoham and Sachs, 1974; Smets and De Ley, 1974). The finding of an increased expression of *H-2* antigens (Lindahl et al., 1973) and Con A binding sites (Huet et al., 1974) on L1210 mouse leukemia cells after treatment with interferon, emphasizes possible similarity in the regulation of the expression of these two cell surface components. However, agglutinability of cells by plant agglutinins and the expression of cell surface antigens seem to be fundamentally different phenomena and extrapolation from one system to the other should be made with caution. In addition, the redistribution of agglutinin binding sites on transformed cells does not *quantitatively* change the agglutinin binding properties of the cells (Cline and Livingston, 1971; Ozanne and Sambrook, 1971; Sela et al., 1971; Nicolson, 1971, 1974).

Gahmberg and Hakomori (1974, 1975) reported that the exposure of ceramide pentasaccharide and ceramide tetrasaccharide on the surface of NIL hamster cells was maximal during the G_1 period although the actual amount of these components associated with the membrane (as determined by direct biochemical assay) did not change appreciably during the cell cycle. The exposure of the so called large, external, transformation sensitive (LETS) glycoprotein is also dependent on the growth state and the cell cycle. The accessibility of LETS protein on the cell surface, as determined by various

TABLE I
SUMMARY OF STUDIES DEALING WITH CELL SURFACE ANTIGENS IN SYNCHRONIZED CULTURES

Cell line	Origin of cell line	Antigens studied	Synchronization method used	Immunologic method used	Maximal antigen expression in	Reference
HeLa	Human cervical carcinoma	Blood group H	Single thymidine block	Mixed hemagglutination	Mitosis	Kuhns and Bramson (1968)
JLS-V9	Mouse bone marrow cells infected in vitro with Moloney leukemia virus (MLV)	H-2 and MLV-determined cell surface antigens (MV-CSA)	Colcemid	Indirect membrane immunofluorescence (IMIF)	G_1	Cikes and Friberg (1971)
P815Y	Mouse mastocytoma	Blood groups B and H	Double thymidine block	IMIF	H in mitosis and G_1; B in S	Thomas (1971)
HR1-K	Burkitt's lymphoma	HL-A	Identification of cells in S by [^{3}H] thymidine at the single cell level	Membrane immunofluorescence	$G_1(?)$ $G_2(?)$	Karb and Goldstein (1971)
V79-753B-3	Chinese hamster	Species specific	Hydroxyurea (HU)	Complement-dependent, antibody-mediated cytotoxicity (CAC)	G_2	Shipley (1971)
P815Y	Mouse mastocytoma	H-2	Velocity sedimentation	CAC	G_1	Pasternak et al. (1971)
YCAB	Mouse lymphoma induced by MLV	H-2 and "viral" antigens	Cell density-dependent arrest in G_1	CAC (1) IMIF (2)	G_1 by CAC; no changes during the cell cycle by IMIF	Lerner et al. (1971)

Cell line	Source	Antigen	Method 1	Method 2	Cell cycle phase	Reference
YAC	Mouse lymphoma induced by MLV	H-2 and MV-CSA	Cell density-dependent arrest in G_1 and HU	Quantitative antibody absorption (QAA)	G_1/S	Cikes and Klein (1972b); Cikes et al. (1972)
WIL_2	Human splenic lymphocytes	HL-A	Cell density-dependent arrest in G_1	QAA	Early S	Pellegrino et al. (1972)
P8154	Mouse mastocytoma	H-2	Velocity sedimentation	CAC, binding of radio labeled antibody and inhibition of CAC	Gradual increase during G_1, thereafter constant	Summer et al. (1973)
RPMI-8866	Human lymphoblastoid cells	HL-A	Velocity sedimentation	QAA and CAC	G_2 by QAA, unclear by CAC	Everson et al. (1974)
WIL_2	Human splenic lymphocytes	HL-A	Cell density-dependent arrest in G_1	CAC, QAA, activation of complement (C') and binding of labeled C' components	No changes during the cell cycle	Ferrone et al. (1973)
LHN_{13}, LHN_6 and DUK	Normal human peripheral lymphocytes	Lymphocyte specific	Double thymidine block	Direct membrane immunofluorescence	G_1	Rosenfeld et al. (1973a,b); Rosenfeld et al. (1973a)
MICH, BEC-11, BRI-7, BRI-8	Human lymphoid cells	"Division antigen"	Velocity sedimentation	IMIF	S and G_2	Thomas (1974)
Daudi	Burkitt's lymphoma	HL-A and IgM	Cell mass and DNA content at single cell level	Cytophotometric measurement of membrane bound immunofluorescent staining	G_2, but no change when related to cell mass	Killander et al. (1974)
T_2	Human neurosarcoma	Sarcoma-associated tumor antigen	Single thymidine block	IMIF	G_1	Burk and Drewinko (1976)

external labeling methods, is maximal in G_1-enriched, stationary phase cultures (Gahmberg et al., 1974; Hynes and Bye, 1974; and review in this volume by Gahmberg, p. 371). These results suggest that cell-to-cell contact and cell cycle-associated changes in structural properties of the cell membrane (Porter et al., 1973; Scott et al., 1973) may influence the topology of the membrane components.

"Masking" substances have been postulated to influence the accessibility of antibody binding sites (Currie, 1967; Sanford, 1967; Currie and Bagshawe, 1968) but subsequent results have shown that histocompatibility antigens cannot be "unmasked" on the cell surface (Ray and Simmons, 1971; Sanford and Codington, 1971; Schlesinger and Gottesfeld, 1971). Friberg and Lilliehöök (1973) found, however, that about 80% of "masked" *H-2* antigens on TA3-Ha mouse adenocarcinoma could be exposed by lyophilization or sonication of the cells.

The "capping" phenomenon (Taylor et al., 1971) also offers a possible mechanism which could affect cell cycle-dependent antigen expression. Indeed, immunofluorescence studies (Cikes, 1970b) have shown that S-enriched logarithmic phase cells predominantly exhibit the sectorial type of immunofluorescent staining whereas G_1-enriched stationary phase cells stain over the entire cell circumference. However, since the G_1-enriched cells yielded greater amounts of extractable *H-2* antigens than the S-enriched cell populations (Cikes and Klein, 1972b), the "capping" mechanism does not appear to play a major role in the changes of antigen expression during the cell cycle. This conclusion would be more convincing if the total antigen content of the G_1- and S-enriched cell populations had been measured. The estimation of the total antigen content would require a quantitative method for solubilization of membrane antigens which is not an easily controlled process.

Assuming constant synthesis of antigenic determinants throughout the entire cell cycle, a cell cycle-dependent degradation (or shedding) of antigens may explain the changes in their expression. Mannick et al. (1964) have demonstrated the presence of transplantation antigens in culture medium of rabbit spleen cells and Cikes (1971b) showed that the proportion of *H-2* antigen-positive mouse lymphoma cells prelabeled by the indirect immunofluorescent method, was lower in cell populations cultured under conditions which favored cell multiplication.

Finally, cell cycle-dependent antigen expression could result from variation in the rate of antigen synthesis. This hypothesis is amenable to experimental verification by measuring the net antigen synthesis in different cell cycle periods. Considerable data now exist which lend support to this hypothesis and show that the synthesis of specialized cell products and some membrane components by cultured cells bears considerable resemblance to the cell cycle-dependent expression of cell surface antigens discussed above. For example, early in the study of mammalian cells in culture, it was noted that cells ceased to form specialized products when cultured in media which favored a rapid cell multiplication (Doljanski, 1930). The mutual exclusiveness of cell proliferation and formation of specialized cell products was subsequently documented in many experimental systems (for reviews see Fischer, 1946; Levintow and Eagle, 1961; Harris, 1964; Herrmann and Marchok, 1967). Some selected observations pertinent to this discussion will be mentioned here.

The specific activity of β-glucuronidase in cultured cells derived from human skin increases 50 to 100% as cell populations become denser and the growth rate decreases. This increase in the enzyme specific activity can be completely suppressed by puromycin (De Mars, 1964). Similar density-dependent increases of the specific activity of other enzymes have been reported (De Luca, 1966; Eagle, 1968). Ruddle and Rapola (1970) studied the activities of lactate dehydrogenase (LDH) and esterase in cultured PK 65 cells derived from swine kidney. The specific activities of these enzymes decreased during the lag phase and remained low during the logarithmic phase. In stationary phase cells, however, the enzyme specific activities rose 2- to 8-fold over the values of the logarithmic phase cells. Histochemical assays of the enzyme activities showed a similar pattern during the growth cycle.

Cultured neuroblastoma cells have proved particularly suitable for in vitro studies of morphological and biochemical markers characteristic of the differentiated state. Reduction of the serum concentration in the culture medium retards the multiplication of C1300 mouse neuroblastoma cells and markedly increases the proportion of cells with axons (Seeds et al., 1970). The specific activity of acetylcholinesterase in C1300 neuroblastoma cells increases 25-fold in stationary phase cultures (Blume et al., 1970). The choline-O-acetyltransferase activity is 5.7-fold greater in stationary phase cells than in logarithmically growing cells, whereas the activity of thymidylate synthetase, an enzyme involved in DNA synthesis, is greater in logarithmically growing cells. Restriction of cell division is accompanied by expression of three additional properties of neurone differentiation in C1300 cells: axon-dendrite development; formation of electrically excitable plasma membranes; and synthesis of acetylcholine receptors (Rosenberg et al., 1971). The prevention by actinomycin D and cycloheximide of the induction of acetylcholinesterase activity suggests that de novo protein and RNA synthesis are required for the observed increases in enzyme-specific activity (Kates et al., 1971). However, the formation of axon-like processes is not affected by these drugs (Seeds et al., 1970). Inhibition of DNA synthesis by cytosine arabinoside and fluorodeoxyuridine produces accumulation of C1300 cells at the G_1/S boundary and markedly increases their acetylcholinesterase specific activity (Kates et al., 1971) and enhances the formation of axon-like processes (Klebe & Ruddle, 1969). The number of acetylcholine receptors and the activity of acetylcholinesterase increases 5- to 10-fold in stationary phase cultures of C1300 neuroblastoma cells (Simantov and Sachs, 1973). The morphological differentiation of cultured NF mouse neuroblastoma cells is accompanied by the expression of neurone-specific cell surface antigens which is also found in mouse brain (Akerson and Herschman, 1974). Cultured mouse C1300 neuroblastoma cells also possess surface receptors for nerve growth factor (NGF) (Revoltella et al., 1975) and it has been shown that the binding of NGF to synchronized C1300 cells is maximal during the G_1 period (Revoltella et al., 1974a,b).

These results suggest that morphological, antigenic and at least some of the biochemical properties characteristic of the differentiated neuronal state are normally expressed in the pre-synthetic (G_1) period of the cell cycle.

Other examples of cell cycle dependent expression of surface receptors have been described in a variety of cell types cultured in vitro. For example, the susceptibility of JLS-V9 mouse cells to infection with Moloney leukemia virus (presumably a function of the density of cell surface receptors required for the virus adsorption) is maximal during the G_1 and early S period (Gergely et al., 1971). Cell cycle-dependent expression of receptors for melanocyte stimulating hormone (MSH) has been reported by Wong et al. (1974) and Varga et al. (1974). MSH binds to the surface of a mouse melanoma cell line (NCTC 3960, CCL 53) during the G_2 period and initiates a sequence of events leading to a dramatic increase in tyrosinase activity, melanin content and changes in cellular morphology and growth characteristics (Varga et al., 1974; Wong et al., 1974). However, since these authors synchronized the melanoma cells by exposure to colchicine for 36 h, the possibility exists that the cells which reached mitosis early after exposure to colchicine were irreversibly damaged and/or that their capacity to traverse the cell cycle was altered. Other cell synchronization methods are therefore required for confirmation of these findings. Isersky et al. (1975) examined the expression of IgE receptors and basophilic granules in cultured rat basophilic leukemia cells (RBL-1) and found that these two differentiation markers were maximally expressed in stationary phase cultures in which the cells were arrested in the G_1 period. Everhart and Rubin (1974) reported that cytochalasin B inhibits the thymidine uptake in Chinese hamster ovary cells by binding to the cell surface during the G_1 period.

Melanin was one of the first specialized cell products whose production in cultured cell was studied extensively. Explanted chick iris epithelium ceases to produce melanin when placed in medium which favours rapid cell multiplication (Doljanski, 1930). Production of melanin by cultured chick retinal pigment cells is related to the inhibition of cell growth brought about by increase in cell density (Whittaker, 1963). Density-dependent increase of melanin synthesis is not confined merely to cells of non-neoplastic origin. Cultured Syrian hamster melanoma cells have only a few pigment granules in the logarithmic phase of growth but a decrease in cell growth rate is accompanied by an increase in pigment production (Moore, 1964). The inverse relationship between cell growth rate and the production of melanin has also been observed in two mouse melanoma lines, the Hardy-Passey line (Schachtschnabel et al., 1970; Schachtschnabel, 1971) and the sublines of the HFH-8 melanoma (Kitano and Hu, 1971). The rate of [^{14}C]tyrosine incorporation into melanin was also found to be dependent on the growth phase, with maximum incorporation occurring during the stationary phase (Kitano and Hu, 1971). This suggests that changes in melanogenesis during the growth cycle of mouse melanoma cells are due to de novo synthesis of melanin. Cell density-dependent synthesis of melanin in cultured B16 mouse melanoma cells has been demonstrated recently by Kreider and Schmoyer (1975). Similar findings have been made in vivo where melanogenesis in a golden hamster melanoma was found to be inversely proportional to the growth rate (Gray and Pierce, 1964).

Goldberg and Green (1964) studied collagen synthesis in mouse fibroblast lines (3T6, 3T12A and P-3T3-1A) by electron microscopy and hydroxyproline determination. With both methods it was shown that collagen synthesis was not detectable during the logarithmic phase of growth whereas in stationary phase cells collagen accumulated and the hydroxyproline concentration progressively increased. Manner (1971) found, however, that collagen synthesis was more rapid in proliferating than in stationary cultures of human fibroblasts and that the higher ratio of collagen to total proteins synthesized in stationary phase cultures was due not to stimulation of collagen synthesis but because synthesis of other proteins was more repressed than collagen synthesis.

Myogenesis in cultured cells derived from embryonic skeletal muscles closely simulates myogenesis in vivo. Both in vivo and in vitro studies have revealed that morphological markers of muscle cells (cross-striated muscle fibers, myosin and production of creatine kinase) are present only in multinucleated cells (myotubes) which do not synthesize DNA (Stockdale and Holtzer, 1961; Okazaki and Holtzer, 1965; Coleman and Coleman, 1968; Nameroff et al., 1968) and that fusion of mononucleated myoblasts into multinucleated cells occurs about 5 h following the entrance of the cells into the G_1 period (Bischoff and Holtzer, 1969). Merlie et al. (1975) showed that the differentiation markers (actomyosin and creatin kinase) as well as the de novo synthesis of acetylcholine receptors in membranes of cultured myogenic cells from foetal calf increased in parallel with the course of cell fusion.

Cultured chondrocytes from chick vertebrae cease to divide and synthesize chrondroitin sulphate when the cells attain a critical density but detectable synthesis of chondroitin sulphate does not occur in rapidly dividing chondrocytes (Abbot and Holtzer, 1966). Another example of cell density-dependent formation of specialized cell products is the study of Pfeiffer et al. (1970) on the production by C_6 clonal rat glial cell line of a protein unique to nervous system, the "S-100 protein" (named so on the basis of its solubility in 100% saturated ammonium sulphate). Serological determination of S-100 protein indicated that it accumulated only during the stationary phase (when 76% of cells were in the G_1 period). C_6 cells adapted for growth in suspension culture did not accumulate this protein, even when they were maintained in stationary suspension culture for extended periods of time. When C_6 cells were readapted to monolayer growth they regained the ability to form S-100 protein. The changes in the production of S-100 protein paralleled the changes in the amount of an antigen which reacted with antiserum prepared in rabbits against the whole C_6 cells (Pfeiffer et al., 1971).

The morphological differentiation of a mouse myeloma has also been shown to be dependent on cell density. Fully matured plasma cells were observed 8 days after last subculture whereas at earlier stages (2 days after subculture) immature forms of plasma cells predominated (Saunders and Wilder, 1971).

Cell density- and cell cycle-dependent synthesis of cell surface macromolecular components have been studied extensively. Since various aspects of this subject are reviewed elsewhere in this volume only a few representative examples will be mentioned here. Hakomori et al. (1970) found that the chemical quantity of several glycolipids increased in confluent untransformed hamster BHK cells. However, virus-transformed BHK cells had a lower concentration of the glycolipids and no longer exhibited the density-dependent increase of glycolipid concentration found in normal cells. Robbins and Macpherson (1971) studied the glycolipid synthesis in NIL hamster cells by measuring cellular incorporation of [^{14}C]palmitate. A 1.5 to 15.4-fold increase in the incorporation of [^{14}C] from palmitate into various glycolipids was observed in cells from high cell density cultures compared with those from low density cultures. Similar studies were performed by Sakiyama et al. (1972) with a number of clonal cell populations derived from NIL2 hamster cells. In all of the untransformed clones tested, the levels of so called "higher" glycolipids (trihexosyl ceramide, globoside and Forssman antigen) increased by a factor of 2 to 6 when the cells approached saturation density. Kijimoto and Hakomori (1972) observed a similar increase in the net synthesis of Forssman antigen and hematoside in confluent monolayers of NIL hamster cells. This was measured by both the incorporation of [^{14}C]galactose into the glycolipids and by direct chemical assay of the quantities of the specific glycolipids present. A 3- to 5-fold increase in [^{14}C]galactose incorporation was observed in high-density cultures. However, the increase in the quantities of Forssman antigen and hematoside in these cultures was smaller than the incorporation of [^{14}C]galactose, suggesting that both synthesis and degradation were increased on cell contact. Increased binding of radiolabeled anti-Forssman antibody to the surface of crowded NIL cells has been described by Hakomori and Kijimoto (1972). A similar increase in the synthesis of higher glycolipids (ceramide tri-, tetra- and pentahexoside) in dense cultures compared with sparse cultures of NIL2 cells has been reported by Critchley and Macpherson (1973) and Sakiyama and Terasima (1975) described a gradual increase in the concentration of Forssman antigen in monolayer cultures of NIL cells after the cells attained a critical density.

Cell cycle-dependent synthesis of cell surface components has been shown more directly in studies using synchronized cultures. Onodera and Sheinin (1970) synchronized 3T3 cells in the G_1 period by growing them to confluency and then followed the synthesis of a trypsin-labile surface macromolecular glucosamine-containing component. The synthesis of this material began 2 h after plating of freshly trypsinized cells, increased gradually throughout the G_1 period, reached a maximum at about 12 h after plating and declined as the cells entered the S period. An increase of the synthesis of this material was observed during the G_1 period of the next cycle. Similar observations were made by Gerner et al. (1971). These investigators synchronized KB cells by the double thymidine method and measured the incorporation of [^{14}C]leucine, [^{14}C]D-glucosamine and [^{14}C]choline into surface membranes, cell particulates and soluble proteins during synchronous growth. They found a marked increase in the rate of incorporation of the three labelled compounds into surface membranes just after mitosis. Nowakowski et al. (1972) observed maximum incorporation of [^{3}H]leucine into plasma membranes of HeLa cells during the late S phase. The incorporation of [^{14}C]galactose into glycosphingolipids of synchronized KB cells, as well as the levels of gangliosides and total neutral glycosphingolipids, were maximal during the M and G_1 periods (Chatterjee et al., 1973). Wolf and Robbins (1974) monitored the incorporation of radio-labeled palmitate, glucosamine and galactose into phospholipids and glycolipids of NIL cells and showed that incorporation into phospholipids increased gradually as the cells traversed the cell cycle, while maximum labeling of ceramide trihexoside, tetrahexoside and pentahexoside (the Forssman antigen) occurred during the G_1 period. Incorporation of the labels into ceramide monohexoside, tetrahexoside and hematoside did not change during the cell cycle.

The synthesis of a variety of cell products has been studied in synchronized cultures. Buell and Fahey (1969) found maximal synthesis of IgG and IgM in human lymphoblastoid cell lines during the late G_1 and early S period. Byars and Kidson (1970) studied the production of γG_{2a} immunoglobulins in synchronized mouse myeloma cells and found that maximum synthesis occurred during the early S period. In Chang liver cells the levels of catecholamine receptors is low during the S period and reaches maximal values during mitosis (Makman and Klein, 1972).

Virus production has also been found to be dependent on the cell cycle. In chick embryo fibroblast transformed by the Schmidt-Ruppin strain of Rous sarcoma virus, virus release and the synthesis of virus-specific RNA and proteins was restricted to the G_1 and early S periods, respectively (Leong et al., 1972). The authors postulated that a substance was produced during the early G_1 period which was essential for the completion of the virion. Significant differences in virus replication in cells at different stages of the cell cycle in synchronised cultures and between actively replicating logarithmic phase cells and non-replicating stationary phase cells have been documented in a wide range of other virus-host cell systems in vitro (Basilico and Marin, 1966; Carp and Gilden, 1966; Pages et al., 1973; Thorne, 1973; Fiszman et al., 1974; Poste et al., 1974; Kilham and Margolis, 1975; Paskinid et al., 1975).

These results suggest that synthesis of specialized cell products and certain cell surface components, including those carrying specific antigenic determinants, is programmed in the cell genome and in most instances synthesis occurs during the G_1 period.

4. Macromolecular synthesis and the expression of cell surface antigens

One of the ways to approach the problem of regulation of cell surface antigen expression is to use inhibitors of nucleic acid and protein synthesis to block or modify antigen production. This method has been applied to study the regulation in cultured cells of enzyme activities (McAuslan, 1963; Garren et al., 1964; Rosen et al., 1964; Peterkofsky and Tomkins, 1967; Moscona et al., 1968; Martin et al., 1969; Gelehrter and Tomkins, 1970; Horowitz et al., 1970; Thompson et al., 1970; Reif-Lehrer, 1971) and the formation of specialized cell products (Dobbs et al., 1968; Harris, 1968; Ambrose, 1969; Vilček et al., 1969; Stejskalova et al., 1970; Tan et al., 1970). Although prone to many objections, this approach has enlarged our understanding of the cellular regulatory mechanisms involved.

Cikes and Klein (1972b) studied the expression of *H-2* and MV-CSA antigens in cultured murine lymphoma cells treated with metabolic inhibitors. Actinomycin-D (AD) added to logarithmically growing YCAB mouse lymphoma cells increased the proportion of antigen-positive cells and their antibody absorptive capacity. This effect was dose-dependent. Concentrations of AD which inhibited [^{3}H]uridine incorporation by more than 95% over untreated control cells still enhanced the proportion of antigen-positive cells compared to untreated cultures. On the other hand a decrease in antigen expression was paralleled by the inhibition of [^{3}H]leucine incorporation in AD-treated cultures. The AD-induced antigen expression was blocked in a dose-response fashion by inhibitors of protein synthesis (cycloheximide and puromycin). However, inhibitors of DNA synthesis (cytosine arabinoside, hydroxyurea and mitomycin C) had no effect.

The finding of an enhanced antigen expression in cells with suppressed RNA synthesis is suggestive of a posttranscriptional control of antigen synthesis. This is reminiscent of the superinduction by AD of the production of enzymes (McAuslan, 1963; Garren et al., 1964; Rosen et al., 1964; Thompson et al., 1964; Horowitz et al., 1970; Thompson et al., 1970; Reif-Lehrer, 1971), antibodies (Dobbs et al., 1968; Harris, 1968; Ambrose, 1969; Stejskalova et al., 1970), and interferon (Vilček et al., 1969; Tan et al., 1970; Myers and Friedman, 1971; Tan et al., 1971; Vilček and Ny, 1971; Ho et al., 1972) by cultured cells. This paradoxical effect of AD has been explained in the model for regulation of gene activity in cells of higher animals proposed by Tomkins et al. (1969). This model postulates the existence of a short-lived repressor which inactivates the specific messenger RNA. AD is envisaged as preferentially inhibiting the short-lived mRNA for repressor(s) of specific mRNAs. The latter RNAs are relatively long-lived and in the absence of their repressor(s) can thus be fully translated.

Cikes and Klein (1972b) also showed that addition of fresh culture medium to "starved" YCAB lymphoma cells enhanced their cell surface antigen expression. However, this "refeeding" effect was inhibited by AD and by inhibitors of protein synthesis (cycloheximide and puromycin) while three inhibitors of DNA synthesis (cytosine arabinoside, mitomycin C and hydroxyurea) were ineffective. Thus, in "starved" cells, in which the intracellular pools of RNA and proteins were presumably exhausted, both RNA and protein synthesis were required for the re-expression of cell surface antigens.

Turner et al. (1972) enzymatically removed the histocompatibility antigens from the surface of intact human peripheral blood lymphocytes and studied their re-expression in the presence of puromycin and AD. The *HL-A.2* antigenic specificity "stripped" from the cells by the enzyme treatment was fully re-expressed during a 6-h incubation of the cells in fresh culture medium. The re-expression was inhibited completely by puromycin (50 μg/ml). This inhibition was reversible, since puromycin treated cells washed and incubated in fresh culture medium resynthesized the *HL-A.2* antigenic specificity within a 6-h incubation in fresh culture medium. On the other hand, AD concentrations which caused a 98% inhibition of [^{3}H]-uridine incorporation, allowed a 70% reexpression of the *HL-A.2* antigenic content. The inhibition of *HL-A.2* antigen expression by puromycin was interpreted as an indication that the antigen was resynthesized de novo. The failure of AD to inhibit the antigen synthesis also suggested that the synthesis had been taking place on pre-existing templates. Similar kinetics for the re-expression of cell surface antigens has been observed on Meth A ascites fibrosarcoma cells following removal of their *H*-2 antigen by papain (Schwartz and Nathenson, 1971) and for the regeneration of surface sialoglycoproteins on TA3 mouse mammary adenocarcinoma cells following treatment with neuraminidase (Hughes et al., 1972).

Ferrone et al. (1972) observed a reduced *HL-A* antigen expression in puromycin treated WIL_2 cultured human lymphoid cells whereas several other inhibitors of macromolecular synthesis had no effect on the expression of these antigens. The same group of investigators found that AD reduced *HL-A* antigen expression on cultured RAJI cells (a lymphoblastoid cell line derived from a patient with Burkitt's lymphoma) but not on WIL_2 and RPMI 8866 cells. Cycloheximide increased the sensitivity of the three lymphoblastoid cell lines to complement-dependent immune lysis but did not affect their antibody absorptive capacity whereas puromycin reduced *HL-A* antigen expression in these cells (Ferrone et al., 1974). Rubens and Dulbecco (1974) have reported that the cytotoxic effect of chlorambucil on polyoma transformed BK cells (J_1) was potentiated by rabbit antiserum against J_1 cell surface antigens. However, the effect of chlorambucil on the antigen expression and the complement-dependent immune lysis of cells was not examined. Specific inhibitors of macromolecular synthesis and chemotherapeutic drugs used in cancer therapy have been found to increase the susceptibility of cells to complement-dependent immune lysis but this was not due to increased antigen expression or binding of complement components (Segerling et al., 1974, 1975a,b,c).

Since these in vitro findings may have practical applications similar to those of an in vivo model system (Davis et al., 1974), further elucidation of the effects of chemotherapeutic drugs on the sensitivity of cells to specific immune lysis warrants further investigation. The general picture which emerges from the data available is that the physiological state of the cell is an important parameter which determines cellular response to these drugs.

5. Culture- and transformation-induced alterations of cell surface antigen expression

5.1. Alterations of cell surface antigen expression in explanted cells

Histocompatibility antigens are the best defined and most extensively studied of the cell surface antigens. They were found in established cell lines, in some cases as long as 25 years after explantation (Cann and Herzenberg, 1963a; Metzgar et al., 1965; Drysdale et al., 1966; Gangal et al., 1966; Papermaster et al., 1969; Rogentine and Gerber, 1969; Klein et al., 1970) but were absent on certain long established human cancer cell lines (Kersey et al., 1973). However, qualitative rather than quantitative methods were used to demonstrate these antigens and, with the exception of some studies (Papermaster et al., 1969; Rogentine and Gerber, 1969) no comparison was made with an antigenically stable reference cell line.

Cikes et al. (1973) observed a gradual decrease of H-2 antigen content with a concomitant increase of virally determined MV-CSA in three different murine lymphoma lines (YCAB, YAC and WL4) during serial passages in vitro. The cell surface antigen content was estimated by the sensitivity of cells to complement-dependent immune cytolysis, by indirect membrane immunofluorescence and by quantitative antibody absorption tests. Results obtained with the three tests agreed well, and all three tumor lines behaved similarly. The decrease in H-2 and the increase of MV-CSA antigen content was correlated with the number of cell generations in culture. For instance, a long established YCAB lymphoma line (>1000 cell generations in culture) and a recently established YCAB line (60 generations in culture) had 21.3 and 2.8-fold lower H-2 antigen content, respectively, than YCAB ascites cells propagated in the peritoneal cavity of syngeneic A mice. On the other hand, the MV-CSA antigen content increased 3.5-fold on the long established YCAB line compared with a newly established YCAB line which had been propagated in culture for only 50 cell generations. Similar results were obtained with two other murine lymphomas, YAC and WL4. The latter line was a spontaneous lymphoma which arose in an A.SW mouse (H-2^s). After about 100 cell generations in culture, WL4 cells lost the sensitivity to complement dependent lysis with anti-H-2^s (A anti A.SW) serum but acquired a sensitivity to specific immune lysis with anti-MV-CSA serum. The final cell density attained by the three cultured murine lymphoma lines was correlated directly with the number of cell generations in culture, as were the changes in concentration of cell surface antigens discussed above. Thus, there is little doubt that the changes in antigenic properties were correlated with the gradual loss of in vitro growth control mechanisms in the three explanted murine lymphomas. However, the loss of these in vitro growth restraints did not correlate with the tumorigenic potential of these cells in vivo (Cikes, unpublished observations), due most probably to the stronger expression of the virally determined antigens on the cell surface.

Ting and Herberman (1971) have made similar observations on the behaviour of *H-2* and polyoma virus-specific cell surface antigens in polyoma transformed cell lines of C3H/HeN origin. These investigators used the isotopic antiglobulin technique to estimate the amount of the cell surface antigens. They established a polyoma transformed cell line (4198) from the original 4098 line derived from C3H/HeN parotid gland. The 4198V cell line was derived from 4198 cells but was carried for more generations in culture than 4196 cells. Antibody absorption tests showed that 4198V cells had 8.8 times more polyoma specific surface antigens and 4.5 times less *H-2* antigens than 4198 cells. The *H-2* antigen content on the original untransformed 4098 cell line exceeded that on 4198 cells. Hyman et al. (1972) isolated several spontaneously arising mouse myeloma variants in culture but, contrary to the results discussed above, these authors found that variants with reduced concentration of *H-2* antigens (which had also lost the ability to synthesize immunoglobulin and to express plasma cell-specific antigen, PC.1) had reduced amounts of Gross leukemia virus antigen.

Evans et al. (1975) analyzed tumor specific and Forssman antigens on the surface of two diethylnitrosamine-induced guinea pig hepatomas and compared cells propagated in vivo and in vitro with respect to their content of the two types of antigens. Cells propagated in vitro had a greater concentration of both Forssman and tumor-specific antigens than the cells grown in vivo. Jones et al. (1974) found that rat Moloney sarcoma cells (MST) had reduced amounts of rat alloantigens compared with normal syngeneic spleen rat cells and that serial in vitro propagation of these cells was associated with a further reduction of rat alloantigen content and increased expression of Moloney sarcoma virus-specific antigen (Jones and Feldman, 1975). An increase in cell surface antigens reacting with autochthonous patient's serum was detected on cultured lymphoblastoid leukemia cells when compared with fresh patient's lymphoblasts (Belpomme et al., 1969). Klein (1975) has found that cultured methylcholantrene-induced mouse sarcomas express surface tumor specific antigens and that antibodies to these antigens are present in sera of older mice as well as in *H-2*, Ia and Thy-1 typing antisera. Nowinski and Klein (1975) have since shown that the tumor antigens involved are the p15 and gp70 envelope proteins of murine leukemia virus.

Antigenic changes similar to those observed during prolonged in vitro cultivation also occur in tumors during serial propagation in vivo. Haywood and McKhann (1971) studied the antigenic properties of several methylcholantrene-induced sarcomas during 12 to 22 passages in syngeneic C3H mice. The antigenicity of these tumor cells was evaluated by complement-dependent immune cytolysis, quantitative antibody absorption and immunogenicity in syngeneic mice. Quantitative antibody absorption showed that the five different tumors had high, intermediate, or low *H-2* antigen content. The extent of the tumor-specific immune response induced by immunization with tumor cells was inversely proportional to their *H-2* antigen content.

Serial transplantation in syngeneic hosts of IgA-synthesizing plasmocytoma cells from a BALB/c mouse (58-8) was accompanied by gradual loss of $H-2^d$ antigens and cell surface antigens reacting with a rabbit antiserum to 58-8 cells as well as a B lymphocyte surface antigen (Ohno et al., 1975). The authors were able to demonstrate the former two antigens after a proteolytic treatment of the cells.

The reciprocal relationship between the expression of viral and cellular antigens discussed above, might reflect a competition between the two types of antigen for membrane space on the cell surface resulting in topographic exclusion. Alternatively, the competition might take place at a more proximal level, such as in the production and/or assembly of antigenic determinants.

5.2. Reversion of culture-induced antigenic changes by retransplantation of cultured cells into syngeneic hosts

Certain of the antigenic changes induced in tumor cells by cultivation in vitro are reversed when the cells are passaged in syngeneic hosts. The amounts of $H-2$ and MV-CSA in three cultured mouse lymphoma lines (Cikes et al., 1973) showed a tendency to revert to the antigenic pattern characteristic of the corresponding lines propagated in vivo. Cultured WL4-1 lymphoma cells passaged once in the peritoneal cavity of syngeneic A.SW mice expressed levels of $H-2$ and MV-CSA antigens identical to that of WL4 ascites cells. A single passage of cultured YCAB-1 and YAC-1 mouse lymphoma cells in syngeneic mice produced a 10- and a 5-fold increase of $H-2$ antigen content, respectively, while the content of MV-CSA on these in vivo passaged cells was about 10 times lower than that on the corresponding culture lines (Cikes, 1975). The suppression of viral antigens in tumors growing in original hosts has been observed in several experimental systems. Precipitating, cell surface antigens, as well as nuclear (T) and cytoplasmic (C) antigens, were not detectable in transformed hamster cells (TSV_5Cl_2) derived from an SV40-induced hamster tumor when the cells were retransplanted into hamsters. Re-explanted TSV_5Cl_2 cells again became sensitive to the cytotoxic effect of specific antisera and re-expressed both T and C antigens (De Vaux-Saint-Cyr, 1969; 1972).

Alternative passage of Gross leukemia virus transformed cells in vivo and in vitro resulted in "switch off" and "switch on" of gs_1 antigen. Cultured cells were 100% positive for gs_1 antigen whereas the transplanted tumors established in adult rats by inoculation of Gross leukemia virus-transformed cells were negative for this antigen. Re-explanted rat tumor cells became gs_1-positive after 5 to 80 days in culture (Ioachim et al., 1972).

Antibodies to virus specific antigens may be responsible for the reduced expression of viral antigens in tumor cells grown in original hosts but other mechanisms, such as nonimmunologic suppression of the viral gene activity and the selection of pre-existing cells with the antigenic properties found in the in vivo propagated cells cannot at present be excluded.

Explantation of the TA3-Ha mouse ascites adenocarcinoma in suspension culture caused a loss of a high molecular weight cell surface glycoprotein, epiglycanin, which was held to be responsible for transplantability of TA3-Ha cells in allogeneic hosts. The surface epiglycanin reappeared after retransplantation of TA3-Ha cells into syngeneic mice (Miller et al., 1975). Smets and Broekhuysen-Davis (1972) made similar observations in mouse lymphosarcoma cells and demonstrated that a trypsin-sensitive cell surface component which facilitated transplantability in vivo was lost after explantation of the cells. The possibility that the reduced transplantability of the explanted mouse lymphosarcoma cells was due to an increased expression of viral or embryonic antigens in explanted cells was not examined.

5.3. Alterations of cell surface antigen expression in transformed cells

Malignant transformation, both in vitro and in vivo, is often accompanied by quantitative changes in the expression of normal cell surface antigens. Mouse fibroblasts transformed spontaneously by murine sarcoma virus have a higher *H-2* antigen concentration than untransformed parental cells whereas those transformed by SV40 and methylcholantrene show a decreased *H-2* antigen concentration (Tsakraklides et al. 1974). Kersey et al. (1973) were unable to detect any changes in cytotoxic sensitivity to *HL-A* antisera after transformation of human fibroblasts with SV40. Shantz and Lausch (1974) showed by quantitative absorption tests that the expression of Forssman antigen was reduced on virally transformed hamster embryo fibroblasts compared to untransformed parent cells. However, cells of a hamster kidney line (BHK21) which normally have no demonstrable Forssman antigen acquire it after transformation with both RNA- and DNA-containing tumor viruses (Fogel and Sachs, 1962; O'Neill, 1968; Robertson and Black, 1969).

Numerous examples of partial or complete loss of normal cell surface antigens on tumor cells in vivo have been reported. Kay and Wallace (1961) found reduced amounts of blood group antigens on cells from human tumors which arose in the urinary epithelium and demonstrated a partial correlation between the frequency of antigen loss and malignancy of the tumors. Loss of blood group antigens has also been detected on human leukemic cells (Chessin et al., 1965) and on cells of primary and metastatic carcinomas of pancreas (Davidson et al., 1971). EL_4 mouse leukemia cells derived from a chemically induced leukemia and maintained by serial transplantation in syngeneic mice for more than 25 years, lack an antigen present on normal syngeneic lymphoid cells and on the Friend and Rauscher virus induced lymphomas of the same mouse strain (Rubin, 1970). *HL-A* antigens were not detectable on peripheral lymphocytes of a patient with lymphoma during remission of the disease (Seigler et al., 1967, 1971). Baldwin and Glaves (1972) observed the disappearance of a normal liver cell membrane antigen and the concomitant expression of new hepatoma specific antigens in aminoazodyl-induced rat hepatomas. Motta and Bruley (1973) studied *H-2* antigen expression on four chemically induced leukemias in (DBA/2 × C57BL/6) F_1 hybrid mice. In one of these leukemias a reduction of the amounts of all *H-2* specificities was observed. Some specificities were lost on the remaining three leukemias but in one case an increase in some *H-2* specificities was found. There was no correlation between the extent of *H-2* antigen expression and the transplantability on the four leukemias in vivo. Birch et al. (1975) reported decreased *H-2* antigen expression on a series of chemically and irradiation induced mouse leukemias.

Fibroblast cell surface antigen (Wartiovaara et al., 1974; Vaheri et al., 1976) and a large molecular weight cell surface protein (synonyms: LETS protein; galactoprotein A; Z protein; band 1 protein; see review by Hynes, 1976) found on normal cells of a wide variety of species (Hynes, 1973; Wickus and Robbins, 1973; Hogg, 1974; Gahmberg et al., 1974; Stone et al., 1974) are reported to be partially or completely lost in transformed cells (review Hynes, 1976).

Transformation induced acquisition of neoantigens has generally received more attention than the alterations of normal cell components in malignant cells. Tumor specific neoantigens are often functionally inactive as targets for host immunosurveillance response and hence of minor importance in control of tumor growth in vivo. The immunologic approach to the control of neoplasia is further confounded by immune enhancement of tumor growth (Kallis, 1962; Prehn, 1972) and production of "blocking factors" (Sjögren et al., 1971). As shown in this section, malignant transformation is frequently associated with a decrease or loss of normal cell components. Apart from the immediate practical aspects of these findings (e.g. in tissue typing), their relevance to the control of tumor growth in vivo will become more meaningful only when the effects of malignant transformation on the cell components involved in interaction with growth and differentiation regulating factors are elucidated.

6. Conclusions

The expression of cell surface antigens in cultured cells is a dynamic process. In random cultures antigen expression is dependent on the phase of cell growth (lag, log and stationary phase) which, in turn, determines the position of cells within the cell cycle. The cell cycle dependence of antigen expression has been shown convincingly in studies with synchronized cultures. Various mechanisms may be involved in the variations of antigen expression during the cell cycle but several lines of evidence support the hypothesis that changes in the synthesis of antigenic determinants are an important mechanism underlying these variations. Studies with inhibitors of macromolecular synthesis suggest that the regulation of antigen expression may take place at transcriptional or posttranscriptional level, depending on the physiological state of the cell. Finally, malignant transformation is accompanied by qualitative and quantitative changes in cell surface antigenic composition. Decrease or loss of normal cell surface components is frequently observed in transformed cells and this may well prove to be more relevant to the control of tumor growth in vivo than the acquisition of tumor specific neoantigens.

References

Abbot, J. and Holtzer, H. (1966) The loss of phenotypic traits by differentiated cells, III. The reversible behaviour of chondrocytes in primary cultures. J. Cell Biol. 28, 473–487.

Akerson, R. and Herschman, H. (1974) Neural antigens of morphologically differentiated neuroblastoma cells. Nature (London) 249, 620–623.

Ambrose, Ch.T. (1969) Regulation of the secondary antibody response in vitro. Enhancement by Actinomycin D and inhibition by a macromolecular product of stimulated lymph node cultures. J. Exp. Med. 130, 1003–1029.

Baker, J.B. and Humphreys, T. (1971) Serum-stimulated release of cell contacts and the initiation of growth in contact-inhibited chick fibroblasts. Proc. Natl. Acad. Sci. USA 68, 2161–2164.

Baldwin, R.W. and Glaves, D. (1972) Deletion of liver-cell surface membrane components from amino-azodye-induced rat hepatomas. Int. J. Cancer 9, 76–85.

Basch, R.S. (1974) Effects of antigen density and non-complement fixing antibody on cytolysis by alloantisera. J. Immunol. 113, 554–562.

Basilico, C. and Marin, G. (1966) Susceptibility of cells in different stages of the cell cycle to transformation by polyoma virus. Virology 28, 429–437.

Becker, H., Stanners, C.P. and Kudlow, J.E. (1971) Control of macromolecular synthesis in proliferating and resting Syrian hamster cells in monolayer culture, II. Ribosome complement in resting and early G_1 cells. J. Cell Physiol. 77, 43–50.

Becker, Y. and Levitt, J. (1968) Stimulation of macromolecular processes in BSC_1 cells due to medium replenishment. Exp. Cell Res. 51, 27–33.

Bellanger, F., Jullien, M. and Harrel, L. (1970) Inhibition de proximité. Libération par les cellules d'inhibiteur de synthèse du DNA et des protéines. C.R. Acad. Sci. Paris, Sér. D, 270, 2232–2235.

Belpomme, D., Seman, G., Doze, J.-F., Veanut, A-M., Berumen, L., Le Borgne de Kaouel, C. and Mathe, G. (1969) Established cell line (ICI 101) obtained from frozen human lymphoblastic leukemia cells: Comparison with the patient's fresh cells. Eur. J. Cancer 5, 55–59.

Ben-Bassat, H., Inbar, M. and Sachs, L. (1971) Changes in the structural organization of the surface membrane in malignant cell transformation. J. Membrane Biol. 6, 183–194.

Birch, J.M., Moore, M. and Craig, A.W. (1975) Cell surface antigen expression on chemically induced murine leukemias. Br. J. Cancer 31, 630–640.

Bischoff, R. and Holtzer, H. (1969) Mitosis and the process of differentiation of myogenic cells in vitro. J. Cell. Biol. 41, 188–200.

Blume, A., Gilbert, G., Wilson, S., Farber, J., Rosenberg, R. and Nirenberg, M. (1970) Regulation of acetylcholine in neuroblastoma cells. Proc. Natl. Acad. Sci. USA 67, 786–792.

Borsos, T. and Rapp, H.J. (1965) Complement fixation on cell surfaces by 19S and 7S antibodies. Science 150, 505–506.

Boyle, M.D.P., Ohanian, S.A. and Borsos, T. (1975) Lysis of tumor cells by antibody and complement, III. Lack of correlation between antigen movement and cell lysis. J. Immunol. 115, 473–475.

Buell, D.H. and Fahey, J.L. (1969) Limited periods of gene expression in immunoglobulin synthesizing cells. Science, 164, 1524–1525.

Burger, M.M. (1969) A difference in the architecture of the surface membrane of normal and virally transformed cells. Proc. Natl. Acad. Sci. USA 62, 994–1001.

Burger, M.M. and Goldberg, A.R. (1967) Identification of a tumor-specific determinant on neoplastic cell surface. Proc. Natl. Acad. Sci. USA 57, 359–366.

Burk, K.H., Drewinko, B., Lichtiger, B. and Trujillo, J.M. (1976) Cell cycle dependency of human sarcoma associated tumor antigen expression. Cancer Res. 36, 3535–3538.

Byars, N., and Kidson, C. (1970) Programmed synthesis and export of immunoglobulin by synchronized myeloma cells. Nature (London) 226, 648–650.

Cann, H.M. and Herzenberg, L.A. (1963a) In vitro studies of mammalian somatic cell variations, I. Detection of H-2 phenotype in cultured mouse cell lines. J. Exp. Med. 117, 259–265.

Cann, H.M. and Herzenberg, L.A. (1963b) In vitro studies of mammalian somatic cell variation, II. Isoimmune cytotoxicity with a cultured mouse lymphoma and selection of resistant variants. J.

Exp. Med. 117, 267–283.

Carp, R.I. and Gilden, R.V. (1966) A comparison of the replication cycles of simian virus 40 in human diploid and African green monkey kidney cells. Virology 28, 150–162.

Chatterjee, S., Sweeley, Ch.S. and Velicer, L.F. (1973) Biosynthesis of proteins, nucleic acids and glycosphingolipids by synchronized KB cells. Biochim. Biophys. Acta 54, 585–592.

Chessin, L.N., Bramson, S., Kuhns, W.J. and Hirschhorn, K. (1965) Studies of the A, B, O(H) blood groups of human cells in culture. Blood 25, 944–953.

Cikes, M. (1970a) Antigenic expression of a murine lymphoma during growth in vitro. Nature (London) 225, 645–647.

Cikes, M. (1970b) Relationship between growth rate, cell volume, cell cycle kinetics and antigenic properties of cultured murine lymphoma cells. J. Natl. Cancer Inst. 45, 979–988.

Cikes, M. (1971a) Variations in expression of surface antigens on cultured cells. Ann. N.Y. Acad. Sci. 177, 190–200.

Cikes, M. (1971b) Expression of surface antigens on cultured tumor cells in relation to cell cycle. Transplantation Proc. 3, 1161–1166.

Cikes, M. (1975) Antigenic changes in cultured murine lymphomas after retransplantation into syngeneic hosts. J. Natl. Cancer Inst. 54, 903–906.

Cikes, M. and Friberg Jr., S. (1971) Expression of H-2 and Moloney leukemia virus-determined cell-surface antigens in synchronized cultures of a mouse cell line. Proc. Natl. Acad. Sci. USA 68, 566–569.

Cikes, M. and Klein, G. (1972a) Quantitative studies of antigen expression in cultured murine lymphoma cells, I. Cell-surface antigens in "asynchronous" cultures. J. Natl. Cancer Inst. 49, 1599–1606.

Cikes, M. and Klein, G. (1972b) Effects of inhibitors of protein and nucleic acid synthesis on the expression of H-2 and Moloney leukemia virus-determined cell-surface antigens on cultured murine lymphoma cells. J. Natl. Cancer Inst. 48, 509–515.

Cikes, M., Friberg Jr., S. and Klein, G. (1972) Quantitative studies of antigen expression in cultured murine lymphoma cells, II. Cell-surface antigens in synchronized cultures. J. Natl. Cancer Inst. 49, 1607–1611.

Cikes, M., Friberg Jr., S. and Klein, G. (1973) Progressive loss of H-2 antigens with concomitant increase of cell-surface antigen(s) determined by Moloney leukemia virus in cultured murine lymphoma. J. Natl. Cancer Inst. 50, 347–362.

Cline, N.J. and Livingston, D.C. (1971) Binding of ^{3}H-Concanavalin A by normal and transformed cells. Nature (London) 232, 155–156.

Coleman, J.R. and Coleman, A.W. (1968) Muscle differentiation and macromolecular synthesis. J. Cell Physiol. 72, Suppl. 1, 19–34.

Critchley, D.R. and MacPherson, I. (1973) Cell density dependent glycolipids in NIL_2 hamster cells, derived malignant and transformed cell lines. Biochim. Biophys. Acta 296, 145–159.

Currie, G.A. (1967) Masking of antigens on Landschütz ascites tumour. Lancet II, 1336–1338.

Currie, G.A. and Bagshave, K.D. (1968) The effect of neuraminidase on the immunogenicity of the Landschütz ascites tumour. Site and mode of action. Br. J. Cancer 22, 588–594.

Davidson, I., Ni, L.J. and Stejskal, R. (1971) Tissue isoantigens A, B, and H in carcinoma of the pancreas. Cancer Res. 31, 1244–1250.

Davis, D.A.L., Buckham, S. and Manstone, A.J. (1974) Protection of mice against syngeneic lymphomata, II. Collaboration between drugs and antibodies. Br. J. Cancer 30, 305–311.

DeLuca, C. (1966) Effects of mode of culture and nutrient medium on cyclic variations in enzyme activities of mammalian cells cultured in vitro. Exp. Cell Res. 43, 39–50.

De Mars, R. (1964) Some studies of enzymes in cultured human cells. Natl. Cancer Inst. Monogr. 13, 181–195.

De Vaux-Saint-Cyr, Ch. (1969) Variations de l'antigénicité des fibroblastes de Hamster transformés par le SV40 puis clonés (TSV_5 Cl_2) selon qu'ils sont maintenus en culture de tissu ou propagés chez l'animal. C.R. Acad. Sci., Paris, Sér. D, 269, 1148–1150.

De Vaux-Saint-Cyr, Ch. (1972) Modulation antigénique et modulation de synthèse de certains antigènes. Ann. Inst. Pasteur, 122, 603–607.

Dobbs, J., Rivero, I., Sabb, F. and Lee, S.L. (1968) Enhancement of antibody production after treatment with actinomycin-D: Interrelationships between 7S and 19S antibody. Immunology, 14, 213–223.

Doljanski, L. (1930) Sur le rapport entre la prolifération et l'activité pigmentogène dans les cultures d'épithelium de l'iris. C.R. Soc. Biol. 105, 343–345.

Drysdale, R.G., Merchant, D.J., Schreffler, D.C. and Parker, F.R. (1966) Distribution of H-2 specificities within the L-M mouse cell line and derived lines. Proc. Soc. Exp. Biol. Med. 124, 413–418.

Eagle, H. (1968) Growth-regulatory effects of cellular interaction in vitro, and their relevance to cancer. In: Proliferation and Spread of Neoplastic Cells: Twenty-First Annual Symposium on Fundamental Cancer Research, William and Wilkins, Baltimore, pp. 7–20.

Edidin, M. and Henney, C.S. (1973) The effect of capping H-2 antigens on the susceptibility of target cells to humoral and T cell-mediated lysis. Nature (London) New Biol. 246, 47–49.

Enger, M.D. and Tobey, R.A. (1969) RNA synthesis in Chinese hamster cells, II. Increase in rate of RNA synthesis during G_1. J. Cell Biol. 42, 308–315.

Enger, M.D. and Tobey, R.A. (1972) Effects of isoleucine deficiency on nucleic acid and protein metabolism in cultured Chinese hamster cells. Continued ribonucleic acid and protein synthesis in the absence of deoxyribonucleic acid synthesis. Biochemistry, 11, 269–277.

Evans, Ch.H., Ohanian, H. and Cooney, A.M. (1975) Tumor-specific and Forssman antigens of guinea-pig hepatoma cells: Comparison of tumor cells grown in vivo and vitro. Int. J. Cancer 15, 512–521.

Everhart, L.P. and Rubin, R.W. (1974) Cyclic changes in the cell surface, I. Change in thymidine transport and its inhibition by cytochalasin B in Chinese hamster ovary cells. J. Cell Biol. 60, 437–441.

Everson, L.K., Plocinik, B.A. and Rogentine Jr., G.N. (1974) HL-A expression on the G_1, S and G_2 cell-cycle stages of human lymphoid cells. J. Natl. Cancer Inst. 53, 913–920.

Ferrone, S., Del Villano, B., Pellegrino, M.A., Lerner, R.A. and Reisfeld, R.A. (1972) Expression of HL-A antigens on the surface of cultured lymphoid cells: Effects of inhibitors of protein and nucleic acid synthesis. Tissue Antigens 2, 477–453.

Ferrone, S., Cooper, N.R., Pellegrino, M.A. and Reisfeld, R.A. (1973) Interaction of histocompatibility (HL-A) antibodies and complement with synchronized human lymphoid cells in continuous culture. J. Exp. Med. 137, 55–68.

Ferrone, S., Pellegrino, M.A., Dierich, M.P. and Reisfeld, R.A. (1974) Effect of inhibitors of macromolecular synthesis on HL-A antibody mediated lysis of cultured lymphoblasts. Tissue Antigens 4, 275–282.

Fischer, A. (1946) Biology of Tissue Cells. Steckert, New York.

Fiszman, M., Reynier, M., Bucchini, D. and Girard, M. (1974) Retarded growth of poliovirus in contact inhibited cells. J. Gen. Virol. 23, 73–82.

Fogel, M. and Sachs, L. (1962) Studies on the antigenic composition of hamster tumors induced by polyoma virus and of normal hamster tissue in vivo and in vitro. J. Natl. Cancer Inst. 29, 239–252.

Fox, T.O. and Pardee, A.B. (1972) Animal cells: Noncorrelation of length of G_1 phase with size after mitosis. Science 167, 80–82.

Fox, T.O., Sheppard, J.R. and Burger, M.M. (1971) Cyclic membrane changes in animal cells: Transformed cells permanently display a surface architecture detected in normal cells only during mitosis. Proc. Natl. Acad. Sci. USA 68, 244–247.

Franks, D. (1967) Antigenic heterogeneity in cultures of mammalian cells. In Vitro, 2, 74–81.

Franks, D. (1968) Antigens as markers on cultured mammalian cells. Biol. Rev. Camb. Phil. Soc. 43, 17–50.

Franks, D. and Dawson, A. (1966) Variation in the expression of blood group antigen A in clonal cultures of rabbit cells. Exp. Cell Res. 42, 543–561.

Friberg Jr., S. (1972) Comparison of an immunoresistant and an immunosusceptible ascites subline from murine tumor TA3, II. Immunosensitivity and antibody-binding capacity in vitro, and immunogenicity in allogeneic mice. J. Natl. Cancer Inst. 48, 1477–1489.

Friberg Jr., S. and Lilliehöök, B. (1973) Evidence for non-exposed H-2 antigens in immunoresistant murine tumor. Nature (London) New Biol. 241, 112–114.

Gahmberg, C.G. and Hakomori, S. (1974) Organization of glycolipids and glycoproteins in surface membranes: Dependency on cell cycle and on transformation. Biochem. Biophys. Res. Commun. 59, 283–291.

Gahmberg, C.G. and Hakomori, S. (1975) Surface carbohydrates of hamster fibroblasts, I. Chemical characterization of surface-labeled glycosphingolipids and specific ceramide tetrasaccharide for transformants. J. Biol. Chem. 250, 2438–2446.

Gahmberg, C.G., Kiehn, D. and Hakomori, S. (1974) Changes in a surface-labelled galactoprotein and in glycolipid concentration in cells transformed by a temperature-sensitive polyoma virus mutant. Nature (London) 248, 413–415.

Gangal, S.G., Merchant, D.J. and Schreffler, D.C. (1966) Characterization of the H-2 antigens of L-M mouse cells grown in culture. J. Natl. Cancer Inst. 36, 1151–1159.

Garren, L.D., Howell, R.R., Tomkins, G.M. and Grocco, R.M. (1964) A paradoxical effect of Actinomycin D: The mechanism of regulation of enzyme synthesis by hydrocortisone. Proc. Natl. Acad. Sci. USA 52, 1121–1129.

Gelehrter, T.D. and Tomkins, G.M. (1970) Posttranscriptional control of tyrosine aminotransferase synthesis by insulin. Proc. Natl. Acad. Sci. USA 66, 391–397.

Gergely, L., Cikes, M., Klein, E., Fenyö, E.M. and Friberg Jr., S. (1971) Sensitivity of JLS-V9 cells to Moloney leukemia virus in relation to cell cycle. Exp. Cell Res. 64, 230–232.

Gerner, E.W., Glick, M.C. and Warren, L. (1971) Membranes of animal cells, V. Biosynthesis of the surface membrane during the cell cycle. J. Cell Physiol. 75, 275–279.

Glinos, A.D. and Werrlein, R.J. (1972) Density-dependent regulation of growth in suspension culture of L-929 cells. J. Cell Physiol. 79, 79–90.

Goetze, D., Pellegrino, M.A., Ferrone, S. and Reisfeld, R.A. (1972) Expression of H-2 antigens during growth of cultured tumor cells. Immunol. Commun. 1, 533–544.

Goldberg, B. and Green, H. (1964) An analysis of collagen secretion by established mouse fibroblast lines. J. Cell Biol. 22, 227–258.

Goldschneider, I. and Barton, R.W. (1976) Development and differentiation of lymphocytes. In: G. Poste and G.L. Nicolson (Eds.), The Cell Surface in Animal Development, Cell Surface Reviews, Vol. 1, North Holland, Amsterdam, pp. 599–696.

Gray, J.M. and Pierce, G.B. (1964) Relationship between growth rate and differentiation of melanoma in vivo. J. Natl. Cancer Inst. 32, 1201–1211.

Hakomori, S.J. (1970) Cell density-dependent changes of glycolipid concentrations in fibroblasts, and loss of this response in virus-transformed cells. Proc. Natl. Acad. Sci. USA 67, 1741–1747.

Hakomori, S. and Kijimoto, S. (1972) Forssman reactivity and cell contacts in cultured hamster cells. Nature (London) New Biol. 239, 87–88.

Harris, G. (1968) Antibody production in vitro, II. Effects of actinomycin D and puromycin on the secondary response to sheep erythrocytes. J. Exp. Med. 127, 675–691.

Harris, M. (1964) Cell Culture and Somatic Variation. Holt, Rinehart and Winston, New York.

Harris, M. (1968) Phenotypic expression and cell marker systems. Natl. Cancer Inst. Monogr. 29, 1–7.

Haywood, G.D. and McKhann, Ch.F. (1971) Antigenic specificities of murine sarcoma cells. Reciprocal relationship between normal transplantation antigen (H-2) and tumor-specific immunogenicity. J. Exp. Med. 133, 1171–1187.

Herrmann, H., Marchok, A.C. and Baril, E.F. (1967) Growth rate and differentiated functions of cells. Natl. Cancer Inst. Monogr. 26, 303–326.

Hill Jr., R.B., Bensch, K.G., Simbonis, S. and King, D.W. (1959) Variability of content of deoxyribonucleic acid in L-strain fibroblasts. Nature (London) 183, 1818–1819.

Ho, M., Tan, Y.H. and Armstrong, J.A. (1972) Accentuation of production of human interferon by metabolic inhibitors. Proc. Soc. Exp. Biol. Med. 139, 259–262.

Holley, R.W. and Kiernan, J.A. (1968) "Contact inhibition" of cell division in 3T3 cells. Proc. Natl. Acad. Sci. USA 60, 300–304.

Hogg, M.A. (1974) A comparison of membrane proteins of normal and transformed cells by lactoperoxidase labeling. Proc. Natl. Acad. Sci. USA 71, 489–492.

Horowitz, N.H., Feldman, H.M. and Pall, M.L. (1970) Depression of tyrosinase synthesis in neurospora by cycloheximide, actinomycin D, and puromycin. J. Biol. Chem. 245, 2784–2788.

Huet, Ch., Gresser, J., Bandu, M.T. and Lindahl, P. (1974) Increased binding of Concanavalin A to interferon-treated murine leukemia L 1210. Proc. Soc. Exp. Biol. Med. 147, 52–57.

Hughes, R.C., Sanford, B. and Jeanloz, R.W. (1972) Regeneration of the surface glycoproteins of a transplantable mouse tumor cell after treatment with neuraminidase. Proc. Natl. Acad. Sci. USA 69, 942–945.

Humphrey, J.H. and Dourmashkin, R.R. (1965) Electron microscope studies of immune cell lysis. In: Ciba Symposium on Complement, Churchill, London, pp. 175–186.

Hyman, R., Ralph, P. and Sarkar, S. (1972) Cell-specific antigens and immunoglobulin synthesis of murine myeloma cells and their variants. J. Natl. Cancer Inst. 148, 173–184.

Hynes, R.O. (1973) Alteration of cell-surface proteins by viral transformation and proteolysis. Proc. Natl. Acad. Sci. USA 70, 3170–3174.

Hynes, R.O. (1976) Cell surface proteins and malignant transformation. Biochim. Biophys. Acta 458, 73–107.

Hynes, R.O. and Bye, J.M. (1974) Density and cell cycle dependence of cell surface proteins in hamster fibroblasts. Cell 3, 113–120.

Inbar, M. and Sachs, L. (1969a) Structural difference in sites on the surface membrane of normal and transformed cells. Nature (London) 223, 710–712.

Inbar, M. and Sachs, L. (1969b) Interaction of the carbohydrate-binding protein concanavalin A with normal and transformed cells. Proc. Natl. Acad. Sci. USA 63, 1418–1425.

Ioachim, H.L., Dorsett, D., Sabbath, M. and Keller, S. (1972) Loss and recovery of phenotypic expression of Gross leukemia virus. Nature (London) New Biol. 237, 215–218.

Isersky, Ch., Metzger, H. and Buell, D. N. (1975) Cell cycle-associated changes in receptors for IgE during growth and differentiation of a rat basophilic leukemia cell line. J. Exp. Med. 141, 1147–1162.

Ishizaka, T., Fada, T. and Ishizaka, K. (1968) Fixation of C_{1a} by rabbit G- and M-antibodies with particulate and soluble antigens. J. Immunol. 100, 1145–1153.

Jones, J.M. and Feldman, J.D. (1975) Alloantigen expression of a rat Moloney sarcoma. J. Natl. Cancer Inst. 55, 995–999.

Jones, J.M., Jensen, F., Veit, B. and Feldman, J.D. (1974) In vivo growth and antigenic properties of a rat sarcoma induced by Moloney sarcoma virus. J. Natl. Cancer Inst. 52, 1771–1777.

Kallis, N. (1962) The elements of immunologic enhancement. A consideration of mechanisms. Ann. N.Y. Acad. Sci. 101, 64–79.

Karb, K. and Goldstein, G. (1971) Combination autoradiography and membrane fluorescence in studying cell cycle transplantation antigen relationships. Transplantation 11, 569–573.

Kates, J.R., Winteron, R. and Schlesinger, K. (1971) Induction of acetylcholinesterase activity in mouse neuroblastoma tissue culture cells. Nature (London) 229, 345–347.

Kay, H.E.M. and Wallace, D.M. (1971) A and B antigens of tumors arising from urinary epithelium. J. Natl. Cancer Inst. 26, 1349–1365.

Kersey, J.H., Yunis, E.J., Todaro, G.J. and Aaronson, S.A. (1973) HL-A antigens of human tumor derived cell lines and viral transformated fibroblasts in culture. Proc. Soc. Exp. Biol. Med. 143, 453–456.

Kijimoto, S. and Hakomori, S.J. (1972) Contact-dependent enhancement of net synthesis of Forssman glycolipid antigen and hematoside in NIL cells at the early stage of cell-to-cell contact. FEBS Lett. 25, 38–42.

Killander, D. and Zetterberg, A. (1965) A quantitative cytochemical investigation of the relationship between cell mass and initiation of DNA synthesis in mouse fibroblasts in vitro. Exp. Cell Res. 40, 12–20.

Killander, D., Klein, E. and Levine, A. (1974) Expression of membrane-bound IgM and HL-A antigens on lymphoblastoid cells in different stages of the cell cycle. Eur. J. Immunol. 4,

327–332.

Kitano, Y. and Hu, F. (1971) Proliferation and differentiation of pigment cells in vitro. J. Invest. Dermatol. 55, 444–451.

Klebe, R.J. and Ruddle, F.H. (1969) Neuroblastoma: Cell culture analysis of a differentiating stem cell system. J. Cell Biol. 43, 69a.

Klein, D., Merchant, D.J., Klein, J. and Schreffler, D.C. (1970) Persistence of H-2 and some non-H-2 antigens on long-term-cultured mouse cell lines. J. Natl. Cancer Inst. 44, 1149–1160.

Klein, P.A. (1975) Anomalous reaction of mouse alloantisera with cultured tumor cells, I. Demonstration of widespread occurrence using reference typing sera. J. Immunol. 115, 1254–1260.

Kreider, J.W. and Schmoyer, M.E. (1975) Spontaneous maturation and differentiation of B16 melanoma cells in culture. J. Natl. Cancer Inst. 55, 641–647.

Kuhns, W.J. and Bramson, S. (1968) Variable behaviour of blood group H on HeLa cell populations synchronized with thymidine. Nature (London) 219, 938–939.

Leong, J.A., Levinson, W. and Bishop, J.M. (1972) Synchronization of Rous sarcoma virus production in chick embryo cells. Virology 47, 133–141.

Lerner, R.A., Oldstone, M.B. and Cooper, N.R. (1971) Cell cycle-dependent immune lysis of Moloney virus-transformed lymphocytes: Presence of viral antigen, accessibility to antibody, and complement activation. Proc. Natl. Acad. Sci. USA 68, 2584–2588.

Lesley, J., Hyman, R. and Dennert, G. (1974) Effect of antigen density on complement-mediated lysis, T-cell-mediated killing, and antigen modulation. J. Natl. Cancer Inst. 53, 1759–1765.

Levine, E.M., Becker, Y., Boone, Ch.W. and Eagle, H. (1965) Contact inhibition of macromolecular synthesis and polyribosomes in cultured human diploid fibroblasts. Proc. Natl. Acad. Sci. USA 53, 350–356.

Levintow, L. and Eagle, H. (1961) Biochemistry of cultured mammalian cells. Ann. Rev. Biochem. 30, 605–640.

Ley, K.D. and Tobey, R.A. (1970) Regulations of initiation of DNA synthesis in Chinese hamster cells, II. Induction of DNA synthesis and cell division by isoleucine and glutamine in G_1-arrested cells in suspension culture. J. Cell Biol. 47, 453–459.

Lindahl, P., Leary, P. and Gresser, I. (1973) Enhancement by interferon of the expression of surface antigens on murine leukemia L 1210 cells. Proc. Natl. Acad. Sci. USA 70, 2785–2788.

Makman, M.H. and Klein, M.J. (1972) Expression of adenylate cyclase, catecholamine receptor, and cyclic adenosine monophosphate-dependent protein kinase in synchronized cultures. Proc. Natl. Acad. Sci. USA 69, 456–458.

Manner, G. (1971) Cell division and collagen synthesis in cultured fibroblasts. Exp. Cell Res. 65, 49–60.

Mannick, J.A., Graziani, J.T. and Edgahl, R.H. (1964) A transplantation antigen recovered from cell culture medium. Transplantation 2, 321–333.

Martin Jr., D.W., Tomkins, G.M. and Bresler, M.A. (1969) Control of specific gene expression examined by synchronized mammalian cells. Proc. Natl. Acad. Sci. USA 63, 842–849.

McAuslan, B.R. (1963) The induction and repression of thymidine kinase in the Poxvirus-infected HeLa cells. Virology, 21, 383–389.

Merlie, J.P., Sobel, A., Changeux, J.P. and Gros, F. (1975) Synthesis of acetylcholine receptor during differentiation of cultured embryonic muscle cells. Proc. Natl. Acad. Sci. USA 72, 4028–4032.

Metzgar, R.S., Flanagan, J.F. and Zmijewski, C.M. (1965) Detection of tissue isoantigens on primary and serial heteroploid cell lines of human origin by the mixed agglutination technique. J. Immunol. 95, 494–500.

Miller, D.K., Cooper, A.G., Brown, M.C. and Jeanloz, R.W. (1975) Reversible loss in suspension culture of a major cell-surface glycoprotein of the TA3-HA mouse tumor. J. Natl. Cancer Inst. 55, 1249–1252.

Möller, E. and Möller, G. (1962) Quantitative studies of the sensitivity of normal and neoplastic mouse cells to the cytotoxic action of isoantibodies. J. Exp. Med. 115, 527–533.

Moore, G.E. (1964) In vitro cultures of a pigmented hamster melanoma cell line. Exp. Cell Res. 36, 422–423.

Moscona, A.A., Moscona, M.M. and Saenz, N. (1968) Enzyme induction in embryonic retina: The role of transcription and translation. Proc. Natl. Acad. Sci. USA 61, 160–167.

Motta, R. and Bruley, M. (1973) Quantitative study of the histocompatibility antigens on the surface of normal and leukemic cells in mice, I. Variations in the expression of groups of H-2 specificities in four leukemias induced by 7,12-dimethylbenz(a)anthracene. Transplantation 15, 22–30.

Myers, H.W. and Friedman, R.M. (1971) Potentiation of human interferon production by superinduction. J. Natl. Cancer Inst. 47, 757–764.

Nameroff, M.A., Reznik, M., Anderson, P. and Hansen, J.L. (1968) Differentiation and control of mitosis in a skeletal muscle tumor. Cancer Res. 30, 596–600.

Nicolson, G.L. (1971) Difference in topology of normal and tumor cell membranes shown by different surface distribution of ferritin-conjugated concanavalin A. Nature (London) New Biol. 233, 244–246.

Nicolson, G.L. (1974) The interactions of lectins with animal cell surfaces. Int. Rev. Cytol. 39, 89–190.

Nilausen, K. and Green, H. (1965) Reversible arrest of growth in G_1 of an established fibroblast line (3T3). Exp. Cell Res. 40, 166–168.

Noonan, K.D., Levine, A.J. and Burger, M.M. (1973) Cell cycle-dependent changes in the surface membrane as detected with [^{3}H] Concanavalin A. Fed. Proc. 58, 491–497.

Nowakowski, M., Atkinson, P.H. and Summers, D.F. (1972) Incorporation of fucose into HeLa cells plasma membranes during the cell cycle. Biochim. Biophys. Acta 266, 154–160.

Nowinski, R.C. and Klein, P.A. (1975) Anomalous reactions of mouse alloantisera with cultured tumor cells, II. Cytotoxicity is caused by antibodies to leukemia viruses. J. Immunol. 115, 1261–1268.

Ohanian, S.H., Borsos, T. and Rapp, H.J. (1973) Lysis of tumor cells by antibody and complement, I. Lack of correlation between antigen content and lytic susceptibility. J. Natl. Cancer Inst. 50, 1313–1320.

Ohno, S., Natsu-Ume, S. and Migita, S. (1975) Alteration of cell-surface antigenicity of the mouse plasmocytoma, I. Immunologic characterization of surface antigens masked during successive transplantations. J. Natl. Cancer Inst. 55, 569–577.

Okazaki, K. and Holtzer, H. (1965) An analysis of myogenesis in vitro using fluorescein-labeled antimyosin. J. Histochem. Cytochem. 13, 726–739.

O'Neill, C.H. (1968) An association between viral transformation and Forssman antigen detected by immune adherence in cultured BHK 21 cells. J. Cell Sci. 3, 405–419.

Onodera, K. and Sheinin, R. (1970) Macromolecular glucosamine-containing component on surface of cultivated mouse cells. J. Cell. Sci. 7, 337–355.

Ozanne, B. and Sambrook, J. (1971) Binding of radioactively labelled Concanavalin A and wheat germ agglutinin to normal and virus transformed cells. Nature (London) 232, 156–160.

Pages, J., Mantenil, S., Stehelin, D., Fiszman, M., Marx, M. and Girard, M. (1973) Relationship between replication of simian virus 40 DNA and specific events in the host cell cycle. J. Virol. 12, 99–107.

Papermaster, W.M., Papermaster, B.W. and Moore, G.E. (1969) Histocompatibility antigens of human lymphocytes in long term culture. Fed. Proc. 28, 379.

Paskinid, M.P., Weinberg, R.A. and Baltimore, D. (1975) Dependence of Moloney leukemia virus production on cell growth. Virology 67, 242–248.

Pasternak, C.A., Warmsley, A.M.H. and Thomas, D.B. (1971) Structural alterations in the surface membrane during the cell cycle. J. Cell Biol. 50, 562–564.

Pellegrino, M.A., Ferrone, S., Natali, G.G. and Reisfeld, R.A. (1972) Expression of HL-A antigens in synchronized cultures of human lymphocytes. J. Immunol. 108, 573–576.

Pellegrino, M.A., Ferrone, S., Pellegrino, A. and Reisfeld, R.A. (1973) The expression of HL-A antigens during the growth cycle of cultured human lymphoid cells. Clin. Immunol. Immunopathol. 1, 182–189.

Pellegrino, M.A., Ferrone, S., Cooper, N.R., Dierich, M.P. and Reisfeld, R.A. (1974) Variation in susceptibility of a human lymphoid cell line to immune lysis during the cell cycle. Lack of correlation with antigen density and complement binding. J. Exp. Med. 140, 578–590.

Peterkofsky, B. and Tomkins, G.M. (1967) Effects of inhibitors of nucleic acid synthesis on steroid-mediated induction of tyrosine aminotransferase in hepatoma cell cultures. J. Mol. Biol. 30, 49–61.

Pfeiffer, S.E., Herschman, H.R., Lightboy, J. and Sato, G. (1970) Synthesis by a clonal line of rat glial cells of a protein unique to the nervous system. J. Cell Physiol. 75, 329–339.

Pfeiffer, S.E., Herschman, J.E., Lightboy, J.E., Sato, G. and Levine, L. (1971) Modification of cell surface antigenicity as a function of culture conditions. J. Cell. Physiol. 78, 145–152.

Porter, K., Prescott, D. and Frye, J. (1973) Changes in surface morphology of Chinese hamster ovary cells during the cell cycle. J. Cell Biol. 57, 815–836.

Poste, G., Schaeffer, B., Reeve, P. and Alexander, D.J. (1974) Rescue of simian virus 40 (SV40) from SV40-transformed cells by fusion with anucleate monkey cells and variation in the yield of virus rescued by fusion with replicating or non-replicating monkey cells. Virology 60, 85–95.

Prehn, R.T. (1972) The immune reaction as a stimulator of tumor growth. Science 176, 170–171.

Ray, P.K. and Simmons, R.L. (1971) Failure of neuraminidase to unmask allogeneic antigens on cell surface. Proc. Soc. Exp. Biol. Med. 138, 600–604.

Reif-Lehrer, L. (1971) Actinomycin-D enhancement of glutamine synthetase activity in chick embryo retina cultures in the presence of cortisol. J. Cell Biol. 51, 303–311.

Revoltella, R., Bertolini, L. and Pediconi, M. (1974a) Unmasking of nerve growth factor membrane-specific binding sites in synchronized murine C1300 neuroblastoma cells. Exp. Cell Res. 85, 89–94.

Revoltella, R., Bertolini, L., Pediconi, M. and Vigneti, E. (1974b) Specific binding of nerve growth factor (NGF) by murine C1300 neuroblastoma cells. J. Exp. Med. 140, 437–451.

Revoltella, R., Bosman, C. and Bertolini, L. (1975) Detection of nerve growth factor binding sites on neuroblastoma cells by rosette formation. Cancer Res. 35, 890–895.

Robbins, E. and Scharff, M.D. (1967) The absence of a detectable G_1 phase in a cultured strain of Chinese hamster lung cell. J. Cell Biol. 34, 684–686.

Robbins, P.W. and MacPherson, I. (1971) Glycolipid synthesis in normal and transformed cells. Proc. Roy. Soc. Lond. Ser. B 177, 49–58.

Robertson, H.T. and Black, P.H. (1969) Changes in surface antigens of SV40 virus-transformed cells. Proc. Soc. Exp. Biol. Med. 130, 363–370.

Rogentine Jr., G.N. and Gerber, P. (1969) HL-A antigens of human lymphoid cells in long-term tissue culture. Transplantation 8, 28–37.

Rosen, F., Raina, P.M., Milholland, R.J. and Nichol, Ch.A. (1964) Induction of several adaptive enzymes by Actinomycin D. Science 146, 661–663.

Rosenberg, R.N., Vandeventer, L., De Francesco, L. and Friedkin, M.E. (1971) Regulation of the synthesis of choline-O-acetyltransferase and thymidylate synthetase in mouse neuroblastoma in cell culture. Proc. Natl. Acad. Sci. USA 68, 1436–1440.

Rosenfeld, C., Dore, J.F., Choquet, C., Venaut, A.-M., Ajuria, E., Marholev, L. and Vastiaux, J.P. (1973a) Variations in expression of cell membrane antigens by cultured cells. A study of antilymphocytic globulin binding by human lymphoblastoid cells before, during, and after their establishment as culture lines. Transplantation 16, 279–286.

Rosenfeld, C., Dore, J.F., Choquet, C., Venaut, A.-M., Vastiaux, J.-P., Guibout, C. and Pico, J.L. (1973b) Variations au cours du cycle cellulaire de l'expression des antigènes de membranes décelés par des globulines anti-lymphocytes après synchronisation des cellules d'une lignée lymphoblastoïde. C.R. Acad. Sci. Paris Sér. D 277, 2829–2832.

Rubens, R.D. and Dulbecco, R. (1974) Augmentation of cytotoxic drug action by antibodies directed at cell surface. Nature (London) 248, 81–82.

Rubin, D.J. (1970) Antigenic loss in a transplantable, chemically induced leukemia of C57BL/6 mice. J. Natl. Cancer Inst. 44, 975–979.

Ruddle, F.H. and Rapola, J. (1970) Changes in lactate dehydrogenase and esterase specific activities, isozymic patterns, and cellular distribution during the growth cycle of PK cells in vitro. Exp. Cell Res. 59, 399–412.

Sakiyama, H. and Terasima, T. (1975) The synthesis of Forssman glycolipid in clones of Nil 2 hamster fibroblasts grown in monolayer or spinner culture. Cancer Res. 35, 1723–1726.

Sakiyama, H., Gross, S.K. and Robbins, P.W. (1972) Glycolipid synthesis in normal and virus-transformed hamster cell line. Proc. Natl. Acad. Sci. USA 69, 872–876.

Sanford, B.H. (1967) An alteration in tumor histocompatibility induced by neuraminidase. Transplantation 5, 1273–1279.

Sanford, B.H. and Codington, J.F. (1971) Alteration of the tumor cell surface by neuraminidase (Discussion) Transpl. Proc. 3, 1155–1156.

Saunders, G.C. and Wilder, M. (1971) Repetitive maturation cycles in a cultured mouse myeloma. J. Cell Biol. 51, 344–348.

Schachtschnabel, D. (1971) Speziphische Zellfunktionen von Zell- und Gewebekulturen, I. Züchtung von Melanin-bildenden Zellen des Haroling-Passey-Melanomas in Monolayer-Kultur. Virchows Arch. Abt. Zellpathol. 7, 27–36.

Schachtschnabel, D., Fischer, R.-D. und Zilliken, F. (1970) Speziphische Zellfunktionen von Zell- und Gewebekulturen, II. Untersuchungen zur Kontrolle der Melanin-Synthese in Zellkulturen des Harding-Passey-Melanomas. Z. Physiol. Chem. 351, 1402–1410.

Schlesinger, M. and Gottesfeld, S. (1971) The effect of neuraminidase on expression of cellular antigens. Transplantation Proc. 3, 1151–1154.

Schwartz, B.D. and Nathenson, S.G. (1971) Regeneration of transplantation antigens on mouse cells. Transplantation Proc. 3, 180–182.

Scott, R.E., Furcht, L.T. and Kersey, J.H. (1973) Changes in membrane structure associated with cell contact. Proc. Natl. Acad. Sci. USA 73, 3631–3635.

Seeds, H.W., Gilman, A.G., Amano, T. and Nirenberg, N.W. (1970) Regulation of axon formation by clonal lines of a neural tumor. Proc. Natl. Acad. Sci. USA 66, 160–167.

Segerling, M., Ohanian, S.H. and Borsos, T. (1974) Effect of metabolic inhibitors on killing of tumor cells by antibody and complement. J. Natl. Cancer Inst. 53, 1411–1413.

Segerling, M., Ohanian, S.H. and Borsos, T. (1975a) Chemotherapeutic drugs increase killing of tumor cells by antibody and complement. Science 188, 55–57.

Segerling, M., Ohanian, S.H. and Borsos, T. (1975b) Enhancing effect by metabolic inhibitors on the killing of tumor cells by antibody and complement. Cancer Res. 35, 3195–3203.

Segerling, M., Ohanian, S.A. and Borsos, T. (1975c) Effect of metabolic inhibitors on the ability of tumor cells to express antigen and bind complement components C4 and C3. Cancer Res. 35, 3204–3208.

Sela, B., Lis, H., Sharon, N. and Sachs, L. (1970) Different locations of carbohydrate-containing sites in the surface membrane of normal and transformed mammalian cells. J. Membrane Biol. 3, 267–269.

Sela, B., Lis, H., Sharon, N. and Sachs, L. (1971) Quantitation of N-acetylgalactosamine-like sites on the surface membrane of normal and transformed cells. Biochim. Biophys. Acta 249, 564–568.

Shantz, G.D. and Lausch, R.N. (1974) Variation in quantitative expression of Forssman antigen on virus-transformed hamster cells. J. Natl. Cancer Inst. 53, 239–246.

Shipley, W.V. (1971) Immune cytolysis in relation to the growth cycle of Chinese hamster cells. Cancer Res. 31, 925–929.

Shoham, J. and Sachs, L. (1974) Different cyclic changes in the surface membrane of normal and malignant transformed cells. Exp. Cell Res. 85, 8–14.

Simantov, R. and Sachs, L. (1973) Regulation of acetylcholine receptors in relation to acetylcholinesterase in neuroblastoma cells. Proc. Natl. Acad. Sci. USA 70, 2902–2905.

Singer, S.J. and Nicolson, G.L. (1972) The fluid model of the structure of cell membranes. Science 175, 720–731.

Sjögren, H.D., Hellström, I., Bansal, S.C. and Hellström, K.E. (1971) Suggestive evidence that the "blocking antibodies" of tumor-bearing individuals may be antigen-antibody complexes. Proc. Natl. Acad. Sci. USA 68, 1372–1375.

Smets, L.A. (1973) Agglutination with Con A dependent on cell cycle. Nature (London) New Biol. 245, 113–115.

Smets, L.A. and Broekhuysen-Davies, J. (1972) Shielding of antigens and concanavalin A agglutinin sites by a surface coat of transplantable mouse lymphosarcoma cells. Eur. J. Cancer 8, 541–548.

Smets, L.A. and DeLey, L. (1974) Cell cycle dependent modulation of the surface membrane of normal and SV40 virus transformed 3T3 cells. J. Cell. Physiol. 84, 342–348.

Stejskalova, V., Ivanyi, J. and Kara, J. (1970) Enhancement of antibody synthesis in vitro by 6-azauridine, uridine and actinomycin D. Folia Biol. (Praha) 16, 250–258.

Stockdale, F.E. and Holtzer, H. (1961) DNA synthesis and myogenesis. Exp. Cell Res. 24, 508–520.

Stone, K.R., Smith, R.E. and Joklik, W.K. (1974) Changes in membrane polypeptides that occur when chick embryo fibroblasts and NRK cells are transformed with avian sarcoma viruses. Virology 58, 86–100.

Summer, M.C.B., Collin, R.C.L.S. and Pasternak, C.A. (1973) Synthesis and expression of surface antigens during the cell cycle. Tissue Antigens 3, 477–484.

Tan, Y.H., Armstrong, J.A., Ke, Y.H. and Ho, M. (1970) Regulation of cellular interferon production. Enhancement by anti-metabolites. Proc. Natl. Acad. Sci. USA 67, 464–471.

Tan, Y.H., Armstrong, J.A. and Ho, M. (1971) Accentuation of interferon production by metabolic inhibitors and its dependence on protein synthesis. Virology 44, 503–509.

Taylor, R.B., Duffus, P.H., Raff, M.C. and de Petris, S. (1971) Redistribution and pinocytosis of lymphocyte surface immunoglobulin molecules induced by antiimmunoglobulin antibody. Nature (London) New Biol. 233, 225–229.

Thomas, D.B. (1971) Cyclic expression of blood group determinants in murine cells and their relationship to growth control. Nature (London) 233, 317–321.

Thomas, D.B. (1974) The i antigen complex: a new specificity unique to dividing human cells. Eur. J. Immunol. 4, 819–824.

Thomas, D.B. and Phillips, B. (1973) Membrane antigens specific for human lymphoid cells in the dividing phase. J. Exp. Med. 138, 64–70.

Thompson, E.B., Granner, D.K. and Tomkins, G.M. (1970) Superinduction of tyrosine aminotransferase by Actinomycin D in rat hepatoma (HTC) cells. J. Mol. Biol. 54, 159–175.

Thorne, H.V. (1973) Cyclic variation in the susceptibility of BALB/c 3T3 cells to polyoma virus. J. Gen. Virol. 18, 163–169.

Ting, C.C. and Herberman, R.B. (1971) Inverse relationship of polyoma tumor specific cell surface antigen to H-2 histocompatibility antigens. Nature (London) New Biol. 232, 118–120.

Tobey, R.A. and Ley, K.D. (1970) Regulation of initiation of DNA synthesis in Chinese hamster cells, I. Production of stable, reversible G_1-arrested populations in suspension cultures. J. Cell Biol. 46, 151–157.

Todaro, G.J., Lazar, G.K. and Green, H. (1965) The initiation of cell division in a contact-inhibited mammalian cell line. J. Cell. Comp. Physiol. 66, 325–334.

Todo, A., Strife, A., Fried, J. and Clarkson, B.D. (1971) Proliferative kinetics of human hematopoietic cells during different growth phases in vitro. Cancer Res. 31, 1330–1340.

Tomkins, G.M., Gelehrter, T.D., Granner, D., Martin Jr., D., Samuels, H.H. and Thompson, E.B. (1969) Control of specific gene expression in higher organisms. Science 166, 1474–1480.

Tsakraklides, E., Smith, C., Kersey, J.H. and Good, R.A. (1974) Transplantation antigens (H-2) on virally and chemically transformed BALB/3T3 fibroblasts in culture. J. Natl. Cancer Inst. 52, 1499–1504.

Turner, M.J., Strominger, J.L. and Sanderson, A.R. (1972) Enzymic removal and re-expression of histocompatibility antigen, HL-A.2 at the surface of human peripheral lymphocytes. Proc. Natl. Acad. Sci. USA 69, 200–202.

Vaheri, A., Ruoslahti, E., Westermark, B. and Pontén, J. (1976) A common cell-type specific surface antigen in cultured human glial cells and fibroblasts: Loss in malignant cells. J. Exp. Med. 143, 64–72.

Varga, J.M., Dipasquale, A., Pawelek, J., McGuire, J.S. and Lerner, A.B. (1974) Regulation of melanocyte stimulating hormone action at the receptor level: Discontinuous binding of hormone to synchronized mouse melanoma cells during the cell cycle. Proc. Natl. Acad. Sci. USA 71, 1590–1593.

Vilček, J. and Ny, H.M. (1971) Post-transcriptional control of interferon synthesis. Virology 7, 588–594.
Vilček, J., Rossman, T.G. and Varacalli, F. (1969) Differential effects of Actinomycin D and puromycin on the release of interferon induced by double stranded RNA. Nature (London) 222, 682–683.
Warmsley, A.M.H. and Pasternak, C.A. (1970) The use of conventional and zonal centrifugation to study the life cycle of mammalian cells. Biochem. J. 119, 493–499.
Wartiovaara, J., Linder, E., Ruoslahti, E. and Vaheri, A. (1974) Distribution of fibroblast surface antigen. Association with fibrillar structures of normal cells and loss upon viral transformation. J. Exp. Med. 140, 1522–1533.
Whittaker, J.R. (1963) Changes in melanogenesis during the differentiation of chick pigment cells in cell culture. Dev. Biol. 8, 99–127.
Wickus, G.G. and Robbins, P.W. (1973) Plasma membrane proteins of normal and Rous sarcoma virus-transformed chick embryo fibroblasts. Nature (London) New Biol. 245, 65–67.
Wolf, B.A. and Robbins, P.W. (1974) Cell mitotic cycle synthesis of NIL hamster glycolipids including the Forssman antigen. J. Cell Biol. 61, 676–687.
Wong, G., Pawelek, J., Sansone, M. and Morowitz, J. (1974) Response of mouse melanoma cells to melanocyte stimulating hormone. Nature (London) 248, 351–354.
Wright, B.S., O'Brien, P.A., Shibley, G.P. and Mayyasi, S.A. (1967) Infection of an established mouse bone marrow cell line (JLS-V9) with Rauscher and Moloney murine leukemia viruses. Cancer Res. 27, 1672–1677.
Yeh, J. and Fischer, H.W. (1969) A diffusible factor which sustains contact inhibition of replication. J. Cell Biol. 40, 382–388.
Zetterberg, A. (1970) Nuclear and cytoplasmic growth during interphase in mammalian cells. Adv. Cell Biol. 1, 211–232.
Zwerner, R.K. and Acton, R.T. (1975) Growth properties and alloantigenic expression of murine lymphoblastoid cell lines. J. Exp. Med. 142, 378–390.

Somatic genetic analysis of the surface antigens of murine lymphoid tumors

Robert HYMAN*

1. Rationale for the somatic analysis of surface antigens

The differentiation of a complex organism from a single-celled zygote requires the selective activation and inactivation of numerous genes (Bennett et al., 1972; Lengerová, 1972; Deucher, 1975). The aim of developmental biology is to work out the specific genetic program(s) resulting in a particular differentiated cell. One logical approach to achieving this objective is that of identifying molecules which distinguish one cell type from another and then determining the genes which code for these molecules and studying the factors which regulate gene expression. The problem with respect to the cell surface is illustrated in Fig. 1. Here we ask whether, given two "different" cells A and B, we can demonstrate unique cell surface molecules A and B, characteristic for each cell type. If so, then is it possible to identify and locate the structural genes A and B coding for these molecules? Also, is it possible to define other genes X, Y and O, P whose products regulate the expression of structural genes A and B, respectively and to determine the nature of this regulation?

The cells of the immune system provide a useful model for studying differentiation in this way. These cells all originate from a common stem cell in the bone marrow (Wu et al., 1968; Owen et al., 1973; Goldschneider and Barton, 1976). This stem cell, which initially has the potential to differentiate into cells of the erythroid, myeloid, and lymphocytic systems, undergoes progressive restriction in differentiative potential. The lymphoid subline differentiates along two pathways (Fig. 2). One path gives rise to the antigen-sensitive precursor of the antibody-producing plasma cell. The second leads to a variety of cell types derived from cells of the thymus. These include helper cells involved in the triggering of antigen sensitive cells, suppressor cells involved in regulating antibody synthesis, and effector cells involved in cell-mediated killing, graft versus host reactions, and delayed hypersensitivity.

Are there molecules characteristic of cells at given stages along the two pathways? One way to approach this question is immunologically – by at-

---

*Scholar of the Leukemia Society of America.

G. Poste & G.L. Nicolson (eds.) *Dynamic Aspects of Cell Surface Organization*
© Elsevier/North-Holland Biomedical Press, 1977

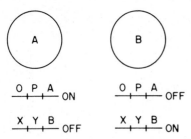

Fig. 1. Diagrammatic representation of cell surface differentiation. A and B represent two different cells bearing distinct surface antigens A and B coded for by structural genes A and B. X, Y and O, P represent regulatory genes whose products regulate the expression of structural genes A and B, respectively.

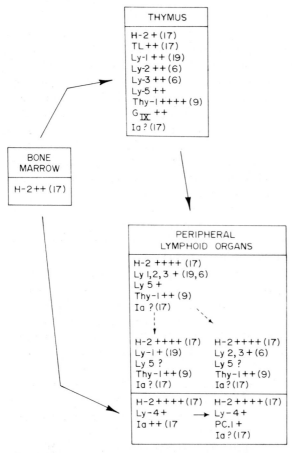

Fig. 2. Surface alloantigens expressed by cells of the lymphoid system. Arrows represent pathways of differentiation; dotted arrows are inferred but not conclusively proven. + to ++++ represent increasing amounts of antigen; ? indicates that the presence of the antigen is uncertain. Numbers in parentheses represent chromosome numbers on which the structural genes for the indicated antigens are borne.

tempting to obtain antisera specific for a given cell type and, more particularly, specific for the *surface* of a given class of cells (Bennett et al., 1972). This strategy has both the advantages and disadvantages of any abstract approach. Thus, it is possible to define cell surface antigens specific for particular subpopulations of the lymphoid system (Boyse and Old, 1969; Boyse, 1973; Cantor and Boyse, 1975) without knowing anything about the molecular structure of the antigens. However, no matter how precise the serology, it is impossible to go much further than descriptive cataloguing without further genetic and biochemical experimentation.

Genetic analysis implies that allelic genes exist which code for molecules possessing phenotypic differences. Heteroimmunization, in which cells from one species are used to immunize a second species, is therefore not as useful as alloimmunization in which the immunization is made between individuals of the same species (see Bennett et al., 1972). This fact has restricted much of the work on the genetics of cell surface antigens to the mouse, where a wide variety of inbred strains are available. Alloimmunization and breeding studies have defined a number of lymphoid cell surface antigens and have localized the structural genes coding for some of these alloantigenic determinants (Fig. 2). Other genes must regulate antigen expression. Also, the products of additional structural genes may contribute to the structure of the molecule bearing the antigenic determinant. For example, genes may code for glycosyltransferases which attach carbohydrate residues to a protein antigen. Obtaining mutants for regulatory genes and for many structural genes in the whole animal may be difficult, since mutations leading to gross alterations in, or nonexpression of, a surface antigen might be expected to be lethal in vivo where the presence of the surface antigen may be necessary for proper cell function or interaction with other cells.

An alternative approach to the further genetic analysis of surface antigens is to study tumor cells growing in vitro and to use somatic genetic approaches to isolate and characterize variants showing altered expression of surface antigens. This approach has the advantage that it may be possible to study mutants which would be lethal in the whole animal due to inability of the cell to carry out its proper function in vivo. It also obviates the possibility of immunological selection against cells with altered surface properties. The use of tumor cells also exploits the ability of these cells to lose or rearrange chromosomes (Ohno, 1971). Such losses and rearrangements may reveal new patterns of antigen expression that would not be seen in the normal counterparts of these cells in vivo. Also, since the tumor cell may be monosomic for all or a portion of a given chromosome (see section 4.2), it may be possible to use appropriate selections to isolate recessive mutations.

Cell hybridization enables further analysis of particular genes. Complementation analysis carried out by fusing independent cell variants with the same antigenic phenotype can be used to define genes. The pattern of antigenic expression in hybrids between variant and parental cells can be used to decide among a number of explanations for a given variant phenotype (Table I).

TABLE I

INTERPRETATIONS OF RESULTS OF HYBRIDS BETWEEN PARENTAL (ANTIGEN-POSITIVE) AND VARIANT (ANTIGEN-NEGATIVE) CELL LINES EACH FROM MOUSE STRAINS EXPRESSING DIFFERENT ALLELIC FORMS OF A GIVEN SURFACE ANTIGENIC DETERMINANT

Result	Interpretation of variant phenotype
(1) Both allelic forms absent	(1A) A dominant regulatory mutation resulting in repression of antigen synthesis in the variant.
	(1B) Activation in the variant of a gene which codes for a product that masks or degrades the antigenic determinant.
(2) Only the parental allelic form is expressed.	(2A) A mutation in the structural gene or a cis acting regulatory mutation affecting the structural gene coding for the allelic determinant characteristic of the strain from which the variant cell was derived.
(3) Both allelic forms present	(3A) A mutation in the structural gene or a cis acting regulatory mutation affecting the structural gene coding for a gene product necessary for the synthesis of the determinant but not comprising the determinant itself (e.g., a precursor of the determinant; a carbohydrate necessary for integration of the molecule bearing the antigenic determinant into the cell membrane).
	(3B) A mutation in a regulatory gene coding for the repressor of a structural gene (normally repressed in the antigen-expressing parent) whose product masks or degrades the determinant.
	(3C) A mutation in a gene coding for a trans-acting activator necessary for expression of the structural gene coding for the determinant.

Difficulties in interpretation may arise because of the possibilities of chromosome loss in hybrids and the inability to perform segregation analyses (Davis and Adelberg, 1973). Nevertheless, these problems can often be minimized by judicious selection of test systems and by coupling genetic experiments with biochemical analysis of the gene product.

Nearly all of the work on the somatic genetic analysis of the surface antigens of tumor cells has been done with cells of the murine lymphoid system. The reasons for concentration on these cells are both intellectual – the pathways of differentiation are reasonably well understood in this system (Fig. 2), and thus a genetic approach is both feasible and interesting – and technical – the cells of this system are particularly well suited for immunogenetic analysis (Boyse and Old, 1969). In what follows, our current knowledge of the serology, biochemistry and in vivo genetics of several antigens which have been used as model antigens for in vitro studies will be briefly summarized. With this background, surface antigen variation in tumor cell populations will then be described, and the question of whether the basis of this variation is genetic will be discussed. The number and nature of the alterations leading to a given variant phenotype will be considered. Finally, the potential application of somatic cell genetic analysis to the study of tumor-associated antigens will be discussed.

2. Characteristics of murine lymphoid cell surface antigens which have been used in somatic genetic studies

2.1. H-2

The major histocompatibility gene complex of the mouse (H-2) actually refers to a number of closely linked genes located on chromosome 17 which code for distinct gene products (J. Klein, 1975). In addition to the genes comprising the H-2 complex, two other genes present on chromosome 17, the T/t locus and Tla, code for gene products which bear biochemical similarities to those coded for by the genes of the H-2 complex (Artzt and Bennett, 1975; Vitetta et al., 1975).

The entire H-2 complex occupies about 0.5 map unit and presumably contains several hundred genes of which only a few marker loci have been identified (J. Klein and Shreffler, 1971; J. Klein et al., 1974; J. Klein, 1975). The two regions at either end of the H-2 complex (K and D) contain the genes coding for the serologically detectable H-2 antigens. The I region comprises a group of genes involved in control of immune responsiveness as well as genes coding for the serologically detectable Ia antigens. The S region contains the structural gene for a protein found in serum which has been identified as murine complement component C4 (Curman et al., 1975; Lachman et al., 1975; Meo et al., 1975). There is also evidence for a fifth region, G, located between S and D that determines a serologically detectable erythrocyte antigen (David et al., 1975; J. Klein et al., 1975).

Biochemically, the serologically detectable H-2 antigens (K and D) are glycoproteins (Muramatsu and Nathenson, 1970; Schwartz and Nathenson, 1971). The basic subunit has a molecular weight of about 45 000 (Schwartz et al., 1973). Each monomer has one or more molecules with a molecular weight of 12–14 000 daltons associated noncovalently with it; the small molecule is assumed to represent the murine analogue of the human β_2-microglobulin (Silver and Hood, 1974; Peterson et al., 1975; Vitetta et al., 1975a,b; Silver and Hood, 1976). It has been suggested that the H-2 antigen is present in the membrane as a four-chain structure (molecular weight ~120 000 daltons) composed of two disulfide bonded 45 000 molecular weight units and two noncovalently associated β_2-microglobulin-like units (Peterson et al., 1975). While other workers also find such structures (Vitetta et al., 1975a), some do not and caution that the apparent disulfide bonding found between monomers may be due to disulfide interchange during antigen purification (Snary et al., 1975). Alkylation with iodiacetamide prior to extraction with detergent markedly reduces the amount of dimer obtained (Henning et al., 1976), also arguing that the antigen may be present as monomers in the cell membrane. The 45 000 molecular weight subunit possesses two carbohydrate chains each with a molecular weight of 3300 daltons but differing in net charge (Muramatsu and Nathenson, 1970).

Double precipitation studies (Cullen et al., 1972a,b) and antigen redistribution studies (Neauport-Sautes et al., 1973; Hauptfeld and Klein, 1975) indicate that each *H-2* haplotype codes for two separate serologically detectable molecules (K and D). Each molecule bears a serological specificity characteristic for the given haplotype (private) and a number of specificities shared among different haplotypes (public). The biochemical studies thus support the genetic results (see J. Klein and Shreffler, 1971; J. Klein, 1975) in which each *H-2* haplotype is found to contain two distinct genes coding for separate molecules (K and D) bearing the serologically detectable *H-2* antigens. The alleles of each gene are defined serologically by a given private specificity for each allele of the K and D loci. Each haplotype, therefore, is characterized by two private specificities. For example, the *H-2*^{k} haplotype has as marker loci the K^k gene, governing private specificity 23, and the D^k gene, governing private specificity 32.

The *H-2* specificities appear to reflect differences in the amino acid sequence of the molecule rather than differences in carbohydrate. *H-2* antigens from different haplotypes have significant differences in their peptide maps (Shimada et al., 1970) and peptide profiles as determined by molecular sieve chromatography (Brown et al., 1974), but little gross difference in carbohydrate (although detailed sequence studies have not been done) (Muramatsu and Nathenson, 1970). Removal of 25% of the glucosamine, 70% of the galactose, and 100% of the sialic acid from the purified *H-2* antigen has no significant effect on antigenic activity (Nathenson and Muramatsu, 1971). Also, glycopeptides isolated from the purified molecule have no inhibitory effect on anti-*H-2* sera (Nathenson and Muramatsu, 1971).

A beginning has been made toward sequencing a number of K and D gene products (Henning et al., 1976; Silver and Hood, 1976; Vitetta et al., 1976). The results show that the products of the K and D loci are clearly related, supporting a common evolutionary origin, although, paradoxically, there is no evidence that K gene products are more closely related to one another than they are to the D gene products (Silver and Hood, 1976). Such a relationship might be expected if K and D genes diverged from one another before mammalian speciation occurred. Such divergence has been thought to have occurred since the major histocompatibility complex of the human

(HL-A) has similar genetic properties to the H-2 complex of the mouse and codes for serologically detectable antigens with similar structural properties (Strominger et al., 1974; Peterson et al., 1975) and amino acid sequence homologies (Terhorst et al., 1976) to the products of the H-2 complex. The allelic products of the K or D genes differ from one another in multiple amino acid residues (Silver and Hood, 1976), supporting the conclusion that the antigenic specificities reflect amino acid sequence differences.

The Ss protein coded by the Ss locus in the S region of the H-2 complex has been reported to have a similar structure to the H-2 antigen (Capra et al., 1975), while the serologically detectable Ia antigens specified by genes in the I region are glycoproteins with molecular weights of about 30 000 daltons (Cullen and Nathenson, 1974; Freed et al., 1975).

2.2. TL

The TL complex of antigens is determined by locus (Tla) in chromosome 17 about one centimorgan distal (with respect to the centromere) to H-$2D$ (Boyse et al., 1965). Four serologically distinguishable specificities have been defined (Boyse et al., 1969). Different Tla haplotypes determine particular patterns of the serologically defined specificities, which are expressed concomitantly on normal thymocytes (Table II). TL negative mouse strains do not express serologically detectable TL specificities, and it is not clear whether or not they have a non-antigenic analogue of the TL molecule. The serologically detectable TL specificities are normally found only on thymocytes. The antigen is not found on the peripheral thymus-derived cells of the lymph nodes and spleen, on thymic precursor cells derived from the bone marrow, nor on non-lymphoid tissues (Boyse et al., 1965; Schlesinger and Hurvitz, 1968; Owen and Raff, 1972). Within the thymus there is a TL negative subpopulation which represents the immunologically competent cells of this organ (Leckband and Boyse, 1971). The TL antigen, therefore, can be considered to be a "differentiation alloantigen" (Boyse and Old, 1969), i.e. an antigen which is expressed only at a particular stage along the pathway of differentiation.

TABLE II
EXPRESSION OF TL ANTIGEN ON NORMAL AND LEUKEMIC MOUSE CELLS (FROM BOYSE et al., 1969)

TL specificities expressed on		
Normal thymocytes	Leukemias	Mouse strains
1, 2, 3	1, 2, 3	A, C58
2	1, 2	BALB/c
2	1, 2, 4	DBA/2
—	1, 2, 4	C57BL/6

Leukemic cells which express TL may show an altered pattern of TL specificities as compared to the TL phenotype of the corresponding normal thymocytes (Table II). Since the leukemias of all strains express TL specificities 1 and 2, it has been argued that the structural genes for at least these two antigens must be present in all mouse strains and that associated "regulatory" genes determine TL expression or non-expression on the normal thymocytes of a given mouse strain (Schlesinger, 1970; Boyse and Old, 1971). Since the regulatory genes determining TL expression or non-expression are the genes which are linked to H-2, it is, in theory, possible that the structural genes for TL are located elsewhere in the mouse genome (Schlesinger, 1970). There are compelling, although not conclusive, arguments, however, that the structural genes for TL are also linked to the H-2 locus (Boyse and Old, 1971).

It is not certain whether the individual TL specificities represent distinct molecules or whether the individual specificities are determinants present on a single molecule. Treatment with monospecific anti-TL3 leads to loss of cytotoxic sensitivity to anti-TL2 or anti-TL1,3 (Old et al., 1968). Loss of cytotoxic sensitivity (antigenic modulation) in the TL system is correlated with capping and endocytosis of TL antigen if divalent antibody is used for modulation (Stackpole et al., 1974; Loor et al., 1975) indicating that modulation can be considered as the correlate of antigen redistribution methods based on fluorescence (see Hauptfeld and Klein, 1975). Thus, if one assumes that antigens which move together in the cell membrane are on the same molecule, the ability of anti-TL3 to modulate TL1 and TL2, argues that all three determinants are on the same molecule. Sequential precipitation experiments using purified TL antigens, as have been done for H-2 (Cullen et al., 1972a,b), will be necessary to settle the question of whether the TL specificities represent one or several molecules.

The molecule(s) bearing the TL specificities is a glycoprotein (Muramatsu et al., 1973; Yu and Cohen, 1974) with many resemblances to H-2. A protein with a molecular weight of 40–50 000 daltons can be precipitated by anti-TL sera from NP40 extracts of cells (Vitetta et al., 1972; Yu and Cohen, 1974). This 40–50 000 molecular weight protein is non-covalently associated with a β_2-microglobulin-like subunit having a molecular weight of ~12–14 000 daltons (Ostberg et al., 1975; Vitetta et al., 1975a). There is evidence that two of the larger chains are disulfide bonded to give a four chain structure with a molecular weight of 120 000 daltons, similar to that proposed for H-2 (Anundi et al., 1975). However, in view of the results of Snary et al. (1975), the same cautions as mentioned in the case of H-2 apply to this model (see section 2.1). A carbohydrate chain with a molecular weight of 4500 daltons can be isolated from the 40–50 000 molecular weight subunit (Muramatsu et al., 1973).

Although the H-2 and TL antigens show common structural characteristics, the two molecules are clearly not identical since anti-TL precipitation leaves all the H-2 in the supernatant (Yu and Cohen, 1974). One possibility is that TL and H-2 arose by gene duplication of a common precursor gene. A second possibility, that TL is a modified H-2 molecule, has not been formally disproven but seems unlikely since there is no serological cross reactivity between the two antigens.

The characteristic expression of different TL specificities on the thymocytes of various mouse strains and the activation of new specificities in a defined pattern on leukemic cells (Table II) raises the question of how the expression of the individual TL specificities is regulated. In the absence of exact knowledge of the molecular nature of the TL specificities and with only a limited amount of genetic data available, it is not possible to develop a precise model for TL antigen regulation. It seems most reasonable to think of the TL antigen as being determined by a series of closely linked regulatory and structural genes. One possible way of looking at things is illustrated in Table III, which is presented as a way of organizing what data are available and not as a formal proposal of the way things really are. It is assumed in Table III that each mouse strain carries the structural genes coding for TL1, TL2 and either TL3, TL4, or a null allele. The associated regulatory genes determine the expression of their cognate structural gene and are considered to be activator genes (the hypothesis could be modified to make the regulatory genes code for repressors of the structural genes). The activator genes are presumed to be expressed in particular combinations, giving rise to the TL antigen patterns of normal thymocytes. Leukemogenesis would involve the activation of all regulatory genes and expression of all possible TL structural genes. The structural genes could represent genes coding for individual glycoproteins (if each TL specificity represents a separate molecule) or, perhaps more likely, could represent genes coding for enzymes which modify a common framework molecule (if all TL specificities are found on the same molecule).

Experimental evidence to support these very general ideas is presently limited. The fact that specificities are activated during leukemogenesis caused by diverse agents argues that the structural genes for TL specificities not normally expressed are present in a "silent form" in the normal mouse

TABLE III

HYPOTHETICAL MODEL FOR STRUCTURAL GENES DETERMINING TL SPECIFICITIES AND ASSOCIATED REGULATORY (ACTIVATOR) GENES[a]

Mouse strain	Normal thymocytes		Leukemic cells	
	Structural genes	Activator genes	Structural genes	Activator genes
A	1 2 3	1 2 3	1 2 3	1 2 3
BALB	1 2 0	1 2 0	1 2 0	1 2 0
C57BL	1 2 4	1 2 4	1 2 4	1 2 4
DBA/2	1 2 4	1 2 4	1 2 4	1 2 4

[a]Underlined genes are expressed while others not underlined are silent.

genome (Boyse et al., 1969; Boyse and Old, 1971). The nature of the regulation of the structural genes coding for the TL specificities is not clear. Thymocytes of TL1,2,3 × TL2 matings express quantities of TL1 and TL3 equal to those of TL1,2,3/TL1,2,3 homozygotes (Boyse et al., 1968) which would be consistent with an activator that could act trans. Crosses between TL⁻ and TL1,2,3 mice, however, have thymocytes with only half the quantity of TL1 and TL3 of TL1,2,3/TL1,2,3 homozygotes (Boyse et al., 1968), an observation which is inconsistent with trans activation. Both of these observations are complicated by the fact that the antigen is measured by cytotoxic absorption assays using intact cells. Since steric factors may play an undefined role in such assays (Boyse et al., 1968), it is not certain how differences in the number of cells necessary to absorb cytotoxic activity relate to actual numbers of molecules expressed by the cells.

There has been speculation that the Tla locus represents a defective virus (Boyse, 1972). There are analogies between TL and the G_{IX} antigen which is clearly a leukemia virus gene product (Stockert et al., 1971; Obata et al., 1975; Tung et al., 1975). Both antigens are inherited in a Mendelian manner, and both are activated in leukemias of strains not normally expressing the antigen. There is, however, no evidence of virus production consequent to the activation of TL, so if the Tla locus represents a viral genome, it must be so defective that antigen production (and leukemic conversion?) are its only detectable functions (Boyse, 1972).

2.3. Thy-1

The Thy-1 alloantigen (formerly known as theta, θ) possesses two alleles: Thy 1.1 (θ-AKR) found in AKR/J mice and a few other strains and Thy 1.2 (θ-C3H) found in most strains of mice, including certain substrains of AKR (Reif and Allen, 1964; Acton et al., 1973). The Thy-1 antigen shows a restricted tissue distribution within the mouse, being present on thymocytes and thymus-derived cells of the peripheral lymphoid tissue (Reif and Allen, 1964), on brain (Reif and Allen, 1964), on epidermal cells (Scheid et al., 1972), on fibroblasts (Stern, 1973) and on mammary tissue (Hilgers et al., 1975). Within the lymphoid system the alloantigen is a marker of thymus-derived lymphocytes, being depleted in neonatally thymectomized animals and in nude mice which have a congenital absence of the thymus (Raff, 1969; Raff and Wortis, 1970; see also Loor and Roelants, 1974). The immunologically incompetent cells of the thymic cortex have a high density of Thy-1 (and a low density of H-2), while the immunologically competent subpopulation in the thymus and the thymus-derived cells of the periphery have a lower density of Thy-1 and a higher density of H-2 with respect to the cortical cells (Aoki et al., 1969; Hämmerling and Eggers, 1970; Stobo, 1972). Studies using thymic grafts (Schlesinger and Hurvitz, 1968; Owen and Raff, 1970) give evidence for circulation of a Thy-1 negative bone marrow derived stem cell (Micklem et al., 1966; Wu et al., 1968) to the thymus where it acquires TL and Thy-1 antigens. The nature of the inductive influence is still controversial (Davies, 1975); however, it has been shown recently that the antigens can be induced in bone marrow stem cells incubated in vitro for several hours with thymopoeitin (Basch and Goldstein, 1974), a molecule with a molecular weight of 7000 daltons purified from bovine thymus (Goldstein, 1975). In contrast to other reported inducers, thymopoeitin induces these thymic antigens specifically and does not induce B cell precursors to form complement receptors (Scheid et al., 1975). Since agents which raise cyclic AMP levels induce antigen differentiation (Scheid et al., 1975), the adenyl cyclase system has been implicated as a mediator of these effects.

The structural gene for the Thy-1 determinant is on chromosome 9 of the mouse (Blankenhorn and Douglas, 1972; Itakura et al., 1972). A determinant which appears identical to the mouse Thy 1.1 antigen is present in the rat (Douglas, 1972) where it is present on thymocytes and brain and, in low concentrations, on what seems to be a subpopulation of peripheral lymphoid thymus-derived cells (Douglas, 1973; Acton et al., 1974). The antigen is also found on rat fibroblasts (Stern, 1973). In the mouse brain, the antigen increases in amount during the first five weeks after birth (Reif and Allen, 1964; Schachner and Hämmerling, 1974) and is found on neuronal-like cells and fibroblastic cells but not on glial cells (Mirsky and Thompson, 1975).

Williams and his co-workers have purified the molecule bearing the Thy-1 determinant from deoxycholate solubilized membranes of rat brain and thymus, using the inhibition of antiserum binding to glutaraldehyde-fixed thymocytes as an assay for antigen. The determinant is present on a glycoprotein with a molecular weight of 25–30 000 daltons (Letarte-Muirhead et al., 1974, 1975; Barclay et al., 1975). A similar conclusion, using different methodology and the anti-purified rat brain Thy-1 serum prepared by Barclay et al. (1975), has been reached for the mouse by Trowbridge et al. (1975b; and Trowbridge and Mazauskas, 1976). Other workers (Vitetta et al., 1973; Esselman and Miller, 1974) have presented evidence consistent with the Thy-1 determinant being a glycolipid, possibly G_{M1} ganglioside (Esselman and Miller, 1974). The apparent discrepancy could be resolved if the Thy-1 determinant was present on both glycolipids and glycoproteins, in similar fashion to the ABO blood group determinants (Watkins, 1966), but direct evidence on this point is not yet available.

Immunization of rabbits with mouse or rat brain gives antisera which, after absorption with mouse liver and red cells, reacts on thymocytes and behaves similarly to anti-Thy-1 (Golub, 1971). Serological analysis (Thiele et al., 1972; Clagett et al., 1973; Morris et al., 1975) indicates that these antisera recognize at least three distinct determinants: the Thy-1 allo-antigenic determinant, a species-specific determinant, and a determinant common to both mice and rats. Since all these determinants co-cap (Thiele and Stark, 1974), and since absorption of the purified 25–30 000 molecular weight glycoprotein on an antibody-affinity column made of antibody to one determinant depletes all determinants (Letarte-Muirhead et al., 1975; Morris et al., 1975), the three determinants must be on the same molecule.

3. Surface antigen variation in tumor cell populations

With the information in section 2 as background, we will now discuss whether populations of tumor cells exhibit variation in their expression of surface antigens. If variants expressing an altered surface antigen pattern exist, then can they be isolated so that they can be subjected to genetic analysis? If so, do the variants represent mutants or epigenetic alterations in the expression of existing genes?

3.1. Selection of tumor cell surface antigen variants

3.1.1. Selection of H-2 antigen variants in vivo

The original protocol for selection of surface antigen variants was developed by G. and E. Klein and their co-workers. A tumor of a mouse strain heterozygous for the *H-2* antigen (e.g., H-$2^a/H$-2^b) should grow in F_1 mice but be rejected by either homozygous parental strain. A rare variant which no longer expressed one allele (e.g., H-$2^a/H$-$2^b \rightarrow H$-$2^a/\bar{}$) or which underwent a (hypothetical) mutational change from heterozygous to homozygous (e.g., H-$2^a/H$-$2^b \rightarrow H$-$2^a/H$-2^a) would be able to grow in one parent (H-2^a), and in the F_1, but would not be able to grow in the other parental strain. It should be possible to select for such a variant by inoculating heterozygous tumor cells into parental strains and recovering any tumor which was able to withstand the host's immune response.

A requirement of such a protocol is that the variant cell must not be killed nonspecifically as an "innocent bystander" during the immune response against the large number of nonvariant cells injected. Control reconstruction experiments, in which small numbers of homozygous parental tumor cells were injected together with F_1 tumor cells into the parental strain, showed that such nonspecific killing did not occur. Homozygous parental tumors could be recovered from parental mice, even if as few as 20 parental cells were injected in a mixture of $5 \cdot 10^7$ F_1 cells (G. Klein and E. Klein, 1956).

Selection was then carried out against an F_1 H-$2^a/H$-2^s tumor by passaging it in homozygous parental H-$2^a/H$-2^a and H-$2^s/H$-2^s mice. Variants could be isolated from 24% of H-$2^s/H$-2^s mice injected with F_1 cells and from 2% of H-$2^s/H$-2^s mice pre-immunized with H-$2^a/H$-2^a tissue and then given F_1 cells (Bayreuther and Klein, 1958). Although it was not possible to isolate variants compatible with the H-$2^a/H$-2^a parent, this was an artifact due probably in part to residual heterozygosis of the strains used (Linder and Klein, 1960; J. Klein, 1975), and in part to peculiarities of the tumor used, since variant lines of an H-$2^a/H$-2^s murine lymphoma which were compatible with either parental strain could be isolated (Hellström, 1960).

Immunization with the variant cells (E. Klein et al., 1957; Bayreuther and Klein, 1958), absorption studies (E. Klein et al., 1957; E. Klein, 1961), and studies of the sensitivity of the cells to cytotoxic antibody (Hellström, 1960) all showed that the variant cells had lost the *H-2* determinants selected against but had retained the determinants of the selective host (e.g., $H\text{-}2^a/H\text{-}2^s \to -/H\text{-}2^s$ if selected in an $H\text{-}2^s/H\text{-}2^s$ parental host).

Attempts have been made to select in vivo against the antigens coded by the complex of a homozygous tumor or to obtain "zero" variants by passaging a variant derived from an F_1 tumor and compatible with one parental strain in the opposite parent (Hellström, 1960; Möller, 1964; Bjaring and Klein, 1968). These attempts have all been unsuccessful. Either the tumors retained their antigenic characteristics or they became "nonspecific", able to grow in a wide variety of strains and showing a reduced, but still measurable, quantity of serologically detectable $H\text{-}2$ specificities. These nonspecific variants presumably grew by inducing a state of tumor enhancement such that the host's immune response was rendered ineffective (Möller, 1964; Bjaring and Klein, 1968). The inability to obtain complete loss variants for the $H\text{-}2$ complex by in vivo selection led to the suggestion that one or more products of the complex, perhaps the serologically detectable $H\text{-}2$ antigen itself, plays a vital role in the physiology of the cell such that its loss would be lethal (Bjaring and Klein, 1968). However, later experiments (see section 3.1.3) have raised questions concerning this hypothesis.

3.1.2. Selection of H-2 and HL-A variants in vitro
Selections against the antigens coded for by the $H\text{-}2$ complex have been carried out in vitro by treating cultured lymphomas with antiserum and complement and then growing out the surviving population. This approach has the advantage that the selection conditions can be controlled much more closely than with in vivo selection and has the potential of allowing the isolation of variants which might be selected against in the whole animal (e.g., because they might be gain mutations and thus be immunogenic). Papermaster and Herzenberg (1966) isolated a stable variant from an $H\text{-}2^d/H\text{-}2^k$ hybrid lymphoma by selection against the antigens coded for by the $H\text{-}2^d$ haplotype. The variant lost a private specificity characteristic of $H\text{-}2^d$ but retained a private specificity characteristic of $H\text{-}2^k$. As with in vivo selection, attempts to select against the antigens coded by the remaining $H\text{-}2^k$ haplotype were unsuccessful. It was also not possible to select in vitro a variant lacking all $H\text{-}2$ antigens from a homozygous lymphoma, although variants which expressed less antigen than the parental line could be obtained (Cann and Herzenberg, 1963).

Analogous stable variants have been selected in vitro from a human lymphoid cell line heterozygous for the $HL\text{-}A$ complex (Pious et al., 1973). Certain of these variants react with rabbit sera made against the variant and absorbed with the parental line (Pious and Soderland, 1974). This result might suggest that the variants are "gain mutations" which have new $HL\text{-}A$ specificities. In the absence of direct evidence that the new specificities are the result of mutations in the gene coding for the $HL\text{-}A$ antigen selected against, however, this interpretation can only be considered as one of several alternatives. It is possible that genes coding for antigens other than $HL\text{-}A$ have been activated in these variants. Alternatively, the absence of the *HL-A* specificities selected against may lead to abnormal interaction of remaining membrane components resulting in the formation of new antigenic configuations on the cell surface. Thus, both in vivo and in vitro selections have so far given no clear evidence of mutations which result in the appearance of new specificities on the molecule selected against. Only variation leading to losses of the antigen selected against is known.

3.1.3. Selection of variants for Thy-1 and TL from homozygous tumor cells

In contrast to *H-2*, it has been possible to select variants for Thy-1 and TL from mouse lymphomas which are homozygous for these antigens.

Hyman (1973) and Hyman and Stallings (1974) have isolated four independent variants for the Thy-1 antigen from mouse lymphomas growing in vitro. The variants were stable in the absence of additional selection and by quantitative absorption analysis expressed undetectable Thy-1 antigen on their surface ($<1\%-<0.1\%$ of the parental amounts). There was little or no alteration in other surface antigens or in cellular sensitivity to drugs (Hyman, 1973; Ralph et al., 1973). Analogous variants have been isolated from another cell line by Buxbaum et al. (1976).

A variant which did not express detectable TL antigen ($<1\%$ of parental amount) was isolated from a homozygous TL1,2,3 positive lymphoma (Hyman and Stallings, 1976). This variant was stable and showed no changes in expression of Thy-1 antigen or in Gross virus related specificities. Both the D^k and K^k private specificities coded for by the genes of the H-2^k complex were not detectable, however, with quantitative absorption experiments showing $<2-5\%$ of the parental amount of these antigens. Since the *H-2* complex and the Tla locus are linked, this result might suggest deletion of all or a portion of chromosome 17 in the variant (Hyman and Stallings, 1976); however, cell hybridization experiments (see section 4.2) rule out this possibility. The loss of serologically detectable *H-2* on the surface of this variant indicates that expression of the *H-2* antigen on the cell surface is not necessary for cell viability in vitro (see section 3.1.2), although it is difficult to rule out the presence of minute quantities of antigen on the variant below that detectable by the quantitative antibody absorption method ($<1\%$ of parental amount). The behavior of the variant in vivo indicates that for some unknown reason the variant cell is at a disadvantage in the animal. Although the variant can grow in both the strain of origin (C58) and strains bearing other *H-2* haplotypes, the frequency of takes is much reduced in the C58 strain compared to the parental line (Hyman, unpublished observations). This behavior may partially explain why it was not possible to isolate such variants by in vivo selection. Other factors which might have allowed isolation of an *H-2* negative variant from a homozygous tumor in this case, but not in the others (see sections 3.1.1 and 3.1.2), include the use of mutagenesis before selection was begun and the use of dinitrophenol to block antigenic modulation (Hyman and Stallings, 1976).

3.2. Are surface antigen variants of tumor cell mutants?

There are two alternative ways in which surface antigen variants might arise. The variants might be mutants for structural or regulatory genes concerned with surface antigen synthesis, or they might be the result of changes in the expression of existing genes. The distinction is between models in which a change in genetic information has occurred (alterations in DNA sequences or loss of DNA sequences) and models in which there has been no change of information, but rather a change in its expression (epigenetic models). These latter models comprise cases in which genes that were formerly expressed are now not expressed or vice versa, and cases in which the alteration is not at the level of the genes at all (at least not directly), but rather at the level of the membrane. There are precedents in protozoa both for environmentally induced metastable changes in expression of the genes determining surface antigenicity – serotype transformation (Sommerville, 1970) – and for perpetuation of induced alterations of the cell membrane pattern through subsequent generations – cortical mutants (Sonneborn, 1970). It is possible to conceive of models in which the events occurring during immunoselection would alter the distribution of antigen in the membrane, with this altered pattern then being maintained in a stable state without the necessity for further immunoselection. For example, if it were necessary to have antigen in the membrane as a "primer" to put more antigen in the membrane, then antibody-induced capping (de Petris and Raff, 1973) might lead to aggregation of antigen at one pole of the cell. Division in the correct plane might then result in one daughter cell lacking membrane primer. This cell would remain as a stable negative (see Hyman and Stallings, 1974). Clearly any such postulated epigenetic change at the level of the membrane must be very stable, since the variant cells maintain their negative phenotype for many generations in the absence of further selection (Hellström, 1960; Hyman, 1973; Hyman and Stallings, 1976). This stability sharply distinguishes antigenic variants obtained by immunoselection from cells undergoing antigenic modulation, i.e. an antibody-induced loss of surface antigen mediated at the level of the membrane which is reversible within hours after the antibody is removed (Old et al., 1968; Lesley and Hyman, 1974).

The latter sort of epigenetic model, in which variants are induced by antibody-mediated capping, can be distinguished clearly from models based on mutation and epigenetic models based on "spontaneously occurring" shifts in gene expression in that variants induced by antibody-mediated capping (as might occur during antibody-complement selection) would not pre-exist in the population before the selection was carried out but rather would be induced during the process of selection itself.

In the Thy-1 system it has been possible to study whether the variants pre-exist in the population before selection (Hyman and Stallings, 1974). By carrying out a sib-selection experiment (Cavalli-Sforza and Lederberg, 1956), it was possible to isolate variants which had 1–4% of the parental amount of antigen by indirect selection without exposing the cultures at any time to antibody and complement. This experiment was carried out by setting up a series of replicate cultures, each derived from a small inoculum, and then testing a portion of each culture for the number of cells surviving exposure to a given concentration of antibody and complement. The culture with the highest frequency of surviving cells was chosen, and a new series of cultures was set up from the portion of the culture not used for testing. The inoculum size was chosen so that each culture dish would receive, on the average, less than one variant. Since the variant cell cannot be divided, it was expected that if stable variants pre-exist in the parental population, the dish receiving a variant cell would be enriched for resistant cells at the next testing (assuming roughly equal growth rates of parent and variant cells). This result was obtained, and after a series of enrichment cycles, the variant cells could be cloned out, proving that variants having less than the parental amount of antigen pre-exist in the parental population and do not require the presence of antibody for their induction.

The variants obtained in the sib-selection experiment still had detectable Thy-1 antigen on their surface, although in much reduced amount compared to parental cells. Antibody-complement-mediated selection carried out on the parental population using the same concentrations of antibody and complement as those used in the screening assays of the sib selection experiments gave variants with <0.2–10% of the parental amount of antigen (Hyman and Stallings, 1974). The most reasonable conclusion from these results is that a spectrum of variants having various amounts of antigen pre-exist in the original tumor cell population and that a given selection condition selects for variants with a threshold amount of antigen or less (Hyman and Stallings, 1974). For the lymphoid cell lines being considered here, sensitivity to antibody and complement is proportional to antigen-density, with the exact threshold amount of antigen at which killing occurs subject to variation depending on the nature of the antibody and complement source used (Möller and Möller, 1964; Haughton and McGhee, 1969; Linscott, 1970; Lesley et al., 1974). If rigorous enough selection conditions are chosen, then the variants isolated will appear "negative" for the antigen selected against – all assay systems having a finite limit to the amount of antigen they can detect. If less rigorous selection conditions are chosen (as in the testing assays used for the sib-selection experiment described above), then some variants with a low but detectable amount of antigen will also survive.

It was not possible to conclude from the experiments of Hyman and Stallings that no variants are ever induced by antibody-induced mechanisms but only that, since variants pre-exist in the population, there is no necessity for induction to obtain variants. Besides "capping" as a mechanism for variant induction, it is conceivable that various agents, including antibody, might alter expression of genes involved in surface antigen synthesis in a manner which would lead to a new stable state that would be maintained when the inductive influence was removed. The closest analogy for this model is the apparently irreversible loss of surface immunoglobulin occurring in mouse bone marrow and fetal liver cells cultured in vitro for 48 h with a pulse of rabbit anti-immunoglobulin (Raff et al., 1975). The sib-selection experiments, however, argue that this type of induced change in gene expression is not necessary to obtain variants. The experiments do not rule out that such changes in gene expression might occur spontaneously at low frequency. Such epigenetic variants would pre-exist in the parental population but would not be mutants.

Evidence that the apparently pre-existing stable variants are mutants is limited. Attempts to increase the frequency of heterozygous loss variants for H-2 with mutagens have given inconclusive results. E. Klein et al. (1960) found no effect of X-rays in some experiments and a modest effect in others (E. Klein et al., 1957). Slight increases in the frequency of variants was seen by Dhaliwal (1961) after pretreatment of the inoculum with X-rays or the alkylating agent triethylene-melamine. J. Klein (1967) found insignificant changes or very slight increases in variant frequency after pretreatment of the inoculum with nitrogen mustard. The effect of mutagens on the frequency of variants isolated in vitro has not been studied extensively. A thirty-fold increase in variants resistant to anti-Thy-1 and complement has been found after pre-treatment with the alkylating agent N-methyl-N'-nitronitrosoguanidine (Hyman and Trowbridge, 1976). The frequency of HL-A loss variants derived from heterozygous tumors has also been reported to be increased by ethyl methanesulfonate (an alkylating agent) and the frameshift mutagen ICR 191 (Pious and Soderland, 1974).

Given the fact that stable antigen loss variants pre-exist in the population and that, in at least some instances, there is a modest increase in their frequency after treatment with mutagens, it seems reasonable to assume that they are mutants in genes affecting surface antigen expression. While this assumption will be made here, it is obvious that more evidence on this point is desirable. The nature of the possible mutation will be discussed below (section 4).

Estimates of variant frequency have been made for HL-A variants derived from a heterozygous line and selected in vitro for loss of a single HL-A specificity. Variants occur with a frequency of 10^{-5} to 10^{-6} (Pious et al., 1973). Estimates of the frequency of variants in the Thy-1 system are also of the order of 10^{-5} to 10^{-7} (Hyman and Stallings, 1974; Bauxbaum et al., 1976). Several facts complicate the interpretation of these experiments. Since there are variants showing different degrees of reduction in the amount of surface antigen compared to the parental population, the frequency of "negative" variants may differ depending on the strength of the selection conditions imposed and the amount of antigen below which negativity is assumed (Hyman and Stallings, 1974). Also, it is not necessarily the case that the antigen-negative phenotype always has the same genetic basis. In the Thy-1 system three complementary defects have been found, each giving rise to the Thy-1 negative phenotype (Hyman and Stallings, 1974). Thus, in the absence of further genetic analysis, estimates of "mutation" frequencies based on the frequency of negative phenotypes can only be considered to be approximate estimates. Nevertheless, it is noteworthy that the frequency of variants which have lost antigens coded for by one *HL-A* haplotype in an *HL-A* heterozygote is only about 10- to 100-fold higher than the frequency for Thy-1 variants in homozygous lymphomas. This observation raises the question as to the nature of the genetic changes leading to the variant phenotype – whether recessive or dominant, whether due to recombination or mutation, and whether occurring in one or several steps.

4. Genetic behavior of antigen loss variants

4.1. Antigen loss variants derived from heterozygous tumors

The initial evidence from studies on variants derived from heterozygous tumors favored the idea that variants arose by mitotic recombination. The evidence in favor of this model came from studies in which a tumor was used where one *H-2* haplotype was derived from a recombinant that had the K^k allele of the K region and the D^d allele of the D region. Selections were carried out against the antigens coded by either the K region or the D region of this recombinant *H-2* haplotype or against the antigens coded by the entire *H-2* haplotype (both K and D regions). Schematically, selection was carried out on a tumor which was K^kD^d/K^xD^x in selective hosts which, for example, were K^kD^k/K^xD^x (to select against the antigen coded by the D^d regions), or K^dD^d/K^xD^x (to select against the antigens coded by the K^k region) or K^yD^y/K^xD^x (to select against the antigens coded by both K^k and D^d regions). When selection was carried out against the entire K^kD^d haplotype, all the variants obtained were negative for the antigens coded by both the K and D regions (K^-D^-). When selection was carried out against the K region antigens only, both K^-D^- and K^-D^d variants were obtained. However, when selection was carried out against the D region antigens, only K^-D^- variants were obtained; K^kD^- variants were never seen (Hellström, 1961; E. Klein and G. Klein, 1964; E. Klein, 1969).

Mitotic recombination was favored as the hypothesis to explain these results (E. Klein and G. Klein, 1964). It was assumed that the gene order of the *H-2* regions in chromosome 17 was centromere-D-K. Mitotic recombination with single crossovers between the centromere and D or between D and K can then give only the outcomes obtained when the various selections are applied (Table IVa). A serious problem with the mitotic recombination hypothesis, however, is the fact that the correct gene order has subsequently been shown to be centromere-K-D (Lyon et al., 1968) which is not compatible with the observations (Table IVb). The results could be rationalized if a third histocompatibility locus is postulated to the left of the K region against which selection is also directed (J. Klein, 1972), but there is no evidence for such a locus.

If mitotic recombination were the correct explanation for the variants arising in heterozygous tumors, then a homozygous, rather than a hemizygous, amount of the antigen coded for by the remaining *H-2* haplotype (the "X" allele) might be expected since this haplotype would be present in double dose.

While such an increase in the amount of the antigen coded for by the remaining haplotype was seen for one set of variants (Hellström and Bjaring, 1966), other variants for *H-2* (Boyse et al., 1970) and for *HL-A* (Pious et al., 1974) still express the hemizygous amount of the histocompatibility antigen. While these exceptions argue against mitotic recombination as a general explanation for these variants, it should be pointed out that interpretations based on the quantity of antigen expressed by the variant lines are by no means conclusive. In the variant described by Boyse et al. (1970), selection was carried out against the antigens of the $H-2^a$ haplotype of a lymphoma which was $H-2^a$ TL1,2,3/$H-2^b$ TL1,2,4. Both the $H-2^a$ antigens and the unselected TL3 specificity were lost by the variant. The $H-2^b$ specificity was expressed in hemizygous amount, and the amounts of TL1 and TL2 were reduced by about half, consistent with loss of the TL1 and TL2 genes contributed by the $H-2^a$ parent.

TABLE IV
EXPECTED PHENOTYPES OF VARIANTS OF A HETEROZYGOUS K^kD^d/K^xD^x TUMOR, ASSUMING THAT THE VARIANTS AROSE BY MITOTIC RECOMBINATION AND THAT THE CORRECT GENE ORDER IS (a) CENTROMERE-D-K OR (b) CENTROMERE-K-D

(a) Gene order: centromere-D-K Recombinant genotype		Survivors if select against K^k	D^d
D^dK^k D^xK^k	D^dK^k D^xK^x	D^dK^x D^xK^x	none
D^dK^k D^xK^x	D^dK^x D^xK^k		
D^dK^k D^dK^k	D^xK^x D^xK^x	D^xK^x D^xK^x	D^xK^x D^xK^x
D^dK^k D^xK^x	D^xK^x D^dK^k		

(b) Gene order: centromere-K-D Recombinant genotype		Survivors if select against K^k	D^d
K^kD^d K^kD^d	K^kD^x K^xD^x	none	K^kD^x K^xD^x
K^kD^d K^xD^x	K^kD^x K^xD^d		
K^kD^d K^kD^d	K^xD^x K^xD^x	K^xD^x K^xD^x	K^xD^x K^xD^x
K^kD^d K^xD^x	K^xD^x K^kD^d		

The amount of TL4 antigen, however, was increased twelve-fold. The explanation for the increase is unknown; steric changes leading to increased accessibility of the TL4 antigen to antibody are possible, but regulatory changes which might be primary or secondary consequences of the event giving rise to the loss of the H-2^a antigen are equally possible. Whatever the explanation, this result makes genetic interpretations based on the amount of antigen expressed somewhat difficult.

The fact that selection against the antigens coded by the H-2 haplotype can result in the loss of TL specificities determined by a locus one centimorgan from H-2 suggests that the event leading to H-2 variant formation in heterozygous tumors is a deletion of all or a part of one copy of chromosome 17. It is clear that the entire chromosome bearing the haplotype selected against is not lost in these variants. Chromosome banding studies show that both chromosomes 17 are present in the heterozygous variants lacking one H-2 or H-2 and Tla haplotype and show also that there are no obvious changes in banding patterns of chromosome 17 (Wiener et al., 1974, 1975; Hauschka et al., 1975). These studies suggest that if the genetic event responsible for the antigen loss phenotype in the heterozygous variants for H-2 is occurring in chromosome 17 (which is probable but not certain), it must be subchromosomal and presumably due to a point mutation or to a small deletion. In the HL-A system the linked phosphoglucomutase locus is not lost in the heterozygous variant cells which have lost a single HL-A specificity (Pious et al., 1973), indicating that here also the genetic change responsible for the variant is subchromosomal. The alterations may be in structural or regulatory genes, but if the genes affected are regulatory, they must act cis only since the antigens of the unselected haplotype are still expressed in the variants. It would be most interesting to know whether the variants are revertible. To approach this question requires a good method of positive selection for cells which have regained the antigens of the missing haplotype. Possibly fluorescent activated cell sorting techniques will provide the methods to answer this question.

4.2. Antigen loss variants derived from homozygous tumors

The usefulness of combined somatic cell genetic and biochemical studies to study antigen loss variants can be illustrated by the Thy-1 system. Three classes of loss variants have been defined, each of which complements the other in cell hybridization experiments (Hyman, 1973; Hyman and Stallings, 1974). When each of these classes of variants was fused to a parental cell line expressing the alternate Thy-1 allele, both Thy-1 alleles were expressed in the hybrid (Hyman, 1973; Hyman and Stallings, 1974). That is, the variant phenotype was recessive in these hybrids, and so the genetic information for the Thy-1 determinant must have been present in the variant, although the determinant was not expressed on the cell surface. It is unlikely that the recessive phenotype of the hybrid was a consequence of the loss from the hybrid of a chromosome derived from the variant and bearing a regulatory gene whose product represses Thy-1 or a structural gene whose product masks Thy-1 (Hyman, 1974). All of over 30 hybrids examined express Thy-1, and any postulated chromosome loss must have occurred in all, which seems unlikely. Also, selection against Thy-1 in the hybrid can give cells which again lack both Thy-1 alleles. If appearance of Thy-1 in the hybrid was a consequence of chromosome loss, the Thy-1 positive phenotype should behave as a stable deletion and should not be readily lost upon reselection of the hybrid (see Hyman, 1974).

The hybridization results were most consistent with a model in which the Thy-1 variants were considered to be mutants in structural genes coding for some portion of the 25 000 molecular weight glycoprotein bearing the Thy-1 determinant (see section 1.3). The mutation was not in the gene coding for the Thy-1 determinant itself, since both Thy-1 alleles were expressed when complementary variants derived from tumors bearing alternative alleles, respectively, were fused. Rather the mutation was best interpreted as affecting a structural gene coding for a portion of the glycoprotein other than the Thy-1 determinant. An alternative interpretation assumed that the variant phenotype was due to a defect in a gene coding for a repressor for a gene whose product masked the Thy-1 determinant on the cell surface. The variant would then be constitutive for synthesis of the masking substance, since no functional repressor would be made. If the latter alternative were true for all variants, it would be necessary to postulate three complementary defects in what is presumably a single repressor, which a priori seems unlikely (Hyman and Stallings, 1974; Trowbridge and Hyman, 1975).

Biochemical studies (Trowbridge and Hyman, 1975; Hyman and Trowbridge, 1977) have confirmed and extended the serological and genetic results. Lactoperoxidase labeling and gel electrophoresis showed that the variant cell lines did not have the 25 000 molecular weight molecule bearing the Thy-1 determinant (T25) on the surface. This result renders unlikely those hypotheses in which the variant is suggested to result from masking of the Thy-1 determinant. Hybrids between complementary variants expressed T25 while hybrids between noncomplementary variants did not. Biosynthetic labeling of the variant cells with [^{3}H]2-mannose revealed that at least three variants synthesized a mutant molecule with a mobility on gels slightly faster than T25 but which reacted with rabbit antisera to it. These results, together with the genetic experiments, suggested a working model (Trowbridge and Hyman, 1975) in which the T25 molecule was assumed to require the addition of specific carbohydrate residues in order to be stably integrated into the cell membrane in a manner such that the molecule is accessible to lactoperoxidase and to antibody. The variant cells were assumed to be mutants in genes coding for glycosyltransferases responsible for addition of the relevant carbohydrate residues. The three complementation classes were considered to represent mutations affecting different structural genes coding for their respective glycosyltransferases. It was assumed that the Thy-1 determinant was a carbohydrate residue distal to those missing in the variant molecule and that the gene responsible for the determinant coded for a different glycosyltransferase for each of the alternate alleles of Thy-1. While this last assumption is a logical extension of the model, it is still possible that the Thy-1 determinant is protein in nature and that the sugars are only required for integration of T25 into the membrane. It is also possible that one of the variant classes is due to a defect in the protein portion of T25, but it is not likely that they all are since at least three complementary classes of variants exist (Trowbridge and Hyman, 1975). The fate of the variant T25 molecule is not known. It may be excreted from the cell without being integrated into the membrane, diverted to some other product, or (perhaps most likely) degraded inside the cell as with the non-secreted immunoglobulin produced by mutant myeloma cells with blocks in immunoglobulin assembly (Schubert and Cohn, 1968).

The biochemical data necessary to prove the model suggested by the results so far in the Thy-1 variants are not yet available. It is worth mentioning, however, that lectin-resistant variants derived from Chinese hamster cells show a striking similarity to the Thy-1 variants. The lectin resistant phenotype behaves as recessive in fusions with parental cells (Stanley et al., 1975a) and complementation classes occur (Stanley et al., 1975b). There is biochemical evidence that one lectin-resistant variant is due to a defect in the gene coding for UDP-N-acetylglucosamine-glycoprotein N-acetylglucosaminyl transferase (Stanley et al., 1975c).

The proposal that the allelic Thy-1 determinants are due to differences in glycosyltransferases has precedent in the ABO blood group system (Boettcher, 1966). Both the ABO system and the proposed model for Thy-1 raise the question of how allelic differences which code for separate glycosyltransferases originate. One possibility, proposed for the ABO system, is that the "alleles" may actually represent complex loci with regulation such that only one or the other gene is expressed (Boettcher, 1966; Bodmer, 1973). The other possibility assumes that the genes are, in fact, alleles. There is evidence from bacterial systems that a limited number of mutations can give rise to genes coding for enzymes with an altered substrate specificity (Hegeman and Rosenberg, 1970; Betz and Clarke, 1972). Thus, the proposal that Thy-1 alleles code for alternative glycosyltransferases does not seem to present insuperable evolutionary difficulties.

The recessive behavior of the variant phenotype seen with the Thy-1 variants also occurs in other systems. G. Klein et al. (1972) found that the TA3-Ha mammary carcinoma, which expresses some fifty-fold less H-2^a antigen than another subline of the same tumor (TA3-St), expresses the full quantity of antigen when fused to another H-2 bearing line. Remembering that the Thy-1 loss variants have a cross-reactive molecule inside the cell which precipitates with antisera against T25, it is noteworthy that material reacting with anti-H-2 sera can be found in lysates of the TA3-Ha tumor cells (Friberg and Lilliehöök, 1973). The findings on the TA3-Ha variant line are complicated by the fact that these cells show altered agglutinability to plant lectins (Friberg et al., 1974) and have a high molecular weight surface glycoprotein, epiglycanin (Codington et al., 1975). However, in hybrids between the TA3-Ha cells and fibroblasts, concanavalin A agglutinability was low, as in the TA3-Ha parent, but expression of H-2^a antigen was similar to the nonvariant TA3-St line, indicating that the alterations of these two characters in the TA3-Ha line are independent one from the other (Friberg et al., 1973).

The homozygous variant isolated by Hyman and Stallings (1976) which lost both TL and H-2 when only TL was selected against (section 3.1.3) has also been studied by cell fusion, and here as well the variant phenotype is recessive in hybrids (Hyman and Stallings, 1977). The particular lines used in this hybridization were the variant line and EL4, which expresses H-2^b but is TL negative. Nevertheless, both the TL3 specificity and the H-2^k haplotype characteristic of the C58 tumor from which the variant was derived were expressed in the hybrid. This result indicates that the variant cell line did not lose the structural genes for H-2 and TL in chromosome 17. Rather the defect appears to be in a gene coding for a component common to both antigens. The β_2-microglobulin-like subunit found in both H-2 and TL is one candidate, and a (carbohydrate?) component common to both antigens and necessary for their expression in the membrane is another. Immunoprecipitation analysis of iodinated cells shows that β_2-microglobulin is not detectable on the cell surface (Hyman and Trowbridge, 1976), a finding consistent with the first alternative.

The recessive behavior in hybrids of all of the variants discussed so far raises the problem of how such recessive variants arise in what are presumably diploid cells. The probability that the variants represent mutations of both copies of the relevant gene seems remote, since the frequency of mutation would have to be $1 \cdot 10^{-3}$–$1 \cdot 10^{-4}$ to allow such variants to occur with frequencies of $1 \cdot 10^{-6}$–$1 \cdot 10^{-7}$. A reasonable hypothesis to explain the recessive nature of the defects observed is that the tumor cells used for selection are not diploid for all genes but rather are monosomic for some genes as a consequence of chromosome rearrangements and deletions.

There is extensive evidence for chromosomal rearrangement in tumor cell lines (Ohno, 1971; Deaven and Peterson, 1973). While there is no direct evidence that any variant cell line is, in fact, monosomic for the gene whose alteration is responsible for the variant phenotype, circumstantial supporting evidence has been presented by Chasin. Thus, the frequency of mutation to 6-thioguanine resistance in CHO cells was reduced only 25-fold when diploid and tetraploid lines were compared, the relative rates being consistent with the rate of accumulation of chromosome segregants from the tetraploid line (Chasin, 1973). Also, the frequency of ethyl methanesulfonate induced mutations to diaminopurine resistance at the adenine phosphoribosyltransferase locus was found to be $2 \cdot 10^{-7}$ when "diploid" cells were used. However, when a drug sensitive revertant, which was derived from a diaminopurine resistant line and presumably was carrying only one functional phosphoribosyltransferase gene, was used, a mutation frequency of $2 \cdot 10^{-4}$ was obtained (Chasin, 1974). Finally, 50% of tetraploid hybrids, homozygous at the hypoxanthine phosphoribosyltransferase locus but heterozygous at the linked glucose-6-phosphate dehydrogenase locus, lost the glucose-6-phosphate dehydrogenase locus when selection was carried out against the hypoxanthine phosphoribosyl transferase locus, arguing that chromosome segregation was occurring so that only one copy of the relevant chromosome remained in the mutant cells (Chasin and Urlaub, 1975).

The assumption is that similar segregation is occurring in the tumor lines used to derive surface antigen variants and that variants can be selected from cells which have (a) become monosomic for all or a portion of the relevant chromosome and (b) become mutant at the relevant locus. Consistent with this idea is the fact that the TL-negative, H-2 negative murine lymphoma variant isolated by Hyman and Stallings (1976) has lost a chromosome. There is as yet no direct evidence that the chromosome which has been rendered monosomic bears the mutant gene, although this question might be approachable by banding analysis if a number of additional noncomplementary variants could be isolated.

Attempts to relate the variant cell lines to precise models of genetic regulation based on bacterial systems must be considered highly speculative (see Davis and Adelberg, 1973; Beckwith and Rossow, 1974 for discussion of possible pitfalls). The recessive nature of the variant defects, the apparent necessity for monosomy for variants arising in homozygous tumors, and the cis nature of the defects in variants of heterozygous tumors all suggest some type of intrachromosomal mechanism. It is possible that the relevant genes may have become translocated with respect to some heterochromatin region resulting in an ill-defined localized repression (Brown, 1966; Markert and Ursprung, 1971). Another, somewhat more precise, model would consider the variants to be mutations which affect the rate of gene transcription. Such a mutation would account for the occurrence of "quantitative variants" which show reduced, but detectable, levels of antigen compared to parental lines. A mutation in a gene coding for an "activator" which reduced the quantity of activator made, or lowered the affinity of the activator for its binding site in the genome, would also give the observed results. If mutation in the gene coding for an activator is the explanation for variants of heterozygous tumors, however, then the activator must act cis only, since the unselected marker is expressed and the selected marker is not.

Structural gene mutations are also possible (Stocker and Mäkelä, 1971). Quantitative variants might be accounted for by mutations which alter base sequence, giving a partially functional molecule. Such changes could account for the partial losses of H-2 specificities seen in certain H-2 variants (Bjaring and Klein, 1968). "Mutant molecules" may not integrate into the membrane effectively and/or be degraded, resulting in a quantitative loss of antigen expression on the cell surface. Extreme antigen-loss variants with undetectable amounts of antigen on the surface may be due to mutations resulting in chain termination or to deletion of the structural gene.

4.3. Dominant "suppression" of antigen expression

Cell fusion studies have given evidence for genes which act by suppressing antigen expression. By "suppression" is simply meant the inability to detect surface antigen in a hybrid between a cell expressing the antigen in question and a cell not expressing it. The nonexpression might be due to repression of the gene coding for the antigen or for a precursor of the antigen, to masking of the antigen by a gene product contributed by the non-expressing parent, or to destruction or diversion of the antigen so that it is not present on the cell surface.

Several workers have reported that hybrids between mouse cells and either normal mouse thymocytes or murine lymphomas which bear the Thy-1 antigen do not express detectable Thy-1 (Parkman and Merler, 1973; Hyman and Kelleher, 1975; Liang and Cohen, 1975). The suppression is specific since other surface antigens such as H-2 and TL are expressed, and probably is not due to loss of the chromosomes bearing the genes involved in Thy-1 synthesis from the hybrid (Hyman and Kelleher, 1975), though this point is somewhat uncertain since no revertants expressing Thy-1 have been isolated. The nature of the suppression is totally unclear, although biochemical experiments similar to those carried out for the Thy-1 variants may help to resolve this point.

An interesting question is whether the failure of the hybrids to express Thy-1 is an example of the "extinguishing" of differentiated traits by the less differentiated L cell (cf. Davidson, 1974). As discussed by Hyman and Kelleher (1975), this point is difficult to establish clearly. Certainly the fact that TL antigen is still expressed in such hybrids (Liang and Cohen, 1975) indicates that the L cell does not shut off all differentiated traits, although it might be argued that since the TL bearing parent is a leukemic cell, regulation of this antigen has been "altered". Also, certain "undifferentiated" 3T3 cell lines express Thy-1 while others do not (Stern, 1973; Hyman, unpublished observations) making it difficult to correlate the presence or the absence of Thy-1 with the state of differentiation of the cell expressing it. The possible role of the virus carried by the L cell in altering regulation of Thy-1 expression must also be considered (Hyman and Kelleher, 1975).

In contrast to the suppression of Thy-1 consequent to the hybridization of Thy-1 bearing lines with L cells is the suppression of a variety of L cell antigens after hybridization of these cells with Ehrlich ascites cells. Ehrlich ascites cells have been passaged for years in a variety of mouse strains and express little detectable H-2 antigen (Hauschka and Amos, 1957). When Ehrlich cells were hybridized with L cells, the H-2^k, the Moloney virus related, and the L cell virus related antigens of the L cell were suppressed to varying degrees (Klein et al., 1970; Grundner et al., 1971). Revertants could be isolated which had lost a number of chromosomes and re-expressed one or more of the L cell antigens (Grundner et al., 1971). The suppression of each surface antigen, therefore, was mediated independently of the others, indicating that the suppression was due to the action of a number of independent genes.

5. Future prospects

5.1. Analysis of differentiation alloantigens

From what has been said above, it is apparent that it is possible to obtain variant tumor cell lines which no longer express one or more known cell surface antigens and to carry out some genetic analyses of these variants using cell hybridization techniques. In favorable systems, such as Thy-1, new genes which have not been detected by whole animal studies, can be defined and analyzed by complementation analysis, and the action of these genes can be studied biochemically. It is apparent, however, that no variant is yet thoroughly understood in both genetic and biochemical terms. It seems likely that, for the immediate future, the most profitable approach will probably be to carry out a more exhaustive analysis of the existing systems, all of which have the advantage that the chemistry of the relevant surface antigen is at least partially understood. One tack would be to further characterize "mutant molecules" such as appear to exist in the Thy-1 system (Trowbridge and Hyman, 1975) with the aim of identifying specific biochemical alterations in the variant cells. Another tack would be to isolate additional variants for a given system and to characterize the variants by somatic genetic methods. Although the analysis is tedious, only in this way will it be possible to identify additional genes and to find new classes of variants, in particular variants with regulatory defects, for which there is as yet no clear evidence. The ultimate aim is to enumerate the genes responsible for the expression of at least one antigen and to understand the mechanisms by which normal antigen expression is regulated.

5.2. Tumor-associated antigens

The use of somatic cell genetic methods also may allow a study of the role of host cell genes in the expression of tumor-associated antigens. Tumor-associated antigens include a number of distinct classes of antigens which may be more or less amenable to genetic analysis at the level of the tumor cell. Certain antigens are clearly structural components of tumor-associated viruses (Bauer, 1974) and as such may not, a priori, seem amenable to somatic genetic analysis. Nevertheless, there are a number of reports of variant tumor cell lines which no longer express or express reduced quantities of one or another murine leukemia virus related surface antigen (Fenyö et al., 1968; Ferrer and Gibbs, 1969; Hyman et al., 1972; Ioachim et al., 1974). Genetic studies have not been carried out on these variants with the exception of a Moloney virus induced lymphoma which expressed a reduced (nonzero) amount of Moloney surface antigen (Fenyö et al., 1968). When this variant was fused to L cells, the full (parental) amount of the Moloney surface antigen was expressed in the hybrid, although, presumably due to an incompatibility at the Fv-1 locus (Fenyö et al., 1973), infectious virus was not produced in the hybrid (Fenyö, 1971). The basis of the defect in the variant is unknown and may not be cellular since passage of the variant cells in Moloney virus-infected mice restores surface antigen expression (Fenyö et al., 1969). It is clear that the antigen which is reduced in quantity in the variant cell line is not one of the known viral proteins (Fenyö and Klein, 1976), although it is presumably virus-coded. Studies on the genetic basis of other variants may reveal host genes which modify or regulate virus antigen expression at the cell surface.

There are cases where genes coding for virus structural components seem to be integrated into the chromosome of normal host cells and to be inherited in a Mendelian manner. The clearest example of this situation is G_{IX} (Stockert et al., 1971) which is a type-specific determinant found on the murine viral glycoprotein with a molecular weight of about 70 000 daltons (Obata et al., 1975; Tung et al., 1975). As discussed above (section 2.2), the TL alloantigen is similar to G_{IX} in that it shows Mendelian inheritance and is expressed in the leukemic cells of strains of mice which do not bear the antigen on their normal thymocytes (see Table II). However, there is no direct evidence that the TL specificities are coded for by viral genes. In the cases of both TL and G_{IX} it might be presumed that host cell genes are involved in regulating antigen expression (see 2.2). Cell hybridization experiments might indicate whether leukemic cells have "trans-acting" genes whose products can activate antigen expression, or, conversely, whether normal cells possess genes which can alter expression of tumor-associated antigens (e.g., by repression or masking). Experiments in the TL system either between thymocytes and spleen cells (Allison et al., 1975) or between lymphomas (Hyman and Stallings, 1976b) show that such analysis is feasible, although in neither case examined in this system was there evidence for intergenomic effects on TL antigen expression.

Another class of tumor-associated antigens comprises the unique specificities found in chemical carcinogen induced tumors (Baldwin, 1973). These antigens present an interesting genetic problem, since tumors of independent origin possess non-crossreacting antigens as determined by transplantation tests, even when the tumors are induced in cells derived from the same cloned cell line (Basombrio and Prehn, 1972; Embleton and Heidelberger, 1972). It has been suggested that such unique tumor-specific antigens represent modified normal cell surface components. The original basis of this suggestion was an apparent reciprocal relationship between the amount of *H-2* antigen present on a series of tumors and the amount of tumor-specific antigen as measured by ability to induce tumor immunity (Haywood and McKhann, 1971). In the case of a chemically induced rat hepatoma there is also chemical evidence for the tumor-specific antigen being on the same molecule as the major transplantation antigen (Bowen and Baldwin, 1975). Such modifications of normal antigens, if they exist, could be regulatory changes resulting in "derepression" of silent genes (Bodmer, 1973) or mutational alterations in already expressed genes coding for membrane components (Huberman and Sachs, 1976). The question of whether tumor-specific determinants in a methylcholanthrene induced sarcoma are modified *H-2* antigens has been approached by selecting against the *H-2* antigens of the methylcholanthreme sarcoma in a hybrid between the sarcoma and an ascites carcinoma (G. Klein and E. Klein, 1975). Loss of the chromsomes bearing the appropriate *H-2* genes did not lead to complete loss of the tumor-specific antigen (although there may have been some quantitative effects), arguing that the tumor-specific antigens were not simply modified *H-2* antigens. This experiment of the Kleins shows the usefulness of somatic genetic analysis in studying the nature and linkage of tumor-specific antigens. Future experiments using a similar approach could actually establish the linkage of the tumor antigens of chemical carcinogen-induced tumors (of particular interest would be to study a series of independently derived chemical carcinogen-induced transformants of the same cloned cell line). Complementation analysis of a number of antigen loss variants derived from such a series might also provide useful information on the nature of the genetic changes giving rise to the tumor-specific transplantation antigens of chemical carcinogen-induced tumors.

Serological studies of the antigens of human tumors have been slow, since xenogeneic antisera must usually be used to define antigens on human cells and the serological analysis of such antisera is difficult. One possible approach to this problem is the use of hybrids between human cells and rodent cells which have lost all but one human chromosome (Kao et al., 1976). Immunization with hybrids retaining genes coding for only a limited number of human surface antigens may allow the production of much simpler antisera which are amenable to serological analysis, particularly when this approach is coupled with biochemical techniques that enable the rational analysis of complex xenoantisera (Trowbridge et al., 1975a,b). Such hybrids will also allow the linkage of particular surface components to be clearly established (Jones et al., 1975; Kao et al., 1976).

The existence of antigenic variation in tumor cells and the study of antigen expression in hybrids is thus of interest both for the knowledge it can provide about the genetic mechanisms by which expression of tumor-associated antigens are controlled and, more practically, as a potential means for defining tumor-associated antigens in outbred systems such as man. One further ramification of somatic cell genetic systems based on tumor cells relates to an understanding of how tumor cells evade the action of the immune response of the host. The existence of antigenic variants which are able to grow in the face of antibody-complement selective pressures must be taken into account in considering the potential limitations of immunotherapy (Hyman, 1974). The possibility that variants exist which may escape from cell-mediated killing (Knowles and Swift, 1975) and the fact that many spontaneous tumors are relatively nonantigenic (Baldwin et al., 1974) suggests that antigenic variation and immunoselection may be as important in the evolution of successful tumors as are other mechanisms which operate at the level of immune response of the host.

Acknowledgments

I would like to thank M. Bevan, M. Cohn, G. Dennert, G.L. Nicolson and I. Trowbridge for their comments on an early draft. M. Cohn gave me the idea for Table III. My own studies have been supported by grants CA-13287 and CA-17020 from the National Cancer Institute, United States National Institutes of Health and by a Scholar award from the Leukemia Society of America.

References

Abelson, H. and Rabstein, L. (1970) Lymphosarcoma: Virus-induced thymic independent disease in mice. Cancer Res. 30, 2213–2222.

Acton, R., Blankenhorn, E., Douglas, T., Owen, R., Hilgers, J., Hoffman, H. and Boyse, E. (1973) Variations among sublines of inbred AKR mice. Nature (London) New Biol. 245, 8–10.

Acton, R., Morris, R. and Williams, A. (1974) Estimation of the amount and tissue distribution of rat Thy 1.1 antigen. Eur. J. Immunol. 4, 598–602.

Allison, D., Meier, P., Majeune, M. and Cohen, E. (1975) Antigen expression of mouse spleen x thymocyte heterokaryons. Cell 6, 521–527.

Anundi, N., Rask, L., Ostberg, L. and Peterson, P. (1975) The subunit structure of thymus-leukemia antigens. Biochemistry 14, 5046–5054.

Aoki, T., Hämmerling, U., de Harven, E., Boyse, E. and Old, L. (1969) Antigenic structure of cell surfaces. An immunoferritin study of the occurrence and topography of H-2, θ, and TL alloantigens on mouse cells. J. Exp. Med. 130, 979–1001.

Artzt, K. and Bennett, D. (1975) Analogies between embryonic (T/t) antigens and adult major histocompatibility antigens. Nature (London) 256, 545–547.

Baldwin, R. (1973) Immunological aspects of chemical carcinogenesis. Adv. Cancer Res. 18, 1–75.

Baldwin, R., Embleton, M., Price, R. and Vose, B. (1974) Embryonic antigen expression on experimental rat tumors. Transplant. Rev. 20, 77–99.

Barclay, A., Letarte-Muirhead, M. and Williams, A. (1975) Purification of the Thy-1 molecule from rat brain. Biochem. J. 151, 699–706.

Basch, R. and Goldstein, G. (1974) Induction of T-cell differentiation in vitro by thymin, a purified polypeptide hormone of the thymus. Proc. Natl. Acad. Sci. USA 71, 1474–1478.

Basombrio, M. and Prehn, R. (1972) Antigenic diversity of tumors chemically induced within the progeny of a single cell. Int. J. Cancer 10, 1–8.

Bauxbaum, J., Basch, R. and Szabadi, R. (1976) Isolation and characterization of murine-T lymphoma cell variants selected for loss of the Thy-1 antigen. Fed. Proc. 35, 514.

Bayreuther, K. and Klein, E. (1958) Cytogenetic, serologic, and transplantation studies on a heterozygous tumor and its derived variant sublines. J. Natl. Cancer Inst. 21, 885–923.

Beckwith, J. and Rossow, P. (1974) Analysis of genetic regulatory mechanisms. Ann. Rev. Genet. 8, 1–13.

Bennett, D., Boyse, E. and Old, L. (1972) Cell surface immunogenetics in the study of morphogenesis. In: L. Silvestri (Ed.), Cell Interactions, Third Lepetit Colloquium, North-Holland, Amsterdam, pp. 247–263.

Betz, J. and Clarke, P. (1972) Selective evolution of phenylacetamide-utilizing strains of *Pseudomonas aeruginosa*. J. Gen. Microbiol. 73, 161–174.

Bjaring, B. and Klein, G. (1968) Antigenic characterization of heterozygous mouse lymphomas after immunoselection in vivo. J. Natl. Cancer Inst. 41, 1411–1429.

Blankenhorn, E. and Douglas, T. (1972) Location of the gene for theta antigen in the mouse. J. Hered. 63, 259–263.

Bodmer, W. (1973) A new genetic model for allelism at histocompatibility and other complex loci: Polymorphism for control of gene expression. Transplant. Proc. 5, 1471–1475.

Boettcher, B. (1966) Modification of Bernstein's multiple allele theory for the inheritance of the ABO blood groups in the light of modern genetical concepts. Vox Sang. 11, 129–136.

Bowen, J. and Baldwin, R. (1975) Tumor-specific antigen related to rat histocompatibility antigens. Nature (London) 258, 75–76.

Boyse, E. (1973) Immunogenetics in the study of cell surfaces: Some implications for morphogenesis and cancer. In: C. Anfinsen, M. Potter and A. Schechter (Eds.), Current Research in Oncology. Academic Press, New York, pp. 57–94.

Boyse, E. and Old, L. (1969) Some aspects of normal and abnormal cell surface genetics. Ann. Rev. Genet. 3, 269–290.

Boyse, E. and Old, L. (1971) A comment on the genetic data relating to expression of TL antigens. Transplantation 11, 561–562.

Boyse, E., Old, L. and Stockert, E. (1965) The TL (thymus leukemia) antigen: A Review. In: P. Grabar and P. Miescher (Eds.), Immunopathology, IVth Int. Symp. Schwabe, Basel, pp. 23–40.

Boyse, E., Stockert, E. and Old, L. (1968) Isoantigens of the H-2 and Tla loci of the mouse. Interactions affecting their representation on thymocytes. J. Exp. Med. 128, 85–95.

Boyse, E., Stockert, E. and Old, L. (1969) Properties of four antigens specified by the Tla locus. Similarities and differences. In: N. Rose and F. Milgrom (Eds.), International Convocation on Immunology. Karger, Basel, pp. 353–357.

Boyse, E., Stockert, E., Iritani, C. and Old, L. (1970) Implications of TL phenotype changes in an H-2-loss variant of a transplanted $H-2^b/H-2^a$ leukemia. Proc. Natl. Acad. Sci. USA 65, 933–938.

Brown, J., Kato, K., Silver, J. and Nathenson, S. (1974) Notable diversity in peptide composition of murine H-2K and H-2D alloantigens. Biochemistry 13, 3174–3178.

Brown, S. (1966) Heterochromatin. Science 151, 417–426.

Cappra, J., Vitetta, E. and Klein, J. (1975) Studies on the murine Ss protein, I. Purification, molecular weight, and subunit structure. J. Exp. Med. 142, 664–672.

Cann, H. and Herzenberg, L. (1963) In vitro studies of mammalian somatic cell variation, II. Isoimmune cytotoxicity with a cultured mouse lymphoma and selection of resistant variants. J. Exp. Med. 117, 267–283.

Cantor, H. and Boyse, E. (1975) Functional subclasses of T lymphocytes bearing different Ly antigens, I. The generation of functionally distinct T-cell subclasses is a differentiative process independent of antigen. J. Exp. Med. 141, 1376–1389.

Cavalli-Sforza, L. and Lederberg, J. (1956) Isolation of pre-adaptive mutants in bacteria by sib selection. Genetics 41, 367–381.

Chasin, L. (1973) The effect of ploidy on chemical mutagenesis in cultured Chinese hamster cells. J. Cell Physiol. 82, 299–308.

Chasin, L. (1974) Mutations affecting adenine phosphoribosyl transferase activity in Chinese hamster cells. Cell 2, 37–41.

Chasin, L. and Urlaub, G. (1975) Chromosome-wide event accompanies the expression of recessive mutations in tetraploid cells. Science 187, 1091–1093.

Clagett, J., Peter, H., Feldman, J. and Weigle, W. (1973) Rabbit antiserum to brain-associated thymus antigens of mouse and rat, II. Analysis of species-specific and cross-reacting antibodies. J. Immunol. 110, 1085–1089.

Codington, J., Linsley, K., Jeanloz, R., Irimura, T. and Osawa, T. (1975) Immunochemical and chemical investigations of the structure of glycoprotein fragments obtained from epiglycanin, a glycoprotein at the surface of the TA3-Ha cancer cell. Carbohydrate Res. 40, 171–182.

Cullen, S. and Nathenson, S. (1974) Further characterization of Ia (immune response region associated) antigen molecules. In: E. Sercarz, A. Williamson and C.F. Fox (Eds.), The Immune System: Genes, Receptors, Signals. Academic Press, New York, pp. 191–200.

Cullen, S., Schwartz, B. and Nathenson, S. (1972a) The distribution of alloantigenic specificities of native H-2 products. J. Immunol. 108, 596–600.

Cullen, S., Schwartz, B., Nathenson, S. and Cherry, M. (1972b) The molecular basis of codominant expression of the histocompatibility-2 genetic region. Proc. Natl. Acad. Sci. USA 69, 1394–1397.

Curman, B., Östberg, L., Sandberg, L., Malmheden-Eriksson, I., Stadenheim, G., Rask, L. and

Peterson, P. (1975) H-2 linked Ss protein is C4 component of complement. Nature (London) 258, 243–245.
David, C., Stimpfling, J. and Shreffler, D. (1975) Identification of specificity H2.7 as an erythrocyte antigen. Control by an independent locus, H-2G, between the S and D regions. Immunogenetics 2, 131–139.
Davidson, R. (1974) Gene expression in somatic cell hybrids. Ann. Rev. Genet. 8, 195–218.
Davies, A. (1975) Thymic hormones? Ann. N.Y. Acad. Sci. 249, 61–67.
Davis, F. and Adelberg, E. (1973) Use of somatic cell hybrids for analysis of the differentiated state. Bacteriol. Rev. 37, 197–214.
Deaven, L. and Petersen, D. (1973) The chromosomes of CHO, an aneuploid Chinese hamster cell line: G-band and autoradiographic analysis. Chromosoma 41, 129–144.
de Petris, S. and Raff, M. (1973) Normal distribution, patching, and capping of lymphocyte surface immunoglobulin studied by electron microscopy. Nature (London) New Biol. 241, 257–259.
Deuchar, E. (1975) Cellular Interactions in Animal Development. Chapman and Hall, London, 298 pp.
Dhaliwal, S. (1961) Studies on histocompatibility mutations in mouse tumor cells using isogenic strains of mice. Genet. Res. 2, 309–332.
Douglas, T. (1972) Occurrence of a theta-like antigen in rats. J. Exp. Med. 136, 1054–1062.
Douglas, T. (1973) A rat analogue of the mouse theta antigen. Transplant. Proc. 5, 79–82.
Embleton, M. and Heidelberger, C. (1972) Antigenicity of clones of mouse prostate cells transformed in vitro. Int. J. Cancer 9, 8–18.
Esselman, W. and Miller, H. (1974) The ganglioside nature of θ antigens. Fed. Proc. 33, 771.
Fenyö, E. and Klein, G. (1976) Independence of Moloney virus-induced cell-surface antigen and membrane-associated virion antigens in immunoselected lymphoma sublines. Nature (London) 260, 355–357.
Fenyö, E., Klein, E., Klein, G. and Swiech, K. (1968) Selection of an immunoresistant lymphoma subline with decreased concentration of tumor-specific surface antigens. J. Natl. Cancer Inst. 40, 69–89.
Fenyö, E., Biberfeld, P. and Klein, G. (1969) Studies on the relations between virus release and cellular immunosensitivity in Moloney lymphomas. J. Natl. Cancer Inst. 42, 837–849.
Fenyö, E., Grundner, G., Klein, G., Klein, E. and Harris, H. (1971) Surface antigens and release of virus in hybrid cells produced by the fusion of A9 fibroblasts with Moloney lymphoma cells. Exp. Cell Res. 68, 323–331.
Fenyö, E., Grundner, G., Wiener, F., Klein, E., Klein, G. and Harris, H. (1973) The influence of the partner cell on the production of L virus and the expression of viral surface antigen in hybrid cells. J. Exp. Med. 137, 1240–1255.
Ferrer, J. and Gibbs, F. (1969) Concomitant loss of specific cell-surface antigen and demonstrable type-C virus particles in lymphomas induced by radiation leukemia virus in rats. J. Natl. Cancer Inst. 43, 1317–1330.
Freed, J., Brown, J. and Nathenson, S. (1975) Studies on the carbohydrate structure of Ia alloantigens: Comparison with H-2K and H-2D gene products. In: M. Seligman, J. Preud'homme and R. Kourilsky (Eds.), Membrane Receptors of Lymphocytes. Elsevier, New York, pp. 241–246.
Friberg, S. and Lilliehöök, B. (1973) Evidence for non-exposed H-2 antigens in immunoresistant murine tumor. Nature (London) New Biol. 241, 112–114.
Friberg, S., Klein, G., Wiener, F. and Harris, H. (1973) Hybrid cells derived from fusions of TA3-Ha carcinoma with normal fibroblasts, II. Characterization of isoantigenic variant sublines. J. Natl. Cancer Inst. 50, 1269–1286.
Friberg, S., Molnar, J. and Pardoe, G. (1974) Tumor cell-surface organization: Differences between two TA3 sublines. J. Natl. Cancer Inst. 52, 85–93.
Goldschneider, I. and Barton, R.W. (1976) Development and differentiation of lymphocytes. In: G. Poste and G.L. Nicolson (Eds.), The Cell Surface in Animal Embryogenesis and Development, Cell Surface Reviews, Vol. 1. North-Holland, Amsterdam, pp. 599–668.
Goldstein, G. (1975) The isolation of thymopoeitin (thymin). Ann. N.Y. Acad. Sci. 249, 177–185.

Golub, E. (1971) Brain-associated θ-antigen: Reactivity of rabbit anti-mouse brain with mouse lymphoid cells. Cell. Immunol. 2, 353–361.

Grundner, G., Fenyö, E., Klein, G., Klein, E., Bregula, U. and Harris, H. (1971) Surface antigen expression in malignant sublines derived from hybrid cells of low malignancy. Exp. Cell Res. 68, 315–322.

Hämmerling, U. and Eggers, H. (1970) Quantitative measurement of uptake of alloantibody on mouse lymphocytes. Eur. J. Biochem. 17, 95–99.

Haughton, G. and McGhee, P. (1969) Cytolysis of mouse lymph node cells by alloantibody: A comparison of guinea pig and rabbit complements. Immunology 16, 447–461.

Hauptfeld, V. and Klein, J. (1975) Molecular relationship between private and public H-2 antigens as determined by antigen redistribution method. J. Exp. Med. 142, 288–298.

Hauschka, T. and Amos, D. (1957) Cytogenetic aspects of compatibility. Ann. N.Y. Acad. Sci. 69, 561–579.

Hauschka, T., Kitt, S., Zumpft, M., Shows, T. and Boyse, E. (1975) Immunoselective loss of parental H antigens by somatic reduction in an $H-2^a/H-2^b$ hybrid mouse leukemia. Transplant. Proc. 7, 165–171.

Hegeman, G. and Rosenberg, S. (1970) The evolution of bacterial enzyme systems. Ann. Rev. Microbiol. 24, 429–462.

Hellström, K. (1960) Studies on isoantigenic variation in mouse lymphomas. J. Natl. Cancer Inst. 25, 237–269.

Hellström, K. (1961) Studies on the mechanism of variant formation in heterozygous mouse tumors, II. Behavior of H-2 antigens D and K: Cytotoxic tests on mouse lymphomas. J. Natl. Cancer Inst. 27, 1095–1105.

Hellström, K. and Bjaring, B. (1966) Studies on the mechanism of isoantigenic variant formation in heterozygous mouse tumors, V. Quantitative studies of residual H-2 antigens in isoantigenic variant sublines isolated from a mouse lymphoma of F_1 hybrid origin. J. Natl. Cancer Inst. 36, 947–952.

Henning, R., Milner, R., Reske, K., Cunningham, B. and Edelman, G. (1976) Subunit structure, cell surface orientation, and partial amino-acid sequences of murine histocompatibility antigens. Proc. Natl. Acad. Sci. USA 73, 118–122.

Hilgers, J., Haverman, J., Nusse, R., van Blitterswijk, W., Clefton, F., Hageman, Ph., van Nie, R. and Calafat, J. (1975) Immunologic, virologic, and genetic aspects of mammary tumor virus-induced cell surface antigens: Presence of these antigens and the Thy 1.2 antigen on murine mammary gland and tumor cells. J. Natl. Cancer Inst. 54, 1323–1333.

Huberman, E. and Sachs, L. (1976) Mutability of different genetic loci in mammalian cells by metabolically activated carcinogenic polycyclic hydrocarbons. Proc. Natl. Acad. Sci. USA 73, 188–192.

Hyman, R. (1973) Studies on surface antigen variants. Isolation of two complementary variants for Thy 1.2. J. Natl. Cancer Inst. 50, 415–422.

Hyman, R. (1974) Genetic alterations in tumor cells resulting in their escape from the immune response. Cancer Chemother. Rept. 58, 431–439.

Hyman, R. and Kelleher, R. (1975) Absence of Thy-1 antigen in L-cell x mouse lymphoma hybrids. Somatic Cell Genet. 1, 335–343.

Hyman, R. and Stallings, V. (1974) Complementation patterns of Thy-l variants and evidence that antigen loss variants "pre-exist" in the parental population. J. Natl. Cancer Inst, 52, 429–436.

Hyman, R. and Stallings, V. (1976) Characterization of a TL$^-$ variant of a homozygous TL$^+$ mouse lymphoma. Immunogenetics 3, 75–84.

Hyman, R. and Stallings, V. (1977) Analysis of hybrids between an $H-2^+$, TL$^-$ lymphoma and an $H-2^+$, TL$^+$ lymphoma and its $H-2^-$, TL$^-$ variant subline. Immunogenetics 4, 171–181.

Hyman, R. and Trowbridge, I. (1977) Analysis of lymphocyte surface antigen expression of the use of variant cell lines. Cold Spring Harbor Symp. Quant. Biol., in press.

Hyman, R., Ralph, P. and Sarkar, S. (1972) Cell-specific antigens and immunoglobulin synthesis of murine myeloma cells and their variants. J. Natl. Cancer Inst. 48, 173–184.

Ioachim, H., Keller, S., Dorsett, B. and Pearse, A. (1974) Induction of partial immunologic tolerance in rats and progressive loss of cellular antigenicity in Gross virus lymphoma. J. Exp. Med. 139, 1382–1394.

Itakura, K., Hutton, J., Boyse, E. and Old, L. (1972) Genetic linkage relationships of loci specifying differentiation alloantigens in the mouse. Transplantation 13, 239–243.

Jones, C., Wuthier, P. and Puck, T. (1975) Genetics of somatic cell surface antigens, III. Further analysis of the A_L marker. Somatic Cell Genet. 1, 235–246.

Kao, F., Jones, C. and Puck, T. (1976) Genetics of somatic mammalian cells: Genetic, immunologic, and biochemical analysis with Chinese hamster cell hybrids containing selected human chromosomes. Proc. Natl. Acad. Sci. USA 73, 193–197.

Klein, E. (1961) Studies on the mechanism of isoantigenic variant formation in heterozygous mouse tumors, I. Behavior of H-2 antigens D and K: Quantitative absorption tests on mouse sarcomas. J. Natl. Cancer Inst. 27, 1069–1093.

Klein, E. and Klein, G. (1964) Studies on the mechanism of isoantigenic variant formation in heterozygous mouse tumors, III. Behavior of H-2 antigens D and K when located in the trans position. J. Natl. Cancer Inst. 32, 569–578.

Klein, E., Klein, G. and Hellström, K. (1960) Further studies on isoantigenic variation in mouse carcinomas and sarcomas. J. Natl. Cancer Inst. 25, 271–294.

Klein, G. and Klein, E. (1956) Genetic studies of the relationship of tumor-host cells. Detection of an allelic difference at a single gene locus in a small fraction of a large tumor-cell population. Nature (London) 178, 1389–1391.

Klein, G. and Klein, E. (1975) Are methylcholanthrene-induced sarcoma-associated, rejection-inducing (TSTA) antigens, modified forms of H-2 or linked determinants? Int. J. Cancer 15, 879–887.

Klein, G., Gars, U. and Harris, H. (1970) Isoantigen expression in hybrid mouse cells. Exp. Cell Res. 62, 149–160.

Klein, G., Friberg, S. and Harris, H. (1972) Two kinds of antigen suppression in tumor cells revealed by cell fusion. J. Exp. Med. 135, 839–849.

Klein, J. (1967) Further evidence on the origin of alloantigenic variants of F_1 tumors grown in semi-isogeneic F_1 hosts. In: J. Klein, M. Vojtǐsková and V. Zeleny (Eds.), Genetic Variations in Somatic Cells. Academia, Prague, pp. 385–392.

Klein, J. (1972) Is the H-2K locus of the mouse stronger than the H-2D locus? Tissue Antigens 2, 262–266.

Klein, J. (1975) Biology of the Mouse Histocompatibility-2 Complex. Springer, New York, 620 pp.

Klein, J. and Shreffler, D. (1971) The H-2 model for the major histocompatibility systems. Transplant. Rev. 6, 3–29.

Klein, J., Bach, F., Festenstein, H., McDevitt, H., Shreffler, D., Snell, G. and Stimpfling, J. (1974) Genetic nomenclature for the H-2 complex of the mouse. Immunogenetics 1, 184–188.

Klein, J., Hauptfeld, V. and Hauptfeld, M. (1975) Evidence for a fifth (G) region in the H-2 complex of the mouse. Immunogenetics 2, 141–150.

Knowles, B. and Swift, K. (1975) Cell-mediated immunoselection against cell-surface antigens of somatic cell hybrids. Somatic Cell Genet. 1, 123–136.

Lachman, P., Grennan, D., Martin, A. and Demant, P. (1975) Identification of Ss protein as murine C4. Nature (London) 258, 242–243.

Leckband, E. and Boyse, E. (1971) Immunocompetent cells among mouse thymocytes: A minor population. Science 172, 1258–1260.

Lengerova, A. (1972) The expression of normal histocompatibility antigens in tumor cells. Adv. Cancer Res. 16, 235–271.

Lesley, J. and Hyman, R. (1974) Antibody-induced changes in expression of the H-2 antigen. Eur. J. Immunol. 4, 732–739.

Lesley, J., Hyman, R. and Dennert, G. (1974) Effect of antigen density on complement mediated lysis, T-cell-mediated killing and antigenic modulation. J. Natl. Cancer Inst. 53, 1759–1765.

Letarte-Muirhead, M., Acton, R. and Williams, A. (1974) Preliminary characterization of Thy 1.1 and Ag-B antigens from rat tissues solubilized in detergents. Biochem. J. 143, 51–61.

Letarte-Muirhead, M., Barclay, A. and Williams, A. (1975) Purification of the Thy-1 molecule – A major cell surface glycoprotein of rat thymocytes. Biochem. J. 151, 685–697.

Liang, W. and Cohen, E. (1975) Somatic hybrid of thymus leukemia (+) and (−) cells forms thymus leukemia antigens but fails to undergo modulation. Proc. Natl. Acad. Sci. USA 72, 1873–1877.

Linder, O. and Klein, E. (1960) Skin and tumor grafting in coisogenic resistant lines of mice and their hybrids. J. Natl. Cancer Inst. 24, 707–720.

Linscott, W. (1970) An antigen density effect on the hemolytic efficiency of complement. J. Immunol. 104, 1307–1309.

Loor, F. and Roelants, G. (1974) High frequency of T lineage lymphocytes in nude mouse spleen. Nature (London) 251, 229–230.

Loor, F., Block, N. and Little, J. (1975) Dynamics of the TL antigens on thymus and leukemia cells. Cell. Immunol. 17, 361–365.

Lyon, M., Butler, J. and Kemp, R. (1968) The position of the centromeres in linkage groups II and IX of the mouse. Genet. Res. 11, 193–199.

Markert, C. and Ursprung, H. (1971) Developmental Genetics. Prentice-Hall, Englewood Cliffs, N.J., pp. 64–82.

Meo, T., Krasteff, T. and Shreffler, D. (1975) Immunochemical characterization of murine H-2 controlled Ss (serum substance) protein through identification of its human homologue as the fourth component of complement. Proc. Natl. Acad. Sci. USA 72, 4536–4540.

Micklem, H., Ford, C., Evans, E. and Gray, J. (1966) Interrelationships of myeloid and lymphoid cells: Studies with chromosome-marked cells transfused into lethally irradiated mice. Proc. Roy. Soc. Lond. Ser. B. 165, 78–102.

Mirsky, R. and Thompson, E. (1975) Thy-1 (Theta) antigen on the surface of morphologically distinct brain types. Cell 4, 95–101.

Möller, E. (1964) Isoantigenic properties of tumors transgressing histocompatibility barriers of the H-2 system. J. Natl. Cancer Inst. 33, 979–989.

Möller, E. and Möller, G. (1962) Quantitative studies of the sensitivity of normal and neoplastic mouse cells to the cytotoxic action of isoantibodies. J. Exp. Med. 115, 527–533.

Morris, R., Letarte-Muirhead, M. and Williams, A. (1975) Analysis in deoxycholate of three antigenic specificities associated with the rat Thy-1 molecule. Eur. J. Immunol. 5, 282–285.

Muramatsu, T. and Nathenson, S. (1970) Studies on the carbohydrate portion of membrane-located mouse H-2 alloantigens. Biochemistry 9, 4875–4883.

Muramatsu, T., Nathenson, S., Boyse, E. and Old, L. (1973) Some biochemical properties of thymus leukemia antigens solubilized from cell membranes by papain digestion. J. Exp. Med. 137, 1256–1272.

Nathenson, S. and Muramatsu, T. (1971) Properties of the carbohydrate portion of mouse H-2 alloantigen glycoproteins. In: G. Jamieson and T. Greenwalt (Eds.), Glycoproteins of Blood Cells and Plasma. Lippincott, Philadelphia, pp. 245–262.

Neauport-Sautes, C., Lilly, F., Silvestre, D. and Kourilsky, F. (1973) Independence of H-2K and H-2D antigenic determinants on the surface of mouse lymphocytes. J. Exp. Med. 137, 511–526.

Obata, Y., Ikeda, H., Stockert, E. and Boyse, E. (1975) Relation of G_{ix} antigen of thymocytes to envelope glycoprotein of murine leukemia virus. J. Exp. Med. 141, 188–197.

Old, L., Stockert, E., Boyse, E. and Kim, J. (1968) Antigenic modulation. Loss of TL antigen from cells exposed to TL antibody. Study of the phenomenon in vitro. J. Exp. Med. 127, 523–539.

Ohno, S. (1971) Genetic implication of karyological instability of malignant somatic cells. Physiol. Rev. 51, 496–526.

Ostberg, L., Rask, L., Wigzell, H. and Peterson, P. (1975) Thymus leukemia antigen contains β_2-microglobulin. Nature (London) 253, 735–737.

Owen, J. and Raff, M. (1970) Studies on the differentiation of thymus-derived lymphocytes. J. Exp. Med. 132, 1216–1232.

Papermaster, B. and Herzenberg, L. (1966) Isolation and characterization of an isoantigenic variant from a heterozygous mouse lymphoma in culture. J. Cell. Physiol. 67, 407–420.

Parkman, R. and Merler, E. (1973) Thymus-dependent functions of mouse thymocyte-fibroblast hybrid cells. Nature (London) New Biol. 245, 14–16.

Peterson, P., Rask, L., Sege, K., Klareskog, L., Anundi, H. and Östberg, L. (1975) Evolutionary relationship between immunoglobulins and transplantation antigens. Proc. Natl. Acad. Sci. USA 72, 1612–1616.

Pious, D. and Soderland, C. (1974) Expression of new antigens by HL-A variants of cultured lymphoid cells. J. Immunol. 113, 1399–1404.

Pious, D., Bodmer, J. and Bodmer, W. (1974) Antigenic expression and cross reactions in HL-A variants of lymphoid cell lines. Tissue Antigens 4, 247–256.

Pious, D., Hawley, P. and Forrest, G. (1973) Isolation and characterization of HL-A variants in cultured human lymphoid cells. Proc. Natl. Acad. Sci. USA 70, 1397–1400.

Raff, M., Owen, J., Cooper, M., Lawton, A., Myron, M. and Gathings, W. (1975) Differences in susceptibility of mature and immature mouse B lymphocytes to anti-immunoglobulin-induced immunoglobulin suppression in vitro. Possible implications for B-cell tolerance to self. J. Exp. Med. 142, 1052–1064.

Ralph, P., Hyman, R., Epstein, R., Nakoinz, I. and Cohn, M. (1973) Independence of θ and TL surface antigens and killing by thymidine, cortisol, phytohemagglutinin, and cyclic AMP in a murine lymphoma. Biochem. Biophys. Res. Commun. 55, 1085–1091.

Reif, A. and Allen, J. (1964) The AKR thymic antigen and its distribution in leukemias and nervous tissues. J. Exp. Med. 120, 413–433.

Schachner, M. and Hämmerling, U. (1974) The postnatal development of antigens on mouse brain cell surfaces. Brain Res. 73, 362–371.

Scheid, M., Boyse, E., Carswell, E. and Old, L. (1972) Serologically demonstrable alloantigens of mouse epidermal cells. J. Exp. Med. 135, 938–955.

Scheid, M., Goldstein, G., Hämmerling, U. and Boyse, E. (1975) Lymphocyte differentiation from precursor cells in vitro. Ann. N.Y. Acad. Sci. 249, 531–540.

Schlesinger, M. (1970) How cells acquire antigens. Prog. Exp. Tumor Res. 13, 28–83.

Schlesinger, M. and Hurvitz, D. (1968) Serological analysis of thymus and spleen grafts. J. Exp. Med. 127, 1127–1137.

Schubert, D. and Cohn, M. (1968) Immunoglobulin biosynthesis, III. Blocks in defective synthesis. J. Mol. Biol. 38, 273–288.

Schwartz, B. and Nathenson, S. (1971) Isolation of the H-2 alloantigens solubilized by the detergent NP-40. J. Immunol. 107, 1363–1367.

Schwartz, B., Kato, K., Cullen, S. and Nathenson, S. (1973) H-2 histocompatibility alloantigens. Some biochemical properties of the molecules solubilized by NP-40 detergent. Biochemistry 12, 2157–2164.

Shimada, A. and Nathenson, S. (1971) Removal of neuraminic acid from H-2 alloantigens without effect on antigenic reactivity. J. Immunol. 107, 1197–1199.

Shimada, A., Yamane, K. and Nathenson, S. (1970) Comparison of the peptide composition of two histocompatibility-2 alloantigens. Proc. Natl. Acad. Sci. USA 65, 691–696.

Silver, J. and Hood, L. (1974) Detergent-solubilized H-2 alloantigen is associated with a small molecular weight polypeptide. Nature (London) 249, 764–765.

Silver, J. and Hood, L. (1976) Structure and evolution of transplantation antigens: Partial amino-acid sequences of H-2K and H-2D alloantigens. Proc. Natl. Acad. Sci. USA 73, 599–603.

Snary, D., Goodfellow, P., Bodmer, W. and Crumpton, M. (1975) Evidence against a dimeric structure for membrane-bound HLA antigens. Nature (London) 258, 240–242.

Sommerville, J. (1970) Serotype expression in Paramecium. Adv. Micro Physiol. 4, 131–178.

Sonneborn, T. (1970) Gene action in development. Proc. Roy. Soc. Lond. Ser. B. 176, 347–366.

Stackpole, C., Jacobson, J. and Lardis, M. (1974) Antigenic modulation in vitro, I. Fate of thymus-leukemia (TL) antigen-antibody complexes following modulation of TL antigenicity from the surfaces of mouse leukemia cells and thymocytes. J. Exp. Med. 140, 939–957.

Stanley, P., Caillibot, V. and Siminovitch, L. (1975a) Stable alterations at the cell membrane of Chinese hamster ovary cells resistant to the cytotoxicity of phytohemagglutinin. Somatic Cell Genet. 1, 3–26.

Stanley, P., Caillibot, V. and Siminovitch, L. (1975b) Selection and characterization of eight phenotypically distinct lines of lectin-resistant Chinese hamster ovary cells. Cell 6, 121–128.

Stanley, P., Narashimihas, S., Siminovitch, L. and Schachter, H. (1975c) Chinese hamster ovary cells selected for resistance to the cytotoxicity of phytohemagglutinin are deficient in a UDP-N-acetylglucosamine-glycoprotein N-acetylglucosaminyltransferase activity. Proc. Natl. Acad. Sci. USA 72, 3323–3327.

Stern, P. (1973) θ alloantigen on mouse and rat fibroblasts. Nature (London) New Biol. 246, 76–78.

Stobo, J. (1972) Phytohemagglutinin and concanavalin A: Probes for murine 'T' cell activation and differentiation. Transplant. Rev. 11, 60–86.

Stocker, B. and Mäkelä, P. (1971) Genetic aspects of biosynthesis and structure of *Salmonella* lipopolysaccharide. In: G. Weinbaum, S. Kadis and S. Ajl (Eds.), Microbial Toxins, Vol. 4. Academic Press, New York, pp. 369–438.

Stockert, E., Old, L. and Boyse, E. (1971) The G_{IX} system. A cell surface alloantigen associated with murine leukemia virus; implications regarding chromosomal integration of the viral genome. J. Exp. Med. 133, 1334–1355.

Strominger, J., Cresswell, P., Grey, H., Humphreys, R., Mann, D., McCune, J., Parkam, P., Robb, R., Sanderson, A., Springer, T., Terhorst, C. and Turner, J. (1974) The immunoglobulin-like structure of human histocompatibility antigens. Transplant. Rev. 21, 126–143.

Terhorst, C., Parham, P., Mann, D. and Strominger, J. (1976) Structure of HL-A antigens: Amino-acid and carbohydrate compositions and NH_2-terminal sequences of four antigen preparations. Proc. Natl. Acad. Sci. USA 73, 910–914.

Thiele, H. and Stark, R. (1974) Common antibody-induced redistribution on thymocytes of strain-, species- and non-species-specific antigenic determinants shared by brain and thymocytes of mice. Immunology 27, 807–813.

Thiele, H., Stark, R. and Keeser, D. (1972) Antigenic correlations between brain and thymus. I. Common antigenic structures in rat and mouse brain tissue and thymocytes. Eur. J. Immunol. 2, 424–429.

Trowbridge, I. and Hyman, R. (1975) Thy-1 variants of mouse lymphomas: Biochemical characterization of the genetic defect. Cell 6, 279–287.

Trowbridge, I. and Mazauskas, C. (1976) The immunological properties of murine thymus-dependent lymphocyte surface glycoproteins. Eur. J. Immunol. 6, 777–782.

Trowbridge, I., Ralph, P. and Bevan, M. (1975a) Differences in the surface proteins of mouse B and T cells. Proc. Natl. Acad. Sci. USA 72, 157–161.

Trowbridge, I., Weissman, I. and Bevan, M. (1975b) Mouse T cell surface glycoprotein recognized by heterologous anti-thymocyte sera and its relationship to the Thy-1 antigen. Nature (London) 256, 652–654.

Tung, J., Vitetta, E., Fleissner, E. and Boyse, E. (1975) Biochemical evidence linking the G_{IX} thymocyte surface antigen to the gp 69/71 envelope glycoprotein of murine leukemia virus. J. Exp. Med. 141, 198–205.

Vitetta, E., Uhr, J. and Boyse, E. (1972) Isolation and characterization of H-2 and TL alloantigens from the surface of mouse lymphocytes. Cell. Immunol. 4, 187–191.

Vitetta, E., Boyse, E. and Uhr, J. (1973) Isolation and characterization of a molecular complex containing Thy-1 antigen from the surface of murine thymocytes and T cells. Eur. J. Immunol. 3, 446–453.

Vitetta, E., Uhr, J. and Boyse, E. (1975a) Association of a β_2-microglobulinlike subunit with H-2 and TL alloantigens on murine thymocytes. J. Immunol. 114, 252–254.

Vitetta, E., Artzt, K., Bennett, D., Boyse, E. and Jacob, F. (1975b) Structural similarities between a product of the T/t locus isolated from sperm and teratoma cells and H-2 antigens isolated from splenocytes. Proc. Natl. Acad. Sci. USA 72, 3215–3219.

Vitetta, E., Capra, J., Klapper, D., Klein, J. and Uhr, J. (1976) The partial amino acid sequence of an H-2K molecule. Proc. Natl. Acad. Sci. USA 73, 905–909.

Watkins, W. (1966) Blood group substances. Science 152, 172–181.

Wiener, F., Dalianis, T. and Klein, G. (1975) Cytogenetic studies on H-2 alloantigenic loss variants selected from heterozygous tumors. Immunogenetics 2, 63–72.

Wiener, F., Dalianis, T., Klein, G. and Harris, H. (1974) Cytogenetic studies on the mechanism of formation of isoantigenic variants in somatic cell hybrids, I. Banding analyses of isoantigenic variant sublines derived from the fusion of TA3Ha carcinoma with MSWBS sarcoma cells. J. Natl. Cancer Inst. 52, 1779–1796.

Wu, A., Till, J., Siminovitch, L. and McCulloch, E. (1968) Cytological evidence for a relationship between normal hematopoeitic colony-forming cells and cells of the lymphoid system. J. Exp. Med. 127, 455–463.

Yu, A. and Cohen, E. (1974) Studies on the effect of specific antisera on the metabolism of cellular antigens, I. Isolation of thymus leukemia antigens. J. Immunol. 112, 1285–1295.

Dynamics of antibody binding and complement interactions at the cell surface

Karl-Gösta SUNDQVIST

1. Introduction

Information obtained in studies of the interaction of antibodies with different components of the animal cell plasma membrane has contributed significantly to our present understanding of membrane structure and function. Experimental observations showing that multivalent antibody molecules can induce topographic rearrangements of cell surface components have played a major part in the emergence and subsequent consolidation of the concept that the plasma membrane is a dynamic structure in which the various components are capable of undergoing rapid and reversible topographic rearrangements. Antibodies, by virtue of their specificity, provide a powerful tool for the experimental study of membrane structure and the analysis of the functional properties of individual membrane components. The combination of antibody with antigen is a fundamental process in immunology which has been analysed extensively in regard to antibody reactions with simple haptens and with soluble macromolecules. Antigen–antibody interactions at the surface of living nucleated cells may be considered as more complex, since they are influenced by the fact that the antigen molecules belong to a metabolically active organism. Various processes and perturbations at the surface membrane of cells, referred to collectively as membrane dynamics, may markedly influence the interaction of antibodies with the cell surface. Such phenomena can alter and even circumvent the detection of antigen–antibody interactions at the cell surface. However, an important formal similarity between cell surface antigens and soluble macromolecular antigens is that their interaction with antibodies is limited in both cases to small areas on the antigenic surface. Therefore, the laws governing the interaction between macromolecular antigens and antibodies in solution should also apply, with certain exceptions, to the interaction between surface antigens of living cells and antibodies. Analysis of antigen–antibody interactions at the surface of living cells must therefore consider two main components: (1) the behaviour of the cell membrane elements; and (2) the molecular interactions between the ligands involved.

Another group of protein molecules involved in dynamic interactions at the surface membrane of cells are the various components of the complement system. Complement proteins interact with both membrane-bound antibodies and cell surface receptors. There is some evidence that complement components modulate the behaviour of membrane-bound antibodies and may also affect the cell surface membrane directly. Activation of complement by antibody at the surface of sensitized cells provokes a transfer of complement factors from the soluble form to the solid phase of the cell surface, possibly involving final association with the plasma membrane as peripheral or integral membrane components.

The primary object of this review is to discuss antibody lattice formation at the cell surface and its bearing on the behaviour and fate of membrane-bound antibodies. Attention will also be given to aspects of the dynamics of complement interaction with antibodies at the cell surface.

2. Cell surface molecules as antibody-binding structures

The primary aim of this section is to consider the various structural features which influence the capacity of cell surface antigen molecules to interact with the combining sites of specific antibodies, namely: (1) accessibility; (2) valency; (3) density; and (4) topographic distribution on the membrane.

Binding of antibodies to Fc-receptors will not be considered although it should be regarded as a potential source of error in studies of specific antibody binding to antigenic determinants at a cell surface. This contention is supported by results of lymphocyte classification using anti-immunoglobulin sera (Lobo et al., 1975; Winchester et al., 1975).

2.1. Accessibility

There is evidence that not all cell surface antigens which could potentially combine with antibodies in a particular antiserum are accessible to antibody. Several reports have shown that enzyme treatment of viable normal cells exposes previously undetectable sites. For example, previously undetected *HL-A* specificities appear on normal human lymphocytes after trypsin treatment (Gibovski and Terasaki, 1972). Trypsin treatment also increases the number of detectable Gross leukemia virus associated surface antigens on AKR virus-induced rat lymphoma cells (Brandschaft and Boone, 1974). A variety of enzymes, *Vibrio cholerae* neuraminidase, trypsin, chymotrypsin, papain, L-asparaginase and lysozyme have been shown to increase the sensitivity of mouse lymphoid cells to lysis by alloantibody and complement (Ray and Simmons, 1973).

The increased number of surface antigens found on certain cells after enzyme treatment could result from either the removal of surface components which sterically block access of antibody to the antigens or from the rearrangement of cell surface architecture, or both. For example, sialic acid may influence antigen expression on tumor cells, since its removal by neuraminidase exposes previously undetectable antigenic sites (Currie, 1967; Currie and Bagshawe, 1967; Sanford, 1967; Bagshawe and Currie, 1969; Kassulke et al., 1969; Bekesi et al., 1971; Simmons et al., 1971). The increased agglutinability of cells by plant lectins induced by mild proteolysis (Burger, 1969; Inbar and Sachs, 1969) has also been proposed as resulting from the exposure of additional cryptic sites for lectin binding in normal cell membranes (Burger, 1969), though binding studies in many laboratories using radiolabeled lectins have failed to confirm this proposal (review, Nicolson, 1974).

Alternative mechanisms which may account for variations in the binding of antibodies and other ligands to the cell surface and possibly also for variations in antigen expression focus on the fact that surface membrane molecules can be redistributed within the plane of the membrane. Mild proteolysis has been shown to induce clustering of lectin binding sites at the surface of 3T3 cells (Marchesi et al., 1972; Nicolson, 1972). It is likely that the increased local density of lectin binding sites in these clusters increases the probability of lectin-mediated crossbridging of receptors on separate cells, thus enhancing the opportunities for agglutination. In accordance with this line of reasoning, the probability of antibody crosslinking of antigen determinants on different surface antigen molecules and lattice formation would also be expected to increase following redistribution of antigen molecules. It is also possible that multivalent ligands such as antibodies on interacting with cell surfaces to redistribute antigen molecules may simultaneously cause rearrangements of surface molecular conformation exposing hidden sites. However, data on molecular conformation and antigenicity strongly indicate that it is the surface residues which comprise the antigenic determinants in proteins (Prager and Wilson, 1971; Reichlin, 1972; Atassi et al., 1973). Therefore, the latter possibility seems less likely.

The accessibility of antigen molecules is probably influenced by their localization in the plasma membrane. This concept is illustrated by immunoglobulin molecules. For example, it would be predicted that these molecules expose Fab-portions to the exterior while their Fc-portion is attached to the plasma membrane and would probably be less accessible. Consistent with this idea is the finding that certain antisera to the Fc-portion did not stain cell surface immunoglobulin (Pernis et al., 1970; Froland and Natvig, 1972). Further evidence in support of this concept is that certain antisera to IgM are unable to react with surface IgM on human leukemia cells (Fu and Kunkel, 1974). These antisera were directed against antigen determinants in the last 50 amino acids of the carboxy-terminal end of the μ-chain. This suggests that the μ-chain is buried in the plasma membrane. Additional support for the concept that the Fc-part and especially the μ-chain is buried in, or hidden by, the plasma membrane is provided by the finding that antibodies to the L-chain detect 90% of radioiodinated cell surface immunoglobulin on mouse spleen cells whereas μ-chains bind approximately 50% of anti μ-chain antibodies (Vitetta and Uhr, 1974; 1975).

One type of cell surface associated antigen which is normally not accessible to antibody at all is found in the submembranous microfilament system. Microfilament systems can be visualized close to the plasma membrane in the cytoplasm of thyroid epithelial cells (G. Biberfeld et al., 1974), liver cells (Farrow et al., 1971), and thymus cells (Fagraeus et al., 1973) by the use of smooth muscle antibodies (SMA) obtained from patients with active chronic hepatitis. Detection of these structures is only possible when the cells have been acetone-fixed or sectioned. Smears of suspended cells should also be treated with a chelating agent in order to expose the structures to antibodies (Fagraeus et al., 1975a,b). The submembranous antigen detected by SMA has been identified as actin (Gabbiani et al., 1973; Lidman et al., 1976). SMA will certainly prove to be a useful tool in studies of the role of cell-surface associated microfilaments in cell membrane dynamics (see also section 5).

2.2. Valence

A prerequisite for antibody lattice formation on the cell surface is that the interacting antibodies and antigen molecules are multivalent. Consequently, in order to be precipitated by antibody membrane molecules must consist of several antigenic determinants (see also section 3.2).

Virtually no information is presently available on the epitope valence of surface membrane molecules in situ. Existing data derive from experiments with solubilized membrane components. The strong histocompatibility antigens of mouse in soluble form were not precipitated by alloantisera using the gel diffusion technique (Sanderson and Welsh, 1974). This failure to obtain precipitation probably excludes that the same epitope is represented several times per molecule. This inference is also supported by the finding that solubilized histocompatibility antigens do not activate complement, a process requiring binding of two antibody molecules per antigen molecule. Results obtained with papain-solubilized *HL-A2* and *HL-A7* molecules and monospecific antisera against *HL-A2* and *HL-A7* indicate that these antigens occur only once per molecule. This conclusion was reached on the basis of the size of the antigen-antibody complexes formed using Fab-fragments or intact antibodies (Sanderson and Welsh, 1974).

There is also evidence that the major histocompatibility determinants are expressed on separate molecules. Support for this contention was obtained by independent cap formation (Preud'homme et al., 1972; Neauport-Sautes et al., 1973) and from precipitation tests using solubilized antigens (Sanderson, 1968; Cullen et al., 1972; Sanderson and Welsh, 1974). The fact that determinants of certain antigenic specificities are expressed on separate molecules and only once per molecule has obvious implications to every theory concerning antibody lattice-formation at the cell surface (see also section 3).

2.3. Number of antigenic sites per cell and antigenic density

Antigen determinants at a cell surface can be quantitated by measurements of the binding of a labeled antibody molecule to the cells. It is important to realize that calculations of the amount of antigen determinants per cell give minimum values only, and that the number of antigenic sites calculated per cell is not necessarily the same as the number of antigen molecules. Each antigen molecule may contain from one to a cluster of sites. The values calculated for antigenic density generally also do not pay attention to the fact that cell surfaces are not perfectly smooth. Calculations of this type should therefore be regarded with caution.

Quantitative measurements of the number of cell surface antigens may yield interesting information as to antibody binding, the crosslinking of different sites located on individual molecules and data on antigenic distribution. There is now good evidence that the antigenic distribution at long range is uniform (see section 2.4). So assuming a random distribution of cell surface antigens one may obtain approximative values of antigenic density. Calculations of the number of surface antigens per cell show large variation between different antigen specificities. For example, the relatively low value of 10^4 Rh(D) sites per red blood cell (Singer, 1974) means that the average distance of separation of the molecules is more than 1000 Å which, in turn, implies that crosslinking of these sites by IgG antibodies (with a span of 140 Å) is impossible. Corresponding values for other antigens are: immunoglobulin 10^5 molecules (Abbas et al., 1975); Forssman antigen, $6 \cdot 10^5$ molecules (Singer, 1974); and *HL-A* 7000 and 30 000 sites, respectively, for heterozygously and homozygously expressed specificity (Sanderson and Welsh, 1974).

2.4. Distribution

The distribution of antigen molecules at the cell surface has been and is still a controversial subject. A few years ago cell surface antigens were considered to be essentially continuously or discontinuously distributed (Cerrotini and Brunner, 1967; Aoki et al., 1969; Fagraeus and Jonsson, 1970; Kourilsky et al., 1971; Stackpole et al., 1971). It was clearly demonstrated, however, that the indirect labelling techniques used to detect surface membrane antigens induced clustering of antigen-antibody complexes at the cell surface and the clustering was responsible for the discontinuous distribution of cell surface antigens (Davis and Silverman, 1968; Davis, 1972; Davis et al., 1972). However, by using Fab fragments, binding of antibodies to the cells at low temperature or prefixation of the cells, surface membrane antigens were found to be essentially uniformly distributed. This pattern has been demonstrated for the Rh (D) antigen (Nicolson et al., 1971a), alloantigens (Nicolson, 1972; Sundqvist, 1972; Davis, 1972; de Petris and Raff, 1974; Loor et al., 1975), blood group substance A (Sundqvist, 1972), and for immunoglobulin (Ig) molecules on mouse B-cells (Taylor et al., 1971; de Petris and Raff, 1972, 1973; Loor et al., 1975). Lectin receptors have also been found to have a continuous random distribution at the cell surface (Karnovsky et al., 1972; Yahara et al., 1972; de Petris and Raff, 1974; Loor, 1974; Nicolson and Singer, 1974; Weller, 1974). By contrast the θ-antigen on mouse thymocytes was concentrated to a uropod (de Petris and Raff, 1974).

The concept of a random distribution of cell surface antigen molecules has been established by immunofluorescence techniques at the light microscope level and by electron microscopic examination of thin sections of cells. However, studies of thin sections reveal the distribution of markers in one plane of the cells only and the immunofluorescence technique has a low resolution. Taking advantage of the fact that freeze-fracturing gives a two dimensional map of the cell membrane, it has been demonstrated that immunoglobulin on murine lymphocytes (Abbas et al., 1975) was located in microclusters with interconnecting networks. In another case, combining electron microscopy and an immunolatex method, the surface immunoglobulin on lymphocytes was found in a similar clustered pattern (Linthicum and Sell, 1975).

Surface membrane components of different kinds have been demonstrated to redistribute spontaneously without any influence of external ligands (Frye and Edidin, 1970; Poo and Cone, 1974; Ehrnst and Sundqvist, 1975; McDonough and Lilien, 1975). Such non-ligand induced redistributions could possibly account for some deviation from a random distribution of cell surface components.

The organization of molecules in the plasma membrane has obvious biological importance in determining the functional properties of the cell surface. Clustered antigen receptors, for example, could be predicted to bind antigens and antibodies more effectively than more segregated receptors. The probability of antibody crossbridging of clustered antigen molecules would be higher than crossbridging of isolated randomly distributed ones. Differences in native clustering of different cell surface antigens may therefore account for variations in antibody binding to cells, and possibly changes in antigen expression (see section 6.2), and capping tendency in different systems (see section 5.3.2).

3. The precipitin reaction at the cell surface

The lattice theory (Marrack, 1938; Heidelberger, 1956) has been applied to describe the formation of complexes of multivalent antigens and antibodies in solution. There is also evidence, however, that binding of antibodies to antigens on a cell membrane leads to formation of macromolecular aggregates, governed by similar principles to the precipitation of antigen by antibodies in solution. The primary purpose of this section is to discuss the precipitin reaction at the cell surface in order to provide a background for understanding the events, that accompany interactions between surface antigens and antibodies discussed later in sections 4 and 5.

3.1. General considerations of antibody structure and its implications for lattice formation

From a functional point of view antibody molecules consist of two regions, Fab and Fc. Fab is responsible for the specific reaction with antigen or the primary function of the antibody molecule. Fc on the other hand determines to a great extent those biological activities that follow antigen-antibody reactions, the so called secondary functions. Antibodies interact with cell surfaces both by virtue of their Fab and Fc-portions. However, there is probably no energetic role for the Fc-fragment in specific Fab-mediated binding of antibody to antigens (Hornick and Karush, 1972) and accordingly no function for Fc in lattice formation.

Lattice formation depends on crossbridging of determinants on different antigen molecules by antibody. Structural features of antibodies which are crucial for their capability to crosslink antigens are valency, flexibility, and span of the antigen-binding region.

3.1.1. Multivalent interaction and affinity
Multivalency is a general property of antibodies, the "energetic advantage" of which probably is of great biological importance (Metzger, 1970). Multivalent interaction increases the affinity of antibodies significantly compared with the affinity offered by single combining sites and multivalent molecules are much more effective in mediating immunity compared to monovalent fragments.

The important role of the multivalency of antibodies for their biological function came from early studies on the neutralization of viruses and haptens by antibody (Kalmanson and Bronfenbrenner, 1943; Lafferty, 1963; Blank et al., 1971). The importance of multivalency in antibody function has been analyzed extensively in recent years (Hornick and Karush, 1969; 1972). The importance of the multivalency of antibody molecules for their binding to cell surfaces is also well documented. One report compared the binding to red cells of bivalent 7 S and monovalent 3.5 S fragments (Greenbury et al., 1965). Binding constants of the bivalent species was 150–450 times larger. IgM and IgG antibodies with similar binding constants for monovalent hapten were compared in haemagglutination and haemolytic tests. IgM was approximately 100-fold more effective. Solheim (1972) found differences in association constants of 1900–5800 when comparing the binding to red cells of divalent fragments of anti-blood group I with intact IgM. Monovalent fragments exhibited K-values which were 12 500–38 500 times lower than intact antibody. Essentially the same results have been obtained with *Helix pomatia* A haemagglutinin and human erythrocytes (Hammarström, 1973). A clear demonstration of the more effective binding of IgM compared to IgG is also provided by the finding that IgM can almost completely displace IgG from the surface of sheep red cells (Humphrey and Dourmashkin, 1965). Agglutination of cells by antibodies is another example of a function where IgM due to its higher valency is more effective than IgG. The increased binding efficiency of multipoint versus single-point interactions means that the half-life of an antibody bound to a cell by monovalent binding is a matter of seconds whereas bivalent binding expands the half-life to 30 min and binding by three combining sites expands the half-life to hours.

Because of their multivalent nature, antibodies can crosslink separate molecules in the cell membrane. This is a prerequisite for such dynamic membrane events as the patching and capping of plasma membrane components.

3.1.2. Span and flexibility

Immunoglobulin molecules have been shown by electron microscopy to be Y-shaped structures (Noelken et al., 1965). The angle between the arms of the Fab_2-fragment and the degree of mobility of the arms could be anticipated to be critical for binding of antibodies to antigen and for the crosslinking of spatially separated sites, perhaps unfavorably disposed on antigen molecules.

Binding of antibody to the cell surface is accompanied by pronounced alterations in the distribution of antigen-antibody complexes leading to increased local densities of molecules. Maintenance of the union between antigen and antibody under these conditions is probably to a great extent dependent on the flexibility of the antibody molecules which crosslink cell surface antigens.

Values of the inter-Fab angle have been calculated based on examination of immunoglobulin molecules by electron microscopy. For rabbit IgG the angle is 60° (Green, 1969), for human IgG, 120° (Pilz et al., 1970), and for IgM, 55–70° (Feinstein et al., 1971). Examination of the relevant literature seems to indicate that immunoglobulin molecules possess limited flexibility in regard to freedom of motion of their Fab regions (Yguerabide et al., 1970; Werner et al., 1973). Support for the notion that immunoglobin molecules display limited flexibility is also provided by the finding that IgM molecules exhibit a functional pentavalency, in spite of the fact that this molecule has the expected number of combining sites when isolated Fab regions are examined separately (Metzger, 1970). There are also data which suggest that the effective valency of IgG in great antigen excess, where cooperative binding plays little role, is one (Arend et al., 1972). It could be expected in view of this that membrane events depending on crosslinking of antigens would have little tendency to occur at low antibody concentrations.

3.2. Lattice formation by a single layer of antibody

3.2.1. Model systems

In accordance with the reaction between antibody molecules and soluble antigens, certain predictions can be made about the precipitin reaction involving cell surface antigens and antibodies. Multivalency of the reactants is required. At extreme antibody excess crosslinking of antigen molecules would be unusual or absent (Fig. 1A). At moderate excess of antibody relative to antigen, crosslinking and lattice formation are presumably maximal (Fig. 1B). The lower the concentration of antibody relative to antigen the less the tendency to crosslinking (Figs. 1C and D).

The number of determinants per antigen molecule exerts a critical influence on lattice formation. Consider first membrane antigen molecules with one

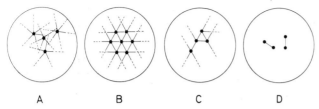

Fig. 1. Schematic representation of the formation of precipitates between cell surface antigens (●) and bivalent antibodies (——) according to the lattice theory. Dotted lines indicate that the complexes may continue to extend. The formation of precipitates will vary depending on the concentration of the reactants: (A) extreme antibody excess; (B) moderate antibody excess; (C) the equivalence zone; and (D) antigen excess zone.

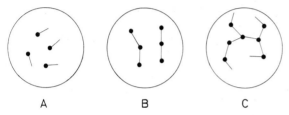

Fig. 2. Schematic representation of the formation of complexes between cell surface antigens (●) and bivalent antibodies (——) when the antigen molecule possesses one (A), two (B), or three (C) determinants.

antigenic site only (Fig. 2A). Although crossbridging of two antigen molecules may occur, lattice formation is impossible. If antigen molecules have two sites, lattice formation is possible but the lattices formed will consist of linear aggregates only (Fig. 2B). With three or more determinants per antigen molecule increasingly larger precipitates are formed (Fig. 2C) (see also section 2.2).

Crosslinking is impossible when the distance between identical determinants on separate antigen molecules exceeds the span of the Fab_2-region of a maximally open IgG-antibody or approximately 140 Å.* Obviously IgM, due to its size, can be postulated to be a more efficient "crosslinker" than IgG, especially if the antigenic density is low. The spatial relationship of antigen determinants on separate antigen molecules could also be unsuitable for crosslinking. For example, the determinants may be buried in the antigen molecule.

Antigen molecules are able to diffuse laterally within the plane of the plasma membrane. This movement can be both "spontaneous" (Frye and Edidin, 1970; McDonough and Lilien, 1975; Schlesinger et al., 1976; Poo and Cone, 1974; Ehrnst and Sundqvist, 1975) and ligand-induced (Taylor et al., 1971; Sundqvist, 1972; Edidin and Weiss, 1972; Doyle et al., 1974). Movement of antigens may influence

---

*A double-layer of antibody may crosslink sites even above this distance but not more than 400 Å apart.

the conditions for lattice formation. Antigen molecules which are ordinarily distributed too far apart from each other for crosslinking to occur could perhaps redistribute by diffusion to permit crosslinking. Cell membrane mobility may cause increasing aggregation of an antibody lattice at the cell surface. This aggregation takes place even at low temperature but is then limited to formation of microaggregates or patching (de Petris and Raff, 1973). In contrast, at 37°C increasingly larger aggregates with a tendency to polar localization on the cell membrane are formed (Taylor et al., 1971). The aggregation presumably affects the lattice structure, but does not disrupt it, since there is little loss of antibody molecules from the lattice during redistribution, at least under conditions when crosslinking is optimal (Sundqvist, 1973b; 1974a).

If the same surface membrane molecule possesses two or more identical sites in positions which permit association to both combining sites of a single IgG antibody this will probably decrease the probability of the antibody crosslinking separate antigen molecules on the cell membrane. This conclusion is supported by calculated predictions which state that "where multisite adherence to a single particle and crosslinking of discrete particles are both possible, the former is predicted to predominate strongly" (Crothers and Metzger, 1972). Protein antigens containing several identical sites probably consist of several at least partially identical polypeptide chains, since groups of the same amino acid residues do not recur as repetitive sequences in a given polypeptide chain.

3.2.2. Experimental systems

One experimental approach which has provided information concerning antibody lattice formation at the cell surface involves determination of the amount of membrane-bound antibody and the degree of crosslinking of membrane molecules obtained with this antibody at different antibody concentrations (Sundqvist, 1973a,b; 1974a). The amount of membrane-bound antibody can be measured by quantitative immunofluorescence at the single cell level. This enables determination of the optimal antibody concentration for crosslinking and lattice formation. The microfluorometric measurements are performed on both a per cell membrane unit and a per cell basis.

Because crosslinking is necessary for redistribution of surface membrane molecules (Taylor et al., 1971), and redistribution can be readily observed visually by immunofluorescence or immunoelectronmicroscopy, redistribution provides a convenient parameter for the assay of antibody-induced crosslinking. Accordingly, data from immunofluorescence studies of antibody-induced redistribution at different antibody concentrations in several test systems can be divided into two categories: (1) test systems that exhibit crosslinking; and (2) test systems that do not exhibit crosslinking (Table I). Figure 3 demonstrates results representative of a test system where crosslinking was very effective as demonstrated by rapid and pronounced capping of surface immunoglobulins (Ig) on mouse lymphocytes. It should be emphasized, however, that this does not mean that those test systems which do not exhibit redistribution/crosslinking with a single-layer of antibody are incapable of redistribution, since this can

TABLE I

EVIDENCE FOR CROSSLINKING OF CELL SURFACE ANTIGENS BY ANTIBODY IN DIFFERENT TEST SYSTEMS AS REVEALED BY ANTIBODY-INDUCED REDISTRIBUTION OF SURFACE ANTIGENS

Cell type	Antigen specificity	Cross-linking and redistribution	
		Single-layer of antibody	Double-layer of antibody
Mouse spleen	Immunoglobulin	Yes	Yes
Mouse spleen	H-2	No	Yes
Mouse spleen	Species antigens	No	Yes
Human blood lymphocytes	Species antigens	No	Yes
Monkey kidney	Blood group A	No	Yes
Mouse L-cells	H-2	No	Yes
Entamoeba histolytica	Surface antigens (not defined)	No	Yes
Lu 106	β_2m	No	Yes
Lu 106	Species antigens	No	Yes
Lu 106 (measles virus infected)	Measles virus hemagglutinin	Yes	Yes

usually be induced by a double-layer of antibody (see section 3.3).

Possible interpretations of these results are: (1) that test systems showing evidence of cross-linking (i.e. antigen redistribution occurred) were the only ones with multiple determinants per antigen molecule; or (2) that the distance was too great or the spatial relationship between separate antigen molecules was unsuitable in all of the test systems that did not exhibit crosslinking; or (3) that antibody binding to identical sites on the same antigen molecule predominated in those test systems where no evidence for crosslinking was noted.

At high antibody concentrations in the Ig-mouse lymphocyte system a plateau was reached, above which further antibody-binding to the cells was limited (Fig. 3). It is interesting to note that antibody-induced redistribution of cell surface antigens occurred at all other antibody concentrations except those corresponding to this plateau. There was no crosslinking at the high antibody concentrations corresponding to the plateau and it is logical to assume that this mimicked the conditions of extreme antibody excess illustrated in Fig. 1A, so that antibodies were bound univalently to the cell surface. This probably implies that the affinity of binding at extreme antibody excess was several orders of magnitude lower than in other regions of the reaction curve. Such low-affinity univalent binding probably causes a relatively higher dissociation of antibody from the cells compared with other antibody concentrations in which crosslinking is more pronounced.

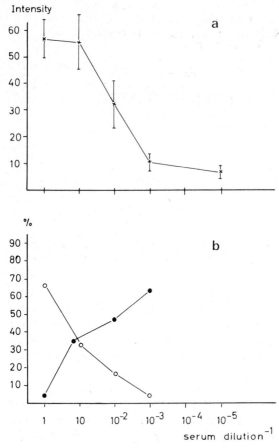

Fig. 3. Simultaneous determination of antibody binding (a) and the degree of antibody-induced redistribution of cell surface antigens (b) at different antibody concentrations using a single layer of antibody. Mouse spleen cells were reacted with fluorescein-conjugated anti-immunoglobulin, which undiluted had a protein concentration of 19.2 mg/ml during 30 min at 0°C. The cells were then washed, incubated for 10 min at 37°C and subsequently fixed in 4% paraformaldehyde. Antibody binding was determined by microfluorometry and the reaction intensity per cell is given in arbitrary units. The distribution of antigen is classified as being either predominantly random (○) or capped (●). The ordinate in B represents capped cells as a percentage of the total number of stained cells.

3.3. Lattice formation by a double layer of antibody

3.3.1. Model systems

In contrast to a lattice consisting of a single antibody layer, the formation of a double-layer lattice is not limited by the number of determinants per antigen molecule at the cell surface since the antibodies applied in the first step can serve as antigens for antiglobulin antibodies applied in a second step. Assuming that the antiglobulin antibodies are IgG, then these will bind one or two of the membrane-bound antibodies. One membrane-bound antibody can then be crosslinked with other membrane-bound antibodies by one or several antiglobulin antibodies. The amount of antiglobulin antibodies attached to each membrane-bound antibody is maximized by the number of sites per molecule of these, probably seven. Effective crosslinking and a stable lattice formation can be predicted to require a certain amount of antiglobulin molecules relative to membrane-bound antibody.

3.3.2. Experimental systems

The approach used to study antibody lattice formation using a double layer of antibody is the same as with a single layer, i.e. antibody binding to cells at different antibody concentrations is compared with the tendency of the antibody to induce redistribution of surface determinants within the plane of the membrane (Sundqvist, 1973a,b; 1974a). The tendency to redistribution is used as an indication of crosslinking of membrane-bound antibodies by antiglobulin-molecules which thus provides indirect information on the behavior of the surface antigens per se. The method used to determine the amount of antibody at the cell surface is the same as when lattice formation of a single antibody layer is studied, i.e. quantitative immunofluorescence. In addition, two other quantitative methods can be employed, the isotope-antiglobulin technique (Harder and McKhann, 1968; Sparks et al., 1969), and the isotope modification of the mixed haemadsorption technique (Sundqvist and Fagraeus, 1972; Sundqvist, 1974b).

The fact that most cell surface antigens do not cap with a single layer of antibody but require a double layer may contain an important message. It has been proposed (Karnovsky et al., 1972) that surface antigens on lymphocytes which were not capped by a single antibody layer, e.g. histocompatibility antigens and species antigens, have a lower cell surface density than immunoglobulins, which cap with a single layer of antibody. However, as mentioned earlier, another explanation may be that membrane molecules which require a double antibody layer in order to cap possess only one antigenic determinant.

When a double layer of antibody is allowed to react with the cell surface the reaction curves obtained in most test systems (Table I) consisted of a prozone followed by a peak or a plateau and a decreasing part of the curve (Fig. 4). This is in contrast to the direct techniques, where a clear plateau of binding is not usually observed (Sundqvist, 1973a). The prozone effect, which is very pronounced in many test systems, will be discussed in detail later in section 4. Maximal crosslinking/redistribution is always obtained in regions corresponding to the decreasing part of the reaction curves, i.e. at a high concentration of antiglobulin molecules relative to membrane-bound antibody (Fig. 4) (Sundqvist, 1973a,b; 1974a). In contrast, at antibody concentrations corresponding to the equivalency zone, and especially to the prozone, of the reaction curves, crosslinking/redistribution was less effective.

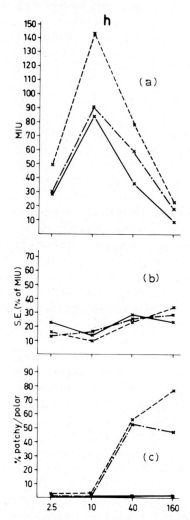

Fig. 4. Simultaneous determination of antibody binding (a) and the degree of antibody-induced redistribution of surface antigens (c) at different antibody concentrations using double-layer technique. Monkey kidney cells (blood group A positive) were reacted with anti-A antiserum at different dilutions (see abscissa) followed by incubation with fluorescein conjugated antiglobulin at 0°C (×——×); 37°C (×---×); and at 37°C in the presence of 10^{-2} M NaN_3 (—·—·—). The reaction intensity (a) was quantitated per membrane unit by microfluorometry (ordinate). The degree of antibody induced redistribution is expressed as patchy-polar cells as a percentage of the total number of stained cells (c), and (b) shows the standard percent error of the mean intensity per membrane unit (MIU).

4. Prozone effects

4.1. Observations of prozone effects

Prozone effect means an absence of detectable antibody reactivity to antigen at high antibody concentration, but not at lower ones. The prozone is the part of an antibody titration curve, showing weak reactivity, or false negative results, despite high antibody concentration. According to text books of Immunology, prozone reactions are well known in agglutination tests but are not mentioned as occurring in other immunological assays (Davis et al., 1967). It is also stated that the molecular basis for this anomalous behavior is not understood. Prozone in agglutination reactions has tentatively been attributed to blocking antibodies. It seems, however, that prozone effects exist not only in agglutination tests but are also frequently obtained with all indirect immunological methods, i.e. those involving a double layer of antibody (Sundqvist, 1972; 1973a; 1974a,b). In a study of virus antigen-antibody systems a prozone was noted but according to the authors "no valid explanation for the prozone could be given" (Espmark et al., 1971). Titration of anti-complementary human sera, containing antibodies to Epstein-Barr virus-associated nuclear antigen, also revealed a prozone effect in anti-complementary immunofluorescence (Henle et al., 1974). The anti-complementarity suggests that the prozone was due to presence of antigen-antibody complexes in the sera. A tendency to prozone and a definite plateau of binding was also observed in studies using an indirect "radioimmunolabeling technique" for detection of lymphocyte surface receptors (Nossal et al., 1972).

The evidence that indirect immunological methods generally exhibited prozone effect (Sundqvist, 1972, 1973a, 1974a,b) may seem in contradiction to other investigators who obtained linear or almost linear relationships between reaction intensity and antibody concentration (Nairn et al., 1969; Strom and Klein, 1969; Killander et al., 1970; Wahren, 1971; Levin et al., 1971). Furthermore, despite the fact that indirect methods are generally employed, and have been for a long time, the immunological literature contains few reports of prozone effects. Two likely explanations of this discrepancy are that authors reporting a linear relationship used concentrations of serum reagents corresponding to the falling phase of the intensity curves, and that quantitative methods, which facilitate detection of prozone effects, have been used relatively little. It is also possible that prozone effects may account for many failures to demonstrate antibody reactivity by immunological methods.

4.2. Studies of the mechanism

By the use of three independent double-layer techniques, indirect immunofluorescence (Sundqvist, 1973a), the isotope antiglobulin technique (Sundqvist, 1974a), and mixed hemadsorption (Sundqvist and Fagraeus, 1972; Sundqvist, 1974b) the prozone effect has been detected in several different antigen-antibody systems (Fig. 4). Initially a number of explanations seemed possible to account for the prozone effect. A series of experiments, some of which are illustrated in Fig. 5 and Table II, were therefore undertaken to analyze the prozone mechanism. These experiments, performed with three different methods, gave similar results, suggesting that the prozone effect observed with the three methods probably involved a similar underlying mechanism. The data from these studies showed that:

(a) the reaction curves were essentially identical;

(b) the course and prozone tendency of the reaction curves were temperature independent but varied with the relative concentrations of antibody in the two steps of the double layer, indicating that lattice formation was optimal at certain concentrations of the reactants, the immunoglobulin (step 1) and the antiglobulin (step 2);

(c) a high ratio of antiglobulin to membrane-bound antibody, chemical fixation of antibody bound to the cells, and extensive washing of these cells shifted the peak of the reaction curves towards an excess of membrane-bound

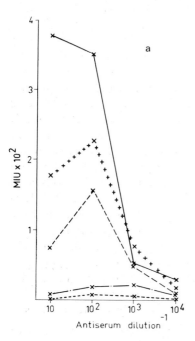

Fig. 5a. Analysis of the mechanism of the prozone effect obtained by indirect labeling techniques. Intensity curves obtained with the indirect immunofluorescence method illustrating the dependence of the intensity curves on the concentration ratio of antibodies in the two steps of the double layer. The concentration of antibodies in the first step was kept constant according to the units on the abscissa, whereas the concentration of the fluorescent conjugate was varied. The dilutions of this were: 1:5 (×——×); 1:10 (×+++×); 1:20 (×————×); 1:80 (×—·—·—·×); and 1:160 (×---×). Background intensity of controls containing only fluorescent conjugate did not increase when the concentration of fluorescent conjugate was increased. Test system: species antigens on human lymphocytes. MIU, mean intensity per membrane unit.

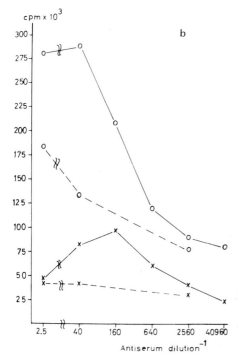

Fig. 5b. Analysis of the mechanism of the prozone effect obtained by indirect labeling techniques. Experiment showing that membrane-bound antibody can be detected at antibody concentrations corresponding to the prozone, which indicates that this is not caused by the absence of membrane-bound antibody. The cells were reacted with antiserum (abscissa) followed by antiglobulin conjugate at a dilution 1/5 (×——×) leading to prozone. Subsequently these cells were exposed to [^{125}I]antiglobulin conjugate diluted 1:2.5 (○——○). The same symbols with dashed lines indicate the intensity obtained with normal control serum. The cells were washed once after step 1. Test system: H-2 antigens on L-cells.

antibody, decreasing the prozone effect and increasing the amplitude of the reaction curves;

(d) the prozone was not due to an absence but rather to a relatively higher amount of membrane-bound antibody on the cell membrane at antibody concentrations corresponding to the prozone;

(e) the prozone was also obtained with purified IgG antibodies;

(f) the disappearance of "prozone factors", presumably low-affinity antibodies, from the cells abolished the prozone. When transferred to antibody-coated cells, "the prozone factors" abrogated the reaction of antiglobulin to this antibody;

(g) the number of membrane-bound antibody molecules disappearing from the cell surface per time unit was probably proportional to antibody concentration, and therefore maximal in regions corresponding to the prozone. Of interest in this context is that C1q, which is bound to membrane-bound antibody with an ionic bond (Laporte et al., 1957; Borsos et al., 1964), dis-

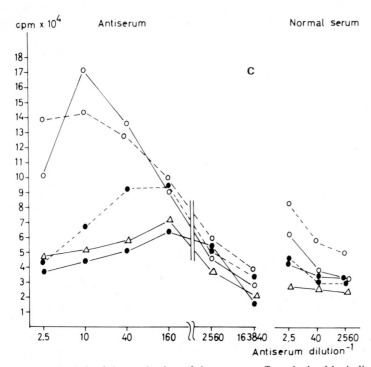

Fig. 5c. Analysis of the mechanism of the prozone effect obtained by indirect labeling techniques. The effect on the reaction curves of fixation of membrane-bound antibody and target cells. Test systems: filled lines, species antigens on human lymphocytes; dashed lines, *H-2* antigens on L-cells: (—●—) unfixed cells; (—○—) cells fixed with paraformaldehyde after antibody coating; (—△—) cells fixed with glutaraldehyde before reaction with antibody. The cells were washed once after step 1. The results are based on data from five experiments.

appeared rapidly from the cell surface, and detection of this molecule implied a very pronounced prozone (Sundqvist et al., 1974). However, since the Clq-protein most likely possesses considerable intramolecular flexibility (Svehag et al., 1972), this might also have contributed to the prozone; and

(h) the more multivalent the binding of the antibody complex to the cell membrane the less the rate of antibody disappearance. The experimental data indicated that multipoint binding was not achieved in the prozone region. Furthermore, redistribution of membrane-bound antibody requiring multivalent attachment to the cell surface, was never observed at antibody concentrations corresponding to the prozone.

The ring zone effect, which was considered as possibly equivalent to the prozone in IMHT, was in fact attributed to a steric hindrance effect produced by a high density of membrane-bound antibody to the indicator cells, resulting in detachment of these cells in the empty region of the zone (Jonsson and Fagraeus, 1969; 1973).

Certain factors of possible importance to the prozone effect are less accessible to experimental analysis, including such questions as: the distance between

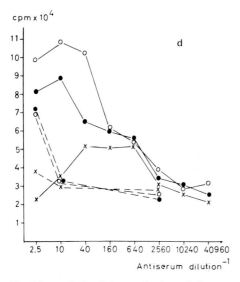

Fig. 5d. Analysis of the mechanism of the prozone effect obtained by indirect labeling techniques. The influence on the intensity curves of the washing procedure after incubation of the target cells with antiserum. Number of washes: 1 (×——×); 6 (●——●) and 9 (○——○). Dashed lines show the corresponding reactivity of normal control serum. Test system: *H-2* antigens on L-cells. The [^{125}I]antiglobulin was diluted 1/10. The results are based on data from six experiments.

Fig. 5e. This experiment is illustrated in Fig. 4, showing that the reaction intensity, but not the prozone effect, is temperature dependent.

different antigenic sites; the accessibility of membrane-bound antibody to crosslinking antiglobulin molecules; and the spatial arrangement of antigen molecules at the cell surface (see also section 2).

Based on the data outlined above the following explanations can be advanced to account for the prozone effect: (1) disappearance of membrane-bound antibody (ligand) of low affinity; (2) instability of membrane-bound antibody complexes; and (3) steric hindrance.

According to the hypotheses 1 and 2 the antiglobulin (conjugate) molecules prevent disappearance of membrane-bound antibody by crosslinking it, thereby increasing the avidity of the antibody binding to the cell. This exerts a stabilizing effect on the molecular lattice being formed at the cell surface. The steric hindrance hypothesis (3) does not account for the increased reactivity and decreased prozone noted when increasing the antiglobulin concentration. If steric hindrance caused the prozone an even higher concentration of antiglobulin should increase rather than decrease the prozone tendency. The possibility exists, however, that fixation of membrane-bound antibody by inhibiting distributional membrane changes decreases the tendency to steric hindrance by antibody clustering and therefore prevents the prozone effect.

TABLE II
THE EFFECT OF TRANSFERRED WASHING FLUIDS ON THE DETECTABILITY OF MEMBRANE-BOUND ANTIBODY

Cells were reacted with antibody for 30 min at 0°C, washed 6 times, and then treated with the different concentrated washing fluids for 30 min at 0°C. These washing fluids had been obtained from cells incubated with antibody in excess (diluted 1:2.5), i.e. corresponding to the prozone of the reaction curves. The cells were then washed once after which membrane-bound antibody was determined using ^{125}I-antiglobulin diluted 1/10 in the case of S-HLy and 1/5 in the case of H-2-L. Each value is the mean of triplicate tubes and the results are based on two experiments.

Test system	Antiserum dilution^{-1}	Uncoated and antibody-coated cells treated with washing fluid, No.				Cells reacted with antiserum; no washing fluid added	Cells reacted with normal control serum; no washing fluid added	Cells reacted only with ^{125}I-antiglobulin
		1–3	4–6	7–9	10–12			
S-HLy	2.5	34790	61619	112998	136710	188095	88551	
	160	37272	57595	54478	80868	124951	26894	35815
	—	43746	59584	51000	60336	—	—	
H-2-L	10	54037	72306	108159	127842	98671	42507	
	40	56413	71344	93836	105273	101398	—	35511
	—	40402	56818	42906	43124	—	—	

However, this is contradicted by the fact that distributional changes were most infrequent in the region corresponding to the prozone.

It is concluded therefore that the prozone effect is caused by release of low-affinity antibodies from the cell surface and this leads to simultaneous loss of the antiglobulin antibody attached to them.

5. Behaviour and fate of antibody following binding to the cell surface

In the preceding sections special attention was given to the structural features of importance in determining the formation of an antibody lattice on the cell surface. The present section will deal with the influence of various cell functions on the behaviour and fate of an antibody lattice at the cell surface.

It is reasonable to assume that the interaction between external ligands, such as antibodies, and the surface membrane of living nucleated cells is the net product of a number of different membrane functions and the influence the ligands exert on these processes. Therefore, in order to interpret the behaviour of antibodies at the cell surface, it would seem essential that we understand the membrane dynamics of the unperturbed cell in the absence of ligands.

The cell membrane has been reported to be in a continuously active metabolic state, with continuous synthesis and incorporation of new material and a high turnover rate of membrane components (Marcus, 1962; Kraemer, 1966; Harris et al., 1969; Warren, 1969; Manuel et al., 1970; Cone et al., 1971; Gordon and Cohn, 1971; Schwartz and Nathenson, 1971; Baker and Humphreys, 1972). Certain cell surface components appear to be shed continuously (Ben-Or and Doljanski, 1960; Marcus, 1962; Weiss and Coombs, 1963; Molnar et al., 1965; Kornfeld and Ginsburg, 1966; Kraemer, 1966; Hayden et al., 1970; Cone et al., 1971; Kapeller et al., 1973), but the rate of this process may be slower than previously thought (Melchers and Andersson, 1973; Werner et al., 1973). Studies of turnover of membrane components using iodinated cells in culture also indicate that the process is very slow. In contrast, experiments where reappearance of antigens after enzymatic digestion from the cell surface was studied showed a more rapid membrane turnover (Vitetta and Uhr, 1975). There is no evidence for the internalization of membrane components such as immunoglobulins (Vitetta and Uhr, 1974, 1975), which suggests that turnover of the cell surface membrane may be accounted for entirely by shedding. Another cell surface process, which has been reported to occur spontaneously in the absence of ligands involves the redistribution of molecules within the plane of the membrane (Frye and Edidin, 1970; Poo and Cone, 1974; Ehrnst and Sundqvist, 1975; McDonough and Lilien, 1975).

Following binding of antibody to cell surface components, the antibody may either dissociate from its receptor, be released complexed with the receptor, be redistributed within the plane of the membrane and/or may be endocytosed and possibly degraded by the metabolic machinery of the cell. The important questions to be solved are the relative magnitude of these processes in different cell types and the identity of the factors influencing which of the alternative fates of the antibody takes place.

5.1. Dissociation of antibodies from antigenic determinants

Viable cells react with antibody in excess and spontaneously release appreciable amounts of this antibody (Chang et al., 1971; Sundqvist, 1974a). It has been calculated that approximately one-third of the initially bound antibody disappears from the cell surface in this way (Chang et al., 1971). This kind of rapid disappearance is probably due partially to dissociation of antibody from the cell surface.

Dissociation, as well as other kinds of loss of antibody from the cell surface, is inhibited by fixation of the antibody-coated cells with paraformaldehyde (Sundqvist, 1973b; G. Biberfeld et al., 1974). Dissociation can also be reduced significantly by extensive washing of antibody-coated cells which probably selectively removes low-affinity antibodies (Sundqvist, 1974a).

5.2. Release of antigen-antibody complexes from the cell surface

5.2.1. Demonstration of release and its differentiation from dissociation

The problems in this context are whether, and to what extent, antibodies disappear from the cell surface to the external milieu of the cell and whether the lost antibody is coupled to cell surface antigen. In other words, it is necessary to distinguish dissociation of antibody alone from the release of antigen-antibody complexes.

By using isotopic or fluorescein-labeled antibodies to different antigen specificities at the cell surface it has been shown that antibody-coated cells in tissue culture release antibodies (Wilson et al., 1972; Engers and Unanue, 1973; Sundqvist, 1973b, 1974a; Antoine and Avrameas, 1974; Fish et al., 1974; Ran et al., 1974). This process involves direct release of labeled material into the medium but may also be preceded by endocytosis and intracellular degradation by lysosomal enzymes. Both the magnitude of the release and its pathway seem to depend on the test system studied (see also section 5.2.3).

There is substantial evidence that the release of antibodies from cells is an active cellular process. First, the binding of antibody to antigens chemically coupled to latex particles or erythrocytes results in very little disappearance compared with that from living cells (Cone et al., 1971; Wilson et al., 1972; Sundqvist, 1973b). This argues against simple dissociation as the major cause of disappearance. Second, identification and analysis of material released into the medium from antibody-coated cells suggests that this material consists either of antibody complexed with antigen (Antoine and Avrameas, 1974) or of antibody in a degraded state (Engers and Unanue, 1973; Fish et al., 1974).

The remarkable differences in the rate of disappearance of antibody from cells in different test systems also suggest that a metabolically dependent release reaction is responsible for the disappearance (Cone et al., 1971; Wilson et al., 1972; Sundqvist, 1973b). There is no reason why high affinity antibodies of certain specificities should be more prone to dissociate than others. It cannot be excluded, however, that certain membrane components are more sensitive than others to extrinsic influences. It is also necessary to consider the reverse question of whether antibody binding to cells counteracts the release of membrane constituents by crosslinking these components to more firmly attached membrane components.

5.2.2. Metabolic dependence of antibody release

The disappearance of antibodies from the cell surface, under conditions that appear to exclude dissociation (see section 5.1), is a temperature-dependent reaction, occurring at 37°C but not at 4°C (Wilson et al., 1972; Sundqvist, 1973b, 1974a). However, when this experiment was performed under similar conditions using cells which had been fixed before coating with antibody, the disappearance of antibodies from the cell surface was very small (Sundqvist, 1974a). This again argues against the possibility that uncoupling of antibody from antigen contributed significantly to the disappearance of antibodies from cells. Additional support for the concept that release of antibody from the cell surface is due to true metabolic turnover of membrane components is provided by the finding that antibody disappearance is counteracted in the presence of metabolic inhibitors (Wilson et al., 1972).

5.2.3. Kinetics of release of antibodies from cells

The kinetics of disappearance of antibodies from cells generally consist of a rapid initial phase followed by a slow phase, though the overall rate of disappearance is influenced by the type of antigen studied (Wilson et al., 1972; Sundqvist, 1973b; Antoine and Avrameas, 1974; Aust-Kettis and Sundqvist, 1977b). From 3–50% of the initially bound antibody was reported to disappear during the rapid phase, and this probably reflected dissociation to some extent. The disappearance of alloantibodies or antibodies to species antigens was a very slow process compared with that of antibodies to μ- and K-chains of splenocytes. In the latter case 90–95% of the original labeled material was reported to disappear within 20 h (Wilson et al., 1972). However, this high value is not consistent with the extremely slow shedding obtained by other methods (Melchers and Andersson, 1973).

It is likely that the release of complexes of immunoglobulins and antibodies in lymphocytes (Wilson et al., 1972) is influenced by capping and that the main fate of anti-immunoglobulin antibodies at the lymphocyte surface is capping followed by endocytosis of the capped antigen-antibody complexes (Engers and Unanue, 1973; Sundqvist, 1973b). The study of Engers and Unanue (1973) also showed that the lymphocyte was able to degrade at least part of the internalized antibodies. Mouse mammary carcinoma cells also exhibit this property (Fish et al., 1974).

Release of antigen-antibody complexes in certain other test systems under single-layer conditions is not influenced by capping and endocytosis (Sundqvist,

1973b). In these systems both the release and the redistribution of antibodies at the cell surface occurred significantly slower than with surface immunoglobulins in mouse lymphocytes.

By enzymatic iodination of surface proteins it is possible to selectively label the surface membrane of cells with isotope (see chapter in this volume by Gahmberg, p. 371) and to study subsequent shedding of labeled cell surface molecules (Cone et al., 1971; Vitetta and Uhr, 1972). Shedding, as revealed by this method, is dependent on cellular respiration and protein synthesis, and the kinetics of the shedding process correlate well with that found in studies using radiolabeled antibodies to surface antigens as indicators of shedding (Cone et al., 1971; Vitetta and Uhr, 1972; Wilson et al., 1972; Sundqvist, 1973b, 1974a). However, as mentioned earlier, these observations showing substantial shedding of membrane components are not consistent with certain other results (Melchers and Andersson, 1973).

5.3. Redistribution of plasma membrane components within the plane of the membrane

This book contains a separate review on capping of cell surface antigens in lymphocytes where the phenomenon has been studied extensively (see chapter by de Petris, p. 643). In this review I will limit my comments to a discussion of the different fates of antibodies and also discuss ligand-induced redistribution of cell surface antigens as a general cellular phenomenon.

The predominating opinion five years ago was that the plasma membrane was a static structure and the various membrane components occupied fixed positions, being either clustered or distributed randomly (Cerrotini and Brunner, 1967; Aoki et al., 1969; Fagraeus and Jonsson, 1970; Kourilsky et al., 1971; Stackpole et al., 1971). By now it is well known that this concept was derived from the fact that the indirect labeling techniques used at that time to map the cell surface caused aggregation of membrane-bound antibodies leading to a discontinuous distribution of cell surface antigens (Davis and Silverman, 1968; Davis, 1972; Davis et al., 1972).

The recognition that the lipid bilayer of the plasma membrane in animal cells is a "fluid" matrix in which other membrane components such as proteins were able to move and undergo topographic rearrangement represented a breakthrough in membrane biology. The demonstration of rapid temperature-dependent intermixing of cell surface antigens after heterokaryon formation (Frye and Edidin, 1970), together with the demonstration of so-called antigenic modulation (Old et al., 1968), provided the first strong arguments in favor of a dynamic state of membrane organization. In 1971 Taylor and his colleagues showed that anti-immunoglobulin antibody was able to induce redistribution and aggregation of mouse lymphocyte immunoglobulin receptors over one pole of the cell (Taylor et al., 1971). This type of ligand-induced redistribution of surface receptors is now referred to as capping. The capping phenomenon, as well as the other experimental evidence described in this volume for the mobility of components in the plane of the cell surface membrane, has provided valuable confirmation of the major proposals made in the fluid mosaic model of membrane structure (Singer and Nicolson, 1972). Most importantly, it has been shown that capping requires bivalent antibodies, thus providing strong evidence that this phenomenon depends on the formation of a lattice of antibody and antigen at the cell surface (Taylor et al., 1971). The observations of Taylor et al. on capping were soon extended to other antigenic specificities on lymphocytes (Karnovsky et al., 1972; Loor et al., 1972; Santer, 1972; Unanue et al., 1972) and also other cell types (Edidin and Weiss, 1972; Sundqvist, 1972, 1973a,b). Furthermore, surface components on non-mammalian cells such as Leishmania (Doyle et al., 1974) and amoeba (Aust-Kettis and Sundqvist, 1975; Pinto da Silva et al., 1975) have also been shown to cap. It therefore seems that the capacity of cells to actively move their membrane components is a fundamental property found in most, if not all, cells.

It is worth recalling that capping and endocytosis of membrane-bound antibody and other related events requiring mobility and redistribution of membrane components have probably been noted many times in the past. A high degree of mobility of erythrocytes bound to the surface of myxovirus-infected cells was described more than ten years ago by Marcus (1962). Similar examples include ligand-induced disappearance of cell surface components from Paramecia (Beale, 1957), the stimulation of pinocytosis in amoeba by anti-amoeba antibodies (Wolpert and O'Neill, 1962), the effect of antibody on pinocytosis in Krebs ascites tumor cells (Easton et al., 1962) and the interaction of antibodies to blood group substance A with human erythrocytes in newborns (Blanton et al., 1968).

There are several factors which influence the redistribution process. The general characteristics of ligand-induced redistribution of cell surface components and most of the prerequisites with respect to suitable ligands and experimental conditions are now well recognized. However, the cellular mechanisms responsible for the phenomenon are not understood.

5.3.1. *Characteristics of ligand-induced redistribution of surface antigens*

Ligand-induced capping of cell surface antigens consists of two separate stages. The first phase involves complexing and aggregation of membrane antigens by the ligand. These initial events are then followed by the second stage in which the antigen-antibody complexes move to one pole of the cell. Initial aggregation is a passive process, but the later capping phase is an active process with respect to its dependence on cell metabolism. There is probably a critical temperature (16°C) which distinguishes these stages and above which distributional changes are accelerated (Loor et al., 1972).

A relatively slow aggregation of surface antigen-antibody complexes at 0–4°C probably results from the increased resistance to movement afforded by the more dense, viscous cell membrane. The low tendency of the ligand to induce capping at 0°C compared with 37°C can also probably be attributed in part to the inhibitory effects of low temperature on cell metabolism and also possibly to microtubule polymerization.

In order to be redistributed, cell surface antigens must be crosslinked by multivalent molecules. Redistribution can be induced by a wide variety of multivalent ligands, including: bivalent antibodies (but not by Fab-fragments) (Taylor et al., 1971); multivalent haptens, tobacco mosaic virus, cellular antigens on sheep red cells (Taylor et al., 1971; Loor et al., 1972; Ashman, 1973); the mitogens concanavalin A and phytohemagglutinin (Greaves et al., 1972; Loor, 1973; Nicolson, 1973); and by the complement protein Clq (Sundqvist et al., 1974).

5.3.2. Capping is influenced by the kind of antigen and the cell type
There are marked differences between capping in different antigen-antibody systems. A limited number of surface components can be induced to cap with a single antibody-layer, but most require a double antibody-layer (as discussed earlier in section 3 in connection with antibody-lattice formation).

Another difference in capping characteristics in various antigen-antibody systems concerns the rate of redistribution and the final pattern of the redistributed antigen-antibody complexes (Sundqvist, 1973b; Menne and Flad, 1974; Stackpole et al., 1974; Yefenof and Klein, 1974). As shown in Fig. 6, the rate of redistribution is influenced both by the kind of antigen and the cell type. The rate of capping is particularly high in mouse B cells and *Entamoeba histolytica* (Aust-Kettis and Sundqvist, 1975, 1977a,b). Surface immunoglobulins on mouse B cells cap significantly faster than H-2 antigens and species-specific antigens on the same cell (Unanue et al., 1972; Sundqvist, 1973b). It is interesting that both mouse B cells and *E. histolytica* exhibit pronounced endocytosis of capped material compared with other cells, suggesting a possible correlation between the facility of capping and endocytosis (for further discussion, see section 5.4 below). The reason why some cell types cap very slowly or not at all (Sundqvist, 1973b; Menne and Flad, 1972; Yefenof and Klein, 1974) is not clear.

There is an interesting correlation between the quantity of microfilaments of different cell types, as detected by specific antibodies, and cell motility (Fagraeus et al., 1975a,b). This correlation could possibly apply also to capping (see also section 5.5), but firm evidence to support this possibility is lacking.

5.4. Endocytosis of ligand-receptor complexes
The capping of surface immunoglobulins in murine lymphocytes by anti-immunoglobulin antibodies (Taylor et al., 1971) and surface antigens and concanavalin A receptors in *E. histolytica* (Aust-Kettis and Sundqvist, 1975, 1977a,b) is followed rapidly by endocytosis of the ligand-receptor complex. In contrast, most other test systems (e.g., alloantigens on mouse L cells and all of the test systems shown in Fig. 6) do not exhibit any marked endocytosis following capping (Sundqvist, 1973b).

Endocytosis of ligand-receptor complexes is an energy-dependent process requiring an active cell metabolism (Taylor et al., 1971; Loor et al., 1972). It is followed by degradation of the internalized antibodies both in mouse lymphocytes (Engers and Unanue, 1973) and in *E. histolytica* (Aust-Kettis and Sundqvist, 1977b).

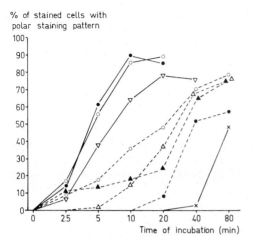

Fig. 6. Rate of appearance of polar staining pattern or marked patch formation in different test systems as recorded by visual observation. The cells were stained at 0°C and afterwards transferred to 37°C and incubated in culture medium or Tyrode's solution supplemented with 1% normal sheep serum for various times. Test systems: ●——●, immunoglobulin determinants on mouse lymphocytes A/sn strain, direct immunofluorescence, antiserum diluted 12.5; ○——○, immunoglobulin determinants on guinea pig lymphocytes, direct immunofluorescence, antiserum diluted 1:2.5; ▽——▽, immunoglobulin determinants on human lymphocytes, indirect immunofluorescence, antiserum diluted 1/10; ○---○, H-2 antigens on mouse lymphocytes, indirect immunofluorescence, step 1 antibody diluted 1/40 and step 2 antibody diluted 1/5; △---△, species antigens on mouse lymphocytes, indirect immunofluorescence, step 1 antibody diluted 1/40 and step 2 antibody diluted 1/5; ▲---▲, species antigens on cells of human lymphoid cell line (Robinson), indirect immunofluorescence, step 1 antibody diluted 1/1000 and step 2 antibody diluted 1/10; ●--●, H-2 antigens on mouse L cells (C3H/K), indirect immunofluorescence, step 1 antibody diluted 1/40 and step 2 antibody diluted 1/5; and ×——×, blood group A on monkey kidney cells, indirect immunofluorescence, step 1 antibody diluted 1/160 and step 2 antibody diluted 1/4 (reproduced with permission from Scand. J. Immunol. 2 (1973) 495.

Capping is not, however, a prerequisite for endocytosis of membrane-bound ligand. Even a monovalent antibody, such as Fab-anti-immunoglobulin, which does not induce capping, is internalized in pinocytotic vesicles (de Petris and Raff, 1973; Antoine and Avrameas, 1974). On the other hand, both the most rapid capping as well as the most complete endocytosis have been observed in mouse lymphocytes (capped immunoglobulins) and *E. histolytica* (capped surface antigens and concanavalin A receptors). This suggests that the two processes may well be coupled.

5.5. Mechanisms for the control of distribution and mobility of cell surface components and bound ligands

Capping, which represents the most pronounced form of ligand-induced redistribution of surface components, can be distinguished from ligand-induced patching of receptors by several criteria. Capping is temperature-dependent and requires a metabolically active cell (Loor et al., 1972; de Petris and Raff, 1973). Inhibitors of glycolysis or of oxidative phosphorylation prevent capping, but inhibitors of protein synthesis and dibutyryl cyclic AMP do not interfere or interfere only slightly with capping (Unanue et al., 1973). Patching, however, is unaffected by any of these treatments.

There is some evidence as to the cellular mechanisms that generate the movement of cell surface components leading to cap formation. Cytochalasin B inhibits ligand-induced movement and redistribution of surface membrane components, almost completely in some cases (de Petris, 1974, 1975) and only partially in others (Taylor et al., 1971; Unanue et al., 1973; Ehrnst and Sundqvist, 1975). This compound can also reverse the capping process (de Petris, 1974, 1975; Poste et al., 1975). Similar effects are produced by tertiary amine local anaesthetics (Ryan et al., 1974; Poste et al., 1975).

Cytochalasin B probably affects an actomyosin-containing microfilament system which has been shown by transmission electron microscopy to occur in bundles near membranes, particularly in association with regions of contact with other cells or at cell extensions. Several types of mechanical behaviour of cell surface membranes (i.e., cell motility, cell division, phagocytosis and ruffling) appear to be under control of this microfilament system (Copeland, 1974; Miranda et al., 1974). One may hypothesize that surface membrane molecules are anchored in an as yet unknown fashion to the actin filaments in the cytoplasm.

Movement of cell surface components does not require extracellular calcium. Interestingly, however, there is evidence that transmembrane passage of calcium, induced by the calcium ionophore A23187, can modify cap formation and even disrupt formed caps (Poste and Nicolson, 1976; Schreiner and Unanue, 1976) and also modify lectin-induced clustering of receptors on fibroblasts (Poste and Nicolson, 1976).

There are also data linking microtubule activity with the topography of membrane components. For example, the microtubule-reactive drugs, colchicine and vinblastine, modify ligand-induced redistribution of surface receptors (see Edelman, 1976). In the case of capping, colchicine and vinblastine alone do not affect this phenomenon in lymphocytes (de Petris, 1974), but these drugs, when used together with cytochalasin B, exert a synergistic inhibitory influence on capping in both mouse lymphocytes (de Petris, 1974, 1975; Unanue and Karnovsky, 1974; Poste et al., 1975) and in *E. histolytica* (Aust-Kettis and Sundqvist, 1975, 1977a). Concanavalin A at high doses inhibits the capping of immunoglobulins (Ig) on mouse B cells induced by anti-Ig antibodies. This inhibition of anti-Ig capping by concanavalin A is abolished by colchicine (Yahara and Edelman, 1973, 1975; Unanue and Karnovsky, 1974) or by lowering of the temperature to +4°C. It has been suggested that surface molecules, such as immunoglobulin and concanavalin A receptors, exist in two "interconvertible states", a free state where they are free to diffuse within the membrane and an "anchored" state in which they interact directly or indirectly with microtubular structures inside the cell which limits their mobility within the membrane

(Edelman, 1976). Following crosslinking by concanavalin A the microtubule system would control or modulate the capping of Ig which would normally occur without microtubule involvement (see Edelman, 1976).

Another explanation for concanavalin A-induced inhibition of anti-Ig capping is that concanavalin A may directly crosslink immunoglobulin molecules and α-methyl-D-mannoside residues on other membrane molecules (de Petris, 1975). On the other hand, there are results (Yahara and Edelman, 1975) which argue against this rather plausible explanation of a direct complexing of immunoglobulin and concanavalin A receptors at the cell surface. This study took advantage of the fact that a small proportion of all concanavalin A receptors on the lymphocyte surface can be complexed with concanavalin A bound to a small particle. In this particular experiment concanavalin A bound to platelets was used. Complexing of only a small proportion of the total concanavalin A receptors at the lymphocyte surface by concanavalin A bound to platelets still prevented movement of the immunoglobulin. However, in colchicine-treated cells exposed to platelet-bound concanavalin A no restriction of immunoglobulin movement was observed. Co-capping of immunoglobulin and concanavalin A receptors on the colchicine-treated cells was not observed (i.e., the motion of concanavalin A and the immunoglobulin molecules was independent). This experiment thus provides strong support for the contention that concanavalin A-induced immunoglobulin movement is influenced by microtubules and is not an artefact generated by simple crosslinking of immunoglobulin molecules and concanavalin A receptors at the cell surface.

On the basis of the evidence indicating that microfilaments and microtubules control the lateral mobility of surface membrane components, it is now an important experimental task to find evidence for or against an interaction between microfilaments, microtubules and various plasma membrane components. One experimental approach would be to induce redistribution of surface antigens by antibody and to then examine whether this leads to a coincident redistribution of microfilaments and/or microtubules. The failure to do this (Fig. 7), in spite of several attempts, may be attributed to the fact that crosslinking of cell surface antigens by antibody probably involves only a minor proportion of the total microfilament population, i.e., only those which interact with the antigenic specificity under study.

5.6. Factors determining the behaviour and fate of membrane-bound antibodies

What factors determine the behaviour and fate of antibody at the cell surface membrane? The available information relevant to this problem may be summarized as follows.

5.6.1. The ligand
The redistribution of ligand-receptor complexes and their subsequent endocytosis are favored by conditions of antibody binding which achieve optimal crosslinking of determinants on separate antigen molecules (see also section 3).

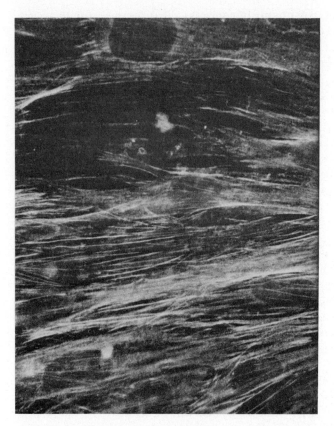

Fig. 7. Demonstration of actin-containing microfilaments in acetone-fixed human fibroblasts by indirect immunofluorescence staining using human antibodies to actin and a fluorescein conjugated sheep anti-human immunoglobulin. The anti-actin-antibodies were kindly provided by Professor Astrid Fagraeus. Capping of the surface antigens in these cells did not alter this staining pattern.

This generally corresponds to the falling phase of a curve showing reaction intensity versus antibody concentration (Figs. 3 and 4). Under these conditions the disappearance of antibodies from cells is relatively small (Fig. 8).

The release and dissociation of ligand or ligand-receptor complexes are favored by conditions of antibody binding when crosslinking of determinants on separate antigen molecules is relatively limited (see section 3.2 and 3.3) or when antibodies are loosely bound (see section 5.1). Crosslinking is limited at antibody concentrations corresponding to the prozone or the plateau in the reaction curves showing reaction intensity versus antibody concentration (Figs. 3 and 4).

Comparison of the kinetics of disappearance of a double-layer of antibody and a corresponding single-layer (Fig. 8) indicates that the higher the concentration of crosslinking antiglobulin relative to antibody in the first step, the less

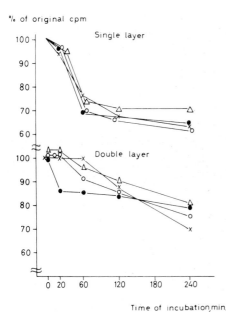

Fig. 8. The fate of a single (A) and a double layer (B) of membrane-bound antibody, respectively, as revealed by the kinetics of disappearance from the cell surface. Test system: species antigens on cells of a human B-cell line. The cells were reacted with antiserum at the following dilutions for 30 min at 0°C: 1/10 ●——●; 1/40 ×——×; 1/160 ○——○ and 1/640 △——△. In A the cells were then washed six times, incubated at 37°C for different times (abscissa) and fixed with paraformaldehyde after which the amount of membrane-bound antibody was determined by measuring the binding of [^{125}I]antiglobulin (diluted 1:5) after 30 min at 0°C. In B the cells were reacted with antiserum in the same way as in A, but cultivated first after the binding of [^{125}I]antiglobulin to the cells. These cells were then fixed with paraformaldehyde at different times of cultivation (abscissa).

the disappearance of antibody from the cells. With a Fab-fragment, the disappearance from cells was more pronounced and endocytosis less pronounced than in comparable studies with whole antibody (Antoine and Avrameas, 1974).

In conclusion, the fate of antibody bound to the cell surface seems to be determined by the degree of crosslinking and/or by how many membrane components or mobile units are united by the antibody lattice. Presumably this reflects differences in the avidity of antibody binding and/or inhibition of membrane turnover. The more multivalent binding that occurs, the less the release of the antigen-antibody complexes and the more pronounced the redistribution of these complexes.

5.6.2. The cell

It is evident from the earlier parts of this section that the fate and behavior of antibodies at the cell surface is dependent on the cell type used and the particular antigen to which the antibodies are bound. For instance, capping and endocytosis are comparatively very pronounced in *E. histolytica* (Aust-Kettis and Sundqvist, 1975, 1977a) and in mouse splenocytes (Taylor et al., 1971; Sundqvist, 1973a). In addition, as mentioned earlier, certain antigens are redistributed by a single antibody layer, whereas others are not.

6. *The interaction of complement with antibodies at the cell surface*

The activation of complement by antibodies is a dynamic process which involves eleven glycoproteins reacting sequentially. This so-called classical pathway of complement activation consists of a recognition unit, C1, an activation units, C2–4, and a membrane attack system, C5–C9 (Müller-Eberhard, 1975). The C1q-protein of the recognition unit interacts with membrane-bound antibodies. Factors of the activation unit, and of the membrane attack mechanism, when activated close to a cell surface, may interact with cell surface receptors for activated complement factors. It seems likely that these interactions influence the behavior of both membrane-bound antibodies and cell surface structures. This concept is supported by data concerning the molecular properties of the complement proteins (for review see Müller-Eberhard, 1975) and by direct experimental evidence. Conversely, ligand-induced cell surface perturbations have been demonstrated to influence complement activation and/or the cytolytic susceptibility of target cells to antibodies and complement.

C1q, the largest complement protein, having a molecular weight of 400 000 daltons, is a collagen-like molecule (Calcott and Müller-Eberhard, 1972). C1q has a recognition function and is the functional link between the antibody and complement systems. To recognize immune complexes C1q has six binding sites for IgG per molecule (Schumaker et al., 1975). C1q probably consists of six subunits of similar size, connected by fibril-like strands to a common central portion, probably with the binding sites located to the each peripheral subunit (Shelton et al., 1972; Svehag et al., 1972). The structure of C1q probably implies a considerable intramolecular flexibility. Therefore, this molecule may be an excellent crosslinker of membrane-bound antibody (see also sections 3, 4 and 5).

C3, C4, C5, C6, C7 and probably C8 (Müller-Eberhard et al., 1966; Arroyave and Müller-Eberhard, 1973) possess the potential capacity for firm attachment to membrane structures. This attachment is achieved by enzymatic activation and exposure of previously concealed binding sites. The capacity of the activated membrane binding sites to bind to their membrane receptors is highly transient (Kolb and Müller-Eberhard, 1973). If an activated complement molecule does not reach its membrane receptor within less than 0.1 sec the binding site is lost and the molecule becomes inactive.

It has been proposed that complement cytolysis is generated by insertion of terminal complement factors into the cell membrane (Mayer, 1972). This concept has received experimental support from the demonstration that C5b and C7 become inserted into the erythrocyte membrane (Hammer et al., 1975).

6.1. Factors influencing the susceptibility of target cells to lysis by antibodies and complement

Despite the presence of a high concentration of antibodies directed against cell surface antigens and an active complement source it is not always possible to lyse target cells. The reason for this lack of cytotoxicity, despite seemingly optimal conditions, may reside both at the target cell level and in the antibody reagent utilized.

In general, IgM antibodies are more effective in inducing lysis of target cells than antibodies of the IgG class. This difference has been explained by the proposal that at least two adjacent IgG molecules, but only one IgM molecule, are required in order to activate complement (Humphrey and Dourmashkin, 1965; Borsos and Rapp, 1965). This property is probably related to the larger molecular size and higher valency of binding to the cell surface of the IgM molecule.

The complement-binding capacity of different IgG subclasses also varies. The binding affinity of C1q for different IgG subclasses is in the order: IgG3 > IgG1 > IgG2 > IgG4 (Schumaker et al., 1975). Binding of C1q to IgG is necessary for activation of complement but does not necessarily lead to activation because IgG4 binds C1q but does not activate the rest of the complement system (Schumaker et al., 1975).

Antibodies directed against histocompatibility antigens sometimes give limited cytolysis and require relatively high complement quantities in order to kill cells (Boyse et al., 1968; Miller and DeWitt, 1972). This is generally attributed to non-complement fixing antibodies present in the sera used (Harris and Harris, 1972; Kinsky et al., 1972; Miller and De Witt, 1972). The elicitation of antibodies with such an aberrant behavior in cytotoxicity tests was attributed to weak histocompatibility antigens. However, even strong histocompatibility antigens seem to give rise to non-complement fixing antibodies (Basch, 1974). During prolonged immunization with strong histocompatibility antigens early antisera were effective in inducing cytolysis whereas late antisera frequently exhibited a "prozone" effect and failed to lyse target cells to the same extent as the early sera. A fraction of non-complement fixing antibody was shown to be responsible for the lower cytotoxic efficiency of the late antisera. Interestingly, these non-complement fixing antibodies reduced the cytotoxic efficiency of the early sera. Despite these results, it was concluded that the decreased cytolytic sensitivity observed was due to a decrease in the effective concentration of antigens on the target cells. This decrease was attributed directly to the presence of the non-complement fixing antibodies.

It has been suggested that non-complement fixing antibodies could be responsible for immunologic enhancement by producing inhibition of cell-mediated immune reactions (Snell et al., 1960; Möller, 1965; Brunner et al., 1968; Smith, 1968; Hellström et al., 1969). There is an inverse relationship between cellular susceptibility to lysis by antibody and complement and ease of enhancement (Möller, 1963; Kaliss, 1966; Smith, 1968). However, antigen density is probably more important to the enhancement phenomenon than any particular property of the antibody, because cytotoxic antibodies function as enhancing antibodies provided the experimental conditions favor this (Linscott, 1970).

The results of several studies in different antigen-antibody systems indicate that there is a direct relationship between the concentration of antigens at the cell surface and the susceptibility of cells to cytotoxic antibody and effector cells (Möller and Möller, 1962, 1967; Winn, 1962; Cann and Herzenberg, 1963, Fenyö et al., 1968; Friberg, 1973; and discussion in this volume by Cikes and Friberg (p. 473). As mentioned earlier, complement activation can be achieved by a single IgM molecule whereas two IgG molecules disposed at a certain minimal distance at the cell surface are required. A reduction in antigen density would therefore be expected to decrease the probability of IgG-dimer formation. When the antigen density is low, as in the case of weak histocompatibility antigens and certain tumor neoantigens, this may critically influence the outcome of cytotoxicity tests. This is illustrated by the decreased effective concentration of isoantigens achieved by non-complement fixing antibodies discussed earlier. Also of interest in this context is the finding that the differing efficiency of IgM and IgG in cytotoxic tests is especially pronounced when the test cells have a low *H-2* antigen concentration on their plasma membranes (Rubio, 1974). Linscott (1970), studying a model system of sheep erythrocytes and covalently linked arsenilate residues, demonstrated that changes in antigen density alone were sufficient to account for major reductions in cellular susceptibility to cytotoxic antisera.

6.2. Cell cycle-dependent variation in cellular susceptibility to lysis by antibodies and complement

It has been shown for several cell lines that their susceptibility to lysis by complement and antibodies directed against membrane antigens changes during the cell growth cycle (Cikes, 1970; Lerner et al., 1971; Shipley, 1971; Cikes et al., 1972; Götze et al., 1972). The times of maximal and minimal susceptibility to complement-mediated lytic damage corresponded to the G_1 and S phases of the cell cycle, respectively. A correlation was established between the period of maximal lytic susceptibility and the maximum expression of antigens at the cell surface as revealed by quantitative antibody absorption assays. However, despite the variations in antigenic expression there was no variation in the capability of the cells to activate complement or to bind late complement components (Lerner et al., 1971; Shipley, 1971).

There are also cell lines which exhibit variations in their susceptibility to lysis but which do not show any marked variation in the expression of surface antigens (Pellegrino et al., 1974). This conclusion was reached by three independent tests: microabsorption assays, isotope antiglobulin technique and the yields of soluble extractable *HL-A* antigens. It was further demonstrated that the ability of the cells after sensitization with various antisera to activate the complement system or to bind radiolabelled C3, C4 and C8 was independent of the stage in the growth cycle at which the tests were performed. Nor did the pathway of complement activation change during the cell cycle. Other data suggested that there was no direct relationship between the cytolytic susceptibility of the target cells and their ability to activate complement or to bind activated complement components. In view of these results the differential susceptibility to lysis was interpreted as resulting from a change in the properties of the cell surface membrane during the cell cycle. One possible explanation is that the membrane possesses the capacity to repair complement-induced damage only during certain periods of the cell cycle. Several structural and functional properties of the cell surface membrane have been shown to vary during the growth cycle. These include synthesis of immunoglobin (Buell and Fahey, 1969; Takahashi et al., 1969) and membrane associated enzymes (see Cikes and Friberg, this volume, p. 473), susceptibility to virus infection (Basilico and Marin, 1966) and to oncogenic transformation by chemical carcinogens (Bertram and Heidelberger, 1974), and the turnover of membrane components (Warren and Glick, 1968). Additional support for the concept of a change in the structure of the cell-surface membrane during the growth cycle is the demonstration that the fragility of cells to cytolysis caused by detergents, hypotonicity, or freeze thawing shows the same variation as the expression of *H-2* antigens as measured by immune cytolysis (Sumner et al., 1973). As will be discussed later (section 6.3), cellular susceptibility to lysis by antibody and complement is influenced by movement of cell surface molecules in the plane of the membrane. Variations in cytolytic susceptibility may therefore be due to variations in mobility of different membrane components during the cell cycle, but results which contradict this hypothesis have also been presented (Kerbel et al., 1975; Killander et al., 1974).

6.3. Influence of the mobility of membrane components on cellular susceptibility to lysis by antibodies and complement

During the course of an antibody-induced redistribution of cell surface antigens the concentration of antigen-antibody complexes per membrane unit will gradually increase. As mentioned earlier in section 6.1, a high antigenic density is correlated with a relatively high cellular susceptibility to lysis by antibodies and complement. There is experimental evidence that the sensitivy of cells to antibody-mediated complement-dependent cytolysis also varies with the topographic distribution of membrane-bound antibody at the target cell surface. Mouse leukemia bone-marrow cells sensitized with *H-2* antibodies were rendered cytotoxicity-resistant when incubated, before exposure to complement, with rabbit anti-mouse immunoglobulin at 37°C, but not at 0°C (Lengerová et al., 1972). This resistance to cytotoxicity at 37°C was associated with a high incidence of cells with a polar distribution of immunofluorescence. Capping by antibody of *H-2* antigens on mouse mastocytoma cells (Edidin and Henney, 1973) and of *HL-A* antigens on human lymphocytes (Bernoco et al., 1972) decreased their cytotoxic susceptibility to antibody and complement. It is interesting, however, that capped cells were as sensitive to lysis by T-cells as were uncapped control cells (Edidin and Henney, 1973).

Addition of anti-Ig to cells sensitized by alloantibodies has been shown to either augment or reduce the cytotoxic effect of these antibodies (Fass and Herberman, 1969; Motta, 1970; Takahashi, 1971; Woodruff and Inchley, 1971). The concentration of anti-Ig antibodies seemed to be the factor determining which of these effects occurred. It is conceivable that the anti-Ig added to the cells induced a redistribution of antigen and antibodies at the cell surface and that the degree of crosslinking and redistribution varied with the concentration of the anti-Ig antibodies (cf. Sundqvist 1972, 1973a). This may have been responsible for the augmentation or reduction in cytotoxicity.

In another study the distribution of antibody at the cell surface was compared with the cytolytic susceptibility of the target cells using antisera to alloantigens on mouse fibroblasts and an anti-thymocyte IgG and human lymphocytes (Sundqvist et al., 1974). Using these test systems the sensitivity of target cells to antibody-mediated complement-dependent cytolysis was found to vary with the distribution of membrane-bound antibody on the target cell surface (Fig. 9). Using one anti-*H-2* serum (ACA anti-A) the relatively low initial cytotoxic sensitivity increased with redistribution of the surface antigen-antibody complexes, with maximum sensitivity corresponding to a predominantly patchy antibody distribution. Subsequently, the cytotoxic sensitivity decreased again to a minimum, at which point the staining pattern consisted of capped cells. The test system consisting of horse anti-human thymocyte IgG and human lymphocytes showed similar variation of cytolytic sensitivity in relation to antibody redistribution. With another anti-*H-2* serum (C57Bl anti-C3H/K) the cytotoxic sensitivity was highest initially and then decreased gradually in relation to the redistribution.

ACA anti-A antiserum is reported to contain relatively more IgG than C57BL anti-C3H/K, and the latter relatively more IgM than ACA anti-A (Wigzell, 1967). The different cytotoxic sensitivity profiles of these two sera in relation to antibody redistribution outlined above might possibly be explained therefore by the hypothesis that an increased density of the IgG antibodies in the ACA anti-A serum enhanced their complement fixing ability, whereas the IgM antibodies in the other serum were efficient complement fixers even without such a distributional change.

The altered sensitivity of target cells to antibody-mediated cytolysis might also be explained by differences in the amount of membrane-bound antibody. However, this explanation receives no support from experimental measurements of antibody binding either at the single-cell level (Fig. 10) or at the cell population level (Sundqvist, 1973b, Sundqvist et al., 1974).

The binding of C1q to membrane-bound antibody is independent of the cell surface distribution of the antibody. It seems unlikely therefore that the altered cytotoxic sensitivity is due to a different accessibility of membrane-bound antibody to complement during the redistribution process.

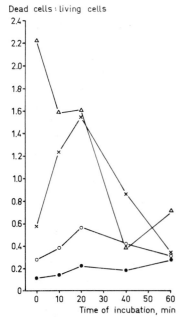

Fig. 9. Cytolytic sensitivity of mouse L-cells at different times in relation to the redistribution of membrane-bound antibody. Cells were reacted with antiserum and antiglobulin at 0°C and subsequently incubated at 37°C for the times shown on the abscissa. After this incubation the cytotoxic sensitivity of the target cells was determined. Sera: ×——×, ACA anti-A/Sn diluted 1:40 + fluorescein isothiocyanate (FITC)-sheep anti-mouse IgG diluted 1:5; ○——○, ACA anti-A (Sn) diluted 1:80 + FITC-sheep anti-mouse IgG diluted 1:5; △——△, C57BL anti-C3H/K diluted 1:40 + FITC-sheep anti-mouse IgG diluted 1:5; and ●——●, normal mouse serum diluted 1:40 + FITC-sheep anti-mouse IgG diluted 1:5. Four experiments were performed using ACA anti-A/Sn and two experiments using C57BL anti-C3H/K with consistent results. Each value is based on counts of 300 to 600 cells. The maximal number of dead cells with the sera used were 64% with ACA anti-A/Sn, 70% with C57BL anti-C3H/K, and 20% with normal mouse serum. Antiserum treated cells in the absence of complement showed lower or approximately the same ratios as cells treated with normal mouse serum (reproduced with permission from Scand. J. Immunol. 3 (1974) 237.

There are some experimental data which could account for the decreased cytotoxic sensitivity found in association with a high frequency of capped cells. The lytic action of complement decreases with the distance between the site of activation and the membrane (Humphrey and Dourmashkin, 1965; Rosse et al., 1966; Rowley and Turner, 1968). Perhaps complement activation over the capped antigen-antibody complex occurs at too great a distance from the cell surface to generate membrane damage. Other possibilities might be that the capping process alters the membrane receptor sites for activated complement factors, or that such receptors, if not involved in the capping process, are not available in sufficient amounts in the region of the cell surface corresponding to the capped antigen-antibody complex.

These results indicate that antibody can increase or decrease the cytotoxic

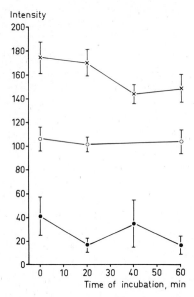

Fig. 10. Mean immunofluorescence intensity per cell determined by microfluorometry at different times in relation to redistribution of membrane-bound antibody. Mouse L-cells were incubated with anti-*H-2* serum and antiglobulin conjugate at 0°C and then cultivated at 37°C for the time shown on the abscissa before final fixation with paraformaldehyde and measurement of the fluorescence intensity. ×——×, ACA anti-A/Sn, 1:40 + fluorescein isothiocyanate (FITC)-sheep anti-mouse IgG, 1:5; ●——●, complement was added to cells treated according to ×——× above and dead cells were selected by phase contrast and their intensity measured; and ○——○, ACA anti-A/Sn, 1:80 + FITC-sheep antimouse IgG, 1:5. Each value is based on determinations of the staining intensity for 15 cells. The data are from one representative individual experiment of a total of about twenty (reproduced with permission from Scand. J. Immunol. 3 (1974) 237.

sensitivity of cells. However, the double-layer conditions required to achieve redistribution of membrane-bound antibody in many test systems raise questions concerning their biological relevance. Systems in which redistribution of surface antigens and cytotoxicity can be generated using a single antibody layer may thus be better systems for experimental analysis. Apart from the capping of surface immunoglobulin on lymphocytes which can be induced by a single antibody layer, certain viral antigens also cap after exposure to a single antibody-layer. Measles virus hemagglutinin is redistributed into a polar staining pattern by a single antibody-layer but the capped antigen-antibody complexes undergo little endocytosis compared with capped immunoglobulins on mouse lymphocytes (Ehrnst and Sundqvist, 1975). Similarly, cold agglutinins cap surface antigens on human lymphocytes without the requirement of a second antibody-layer (P. Biberfeld et al., 1974). Also, rheumatoid factor (R.F.), in similarity with other anti-antibodies, induces redistribution of antigen and membrane-bound antibody at the cell surface (Sundqvist et al., unpublished observations). There was also a change in the cytolytic susceptibility of these cells following redistribution of surface components by rheumatoid factor. These examples point to the possibility that capping of surface antigens may exert an important effect in determining cellular susceptibility to lysis by antibodies and complement in vivo.

The complement protein C1q also redistributes membrane-bound antibodies (Sundqvist et al., 1974). This redistribution is pronounced if C1q is allowed to react with the cells simultaneously with the antibody at 37°C. However, when the reactants are allowed to interact with the cells one after the other, or at 4°C, there is only a partial capping effect (Sundqvist, unpublished observations). The fact that C1q has the capacity to increase the local density of antibodies at the cell surface by redistribution may also be of biological importance in vivo. This might represent a mechanism by which the complement system can directly increase the cytotoxic susceptibility of target cells.

6.4. Complement modulates the binding of antibodies and immune complexes to cells

There is evidence that the presence of complement enhances the attachment of antibodies to the cell surface. Fresh serum increased antibody binding to human red cells and this effect was abrogated by heating or treatment of the serum with EDTA (Stratton, 1960; Polley and Mollison, 1961). C1 has been reported to enhance the attachment of cold-agglutinating antibody to red cells (Rosse et al., 1968) and virus neutralization by antibodies is increased in the presence of complement (Sabin, 1950; Barlow et al., 1958). Further evidence for a role of complement in antibody binding to cells has been obtained using antisera to tumours (Drake and Mardiney, 1975). This study demonstrated that a tumour antiserum with specificity for a particular tumour cell manifested additional previously unrelated specificities in the presence of complement. This complement-mediated influence on antibody specificity required the whole complement sequence and not only C1, since C4- or C6-deficient complement was ineffective. The mechanism of the increased binding of anti-tumour antibodies was explained as resulting from the presence of determinants homologous to the tumour cell membrane but unique in conformation. Complement was suggested to change the conformation of the antigen and/or antibody allowing for a firmer union. As an alternative possibility, it was also proposed that complement induced lysis of cells sensitized with anti-tumour antibody. This would cause release of additional antigen of the same specificity or of different specificity which would neutralize the remaining anti-tumour antibody and thus account for the increased absorption of antibody in the presence of complement.

Complement, or some other heat-labile factor in serum, has also been shown to prevent the detection of antibodies to antigens in tissue sections by indirect immunofluorescence (P. Biberfeld et al., 1974).

Capping is another means by which the complement system, especially C1q, may influence antibody binding to cells (Sundqvist et al., 1974; and unpublished observations).

The data discussed above indicate that complement may modulate antibody binding to cells by several mechanisms involving either only C1 or, in some cases, all of the complement factors. Possibly this effect is achieved by cross-linking of antibodies possessing weak binding to the cell surface, thereby increasing the avidity of the binding. It may be relevant in this context that C1q under certain conditions seems to stabilize antibody binding to the cell surface (Sundqvist et al., 1974). There is strong evidence that complement both enhances the binding of soluble immune complexes to lymphocytes and releases immune complexes bound to the lymphocyte surface. Soluble immune complexes, in the presence of complement, bind to lymphocytes by means of membrane receptors for complement factors (Eden et al., 1973). It is uncertain which complement factors are responsible but a split-product of C3 is considered likely to participate.

Immune complexes can also be released from the cell surface by a complement-dependent mechanism (Gewurtz et al., 1968; Sandberg et al., 1970; Götze and Müller–Eberhard, 1971; Marcus et al., 1971). This probably involves the alternative pathway of complement activation since C3 alone, in the absence of serum, does not release the complexes, whereas a C4-deficient serum did. In addition, treatment of the serum at 50°C for 30 minutes abolished the releasing activity and releasing activity was Mg^2-dependent but not Ca^{2+}-dependent (Miller et al., 1973).

Complement components may thus exert an influence both on the binding of and on the release of immune complexes and antibodies to the cell surface.

Acknowledgements

My own work cited in this article was supported by the Swedish Medical Research Council (project No B76-16X-04774-01).

References

Abbas, A.K., Ault, K.A., Karnovsky, M.J. and Unanue, E.R. (1975) Non-random distribution of surface immunoglobulin on murine B lymphocytes. J. Immunol. 114, 1197–1203.
Antoine, J.-C. and Avrameas, S. (1974) Surface immunoglobulins of rat immunocytes: quantitation and fate of cellbound peroxidase-labelled antibody and Fab fragment. Eur. J. Immunol. 4, 468–474.
Aoki, T., Hämmerling, U., de Harven, E., Boyse, E.A. and Old, L.J. (1969) Antigenic structure of cell surfaces: an immunoferritin study of the occurrence and topology of H-2, ϑ, and TL alloantigens on mouse cells. J. Exp. Med. 130, 979–1001.
Arend, W.P., Teller, D.C. and Mannik, M. (1972) Molecular composition and sedimentation characteristics of soluble antigen-antibody complexes. Biochemistry 11, 4063–4072.
Arroyave, C.M. and Müller-Eberhard, H.J. (1973) Interactions between human C5, C6 and C7 and their functional significance in complement-dependent cytolysis. J. Immunol. 111, 536–545.
Ashman, R.F. (1973) Lymphocyte receptor movement induced by sheep erythrocyte binding. J. Immunol. 111, 212–220.
Atassi, M.Z., Perlstein, M.T. and Staub, D.J. (1973) Conformation and immunochemistry of derivatives modified at lysines 98, 140 and 145 by reaction with 3,3-tetramethyleneglutaric anhydrine. Biochim. Biophys. Acta 328, 278–288.
Aust-Kettis, A. and Sundqvist, K.G. (1975) Capping in *Entamoeba histolytica*. Acceleration and

inhibition by experimental procedures. Proceedings of the 7th international Congress of Amebiasis.

Aust-Kettis, A. and Sundqvist, K.A. (1977a) Dynamics of the interaction between *Entamoeba histolytica* and components of the immune response, I. Capping and endocytosis. Influence of inhibiting and accelerating factors. Submitted.

Aust-Kettis, A. and Sundqvist, K.G. (1977b) Dynamics of the interaction between *Entamoeba histolytica* and components of the immune response, II. Fate of antibodies following binding to the cell surface. Submitted.

Bagshawe, K.D. and Currie, G.A. (1968) Immunogenicity of L 1210 murine leukaemia cells after treatment with neuraminidase. Nature (London) 218, 1254–1255.

Baker, J.B. and Humphreys, T. (1972) Turnover of molecules which maintain the normal surface of contact-inhibited cells. Science 175, 905–906.

Barlow, J., van Vunakis, H. and Levine, L. (1958) Studies of the inactivation of phage by the properdin system, I. Evidence for complement, properdin and magnesium requirements. J. Immunol. 80, 339–355.

Basch, R.S. (1974) Effects of antigen density and noncomplement fixing antibody on cytolysis by alloantisera. J. Immunol. 113, 554–561.

Basilico, C. and Marin, G. (1966) Susceptibility of cells in different stages of the mitotic cycle to transformation by polyma virus. Virology 28, 429–437.

Beale, G.H. (1957) The antigen system of *Paramecium aurelia*. Int. Rev. Cytol. 6, 1–23.

Bekesi, J.G., St-Arneault, G. and Holland, J.F. (1971) Increase of leukemia L 1210 immunogenicity by *Vibrio cholerae* neuraminidase treatment. Cancer Res. 31, 2130–2132.

Ben-Or, S. and Doljanski, F. (1960) Single cell suspensions at tissue antigens. Exp. Cell Res. 20, 64.

Bernoco, D., Cullen, S., Scudeller, G., Trinchieri, G. and Ceppellini, R. (1972) HL-A molecules at the cell surface. In: J. Dausset and J. Colombani (Eds.), Histocompatibility Testing, Munksgaard, Copenhagen, pp. 527–537.

Bertram, J.S. and Heidelberger, C. (1974) Cell cycle dependency of oncogenic transformation induced by N-methyl-N'-nitronitrosoguanidine in culture. Cancer Res. 34, 526–537.

Biberfeld, G., Fagraeus, A. and Lenkel, L. (1974) Reaction of human smooth muscle antibody with thyroid cells. Clin. Exp. Immunol. 18, 371–377.

Biberfeld, P., Biberfeld, G., Molnar, Z. and Fagraeus, A. (1974) Fixation of cell-bound antibody in the membrane immunofluorescence test. J. Immunol. Methods 4, 135–148.

Blank, S.E., Gerrie, A.L. and Clem, L.W. (1971) Antibody affinity and valence in viral neutralization. J. Immunol. 108, 665–673.

Blanton, P.L., Martin, J. and Haberman, S. (1968) Pinocytotic response of circulating erythrocytes to specific blood grouping antibodies. J. Cell Biol. 37, 716–728.

Borsos, T. and Rapp, H.J. (1965) Hemolysin titration based on fixation of the activated first component of complement: evidence that one molecule of hemolysin suffices to sensitize an erythrocyte. J. Immunol. 95, 559–566.

Borsos, T., Rapp, H.J. and Waltz, U. (1964) Action of the first component of complement. Activation of C'1 to C'1a in the hemolytic system. J. Immunol. 92, 108–112.

Boyse, E.A., Miyazawa, M., Aoki, T. and Old, L.J. (1968) Ly-A and Ly-B: two systems of lymphocyte isoantigens in the mouse. Proc. Roy. Soc. London Ser. B. 170, 175–193.

Brandschaft, P.B. and Boone, C.W. (1974) Increase in gross (G) antigen sites on the surface of AKR-virus-induced rat lymphoma cells after treatment with trypsin. J. Immunol. 113, 94–101.

Brunner, K.T., Mauel, J., Cerottini, J.-C. and Chapius, B. (1968) Quantitative assay of the lytic action of immune lymphoid cells on ^{51}Cr-labelled allogenic target cells in vitro; inhibition by isoantibody and by drugs. Immunology 14, 181–196.

Buell, D.N. and Fahey, J.L. (1969) Limited periods of gene expressions in immunoglobulin synthesizing cells. Science 164, 1524–1525.

Burger, M.M. (1969) A difference in the architecture of the surface membrane of normal and virally transformed cells. Proc. Natl. Acad. Sci. USA 62, 994–1001.

Calcott, M.A. and Müller-Eberhard, H.J. (1972) C1q protein of human complement. Biochemistry 11, 3443–3450.

Cann, H.M. and Herzenberg, L.A. (1963) In vitro studies of mammalian somatic cell variation, II. Isoimmune cytotoxicity with a cultured mouse lymphoma and selection of resistant variants. J. Exp. Med. 117, 256–284.

Cerottini, J.C. and Brunner, K.T. (1967) Localization of mouse isoantigens on the cell surface as revealed by immunofluorescence. Immunology 13, 395–403.

Chang, S., Stockert, E., Boyse, E.A., Hämmerling, U. and Old, L.J. (1971) Spontaneous release of cytotoxic alloantibody. Immunology, 21, 829–838.

Cikes, M. (1970) Relationship between growth rate, cell volume, cell cycle kinetic and antigenic properties of cultured murine lymphoma cells. J. Natl. Cancer Inst. 45, 979–988.

Cikes, M., Friberg Jr., S. and Klein, G. (1972) Quantitative studies of antigen expression in cultured murine lymphoma cell. II. Cell-surface antigens in synchronized cultures. J. Natl. Cancer Inst. 49, 1607–1611.

Cone, R.E., Marchalonis, J.J. and Rolley, R.T. (1971) Lymphocyte membrane dynamics. J. Exp. Med. 134, 1373–1384.

Copeland, M. (1974) The cellular response to cytochalasin B: A critical overview. Cytologia 39, 709–727.

Crothers, D.M. and Metzger, H. (1972) The influence of polyvalency on the binding properties of antibodies. Immunochemistry, 9, 341–357.

Cullen, S.E., Schwartz, B. and Nathenson, S.G. (1972) The molecular basis of codominant expression of the histocompatibility-2 genetic region. Proc. Natl. Acad. Sci. USA 69, 1394–1397.

Currie, G.A. and Bagshawe, K.D. (1967) The masking of antigens on trophoblast and cancer cells. Lancet 1, 708–710.

Currie, G.A. and Bagshawe, K.D. (1968) The role of sialic acid in antigenic expression: Further studies of the Landschütz ascites tumour. Brit. J. Cancer, 22, 843–853.

Currie, G.A. and Bagshawe, K.D. (1969) Tumour specific immunogenicity of methylcholanthrene induced sarcoma cells after incubation in neuraminidase. Br. J. Cancer, 23, 141–149.

Currie, G.A. and Lond, M.B. (1967) Masking of antigens on the Landschütz ascites tumour. Lancet 2, 1336–1338.

Davis, B.D., Dulbecco, R., Eisen, H.N., Ginsberg, H.S. and Wood Jr., W.B. (1967) Microbiology. Harper and Row, New York.

Davis, W.C. (1972) H-2 antigen on cell membranes: an explanation for the alteration of distribution by indirect labelling techniques. Science 175, 1006–1008.

Davis, W.C. and Silverman, L. (1968) Localization of mouse H-2 histocompatibility antigen with ferritin-labelled antibody. Transplantation 6, 535.

Davis, W.C., Alspaugh, M.A., Stimpfling, J.H. and Walford, R.L. (1971) Cellular surface distribution of transplantation antigens: discrepancy between direct and indirect labelling techniques. Tissue Antigens 1, 89.

Doyle, J.J., Behin, R., Mauel, J. and Rowe, D.S. (1974) Antibody-induced movement of membrane components of Leishmania enriettii. J. Exp. Med. 139, 1061–1069.

Drake, W.P. and Mardiney Jr., M.R. (1975) Complement-mediated alteration of antibody specificity in vivo. J. Immunol. 114, 1052–1057.

Easton, J.M., Goldberg, B. and Green, H. (1962) Demonstration of surface antigens and pinocytosis in mammalian cells with ferritin antibody conjugates. J. Cell Biol. 12, 437–443.

Edelman, G.M. (1976) Surface modulation in cell recognition and cell growth. Science 192, 218–226.

Edelman, G.M., Yahara, I. and Wang, J.L. (1973) Receptor mobility and receptor-cytoplasmic reactions in lymphocytes. Proc. Natl. Acad. Sci. USA 70, 1442–1446.

Eden, A., Bianco, C. and Nussenzweig, V. (1973) Mechanism of binding of soluble immune complexes to lymphocytes. Cell. Immunol. 7, 459–473.

Edidin, M. and Henney, C.S. (1973) The effect of capping H-2 antigens on the susceptibility of target cells to humoral and t cell-mediated lysis. Nature (London) New Biol. 246, 47–49.

Edidin, M. and Weiss, A. (1972) Antigen cap formation in cultured fibrolasts: A reflection of membrane fluidity and of cell mobility. Proc. Natl. Acad. Sci. USA 69, 2456–2459.

Ehrnst, A. and Sundqvist, K.G. (1975) Polar appearance and nonligand induced spreading of measles virus hemagglutinin at the surface of chronically infected cells. Cell 5, 351–359.

Engers, H.D. and Unanue, R.R. (1973) The fate of anti-Ig surface Ig complexes on B lymphocytes. J. Immunol. 110, 465–475.
Espmark, J.Å., Grandien, M. and Bergquist, N.R. (1971) Quantitative evaluation of direct and indirect immunofluorescence tests for viral antigens. Ann. N.Y. Acad. Sci. 177, 98–110.
Fagraeus, A. and Jonsson, J. (1970) Distribution of organ antigens over the surface of thyroid cells as examined by the immunofluorescence test. Immunology 18, 413–416.
Fagraeus, A., The, H. and Biberfeld, G. (1973) Reaction of human smooth muscle antibody with thymus medullary cells. Nature (London) New Biol. 246, 113–115.
Fagraeus, A., Nilsson, K., Lidman, K. and Norberg, R. (1975a) Reactivity of smooth-muscle antibodies, surface ultrastructure and mobility in cells of human hematopoietic cell lines. J. Natl. Cancer Inst. 55, 783–789.
Fagraeus, A., Lidman, K. and Norberg, R. (1975b) Indirect immunofluorescence staining of contractile proteins in smeared cells by smooth muscle antibodies. Clin. Exp. Immunol. 20, 469–477.
Farrow, L.J., Holborow, E.J. and Brighton, W.D. (1971) Reaction of human smooth muscle antibody with liver cells. Nature (London) New Biol. 232, 186–187.
Fass, L. and Herberman, R.B. (1969) A cytotoxic antiglobulin technique for assay of antibodies to histocompatibility antigens. J. Immunol. 102, 140–144.
Feinstein, A., Munn, E.A. and Richardson, N.E. (1971) The three-dimensional conformation of γM and γA globulin molecules. Ann. N.Y. Acad. Sci. 190, 104–121.
Fenyö, E.M., Klein, E., Klein, G. and Swiech, K. (1968) Selection of an immunoresistant Moloney lymphoma subline with decreased concentration of tumour specific surface antigens. J. Natl. Cancer Inst. 40, 69–89.
Fish, F., Witz, I.P. and Klein, G. (1974) Tumour-bound immunoglobulin. The rate of immunoglobulin disappearing from the new surface of coated tumour cells. Clin. Exp. Immunol. 16, 355–365.
Friberg, S. (1973) Comparison of an immunoresistant and an immunosusceptible ascites subline from the murine tumour TA3. Thesis, Stockholm.
Fröland, S.S. and Natvig, J.B. (1972) Class, subclass and allelic exclusion of membrane-bound Ig of human B lymphocytes. J. Exp. Med. 136, 409–414.
Frye, L.D. and Edidin, M. (1970) The rapid intermixing of cell surface antigens after formation of mouse–human heterokaryons. J. Cell Sci. 7, 319–335.
Fu, S.M. and Kunkel, H.G. (1974) Membrane immunoglobulins of B lymphocytes. Inability to detect certain characteristic IgM and IgD antigens. J. Exp. Med. 140, 895–903.
Gabbiani, G., Ryan, G.B., Lamelin, J.-P., Vassalli, P., Majno, G., Bouvier, C.A., Cruchaud, A. and Lüscher, E.F. (1973) Human smooth muscle autoantibody. Am. J. Path. 72, 473–488.
Gewurtz, H., Shin, H.S. and Mergenhagen, S.E. (1968) Interactions of the complement system with endotoxic lipopolysaccharide: consumption of each of the six terminal complement components. J. Exp. Med. 128, 1049–1057.
Gibowsky, A. and Terasaki, P.I. (1972) Trypsinization of lymphocytes for HL-A typing. Transplantation 13, 192–194.
Gordon, S. and Cohn, Z. (1971) Macrophage-melanoma cell-heterokaryons, IV. Unmasking the macrophage-specific membrane receptor. J. Exp. Med. 134, 947–962.
Greaves, M.F., Bauminger, S. and Janossy, G. (1972) Lymphocyte activation, III. Binding sites for phytomitogens on lymphocyte subpopulation. Clin. Exp. Immunol. 10, 537–554.
Green, N.M. (1969) Electron microscopy of the immunoglobulins. Adv. Immunol. 11, 1–28.
Greenbury, C.L., Moore, D.H. and Nunn, L.A.C. (1965) The reaction with red cells of 7S rabbit antibody, its subunits and their recombinants. Immunology 8, 420–431.
Götze, O. and Müller-Eberhard, H.J. (1971) The C3 activator system: an alternate pathway of complement activation. J. Exp. Med. 134, 90s–108s.
Götze, D., Pellegrino, M.A., Ferrone, S. and Reisfeld, R.A. (1972) Expression of H-2 antigens during the growth cycle of cultured tumour cells. Immunol. Commun. 1, 533–544.
Hammer, C.H., Nicholson, A. and Mayer, M.M. (1975) On the mechanism of cytolysis by complement: Evidence on insertion of C5b and C7 subunits of the C5b,6,7 complex into

phospholipid bilayers of erythrocyte membranes. Proc. Natl. Acad. Sci. USA 72, 5076–5080.

Hammerström, S. (1973) Binding of *Helix pomatia* A hemagglutinin to human erythrocytes and other cells. Influence of multivalent interaction on affinity. Scand. J. Immunol. 2, 53–66.

Harder, P.H. and McKhann, C.F. (1968) Demonstration of cellular antigens on sarcoma cells by an indirect ^{125}I-labelled antibody technique. J. Natl. Cancer Inst. 40, 231–241.

Harris, H., Sidebottom, E., Grace, D.M. and Bramwell, M.E. (1969) The expression of genetic information: A study with hybrid animal cells. J. Cell Sci. 4, 499–525.

Harris, T.N. and Harris, S. (1972) Effect of anti-mouse IgG on some complement-requiring alloantibodies of the mouse. J. Immunol. 109, 1096–1108.

Hayden, G.A., Crowley, G.M. and Jamieson, G.A. (1970) Studies on glycoproteins, V. Incorporation of glucosamine into membrane glycoproteins of phytohemagglutinin-stimulated lymphocytes. J. Biol. Chem. 245, 5827–5832.

Heidelberger, M. (1956) Lectures in Immunochemistry. Academic Press, New York.

Hellström, I., Hellström, K.E., Heppner, G.H., Pierce, G.E. and Yang, J.P.S. (1969) Serum-mediated protection of neoplastic cells from inhibition by lymphocytes immune to their tumour-specific antigens. Proc. Natl. Acad. Sci. USA 62, 362–368.

Henle, W., Guerra, A. and Henle, G. (1974) False negative and prozone reactions in tests for antibodies to Epstein-Barr virus-associated nuclear antigen. Int. J. Cancer, 13, 751–754.

Hornick, C.L. and Karush, F. (1969) The interaction of hapten-coupled bacteriophage IX174 with anti-hapten antibody. Israel J. Med. Sci. 5, 163–170.

Hornick, C.L. and Karush, F. (1972) Antibody affinity, III. The role of multivalence. Immunochemistry 9, 325–349.

Humphrey, J.H. and Dourmashkin, R.R. (1965) Electron microscope studies of immune cell lysis. In: G.E.W. Wolstenholme and J. Knight (Eds.), Complement. Ciba Foundation Symposium, Churchill, London, pp. 175–186.

Humphrey, J.H. and Dourmashkin, R.R. (1969) The lesions in cell membranes caused by complement. Adv. Immunol. 11, 75–115.

Inbar, M. and Sachs, L. (1969) Interaction of the carbohydrate-binding protein concanavalin A with normal and transformed cells. Proc. Natl. Acad. Sci. USA 63, 1418–1425.

Jonsson, J. and Fagraeus, A. (1969) On the mechanism of the ring zone effect obtained with the mixed hemadsorption technique. Immunology 17, 387–411.

Jonsson, J. and Fagraeus, A. (1973) Studies with human anti-thyroid sera reacting with thyroid monolayer cultures. Acta Pathol. Microbiol. Scand. Section B 81, 165–175.

Kaliss, N. (1966) Immunological enhancement: Conditions for its expression and its relevance for grafts of normal tissues. Ann. N.Y. Acad. Sci. 129, 155–163.

Kalmanson, G.M. and Bronfenbrenner, J. (1943) Restoration of activity of neutralized biologic agents by removal of the antibody with papain. J. Immunol. 47, 387–407.

Kapeller, M., Gal-Oz, R., Grover, N.B. and Doljanski, F. (1973) Natural shedding of carbohydrate-containing macromolecules from cell surfaces. Exp. Cell Res. 79, 152–158.

Karnovsky, M.J., Unanue, E.R. and Leventhal, M. (1972) Ligand-induced movement of lymphocytic membrane macromolecules, II. Mapping of surface moieties. J. Exp. Med. 136, 907–930.

Kassulke, J.T., Stutman, O. and Yunis, E.J. (1971) Blood-group isoantigens in leukemic cells: Reversibility of isoantigenic changes by neuraminidase. J. Natl. Cancer Inst. 46, 1201–1208.

Kerbel, R.S., Birbeck, M.S.C., Robertson, D. and Cartwright, P. (1975) Ultrastructural and serological studies on the resistance of activated B cells to the cytotoxic effects of anti-immunoglobulin serum. Clin. Exp. Immunol. 20, 161–177.

Killander, D., Levin, A., Inoue, M. and Klein, E. (1970) Quantification of immunofluorescence on individual erythrocytes coated with varying amounts of antigen. Immunology 19, 151–156.

Killander, D., Yefenof, E. and Klein, E. (1975) Antibody induced redistribution of surface antigens is independent of the amount of antigen and the cell cycle position. Exp. Cell Res. 89, 413–415.

Kinsky, R.G., Voisin, G.A. and Duc, H.T. (1972) Biological properties of transplantation immune sera, III. Relationship between transplantation (facilitation or inhibition) and serological (anaphylaxis or cytolysis) activities. Transplantation 13, 452–466.

Kolb, W.P. and Müller-Eberhard, H.J. (1973) The membrane attack mechanism of complement: Verification of a stable C5-9 complex in free solution. J. Exp. Med. 138, 438–451.

Kornfeld, S. and Ginsburg, V. (1966) The metabolism of glucosamine by tissue culture cells. Exp. Cell Res. 41, 592–600.

Kourilsky, F.M., Silvestre, D., Levy, J.P., Dausset, J., Nicolai, M.G. and Senik, A. (1971) Immunoferritin study of the distribution of HL-A antigens on human blood cells. J. Immunol. 106, 454–466.

Kraemer, P.M. (1966) Regeneration of sialic acid on the surface of Chinese hamster cells in culture, I. General characteristics of the replacement process. J. Cell Physiol. 68, 85–90.

Lafferty, K.J. (1963) The interaction between virus and antibody, I–II. Virology 21, 61–90.

Laporte, R., Hardre de Looze, L. and Sillard, R. (1957) Contribution à l'étude du complément, II. Premiers stades de l'action hémolytique du complément. Role particulier du premier composant. Ann. Inst. Pasteur. 92, 15–42.

Lengerová, A., Pokorná, Z., Viklický, V. and Zelený, V. (1972) Phenotypic suppressions of H-2 antigens and topography of the cell surface. Tissue Antigens 2, 332–340.

Lerner, R.A., Oldstone, M.B.A. and Cooper, N.R. (1971) Cell cycle-dependent immune lysis of Moloney virus-transformed lymphocytes: presence of viral antigen, accessibility to antibody and complement activation. Proc. Natl. Acad. Sci. USA 68, 2584–2588.

Levin, A., Killander, D., Klein, E., Nordenskjöld, B. and Inoue, M. (1971) Applications of microspectrofluorometry in quantitation of immunofluorescence on single cells. Ann. N.Y. Acad. Sci. 177, 481–489.

Lidman, K., Biberfeld, G., Fagraeus, A., Norberg, R., Torstensson, R. and Utter, G. (1976) Anti-actin specificity of human smooth muscle antibodies in chronic hepatitis. Clin. Exp. Immunol. 24, 1–7.

Lindenmann, J. and Klein, P.A. (1967) Immunological aspects of viral oncolysis. In: P. Rentchnick (Ed.), Recent Results in Cancer Research, Vol. 9. Springer, New York.

Linscott, W.D. (1970) Complement fixation: The effects of IgG and IgM antibody concentration on C1-binding affinity. J. Immunol. 105, 1013–1023.

Linthicum, D.S. and Sell, S. (1975) Topography of lymphocyte surface immunoglobulin using scanning immunoelectron microscopy. J. Ultrastruct. Res. 51, 55–68.

Lobo, P.I., Westervelt, F.B. and Horwitz, D.A. (1975) Identification of two populations of immunoglobulin bearing lymphocytes in man. J. Immunol. 114, 116–119.

Loor, F. (1973) Lymphocyte membrane particle redistribution induced by a mitogenic/capping dose of the phytohemagglutinin of Phaseolus vulgaris. Eur. J. Immunol. 3, 112–116.

Loor, F. (1974) Binding and redistribution of lectins on lymphocyte membrane. Eur. J. Immunol. 4, 210–220.

Loor, F., Forni, L. and Pernis, B. (1972) The dynamic state of the lymphocyte membrane. Factors affecting the distribution and turnover of surface immunoglobulins. Eur. J. Immunol. 2, 203–212.

Loor, F., Naomi, B. and Russell Little, J. (1975) Dynamics of the TL antigens on thymus and leukemia cells. Cell Immunol. 17, 351–365.

Manuel, J., Rudolf, H., Chapius, B. and Brunner, K.T. (1970) Studies of allograft immunity in mice, II. Mechanism of target cell inactivation in vitro by sensitized lymphocytes. Immunology 18, 517–535.

Marchesi, V.T., Tillack, T.W., Jackson, R.L., Segnest, J.P. and Scott, R.E. (1972) Chemical characterization and surface orientation of the major glycoprotein of the human erythrocyte membrane. Proc. Natl. Acad. Sci. USA 69, 1445–1449.

Marcus, P.I. (1962) Dynamics of surface modification in myxovirus-infected cells. Cold Spring Harbor Symp. Quant. Biol. 27, 351–365.

Marcus, R.L., Shin, H.S. and Mayer, N.M. (1971) Analternate complement pathway: C-3 cleaving activity, not due to C4, 2a on endotoxic lipopolysaccharide after treatment with Guinea pig serum: Relation to properdin. Proc. Natl. Acad. Sci. USA 68, 1351–1354.

Marrack, J.R. (1938) The chemistry of antigens and antibodies. H.M. Stationary Office, London.

Mayer, M. (1972) Mechanism of cytolysis by complement. Proc. Natl. Acad. Sci. USA 69, 2954–2958.

McDonough, J. and Lilien, J. (1975) Spontaneous and lectin-induced redistribution of cell surface receptors on embryonic chick neural retina cells. J. Cell Sci. 19, 357–368.

Melchers, F. and Andersson, J. (1973) Synthesis, surface deposition and secretion of immunoglobulin M in bone marrow-derived lymphocytes before and after mitogenic stimulation. Transplantation 14, 76–130.

Menne, H.D. and Flad, H.D. (1973) Membrane dynamics of HL-A-anti-HL-A complexes of human normal and leukaemic lymphocytes. Clin. Exp. Immunol. 14, 57–67.

Metzger, H. (1970) Structure and function of γM macroglobulins. Adv. Immunol. 12, 57–116.

Miller, C. and DeWitt, C. (1972) Rat alloantibody responses against strong and weak histocompatibility antigens. J. Immunol. 109, 919–926.

Miller, G.W., Slauk, P.H. and Nussenzweig, V. (1973) Complement-dependent release of immunecomplexes from the lymphocyte membrane. J. Exp. Med. 138, 495–507.

Miranda, A.F., Godman, G.C., Deitch, A.D. and Tanenbaum, S.W. (1974) Action of cytochalasin D on cells of established lines. J. Cell Biol. 61, 481–500.

Möller, E. (1965) Antagonistic effects of humoral isoantibodies on the in vitro cytotoxicity of immune lymphoid cells. J. Exp. Med. 122, 11–23.

Möller, G. (1963) Studies on the mechanism of immunological enhancement of tumour homografts, I. Specificity of immunological enhancement. J. Natl. Cancer Inst. 30, 1153–1175.

Möller, E. and Möller, G. (1962) Quantitative studies of the sensitivity of normal and neoplastic mouse cells to the cytotoxic action of isoantibodies. J. Exp. Med. 115, 527–553.

Möller, G. and Möller, E. (1967) Immune cytotoxicity and immunological enhancement in tissue transplantation. In: B. Cinader (Ed.), Antibodies to Biologically Active Molecules. Pergamon, Oxford, pp. 349–407.

Molnar, J., Reegraden, D.W. and Winzler, R.J. (1965) The biosynthesis of glycoprotein, VI. Production of extracellular radioactive macromolecules by Ehrlich ascites carcinoma cells during incubation with glucosamine-^{14}C. Cancer Res. 25, 1860–1866.

Motta, R. (1970) An indirect cytotoxicity test with increased sensitivity against the normal and leukaemic transplantation antigens. Eur. J. Clin. Biol. Res. 15, 510–514.

Müller-Eberhard, H.J. (1975) Complement. Annu. Rev. Biochem. 44, 697–724.

Müller-Eberhard, H.J., Dalmasso, A.P. and Calcott, M.A. (1966) The reaction mechanism of B_{1c}-globulin (C'3) in immune hemolysis. J. Exp. Med. 123, 33–54.

Nairn, R.C., Herzog, F., Ward, H.A. and de Boer, W.G.R.M. (1969) Microphotometry in immunofluorescence. Clin. Exp. Immunol. 4, 687–705.

Neauport-Sautes, C., Silvestre, D. and Lilly, F. (1973) Independence of H-2k and H-2D antigenic determinants on the surface of mouse lymphocytes. Transplant Proc. 5, 443–446.

Nicolson, G.L. (1972) Topography of membrane concanavalin A sites modified by proteolysis. Nature (London) New Biol. 239, 193–197.

Nicolson, G.L. (1973) Temperature-dependent mobility of concanavalin A sites on tumour cell surfaces. Nature (London) New Biol. 23, 218–220.

Nicolson, G.L. (1974) The interactions of lectins with animal cell surfaces. Int. Rev. Cytol. 39, 89–190.

Nicolson, G.L. and Hyman, R. (1971) The two-dimensional topographic distribution of H-2 histocompatibility alloantigens on mouse red blood cell membranes. J. Cell Biol. 50, 905–910.

Nicolson, G.L. and Singer, S.J. (1974) The distribution and asymmetry of mammalian cell surface saccharides utilizing ferritin-conjugated plant agglutinins as specific saccharide stains. J. Cell Biol. 60, 236–248.

Nicolson, G.L., Masouredis, S.P. and Singer, S.J. (1971) Quantitative two-dimensional ultrastructural distribution of Rh_0 (D) antigenic sites on human erythrocyte membranes. Proc. Natl. Acad. Sci. USA 68, 1416–1420.

Noelken, M.E., Nelson, C.A., Buckley, C.E. III and Tanford, C. (1965) Gross conformation of rabbit 7Sγ-immunoglobulin and its papain-cleaved fragments. J. Biol. Chem. 240, 218–223.

Nossal, G.J.V., Warner, N.L., Lewis, H. and Sprent, J. (1972) Quantitative features of a sandwich radioimmunolabeling technique for lymphocyte surface receptors. J. Exp. Med. 135, 405–428.

Old, L.J., Stockert, E., Boyse, E.A. and Kim, J.H. (1968) Antigenic modulation. Loss of TL antigen from cells exposed to TL antibody. Study of the phenomena in vitro. J. Exp. Med. 127, 523–532.

Pellegrino, M.A., Ferrone, S., Cooper, N.R., Dierich, M.P. and Reisfeld, R.A. (1974) Variation in susceptibility of human lymphoid cell line to immune lysis during the cell cycle. Lack of correlation with antigen density and complement binding. J. Exp. Med. 140, 578–588.

Pernis, B., Forni, L. and Amante, L. (1970) Immunoglobulin spots on the surface of rabbit lymphocytes. J. Exp. Med. 132, 1001–1019.

Petris, S. de (1974) Inhibition and reversal of capping by cytochalasin B, vinblastine and colchicine. Nature (London) 250, 54–55.

Petris, S. de (1975) Concanavalin A receptors, immunoglobulins and θ antigen of the lymphocyte surface. J. Cell Biol. 65, 123–146.

Petris, S. de and Raff, M.C. (1972) Distribution of immunoglobulin on the surface of mouse lymphoid cells as determined by immunoferritin electron microscopy. Antibody-induced, temperature-dependent redistribution and its implications for membrane structure. Eur. J. Immunol. 2, 523–535.

Petris, S. de and Raff, M.C. (1973) Normal distribution, patching and capping of lymphocyte surface immunoglobulin studied by electron microscopy. Nature (London) New Biol. 241, 257–259.

Petris, S. de and Raff, M.C. (1974) Ultrastructural distribution and redistribution of alloantigens and concanavalin A receptors on the surface of mouse lymphocytes. Eur. J. Immunol. 4, 130–137.

Pilz, I., Puchwein, G., Krathky, O., Herbst, M., Haager, O., Gall, W.E. and Edelman, G.M. (1970) Small angle X-ray scattering of a homogeneous γG1 immunoglobulin. Biochemistry 9, 211–219.

Pinto da Silva, P., Martínez-Palomo, A. and Gonzáles-Robles, A. (1975) Membrane structure and surface coat of *Entamoeba histolytica*. Topochemistry and dynamics of the cell surface: Cap formation and microexudate. J. Cell Biol. 54, 538–550.

Polley, M.J. and Mollison, P.L. (1961) The role of complement in the detection of blood group antibodies: special reference to the antiblobulin test. Transfusion 1, 9–22.

Poo, M. and Cone, R.A. (1974) Lateral diffusion of rhodopsin in the photoreceptor membrane. Nature (London) 247, 438–441.

Poste, G. and Nicolson, G.L. (1976) Calcium ionophores A23187 and X537A affect cell agglutination by lectins and capping of lymphocyte surface immunoglobulins. Biochim. Biophys. Acta 426, 148–155.

Poste, G., Papahadjopoulos, D. and Nicolson, G.L. (1975) Local anesthetics affect transmembrane cytoskeletal control of mobility and distribution of cell surface receptors. Proc. Natl. Acad. Sci. USA 72, 4430–4434.

Prager, E.M. and Wilson, A.C. (1971) The dependence of immunological cross-reactivity upon sequence resemblance among lysozymes. Comparison of precipitin and micro-complement fixation studies. J. Biol. Chemistry 246, 5978–5989.

Preud'homme, J.L., Neauport-Sautes, C., Piat, S., Silvestre, D. and Kourilsky, F.M. (1972) Independence of HL-A antigens and immunoglobulin determinants on the surface of human lymphoid cells. Eur. J. Immunol. 3, 297–300.

Ran, M., Fish, F., Witz, I.P. and Klein, G. (1974) Tumour-bound immunoglobulins. The in vitro disappearance of immunoglobulin from the surface of coated tumour cells, and some properties of released components. Clin. Exp. Immunol. 16, 335–353.

Ray, P.K. and Simmons, R.L. (1973) Serological studies of enzyme-treated murine lymphoid cells. Proc. Soc. Exp. Biol. Med. 142, 846–852.

Reichlin, M. (1972) Localizing antigenic determinants in human haemoglobin with mutants: molecular correlations of immunological tolerance. J. Mol. Biol. 64, 485–496.

Rosse, W.F., Dourmashkin, R.R. and Humphrey, J.H. (1966) Immune lysis of normal human and paroxysmal nocturnal hemoglobinuria (PNH) red blood cells, III. The membrane defects caused by complement lysis. J. Exp. Med. 123, 969–984.

Rosse, W.F., Borsos, R. and Rapp, H.J. (1968) Cold-reacting antibodies. The enhancement of antibody fixation by the first component complement (C'1a) J. Immunol. 100, 259–265.

Rowley, D. and Turner, K.J. (1968) Passive sensitization of *Salmonella adelaide* to the bacterial action of antibody and complement. Nature (London) 217, 657–658.

Rubin, H. (1966) Fact and theory about the cell surface in carcinogenesis. In: M. Locke (Ed.), 25th symposium of the Society for Developmental Biology, Academic Press, New York, pp. 317–332.

Rubio, N. (1974) Surface H-2 antigen concentration requirement of somatic hybrid cells for IgM-mediated cytotoxicity. Nature (London), 249, 461–463.

Ryan, G.B., Unanue, E.R. and Karnovsky, M.J. (1974) Inhibition of capping of surface macromolecules by local anesthetics and tranquillizers. Nature (London) 250, 56–57.

Sabin, A.B. (1950) The Dengue group of viruses and its family relationships. Bacteriol. Rev. 14–15, 225–232.

Sandberg, A.L., Osler, A.G., Shin, H.S. and Oliveira, B. (1970) The biologic activities of guinea pig antibodies, II. Modes of complement interaction with γ1 and γ2 immunoglobulins. J. Immunol. 104, 329–334.

Sanderson, A.R. (1968) HL-A substances from human spleens. Nature (London) 220, 192–195.

Sanderson, A.R. and Welsh, K.I. (1974) Properties of histocompatibility (HL-A) determinants. Transplantation 17, 281–289.

Sanford, B.H. (1967) An alteration in tumour histocompatibility induced by neuraminidase. Transplantation 4, 1273–1279.

Santer, H.P. (1972) Ultrastructural distribution of surface immunoglobulin determinants on mouse lymphoid cells. Exp. Cell Res. 72, 377–386.

Schlesinger, J., Koppel, D.E., Axelrod, D., Jacobson, K., Webb, W.W. and Elson, E.L. (1976) Lateral transport of cell membranes: Mobility of concanavalin A receptors on myoblasts. Proc. Natl. Acad. Sci. USA 73, 2409–2413.

Schreiner, G.P. and Unanue, E.R. (1976) Calcium-sensitive modulation of Ig capping: Evidence supporting a cytoplasmic control of ligand-receptor complexes. J. Exp. Med. 143, 15–31.

Schumaker, V.N., Calcott, M.A., Spiegelberg, H. and Müller-Eberhard, H.J. (1975) Human myeloma IgG half-molecules. Structural and antigenic analyses. Biochemistry 10, 2157–2163.

Schwartz, B.D. and Nathenson, S.G. (1971) Regeneration of transplantation antigens on mouse cells. Transplant. Proc., 3, 180–182.

Shelton, E., Yonemasu, K. and Stroud, R.M. (1972) Ultrastructure of human complement component, C1q. Proc. Natl. Acad. Sci. USA 69, 65–68.

Shipley, W.U. (1971) Immune cytolysis in relation to the growth cycle of Chinese hamster cells. Cancer Res. 31, 925–929.

Simmons, R.L., Rios, R., Ray, P.K. and Lundgren, G. (1971) Effect of neuraminidase on growth of a 3-methylcholanthrene-induced fibrosarcoma in normal and immunosuppressed syngeneic mice. J. Natl. Cancer Inst. 47, 1087–1094.

Singer, S.J. (1974) Molecular biology of cellular membranes with applications to immunology. Adv. Immunol. 19, 1–66.

Singer, S.J. and Nicolson, G.L. (1972) The fluid mosaic model of the structure of cell membranes. Science, 175, 720–730.

Smith, R.T.N. (1968) Tumour-specific immune mechanisms (continued). New Eng. J. Med. 278, 1268–1275.

Snell, G.D., Winn, H.J., Stimfling, J.H. and Parker, S.J.J. (1960) Depression by antibody of the immune response to homografts and its role in immunological enhancement. J. Exp. Med. 112, 293–314.

Solheim, B.G. (1972) Kinetics of the reaction between a monoclonal IgM anti-I and rabbit red cells. Scand. J. Immunol. 1, 179–191.

Sparks, F.C., Ting, C.C., Hammond, W.G. and Herberman, R.B. (1969) An isotopic antiglobulin technique for measuring antibodies to cell-surface antigens. J. Immunol. 102, 842–847.

Stackpole, C.W., Aoki, T., Boyse, E.A., Old, L.J., Lymley-Frank, J. and de Harven, E. (1971) Cell surface antigens: Serial sectioning of single cells as an approach to topographical analyses. Science 172, 472–474.

Stackpole, C.W., Jacobson, J.B. and Lardis, M.P. (1974) Two distinct types of capping of surface receptors on mouse lymphoid cells. Nature (London) 248, 232–234.

Stratton, F. (1960) Some factors involved in the demonstration of complement fixing blood group antibodies using the antiglobulin test. Vox Sang. 5, 201–223.

Strom, R. and Klein, E. (1969) Fluorometric quantitation of fluorrescein-coupled antibodies attached to the cell membrane. Proc. Natl. Acad. Sci. USA 63, 1157–1163.

Sumner, M.C.B., Collin, R.C.L.S. and Pasternak, C.A. (1973) Synthesis and expression of surface antigens during the cell cycle. Tissue Antigens 3, 477–484.

Sundqvist, K.G. (1972) Redistribution of surface antigens – a general property of animal cells? Nature (London) New Biol. 239, 147–149.

Sundqvist, K.G. (1973a) Quantitation of cell membrane antigens at the single cell level, I. Microfluorometric analysis of reaction patterns in various test systems, with special reference to the mobility of the cell membrane. Scand. J. Immunol. 2, 479–494.

Sundqvist, K.G. (1973b) Quantitation of cell membrane antigens at the single cell level, II. On the mechanisms of the dynamic interaction between cell surface and bound ligands. Scand. J. Immunol. 2, 495–509.

Sundqvist, K.G. (1974a) Further studies on dynamic aspects of antigen-antibody interaction at the surface of animal cells by isotope antiglobulin technique. Thesis, Karolinska Institutet, Stockholm, Chapter I.

Sundqvist, K.G. (1974b) Further studies on the isotope modification of the mixed haemadsorption technique. Factors influencing the prozone effect. Scand. J. Immunol. 3, 251–260.

Sundqvist, K.G. and Fagraeus, A. (1972) A sensitive isotope modification of the mixed haemadsorption test applicable to the study of prozone effects. Immunology 22, 371–380.

Sundqvist, K.G., Svehag, S.E. and Thorstensson, R. (1974) Dynamic aspects on the interaction between antibodies and complement at the cell surface. Scand. J. Immunol. 3, 237–250.

Svehag, S.-E., Manhem, L. and Bloth, B. (1972) Ultrastructure of human Clq protein. Nature (London) New Biology, 238, 117–118.

Takahashi, M., Yagi, Y., Moore, G.E. and Pressman, D. (1969) Immunoglobulin production in synchronized cultures of human hematopoietic cell lines. J. Immunol. 103, 834–843.

Takahashi, T. (1971) Antigenic expression as an escape route from immunological rejection. Possible examples of antigenic modulation affecting H-2 antigens and cell surface immunoglobulins. Transplant. Proc. 3, 1217–1220.

Taylor, R.B., Duffus, P.H., Raff, M.C. and de Petris, S. (1971) Redistribution and pinocytosis of lymphocyte surface immunoglobulin molecules induced by anti-immunoglobulin antibody. Nature (London) New Biol. 233, 225–229.

Unanue, E.R. and Karnovsky, K.J. (1974) Ligand-induced movement of lymphocyte membrane macromolecules. J. Exp. Med. 140, 1207–1220.

Unanue, E.R., Perkins, W.D. and Karnovsky, M.J. (1972) Ligand-induced movement of lymphocytes membrane macromolecules, I. Analysis by immunofluorescence and ultrastructural radioautography. J. Exp. Med. 136, 885–906.

Unanue, E.R., Karnovsky, M.J. and Engers, H.D. (1973) Ligand-induced movement of lymphocyte surface macromolecules, III. Relationship between the formation and fate of anti-Ig-surface Ig complexes and cell metabolism. J. Exp. Med. 137, 675–680.

Vitetta, E.S. and Uhr, J.W. (1972) Cell surface Ig, V. Release from murine splenic lymphocytes. J. Exp. Med. 136, 676–696.

Vitetta, E.S and Uhr, J.W. (1974) Cell surface immunoglobulin. A new method for the study of synthesis, intracellular transport, and exteriorization in murine splenocytes. J. Exp. Med. 139, 1599–1620.

Vitetta, E.S. and Uhr, J.W. (1975) Immunoglobulin and alloantigens on the surface of lymphoid cells. Biochim. Biophys. Acta 415, 253–271.

Wahren, B. (1971) Quantitative immunofluorescence of EB virus-infected cells. Ann. N.Y. Acad. Sci. 177, 113–120.

Warren, L. (1969) The biological significance of turnover of the surface membrane of animal cells. In: A.A. Moscona and A. Monroy (Eds.), Current Topics in Developmental Biology, Vol. 4, Academic Press, New York, pp. 197–222.

Warren, L. and Glick, M.C. (1968) Membranes of animal cells, II. The metabolism and turnover of the surface membrane. J. Cell. Biol. 37, 729–746.

Weiss, L. and Coombs, R.R.A. (1963) The demonstration of rupture of cell surfaces by an immunological technique. Exp. Cell Res. 30, 331–338.

Weller, N. (1974) Visualization of Concanavalin A-binding sites with scanning electron microscopy. J. Cell Biol. 63, 699–707.

Werner, D., Vitetta, E.S., Boyse, E.A. and Uhr, J.W. (1973) Synthesis, intracellular distribution and secretion of Ig and H-2 antigen from splenocytes of unimmunized mice. J. Exp. Med. 138, 847–857.

Werner, T.C., Bunting, J.R. and Cathon, R.E. (1972) The shape of immunoglobulin G molecules in solution. Proc. Natl. Acad. Sci. USA 69, 795–799.

Wigzell, H. (1967) Studies on some factors regulating antibody synthesis. Thesis, Karolinska Institutet, Stockholm.

Wilson, J.D., Nossal, G.J.V. and Lewis, H. (1972) Metabolic characteristics of lymphocyte surface immunoglobulins. Eur. J. Immunol. 2, 225–232.

Winchester, R.J., Fu, S.M., Hoffman, T. and Kunkel, H.G. (1975) IgG on lymphocyte surfaces: Technical problems and the significance of a third cell population. J. Immunol. 114, 1210–1212.

Winn, H.J. (1962) The participation of complement in isoimmune reactions. Ann. N.Y. Acad. Sci. 101, 23–45.

Witz, I.P., Kinamon, S., Ran, M. and Klein, G. (1974) Tumour-bound immunoglobulins. Fixation of anti-Ig reagents by tumour cells. Clin. Exp. Immunol. 16, 321–333.

Wolpert, L. and O'Neill, C.H. (1962) Dynamics of the membrane of amoeba proteus studied with labelled specific antibody. Nature 196, 1261–1266.

Woodruff, M.F.A. and Inchley, M.P. (1971) Cytolytic efficiency of rabbit anti-mouse antilymphocytic globulin and its augmentation by antiglobulin. Clin. Exp. Immunol. 9, 839–851.

Yahara, I. and Edelman, G.M. (1972) Restriction of the mobility of lymphocyte immunoglobulin receptors by concanavalin A. Proc. Natl. Acad. Sci. USA 69, 608–612.

Yahara, I. and Edelman, G.M. (1973) Modulation of lymphocyte receptor redistribution by concanavalin A, anti-mitotic agents and alterations of pH. Nature (London), 236, 152–154.

Yahara, I. and Edelman, G.M. (1975) Electron microscopic analysis of the modulation of lymphocyte receptor mobility. Exp. Cell Res. 91, 125–142.

Yefenof, E. and Klein, G. (1974) Antibody-induced redistribution of normal and tumour associated surface antigen. Exp. Cell Res. 88, 217–224.

Yguerabide, J., Epstein, H.F. and Stryer, L. (1970) Segmental flexibility in an antibody molecule. J. Mol. Biol. 51, 573–590.

Mitogen stimulation of B lymphocytes. A mitogen receptor complex which influences reactions leading to proliferation and differentiation 12

Jan ANDERSSON and Fritz MELCHERS

1. Introduction

An immune response is the result of the interaction between antigen and lymphocytes of the immune system (Ehrlich, 1900; Gowans and McGregor, 1965; Mitchison, 1967). The first step in this reaction is the binding of antigen to specific receptors located on the surface membrane of lymphocytes (Naor and Sulitzeanu, 1967; Byrt and Ada, 1969).

A humoral immune response is governed by the ability of an antigen to induce clonal proliferation of immunocompetent bone-marrow derived lymphocytes and subsequent maturation of these cells into high rate antibody producing plasma cells (Miller and Mitchell, 1968). Therefore, successful immune induction is recorded as an increased concentration of specific antibody in the circulation or by enumeration of the number of high rate antibody producing cells by means of the haemolytic plaque assay developed by Jerne et al. (1963).

Depending on the nature of an antigen a humoral immune response may either be thymus-dependent or thymus-independent. Thymus-dependent responses engage the two major classes of lymphocytes, thymus-derived (T) and bone marrow-derived (B) cells, which cooperate in such responses (Claman et al., 1966). Two determinants are needed on an antigen to become immunogenic (Rajewsky et al., 1969). One, the haptenic determinant, is recognized by B cells while the second, the carrier determinant (in the case of T-dependent antigens) is recognized by T cells. Since both T and B cells recognize antigen specifically, both must possess antigen-specific receptor structures. On B cells these receptors are immunoglobulin (Ig) molecules (Pernis et al., 1971; Rabellino et al., 1971; Vitetta and Uhr, 1972), predominantly of the μ-heavy chain class (Greaves and Hogg, 1971). The nature of the receptor molecules on T cells is not known at present.

A thymus-independent response is induced equally well in the presence as well as in the absence of T cells. In this case recognition of carrier-determinants by T cells appears to be unnecessary for induction of hapten-specific responses in B cells.

Two hypotheses have been advanced to explain activation of B cells by antigen. In one model (Bretscher and Cohn, 1969), binding of a haptenic determinant to the corresponding binding site of the Ig molecules situated on the surface of antigen sensitive B cells, gives a signal (Signal 1) to the B cells. Signal 1 alone, it is postulated, will induce unresponsiveness (tolerance) in B cells. If, however, antigen binding via its haptenic determinants to Ig on B cells is accompanied by the binding of "associative antibody" produced by T cells in response to carrier-determinant recognition, then a second signal (Signal 2) is given to the B cells via an "associative antibody"-specific receptor site. Signal 1 plus Signal 2 stimulate B cells to grow and differentiate. The second hypothesis (Coutinho and Möller, 1974) assumes only one, antigen-nonspecific signal will suffice for B-cell stimulation. This signal is received on B cells by non-clonally distributed surface structures other than Ig molecules. Binding of antigen to Ig by itself constitutes no signal. Therefore, antigens must carry structures which fit these non-clonally distributed receptors on B cells and which have the property of activating B cells, which means that these structures are mitogenic. It is postulated in this second hypothesis that T-independent antigens have inherent mitogenic properties, while T-dependent antigens have none.

T cell-dependent antigens are thought to be first recognized by T cells. These T cells then supply molecules, which, together with antigen, are stimulating (mitogenic) for the B cells. For both T-independent and T-dependent antigens, Ig molecules on the surface of B cells focus the haptenic determinants on the correct B cells with high affinities for the hapten and thus concentrate amounts of mitogen sufficient to activate them.

One important distinction in these two hypotheses is the role of surface Ig molecules in triggering of B cells by antigen. The second hypothesis postulates a purely passive role for focussing antigen to the right cell, while the first hypothesis states that occupance of surface Ig by antigen alone induces a reaction in the B cells leading to unresponsiveness. We think that both hypotheses are not compatible with the findings that binding of anti-Ig antibodies to the surface of B cells will lead to inhibition of B-cell development into Ig-secreting cells upon subsequent stimulation (Andersson et al., 1974b). Our experiments outlined below suggest that:
(1) surface Ig on B cells modulates the reactions of B cells leading to either stimulation or unresponsiveness; and
(2) polyclonally active, mitogenic molecules must give a signal to induce both activation and suppression of B cells.

T cell-independent antigens induce a large proportion of all B cells because of the mitogenic principle of these particular antigens (Andersson et al., 1972a). This activation is achieved by simply increasing the concentration so that binding of the mitogenic part of the molecule to the mitogen receptor on B cells becomes efficient enough to polyclonally activate all B cells. This overrides the antigen focussing, hapten-determinant-binding step to the Ig receptors.

Such polyclonal activation will mimic the action of antigen in that B cells are induced to synthesize DNA, to proliferate and to differentiate into plasma cells, secreting large amounts of immunoglobulins (Melchers and Andersson, 1973). Mitogen-stimulated lymphocytes are, therefore, useful for biochemical studies of the cellular and molecular events connected with lymphocyte stimulation since a large number of cells in a population enable biochemical monitoring of such changes.

It is the purpose of this article to describe B lymphocytes before and after activation and to discuss some of the molecular processes involved in this onset of proliferation and differentiation.

2. Lymphocyte heterogeneity

Practically every lymphoid organ at almost all times of prenatal and postnatal life consists of a mixture of lymphocytes at different stages in their differentiation. Certain organs, at specific times in ontogeny, are sites where particular subpopulations may be enriched. Thus, the neonatal or adult thymus appears to contain very few, if any, B cells and is therefore regarded as a source of T cells. Spleen and lymph nodes from normal mice constitute a mixture of T and B cells. Bone marrow, in the adult mouse, appears to be the major site for lymphocyte generation. It therefore contains stem cells, precursors for B and T cells, and the earliest committed stages of Ig-positive B cells, but very few differentiated theta-bearing T cells. Fetal liver serves a similar function during embryonic development until birth. Adult spleen is one of the most heterogeneous lymphoid organs, since it contains stem cells, precursor cells, and early and late differentiated T and B cells. Genetically athymic "nude" mice lack the thymus and are devoid of the differentiated, functional forms of T cells. In such mice spleen and lymph nodes consist of the different developmental stages of B cells and possibly of precursor T cells. Different stages in B-cell development appear to be stimulated by different mitogens. During life the compositions of the lymphoid organs, i.e. their content of cells in the various stages of B-cell development, change. Mitogen-reactive B cells are left in the spleen (Melchers, unpublished observations).

It is obvious that at least partial purification of different lymphocyte populations is necessary in order to perform biochemical experiments. In the results to be reported herein, we have used purified small B lymphocytes from the spleen of nude mice.

3. The small B lymphocyte

Small lymphocytes which carry immunoglobulin of μ-class on the surface membrane are classified as B cells. They are called quiescent B cells because they synthesize very little, if any, DNA and do not divide. Only around 30–40%

of the total number of lymphoid cells in the spleen from normal mice have these characteristics. We call them mitogen-sensitive B cells because they can be induced to proliferate and differentiate to Ig-synthesizing and secreting plasma cells by a suitable dose of most B-cell mitogens.

Small, quiescent Ig-bearing B lymphocytes synthesize RNA and protein. Among the proteins synthesized only 1–3% is immunoglobulin and the only class of newly synthesized Ig which can be detected in these cells is IgM (Parkhouse et al., 1972; Andersson and Melchers, 1973). It was calculated that one small, resting B cell synthesizes around $2.5-5 \cdot 10^2$ IgM 7–8S subunit molecules per hour (Melchers and Andersson, 1974a). The main cellular pool of these IgM molecules is the surface membrane into which the molecules are inserted as 7–8S subunits and as such may serve receptor functions for haptenic determinants of antigen. From measurements on the total number of IgM molecules located in the surface membrane ($5-10 \cdot 10^4$ [Rabellino et al., 1971]) we estimate that one small B cell synthesizes half of its total cellular pool of IgM within a day. Turnover of surface membrane-located IgM is therefore slow ($t_{1/2} \sim 20$ h). The IgM molecules which turn over are shed from the surface membrane as 7–8S subunits into the extracellular fluid. Such molecules, whether located on the membrane or shed, contain the "core" sugars glucosamine and mannose, but not the penultimate galactose or terminal fucose in carbohydrate moieties attached to the μ-heavy chains (Andersson et al., 1974a).

4. The mitogens

Until recently, stimulation of B lymphocytes into Ig-producing and -secreting B plasma cells could only be studied by functional tests involving the stimulating antigen (Jerne et al., 1963; Dutton and Mishell, 1967) or as morphological changes in individual cells in the population. Biochemical analysis of changes in B cells after stimulation were not possible since any given antigen stimulates only less than 1% of a population of resting B lymphocytes. Recently, however, substances have become known which stimulate a large portion of all B lymphocytes (see G. Möller, 1972). They therefore induce *poly*clonal growth and differentiation into secreting cells and thereby circumvent the hapten-binding step to the surface Ig-receptor molecule. Although most B-cell mitogens are thymus-independent antigens, there is no apparent structural characteristic common to the wide variety of polyclonal B cell activators, although the molecular structure of some of them has been elucidated. The commonly studied B cell mitogens include glycolipids such as lipopolysaccharide from Gram-negative bacteria (LPS) (Andersson et al., 1972a), proteins such as "purified protein derivative from Tubercle bacilli" (PPD) (Nilsson et al., 1973) and structurally unknown components of fetal calf serum (FCS) (Coutinho et al., 1973), polysaccharides such as pneumococcal polysaccharide SIII (Coutinho and Möller, 1973a) or dextran and dextran sulfate (Coutinho et al., 1974b) and lipoproteins such as that from the outer membrane of *Escherichia coli* (Melchers et al., 1975a).

Consequently, and unlike with lectins and their stimulation of T cells, no membrane structure on B cells is evident which could serve as receptor for all these different mitogens. In fact, receptors for B-cell mitogens have, so far, not been isolated and this makes it difficult to study the initial events of ligand-receptor interaction in the induction of B cells. It is not even clear at present whether conventional methods of affinity chromatography could be used to purify and identify mitogen receptors, since it may be suspected that affinities of mitogens for their receptors may be low and that rates of association of mitogen with receptor may be slow.

Most lectins which are mitogens for lymphocytes are selectively inducing proliferation only in T lymphocytes (see G. Möller, 1972). It is of considerable interest to note that, in general, lectins are not mitogenic for B lymphocytes although these cells show the same number of lectin receptors on the surface as do T cells (Andersson et al., 1972b; Stobo et al., 1972). Although the lymphocyte receptors for soluble lectins like Concanavalin A (Con A) or phytohaemagglutinin (PHA) are heterogenous (Krug et al., 1973), these lectins display identical apparent affinities for the receptors on both T and B cells (Stobo et al., 1972). But only T cells are induced indicating that binding of mitogens to cells is not the only step in the interaction of mitogens with lymphocytes leading to induction of growth and differentiation.

It was shown that lectins like Con A or PHA could be rendered stimulatory for B cells provided that they were insolubilized on Sepharose® (Greaves and Bauminger, 1972; Andersson and Melchers, 1973) or tissue culture petri dishes (Andersson et al., 1972c). In two of these reports the evidence for insolubilized lectins acting as direct B-cell mitogens was insufficient, since only thymidine uptake measurements were done, and no direct B-cell function (such as immunoglobulin synthesis) was recorded. In one report it was shown that Con A-Sepharose beads could activate B cells to increase immunoglobulin synthesis and secretion (Andersson and Melchers, 1973).

In the light of recent experiments with more defined B-cell mitogens it is not clear whether stimulation of B cells by insolubilized Con A is the result of a direct or an indirect action of insolubilized lectin on B cells. It may be that insolubilized Con A renders B cells more susceptible to factors present in serum or produced by other cells, or to the agarose or dextran backbone of the bead material, all of which may act as a subsequent mitogenic signal. Taken together, these considerations indicate that for the present insolubilized Con A is a less suitable "B-cell" mitogen.

For studies on the stimulation of B lymphocytes it is recommended that mitogens of more defined action on B cells be used. It was found by Schumann et al. (1973) that the fucose binding protein from *Ulex europeus* was mitogenic for B cells. Again, only increases in thymidine uptake were used as an assay and no specific B-cell functions were recorded. Because of the obvious advantages of a defined soluble lectin with known binding characteristics (Matsumo and Osawa, 1969) as a selective B-cell mitogen, we repeated these experiments. We used fucose binding proteins from both *Ulex europeus* and

Lotus tetragonolobus. The lectins were purified by conventional chemical techniques or by affinity chromatography on fucosyl-epsilon amino caproyl-agarose (Yariv et al., 1967) and subsequent elution of the binding proteins by L-fucose. Several batches of chemically purified fucose binding proteins from either *U. europeus* or *L. tetragonolobus* were mitogenic for B lymphocytes as revealed by increased thymidine uptake and increased rates of immunoglobulin synthesis and secretion. *All* preparations of affinity chromatography purified fucose binding proteins were non-mitogenic for both T- and B-lymphocytes. There was no difference between mitogenic and non-mitogenic preparations in their ability to agglutinate human red blood cells of type O, and there was no difference in the concentration of L-fucose required to inhibit haemagglutination. Lymphocyte mitogenesis by the fucose binding protein preparations was not prevented by concentrations of L-fucose which completely abolished the anti-blood group O haemagglutinating activity. Similarly the lymphocyte mitogenic activity, unlike the fucose binding activity, was not heat labile.

Thus, we conclude that the mitogenic principle of fucose binding protein preparations from *Ulex europeus* or *Lotus tetragonolobus* is not associated with the fucose binding properties of these preparations.

5. The induction of B lymphocytes by mitogens

Polyclonal activation of B cells by mitogens has been used to delineate cellular and molecular changes which occur after activation. Mitogenic activation of B cells appears not to require T cells or other accessory cells, since a population of cells consisting of more than 95% Ig bearing small lymphocytes can be stimulated.

Upon treatment with mitogen (1 μM–10 μM) B cells respond by increased rates of DNA synthesis after 14–16 h of culture. This lag period observed for induction of DNA synthesis seems common for all mammalian cells where a transition from G_0 to the S-phase of the cell cycle takes place (Baserga, 1969; Clarkson and Baserga, 1974).

After stimulation with mitogen, morphological changes occur in the B lymphocytes. From small cells with little cytoplasm blast cells form with more cytoplasm and then plasma cells with even more cytoplasm develop. The cytoplasm of these plasma cells contains large amounts of membraneous structures such as the rough and smooth endoplasmic reticulum and Golgi apparatus (de Petris et al., 1963). These cells now synthesize 10–100 times more IgM-molecules than small B lymphocytes and actively secrete them (Melchers and Andersson, 1974a). Consequently, induction of B cells to plasma cells leads to changes in the regulatory states of IgM synthesis, transport and secretion. We call this amplification of the phenotypic expression of Ig-genes governing B-cell maturation. Some of the possible molecular processes leading to this maturation may be understood by describing the changes in synthesis, turnover, carbohydrate composition, surface deposition and active secretion of IgM which occur early and late in the response of small B lymphocytes to mitogenic stimulation.

An increased rate of IgM synthesis occurs within the first hour of mitogenic stimulation. This stimulation, which may be obtained with LPS as a B-cell mitogen, is 2- to 3-fold and furthermore is dose-dependent. IgM molecules synthesized at this increased rate are actively secreted from cells with a medium disappearance time of 2 to 4 h and are no longer shed from the cells as 7–8S subunits but are actively secreted as 19S pentamers. In the secreted 19S form IgM molecules contain the "branch" sugars galactose and fucose attached to the Hμ-chains. With increasing time of stimulation the rate of synthesis of the glycoprotein IgM increases over the rates of synthesis of other proteins and carbohydrate containing macromolecules. At 70 h of mitogenic stimulation cells have developed which devote more than 10% of their protein synthetic capacity to the synthesis of IgM and which secrete more than 70% of their proteins as IgM molecules. Since in a given B-cell population the number of IgM-secreting cells increases with time of stimulation, the selective increase of IgM synthesis and secretion indicates that more and more cells within the population change from non-secretors to secretors and that this change occurs in a short time (hours) compared to the total time (days) required for the development of a B-cell clone after immune induction.

The initial change in rate and type of IgM synthesis after mitogenic stimulation can also be observed in the presence of actinomycin D (AMD) at doses which completely suppress DNA-dependent RNA synthesis. IgM synthesis, which is more sensitive to inhibition by AMD than is general protein synthesis in small B cells, is rendered more resistant immediately after mitogenic stimulation. Induction of small resting B lymphocytes by mitogens, in the presence or absence of AMD, leads to a redistribution of ribosomes from monoribosomes to polyribosomes within the first hour of stimulation. It remains to be established whether the newly formed polyribosomes contain mRNAs for μ- and L-chains of IgM (Melchers and Andersson, 1974c).

The central problem of B-lymphocyte stimulation remains how the cell achieves selective increase of phenotypic expression of the Ig-genes. It is tempting to speculate that stimulation of B cells by mitogen stabilizes RNA synthesis-dependent components of IgM synthesis, such as mRNAs for the polypeptide chains of IgM, from degradation through the formation of polyribosomes. This would then account for the reprogramming of IgM synthesis to active secretion and the selective expression of the Ig-genes over other genes. Studies on the synthesis, processing and fate of mRNAs for Ig in lymphocytes before and after induction may reveal the detailed mechanisms leading to maturation.

6. Proliferation and maturation of B lymphocytes

Stimulation of B lymphocytes by a mitogen such as LPS leads to clonal proliferation and also to maturation of B cells to high rate Ig-secreting plasma cells. A response of a lymphocyte population can therefore be expressed by the number of divisions and the efficiency of maturation which a clone of B cells undergoes after stimulation. We, and most of the other investigators in this field, have used increases in [^{3}H]thymidine uptake as a measure for *proliferation*. The inherent difficulties with this technique and the drawback that it at most only allows qualitative statements regarding B-cell proliferation have been discussed in detail elsewhere (Andersson and Melchers, 1976).

Recently, culture conditions have been developed (Melchers et al., 1975c) which allow extensive B-cell proliferation with division times of around 18 h so that the the future proliferation can be quantitated. *Maturation* of B cells after mitogenic stimulation can be measured by serological analysis of the immunoglobulin biosynthetically labelled with radioactive precursors, or, most frequently, B-cell maturation is measured as an increase in the numbers of cells secreting immunoglobulin with affinities for determinants on erythrocyte membranes by means of a modified (Bullock and Möller, 1972) haemolytic plaque assay (Jerne et al., 1963). This latter assay is a reliable and quantitative technique but measures, unfortunately, only the end stage of a number of events between binding of ligand to the cell surface and development of plasma cells and has, thus far, only detected IgM-secreting cells.

A number of experimental approaches have been used to see whether B lymphocytes after stimulation balance their response between proliferation and maturation. First, B cells will normally respond to stimulation by dividing before they mature to plasma cells. Thus, in time, the peak of proliferation always precedes that of maturation (Melchers et al., 1975c). A "hot pulse" (Dutton and Mishell, 1967) of high specific activity thymidine abolishes subsequent mitogen-induced maturation into IgM-secreting cells, when such pulses are given at the time of induction of DNA-synthesis in B cells (from 18 h after stimulation onwards) (Melchers and Andersson, 1974d).

We clearly need to monitor early changes in B cells after stimulation which can easily be measured and which can be definitely normalized to physiologically relevant B-cell changes such as growth and Ig-secretion.

From such experiments it can be calculated that more than 90% of the B cells stimulated to mature by mitogen also incorporate thymidine into DNA and proliferate. When B cells are induced to proliferate by LPS first, then treated with a "hot pulse" and subsequently restimulated with a second B-cell mitogen, like PPD or FCS, maturation to PFC is also abolished. These experiments indicate that one and the same population of cells respond to LPS and to PPD to go through at least one round of division before they mature to PFC.

Second, stimulation of B cells by mitogen in the presence of inhibitors of DNA-synthesis, such as cytosine arabinoside or hydroxyurea (Krakoff et al., 1968) will still lead to maturation into 19S IgM-secreting, plaque-forming B cells

(Andersson and Melchers, 1974). The change from synthesis of membrane-bound IgM (7–8S IgM, slow turnover, AMD-sensitive) after stimulation (Andersson et al., 1974a) to synthesis of secreted IgM (19S IgM, fast turnover, AMD-resistant (Melchers and Andersson, 1974a) occurs within the first hour, some 10–15 h before initiation of DNA synthesis. These experiments indicate in B cells that molecular processes leading to maturation can take place in the absence of proliferation (Andersson and Melchers, 1974). A comparison of the maturation of proliferation-inhibited, stimulated B cells with non-inhibited, proliferating cells shows that two stages of B-cell maturation and differentiation are distinguishable. In the absence of DNA synthesis the first-stage immature plasmablasts develop mostly surface-located and little intracytoplasmic Ig, while a second or later stage of B-cell differentiation to mature plasma cells requiring DNA synthesis ensues with abundant intracytoplasmic Ig and well developed rough endoplasmic reticulum.

Third, a series of plasma cell-like mouse myeloma tumours, all secreting IgM, have been compared with normal, LPS-stimulated B cells for their degree of differentiation into mature plasma cells by morphological and biochemical parameters (Andersson et al., 1974c). Different tumour cell lines resemble normal cells at different stages of plasma cell development. Each tumour line, in all parameters tested, appears to be equivalent to B cells arrested at a particular time after mitogenic stimulation. These tumours, therefore, appear to differently balance between proliferation and maturation.

The above three cases indicate that a balance may exist in B lymphocytes between reactions leading to proliferation and those leading to maturation. In a sense they appear to act antagonistically to each other to control onset and limitation of B-cell growth. In the case of B-derived tumour cells this balance of proliferation and maturation has been upset.

Some of the intriguing problems in growth control of lymphocytes are to understand: which reactions lead to B-cell proliferation and maturation; how B cells regulate these reactions; when they occur during the cell cycle; and how they are influenced from outside of the cells by stimulatory or inhibitory factors.

7. The role of surface membrane-bound Ig in the induction of B cells to growth and differentiation

During an immune response to antigen, surface Ig acts to focus the "right" haptenic determinant of antigen (with high binding capacity) to a B cell. It has been postulated that this binding by itself has either no signalling effect to the cell (Coutinho and Möller, 1974) or will induce unresponsiveness to subsequent stimulation. Growth and differentiation of B cells is induced by mitogen (Andersson et al., 1972a) or associative antibody (Bretscher and Cohn, 1969; Cohn, 1973) which is supplied, at least in the case of T cell-dependent antigens, by T cells and/or accessory cells.

Antibody against Ig molecules can be used as probes to study the involvement of surface-located Ig molecules in polyclonally, mitogen-stimulated B-cell populations, since they will bind to all clones of B cells expressing Ig on their surface. When small, surface Ig-positive cells are loaded with anti-Ig antibodies prior to mitogenic stimulation by LPS, inhibition of the increase in Ig synthesis and secretion and therefore inhibition of maturation to PFC is observed (Andersson et al., 1974b). The inhibition is specific for the binding of anti-Ig antibodies to surface Ig molecules. Thus, (Fab)$_2$-fragments of the same anti-Ig antibodies also mediate inhibition. Non-specific antibodies, like anti-ovalbumin antibodies, are ineffective. This indicates that the binding of F_c-portions of antibodies to F_c-receptors on B cells (Möller and Coutinho, 1975) does not mediate this inhibition.

The inhibition of PFC maturation can be abrogated by proteolytic treatment of the cells or by dissociation of the anti-Ig antibodies from the cells, possibly through turnover, but only prior to the addition of mitogen to the cells. It is important to emphasize that this inhibition can be reversed by taking off the bound anti-Ig molecules from the surface-bound Ig, but only if mitogen has not been added. This indicates that anti-Ig binding to surface Ig modulates in some way the cell surface to react to mitogen in a *negative* way, i.e. to become suppressed, and stresses that the sole important role of mitogens as agents is to induce *irreversible* changes in B cells leading to either response or unresponsiveness.

Interestingly, if LPS is added prior to anti-Ig antibodies to B cells, a very rapid change in response is observed. Cells stimulated for 6 h by LPS can no longer be inhibited from differentiating to PFC by anti-Ig antibodies. This indicates that changes take place in the mitogen-stimulated cells which may involve surface-bound Ig molecules. Earlier observations (Melchers and Andersson, 1973, 1974a) had also led us to think that surface-bound Ig molecules might be involved in reactions stimulated by mitogens such as LPS, PPD or FCS leading to B-cell differentiation into plasma cells which actively secrete IgM. Surface-located Ig molecules on small lymphocytes were found to aggregate into large, detergent insoluble complexes within 30 to 60 min after LPS has been added to the cells. In the same time period the capacity of these LPS-exposed lymphocytes to bind ^{125}I-labelled anti-Ig antibodies or Fab antibody fragments decreased drastically. Thus aggregated surface-located Ig molecules are being progressively degraded by membrane bound proteases. The kinetics for maximal aggregation of surface-Ig and for disappearance of binding sites for ^{125}I-labelled anti-Ig antibodies on the cells are strikingly similar to the time in which LPS changes splenic lymphocytes to become inaccessible to the inhibitory action of anti-Ig antibodies.

The inhibitory action of anti-Ig antibodies is usually exerted on a heterogeneous population of B lymphocytes. The original observation, that anti-Ig antibodies were inhibiting maturation to PFC but not the proliferation of B cells, was confirmed with B-lymphocyte subpopulations from different lymphoid organs. Recirculating B cells from the ductus thoracicus (TDL) were found to be inhibited both for induction of PFC formation and for thymidine uptake. A similar type of B cell could be found in spleen. When such spleen cells were separated by free-flow electrophoresis, the least-migrating fraction of cells were inhibited for both PFC formation and thymidine uptake similar to TDL-B-cells. Another population of cells, deflected more towards the anode, was only inhibited for PFC formation, but not (or only partially) for thymidine uptake. Other, even more deflected, fractions of cells did not yield any plaques in the mitogen-stimulated cultures, but were stimulated to take up thymidine. Some could while others could not be inhibited for uptake of thymidine by anti-Ig antibodies. This indicated that spleen is a very heterogeneous mixture of cells. When anti-Ig inhibited thymidine uptake and/or PFC formation, it appeared reasonable to assume that this inhibition was exerted through the binding of anti-Ig to surface-bound Ig molecules which are characteristically on B cells. From these experiments it was not clear whether those cells which were stimulated by the B-cell mitogen LPS to undergo thymidine uptake in the presence of anti-Ig antibodies were surface-Ig-containing B cells. However, passage of spleen cells over columns of anti-Ig-coated Sephadex resulted in retention of all cells which could be stimulated by LPS (either PFC-induction or thymidine uptake), while these same cells passed through Sephadex columns coated with nonspecific antibodies (Table I). This indicates that the cells which take up thymidine after exposure to anti-Ig antibodies and mitogen contain surface-bound Ig, and are therefore B cells.

There are three explanations for the nature of such B cells:

(1) there is one population of surface Ig-containing B cells which can balance

TABLE I

PASSAGE OF SPLEEN CELLS OVER ANTI-IMMUNOGLOBULIN COLUMNS REMOVES *ALL* CELLS RESPONDING BY PROLIFERATION AND MATURATION AFTER LPS TREATMENT

Normal mouse spleen cells were passed over Sephadex G100 columns coupled with normal rabbit immunoglobulin (NRIg = control) or rabbit anti mouse immunoglobulin (Ranti-MIg). The passed or non-passed cells were cultured at 10^6 cells/ml in the presence of Con A (5 µg/ml) or LPS (50 µg/ml) and [^{3}H] thymidine uptake determined after 2 days or in the response to LPS the number of antibody secreting plaque-forming cells to heavily haptenated (trinitrophenyl) sheep erythrocytes was determined after 3 days of culture.

Treatment of cells	Response to Con A ("T-cells")	Response to LPS LPS ("B-cells")	
	[^{3}H] thymidine uptake (cpm/10^6 cells)	[^{3}H] thymidine uptake (cpm/10^6 cells)	Plaque-forming cells (PFC)
Nonpassed	72 000	18 000	2200
Nonpassed + anti-Ig	70 000	21 000	10
Passed NRIg	65 000	22 000	2300
Passed NRIg + anti-Ig	62 000	26 000	15
Passed Ranti-MIg	68 000	1 200	2

between proliferation and maturation to PFC. Anti-Ig antibodies inhibit maturation, but not mitogen-stimulated proliferation;

(2) there are two populations of B cells and one is stimulated by LPS to PFC formation but has little thymidine uptake, while the other is induced to only incorporate thymidine. Only the former but not the latter can be inhibited by anti-Ig antibodies; and

(3) there is one population of cells and this is initially only induced by mitogen to uptake of thymidine but not to form PFC. After a period in tissue culture this population develops into a second type which cannot be inhibited by anti-Ig antibodies in the presence of mitogen. In this case the inhibitory properties but not surface-Ig develop.

Only by studying single cells or clones of cells, will we finally be able to decide between these possibilities.

8. Mitogen-receptors on B cells

B cells must contain (polyclonally) receptors for mitogens such as LPS, lipoprotein, PPD etc. or natural mitogens supplied by T- or accessory cells for the following reasons:

(1) the stimulation of splenic lymphocytes by either LPS or PPD involves the majority (at least 70% to 80%) of all Ig-positive lymphocytes and leads at day 3 of the response after an average of two cell divisions to cell populations which in IgM-synthetic and secretory capacities resemble cell suspensions from mature plasma cell tumours synthesizing and secreting large amounts of Ig (Andersson and Melchers, 1973);

(2) different mitogens – LPS, PPD and FCS – stimulate the same population of splenic lymphocytes to undergo clonal development to IgM producing plaque-forming cells through proliferation and differentiation (Melchers and Andersson, 1974d). Thus, the involvement of surface-membrane-located Ig molecules in the mitogenic stimulation of lymphocytes is likely not to occur via the antigen-recognizing combining site of that molecule (Andersson et al., 1972b; Andersson et al., 1973). If mitogens are in fact antigens effecting stimulation of B cells via binding to the variable, antigen-binding portion of the Ig-receptor molecules, then (a) LPS, PPD and FCS should contain the same antigenic structural principle, and (b) our nude splenic lymphocytes must contain an extremely high number of cells committed to these structures; and

(3) splenic lymphocytes of certain inbred strains of mice are high responders for one mitogen, but low responders for another (i.e., C3H/HeJ are high for PPD, low for LPS [Sultzer and Nilsson, 1972]). Responsiveness to LPS and PPD is inherited in a mendelian way as a single autosomal gene (Watson and Riblet, 1974). This argues that receptors for LPS and for PPD are separate structures on the surface of B cells.

These observations suggest that a complex of structures on the surface of Ig-positive, small lymphocytes is involved in lymphocyte stimulation. The complex, probably exposed to the outside of the cell on the surface membrane consists of Ig and mitogen receptors. It is postulated that this complex can exist in different conformations in which it has increased tendencies to associate either with a series of molecules constituting chains of reactions leading to response (growth) (go-signals) or with another series of molecules constituting chains of reactions leading to suppression of further stimulation (stop-signals) (Fig. 1).

While the complex can change back and forth between the two conformations in a monomolecular reaction, ligand-binding to Ig or to mitogen-receptor will fix one of the two conformations of the complex. Mitogen-binding to mitogen receptors irreversibly modifies receptor conformations and initiates go- or stop-reactions, while antigen or mitogen binding to Ig molecules fixes conformations but does not induce the go- or stop-reactions.

At present there is no good evidence for the molecular nature of initiating go-reactions. Stop-reactions appear to be initiated by raising levels of cyclic AMP in B cells as a result of an increased adenyl-cyclase activity and/or a decreased phosphodiesterase activity. Thus, mitogenic stimulation of splenic lymphocytes to the development of PFC is inhibited by incubation of the cells with cholera toxin prior to exposure to the mitogens LPS or PPD. This inhibition by cholera toxin (F. Melchers and P. Cuatrecasas, unpublished observations) later strikingly resembles the inhibition by anti-Ig antibodies described in this paper, although the structures to which cholera toxin initially binds on the surface of lymphocytes is probably not Ig, but a glycolipid, ganglioside GM_1 (Holmgren et al., 1974).

Two interesting examples of changes in this postulated receptor complex appear to occur during ontogeny of B cells (Melchers et al., 1975b):

(1) Early in foetal development, between day 12 and 15 of gestation, IgM is synthesized by large, pre-B-like cells in foetal liver (Melchers et al., 1975). This 7–8S IgM can be radioiodinated by the lactoperoxidase-catalyzed radioiodination reaction, thus it appears to exist on the surface of the cells. Its turnover is, however, very rapid ($t_{1/2} = 45$ min) indicating that its median time on the surface is very short. The dynamic state of lymphocyte membranes, i.e. fluidity and rearrangement of surface-located components *within* the plane of the

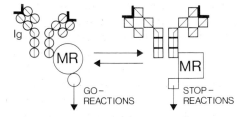

Fig. 1. The possible association between immunoglobulin receptors (Ig) and mitogen receptors (MR) in the surface membrane of B-lymphocytes. Circles and squares indicate different conformations in the mitogen-Ig-receptor complex which favour either stimulatory or inhibitory reactions.

membrane, is thought to influence the reactivity of lymphocytes to external stimuli. It appears necessary to include rates of turnover of surface components, maybe of a receptor complex, as well as rates of movement *across* the plane of the surface membrane in considerations of lymphocyte reactivity. Reactions leading to either growth, suppression or elimination of antigen-binding B cells may very well depend on turnover rates of the antigen-binding Ig molecules on the cell surface;

(2) Later, in foetal development, between day 16 and 19 (birth), small surface Ig-positive B cells exist in foetal liver, which show the slow turnover rate of surface IgM ($t_{1/2} = 20$ h) characteristic of mitogen-reactive small B cells. They are, however, not yet capable of responding to LPS by growth and maturation to PFC, but acquire this capacity around birth. It appears possible that they do not yet express the putative LPS-receptor (see above).

B-cell reactivity to growth induction may depend, therefore, on the existence of an intact receptor complex of slowly turning-over IgM and a mitogen receptor on the surface of the cell.

Many observations remain ill-understood. One small B lymphocyte may contain as many as 10^5 Ig molecules on its surface. Are all of them associated in receptor complexes, and how do 10^5 receptors communicate with each other? Crosslinking of Ig receptors in the plasma membrane has often been claimed to be mandatory for stimulatory signals to B cells (see Möller, 1975). Nothing is known concerning the structural basis of mitogen receptors, and a series of surface receptors and antigens on B cells (F_c-receptor, C'-receptor, MBLA-antigen) have yet to be associated with any known B-cell function. Binding of anti-Ig antibodies to surface Ig of murine B cells and subsequent stimulation by mitogen results in suppression. However, rabbit peripheral lymphocytes are stimulated by anti-Ig antibodies, although it is less clear in this case that the stimulated cells are B cells and it cannot be ascertained if mitogens play a role in this stimulation (Sell and Gell, 1965). It also remains puzzling that while anti-Ig binding causes murine B-cell suppression binding of hapten-mitogen-conjugates (again binding to surface Ig via hapten) causes growth and maturation of B cells (Coutinho et al., 1974a). Furthermore, hapten-protein conjugates with no intrinsic mitogenic activity cause hapten-specific B-cell suppression of subsequent mitogen-induced stimulation (Bullock and Andersson, 1973).

Future experiments will have to consider possible conformational changes in the surface-bound Ig molecule (Huber, 1976; Pecht, 1976), as well as the effects of hapten and different populations of anti-Ig antibodies on conformations of a possible receptor complex between surface-bound Ig and mitogen receptors.

References

Andersson, J. and Melchers, F. (1973) Induction of immunoglobulin M synthesis and secretion in bone marrow-derived lymphocytes by locally concentrated Concanavalin A. Proc. Natl. Acad. Sci USA 70, 416–420.

Andersson, J. and Melchers, F. (1974) Maturation of mitogen-activated bone marrow-derived lymphocytes in the absence of proliferation. Eur. J. Immunol. 4, 533–539.

Andersson, J. and Melchers, F. (1976) Lymphocyte stimulation by Con A. In: H. Bittiger and H.P. Schnebli (Eds.), Concanavalin A as a Tool. Wiley, New York, p. 505.

Andersson, J., Sjöberg, O. and Möller, G. (1972a) Induction of immunoglobulin and antibody synthesis in vitro by lipopolysaccharides. Eur. J. Immunol. 2, 349–353.

Andersson, J., Sjöberg, O. and Möller, G. (1972b) Mitogens as probes for immunocyte activation and cellular co-operation. Transplant. Rev. 11, 131–177.

Andersson, J., Edelman, G.M., Möller, G. and Sjöberg, O. (1972c) Activation of B lymphocytes by locally concentrated Concanavalin A. Eur. J. Immunol. 2, 233–235.

Andersson, J., Melchers, F., Galanos, C. and Lüderitz, O. (1973) The mitogenic effect of lipopolysaccharide in bone-marrow derived mouse lymphocytes. Lipid A as the mitogenic part of the molecule. J. Exp. Med. 137, 943–953.

Andersson, J., Lafleur, L. and Melchers, F. (1974a) Immunoglobulin M in bone marrow-derived lymphocytes. Synthesis, surface deposition turnover and carbohydrate composition in unstimulated mouse B-cells. Eur. J. Immunol. 4, 170–180.

Andersson, J., Bullock, W.W. and Melchers, F. (1974b) Inhibition of mitogenic stimulation of mouse lymphocytes by anti-mouse immunoglobulin antibodies, I. Mode of action. Eur. J. Immunol. 4, 715–722.

Andersson, J., Buxbaum, J., Citronbaum, R., Douglas, S., Forni, L., Melchers, F., Pernis, B. and Stott, D. (1974c) IgM producing tumors in the Balb/C mouse. A model for B-cell maturation. J. Exp. Med. 140, 742–763.

Baserga, R. (1969) Ed. Biochemistry of Cell Division, C.C. Thomas, Springfield, Illinois.

Blumberg, S., Hildesheim, J., Yariv, J. and Wilson, K.J. (1972) The use of 1-amino-L-fucose bound to Sepharose in the isolation of L-fucose-binding proteins. Biochim. Biophys. Acta 264, 171–176.

Bretscher, P. and Cohn, M. (1970) A theory of self-non-self discrimination. Science 169, 1042–1049.

Bullock, W.W. and Andersson, J. (1973) Mitogens as probes for immunocyte regulation specific and non-specific suppression of B cell mitogenesis. In: G.E.W. Wolstenholme and J. Knight (Eds.), Ciba Foundation Symposium on Immunpotentiation. Excerpta Medica, Amsterdam, pp. 173–188.

Bullock, W.W. and Möller, E. (1972) Spontaneous B-cell activation due to loss of normal mouse serum suppressor. Eur. J. Immunol. 2, 514–517.

Byrt, P. and Ada, G. (1969) An in vitro reaction between labelled flagellin or haemocyanin and lymphocyte-like cells from normal animals. Immunology 17, 503–516.

Claman, H.N., Chaperon, E.A. and Triplett, R.F. (1966) Thymus marrow cell combinations. Synergy in antibody production. Proc. Soc. Exp. Biol. Med. 122, 1167–1172.

Clarkson, B. and Baserga, R. (1974) (Eds.) Control of Proliferation in Animal Cells. Cold Spring Harbor Laboratory, New York, pp. 833–1007.

Cohn, M. (1973) Conference evaluation and commentary. In: H.O. McDevitt and M. Landy (Eds.), Genetic Control of Immune Responsiveness. Academic Press, New York, pp. 367–448.

Cosenza, H. and Köhler, H. (1972) Specific suppression of the antibody response by antibodies to receptors. Proc. Natl. Acad. Sci. USA 69, 2701–2705.

Coutinho, A. and Möller, G. (1973a) B cell mitogenic properties of thymus-independent antigens. Nature (London) New Biol. 245, 12–14.

Coutinho, A. and Möller, G. (1973b) Mitogenic properties of the Tl antigen pneumococcal polysaccharide S111. Eur. J. Immunol. 3, 608–613.

Coutinho, A. and Möller, G. (1974) Immune activation of cells: evidence for "one non-specific triggering signal" not delivered by the Ig receptors. Scand. J. Immunol. 3, 133–146.

Coutinho, A., Möller, G., Andersson, J. and Bullock, W.W. (1973) In vitro-activation of mouse lymphocytes in serum-free medium. Effect of T and B cell mitogens on proliferation and immunoglobulin synthesis. Eur. J. Immunol. 3, 299–306.

Coutinho, A., Gronowicz, E., Bullock, W.W. and Möller, G. (1974a) Mechanism of thymus-independent immunocyte triggering. Mitogenic activation of B cells results in specific immune responses. J. Exp. Med. 139, 74–92.

Coutinho, A., Möller, G. and Richter, W. (1974b) Molecular basis of B cell activation, I. Mitogenicity of native and substituted dextrans. Scand. J. Immunol. 3, 321–338.

de Petris, S., Karlsbad, G. and Pernis, B. (1963) Localization of antibodies in plasma cells by electron microscopy. J. Exp. Med. 117, 849–862.

Dutton, R.W. and Hunter, P. (1974) The effect of mitogen-stimulated T cells on the response of B cell to antigen and the mechanism of T cell stimulation of the B cell response. In: G.M. Edelman (Ed.), Cellular Selection and Regulation in the Immune Response. Raven Press, New York, pp. 199–215.

Dutton, R.W. and Mishell, R.I. (1967) Cell population and cell proliferation in the in vitro response of normal mouse spleen to heterologous erythrocytes. Analysis by the hot pulse technique. J. Exp. Med. 126, 443–454.

Dutton, R.W., Falkoff, R., Hirst, J.A., Hoffmann, M., Kappler, J.W., Kettman, J.R., Lesley, J.F. and Vann, D. (1971) Is there evidence for a non-antigen specific diffusable chemical mediator from the thymus-derived cell in the initiation of the immune response? Progr. Immunol. 1, 355–368.

Ehrlich, P. (1900) Croonian lecture: On immunity with special reference to cell life. Proc. Roy. Soc. Lond. Ser. B, 66, 424–448.

Fuji, H. and Jerne, N.K. (1969) Primary immune response in vitro: Reversible suppression by anti-globulin antibodies. Ann. Inst. Pasteur 117, 801–805.

Gowans, J.L. and McGregor, D.D. (1965) The immunological activities of lymphocytes. Progr. Allergy 9, 1–78.

Greaves, M.F. and Bauminger, S. (1972) Activation of T and B cells by insoluble phytomitogens. Nature (London) 235, 67–70.

Greaves, M.F. and Hogg, N.M. (1971) Immunoglobulin determinants on the surface of antigen binding T and B lymphocytes in mice. Progr. Immunol. 1, 111–126.

Gronowicz, E. and Coutinho, A. (1974) Selective triggering of B cell subpopulations by mitogens. Eur. J. Immunol. 4, 771–776.

Holmgren, J., Lindholm, L. and Lönnroth, I. (1974) Interaction of cholera toxin and toxin derivatives with lymphocytes, I. Binding properties and interference with lectin-induced cellular stimulation. J. Exp. Med. 139, 801–819.

Huber, R. (1976) 27th Mosbach Kolloquium on "The Immune System", p. 26.

Hünig, T., Schimpl, A. and Wecker, E. (1974) Autoradiographic studies on the proliferation of antibody-producing cells in vitro. J. Exp. Med. 139, 754–760.

Jerne, N.K., Nordin, A.H. and Henry, C. (1963) The agar plaque technique for recognizing antibody-producing cells. In: B. Amos and H. Koprowski (Eds.), Cell-bound Antibodies. Wistar Institute Press, Philadelphia, pp. 109–125.

Kagnoff, M.F., Billings, P. and Cohn, M. (1974) Functional characteristics of Peyer's patch lymphoid cells, II. Lipopolysaccharide is thymus dependent. J. Exp. Med. 139, 407–413.

Kishimoto, T. and Ishiyaka, K. (1971) Regulation of antibody response in vitro, I. Suppression of secondary response by anti-immunoglobulin heavy chains. J. Immunol. 107, 1567–1575.

Krakoff, I.H., Brown, N.C. and Reichard, P. (1968) Inhibition of ribonucleoside diphosphate reductase by hydroxyurea. Cancer Res. 28, 1559–1565.

Krug, U., Hollenberg, M.D. and Cuatrecasas, P. (1973) Changes in the binding of concanavalin A and wheat germ agglutinin to human lymphocytes during in vitro transformation. Biochem. Biophys. Res. Commun. 52, 305–311.

Matsumoto, I. and Osawa, T. (1969) Purification and characterization of an anti-H (O) phytohaemagglutinin of *Ulex europeus*. Biochim. Biophys. Acta 194, 180–189.

Melchers, F. and Andersson, J. (1973) Synthesis, surface deposition and secretion of immunoglobulin M in bone marrow-derived lymphocytes before and after mitogenic stimulation. Transplant. Rev. 14, 76–130.

Melchers, F. and Andersson, J. (1974a) Immunoglobulin M in bone marrow-derived lymphocytes. Changes in synthesis, turnover and secretion, and in number of molecules on the surface of B-cells after mitogenic stimulation. Eur. J. Immunol. 4, 181–188.

Melchers, F. and Andersson, J. (1974b) Proliferation and maturation of bone marrow-derived lymphocytes. In: G.M. Edelman (Ed.), Cellular Selection and Regulation in the Immune Response. Raven Press, New York, pp. 217–231.

Melchers, F. and Andersson, J. (1974c) Early changes in immunoglobulin M synthesis after mitogenic stimulation of bone marrow-derived lymphocytes. Biochemistry 13, 4645–4653.

Melchers, F. and Andersson, J. (1974d) The kinetics of proliferation and maturation of mitogen-activated bone marrow-derived lymphocytes. Eur. J. Immunol. 4, 687–691.

Melchers, F., Braun, V. and Galanos, C. (1975a) The lipoprotein of the outer membrane of *Escherichia coli*: A B-lymphocyte mitogen. J. Exp. Med. 142, 473–482.

Melchers, F., von Boehmer, H. and Phillips, R.A. (1975b) B-lymphocyte subpopulations in the mouse. Organ distribution and ontogeny of immunoglobulins synthesizing and of mitogen-sensitive cells. Transplant. Rev. 25, 26–58.

Melchers, F., Coutinho, A., Heinrich, G. and Andersson, J. (1975c) Continuous growth of mitogen-reactive B-lymphocytes. Scand. J. Immunol. 4, 853–858.

Miller, J.F.A.P. and Mitchell, G.F. (1968) Cell to cell interaction in the immune response. I. Hemolysin-forming cells in neonatally thymectomized mice reconstituted with thymus or thoracic duct lymphocytes. J. Exp. Med. 128, 431–439.

Mitchison, N.A. (1967) Antigen recognition responsible for the induction in vitro of the secondary response. Cold Spring Harbor Symp. Quant. Biol. 32, 431–439.

Möller, G. (1969) Ed. Antigen sensitive cells. Their source and differentiation. Transplant. Rev. 1.

Möller, G. (1972) Ed. Lymphocyte activation by mitogens. Transplant. Rev. 11.

Möller, G. (1975) Ed. Concepts of B lymphocyte activation. Transplant Rev. 23.

Möller, G. and Coutinho, A. (1975) Role of C'3 and F_c receptors in B-lymphocyte activation. J. Exp. Med. 141, 647–663.

Mosier, D.E. (1967) A requirement for two cell types for antibody formation in vitro. Science 158, 1573–1575.

Naor, D. and Sulitzeanu, D. (1967) Binding of radioiodinated bovine serum albumin to mouse spleen cells. Nature (London) 214, 687–688.

Nilsson, B.S., Sultzer, B.M. and Bullock, W.W. (1973) PPD tuberculin induces immunoglobulin production in normal mouse spleen cells. J. Exp. Med. 137, 127–136.

Parkhouse, R.M.E., Janossy, G. and Greaves, M.F. (1972) Selective stimulation of IgM synthesis in mouse B-lymphocytes by pokeweed mitogen. Nature (London) 235, 21–23.

Pecht, I. (1976) 27th Mosbach Kolloquium on "The Immune System" p. 41.

Pernis, B. Forni, L. and Amante, L. (1971) Immunoglobulins as cell receptors. Ann. N.Y. Acad. Sci. 190, 420–429.

Rabellino, E., Colon, D., Grey, H.M. and Unanue, E.R. (1971) Immunoglobulins on the surface of lymphocytes, I. Distribution and quantitation. J. Exp. Med. 133, 156–167.

Raff, M.C., Feldman, M. and de Petris, S. (1973) Monospecificity of bone marrow-derived lymphocytes. J. Exp. Med. 137, 1024–1030.

Rajewsky, K., Schirrmacher, V., Nase, S. and Jerne, N.K. (1969) The requirements of more than one antigenic determinant for immunogenicity. J. Exp. Med. 129, 1131–1143.

Schlossman, C.F. and Hudson, L. (1973) Specific purification of lymphocyte populations on a digestable immunoadsorbant. J. Immunol. 110, 313–315.

Schumann, G., Schnebli, H.P. and Dukor, P. (1973) Selective stimulation of mouse lymphocyte populations by lectins. Int. Arch. Allergy 45, 331–340.

Sell, S. and Gell, P.G.M. (1965) Studies on rabbit lymphocytes in vitro, I. Stimulation of blast transformation with an antiallotype serum. J. Exp. Med. 122, 423–440.

Stobo, J.D., Rosenthal, A.S. and Paul, W.E. (1972) Functional heterogeneity of murine lymphoid cells, I. Responsiveness to and surface binding of concanavalin A and phytohemagglutinin. J. Immunol. 108, 1–17.

Sultzer, B.M. and Nilsson, B.S. (1972) PPD tuberculin – a B-cell mitogen. Nature (London) New Biol. 240, 198–200.

Vitetta, E.S. and Uhr, J.W. (1972) Cell surface immunoglobulin, V. Release from murine splenic lymphocytes. J. Exp. Med. 136, 676–696.

Watson, J. and Riblet, R. (1974) Genetic control of responses to bacterial lipopolysaccharides in mice, I. Evidence for a single gene that influences mitogenic and immunogenic responses to lipopolysaccharides. J. Exp. Med. 140, 1147–1161.

Yariv, J., Kalb, A.J. and Katchalski, E. (1967) Isolation of an L-fucose binding protein from Lotus tetragonolobus seed. Nature (London) 215, 890–891.

Structure and function of surface immunoglobulin of lymphocytes

Emil R. UNANUE and George F. SCHREINER

1. Introduction

B lymphocytes, the progenitors of antibody-forming cells, represent one of the two major cellular branches of the immune tissues. The other cells are the thymic-derived cells involved in a series of close regulatory interactions with the B cells. B cells recognize antigen molecules by way of receptor molecules on their plasma membranes. These receptor molecules are immunoglobulin molecules (Ig). Ig molecules are made up of two identical pairs of polypeptide chains, the heavy (H) and light (L) chains, each with an N terminal variable region and a C terminal constant region. The constant region of the heavy chains establishes the class of Ig molecules, whether IgM, IgG, IgA, IgE, or IgD. The variable region of each of the heavy and light chains together establishes the site that combines with antigen molecules. Ig molecules are unique, each H and L polypeptide chain is coded by two distinct genes – the genes for the variable and for the constant portion.

Interaction of antigen with surface Ig produces a stimulation of the B cell provided that there follows a second, ill-defined interaction with T cells and/or macrophages. The B cells proliferate, expanding the clone, and differentiation takes place. The differentiated cells become actively secretory with abundant endoplasmic reticulum. The secretory cell – the plasma cell – makes Ig molecules having a combining site for antigen identical to that found in the receptor Ig of the progenitor B cell. The process of differentiation, its control, and the genetics of Ig synthesis are fascinating aspects of the B cell physiology.

This chapter reviews some of the characteristics of Ig as a receptor protein on the plasma membrane of B cells. The rapidly expanding knowledge of the structure and function of surface Ig has provided a conceptual focus for current attempts to understand the cellular basis of the B cell response to antigens and to other cells of the immune system. It has also provided a fascinating model for the more general biological question of the interplay between membrane and cytoplasm in a cellular response evoked by a ligand-membrane-receptor interaction. In this review we have not made an extensive review of the literature but have cited only pertinent or key references.

Extensive reviews on various aspects of Ig and B cell physiology have appeared recently (Warner, 1974; Schreiner and Unanue, 1976c).

2. *Detection of surface Ig*

In this section we will provide a brief explanation of the methodologies used in studying surface Ig. Surface Ig can be detected on B cells either by biochemical or cytochemical methods. In both assays the detection system involves the use of an antibody directed against Ig isolated from serum. Surface and serum Ig share most antigenic determinants. The biochemical method used most extensively is the radioiodination of surface molecules using the procedure developed by Phillips and Morrison (1970). Surface proteins are radiolabeled with ^{125}I, using lactoperoxidase-catalyzed iodination. The cell is then lysed usually with non-ionic detergents; the radiolabeled proteins are precipitated with anti-Ig antibodies; the precipitates are then washed, dissolved, and examined by polyacrylamide gel electrophoresis using sodium dodecylsulphate (SDS) (Baur et al., 1971; Vitetta et al., 1971; Marchalonis and Cone, 1973). Controls in these studies involve the use of non-specific antigen-antibody complexes in order to determine whether any of the labeled products may bind non-specifically to immune complexes. The presence of dead cells or antibody-secreting cells may introduce false results and should be minimized.

Cytochemical methods employ anti-Ig antibodies conjugated to a suitable visual marker. All methods employ live lymphocytes, which are incubated with the antibody, washed, and examined fresh or after fixation. The most commonly used method is immunofluorescence. Surface Ig is identified as small, discrete dots of fluorescence scattered over the entire cell surface (Pernis et al., 1970; Raff et al., 1970; Rabellino et al., 1971). Autoradiography using ^{125}I-labeled antibodies has also been used. Although more sensitive than fluorescence, it has the disadvantage of being a time-consuming procedure and less precise insofar as localization of the labeled products. Extensive studies have been made at the electronmicroscope level using anti-Ig tagged with ferritin, hemocyanin, peroxidase, or viruses (see, among many, Hämmerling et al., 1968; Aoki et al., 1971; Karnovsky et al., 1972; Antoine et al., 1974).

One of the most serious problems in detecting surface Ig concerns the binding of serum Ig or the labeled reagent to lymphoid cells that have Fc receptors. Lymphoid cells have a surface site which binds specifically to the Fc fragment of Ig, usually IgG. (Fc and Fab fragments of Ig are produced by partial proteolysis of the molecule using papain. The Fc fragments include the C terminal halves of the two H chains held by non-covalent bonds. The Fc fragment mediates many of the biological functions of Ig such as fixation of complement, catabolic rates, transplacental passage, etc. The Fab fragments include the L chain and the N terminal half of the H chain and contain the antibody combining site. For two excellent books on Ig chemistry, see Nisonoff et al. (1975) and Kabat (1976)). The Fc site has been poorly characterized in biochemical terms, although a first attempt indicates that the receptor may be relatively easy to identify. A protein of about 60 000 daltons has been isolated recently which is thought to be the Fc receptor (Rask et al., 1975). The Fc receptor has been well characterized biologically. An immune complex (be it an antibody-coated red cell or a fluorescein-labeled complex) will bind avidly to cells having Fc receptors – this binding can be shown to be specific for the Fc portion of IgG and can only be blocked by addition of aggregated IgG and not by any other protein. In the macrophage, where extensive studies have been made, the Fc receptor has specificity for subclasses of IgG. The presence of this receptor is crucial for the phagocyte. The receptor has the important function of binding antibody-coated microorganisms with great avidity to the phagocyte surface, thus favoring phagocytosis and elimination of the foreign material (reviewed in Cohn, 1968). Up to about three years ago, it was thought that only the phagocytes contained Fc receptors. However, as studies with surface Ig progressed and methodologies were improved and made more sensitive, it became apparent that lymphocytes also contained an Fc receptor (Basten et al., 1972; Dickler and Kunkel, 1972; Paraskevas et al., 1972). It is now accepted that most B cells have Fc receptors, as well as some T cells and a third type of lymphocyte with no known surface markers.

The identification of Fc receptors on B cells raised the question whether some of the Ig found on lymphocyte surfaces was derived from serum Ig or from attachment of the antibody in the assay, rather than being a product of synthesis of the lymphocyte. The following conclusions have been reached: (1) B lymphocytes, the progenitors of antibody-forming cells, have Ig on their membranes which they have made and inserted in their plasma membrane. Besides this Ig, B cells have an Fc receptor which allows them to bind aggregated or complexed Ig. It is possible that under certain circumstances there may be some loosely bound Ig on the B cell membrane acquired from serum. This serum-derived Ig is usually IgG; in contrast, the Ig's made by the B cell, at least in man, are IgM and IgD. Cells cultured for several hours keep on synthesizing their Ig receptors and loose the Ig acquired from serum; (2) the Fc receptor is independent of surface Ig molecules (see below); and (3) a class of lymphocytes has been found, characterized by the presence on its membrane of an Fc receptor of high avidity, some of which is bound to serum IgG (Kurnick and Grey, 1975; Lobo et al., 1975; Winchester et al., 1975). With cultivation in vitro, the IgG is lost with time. These lymphocytes that bear IgG on their membranes may thus be wrongly classified as B cells. However, these cells are not progenitors of antibody-forming cells since no synthesis of Ig takes place in them. Their role in immunity is not clear – they may represent an early monocyte or a new class of lymphocytes.

3. Class and antigen specificity of Ig

3.1. Class

Extensive studies have been carried out with human and murine B cells using class-specific antibody (i.e., antibodies reacting with only a particular class of Ig, i.e., IgM, IgG, IgA, IgE, or IgD; antibodies to Fab determinants or to light chains have also been used). In man and mouse, as well as in most other species, IgM is one of the major immunoglobulins on the cell surface (Vitetta et al., 1971, 1974; Marchalonis and Cone, 1973; Andersson et al., 1974; Vitetta and Uhr, 1974). The IgM on the membrane is in monomer form with a molecular weight of about 70 000 daltons instead of the pentameric form of about 10^6 daltons found in serum. van Boxel et al. (1972) made the unexpected observation that a high percentage of human blood lymphocytes carried IgD on their membrane. IgD is an Ig found in trace amounts in serum at concentrations of about 50 μg per ml. This Ig was only discovered because of the availability of a myeloma monoclonal Ig protein found to be unreactive with the available class-specific antibodies and, therefore, identified as a new class of Ig (Rowe and Fahey, 1965). Antibodies to this myeloma protein were used to identify a small amount of the protein in normal serum.

The finding of van Boxel et al., was rapidly confirmed by Rowe et al. (1973), who found that IgD was present in most B cells from newborns. Using antibodies to μ and δ chains, it has now been clearly established that about three-fourths of B cells in human peripheral blood bear both IgM and IgD.

The question concerning the presence or absence of IgG as an antigen receptor on human lymphocytes, which has been of some controversy, has been settled to a great extent. As mentioned before, IgG-bearing cells are found, but the IgG in most of these cells is passively acquired from serum onto Fc receptors and is not made by the lymphoid cells. The number of true IgG-synthesizing B cells in man is quite low.

Is IgD a membrane Ig in other species besides man? One problem in trying to answer this question is that there are no available antibodies to δ chains in any other species. Despite the availability of myelomas in mice, there has been no instance of a myeloma protein analogous to the human IgD. (An analogue of IgD has been found in serum of monkeys using anti-human IgD antibodies.) The question of IgD as a membrane receptor in other species has been tackled by examining radioiodinated surface Ig. The surface Ig, after radioiodination is precipitated by a polyvalent antibody (one recognizing Fab determinants and thus precipitating all Ig molecules regardless of their class). The precipitates are dissolved, the Ig reduced and alkylated, and examined by SDS polyacrylamide gel electrophoresis. Studies done in this way in the mouse indicate the presence of two Ig's in the membrane: one being IgM; the other unidentifiable with class-specific antibodies (Abney and Parkhouse, 1974; Melcher et al., 1974). The radiolabeled heavy chains were two – one having the same position as purified μ chain (i.e., about 70 000 daltons), the other sedimenting between μ

and γ chains. This last Ig was sensitive to proteolysis and interpreted by its discoverers to be the murine equivalent of the human IgD.

The IgD-like molecule (for convenience, we will use the term IgD to describe it, acknowledging that final proof must await serological identification) is indeed synthesized by the spleen cells. It can be stripped off the membrane by proteolysis, and, upon culture, the lymphocytes can reexpress it on their membrane. It has also been found that the IgD appears at a certain stage in the maturation of the mouse. The neonatal mouse is known to be born very immature immunologically and, its B cells lack some characteristics of mature B cells (Gelfand et al., 1974; Sidman and Unanue, 1975). The immature B cells were found by Vitetta et al. (1975) to have only IgM and to lack IgD. However, at about two weeks of life, IgD appears on the surface of the B cells, precisely at the time the mouse becomes fully competent. In adults, IgD is the major surface component. IgD also appears in the maturation of B cells from nude athymic mice, thus ruling out a role for the thymus and/or thymus-derived cells in this phenomenon. It has been possible to determine, using antibodies to μ chains and to Fab determinants, that about one-half to three-fourths of murine splenic B cells have both Ig's.

3.2. Antibody specificity

All the receptors for antigen in B lymphocytes appear to have the same combining site; that is to say, one lymphocyte equals recognition of one unique antigenic determinant. Many of the studies on antigenic specificity involve analysis of antigen binding to B cells. Mose commonly, protein antigens heavily radioiodinated, have been incubated with lymphocytes and the binding estimated by autoradiography (Naor and Sulitzeanu, 1967, 1970; Byrt and Ada, 1969; Unanue, 1971; Davie and Paul, 1972).

The interaction of antigen molecules with B cells has been difficult to study because of the number of cells binding antigen molecules is very sparse, on the order of 10 to 50 per 10^5 lymphocytes examined. The binding of antigen was markedly inhibited by prior treatment of the cells with anti-Ig antibodies (Warner and Byrt, 1970). That the antigen-binding B cells were indeed the cells responsible for the secretion of antibody molecules was established by experiments in which the cells were depleted (or killed) following exposure to the antigen. Wigzell and Andersson (1969) passed suspensions of lymphocytes through columns containing antigen bound to glass beads. The cells binding the antigen were retained in the column. The cells passing through were unable to respond to the antigen in question but retained their capacity to make an immune response to other unrelated antigens. In a different approach, Ada and Byrt (1969) and Humphrey and Keller (1970) briefly incubated lymphocytes with highly radioactive protein antigens (about 0.2 to 0.5 mCi per μg) and then tested for biological activity. They found that the capacity of lymphocytes to respond to the antigen was abrogated – the response to unrelated antigen, however, was not impaired. Apparently the interaction with highly radioactive antigen killed those cells binding the antigen therefore depleting the population of cells competent to make an immune response. These two approaches have not only indicated that the cells binding to antigen are indispensible for immune expression but they strongly indicate that each response is only carried about by a specific clone of cells. (The inverse approach, that of enrichment, has now been done using cells isolated by fluorescent-activated cell sorter [Julius et al., 1972].)

Direct cytochemical determination of the antigen specificity of all surface Ig molecules has been more difficult to determine. An ingenious experiment of Raff et al. (1973), made use of the capping phenomenon (see section 4). Interaction of anti-Ig antibodies with surface Ig leads to a redistribution of the complexes within the plane of the membranes, all coalescing into a small mass at one pole of the cell – the cap of the complexes. B cells were exposed to the antigen flagellin, a polymeric protein obtained from the flagella of *Salmonella* organisms. The flagellin-surface Ig's were capped entirely, leaving no surface Ig outside the cap. Had the surface Ig shown specificity to antigens unrelated to flagellin, these would have not redistributed, remaining diffusely scattered over the cell surface. These results were later confirmed by Nossal and Layton (1976). Since the B cells of the above experiments were from adult mice known to bear both IgD and IgM, one can conclude that the combining site of both Igs is identical. This point was also tested directly by means of idiotypic antibodies. Idiotypic antibodies are antibodies directed to an antigenic conformation associated with the combining site of an Ig molecule. It has been found in cases of abnormal lymphocyte proliferation in man (where a single species of Ig is secreted) that idiotypic antibodies made against the serum monoclonal Ig would bind to all the Ig's found on the membrane of the abnormal cells (Wernet et al., 1972; Schroer et al., 1974). Fu et al. (1975) have determined that IgD and IgM carried the same idiotype and thus identical antigen-recognition sites.

Several characteristics of antigen-binding B cells are worth analyzing. The B cells binding a given antigen molecule are found in an individual prior to antigen stimulation but increase in number and affinity following exposure to antigen. The immune response is thus viewed as an antigen selection process and antigen selects those cells that bear receptors which best fit with it. The results analyzed above support the clonal selection theory of Burnet (1959). Burnet envisioned that during development B cells would arise with receptors for a given antigen, a copy of which was found in serum secreted as antibody. Antigen would select the clone of antibody-forming cell progenitors. During development, the clone of B cells having specificity to autologous antigen would be depleted – or inactivated – explaining the inability of an individual to respond to self.

3.3. Allelic exclusion

Ig molecules – either the H or L chains – contain antigenic determinants that differ within the members of the same species. These antigenic determinants are recognized by immunizing within the species, one individual with Ig of another individual. These antigenic determinants are explained, in many cases, by a single amino acid substitution. The term allotype was coined by Oudin to denote those antigenic markers found only in some individuals of an animal species. Allotypes have been studied extensively in man, rabbit, and mouse Ig and found in L chains as well as in different portions of the H chains. Ig allotypes are inherited as an autosomal mendelian trait. Indeed, some of the allotypic determinants that have been recognized are inherited as allelic traits in a codominant way. Thus, a heterozygote individual for a given trait will have serum Ig molecules of both allelic forms. In the middle 1960s, it was found by immunofluorescence that individual plasma cells, the differentiated product of the B cell, contained Ig only of a given class and allotype, and with the same specificity for antigen (Pernis et al., 1965; Cebra et al., 1966). In other words, Ig found in plasma cells of an individual heterozygote for a given allotypic trait contained only one of the allelic forms. This was a clear example of allelic exclusion of an autosomally inherited trait.

Studies have been made in B cells with regard to the question of allelic exclusion. Does the Ig represented on B cell membranes also show allelic exclusion, or does this phenomenon develop only as the cell undergoes differentiation? The answer appears to be that B cells also show allelic exclusion, although arguments against this interpretation have been made. In early studies, Gell and Sell and their associates found that treatment of rabbit lymphocytes with antibodies to Ig allotypes induced the cell to undergo DNA synthesis. Using cells from an individual heterozygote for a given trait and testing in this biological assay, Gell and Sell (1965) (reviewed in Sell and Asofsky, 1968) concluded that B cells contained both allelic forms. A cytochemical analysis by Wolf et al. (1971) confirmed this finding. Other studies, however, have shown allelic exclusion (Davie et al., 1971). Jones et al. (1973, 1974a), concluded from a series of detailed studies in the rabbit that B cells showed allelic exclusion and explained the earlier findings by noting that some *serum* Ig could also be found bound to lymphocytes. This acquired serum Ig which is not made by the cell could clearly explain the findings of more than one allotype class of Ig on the cells. Jones et al., studied the b marker localized in the kappa L chain of rabbit Ig. Heterozygous rabbits (b^5b^9, for example) had large numbers of B cells with surface Ig of both allotypes. The surface Ig was then removed by pronase treatment, and the cells were allowed to regenerate their surface Ig in culture in the absence of rabbit serum in the medium. The cells were found to regenerate only one of the allotypes. Jones et al. (1973) further showed that rabbit B cells could indeed acquire some Ig if cultured in rabbit serum. The exact amounts have not been determined, but it appears sufficient to be detected by sensitive assays. Allelic exclusion has also been reported in man (Froland and Natvig, 1972).

In summary, therefore, the expression of receptor Ig is generally restricted in the B cell. Of the apparently extensive library of variable genes, cells are committed to read only one set in H chain and one in the L chain. There is, likewise, restriction of the expression of constant heavy chain genes in that only μ and δ chains are found, both having the same combining site. Finally, the cell reads the genetic information for Ig genes of only one chromosome. The mechanisms restricting the translation of genetic information are unknown.

4. Topography and redistribution

How is surface Ig anchored in the plasma membrane, how is it distributed, and what is its relationship to other surface molecules?

Ig molecules are unique in that they function both as soluble molecules, in serum and extracellular fluids, and as integral membrane proteins which serve as antigen receptors on B cells. Secreted Ig is derived from intracellular sources apparently having no relationship with the surface Ig. Surface Ig appears to be an integral membrane protein: it cannot be solubilized from the membrane by treatment with various salts (Kennel and Lerner, 1973; Ault and Unanue, 1974). Furthermore, surface Ig solubilized by non-ionic detergents becomes insoluble when attempts are made to remove the detergent (Melcher et al., 1975). There is little understanding of how surface Ig is actually fixed to the membrane. The C terminal region of the μ chains is somewhat rich in hydrophobic amino acids, although not particularly long sequences are found (Putnam et al., 1973). It is possible that, in the surface-bound form, this portion of the H polypeptide chain could be unfolded and partially buried inside the bilipid layer. Several reports have indicated that surface Ig lacks certain antigenic determinants found in serum Ig, suggesting strongly that part of the Fc fragment of the molecule could be hidden (Froland and Natvig, 1973; Fu and Kunkel, 1974; Jones et al., 1974b). In the study of Fu and Kunkel it was found that B cells were incapable of absorbing all the activity of an anti-μ chain antiserum raised against serum μ chains. In one of their experiments a fluorescein-conjugated anti-μ chain antibody stained surface IgM of neoplastic B cells as well as cytoplasmic IgM. (For surface staining, live cells are incubated with the antibody; for cytoplasmic staining, the cells are smeared, fixed in alcohols or acetone making the membrane permeable, and then stained with the labeled antibody.) The absorption of the anti-μ antibody with B cells abolished the reaction with surface IgM but not with cytoplasmic IgM. Clearly, there were antigenic specificities found in secretory IgM and not in surface IgM. The interpretation of these experiments is that the lack of expression of all the antigen of μ chain is because some antigens are either hidden inside the surface or in some way changed or sterically covered in the membrane.

Biochemical analysis of radioiodinated surface Ig of normal B cells has not yielded any particular information within the limitations of the method. Kennel and Lerner (1973) found that μ chain in transformed B cells was somewhat larger than that of serum IgM, but such differences have not been observed with normal human or mouse B cells. (Also, in the study of Milstein et al. (1972) it was found that intracellular L chains were about 1000 daltons larger than the

L chains in serum Ig. The authors speculated on a possible role of an extra piece involved in the secretory process. As with μ chains, the L chains found in surface Ig are of the same size as that in serum Ig.)

The possibility that surface Ig is associated somehow with an anchoring peptide has been raised. The method of examination of surface Ig using iodination and denaturing conditions, however, may not disclose such an association. In this regard, attention has been given to the Fc receptor as a possible anchoring structure (Ramasamy et al., 1974). By differential redistribution studies described below, it has been shown that the Fc receptor on the B cell surface is independent of surface Ig molecules (Abbas and Unanue, 1975). The conclusion is that there are Fc receptors with free available sites independent of surface Ig; it does not rule out that there may be other Fc receptors saturated with surface Ig. There is an interesting association between surface Ig and the receptor when the former is linked by antibody (this point is discussed later).

As expected from the fluid mosaic model of membrane structure (Singer and Nicolson, 1972), surface Ig is free to diffuse within the plane of the membrane. The induced redistribution of surface Ig has been studied quite extensively. Capping of surface Ig by B cells is probably one of the best studied systems, paralleled only by the studies of lectin redistribution in fibroblasts. One point to consider before briefly analyzing capping of surface Ig is its distribution on the cell surface before any interaction with ligands. Several methods have been employed in studying topography of surface Ig – or other surface components – of the lymphocyte surface. Thin sections of cells exposed to anti-Ig antibodies bound to ferritin or other electronmicroscopic markers convey information of limited value since they detect the molecules only in the plane of sectioning. Freeze-etching procedures, on the other hand, give a flat two-dimensional appraisal of the cell surface, although the distribution of the marker along the villi cannot be determined. Few studies on topographical distribution have been made using scanning electron microscopy because of its relatively low resolution. Topographical studies require the use of monovalent antibodies that do not cross-link the components (Davis et al., 1971; de Petris and Raff, 1972). Bivalent antibodies might approximate adjacent sites and distort the normal architecture. An alternative approach is to do the labeling on prefixed cells in conditions where there is no mobility of the surface proteins. (Fixation, however, may alter the antigenic determinants of a particular membrane component.) In a study on distribution of surface Ig using monovalent antibodies, de Petris and Raff (1972) found isolated molecules in no particular pattern. Similar results were obtained by Reyes et al. (1972), using peroxidase-labeled antibody. In the latter study, Ig was found also along the surface of microvilli (depending on culture conditions, B cells contain a number of thin villi dispersed throughout their surface). Studies of the topography were made by Abbas et al. (1975) using the freeze-etching technique. They used a monovalent anti-Ig antibody coupled to the hapten fluorescein. (Fluorescein isothiocyanate used for fluorescence microscopy is an excellent immunogenic hapten if bound to a protein.) A second step used anti-fluorescein antibodies, also monovalent, conjugated to ferritin. Abbas et al., found surface Ig to be distributed throughout the cell surface in no particular pattern except for a definite tendency to form small microclusters of four to eight ferritin grains. These patterns, when subjected to mathematical analysis, did not correspond to a random distribution of molecules since there was too much membrane devoid of molecules. The patterns did not suggest, however, any apparent long-range organization. This random microclustering was found under various experimental conditions – using monovalent antibodies at various hapten:protein ratios or ferritin:protein ratios, in cells prefixed or in cells handled in the cold. The particular reasons for the microclustering of surface Ig, if indeed the true pattern on the membrane, has not been determined. Attempts to disrupt or change the pattern by changing the extracellular Ca^{2+}, for example, or perturbing intracellular microtubules (by treatment with colchicine) were unsuccessful. It is possible that the microclustering may reflect a variable degree of association of surface Ig with a multivalent anchoring peptide, or possibly the interaction of surface Ig with other proteins or lipids in the membrane.

A point to note at this juncture concerns the amount of Ig on the cell surface. There is heterogeneity in the content of surface Ig on B cells with figures varying from 50 000 to 150 000 molecules per B cell (Rabellino et al., 1971; see also Stobo et al., 1972). The number of surface Ig sites needed for stimulation by antigen have not been determined.

Regardless of its mode of insertion and its true topography, surface Ig can readily diffuse upon interaction with cross-linking ligands, producing a cap. Capping has been extensively reviewed recently (Schreiner and Unanue, 1976c) and will only be summarized here. Capping involves a segregation of Ig-ligand complexes to one small area of the plasma membrane of B cells (Taylor et al., 1971; Loor et al., 1972; Unanue et al., 1972). It involves a perfectly coordinated flow of complexes within the plane of the membrane dependent on energy metabolism and is influenced by manipulations of the microfilament-microtubular system of the cell. Within minutes of interaction with ligands (anti-Ig or multivalent antigens), small diffuse clusters of complexes rapidly flow to one pole of the cell where they coalesce. The formation of the Ig-ligand cap does not involve a progressive coalescence of increasingly large patches of complexes but rather takes place immediately as an integrated flow of the entire lattice of complexes to a single site on the membrane. Kenneth A. Ault in our department has estimated the diffusion rate of the complexes of surface Ig at the time of cap formation to be about $10^{-8}\,cm^2\,sec^{-1}$. These figures are about the diffusion rate of phospholipids in natural membranes (i.e., about 0.5 to $11 \cdot 10^{-8}\,cm^2\,sec^{-1}$). They are, however, faster than the spontaneous rate of diffusion of various proteins on the cell surface; for example, rhodopsin diffuses at a rate of 3.5 to $3.9 \cdot 10^{-9}\,cm^2\,sec^{-1}$ (Poo and Cone, 1974); proteins on muscle fibers at 1 to $2 \cdot 10^{-9}\,cm^2\,sec^{-1}$ (Edidin and Fambrough, 1973) and on L cells at $2.6 \cdot 10^{-10}\,cm^2\,sec^{-1}$ (Edidin et al., 1976). Thus, capping of surface Ig cannot be explained on the basis of agglutination of complexes diffusing at random. Instead, the development of capping, its rate of formation, and energy requirements imply that its operation is associated somehow with a motile force, most likely involving elements of the cell cortex.

Following the formation of the cap, a contractile ring develops just below the cap area, the cytoplasm is displaced forward opposite the cap, and the cell undergoes translatory motion (Schreiner and Unanue, 1976a). The cell is perfectly polarized at this time – with the capped receptors at one end, followed by an adjacent zone of dense, contracted microfilaments, then by the cytoplasm containing organelles, the nucleus, and a front portion of cytoplasm streaming forward. We have proposed that one function of Ig capping in the B cell may be a means to stimulate and orient the normally sessile lymphocyte to undergo a motile response (Schreiner and Unanue, 1975, 1976a, 1976b). Subsequent to the formation of the cap, the complexes of Ig-ligand are interiorized in vesicles and suffer intracellular digestion. A small percentage of the complexes are eliminated into the extracellular milieu. At the end of the process, the B cell surface has been cleared of its surface Ig.

Of interest are the observations that capping of other surface macromolecules has different characteristics, suggesting that different mechanisms may be operating in an apparently similar phenomenon. For example, capping of transplantation antigens is a slow, disorganized process, requiring two antibodies and never involving the stimulation of motility (Unanue, 1976). Capping of concanavalin A (Con A) is strictly dependent on translatory motion (de Petris, 1975; Unanue and Karnovsky, 1975).

The role of microfilaments and microtubules has been extensively investigated and debated. It is our feeling that Ig capping may involve the microfilament system which actively engages in the transport of the surface complexes through the membrane. We postulate that the Ig-ligand complex is somehow linked to the actin-myosin network in the cell cortex, probably by a trans-membrane protein, and the contractile system provides the force and the orientation for the flow of complexes. This postulate is based on results of manipulations that presumably affect contractile elements and which have an effect on capping. One of the drugs used most frequently has been the cytochalasins which, at high doses, can reduce capping (Taylor et al., 1971). Another approach is to manipulate cellular Ca^{2+} by the use of Ca^{2+} ionophore A23187 or by local anesthetics (Poste et al., 1975; Schreiner and Unanue, 1976a, 1976b). Introduction of Ca^{2+} into the cell prevents the cell from capping. Also, introduction of Ca^{2+} to cells with already formed Ig caps produces a striking disruption of the cap, which then disperses over the entire cell surface. The disruption of the cap by the introduction of Ca^{2+} requires metabolic energy, as does capping itself. Our interpretation of this phenomenon produced by Ca^{2+} entry is that it is associated with an activation of the entire contractile elements of the cell cortex. In the instances where this contraction takes place prior to capping, it becomes impossible for those contractile elements associated with the surface complexes to actively displace them. The disruption of the cap is viewed as being produced by an effect of contraction of the cortex on the filaments associated with the cap. Not only is Ca^{2+} involved, presumably in the activation of the actin-myosin system, but it may be an element in the putative link between surface complexes and the contractile elements. An experiment suggesting this includes the use of local anesthetics, some of which displace Ca^{2+} from the inner leaflet of the plasma membrane. Chlorpromazine treatment stops capping (Ryan et al., 1974) and also disrupts the caps already formed (Poste et al., 1975; Schreiner and Unanue, 1976b). This disruption of the cap, in contrast to that produced by the Ca^{2+} ionophore A23187, is passive, not requiring energy metabolism. The effects of local anesthetics on capping can be minimized to great extent by increasing the concentration of extracellular Ca^{2+}. Finally, support for an association between the contractile network and the surface complexes is provided by two recent observations – first that the area of the cap is associated with a marked condensation of microfilaments which, upon depletion of ATP, hypercontract; and second, and more important, there appears to be an association between the surface patches and increased local concentration of myosin, using anti-myosin antibodies (Schreiner and Fujiwara, in preparation). Two crucial points in all these studies are to determine whether indeed the surface complexes are linked physically to actin-myosin filaments (and, if so, how) and the mechanisms leading to the activation of these contractile elements.

With respect to membrane-associated microtubules, it has been hypothesized that these elements may have a modulatory role not only in cellular movement and changes in shape but also in the surface disposition of ligand-receptor complexes (Ukena and Berlin, 1972; Berlin et al., 1974; also see chapter 1 of this volume). We will concern ourselves here mainly with a discussion of capping of Ig-ligand complexes. Previous reviews and discussion on capping have focused on this problem in a more extensive way (Yahara and Edelman, 1973a,b; de Petris, 1975). Capping is not affected by treatment with colchicine, a drug which results in increased number of disassembled microtubules (Taylor et al., 1971). Furthermore, in a number of instances, colchicine treatment results in an enhancement of lymphocyte capping and/or motility. The best example of this is the inhibition of Ig capping produced by Con A (Yahara and Edelman, 1973a,b). Lymphocytes exposed to Con A are found to cap Ig poorly unless treated with colchicine. Yahara and Edelman interpret this finding as a reflection of increased microtubular activity produced by Con A which leads to a restriction in the motility of surface complexes. In contrast to Ig capping in normal cells, we believe that capping of Ig in Con A-treated cells may be strictly dependent on cell motility, and that this is the reason for the effects of colchicine (Unanue and Karnovsky, 1974). In some way, perhaps by extensively cross-linking surface glycoproteins, Con A may inhibit their surface motility through the creation of an externally imposed lattice; however, when the cells move, they put forward uninvolved areas of membrane, leaving behind the fixed surface matrix of complexes. Capping in this case represents a countercurrent effect of new membrane relative to the anchored stationary areas (see, de Petris, 1975). As would be expected, stopping cell motility stops Con A-induced capping in the presence of colchicine (Unanue and Karnovsky, 1974; de Petris, 1975). In essence, the role of microtubules is still not apparent. Perhaps these organelles may play some role in offering direction and orientation to the cell. The nature of the interaction of the tubules with the membrane is not clear.

Finally, what is the function of Ig Capping? Capping of antigen receptors can be visualized as serving several purposes. First, it enhances the binding of antigen by establishing stable complexes of multipoint binding sites. Secondly, as a prelude to endocytosis, it may function to eventually clear the membrane in the absence of additional stimulatory signals from thymus-derived lymphocytes. This would prevent the cell from spontaneously dividing and differentiating outside of the recognition controls imposed by T lymphocytes. Finally, as noted earlier, it permits the polarization of a general membrane stimulus to allow the induction of directional motility.

One last point to analyze concerning the distribution of surface Ig is its relationship to other membrane proteins and to intramembranous particles. The topographic interrelationship between various surface components has been studied quite extensively, making use of the redistribution phenomenon detailed above. Assume two receptors, (a) and (b), are situated on independent surface molecules. If (a) is capped with an appropriate ligand, (b) will not be changed in its original distribution. If, on the other hand, (a) and (b) are linked together forming a single unit, capping of (a) will produce co-capping of (b). By following this approach, one can determine whether surface Ig is associated with other membrane components. Using this approach it has been possible to show that : (1) IgM and IgD are independent of each other (Knapp et al., 1973; Rowe et al., 1973); (2) all Ig molecules have the same Ig combining site (Raff et al., 1973); and (3) surface Ig molecules are independent of transplantation antigens (Preud'homme et al., 1972; Unanue et al., 1974) and of Fc receptor (Abbas and Unanue, 1975; Forni and Pernis, 1975).

The relationship between surface Ig and Fc receptor molecules is of particular interest. Capping of immune complexes bound to Fc receptors does not co-cap surface Ig, which remains diffuse. However, when doing the reaction in reverse, one finds the contrary; capping of anti-Ig-Ig complexes produces co-capping of Fc receptors (Abbas and Unanue, 1975; Forni and Pernis, 1976). Thus, there is an apparent association of Ig in complexed form to the Fc receptor, which may result from increased affinity of surface Ig in aggregated form to this structure.

Surface Ig bears no relationship to the intramembranous particles observed by freeze cleavage of the plasma membrane since the capping of Ig does not alter the distribution of the particles (Karnovsky and Unanue, 1973).

5. Synthesis and dynamics of surface Ig

The metabolism of surface Ig varies among cells of the B lineage. The B lymphocyte is derived from a stem cell through a maturation process that is independent of antigen stimulation. Before birth, progenitors of murine B cells are found in the liver and in the spleen; after birth, the bone marrow becomes the major source. Different stages of maturation can be identified by the use of surface markers or by studying their reactivity to antigen. Thus, the B cells appear to evolve from stem cell to an early, intermediate stage to a late, mature cell. The mature, competent B cell differentiates upon antigen stimulation to an antibody-secreting cell. The end stage of this antigen-driven differentiation process is a plasma cell. Plasma cells contain abundant cytoplasm, rich in endoplasmic reticulum and have a small nucleus. Plasma cells do not divide, having a life of about two to three days. Between the end-stage plasma cell and the B cell, there are intermediate stages recognized as cells of large size, with more cytoplasm, and active in antibody secretion.

The different functional stages of B cells have been studied by examining lymphoid tissues from experimental animals either non-stimulated or undergoing immunization. Especially useful has been the property of certain compounds to non-specifically stimulate a large number of B cells. These, most notably the lipopolysaccharides from Gram-negative bacteria, stimulate B cells to undergo division and differentiation to secretory cells (reviewed by Coutinho and Möller, 1975; and Andersson and Melchers in this volume, p. 601).

We will limit ourselves to an analysis of the metabolism of surface Ig in relation to the activity of B cells. Discussion of the fascinating process of antigen activation have just been summarized in an issue of Transplantation Reviews, edited by Möller (1975) and will not be discussed at length here. The binding of most ligands to the Ig receptor – be it an anti-receptor like anti-Ig antibody or antigen – results in very little stimulation of the differentiation process unless this interaction is followed by a second one with interacting cells like the thymus-derived lymphocyte. Thus, B cell stimulation, in general, is regarded as a multisignal process, requiring more than one surface stimulus.

In examining lymphoid tissues for metabolic studies, one important point to consider is the heterogeneity of B cells. Lymphoid tissues, including those from non-stimulated animals, contain B cells at all stages of differentiation. One contaminating plasma cell among several hundred B cells is more than enough to produce false results with respect to rates of secretion or turnover rates (Vitetta et al., 1974). Melchers and associates, as well as Vitetta, Uhr and their colleagues, have undertaken, using semipurified populations of B cells, a detailed study of the various parameters and the reader more interested in detail should consult these publications (Andersson et al., 1974; Melchers and Andersson, 1974; Vitetta and Uhr, 1974; Vitetta et al., 1974; Melchers and Cone, 1975; Melchers et al., 1975). In the studies of Melchers and associates, the cells were harvested from different tissues and fractionated according to density, size, and charge. The behavior of surface Ig was quite different in different classes of cell, reflecting the stage of B cell activation.

The small resting B lymphocyte is characterized by having most of the Ig that is synthesized transported to its surface to serve as receptor protein. Melchers and Andersson (1974) estimated that about 90% of the synthesized Ig remains surface bound after a 4-h period of labeling with radioactive leucine. The half-life of the cell-associated surface Ig has been estimated to be about 20 to 30 h (Andersson et al., 1974; Melchers and Cone, 1975). Results published earlier had indicated a much shorter life, but these can now be explained as a result of heterogeneity of the B cell population. The turnover rate has been estimated by labeling surface Ig with ^{125}I and then determining the fall of cell-associated radioactivity; alternatively, turnover rates had also been measured using internally labeled Ig.

The B cell secretes small amounts of IgM into the medium, but this secreted material has a size of 7S to 8S instead of the 19S size of serum IgM (Andersson et al., 1974). The secreted IgM is presumed to derive from membrane IgM. Indeed, cells that have their surface proteins radiolabeled with ^{125}I release radioactive monomeric IgM into the medium. The mechanisms involved in the spontaneous release, or shedding, of IgM must lead to a change whereby Ig, as an integral membrane protein insoluble in aqueous milieu, now becomes soluble in serum. Perhaps this may result from some limited proteolysis at the cell surface. Vitetta and Uhr (1972) made a series of observations suggesting that the ^{125}I-labeled Ig secreted into the medium was non-covalently bound to a plasma membrane lipid. No further studies have been made of this very interesting point.

The mechanisms of assembly and transport of Ig from sites of synthesis in the cytoplasm to the plasma membrane are not known. In analogy to the process found in plasma cells, it is thought that lymphocytes make Ig in their endoplasmic reticulum and from there it is carried into the Golgi area and then to the plasma membranes via vesicles which fuse with the plasma membrane. Small lymphocytes, however, show very few profiles of endoplasmic reticulum. However, Andersson et al. (1974), have calculated that very few ribosomes would be needed to manufacture the amount of Ig found in B cell membrane. An alternative site of synthesis could be the ribosomes associated with the plasma membrane of the cell. In an attempt to distinguish between synthesis in rough endoplasmic reticulum or on polyribosomes close to the plasma membrane, Vitetta and Uhr (1974) pulsed cells with radioactive amino acids and sugars and found an interval of about two hours between the time intracellular Ig was labeled to the time of appearance of radiolabeled material. They interpreted this latency period to reflect the time required for synthesis, assembly, and packaging of the Ig in the endoplasmic reticulum. Their studies also indicated that surface Ig was glycosylated (Ig molecules are glycoproteins; each Ig class has a variable amount of carbohydrate, usually bound to the H chain). The experiments of Vitetta and Uhr revealed that there was a sequence of incorporation of the different sugars. Previous studies in plasma cells had shown that glucosamine was added into nascent chains and that additional sugars – galactose, fucose – were incorporated into molecules assembled in the Golgi area (reviewed in Uhr, 1970). A similar sequence was suggested for B cells by calculating the percentage of surface Ig containing different radioactive sugars relative to radioactive amino acids. Andersson et al. (1974) had previously found that both cell-associated and released Ig were glycosylated.

As described earlier, B cells that interact with anti-Ig antibodies or with Ig antigen (in circumstances not leading to stimulation of differentiation), clear the Ig-ligand complexes from the membrane. Such B cells will reexpress surface Ig within a few hours of culture (Elson et al., 1973; Ault and Unanue, 1974). The amounts of Ig reexpressed on the membrane are similar to those found in the cell prior to interaction with the ligand (Ault and Unanue, 1974). In unpublished studies we have found that the reexpression of surface Ig is, in great part, curtailed by treatment of cells with colchicine, suggesting a role for microtubules for the putative transport from intracellular vesicles to the membrane.

Stimulation by antigen, together with some signal from T cells, results in a dramatic alteration of B cell physiology, inducing cell division and differentiation. Differentiation entails the development of endoplasmic reticulum, increased biosynthesis of Ig, secretion of Ig at a high rate, a change in the Ig secreted by the cell, and a high turnover of surface IgM. These changes are not only found in plasma cells but also in large lymphocytes that develop from the small B cells. The turnover of surface Ig markedly changes to a half-life of 2 to 4 h, about one-tenth of that found in the resting cell (Andersson et al., 1974; Melchers and Cone, 1975). The Ig that is secreted by the stimulated B cell or its derivatives is now 19S IgM or the other Ig classes. This IgM secreted by the cell cannot be radioiodinated at the cell membrane, clearly indicating that the pathway of secretion is independent from that of transport of Ig to the membrane (Vitetta and Uhr, 1974; Melchers et al., 1975). Differences in turnover rates between secreted Ig and surface Ig were first recognized by Lerner et al. (1972), when studying a transformed line of B cells which had surface-bound 7S IgM and also secreted large amounts of 19S IgM.

As noted earlier, the bulk of surface Ig receptors belong to the IgM or IgD class. Yet, when differentiation takes place, there is secretion of IgG or other classes, suggesting that differentiation entails preservation of the recognition site to which the cell is precommitted, together with profound structural changes as the secreted Ig evolves into different structures and classes. Thus, the cell, with a 7S monomeric IgM on its surface, begins to secrete a 19S soluble IgM or an IgA. The change in expression of H chain implies a switch in the genes coding for the different H chains while maintaining its previous reading of H genes coding for the variable portion of the *same polypeptide chain*. There are several theories based on some kind of switch mechanism; they remain intriguing and unproven (see, Gally and Edelman, 1970; Sledge et al., 1976).

An alternative explanation is that each cell that secretes a given Ig class derives from a precursor bearing the same H chain. This has become less tenable with the recent evidence discussed earlier that the number of cells truly bearing IgG is very low. Evidence in favor of a genetic switch comes from a number of observations. Pierce and associates studied the immune response in vitro of murine spleen cells to foreign red cells in the presence of an excess amount of various anti-Ig antibodies (Pierce et al., 1972a,b). The antibodies, by reacting with the B cells and covering and/or clearing surface Ig, would stop the cell from interacting with antigen. They found that exposure to anti-IgM antibodies stopped cells from making not only antibodies of the IgM class but also antibodies of the IgG or IgA class. Thus, the indications were that the precursor cells of plasma cells secreting γ or α chain Ig must have derived from the IgM-bearing B cells. In other experiments it was found that in *immune* animals (that is to say, in spleen cells of animals that had been previously immunized) IgM was not as effective in stopping IgG production, suggesting that some of the precursor B cells – which had immunological "memory" – must have had IgG-bearing receptors. This observation has just been confirmed in another system in which murine spleen B cells incubated with lipopolysaccharide are stimulated to differentiate to secreting IgM or IgG antibodies (Kearney et al., 1976). It was known that exposure of these B cells to anti-Ig can shut off this differentiation process by ways not yet clear. In the experiment of Kearney et al. (1976) it was found that anti-μ chain antibodies stopped the differentiation to both IgM and IgG secretion.

Essentially similar results have been obtained in vivo. Kincade and associates injected antibodies to IgM into newborn chickens coupled with bursectomy and found that the development of plasma cells bearing IgM and IgG was prevented (Kincade et al., 1970). Similar results were obtained by Lawton et al. (1972) in the mouse (newborn B cells are especially sensitive to the effects of anti-Ig antibodies or antigen and rapidly become inactivated). The explanation of Lawton, Kincade, and their associates is that they are stopping an antigen-independent maturation stage which goes from IgM to IgG to IgA. However, the in vitro experiments discussed before argue strongly in favor of an additional substantial switch in class of Ig by an antigen-driven process (such a switch has not been observed to a major extent in the rabbit [Jones et al., 1974a,b]).

A second line of evidence in favor of class switch comes from cytological studies in which plasma cells have been found to contain IgG in their cytoplasm while exhibiting IgM on their membranes. Pernis et al. (1971) found about 15% of the IgG-secreting plasma cells to have surface IgM while Jones et al. found between 4% and 10%.

6. Summary

We have analyzed in this review the main properties of surface Ig as a receptor protein. The biological structure, topography, and surface redistribution of Ig molecules have been well defined, as well as the general change exhibited as a result of an encounter with antigen. B lymphocytes are unique cells, particularly suited for analysis because of the availability of a number of antibodies that can be used as a probe. Also, the end result of differentiation is well defined and easy to establish and a number of in vitro assays for B cell function are available. There are still important gaps in our understanding of surface Ig physiology – its attachment to the membrane, its mode of interaction with the cytoplasm, its biosynthetic pathway, and its assembly in the membrane. All of these are important questions whose answers will be emerging in the near future.

References

Abbas, A.K. and Unanue, E.R. (1975) Interrelationship of surface immunoglobulin and Fc receptors on mouse B lymphocytes. J. Immunol. 115, 1665–1671.

Abbas, A.K., Ault, K.A., Karnovsky, M.J. and Unanue, E.R. (1975) Non-random distribution of surface immunoglobulin on murine B lymphocytes. J. Immunol. 114, 1197–1204.

Abney, E.R. and Parkhouse, R.M.E. (1974) Candidate for immunoglobulin D present on murine B lymphocytes. Nature (London) 252, 600–602.

Ada, G.L. and Byrt, P. (1969) Specific inactivation of antigen-reactive cells with ^{125}I-labelled antigen. Nature (London) 222, 1291–1292.

Andersson, J., LaFleur, L. and Melchers, F. (1974) IgM in bone marrow-derived lymphocytes. Synthesis, surface deposition, turnover and carbohydrate composition in unstimulated mouse B cells. Eur. J. Immunol. 4, 170–180.

Antoine, J.-C., Avrameas, S., Gonatas, N.K., Stieber, A. and Gonatas, J.O. (1974) Plasma membrane and internalized immunoglobulins of lymph node cells studied with conjugates of antibody or its Fab fragments with horseradish peroxidase. J. Cell Biol. 63, 12–23.

Aoki, T.H., Wood, H.A., Old, L.J., Boyse, E.A., de Harven, E., Lardis, M.P. and Stackpole, C.W. (1971) Another visual marker of antibody for electron microscopy. Virology 65, 858–862.

Ault, K.A. and Unanue, E.R. (1974) Events after the binding of antigen to lymphocytes: removal and regeneration of the antigen receptor. J. Exp. Med. 139, 1110–1124.

Basten, A., Miller, J.F.A.P., Sprent, J. and Pye, J. (1972) A receptor for antibody on B lymphocytes. I. Method of detection and functional significance. J. Exp. Med. 135, 610–626.

Baur, S., Vitetta, E.S., Sherr, C.J., Schenkein, I. and Uhr, J.W. (1971) Isolation of heavy and light chains of immunoglobulin from the surfaces of lymphoid cells. J. Immunol. 106, 1133–1135.

Berlin, R.D., Oliver, J.M., Ukena, T.E. and Yin, H.H. (1974) Control of cell surface topography. Nature (London) 247, 247, 45–46.

Burnet, F.M. (1959) The Clonal Selection Theory of Acquired Immunity. Cambridge University Press, Cambridge.

Byrt, P. and Ada, G.L. (1969) An in vitro reaction between labelled flagellin or haemocyanin and lymphocyte-like cells from normal animals. Immunology 17, 503–521.

Cebra, J.J., Colberg, J.E. and Dray, S. (1966) Rabbit lymphoid cells differentiated with respect to α, γ, and μ-heavy polypeptide chains to allotype markers Aa1 and Aa2. J. Exp. Med. 123, 547–561.

Cohn, Z.A. (1968) The structure and function of monocytes and macrophages. Adv. Immunol. 9, 163–214.

Coutinho, A. and Moller, G. (1975) Thymus-independent B cell induction and paralysis. Adv. Immunol. 21, 113–133.

Davie, J.M. and Paul, W.E. (1972) Receptors on immunocompetent cells, IV. Direct measurement of avidity of cell receptors and cooperative binding of multivalent ligands. J. Exp. Med. 135, 643–674.

Davie, J.M., Paul, W.E., Mage, R.G. and Goldman, M.B. (1971) Membrane-associated immunoglobulin of rabbit peripheral blood lymphocytes: allelic exclusion at the *b* locus. Proc. Natl. Acad. Sci. USA 68, 430–441.

Davis, W.C., Alspaugh, M.A., Stimpfling, J.H. and Walford, R.L. (1971) Cellular surface distribution of transplantation antigens: discrepancy between direct and indirect labeling techniques. Tissue Antigens 1, 89–96.

de Petris, S. (1975) Concanavalin A receptors, immunoglobulins, and θ antigens of the lymphocyte surface. J. Cell Biol. 65, 123–146.

de Petris, S. and Raff, M.D. (1972) Distribution of immunoglobulin on the surface of mouse lymphoid cells as determined by immunoferritin electron microscopy. Antibody-induced, temperature-dependent redistribution and its implications for membrane structure. Eur. J. Immunol. 2, 523–535.

Dickler, H.B. and Kunkel, H. (1972) Interaction of aggregated γ-globulin with B lymphocytes. J. Exp. Med. 136, 191–199.

Edidin, M. and Fambrough, D. (1973) Fluidity of the surface of cultured muscle fibers. Rapid lateral diffusion of marked surface antigens. J. Cell Biol. 57, 27–37.

Edidin, M., Yagyansky, Y. and Lardner, T.J. (1976) Measurement of membrane protein lateral diffusion in single cells. Science, 191, 466–469.

Elson, C.J., Singh, J. and Taylor, R.B. (1973) The effect of capping by anti-immunoglobulin antibody on the expression of cell surface immunoglobulin and on lymphocyte activation. Scand. J. Immunol. 2, 143–149.

Forni, L. and Pernis, B. (1975) Interactions between Fc receptors and membrane immunoglobulins on B lymphocytes. In: M. Seligmann, J.L. Preud'homme and F.M. Kourilsky (Eds.), Membrane Receptors of Lymphocytes. North-Holland, Amsterdam, pp. 193–201.

Froland, S.S. and Natvig, J.B. (1972) Class, subclass and allelic exclusion of membrane-bound Ig of human B lymphocyte. J. Exp. Med. 136, 409–414.

Fu, S.M. and Kunkel, H. (1974) Membrane immunoglobulins of B lymphocytes. Inability to detect certain characteristic IgM and IgD antigens. J. Exp. Med. 140, 895–903.

Fu, S.M., Winchester, R.J. and Kunkel, H. (1975) Similar idiotypic specificity for the membrane IgD and IgM of human B lymphocytes. J. Immunol. 114, 250–252.

Gally, J.A. and Edelman, G.M. (1970) Somatic translocation of antibody genes. Nature (London) 227, 341–348.

Gelfand, M.C., Elfenbein, G.F., Frank, M.M. and Paul, W.E. (1974) Ontogeny of B lymphocytes, II. Relative rates of appearance of lymphocytes bearing surface immunoglobulin and complement receptors. J. Exp. Med. 139, 1125–1153.

Gell, P.G.H. and Sell, S. (1965) Studies on rabbit lymphocytes in vitro, II. Induction of blast transformation with antisera to six IgG allotypes and summation with mixtures of antisera to different allotypes. J. Exp. Med. 122, 813–821.

Hämmerling, U., Aoki, T., de Harven, E., Boyse, E.A. and Old, L.J. (1968) Use of hybrid antibody with anti-γG and anti-ferritin specificities in locating cell surface antigens by electron microscopy. J. Exp. Med. 128, 1461–1469.

Humphrey, J.H. and Keller, H.U. (1970) Some evidence for specific interaction between immunologically competent cells and antigens. In: J. Sterzl and I. Rina (Eds.), Symposium of Developmental Aspects of Antibody Formation and Structures. Publishing House of the Czechoslovak Academy of Science, Prague, pp. 485–502.

Jones, P.P., Cebra, J.J. and Herzenberg, L.A. (1973) Immunoglobulin (Ig) allotype markers on rabbit lymphocytes: separation of cells bearing different allotypes and demonstration of the binding of Ig to lymphoid cell membranes. J. Immunol. 111, 1334–1348.

Jones, P.P., Cebra, J.J. and Herzenberg, L.A. (1974a) Restriction of gene expression in B lymphocytes and their progeny, I. Commitment to immunoglobulin allotype. J. Exp. Med. 139, 581–599.

Jones, P.P., Craig, S.W., Cebra, J.J. and Herzenberg, L.A. (1974b) Restriction of gene expression in B lymphocytes and their progeny, II. Commitment to immunoglobulin heavy chain isotype. J. Exp. Med. 140, 452–469.

Julius, M.H., Masuda, T. and Herzenberg, L.A. (1972) Demonstration that antigen-binding cells are precursors of antibody-producing cells after purification with a fluorescence-activated cell sorter. Proc. Natl. Acad. Sci. USA 69, 1934–1938.

Kabat, E.A. (1976) Structural Concepts in Immunology and Immunochemistry. Holt, Rinehart and Winston, New York.

Karnovsky, M.J. and Unanue, E.R. (1973) Mapping and migration of lymphocyte surface macromolecules. Fed. Proc. 32, 55–59.

Karnovsky, M.J., Unanue, E.R. and Leventhal, M. (1972) Ligand-induced movement of lymphocyte membrane macromolecules, II. Mapping of surface moieties. J. Exp. Med. 136, 907–930.

Kearney, J.F., Cooper, M.D. and Lawton, A.R. (1976) B lymphocyte differentiation induced by lipopolysaccharide, III. Suppression of B cell maturation by anti-mouse immunoglobulin antibodies. J. Immunol. 116, 1664–1668.

Kennel, S.F. and Lerner, R.A. (1973) Isolation and characterization of plasma membrane associated immunoglobulin from culture human diploid lymphocytes. J. Mol. Biol. 76, 485–502.

Kincade, P.W., Lawton, A.R., Bockman, D.E. and Cooper, M.D. (1970) Suppression of immunoglobulin G synthesis as a result of antibody-mediated suppression of immunoglobulin M synthesis in chickens. Proc. Natl. Acad. Sci. USA 67, 1918–1925.

Knapp, W., Bolhuis, R.L.H., Radl, J. and Hijmans, W. (1973) Independent movement of IgD and IgM molecules on the surface of individual lymphocytes. J. Immunol. 111, 1295–1298.

Kourilsky, F.M., Silvestre, D., Neauport-Sautes, C., Loosfelt, Y. and Dausset, J. (1972) Antibody-induced redistribution of HL.A antigens at the cell surface. Eur. J. Immunol. 2, 249–257.

Kurnick, J.T. and Grey, H.M. (1975) Relationship between Ig-bearing lymphocytes and cells reactive with sensitized human erythrocytes. J. Immunol. 115, 305–309.

Lawton, A.R., Asofsky, R., Hylton, M.B. and Cooper, M.D. (1972) Suppression of immunoglobulin class synthesis in mice, I. Effects of treatment with antibody to μ chain. J. Exp. Med. 135, 277–297.

Lerner, R.A., McConahey, P.J., Jansen, I. and Dixon, F.J. (1972) Synthesis of plasma membrane-associated and secretory immunoglobulin in diploid lymphocytes. J. Exp. Med. 135, 136–149.

Lobo, P.I., Westervelt, F.B. and Horwitz, D.A. (1975) Identification of two populations of immunoglobulin-bearing lymphocytes in man. J. Immunol. 114, 116–119.

Loor, F., Forni, L. and Pernis, B. (1972) The dynamic state of the lymphocyte membrane. Factors affecting the distribution and turnover of surface immunoglobulins. Eur. J. Immunol. 2, 203–211.

Marchalonis, J.L. and Cone, R.E. (1973) Biochemical and biological characteristics of lymphocyte surface immunoglobulin. Transplant. Rev. 14, 3–49.

Melcher, U., Vitetta, E.S., McWilliams, M., Lamm, M.E., Phillips-Quagliata, J.M. and Uhr, J.W. (1974) Cell surface immunoglobulin, X. Identification of an IgD-like molecule on the surface of murine splenocytes. J. Exp. Med. 140, 1427–1431.

Melcher, U., Eidels, L. and Uhr, J.W. (1975) Are immunoglobulins integral membrane proteins? Nature (London) 258, 434–437.

Melchers, F. and Andersson, J. (1974) IgM in bone marrow-derived lymphocytes. Changes in synthesis, turnover and secretion, and in numbers of molecules on the surface of B cells after mitogenic stimulation. Eur. J. Immunol. 4, 181–188.

Melchers, F. and Cone, R.E. (1975) Turnover of radioiodinated and of leucine-labeled immunoglobulin M in murine splenic lymphocytes. Eur. J. Immunol. 5, 234–240.

Melchers, F., Cone, R.E., von Boehmer, H. and Sprent, J. (1975) Immunoglobulin turnover in B lymphocyte subpopulations. Eur. J. Immunol. 5, 382–388.

Milstein, C., Brownlee, G.G., Harrison, T.M. and Mathews, M.B. (1972) A possible precursor of immunoglobulin light chains. Nature (London) New Biol. 238, 117–120.

Möller, G. (1975) Concepts of B lymphocyte activation. Transplant. Rev. 23, 5–265.

Naor, D. and Sulitzeanu, D. (1967) Binding of radioiodinated bovine serum albumin to mouse spleen cells. Nature (London) 214, 687–689.

Naor, D. and Sulitzeanu, D. (1970) Affinity of radioiodinated bovine serum albumin for lymphoid cells, III. Further experiments with cells of normal animals. Israel J. Med. Sci. 6, 519–529.

Nicolson, G.L. (1976) Trans-membrane control of the receptors on normal and tumor cells, I. Cytoplasmic influence over cell surface components. Biochim. Biophys. Acta 457, 57–108.

Nisonoff, A., Hopper, J.E. and Spring, S.B. (1975) The Antibody Molecule. Academic Press, New York.

Nossal, G.J.V. and Layton, J.E. (1976) Antigen-induced aggregation and modulation of receptors on hapten-specific B lymphocytes. J. Exp. Med. 143, 511–528.

Paraskevas, F., Lee, S.-T., Orr, K.B. and Israels, L.G. (1972) A receptor for Fc on mouse B lymphocytes. J. Immunol. 108, 1319–1327.

Pernis, B., Chiappino, G., Kelus, A.S. and Gell, P.G.H. (1965) Cellular localization of immunoglobulins with different allotypic specificities rabbit lymphoid tissues. J. Exp. Med. 122, 853–876.

Pernis, B., Forni, L. and Amante, L. (1970) Immunoglobulin spots on the surface of rabbit lymphocytes. J. Exp. Med. 132, 1001–1018.

Pernis, B., Ferrarini, M., Forni, L. and Amante, L. (1971) In: B. Amos (Ed.), Progress in Immunology, Academic Press, New York, pp. 95–106.

Phillips, D.R. and Morrison, M. (1970) The arrangement of proteins in the human erythrocyte membrane. Biochem. Biophys. Res. Commun. 40, 284–289.

Pierce, C.W., Solliday, S.M. and Asofsky, R. (1972a) Immune responses in vitro, IV. Suppression of primary γM, γG and γA plaque-forming cell responses in mouse spleen cell cultures by class-specific antibody to mouse immunoglobulins. J. Exp. Med. 135, 675–697.

Pierce, C.W., Solliday, S.M. and Asofsky, R. (1972b) Immune responses in vitro, V. Suppression of γM, γG and γA plaque-forming cell responses in cultures of primed mouse spleen cells by class-specific antibody to mouse immunoglobulin. J. Exp. Med. 135, 698–710.

Poo, M. and Cone, R.A. (1974) Lateral diffusion of rhodopsin in the photoreceptor membrane. Nature (London) 247, 438–441.

Poste, G. and Nicolson, G.L. (1976) Calcium Ionophores A23187 and X537A affect cell agglutination by lectins and capping of lymphocyte surface immunoglobulin. Biochim. Biophys. Acta 426, 148–155.

Poste, G., Papahadjopoulos, D. and Nicolson, G.L. (1975) Local anesthetics affect transmembrane cytoskeletal control of mobility and distribution of cell surface receptors. Proc. Natl. Acad. Sci. USA 72, 4430–4434.

Preud'homme, J.L., Neauport-Sautes, C., Piat, S., Silvestre, D. and Kourilsky, F.M. (1972) Independence of HL-A antigens and immunoglobulin determinants on the surface of human lymphoid cells. Eur. J. Immunol. 2, 297–300.

Putnam, F.W., Florent, G., Paul, C., Shinoda, T. and Shimiza, A. (1973) Complete amino acid sequence of the μ heavy chain of a human IgM immunoglobulin. Science 182, 287–291.

Rabellino, E., Colon, S., Grey, H.M. and Unanue, E.R. (1971) Immunoglobulins on the surface of lymphocytes, I. Distribution and quantitation. J. Exp. Med. 133, 156–167.

Raff, M.C., Sternberg, M. and Taylor, R.B. (1970) Immunoglobulin determinants on the surface of mouse lymphoid cells. Nature (London) 225, 553–554.

Raff, M.C., Feldmann, M. and de Petris, S. (1973) Monospecificity of bone marrow-derived lymphocytes. J. Exp. Med. 137, 1024–1030.

Ramasamy, R., Munro, A. and Milstein, C. (1974) Possible role for the Fc receptor on B lymphocytes. Nature (London) 249, 573–574.

Rask, L., Klareskog, L., Ostberg, L. and Peterson, P.A. (1975) Isolation and properties of a murine spleen cell Fc receptor. Nature (London) 257, 231–233.

Reyes, F., Lejonc, J.L., Goudin, M.F., Mannoni, P. and Dreyfus, B. (1975) The surface morphology of human B lymphocytes as revealed by immunoelectron microscopy. J. Exp. Med. 141, 392–410.

Rowe, D.S. and Fahey, J.L. (1965) A new class of human immunoglobulins, I. A unique myeloma protein. J. Exp. Med. 121, 171–199.

Rowe, D.S., Hug, K., Forni, L. and Pernis, B. (1973) Immunoglobulin D as a lymphocyte receptor. J. Exp. Med. 138, 965–972.

Ryan, G.B., Unanue, E.R. and Karnovsky, M.J. (1974) Inhibition of surface capping macromolecules by local anesthetics and tranquilizers. Nature (London) 250, 56–57.

Schreiner, G.F. and Unanue, E.R. (1975) The modulation of spontaneous and anti-Ig stimulated motility of lymphocytes by cyclic nucleotides and adrenergic and cholinergic agents. J. Immunol. 114, 802–808.

Schreiner, G.F. and Unanue, E.R. (1976a) Calcium-sensitive modulation of Ig capping – evidence supporting a cytoplasmic control of surface-receptor complexes. J. Exp. Med. 143, 15–31.

Schreiner, G.F. and Unanue, E.R. (1976b) The disruption of immunoglobulin caps by local anesthetics. Clin. Immunol. Immunopathol. 6, 264–269.

Schreiner, G.F. and Unanue, E.R. (1976c) Membrane and cytoplasmic changes in B lymphocytes induced by ligand-surface immunoglobulin interaction. Adv. Immunol. 24, 37–165.

Schroer, K.R., Briles, D.E., van Boxel, J.A. and Dane, J.H. (1974) Idiotypic uniformity of cell surface immunoglobulin in chronic lymphocyte leukemia. J. Exp. Med. 140, 1416–1420.

Sell, S. and Asofsky, R. (1968) Lymphocytes and immunoglobulins. Prog. Allergy 12, 86–106.

Sidman, C.L. and Unanue, E.R. (1975) Development of B lymphocytes, I. Cell populations and a critical event during ontogeny. J. Immunol. 114, 1730–1735.

Singer, S.J. and Nicolson, G.L. (1972) The fluid mosaic model of the structure of cell membranes. Science 175, 720–731.

Sledge, C., Fair, D.S., Black, B., Kreuger, R.G. and Hood, L. (1976) Antibody differentiation: apparent sequence identity between variable regions shared by IgA and IgG immunoglobulins. Proc. Natl. Acad. Sci. USA 73, 923–927.

Smith, R.S., Longmire, R.L., Reid, R.T. and Farr, R.S. (1970) The measurement of immunoglobulin associated with human peripheral lymphocytes. J. Immunol. 104, 367–376.

Stobo, J.D., Rosenthal, A.S. and Paul, W.E. (1972) Functional heterogeneity of murine lymphoid cells, I. Responsiveness to and surface binding of concanavalin A and phytohemagglutinin. J. Immunol. 108, 1–17.

Taylor, R.B., Duffus, P.H., Raff, M.C. and de Petris, S. (1971) Redistribution and pinocytosis of lymphocyte surface immunoglobulin molecules induced by anti-immunoglobulin antibody. Nature (London) New Biol. 233, 225–229.

Uhr, J.W. (1970) Intracellular events underlying synthesis and secretion of immunoglobulin. Cell. Immunol. 1, 228–244.

Ukena, T.E. and Berlin, R.D. (1972) Effect of colchicine and vinblastine on the topographical separation of membrane functions. J. Exp. Med. 136, 1–7.

Unanue, E.R. (1971) Antigen binding cells, II. Effect of highly radioactive antigen on the immunological function of bone marrow cells. J. Immunol. 107, 1663–1665.

Unanue, E.R. (1976) Cytological analysis of histocompatibility molecules. In: D.H. Katz and B. Benacerraf (Eds.), The Role of the Products of the Histocompatibility Gene Complex in Immune Response. Academic Press, New York, pp. 603–642.

Unanue, E.R. and Karnovsky, M.J. (1974) Ligand-induced movement of lymphocyte membrane macromolecules, V. Capping, cell movement, and microtubular function in normal and lectin-treated lymphocytes. J. Exp. Med. 140, 1207–1220.

Unanue, E.R., Perkins, W.D. and Karnovsky, M.J. (1972) Ligand-induced movement of lymphocyte membrane macromolecules, I. Analysis of immunofluorescence and ultrastructural autoradiography. J. Exp. Med. 136, 885–906.

Unanue, E.R., Dorf, M.E., David, C.S. and Benacerraf, B. (1974) The presence of I-region-associated antigens on B cells in molecules distinct from immunoglobulin and H-2K and H-2D. Proc. Natl. Acad. Sci. USA 71, 5014–5016.

van Boxel, J.A., Paul, W.E., Terry, W.D. and Green, I. (1972) IgD-bearing human lymphocytes. J. Immunol. 109, 648–651.

Vitetta, E.S. and Uhr, J.W. (1972) Cell surface immunoglobulin, V. Release from murine splenic lymphocytes. J. Exp. Med. 136, 676–696.

Vitetta, E.S. and Uhr, J.W. (1974) Cell surface immunoglobulin, IX. A new method for the study of synthesis, intracellular transport, and exteriorization in murine splenocytes. J. Exp. Med. 139, 1599–1620.

Vitetta, E.S., Baur, S. and Uhr, J.W. (1971) Cell surface immunoglobulin, II. Isolation and characterization of immunoglobulin from mouse splenic lymphocytes. J. Exp. Med. 134, 242–264.

Vitetta, E.S., Gruncke-Igbal, I., Holmes, K.V. and Uhr, J.W. (1974) Cell surface immunoglobulin, VII. Synthesis, shedding and secretion of immunoglobulin by lymphoid cells of germ-free mice. J. Exp. Med. 139, 862–876.

Vitetta, E.S., Melcher, U., McWilliams, M., Lamm, M.E., Phillips-Quagliata, J.M. and Uhr, J.W. (1975) Cell surface immunoglobulin, XI. The appearance of an IgD-like molecule on murine lymphoid cells during ontogeny. J. Exp. Med. 141, 206–215.

Warner, N.L. (1974) Membrane immunoglobulins and antigen receptors on B and T lymphocytes. Adv. Immunol. 19, 67–216.

Warner, N.L. and Byrt, P. (1970) Blocking of the lymphocyte antigen receptor site with anti-immunoglobulin sera in vitro. Nature (London) 226, 942–943.

Wernet, P., Feizi, T. and Kunkel, H. (1972) Idiotypic determinants of immunoglobulin M detected on the surface of human lymphocytes by cytotoxic assays. J. Exp. Med. 136, 650–655.

Wigzell, H. and Andersson, B. (1969) Cell separation on antigen-coated columns. Elimination of high rate antibody-forming cells and immunological memory cells. J. Exp. Med. 129, 23–36.

Winchester, R.J., Fu, S.M., Hoffman, T. and Kunkel, H. (1975) IgG on lymphocyte surfaces: technical problems and the significance of a third cell population. J. Immunol. 114, 1210–1212.

Wolf, B., Janeway, Jr., C.A., Coombs, R.R.A., Catty, D., Gell, P.G.H. and Kelus, A.S. (1971) Immunoglobulin determinants on the lymphocytes of normal rabbits, III. As4 and As6 determinants of individual lymphocytes and the concept of allelic exclusion. Immunology 20, 931–944.

Yahara, I. and Edelman, G.M. (1973a) Modulation of lymphocyte receptor redistribution by concanavalin A, anti-mitotic agents and alterations of pH. Nature (London) 236, 152–154.

Yahara, I. and Edelman, G.M. (1973b) The effects of concanavalin A on the mobility of lymphocyte surface receptors. Exp. Cell Res. 81, 143–155.

Distribution and mobility of plasma membrane components on lymphocytes

14

S. de PETRIS

1. Introduction

The studies carried out in recent years on the distribution and dynamic behaviour of lymphocyte plasma membrane components have contributed considerably to our present understanding of the general characteristics of biological membranes, and of the functional relationships between plasma membrane components and cytoplasmic structures. These studies have been mainly stimulated by the discovery of the phenomena of ligand-induced redistribution of lymphocyte surface components, and in particular the phenomenon of capping (Taylor et al., 1971). These phenomena have provided an additional proof of the mobility of the macromolecular components in the plane of the plasma membrane first demonstrated by the experiments of Frye and Edidin on the diffusion of membrane antigens of heterokaryons (Frye and Edidin, 1970); in addition they have shown that this mobility is also under the control of cytoplasmic structures. The results of these investigations are reviewed in this article. Although several basic features of the redistribution phenomena appear to be clearly established, the general picture is still largely incomplete, and the interpretation of the experimental observations often remains highly speculative, and sometimes controversial, especially with regard to the mechanical and biochemical mechanisms which control redistribution. Despite these difficulties, this review has concentrated more on the general aspects of redistribution phenomena in lymphocytes rather than examining in detail the particular redistribution characteristics of the individual surface components of these cells, because this approach, although provisional and incomplete, may be ultimately more useful for an appreciation of the significance of the various observations. The possible immunological significance of redistribution phenomena has been only considered briefly; because the general subject of membrane phenomena in mitogenic stimulation is reviewed elsewhere in this volume (see chapter by J. Andersson and F. Melchers [p. 601]), and the relatively few well-studied cases of antigen-induced redistribution in B lymphocytes can be considered, from the point of view of the underlying mechanisms, as special cases of surface Ig redistribution. Moreover, the immunological aspects of these phenomena have been recently reviewed in considerable detail by Schreiner and Unanue (1976c).

2. General characteristics of redistribution of lymphocyte surface components

2.1. Lymphocyte surface components

Redistribution phenomena have been demonstrated to occur, albeit with different modalities, in virtually all types of animal cells thus far examined; however, the lymphocyte still remains one of the preferred subjects for this type of study. This preference can be attributed not only to the intrinsic importance of this type of cell as a key cell in immunological phenomena, and as a cellular model for the study of mitogenic processes in vitro (see Andersson and Melchers, this volume, p. 601), but also to the fact that redistribution phenomena, and in particular capping, can be induced more readily in lymphocytes than in many other cell types. Moreover, a number of different antigens (more than ten) have been immunologically characterized on the surfaces of lymphocytes, providing a repertoire of cell membrane markers (antigens) and specific ligands (antibodies) larger than that presently available for any other cell type. Several of these antigens are common to all lymphocytes and to other cells as well (e.g. the histocompatibility antigens), while others are specific for lymphocytes and are variously represented on different lymphocyte subpopulations. The latter antigens are often used as specific markers for the morphological and functional identification of these subpopulations. The term "lymphocyte" is used in fact to designate a heterogeneous class of cells characterized by certain basic morphological and functional properties (e.g. properties of a mobile cell showing no contact inhibition of movement, a high nuclear-cytoplasmic ratio, etc.), origin, anatomical localization which includes several subpopulations which differ in their functional, and probably also in their structural characteristics. The great majority of lymphocytes in all vertebrates are small non-proliferating cells ("small lymphocytes") of 5–6 μm diameter. Larger cells ("medium" and "large" lymphocytes) are, in general, proliferating cells which are derived from, or are precursors of, small lymphocytes. The two major lymphocyte subpopulations are the B lymphocytes ("bone-marrow-derived" lymphocytes in mammals or "bursa-dependent" lymphocytes in birds) and the T ("thymus-dependent") lymphocytes (Claman and Claperon, 1969; Miller and Mitchell, 1969; see Greaves et al., 1976). The B lymphocytes are precursors of the antibody-producing cells (e.g., plasma cells) and are characterized by the presence on their plasma membranes of membrane-bound immunoglobulin (surface Ig) which is easily detectable by common immunological, immunoradiochemical or immuno-optical labelling methods (Pernis et al., 1970; Raff et al., 1970; Rabellino et al., 1971; Warner, 1974; Vitetta and Uhr, 1975; Greaves et al., 1976). Surface Ig consists of Ig molecules of different classes (IgM, IgG, IgD, etc.) which probably function as specific receptors (antibody) for the various antigens; molecules of different classes can sometimes exist in combination on the same cell (Pernis et al., 1970; Rabellino et al., 1971; von Boxel et al., 1972; Knapp et al., 1973; Rowe et al., 1973; Vitetta and Uhr, 1975). The

T-lymphocytes are a lymphocyte subclass which includes cells responsible for the so-called "cell-mediated immunity" reactions and they may carry receptors for antigens but do not carry surface Ig molecules detectable by standard immunofluorescence methods. In the mouse these cells are characterized by the presence on their plasma membranes of the Thy. 1 (theta, θ) antigen (Raff, 1969; Schlesinger and Yron, 1969; Greaves et al., 1976) which is present also in virtually all thymic lymphocytes, and other cells as well (brain cells, fibroblasts) (Greaves et al., 1976). A minor subpopulation of mononuclear (lymphoid) cells do not belong to any of the two major subclasses, and do not carry their typical surface markers ("null" lymphocytes) (Froland and Natvig, 1973). However, like the majority of B lymphocyte and macrophages, null lymphocytes carry receptors for the Fc part of the gamma-globulin (IgG) molecule (Fc receptors), and through these receptors they can passively absorb soluble IgG molecules (which are weakly bound and easily removed) from the surrounding medium (Froland et al., 1974; Kurnick and Grey, 1975; Lobo et al., 1975). The topography and redistribution of Fc receptors in B lymphocytes and in other mononuclear cells are generally studied using labelled, aggregated immunoglobulin (Dickler and Kunkel, 1972; Dickler, 1974) or antigen-antibody complexes (Basten et al., 1972; Anderson and Grey, 1974; Abbas and Unanue, 1975; Forni and Pernis, 1975).

The dynamic (rheological) behaviour of surface Ig, θ antigens, and several other antigens and surface components (e.g. lectin receptors, Fc receptors, etc.) which follows the binding of an appropriate specific ligand (antibody, lectin, markers for Fc receptors, etc.) has been studied on lymphoid cells of different species (Table I, p. 661). The majority of these distribution and redistribution studies have been carried out on normal spleen, lymph node, and thymus cells of mammals (mouse, rabbit, man) and on some murine and human neoplastic (leukemic) cells and lymphoid cell lines. The molecular characteristics of the surface components so far studied are only incompletely known (cf. Vitetta and Uhr, 1975). These components are probably all, or almost all, glycoproteins which are synthesized by the same cell which carries them on their surface, although this point has not been proved for all cases (cf. Vitetta and Uhr, 1975) [an exception such as the non-glycoprotein β_2-microglobulin is only apparent, because this molecule is a non-covalently linked 12 000 dalton subunit of the HL-A, H-2 and TL glycoprotein antigens (cf. Vitetta and Uhr, 1975; Vitetta et al., 1976)]. Surface immunoglobulins have the well known covalently linked dimeric structure, constituted by two heavy and two light chains (mol. wt. 150 000–180 000 daltons). They are similar to serum Ig, but have some hydrophobic characteristics, probably due to the presence of a peptide with hydrophobic side chains through which they are perhaps inserted into the membrane (Melcher et al., 1975; reviewed in Vitetta and Uhr, 1975). HL-A, H-2 and TL antigens are constituted of two dissimilar non-covalently linked subunits of about 45 000 and 12 000 daltons, respectively, the latter corresponding to the β_2-microglobulin (reviewed in Vitetta and Uhr, 1975). Ia antigens are also constituted of two subunits, non-identical to those of the H-2 or TL antigens (Cullen et al., 1974; Schwartz et al., 1976). Although these and other lymphocyte molecules require detergents for their isolation, the precise way in which they are inserted into, or attached to the membrane is still unknown (cf. Vitetta and Uhr, 1975). However, with respect to their mobile behaviour all of these components, independently of their origin or particular molecular characteristics, can be considered as intrinsic or integral membrane components, either because they are themselves intrinsic membrane proteins (Singer, 1971; Singer and Nicolson, 1972), with part of their polypeptide chain(s) inserted in the lipid bilayer (e.g., Melcher et al., 1975; Vitetta et al., 1976), or because they are stably linked to such proteins (it is not known how many of these proteins, if any, have an hydrophobic part of their polypeptide chain completely spanning the membrane like the major glycoproteins of the erythrocyte [reviewed in Nicolson, 1976a]). This is consistent with the fact that the basic features of the redistribution of all these membrane components appear to be essentially the same, although quantitative and sometimes also important qualitative differences have been observed in the different experimental systems. Many of these differences are probably related to the characteristics of the individual membrane components and their relative ligands, and to the characteristics of the interactions of the components with internal cellular structures (see Edelman, 1974; Nicolson, 1976a; section 5.6). Other differences probably reflect differences in the general structural and functional characteristics of the cells of the particular subpopulations (or lines) on which the components are present, and in particular differences in the cellular mechanical system which controls redistribution.

2.2. Experimental methods

Almost all distribution and redistribution studies have been carried out using immuno-optical techniques (immunofluorescence, autoradiography and immunoelectron microscopy) which detect patterns of distribution of a suitably tagged ligand over the cell surface (for reviews and references on these techniques see Abbas et al., 1976b; de Petris, 1977). These methods in general do not give direct information about the mobility of the individual labelled components, but only information about the overall rearrangement of these components as a function of time. This depends on several other factors in addition to the mobility of the individual components (see below). Quantitative experiments based on collision rates of individual macromolecular components have not yet been performed, but in a few cases changes in parameters related to general membrane physical properties during application of a ligand have been determined, such as changes in membrane fluidity measured by electron paramagnetic resonance during lectin binding (Barnett et al., 1972; Dodd, 1975) or rotational diffusion of membrane bound ligands (concanavalin A) measured by fluorescence polarization (Inbar et al., 1973a,b). The interpretation of some of these data is, however, still uncertain (Shinitzky and Inbar, 1974; Dodd, 1975) (see also section 4).

2.3. Early distribution studies of lymphocyte surface components

Early distribution studies with immunofluorescence (Möller, 1961; Pernis et al., 1970; Raff et al., 1970; Rabellino et al., 1971) or electron microscopy (Aoki et al., 1969; Mandel et al., 1969; Hammond, 1970; Jones et al., 1970) using antibodies against alloantigens, surface Ig and in a few cases lectin molecules (Osunkoya et al., 1970; Smith and Hollers, 1970) had led to the conclusion that these surface components were distributed in discrete areas ("patches", "crescents", "caps", "spots") on the surface, although unexplained discrepancies were sometimes noted between the results from different laboratories. For example, surface Ig's were detected as multiple "spots" by some authors (Pernis et al., 1970) or collected in a single "cap" by others (Raff et al., 1970). Based on the immunoelectronmicroscopy studies of Aoki et al. (1969) on the distribution of H-2, θ and TL antigens on the surface of mouse splenocytes and thymocytes, the concept was developed that antigens on the lymphocyte surface were arranged in a mosaic of discrete patches of variable sizes and patterns characteristic for each antigen. In addition Boyse et al. (1968) found evidence for the existence of a fixed *short-range* mosaic for various antigens on the lymphocyte surface and attempted to construct two-dimensional molecular "maps" of the relative antigen positions in the mosaic by measuring the degree of inhibition that the binding of an antibody directed against one particular antigen had on the binding of an antibody directed against another antigen. Although this was not mentioned at that time, this "short-range" mosaic pattern was difficult to reconcile with the "long-range" mosaic pattern derived by immunofluorescence and immunoelectronmicroscopy.

In all of these studies, as well as in most of the subsequent studies, the antibodies were applied to living (i.e., unfixed) cells, because fixation often alters or destroys surface antigens. The distribution patterns obtained under these conditions are now known to be the result of ligand-induced redistribution.

2.4. Basic redistribution phenomena

The interpretation of most of the previous observations and the explanation of the discrepancies became clear when it was realized that surface antigens are mobile in the plane of the membrane and that the pattern observed by immuno-optical methods was a redistribution pattern induced by the ligand (antibody), dependent on ligand characteristics (mainly the valency), and on other general experimental conditions such as temperature. This was discovered independently by two groups studying the distribution of mouse surface Ig by immunofluorescence and electron microscopy (Taylor et al., 1971; de Petris and Raff, 1972; Raff and de Petris, 1972), and was soon confirmed in other laboratories (Loor et al., 1972; Karnovsky et al., 1972; Kourilsky et al., 1972; Unanue et al., 1972; Yahara and Edelman, 1972) and found to apply also to other antigens (Taylor et al., 1971; Kourilsky et al., 1972) and surface components of lymphoid (Greaves et al., 1972; Yahara and Edelman, 1972) and non-lymphoid

(Edidin and Weiss, 1972; Sundqvist, 1972) cells. In the original studies on surface Ig three basic patterns of distribution of the labelled component were observed: dispersed, patches and caps (Fig. 1). These could eventually be accompanied by endocytosis of the label. These results can be summarized as follows:

(1) When Ig-carrying lymphocytes were labelled with fluorescent *monovalent* (Fab) anti-Ig antibody the fluorescent antibody remained uniformly distributed over the surface, appearing as a continuous *uniform ring* around the cell contour at all temperatures (from 0° to 37°C). Electron microscopy later confirmed that the distribution of surface Ig was essentially dispersed under these conditions (see section 3).

(2) If cells were labelled with a *divalent* fluorescent anti-Ig antibody at low temperature (0°–4°C), or at any temperature in the presence of metabolic inhibitors, the label appeared to be concentrated in discrete *"spots"* or *"patches"*, separated by unlabelled areas ("spotty" or broken "rings").

(3) If cells labelled at 0°–4°C were brought to higher temperatures or were directly labelled at high temperature (approx. 15° to 37°C) under normal metabolic conditions, all the label (e.g. the previously dispersed patches) coalesced rapidly (in 1–5 min) into a single *"cap"* at one pole of the cell. Cells labelled with monovalent Fab antibody did not cap, but were able to do so if subsequently incubated (and crosslinked) with a divalent anti-Fab antibody.

(4) "Capping" (and more generally surface antigen labelling, see section 6) was usually accompanied by *endocytosis* of the areas of membrane containing the label resulting in the progressive disappearance of the label from the surface.

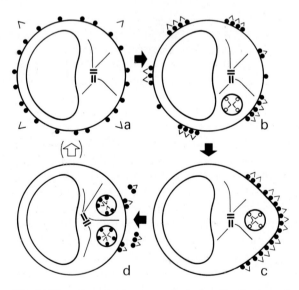

Fig. 1. Diagrammatic representation of redistribution phenomena in lymphocytes. (a) normal dispersed distribution; (b) patch formation on addition of a multivalent ligand; (c) cap formation in a metabolically active cell; (d) disappearance of the crosslinked complexes by pinocytosis, or by shedding.

These experiments demonstrated that surface Ig (and other molecules as well) change their relative position in the plane of the membrane, providing additional evidence for a concept of a fluid membrane suggested by the data of Frye and Edidin (1970). The observations were interpreted (Taylor et al., 1971; de Petris and Raff, 1972; Loor et al., 1972; Raff and de Petris, 1972) on the basis of those data and of the then prevailing concepts on the structural make-up of plasma membranes (as mosaics of proteins and lipids with the lipid mainly arranged in an uninterrupted bilayer and the proteins partially inserted into the bilayer), concepts which were shortly thereafter expressed in detail in the "fluid mosaic model" of Singer and Nicolson (1972), and which are now supported by considerable additional experimental evidence (see Nicolson et al., this volume, p. 1). According to this interpretation, surface Ig molecules were uniformly distributed and free to diffuse in the plane of the plasma membrane. Monovalent ligand did not disturb this distribution, but bivalent (or in general, multivalent) ligands were able to crosslink and aggregate receptors into discrete patches. In metabolically active cells an active process was superimposed upon these metabolically passive redistribution phenomena, and this latter process led to a polarized movement and coalescence of the crosslinked patches into a single *cap*, and/or to their interiorization by pinocytosis. This scheme is now supported by considerable evidence, which will be examined in detail in the following sections.

3. Normal distribution of membrane components

Uniform "ring" staining is usually obtained using monovalent fluorescent ligands, e.g. fluorescent labelled monovalent Fab anti-Ig antibody fragments (Taylor et al., 1971; Karnovsky et al., 1972; Loor et al., 1972) or monovalent fragments of staphylococcal protein A (Ghetie et al., 1974). Similar staining is obtained with Fab antibody fragments against other antigens (e.g., in lower organisms [Beug et al., 1973]) or with lectins at the light microscopical level (Yahara and Edelman, 1972; Nicolson, 1973). This uniform staining indicates that the distribution of surface Ig as well as that of other antigens in an unperturbed membrane is uniform within the limits of resolution of the optical microscope (0.2–0.3 μm). The nature of the distribution at the molecular level is, however, less obvious. In fact, the distributions of independent molecules in a perfectly fluid plasma membrane would be expected to be randomly dispersed, but interactions between membrane molecules, or between membrane molecules and cellular structures, which must occur to at least some extent in any living cell, would necessarily induce some deviations from complete randomness. Experimental determination of the true distribution at the ultrastructural level is based on immunoelectronmicroscopy techniques, but the interpretation of these observations presents some difficulty because, due to the fluid nature of the membrane, the binding of any label could perturb the distribution. This difficulty is minimized, but not completely overcome, by labelling the surface components with monovalent ligands (Fab-antibody fragments which are free of aggregates (Taylor et al., 1971; Loor et al., 1972; de Petris and Raff, 1973a), or by immobilizing the distribution by prefixation with aldehydes (Comoglio and Guglielmone, 1972; de Petris et al., 1973; Inbar et al., 1973a; Nicolson, 1973; Parr and Oei, 1973a,b; de Petris and Raff, 1974). The second procedure is particularly useful for the examination of the carbohydrate lectin receptors (Comoglio and Guglielmone, 1972; de Petris et al., 1973; Inbar et al., 1973a; Nicolson, 1973) which are not modified by the aldehydes, but is not as successful for the immobilization and preservation of protein antigens (Parr and Oei, 1973a,b; de Petris and Raff, 1974). In this case a compromise has to be sought between the preservation of the antigenic structure and the complete immobilization of the molecule (see de Petris, 1977). The conclusion derived from the application of these techniques to the study of the distribution of several plasma membrane components of lymphocytes and other cells (mainly fibroblasts) is that the normal distributions of protein molecules in the plasma membrane (e.g., surface Ig [de Petris and Raff, 1973a], H-2 histocompatibility antigens [Davis, 1972; Parr and Oei, 1973a,b], θ antigens [de Petris and Raff, 1974], Con A receptors [de Petris and Raff, 1973b; de Petris et al., 1973; Rosenblith et al., 1973], and probably TL antigens [de Petris and Raff, 1974]) are *essentially* uniform and dispersed and do not present any appreciable short-range regularity also at the ultrastructural level in the temperature range between 0° and 37°C. These results have been generally interpreted (de Petris and Raff, 1972, 1973a) as indicating that these components are distributed as independent "mobile units" (i.e. as individual molecules with negligible, or weak and reversible, mutual interactions) which are probably constituted by one, or a few polypeptide chains identical to those which can be isolated after solubilization of the membrane with detergents (cf. Vitetta and Uhr, 1975).

Some uncertainty and controversy still remains, however, concerning the short range distribution of these molecules, in particular surface Ig (Ault et al., 1973; Abbas et al., 1975, 1976a). For example, in the studies quoted above, in addition to the dispersed distribution of surface labelling molecules clusters of the latter were also sometimes observed, but these were generally interpreted as corresponding to ligand (antibody) molecules bound to different sites of the same membrane molecule, and/or to small aggregates present in the labelling reagents (de Petris and Raff, 1973a). The concept that membrane antigens are actually dispersed in the plane of the membrane was challenged by Karnovsky and Unanue (Karnovsky and Unanue, 1973; Unanue and Karnovsky, 1973), originally on the inaccurate assumption (cf. de Petris and Raff, 1973a; Raff and de Petris, 1973) that no redistribution occurred at 4°C in the lymphocyte membrane. Therefore, they reasoned that the pattern observed at that temperature corresponded to the pattern originally present in the membrane. Similarly, the patchy distribution of surface Ig observed by Ault et al. (1973) at 0° to 4°C on human peripheral blood lymphocytes was subsequently recognized by the same authors (Abbas et al., 1975; see also Schreiner and Unanue, 1976c) to have been due to redistribution induced by the divalent antibody, and, probably, to Ig molecules secondarily absorbed on the Fc receptors of the cells. In a more recent study, however, Abbas et al. (1975, 1976a) have shown that the distribution of surface Ig (Abbas et al., 1975), and that of Ia antigens (Abbas et al., 1976a) on the surface replicas of fixed and unfixed B lymphocytes labelled with two layers of Fab-antibody (e.g. an haptenated anti-Ig Fab followed by an anti-hapten Fab-ferritin conjugate), though disordered and essentially dispersed, was not random in the sense that it did not follow a Poisson's distribution. On the contrary, the distribution appeared to be loosely microclustered, both in the cold and at room temperature. The meaning of these observations is uncertain, however, because the two-layer labelling system used by Abbas et al. (1975) is prone to give aggregates and the mild fixation used in some of their experiments is probably insufficient to prevent local movement and clustering as suggested by the inability of a comparable treatment to stop diffusion of haptenated molecules on fibroblasts (Edidin et al., 1976). The real distribution, therefore, is probably less clustered than described by Abbas et al. (1975). Thus, these observations probably do not affect the main conclusions of previous studies, i.e. that (a) there is no definite short-range order in the distribution of receptors on lymphocytes, and (b) surface Ig as well as the other molecules so far examined are essentially mutually independent. Abbas et al. (1975, 1976a) are probably correct, however, in concluding that the distribution of these molecules is not strictly random. Such non-randomness could be derived from several different causes:

(a) preferential aggregation as the result of interactions between different membrane molecules in the plane of the membrane (e.g. *cis*-membrane interactions; cf. Nicolson, 1976a). To account for the relatively small effects observed in the systems so far examined, these interactions could only be weak (and reversible);

(b) aggregation (and/or immobilization) induced by interactions between membrane molecules and cytoplasmic structures (*trans*-membrane interactions; cf. Nicolson, 1976a). An example of this type of effect is possibly provided by the localization of a specific glycoprotein in distinct areas of the surface of fibroblasts and glial cells spread on a solid surface, which in turn might be topographically correlated with the presence of submembranous cytoplasmic structures (Wartiovaara et al., 1974; Pollack and Rifkin, 1975; Vaheri et al., 1976). There is no evidence that any of the molecules *so far examined* on the surface of lymphocytes presents similar characteristics; or

(c) exclusion of the molecules under consideration from areas of the membrane occupied by other molecules or molecular aggregates. This could lead not only to "short-range" deviations but also to "long-range" deviations from uniformity, although in principle the excluded molecules could still be distributed at random inside the accessible areas.

Evidence has been recently obtained in lymphocytes for the existence of "long-range" deviations from an average uniform distribution which may have arisen by the third mechanism mentioned above. These deviations have been observed in lymphocytes in which a temporary alteration of cell shape was taking place as a result of contraction of cytoplasmic structures (de Petris and Raff, 1974, and unpublished results). Thus, on examining the "long-range" distribution of θ

antigen (de Petris and Raff, 1974) and of concanavalin A receptors [unpublished data] on thymocytes, the present author found that on prefixed spherical cells this distribution was essentially uniform, as expected, while on prefixed cells which had spontaneously formed "uropods" (probably as a result of contact with a solid substrate) the distribution was markedly polarized. In these latter cells the area of the postnuclear constriction (= constriction ring [Norberg, 1971]) and to a lesser extent that of the uropod itself were depleted of θ antigen and of concanavalin A receptors. These components tended instead to accumulate on the anterior spherical part of the cell (Figs. 2 and 3). This spontaneous redistribution was observed also in the absence of microtubules [unpublished observation]. The interpretation for this phenomenon is that unidentified membrane molecules which are connected with the cytoplasmic structures responsible for the change in cell shape, the formation of the uropod and of the post-nuclear constriction, become concentrated in the "contracted" areas of the membrane, whereas other molecules which are not connected with these structures (including θ antigens, concanavalin A receptors, or at least most of the latter) are displaced from the same areas into the "relaxed" regions of the surface (i.e. into the anterior region of the cell) (Fig. 3). These observations suggest that not all surface molecules are necessarily connected (permanently or transiently) with cytoplasmic structures, or, alternatively, that different molecules are connected with different types of

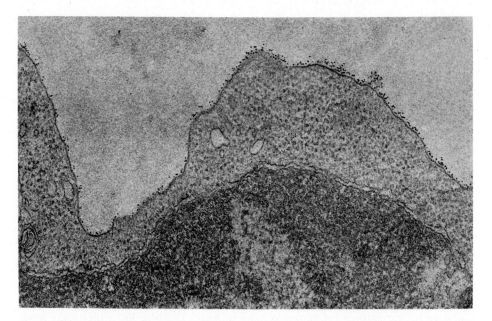

Fig. 2. Non-uniform distribution of concanavalin A receptors on a mouse uropod-forming thymocyte prefixed with glutaraldehyde, and labelled with a ferritin-lectin conjugate. The receptors are more concentrated around the nucleus (right) than on the post-nuclear constriction and on the uropod (left). ×56 000. In these cells the labelled receptors were 2 to 5 times more concentrated on the anterior than on the posterior part of the cell.

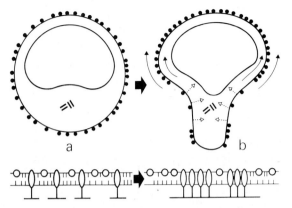

Fig. 3. Diagrammatic representation of the spontaneous redistribution of surface components (concanavalin A receptors, θ antigen) in thymocytes during uropod formation (top), and of its possible origin by exclusion from "contracted" areas (bottom). For the sake of clarity in this and in the following schematic membrane drawings only the molecules directly interacting with some cytoplasmic structures are drawn as spanning the membrane, although this characteristic could be common also to other membrane proteins.

cytoplasmic structures which probably have different functions (mechanical or other).

An inhomogeneous ("cap-like") distribution of surface Ig molecules has also been observed in chicken bursa lymphocytes stained with fluorescent monovalent Fab-anti-Ig antibody (Tao-Wiedmann et al., 1975). The significance of these observations, however, is somewhat uncertain because of the lower resolution of the technique. A clustered distribution of surface Ig with more label concentrated on microvilli was reported by Linthicum and Sell (1975) using scanning electron microscopy and immunolatex spheres as a ligand, although the results were not confirmed by Molday et al. (1975) using a similar technique. Due to technical limitations the significance of these observations remained uncertain. Recently, however, the present author, using an immunoferritin technique, has clearly observed a higher density of surface Ig on lymphocyte microvilli compared to that on the cell body, whereas he has failed to observe any obvious difference in the distribution of concanavalin A receptors (unpublished data). It seems likely, therefore, that some surface components may display such differences, and that for the reasons mentioned above even mobile membrane molecules which are indirectly restrained may at times be non-uniformly distributed.

4. Metabolically independent redistribution: patching

The phenomenon of patching in lymphocytes and in other cells consists in the segregation of a surface component into discrete aggregates separated by membrane areas depleted of that component and occurs when the component is crosslinked into a two-dimensional lattice by a bivalent antibody or, in general, by a multivalent ligand (de Petris and Raff, 1972, 1973a; Raff and de Petris, 1972; Karnovsky et al., 1972; Loor et al., 1972; Unanue et al., 1972; Yahara and Edelman, 1972). Since crosslinking of membrane molecules can be considered the result of collision between diffusing molecules in the plane of the membrane, patching can be considered the reverse of the phenomenon of diffusion and intermixing of previously separated antigens described by Frye and Edidin (1970). In some systems (e.g., the formation of mouse surface Ig-anti-mouse surface Ig antibody complexes) patching can be observed only when the cells are metabolically inactive (at low temperature, or in the presence of metabolic energy inhibitors), i.e. under conditions in which the metabolically dependent superimposed phenomenon of capping is prevented (see next section) (Taylor et al., 1971; Loor et al., 1972). In many other systems in which the ligand does not induce capping, patching can also be observed in metabolically active cells. Patches can be best studied by electron microscopy (de Petris and Raff, 1972; Karnovsky et al., 1972; Unanue et al., 1972), especially by replica techniques (Karnovsky et al., 1972), and can also be analysed by immunofluorescence (Taylor et al., 1971; Loor et al., 1972) if the clusters and patches are separated by sufficiently large gaps of unlabelled membrane ($\geqslant 0.2$–0.3 μm).

The number and size of the patches formed by ligand-surface component complexes vary in different systems and under different experimental conditions. There appear to be three basic factors which determine or influence the patching pattern:

(1) the rates of diffusion of the individual components, and of the ligand-surface component complexes in the plane of the membrane which depend on the temperature and membrane viscosity;

(2) the degree of crosslinking of the membrane components by the ligand which, in turn, is determined by (a) the valence of the reactants and (b) their concentration;

(3) the possible contribution to patching of metabolically-dependent non-diffusional processes.

Very little is known about the actual rates of diffusion of protein molecules in the lymphocyte membrane and their dependence on the temperature. This rate is expected to be mainly determined by the physical state and the viscosity of the lipid milieu, but could be also affected by *cis*-, or *trans*-membrane interactions of mobile membrane proteins with other membrane, or cytoplasmic structures (see Nicolson et al., this volume, p. 1). Some measurements of the rotational diffusion of lectins bound to lymphocyte surface receptors at room temperature have been carried out (Inbar and Shinitzky, 1973a,b), but they are not very informative about the translational mobility of the receptors in the plane of the membrane (cf. Shinitzky and Inbar, 1974). The existence of apparent transitions in the lipid physical state of the membrane of B and T lymphocytes between about 18° and 25°C measured by electron spin resonance has been reported (Cone, 1976), but their significance in terms of membrane protein mobility is unclear, and it is difficult to correlate them with ligand-induced redistribution phenomena which occur both above and below this temperature range. The same can be said for the changes in the fluidity of the lipid environment of molecular probes observed after the binding of lectins (Barnett et al., 1974; Dodd, 1975). On the contrary, an invariant phase in the bulk lipids of the plasma membranes of rat and mouse lymphocyte was observed by fluorescence polarization between 0° and 38°C by Shinitzky and Inbar (1974). They found that the lipid microviscosities varied between 1.7 poise at 37°C and 7.5 poise at 4°C (Shinitzky and Inbar, 1974).

Relatively minor variations in lipid viscosity and in the rate of diffusion of lymphocyte membrane proteins are probably unimportant in determining the pattern of patching, at least at physiological temperatures. By analogy with the values found for the translational diffusion coefficients of proteins in the plasma membranes of mammalian cells (Edidin and Fambrough, 1973; Edidin et al., 1976; Jacobson et al., 1976), diffusion coefficients of lymphocyte membrane proteins would be expected to be of the order of 10^{-10} to 10^{-12} cm^2 sec^{-1} at room temperature. Even with these low diffusion coefficients the influence of the rate of diffusion in determining the observed pattern of patching would not be expected to be very critical in relatively long-term ($\geqslant 20$ min) immuno-optical experiments at room temperature, or above, as an extensive degree of crosslinking would be obtained within a few minutes*. This influence could only become appreciable at low temperature, if there were a drastic decrease of: (a) bulk lipid viscosity; (b) mobility of the proteins as a result of local changes in the physical state of the lipid in the environment of the proteins (phase transitions, lateral phase separations: see Nicolson et al., this volume, p. 1; or (c) the result of formation of protein-protein links. If, however, the bulk lipids were to remain fluid even at low temperatures and no direct constraints were imposed on the proteins, the effect of the various factors even at 0°C would probably be to delay the coalescence of clusters and patches into larger aggregates rather than to prevent the initial formation of clusters during the time of the experiment. This conclusion is supported by the experimental evidence available for lymphocytes. Well defined patches of surface Ig-anti-Ig are formed at 4°C (de Petris and Raff, 1972, 1973a; Karnovsky et al., 1972; Antoine et al., 1974) and patches of several alloantigens with their respective alloantibodies plus an anti-Ig conjugate, and of concanavalin A receptors with concanavalin A were formed at 6°C (de Petris and Raff, 1974), or at 4°C (Karnovsky et al., 1972) in immunoelectronmicroscopy experiments (although in some of these experiments [Karnovsky et al., 1972] the patches were initially considered to be preexistent to the labelling, and not due to redistribution at low temperature). Formation of patches was observed also at 0°C (Loor et al.,

---

*On the basis of the known relationship $d^2 = t/(4D)$ between the mean square displacement d^2 of a particle in a time t and the diffusion coefficient D, one can estimate that the time required for molecules for which $D = 10^{-9}$ to 10^{-11} cm^2 sec^{-1} to have a root mean square displacement of about 40 nm (equal to approximate distance between surface Ig molecules on a B lymphocyte) is about 0.004–0.4 sec. Although the time necessary for a successful collision (i.e. a collision leading to formation of a chemical bond) could be 100-fold longer than this interval, it would still be much shorter than the time of completion of a standard immunofluorescence or immunoelectronmicroscopy labelling experiment. These times could become comparable only for much lower values of D, such as those characteristic of large patches at low temperature.

1972; de Petris and Raff, 1973a) (or even at −5°C [Stackpole et al., 1974a]), although it proceeded more slowly than at 4°C. The coalescence of well defined patches into a larger coarse network did not proceed appreciably, however, during 4 h at 4°C, while it occurred readily at 37°C in the presence of 5 mM iodoacetamide (Unanue et al., 1973). Thus, although the rate of aggregation of the complexes is decreased at low temperature, the lymphocyte membrane, and presumably the bulk of its membrane lipids, are not "frozen" even at 0°C (de Petris and Raff, 1973a).

The second group of factors are those determining the degree of crosslinking. Among these, in particular, the valence of the surface component with respect to the antibody (or, more correctly, to the particular spectrum of antibodies present in the antiserum), or with respect to any particular ligand, is probably the most important element in determining the pattern of distribution (size, number) of the surface component-ligand complexes (de Petris and Raff, 1974; de Petris, 1977). Thus, with divalent antibodies as ligands univalent antigens would give a diffuse distribution of small complexes composed of one antibody and at most two antigen molecules; bivalent antigens would form linear chains of alternated antigen and antibody molecules; and three- or multi-valent antigens would yield two dimensional lattices of increasing complexity. The importance of this factor is clearly indicated by the variations in the pattern of distribution which are observed when the degree of crosslinking is varied under otherwise comparable experimental conditions (Davis, 1972; Unanue et al., 1972; de Petris and Raff, 1973a, 1974). Thus, surface Ig molecules, which are presumably multivalent, form large patches, whereas alloantigens of restricted specificity (i.e. low valency [cf. Sanderson and Welsh, 1974]) form only small and loose patches and clusters. Some alloantigens (the θ alloantigens) which are in all probability practically monovalent towards their particular alloantiserum, fail even to form clearly recognizable patches (de Petris and Raff, 1974). The effect of reagent concentration on the size of the patches is also obvious. As the molar ratio of bound ligand to total antigen is progressively decreased below 1, the patches are expected to become smaller and the distribution more diffuse, the effect being more marked the lower the valence of the reagents. This is consistent with the more uniform, but weaker, staining of cells at decreasing concentrations of fluorescent antibody (Unanue et al., 1972). Inhibition of patching would also occur in ligand excess ("prozone" effect), because in this case most of the ligand molecules would remain monovalently bound during the period of incubation and would be largely released during washing (Chang et al., 1971; unpublished observations of S. de Petris and M.C. Raff). Patching of surface Ig and other antigens is also inhibited by the presence of high concentrations of concanavalin A (Loor et al., 1972; Yahara and Edelman, 1972), but the mechanism of inhibition is not related to a normal "prozone" effect and will be discussed in section 5.6. The role of "non-specific" interactions such as changes in the electrostatic repulsion between protein molecules (Singer and Nicolson, 1972) which could operate in addition to specific crosslinking to induce aggregation of surface molecules seems to be generally negligible in ligand-induced redistribution phenomena in lymphocytes, although a minor role of these non-specific interactions cannot be completely excluded in certain cases (Stackpole et al., 1974a). For example, these interactions could perhaps be responsible for the failure of patches of surface Ig and F(ab)$'_2$ anti-Ig divalent fragments to re-disperse when the F(ab)$'_2$ are reduced into monovalent fragments by dithiothreithol (Loor et al., 1972). This failure, however, could also have been due to other reasons such as, for example, the fact that the surface IgM molecules might have been additionally crosslinked in the plane of the membrane by membrane Fc receptor molecules. These receptors are apparently able to bind to surface IgM (but not surface IgD) when the latter are aggregated by an antibody (Abbas and Unanue, 1975; Forni and Pernis, 1975). An important case in which non-specific aggregation could play a major role is in the spontaneous aggregation of surface proteins into patches and even caps which is sometimes observed when monovalent chemical groups (haptens) are conjugated to surface proteins (Ryan et al., 1974b; F. Loor, personal communication). This point, however, will require further investigation (see also sections 5.2 and 5.5).

The possibility that residual metabolically dependent non-diffusional membrane movements might contribute to the formation of large patches by active aggregation of smaller ones, even in the cases in which capping does not occur, is more difficult to ascertain, but cannot be excluded. In the presence of certain drugs the size of the patches is smaller than that observed in their absence and, as discussed before, the effect would be difficult to explain only in terms of drug-induced modifications of lipid viscosity. Patches of surface Ig-anti Ig antibody smaller than those observed in metabolically inhibited cells were present in mouse lymphocytes treated with xylocaine and other local anaesthetics at 37°C (Ryan et al., 1974a). Similarly, in cells in which capping was partially or completely blocked by 1–$2 \cdot 10^{-4}$ M of tosyl-lysyl-chloromethylketone (TLCK) which is a serine hydrolase inhibitor (Schnebli and Burger, 1972), but also a selective sulphydryl alkylating agent (Rossman et al., 1974) and an inhibitor of other biochemical processes (Pong et al., 1975), the patches of surface Ig-anti Ig or anti Ig-ferritin complexes formed at 37°C were much smaller than those observed on cells in which capping was blocked by drugs such as sodium azide or 10^{-6} M valinomycin (Fig. 4) (de Petris, unpublished results). Local anaesthetics such as xylocaine are thought to be able to disrupt the cortical microfilament-microtubular system of the cell blocking the non-diffusional movement of membrane components (Poste et al., 1975a; Nicolson et al., 1976). The electronmicroscopic morphological evidence is not incompatible with a somewhat similar effect of TLCK under these particular experimental conditions. On the contrary, metabolic inhibitors at the concentrations employed to block capping may still leave enough ATP in the cell to induce some active movement of the membrane which may favour the coalescence of patches formed by diffusion into larger aggregates. Prolonged incubation with NaN_3 for instance, was noted to reduce the number of nylon-fiber-bound cells which formed light microscopic visible patches when stained by fluorescent anti-surface Ig antibody, and to increase the number of the cells which were stained uniformly (Rutishauser et al., 1974).

As it is an essentially diffusion-dependent redistribution process, patching is difficult to inhibit by modifying the experimental conditions. In addition to being

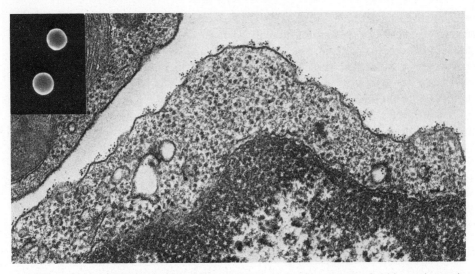

Fig. 4. Distribution in small patches of surface Ig labelled with an anti-Ig-ferritin conjugate in a mouse B lymphocyte incubated at 37°C with $2 \cdot 10^{-4}$ M TLCK. ×52 000. Inset: Cells from the same preparation fixed with glutaraldehyde and labelled with fluorescent anti-ferritin antibodies. Note the almost uniform distribution; pinocytosis is not detected by this method.

inhibited by concanavalin A (see, however, section 5.6), and to a more limited extent by low temperature, patching can be inhibited (at the light microscopic level) by low pH (<5.2–5.5) (Yahara and Edelman, 1973a), and by high concentrations of sugars such as 0.5 M mannose (Loor et al., 1972). The effect of acidic pH may be related to changes in electrostatic interactions between molecules (proteins, phospholipid polar groups) or to an effect on sub-membraneous cytoskeletal structures (Yahara and Edelman, 1973a). The effect of sugars is unclear, and possibly related to the hypertonicity of the medium. Other substances like $LaCl_3$ can also block patching (Loor et al., 1972), but they apparently "fix" the membrane and kill the cells (Yahara and Edelman, 1973a).

A phenomenon which has been related to the ability of the cells to form clusters and patches of lectin receptors crosslinked by the specific ligand is cell agglutination by lectins (concanavalin A, wheat germ agglutinin, etc.) (Nicolson, 1971, 1974a,b; Poste et al., 1975b), although other factors in addition to clustering are probably involved in the phenomenon (Nicolson, 1974a,b; Raff et al., 1974; Nicolson, 1976b). It is uncertain whether formation of patches can account for agglutination in lymphocytes (Loor, 1973a). Lymphocytes are readily agglutinated by high doses of lectin, regardless of the metabolic state of the cell and of the presence or absence of microvilli on their surface (Loor, 1973a). At low doses, however, agglutination does not occur readily and requires some metabolic activity (Loor, 1973a). Capping in itself may decrease agglutination, conceivably by reducing the amount of surface area covered by the lectin and hence the probability of establishing a successful cell-to-cell contact (Ukena et al., 1974). Agglutination by concanavalin A, however, is drastically decreased by doses of methylketones (TLCK: $1-2 \cdot 10^{-4}$ M) which inhibit or delay capping (S. de Petris and G. Trinchieri, unpublished results). The distribution of concanavalin A-ferritin in capping- and agglutination-inhibited cells does not appear obviously different from that on control cells. However, as mentioned above, patching of surface Ig, which can be observed more easily because of the lower surface concentration of the molecules, is somewhat reduced, in the sense that more numerous and smaller patches are formed. Moreover, the inhibited cells appear to be almost perfectly spherical. The meaning of these observations in the understanding of lymphocyte agglutination is not clear and a full discussion of this problem is outside the scope of this review.

5. Metabolically dependent redistribution: capping

In contrast to the essentially diffusional nature of the patching process, the phenomenon of capping represents a non-diffusional, polarized and metabolically dependent process consisting in the displacement of the surface components crosslinked by a ligand to one pole of the cell (Fig. 1; Fig. 5) (Taylor et al., 1971). Capping is the most interesting of the redistribution phenomena, because it provides an experimental tool for investigating the mechanisms by which energy-dependent cytoplasmic systems control the movement of membrane components. Under suitable conditions capping can be readily induced in lymphocytes, and also in other cells such as polymorphonuclear leukocytes (Ryan et al., 1974b; Oliver et al., 1975), fibroblasts (Sundqvist, 1972; Edidin and Weiss, 1972), amoebae (Beug et al., 1973; Pinto da Silva et al., 1975). The mechanism of capping is believed to be basically the same in lymphoid and non-lymphoid cells, although there are considerable differences in the details of the process. These are probably related to differences in the organization of the internal mechanical structures and in the nature of the surface molecules, and sometimes to the anisotropy imposed on the cells by their attachment to a solid substrate.

The capping ability and in general the ligand-induced redistribution properties of at least fifteen lymphocyte surface components have been studied in normal lymphoid cells or in lymphoid tumour cells of different species (Table I). The two components studied in the most detail, and on which most of our present understanding of redistribution phenomena is based, are *surface Ig* (with its various subclasses) (cf. Vitetta and Uhr, 1975), and *concanavalin A receptors*, namely that class of surface molecules which carry receptors for the lectin concanavalin A. The latter is molecularly heterogeneous (Allan et al., 1972; Krug et al., 1973; Hunt and Marchalonis, 1974; cf. de Petris, 1975), and probably corresponds to a large proportion of the membrane glycoproteins (Allan et al., 1972; Hunt and Marchalonis, 1974) including the surface Ig molecules themselves (Loor, 1974; Hunt and Marchalonis, 1974; de Petris, 1975) (see below). The ability of the different ligand-surface component complexes to cap varies greatly,

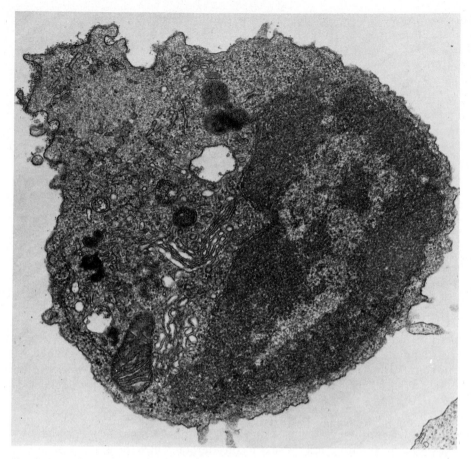

Fig. 5. Cap of surface Ig-anti Ig antibody complexes on mouse spleen B lymphocyte after incubation at 26°C for 30 min in the presence of $5 \cdot 10^{-5}$ M colcemid. Microtubules are absent, except for an occasional microtubule remaining near the centrioles. Note the small uropod, the preserved polarity, and the presence of light and dense endocytic vesicles. Most of the cytoplasm in the uropod area is filled with a dense network of microfilaments. ×29 000.

although each of the antigens or lectin receptors so far examined has been induced to cap in at least some cells under the proper experimental conditions.

5.1. Dependence on cell metabolism

In lymphocytes, as well as in other cells, capping occurs only in metabolically active cells at relatively high ("physiological") temperatures. Mammalian cells generally cap at temperatures higher than 12–15°C (Taylor et al., 1971; Loor et al., 1972; Yahara and Edelman, 1972; Hilgers et al., 1975; Hinuma et al., 1975). The temperature limit is not absolute, and capping of some exceptional components can take place also at lower temperatures. For example, capping of

TABLE I
LIGAND-INDUCED REDISTRIBUTION OF SURFACE COMPONENTS IN LYMPHOID CELLS

Surface component	Species	Cell type[a,b]	References
Surface Ig (and B lymphocyte antigen receptors)	Mouse	Spleen[a]	Taylor et al., 1971; de Petris and Raff, 1972, 1973a; Diener and Petkau, 1972; Dunham et al., 1972; Karnovsky et al., 1972; Loor et al., 1972; Unanue et al., 1972, 1973, 1974a; Yahara and Edelman, 1972; Ashman, 1973; Kiefer, 1973; Sundqvist, 1973a; Elson et al., 1973; de Petris, 1974, 1975; Wofsy, 1974; Antoine et al., 1974; Ault et al., 1974; Stackpole et al., 1974a,b,c; Poste et al., 1975a,b; Poste and Nicolson, 1976; Schreiner and Unanue, 1975a,b, 1976a,b; Diener et al., 1976; Nossal and Layton, 1976
	Rabbit	Spleen	Loor et al., 1972; Sundqvist, 1973a; Linthicum and Sell, 1974b, Lustig et al., 1975
		Blood	Linthicum and Sell, 1974a
	Guinea pig	Lymph node	Rosenthal et al., 1973; Sundqvist, 1973a
	Chicken	Bursa	Albini and Wick, 1973; Tao-Wiedman et al., 1975
	Fish (skate *Raja naevus*)	Spleen, thymus	Ellis and Parkhouse, 1975
	Man	Blood	Preud'homme et al., 1972; Ault et al., 1973; Rowe et al., 1973; Yefenov and Klein, 1974; Cohen and Gilbertsen, 1975
		Leukaemia	Yefenov and Klein, 1974; Cohen and Gilbertsen, 1975
Antigen-receptors on T lymphocytes	Mouse	Spleen	Ashman and Raff, 1973; Roelants et al., 1973, 1974
Histocompatibility antigens (H-2, HL-A)	Mouse (H-2)	Spleen and thymus	Davis, 1972; Lengerová et al., 1972; Sundqvist, 1973a,b; Neauport-Sautes et al., 1973; de Petris and Raff, 1974; Stackpole et al., 1974a,b,c; Yefenov and Klein, 1974
		Leukaemia[b]	Yefenov and Klein, 1974; Schrader et al., 1975
	Man (HL-A)	Blood	Bernoco et al., 1972; Kourilsky et al., 1972; Preud'homme et al., 1972; Cullen et al., 1973; Menne and Flad, 1973; Poulik et al., 1973; Neauport-Sautes et al., 1974; Solheim and Thorsby, 1974

TABLE I (cont'd)

Surface component	Species	Cell type[a,b]	References
		Leukaemia	Menne and Flad, 1973
β_2 microglobulin	Man	Blood	Poulik et al., 1973; Bismuth et al., 1974: Neauport-Sautes et al., 1974; Solheim and Thorsby, 1974
Ia antigens	Mouse	Spleen	Unanue et al., 1974b; Hauptfeld et al., 1975; Schultz et al., 1975
TL antigens	Mouse	Thymus	de Petris and Raff, 1974; Stackpole et al., 1974a–d; Loor et al., 1975
		Leukaemia (thymus)	Stackpole et al., 1974d; Yu and Cohen, 1974;; Loor et al., 1975
Thy-1 (θ) antigen	Mouse	Thymus	Taylor et al., 1971; de Petris and Raff, 1974; Matter and Bonnet, 1974; Stackpole et al., 1974a–c; Hilgers et al., 1975; Loor et al., 1975
		Leukaemia (thymus)	Hilgers et al., 1975
Fc receptor	Mouse	Spleen	Eden et al., 1973; Anderson and Grey, 1974; Abbas and Unanue, 1975; Forni and Pernis, 1975; Ramasamy and Lawson, 1975
	Rat	Thoracic duct	Parish and Hayward, 1974
	Man	Blood	Dickler, 1974; Forni and Pernis, 1975
C3 (complement) receptor	Mouse	Lymph node, spleen	Eden et al., 1973; Dickler, 1974; Nussenzweig, 1974; Gormus and Shands, 1975
	Man	Blood	Abrahamsohn et al., 1974
Sheep red blood cell receptors on human T cells	Man	Blood	Yu (1974)
Virus- and tumour-dependent antigens	Mouse	Leukaemia	Yefenov and Klein, 1974; Hilgers et al., 1975; Schrader et al., 1975
	Man	Leukaemia	Yefenov and Klein, 1974
Virus receptors (EBV)	Man	Leukaemia	Hinuma et al., 1975
Cholera toxin receptor (G_{M1} ganglioside)	Rat, mouse, man	Spleen, blood	Craig and Cuatrecasas, 1975; Révész and Greaves, 1975

TABLE I (cont'd)

Surface component	Species	Cell type[a,b]	References
Concanavalin A receptors	Mouse	Spleen, thymus	Smith and Hollers, 1970; Comoglio and Guglielmone, 1972; de Petris and Raff, 1972, 1974; Unanue et al., 1972; Edelman et al., 1973; Gunther et al., 1973; Inbar and Sachs, 1973; Sundqvist, 1973a,b; Yahara and Edelman, 1973a,b; de Petris, 1974, 1975; Rutishauser et al., 1974; Stackpole et al., 1974b,c; Unanue and Karnovsky, 1974; Wofsy, 1974; Loor, 1974; Yahara and Edelman, 1975a,b
		Lymphoma	Inbar and Sachs, 1973
	Chicken	Thymus, bursa	Sällström and Alm, 1972
	Man	Blood	Barat and Avrameas, 1973
Receptors for other lectins (phytohaemagglutinin, pokeweed mitogen)	Mouse	Spleen	Loor, 1974
Polylysine-binding sites	Mouse	Spleen	Loor, 1974

[a]Spleen: occasionally including lymph node cells. [b]Leukaemia: occasionally including other lymphoid tumours.

polylysine-binding sites occurs at 4°C (Loor, 1974), and in poikolotherm animals living in cold waters (such as in the skate *Raja naevus*) also surface Ig capping occurs at this temperature (Ellis and Parkhouse, 1975). The possible structural or biochemical reasons for these differences are not known. Inhibition of capping by low temperature (Taylor et al., 1971; de Petris and Raff, 1972; Sällström and Alm, 1972; Loor et al., 1972; Unanue et al., 1972, 1973) is virtually immediate, and probably depends only partially on a restriction in ATP production, because ATP levels are not immediately lowered by rapid chilling (S. de Petris and G. Trinchieri, unpublished data). It may be related to the effect of temperature on sensitive mechanical structures or to physical changes in some membrane lipids, although as discussed above some bulk fluidity of the lymphocyte membrane persists even at low temperatures (de Petris and Raff, 1973a). However, a physical change in certain membrane lipids would be consistent with the observation that lymphoid cells which cannot spontaneously cap complexes of the Thy-1 (θ) antigen, or a tumour antigen, with the respective antibodies can cap them if subjected to upward or downward temperature shifts around 13–15°C (Hilgers et al., 1975; cf. also Hinuma et al., 1975). Sharp changes in antigen mobility (Petit and Edidin, 1974) and in other membrane properties which are probably related to changes in the physical state of the lipids have been observed in fibroblasts at about these same temperatures (Noonan and Burger, 1973), but it is not known whether they have any relationship to the capping phenomena observed in lymphocytes.

Capping at physiological temperatures is inhibited by drugs which inhibit ATP production, such as sodium azide ($1 \cdot 10^{-3}$–$3 \cdot 10^{-2}$ M) (Taylor et al., 1971; Loor et al., 1972; Sällström and Alm, 1972; Yahara and Edelman, 1972; de Petris and Raff, 1973a; Unanue et al., 1973; Stackpole et al., 1974b), 2,4-dinitrophenol (10^{-3}–10^{-2} M) (Taylor et al., 1971; Loor et al., 1972; Unanue et al., 1973), oligomycin (100 μg/ml) (Unanue et al., 1973), sodium (Unanue et al., 1973; Ryan et al., 1974b) and potassium (Cohen and Gilbertsen, 1975) cyanide (10^{-4}–10^{-3} M) and other cyanide compounds (Loor et al., 1972) and iodoacetamide (10^{-3} M) (Sällström and Alm, 1972; Unanue et al., 1973; Ryan et al., 1974b). Surface Ig capping is also inhibited by 10^{-6} M valinomycin (Yahara and Edelman, 1975b), although this inhibition has not always been observed (Poste and Nicolson, 1976). The inhibition of concanavalin A receptor or surface Ig capping by 10 mM azide in mouse splenocytes is not reversed, or reversed only partially, by 10 mM glucose (de Petris, unpublished data). Other substances inhibit capping by mechanisms not fully elucidated, but probably related to alterations in membrane structure and function (Loor et al., 1972). Also, the effect of dinitrophenol could in part be due to a direct effect of the drug on the membrane bilayer, because a similar effect was observed with 2,4,6-trinitrophenol, which is not an uncoupler of oxidative phosphorylation (Sheetz et al., 1976).

Surface Ig capping is inhibited by alkaline (pH > 8) and acidic media (pH < 5.5) (Yahara and Edelman, 1973a; Hinuma et al., 1975). At low pH patching is also inhibited (Yahara and Edelman, 1973a). Inhibition of capping at high pH is probably due to general cell damage and alteration of cell metabolism, while that at low pH could be related to alterations in the organization of charged membrane components such as phospholipids or proteins. Like patching (Loor et al., 1972), capping can also be inhibited by high concentrations of some sugars (glucose, mannose, mannitol) (Hinuma et al., 1975). Capping is not affected by inhibitors of DNA, RNA (Kourilsky et al., 1972; Unanue et al., 1973) and protein synthesis (puromycin [Kourilsky et al., 1972; Unanue et al., 1973; Cohen and Gilbertsen, 1975] or cycloheximide [Kourilsky et al., 1972; Poulik et al., 1973; Unanue et al., 1973]) under conditions in which appearance of new antigen on the membrane is prevented (Poulik et al., 1973) suggesting that capping is not dependent on synthesis of new membrane (de Petris and Raff, 1972).

5.2. Dependence on crosslinking

As a rule capping of one particular surface component occurs only if the molecules of this component are aggregated, typically if they are specifically crosslinked by a multivalent ligand (Taylor et al., 1971; Karnovsky et al., 1972;

Loor et al., 1972; Unanue et al., 1972; de Petris and Raff, 1973a). Monovalent ligands (Fab-antibody fragments [Taylor et al., 1971; Loor et al., 1972; Unanue et al., 1972; de Petris and Raff, 1973a] or monovalent protein A fragments from *Staphylococcus aureus* [Ghetie et al., 1974]) as a rule do not induce capping. A non-uniform cap-like distribution of Fab anti-surface Ig fluorescent-labelled (Tao-Wiedmann et al., 1975) or peroxidase-coupled (Antoine et al., 1974) antibody has been observed occasionally on lymphoid cells, but it is likely that this pattern either corresponded to a pre-existent non-uniform distribution such as that of θ antigens and concanavalin A receptors observed on thymocytes (see section 3.), or was due to pinocytosis (Antoine et al., 1974). It is probable that non-specific aggregation of surface molecules due to causes other than specific cross-linking (e.g., changes in repulsive electrostatic forces between charged molecules [Singer and Nicolson, 1972]) can also induce capping in particular cases, such as in the spontaneous capping of concanavalin A receptors on chick embryo retinal cells after trypsinization (McDonough and Lilien, 1975) or in the non-specific accumulation of debris (aggregated or particulated material) to one pole of the lymphoid cell (McFarland et al., 1966; Biberfeld, 1971a; de Petris and Raff, 1972; Rosenthal et al., 1973). In the reviewer's opinion, capping of surface proteins labelled with the monovalent hapten fluorescein isothiocyanate as observed by Ryan et al. on mobile polymorphonuclear leukocytes (Ryan et al., 1974b) is due to non-specific aggregation, an interpretation reinforced by the fact that the labelled material is eventually shed (see also section 5.5.). Similar patching and capping are also induced by labelling surface proteins of living cells with other monovalent haptens which presumably alter the proteins (F. Loor, personal communication). Other experimental manipulations such as protein iodination (see the uneven distribution of autoradiographic grains in Plate I of Marchalonis et al., 1971) might also perturb the distribution of surface proteins. In the majority of cases in which capping follows the binding of a hydrophilic multivalent protein ligand, such as an antibody, non-specific aggregation is probably of negligible importance, and capping can be attributed to specific cross-linking.

Multivalency of the ligand, though necessary, is not by itself sufficient to induce capping. A relatively high degree of cross-linking appears to be required, although the exact requirements are not known. This requirement has been generally expressed by saying that capping requires the prior formation of patches (de Petris and Raff, 1972; Loor et al., 1972; Raff and de Petris, 1972; Yahara and Edelman, 1972). This does not mean that large discrete patches must necessarily form before capping can start. In fact, in several cases (e.g. concanavalin A [Yahara and Edelman, 1972; de Petris, 1975] and TL antigen capping [Loor et al., 1975]) the level of crosslinking is apparently sufficient to start the polar movement of the complexes before discrete patches become visible under the light microscope*. The dependence of capping on cross-linking

---

*The need for a previous formation of patches as a prerequisite for capping has been sometimes questioned on the basis of this failure to detect distinct patches before, or during, capping of some

implies that the phenomenon is broadly dependent on the same factors which favour formation of large patches and, in particular, that it is dependent on the valence of the reactants (de Petris and Raff, 1972; 1974; see also de Petris, 1977) and on the concentration of the ligand. Thus, excess ligand ("prozone" or "antigen excess") or too little ligand ("antigen excess"), both leading to the formation of smaller patches, were less effective in inducing capping (Taylor et al., 1971; Unanue et al., 1972; Sundqvist, 1973a,b; Stackpole et al., 1974b). Some of the "prozone" effects, however, in particular those occurring with surface components present at high concentration on the cell surface (Sundqvist, 1973a,b; Stackpole et al., 1974b) such as concanavalin A receptors (Yahara and Edelman, 1972; Edelman et al., 1973), may not be due to an excess of monovalent binding, but to other inhibitory mechanisms which involve a high level of cross-linking (see below). Cross-linking of two or a few surface molecules and formation of small clusters (such as complexes of mouse θ, H-2 [Davis, 1972; de Petris and Raff, 1974] and human HL-A [Bernoco et al., 1972] antigens with the respective alloantibodies [cf. Sanderson and Welsh, 1974] or concanavalin A receptors cross-linked by divalent concanavalin A [Gunther et al., 1973]) appear to be insufficient to induce capping. Thus, with the exception of mouse B lymphocyte surface Ig, thymocyte TL antigen and, in part, splenocyte concanavalin A receptors which are capped by antibodies or tetravalent lectin, respectively, most of the other mouse antigens form patches and do not cap with their respective antibodies unless these are further crosslinked by a second anti-Ig antibody (Taylor et al., 1971; Bernoco et al., 1972; Unanue et al., 1972; Sundqvist, 1973a,b; Stackpole et al., 1974b). However, even in this case capping does not occur in all cells even after prolonged incubation. This suggests that capping does not correspond to a mechanical process going on continuously in the cell, even in the absence of the ligand, which leads to a progressive accumulation of patches to a pole of the cell as they grow in size, but is instead a process which is "triggered" into action in each individual cell when the degree of crosslinking on the surface reaches some critical level, probably dependent on the nature of the crosslinked molecules as well as on quantitative and qualitative characteristics of the crosslinked lattice. This concept is also consistent with the kinetics of capping (see next section). With regard to the required characteristics of the ligand-surface component lattice, it was originally suggested by Karnovsky and Unanue (Karnovsky et al., 1972; Unanue and Karnovsky, 1973) that the formation of a continuous network of interconnected patches (e.g., formed by surface Ig-anti-Ig or Con A-receptors-Con A-haemocyanin complexes at 4°C on mouse splenocytes [Karnovsky et al., 1972]), or of patches so close to each other than they can be spanned by an antibody, is a prerequisite for capping. The

---

surface components (concanavalin A receptors, TL antigens) by immunofluorescence (Loor et al., 1975; Yahara and Edelman, 1975b). Considering the low resolution of the technique and, in the case of the lectin receptors, the high concentration of the receptors, the objection is not conclusive. Using different antisera, or different experimental conditions, patches of both TL (Aoki et al., 1969; de Petris and Raff, 1974; Stackpole et al., 1974a) and concanavalin A receptors (de Petris, 1975) have been observed.

formation of an extended network of clusters is probably important, although the formation of one *single* interconnected network does not seem to be required, because the phenomenon of cap reversal suggests that caps can be composed of separate patches (de Petris, 1974, 1975; Poste et al., 1975a; cf. Schreiner and Unanue, 1976c). The second condition, closeness of clusters, may be a prerequisite for the formation of large patches when the cells are labelled and washed in the cold, but is probably not necessary when the cells are labelled at physiological temperatures where the patches can move and be further crosslinked to each other by the ligand still present in the medium. This is possibly the explanation for the different capping behaviour between TL antigen labelled in the cold and at 37°C (Loor et al., 1975). "Tightness" of crosslinking or packing, closeness of the crosslinked molecules combined with stability of the chemical links (hence rigidity of the network) are also probably important. In this respect the constraints imposed on the antigen molecule by several ligand molecules bound directly to its antigenic sites are not necessarily identical to those imposed by the binding of fewer molecules which are then further crosslinked by a second antibody. Tightness of crosslinking could be important in relation to the mechanism of transmission of a directional movement from cytoplasmic structures to the surface molecules (see sections 5.6 and 5.7), and possibly also in the cap triggering process. A triggering signal could be of relatively non-specific nature, corresponding to some physical perturbation in membrane structure (for example leading to critical ion permeability changes, etc.), and this perturbation could in general be expected to be higher if stricter constraints were imposed on the molecule. The delivery of a triggering signal could be dependent also on the individual nature of the crosslinked component, and on the nature of its relationship to cytoplasmic structures. In this sense it has been suggested that molecules such as surface Ig might have unique characteristics (Schreiner and Unanue, 1976c). Thus, both quantitative factors related to the degree of crosslinking, and qualitative factors related to the individual nature of the surface molecules are probably important in determining the onset of capping in individual cells, although their precise role remains to be determined.

5.3. Kinetics of capping

Cap formation can be very rapid. When well crosslinked patches of surface Ig (Taylor et al., 1971), Con A receptors (Stackpole et al., 1974b), or alloantigens crosslinked by more than one layer of antibody (Stackpole et al., 1974b) are pre-formed in the cold, and the cells are then rapidly warmed to 20 or 37°C, a cap can form in a few minutes (less than 1 min in several individual cells) corresponding to relative rates of displacement of surface molecules in the plane of the membrane of at least 1–5 μm/min. For surface Ig the rate is 2–3 times faster at 37°C than at 20°C (Taylor et al., 1971). High rates of surface movement are also generally observed when the label is added directly at high temperature, although detailed measurements have not been performed. Slower rates were reported for movement of surface Ig, and more precisely for the movement of specific anti-sheep-erythrocyte antibodies crosslinked by the massive red cells themselves on the surface of sensitized mouse B lymphocytes. Even in that case the movement was completed within 10 min at 23°C (Ashman, 1973). The actual speed of movement of the complexes on the surface of the individual cells may have little relation to the "overall rate of capping" as measured by the percentage of capped cells at different times after the addition of the ligand (Ashman, 1973). This overall rate, in fact, depends also on the variable lag time which occurs between the addition of the ligand and the onset of the actual surface movement. The rate may vary considerably for different components on the same cells, possibly reflecting the time required to form patches of adequate size and adequate degree of crosslinking. This is consistent with the concept that capping is triggered only when the distribution of the crosslinked component on the surface reaches some critical state. In systems such as surface Ig/anti Ig antibody, or concanavalin A receptors/concanavalin A, capping occurs very rapidly with only a slight lag of one or a few minutes; in other systems such as H-2/anti-H-2 and β_2 microglobulin/anti-β_2 microglobulin capping is considerably slower and usually restricted to only part of the cell population (Bismuth et al., 1974; Poulik et al., 1973; Sundqvist, 1973b).

There are often considerable variations in the overall rate and in other characteristics of capping of the same membrane component in different cell types. For example, capping of surface Ig (Cohen and Gilbertsen, 1975) and HL-A antigens (Menne and Flad, 1973) in cells of human chronic lymphocitic leukemia is impaired in comparison to that on normal lymphocytes. Variations in phytohemagglutinin receptor capping (although not described as such) in different Burkitt lymphoma cell lines (Osunkoya et al., 1970) or in H-2 antigen capping in fibroblasts (Edidin and Weiss, 1972) have been reported. In this case capping was observed on the more mobile cells. Concanavalin A was found to cap less readily in lymphoma cells than in normal lymphocytes (Inbar and Sachs, 1973); however, when the capping ability of different antigens was compared, the consistent generalization that tumour cells cap less efficiently than normal cells (Inbar and Sachs, 1973) cannot be made (Yefenov and Klein, 1974; Hilgers et al., 1975)*. These differences may be related to quantitative or qualitative (e.g. presence of different subpopulations of lectin receptors) differences in the expression of the same component on the surface of the various cells types. It is likely, however, that they also reflect differences in the organization and function of the cytoplasmic mechanical systems responsible for capping (see below), and in the interaction of these systems with surface components. Alteration of an internal actin-like network of molecules following viral transformation has been demonstrated in fibroblasts (McNutt et al., 1973;

---

*Comparison of capping versus patching of a surface component in different cells has been sometimes taken as an index of the relative mobility of the component in the membrane plane, or even of membrane "fluidity" (Inbar and Sachs, 1973). The comparison of mobilities estimated on the basis of diffusion phenomena with those estimated on the basis of non-diffusional phenomena such as capping is unjustified.

Pollack et al., 1975). Differences between lymphoma cells and normal lymphocytes in regard to their ability to cap lectin receptors has been tentatively attributed to differences in membrane fluidity (Inbar and Sachs, 1973), although the existence of a possible correlation between these physical properties and biological phenomena remains uncertain (Inbar et al., 1973a). An increase in the overall rate and extent of Ig capping in lymphocytes (Kosower et al., 1974) and wheat germ agglutinin receptors in a mastocytoma cell line (Lustig et al., 1975) was obtained using cyclopropane unsaturated fatty acid esters as membrane fluidity increasing agents. For the reasons discussed in section 4, however, it seems doubtful whether capping as detected in relatively long-term experiments can be *directly* sensitive to relatively minor viscosity changes. It seems more likely that the changes in the lipid milieu affect capping by some indirect mechanism, for example, by altering critical interactions between lipids and specific membrane proteins, by modifying the conformation or reactivity of membrane proteins interacting with the cytoplasmic structures, or by a similar indirect process.

5.4. Movement of surface molecules during capping

5.4.1. The polar movement
Capping in lymphocytes normally occurs by an apparently coordinated, polarized movement of the crosslinked complexes towards that pole of the cell which is facing the nuclear indentation and contains most of the lymphocyte cytoplasm including centrioles, Golgi apparatus and most cellular organelles (Smith and Hollers, 1970; Taylor et al., 1971; de Petris and Raff, 1972; Greaves et al., 1972; Unanue et al., 1972). This pole corresponds to the trailing part ("tail") in locomoting cells (Bessis, 1973). The movement of the clusters appears to occur simultaneously over the entire surface. This process is often, but not invariably, accompanied or followed (see below) by a change in cell shape. This consists of elongation of the cell and formation of a cytoplasmic protrusion, the "uropod", on which the cap is located (Taylor et al., 1971; de Petris and Raff, 1972, 1973b; Unanue and Karnovsky, 1973; de Petris, 1975). When pre-formed clusters and patches (e.g. patches of surface Ig formed at 0°–4°C or in the presence of metabolic inhibitors) initially distributed all over the surface of the cell with gaps of unlabelled membrane interspersed between them move towards the pole, they appear to coalesce in an almost continuous layer with partial or complete obliteration of the unlabelled gaps (de Petris and Raff, 1972; Karnovsky et al., 1972). Only the crosslinked component (and other molecules physically linked to them) appears to move to the cap, while physically independent molecules which are either unlabelled (Taylor et al., 1971; Preud'homme et al., 1972; Neauport-Sautes et al., 1973) or insufficiently crosslinked (Loor et al., 1975) remain dispersed over the entire surface. For example, if TL and θ antigens on mouse thymocytes are labelled with mouse anti-TL antiserum followed by monovalent fluorescein-conjugated rabbit anti-

mouse Ig and with rhodamine-conjugated mouse anti-θ alloantibody, respectively, and are then allowed to cap, the TL antigen caps, but the θ antigen which is insufficiently crosslinked (cf., de Petris and Raff, 1974) does not cap and remains dispersed over the cell surface (Loor et al., 1975).

These observations provide one of the most direct demonstrations of the relative movement of surface components with respect to each other (Taylor et al., 1971), and hence of the fluid character of the membrane (de Petris and Raff, 1972; Raff and de Petris, 1972; Yahara and Edelman, 1972; Loor et al., 1972; Karnovsky et al., 1972; Kourilsky et al., 1972; Unanue et al., 1972). Moreover, they demonstrate that capping does not occur by overall backwards flow of the membrane as a whole, but by selective movement of certain components, the coordinate nature of which suggests that it depends on the activity of an organized mechanical system operating over the entire surface (de Petris and Raff, 1972, 1973b) (see below).

5.4.2. Independence of membrane components

The property of independence of surface molecule movement has been often exploited for examining the physical relationship between different molecules on the lymphocyte surface. In this type of experiment one component (labelled with one marker) is capped completely, and the distribution of a second component (labelled with a second marker) is examined under non-capping conditions (Taylor, 1971; Bernoco et al., 1972; Kourilsky et al., 1972; Preud'homme et al., 1972). If, alternatively, the second component is also allowed to cap, normally the complexes move to the same pole where the first cap is located (Preud'homme et al., 1972). Variations of this method have been used where the first component is completely removed from the surface or "modulated" (cf. section 6), and the display of the second antigen on the surface is then tested by immunofluorescence, radiolabelling or complement-mediated cytotoxicity (e.g., Bernoco et al., 1972; Knopf et al., 1973; Lesley and Hyman, 1974; Hauptfeld et al., 1975). By using these methods the mutual independence of several lymphocyte surface components has been demonstrated: surface Ig and murine H2 (Taylor et al., 1971), human HL-A (Preud'homme et al., 1972) and murine Ia (Unanue et al., 1974b; Hauptfeld et al., 1975) antigens; surface Ig and C3 (Abrahamsohn et al., 1974; Nussenzweig, 1974; Parish and Hayward, 1974) and Fc receptors (Abrahamsohn et al., 1974; Abbas and Unanue, 1975; Forni and Pernis, 1975; Ramasamy and Lawson, 1975) (however, see below); surface Ig and determinants recognized by an anti-lymphocyte antiserum (Taylor et al., 1971); surface Ig molecules of different classes (e.g., IgM and IgD) (Knapp et al., 1973; Rowe et al., 1973), H-2 and Ia antigens (Unanue et al., 1974b); HL-A antigens coded by different genetic loci (Bernoco et al., 1972; Lengerová et al., 1972; Neauport-Sautes et al., 1973; Pierres et al., 1975); TL and H-2, and TL and θ antigens on thymocytes (Loor et al., 1975). Other surface components have been found to be non-independent because capping of one of them induces the capping of the other*. This non-independence could be due to several factors:

(a) the two components are essentially identical, or are parts of the same mobile molecular unit, such as two antigens present on the same polypeptide chain (e.g., some H-2 or HL-A antigens [Bernoco et al.,1972]), or present on two linked polypeptide chains (e.g., β_2 microglobulin and human HL-A [Poulik et al., 1973; Solheim and Thorsby, 1974; Neauport-Sautes et al., 1974; Osterberg et al., 1974]). Another example which has immunological relevance is the demonstration that capping of the receptors for a specific antigen induced by the antigen on a B lymphocyte results in the simultaneous capping of *all* the detectable surface Ig (Raff et al., 1973), suggesting that all the Ig molecules have the same specificity. This is presumably because only one of the possible alleles of the genetic loci, coding for the variable regions of the Ig molecule(s) is expressed on the Ig molecules of each individual cell (Raff et al., 1973; Goding and Layton, 1976; Nossal and Layton, 1976);

(b) the two mobile components are independent, but they carry the same antigenic or receptor site, i.e. they are "cross-reacting". For example, Con A receptors and surface Ig, H-2 and θ antigens, Fc receptors and some receptors for other lectins probably all carry mannopyranosyl and similar carbohydrate receptors (e.g. Goldstein et al., 1973) specific for concanavalin A (Allan et al., 1972; Loor, 1974; Hunt and Marchalonis, 1974; de Petris, 1975; Yahara and Edelman, 1975b);

(c) the two components are normally independent, but become secondarily associated as the result of some change in the configuration or state of aggregation of one of the two components due to labelling and crosslinking by the ligand. This seems to be the case for surface IgM and Fc receptors of B lymphocytes, because the former, when crosslinked by anti-Ig antibody, can "co-cap" the latter, whereas the converse does not occur (Abbas and Unanue, 1975; Forni and Pernis, 1975). Other authors, however, have failed to detect such an association (Parish and Hayward, 1974; Ramasamy and Lawson, 1975), perhaps because of differences in the techniques employed.

---

*Artefactual results may sometimes occur with this technique, despite its simplicity. Possibly the most serious artefacts are derived from the undetected presence in the antisera of antibodies against other known or unknown surface components. Moreover, it has been observed that capping of one component might induce capping of another independent component which is normally unable to cap, if both components are labelled and cross-linked simultaneously by different ligands (Ahmann and Sage, 1974). An interesting question is whether a similar effect might not have been partially responsible for some recent findings of Gonatas et al. (1976). These authors radioiodinated mouse lymphocyte surface proteins with lactoperoxidase and then capped surface Ig with an anti-Ig antibody. They observed that about 64% of the total radiolabel was present in the cap area, which means that surface Ig had "co-capped" 25–30% of labelled protein initially present outside the cap area (as estimated from their published data). Part of these co-capped proteins could have corresponded to Fc receptors (see text) and other surface molecules specifically linked to surface Ig, but some additional nonspecific "co-capping" of labelled proteins could have occurred if radioiodination had caused alteration and non-covalent aggregation of some of these proteins (see also discussion in section 5.5.2).

The "co-capping" methods would probably fail to detect preferential association due to weak mutual interactions of essentially independent molecules, especially if these were altered and weakened by the binding of the ligand. It is unclear if these interactions could account for some of the short-range preferential association of H-2, θ and TL antigens detected in the antibody-binding inhibition studies of Boyse et al. (1968).

Conceptually similar methods have been used to study the interdependence between various surface components and the intramembraneous particles (IMP) exposed on the surface of freeze-fractured membranes (Scott and Marchesi, 1972; McIntyre et al., 1974). The number of such particles in the plasma membrane of lymphoid cells is relatively low (200–500 particles/μm^2); they appear to be of different size and are of unknown composition (Scott and Marchesi, 1972; Matter and Bonnet, 1974; McIntyre et al., 1974). No correlation has been found between surface components and IMPs in lymphocytes, or in general in nucleated cells, similar to that demonstrated in erythrocytes (Tillack et al., 1972; Pinto da Silva and Nicolson, 1974). The IMPs do not appear to be involved in capping of lymphocyte surface components. Contrary to an earlier report suggesting a correlation between redistribution of IMPs and capping of phytohaemagglutinin receptors (Loor, 1973b), it has been found that capping of surface Ig, Con A and phytohaemagglutinin receptors, and θ antigen does not induce redistribution of the IMPs (Karnovsky and Unanue, 1973; Matter and Bonnet, 1974; Yahara and Edelman, 1975a). Only in the case of θ antigen capping (Matter and Bonnet, 1974) was a statistically significant decrease in the number of IMPs noted, especially on the outer fracture face of the plasma membrane. At present the meaning of this observation remains unclear.

5.4.3. Mechanism of segregation of membrane proteins

The observation that individual crosslinked surface components move selectively and independently of the other molecules towards one pole of the cell has important consequences for the understanding of the capping mechanism. As already noted, this indicates that capping is not due to a backward movement of the membrane as a whole, similar to the mechanism postulated by others (e.g. Goldacre, 1961; Shaffer, 1962; Abercrombie et al. 1970) to explain the movement of particles attached to the surface of amoebae or tissue cells during cell locomotion. It also raises some doubts about the validity of the mechanism as originally envisaged for explaining the latter phenomena (de Petris and Raff, 1973b). According to that model the backwards-moving membrane would be continuously "resorbed" at the rear (or at the center) of the cell and assembled de novo at the cell front, probably via intracellular recirculation of the "resorbed" membrane (Goldacre, 1961; Shaffer, 1962; Abercrombie et al., 1970). A similar model has been recently re-proposed by Bretscher in a modified version to explain capping (Bretscher, 1976). This model takes into account the fluidity of the membrane and the independent movement of the surface molecules, but retains the same concept of a *continuous* backwards flow of membrane on the cell surface coupled to an intracellular forwards transport of "resorbed" membrane as a source of movement (Fig. 6a). The recirculation would be limited, however, to the membrane lipids and to some specific protein molecules. The majority of the other proteins would remain dispersed on the surface and would be passively carried towards the rear of the cell by lipid flow, a movement counteracted by their diffusion. As a result a shallow positive gradient towards the rear of the cell would be established. As soon as these proteins were crosslinked into large patches, their diffusion coefficient would drop, and the flow would lead to an accumulation of patches at the rear of the cell. This model depends critically on the diffusion coefficient of the proteins and might be useful to explain capping on immobile cells (see next section), but it does not explain other redistribution phenomena such as: (a) the formation of gradients in a direction opposite to that postulated by the model (those of θ antigens and concanavalin A receptors described in section 3); (b) the formation of "reverse" caps (see below); and (c) capping in the absence of microtubules. These are all processes for which additional mechanisms must be postulated. Moreover, there is little or no morphological evidence for the postulated intracellular transport of membrane towards the front of the cell; and if the flow were to occur, it would have to be intermittent or triggered by the ligand. For example, in several systems large patches are formed, but they do not cap.

An alternative model of surface movement and capping has been proposed by de Petris and Raff (1972, 1973b) who suggest that surface molecules segregate from each other by a countercurrent movement. This alternative model represents a development of earlier "fluid" or "plastic" models of membrane movement during cell locomotion (for references see: Wolpert and

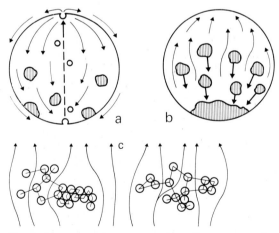

Fig. 6. Diagrammatic representation of two suggested types of surface movements leading to cap formation: (a) unidirectional overall membrane (lipid) flow (Bretscher, 1976); (b), (c) countercurrent movement (de Petris and Raff, 1973b). Fig. c indicates schematically that countercurrent flow could occur on a molecular scale, with "free" molecules flowing around, and in part seeping through, the crosslinked patches.

Gingell, 1968; de Petris and Raff, 1973b) and assumes that the membrane is essentially conserved during capping. Thus, any intracellular recirculation during the actual process is negligible, or irrelevant, so that accumulation of molecules of one component to one pole of the cell must be accompanied by displacement of molecules of other components in the opposite direction (Fig. 6b). In this sense capping is a process which involves the entire membrane, not only the molecules which are capped. The countercurrent flow could involve different domains of the membrane, or it could occur on a molecular scale with the "free" molecules seeping through the lattice formed by the "immobilized" components (Fig. 6c). The segregation of patches of crosslinked molecules is attributed to the direct application to these molecules of forces generated by a cytoplasmic mechanical system present underneath the membrane (see also sections 5.6. and 5.7.). According to this mechanism, in order for a surface component to separate from the others and to move to one pole of the cell this component (or the others) must be directly or indirectly connected with an internal cellular framework and be either displaced with respect to this framework or immobilized with respect to it while other surface constituents are moved in the opposite direction (de Petris and Raff, 1972, 1973b). The most obvious hypothesis is that the cross-linked complexes moving towards the Golgi area are the components which are connected to the cellular framework, and the molecules moving in the opposite direction are free (de Petris and Raff, 1972, 1973b). This concept of a countercurrent movement is consistent with the observation that in addition to the usual or "normal" caps formed around the Golgi area "reversed" caps can sometimes be formed with certain antigens at the opposite pole of the cell. In certain systems the two types of caps may be present on the same cell (D. Bernoco, personal communication). "Reverse" caps have been described by Stackpole et al. (1974c,d) who found by electron microscopy that H-2 and θ antigens of mouse thymocytes labelled by specific alloantibodies followed by an anti-Ig/anti-ferritin hybrid antibody and finally ferritin capped on the pole *opposite* to the one containing the Golgi-centrosomal apparatus, i.e. at the front of the cell. This reviewer tends to attribute the formation of these "reverse" caps to a failure to reach a critical level of crosslinking for the triggering of the "normal" process of capping, since the same θ and H-2 thymocyte antigens were found to cap "normally" (toward the Golgi area) in practically all cells when the antigens were labelled with the alloantibody followed by a divalent anti-Ig-ferritin conjugate (S. de Petris and M.C. Raff, unpublished results). This two-layer system is expected to crosslink the antigen more efficiently than the hybrid antibody method used by Stackpole et al. (1974c,d). The "reverse" movement of poorly crosslinked molecules would then correspond to a spontaneous forward movement of "unlocked" complexes similar to that of the unlabelled θ antigen and concanavalin A receptor molecules observed in thymocytes forming a uropod (see section 3), and the "reverse" cap would correspond to a localized "patch" rather than to a real cap.

The countercurrent model postulates the existence of some trans-membrane links (cf. Nicolson, 1976a) between surface components and cytoplasmic constituents. No definite information is presently available about the nature and number of the molecules constituting these putative links, except the negative one that they do not appear to correspond to the intramembranous particles seen on the freeze-fractured surface of lymphocyte plasma membranes, since, as mentioned above, the distribution of the latter appears to be independent of the movement of all surface components tested so far.

5.5. Cell movement, cell morphology and capping

5.5.1. Stimulation of motility and changes of cell shape

Cap formation is often (but not always) accompanied by stimulation of cell motility and a change in cell shape which consists of cell elongation and formation of a more or less pronounced protrusion called a "uropod" (nomenclature of McFarland et al., 1966; McFarland and Schechter, 1970) on which the cap is located*. Microvilli or microspikes are often associated with the uropod, and these may move to this region during capping, whereas the rest of the cell is relatively smooth or undulated (de Petris and Raff, 1973b; Karnovsky et al., 1975; Kay, 1975; Loor and Hägg, 1975; Schreiner and Unanue, 1976c). The overall shape of cells with uropods is similar to the "hand-mirror" shape of lymphocytes moving on substrates (Lewis, 1931; de Bruyn, 1946; Marshall and Roberts, 1965; Bessis, 1973) or interacting with other cells (McFarland et al., 1966; Astrom et al., 1968; McFarland and Schechter, 1970; Biberfeld, 1971a). Elongation and formation of a "uropod" are accompanied by the formation of small protrusions and membrane ruffles at the front of the cell (Karnovsky et al., 1975; Kay, 1975; Schreiner and Unanue, 1976a) which may be taken as evidence for the stimulation of some cytoplasmic streaming. Changes in cell shape correlate with stimulation of cell motility and, in fact, can be considered a manifestation of the increased motility, whether or not they are modified by the contact of the cell with a solid substrate. These phenomena have been studied in more detail with respect to surface Ig (Unanue et al., 1974a; Schreiner and Unanue, 1975b) and concanavalin A receptor systems (Unanue and Karnovsky, 1974; de Petris, 1975). The stimulation of cell motility by anti-surface Ig antibodies is of general nature and is random, and not chemotactically directed (Unanue et al., 1974a; Schreiner and Unanue, 1975b), (however, see unpublished data quoted in Schreiner and Unanue, 1976c). Both changes in cell shape and stimulation of motility depend on multivalent cross-linking of a surface component by a ligand (Unanue and Karnovsky, 1974; Unanue et al., 1974a; see also discussion in de Petris and Raff, 1973b) and are apparently stimulated by the same cross-linking event which triggers capping. The signal for stimulation of motility does not seem to be identical to that for

---

*The "uropod" formed during capping on B lymphocytes, and probably also T lymphocytes, may not be morphologically identical to the uropod which forms *spontaneously* on thymocytes and T lymphocytes (Rosenthal et al., 1973).

cap formation, but to occur in addition to the former signal. In fact, the manifestations of increased motility are observed only in cells which cap, but not in all of them (see below). Thus, in the surface Ig system shape change is not induced by the binding of monovalent Fab antibody, whereas elongation and/or uropod formation is induced in a considerable fraction of cells (typically 30%) by the binding and capping of bivalent antibody (see Unanue et al., 1974a; de Petris, unpublished observations; discussion in de Petris and Raff, 1973b). Shape changes are even more accentuated during concanavalin A capping (Smith and Hollers, 1970; Greaves et al., 1972; Unanue and Karnovsky, 1974; de Petris, 1975), a fact which has led to the suggestion that concanavalin A capping requires cell movement (Schreiner and Unanue, 1976a,c) (however, see below). Changes in cell shape during surface Ig (Unanue et al., 1974a) and concanavalin A capping (S. de Petris, unpublished observations; cf. Unanue and Karnovsky, 1974) can take place also in suspension, although in the latter case alterations are less frequent and marked than when the cells are substrate-attached. These observations are consistent with the hypothesis that a motility-triggering signal can be delivered by surface crosslinking equally well to attached or suspended cells, but the effects are clearer when the cells are in contact with a substrate. It is not inconceivable that tension is generated inside the cell during cap formation which would tend to elongate it rendering the cell more "stretched", if its surface were anchored to a fixed point on the substrate.

The change in cell shape is transient during surface Ig capping and is apparently reversed slowly as the antigen-antibody complexes are cleared from the surface (Unanue et al., 1974a). The reversal is complete within 20–30 minutes suggesting that the stimulation might persist until some critical number of complexes remain on the cell surface*. Alternatively, the perturbation might have a programmed duration determined by the characteristics of the cellular mechanisms (Schreiner and Unanue, 1976c). The uropods induced by concanavalin A are often very extended with a convoluted surface (an abortive attempt at pinocytosis?) and remain "frozen" in a stretched, contracted state for a considerable time (Unanue and Karnovsky, 1974; de Petris, 1975). This is possibly related to the larger area covered by the cap, and the permanence of receptor-ligand aggregates on the surface. In "mobile" cells, i.e., cells actually moving or showing a definite morphological polarization, the caps become smaller and more concentrated as in surface Ig capping on lymphocytes (Unanue and Karnovsky, 1974), or are displaced from a central position to the trailing part of the cell as in concanavalin A capping in granulocytes (Ryan et al., 1974b).

---

*The continuous presence of complexes on the surface could be necessary though insufficient for maintaining the stimulus for alteration of cell shape. An example of a ligand-induced biochemical event which is possibly connected with pinocytosis and persists only as long as the ligand remains bound to the surface is the concanavalin A-stimulated increase in the rate of respiration and glucose oxidation by way of the hexose monophosphate shunt in polymorphonuclear leukocytes and macrophages (Romeo et al., 1973).

The motility and shape alterations observed during surface Ig and concanavalin A receptor capping do not occur readily in other systems. Cell motility and uropod formation are stimulated by insolubilized antigen-antibody complexes on a plastic surface to which the lymphocytes are adhering (Alexander and Henkart, 1976). On the contrary, Schreiner and Unanue (1975b) observed that bivalent anti-θ, anti-H-2, or "anti-lymphocyte" antibodies (which normally do not cap) did not change cell shape, even when the latter antibodies were further crosslinked by an anti-Ig antibody (a condition under which they would normally cap) (Schreiner and Unanue, 1975b). The present author has observed, however, that small uropods or cell elongations often accompanied capping of θ antigen when the alloantibody was further crosslinked by anti-IgG (unpublished data).

5.5.2. Relationship between cell movement and capping
The lack of a consistent correlation between capping and cell motility in the systems discussed above and the variable occurrence of motility stimulation in other systems raises the question of whether stimulation of cell motility by surface receptor crosslinking plays any role in facilitating capping in some cells in which it would not otherwise occur. This question is related to the more general problem of whether there is any relationship between capping and cell movement (spontaneous or induced).

Partial answers have been found to these questions. With regard to the relationship between capping and cell locomotion, the experimental evidence clearly indicates that capping can also occur in cells which do not move or which are not in contact with a solid substrate. These points were recognized by Taylor et al. (1971) and de Petris and Raff (1973b) who found that cells kept in suspension by continuous stirring were able to cap surface Ig-anti-Ig complexes equally as well as stationary cells, and these observations were confirmed by other authors studying capping of surface Ig on cells kept either in suspension (Unanue et al., 1974a) or attached to nylon fibers (Rutishauser et al., 1974). This property is not restricted to capping of surface Ig, but seems to be general, because concanavalin A receptors can cap on cells in suspension under conditions in which there is practically no cell-to-cell or cell-to-substrate contact (de Petris, unpublished). The latter point was also indirectly confirmed by the experiments of Unanue and Karnovsky (1974) who observed capping of surface Ig on cells in suspension which had been pre-incubated with (unlabelled) concanavalin A. Under those conditions surface Ig capping reflects the behaviour of concanavalin A capping (de Petris, 1975). The possibility that actual cell movement is required for capping of some membrane components (concanavalin A receptors) but not for others (surface Ig) (Unanue and Karnovsky, 1974; Schreiner and Unanue, 1976a) seems to be remote. Thus, there is no evidence that actual cell locomotion is necessary for capping any of the multivalent ligand-surface component systems so far analyzed, although the possibility cannot be completely excluded.

Cell movement might be required for other special systems. An interesting case was reported by Ryan et al. (1974b) who examined the behaviour of the monovalent fluorescent label fluorescein-isothiocyanate (FITC) conjugated directly to surface molecules (presumably protein) of human polymorphonuclear leukocytes in the absence of any added ligand (Ryan et al., 1974b). If the cells were allowed to move normally at 37°C, a bright FITC cap formed on the tail of the cell which was later shed into the medium, although a small amount of label remained dispersed over the cell body. If the cells were immobilized (by incubation at 18°C in the absence of serum, or by addition of cytochalasin B), the label remained generally diffuse, although it formed a central cap when the labelled components were cross-linked by anti-FITC antibody. These experiments were interpreted by the authors as evidence for a passive backward flow of *non-crosslinked* components occurring on locomoting but not on stationary cells. This was contrasted with the active flow of *crosslinked* components triggered by a multivalent ligand occurring on both types of cells. This reviewer finds it difficult to understand how isolated molecules could cap by *any* mechanism *if the membrane is fluid*. A more likely possibility is that the surface molecules treated with the monovalent reagent were actually altered, and they subsequently aggregated as a consequence of the modification (similarly to what sometimes occurs during conjugation of FITC to soluble serum proteins) but that the aggregation was probably insufficient to signal capping and set into motion the internal cytoskeletal system. This signal could be delivered only if the labelled molecules were further crosslinked by an antibody. In spontaneously moving cells, however, the same system and similar surface movement might be independently activated by a different stimulus. This interpretation is based on the fundamental similarity between the mechanism of capping and that of cell movement as suggested by de Petris and Raff (1972, 1973b). According to this concept, the cytoplasmic structures (e.g., microfilaments, microtubules) and the basic mechanisms which are responsible for the movement of crosslinked complexes during capping (with or without cell movement) are essentially the same as those which operate during cell motility in the absence of the ligand, despite the possible existence of secondary mechanical and biochemical difference between the two processes*. One of these differences is that locomotion always presumably involves cytoplasmic streaming, whereas capping may occur with or without cytoplasmic streaming. Thus, in both cases the basic process would consist in active backwards displacement of a "patch" of crosslinked surface material with respect to an internal

---

*Some misunderstanding about the distinction between "dependence of capping on cell movement" and "dependence of capping on the mechanism responsible for cell movement" was probably at the origin of an apparent controversy (Unanue et al., 1974a) about the need of cell movement for capping. This controversy was actually non-existent. Since the first experiments on capping it was clear to the present author and to his colleagues that cells could cap surface Ig without actually moving or being in contact with the substrate, as this point was intentionally investigated (Taylor et al., 1971; de Petris and Raff, 1972, 1973b). However, some diverging opinions still remain about the role of cell movement in capping of different surface components, and in general on the mechanism of capping, but on these subjects the experimental evidence is not yet conclusive (see text).

framework (de Petris and Raff, 1973b). In the case of cell locomotion the "patch" would consist of the surface molecules adhering to, and "crosslinked" by, the substrate. Because this patch is fixed to the substrate, its *relative* backward movement with respect to the internal framework would lead to a forward displacement of the framework itself, and of the cell as a whole (de Petris and Raff, 1973b). According to this hypothesis spontaneous cell movement could induce capping of material unable to stimulate polar surface movements by itself, but only if the material were already aggregated. Thus, in the case of labelled granulocytes, a truly monovalent ligand would not be able to cap even on moving cells, a point which unfortunately was not explicitly tested (Ryan et al., 1974b).

The hypothesis that the cell mechanical systems involved in capping and cell movement are essentially the same, despite the apparent independence of the two phenomena, is consistent with the frequently reported observation that capping of a particular component (e.g. surface Ig, H-2) on different cell lines occurs more readily on those cells which are more mobile (Osunkoya et al., 1970; Edidin and Weiss, 1972). Thus, cells in which the locomoting systems are more active or more easily triggered would also be more active in capping.

Evidence against a role in capping not only for actual cell locomotion but also for cell motility phenomena stimulated by crosslinking (which can probably operate in the absence of actual cell locomotion) has been provided by Unanue and Karnovsky (1974) who demonstrated that ligand-induced motility and cell shape changes can be dissociated from capping in the surface Ig system. These authors have shown that uropod formation in this system does not occur simultaneously with cap formation, but it follows the latter after a short lag of 30 sec or longer. The number of cells forming uropods reaches a peak about 10 min after the addition of the antibody. More significantly, both increases in motility and changes in cell shape could be prevented by incubating the cells either with the serine esterase inhibitor diisopropylfluorophosphate (DFP) (Unanue et al., 1974a), or with agents that increase cellular cyclic AMP levels (Schreiner and Unanue, 1975a), or with antilymphocyte antiserum (ALS) (Schreiner and Unanue, 1975b). This occurred without an impairment in the ability of the cells to patch, cap, or endocytose the labelled antigen-antibody complexes. Cyclic GMP and cholinergic drugs, on the other hand, had a stimulatory effect on the *spontaneous* motility of lymphocytes, especially of B cells (Schreiner and Unanue, 1975b; cf. Estensen et al., 1973). The inhibitory effect of ALS and cyclic AMP has been tentatively attributed to "stabilization" of microtubules, because inhibition could be reversed by colchicine (Schreiner and Unanue, 1975a,b) (see also section 5.6.3). The hypothesis that cyclic AMP favours microtubule polymerization is consistent with some morphological (e.g. Porter et al., 1974; Brinkley et al., 1975) and biochemical (Sloboda et al., 1975) evidence, although the evidence is not yet conclusive (cf. Oliver et al., 1975; see also section 5.6.3). The activation of cell motility appears to be dependent on the activation of a cullular esterase which is inactive before the addition of the ligand (Unanue et al., 1974a; Becker and Unanue, 1976).

The role of ligand-stimulated motility in other systems is more uncertain. On the basis of differences in the susceptibility of capping and its reversal to drugs (mainly cytochalasin B) Unanue, Karnovsky and Schreiner have suggested that cell motility may be essential for capping concanavalin A receptors (Unanue and Karnovsky, 1974; Schreiner and Unanue, 1976a,c). However, also in this system it appears that cells can normally cap without appreciably altering their spherical shape (Greaves et al., 1972; S. de Petris, unpublished observations), a conclusion in agreement with the fact that concanavalin A-capping-inhibited cells bound to concanavalin A-derivatized nylon fibres remain spherical when the inhibition is relieved by colchicine (cf. Table 3 in Rutishauser et al., 1974). Thus, if cell motility has any effect in this system, the effect is probably only partial. Motility and cell shape alterations are inhibited in all surface systems by cytochalasin B (e.g. Rutishauser et al., 1974; de Petris, 1975), but capping is also affected to a variable extent (see next section).

In summary, capping of surface receptors can take place either on substantially immobile cells without an appreciable change of cell shape, or on motile but not necessarily locomoting cells with induction of cytoplasmic streaming accompanied by formation of pseudopods and cytoplasmic projections such as those often observed in concanavalin A-capped cells (Loor, 1974; Unanue and Karnovsky, 1974; de Petris, 1975). These two modes of capping may correspond to two different mechanisms of capping used differentially by the various surface systems such as surface Ig and concanavalin A receptors, as suggested by Schreiner and Unanue (1976a,c). Alternatively, the same process as that taking place on spherical cells could represent the basic polar mechanism of capping present in all systems, including the concanavalin A receptor system. The additional phenomena of local contraction, increase in internal pressure and generation of cytoplasmic streaming would be superimposed with variable frequency in the different systems (Fig. 7). In the cases where cell motility is stimulated by the same crosslinking event which stimulates capping, such as normally happens in lymphocytes incubated with a multivalent ligand, the second set of events would probably not increase the efficiency of capping, although it could alter its morphology. In the cases where motility is stimulated by an independent event, such as in the mobile FITC-treated polymorphonuclear leukocytes of Ryan et al. (1974b), cell motility could also induce capping of surface components which might be able to form aggregates but not deliver the triggering signal for setting in motion the normal capping process. In the reviewer's opinion the experimental evidence is more consistent with the second alternative, although it is still difficult to confirm or rule out either possibility.

5.5.3. Capping on cells bound to a solid substrate
An interesting case illustrating the effects that the presence or absence of cell motility and cytoplasmic streaming may have on the localization of the cap is represented by the observations of Rutishauser et al. (1974). These authors studied cap behaviour during surface Ig-capping on B lymphocytes which

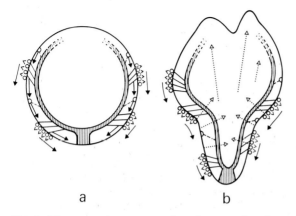

Fig. 7. Diagrammatic representation of two modes of capping, i.e. in the presence or absence of cytoplasmic streaming. Under the hypothesis that a shearing force is applied directly by a cellular framework to membrane molecules, capping occurs by a backwards movement of the crosslinked patches with respect to the framework (a). On this movement a second process may be superimposed, represented by contraction and deformation of the framework with generation of cytoplasmic streaming. This pushes forwards cellular elements not connected to the framework (b). The possible permanent connection of the framework with some membrane components is schematically indicated at the rear of the cell.

carry specific anti-dinitrophenyl surface antibodies and are bound through these antibodies to nylon fibres covered with dinitrophenylated bovine serum albumin (DNP-BSA). Also in this system dissociation of capping and cell movement was observed, because the redistributable Ig molecules remaining on the cell surface (i.e. those not bound to the fibre) could be capped by anti-Ig antibody both on round immobile cells and on cells displaying motile activity. An important difference, however, was noted in the two cases. In round immobile cells the caps were distributed at random with respect to the point of cell attachment to the fibre. On the contrary, on cells showing spontaneous or induced motile activity the caps moved to a distal position with respect to the fibre, although the cell did not move along the fibre. A possible interpretation of this behaviour is in terms of the counter-current model of capping described in the previous sections (which assumes that the capping molecules are connected to, and move backwards with respect to, an internal framework) combined with the notion that capping can occur either in the presence or absence of large-scale cytoplasmic streaming (Fig. 8). Thus in immobile round cells there would be negligible cytoplasmic streaming, and therefore no appreciable force or pressure applied to the point of contact with the fibre (Fig. 8a). On the contrary, in motile cells vigorous forward cytoplasmic and membrane streaming of material not connected with the framework would be generated, for example by contraction of framework structures at the back of the cell. Streaming would not affect the capping complexes as these remain locked to the framework. If the forward streams encounter some resistance at the cortical layer of the cell immobilized by contact with the fibre, a reaction

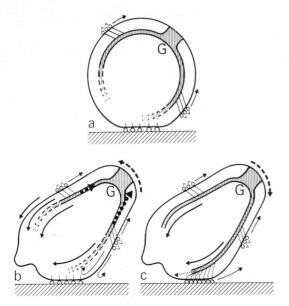

Fig. 8. Diagrammatic illustration of the possible mechanism of formation of "normal" (a,c) and "reverse" (b) caps on cells bound to nylon fibres. In (a) the cell is immobile, no streaming is generated and the fibre-immobilized surface molecules are not connected to the framework. Instead the framework is connected to the patches of ligand-crosslinked molecules, and the latter move backwards with respect to the immobile framework. In (b) the situation is identical, but cytoplasmic and membrane streaming are generated by framework structures (e.g. by their contraction). The unequal resistance encountered on different sides of the cell by the streaming at the site of attachment of the cell to the fibre generates by reaction a torque which causes rotation of the framework. The backwards-moving ligand-crosslinked molecules connected with the framework move together with the framework to a symmetrical equilibrium position opposite to the fibre. The fibre-bound molecules remain at the front of the cell ("reverse" cap). In (c) the surface molecules attached to the fibre become connected to the cellular framework and move backwards with respect to the latter, similarly to the patches of ligand-crosslinked molecules (independently of the streaming). This causes a rotation of the framework and of all the cellular structures until the cap reaches the point of cell contact with the fibre. Cellular structures are not indicated in the drawing, except for the approximate position of the centriolar-Golgi complex (G) with respect to the framework. It should be noted that a unidirectional backwards flow on the surface (such as that in Fig. 6a) in principle would always cause by reaction a rotation of the cap towards the fibre.

applied to the stream-generating framework would be produced, causing a rotation of the framework itself until the latter, along with the attached cellular structures, would reach a position of symmetrical equilibrium with respect to the fibre, i.e. with the rear of the cell and the cap in the distal position from the fibre (Fig. 8b). This interpretation is based on the assumption (suggested also by Rutishauser et al. [1974]) that surface Ig caps are formed on fibre-attached cells around the centrosomal Golgi area, a point amenable to direct experimentation.

On cells attached to a fibre a second "cap" can be thought of as present at the point of cell contact with the fibre, consisting of the surface Ig molecules immobilized by the fibre-bound antigen. Therefore, during rotation of the cell this second "cap" would move to a position equivalent to that of the "reverse" caps found by Stackpole et al. (1974c) which, as discussed above, are probably formed by molecules which do not reach the critical level of crosslinking necessary to trigger actual capping. In contrast with the observations made on cells attached to DNP-BSA-derivatized nylon fibres, Loor (1974) and Kiefer (1973) examined cells attached to lectin- and mitogen-derivatized Sepharose and Sephadex beads and DNP-derivatized nylon fibres, and found that the caps moved to the point of cell attachment to the substrate. Similar observations were also made by Yahara and Edelman using concanavalin A-derivatized platelets and latex beads (Yahara and Edelman, 1975b). A possible explanation of this discrepant behaviour is that in the second set of experiments the degree of cross-linking reached a critical level and triggered capping (for unknown reasons which may be related to the distribution of the immobilized ligand on the bead or fibre), so that all molecules became locked to the internal cellular framework and therefore moved toward the back of the cell. However, since some of the patches were fixed to the bead, the back of the cell would move toward the bead (Fig. 8c). This interpretation based on the relative movements of surface molecules with respect to a cellular framework was originally proposed by de Petris and Raff (1972) to explain why lymphocytes interacting with other cells (target cells, macrophages, etc.) end up by contacting the cells through the uropod (McFarland et al., 1966; Ax et al., 1968; Smith and Goldman, 1970; Biberfeld, 1971a).

5.6. Role of intracellular structures in capping

5.6.1. Structural cellular components (microfilaments, microtubules)
As discussed in the previous section, the movement characteristics of the crosslinked molecules during capping, in particular its non-diffusional character, polarity, and dependence on metabolic energy, indicate that such movement has to be considered in relation to a cellular framework located in the cytoplasm which somehow generates the motive force for the movement. This framework should be conserved as a whole during the process, and its component parts should remain interconnected, although relative displacements of its various parts could take place in concurrence with movements on the surface. For example, some of its components could be depolymerized in certain regions of the framework and polymerized in others. Very little is known about the precise nature and organization of this structure, but putative structural elements in lymphocytes as well as in other cells are the microfilaments (Wessells et al., 1971, 1973; Komnick et al., 1973; McNutt et al., 1973; Pollard and Weihing, 1974; Goldman et al., 1975) and the microtubules (Olmsted and Borisy, 1973; Wilson and Bryan, 1974; Weber et al., 1975). Considerable, though indirect, experimental evidence suggests that both these types of structures play a role in capping.

Microfilaments (5–6 nm diameter in thin sections) are probably actin-containing filaments (Komnick et al., 1973; Wessells et al., 1973; Fagraeus et al., 1974; Pollard and Weihing, 1974) which in different cells are variously arranged in parallel bundles, sheets, and irregular networks. They are probably intermixed and interacting with myosin-like molecules, although the latter are not usually recognizable as independent structures. Myosin and actin molecules are closely associated with mammalian membranes (Kemp et al., 1973; Pollard and Korn, 1973; Gruenstein et al., 1975; Painter et al., 1975), but their precise relationship with the latter is still unknown. Leukocytes in general and lymphocytes in particular contain a considerable amount of actin and myosin (Stossel and Pollard, 1973; Fagraeus et al., 1974; Shibata et al., 1975) associated in part with the plasma membrane (Barber and Crumpton, 1976). However, these do not appear to be organized in any characteristic, regular structure which is easily identifiable by electron microscopy such as that present in skeletal muscle. It is likely that only part of the cell actin is polymerized into filaments, the remaining being distributed sparsely in the cytoplasm in a globular or partially polymerized form (Pollard and Weihing, 1974). Under particular conditions, e.g., under ATP deprivation, these molecules can polymerize and form large masses of microfilaments located mainly at the cell periphery, in microvilli and in the perinuclear area (Fig. 9) (de Petris, unpublished data). The most obvious localization of these filaments is underneath the plasma membrane where they form a filamentous network of variable thickness (usually less than 0.1 μm) which separates the plasma membrane from the bulk of the cytoplasm where many ribosomes are located (Fig. 10). This network probably is largely in-

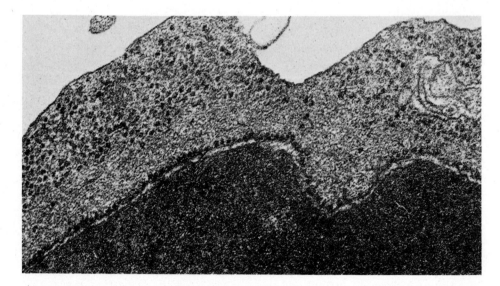

Fig. 9. ATP deprivation by incubation with sodium azide appears to induce polymerization of a large number of actin-like microfilaments in the cytoplasm of mouse lymphocytes. Masses of filaments accumulate, especially in the perinuclear area (illustrated here), and in the cortical area of the cell, mainly in microvilli (not shown). ×79 000.

terconnected, since it appears to be "impermeable" to ribosomes. It is probably also physically linked to the membrane (cf., Barber and Crumpton, 1976), since it remains adjacent to it despite thermal agitation. The highest concentration of the microfilaments is usually found in the posterior part of the cell, particularly in those cells which have a uropod (cf., Fig. 5). Microfilaments can be easily identified after glycerol extraction of the cell and decoration with heavy meromyosin to demonstrate their actin-like character according to the technique of Ishikawa et al. (1969). When this technique is used, the filaments become fuzzy, although regular arrowheads cannot be easily seen (Fig. 11). Although in the cytoplasm underneath the cap and in particular in the uropods, a thick network of microfilaments is often present, it is difficult to establish whether this appearance is derived from contraction of a pre-existent microfilament layer accompanied by expulsion of non-filamentous cytoplasm, from transport of filaments into the area from other areas of the cell, or from some other process. There is no obvious difference between the appearance of the network present in the area underlying a patch of surface Ig-anti-Ig complexes, and that present in areas free of complexes (de Petris and Raff, 1972). Individual microfilaments running parallel to and under the membrane at a distance of about 15–20 nm are occasionally visible. In some areas, such as in the area of constriction of cells forming a uropod, the submembranous layer of microfilaments often appears thickened, and the filaments are prevalently oriented in a parallel configuration. A clear distinction between an organization of microfilaments either in a "random network", or in oriented "sheaths" or "bundles" has been described in other cell types (e.g. Wessells et al., 1971, 1973; Reaven and Axline, 1973) but is not clearly detectable in lymphocytes, except for the presence of parallel bundles of filaments in microvilli (Fig. 10c). However, it is conceivable that despite the lack of clear morphological discrimination the lymphocyte "microfilaments" are actually organized, either alone or in combination with other structures, in more than one structural system with differing or complementary mechanical functions. Future immunofluorescence and serial section electronmicroscopy studies (cf., Reaven and Axline, 1973; Axline and Reaven, 1974) may help unravel this organization. In addition to actin-like microfilaments, "100 Å *filaments*" (e.g., Ishikawa et al., 1969; McNutt et al., 1973) of unknown function are present in lymphocytes. In the mouse they are fairly common in T-lymphocytes and thymocytes, but rare or absent in B cells. Similar filaments connected in regular arrays by side-arms distributed along their axis with a period of about 75 Å are present in thymocytes. It is not known whether these filaments correspond to a particular arrangement of the normal smooth 100 Å filaments, or to a completely different class of filaments (S. de Petris, unpublished observations).

Microtubules are quite numerous in lymphocytes at 37°C and are detectable also at lower temperatures (down to 10–15°C). Most or perhaps all of them are connected to the centrioles (Fig. 12a) from which they radiate. They usually extend along the nuclear invagination, the perinuclear membrane and the sides of the cell forming an extensive interconnected network. Micro-

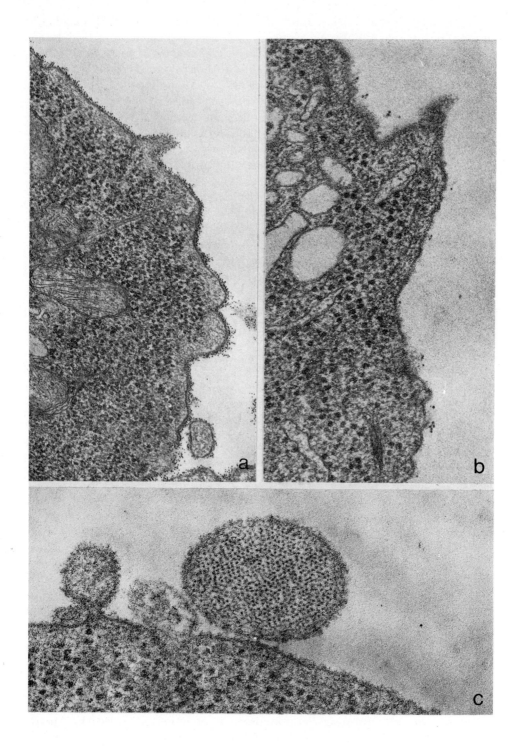

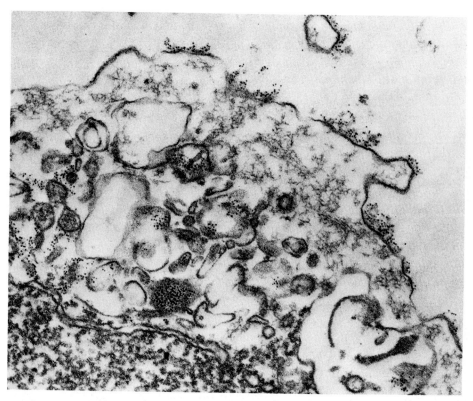

Fig. 11. Mouse lymphocyte with a cap of surface Ig-anti-Ig antibody complexes that was extracted with glycerol and incubated with heavy meromyosin. A layer of microfilaments present between the labelled membrane and the organelles of the centrosomal area which include several ferritin-laden pinocytic vesicles acquires a typical fuzzy appearance. ×62 000.

tubules often run parallel to the plasma membrane from which they are usually separated by the layer of microfilaments; however, some start at the centrioles and terminate at the plasma membrane or close to it (Fig. 12). The total number of these is unlikely to be more than one hundred in the entire cell. Increased concentrations of microtubules correlated with the formation of a concanavalin A cap in the central area of the cell have been reported in ovarian granulosa cells (Albertini and Clark, 1975). However, nothing definite

---

Fig. 10. Microfilament structures in lymphocytes. (a) Cortical microfilament layer (network) separating the lymphocyte plasma membrane from the ribosome-studded cytoplasm. Mouse thymocytes labelled with ferritin-conjugated concanavalin A (detail of a cap). A microtubule is visible which probably terminates near the membrane. ×49 000. (b) Mouse thymocyte lightly labelled with anti-θ antibody showing the cortical microfilament layer at higher magnification. ×85 000. (c) Conspicuous parallel bundle of actin-like microfilaments in a transversely cut microvillus of a rabbit popliteal lymph node cell which is probably connected to the lymphocyte at the bottom of the figure. Longitudinal sections of other microvilli in this preparation demonstrated that they were connected either to lymphocytes, or to plasma cells. ×100 000.

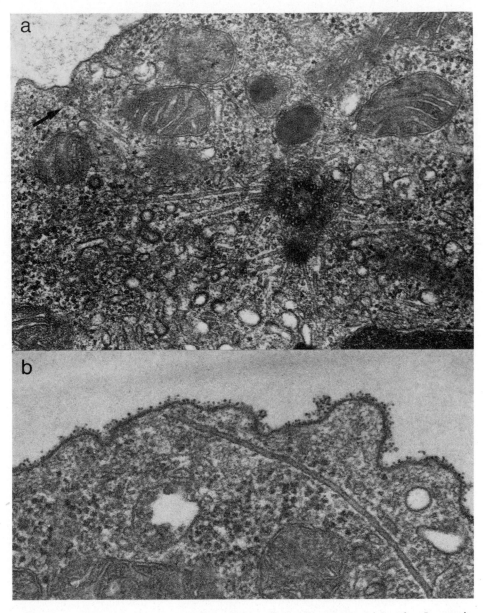

Fig. 12. Lymphocyte microtubules. (a) Microtubules radiating from the centrioles of an Ig-negative mouse spleen small lymphocyte. One microtubule probably ends at the membrane (arrow). ×57 000. (b) Microtubule ending near the membrane of a small mouse thymocyte. No well defined structures are visible near its end. Note the microfilament network separating the membrane from the area of the cytoplasm containing the ribosomes. The demonstration that microtubules can end near the membrane has been obtained in several cases by serial sectioning. ×92 000.

is known about the behaviour of microtubules during the capping process in lymphocytes, apart from the observation that some of them are present in the cap area, and often run more or less parallel to the axis of the uropod, when this is present.

5.6.2. Evidence for a role of microfilaments

Experimental evidence for a functional role of a contractile microfilament system in the capping process is based mainly on studies using the drug cytochalasin B. The cytochalasins are mold metabolites with multiple inhibitory effects on mammalian cells. At low doses (of the order of 10^{-7}–10^{-6} M) cytochalasin B inhibits hexose transport (Estensen and Plagemann, 1972; Mizel and Wilson, 1972; Kletzien and Perdue, 1973; Axline and Reaven, 1974), whereas at higher doses it inhibits cell movement, cytoplasmic streaming and membrane movements (Carter, 1967; Williamson, 1972; Allison, 1973; Bradley, 1973). These latter effects probably occur by functional and in part structural impairment of contractile microfilament structures (Schroeder, 1970; Wessells et al., 1973; Axline and Reaven, 1974; Pollard and Weihing, 1974). Cytochalasin B binds with low affinity to actin (Lin and Spudich, 1974), but its mechanism of action is obscure. Cytochalasin D induces a state of hypercontraction and clumping of microfilaments and possibly alters their interaction with the plasma membrane (Miranda et al., 1974). Cytochalasin B probably causes similar effects, because at concentrations of about 10 μg/ml ($=2.1\cdot 10^{-5}$ M) it inhibits almost completely both ligand-induced modifications in cell shape (de Petris, 1974, 1975; Rutishauser et al., 1974) and motility (de Petris, 1974, 1975; Rutishauser et al., 1974; Unanue et al., 1974a). These changes are presumably dependent on the activity of an actomyosin system and on cytoplasmic streaming.

Capping is inhibited by the cytochalasins, but to a variable degree in different systems. The inhibition of concanavalin A receptor capping in splenocytes by cytochalasin B is virtually complete (de Petris, 1974, 1975; Unanue and Karnovsky, 1974), while surface Ig (Taylor et al., 1971; Stackpole et al., 1974c; Unanue et al., 1974a; de Petris, 1975), θ and H-2 antigen capping (Stackpole et al., 1974c) in splenocytes is only partially inhibited. Surface Ig capping is more strongly inhibited by cytochalasin D (Unanue and Goldman, quoted in: Schreiner and Unanue, 1976c). The effect of cytochalasin B on θ, H-2 and TL antigen capping on thymocytes is practically negligible (Stackpole et al., 1974c; de Petris, 1975). The partial inhibition of surface Ig capping is not enhanced by increasing the dose above 5–10 μm/ml (Taylor et al., 1971). This could indicate that even at saturation of the drug-sensitive sites the inhibition remains incomplete, or it may suggest that the polarized translocation of crosslinked components operates by more than one mechanism, and that each mechanism is differentially affected by cytochalasin B and used to differing extents in each cell type and surface component system.

Cytochalasin-sensitive structures are required for maintaining the cap once this has been formed (de Petris, 1974, 1975). In fact, concanavalin A and to a great extent surface Ig caps can be readily reversed by addition of 10–20 μg/ml of cytochalasin B (de Petris, 1974, 1975; Schreiner and Unanue, 1976a) or 5 μg/ml of cytochalasin D (Schreiner and Unanue, 1976a). The complexes which have not been pinocytosed redistribute rapidly over the entire surface (de Petris, 1974, 1975; Schreiner and Unanue, 1976a). Reversal of surface Ig caps is not enhanced by microtubule-disrupting drugs alone (Schreiner and Unanue, 1976a), although the phenomenon is more clearly detected in cells preincubated with colchicine. This could be due to the delaying effect of the drug on the pinocytosis of the complexes (de Petris, 1975). Concanavalin A caps are also partially reversed by addition of sodium azide (Sällström and Alm, 1972; de Petris, 1975) suggesting that the maintenance of the cap requires metabolic energy. However, metabolic inhibitors used at concentrations which inhibit capping do not reverse surface Ig caps (Schreiner and Unanue, 1976a), and in fact these are often examined at room temperature in the presence of sodium azide (Taylor et al., 1971). The variable effect of azide may be related to the presence of some residual ATP in the cell which is too low to allow cap formation, but may be sufficient to prevent its disruption in some, but not other, surface component systems. Other authors, however, have considered the differences to be of more fundamental nature (see below) (Schreiner and Unanue, 1976a,c). The demonstration of cap reversal suggests that separated patches can be kept together in a cap by intracellular structures present on the inner side of the membrane even if the patches are not crosslinked in a single continuous complex by the ligand (de Petris, 1975). The reversal could be interpreted as a result of relaxation (or disruption) of an activated actomyosin system which counteracts a spontaneous tendency to disperse (de Petris, 1975), although other interpretations (hypercontraction) have been suggested (Schreiner and Unanue, 1976a). The rapidity of the reversal (de Petris, 1975), and other indirect observations (Schreiner and Unanue, 1976a), have suggested the possibility that the dispersal of the patches when the cap is disrupted by cytochalasin B may be helped by some metabolically dependent process (Schreiner and Unanue, 1976a).

The variable effect of the cytochalasins and azide on the inhibition and reversal of capping of different surface components has been taken by some authors as evidence for the existence of independent capping mechanisms for different surface components (Schreiner and Unanue, 1976a,c). In particular, Schreiner and Unanue have suggested that surface Ig capping is an active process in the sense that the molecules might be directly connected to, and displaced by the cytoplasmic mechanical structures (microfilaments), whereas capping of concanavalin A receptors is passive, and occurs as result of cell movement (Schreiner and Unanue, 1976a,c). For reasons discussed in sections 5.5 and 5.6.3, the present reviewer feels that clear-cut evidence for the existence of a qualitatively different capping mechanism for surface Ig and concanavalin A is still lacking. It seems more likely that the mechanism of capping is essentially the same for all the crosslinked surface components systems, but that the forces generated by the mechanical structures have to overcome variable "resistances" in the different systems. Therefore, the partial functional impairment of the mechanical structures by drugs would have a variable effect according to the "resistance" offered by each system. A precise definition of "resistance" in detailed mechanical terms would be very difficult to give at present, although on general grounds a high "resistance" could be broadly and tentatively correlated with the occupation of a large fraction of the cell surface by crosslinked components, combined with tightness of the resulting lattice. According to this view, the individual characteristics of the components would be more important in determining the ability of the aggregated molecules to deliver a triggering signal for capping and in determining the "resistance" to movement, than in affecting the intrinsic mechanism of movement (see also section 5.6.3). Some authors have proposed that the microtubule system may provide this resistance (Edelman et al., 1973; Ukena et al., 1974; Poste et al., 1975a; discussed in Chapter 1, this volume, p. 40 and section 5.6.3).

Calcium is important in the control of muscle cell contraction and is probably also important in contractile events in non-muscle cells, in particular during cell locomotion (e.g., Norberg, 1971; Gail et al., 1973; Huxley, 1973; Pollard and Weihing, 1974; Taylor et al., 1976) and cap formation (Taylor et al., 1971; Schreiner and Unanue, 1976a). Capping can occur either in the presence or in the absence of extracellular calcium (Taylor et al., 1971; Becker et al., 1973; Schreiner and Unanue, 1976a) and extracellular magnesium (Taylor et al., 1971). Possible quantitative effects of prolonged Ca^{2+}-depletion have not been studied in lymphocytes, as they have in fibroblasts (Gail et al., 1973). This does not exclude a role of Ca^{2+} in the intracellular mechanism of these phenomena, but it may suggest that they depend on mobilization of intracellular Ca^{2+} sources. Introduction of Ca^{2+} into the cells by the use of a Ca^{2+} ionophore has been reported to inhibit capping of surface Ig, or to cause a reversal of the preformed caps (Schreiner and Unanue, 1976a). These effects have been interpreted as the result of a generalized state of contraction of the membrane-associated contractile system which under normal conditions would operate according to an organized spatial and temporal sequence of contractions and relaxations (Schreiner and Unanue, 1976a). Reversal of capping by Ca^{2+} did not occur in cells depleted of ATP, suggesting that in this case the structures were immobilized in a contracted state similar to "rigor mortis" (Schreiner and Unanue, 1976a). However, according to other authors, introduction of Ca^{2+} by calcium ionophores had no effect on surface Ig capping, although it was able to release the inhibition of capping of surface Ig induced by concanavalin A (see below) (Poste and Nicolson, 1976). This calcium effect, similar in result to that induced by colchicine, was attributed to the disruption of microtubules by high intracellular Ca^{2+} concentrations (Weisenberg, 1972; Gallin and Rosenthal, 1974; Kirschner and Williams, 1974; Olmsted et al., 1974; Rosenfeld et al., 1976). Ca^{2+} introduced by an ionophore into the cell can in fact substitute for colchicine in the synergistic action displayed by combinations of colchicine and cytochalasin B on the concanavalin A-induced agglutination of 3T3 fibroblasts (Poste and Nicolson, 1976). Displacement of Ca^{2+} from cellular membranes resulting in modifications of membrane structure and increases in free cytoplasmic Ca^{2+} concentrations which in turn result in alterations of microtubules and microfilaments, and of their membrane relationship, has also been suggested (Poste et al., 1975a,b; Nicolson and Poste, 1976a) as a possible mechanism for the cellular effects of tertiary amine local anaesthetics (Papahadjopoulos, 1972; Seeman, 1974; Poste et al., 1975c), and in particular their ability to inhibit or reverse capping (Ryan et al., 1974a; Poste et al., 1975a,b; Schreiner and Unanue, 1976b) (see also next section). This view is supported by the fact that high levels of extracellular Ca^{2+} counteract cap disruption (Schreiner and Unanue, 1976b) and other cellular effects of local anaesthetics (Poste et al., 1975c). Therefore, Ca^{2+} ions could be involved in the control of capping and in general surface movements, not only by activating the actomyosin system, but also by linking some of its elements to membrane structures (Nicolson and Poste, 1976a; Schreiner and Unanue, 1976c). Thus, Schreiner and Unanue (1976c) have suggested that introduction of Ca^{2+} with a calcium-ionophore can inhibit capping (but not pinocytosis) by blocking the actomyosin system, whereas local anaesthetics can inhibit both capping and pinocytosis by displacing membrane Ca^{2+} and disrupting the microfilament-membrane nexus. In addition, Ca^{2+} could have a destabilizing effect on microtubules (Poste et al., 1975b; Nicolson and Poste, 1976a), although it is not known whether this effect would be appreciable at physiological concentrations of free intracellular Ca^{2+}.

Capping is insensitive to extracellular divalent ions and also to other modifications of the extracellular medium. Absence of K^+ from the medium (Taylor et al., 1971) or incubation in the presence of ouabain (Unanue et al., 1973), has no effect on capping. Depolarization of the membrane by substituting part of the NaCl (about 100 mM) with KCl does not affect capping (data quoted in Raff and de Petris [1973] and unpublished experiments). This result is consistent with the similar observation that membrane depolarization by K^+ fails to affect cell movement also in fibroblasts (Gail et al., 1973), without excluding the possibility that trans-membrane polarization changes occur during capping. Capping is inhibited by the K^+-ionophore valinomycin (Yahara and Edelman, 1975b) (although no inhibition was reported by Poste and Nicolson [1976]). The inhibition probably does not depend on plasma membrane depolarization.

5.6.3. Evidence for a role of microtubules
The role of microtubules in the capping process is also complex, as the experimental evidence suggests that these structures can either take an active part in the process and even be necessary for its occurrence, or have instead a hindering effect. Studies on the role of microtubules in surface movements are based on the use of microtubule-disrupting drugs, such as vinblastine, colchicine, podophyllotoxin, and others (Borisy and Taylor, 1967; Weisenberg and Timasheff, 1970; Wilson, 1970; Owellen et al., 1972). Although these drugs can have other secondary effects on various cellular functions (cf., Allison, 1973), in particular on membrane functions (Jacob et al., 1972; Seeman et al., 1973), it seems very likely that their effect on capping (at least at moderate

concentrations) is actually due to their disruptive effect on microtubules (Yahara and Edelman, 1973a). For example, colchicine must penetrate into the cytoplasm in order to affect concanavalin A capping in Chinese hamster ovary cells (Aubin et al., 1975), and lumicolchicine, a structural isomer of colchicine which does not affect microtubular structure, does not modify capping (Yahara and Edelman, 1973a). In general, the activity of these chemically different drugs is correlated with their effect on microtubules (Yahara and Edelman, 1973a); however, the possibility that these drugs affect some protein with microtubule-like characteristics on the inner side of the membrane cannot be ruled out.

Since the earliest studies on capping, it was established that microtubule-disrupting drugs, when used alone, had no inhibitory effect on lymphocyte capping (or they had an enhancing effect) suggesting that microtubules are not required for capping (Taylor et al., 1971; Edelman et al., 1973; Unanue et al., 1973; de Petris, 1975). This was a somewhat unexpected result considering the marked polarity of the phenomenon (cf., section 5.6.4). However, when vinblastine or colchicine were used in combination with cytochalasin B, the two groups of drugs together appeared to exert a considerable dose-dependent synergistic inhibitory effect on capping (de Petris, 1974, 1975; Poste et al., 1975a; Nicolson and Poste, 1976a; Schreiner and Unanue, 1976a). Thus capping of mouse surface Ig, which is only partially inhibited by cytochalasin B, was virtually completely inhibited when vinblastine or colchicine were additionally added (de Petris, 1974, 1975; Poste et al., 1975a). The combination of these two drugs had, however, only a partial inhibitory effect on capping of θ-anti-θ-rabbit anti-mouse IgG complexes on mouse thymocytes, and this may reflect the low sensitivity of these cells to cytochalasin B (cf., de Petris, 1975) observed when this drug is used alone (Stackpole et al., 1974c; de Petris, unpublished results). These observations suggest that capping in lymphocytes is mainly dependent on the intactness of a cytochalasin-sensitive system. When this system is impaired, the support of microtubules may become essential. In line with the suggestions made above, this could mean that microtubules prevent further disorganization of a mechanism already partially impaired, or it could indicate the existence of at least two mechanisms contributing to the displacement of surface molecules in lymphocytes – the most important one of which would be cytochalasin-sensitive, and the other microtubule-dependent and relatively cytochalasin-resistant. Indirect evidence for the existence of capping mechanisms which rely either on the intactness of microtubules or on cytochalasin-sensitive structures, or both, has also been obtained for other cell types, although the relative importance of the various structures seems to vary in the different cells. For example, cytochalasin B and colchicine were found to inhibit synergistically capping of concanavalin A receptors on polymorphonuclear leukocytes resulting in complete inhibition at high lectin concentrations, and partial inhibition at low lectin concentrations (Ryan et al., 1974b). Cytochalasin B but not colchicine inhibited capping of concanavalin A receptors on macrophages (Bishey and Mehta, 1975), and on virus-transformed fibroblasts (Ukena et al., 1974), whereas colchine alone was sufficient to inhibit capping of H-2 antigens on mouse embryo fibroblasts (Edidin and Weiss, 1972), and of concanavalin A receptors on Chinese hamster ovary cells (Aubin et al., 1975).

Simultaneous impairment of microfilament and microtubule structures analogous to that caused by cytochalasin B plus microtubule-disrupting drugs has also been attributed (Poste et al., 1975a) to the inhibitory effect of tertiary amine local anaesthetics on surface Ig capping (Ryan et al., 1974a; Poste et al., 1975a; Nicolson and Poste, 1976a). This interpretation is also supported by the fact that anaesthetics can reverse preformed surface Ig caps as does cytochalasin B alone or in combination with colchicine or vinblastine (Poste et al., 1975a). As discussed in the previous section, anaesthetics might affect microfilaments also by altering their linkage with membrane proteins (see Chapter by Nicolson et al., p. 45).

The opposite role of the microtubules as elements controlling and restraining the movement of crosslinked surface molecules has been demonstrated by experiments on the inhibition of surface receptor redistribution by concanavalin A (Edelman et al., 1973). Loor et al. (1972) and later Yahara and Edelman (Yahara and Edelman, 1973b; Edelman et al., 1973) noted that pre-incubation of lymphocytes at 20 or 37°C with relatively high doses of tetravalent concanavalin A (>10 μg/ml, generally $\geqslant 50$ μg/ml) prevented subsequent capping (and also patching at the light microscopical level) of surface Ig. At high lectin doses, capping of the lectin receptors themselves was also impaired (Greaves et al., 1972; Yahara and Edelman, 1972, 1973b). These inhibitory effects were confirmed by others (Loor, 1974; Unanue and Karnovsky, 1974; de Petris, 1975; Oliver et al., 1975; Poste and Nicolson, 1976), although in many cases the inhibition appeared to be only partial. The inhibition was subsequently found to be of general character, because in addition to capping of surface Ig, the lectin was also able to inhibit capping of θ antigens, H-2 antigens, β_2 microglobulin, Fc receptors and receptors for certain other lectins (Yahara and Edelman, 1973b, 1975b). The inhibition appeared to depend on a high level of crosslinking of the receptors by the lectin, since divalent succinylated concanavalin A or low doses of tetravalent concanavalin A failed to inhibit surface Ig capping (Gunther et al., 1973). In cells where surface Ig capping was inhibited, concanavalin A itself failed to cap its own receptors (Edelman et al., 1973; Yahara and Edelman, 1973). On the contrary, in cells where capping of concanavalin A receptors occurred (spontaneously or in the presence of microtubule-disrupting drugs, see below), concanavalin A was able to "co-cap" the inhibited component; however, the reverse did not occur (Loor, 1974; de Petris, 1975; Yahara and Edelman, 1975b). The simplest explanation for these observations is that the inhibited membrane components correspond to subpopulations of concanavalin A receptors, i.e. to glycoproteins which carry carbohydrate groups specific for the lectin (Loor, 1974; de Petris, 1975). This conclusion is supported by biochemical evidence for at least surface Ig, H-2 and Ia antigens (Hunt and Marchalonis, 1974). At high levels of crosslinking by concanavalin A these glycoproteins are therefore crosslinked and dispersed among the other receptors, and they fail to cap as a result of the failure of the latter to cap (Loor, 1974; de Petris, 1975), whereas at low levels of crosslinking they can move and cap independently of the majority of the other receptors. The failure of tetravalent concanavalin A to cap its own receptors cannot be attributed to insufficient crosslinking, because at least in a fraction of the cells the inhibition can be relieved if the cells are incubated before, during, or after addition of the lectin with microtubule-disrupting agents (Edelman et al., 1973; Yahara and Edelman, 1973b, 1975a). This strongly suggests that concanavalin A somehow blocks (or fails to activate) the cytoplasmic mechanical system which caps the receptors, and that disruption of microtubules releases this block (Edelman et al., 1973; Yahara and Edelman, 1973b, 1975a). To explain this phenomenon, Yahara and Edelman (1973b, 1975a; Edelman et al., 1973), developing on an idea previously advanced in another context by Ukena and Berlin (1972), have suggested that the concanavalin A receptors (and the other surface molecules the movement of which is restricted by the lectin) are physically connected to an intracellular microtubular or microtubule-like system that controls or "modulates" their movement. The connection could be direct, or, as subsequently proposed, indirect, and mediated by a layer of microfilaments (Berlin et al., 1974; Edelman, 1974; Yahara and Edelman, 1975a,b). Extensive receptor crosslinking would somehow paralyze the microtubular system inhibiting the movement of all the surface molecules connected to it, while disruption of microtubules would release the restrained molecules (Edelman et al., 1973; Yahara and Edelman, 1973b, 1975a). Supporting this concept is the observation that surface Ig and concanavalin A capping are not inhibited if the cells are preincubated with the lectin at 0°C (where microtubules are disrupted), washed, and then incubated at room temperature or at 37°C. The significance of this last finding is, however, somewhat uncertain as the binding of concanavalin A at 0°–4°C is quantitatively (Ryan et al., 1974b; Yahara and Edelman, 1975a) and perhaps also qualitatively (Gordon and Marquardt, 1974; Huet, 1975) different from that at higher temperatures. In other cell types such as human polymorphonuclear leukocytes (Ryan et al., 1974b) and ovarian granulosa cells (Albertini and Clark, 1975) concanavalin A capping can occur after labelling the cells either at 0–4°C or at 37°C, but the cellular location of the cap may be different (Ryan et al., 1974b). In agreement with the hypothesis of Yahara and Edelman (1973a,b, 1975a,b), Oliver et al. (1975) found that normal mouse polymorphonuclear leukocytes capped only limitedly with 5 μg/ml of concanavalin A, but that capping was dramatically increased in the presence of colchicine. On the contrary, leukocytes from beige CH mice (an analogue of human Chediak-Higashi syndrome characterized by the presence of abnormal granules, defective degranulation and impaired chemotaxis, possibly due to microtubule or membrane defects) capped much more effectively under the same conditions, and in these cells capping was not enhanced by colchicine. In both colchicine-treated normal cells and in untreated beige CH cells capping was greatly reduced by agents which decreased the intracellular level of cyclic GMP (Oliver et al., 1975). This effect was interpreted as due to stabilization of microtubules or of their interaction with the membrane (Oliver, 1975; Oliver et al., 1975, 1976). On the other hand, cyclic AMP to which similar microtubule-stabilizing properties have been attributed by others (cf., Porter et al., 1974; Brinkley et al., 1975; Schreiner and Unanue, 1975a,b) had no effect.

The strongest argument in favour of intracellular control of the movement of cell surface components through the microtubular system comes from capping experiments where cells are bound to concanavalin A-derivatized nylon fibers (Rutishauser et al., 1974), latex beads, or platelets (Yahara and Edelman, 1975b). In these experiments the cells are in contact with the lectin only over the limited part of their surface area which is in contact with the fibre, or with the bead. However, this local contact is sufficient to appreciably inhibit surface Ig capping on the part of the cell not in contact with the immobilized lectin when soluble anti-Ig antibodies are added to the cells. The inhibition is relieved by colchicine or vinblastine*. As in the case of inhibition by soluble lectin, a minimum amount of surface area must be covered by the lectin, and/or a minimum number of lectin receptors must be bound and immobilized to obtain inhibition (Yahara and Edelman, 1975b). For example, more than 10 platelets, each carrying an average of $4 \cdot 10^5$ molecules, must be bound to a lymphocyte in order to inhibit capping; thus, the minimum number of bound lectin molecules is probably of the order of $4 \cdot 10^5$ if it is assumed that each platelet is in contact with a lymphocyte with at least one tenth of its surface area.

The phenomena of capping-inhibition and capping-enhancement by antimitotic drugs do not seem to be restricted to the concanavalin A receptor system, but probably occur with other surface molecules, in particular with those which are present at high concentrations on the surface. Several of these molecules may in fact correspond to subclasses of concanavalin A receptors. Thus, some of the capping "prozone" effects observed with antigens present at high concentrations on the cell surface could represent inhibition phenomena (Stackpole et al., 1974b). A particular example could be the inhibition of capping of the θ antigen-anti-θ alloantibody complexes further crosslinked by an anti-mouse Ig antibody which was observed to occur with increasing concentrations of anti-θ antibody at room temperature by Taylor et al. (1971). It can be predicted that in these cases microtubule-disrupting agents would partially or totally release the inhibition. It is interesting to note that the relatively low capping ability of surface Ig on B lymphocytes bound to DNP-BSA derivatized nylon fibres in the absence of concanavalin A was considerably

---

*An alternative explanation of the failure of surface Ig to cap in the presence of *soluble* concanavalin A would be that the surface Ig molecules which are dispersed among the lectin molecules are unable to deliver a suitable stimulatory signal to the cell (de Petris, 1975). However, this could not explain capping inhibition by locally bound immobilized concanavalin A, unless a sizeable fraction of surface Ig were sequestered in the area of cell contact to the derivatized bead or fibre so that the number of molecules remaining free on the surface would remain below the threshold number required for triggering. Although this possibility has not been ruled out, the paralysis of an internal mechanical system seems more likely.

enhanced when vinblastine was added (see Table 3 in Rutishauser et al., 1974). A marginal effect of vinblastine was sometimes noticeable also during surface Ig capping in cell suspensions (de Petris, 1975). More recently Fram et al. have shown that colchicine accelerates the rate of surface Ig capping (in the absence of concanavalin A) in lymphocytes from mouse strains in which this rate is genetically programmed to be low (Fram et al., 1976).

5.6.4. Mechanisms of membrane–cytoplasm interaction
The suggestion of Yahara and Edelman (Edelman et al., 1973; Yahara and Edelman, 1973b; Edelman, 1974) that the movement of a variety of surface molecules is under the control of microtubules or microtubule-like proteins appears to be supported by considerable experimental evidence. How this control is exerted, however, and in particular whether it is exerted through a direct "anchoring" of the molecules to cytoplasmic structures remains to be established. For numerical reasons microtubules cannot bind to all of the "controlled" molecules but could do so indirectly by controlling other structures (e.g. microfilaments) connected to the surface molecules, as suggested by Yahara and Edelman (Edelman et al., 1973; Edelman, 1974; Yahara and Edelman, 1975a) and Berlin et al. (1974) (see Nicolson et al., this volume, p. 48). This direct control would imply the existence of either direct *trans*-membrane interactions with the surface molecules, or interactions mediated by intercalated membrane proteins (cf. Edelman, 1974; Nicolson, 1976a). In either case the hypothesis of a direct control of surface components faces the difficulty that a number of structurally different and presumably functionally unrelated surface molecules would have to be able to interact with one, or a few, types of proteins, the function of which is probably very specific, i.e. essentially mechanical. Although this cannot be excluded, a more appealing possibility is that the interaction with microtubules and microfilaments is restricted to only a few specific membrane molecules, which are present on the cytoplasmic side of the membrane and probably span the latter, and which may, or may not, correspond to any of the surface molecules so far analyzed in capping inhibition experiments. By controlling the movement of these molecules the cytoplasmic structures would be able to control indirectly the *active* movement of all other surface molecules, and of the membrane as a whole (de Petris, 1975). Similar concepts have been recently expressed by Nicolson (1976a), Nicolson and Poste (1976b) and Oliver et al. (1975). In support of these concepts Oliver et al. (1975) have suggested that an arginine-rich protein (or proteins) could actually be the protein responsible for controlling the deformability of the membrane and the mobility of surface molecules. Oliver et al. (1975) demonstrated that the protein-specific reagent phenylglyoxal, which reacts with arginine residues and with the protein α-amino terminal group, is able to inhibit concanavalin A capping, patching, endocytosis, and modifications of cell shape in polymorphonuclear leukocytes and macrophages (Oliver et al., 1976). The actual mechanism of this inhibition, however, is still unclear.

If the active movement of surface molecules were controlled only by one or a few specific proteins and most of the surface molecules were not connected to these proteins, the problem arises of how these molecules would be able to cap, and/or inhibit the capping of other molecules. The simplest and perhaps the only possible mechanism by which this could occur would be by forming a sufficiently tight and extensive lattice around the cytoplasm-driven molecules, so that the aggregated molecules could either be dragged by the latter or be able to hinder their movement (Fig. 13). This would be in agreement with the broad requirements for a high level of crosslinking during capping. As suggested by some co-capping experiments (Abbas and Unanue, 1975; Forni and Pernis, 1975), in addition to or as a consequence of external ligand-induced crosslinking, secondary links between surface molecules or between the cytoplasma-driven proteins could be possibly established inside the membrane. It is in this process that Ca^{2+} ions could have an important role as suggested by Schreiner and Unanue (1976c). An implication of the hypothesis on the independence of most surface molecules from cytoplasmic constraints is that only *active* movements of crosslinked surface molecules would be under the control of these structures. The passive, diffusion-dependent formation of clusters and patches would not be impeded, although it is possible that the further coalescence of patches into very large aggregates would be seriously hindered either by the inhibition of active movements, and/or by the diffusional hindrance represented by the presence of the immobilized molecules (possibly trapped inside the patches) spanning the membrane. This possibility is not apparently supported by the observation that concanavalin A is also able to inhibit patch formation of surface Ig (Yahara and Edelman, 1973b) and other antigens (de Petris, 1975). However, considering the technical limitations of the techniques used to establish this point (mainly immunofluorescence) and the fact that the antigen molecules are dispersed among other concanavalin A receptors, the author feels that lectin-mediated inhibition of diffusional motility of molecules not directly crosslinked by the ligand remains to be demonstrated*. Another consequence of the hypothesis would be that non-

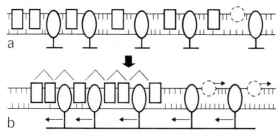

Fig. 13. Postulated mechanism of cytoplasm-driven displacement of surface molecules unconnected to cytoplasmic structures (a) by formation of a tight ligand-induced lattice around independent cytoplasm-driven specialized molecules (b). The efficiency of motion transmission between the internal system and different sets of surface molecules would depend quantitatively on the degree of crosslinking and the physical characteristics of the surface molecules.

---

*This point can be tested experimentally. Gaps in the distribution of uncapped, unlabelled concanavalin A were observed by electron microscopy (de Petris, 1975), but these observations were inconclusive, as they were made on samples in which inhibition of capping was incomplete.

protein molecules (e.g. membrane lipids) could in principle be able to cap if tightly crosslinked by two or more layers of antibody or other suitable ligands. This would be consistent with the recent demonstration that the ganglioside G_{M1}, which is a receptor for cholera toxin and cholerogenoid, can be capped by these (multivalent) proteins (Craig and Cuatrecasas, 1975; Révész and Greaves, 1975). In this particular case, however, it is possible that the ganglioside molecules are or become tightly associated with some specific membrane protein, a hypothesis which would more easily account for the biological effect of toxin binding and would provide an alternative explanation for the capping of gangliosides (Craig and Cuatrecasas, 1975).

The experimental evidence summarized in this and in the previous sections clearly points to an involvement of both an actomyosin-like microfilament system and the microtubular system in the generation and control of the active displacement of surface molecules during capping. However, it provides few clues about the organization of these systems, and the ways the forces they generate are used to obtain directional displacement of membrane molecules. With the exception of the model adopted by Bretscher (1976) where these forces are proposed to be used only for the forward intracellular transport of membrane elements*, all the other models of active surface movements considered by the various authors have proposed that membrane forces are applied directly to one or more classes of membrane molecules and transmitted directly or indirectly to the surface receptors by a *trans*-membrane interaction mechanism (cf., Nicolson, 1976a). Although this point has not been conclusively proven, these forces are probably generated by an actomyosin system with the microtubules playing only a minor role, if any, in this respect. The microtubules probably provide a framework which is connected to the microfilament-filament system and which organizes and coordinates the pattern of the surface and cytoplasm movements generated by this system. The microtubules cannot be considered simply as elements "antagonizing" these movements (cf. Poste et al., 1975a,b; Schreiner and Unanue, 1976c), because in some cases their integrity appears to be necessary for capping (Edidin and Weiss, 1972; de Petris, 1974, 1975; Aubin et al., 1975; Poste and Nicolson, 1976; Schreiner and Unanue, 1976a). This apparent contradiction may perhaps be interpreted in the sense that the microtubules, by imposing some static or dynamic constraints on the orientation and internal movements of the actomyosin system are limiting and restricting the movements of the latter system, or the movements induced by it. At the same time they accentuate the polarized character which is necessary for capping. Thus, when the microfilament system is partially impaired by cytochalasin B, addition of vinblastine or colchicine never enhances, but on the contrary always depresses capping independently of

---

*The evidence provided by the effects of microtubules and microfilaments inhibitors does not rule out capping mechanisms based entirely on membrane flow and intracellular membrane recirculation (Bretscher, 1976), although models based on these mechanisms face the difficulty of explaining how in lymphocytes this intracellular transport remains unaffected by microtubule disruption at concentrations which affect vesicle or granule transport in several other systems.

the presence of concanavalin A (de Petris, 1974, 1975; Unanue and Karnovsky, 1974). This suggests that a coordinate polarized movement of membrane and cytoplasm occurs more efficiently in the presence of microtubules. The role of microtubules in capping appears to be important also in the absence of appreciable cytoplasmic streaming, when streaming is practically abolished by cytochalasin B (de Petris, 1974, 1975). However, it is also clear that polarity of surface movements cannot be attributed entirely to the presence of microtubules (see below). The complex nature of the microfilament–microtubule relationship and the almost complete lack of experimental data on the subject make it impossible at present to understand how the inhibition of active surface movements by concanavalin A occurs. It is not known, for example, whether the lectin-inhibited microtubular system acts simply by mechanically opposing an actomyosin system in its normal state of activity, or whether changes are additionally induced in the organization and activity of the actomyosin system, a possibility not inconsistent with the spherical shape and motionless appearance of concanavalin A-inhibited cells. Conversely, it is not known whether the release from concanavalin A inhibition involves simply the release or bypass of certain constraints from an activated actomyosin system, or whether the release is accompanied by stimulation of a previously inactivated system.

Similar problems arise when one considers the possible ways in which a directional movement of surface molecules is generated. Although it seems practically certain that the mechanical work produced during capping is provided by a microfilament-filament system of the actomyosin type, the scarcity of information on its structural organization has not yet made possible the development of a detailed model of its mechanism of action. However, if the forces were applied directly to surface molecules and not used to generate a membrane flow, the characteristics of the phenomenon would impose some general limitations on the possible ways in which the system could be organized and operated. Actomyosin mechanical systems appear to be ubiquitous in eukaryotic muscle and non-muscle cells and their basic molecular constituents and biochemical characteristics appear to be essentially conserved in different species, despite the relatively minor variations on the basic molecular patterns of the various constituents, and the considerable differences in their supramolecular organization in different cells and organisms. This basic similarity suggests that all these systems, whether or not differently organized, may operate essentially on the basis of the same common principle. This common feature is probably the "sliding filament" mechanism which has been developed to a high level of specialization in skeletal muscle (cf., Jahn and Bovee, 1969; Huxley, 1973; Komnick et al., 1973; Pollard and Weihing, 1974), although the occurrence of additional mechanisms cannot yet be excluded. It is well known that this mechanism operates by generation of shearing forces between two parallel filament systems resulting in the relative displacement of the two interacting systems in opposite directions (Jahn and Bovee, 1969; Huxley, 1973; Pollard and Weihing, 1974). (The reader should consult specialized reviews for a detailed description.) To obtain displacement the two interacting filament systems must have an intrinsic polarity and be properly oriented with respect to each other (Jahn and Bovee, 1969; Huxley, 1973).

The basic mechanism of displacement by an interacting filament system could be used to induce polarized displacements of surface components via a structure which could be organized in various ways. For example, surface molecules or other structures connected to them could be forced to "slide" over a set of fixed filaments (a fixed framework) parallel to the membrane. Alternatively, surface molecules attached to cortical cytoplasmic structures (the microfilament "network"?) could slide over parallel structures situated more internally in the cell. This sliding process would be limited to a superficial cell layer, and generating forces parallel to the surface. In addition, it would have the appealing feature that although it would induce some passive counterflow of "fluid" membrane components, it would not necessarily generate either appreciable internal pressure in the fluid cytoplasm or forced cytoplasmic streaming, and it could therefore be well suited to account for capping in motionless round cells (cf. Fig. 7a). Another possibility might be represented by the displacement of patches induced by a shortening of the cables connecting them to internal framework structures situated at the rear of the cell; this would be equivalent in substance to a contraction and shortening of (part of) the framework. The important point is that independently from the particular mechanism of displacement, the fact that the polarity of movement with respect to internal cellular structures is maintained during capping implies also that the fibrous structures operating this movement must maintain the appropriate orientation with respect to the overall cellular framework. This conclusion is consistent with the observation that capping can occur with unchanged polarity even in the absence of microtubules. In some cells polarized capping occurs even after 23 hours in the continuous presence of 1×10^{-5} M colchicine (unpublished observations) indicating that cell polarity must also reside in structures other than microtubules, either in the microfilament-filament system itself and/or in its connection with the plasma membrane (de Petris and Raff, 1973b; de Petris, 1975). This does not imply that the filamentous framework is a permanent structure, but it suggests that when its parts are disassembled, they are re-assembled following a definite oriented pattern. The need to maintain some polarity during movement could also impose limitations on the arrangement of the membrane molecules interacting with cytoplasmic structures. For example, if the former were to interact *directly* with the latter through a cyclic actin–myosin-like type of interaction (Huxley, 1973), the possibility would arise that these components have to be arranged in pauci- or multi-molecular complexes rather than to be isolated and freely floating as in an ideal fluid membrane. It is clear that if an intermittent force were applied to isolated protein molecules in a fluid membrane, very little net displacement of the latter could be expected, because any oriented displacement would be rapidly cancelled by random diffusion of the molecule released at the end of a cycle of a myosin–actin-like interaction. Net movement could only result if the interacting membrane molecules were connected as in a "patch" with at least a fraction of the molecules presenting the correct orientation for the interaction with an oriented sub-membrane framework (cf., Bray, 1973). This "patch" could be formed by external ligand-induced crosslinking, although in this case only a few of the crosslinked molecules would be expected to have the correct orientation at any time. However, if, as suggested previously, the framework–membrane interactions were limited only to certain specific components present on the cytoplasmic side of the membrane which they probably span, these could be organized transiently or semi-permanently in suitably oriented patches or networks that would result from linkages established between the membrane molecules themselves or between the molecules and underlying cytoplasmic structures. This would insure a better continuity of interaction when the patch is suitably oriented. An alternative, of course, would be that these type of interactions do not occur at the level of the plasma membrane but more internally between oriented cytoplasmic structures to which the plasma membrane molecules are connected. It could be further speculated that the formation of integral inner surface patches which would be required for capping any of the surface components, could be part of the triggering signal for the initiation of translational movement, as suggested by Schreiner and Unanue (1976c).

The tentative scheme presented above represents only one possible attempt at reconciling the "local" aspects of the polarized movement during capping, which is restricted to the crosslinked patches, with its general character of coordinate process simultaneously induced over the entire cell surface. The arrangement and movement of specialized surface molecules composing the "inner surface patches" could in fact be coordinated over the entire membrane, because the connected set of molecules could move in a "comb-like" fashion inside the membrane bilayer without affecting the mobility of other unconnected molecules, unless the latter become extensively crosslinked and "locked" into the "patches". The concept that the mobility of some classes of molecules present on the inner surface of the membrane of lymphocytes and other cells may be more restricted than that of the majority of the molecules on the outer face of the membrane, though still purely speculative, is not inconsistent with the demonstration that an extensive spectrin-actin network exists in the erythrocyte membrane (cf. Nicolson, 1976a; Elgsaeter et al., 1976), and with the demonstration that a myosin-containing fuzzy layer of material is attached to the inner surface of the plasma membrane of mammalian cells (Painter et al., 1975).

5.7. Biological significance of capping

Capping is normally studied as an in vitro phenomenon, but it can occur also in vivo if a sufficient amount of ligand (antibody) is injected into an animal such as anti-γ_M or anti-γ_D antibody in monkeys (B. Pernis and L. Forni, personal communication). However, a specific physiological function for capping has not yet been established. An obvious function would be that of concentrating surface component-ligand complexes, or in general unwanted membrane-attached material, into a particular area of the cell surface where pinocytosis or shedding would be facilitated. Since the latter processes, however, do not strictly require capping, the usefulness of capping for the removal of unwanted material, though not excluded, remains to be conclusively proven (see section 6). Multivalent binding and ligand-induced clustering of the molecules of one particular membrane component could be the starting event for a number of biological phenomena, since clustering could be followed by conformational changes, additional binding of unrelated molecules, activation of membrane-bound enzymes, etc., which could initiate a particular series of biochemical events inside the cell. However, ligand-induced clustering and patching in principle can be obtained purely by diffusional mechanisms without the necessary participation of non-diffusional surface movements.

Several authors have investigated the possibility of a correlation between capping and mitogenic stimulation of lymphoid cells, but the lack of a definite relationship between surface redistribution and stimulation by antigen (Dunham et al., 1972; Diener et al., 1976), or antibody (Elson et al., 1973) has been observed in several of such studies. This, in addition to the demonstration that cells can be stimulated by divalent concanavalin A which is unable to cap Gunther et al., 1973, has shown that capping is not necessarily required for stimulation of cell proliferation (see also Schreiner and Unanue [1976c]). The apparent correlation sometimes observed between capping and some of the above phenomena is probably due to the fact that stimulatory reagents are usually multivalent, and therefore potentially capable of capping their specific receptors (Schechter et al., 1976). Moreover, capping might constitute an important step (the removal of receptors from the surface [Ault et al., 1974; Nossal and Layton, 1976]) in a chain of events leading to stimulation, without being itself one of the critical triggering steps. Similar considerations are probably valid for the role of capping (or lack of capping) in the induction of immunological tolerance (Diener et al., 1976) (see also section 6.4).

Although capping, together with other redistribution phenomena, might have been adapted to serve some immunological functions, its primary origin and function are certainly not immunological. Similar phenomena occur also in other tissue cells, and in unicellular organisms such as protozoa (Doyle et al., 1974; Pinto da Silva et al., 1975) and slime mold amoebae (Beug et al., 1973) which are clearly devoid of any immunological functions.

The lack of a recognizable specific function renders it unlikely that capping occurs by a mechanism developed ad hoc by the cell. On the other hand, its widespread occurrence in cells of primitive as well as highly developed organisms suggests that it reflects the activity of a mechanism which performs a basic function common to all these cells. As discussed in section 5.4, the characteristics of capping suggest that this function is essentially mechanical and that the structures involved in the process are those operating during cell movement (de Petris and Raff, 1972, 1973b). The mechanical analogy between the two processes is that in both cases a patch of cross-linked surface components is displaced backwards with respect to an internal framework (de Petris and Raff, 1973b). Therefore, despite the fact that important biochemical and mechanical differences probably exist between the two processes such as in the composition of the "patches", their topography, the characteristics of their movement (i.e., transient and essentially irreversible in capping, continuous or more prolonged during cell movement), the nature of the biochemical signals originated by the formation of "patches", etc., the general information gained from studying the mechanism of capping will probably be of great value for understanding the mechanism of cell movement, and vice versa.

6. Fate of labelled material

6.1. Pinocytosis

Binding of a macromolecular ligand (antibody, lectin, etc.) to a lymphocyte membrane component often triggers endocytosis of the labelled membrane (Biberfeld, 1971a; Taylor et al., 1971; de Petris and Raff, 1972; Unanue et al., 1972; Barat and Avrameas, 1973; Rosenthal et al., 1973; Linthicum and Sell, 1974a; Antoine et al., 1974; Nicolson, 1976b). This phenomenon also occurs in non-lymphoid cells. For example, it was described by Easton et al. (1962) in Krebs ascites tumor cells using ferritin-conjugated antibody against membrane antigens and has been often observed by immunofluorescence or immunoelectron microscopy in lymphocytes labelled with lectins (Hirsschorn et al., 1965; Packer et al., 1965; Biberfeld, 1971a, and others). Endocytosis of the labelled membrane often follows capping (Taylor et al., 1971; de Petris and Raff, 1972; Unanue et al., 1972), although it appears to be essentially independent of, and not correlated with the latter, as it can occur in its absence (de Petris and Raff, 1972, 1973a; Unanue et al., 1972; Rosenthal et al., 1973; Linthicum and Sell, 1974a; Antoine et al., 1974). The endocytosis process is initiated by an invagination of the labelled membrane which may include unlabelled areas as well, and is followed by pinching off of the vesicle from the membrane as in the typical process of pinocytosis (cf., Allison, 1973). In contrast to capping, a phenomenon, which involves the entire membrane, pinocytosis is a *local* phenomenon with presumably "segmental" characteristics similar to those observed in phagocytosis (Griffin and Silverstein, 1974), which involves a limited region of the membrane, with perhaps little lateral movement of the labelled molecules during the inward flow of the membrane.

The energetic requirements for pinocytosis seem to be less stringent than those for capping. Pinocytosis can occur in the presence of sodium azide under conditions which not only inhibit capping, but also transport of the pinocytic vesicles to the Golgi area (Loor and Hägg, 1975). Similarly, capping of surface Ig complexes is completely inhibited at 4°C in Ig-positive bone marrow lymphocytes (S. de Petris, M.C. Raff and M.D. Cooper, unpublished data), while pinocytosis, or at least the formation of deep membrane invaginations, can still occur.

The pinocytic vesicles can be quite large (0.2–0.4 μm) (de Petris and Raff, 1972) and probably fuse to form larger vesicles in the Golgi area (Biberfeld, 1971a; de Petris and Raff, 1972; Unanue et al., 1972; Rosenthal et al., 1973; Santer, 1974). Marker molecules such as ferritin which are initially bound along the inner surface of the otherwise empty vesicles (Fig. 14), appear to detach subsequently from the membrane and fill the entire volume of the vesicle, which appears to condense (Fig. 15). This probably indicates the beginning of a process of hydrolytic degradation of the enclosed molecules resulting from the fusion of the pinocytic vesicles with lysosomes (about ten lysosomal acid hydrolases have been identified in lymphocytes [Bowers and de Duve, 1967; Bowers, 1972]). Pinocytosis and the final fate of the vesicles labelled with ferritin or ferritin-labelled antibody in phytohaemagglutinin-stimulated cells was studied in detail by Biberfeld who demonstrated the presence of hydrolytic enzymes in the vesicles (Biberfeld, 1971a,b; cf. also Packer et al., 1965).

The extent and rate of ligand-induced pinocytosis vary considerably in the different membrane component/ligand systems (e.g. Karnovsky et al., 1972; Unanue et al., 1972) and in different lymphoid cells (e.g. thymocytes versus splenocytes). This may possibly reflect differences in the (unknown) stimulus for pinocytosis given by the different complexes, or differences in the structural

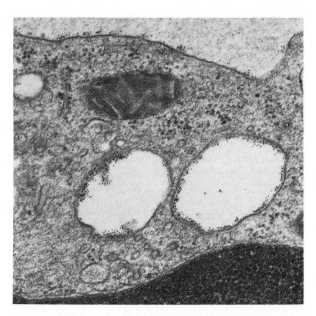

Fig. 14. Pinocytosis of labelled membrane in mouse B lymphocyte labelled with an antibody ferritin conjugate against surface Ig. Initial formation of large vesicles with labelled complexes partially, or completely, lining the internal surface of the vesicles. ×45 000.

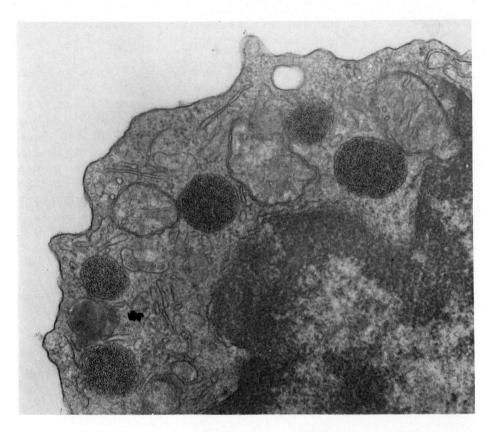

Fig. 15. Later stage of pinocytosis of labelled membrane. Detachment of the labelling molecules from the wall of the pinocytic vesicles, and condensation of the latter. Mouse B lymphocyte incubated with ferritin-conjugated anti-Ig antibody for 2 h at 37°C and reincubated with fresh conjugate at 0°C: almost no antigen can be relabelled on the surface ("antigenic modulation"). ×40 000.

characteristics of the cells. In the surface Ig system ingestion of the complexes is very rapid and can be practically completed in 10–15 min (Antoine and Avrameas, 1974; de Petris, 1975), and is followed by slower degradation of the ingested labelled antibody (Wilson et al., 1972; Engers and Unanue, 1973; Knopf et al., 1973). The process of ingestion and subsequent degradation of the complexes, together with the simultaneous process of shedding (see below), can lead to virtually complete disappearance of the label and of the labelled membrane component from the surface of a large proportion of cells (Fig. 15) (Taylor et al., 1971; Loor et al., 1972; Ault and Unanue, 1974). Similar observations were made in experiments with labelled multivalent soluble antigen reacting with specific surface Ig antibody (Ault and Unanue, 1974; Ault et al., 1974; Nossal and Layton, 1976). On the contrary, no ingestion was observed with a particulated antigen (red cells), in agreement with the known inability of lymphocytes to phagocytize large particles. The red cell-antibody complexes were eventually shed (see section 6.2) (Ashman, 1973; cf., also Yu, 1974).

Endocytosis of histocompatibility antigen complexes (H-2, HL-A) with their antibody is relatively slow, and incomplete (Kourilsky et al., 1972; Unanue et al., 1972; Cullen et al., 1973; Menne and Flad, 1973; Lesley and Hyman, 1974), although it is accelerated by crosslinking of the complexes by a second (anti-IgG) antibody (Unanue et al., 1972; Menne and Flad, 1973). Complexes formed by surface components with anti-lymphocyte serum (ALS) are poorly endocytised, even after crosslinking with a second antibody (Unanue et al., 1972). Therefore, induction of pinocytosis may be even more strongly dependent on the individual molecular characteristics of the labelled component than the process of capping, with the complexes of surface Ig being particularly efficient in inducing the process (see also the review by Schreiner and Unanue [1976c]).

The crosslinking requirements for stimulation of ligand-induced pinocytosis remain somewhat uncertain, although in general they appear less stringent than those for capping. Sparse clusters or patches of surface complexes are sufficient to induce pinocytosis. Pinocytosis can also be induced by the binding of monovalent (Fab) anti-Ig (and probably anti-TL) antibodies, monovalent antigens, or their immunofluorescent derivatives (de Petris and Raff, 1973a; Antoine and Avrameas, 1974; Antoine et al., 1974; Loor and Hägg, 1975), although in these experiments it is difficult to rule out completely the possibility that pinocytosis is stimulated by the presence of traces of dimeric or polymeric aggregates in the reagents which could bind with a 10^2–10^3-fold higher affinity to the membrane (cf., Becker et al., 1973; Bystryn et al., 1973). Monovalent ligands seem to be less efficient than polyvalent ligands in inducing pinocytosis, possibly because of their higher probability of dissociation from the membrane (Bystryn et al., 1973; Antoine and Avrameas, 1974; Ault and Unanue, 1974), although there is some indication that a continuous layer of well crosslinked molecules such as the caps formed by Ig-anti-Ig complexes may be pinocytosed more slowly than looser aggregates or isolated molecules (Antoine et al., 1974). A continuous compact layer of molecules similar to those formed by concanavalin A caps sometimes appears to induce undulations and invaginations in the membrane (de Petris, 1975) without the formation of true pinocytic vesicles. This may perhaps be due to interference with the fusion of the opposed membranes due to the presence of an almost continuous layer of concanavalin A. Barat and Avrameas (1973) observed that some pinocytic vacuoles containing concanavalin A-peroxidase complexes were still open to the cell exterior after 3 h at 37°C, and concanavalin A receptors (which include surface Ig [Hunt and Marchalonis, 1974; de Petris, 1975]) are generally pinocytosed more slowly than surface Ig-anti-Ig antibody complexes (e.g., Greaves et al., 1972; de Petris, 1975). These effects have not been studied in sufficient detail to draw any firm conclusions, but they could perhaps bear some relation to the inhibition of Sendai virus induced cell fusion by concanavalin A (Bächi et al., 1973) and the inhibition of fusion of peritoneal macrophage lysosomes with pinosomes. The latter event, however, occurs when concanavalin A is bound to the inner surface of the pinosomes (Edelson and Cohn, 1974), and the fusion does not involve the extracellular, but the cytoplasmic side of the vesicles. A point of some potential interest is whether there is any release of hydrolytic enzymes into the medium from lymphocytes or other cells during pinocytosis, similar to that observed during phagocytosis in macrophages (cf. Cohn, 1975) and granulocytes (e.g. Henson, 1971; Becker and Henson, 1973). This last point could be potentially important, since the released substances could have secondary effects on the lymphocyte cell surfaces. The limited evidence available suggests that an appreciable release of hydrolytic enzymes may occur from non-lymphoid cells present in lymphoid cell suspensions (phagocytes, "adherent cells"), but not from the lymphocytes themselves (Jacot-Guillarmod et al., 1975; Trinchieri et al., 1975).

Pinocytic vesicles normally accumulate in the Golgi area of the cell (Taylor et al., 1971; de Petris and Raff, 1972; Unanue et al., 1972). Since pinocytosis can take place at different points of the surface, this accumulation could be due either to a rapid intracellular transport of the vesicles to the Golgi area, or to a preferential formation of vesicles in the posterior (post-nuclear) region of the cell (this second process, of course, takes place preferentially when a cap has previously formed in this area). There is little information on the relative importance of these two pathways of vesicle accumulation in lymphocytes, although the second seems to be the major one. Thus, Biberfeld has observed that in phytohaemagglutinin-stimulated cells the vesicles formed mainly in the uropod region (Biberfeld, 1971a).

Drugs such as colchicine and vinblastine at doses sufficient to disrupt microtubules do not inhibit ligand-induced pinocytosis of surface Ig or concanavalin A receptors (de Petris, 1974, 1975; Loor and Hägg, 1975) in lymphocytes, although they appear to slow down its rate (de Petris, 1974, 1975), whereas at the same time they favour the formation of larger caps. Vinblastine (but not colchicine) at very high doses (10^{-4} M) blocks pinocytosis (Loor and Hägg, 1975) perhaps by a non-microtubule membrane effect. It has been suggested that disruption of microtubules may alter the topography of the membrane (Ukena and Berlin, 1972) or alter microtubule-directed movement of the membrane during interiorization (Oliver et al., 1974). These experiments have shown that particular membrane components are selectively ingested during phagocytosis in rabbit polymorphonuclear leukocytes and that the selectivity is lost after microtubule disruption by colchicine (Ukena and Berlin, 1972; Oliver et al., 1974); however, others have been unable to observe similar colchicine effects in human leukocytes (Dunham et al., 1974). There are no data concerning similar microtubule drug effects on pinocytosis in lymphocytes. Cytochalasin B is unable to block ligand-induced pinocytosis (de Petris, 1975; Loor and Hägg, 1975) in lymphocytes, although it is unclear whether or not the latter can be delayed. In this respect pinocytosis of large labelled vesicles by lymphocytes has the characteristics of the "small vesicle" pinocytosis in macrophages (Wills et al., 1972) which is also not inhibited by cytochalasin B. On the other hand, "macropinocytosis" and phagocytosis of particulated material are inhibited by cytochalasin B (Allison et al., 1971; Wagner et al., 1971). Combinations of cytochalasin B with microtubule-disrupting agents do not appear to be effective in inhibiting completely pinocytosis in mouse spleen cells, although it appears to inhibit the intracellular transport of the pinocytotic vesicles toward the Golgi area (de Petris, 1974, 1975). This suggests that the transport of the vesicles, which also requires metabolic energy (Antoine et al., 1974; Loor and Hägg, 1975), occurs via a microtubule-microfilament system similar to that involved in capping. It is therefore possible that both capping and polarized vesicle transport utilize the same mechanical system, whereas membrane invagination and vesicle formation depend on a different mechanism (cf. Wills et al., 1972).

6.2. Shedding

Another process by which ligand-surface complexes can disappear from the cell surface is by *shedding* which consists of the release of the complexes into the medium as intact or partially degraded molecules (Karnovsky et al., 1972; Miyajima et al., 1972; Wilson et al., 1972; Jones, 1973; Engers and Unanue, 1973; Cullen et al., 1973; Antoine and Avrameas, 1974; Ault and Unanue, 1974; Cone et al., 1975; Jacot-Guillarmod et al., 1975; Yefenov et al., 1975). This occurs in addition to the normal (reversible) dissociation of some ligand molecules from the membrane components (Amos et al., 1970; Wilson et al., 1972; Cullen et al., 1973; Antoine and Avrameas, 1974; Ault and Unanue, 1974). Particulated complexes like those formed with red cells or bacteria are generally shed from cells rather than ingested (Ashman, 1973; Yu, 1974; Gormus and Shands, 1975). *Spontaneous shedding*, i.e., shedding of a membrane component in the absence of a specific ligand, apparently occurs with some antigens such as surface Ig, but not with others (e.g., H-2 or HL-A antigens) (Vitetta and Uhr, 1972), and the binding of a specific antibody can initiate or accentuate shedding (e.g., Cullen et al., 1973; Cone et al., 1975). The mechanism of shedding and the initiation processes remain uncertain. It is possible that different mechanisms operate in different systems, or at different steps of the process. In the HL-A system shedding is temperature-dependent and requires metabolic energy (Cone et al., 1971; Jacot-Guillarmod et al., 1975). Moreover, it is inhibited by drugs affecting the movement of surface molecules and of the whole cell such as cytochalasin B (alone or in combination with colchicine) with characteristics similar to the inhibition of capping (H. Jacot-Guillarmod, private communication). The mechanism of shedding inhibition is, however, still obscure. In the same system degradation of the shed product is apparently related to the presence of monocytes in the cell suspension and is susceptible to blocking by protease inhibitors, whereas the actual detachment of the complexes from the cell surface is unaffected by the latter (Jacot-Guillarmod et al., 1975). On the contrary, no effect of metabolic inhibitors or cytochalasin B has been reported in the murine surface Ig-anti-Ig system (Unanue et al., 1973, 1974a).

6.3. Antigenic modulation

Redistribution, pinocytosis and shedding are presumably responsible for the so-called phenomenon of "antigenic modulation" in lymphoid cells, which consists in the functional (and possibly physical) disappearance of an antigen from the cell surface following the binding of a specific antibody. This phenomenon could be important in the development in vivo of resistance of tumour cells to immune killing. The phenomenon was originally discovered and studied in *Paramecium* (Beale, 1957) and was first described in lymphoid cells by Boyse, Old and their colleagues while studying the TL antigen of mouse leukaemia cells (Old et al., 1968; Boyse and Old, 1969). It was subsequently demonstrated in several other lymphocyte surface systems: mouse H-2 (Amos et al., 1970; Takahashi, 1971; Schlesinger and Chaouat, 1972, 1973; Lesley and Hyman, 1974; Stackpole et al., 1974d), human HL-A (Bernoco et al., 1972; Menne and Flad, 1973; Cullen et al., 1973), surface Ig (Takahashi, 1971; Knopf et al., 1973), mouse viral antigens (Aoki and Johnson, 1972) and others.

Old et al. (1968) found that if thymoma cells carrying surface TL antigen(s) characteristic of thymus cells from which the leukaemic cells were derived were incubated at room temperature or at 37°C (but not at 4°C) with mouse anti-TL antibody for variable periods of time, they became progressively resistant to cytotoxic lysis on subsequent addition of the alloantibody plus guinea pig complement. The resistant cells were unable to absorb anti-TL antibodies suggesting that most of the antigen had disappeared from the surface (or had become inaccessible to the antibody). Subsequent observations in the same (Stackpole et al., 1974d) and other systems (e.g. H-2 [Lengerová et al., 1972; Lesley and Hyman, 1974] and HL-A antigens [Bernoco et al., 1972; Cullen et al., 1973]) have indicated that at least part of the antigen-antibody complexes may still be present on the surface of the "modulated" cells, although these are no longer lysed by complement. "Modulation" of one antigen is specific, i.e. it does not alter the sensitivity to complement-mediated killing by antibodies against molecularly independent antigens, although it sometimes causes minor quantitative changes (Old et al., 1968). This indicates that the resistance does not depend on some general change of membrane or cell properties, but is due to some modification of the characteristics of the antigen being modulated. Some antigens (e.g., TL [Old et al., 1968], surface Ig [Takahashi, 1971; Knopf et al., 1973]) can be modulated by incubation with the specific antiserum alone, while others (e.g., H-2, HL-A [Takahashi, 1971; Lengerová et al., 1972; Cullen et al., 1973; Stackpole et al., 1974d]) require the use of the specific antiserum followed by anti-Ig antibodies directed against the first antibody. This suggests that the degree of crosslinking, and therefore redistribution of the antigen, are important in the phenomenon. In some cases, such as in that of TL antigen, also monovalent (Fab') antibody appears to be able to induce modulation (Lamm et al., 1968; Stackpole et al., 1974d). The precise mechanism of antigen modulation is therefore still uncertain, and several factors could contribute to a variable extent in different systems to the induction of "antigenic modulation" as functionally measured by resistance to complement-mediated lysis. The most important of these factors are probably the following (see also Nicolson, 1976b):

(a) *Pinocytosis* of the antigen-antibody complexes and their subsequent intracellular degradation which could considerably reduce the amount of antigen remaining on the surface (Taylor et al., 1971). When pinocytosis is rapid and massive as in the case of modulation of surface Ig on B lymphocytes (Takahashi, 1971; Taylor et al., 1971; Knopf et al., 1973), and possibly also in the case of TL antigen on thymus leukaemia cells (Yu and Cohen, 1974), this could be the most important factor which would lead to the virtual disappearance of the antigen from the surface of a considerable proportion of cells. In these systems pinocytosis could possibly also occur to a significant extent using monovalent antibody (see also section 6.1.);

(b) *Shedding* of the intact or partially degraded antigen-antibody complexes into the medium (Cullen et al., 1973; Cone et al., 1975; Jacot-Guillarmod et al., 1975). In one study of TL modulation, shedding of the antigen was not increased by the binding of the antibody (Yu and Cohen, 1974), and in general this factor alone is unlikely to account for modulation in lymphocytes;

(c) Redistribution of the complexes into patches and caps (Bernoco et al., 1972; Lengerová et al., 1972; Cullen et al., 1973) which could prevent the series of complement components to bind to the complexes because of steric configuration that would not allow them to be activated and/or to react with the membrane. This could perhaps be the most important factor, and this is supported by the demonstration that patches and caps are present on the surface of "modulated" cells (Bernoco et al., 1972; Lengerová et al., 1972; Stackpole et al., 1974d) and by the fact that a second layer of antibody is required to make the cells carrying H-2 or HL-A antigens resistant to lysis (Takahashi, 1971; Bernoco et al., 1972; Lengerová et al., 1972; Stackpole et al., 1974c).

These observations suggest that modulation is favoured by conditions which cause large-scale redistribution, e.g., higher antigen valency, high temperature, etc. However, efficient binding of complement (and more precisely of the C1' complement component) to IgG molecules requires the prior formation of some antigen-antibody complexes (e.g., Cohen, 1968; Hyslop et al., 1970), and therefore, probably depends on a minimum degree of redistribution. For example, complexes of monovalent antigens with bivalent antibodies might be unable to fix, or activate, complement. Thus the antibody-dependent cytotoxic activity of complement seems to be optimal when the aggregates are numerous and small, i.e. closely spaced. One possible way of understanding this requirement would be to suppose that a successful lytic "hit" on the membrane could occur only if the complement complex were able to bind to adjacent antigen-antibody complexes in such a way as to come sufficiently close to the membrane in an appropriate configuration and orientation (cf., Müller-Eberhard, 1969; Hammer et al., 1976). For example, this could occur with high probability if the complement complex were located in the "gap" between adjacent antigen-antibody aggregates, and not on top of them. This probability would drop rapidly if the number of the antigen molecules (and hence of the antigen-antibody aggregates) were reduced by pinocytosis or shedding, or if the gaps were obliterated by coalescence of the aggregates into large patches. Reduction of the number of antigen molecules by pinocytosis could also play a role in the modulation of TL antigen by monovalent antibodies, although the antigen is not removed completely from the surface (Stackpole et al., 1974d). Perhaps this factor, combined with the inhibition of the binding of the bivalent antibody used to test the resistance to lysis as the result of the partial occupation of the sites by the residual monovalent antibody, may delay cytolysis until the redistribution induced by the second (testing) antibody can establish a permanent resistance (Stackpole et al., 1974d). However, the mechanism of resistance to lysis induced by monovalent anti-TL antibody remains obscure. Another point which remains obscure is the mechanism of inhibition of TL-antigen modulation by actinomycin D (Lamm et al., 1968; Old et al., 1968).

6.4. Resynthesis of surface components (recovery from modulation)

In most mammalian systems antigenic modulation is only transitory, because new antigen is resynthesised continuously (Loor et al., 1972, 1975; Miyajima et al., 1972; Cullen et al., 1973; Rowe et al., 1973; Lesley and Hyman, 1974). In capped or modulated cells the antigen is fully re-expressed after 4–24 h in different systems and under different experimental conditions, in an amount similar to that present before modulation (Ault and Unanue, 1973; Cullen et al., 1973; Elson et al., 1973; Rowe et al., 1973; Roelants et al., 1974; Loor et al., 1975). However, in some cases a prolonged or permanent modulation has been obtained. Thus Loor et al. (1975) found that TL antigen on mouse thymocytes was not resynthesised after modulation, whereas it reappeared on modulated thymoma cells. The lack of resynthesis in normal thymocytes is presumably connected with the fact that these cells spontaneously lose the antigen during maturation before leaving the thymus, so that the synthesis of the antigen may be already turned off in the cells undergoing modulation. Similarly, surface Ig removed from foetal mouse B lymphocytes by capping and pinocytosis with bivalent antibody is not resynthesised for several days, whereas the same antigen reappears readily in adult lymphocytes (Raff et al., 1975; Sidman and Unanue, 1975). The delayed reappearance of surface Ig in foetal lymphocytes may play a role in the prenatal induction of B cell tolerance to antigens and in the phenomenon of allotype suppression, i.e. in the prolonged disappearance and functional impairment of cells carrying surface Ig of a particular allotype after neonatal exposure to antibodies against that allotype (Dray, 1962; reviewed in Warner, 1974). Of potential immunological significance is also the observation by Ault et al. (1974) that lymphocytes exposed to a tolerogenic dose of a dinitrophenylated (DNP) D-glutamic acid-lysine copolymer eliminated the complexes very poorly and showed an impaired ability to re-express new receptors (surface Ig), whereas after incubation with the immunogenic DNP-L-copolymer they were able to endocytose the complexes and readily re-express new receptors. Similarly, Nossal and Layton (1976) found that both immunogenic (non-receptor-saturating) and tolerogenic (receptor-saturating), doses of polymerised and derivatised flagellin were largely, though incompletely, cleared from the surface of specific receptor-carrying lymphocytes, but in the case of tolerogenic doses the re-expression of receptors was impaired. However, other studies suggest that the induction of a tolerant state cannot be simply ascribed to inhibition of redistribution and clearance of the complexes and failure to resynthesize new receptors (Diener et al., 1976). Although these phenomena in many instances are correlated with the establishment of a tolerant state, the two can be dissociated by experimental manipulations (Diener et al., 1976). In general, it should be noted that receptor modulation may correspond to only one mechanism of immunological tolerance. Other mechanisms appear to exist which may also depend on interactions between different cells.

7. Concluding remarks

Despite the considerable progress made in recent years, great gaps still exist in our knowledge concerning the distribution and motility of the surface components of lymphocytes (and other cells), and especially of the factors which determine and control their motility. Some of these gaps will be undoubtedly filled in the next few years by studying the membrane itself, i.e. by the analysis of the detailed biochemical and biophysical characteristics of the individual components of the plasma membrane of these (and other) cells, and of their interrelationship. From the discussion in this review it is clear, however, that from the structural and functional point of view the plasma membrane cannot be considered as an isolated structure independent from the rest of the cell. In particular with regard to its mechanical properties, the plasma membrane, through specific membrane components, appears to be strictly connected, physically and functionally, to a complex mechanical system located in the cytoplasm. A satisfactory description and interpretation of the motility phenomena on the plasma membrane of lymphocytes, and in general of animal cells, will not be obtained until at least the general constitution and the physical and functional supramolecular organization of this cytoplasmic system is unravelled. Until that moment many conclusions will necessarily remain speculative.

Acknowledgments

The author is indebted to Drs. E.R. Unanue, G.F. Schreiner and G.L. Nicolson for making available to him the text of their review articles before publication.

References

Abbas, A.L. and Unanue, E.R. (1975) Interrelationship of surface immunoglobulin and Fc receptors on mouse B lymphocytes. J. Immunol. 115, 1665–1671.

Abbas, A.K., Ault, K.A., Karnovsky, M.J. and Unanue, E.R. (1975) Non-random distribution of surface immunoglobulin on murine B lymphocytes. J. Immunol. 114, 1197–1204.

Abbas, A.K., Dorf, M.E., Karnovsky, M.J. and Unanue, E.R. (1976a) The distribution of Ia antigens on the surfaces of lymphocytes. J. Immunol. 116, 371–378.

Abbas, A.K., Unanue, E.R. and Karnovsky, M.J. (1976b) Current techniques for the detection of surface immunoglobulin on lymphocytes. Tech. Biochem. Biophys. Morphol., vol. 3, in press.

Abercrombie, M., Heaysman, J.E.M. and Pegrum, S.M. (1970) The locomotion of fibroblasts in culture, III. Movements of particles on the dorsal surface of the leading lamella. Exp. Cell Res. 62, 389–398.

Abrahamsohn, I., Nilsson, U.R. and Abdou, N.I. (1974) Relationship of immunoglobulin to complement receptors on human B cells. J. Immunol. 112, 1931–1938.

Ahmann, G.B. and Sage, H.J. (1974) Binding of purified lectins to guinea pig lymphocytes. Studies of the number, binding constant, and distribution of *Lens culinaris* lectin A and *Agaricus bisporus* molecules on lymphocyte surfaces. Cell. Immunol. 13, 407–415.

Albertini, D.F. and Clark, J.I. (1975) Membrane–microtubule interactions: concanavalin A capping induced redistribution of cytoplasmic microtubules and colchicine-binding proteins. Proc. Natl. Acad. Sci. USA 72, 4976–4980.

Albini, B. and Wick, G. (1973) Immunoglobulin determinants on the surface of chicken lymphoid cells. Int. Arch. Allergy Appl. Immunol. 44, 804–822.

Alexander, E. and Henkart, P. (1976) The adherence of human Fc receptor-bearing lymphocytes to antigen-antibody complexes, II. Morphological alterations induced by the substrate. J. Exp. Med. 143, 329–347.

Allan, D., Auger, J. and Crumpton, M.J. (1972) Glycoprotein receptors for concanavalin A isolated from pig lymphocyte plasma membrane by affinity chromatography in sodium deoxycholate. Nature (London) New Biol. 236, 23–25.

Allison, A.C. (1973) The role of microfilaments and microtubules in cell movement, endocytosis and exocytosis. In: R. Porter and D.W. Fitzsimmons (Eds.), Locomotion of Tissue Cells (CIBA Symp. No. 14), Excerpta Medica, Amsterdam, pp. 109–143.

Allison, A.C., Davies, P. and de Petris, S. (1971) Role of contractile microfilaments in macrophage movement and endocytosis. Nature (London) New Biol. 232, 153–155.

Amos, D.B., Cohen, I. and Klein Jr., W.R. (1970) Mechanisms of immunological enhancement. Transplant. Proc. 2, 68–75.

Anderson, C.L. and Grey, H.M. (1974) Receptors for aggregated IgG on mouse lymphocytes. Their presence on thymocytes, thymus-derived, and bone marrow-derived lymphocytes. J. Exp. Med. 139, 1175–1188.

Antoine, J.-C. and Avrameas, S. (1974) Surface immunoglobulins of rat immunocytes: quantitation and fate of cell-bound peroxidase-labeled antibody and Fab fragment. Eur. J. Immunol. 4, 468–474.

Antoine, J.-C., Avrameas, S., Gonatas, N.K., Stieber, A. and Gonatas, J.O. (1974) Plasma membrane and internalized immunoglobulins of lymph node cells studied with conjugates of antibody or its Fab fragments with horse-radish peroxidase. J. Cell Biol. 63, 12–23.

Aoki, T. and Johnson, P.A. (1972) Suppression of Gross leukemia cell-surface antigens: a kind of antigenic modulation. J. Natl. Cancer Inst., 49, 183–189.

Aoki, T., Hämmerling, U., de Harven, E., Boyse, E.A. and Old, L.J. (1969) Antigenic structure of cell surfaces. An immunoferritin study of the occurrence and topography of H2, θ, and TL alloantigens on mouse cells. J. Exp. Med. 130, 979–1001.

Ashman, R.F. (1973) Lymphocyte receptor movement induced by sheep erythrocyte binding. J. Immunol. 111, 212–220.

Ashman, R.F. and Raff, M.C. (1973) Direct demonstration of theta-positive antigen-binding cells, with antigen-induced movement of thymus-dependent cell receptors. J. Exp. Med. 137, 69–84.

Astrom, K.E., Webster, H.F. and Arnason, B.G. (1968) The initial lesion in experimental neuritis. A phase and electron microscopic study. J. Exp. Med. 128, 469–495.

Aubin, J.E., Carlsen, S.A. and Ling, V. (1975) Colchicine permeation is required for inhibition of concanavalin A capping in Chinese hamster ovary cells. Proc. Natl. Acad. Sci. USA 72, 4516–4520.

Ault, K.A. and Unanue, E.R. (1974) Events after the binding of antigen to lymphocytes. Removal and regeneration of the antigen receptor. J. Exp. Med. 139, 1110–1124.
Ault, K.A., Karnovsky, M.J. and Unanue, E.R. (1973) Studies on the distribution of surface immunoglobulins on human B-lymphocytes. J. Clin. Invest. 52, 2507–2516.
Ault, K.A., Unanue, E.R., Katz, D.H. and Benacerraf, B. (1974) Failure of lymphocytes to reexpress antigen receptors after brief interaction with a tolerogenic D-amino acid copolymer. Proc. Natl. Acad. Sci. USA 71, 3111–3114.
Ax, W., Malchow, H., Zeiss, I. and Fischer, H. (1968) The behaviour of lymphocytes in the process of target cell destruction in vitro. Exp. Cell Res. 53, 108–116.
Axline, S.G. and Reaven, E.P. (1974) Inhibition of phagocytosis and plasma membrane mobility of the cultivated macrophage by cytochalasin B. Role of subplasmalemmal microfilaments. J. Cell Biol. 62, 647–659.
Bächi, T., Aguet, M. and Howe, C. (1973) Fusion of erythrocytes by Sendai virus studied by immuno-freeze-etching. J. Virol. 11, 1004–1012.
Barat, N. and Avrameas, S. (1973) Surface and intracellular localization of concanavalin A in human lymphocytes. Exp. Cell Res. 76, 451–455.
Barber, B.H. and Crumpton, M.J. (1976) Actin associated with purified lymphocyte plasma membrane. FEBS Lett. 66, 215–220.
Barnett, R.E., Scott, R.E., Furcht, L.T. and Kersey, J.H. (1974) Evidence that mitogenic lectins induce changes in lymphocyte membrane fluidity. Nature (London) 249, 465–466.
Basten, A., Warner, N.L. and Mandel, T. (1972) A receptor for antibody on B lymphocytes, II. Immunochemical and electron microscopy characteristics. J. Exp. Med. 135, 627–642.
Beale, G.H. (1957). The antigen system of *Paramecium aurelia*. Int. Rev. Cytol. 6, 1–23.
Becker, E.L. and Henson, P.M. (1973) In vitro studies of immunologically induced secretion of mediators from cells and related phenomena. Adv. Immunol. 17, 93–193.
Becker, E.L. and Unanue, E.R. (1976) The requirement for esterase activation in the anti-immunoglobulin-triggered movement of B lymphocytes. J. Immunol. 117, 27–32.
Becker, K.E., Ishizaka, T., Metzger, H., Ishizaka, K. and Grimley, P.M. (1973) Surface IgE on human basophils during histamine release. J. Exp. Med. 138, 394–409.
Berlin, R.D., Oliver, J.M., Ukena, T.E. and Yin, H.H. (1974) Control of cell surface topography. Nature (London) 247, 45–46.
Bernoco, D., Cullen, S., Scudeller, G., Trinchieri, G. and Ceppellini, R. (1972) HL-A molecules at the cell surface. In: J. Dausset and J. Colombani (Eds.), Histocompatibility Testing 1972. Munksgaard, Copenhagen, pp. 527–537.
Bessis, M. (1973) Living Blood Cells and Their Ultrastructure. Springer, Berlin, pp. 424–427.
Beug, H., Katz, F.E., Stein, A. and Gerisch, G. (1973) Quantitation of membrane sites in aggregating *Dictyostelium* cells by use of tritiated univalent antibody. Proc. Natl. Acad. Sci. USA 70, 3150–3154.
Biberfeld, P. (1971a) Uropod formation in phytohaemagglutinin (PHA) stimulated lymphocytes. Exp. Cell Res. 66, 433–445.
Biberfeld, P. (1971b) Endocytosis and lysosome formation in blood lymphocytes transformed by phytohemagglutinin. J. Ultrastruct. Res. 37, 41–68.
Bishey, A.N. and Mehta, N.G. (1975) The effect of microtubular inhibitors and cytochalasin B on aggregation of concanavalin A receptors in mouse peritoneal macrophages. In: M. Borges and M. de Brabander (Eds.), Microtubules and Microtubule Inhibitors, North Holland, Amsterdam, pp. 355–366.
Bismuth, A., Neauport-Sautes, C., Kourilsky, F.M., Manuel, Y., Greenland, T. and Silvestre, D. (1974) Distribution and mobility of β_2 microglobulin on the human lymphocyte membrane: immunofluorescence and immunoferritin studies. J. Immunol. 112, 2036–2046.
Borisy, G.G. and Taylor, E.W. (1967) The mechanism of action of colchicine. Binding of colchicine-^{3}H to cellular proteins. J. Cell Biol. 34, 525–534.
Bowers, W.E. (1972) Lysosomes in rat thoracic duct lymphocytes. J. Exp. Med. 136, 1394–1403.
Bowers, W.E. and de Duve, C. (1967) Lysosomes in lymphoid tissue, II. Intracellular distribution of acid hydrolases. J. Cell Biol. 32, 339–348.

Boyse, E.A. and Old, L.J. (1969) Some aspects of normal and abnormal surface genetics. Ann. Rev. Genet. 3, 269–290.

Boyse, E.A., Old, L.J. and Stockert, E. (1968) An approach to the mapping of antigens on the cell surface. Proc. Natl. Acad. Sci. USA 60, 886–893.

Bradley, M.O. (1973) Microfilaments and cytoplasmic streaming: inhibition of streaming with cytochalasin. J. Cell. Sci. 12, 327–343.

Bray, D. (1973) Model for membrane movements in the neural growth cone. Nature (London) 244, 93–96.

Bretscher, M.S. (1976) Directed lipid flow in cell membranes. Nature (London) 260, 21–23.

Brinkley, B.R., Fuller, G.M. and Highfield, D.P. (1975) Cytoplasmic microtubules in normal and transformed cells in culture: analysis by tubulin antibody immunofluorescence. Proc. Natl. Acad. Sci. USA 72, 4981–4985.

Bystryn, J.-C., Siskind, G.W. and Uhr, J. (1973) Binding of antigen by immunocytes. I. Effect of ligand valence on binding affinity of MOPC 315 cells for DNP conjugates. J. Exp. Med. 137, 301–316.

Carter, S.B. (1967) Effects of cytochalasins on mammalian cells. Nature (London) 213, 261–264.

Chang, S., Stockert, E., Boyse, E.A., Hämmerling, U. and Old, L.J. (1971) Spontaneous release of cytotoxic antibody from viable cells sensitized in excess antibody. Immunology 21, 829–838.

Claman, H.N. and Claperon, E.A. (1969) Immunological complementation between thymus and marrow cells. A model for two-cell theory of immunocompetence. Transplant. Rev. 1, 92–113.

Cohen, H.J. and Gilbertsen, B.B. (1975) Human lymphocyte surface immunoglobulin capping. J. Clin. Invest. 55, 84–93.

Cohen, S. (1968) The requirement for the association of two adjacent rabbit γG-antibody molecules in the fixation of complement by immune complexes. J. Immunol. 100, 407–413.

Cohn, Z. (1975) Macrophage physiology. Fed. Proc. 34, 1725–1729.

Comoglio, P.M. and Guglielmone, R. (1972) Two dimensional distribution of concanavalin-A receptor molecules on fibroblasts and lymphocyte plasma membranes. FEBS Lett. 27, 256–258.

Cone, R.E. (1976) Dynamics of the lymphocyte membrane. In: J.J. Marchalonis (Ed.), The Lymphocyte: Structure and Function, M. Dekker, New York (in press).

Cone, R.E., Marchalonis, J.J. and Rolley, R.T. (1971) Lymphocyte membrane dynamics. Metabolic release of cell surface proteins. J. Exp. Med. 134, 1373–1384.

Cone, R.E., Bernoco, D., Ceppellini, R., Dorval, G. and Jacot-Guillarmod, H. (1975) Reversible modifications of cell membrane structures induced by antibodies against transplantation antigens of man. In: V.P. Eijsvogel, D. Roos and W.P. Zeijlemaker (Eds.), Leukocyte-Membrane Determinants Regulating Immune Reactivity (Proc. 10th Leukocyte Conf.), Academic Press, New York, pp. 33–43.

Craig, S.W. and Cuatrecasas, P. (1975) Mobility of cholera toxin receptors on rat lymphocyte membranes. Proc. Natl. Acad. Sci. USA 72, 3844–3848.

Cullen, S.E., Bernoco, D., Carbonara, A.O., Jacot-Guillarmod, H., Trinchieri, G. and Ceppellini, R. (1973) Fate of HL-A antigens and antibodies at the lymphocyte surface. Transplant. Proc. 5, 1835–1847.

Cullen, S.E., David, C.S., Shreffler, D.C. and Nathenson, S.G. (1974) Membrane molecules determined by the H-2 associated immune response region: isolation and some properties. Proc. Natl. Acad. Sci. USA 71, 648–652.

Davis, W.C. (1972) H-2 antigen on cell membranes: an explanation for the alteration of distribution by indirect labeling technique. Science 175, 1006–1008.

de Bruyn, P.P.H. (1946) Ameboid movement of leukocytes. Anat. Rec. 95, 177–191.

de Petris, S. (1974) Inhibition and reversal of capping by cytochalasin B, vinblastine and colchicine. Nature (London) 250, 54–56.

de Petris, S. (1975) Concanavalin A receptors, immunoglobulins and θ antigen on the lymphocyte surface: interactions with concanavalin A and with cytoplasmic structures. J. Cell Biol. 65, 123–146.

de Petris, S. (1977) Immunoelectron microscopy and immunofluorescence in membrane biology. In: E.D. Korn (Ed.) Methods in Membrane Biology (in press).

de Petris, S. and Raff, M.C. (1972) Distribution of immunoglobulin on the surface of mouse lymphoid cells as determined by immunoferritin electron microscopy. Antibody-induced, temperature-dependent redistribution and its implications for membrane structure. Eur. J. Immunol. 2, 523–535.

de Petris, S. and Raff, M.C. (1973a) Normal distribution, patching and capping of lymphocyte surface immunoglobulin studied by electron microscopy. Nature (London) New Biol. 241, 257–259.

de Petris, S. and Raff, M.C. (1973b) Fluidity of the plasma membrane and its implications for cell movement. In: R. Porter and D.W. Fitzsimmons (Eds.), Locomotion of Tissue Cells (CIBA Found. Symp. No. 14). Excerpta Medica, Amsterdam, pp. 27–41.

de Petris, S. and Raff, M.C. (1974) Ultrastructural distribution of alloantigens and concanavalin A receptors on the surface of mouse lymphocytes. Eur. J. Immunol. 4, 130–137.

de Petris, S., Raff, M.C. and Mallucci, L. (1973) Ligand-induced redistribution of concanavalin A receptors on normal, trypsinized and transformed fibroblasts. Nature (London) New Biol. 244, 275–278.

Dickler, H.B. (1974) Studies of the human lymphocyte receptor for heat-aggregated or antigen-complexed immunoglobulin. J. Exp. Med. 140, 508–522.

Dickler, H.B. and Kunkel, H.G. (1972) Interaction of aggregated γ-globulin with B lymphocytes. J. Exp. Med. 136, 191–196.

Diener, E. and Petkau, V.H. (1972) Antigen recognition: early surface-receptor phenomena induced by binding of a tritium-labeled antigen. Proc. Natl. Acad. Sci. USA 69, 2364–2368.

Diener, E., Kraft, N., Lee, K.-C. and Shiozawa, C. (1976) Antigen recognition, IV. Discrimination by antigen-binding immunocompetent B cells between immunity and tolerance is determined by adherent cells. J. Exp. Med. 143, 805–821.

Dodd, N.J.F. (1975) PHA and lymphocyte membrane fluidity. Nature (London) 257, 827–828.

Doyle, J.J., Behin, R., Mauel, J. and Rowe, D.S. (1974) Antibody-induced movement of membrane components of *Leishmania henriettii*. J. Exp. Med. 139, 1061–1069.

Dray, S. (1962) Effect of maternal isoantibodies on the quantitative expression of two allelic genes controlling γ-globulin allotypic specificities. Nature (London) 195, 677–680.

Dunham, E.K., Unanue, E.R. and Benacerraf, B. (1972) Antigen binding and capping by lymphocytes of genetic non-responder mice. J. Exp. Med. 136, 403–408.

Dunham, P.B., Goldstein, I.M. and Weissman, G. (1974) Potassium and amino acid transport in human leukocytes exposed to phagocytic stimuli. J. Cell Biol. 63, 215–226.

Easton, J.M., Goldberg, B. and Green, H. (1962) Demonstration of surface antigens and pinocytosis in mammalian cells with ferritin-antibody conjugates. J. Cell Biol. 12, 437–443.

Edelman, G.M. (1974) Surface alterations and mitogenesis in lymphocytes. In: B. Clarkson and R. Baserga (Eds.), Control of Proliferation in Animal Cells. Cold Spring Harbor Laboratory, New York, pp. 357–377.

Edelman, G.M., Yahara, I. and Wang, J.L. (1973) Receptor mobility and receptor-cytoplasmic interactions in lymphocytes. Proc. Natl. Acad. Sci. USA 70, 1442–1446.

Edelson, P.J. and Cohn, Z.A. (1974) Effects of concanavalin A on mouse peritoneal macrophages, I. Stimulation of endocytic activity and inhibition of phago-lysosome formation. J. Exp. Med. 140, 1387–1403.

Eden, A., Bianco, C., Bogart, B. and Nussenzweig, V. (1973) Interaction and release of soluble immune complexes from mouse B lymphocytes. Cell. Immunol. 7, 474–483.

Edidin, M. and Fambrough, D. (1973) Fluidity of the surface of cultured muscle fibers. Rapid lateral diffusion of marked surface antigens. J. Cell Biol. 57, 27–37.

Edidin, M. and Weiss, A. (1972) Antigen cap formation in cultured fibroblasts: a reflection of membrane fluidity and cell mobility. Proc. Natl. Acad. Sci. USA 69, 2456–2459.

Edidin, M., Zagyansky, Y. and Lardner, T.J. (1976) Measurement of membrane protein lateral diffusion in single cells. Science 191, 466–468.

Elgsaeter, A., Shotton, D.M. and Branton, D. (1976) Intramembrane particle aggregation in erythrocyte ghosts, II. The influence of spectrin on aggregation. Biochim. Biophys. Acta 426, 101–122.

Ellis, A.E. and Parkhouse, R.M.E. (1975) Surface immunoglobulins on the thymocytes of the skate *Raja naevus.* Eur. J. Immunol. 5, 726–728.

Elson, C.J., Singh, J. and Taylor, R.B. (1973) The effect of capping by anti-immunoglobulin antibody on the expression of cell surface immunoglobulin and on lymphocyte activation. Scand. J. Immunol. 2, 143–149.

Engers, H.D. and Unanue, E.R. (1973) The fate of anti-Ig-surface Ig complexes on B lymphocytes. J. Immunol. 110, 465–475.

Estensen, R.D. and Plagemann, P.G.W. (1972) Cytochalasin B: inhibition of glucose and glucosamine transport. Proc. Natl. Acad. Sci. USA 69, 1430–1434.

Estensen, R.D., Hill, H.R., Quie, P.G., Hogan, N. and Goldberg, N.D. (1973) Cyclic GMP and cell movement. Nature (London) 245, 458–460.

Fagraeus, A., Lidman, K. and Biberfeld, G. (1974) Reaction of human smooth muscle antibodies with human blood lymphocytes and lymphoid cell lines. Nature (London) 252, 246–247.

Forni, L. and Pernis, B. (1975) Interactions between Fc receptors and membrane immunoglobulins on B lymphocytes. In: M. Seligmann, J.L. Preud'homme and F.M. Kourilsky (Eds.), Membrane Receptors of Lymphocytes. North-Holland, Amsterdam, pp. 193–201.

Fram, R.J., Sidman, C.L. and Unanue, E.R. (1976) Genetic control of ligand-induced events in B lymphocytes. J. Immunol. 117, 1456–1463.

Froland, S.S. and Natvig, J.B. (1973) Identification of three different lymphocyte populations by surface markers. Transplant. Rev. 16, 114–162.

Froland, S.S., Michaelsen, T.E., Wisloff, F. and Natvig, J.B. (1974) Specificity of receptors for IgG on human lymphocyte-like cells. Scand. J. Immunol. 3, 509–520.

Frye, L.D. and Edidin, M. (1970) The rapid intermixing of cell surface antigens after formation of mouse-human heterokaryons. J. Cell Sci. 7, 319–335.

Gail, M.H., Boone, Ch.W. and Thompson, C.S. (1973) A calcium requirement for fibroblast mobility and proliferation. Exp. Cell. Res. 79, 386–390.

Gallin, J.I. and Rosenthal, A.S. (1974) The regulatory role of divalent cations in human granulocyte chemotaxis. Evidence for an association between calcium exchanges and microtubule assembly. J. Cell Biol. 62, 594–609.

Ghetie, V., Fabricius, H.Å., Nilsson, K. and Sjöquist, J. (1974) Movement of IgG receptors on the lymphocyte surface induced by protein A of *Staphylococcus aureus.* Immunology 26, 1081–1091.

Goding, J.W. and Layton, J.E. (1976) Antigen-induced cocapping of IgM and IgD-like receptors on murine B cells. J. Exp. Med. 144, 852–857.

Goldacre, R.J. (1961) The role of the cell membrane in the locomotion of amoebae, and the source of the motive force and its control by feedback. Exp. Cell Res. 8 (Suppl.), 1–16.

Goldman, R.D., Lazarides, E., Pollack, R. and Weber, K. (1975) The distribution of actin in non-muscle cells. The use of actin antibody in the localization of actin within the microfilament bundles of mouse 3T3 cells. Exp. Cell Res. 90, 333–350.

Goldstein, I.J., Reichert, C.M., Misaki, A. and Gorin, P.A.J. (1973) An "extension" of the carbohydrate binding specificity of concanavalin A. Biochim. Biophys. Acta 317, 500–504.

Gonatas, N.K., Gonatas, J.O., Stieber, A., Antoine, J.C. and Avrameas, S. (1976) Quantitative ultrastructural autoradiographic studies of iodinated plasma membranes of lymphocytes during segregation and internalization of surface immunoglobulins. J. Cell Biol. 70, 477–493.

Gordon, J.A. and Marquardt, M.D. (1974) Factors affecting hemagglutination by concanavalin A and soybean agglutinin. Biochim. Biophys. Acta 332, 136–144.

Gormus, B.J. and Shands Jr., J.W. (1975) Capping of the lymphocyte C3 receptor and temperature-dependent loss of C3 rosettes. J. Immunol. 114, 1221–1225.

Greaves, M., Bauminger, S. and Janossy, G. (1972) Lymphocyte activation, III. Binding sites for phytomitogens on lymphocyte subpopulations. Clin. Exp. Immunol. 10, 537–554.

Greaves, M.F., Owen, J.J.T. and Raff, M.C. (1976) T and B Lymphocytes: Origins, Properties and Roles in Immune Response. 2nd Ed., Excerpta Medica, Amsterdam.

Griffin Jr., F.M. and Silverstein, S.C. (1974) Segmental response of the macrophage plasma membrane to a phagocytic stimulus. J. Exp. Med. 139, 323–336.

Gruenstein, E., Rich, A. and Weihing, R.R. (1975) Actin associated with membranes from 3T3 mouse fibroblast and HeLa cells. J. Cell Biol. 64, 223–234.

Gunther, G.G., Wang, J.L., Yahara, I., Cunningham, B.A. and Edelman, G.M. (1973) Concanavalin A derivatives with altered biological activities. Proc. Natl. Acad. Sci. USA 70, 1012–1016.

Hammer, C.H., Nicholson, A. and Mayer, M.M. (1976) On the mechanism of cytolysis by complement: evidence on insertion of C5b and C7 subunits of the C5b,6,7 complex into phospholipid bilayers of erythrocyte membranes. Proc. Natl. Acad. Sci. USA 72, 5076–5080.

Hammond, E. (1970) Ultrastructural characteristics of surface IgM reactive malignant lymphoid cells. Exp. Cell Res. 59, 359–370.

Hauptfeld, V., Hauptfeld, M. and Klein, J. (1975) Induction of resistance to antibody-mediated cytotoxicity. H-2, Ia, and Ig antigens are independent entities in the membrane of mouse lymphocytes. J. Exp. Med. 141, 1047–1056.

Henson, P. (1971) The immunological release of constituents from neutrophil leukocytes, I. The role of antibody and complement on nonphagocytosable surfaces or phagocytosable particles. J. Immunol. 107, 1535–1546.

Hilgers, J., van Blitterswijk, W., Bont, W.S., Theuns, G.J., Nusse, R., Haverman, J. and Emmelot, P. (1975) Distribution and antibody-induced redistribution of mammary tumor virus-induced and normal antigen on the surface of mouse leukemia cells. J. Natl. Cancer Inst. 54, 1335–1342.

Hinuma, Y., Suzuki, M. and Sairenji, T. (1975) Epstein-Barr virus-induced cap formation in human lymphoblastoid cells. Int. J. Cancer 15, 799–805.

Hirsschorn, R., Kaplan, J.M., Goldberg, A.F., Hirsschorn, K.E. and Weissman, G. (1965) Acid phosphatase-rich granules in human lymphocytes induced by phytohemagglutinin. Science 147, 55–57.

Huet, M. (1975) Factors affecting the molecular structure and the agglutinating ability of concanavalin A and other lectins. Eur. J. Biochem. 59, 627–632.

Hunt, S.M. and Marchalonis, J.J. (1974) Radioiodinated lymphocyte surface glycoproteins: concanavalin A binding proteins include surface immunoglobulin. Biochem. Biophys. Res. Commun. 61, 1227–1233.

Huxley, H. (1973) Muscular contraction and cell motility. Nature (London) 243, 445–449.

Hyslop Jr., N.E., Dourmashkin, R.R., Green, M. and Porter, R.R. (1970) The fixation of complement and the activated first component (CĪ) of complement by complexes formed between antibody and divalent hapten. J. Exp. Med. 131, 783–802.

Inbar, M. and Sachs, L. (1973) Mobility of carbohydrate containing sites on the surface membrane in relation to the control of cell growth. FEBS Lett. 32, 124–128.

Inbar, M., Shinitzky, M. and Sachs, L. (1973a) Rotational relaxation time of concanavalin A bound to the surface membrane of normal and malignant transformed cells. J. Mol. Biol. 81, 245–253.

Inbar, M., Shinitzky, M. and Sachs, L. (1973b) Rotational diffusion of lectins bound to the surface membrane of normal lymphocytes. FEBS Lett. 34, 247–250.

Inbar, M., Huet, Ch., Oseroff, A.R., Ben-Bassat, H. and Sachs, L. (1973c) Inhibition of lectin agglutinability by fixation of the cell surface membrane. Biochim. Biophys. Acta 311, 594–599.

Ishikawa, S., Bishop, R. and Holtzer, H. (1969) Formation of arrow-head complexes with heavy meromyosin in a variety of cell types. J. Cell Biol. 43, 312–328.

Jacob, H., Amsden, T. and White, J. (1972) Membrane microfilaments of erythrocytes: alteration in intact cells reproduces the hereditary spherocytosis syndrome. Proc. Natl. Acad. Sci. USA 69, 471–474.

Jacobson, K., Wu, E. and Poste, G. (1976) Measurement of the translational mobility of concanavalin A in glycerol-saline solutions and on the cell surface by fluorescence recovery after photobleaching. Biochim. Biophys. Acta 433, 215–222.

Jacot-Guillarmod, H., Buzzi, G., Carbonara, A.O., Cone, R. and Ceppellini, R. (1975) Shedding of HL-A antigens and antibodies from lymphocyte surface. In: F. Kissmeyer-Nielson (Ed.), Histocompatibility Testing 1975. Munksgaard, Copenhagen, pp. 753–760.

Jahn, T.L. and Bovee, E.C. (1969) Protoplasmic movements within cells. Physiol. Rev. 49, 793–862.

Jones, G. (1973) Release of surface receptors from lymphocytes. J. Immunol. 110, 1526–1531.

Jones, G., Marcuson, E.C. and Roitt, I.M. (1970) Immunoglobulin allotypic determinants on rabbit lymphocytes. Nature (London) 227, 1051–1053.

Karnovsky, M.J. and Unanue, E.R. (1973) Mapping and migration of lymphocyte surface macromolecules. Fed. Proc. 32, 55–59.

Karnovsky, M.J., Unanue, E.R. and Leventhal, M. (1972) Ligand-induced movement of lymphocyte membrane macromolecules, II. Mapping of surface moieties. J. Exp. Med. 136, 907–930.

Karnovsky, M.J., Leventhal, M. and Unanue, E.R. (1975) Correlated fluorescent and scanning microscopy of immunoglobulin-induced capping and movement of B lymphocytes. J. Cell Biol. 67, 201a.

Kay, M.M.B. (1975) Multiple labelling technique used for kinetic studies of activated human B lymphocytes. Nature (London) 254, 424–426.

Kemp, R.B., Jones, B.M. and Gröschel-Stewart, U. (1973) Abolition by myosin and heavy meromyosin of the inhibitory effect of smooth-muscle actomyosin antibodies on cell aggregation in vitro. J. Cell. Sci. 12, 631–639.

Kiefer, H. (1973) Binding and release of lymphocytes by hapten-derivatized nylon fibers. Eur. J. Immunol. 3, 181–183.

Kirschner, M.W. and Williams, R.C. (1974) The mechanism of microtubule assembly in vitro. J. Supramol. Struct. 2, 412–428.

Kletzien, R.F. and Perdue, J.F. (1973) The inhibition of sugar transport in chick embryo fibroblasts by cytochalasin B. Evidence for a specific membrane effect. J. Biol. Chem. 248, 711–719.

Knapp, W., Bolhuis, R.L.H., Radl, J. and Hijmans, W. (1973) Independent movement of IgD and IgM molecules on the surface of individual lymphocytes. J. Immunol. 111, 1295–1298.

Knopf, P.M., Destree, A. and Hyman, R. (1973) Antibody-induced changes in expression of an immunoglobulin surface antigen. Eur. J. Immunol. 3, 251–259.

Komnick, H., Stockem, W. and Wohlfarth-Bottermann, K.E. (1973) Cell motility: mechanisms in protoplasmic streaming and ameboid movement. Int. Rev. Cytol. 34, 169–249.

Kosower, E.M., Kosower, N.S., Faltin, Z., Diver, A., Saltoun, G. and Frensdorff, A. (1974) Membrane mobility agents. A new class of biologically active molecules. Biochim. Biophys. Acta 363, 261–266.

Kourilsky, F.M., Silvestre, C., Neauport-Sautes, C., Loosfelt, Y. and Dausset, J. (1972) Antibody-induced redistribution of HL-A antigens at the cell surface. Eur. J. Immunol. 2, 249–257.

Krug, U., Hollenberg, M.D. and Cuatrecasas, P. (1973) Changes in the binding of concanavalin A and wheat germ agglutinin to human lymphocytes during in vitro transformation. Biochem. Biophys. Res. Commun. 52, 305–312.

Kurnick, J.T. and Grey, H.M. (1975) Relationship between immunoglobulin-bearing lymphocytes and cells reactive with sensitized human erythrocytes. J. Immunol. 115, 305–307.

Lamm, M.E., Boyse, E.A., Old, L.J., Lisowska-Bernstein, B. and Stockert, E. (1968) Modulation of TL (thymus-leukemia) antigens by Fab-fragments of TL antibody. J. Immunol. 101, 99–103.

Lengerová, A., Pokorná, Z., Viklický, V. and Zelený, V. (1972) Phenotypic suppression of H-2 antigens and topography of the cell surface. Tissue Antigens 2, 332–340.

Lesley, J. and Hyman, R. (1974) Antibody-induced changes in expression of the H-2 antigen. Eur. J. Immunol. 4, 732–739.

Lewis, W.H. (1931) Locomotion of lymphocytes. Bull. John Hopkins Hosp. 49, 29–36.

Lin, S. and Spudich, A. (1974) On the molecular basis of action of cytochalasin B. J. Supramol. Struct. 2, 728–736.

Linthicum, D.S. and Sell, S. (1974a) Surface immunoglobulin of rabbit lymphoid cells, I. Ultrastructural distribution and endocytosis of b4 allotypic determinants on peripheral blood lymphocytes. Cell. Immunol. 12, 443–458.

Linthicum, D.S. and Sell, S. (1974b) Surface immunoglobulin on rabbit lymphoid cells, II. Ultrastructural distribution of b4 allotypic determinants and morphology of spleen lymphocytes. Cell. Immunol. 12, 459–471.

Linthicum, D.S. and Sell, S. (1975) Topography of lymphocyte surface immunoglobulin using scanning immunoelectron microscopy. J. Ultrastruct. Res. 51, 55–68.

Lobo, P.I., Westervelt, F.B. and Horwitz, D.A. (1975) Identification of two populations of immunoglobulin-bearing lymphocytes in man. J. Immunol. 114, 116–119.

Loor, F. (1973a) Lectin-induced lymphocyte agglutination. An active cellular process? Exp. Cell. Res.

82, 415–425.

Loor, F. (1973b) Lymphocyte membrane particle redistribution induced by a mitogenic/capping dose of phytohemagglutinin of *Phaseolus vulgaris*. Eur. J. Immunol. 3, 112–116.

Loor, F. (1974) Binding and redistribution of lectins on lymphocyte membrane. Eur. J. Immunol. 4, 210–220.

Loor, F. and Hägg, L.-B. (1975) The modulation of microprojections on the lymphocyte membrane and the redistribution of membrane-bound ligands, a correlation. Eur. J. Immunol. 5, 854–865.

Loor, F., Forni, L. and Pernis, B. (1972) The dynamic state of the lymphocyte membrane. Factors affecting the distribution and turnover of surface immunoglobulins. Eur. J. Immunol. 2, 203–212.

Loor, F., Block, N. and Little, R.J. (1975) Dynamics of the TL antigens on thymus and leukemia cells. Cell. Immunol. 17, 351–365.

Lustig, S., Pluznik, D.H., Kosower, N.S. and Kosower, E.M. (1975) Membrane mobility agent alters the consequences of lectin-cell interaction in a malignant cell membrane. Biochim. Biophys. Acta 401, 458–467.

Mandel, T., Byrt, P. and Ada, G.L. (1969) A morphological examination of antigen reactive cells from mouse spleen and peritoneal cavity. Exp. Cell. Res. 58, 179–182.

Marchalonis, J.J., Cone, R.E. and Santer, V. (1971) Enzymic iodination. A probe for accessible surface proteins of normal and neoplastic lymphocytes. Biochem. J. 124, 921–927.

Marshall, W.H. and Roberts, K.B. (1965) Continuous cinematography of human lymphocytes cultured with a phytohaemgglutinin including observations on cell division and interphase. Quart. J. Exp. Physiol. 50, 361–374.

Matter, A. and Bonnet, C. (1974) Effect of capping on the distribution of membrane-particles in thymocyte membranes. Eur. J. Immunol. 4, 704–707.

McDonough, J. and Lilien, J. (1975) Spontaneous and lectin-induced redistribution of cell surface receptors on embryonic chick neural retina cells. J. Cell. Sci. 19, 357–368.

McFarland, W. and Schechter, G.P. (1970) The lymphocyte in immunological reactions in vitro: ultrastructural studies. Blood 35, 683–688.

McFarland, W., Heilman, D. and Moorhead, J.F. (1966) Functional anatomy of the lymphocyte in immunological reactions in vitro. J. Exp. Med. 124, 851–858.

McIntyre, J.A., Gilula, N.B. and Karnovsky, M.J. (1974) Cryoprotectant-induced redistribution of intra-membraneous particles in mouse lymphocytes. J. Cell Biol. 60, 192–203.

McNutt, N.S., Culp, L.A. and Black, P.H. (1973) Contact-inhibited revertant cell lines isolated from SV40-transformed cells, IV. Microfilament distribution and cell shape in untransformed, transformed, and revertant Balb/c 3T3 cells. J. Cell Biol. 56, 412–428.

Melcher, U., Eidels, L. and Uhr, J.W. (1975) Are immunoglobulins integral membrane proteins? Nature (London) 258, 434–435.

Menne, H.D. and Flad, H.D. (1973) Membrane dynamics of HL-A-anti-HLA complexes of human normal and leukaemic lymphocytes. Clin. Exp. Immunol. 14, 57–67.

Miller, J.F.A.P. and Mitchell, G.F. (1969) Thymus and antigen-reactive cells. Transplant. Rev. 1, 3–43.

Miranda, A., Godman, G.C. and Tanenbaum, S.W. (1974) Action of cytochalasin D on cells of established lines, II. Cortex and microfilaments. J. Cell Biol. 62, 406–423.

Miyajima, T., Hirata, A.A. and Terasaki, P.I. (1972) Escape from sensitization to HL-A antibodies. Tissue Antigens 2, 64–73.

Mizel, S.B. and Wilson, L. (1972) Inhibition of the transport of several hexoses in mammalian cells by cytochalasin B. J. Biol. Chem. 247, 4102–4105.

Molday, R.S., Dreyer, W.J., Rembaum, A. and Yen, S.P.S. (1975) New immunolatex spheres: visual markers of antigens on lymphocytes for scanning electron microscopy. J. Cell Biol. 64, 75–88.

Möller, G. (1961) Demonstration of mouse isoantigens at the cellular level by the fluorescent antibody technique. J. Exp, Med. 114, 415–434.

Müller-Eberhard, H.J. (1969) Complement. Annu. Rev. Biochem. 38, 389–414.

Neauport-Sautes, C., Lilly, F., Silvestre, D. and Kourilsky, F.M. (1973) Independence of H-2K and H-2D antigenic determinants on the surface of mouse lymphocytes. J. Exp. Med. 137, 511–526.

Neauport-Sautes, C., Bismuth, A., Kourilsky, F.M. and Manuel, Y. (1974) Relationship between HL-A antigens and β2-microglobulin as studied by immunofluorescence on the lymphocyte membrane. J. Exp. Med 139, 957–968.

Nicolson, G.L. (1971) Differences in topology of normal and tumour cell membranes shown by different surface distributions of ferritin-conjugated concanavalin A. Nature (London) New Biol. 233, 244–246.

Nicolson, G.L. (1973) Temperature-dependent mobility of concanavalin A sites on tumour cell surfaces. Nature (London) New Biol. 243, 218–220.

Nicolson, G.L. (1974a) The interactions of lectins with animal cell surfaces. Int. Rev. Cytol. 39, 89–190.

Nicolson, G.L. (1974b) Factors influencing the dynamic display of lectin-binding sites on normal and transformed cell surfaces. In: B. Clarkson and R. Baserga (Eds.), Control of Proliferation in Animal Cells. Cold Spring Harbor Laboratory, New York, pp. 251–270.

Nicolson, G.L. (1976a) Transmembrane control of the receptors on normal and tumor cells, I. Cytoplasmic influence over cell surface components. Biochim. Biophys. Acta 457, 57–108.

Nicolson, G.L. (1976b) Transmembrane control of the receptors on normal and tumor cells, II. Surface changes associated with transformation and malignancy. Biochim. Biophys. Acta 458, 1–72.

Nicolson, G.L. and Poste, G. (1976a) Cell shape changes and transmembrane receptor uncoupling induced by tertiary amine local anesthetics. J. Supramol. Struct. 5, 65–72.

Nicolson, G.L. and Poste, G. (1976b) The cancer cell: Dynamic aspects and modifications in cell-surface organization. [Part 1]. New Engl. J. Med. 295, 197–203; [Part 2]. New Engl. J. Med. 295, 253–258.

Nicolson, G.L., Smith, J.R. and Poste, G. (1976) Effects of local anesthetics on cell morphology and membrane-associated cytoskeletal organization in BALB/3T3 cells. J. Cell Biol. 68, 395–402.

Noonan, K.D. and Burger, M.M. (1973) The relationship of concanavalin A binding to lectin-initiated cell agglutination. J. Cell Biol. 59, 134–142.

Norberg, B. (1971) The formation of an ATP-induced constriction ring in glycerinated lymphocytes during migration. Scand. J. Haematol. 8, 75–80.

Nossal, G.J.V. and Layton, J.E. (1976) Antigen-induced aggregation and modulation of receptors on hapten-specific B lymphocytes. J. Exp. Med. 143, 511–528.

Nussenzweig, V. (1974) Receptors for immune complexes on lymphocytes. Adv. Immunol. 19, 217–258.

Old, L.J., Stockert, E., Boyse, E.A. and Kim, J.H. (1968) Antigenic modulation. Loss of TL antigen from cells exposed to TL antibody. Study of the phenomenon in vitro. J. Exp. Med. 127, 523–539.

Oliver, J.M. (1975) Defects in cyclic GMP generation and microtubule assembly in Chediak-Higashi and malignant cells. In: M. Borgers and M. de Brabander (Eds.), Microtubules and Microtubule Inhibitors. North-Holland, Amsterdam, pp. 341–354.

Oliver, J.M., Ukena, T.E. and Berlin, R.D. (1974) Effects of phagocytosis and colchicine on the distribution of lectin-binding sites on cell surfaces. Proc.Natl. Acad. Sci. USA 71, 394–398.

Oliver, J.M., Zurier, R.B., and Berlin, R.D. (1975) Concanavalin A cap formation on polymorphonuclear leukocytes of normal and beige (Chediak-Higashi) mice. Nature (London) 253, 471–473.

Oliver, J.M., Yin, H.H. and Berlin, R.D. (1976) Control of lateral mobility of membrane proteins. In: V.R. Eijsvogel, D. Roos and W.P. Zeijlemaker (Eds.), Leucocyte Membrane Determinants Regulating Immune Reactivity (Proc. 10th Leucocyte Culture Conference, Amsterdam 1975). Academic Press, New York, pp. 3–14.

Olmsted, J.B. and Borisy, G.G. (1973) Microtubules. Ann. Rev. Biochem. 42, 507–540.

Olmsted, J.B., Marcum, J.M., Johnson, K.A., Allen, C. and Borisy, G.C. (1974) Microtubule assembly: some possible regulatory mechanisms. J. Supramol. Struct. 2, 429–450.

Osterberg, L., Lindblom, J.B. and Peterson, P.A. (1974) Subunit structure of HL-A antigens on cell surface. Nature (London) 249, 463–465.

Osunkoya, B.O., Williams, A.I.O., Adler, W.H. and Smith, R.T. (1970) Studies on the interaction of phytomitogens with lymphoid cells. I. Binding of phytohaemagglutinin to cell membrane receptors of cultured Burkitt's lymphoma and infectious mononucleosis cells. West Afr. J. Med. Sci. 1, 3–16.

Owellen, R.J., Owens Jr., A.H. and Donigian, C. (1972) The binding of vincristine, vinblastine and colchicine to tubulin. Biochem. Biophys. Res. Commun. 47, 685–691.

Packer, J.W., Wakasa, H. and Lukas, R.J. (1965) The morphologic and cytochemical demonstration of lysosomes in lymphocytes incubated with phytohemagglutinin by electron microscopy. Lab. Invest. 14, 1736–1743.

Painter, R.G., Sheetz, M. and Singer, S.J. (1975) Detection and ultrastructural localization of human smooth muscle myosin-like molecules in human non-muscle cells by specific antibodies. Proc. Natl. Acad. Sci. USA 72, 1359–1363.

Papahadjopoulos, D. (1972) Studies on the mechanism of action of local anesthetics with phospholipid model membranes. Biochim. Biophys. Acta 265, 169–186.

Parish, C.R. and Hayward, J.A. (1974) The lymphocyte surface, I. Relation between Fc receptors, C'3 receptors and surface immunoglobulin. Proc. Roy Soc. (Lond.) Ser. B 187, 47–63.

Parr, E.L., and Oei, J.S. (1973a) Paraformaldehyde fixation of mouse cells with preservation of antibody-binding by the H-2 locus. Tissue Antigens 3, 99–107.

Parr, E.L. and Oei, J.S. (1973b) Immobilization of membrane H-2 antigens by paraformaldehyde fixation. J. Cell Biol. 59, 537–542.

Pernis, B., Forni, L. and Amante, L. (1970) Immunoglobulin spots on the surface of rabbit lymphocytes. J. Exp. Med. 132, 1001–1018.

Petit, V.A. and Edidin, M. (1974) Lateral phase separation of lipids in plasma membranes: effect of temperature on the mobility of membrane antigens. Science 184, 1183–1184.

Pierres, M., Frodelizi, D., Neauport-Sautes, C., and Dausset, J. (1975) Third HL-A segregant series: genetic analysis and molecular independence on lymphocyte surface. Tissue Antigens 5, 266–279.

Pinto da Silva, P. and Nicolson, G.L. (1974) Freeze-etch localization of concanavalin A receptors to the membrane intercalated particles of human erythrocyte ghost membranes. Biochim. Biophys. Acta 363, 311–319.

Pinto da Silva, P., Martinez-Palomo, A. and Gonzales-Robles, A. (1975) Membrane structure and surface coat of *Entamoeba histolytica*. Topochemistry and dynamics of the cell surface: cap formation and microexudate. J. Cell Biol. 64, 538–550.

Pollack, R. and Rifkin, D. (1975) Actin-containing cables within anchorage-dependent rat embryo cells are dissociated by plasmin and trypsin. Cell 6, 495–506.

Pollack, R., Osborn, M. and Weber, K. (1975) Patterns of organization of actin and myosin in normal and transformed cultured cells. Proc. Natl. Acad. Sci. USA 72, 994–998

Pollard, T.D. and Korn, E.D. (1973) Electron microscopic identification of actin associated with isolated amoeba plasma membranes. J. Biol. Chem. 248, 448–450.

Pollard, T.D. and Weihing, R.R. (1974) Actin and myosin and cell movement. CRC Critical Rev. Biochem. 2, 1–65.

Pong, S.S., Nuss, D.L. and Koch, G. (1975) Inhibition of initiation of protein synthesis in mammalian tissue culture cells by L-1-tosylamido-2-phenylethyl chloromethylketone. J. Biol. Chem. 250, 240–245.

Porter, K.R., Puck, T.T., Hsie, A.W. and Kelley, D. (1974) An electron microscope study of the effects of dibutyryl cyclic AMP on Chinese hamster ovary cells. Cell 2, 145–162.

Poste, G. and Nicolson, G.L. (1976) Calcium ionophores A23187 and X537A affect cell agglutination by lectins and capping of lymphocyte surface immunoglobulins. Biochim. Biophys. Acta 426, 148–155.

Poste, G., Papahadjopoulos, D. and Nicolson, G.L. (1975a) Local anesthetics affect transmembrane cytoskeletal control of mobility and distribution of cell surface receptors. Proc. Natl. Acad. Sci. USA 72, 4430–4434.

Poste, G., Papahadjopoulos, D., Jacobson, K. and Vail, W.J. (1975b) Effects of local anesthetics on membrane properties. II. Enhancement of the susceptibility of mammalian cells to agglutination by plant lectins. Biochim. Biophys. Acta 394, 520–539.

Poste, G., Papahadjopoulos, D., Jacobson, K. and Vail, W.J. (1975c) Local anesthetics increase susceptibility of untransformed cells to agglutination by concanavalin A. Nature (London) 253,

552–554.

Poulik, M.D., Bernoco, M., Bernoco, D. and Ceppellini, R. (1973) Aggregation of HL-A antigens at the lymphocyte surface induced by antiserum to β2-microglobulin. Science 182, 1352–1355.

Preud'homme, J.L., Neauport-Sautes, C., Piat, S., Silvestre, D. and Kourilsky, F.M. (1972) Independence of HL-A antigens and immunoglobulin determinants on the surface of human lymphoid cells. Eur. J. Immunol. 297–330.

Rabellino, E., Colon, S., Grey, H.M. and Unanue, E.R. (1971) Immunoglobulins on the surface of lymphocytes, I. Distribution and quantitation. J. Exp. Med. 133, 156–167.

Raff, M.C. (1969) Theta isoantigen as a marker of thymus-derived lymphocytes in mice. Nature (London) 224, 378–379.

Raff, M.C. and Petris, S. de (1972) Antibody-antigen reactions at the lymphocyte surface: implications for membrane structure, lymphocyte activation and tolerance induction. In: L.G. Silvestri (Ed.), Cell Interactions. North Holland, Amsterdam, pp. 237–243.

Raff, M.C. and Petris, S. de (1973) Movement of lymphocyte surface antigens and receptors: the fluid nature of the lymphocyte plasma membrane and its immunological significance. Fed. Proc. 32, 48–54.

Raff, M.C., Sternberg, M. and Taylor, R.B.(1970) Immunoglobulin determinants on the surface of mouse lymphoid cells. Nature (London) 225, 553–554.

Raff, M.C., Feldmann, M. and Petris, S. de (1973) Monospecificity of bone marrow-derived lymphocytes. J. Exp. Med. 137, 1024–1030.

Raff, M.C., Petris, S. de and Mallucci, L. (1974) Distribution and mobility of membrane macromolecules: ligand-induced redistribution of concanavalin A receptors and its relationship to cell agglutination. In: B. Clarkson and R. Baserga (Eds.), Control of Proliferation of Animal Cells. Cold Spring Harbor Laboratory, New York, pp. 271–281.

Raff, M.C., Owen, J.J.T., Cooper, M.D., Lawton, A.R.III, Megson, M. and Gathings, W.E. (1975) Differences in susceptibility of mature and immature mouse B lymphocytes to anti-immunoglobulin-induced immunoglobulin suppression in vitro. Possible implications for B-cell tolerance to self. J. Exp. Med. 142, 1052–1064.

Ramasamy, R. and Lawson, Y. (1975) Independent movement of surface immunoglobulin from Fc receptors on lymphocyte membranes. Immunology 28, 301–304.

Reaven, E.P. and Axline, S.G. (1973) Subplasmalemmal microfilaments and microtubules in resting and phagocytizing cultivated macrophages. J. Cell Biol. 59, 12–27.

Révész, T. and Greaves, M. (1975) Ligand-induced redistribution of lymphocyte membrane ganglioside GM1. Nature (London) 257, 103–106.

Roelants, G.E., Forni, L. and Pernis, B. (1973) Blocking and redistribution (capping) of antigen receptors on T and B lymphocytes by anti-immunoglobulin antibody. J. Exp. Med. 137, 1060–1077.

Roelants, G.E., Rydén, A., Hägg, L.-B. and Loor, F. (1974) Active synthesis of immunoglobulin receptors for antigen by T lymphocytes. Nature (London) 247, 106–108.

Romeo, D., Zabucchi, G. and Rossi, F. (1973) Reversible metabolic stimulation of polymorphonuclear leukocytes and macrophages by concanavalin A. Nature (London) New Biol. 243, 111–112.

Rosenblith, J.Z., Ukena, T.E., Yin, H.H., Berlin, R.D. and Karnovsky, M.J. (1973) A comparative evaluation of the distribution of concanavalin-binding sites on the surface of normal, virally-transformed, and protease-treated fibroblasts. Proc. Natl. Acad. Sci. USA 70, 1625–1629.

Rosenfeld, A.C., Zackroff, R.V. and Weisenberg, R.C. (1976) Magnesium stimulation of calcium binding to tubulin and calcium induced depolymerization of microtubules. FEBS Lett. 65, 144–147.

Rosenstreich, D.L., Shevach, E., Green, I. and Rosenthal, A.S. (1972) The uropod-bearing lymphocyte of the guinea pig. Evidence for thymic origin. J. Exp. Med. 135, 1037–1048.

Rosenthal, A.S., Davie, J.M., Rosenstreich, D.L. and Cehrs, K.U. (1973) Antibody-mediated internalization of B lymphocyte surface membrane immunoglobulin. Exp. Cell Res. 81, 317–329.

Rossman, R., Norris, C. and Troll, W. (1974) Inhibition of macromolecular synthesis in *Escherichia coli* by protease inhibitors. Specific reversal by glutathione of the effects of chloromethylketones. J. Biol. Chem. 249, 3412–3417.

Rowe, D.S., Hug, K., Forni, L. and Pernis, B. (1973) Immunoglobulin D as lymphocyte receptor. J.

Exp. Med. 138, 965–972.

Rutishauser, U., Yahara, I. and Edelman, G.M. (1974) Morphology, mobility and surface behaviour of lymphocytes bound to nylon fibers. Proc. Natl. Acad. Sci.USA 71, 1149–1153.

Ryan, G.B., Unanue, E.R. and Karnovsky, M.J. (1974a) Inhibition of surface capping of macromolecules by local anesthetics and tranquillisers. Nature (London) 250, 56–57.

Ryan, G.B., Borysenko, J.Z. and Karnovsky, M.J. (1974b) Factors affecting the redistribution of surface-bound concanavalin A on human polymorphonuclear leukocytes. J. Cell Biol. 62, 351–365.

Sällström, J.F. and Alm, G.V. (1972) Binding of concanavalin A to thymic and bursal chicken lymphoid cells. Exp. Cell. Res. 75, 63–72.

Santer, V. (1974) Ultrastructural radioautographic studies on capping and endocytosis of anti-immunoglobulin by mouse spleen lymphocytes. Aust. J. Exp. Biol. Med. Sci. 52, 241–251.

Sanderson, A.R. and Welsh, K.I. (1974) Properties of histocompatibility (HL-A) determinants, III. The independent precipitation and epitope valence of soluble HL-A determinants. Transplantation 18, 197–205.

Schechter, B., Lis, H., Lotan, R., Novogrodsky, A. and Sharon, N. (1976) The requirement for tetravalency of soybean agglutinin for induction of mitogenic stimulation of lymphocytes. Eur. J. Immunol. 6, 145–149.

Schlesinger, M. and Chaouat, M. (1972) Modulation of the H-2 antigenicity on the surface of murine peritoneal cells. Tissue Antigens 2, 427–435.

Schlesinger, M. and Chaouat, M. (1973) Antibody-induced alteration in the expression of the H-2 antigenicity on murine peritoneal cells: the effect of metabolic inhibitors on antigenic modulation and antigen recovery. Transplant. Proc. 5, 105–110.

Schlesinger, M. and Yron, I. (1969) Antigenic changes in lymph node cells after administration of antiserum to thymus cells. Science 164, 1412–1413.

Schnebli, H.P. and Burger, M.M. (1972) Selective inhibition of growth of transformed cells by protease inhibitors. Proc. Natl. Acad. Sci. USA 69, 3825–3827.

Schrader, J.W., Cunningham, B.A. and Edelman, G.M. (1975) Functional interactions of viral and histocompatibility antigens at tumor cell surfaces. Proc. Natl. Acad. Sci. USA 72, 5066–5070.

Schreiner, G.F. and Unanue, E.R. (1975a) The modulation of spontaneous and anti-Ig-stimulated mobility of lymphocytes by cyclic nucleotides and adrenergic and cholinergic agents. J. Immunol. 114, 802–808.

Schreiner, G.F. and Unanue, E.R. (1975b) Anti-Ig-triggered movement of lymphocytes. Specificity and lack of evidence for directional migration. J. Immunol. 114, 809–814.

Schreiner, G.F. and Unanue, E.R. (1976a) Calcium-sensitive modulation of Ig capping: evidence supporting a cytoplasmic control of ligand-receptor complexes. J. Exp. Med. 143, 15–31.

Schreiner, G.F. and Unanue, E.R. (1976b) The disruption of immunoglobulin caps by local anesthetics. Clin. Immunol. Immunopathol. 6, 264–269.

Schreiner, G.F. and Unanue, E.R. (1976c) Membrane and cytoplasmic changes in B lymphocytes induced by ligand–surface immunoglobulin interaction. Adv. Immunol. 24, 38–165.

Schroeder, T.E. (1970) The contractile ring, I. Fine structure of dividing mammalian (HeLa) cells and the effects of cytochalasin B. Z. Zellforsch. Mikrosk. Anat. 109, 431–449.

Schultz, J.S., Frelinger, J.A., Kim, S.-K. and Shreffler, D.C. (1975) The distribution of Ia antigens of the H-2 complex on lymphnode cells by immunoferritin labelling. Cell. Immunol. 16, 125–134.

Schwartz, B.D., Kask, A.M., Paul, W.E. and Shevach, E.M. (1976) Structural characteristics of the alloantigens determined by the major histocompatibility complex of the guinea pig. J. Exp. Med. 143, 541–558.

Scott, R.E. and Marchesi, V.T. (1972) Structural changes in membranes of transformed lymphocytes demonstrated by freeze-etching. Cell. Immunol. 3, 301–317.

Seeman, P. (1974) The membrane actions of anesthetics and tranquilizers. Pharmacol. Rev. 24, 583–655.

Seeman, P., Chau-Wong, M. and Moyyen, S. (1973) Membrane expansion by vinblastine and strychnine. Nature (London) New Biol. 241, 22.

Shaffer, B.M. (1962) The Acrasina. Adv. Morphogenesis 2, 109–182.

Sheetz, M.P., Painter, R.G. and Singer, S.J. (1976) Biological membranes as bilayer couples, III. Compensatory shape changes induced in membranes. J. Cell Biol. 70, 193–203.

Shibata, N., Tatsumi, N., Tanaka, K., Okamura, Y. and Senda, N. (1975) Leucocyte myosin and its location in the cell. Biochim. Biophys. Acta 400, 222–243.

Shinitzky, M. and Inbar, M. (1974) Differences in microviscosity induced by different cholesterol levels in the surface membrane lipid layer of normal lymphocytes and malignant lymphoma cells. J. Mol. Biol. 85, 603–615.

Sidman, C.L., and Unanue, E.R. (1975) Receptor-mediated inactivation of early B lymphocytes. Nature (London) 257, 149–151.

Singer, S.J. (1971) The molecular organization of biological membranes. In: L.I. Rothfield (Ed.), Structure and Function of Biological Membranes, Academic Press, New York, pp. 145–222.

Singer, S.J. and Nicolson, G.L. (1972) The fluid mosaic model of the structure of cell membranes. Science 175, 720–731.

Sloboda, R.D., Rudolph, S.A., Rosenbaum, J.L. and Greengard, P. (1975) Cyclic AMP-dependent endogenous phosphorylation of a microtubule-associated protein. Proc. Natl. Acad. Sci. USA 72, 177–181.

Smith, C.W. and Goldman, A.S. (1970) Interactions of lymphocytes and macrophages from human colostrum: characteristics of the interacting lymphocytes. J. Reticuloendothel. Soc. 8, 91–104.

Smith, C.W. and Hollers, J.C. (1970) The pattern of binding of fluorescein-labeled concanavalin A to the motile lymphocyte. J. Reticuloendothel. Soc. 8, 458–464.

Solheim, B.G. and Thorsby, E. (1974) β-2-microglobulin. Part of the HL-A molecule in the cell membrane. Tissue Antigens 4, 83–94.

Stackpole, C.W., De Milio, L.T., Hämmerling, U., Jacobson, J.B. and Lardis, M.P. (1974a) Hybrid antibody-induced topographical redistribution of surface immunoglobulins, alloantigens, and concanavalin A receptors on mouse lymphoid cells. Proc. Natl. Acad. Sci. USA 71, 932–936.

Stackpole, C.W., De Milio, L.T., Jacobson, J.B., Hämmerling, U. and Lardis, M.P. (1974b) A comparison of ligand-induced redistribution of surface immunoglobulin, alloantigens, and concanavalin A receptors on mouse lymphoid cells. J. Cell Physiol. 83, 441–448.

Stackpole, C.W., Jacobson, J.B. and Lardis, M.P. (1974c) Two distinct types of capping of surface receptors on mouse lymphoid cells. Nature (London) 248, 232–234.

Stackpole, C.W., Jacobson, J.B. and Lardis, M.P. (1974d) Antigenic modulation in vitro, I. Fate of thymus-leukemia (TL) antigen-antibody complexes following modulation of TL antigenicity from the surface of mouse leukemia cells and thymocytes. J. Exp. Med. 140, 939–953.

Stossel, T.P. and Pollard, T.D. (1973) Myosin in polymorphonuclear leukocytes. J. Biol. Chem. 248, 8288–8294.

Sundqvist, K.G. (1972) Redistribution of surface antigens – a general property of animal cells? Nature (London) New Biol. 239, 147–149.

Sundqvist, K.G. (1973a) Quantitation of cell membrane antigens at the single cell level, I. Microfluorometric analysis of reaction patterns in various test systems, with special reference to the mobility of the cell membrane. Scand. J. Immunol. 2, 479–494.

Sundqvist, K.G. (1973b) Quantitation of cell membrane antigens at the single cell level, II. On the mechanisms of the dynamic interaction between cell surface and bound ligands. Scand. J. Immunol. 2, 495–509.

Takahashi, T. (1971) Possible examples of antigenic modulation affecting H-2 antigens and cell surface immunoglobulins. Transplant. Proc. 3, 1217–1220.

Tao-Wiedmann, T.-W., Loor, F. and Hägg, L.-B. (1975) Development of surface immunoglobulins in the chicken. Immunology 28, 821–830.

Taylor, D.L., Moore, P.L., Condeelis, J.S. and Allen, R.D. (1976) The mechanochemical basis of amoeboid movement, I. Ionic requirements for maintaining viscoelasticity and contractility of amoeba cytoplasm. Exp. Cell Res. 101, 127–133.

Taylor, R.B., Duffus, W.P.H., Raff, M.C. and de Petris, S. (1971) Redistribution and pinocytosis of lymphocyte surface immunoglobulin molecules induced by anti-immunoglobulin antibody. Nature (London) New Biol. 233, 225–229.

Tillack, T.W., Scott, R.E. and Marchesi, V.T. (1972) The structure of erythrocyte membranes studied by freeze-etching, II. Localization of the receptors for phytohemagglutinin and influenza virus to the intramembranous particles. J. Exp. Med. 135, 1209–1227.

Trinchieri, G., Baumann, P., de Marchi, M. and Tökés, Z. (1975) Antibody-dependent cell-mediated cytotoxicity in humans, I. Characterization of the effector cell. J. Immunol. 115, 249–255.

Ukena, T.E. and Berlin, R.D. (1972) Effect of colchicine and vinblastine on the topographical separation of membrane functions. J. Exp. Med. 136, 1–7.

Ukena, T.E., Borysenko, J.Z., Karnovsky, M.J. and Berlin, R.D. (1974) Effects of colchicine, cytochalasin B, and deoxyglucose on the topographical organization of surface-bound concanavalin A in normal and transformed fibroblasts. J. Cell Biol. 61, 70–82.

Unanue, E.R. and Karnovsky, M.J. (1973) Redistribution and fate of Ig complexes on the surface of B lymphocytes: functional implications and mechanism. Transplant. Rev. 14, 184–210.

Unanue, E.R. and Karnovsky, M.J. (1974) Ligand-induced movement of lymphocyte membrane macromolecules, V. Capping, cell movement, and microtubular function in normal and lectin-treated lymphocytes. J. Exp. Med. 140, 1207–1220.

Unanue, E.R., Perkins, W.D. and Karnovsky, M.J. (1972) Ligand-induced movement of lymphocyte membrane macromolecules, I. Analysis by immunofluorescence and ultrastructural radioautography. J. Exp. Med. 136, 885–906.

Unanue, E.R., Karnovsky, M.J. and Engers, H.D. (1973) Ligand-induced movement of lymphocyte membrane macromolecules, III. Relationship between the formation and fate of anti-Ig-surface Ig complexes and cell metabolism. J. Exp. Med. 137, 675–689.

Unanue, E.R., Ault, K.A. and Karnovsky, M.J. (1974a) Ligand-induced movement of lymphocyte membrane macromolecules, IV. Stimulation of cell movement by anti-Ig and lack of relationship to capping. J. Exp. Med. 139, 295–312.

Unanue, E.R., Dorf, M.E., David, C.S. and Benacerraf, B. (1974b) The presence of I-region-associated antigens on B cells in molecules distinct from immunoglobulin and H-2K and H-2D Proc. Natl. Acad. Sci. USA 71, 5014–5016.

Vaheri, A., Ruoslahti, E., Westermark, B. and Pontén, J. (1976) A common cell-type specific surface antigen in cultured human glial cells and fibroblasts: loss in malignant cells. J. Exp. Med. 143, 64–72.

van Boxel, J.A., Paul, W.E., Terry, W.D. and Green, I. (1972) IgD-bearing human lymphocytes. J. Immunol. 109, 648–651.

Vitetta, E.S. and Uhr, J.W. (1972) Cell surface immunoglobulin, V. Release from murine splenic lymphocytes. J. Exp. Med. 136, 676–696.

Vitetta, E.S. and Uhr, J.W. (1975) Immunoglobulins and alloantigens on the surface of lymphoid cells. Biochim. Biophys. Acta 415, 253–271.

Vitetta, E.S., Poulik, M.D., Klein, J. and Uhr, J.W. (1976) Beta-2-microglobulin is selectively associated with H-2 and TL alloantigens on murine lymphoid cells. J. Exp. Med. 144, 179–192.

Wagner, R., Rosenberg, M. and Estensen, R. (1971) Endocytosis in Chang liver cells. Quantitation by sucrose-^{3}H uptake and inhibition by cytochalasin B. J. Cell Biol. 50, 804–817.

Warner, N.L. (1974) Membrane immunoglobulins and antigen receptors on B and T lymphocytes. Adv. Immunol. 19, 67–216.

Wartiovaara, J., Linder, E., Ruosolahti, E. and Vaheri, A. (1974) Distribution of fibroblast surface antigens. Association with fibrillar structures of normal cells and loss upon viral transformation. J. Exp. Med. 140, 1522–1533.

Weber, K., Pollack, R. and Bibring, T. (1975) Antibody against tubulin: the specific visualization of cytoplasmic microtubules in tissue culture cells. Proc. Natl. Acad. Sci. USA 72, 459–463.

Weisenberg, R.C. (1972) Microtubule formation in vitro in solutions containing low calcium concentrations. Science 177, 1104–1105.

Weisenberg, R.C. and Timasheff, S.N. (1970) Aggregation of microtubule subunit protein. Effects of divalent cations, colchicine and vinblastine. Biochemistry 9, 4110–4116.

Wessells, N.K., Spooner, B.S., Ash, J.F., Bradley, M.O., Ludueña, M.A., Taylor, E.L., Wrenn, J.T.

and Yamada, K.M. (1971) Microfilaments in cellular and developmental processes. Science 171, 135–143.

Wessels, N.K., Spooner, B.S. and Ludueña, M.A. (1973) Surface movements, microfilaments and cell locomotion. In: R. Porter and D.W. Fitzsimmons (Eds.), Locomotion of Tissue Cells (CIBA Found. Symp. No. 14), Excerpta Medica, Amsterdam, pp. 53–77.

Williamson, R.E. (1972) A light microscope study of the action of cytochalasin B on the cells and isolated cytoplasm of the *Characeae*. J. Cell. Sci. 10, 811–819.

Wills, E.J., Davies, P., Allison, A.C. and Haswell, A.D. (1972) Cytochalasin B fails to inhibit pinocytosis by macrophages. Nature (London) New Biol. 240, 58–60.

Wilson, L. (1970) Properties of colchicine binding protein from chick embryo brain. Interactions with Vinca alkaloids and podophyllotoxin. Biochemistry 9, 4999–5007.

Wilson, L. and Bryan, J. (1974) Biochemical and pharmacological properties of microtubules. Adv. Cell Mol. Biol. 3, 21–72.

Wilson, J.D., Nossal, G.J.V. and Lewis, H. (1972) Metabolic characteristics of lymphocyte surface immunoglobulins. Eur. J. Immunol. 2, 225–232.

Wofsy, L. (1974) Probing lymphocyte membrane organization. In: E.E. Sercarz, A.R. Williamson and C.F. Fox (Eds.), The Immune System: Genes, Receptors, Signals. Academic Press, New York, pp. 259–269.

Wolpert, L. and Gingell, D. (1968) Cell surface membrane and amoeboid movement. In: P.L. Miller (Ed.), Aspects of Cell Mobility (Symp. Soc. Exp. Biol. 22), Cambridge University Press, London, pp. 168–198.

Yahara, I. and Edelman, G.M. (1972) Restriction of the mobility of lymphocyte immunoglobulin receptors by concanavalin A. Proc. Natl. Acad. Sci. USA 69, 608–612.

Yahara, I. and Edelman, G.M. (1973a) Modulation of lymphocyte receptor redistribution by concanavalin A, anti-mitotic agents and alterations of pH. Nature (London) 246, 152–155.

Yahara, I. and Edelman, G.M. (1973b) The effects of concanavalin A on the mobility of lymphocyte surface receptors. Exp. Cell Res. 81, 143–155.

Yahara, I. and Edelman, G.M. (1975a) Electron microscopic analysis of the modulation of lymphocyte receptor mobility. Exp. Cell Res. 91, 125–142.

Yahara, I. and Edelman, G.M. (1975b) Modulation of lymphocyte receptor mobility by locally bound concanavalin A. Proc. Natl. Acad. Sci. USA 72, 1579–1583.

Yefenov, E. and Klein, G. (1974) Antibody-induced redistribution of normal and tumor associated surface antigens. Exp. Cell Res. 88, 217–224.

Yefenov, E., Witz, I.P. and Klein, E. (1975) The fate of IgM and of anti-IgM on antibody-coated Daudi cells. In: V.P. Eijsvogel, D. Roos and W.P. Zeijlemaker (Eds.), Leukocyte Membrane Determinants Regulating Immune Reactivity (Proc. 10th Leukocyte Culture Conf.). Academic Press, New York, p. 109.

Yu, A. and Cohen, E.P. (1974) Studies on the effect of specific antisera on the metabolism of cellular antigens, II. The synthesis and degradation of TL antigens of mouse cells in the presence of TL antiserum. J. Immunol. 112, 1296–1307.

Yu, D.T.Y. (1974) Human lymphocyte receptor movement induced by sheep erythrocyte binding: effect of temperature and neuraminidase treatment. Cell. Immunol. 14, 313–320.

Subject Index

N-Acetylneuraminic acid (see Sialic acid)
Actin (see also α-Actinin, Lymphocyte, Myosin),
- membrane-associated, 40, 44–48, 96, 101, 668, 683–685
- microfilaments, 40, 44–48, 629, 668, 683–685, 689–692, 692–698, 700–702

α-Actinin,
- membrane-associated, 42, 43–44
- microvilli, 49–50

Actomyosin (see Myosin)
Adenyl cyclase (see Cyclic nucleotides)
Adhesion, cell,
- glycosyltransferase, 407, 410
- LETS (fibronectin), 404, 410
- metastasis, 407
- slime mold, 408

Agglutination, cell (see Anesthetics, local; Calcium; Fibroblast; Fibronectin; Lectins; Lipid, alteration; Microvilli; Mobility, lateral)
Aggregation, cell (see Adhesion, cell)
Alkaloids (see Colchicine; Microtubules)
Alloantigens (see Antibody, cell surface binding; Antigen, cell surface; Antigen mobility; Antigen, tumor cell; Endocytosis; Somatic analysis; Antigen, cell surface)
Anchorage (see Microtubules)
Anesthetics, local (see also Filaments, cytoplasmic; Membrane-associated components; Trans-membrane control),
- agglutination, cell, 44–48
- calcium, 47–48
- capping, 45–48, 630, 657, 689–692, 694
- cytoskeletal disruption, 46–48, 630
- fluidity, 45

Anionic sites,
- erythrocytes, 28
- sperm 30

Antibody accessibility (see Neuraminidase; Protease; Trypsin)
Antibody binding (see Antibody, cell surface binding)
Antibody, cell surface binding (see also Antigen mobility; Antigenic modulation; Complement; Cytolysis, complement-mediated; Shedding),
- accessibility, 552–554, 563–571
- antibody flexibility, 558
- capping, 29–31, 44–51, 560–561, 573, 579, 585–590, 645–704
- cell surface, 551–590
- cell surface structures, 552–556
- complement, 552, 567–568
- crosslinking, 558–571, 575–576, 580–581, 653–658, 674, 678
- detachment, 569–574, 580–581
- endocytosis, 571–574, 575–578, 704–707
- enhancement, 583
- Fab, 558, 581, 648–651, 653, 657, 665, 707
- humoral response, 601–602, 606–614
- immunoglobulin, surface, 12, 30, 44–48, 553, 555–556, 574–581, 620–636, 645–704
- induced redistribution, 12, 30, 44–48, 553, 555–556, 559–564, 574–581, 626–631, 645–704
- lattice formation, 474, 558–564, 566–571, 575, 674
- metabolism, 551, 564, 572–574, 575–576
- modulation, 710–712
- prozone effects, 563–571, 580, 583, 656, 666
- shedding, 423–460, 569–574, 709
  turnover, 571–574
- valency, 474, 557–558, 568, 581, 657

Antibody secretion (see Immune response)
Antigen accessibility (see Protease; Trypsin)
Antigen-antibody complex (see Complement; Cytolysis, complement-mediated)
Antigen, cell surface (see also Antibody, cell surface binding; Antigen mobility; Antigen, tumor cell; Antigenic modulation; B-cell; Cell cycle; Lymphocyte; Microvilli; Mobility, lateral; Shedding; Sialic acid; Somatic analysis, cell surface antigen; T-cell),
- ABO, 90, 153, 336, 480–481, 486, 534
- accessibility, 488, 552–554, 563–571
- antibody binding (see Antibody, cell surface binding)
- capping, 29–31, 44–51, 560–561, 573–576, 579,

Antigen, cell surface, (*continued*)
628–631, 640–704
- cell cycle, 353–354, 393–394, 473–493
- density, 554–555, 561–562, 563
- endocytosis, 571–574, 575–578
- Forssman, 492, 497, 499
- glycolipid, 307, 337–338, 523
- H-2, 11, 27, 31, 382, 402, 429–431, 473–477, 485–487, 494–500, 514, 517–520, 524–526, 530–532, 534–537, 539–540, 552, 570, 584–586, 645–647, 666, 668, 670–672, 689, 694
- H-2 structure, 429–431, 517–519
- hemadsorption, 563, 572
- HL-A, 11, 27, 382, 477, 484, 485–487, 525–526, 554, 561, 584, 645, 666, 668, 670–672
- Ia, 497, 514, 645, 651, 670, 694
- immunoglobulin, surface, 12, 30, 44–48, 555–556, 560–564, 573, 577–579, 619–636, 645–704
- MN, 90, 374–376
- mobility, 11–13, 27–29, 44–48, 555–556, 559, 564, 574–587, 624, 628–631, 645–709
- modulation (see Antigenic modulation), 477, 520, 527, 569–574, 704–707, 710–712
- protease, 426, 434–436, 492–493, 498–499, 552–554
- $Rh_0(D)$, 555
- SF, 30
- shedding, 432–460, 484, 569–574, 590, 665, 709, 711
- T-antigen, 498
- Thy-1(θ), 29, 338, 477, 497, 514, 522–523, 526, 532–534, 537–538, 645, 650–651, 656, 665, 669–670, 672, 673, 684, 693–694, 696
- TL, 29, 31, 415, 440–441, 514, 519–522, 526, 531–532, 535–537, 645, 650–651, 665, 666–670, 689
- TL structure, 520–521
- topographic distribution, 484, 553, 560–564
- tumor-associated, 401, 424–460, 473–480, 484, 494, 495–498, 538–540
- valency, 554, 557–558, 654, 656
- variation, 513–540
- virus, 375, 401, 473–493, 496, 497, 498, 561, 565, 575
Antigen loss (see Somatic analysis, cell surface antigen)
Antigen mobility (see also Antigenic modulation; Lymphocyte; Tumor cell),
- ABO, 651
- β_2-microglobulin, 651, 668, 694
- Fc receptor, 627, 670, 671
- H-2, 11, 27, 31, 561, 645–647, 666, 668, 670–674, 689, 694
- HL-A, 11, 27, 554, 561, 666, 668, 670–672

Antigen mobility, (*continued*)
- Ia, 651, 670
- immunoglobulin, surface, 12, 30, 44–48, 555–556, 559–564, 574–581, 688, 601–604, 609–614, 619–636, 643–709
- modulation, 440, 477, 520, 569–574, 704–707, 710–713
- Thy-1(θ), 29, 645–647, 650–652, 656, 665–666, 674, 689, 693–694, 696
- TL, 29, 31, 520, 645, 650–651, 665–666, 667, 689
- tumor-associated antigens (see Antigen, tumor cell)
- virus antigen, 575, 651
Antigen rearrangement (see Somatic analysis, cell surface antigen)
Antigen stimulation (see Immune response; Mitogen)
Antigen suppression (see Somatic analysis, cell surface antigen)
Antigen, tumor cell (see also Tumor cell),
- Alloantigens, 402, 428–429, 430–431, 440, 441, 473–477, 485–500, 513, 540
- blocking factors, 436–437, 440, 447–453, 454–457, 500
- capping (see Capping)
- carcinoembryonic antigen (CEA), 428, 438, 442, 443, 454
- cell cycle, 353–354, 434, 473–493, 500–501
- cell-mediated immunity, 442–453, 457–560
- cholesterol, 431–432
- complement sensitivity, 475–477, 483–484, 495–497, 498
- degradation, 432, 434–435
- delayed hypersensitivity, 425, 426, 428
- derepression, 539
- detection, 435–438, 443, 444, 445, 446, 454
- embryonic antigens, 428, 438, 440–444
- expression, 473–501
- α-fetoprotein (AFP), 428, 443
- Forssman, 499
- glycolipid, 307, 337–338
- H-2 association, 429–431, 437
- immune complexes, 436–439, 442–454, 455
- immune lysis, 436, 440–442, 474, 475
- immunoselection, 524–540
- inhibitory factors, 451–455
- isolation, 424–431, 446, 457
- macrophage migration inhibitor, 425, 446, 447
- metastasis, 436, 441, 457
- β_2-microglobulin, 431
- modulation, 440–442, 527, 704–707, 710–713
- neuraminidase, 401, 404
- papain, 426–428, 429, 431–432, 446, 449
- proteolytic enzymes, cell surface, 435

Antigen, tumor cell, (continued)
- quantity, relative, 475, 476, 483, 498
- rearrangement, 431, 434, 527
- serum factors, 451–460
- shedding, 423, 432–439, 440–451, 451–453
- synthesis, 432, 435
- T antigen, 498
- trypsin, 492–493, 498, 499, 552–554
- tumor-associated, 401, 424–460, 473–480, 484, 495–498, 538–540
- turnover, 432–436, 439
- variants, 524–540
- virus antigens, 401, 473–477, 486, 487, 496, 497, 498, 538

Antigen unmasking (see Neuraminidase; Protease; Trypsin)
Antigenic expression (see Protease; Trypsin)
Antigenic modulation (see also Antigen mobility; Antigen, tumor cell),
- capping, 440
- endocytosis, 704–707, 711–712
- histocompatibility antigens, 520, 710
- immunoglobulin, surface, 440–441, 710–712
- recovery, 712–713
- redistribution, 477, 710–711
- shedding, receptor, 569–574, 590, 711
- tumor antigen, 439, 440

Antigenic selection (see Somatic analysis, cell surface antigen)
Arrhenius plots (see also Enzyme, kinetics; Lipid, alteration),
- enzyme activity, 231–236, 238–239, 240–256, 259–261, 266–271

Asymmetry, membrane (see Cholesterol; Exchange, lipid; Lipids, membrane; Phospholipids)
ATPase (see Calcium; Lipid, alteration; Proteins, plasma membrane)
Autoradiography,
- electron microscopy, 12, 117, 130
- immuno-, 12

B-cell (bone marrow-derived) (see also Immune response; Immunoglobulin, cell surface; Lymphocyte; Mitogen),
- antigen receptor, 602, 603, 634
- capping, 29–31, 44–48, 575–578, 588, 628–631, 645–709
- immunoglobulin secretion, 606–610, 626, 629, 634–635
- immunoglobulin, surface, 44–51, 560, 578–579, 588, 601–604, 609–614, 619–636
- immunoglobulin synthesis, 607–609, 632–635
- lipids, 654

B-cell, (continued)
- maturation, 513–515, 608–614, 634–636
- mitogenesis, 604–607, 610–614, 713
- secretion, 606–610, 632–635
- secretion, immunoglobulin, 605–610, 626, 629, 634–635
- T-cell cooperation, 601–603
- tolerance, 602, 609, 713

Biosynthesis, glycolipid (see Glycolipids)
Biosynthesis, proteins (see Protein synthesis)
Blocking factors (see Antigen, tumor cell)
Blood group substances (see Antigen, cell surface; Antigen mobility)
Brain lipids (see Exchange, lipid)

C1q (see Complement)
Calcium (see also Anesthetics, local; Membrane, plasma; Mobility, lateral; Phase transition),
- agglutination, 47
- ATPase, 248–249
- capping, 44–48, 629–630, 689–692, 698
- fusion, 19
- induced aggregation, 19, 628–630
- ionophores, 47–48, 691
- mobility, lateral, 18, 19, 44–51, 218–224, 267, 628–630, 691–692
- phospholipid binding, 18, 19, 217, 219–224, 267

Calcium ionophore (see Capping, receptor)
Calorimetry (see Differential scanning calorimetry; Liposome)
Cancer, antigens (see Antigens, tumor cell)
- cell surface glycoproteins (see Glycoproteins, transformed cells)

Capping, receptor (see also Anesthetics, local; Antibody, cell surface binding; Antigen, cell surface; Antigen, tumor cell; Antigenic modulation; B-cell; Calcium; Cytochalasins; Cytoplasmic streaming; Concanavalin A; Immune response; Immunoglobulin, cell surface; Metabolism, cell; Microfilaments; Microtubules; Myosin; T-cell; Wheat germ agglutinin),
- amoebae, 659, 703
- anesthetics, local, 45–48, 629, 630, 657, 689–692, 694
- antibody-complement cytolysis, 586–589
- antigen, 29–31, 44–51, 560–561, 573–576, 579, 628–631, 640–704
- calcium effects, 44–48, 629–630, 689–692, 698
- calcium ionophore, 44–48, 629–630, 689–692
- cell shape, 651–653, 675–677, 679–680, 687
- co-capping, 579, 631, 669–672, 694
- colchicine, 44–48, 579, 680, 690, 692–697

Capping, receptor, (*continued*)
- concanavalin A modulation, 45, 579, 630–631, 694–695
- crosslinking, dependency, 26–31, 558–571, 580–581, 653–658, 664–667, 678–679, 683, 690
- cytochalasins, 43–48, 578–579, 629–630, 679, 689–692, 699
- fibroblast, 576, 659
- freeze fracture particles, 672, 675
- immunoglobulin, 44–51, 560, 578–579, 588, 628–631, 645–704
- lectin, 28–31, 44–51, 579–564, 608, 629–631, 640–704
- lymphocyte, 29–31, 44–51, 560, 575–578, 588, 628–631, 640–704
- mechanisms, 44–48, 697–702
- metabolism, 576–578, 629–630, 657–658, 660, 663–664, 690
- motility, cell, 629–631, 675–683, 690
- pH, 658–664
- polymorphonuclear leukocyte, 44, 659, 678, 680, 695, 697
- rate, 667–669
- "reverse" caps, 674, 682–683
- reversibility, 44, 45, 629–630, 667, 676, 689–692
- significance, 528, 702–704
- surface attachment, 676–683, 696–697
- temperature-dependency, 575, 577, 659–669
- toxin receptor, 342
- transplantation antigens, 629, 655, 679, 689
- tumor-associated antigens (see Antigen, tumor cell)
- ultrastructure, 652–653, 660. 669
- virus antigens, 431, 575, 588

Carbohydrate (see Neuraminidase; Sialic acid)
Carcinoembryonic antigen (see Antigen, tumor cell)
Cell contact (see Adhesion, cell)
Cell cycle (see also Antigen, cell surface; Antigen, tumor cell; Cytolysis, complement-mediated; Fibronectin; Fluidity; Growth, cell; Turnover, membrane),
- antigens, 353–354, 434, 473–493, 500–501
- chondroitin sulfate, 491
- collagen, 491
- glycolipid, 326–329, 333–335, 350, 409, 485, 492–493
- glycoproteins, 393–394, 485–488
- hormone receptors, 489–490
- melanin, 490–491
- mitogenesis, 609, 613–614
- myosin, 491

Cell fusion (see also Endycytosis; Somatic analysis, cell surface antigen)
Cell growth (see Growth, cell)
Cell hybridization (see Somatic analysis, cell surface antigen)
Cell-mediated immunity (see Antigen, tumor cell; Immune response; T-cell)
Cell morphology (see Cyclic nucleotides)
Cell motility (see also concanavalin A; Microfilaments),
- anti-Ig-induced, 675–677
- during capping, 629–631, 675–683, 690
- fiber immobilization, 677
- lectin-induced, 675, 676, 677
- microfilaments, 576, 629, 678–683, 685
- tumor cell, 351

Cell movement (see Cell motility; Concanavalin A; Lymphocyte; Microfilaments; Myosin)
Cell organelle (see Golgi; Mitochondria)
Cell shape (see Capping, receptor; Concanavalin A; Lymphocyte; Myosin)
Cell surface charge (see Anionic sites)
Cell surface labeling (see Fibronectin; Galactose oxidase)
Cholera toxin,
- adenyl cyclase, 339–345
- binding specificity, 338–341
- cell actions, 339–345
- cell growth, 339
- glycolipids, 331, 338–345, 373
- hormones, 341
- NAD, 343–345
- receptor, cell surface, 331, 338–345, 573
- structure, 341–342
- temperature-dependency, 353

Cholesterol (see also Antigen, tumor cell; Differential scanning calorimetry; Fluidity, membrane; Lipid, alteration; Lipids, membrane; Liposome; Mobility, lateral; Phase transition; Phospholipids),
- asymmetry, membrane, 19, 269
- enzyme activity, 254–255, 262
- fluidity, membrane, 15–16, 218–224, 254–255, 267
- lipid packing, 219–224, 259, 269
- membrane alteration, 218–219, 296–299
- mobility, receptor, 431–432
- molecular spacing, 16
- phase transition, 25, 26, 217–219, 259, 262–263

Circular dichroism (CD) (see also Spectroscopy),
- membrane proteins, 9

Co-capping (see Capping, receptor)
Colchicine (see also Immunoglobulin, cell surface; Microtubules),
- capping, 44–48, 579, 680, 690, 692–697

Colchicine, (continued)
- cellular effects, 40, 44–48, 490, 680, 692–697

Complement (see also Antibody, cell surface binding; Antigen, tumor cell; Cytolysis, complement-mediated; Mobility, lateral),
- C1q binding, 567, 568, 582–584, 586–589
- C1q-induced redistribution, 576, 582, 589–590
- cell surface binding, 552, 567, 582–585
- cytolysis, 424, 430, 475–477, 483, 484, 495, 582–589, 711–712
- immune-complex release, 590
- immunoglobulin subclass, 583
- modulation, 710–712
- susceptibility, 585–589

Concanavalin A (see also Endocytosis; Immunoglobulin, cell surface; Lymphocyte; Mitogen),
- agglutination, 44–46, 402, 553–658
- capping, receptor, 28–31, 44–48, 629–630, 664–668, 672, 675, 675–704
- cell movement, 675–678, 679–683, 700
- cell shape, 679–683, 700
- endocytosis, 697, 707
- erythrocyte, 28, 39, 90–91, 377
- fibroblast, 45–47, 91
- fluorescent-labeled, 26–31, 44–48, 579, 650, 655–658, 664–671, 673
- lymphocyte, 12, 26–31, 44–48, 579, 605, 652, 655–658, 659, 664–666, 671–704
- mitogenesis, 605–606, 610
- receptor isolation, 381–382, 404
- receptor mobility, 11, 12, 26–31, 44–48, 579, 650, 652, 655–658, 664–666
- inhibition of, 44–45, 579, 630–631, 656, 658, 694–695
- receptor redistribution, 12, 26–31, 44–48, 91, 579, 650, 652, 655–658, 664–666, 671–675, 676–705
- slime mold, 352
- succinyl derivative, 44, 666

Crosslinking (see also Antibody, cell surface binding; Capping, receptor; Membrane, plasma),
- erythrocyte, 39
- lattice formation, 558–564, 566–571, 575
- lymphocyte receptors, 44–45

Cyclic AMP (see Cyclic nucleotides)
Cyclic GMP (see Cyclic nucleotides)
Cyclic nucleotides (see also Microfilaments),
- adenyl cyclase, 339–345, 346–347
- capping, receptor, 43
- cell morphology, 679
- cyclic AMP, 322, 326–327, 339–345, 346–347, 349–351, 400, 426, 613, 679

Cyclic nucleotides, (continued)
- cyclic GMP, 400, 679
- cytoskeletal organization, 43
- glycoproteins, cell surface, 400, 426
- mitogenesis, 400, 613
- phosphorylation, membrane proteins, 426–427
- transformation, 317, 326–327, 400
- transport, 689

Cytochalasins (see also Capping, receptor; Cytoplasmic streaming; Endocytosis; Microfilaments),
- capping, 43–48, 578–579, 629–630, 679, 689–692, 699
- microfilaments, 40, 43, 44–48, 578–579, 689–692

Cytolysis, complement-mediated (see also Capping, receptor; Complement),
- cell cycle, 475–477, 585
- complement, 424, 430, 475–477, 483, 484, 495, 582–589, 711–712
- temperature-dependency, 585–586

Cytoplasmic streaming,
- cytochalasin B, 700
- during capping, 678–679, 681–682, 700

Cytoskeleton (see Actin; α-Actinin; Anesthetics, local; Cyclic nucleotides; Membrane-associated components; Microfilaments; Microtubules; Myosin)

Cytotoxic reactions (see Complement; Cytolysis, complement-mediated)

Degradation, membrane (see Antigen, tumor cell; Membrane, plasma; Turnover, membrane)

Delayed hypersensitivity (see Antigen, tumor cell)

Desmosome (or *Macula adherens*),
- electron microscopy, 101
- freeze fracture, 101–103

Detergent,
- Brij-98, 428
- deoxycholate, 169–170, 250, 257
- isolation, membrane proteins, 380–384
- Lubrol WX, 264, 342
- NP40, 381
- SDS, 381–384
- Triton X-100, 224, 247–248, 381

Diet (see Lipid, alteration)

Differential scanning calorimetry,
- cells, 225
- cholesterol, 26, 218–219
- lipids, 8, 21–22, 26, 211–212, 262

Differential scanning calorimetry, (continued)
- liposome, 22, 26, 211–224, 225
- phase transitions, 8, 22, 26, 212–224, 225, 239, 266

DNA synthesis (see Immune response; mitogen)

Electron magnetic resonance (see also Magnetic resonance; Neutron scattering; Raman spectroscopy; X-ray diffraction),
- acyl chain flexing, 15, 22, 209–228
- calcium effects, 18, 19
- fluidity, 7, 15–16, 21–26, 224–228, 256, 266–269, 654, 655
- lateral diffusion, 7, 15–18, 21–26, 128, 224–228, 266–269, 654, 655
- lipid flip-flop, 20, 21, 128
- phase transition, 8, 21–26, 209–218, 238, 248, 256, 654
- probes, 6, 7–9, 16–18, 216, 225–226, 767
- phospholipids, 7–9, 15–26, 128, 209–228, 239, 267, 301

Electron microscopy (see also α-Actinin; Autoradiography; Desmosome; Electron microscopy, freeze fracture; Endocytosis; Filaments, cytoplasmic; Gap junction; Junctions, cell; Liposome; Membrane-associated components; Membrane, plasma; Microfilaments; Microsome; Microtubules; Microvilli; Mitochondria; Mobility, lateral; Tight junction),
- autoradiography, 12, 117, 130, 481
- freeze-fracture, 13, 23–26, 28, 34–36, 37, 76–117, 216–218
- histochemical, 12
- membrane purification, 380

Electron microscopy, freeze fracture (see also Capping, receptor; Gap junction; Liposome; Mobility, lateral; Particles, intramembranous; Red blood cell; Tight junction),
- antigen localization, 25–28, 38–39, 51, 90–91, 556
- contamination, 81–84
- cryoprotection, 77–81, 100
- endothelium, 105
- epithelial cells, 95–99, 100–117
- erythrocyte, 13, 28, 39, 83–95, 396–397
- interpretation, 80–84, 86–94, 110–111
- junctions, 34–36, 100–117
- lipids, 24–26, 216–218
- lymphocyte, 556, 627–628, 631, 672, 675
- membranes, (see Membrane, plasma)

Electron microscopy, freeze fracture, (continued)
- muscle, 95, 97, 98
- myelin, 88–90, 91
- particle arrays, 36, 97–99, 100–116
- particles (see also Red blood cell), 13, 25–26, 28, 37–39, 76–117
- phase transition, 24–26, 78
- pinocytosis, 95–96
- protozoan, 96–97
- replica techniques, 76–80
- synapse, 36, 40
- tumor cells, 99–100, 115–116

Electron spin resonance (ESR) (see Electron magnetic resonance; Magnetic resonance; Spectroscopy)

Endocytosis (see also Antibody, cell surface binding; Antigen, cell surface; Antigenic modulation; Golgi; Immunoglobulin, cell surface; Turnover, membrane),
- colchicine, 708–709
- concanavalin A, 697, 707
- cytochalasins, 708–709
- electron microscopy, 95–98, 704–709
- freeze fracture, 95–98
- Golgi, 672, 705, 708 (see also Golgi)
- histocompatibility antigens, 707, 711–712
- immunoglobulin, 571–578, 629, 704–709, 711–712
- ligands, 571–574, 575–578, 697, 704–709
- lysosome, 705
- membrane fusion, 707–708
- metabolism, cell, 704
- protozoa, 96, 621
- rate, 705–706
- smooth muscle cell, 96–98
- vesicle transport, 709
- vinblastine, 708–709

Endoplasmic reticulum (see also Microsome; Mitochondria),
- enzymes, 248–249, 267–269
- Golgi links, 32, 33
- membrane biosynthesis, 32–34
- mitogenesis, 606
- transition temperature, 248–249

Enzyme, kinetics (see also Enzymes), 228–236, 238–239
- Arrhenius plots, 231–236, 238–239, 240–256, 259–261, 266–271

Enzymes (see also Endoplasmic reticulum; Tight junction; Transport, cellular),
- ATPase, 224, 233–236, 247–248, 249–256
- galactose oxidase, 385–386
- glycosidase, 236–238, 384
- hydrogenase, 230, 270

Enzymes, (continued)
- lipase, 241, 248, 270–271
- lipid-dependency, 228–236
- membrane, 181, 208, 228
- neuraminidase, 385–386
- oxidase, 133–134, 241, 247
- phosphatase, 133, 224, 233–236, 240, 251, 257
- proteases, 33–34, 106–107, 112–117, 324, 329, 354, 382, 384, 392, 399–400, 407, 494–495, 552–554, 610, 633, 679, 708
- protein synthesis, 494, 495
- reductase, 133, 267–269

Enzymes, membrane-bound (see also Tumor cell),
- ATPase, 224, 233–236, 238, 240, 244, 247–248, 249–256, 259–266, 270–273
- cholesterol, 208, 238, 254–256, 259
- cytochrome, 221–224, 241, 247, 267–269
- dehydrogenase, 260–262
- esterase, 255
- fluidity, 208, 228–240, 241–244, 256–266
- glycosidase, 236–238, 314–315, 323, 330–335
- glycosyltransferase, 248–249, 256, 313–316, 318–320, 323–325, 328–330, 353, 410
- lipid-dependency, 208, 228–231, 231–240, 256–266
- mitochondrial, 244–248
- oxidase, 133–134, 241, 247–248
- phosphatase, 257
- protease, 426, 435
- reductase, 133, 267–269
- transport, 236–238, 240–244

Erythrocyte (see Anionic sites; Concanavalin A; Electron microscopy, freeze fracture; Exchange, lipid; Phospholipids; Red blood cell; Trans-membrane control)

Exchange, lipid (see also Lipids, membrane; Lipoproteins; Liposome; Microsome; Mitochondria; Phospholipids; Red blood cell; Turnover, membrane),
- asymmetry, membrane, 182–187
- brain, 146–147, 151, 158–159
- chylomicrons, 141–142
- electrostatic interactions, 171–175, 175–181
- erythrocyte, 140, 142–143, 150
- fibroblast, 143
- glycolipids, 316–322
- hydrophobic interactions, 168–171
- ions, 171–175, 180
- lipid synthesis, 131–132, 188–191
- lipoproteins, 140–149, 152–178, 178–181, 188–191
- liposome, 149–154, 163–181, 184–224
- microsome, 129–141, 145–148, 149–153, 158–

Exchange, lipid, (continued)
159, 178–181, 183–187, 188–191
- mitochondria, 129–141, 145–148, 149–153, 158–159, 178–181, 183–187, 188–191
- myelin, 134
- phospholipid class, 139, 145–147, 151, 174
- physiological significance, 187–191
- specificity, 139, 145–149, 151, 174
- synaptosome, 134
- temperature-dependency, 135
- tissue, 143–144, 151, 153

Exchange, protein (see Exchange, lipid)

Fc receptor (see Antigen mobility; Lymphocyte)
Fab (see Antibody, cell surface binding)
Fibroblast (see also Capping, receptor; Concanavalin A; Exchange, lipid),
- agglutination, 45–47
- fibronectin, 331–335, 337, 352–354, 391–398, 410
- freeze fracture, 99–101
- glycolipids, 313–316
- lectin receptors, 45–47, 91, 553
- lipid alteration, 297–302

Fibronectin (LETS) (SF) (see also Fibroblast),
- agglutination, 404, 410
- cell cycle expression, 393–394, 485, 488
- labeling, cell surface, 331–335, 337, 352–354, 391–398, 485, 488
- molecular weight, 391–392
- protease sensitivity, 392, 404
- structure, 391–398

Filaments, cytoplasmic (see also Membrane-associated components; Red blood cell),
- 100 Å filaments (intermediate), 43, 685
- anesthetics local, 45–48, 629, 689–692, 694
- microfilaments (see Microfilaments)
- thick filaments, 684

Filaments, intermediate (see Filaments, cytoplasmic)
Filaments, thick (see Filaments, cytoplasmic)
Fixation,
- aldehydes, 650, 651, 653

Fluidity, membrane (see also Anesthetics, local; Cholesterol; Electron magnetic resonance; Lipid alteration; Lipids, membrane; Membrane, plasma; Mobility, lateral; Mobility, perpendicular; Phospholipids),
- calcium, 18–19, 44–51, 218–224
- cell cycle, 335
- cholesterol, 15–16, 218–224, 254–255, 267
- electron spin resonance (see Electron magnetic resonance)

Fluidity, membrane, (*continued*)
- enzymes, 208, 228–244, 254–256, 269,
- lipid alteration, 228–236, 239–244, 256–269, 296–302, 655
- lipids (see Lipids, membrane)
- liposome (see also Liposome), 4–9, 14–19, 21–26, 209–228, 256–266
- mobility, antigen (see Mobility, lateral)
- phospholipids (see Phospholipids)
- polarization, fluorescence (see Fluorescence, polarization)
- redistribution, receptor (see Mobility, lateral)
- transformed cells, 403–404
- transport, 271

Fluorescence (see also Spectroscopy),
- lectins, 26
- microscopy, 4, 8, 11, 12, 26–31, 345, 390–391, 480–481, 560, 620–621, 627, 647–680
- polarization 8, 9, 16, 26, 226–227, 254, 301, 403–404, 655
- photobleaching, 12, 26
- probes, 8, 9, 12, 26
- quantitative microfluorescence, 475–476, 560

Forssman (see Antigen, cell surface; Antigen, tumor cell; Glycolipids)
Freeze fracture (see Electron microscopy)

Galactose oxidase,
- labeling, cell surface, 332–335, 385–386, 391, 396
- glycolipid exposure, 330–335

Gap junction (or nexus),
- assembly, 112–117
- brain, 117
- development, 115
- electron microscopy, 107–110
- epithelial cells, 108–112
- freeze fracture, 107–117
- model, 111
- reviews, 107
- structure, 107–112
- tumor cells, 115–116
- turnover, 112–117

Genetic analysis (see Somatic analysis, cell surface antigen)
Glycolipid exposure (see Galactose oxidase; Neuraminidase; Sialic acid)
Glycolipids (see also Antigen, cell surface; Antigen, tumor cell; Cell cycle; Exchange, lipid; Fibroblast; Growth, cell; Hormone, cell interactions; Lipid, alteration; Lipids, membrane; Liposome; Membrane, plasma; Metabolism, cell; Microsome; Mitochon-

Glycolipids, (*continued*)
dria; Tumor cell),
- biosynthesis, 310–320, 324–327, 335
- cell adhesion, 351–355
- cell contact, 351–355
- cell cycle dependency, 326–329, 333–335, 350, 409, 485, 492–493
- cell density-dependency, 322–325, 327–330, 333
- cellular distribution, 316–317
- cholera toxin, 331, 338–345, 373
- Forssman, 337–338, 492, 497, 499
- glycosidase, 314–315, 327, 330–335
- glycosyltransferase, 308, 313–316, 318–320, 323–325, 329–330
- glycosyltransferase induction, 330
- growth, cell, 321–322, 336, 348–351
- hormone, 345–348, 349
- induction, 312
- interferon, 347–348, 373
- malignancy, 335–338, 350
- normal cells, 322–335
- revertants, 320
- simplification, 310–315, 317–318, 326, 335–337
- structure, 308–313
- temperature-sensitive mutants, 318–319, 321
- toxin receptors, 331, 338–345, 347, 373
- transformation, 307–322, 335–338, 348–351, 409
- transglycosylation, 328–329
- tumor cells, 307–322, 331–334, 335–338
- virus-transformation, 316–324, 328

Glycoproteins, cell surface (see Anionic sites; Cell cycle; Cyclic nucleotides; Growth, cell; Phase transition; Sialic acid; Turnover, membrane)

Glycoproteins, transformed cells,
- endocytosis (see Endocytosis)
- fibroblastic, 393–398, 404–406
- galactoprotein a, 393–394, 399
- glycopeptides, 404–406
- lectin interactions, 44–48, 402–404
- lectin receptors, 44–46, 402–404
- lost glycoproteins, 393–400, 500
- lymphoid, 398
- mobility, 44–48, 403–404
- plasminogen activator, 399–400
- shedding (see Shedding)
- turnover (see Turnover)

Glycosidase (see Enzymes; Enzymes, membrane-bound; Galactose oxidase; Glycolipids; Neuraminidase)
Glycosyltransferase (see Adhesion, cell; Enzymes, membrane-bound; Glycolipids)

Golgi (see also Endocytosis; Endoplasmic reticulum),
- endocytosis, 672, 705, 708
- secretion, 435, 605–610, 682

Growth, cell (see also Cholera toxin; Glycolipids; Lipid, alteration; Tumor cell),
- cyclic nucleotides, 349–350
- density-dependency, 322–325, 327–330
- glycolipids, 321–322, 336, 348–351
- glycoproteins, 371
- lipids, 296–302

H-2 (see Antigen, cell surface; Antigen mobility; Antigen, tumor cell)

Hemagglutinins (see Concanavalin A; Wheat germ agglutinin)

HL-A (see Antigen, cell surface; Antigen mobility)

Hormone, cell interactions (see also Cell cycle; Cholera toxin),
- ACTH, 346
- adenyl cyclase, 345–347, 349, 350
- FSH, 345–346
- glycogen, 345–347
- glycolipid interaction, 345–348, 349
- insulin, 345
- lutenizing hormone, 346
- prolactin, 345
- thyrotropin, 345, 346, 348, 349

Humoral immunity (see Antibody, cell surface binding; Immune response)

Ia (see Antigen, cell surface; Antigen mobility)

Immune complex (see Antigen, tumor cell; Complement)

Immune response (see also Immunoglobulin, cell surface),
- allelic exclusion, 624–626
- antigen stimulation, 601–602, 624–634
- B-cell maturation, 601–604, 606–614
- BCG, 456, 457
- blocking factor (see Antigen, tumor cell)
- capping, 703
- DNA synthesis, 606
- humoral response, 601–604, 606–614, 624
- metastasis, tumor, 441, 457
- mitogen stimulation, 606–614
- secretion, antibody, 605–610, 619
- secretion, immunoglobulin, 607–609
- T-dependent, 601–603, 610, 634

Immune response, (continued)
- T-independent, 601–603, 606–614
- tumor response, 436–439, 440, 442–453, 457–560

Immunity, cell-mediated (see Cell-mediated immunity)

Immunoglobulin, cell surface (see also Antibody, cell surface binding; Antigen, cell surface; Antigenic modulation; B-cell; Capping, receptor; Cell mobility; Complement; Endocytosis; Immune response; Lymphocyte; Proteins, plasma membrane),
- allelic exclusion, 624–626
- antibody (anti-immunoglobulin) 12, 30, 44–48, 555–556, 574–581, 620–636, 645–704
- antigen receptors, 602–603, 623–624, 634–635
- B-cell, 44–51, 555–556, 560, 578–579, 588, 601–604, 609–614, 619–636, 645–704
- capping, 44–51, 560, 578–579, 588, 628–631, 645–704
- class, 622–623, 633–635, 644
- colchicine, 44–48, 690–692, 692–697
- concanavalin A modulation, 44–45, 630–631, 579, 656, 658, 694–695
- detection, 620–621, 645–646
- endocytosis, 571–578, 629, 704–709, 711–712
- isolation, 626–627, 645–646
- nitrogen receptors, 399, 610–614
- mobility, 12, 30, 44–48, 555–556, 559–564, 574–581, 624, 645–704
- modulation, 710–712
- patching, 649–659
- redistribution, 12, 30, 44–48, 555–556, 574–581, 626–631, 645–704
- secretion, 435, 605–610, 626, 629, 634–635
- specificity, 623–624

Immunoselection (see Antigen, tumor cell; Somatic analysis, cell surface antigen)

Infrared spectroscopy (IR) (see also Spectroscopy),
- protein conformation, 10

Ionophores (see Calcium; Calcium ionophore; Capping, receptor)

Junction, gap (see Gap junction)

Junctions, cell (see also Desmosome; Electron microscopy, freeze-fracture; Gap junction; Tight junction),
- assembly, 106–107, 112–117
- asymmetrical, 117
- gap (nexus), 34, 100, 107–117
- intermediate, 100

Junctions, cell, (*continued*)
- low resistance, 351
- *Macula adherens* (desmosome) 34, 100–103
- *Macula occludens* (tight junction), 100, 103–107
- mobility, 36
- models, 111
- nexus (gap), 34, 100, 107–117
- proteolysis, 106–107, 116
- reviews, 34, 100–117
- septate desmosomes (*Macula adherens*) 34, 100–103, 103–107
- solubilization, 36
- tight (*Macula occludens*), 100, 103–107
- zonulae occludentes (tight), 34, 100

Labeling, cell surface (see Fibronectin; Galactose oxidase; Proteins, plasma membrane)
Lattice formation (see Antibody, cell surface binding)
Lectin receptors, cell surface (see Concanavalin A; Glycoproteins, transformed cells; Lectins; Phytohemagglutinin; *Ricinus communis* agglutinin; Wheat germ agglutinin)
Lectins (see also Capping, receptor; Cell motility; Concanavalin A; Endocytosis; *Ricinus communis* agglutinin; Wheat germ agglutinin)
- agglutination, 44–48, 320, 337, 352–353, 402–403, 410, 484–485, 527, 553, 658
- concanavalin A, 12, 26–31, 44–48, 334, 352, 396, 605–611
- ferritin–conjugated, 28, 91–92
- fluorescent labeled, 12, 16–31, 44–48, 403
- *Helix pomatia*, 557
- phytohemagglutinin, 39, 91, 382
- receptor redistribution, 27–31, 44–48, 403
- *Ricinus communis*, 39, 91, 334, 396
- Sepharose-bound, 382
- slime mold, 351–353
- tissues, 353
- wheat germ agglutinin, 27, 320, 402, 669
LETS protein (see Fibronectin)
Ligand-induced redistribution (see Mobility, lateral)
Ligand interactions (see Cholera toxin; Hormone, cell interactions)
Lipid, alteration (see also Lipids, membrane; Lipoproteins),
- Acholeplasma, 238–239, 266
- agglutination, by lectins, and, 300
- Arrhenius plot, 236–238, 244–256, 259–261, 266–267
- ATPase, 240, 260–266
- catabolism, 296–299

Lipid, alteration, (*continued*)
- cholesterol, 296–299
- diet, 240
- *Escherichia coli*, 236–238
- exogenous lipid, 296–302
- fluidity, 239–244, 256–266, 266–269
- glycolipids, 299, 321–322
- growth, cell, 297–302
- mobility lateral, 266–269
- mycoplasma, 238–266
- phase separation, 240–244, 256–266
- reconstitution, membrane, 256–266
- temperature acclimation, 239–240, 270–271
- transport effects, 236–237, 249
- tumorigenicity, 300–302
Lipid packing (see Cholesterol; Lipids, membrane)
Lipid synthesis (see Microsome; Mitochondria)
Lipids, membrane (see also B-cell; Cholesterol; Differential scanning calorimetry; Electron microscopy, freeze fracture; Exchange, lipid; Growth, cell; Lipid, alteration; Lipoproteins; Metabolism, cell; Mobility, lateral; Phase transition; Phospholipids; T-cell; Turnover, membrane),
- asymmetry, 2, 7, 20, 21, 32, 182–187, 207, 373
- calcium binding, 18, 19
- cholesterol, 15, 16, 25, 218–219, 227, 254–255, 296–299, 372–373
- exchange, 127–191, 296–301
- fluidity, 5–10, 15–19, 25–32, 207, 209–224, 246–256, 266–269
- glycolipids, 296, 699
- melting, 22–26
- mobility, 5–10, 15–19, 25–32, 207, 219–228, 240–244, 266–269
- packing, 19–26, 210–218
- phase separation, 21–26, 219–221, 238, 240–244
- phospholipids, 15, 21–26, 32, 207–302
- protein effects, 226–228
- transition temperature, 22–26, 219–221, 240–244, 246–256
- viscosity, 15–19, 266–269
Lipopolysaccharide (see Mitogen)
Lipoproteins (see also Exchange, lipid; Lipids, membrane, Phospholipids, exchange; Proteins, plasma membrane),
- beef brain, 158–159, 178–181
- beef heart, 157–158
- beef liver, 155–157, 168–175, 178–182
- immunological properties, 162
- lipid binding, 165–168, 168–175, 175–181, 181–182

Lipoproteins, (*continued*)
- lipid exchange, 140–149, 152–154, 154–178, 179–182, 184–187, 188–191
- liposome, 152–154, 163–165, 165–181
- molecular weight, 160, 161
- plant, 159–163
- properties, 154–163, 165–168, 175–178, 181–182
- purification, 148, 154–160
- rat intestine, 158
- structure, 147–148, 160–161, 169

Liposomes (see also Endocytosis; Exchange, lipid; Differential scanning calorimetry; Lipoproteins; Microsome; Mitochondria; Phospholipids; Spectroscopy),
- antigen incorporation, 153
- cholesterol, 15, 16, 25, 218–219, 227, 297
- differential scanning colorimetry, 22, 26, 211–224, 225
- enzyme activity in, 256–261
- fluidity, 4–9, 14–19, 21–26, 209–224, 225–228, 256–266
- freeze fracture, 24–26, 216–218, 223, 228
- lipid exchange, 149–154, 163–181, 184–224
- lipid packing, 19–26, 210–218, 259–266
- permeability, 209, 221
- phase transition, 21–26, 209–218, 256–266
- reconstitution, membrane, 256–266
- structure, 209–224

Locomotion, cell (see Cell motility; Membrane-associated components; Microfilaments; Myosin)

Lymphocyte (see also B-cell; Capping, receptor; Electron microscopy, freeze fracture; Immunoglobulin, cell surface; Mitogen; T-cell; Trans-membrane control),
- actin, 684–692
- antigen stimulation, 601–602, 624, 634–635
- anti-lymphocyte effects, 679
- B-cell, 513–515, 601–604, 604–614, 619–636, 643–713
- blocking factors (see Antigen, tumor cell)
- capping, 29–31, 44–51, 560, 575–578, 588, 628–631, 640–704
- cell movement, 669–670, 672–683
- cell shape, 651–653, 669, 675–678
- concanavalin A, 12, 26–31, 44–48, 579, 605, 652, 655–658, 659, 664–658, 659, 664–666, 671–704
- DNA synthesis, 606, 608–609
- Fc receptor, 621, 627, 631
- heterogeneity, 603–604, 612
- immunoglobulin, secretion, 605–610, 634–635

Lymphocyte, (*continued*)
- – surface, 44–51, 560, 578–579, 588, 601–604, 609–614, 619–636
- LPS, 604, 608–614
- maturation, 513–515, 608–614, 634–636
- microvilli, 653, 675
- mitogen receptors, 605–614
- myeloma, 609
- phytohemagglutinin, 605, 672
- redistribution, 29–31, 44–51, 575–578, 647–709
- stimulation, 399, 601–614, 634–635
- T-cell, 513–515, 601–603
- tumor response, 435, 440, 442–453, 456–460
- ultrastructure, 652–653, 660, 669, 679–680
- uropod, 652, 660, 669, 675, 679

Macrophage migration inhibitor (see Antigen, tumor cell)
Macula adherens (see Desmosome)
Macula occludens (see Tight junction)
Magnetic resonance (see also Electron magnetic resonance; Nuclear magnetic resonance),
- ESR, 4, 5, 7–9, 15, 16, 19–26, 128, 210–221, 224–228, 240–244, 249–254
- NMR, 4, 5, 6, 15, 210–221
Maturation, B-cell (see B-cell; Immune response; Mitogen)
Maturation, T-cell (see T-cell)
Membrane antigens (see Antigen, cell surface; tumor cell)
Membrane-associated components,
- actin, 40, 44–48, 96, 101, 668, 683–685
- α-actinin, 42, 43–44
- anesthetic effects, 45–48, 629
- glycoproteins, cell surface, 45–51, 392–394
- intermediate filaments, 43
- locomotion, cell, 576, 629, 678–680
- microfilaments, 40, 42–51, 96, 99, 101, 342, 554, 576–579, 678, 683–685, 689–692
- microtubules, 40, 41, 44–51, 342, 579, 628–630, 678–680, 683, 685–689, 691, 692–704
- myosin, 43–44, 629, 683–685
- tropomyosin, 43
- tubulin, 40
Membrane lipids (see Lipids, membrane)
Membrane, plasma (see also Electron microscopy; Enzymes, membrane-bound; Immunoglobulin, cell surface; Lipids, membrane; Mobility, lateral; Myelin, plasma membrane; Phase transition; Tumor cell),
- asymmetry, 1, 3, 7, 19, 21, 32, 143, 182–187, 207, 228, 373

Membrane, plasma, (*continued*)
- calcium (see also Calcium), 18, 19, 44–48, 217, 219–221, 248
- cholesterol (see Cholesterol)
- cis-control, 34–37, 647, 651, 654
- crosslinking, 39, 44–45, 558–564, 566–571, 575, 678
- degradation, 33–34, 106–107, 112–117
- domains, 37–38, 268, 651, 674
- electron microscopy, 12, 13, 14, 23–31, 47, 75–117, 113, 216–218, 228, 380, 390–396, 475, 481–482, 647–680, 704–707
- enzymes (see Enzymes, membrane bound)
- enzymes, plasma membrane (see Enzymes, membrane bound)
- exchange, 21, 127–191
- fixation, 650–651, 653
- fluidity (see also Mobility, lateral), 4–9, 11, 14–19, 21–26, 128, 209–224, 248–256, 266–271
- glycolipids, 310–355, 373–374
- hormone, cell interactions (see Hormone, cell interactions)
- ion transport, 248–256, 265, 271
- junctions, 34, 100–117
- lateral compression, 25
- lipid alteration (see Lipid alteration)
- lipids (see also Lipids and Phospholipids), 2, 4–10, 14–26, 127–191, 209–224, 372–373, 655
- microviscosity, 8, 9, 16–19, 226–228, 266–269, 654, 655, 669
- mobility, lateral, 14–31, 44–48, 128, 209–224, 225–228, 266–269, 342–343, 555–556, 620–636, 647–704
- phase separation, 21–26, 210–218, 240–244, 256, 655
- phase transition, 21–26, 210–218, 238, 240–244, 248–256, 655
- proteins (see also Proteins, plasma membrane), 3, 7, 26–31, 38–39, 90–91, 373–410, 645–647
- proteolysis, 106–107, 382, 384, 552–554, 679
- purification, 133–134, 378–380, 389–390
- reconstitution, 256–266
- reviews, 2–4, 28, 205–206
- structure, 1–4, 13–15, 48–51, 75–76, 127–128, 374
- synthesis, 31–34, 131–152, 187–191
- thickness, 22
- trans-membrane control, 38–51, 377, 645–646, 654, 675, 685–702
- transport, 248–256, 265–266
- turnover, 14, 31–34, 106–107, 112–117, 187–191, 504, 569–574, 614, 712–713

Membrane, plasma, (*continued*)
- ultrastructure, 13, 79–117, 650–653
- viruses, interaction with, 36, 561, 575
- virus antigen (see Antigen cell surface)

Membrane proteins (see Proteins, plasma membrane)

Metabolism, cell (see also Endocytosis; Mobility, lateral; Shedding),
- during capping, 576–578, 629–630, 657–658, 660, 663–664, 690
- glycolipids, 317–330
- inhibitors, 656
- lipids, 296–299

Metastasis (see Adhesion, cell; Antigen, tumor cell; Immune response)

Microfilaments (see also Actin; Cell motility; Cytochalasins; Membrane-associated components; Microvilli; Trans-membrane control),
- anesthetics, 45–48, 342, 629, 689–692, 694
- ATP, 684, 691
- capping, 44–48, 576–579, 629–630, 683–685, 689–692, 692–697
- cell motility, 576, 629, 678–683, 685
- cyclic nucleotides, 43
- cytochalasins, 40, 43, 44–48, 578–579, 689–692
- membrane-associated, 40, 42–51, 96, 99, 101, 342, 554, 576–579, 678, 683–685, 689–692
- microvilli, 49, 685
- structure, 40, 43–47, 684–685
- uropod, 685

β_2-Microglobulin (see Antigen, cell surface; Antigen mobility; Antigen, tumor cell; Proteins, plasma membrane)

Microsome,
- enzymes, 248–249
- glycolipids, 316–317
- lipid exchange, 129–141, 145–148, 149–153, 158–159, 178–181, 183–187, 188–191
- lipid synthesis, 131, 132, 188–191
- liposome exchange, 184–187

Microtubules (see also Colchicine; Membrane-associated components; Mobility, lateral; Trans-membrane control),
- alkaloids, 40, 44, 692–697
- anchoring, 44, 579, 690–692
- anesthetics, 46–48, 630, 657, 694
- calcium, 40, 44, 689–692
- capping, 44–45, 629–697
- colchicine, 40, 44, 279, 403, 692–697
- cyclic AMP, 696
- membrane-associated, 40, 41, 44–51, 342, 579, 628–630, 678–680, 683, 685–689, 691, 692–704

Microtubules, (continued)
- podophyllotoxin, 692
- structure, 40, 685–689
- vinblastine, 40, 44, 403, 692–697

Microvilli (see also α-Actinin; Lymphocyte),
- agglutination, cell, 658
- antigens, surface, 30, 434, 653, 675
- microfilaments, 49, 685

Microviscosity (see Membrane, plasma)

Mitochondria,
- enzymes, 133, 188, 246–248
- glycolipids, 316–317
- lipid exchange, 129–141, 145–148, 149–153, 158–159, 178–181, 183–187, 188–191
- lipid synthesis, 131–132, 188–191
- liposome exchange, 184–187

Mitogen,
- B-cell stimulators, 399, 604–607, 607–614
- concanavalin A, 399, 605, 611
- DNA synthesis, 606, 608–609
- lipopolysaccharide, 604, 608–614, 632
- maturation of B cells, 608–612
- PHA, 480–481, 605
- PPD, 604, 608–610, 613–613
- receptors, cell, 399, 610–614
- T-cell stimulators, 605

Mitogenesis (see B-cell; Cell cycle; Concanavalin A; Cyclic nucleotides; Endoplasmic reticulum; Immunoglobulin, cell surface)

Mobility, cell surface antigens (see Antibody, cell surface binding; Antigen, cell surface)

Mobility, lateral (see also Calcium; Electron magnetic resonance; Immunoglobulin, cell surface; Lipid, alteration; Membrane, plasma; Nuclear magnetic resonance; Phospholipids; Rhodopsin),
- agglutination, 44–48, 485, 553, 658
- antigens, 11, 12, 27–31, 44–51, 555–556, 559–564, 571–579, 626–631, 643–709
- calcium, 18, 19, 44–51, 218–224, 267, 628–630, 691–692
- capping, 30, 44–51, 560–561, 578–579, 588, 626–631, 643–704
- cholesterol, 15–16, 218–224, 254–255, 267, 431–432
- complement, 585–590
- electron microscopy, 25–28, 216–218, 643–709
- fluidity, 403–404
- lectin receptors, 26–31, 44–49, 652, 655–659, 664–668, 669–704
- ligand-induced, 11, 12, 26–31, 342–343, 555–556, 559–564, 569–574, 626–631, 643–704
- lipid, 5–10, 14–19, 21–26, 128, 207, 209–218,

Mobility, lateral, (continued)
219–228, 266–269
- marginal redistribution, 29–30
- membrane particles, 11, 26–28, 672
- membrane proteins, 4, 9, 11, 12, 25–31, 44–51, 221–224, 228, 555–556, 571–574, 647–704
- metabolism, 551, 564, 572–576, 654
- microtubules, 34–36, 44–51
- particles, 11, 25–28
- particles, intramembranous, 25–28
- phase separation, 21–26, 210–218, 219–221, 238, 240–244
- restraints, 34–36, 227, 630–631, 656, 658
- temperature dependence, 575–576, 648
- toxin receptors, 342–343

Mobility, perpendicular (see also Electron magnetic resonance; Phospholipids),
- chemical labeling, 21
- cholesterol, 19
- erythrocyte, 142–143, 150, 182–183
- exchange, 21, 127–191
- lipid "flip-flop", 19–21, 128, 207, 227–228
- membrane asymmetry, 19–21, 32, 182–187
- microsomes, 183–184
- mitochondria, 183–184
- toxin, cell entry, 344–345

Mobility, receptor (see Cholesterol; Concanavalin A; Immunoglobulin, cell surface; Wheat germ agglutinin)

Modulation, antigenic (see Antigenic modulation)
- lectin (see Capping, receptor)

Mutation, somatic (see Somatic analysis, cell surface antigen)

Mycoplasma,
- enzymes, 261
- membrane, 238–239, 261, 266
- plasma membrane, 266

Myelin, plasma membrane (see also Exchange, lipid; Electron microscopy, freeze fracture),
- basic protein, 222
- particles, 88–90, 91

Myosin (see also Actin; Cell cycle),
- capping, 44–51, 629, 684–692, 700–702
- membrane-associated, 43–44, 629, 683–685

Neuraminidase (see also Antigen, tumor cell; Enzymes),
- antibody accessibility, 552–554
- antigen unmasking, 401, 431, 553
- asialoglycoproteins, 386
- glycolipid exposure, 330–335

Neuraminidase, (continued)
- lectin agglutinability, 553
Neutron scattering, 5
Nexus (see Gap junction)
Nuclear magnetic resonance, 4, 5, 15–18 (see also Magnetic resonance),
- acyl chain flexibility, 5, 15, 210–211
- lateral mobility, 5, 15–18, 209–224
- ^{31}P-resonance, 15
- phospholipids, 5, 16–18, 128, 209–224, 225
- spin-lattice relaxation, 18
- spin-spin relaxation, 18

Oligosaccharides (see Red blood cell)
Optical rotatory dispersion (ORD) (see also Spectroscopy),
- membrane proteins, 9

Particles, intramembranous, 13, 25–26, 28, 38–39, 76–117, 672 (see also Mobility, lateral)
Permeability (see Liposome)
Phase separation (see Lipid, alteration; Lipids, membrane; Membrane, plasma; Mobility, lateral; Phase transition)
Phase transition (see also Cholesterol; Differential scanning calorimetry; Electron magnetic resonance; Membrane, plasma; Phospholipids)
- calcium, 220–221
- cholesterol, 25, 26, 217–219, 259, 262–263
- electron microscopy, freeze fracture, 24–26, 78
- lateral compression, 25
- lipids, 8, 21–26, 209–224
- phase separation, 21–26, 216–224, 244–256
- proteins, 25
Phospholipids (see also Calcium; Electron magnetic resonance; Exchange, lipid; Lipid, alteration; Lipids, membrane; Lipoproteins; Liposome; Phase transition),
- asymmetry, 2, 7, 20, 21, 32, 182–187, 207, 373
- calcium, 18, 19, 217, 219–224, 267
- catabolism, 188–191, 296–299
- cholesterol, 15, 16, 25, 218–219, 298–299, 372–373
- enzyme-dependency, 208, 228–236, 231–240, 256–266
- erythrocyte, 142–143, 372–373
- exchange, 127–130, 132–191, 296–301
- fluidity, 4–9, 14–19, 21–26, 205–214, 244–256
- lateral mobility, 4–9, 14–19, 21–26, 209–214, 266–269, 655

Phospholipids, (continued)
- lipoproteins, 140–144, 154–178
- liposomes, 144, 149–154, 163–165, 165–181, 209–224
- membrane composition, 15, 21–26, 32, 127–134, 188–191
- microsome, 129–141, 145–148, 149–153, 158–159, 183–184, 188–191
- mitochondria, 129–141, 145–148, 149–153, 158–159, 183–184, 188–191
- phase transition, 21–26, 210–218, 219–221, 240–244, 248–256, 259–266, 655
- plasma membrane, 2, 4–10, 14–26, 127–191, 249–256, 295–302, 372–373
- saturation, 211–215
- synthesis, 131–132, 188–191, 296–299
- temperature acclimation, 239–240
- transition temperature, 21–26, 210–221, 240–244, 246–256
Phytohemagglutinin (*Phaseolus vulgaris*) (PHA) (see also Lectins; Lymphocyte; Mitogen),
- erythrocyte receptors, 39, 91
- lymphocyte, 605, 672
- platelet receptors, 91
Pinocytosis (see Endocytosis)
Plasma membrane (see Membrane, plasma)
Polymorphonuclear leukocyte (see Capping, receptor)
Probes, cell surface (see Electron magnetic resonance; Fluorescence; Magnetic resonance; Neutron scattering; Raman spectroscopy; X-ray diffraction)
Protease (see also Antigen, cell surface; Enzymes; Tight junction; Trypsin),
- agglutination, cell, 659
- cholera toxin binding, 340
- effects on cells, 33–34, 106–107, 112–117, 375–376, 384, 389, 399, 610, 633, 657, 665
- glycolipid exposure, 332, 333, 334
- glycoprotein exposure, 384, 397, 431
- inhibitors, 657, 659, 679
- papain, 426–429, 431
- plasminogen activator, 399–400
Protein synthesis,
- antigen expression, 432, 489, 494, 495
- capping, receptor, 664
- diphtheria toxin, 344
- enzyme induction, 330, 494
- glycoproteins, 495
- immunoglobulin, 607–609, 632–635
- membrane proteins, 32–34, 495
- rat liver, 131
Proteins, plasma membrane (see also Fibronectin; Immunoglobulin, cell surface; Mo-

Proteins, plasma membrane, (*continued*)
  bility, lateral; Phase transition; Protein synthesis; Red blood cell; Transport, cellular),
  – actin (see Actin)
  – asymmetry, 1, 3, 7, 19, 21, 32, 143, 182–187, 207–228, 372
  – ATPase, 224, 233–236, 240, 244, 247–256, 259–266, 270–273
  – cell cycle, 393–394
  – cell density, 353–354, 404–406
  – cis-membrane control, 34–37, 647, 651, 654
  – cytochrome, 221–224, 241, 247, 267
  – enzyme accessibility, 332–333
  – enzymes (see Enzymes, membrane bound)
  – epiglycanin, 498–499
  – fibronectin (LETS), 331–335, 337, 352–354, 391–398, 401–402, 406, 410, 500
  – fluidity, membrane (see Fluidity, plasma membrane)
  – galactose oxidase, 385–386
  – gel electrophoresis, 382–388
  – glycophorin, 25–28, 32, 38–39, 51, 90–91, 374–378
  – glycosyltransferase (see Enzymes, membrane-bound)
  – immunoglobulin, cell surface (see Immunoglobulin, cell surface)
  – integral, 3–4, 222–224, 373–377, 424–425
  – labeling, cell surface, 384–388
  – β_2-microglobulin, 401–402, 520, 645, 661, 668, 694
  – myelin basic protein, 222, 227
  – neuraminidase, 385–386
  – peripheral, 3–4, 222–224, 373–377, 424–425
  – proteolysis, 106–107, 382, 384, 392, 399–400, 407, 425, 494–495, 552–554, 679
  – purification, 380–388, 389–390
  – rhodopsin, 9, 17
  – spectrin, 38, 39, 47, 91–94
  – synthesis, 32–34, 495
  – trans-membrane control, 38–51, 377, 645–646, 654, 675, 685–704
  – transport, 236–237, 240–244, 248–256, 265, 271
  – turnover, 14, 31–34, 106–107, 112–117, 187–191, 452, 569–574, 604, 614, 712–713
  – virus, 36, 401, 561, 575
Proteins, tumor cell (see Glycoproteins, transformed cells)
Proteolysis (see Enzymes, membrane-bound; Junctions, cell; Membrane, plasma; Proteins, plasma membrane; Tight junction; Tumor cell)
Proteolytic enzymes (see Antigen, tumor cell;

Proteolytic enzymes, (*continued*)
  Enzymes, membrane-bound; Proteases; Trypsin)
Prozone effects (see Antibody, cell surface binding)
Purified protein derivative, tubercle bacilli (PPD) (see Mitogen)

Raman spectroscopy, 4, 5
Receptor, cell surface (see Anesthetics, local; Capping, receptor; Cholera toxin; Concanavalin A; Lectins; Immunoglobulin, cell surface; Mitogen; Mobility, lateral; Wheat germ agglutinin)
Receptor mobility (see Mobility, receptor)
Reconstitution, membrane (see Lipid, alteration; Liposome; Membrane, plasma)
Red blood cell,
  – band 3 component, 39, 90–91, 376, 377, 389, 406
  – concanavalin A receptors, 39, 90–91, 377
  – filaments, 92
  – freeze fracture, 13, 28, 39, 83–95, 672
  – glycophorin, 25, 27, 28, 32, 38–39, 51, 90–91, 374–378, 381
  – lectin receptors, 39, 90–91, 377
  – lipid exchange, 142–143, 150
  – lipids, 315, 342, 372–373
  – oligosaccharides, 377–378
  – particles, 13, 28, 39, 83–95
  – protease, 426
  – spectrin, 38–39, 47, 91–94, 377, 403, 702
  – trans-membrane control, 38–39, 403
Redistribution, cell surface (see Antigenic modulation; Capping, receptor; Concanavalin A; Immunoglobulin, cell surface; Mobility, lateral; Trypsin)
Rhodopsin,
  – dichroism, 17
  – lateral mobility, 9; 26
  – rotational mobility, 17
Ricinus communis agglutinin (see also Lectins),
  – erythrocyte receptors, 39, 91
  – fibroblast receptors, 396, 397

Secretion (see Antigen, tumor cell; B-cell; Immunoglobulin, cell surface; Turnover, membrane)
Shedding (see also Antibody, cell surface binding; Antigen, cell surface; Antigen, tumor cell; Antigenic modulation; Turnover, membrane),
  – antibody-receptor complex, 484, 569–574, 590, 665, 709

Shedding, (continued)
- antigens, tumor cell surface, 423–460
- metabolism, 709
- proteolytic enzyme, cell surface, 435
- temperature-dependency, 709
- tumor-associated antigens, 423, 432–439, 440–451, 451–453

Sialic acid,
- alloantigen, 552
- cell surface antigens, 552
- glycolipids, 308–313, 330–335, 345
- glycoproteins, membrane, 385–386, 387, 389

Sodium dodecyl sulfate (SDS) (see Detergent)

Somatic analysis, cell surface antigen,
- antigen loss, 515, 524–536, 540
- antigen masking, 533
- antigen rearrangement, 515, 527, 528
- antigen selection, 524–536
- antigen suppression, 536–537
- cell hybridization, 515–516
- chromosomal rearrangement, 535
- chrosomal segregation, 536
- complementation analysis, 529, 530–531, 533–534, 540
- differentiation, 514–517
- H-2, 517–519, 525–526, 529–532, 535–536
- HL-A, 529–530, 532
- immunoselection, 515, 524–540
- lymphoid cells, 516–540
- mutation, 527–530, 536
- rationale, 513–517
- recombination, 530–532
- surface antigen expression, 513–540
- Thy-1 (θ), 522–523, 526–528, 532–534, 538
- TL, 519–522, 526, 531–532, 535–536
- tumor cells, 515, 524–540

Spectroscopy,
- CD, 9
- ESR, 4, 5, 7–8, 15–18, 19–26, 128, 209–225
- fluorescence, 4, 8, 9, 11, 16, 26–31, 226–227, 254, 301
- IR, 10
- NMR, 4–6, 15–18, 128, 209–224
- ORD, 9
- Raman, 4, 5

Structure, membrane (see Membrane, plasma)

Surface antigens (see Antigens, surface membrane)

Synapse (see Trans-membrane control)

Synaptosome (see Exchange, lipid)

T-antigen (see Antigen, cell surface; Antigen, tumor cell)

Thy-1(θ) (see Antigen, cell surface; Antigen mobility)

TL (see Antigen, cell surface; Antigen mobility)

T-cell (thymus-derived) (see also Lymphocyte; Mitogen),
- antigen-specific receptor, 601–603
- cell-mediated immunity, 436, 645
- filaments, intermediate, 685
- lipids, 654
- maturation, 513–515
- "null", 645
- receptor redistribution, 651–653
- surface antigens, 519–523
- T-dependent antigen, 601–603, 634
- T-independent antigen, 601–603
- uropod, 652

Tight junction (zonulae occludentes and *Macula occludens*),
- assembly, 106–107
- electron microscopy, 103–104
- epithelial cells, 103–107
- freeze fracture, 103–105
- proteolysis, 106–107
- turnover, 106–107

Toxin receptors (see Capping, receptor; Cholera toxin; Glycolipids; Mobility, lateral)

Transformation (see Antigen, tumor cell; Cyclic nucleotides; Fibronectin; Glycolipids; Glycoproteins, transformed cells; Transport, cellular; Tumor cell)

Transition temperature (see Endoplasmic reticulum; Lipids, membrane; Phospholipids)

Trans-membrane control (see also Filaments, cytoplasmic; Membrane, plasma; Proteins, plasma membrane; Red blood cell),
- anesthetics local, 45–48, 629, 630, 567
- calcium, 19, 44, 659–692
- erythrocyte, 38–39
- lymphocyte, 29–31, 44–48, 578–579, 628–631, 685–704
- microfilaments, 40–51, 96, 99, 101, 403, 554, 576, 579, 678, 683–685, 689–692, 697–704
- microtubules, 40–51, 403, 628–680, 685, 685–689, 692–697, 697–704
- synapse, 40
- tumor cell, 44–48, 402–404, 431

Transport, cellular (see also Cyclic nucleotides; Membrane, plasma),
- cell growth, 371
- enzymes, membrane bound, 236–238, 240–244
- insulin binding, 407
- neuraminidase effects, 407
- protease effects, 407
- transformed cells, 406–407

Triton X–100 (see Detergent)
Trypsin,
- antibody accessibility, 552–554
- glycolipid accessibility, 332, 334
- glycoprotein accessibility, 332, 333, 334, 354, 396–397
- glycoprotein cleavage, 375–376, 396–397
- leukemia virus antigen, 552
- receptor redistribution, 665

Tumor-associated antigens (see Antigen, cell surface; Antigen, tumor cell)

Tumor cell (see also Cyclic nucleotides; Gap junction; Glycoproteins, transformed cells; Shedding; Somatic analysis, cell surface antigen; Trans-membrane control),
- adhesion, cell, 407–408
- agglutination, lectin (see Lectins)
- antigens (see Antigen, tumor cell)
- capping (see Capping)
- cytoskeletal (see Cytoskeletal)
- freeze fracture electron microscopy, 99–100
- glycolipids, 307–322, 331–334, 335–338
- glycoproteins (see Glycoproteins, transformed cells)
- growth characteristics, 318–320, 321–322
- immunity (see Antigen, tumor cell)
- lipid alteration, 297–302, 321–322
- morphology, 318–319
- motility, 351
- myeloma, 609
- proteolytic enzymes, 434, 435
- shedding (see Antigen, tumor cell)
- surface peptides, 404–406
- temperature-sensitive mutants, 318–319
- transport, 406–407
- virus transformation, 316–322

Tumorigenicity (see Lipid, alteration)

Turnover, membrane (see also Antibody, cell surface binding; Antigen, tumor cell; Golgi; Lymphocyte; Membrane, plasma; Shedding; Tight junction),
- cell cycle, 393–394

Turnover, membrane, (*continued*)
- degradation, 33–34, 106–107, 112–117, 432, 434–435
- endocytosis, 95–98, 571–574, 575–578, 704–709, 711–712
- exchange, 31
- glycolipid, 322–330
- lipid, 127–191, 296–299
- rates, 31–34, 106–107, 112–117, 614
- secretion, 569–574, 607, 609–610, 632–635
- shedding, 432–439, 440–453, 569–574, 665, 709
- synthesis, 31–34, 106–107, 112–117, 712–713

Ultrastructure (see Capping, receptor; Cytoplasmic streaming; Desmosome; Electron microscopy; Endocytosis; Endoplasmic reticulum; Gap junction; Junctions, cell; Lymphocyte; Membrane, plasma; Microsome; Microvilli; Mitochondria; Particles, intramembranous; Tight junction)

Uropod (see Microfilaments; Lymphocyte)

Vinblastine (see Endocytosis; Microtubules)
Virus antigens (see Antigen, cell surface; Antigen mobility; Antigen, tumor cell; Capping, receptor)
Virus transformation (see Transformation; Tumor cell)

Wheat germ agglutinin,
- cell agglutination, 320, 402, 658
- receptor capping, 669
- receptor mobility, 27, 669

X-ray diffraction, 4
- cholesterol, 26
- gap junction, 109–110
- lipid organization, 4, 5, 209, 214, 220, 238, 251–254

Zonulae occludentes (see Tight junction)

165621